The Legacy of *Niels Henrik Abel*

The Abel Bicentennial,

Oslo, 2002

Springer-Verlag Berlin Heidelberg GmbH

Olav Arnfinn Laudal • Ragni Piene (Editors)

The Legacy of Niels Henrik Abel

The Abel Bicentennial,
Oslo, 2002

Springer

Editors:

Olav Arnfinn Laudal
Ragni Piene
University of Oslo
Department of Mathematics
0316 Oslo, Norway
e-mail: arnfinnl|ragnip@math.uio.no
URL: http://folk.uio.no/arnfinnl/
http://folk.uio.no/ragnip/

Cataloging-in-Publication Data applied for

A catalog record for this book is available from the Library of Congress.

Bibliographic information published by Die Deutsche Bibliothek
Die Deutsche Bibliothek lists this publication in the Deutsche Nationalbibliografie; detailed bibliographic data is available in the Internet at http://dnb.ddb.de

Mathematics Subject Classification (2000): 01-XX, 11-XX, 14-XX, 16-XX, 32-XX, 34-XX, 37-XX, 51-XX

Additional material to this book can be downloaded from http:// extras .springer .com .

ISBN 978-3-642-62350-9 ISBN 978-3-642-18908-1 (eBook)
DOI 10.1007/978-3-642-18908-1

springeronline.com

Originally published by Springer-Verlag Berlin Heidelberg New York in 2004
Softcover reprint of the hardcover 1st edition 2004

Cover design: deblik Berlin
Typesetting: Le-TeX Jelonek, Schmidt & Vöckler GbR, Leipzig
Printed on acid-free paper 46/3142db-5 4 3 2 1 0

Contents

List of Contributors

Michel van den Bergh
Departement WNI
Limburgs Universitair Centrum
3590 Diepenbeek, Belgium
vdbergh@luc.ac.be

Jan-Erik Björk
Matematiska institutionen
Stockholms universitet
10691 Stockholm, Sweden
jeb@matematik.su.se

Aldo Brigaglia
Dipartimento di Matematica
Università di Palermo
Via Archirafi 34
90123 Palermo, Italia
brig@math.unipa.it

Fabrizio Catanese
Lehrstuhl Mathematik VIII
Universität Bayreuth
95440 Bayreuth, Germany
Fabrizio.Catanese
@uni-bayreuth.de

Moira Chas
Department of Mathematics
SUNY at Stony Brook
Stony Brook, NY 11794, USA
moira@math.sunysb.edu

Ciro Ciliberto
Dipartimento di Matematica
Università di Roma Tor Vergata
Via della Ricerca Scientifica
00133 Roma, Italia
cilibert
@axp.mat.uniroma2.it

Herbert Clemens
Mathematics Department
Ohio State University
231 W. 18th Ave.
Columbus, OH 43210, USA
clemens@math.ohio-state.edu

Torsten Ekedahl
Department of Mathematics
Stockholm University
10691 Stockholm, Sweden
teke@math.su.se

Günther Frei
Lützelstrasse 36,
8634 Hombrechtikon, Switzerland
g.frei@active.ch

William Fulton
Department of Mathematics
University of Michigan
Ann Arbor, MI 48109-1109, USA
wfulton@umich.edu

Simon Gindikin
Department of Mathematics
Hill Center, Rutgers University
110 Frelinghuysen Road,
Piscataway, NJ 08854-8019, USA
gindikin@math.rutgers.edu

Tom Graber
University of California at Berkeley
Department of Mathematics
970 Evans Hall
Berkeley, CA 94720-3840, USA
graber@math.berkeley.edu

Mark Green
Department of Mathematics
UCLA, Box 951555
Los Angeles, CA 90095-1555, USA
mlg@ipam.ucla.edu

Phillip Griffiths
Institute for Advanced Study
Princeton, NJ 08540, USA
hastings@ias.edu

Joe Harris
Harvard University
Department of Mathematics
One Oxford Street
Cambridge, MA 02138, USA
harris@math.harvard.edu

Håkan Hedenmalm
Department of Mathematics
The Royal Institute of Technology
14400 Stockholm, Sweden
haakanh@math.kth.se

G. M. Henkin
Institut de Mathématiques
UMR 7586 du CNRS, case 247
Université Pierre et Marie Curie
4 place Jussieu
75252, Paris, Cedex 05, France
henkin@math.jussieu.fr

Christian Houzel
11, rue Monticelli
75014, Paris, France

Christian Kassel
Institut de Recherche Mathématique
Avancée C.N.R.S.
Université Louis Pasteur
7 rue René Descartes
67084 Strasbourg Cedex, France
kassel@math.u-strasbg.fr

Steven L. Kleiman
Math Department, 2-278 MIT
77 Mass Ave,
Cambridge MA 02139, USA
Kleiman@math.MIT.edu

Daniel Lazard
LIP6, Université P. et M. Curie
boite 168
4 place Jussieu
75252 Paris Cedex 05, France
Daniel.Lazard@lip6.fr

Yuri I. Manin
Max–Planck–Institut
für Mathematik
Bonn, Germany
and
Northwestern University
Evanston, USA
manin@mpim-bonn.mpg.de

Barry Mazur
Harvard University
Department of Mathematics
One Oxford Street
Cambridge, MA 02138, USA
mazur@math.harvard.edu

V. P. Palamodov
Tel Aviv University
School of Mathematical Sciences
Ramat Aviv, Tel Aviv 69978, Israel
palamodo@post.tau.ac.il

Mikael Passare
Matematiska institutionen
Stockholms universitet
10691 Stockholm, Sweden

Claudio Pedrini
Dipartimento di Matematica
Università di Genova
Via Dodecaneso 35
16146 Genova, Italia
pedrini@dima.unige.it

Birgit Petri
Technische Universität Darmstadt
Fachbereich Mathematik, AG2
Schloßgartenstr. 7
64289 Darmstadt, Germany
petri
@mathematik.tu-darmstadt.de

Norbert Schappacher
Technische Universität Darmstadt
Fachbereich Mathematik, AG2
Schloßgartenstr. 7
64289 Darmstadt, Germany
schappacher
@mathematik.tu-darmstadt.de

Yum-Tong Siu
Department of Mathematics
Harvard University
Cambridge, MA 02138, USA
siu@math.harvard.edu

Jason Starr
Massachusetts Institute of Technology
Department of Mathematics
Room 2-271
Massachusetts Institute of Technology
77 Massachusetts Ave.
Cambridge, MA 02139, USA
jstarr@math.mit.edu

Arild Stubhaug
Department of Mathematics
University of Oslo
P.O. Box 1053 Blindern
0316 Oslo, Norway
arilds@math.uio.no

Dennis Sullivan
City University of New York
The Graduate Center
365 Fifth Avenue, Room 4208
New York, NY 10016-4309
dsullivan@gc.cuny.edu

August Tsikh
Fakultet matematiki i informatiki
Krasnoyarskiĭ gosudarstvennyĭ
universitet
660041 Krasnoyarsk, Russia

The Legacy of Niels Henrik Abel

The collected works of Niels Henrik Abel, edited by Sophus Lie and Ludwig Sylow in 1881, contain 29 papers. The first was written in 1823 when Abel was 21 years old, and the last was jotted down on his deathbed in January and February 1829.

During this short, five-year mathematical lifespan, Abel produced major works on a variety of topics. Most of these works have become part of today's mathematical vernacular.

The *Abel Year 2002*, commemorating the 200th anniversary of Niels Henrik Abel's birth, was, however, not only an international mathematical event, but indeed a national celebration throughout Norway, with important contributions made by universities, Government, local communities, and the media.

Niels Henrik Abel is, in the minds of his countrymen, not only a great mathematician, but also, to most Norwegians, a scientific and romantic icon.

A lasting expression of this national sentiment has now been translated into the Abel Prize, established by the Norwegian Government on the model of the Swedish Nobel Prizes awarded in the other sciences.

Although proposed by the Prime Minister in 2001 and subsequently approved by the Norwegian Parliament, the formal announcement of the creation of the Abel Prize was not made until the opening of the Abel Bicentennial Conference held at the University of Oslo, June 3–8, 2002.

This Conference, supported by the International Mathematical Union and the European Mathematical Society, attracted an impressive number of distinguished mathematicians.

A year or two earlier some of these mathematicians had been approached by us with a request to submit an authoritative work on Abel's legacy in their own fields, thus contributing to an "Abel book."

The response was very positive and, based on this response, we decided to accept an offer from Springer-Verlag to publish the present volume with the rather ambitious title *The Legacy of Niels Henrik Abel.*

The book is therefore not a usual Proceedings from the Abel Bicentennial Conference. Most of the contributors were invited speakers at the Conference, but not all of them. Moreover, and this is important, most of the many and very interesting talks presented at the Conference are not included in this volume.

The purpose of this book, published at the beginning of the third century after Abel's birth, is to present the mathematical community with an overview of some selected fields where Abel's legacy is still operative.

The book starts with His Majesty King Harald V's Opening Address at the Abel Bicentennial Conference, which underlines Norwegian national sentiment towards Niels Henrik Abel. Then Arild Stubhaug, whose Abel biography is now available in five languages, recounts the story of the mathematician's short and tragic life.

Christian Houzel presents us with an impressive historical overview of the work of Abel, and Phillip Griffiths presents a masterful treatment of Abel's famous Paris Memoir, showing how the original ideas of Abel are still effective in modern algebraic and complex geometry.

Abel's work on algebraic equations is the subject matter for several authors. Daniel Lazard's paper explains how to solve the quintic equation by radicals, when Abel's condition for so doing is satisfied. Birgit Petri and Norbert Shappacher tell the story of algebraic equations and the study of Abelian field extensions, from Abel to Kronecker, and Günther Frei introduces us to the fascinating story behind the Artin reciprocity law for Abelian extensions, and class field theory.

As is clear from the papers of Houzel and Griffiths, algebraic geometry is in great debt to Abel. Aldo Briaglia, Ciro Ciliberto, and Claudio Pedrini look at the roots of the Italian school in algebraic geometry, and bring us up to date on the influence of Abel in this magnificent story. Then Fabrizio Catanese introduces us to Abel the algebraist, and to Abelian varieties, and Steven Kleiman completes the overview of the legacy of Abel in algebraic geometry by presenting a thorough discussion of the history of and the mathematics behind the many variants of Abel's theorem in modern algebraic geometry.

These papers are complemented by a paper by Torsten Ekedahl on the moduli stack of Abel's hyperelliptic curves. The paper by Mark Green and Phillip Griffiths brings up the present problems of intermediate Jacobians, and the corresponding problems of formal deformations of Chow groups. The paper by Herbert Clemens on an analogue of Abel's theorem also fits into this picture.

In the same direction, the paper by Tom Graber, Joe Harris, Barry Mazur, and Jason Starr looks at arithmetical questions related to rationally connected varieties.

In a more complex geometric setting, Yum-Tong Siu treats the methods used to study criteria for hyperbolicity of submanifolds of Abelian varieties and complex projective space.

The second paper of the 21-year-old Niels Henrik Abel, written in 1823, introduced the Abel transform. This has later been the basic tool in studying inverse problems. The history of Abel's transform and the present status of this subject is treated in different settings by three papers — by Gennadi M. Henkin, by Simon Gindikin, and by Victor Palamodov. Closely connected with the paper of Henkin is the contribution of Jan-Erik Björk, which studies residues on complex manifolds and D-modules. This paper also contains an interesting history of the relationship between Abel's search for radical solutions to the quintic equations, and the search for analytic or meromorphic solutions. This is also the point of view of Mikael Passare and August Tsikh in their paper on algebraic equations and hypergeometric series. The paper by Håkan Hedenmalm on Dirichlet series and functional analysis ends this section of the book.

The final string of papers treats problems that might be classified as part of non-commutative geometry, a field which promises to be an essential ingredient of future mathematics.

The first of these is Yuri I. Manin's "Alterstraum," on the use of non-commutative geometry in the study of abelian class field theory for real quadratic fields. The second is the paper by William Fulton, on quantum cohomology of homogeneous varieties. The third is a paper by Christian Kassel on quantum principal bundles up to homotopy equivalence, generalizing the notion of Galois extensions to non-commutative algebras. The string is completed by Michel van den Bergh's paper on non-commutative crepant resolutions of singularities, and finally by the work of Moira Chas and Dennis Sullivan on closed string operators in topology leading to Lie bialgebras and higher string algebra.

Springer Verlag has been most helpful in the process of bringing this book into existence; in particular we would like to acknowledge the enthusiastic support of Martin Peters, assisted by Ruth Allewelt and Claudia Kehl.

Oslo, 17 May 2003,

Olav Arnfinn Laudal
Ragni Piene

Opening address
The Abel Bicentennial Conference
University of Oslo, June 3, 2002

King Harald V

Two hundred years ago, our nation was in its childhood. A young nation needs national heroes. Niels Henrik Abel was such a hero. He was young, bright minded, and he became famous in academic circles at home and in Europe. His untimely death was a loss for the young nation, and also for the development of mathematics throughout the world.

This week we shall honour this bright young man, his ideas and achievements. But we shall also honour the mathematics itself, and the development of a beautiful universe of ideas, theories and applications. I sincerely hope that this conference can be another step in this development, that you through discussions and lectures can broaden your experience in your beloved work. I know that you all have a passionate relationship with mathematics, which is of course difficult for me, being a non-mathematician, to understand.

It is a pleasure for me to see so many professional mathematicians from so many parts of the world gathered here in Oslo. I hope you will all have some spare time to explore the capital and its surroundings during your stay. This time of the year is maybe the most beautiful here in the north of Europe, when nature wakes up after a cold winter and the evenings are long and bright.

With these words I declare the Abel Bicentennial Conference 2002 for open.

The Life of Niels Henrik Abel

Arild Stubhaug

The story goes that when Niels Henrik was born on August 5, 1802, he arrived three months early and was so weak that he had to be washed in red wine and swathed in thick layers of cotton in order for his tiny spark of life not be extinguished. There is much to indicate that this birth took place at Nedstrand, not at Finnøy, an island near by (near Stavanger), where the father of the new-born was a parish vicar.

Niels Henrik's father, Søren Georg Abel, himself a vicar's son, had married Anne Marie Simonsen, daughter of the richest man in Risør, a lively and prosperous coastal town. Niels Henrik was the young couple's second child, and when he was two years old the family moved to Gjerstad, not far from Risør. At Gjerstad Niels Henrik's father took over his father's position, and in this small rural community Niels Henrik grew up along with four brothers and one sister.

Vicar Søren Georg Abel was in all ways an outstanding man, and for a long time his parish flock held him in highest esteem. His time as pupil and student in Denmark had coincided with an epoch when the ideals of the Enlightenment and the ideas of the French revolution were felt everywhere. In all his endeavours father Abel would pay homage to the reasonable and useful, and proclaim his rock solid faith in the human being's ability to solve life's mysteries through reason. For his parishioners he arranged a reading group where they could borrow books on how growing of field crops and animal husbandry could be improved, on how vegetable gardens and orchards could produce a higher yield. He vaccinated his parishioners against smallpox, organised care for the poor and schooling for all; he had a granary built, with a rootcellar for potatoes; he constructed a brickworks and a tar-burning furnace. In the years of famine and war from 1809 to 1812, he gave out recipes for how the wild turnip, a weed (Brassica campestris) could be used in cooking of porridge and the baking of flatbread. He persuaded the village folk that it was reprehensible not to eat horsemeat. Father Abel was a full-blooded rationalist of the type that would later come to be called "potato priests," and in this rationalistic spirit, S. G. Abel wrote a catechism, entitled *The Questions of Religion*, with Answers, Adapted to the Comprehension of the Young – 337 questions and answers. The book was first published in 1806, and in the course of the next ten years it saw five printings, and was much used in schools and churches in preparation for confirmation and communion.

Niels Henrik's mother seems to have thrived best at parties and on festive occasions. She had lost her mother when she was seven, and thereafter had grown up with two stepmothers in a house well-appointed with everything that those luxurious times could yield. At the vicarage in Gjerstad she also arranged many a gallant ball. Up to one hundred people could be gathered in the many rooms of the house, and after a dinner of sumptuous proportions there would be dancing until dawn, to the music of violins and clarinets. Mrs. Abel cared less about her children, and there is much to indicate that early on she became an alcoholic.

Niels Henrik and his siblings were educated by their father, and for a time, also by their mother's sister – later by a young man who became the teacher and parish clerk in the community. Their father's published catechism was probably the children's first primer, and in addition, for teaching purposes he had handwritten an instruction book on history, geography, the mother tongue, and mathematics. In mathematics one found not only multiplication and division tables, but also tables for addition and subtraction; in the first line here there stood: $1 + 0 = 0$.

When 1814 – the momentous year for Norway and its independence – came to pass, father Abel was acquainted with the political conditions, and he praised and promoted the independence of Norway. In the autumn of 1814 he was elected to the Extraordinary Parliament as the representative for his district. Again in 1818 Vicar Abel was a Member of Parliament, and on both occasions he drew notice, mostly from a negative direction. In the Parliament during the autumn of 1814 he was criticised for having one evening invited some of the peasants' parliamentary representatives to a great punch-drinking session, in order to get them to follow his lead in an important vote the following day. To discuss these issues in a private gathering in advance was regarded as almost cheating, and to use alcoholic drinks in such a connection was reprehensible. Representative Abel's behaviour was severely repimanded, but most Norwegians were in favour of the issue that was so won. The issue was the the King's right of veto over citizenship matters, and the majority of Norwegians were reluctant about the King, who was Swedish, deciding who could become a Norwegian citizen.

Parliamentary proceedings took place in the buildings of the Christiania Cathedral School, and the school's great auditorium was the city's most representative hall. (The capital, today Oslo, at that time was called Christiania.) In the autumn of 1814, during his stay in the capital, father Abel had become acquainted with the school's teachers, and quite certainly with the grant possibilities of which the school had command. The father considered that as early as the coming autumn, 1815, his eldest son should journey to Christiania. But when the time of his departure came close, the boy became so sad and despondent about abandoning his home, that his father did not dare to send him off, and thus he sent Niels Henrik instead. Thirteen year old Niels Henrik left home, and in October 1815 he appeared at the Christiania Cathedral School. He passed the entrance examination to the first form, and shortly thereafter he wrote home saying that he felt "right in his Element."

After one year at the school, big brother Hans Mathias arrived to keep Niels Henrik company. They were in the same class and for the next few years the older brother did better than Niels Henrik. Being a pupil at the cathedral school was a fulltime occupation. School began at nine o'clock in the morning and continued until one o'clock, at which time there was a break for a couple of hours, and then the day ended around six in the evening. After this there was homework and preparation for the coming lessons, and day in and day out, this is the way things went, six days a week, broken only by an eight-day holiday at Christmas and two weeks of summer vacation. Theatre performances were the biggest change in the everyday school routine. The disciples, as the pupils at the cathedral school were called, were given free tickets to the final dress rehearsals; the theatre was also a meeting place where the boys from the school met the city's youth of both sexes. Even though most people in the Norwegian capital at this time were extremely interested in the theatre, there is much to indicate that Niels Henrik gradually became an specially avid theatre-goer.

With the great school reforms at the turn of the century, around 1800, a system of specialised teachers instructing in one subject began to replace the old classroom teacher system in which one teacher taught his class all subjects. On paper, according to these new regulations, this implied that the living languages (German, French and English), and the sciences, were included within the range of subjects taught. But the classical subjects, with Latin in the lead, still dominated the pupils' lessons. According to the new reforms, the practice of hitting and beating the pupils should also be phased out. Whereas before, there had been corporal punishment to enforce the diligence of the work ethic and the imbibing of knowledge, now the pupils' sense of honour and their reason were to be appealed to, and weight was placed on awakening the interest of the individual.

But the mathematics teacher Hans Peter Bader at the Cathedral School in Christiania was one of those who adhered to the old ways: Copying from the blackboard and beating for those who didn't learn. One day in November 1817 a fatal occurrence took place. Bader beat and thrashed a pupil so hard and at such length that the boy, after being bedridden for a week, died. Mathematics teacher Bader was not, correctly speaking, found legally responsible – the cause of death was said to have been "fever of the nerves," that is, typhus, but the pupils, who had long suffered under Bader's uncontainable rage and blows, now refused to attend their mathematics class unless Bader was removed. Thus the first student strike in Norwegian history became a fact. Bader was suspended (and never came back). With all possible haste, Rector Jacob Rosted had to find a new mathematics teacher, and he achieved this by finding one of his former pupils, Bernt Michael Holmboe (1795–1850), who in addition to the level of mathematical instruction available at the time in Norway, had studied more extensively on his own. Holmboe was inspired by, and took seriously the new pedagogical ideas; he began to give the pupils independent tasks to challenge them. Young Abel's engagement in the solving of mathematical problems surprised and astounded those around him, but the teachers were also disturbed by the one-sidedness of his interest and concentration. Classical languages and cultural refinement were, and became, the ideal. But Holmboe gave Niels Henrik private tutoring, and guided young Abel further into the mathematical literature. The library records reveal how the nature of Abel's book borrowings changed at this time, from fiction to physics and mathematics.

However, the circumstances around Abel's mathematical "awakening" also had another dimension. At the same time that the sixteen year-old Abel got Holmboe as his teacher, his family suffered a dramatic decline of fortune. It began with his father's catechism. Now there was also a considerable upheaval within the religious life of the new Norwegian nation. The rationalism and "potato priest theology" that had dominated teaching and ecclesiastical affairs was about to fall. The greatest authority on every question of learning was now placed in the new University of Christiania, and at the Faculty of Theology, sat Professor S. B. Hersleb and Lecturer S. J. Stenersen, who both stood in personal and reverential relationship to the looming presence of the Nordic countries' defender of the faith of the day, the young N. F. S. Grundtvig in Denmark. Lecturer Stenersen in particular saw it as his foremost task to root out the erroreous and convoluted teachings and great delusions which, in his view, he detected in Norwegian teaching. And the worst example of wrong thinking and mistaken theology, according to Lecturer Stenersen, was found in S. G. Abel's catechism. Stenersen dissected Abel's book, point by point, in "Det Norske Nationalblad", the country's largest press organ, and this occupied more than thirty pages over several issues of the newspaper. The conclusion was clear: the Gjerstad priest was one of the greatest heathens in the country, and so grossly had father Abel missed the truth that he had to take responsibility for at least 10,000 lost Norwegian souls now on their way to certain perdition. (The number 10,000 was arrived at by theologian Stenersen calculating that when the catechism went through five printings, each of 1,000 copies; at least two young souls had read each copy with its fatal spiritual vandalism.) Father Abel replied in his own defense, and some supported him, but it was Stenersen's judgement that prevailed. Father Abel became the most beleaguered theologian in the land, and the subject of much mocking verse in the newspapers. Niels Henrik's visible reaction to all this was, at the beginning, "exaggerated" cheeriness.

Shortly thereafter, in February 1818, father Abel came to the capital as Member of Parliament. And here he created a scandal and was crucified as a politician. From the Parliament lectern he criticised one of the country's most powerful men, Carsten Anker, owner of the Eidsvoll Iron Mills and the host of the Eidsvoll constitutional proceedings that in 1814 had given Norway the most liberal constitution in Europe. Representative Abel claimed that Carsten Anker did not treat his workers well, that he consistently had reneged on promises of pensions and in this Anker had received the help of an attorney who was also a Member of Parliament. These accusations were considered an outrage against the whole Parliament, and since M. P. Abel was not willing to apologise or withdraw what he had said, the issue went so far that impeachment was threatened. In the end, Odelstinget, the parliamentary upper house, found that impeachment proceedings against Pastor Abel would be a waste of resources, and the matter was dismissed. But father Abel returned to Gjerstad in

September 1818, finished as a politician, crushed as a theologian, and compromised as a human being – and he began to drink more than before. Everyone at Gjerstad knew that the vicar and his wife were drinking heavily, and that they drank separately. For father Abel however, it was a rapid downfall, and he died in the spring of 1820, at the age of forty-eight. During the last period, Mrs. Abel had not wanted to look after her husband, and it was said that he would have died in the most dismal and unsanitary condition if it had not been for the good and helpful folk in the community. At father Abel's funeral, under the gaze of the clergyman and public servants, and with the whole vicarage full of guests, Mrs. Abel got openly drunk and went to bed with one of the hired men on the farm.

At the Christiania Cathedral School the two Abel brothers reacted very differently to the family's and their father's gradual downfall. Niels Henrik had gotten his hands on mathematics and dedicated himself to new knowledge with such speed that his teacher Holmboe, in the pupils' report records, had already described Abel as "A remarkable mathematical Genius." But in the other subjects he did badly and he was only advanced to a higher class on probation. On the other hand, his big brother Hans Mathias completely lost his concentration, his marks fell drastically, and he slipped into a depression that was so deep that the rector had to send him home with the clear message that he would never manage to take the graduating examination, the *examen artium*. Big brother Hans Mathias returned home to Gjerstad, and for the rest of his life – more than twenty years – he went around like a half-wit in the folds of his mother's skirts.

Responsibility for the family now fell to Niels Henrik. In the formerly prosperous trading house of Simonsen in Risør there was nobody to call upon for help. The rich Grandfather Simonsen, like so many other trading houses, went bankrupt in the economic downturn that followed the Napoleonic Wars. Grandfather Simonsen also died during that spring of 1820, plucked bare and penniless. As long as Niels Henrik lived, he tried in various ways to help his mother and his younger siblings.

Niels Henrik Abel became a university student in 1821, with mediocre marks, apart from mathematics for which he certainly received the top mark. Even as a new student, Abel probably had greater knowledge of mathematics than anyone else in the country. B. M. Holmboe had taught him all that he could, and Abel had gone further, studying on his own and borrowing mathematical literature from the rather well-equipped university library. Among other books that he borrowed were those of Newton, Euler, Lacroix, Lagrange, Monge, Legendre, Laplace and Gauss, together with contemporary publications and journals from Paris.

Indeed, even before he became a student Abel began what would become his first great mathematical achievement, namely, work on the quintic equation. At first Abel thought he had found the solution to the general quintic equation, the solution to which Europe's mathematicians had been hunting for over 250 years, ever since Girolamo Cardano in his seminal work *Ars Magna*, in 1545, had presented solutions to equations of the third- and fourth degree. The two professors of mathematics in Christiania, Søren Rasmussen (1768–1850) and Christopher Hansteen (1784–1873), found no errors in Abel's formulae, and sent the work on to the leading mathematician of the Nordic countries, Professor Ferdinand Degen (1766–1825) in Copenhagen. He too discovered no errors, but nonetheless doubted that the solution which so many prominent mathematicians had pursued, could really now have been found by an unknown student in far-off Christiania. But nonetheless, Professor Degen noticed Abel's extraordinary ability, and felt that such a talented young man ought not to use his powers on such a "sterile Subject Matter" as the quintic equation, but ought to turn his attention instead to "the elliptic Transcendents", for that is where, Degen writes (and for this he is remembered in mathematical history), he (Abel) would "discover Magellan Voyages to great Regions of one and the same immense analytic Ocean". Abel would come to do both: he would show that the general fifth degree equation could not be solved only with the help of the classical operations of addition, subtraction, multiplication, division, and the extraction of roots. Moreover, the "elliptic Transcendents", that is, elliptic functions and integrals, was the next field in which Abel would do fundamentally new work. Abel truly found the Straits of Magellan to a new analytical ocean.

As a new fresh student, Abel received free lodging at a student residence located on the city property where the university's lectures were also held. Abel took the obligatory secondary examination in 1822, a rather comprehensive preparatory exam, which was really only an elaboration of the school graduation exam, the *examen artium*, which was known as the "first" or "primary" examination. And in the secondary examination as well, Abel received only mediocre marks, apart from the subject of mathematics. After this there was not much more to study at the university for someone who did not choose the usual public service fields of study: theology, medicine, law or philology – no sciences programme was yet to be found at Norway's only university (founded in 1811).

Some of the professors supported Abel economically from their own pockets and the influential Professor Hansteen opened his home to him. Here Abel met she whom he would refer to as his "second Mother," Mrs. Hansteen. Abel borrowed books from the University Library and studied mathematics on his own in the little attic room. Here he now had a younger brother, Peder, living with him. Abel shared his bed with Peder and helped him prepare for the *examen artium*. In addition he would later help his sister Elisabeth to find a position in Christiania.

Abel made his mathematical debut in the spring of 1823 with an article in the country's first science journal, *Magazin for Naturvidenskaberne*, which that year had produced its first issue. Abel later published a couple of other lesser works in this periodical as well. During that spring of 1823 Abel wrote a longer, and probably much more substantial and important mathematical work – in French – on "a general Demonstration of the Possibility of integrating all possible Differential Equations," and sought support from the university to publish this. Professors Rasmussen and Hansteen were positive, but the Academic Collegium was cautious, and in the mountains of paper that passed between the different offices, this work was lost without a trace for all times.

Professors Rasmussen and Hansteen, as well as others who supported Abel, knew he needed to go abroad to learn more mathematics, and Abel himself also wanted very much to go abroad so as to make contact with the European mathematical milieu. The places where mathematical research was undertaken, were Paris – at l'École Polytechnique, l'École Normale Supérieure and l'Université de la Sorbonne – and in Göttingen, around the great C. F. Gauss, who was almost considered an institution in himself. But the lack of public support forced Abel to remain at the university in Christiania for four years, to study – as it was phrased in the writings of the ministries, "the learned Languages and other useful Co-sciences of his major Field of Mathematics."

In the summer of 1823, Professor Rasmussen gave Abel 100 speciedaler from his own pocket so that Abel could at least journey to Copenhagen to visit mathematicians there. Abel travelled and discovered that indeed he knew as much mathematics as anyone in Copenhagen, and that the library at the new university in the Norwegian capital was equally well-furnished with mathematical literature as any of the libraries in Copenhagen. Abel attended the theatre and took part in student life. He read mathematics, worked a little on Fermat's last theorem, and formulated some inequalities showing that, if solutions were to be found, they would have to be enormously large, and he started work on elliptic functions (not to be published until the summer of 1827). Abel's aunt, the sister of his mother, had married a Danish naval officer, Peder Mandrup Tuxen, and Abel stayed in their home at the Christianshavn naval base in Copenhagen. There, at a ball, he met the young girl Christine Kemp. The orchestra played a waltz, at that time a newfangled dance that neither of them mastered. They just stood on the dance floor looking at one another and in great embarrassment they both backed out. Less than a year later Christine Kemp (1804–62) went to Norway and worked as a governess in Son, on the east side of Oslofjord. Abel's engagement to her at Christmas 1824 came as a great surprise to his comrades at the Regentsen student residence – a sense of surprise that perhaps arose from their experience with Niels Henrik's general relationship to the other sex, and most certainly as well from the fact that Abel had no prospects of an economy that would allow for marriage in the near future.

From the beginning of the year 1824 Abel got 200 speciedaler as a state stipend for a period of two years. During the spring of 1824 he had printed the work on the quintic equation at his own expense. He had written it in French and compressed the proof into six small pages (to make the printing as economical as possible), something that made it almost incomprehensible – and as such it failed to become the "admission ticket" to the learned circles of Europe, as Abel had expected it to be.

By the summer 1825 Abel felt that he had stayed long enough at the University of Christiania, and he wrote a personal letter to King Karl Johan, and asked for permission to travel abroad right away. The King gave his permission, and in September 1825 Abel set sail from Christiania by freight boat, on a course to Copenhagen. The plan and conditions for his stipend were to go to Gauss in Göttingen, and so onward to Paris. But when he had reached Copenhagen, he altered his itinerary. At the recommendation of, among others, the Danish mathematician von Schmidten, who had just returned home after a stay in that city, Abel left for Berlin. And travelling to Berlin must be seen as the great good fortune of his life, for in Berlin he met an engineer interested in mathematics, August Leopold Crelle (1780–1855), who built roads and one of the first railways in Germany (the line from Berlin to Potsdam). And now, meeting with Abel, Crelle, who had high status in the Prussian capital, got the courage to go into action with what had long been one of his desires – namely to publish a mathematical periodical in Berlin that could challenge the well-established journals of France. As early as February 1826 the first issue of *Journal für die reine und angewandte Mathematik* was published, and here Abel would publish the majority of what he managed to write, and thanks greatly to Abel's work, Crelle's *Journal* quickly became renowned as one of Europe's leading periodicals.

Abel stayed four months in Berlin and he had an inspiring time with Crelle and other mathematicians in the circle around him; one of them was the Swiss Jacob Steiner. Abel was invited to "salons" and private parties; he went to the theatre; and above all, he worked. The first thing he published in Crelle's *Journal* was an expanded version of the proof that the general fifth degree equation could not be solved by radicals. With great keenness and virtuosity he demonstrated that an eventual solution would have to include qualities that no radical expression possessed. But many special equations of degree greater than five had solutions, and in his further work in equation theory, Abel went a long way toward finding what the conditions were for such solutions.

The mathematical discussions in Berlin certainly dwelt upon many topics, but an on-going theme seems to have been the new demand for logical stringency in mathematical proof. Engaging in discussions about mathematics seems to have sharpened Abel's critical abilities; he became bolder and more self-confident. Some of the things he carried out seem to have been the result of thoughts and questions that had been percolating inside him even before he went abroad; and new ideas came to him as well. He wrote home to Holmboe about infinite series:

> On the Whole, Divergent Series are the Work of the Devil and it's a Shame that one dares base any Demonstration upon them. You can get whatever result you want when you use them, and they have given rise to so many Disasters and so many Paradoxes. Can anything more horrible be conceived than to have the following oozing out at you:
>
> $$0 = 1 - 2^n + 3^n - 4^n + \text{etc.}$$
>
> where n is an integer number?

Abel gained great inspiration from the possibilities of getting his works directly into print in Crelle's *Journal* – he wrote in French and Crelle translated; later Crelle printed Abel's papers in French. In letters home he expressed this, writing:

> Precisely at this Time there are so many Ideas going around in my Head. Pure Mathematics in its purest Sense most certainly ought to become my future Course of Studies. I will engage all my Powers to bring Light into the enormous Darkness that is now unquestioningly to be found in Analysis. It lacks almost all forms of Plan and System such that it is really astounding that

> it can be studied by so many, and what is worst is that it is still not treated with stringency. There are very few Propositions in higher Analysis that are proven with convincing Rigour. Everywhere one finds the unfortunate Method of deducing the General from the Special Occurrence, and it is strange to the extreme that by following this sort of Approach, so few of the so-called Paradoxes are not to be found. It is really extremely interesting to look into the the Foundations of this phenomenon. According to my Thinkng, the reason for this state of affairs lies in the fact that Functions, that until now have occupied the full attention of Analysis, have, for the most part, been expressed by Powers. As soon as other functions come into Consideration – this however not often happens – things usually do not go very well, and this in turn gives rise to false Conclusions wherein a number of interlinked, yet incorrect Propositions arise.

The first volume of Crelle's *Journal* (1826) included seven works by Abel – two lesser commentaries and five treatises. Here in the *Journal* he could in a much more extensive way demonstrate his work about the impossibility of solving by radicals the general algebraic equation of degree greater than four. And he presented, among other results, the proof of the convergence of the binomial series $(1 + x)^m$ when m is not an integer. Abel demonstrates precisely the exactitude and stringency that frequently was lacking in mathematical proofs, and thus becomes (together with Gauss and Cauchy, whose works Abel seemed to have read in Berlin) one of the pioneers of a modern stringent presentation of proof.

During his stay in Berlin, Abel had the company of four other young Norwegian men of science: C. P. B. Boeck, B. M. Keilhau, N. B. Møller, and N. O. Tank, "Our young travelling Scholars" as they were called at home in Norway – the two first-mentioned also later became professors. They lived together in one room at Am Kupfergraben 4 on the Spree, and according to later statements by Boeck, Abel was now in the habit of getting up almost every night, lighting a candle in order to write down the ideas that had awakened him. It was also here that the five Norwegian students celebrated Christmas (1825), so loudly and boisterously, that the philosopher, G. W. F. Hegel, who lived in the same house, thought that there must have been "russische Bären" (Russian bears) rummaging around on the floor below him.

Because these friends of Abel's studied mineralogy and geology, they wanted to go to the south of Germany, to Austria, Switzerland and northern Italy. And Abel, who had complained in several letters home about how melancholy he would become when he was alone, decided to follow his friends on their southward travels in Europe. In one of his letters he explained himself as follows:

> Now it is a Fact that I have been so constituted that I absolutely cannot, or only with the utmost difficulty, be alone. I become so terribly melancholy, and as such I am not exactly in the best Mood to engage in anything.

But during the whole tour with his friends he had a bad conscience regarding the granting authorities back home in Norway for not having gone immediately and directly to the mathematical milieu of Paris. Abel excused himself by seeking out mathematicians and astronomers in the cities they passed through, and he taught himself German by going to the theatre.

Following Berlin, Abel spent a month in Freiberg, where he met the mathematician August Naumann at the mineralogy seminar there. Abel went on to stay in Dresden for a week, followed by a short week in Prague, where he sought out Professors Aloys David and F. J. von Gerstner (Bernard Bolzano was away); during his six weeks in Vienna Abel spent much time with Joseph von Littrow and he also met Adam von Burg. Here he also seems to have read Paolo Ruffini's work about the quintic equation. Otherwise Abel visited galleries, museums, churches, and he went to all the theatres in the city. He wrote home to Holmboe, who had now become professor:

> You may well imagine that it is wrong for me to while away so much Time with travelling; but I do not think it can be called whiling the Time away. One learns many strange Things on such a Tour, Things of which I can find more Use than if I were purely studying Mathematics. Besides, I always need to have what you know a lazy Period, in order to pull myself together again with renewed Energy.

In Vienna the spire of St. Stephens made a powerful impression; in Trieste he enjoyed returning to the sea again and studying all the various ethnic strata of the city. Among other things about his stay in Venice, he reported: "There is a melancholy Spectre invading Venice. Everywhere one sees Signs of former Glory and contemporary Wretchedness. [...] Everything testifies to decay."

Thereafter they turned northward – via Verona to Bolzano and in to the Dolomites, and here the paths of the five Norwegians diverged. But first they had taken a detour to Predazzo and at the inn "Goldenes Schiff" the Norwegian students signed themselves in as what they hoped and desired to be in later life. Abel signed himself in as "Abel, professore della geometria."

Abel carried on to Innsbruck and he wanted very much to see a little of Switzerland – "So why should I not see a little of it. My Lord! I am not devoid of all Sense of Nature's Beauty," he wrote home to those who had granted him money. Abel hiked to the top of Mt. Rigi, and went on to Basel. From there he took the direct route to Paris, where he arrived, by himself, in July 1826, ten months after leaving Norway. Abel reported home: "I have finally arrived at the Focus of all my mathematical Desires," and about his long detour, he explained yet again: "My Desire to see a little was so great, and besides, does one travel only to study the strictly Scientific?"

Although Abel had now begun to publish in Crelle's *Journal* in Berlin, he had saved a larger work, something that he felt represented completely new insights, for the illustrious French Academy, l'Académie des sciences. To be published in Paris would be to make an impression on the government and administration in Norway, and this was also one of the goals for his travels. With the assistance of the Norwegian painter Johan Gørbitz, Abel found accommodation in the city. Gørbitz, who had been living in Paris for almost twenty years, now also painted a portrait of Abel, the only portrait made of Abel during his lifetime.

As soon as he had found a place to live (right near St. Germain-des-Prés), Abel began to work on what would come to be called his Paris memoir. To this day, there have been scarcely any other mathematical theses that have won as many words of praise as Abel's Paris memoir, an addition theorem for algebraic integrals. What is so impressive is the colossal generality in the way Abel poses the problems. He demonstrated mathematical correlations that no one had earlier dreamed of, and he opened areas of research of which new aspects are still being found. Abel's Paris memoir has continued to stand as a milestone in the development of mathematics. Abel's scenario was larger than those of his predecessors, and from his work there would well up a mass of comparisons and results. But where did this audacity come from, and was it reflected anywhere in his personal life?

Abel completed the Paris memoir and delivered it at the end of October, signing it "par N. H. Abel, Norvégien." He stayed on in Paris to the end of the year, and while he waited for an answer he prepared a couple of other pieces of work. He discovered how the lemniscate could be divided into equally great parts with the help of ruler and compass, and he began a larger work about equation theory which he finished a year and a half later and got published in Crelle's *Journal* just some days before his death.

Abel frequented circles and "soirées" where men of science, both young and old, would meet, and Abel's way of characterising some of these personalities perhaps gives a glimpse of his own sense of self. He describes A-M. Legendre, one of the great mathematicians, and also one of Abel's sources of inspiration, as an "exceedingly obliging Man but unhappily as old as a Stone" – Legendre was 74 years old. About S. D. Poisson, a central figure in this circle, Abel says that he had "a pretty little Stomach" and was "somewhat captivated by himself," and carried his "Body with Dignity" – something that also J-B-J. Fourier expressly did. Fourier was the secretary of the Academy, and one of the three men of science whom Napoleon had taken with him on his expedition to Egypt, when he intended to civilise that ancient land. S-F. Lacroix, professor at the Sorbonne and at l'École Polytechnique, was, in Abel's words "awfully scaldic and remarkably Old." Lacroix was 60. And about A. L. Cauchy – a pathfinder in mathematics and he whom Abel recognised as the greatest in Paris, he wrote: "Cauchy is tremendously Catholic and a bigot. A sorely strange Thing for a Mathematician."

Other men of science Abel met in Paris included A. Bouvard (the observatory director, to whom Abel had a letter of introduction from von Littrow in Vienna, and it was Bouvard who introduced Abel to members of the Institute), Baron Ferrusac (the journal publisher, who opened his well-appointed library to young scientists), J-N-P. Hachette (Monge's assistant and editor of the Academy's *Correspondences*), J. F. Saigey (editor of Ferrusac's *Bulletin*, and who got Abel to write some references for the *Bulletin*), J. Liouville (only seventeen in 1826, who many years later said that one of his life's mistakes was meeting Abel without getting to know him), G. P. L. Dirichlet (the Prussian who regarded Abel as his countryman, and about whom Abel reported back home: "He [Dirichlet] is a very sharp Mathematician. Together with Legendre, he has shown the impossibility of solving in whole Numbers the equation $x^5 + y^5 = z^5$ and other neat Things."), and F-V. Raspail (who later as a famous representative of the French Chamber of Deputies used Abel and his fate as an exemple of how the Academy favoured old men of science, while keeping out the young).

Abel's Paris memoir was put aside and forgotten – precisely by Cauchy – and as long as Abel lived he believed that the work had been lost for ever. His stay in Paris was a disappointment; he didn't feel well, he had a fever and a cough, and one way or another – probably from a doctor in the circle of young scientists he moved among – he learned that someone suspected he suffered from tuberculosis – which was a death sentence at that time. He himself would not believe this, and nor was there very much he could have done even if he had believed the diagnosis. Even in Paris the normal cure was still blood-letting. More and more Abel seems to have prefered being (left) alone. On a page in Abel's Paris notebooks, among mathematical deliberations and a drawing of the lemniscate and geometric "doodlings" and snatches of inner dialogue, there comes out: "Our Father who art in Heaven, Give me bread and beer. Listen for once."

After having celebrated Christmas in Paris, Abel, sad and poor, traveled to visit friends in Berlin. In Berlin, Crelle offered him the editorship of Crelle's *Journal*, but Abel declined. He longed for home and in general wanted his science to serve his "Fatherland," but Crelle went on working to get Abel a permanent position in Berlin.

When at the end of May 1827, Abel returned to Norway, his travels abroad were viewed as almost a failure. He had not been published by anyone in Paris, and he had not visited the great Gauss in Göttingen. Of course Abel had published his works in Crelle's *Journal* but what prestige did this new Berlin journal have? Abel did not receive a renewal of his stipend; he advertised in the newspapers as a private tutor, and he took a private loan of 200 speciedaler from Norges Bank, something he never managed to pay back. In addition to his own expenses he wanted very much to clear up the debts that his mother had in Gjerstad, and the debts his brother Peder had left behind in Christiania. When again the Department of Finance refused Abel a renewal application for a stipend, the Academic Collegium members undertook to support him from the university budget.

Abel now had only a year and a half to live, and this time was filled with an impressive series of dissertations that he sent in a steady stream to Crelle in Berlin, who for his part, was unable to publish them as quickly as they came in. Abel worked both with algebraic equations, elliptic functions, and infinite series, and made pioneering contributions in all fields, the majority of these works being sent to Berlin. In an intense struggle over priority, with the German mathematician C. G. J. Jacobi, Abel published an important paper on elliptic functions in July in *Astronomische Nachrichten* of Altona (a journal which was published by Gauss' friend, H. C. Schumacher, whom Abel visited during his travel abroad).

In Christiania, during the spring of 1828, Abel found a little more breathing space economically. Professor Hansteen now left on a scientific expedition to Siberia, and Abel took over some of his teaching tasks at the university. Abel was appointed to the position of docent, with an annual salary of 400 speciedaler, the usual docent salary being 600. But no view materialised towards a permanent position at the Norwegian university. After much doubt, Abel now decided to accept if any position was found for him in Berlin, and during the summer of 1828 it began to look hopeful. For six weeks that summer Abel was together with his fiancée, who was now a governess in southern Norway at the Froland Ironworks, near Arendal, and they looked forward to getting married and setting up house in Berlin. But the position in Berlin seemed to slide away into the blue, and during the autumn of 1828 Abel worked equally intensely in Christiania, and perhaps even more feverishly than before. He had little social life, particularly after Mrs. Hansteen (his "second Mother") in her husband's absence, had gone to her home city of Copenhagen.

Abel was sick and bedridden for several weeks that autumn, and admitted that the equation theory work now sapped his physical energy. When Christmas was approaching he wanted very much to return to his fiancée and his friends at Froland Ironworks. The trip there through winter conditions was long and exhausting. Abel arrived by sled, cold and coughing, but happy at the reunion. But after a Christmas ball, when he wanted to go outside to cool off, he began to cough blood. He was now bedridden at Froland for twelve weeks, in the residence of the ironworks' owner, one of the most stately homes of the period, and he was under the care of the best doctor in the district. For a time, he felt better, and he got a mathematical paper written: two-three pages where he tried again to formulate the introductory thoughts of the extensive Paris memoir.

The sickbed became a deathbed; haemorrhaging became rampant. Abel was only twenty-six years old and felt that it was ghastly that everything would soon be over. He cursed his God and the science of his day that had not managed to get the better of his illness. One of the house's serving girls, whose task it was to nurse him through the nights, would later become emotionally upset whenever there was talk of Abel's deathbed. But Abel's fury turned into apathy, and in calmer moments he seems to have worried that with his departure, his fiancée would have no one to console her. From his friend Keilhau, Abel secured a promise that he, Keilhau, would take care of Christine. A year and a half later, the two were married, and they lived, apparently, happily together for their rest of their lives. But Niels Henrik's lifespan came to its end on April 6, 1829.

Two days later, on April 8th, without any knowledge of what had taken place at Froland, there was correspondence about Abel in both Paris and Berlin. From Paris it was reported that the Paris memoir had finally been found again, and words of praise were beginning to stream in – the Academy's prize of 1,500 francs was awarded the following year for Abel's work, and the money eventually ended up with his alcoholic mother at Gjerstad. From Berlin, Crelle wrote, radiant with joy, on April 8th, that it was now guaranteed that Abel would get a permanent position in that city. He wrote: "You may now regard your future calmly. You belong among us and you will be safe," and he concluded, "You are coming to a good land, to a better climate, closer to science and to real friends who appreciate you and are fond of you."

On April 13th, after a snowstorm greater than anyone could remember so late in the year, Abel was buried in the Froland churchyard. A couple of years later, a monument, funded by his friends, was put up at the grave. The Parisian men of science expressed the desire that Abel's collected works be published, and in 1839, *Oeuvres complètes, avec des notes et développements*, edited by B. M. Holmboe, was published. But the Paris memoir, which had been located in Paris in 1829, again disappeared, and it was not included in Holmboe's edition. The Paris memoir was not tracked down and printed until 1841, in the journal of the Paris Academy (in volume VII of *Mémoires présentés par divers savants Étrangers*). The full, two-volume publication *Oeuvres complètes de Niels Henrik Abel*, (621 pages and 341 pages, in quarto format) was published in 1881, with editing and commentary by Ludvig Sylow and Sophus Lie. When Sylow and Lie were preparing this edition, they attempted to get hold of the manuscript of the Paris memoir, but it was nowhere to be found in the Academy archives. The manuscript had once more disappeared, and it was not found until 1952 by Viggo Brun. Except for eight pages the manuscript was located in the Moreniana Library of Florence. Four of these missing pages were recovered some years later, however this discovery was not made public until this year by Andrea Del Centina. Thanks to him and to the Moreniana Library Abel's manuscript is very nicely reproduced in Florence District Council's editorial series *Cultura e Memoria*. On loan from the Moreniana Library the manuscript highligted the exhibition at the Abel bicentennial conference in Oslo in June 2002.

Acknowledgment. Thanks to Richard Daly and Bibbi Lee for the translation.

The Work of Niels Henrik Abel

Christian Houzel

During his short life, N.-H. Abel devoted himself to several topics characteristic of the mathematics of his time. We note that, after an unsuccessful investigation of the influence of the Moon on the motion of a pendulum, he chose subjects in pure mathematics rather than in mathematical physics. It is possible to classify these subjects in the following way:

1. solution of algebraic equations by radicals;
2. new transcendental functions, in particular elliptic integrals, elliptic functions, abelian integrals;
3. functional equations;
4. integral transforms;
5. theory of series treated in a rigourous way.

The first two topics are the most important and the best known, but we shall see that there are close links between all the subjects in Abel's treatment. As the first published papers are related to subjects 3 and 4, we will begin our study with functional equations and the integral transforms.

1 Functional Equations

In the year 1823, Abel published two norwegian papers in the first issue of *Magasinet for Naturvidenskaberne*, a journal edited in Christiania by Ch. Hansteen. In the first one, titled *Almindelig Methode til at finde Funktioner af een variabel Størrelse, naar en Egenskab af disse Funktioner er udtrykt ved en Ligning mellom to Variable* (*Œuvres*, t. I, p. 1–10), Abel considers a very general type of functional equation: $V(x, y, \varphi\alpha, f\beta, F\gamma, \dots, \varphi'\alpha, f'\beta, F'\gamma, \dots) = 0$, where $\varphi, f, F, \dots$ are unknown functions in one variable and $\alpha, \beta, \gamma, \dots$ are known functions of the two independent variables x, y. His method consists in successive eliminations of the unknown $\varphi, f, F, \dots$ between the given equation $V = 0$ and the equations obtained by differentiating this equation with α constant, then with β constant, etc. If, for instance $\alpha = \text{const}$, there is a relation between x and y, and y may be considered as a function of x and the constant value of α; if n is the highest order of derivative of φ present in V, it is possible to eliminate $\varphi\alpha$ and its derivatives by differentiating V $n+1$ times with α constant. We then eliminate $f\beta$ and its derivative, and so on, until we arrive at a differential equation with only one unknown function of one variable. Naturally, all the functions, known and unknown, are tacitly supposed indefinitely differentiable.

Abel applies this to the particular case $\varphi\alpha = f(x, y, \varphi\beta, \varphi\gamma)$, where f, α, β and γ are given functions and φ is unkown; he gets a first order differential equation with respect to φ. For instance, the functional equation of the logarithm $\log xy = \log x + \log y$ corresponds to the case where $\alpha(x, y) = xy$, $\beta(x, y) = x$, $\gamma(x, y) = y$ and $f(x, y, t, u) = t + u$; differentiating with $xy = \text{const}$, we get $0 = x\varphi'x - y\varphi'y$, from which, with $y = \text{const}$, we get $\varphi'x = \frac{c}{x}$, where $c = y\varphi'y$. In the same way, the functional equation for arctangent,

$$\arctan\frac{x+y}{1-xy} = \arctan x + \arctan y,$$

corresponds to $\alpha(x, y) = \frac{x+y}{1-xy}$, $\beta(x, y) = x$, $\gamma(x, y) = y$ and $f(x, y, t, u) = t + u$; differentiating with α constant gives $0 = (1+x^2)\varphi'x - (1+y^2)\varphi'y$, whence $\varphi'x = \frac{c}{1+x^2}$ if $c = (1+y^2)\varphi'y$.

When $\beta(x, y) = x$, $\gamma(x, y) = y$ and $f(x, y, t, u) = t \cdot u$, we get first

$$\varphi y \cdot \varphi' x \frac{\partial\alpha}{\partial y} - \varphi x \cdot \varphi' y \frac{\partial\alpha}{\partial x} = 0,$$

whence $\frac{\varphi' x}{\varphi x}$ as a known function of x if y is supposed constant. For $\alpha(x, y) = x + y$, this gives $\frac{\varphi' x}{\varphi x} = c = \frac{\varphi' y}{\varphi y}$, so $\log \varphi x = cx$(for $\varphi(0) = 1$) and $\varphi x = e^{cx}$; for $\alpha(x, y) = xy$, $\frac{\varphi' x}{\varphi x} = \frac{c}{x}$, so $\log\varphi x = c\log x$ ($\varphi(1) = 1$) and $\varphi x = x^c$.

All these examples were classical as is the next one, coming from mechanics. The law of composition of two equal forces making an angle $2x$ leads to the functional equation

$$\varphi x \cdot \varphi y = \varphi(x+y) + \varphi(x-y); \tag{1}$$

where φx is the ratio of the resultant force to one of the two equal forces. Differentiating with $y + x = \text{const}$, one gets $\varphi' x \cdot \varphi y - \varphi x \cdot \varphi' y = 2\varphi'(x-y)$; another differentiation, with $x - y = \text{const}$, gives $\varphi'' x \cdot \varphi y - \varphi x \cdot \varphi'' y = 0$. If y is regarded as constant, this gives $\varphi'' x + c\varphi x = 0$ and $\varphi x = \alpha \cos(\beta x + \gamma)$ with α, β and γ constant. From (1), one sees that $\alpha = 2$ and $\gamma = 0$ and the problem imposes $\varphi\left(\frac{\pi}{2}\right) = 0$, so $\beta = 1$ and $\varphi(x) = 2\cos x$.

Here is another case of application of Abel's general method: the equation has the form $\psi\alpha = F(x, y, \varphi x, \varphi' x, \ldots, fy, f'y, \ldots)$, where α is a given function of x and y and φ, f, ψ are unknown functions. By differentiating with α constant, one gets a relation between $x, \varphi x, \varphi' x, \ldots$ and $y, fy, f'y, \ldots$, whence two differential equations, with respect to φ and to f, considering successively y and x as constant; if φ and f are determined, it is easy to determine ψ by the functional equation. In particular, if $\psi(x+y) = \varphi x \cdot f'y + fy \cdot \varphi' x$, so that $\alpha(x, y) = x + y$, the differentiation with α constant gives $\varphi x \cdot f'' y - fy \cdot \varphi'' x = 0$, and $\varphi x = a\sin(bx + c)$, $fy = a' \sin(by + c)$ then $\psi(x+y) = aa'b\sin(b(x+y) + c + c')$ so that $\psi\alpha = aa'b\sin(b\alpha + c + c')$.

In the case of $\psi(x+y) = f(xy) + \varphi(x-y)$, one gets $0 = f'(xy)(y-x) + 2\varphi'(x-y)$. Abel takes $xy = c$ as constant and writes $\varphi'\alpha = k\alpha$, where $\alpha = x - y$ and $k = \frac{f'(c)}{2}$, so $\varphi\alpha = k' + \frac{k}{2}\alpha^2$; then he takes $x - y = c$ constant and writes $f'\beta = c' = \frac{2\varphi' c}{c}$, so $f\beta = c'' + c'\beta$. Finally

$$\psi(x+y) = c'' + c'xy + k' + \frac{k}{2}(x-y)^2$$

or $\psi\alpha = c'' + c'x(\alpha - x) + k' + \frac{k}{2}(2x - \alpha)^2 = c'' + \frac{k}{2}\alpha^2 + k' + x\alpha(c' - 2k) + (2k - c')x^2$ and we see that the condition $c' = 2k$ is necessary; $\psi\alpha = k' + c'' + \frac{k}{2}\alpha^2$.

The third example is $\varphi(x+y) = \varphi x \cdot fy + fx \cdot \varphi y$, which gives

$$0 = \varphi' x \cdot fy - \varphi x \cdot f'y + f'x \cdot \varphi y - fx \cdot \varphi' y; \tag{2}$$

if one supposes that $f(0) = 1$ and $\varphi(0) = 0$, one gets $0 = \varphi' x - \varphi x \cdot c + fx \cdot c'$ by making $y = 0$ ($c = f'(0)$ and $c' = -\varphi'(0)$); so $fx = k\varphi x + k'\varphi' x$ and, substituting this value in (2) and making y constant: $\varphi'' x + a\varphi' x + b\varphi x = 0$ etc.

Abel returned to the study of functional equations in the paper "*Recherche des fonctions de deux quantités variables indépendantes x et y, telles que $f(x, y)$, qui ont la propriété que $f(z, f(x, y))$ est une fonction symétrique de z, x et y*", published in German in the first volume of Crelle's *Journal* in 1826 (*Œuvres*, t. I, p. 61–65). The condition of the title characterises a composition law which is associative and commutative; it may be written as $f(x, y) = f(y, x)$, $f(z, f(x, y)) = f(x, f(y, z)) = f(y, f(z, x))$ or

$$f(z, r) = f(x, v) = f(y, s) \tag{3}$$

if $f(x, y) = r$, $f(y, z) = v$ and $f(z, x) = s$. Differentiating with respect to x, to y and to z and multiplying the results, one gets

$$\frac{\partial r}{\partial x}\frac{\partial v}{\partial y}\frac{\partial s}{\partial z}=\frac{\partial r}{\partial y}\frac{\partial v}{\partial z}\frac{\partial s}{\partial x}. \tag{4}$$

But, by the definition of v, the quotient of $\frac{\partial v}{\partial y}$ by $\frac{\partial v}{\partial z}$ is a function φy when z is regarded as constant; in the same manner, φx is the quotient of $\frac{\partial s}{\partial x}$ by $\frac{\partial s}{\partial z}$, so (4) becomes $\frac{\partial r}{\partial x}\varphi y=\frac{\partial r}{\partial y}\varphi x$ and this gives r as an arbitrary function ψ of $\Phi(x)+\Phi(y)$, where Φ is a primitive of φ. So $f(x,y)=\psi(\Phi(x)+\Phi(y))$; putting this expression in (3) and making $\Phi z=\Phi y=0$ and $\Phi x=p$, one gets $\Phi\psi p=p+c$, where $c=\Phi\psi(0)$, and then $\Phi f(x,y)=\Phi(x)+\Phi(y)+c$ or

$$\Psi f(x,y)=\Psi(x)+\Psi(y) \tag{5}$$

where $\Psi(x)=\Phi(x)+c$. In other words, Abel finds that f is conjugate to the addition law by the function Ψ: he has determined the one-parameter groups.

The second volume of Crelle's *Journal* (1827) (*Œuvres*, t. I, p. 389–398) contains another paper of Abel on a functional equation:

$$\varphi x+\varphi y=\psi(xfy+yfx)=\psi(r), \tag{6}$$

where $r=xfy+yfx$; this equation includes, as particular cases, the laws of addition for log ($fy=\frac{1}{2}y$, $\varphi x=\psi x=\log x$) and for arcsin ($fy=\sqrt{1-y^2}$, $\varphi x=\psi x=\arcsin x$). One has $\varphi' x=\psi' r\cdot\frac{\partial r}{\partial x}$, $\varphi' y=\psi' r\cdot\frac{\partial r}{\partial y}$, so $\varphi' x\cdot\frac{\partial r}{\partial y}=\varphi' y\cdot\frac{\partial r}{\partial x}$ or

$$\varphi' y\cdot(fy+yf'x)=\varphi' x\cdot(fx+xf'y), \tag{7}$$

whence, for $y=0$,

$$a\alpha-\varphi' x\cdot(fx+\alpha' x)=0, \tag{8}$$

where $a=\varphi'(0)$, $\alpha=f(0)$ and $\alpha'=f'(0)$, a differential equation which determines φ if f is known. Substituting in (7), one gets $(fx+\alpha' x)(fy+yf'x)=(fy+\alpha' y)(fx+xf'y)$ or

$$\frac{1}{y}(\alpha' fy-fy\cdot f'y-\alpha' yf'y)=\frac{1}{x}(\alpha' fx-fx\cdot f'x-\alpha' xf'x)=m,$$

necessarily constant. So

$$f'x\cdot(fx+\alpha' x)+(mx-\alpha' fx)=0, \tag{9}$$

which determines f; as this differential equation is homogenenous, it is easily integreted by putting $fx=xz$, in the form $\log c-\log x=\frac{1}{2}\log(z^2-n^2)+\frac{\alpha'}{2n}\log\frac{z-n}{z+n}$, where $m=-n^2$ and c is a constant of integration. One gets

$$c^{2n}=(fx-nx)^{n+\alpha'}(fx+nx)^{n-\alpha'},$$

with $c = \alpha$, then φ by (8) and (6) is verified if $\psi x = \varphi\left(\frac{x}{\alpha}\right) + \varphi(0)$. Abel explicitly treats the case in which $n = \alpha' = \frac{1}{2}$: $fx = \alpha + \frac{1}{2}x$, then $\varphi x = a\alpha \log(\alpha + x) + k$ and $\psi x = 2k + a\alpha \log(\alpha^2 + x)$.

The relation $\alpha^{2n} = (fx - nx)^{n+\alpha'}(fx + nx)^{n-\alpha'}$, which determines f, allows us to express $fx - nx$, and then x and fx, in terms of $fx + nx = v$; turning back to (8), this gives $\varphi x = \frac{a\alpha}{n+\alpha'} \log(cnx + cfx)$. When $n = 0$, the relation which determines f takes the form $e^{\alpha' x} = \left(\frac{fx}{\alpha}\right)^{fx}$ and we have $\varphi x = \frac{a\alpha}{\alpha'} \log c\alpha + \frac{a\alpha x}{fx}$, $\psi x = \frac{2a\alpha}{\alpha'} \log c\alpha + \frac{ax}{f\left(\frac{x}{\alpha}\right)}$. The equation (6) signifies that $\alpha f\left(\frac{xfy + yfx}{\alpha}\right) = fx \cdot fy$ and Abel verifies that it is satisfied. Another particular case is that in which $\alpha' = \infty$. When m is finite, (9) reduces to $xf'x - fx = 0$, so that $fx = cx$; when m is infinite and equals $-p\alpha'$, (9) becomes $xfx - px - fx = 0$ and $fx = px \log cx$. In this last case, one gets by (7) $y\varphi' y - x\varphi' x = 0$, whence $x\varphi' x = k$ constant and $\varphi x = k \log mx$ (a new m) and then $\psi(pv \log c^2 v) = k \log m^2 v$.

A memoir left unpublished by Abel is devoted to the equation $\varphi x + 1 = \varphi(fx)$, where f is given and φ unknown (*Œuvres*, t. II, p. 36–39, mem. VI). Abel introduces a function ψ such that $f\psi y = \psi(y + 1)$; one may take ψ arbitrarily on the interval $[0, 1]$ and define ψ on $[0, +\infty[$ by $\psi(y + n) = f^n(\psi y)$ (and on $]-\infty, 0]$ by $\psi(y - n) = f^{-n}(\psi y)$ if f is bijective). For $x = \psi y$, the functional equation becomes $1 + \varphi\psi y = \varphi\psi(y + 1)$, so that $\varphi\psi y = y + \chi y$ where χ is any periodic function of y with period 1. Denoting the inverse function of ψ by $`\psi$, Abel gets

$$\varphi x = `\psi x + \chi(`\psi x).$$

As an example, he takes $fx = x^n$ and $\psi y = a^{n^y}$, so that $`\psi x = \frac{\log\log x - \log\log a}{\log n}$ and

$$\varphi x = \frac{\log\log x - \log\log a}{\log n} + \chi\left(\frac{\log\log x - \log\log a}{\log n}\right),$$

for instance $\varphi x = \frac{\log\log x}{\log n}$ if $\chi = 0$ and $a = e$.

Abel treats in a similar manner the general equation $F(x, \varphi(fx), \varphi(\psi x)) = 0$, where F, f and ψ are given functions and φ is unknown. Supposing that $fx = y_t$ and $\psi x = y_{t+1}$ or $y_{t+1} = \psi('fy_t)$, one has $F('fy_t, u_t, u_{t+1}) = 0$, where $u_t = \varphi y_t$; this difference equation has a solution $u_t = \theta t$ and $\varphi z = \theta('y_t)$. For instance the equation $(\varphi x)^2 = \varphi(2x) + 2$ leads to $(u_t)^2 = u_{t+1} + 2$ and, if $u_1 = a + \frac{1}{a}$, this gives $u_t = a^{2^{t-1}} + \frac{1}{a^{2^{t-1}}}$; on the other hand $y_{t+1} = 2y_t$, so that $y_t = c \cdot 2^{t-1}$ (c constant) and $2^{t-1} = \frac{x}{c}$. Finally, $\varphi x = b^x + b^{-x}$ ($b = a^{1/c}$). As we see, this type of equations is treated with a method different from the preceeding one, by reduction to finite difference equation.

Another type of functional equation is related to the *dilogarithm*

$$\psi x = x + \frac{x^2}{2^2} + \frac{x^3}{3^2} + \ldots + \frac{x^n}{n^2} + \ldots,$$

which Abel studies in the posthumous memoir XIV (*Œuvres*, t. II, p. 189–193) after Legendre's *Exercices de Calcul intégral*. The study is based on the summation of the series (for $|x| \leq 1$) in the form of an integral

$$\psi x = -\int_0 \frac{dx}{x} \log(1-x) \tag{10}$$

and Abel reproduces several functional equations given by Legendre, as for example

$$\psi x + \psi(1-x) = \frac{\pi^2}{6} - \log x \cdot \log(1-x).$$

But he adds a remarkable new property:

$$\psi\left(\frac{x}{1-x} \cdot \frac{y}{1-y}\right) = \psi\left(\frac{y}{1-x}\right) + \psi\left(\frac{x}{1-y}\right) \\ -\psi y - \psi x - \log(1-y)\log(1-x) \tag{11}$$

for (x, y) in the interior domain of the figure

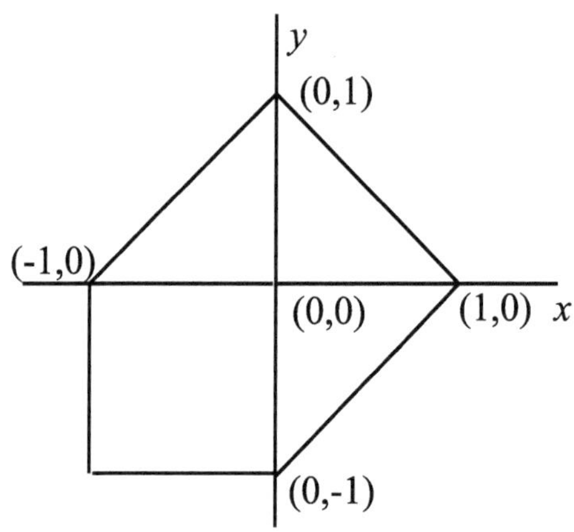

In order to prove (11), Abel substitutes $\frac{a}{1-a} \cdot \frac{y}{1-y}$ for x in (10):

$$\begin{aligned}
\psi\left(\frac{a}{1-a} \cdot \frac{y}{1-y}\right) &= -\int \left(\frac{dy}{y} + \frac{dy}{1-y}\right) \log \frac{1-a-y}{(1-a)(1-y)} \\
&= -\int \frac{dy}{y} \log\left(1 - \frac{y}{1-a}\right) + \int \frac{dy}{y} \log(1-y) \\
&\quad - \int \frac{dy}{1-y} \log\left(1 - \frac{a}{1-y}\right) + \int \frac{dy}{1-y} \log(1-a) \\
&= \psi\left(\frac{y}{1-a}\right) - \psi y - \int \frac{dy}{1-y} \log\left(1 - \frac{a}{1-y}\right) \\
&\quad - \log(1-a)\log(1-y),
\end{aligned}$$

where the remaining integral is computed by taking $z = \frac{a}{1-y}$ as variable:

$$\int \frac{dy}{1-y} \log\left(1 - \frac{a}{1-y}\right) = \int \frac{dz}{z} \log(1-z) = -\psi z = -\psi\left(\frac{a}{1-y}\right) + \text{const.}$$

The constant of integration is determined by taking $y = 0$ and is found to be ψa.

Abel was the first mathematician to give a general and (almost) rigourous proof of Newton's famous binomial formula

$$(1+x)^m = 1 + mx + \frac{m(m-1)}{2}x^2 + \frac{m(m-1)(m-2)}{2\cdot 3}x^3 + \dots \tag{12}$$

He published his demonstration in the first volume of *Crelle's Journal* (1826, *Recherches sur la série* $1 + mx + \frac{m(m-1)}{2}x^2 + \frac{m(m-1)(m-2)}{2\cdot 3}x^3 + \dots$, *Œuvres*, t. I, p. 218–250). He uses an idea of Euler, already exploited by Lagrange and Cauchy: writing $\varphi(m)$ the second member of (12), one proves that

$$\varphi(m+n) = \varphi(m)\varphi(n), \tag{13}$$

so that $\varphi(m) = A^m = (1+x)^m$ for m rational as was observed by Euler. Lagrange extended this proof to every value of m admitting that φ is an analytic function of m. Cauchy used an analogous strategy for m real and $|x| < 1$ using the continuity of φ, for which his proof was unfortunately incomplete. Abel considers the most general case, with x and m complex, with $|x| < 1$ or $|x| = 1$ and $\operatorname{Re} m > -1$ (if $x = -1$, one needs $\operatorname{Re} m > 0$).

For $m = k + k'i$, $\varphi(m) = f(k,k')(\cos\psi(k,k') + i\sin\psi(k,k'))$, with f, ψ continuous functions of k, k' real. The continuity is almost established by Abel in his theorem V, but this theorem is not entirely correct. The concept of uniform convergence did not exist at that time and it was not easy to give a general theorem for the continuity of the sum of a series of continuous functions. The functional equation (13) becomes

$$\begin{aligned} f(k+\ell, k'+\ell') &= f(k,k')f(\ell,\ell'); \\ \psi(k+\ell, k'+\ell') &= 2m\pi + \psi(k,k') + \psi(\ell,\ell'), \end{aligned} \tag{14}$$

where m is an integer, which must be constant because of the continuity of ψ. In a first step, Abel treats the functional equation for ψ; putting $\theta k = \psi(k, k'+\ell') = 2m\pi + \psi(k,k') + \psi(0,\ell')$ he gets

$$\theta k + \theta\ell = a + \theta(k+\ell), \tag{15}$$

with $a = 2m\pi + \psi(0,k') + \psi(0,\ell')$, whence

$$\theta k = ck + a, \tag{16}$$

where c is a function of k', ℓ'. Indeed, taking $\ell = k, 2k, \dots, \rho k$ in (15) and adding the results, Abel gets $\rho\theta k = (\rho - 1)a + \theta(\rho k)$ and $\theta\rho = \rho(\theta(1) - a) + a$ for $k = 1$,

ρ a natural integer; then, for $k = \frac{\mu}{\rho}$ ($\mu, \rho \in \mathbf{N}, \rho \neq 0$), $\rho\theta\left(\frac{\mu}{\rho}\right) = (\rho - 1)a + \theta\mu$ and $\theta\left(\frac{\mu}{\rho}\right) = c\frac{\mu}{\rho} + a$, with $c = \theta(1) - a$. This formula is extended to the negative values of k using $\theta(-k) = 2a - \theta k$ and, by continuity, to every real value of k. So

$$\psi(k, k' + \ell') = ck + 2m\pi + \psi(0, k') + \psi(0, \ell'), \tag{17}$$

where $c = \theta(k', \ell')$, a function of k' and ℓ'. For $k = 0$, this gives

$$\psi(0, k' + \ell') = 2m\pi + \psi(0, k') + \psi(0, \ell'),$$

a functional equation which may be treated as (15) and which has the solution

$$\psi(0, k') = \beta' k' - 2m\pi,$$

with an arbitrary constant β'; then (17) becomes

$$\psi(k, k' + \ell') = \theta(k', \ell') \cdot k + \beta'(k' + \ell') - 2m\pi,$$

also equal to $2m\pi + \psi(k, k') + \psi(0, \ell') = \psi(k, k') + \beta'\ell'$ by (14), so that $\psi(k, k') = Fk' \cdot k + \beta' k' - 2m\pi$, with $Fk' = \theta(k', \ell')$ independent of ℓ' and $F(k' + \ell') = Fk' = F(0) = \beta$ a constant. Finally

$$\psi(k, k') = \beta k + \beta' k' - 2m\pi. \tag{18}$$

To treat the functional equation (14) for f, Abel writes $f(k, k') = e^{F(k,k')}$ and $F(k + \ell, k' + \ell') = F(k, k') + F(\ell, \ell')$, a functional equation analog to that for ψ with $m = 0$, so its solution is of the form $F(k, k') = \delta k + \delta' k'$, with two arbitrary constant δ, δ'. Finally

$$\varphi(k + k'i) = e^{\delta k + \delta' k'}(\cos(\beta k + \beta' k') + i \sin(\beta k + \beta' k')) \tag{19}$$

and it remains to determine the constants β, β', δ and δ'.

For $k = 1$ and $k' = 0$, $\varphi(1) = 1 + x = 1 + \alpha\cos\phi + i\alpha\sin\phi$, where $\alpha = |x| < 1$ and $\phi = \arg x$; this gives $e^{\delta}\cos\beta = 1 + \alpha\cos\phi$ and $e^{\delta}\sin\beta = \alpha\sin\phi$, so that

$$e^{\delta} = (1 + 2\alpha\cos\phi + \alpha^2)^{\frac{1}{2}} \quad \text{and} \quad \tan\beta = \frac{\alpha\sin\phi}{1 + \alpha\cos\phi}, \beta = s + \mu\pi, \tag{20}$$

with $-\frac{\pi}{2} \leq s \leq \frac{\pi}{2}$ and $\mu \in \mathbf{Z}$. Now, for $k' = 0$ and any k, let $p = f\alpha$ and $q = \theta\alpha$ designate the real and the imaginary part of the series $\varphi(k)$, which are continuous functions of α after Abel's theorem IV (which is correct); one has

$$f\alpha = e^{\delta k}\cos ks \cos k\mu\pi - e^{\delta k}\sin ks \sin k\mu\pi,$$
$$\theta\alpha = e^{\delta k}\sin ks \cos k\mu\pi + \varepsilon^{\delta\kappa}\cos ks \sin k\mu\pi$$

and $\cos k\mu\pi = e^{-\delta k}(f\alpha \cdot \cos ks + \theta\alpha \cdot \sin ks)$, $\sin k\mu\pi = e^{-\delta k}(\theta\alpha \cdot \cos ks - f\alpha \cdot \sin ks)$, independent of α by continuity. For $\alpha = 0$, $e^{\delta} = 1$ and $s = 0$ after (19) whereas $f\alpha = 1$ and $\theta\alpha = 0$, so $k\mu\pi = 0$ and

$$f\alpha = (1+2\alpha\cos\phi+\alpha^2)^{\frac{k}{2}}\cos ks, \quad \theta\alpha = (1+2\alpha\cos\phi+\alpha^2)^{\frac{k}{2}}\sin ks; \tag{21}$$

this is Cauchy's result for $f\alpha + i\theta\alpha = |1+x|^k(\cos ks + i\sin ks) = (1+x)^k$.

Abel now considers the case in which $m = in$ is purely imaginary; then the series (12) is convergent for any value of n by d'Alembert's rule (which is Abel's theorem II) and Abel states its continuity as a function of n as a consequence of his theorem V. He writes the real and imaginary parts of the series in the form

$$\begin{aligned} p &= 1 + \lambda_1\alpha\cos\theta_1 + \ldots + \lambda_\mu\alpha^\mu\cos\theta_\mu + \ldots \\ \text{and}\quad q &= \lambda_1\alpha\sin\theta_1 + \ldots + \lambda_\mu\alpha^\mu\sin\theta_\mu + \ldots, \end{aligned}$$

where $\lambda_\mu = \delta_1\delta_2\ldots\delta_\mu$, $\theta_\mu = \mu\phi + \gamma_1 + \gamma_2 + \ldots + \gamma_\mu$ and

$$\frac{ni - \mu + 1}{\mu} = \delta_\mu(\cos\gamma_\mu + i\sin\gamma_\mu).$$

From (19) he knows that $p = e^{\delta' n}\cos\beta' n$ and $q = e^{\delta' n}\sin\beta' n$; in order to determine δ' resp. β', he takes the limits of $\frac{e^{\delta' n}\cos\beta' n - 1}{n}$ resp. $\frac{e^{\delta' n}\sin\beta' n}{n}$ for $n = 0$. As $\delta_\mu \to \frac{\mu-1}{\mu}$ and $\gamma_\mu \to \pi$ ($\mu \geq 2$; for $\mu = 1$, $\gamma_1 = \frac{\pi}{2}$), he gets $\frac{\lambda_\mu}{n} \to \frac{1}{\mu}$ and $\gamma_\mu \to \mu(\phi+\pi) - \frac{\pi}{2}$ so

$$\begin{aligned} \beta' &= \alpha\cos\phi - \frac{1}{2}\alpha^2\cos 2\phi + \frac{1}{3}a\alpha^3\cos 3\phi - \ldots, \\ \delta' &= -\alpha\sin\phi + \frac{1}{2}\alpha^2\sin 2\phi - \frac{1}{3}\alpha^3\sin 3\phi + \ldots \end{aligned}$$

Now, computing in the same manner the limits, for $k = 0$, of $\frac{f\alpha - 1}{k}$ and $\frac{\theta\alpha}{k}$, one gets from (21)

$$\begin{aligned} \delta &= \alpha\cos\phi - \frac{1}{2}\alpha^2\cos 2\phi + \frac{1}{3}\alpha^3\cos 3\phi - \ldots \\ \text{and}\quad \beta &= \alpha\sin\phi - \frac{1}{2}\alpha^2\sin 2\phi + \frac{1}{3}\alpha^3\sin 3\phi - \ldots, \end{aligned} \tag{22}$$

so that $\beta' = \delta$ and $\delta' = -\beta$. The sum (19) of the series (12) for $m = k + k'i$ is

$$e^{\delta k - \beta k'}(\cos(\beta k + \delta k') + i\sin(\beta k + \delta k'))$$

with β and δ as in (20). Let us interpret Abel's result: writing $\delta + i\beta = \log(1+x)$, one gets

$$m\log(1+x) = (k+ik')(\delta+i\beta) = k\delta - k'\beta + i(k\beta + k'\delta),$$

so that $\varphi(m) = (1+x)^m$.

Comparing (20) and (22), Abel gets

$$\frac{1}{2}\log(1+2\alpha\cos\phi+\alpha^2) = \alpha\cos\phi - \frac{1}{2}\alpha^2\cos 2\phi + \frac{1}{3}\alpha^3\cos 3\phi - \ldots$$

and

$$\arctan \frac{\alpha \sin \phi}{1 + \alpha \cos \phi} = \alpha \sin \phi - \frac{1}{2}\alpha^2 \sin 2\phi + \frac{1}{3}\alpha^3 \sin 3\phi - \dots ; \qquad (23)$$

by making α tend toward ± 1, $\frac{1}{2}\log(2 \pm 2\cos\phi) = \pm\cos\phi - \frac{1}{2}\cos 2\phi \pm \frac{1}{3}\cos 3\phi - \dots$ and $\frac{1}{2}\phi = \sin\phi - \frac{1}{2}\sin 2\phi + \frac{1}{3}\sin 3\phi - \dots$ for $-\pi < \phi < \pi$. If $\phi = \frac{\pi}{2}$ and $-1 \leq \alpha \leq 1$ in (23), one gets Gregory's series $\arctan \alpha = \alpha - \frac{1}{3}\alpha^3 + \frac{1}{5}\alpha^5 - \dots$

Taking $x = i \tan \phi$ and m real in the binomial series, Abel's finds

$$\cos m\phi = (\cos\phi)^m \left(1 - \frac{m(m-1)}{1 \cdot 2}(\tan\phi)^2 + \frac{m(m-1)(m-2)(m-3)}{1 \cdot 2 \cdot 3 \cdot 4}(\tan\phi)^4 - \dots\right),$$

$$\sin m\phi = (\cos\phi)^m \left(m \tan\phi - \frac{m(m-1)(m-2)}{1 \cdot 2 \cdot 3}(\tan\phi)^3 + \dots\right)$$

for $-\frac{\pi}{4} \leq \phi \leq \frac{\pi}{4}$ (for $\phi = \pm\frac{\pi}{4}$, m must be > -1).

Now, taking $|x| = 1$ and $m > -1$, he finds as the real part of

$$(1 + x)^m (\cos\alpha - i \sin\alpha) :$$

$$\begin{aligned} &\cos\alpha + \frac{m}{1}\cos(\alpha - \phi) + \frac{m(m-1)}{1 \cdot 2}\cos(\alpha - 2\phi) + \dots \\ &= (2 + 2\cos\phi)^{\frac{m}{2}} \cos\left(\alpha - \frac{m\phi}{2} + m\rho\pi\right) \end{aligned}$$

where ρ is an integer such that $|\phi - 2\rho\pi| \leq \pi$ (with the restriction $m > 0$ in case of equality). The substitutions $\phi = 2x$ and $\alpha = mx, mx + \frac{\pi}{2}, m\left(x + \frac{\pi}{2}\right)$ or $m\left(x + \frac{\pi}{2}\right) - \frac{\pi}{2}$ give Abel various formulae, for instance

$$(2\cos x)^m \cos 2m\rho\pi = \cos mx + \frac{m}{1}\cos(m-2)x + \frac{m(m-1)}{1 \cdot 2}\cos(m-4)x + \dots$$

$$(2\cos x)^m \sin 2m\rho\pi = \sin mx + \frac{m}{1}\sin(m-2)x + \frac{m(m-1)}{1 \cdot 2}\sin(m-4)x + \dots$$

for $2\rho\pi - \frac{\pi}{2} \leq x \leq 2\rho\pi + \frac{\pi}{2}$. Abel was the first to prove rigourously such formulae for m non integer; in a letter to his friend Holmboe (16 January 1826, *Œuvres*, t. II, p. 256), he states his result and alludes to the unsuccessful attempts of Poisson, Poinsot, Plana and Crelle.

Other examples of functional equations in Abel's work may be mentioned, as the famous Abel theorem (see §5), which may be interpreted in this way. In a letter to Crelle (9 August 1826, *Œuvres*, t. II, p. 267), Abel states his theorem for genus 2 in a very explicit manner: he considers the hyperelliptic integral $\varphi(x) = \int \frac{(\alpha+\beta x)dx}{\sqrt{P(x)}}$ where P is a polynomial of degree 6; then Abel's theorem is the functional equation $\varphi(x_1) + \varphi(x_2) + \varphi(x_3) = C - (\varphi(y_1) + \varphi(y_2))$, where x_1, x_2 and x_3 are independant variables, C is a constant and y_1, y_2 are the roots of the equation

$$y^2 - \left(\frac{c_2^2 + 2c_1 - a_4}{2c_2 - a_5} - x_1 - x_2 - x_3\right) y + \frac{\frac{c^2-a}{x_1x_2x_3}}{2c_2 - a_5} = 0,$$

with $P(x) = a + a_1x + a_2x^2 + a_3x^3 + a_4x^4 + a_5x^5 + x^6$ and $c + c_1x_j + c_2x_j^2 + x_j^3 = \sqrt{P(x_j)}$ for $j = 1, 2, 3$. Abel says that this functional equation completely characterises the function φ.

Abel discovered how to express the elliptic functions as quotients of two entire functions of the type of Weierstrass' σ-function; there is an allusion to that in the introduction to his *Précis d'une théorie des fonctions elliptiques*, published in the fourth volume of *Crelle's Journal* (1829, *Œuvres*, t. I, p. 527–528) and in a letter to Legendre (25 November 1828, *Œuvres*, t. II, p. 274–275). The elliptic function $\lambda(\theta)$ is defined by

$$\theta = \int_0^{\lambda\theta} \frac{dx}{\Delta(x,c)}, \qquad \text{where} \quad \Delta(x,c) = \pm\sqrt{(1-x^2)(1-c^2x^2)},$$

and $\lambda\theta = \frac{\varphi\theta}{f\theta}$ where the entire functions φ and f are solutions of the system of functional equations $\varphi(\theta' + \theta) \cdot \varphi(\theta' - \theta) = (\varphi\theta \cdot f\theta')^2 - (\varphi\theta' \cdot f\theta)^2$, $f(\theta' + \theta) \cdot f(\theta' - \theta) = (f\theta \cdot f\theta')^2 - c^2(\varphi\theta \cdot \varphi\theta')^2$. This system is partially solved in a notebook of 1828, with x and y in place of θ' and θ; supposing φ odd and f even and taking the second derivative with respect to x at $x = 0$, Abel finds the equations $f''y + fy - (f'y)^2 = a(fy)^2 - c^2b(\varphi y)^2$ and $-\varphi''y + \varphi y + (\varphi' y)^2 = b(fy)^2 - a(\varphi y)^2$ with $a = f(0) \cdot f''(0)$ and $b = (\varphi' 0)^2$. If it is supposed that $a = 0$ and $b = 1$, this reduces to $(f'y)^2 - f''y \cdot fy = c^2(\varphi y)^2$, $(\varphi' y)^2 - \varphi''y \cdot \varphi y = (fy)^2$. Again differentiating four times at $x = 0$, Abel obtains the derivatives of f up to the 4th order and φ, but his computation, aimed to find differential equations for f and φ, stops here.

Two posthumous papers by Abel are devoted to differential equations of Riccati type. In the first one, *Sur l'équation différentielle* $dy + (p + qy + ry^2)dx = 0$, *où* p, q *et* r *sont des fonctions de* x *seul* (*Œuvres*, t. II, p. 19–25), Abel shows how to transform this equation in another one of the form $dy + (P + Qy^2)dx = 0$. Two methods are proposed. The first one, by putting $y = z + r'$ with $r' = -\frac{q}{2r}$, which gives $dz + (P + Qz^2)dx = 0$ with $P = p - \frac{q^2}{4r} - \frac{dq}{dx}\frac{1}{2r} + \frac{dr}{dx}\frac{q}{2r^2}$ and $Q = r$.

The second one, which is classical, by putting $y = zr'$ with $r' = e^{-\int q dx}$; this gives $P = pe^{\int q dx}$ and $Q = re^{-\int q dx}$. Abel observes that when $pe^{\int q dx} = are^{-\int q dx}$ or $e^{\int q dx} = \sqrt{\frac{ar}{p}}$, the equation, which is written $dy + \left(p + \frac{1}{2}\left(\frac{dr}{rdx} - \frac{dp}{pdx}\right) y + ry^2\right) dx = 0$, may be integrated in finite terms, giving $y = -\sqrt{\frac{p}{r}} \tan\left(\int \sqrt{rp}dx\right)$. For example, the equation $dy + \left(\frac{1}{x} - \frac{y^2}{x}\right) dx = 0$ has a solution of the form $y = \frac{1-cx^2}{1+cx^2}$ and the equation $dy + \left(x^m + \frac{1}{2}(n-m)\frac{y}{x} + x^ny^2\right) dx = 0$ has a solution of the form $y = -x^{\frac{m-n}{2}} \tan\left(c + \frac{2}{m+n+2} x^{\frac{1}{2}(m+n+2)}\right)$; in the case in which $n = -m - 2$, this solution becomes $y = -x^{m+1}\tan(\log k'x)$. Another easy case of integration is given by the relations $\frac{p}{c} = \frac{q}{2a} = r$; in this case $y = -a + \sqrt{a^2 - c}\frac{1+e^{-\frac{2}{c}\sqrt{a^2-c}\int p dx}}{1-e^{-\frac{2}{c}\sqrt{a^2-c}\int p dx}}$.

Abel explains how to solve the equation when a particular solution y' is known. Putting $y = z + y'$, he finds $dz + ((q+2ry')z + rz^2)dx = 0$ and $y = y' + \frac{e^{-\int(q+2ry')dx}}{\int e^{-\int(q+2ry')}rdx}$. For example the equation $dy + \left(\frac{1}{x^2} + \frac{ay}{x} + cy^2\right)dx = 0$ has the particular solution $y' = \left(\frac{1-a}{2c} \pm \sqrt{\left(\frac{1-a}{2c}\right)^2 - \frac{1}{c}}\right)\frac{1}{x}$ and this leads to the general solution

$$y = \left(\frac{1-a}{2c} \pm \sqrt{\left(\frac{1-a}{2c}\right)^2 - \frac{1}{c}}\right)\frac{1}{x} + \frac{kx^{-\left(1\pm\sqrt{(1-a)^2-4c}\right)}}{C \pm \frac{ck}{\sqrt{(1-a)^2-4c}}x^{\mp\sqrt{(1-a)^2-4c}}}.$$

Other cases of integration are found by Euler's method of integrating factor: the expression $zdy + z(p+qy^2)dx$ is a complete differential when $\frac{\partial z}{\partial x} = \frac{\partial(z(p+qy^2))}{\partial y}$ or, if $z = e^r$, when $\frac{\partial r}{\partial x} = (p+qy^2)\frac{\partial r}{\partial y} + 2qy$. Abel tries with $r = a\log(\alpha + \beta y)$ with a constant and α, β functions of x only. He finds the conditions $a\alpha' - a\beta p = a\beta' - 2\alpha q = a\beta q + 2\beta q = 0$, where α', β' are the derivatives of α, β. Thus the equation $dy + \left(\frac{\alpha'}{\beta} - \frac{\beta'}{\alpha}y^2\right)dx = 0$ admits the integrating factor $z = \frac{1}{(\alpha+\beta y)^2}$ and the solution $y = -\frac{\alpha}{\beta} + \frac{1}{\beta^2\left(C - \int\frac{\beta'}{\alpha\beta^2}dx\right)}$.

In the second paper, Abel considers the differential equation

$$(y+s)dy + (p + qy + ry^2)dx = 0,$$

which is reduced to the form $zdz + (P + Qz)dx = 0$ by the substitution $y = \alpha + \beta z$ with $\alpha = -s$ and $\beta = e^{-\int rdx}$. One has $P = (p - qs + rs^2)e^{2\int rdx}$ and $Q = \left(q - 2rs - \frac{ds}{dx}\right)e^{\int rdx}$. If $P = 0$, this equation has the solution $z = \int\left(2rs + \frac{ds}{dx} - q\right)e^{\int rdx}dx$ so that the equation

$$(y+s)dy + (qs - rs^2 + qy + ry^2)dx = 0$$

has for solution $y = -s + e^{-\int rdx}\int\left(2rs + \frac{ds}{dx} - q\right)e^{\int rdx}dx$. When $Q = 0$, the equation in z has the solution $z = \sqrt{2\int(qs - p - rs^2)e^{2\int rdx}dx}$ and the equation

$$(y+s)dy + \left(p + \left(2rs + \frac{ds}{dx}\right)y + ry^2\right)dx = 0$$

has for solution $y = -s + e^{-\int rdx}\sqrt{2\int\left(rs^2 - p + \frac{sds}{dx}\right)e^{2\int rdx}dx}$.

In order that $z = e^r$ be an integrating factor for the equation

$$ydy + (p + qy)dx = 0,$$

we must impose $y\frac{\partial r}{\partial x} - (p+qy)\frac{\partial r}{\partial y} - q = 0$. For $r = \alpha + \beta y$, this gives the conditions $\frac{d\beta}{dx} = \frac{d\alpha}{dx} - q\beta = p\beta + q = 0$, so $\beta = -c$, $\alpha = -c\int qdx$ and $-cp + q = 0$. For $r = \alpha + \beta y + \gamma y^2$, one finds $\gamma = c$, $\beta = 2c\int qdx$, $q + 2cp\int qdx = 0$ and $\alpha = 2c\int qdx\int qdx - \int\frac{qdx}{\int qdx}$. When $q = 1$, we find that the equation $ydy + \left(\frac{1}{c(x+a)} + y\right)dx = 0$ admits the integrating factor $\frac{1}{x+a}e^{-\frac{c}{2}(x+y+a)^2}$. More generally, for $r = \alpha + \alpha_1 y + \alpha_2 y^2 + \ldots + \alpha_n y^n$, one finds $n+2$ conditions $\frac{d\alpha_n}{dx} = 0 = \frac{d\alpha_{n-1}}{dx} - nq\alpha_n = \frac{d\alpha_{n-2}}{dx} - (n-1)q\alpha_{n-1} - np\alpha_n = \ldots = \frac{d\alpha}{dx} - q\alpha_1 - 2p\alpha_2 = q + p\alpha_1 = 0$ for the $n+1$ coefficients α_k; so there is a relation between p and q. For $n = 3$, Abel finds

$$q + 6cp\int qdx\int qdx + 3cp\int pdx = 0.$$

A function $r = \frac{1}{\alpha+\beta y}$ leads to the conditions $\frac{d\beta}{dx} + \beta^2 q = \frac{d\alpha}{dx} - \beta q + 2\alpha\beta q = \alpha^2 q - \beta p = 0$ and the equation $ydy + \left(\left(\frac{C}{(\int qdx)^2} + \frac{1}{2}\right)^2 q\int qdx + qy\right)dx$ admits the integrating factor $e^{\frac{1}{\alpha+\beta y}}$ with $\beta = \frac{1}{\int qdx}$ and $\alpha = \frac{C}{(\int qdx)^2} + \frac{1}{2}$.

Another form tried by Abel is $r = a\log(\alpha + \beta y)$; he finds that $ydy - \left(\frac{a+1}{a^2}q - qy\right)dx = 0$ has the integrating factor $\left(\frac{(a+1)c}{a}\int qdx + cy\right)^a$. More generally $r = a\log(y+\alpha) + a'\log(y+\alpha')$ gives a new form of differential equation integrable by the factor e^r.

2 Integral Transforms and Definite Integrals

The second Norwegian paper of Abel, titled *Opløsning af et Par Opgaver ved Hjelp af bestemte Integraler* (1823, *Œuvres*, t. I, p. 11–27), studies in its first part the integral equation $\psi a = \int\limits_{x=0}^{x=a}\frac{ds}{(a-x)^n}$ where ψ is a given function, s an unknown function of x and $n < 1$.

In the case where $n = \frac{1}{2}$, s is interpreted as the length of a curve to be found, along which the fall of a massive point from the height a takes a time equal to ψa. Let the curve be KCA, the initial position of the falling body be the point C and its initial velocity be 0; when the falling body is in M its velocity is proportional to $\sqrt{a-x}$, where a is the total height AB and x is the height AP. So the fall along an infinitesimal arc MM' takes a time dt proportional to $-\frac{ds}{\sqrt{a-x}}$, where $s = AM$ is the curvilineal abscissa along the curve, and the total duration of the fall is proportional to the integral $\int\limits_{x=0}^{x=a}\frac{ds}{\sqrt{a-x}}$.

Abel's equation is probably the first case of an integral equation in the history of mathematics; before that, Euler had introduced in his *Institutiones Calculi Integralis* the general idea to solve a differential equation by a definite integral, for instance by

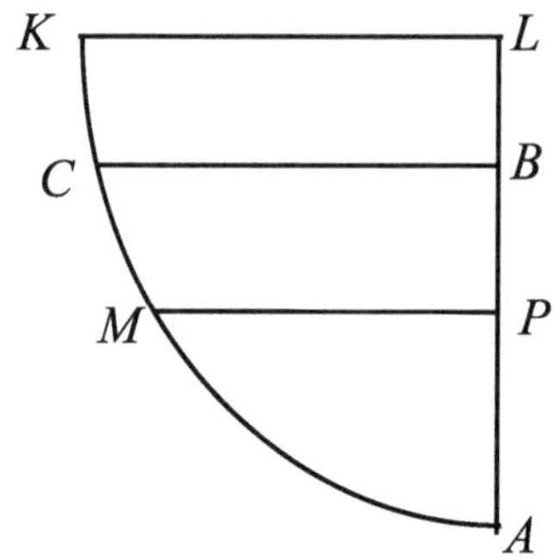

the so called Laplace transform and Fourier (1811) and Cauchy (1817) had studied the Fourier transform and its law of inversion.

Abel supposes that s has a development in power series with respect to x: $s = \sum \alpha^{(m)} x^m$; differentiating and integrating term by term, he obtains $\psi a = \sum m\alpha^{(m)} \int_0^a \frac{x^{m-1}dx}{(a-x)^n}$. One has $m \int_0^a \frac{x^{m-1}dx}{(a-x)^n} = ma^{m-n} \int_0^1 \frac{t^{m-1}dt}{(1-t)^n} = \frac{\Gamma(1-n)\Gamma(m+1)}{\Gamma(m-n+1)} a^{m-n}$, using the Eulerian function Γ, for which Abel refers to Legendre's *Exercices de Calcul intégral*; so

$$\psi a = \Gamma(1-n) \sum \alpha^{(m)} a^{m-n} \frac{\Gamma(m+1)}{\Gamma(m-n+1)}.$$

Let now $\psi a = \sum \beta^{(k)} a^k$ (ψ is implicitly supposed to be analytic); by identification, Abel gets $\alpha^{(n+k)} = \frac{\Gamma(k+1)}{\Gamma(1-n)\Gamma(n+k+1)} \beta^{(k)} = \frac{\beta^{(k)}}{\Gamma n \cdot \Gamma(1-n)} \int_0^1 \frac{t^k dt}{(1-t)^{1-n}}$, so that

$$s = \sum \alpha^{(m)} x^m = \frac{x^n}{\Gamma n \cdot \Gamma(1-n)} \int_0^1 \frac{\sum \beta^{(k)} (xt)^k dt}{(1-t)^{1-n}}$$

$$= \frac{x^n}{\Gamma n \cdot \Gamma(1-n)} \int_0^1 \frac{\psi(xt)dt}{(1-t)^{1-n}} = \frac{x^n \sin n\pi}{\pi} \int_0^1 \frac{\psi(xt)dt}{(1-t)^{1-n}}$$

and, in the particular case where $n = \frac{1}{2}$, $s = \frac{\sqrt{x}}{\pi} \int_0^1 \frac{\psi(xt)dt}{\sqrt{1-t}}$.

Abel applies this result in the case where $\psi a = ca^n$ (c constant, and the exponent n not to be confused with that of $a-x$ in the general problem, which is now $\frac{1}{2}$), in which $s = Cx^{n+\frac{1}{2}}$, with $C = \frac{c}{\pi} \int_0^1 \frac{t^n dt}{\sqrt{1-t}}$; then $dy = \sqrt{ds^2 - dx^2} = dx\sqrt{kx^{2n-1} - 1}$, where $k = \left(n + \frac{1}{2}\right)^2 C^2$, so

$$y = \int dx\sqrt{kx^{2n-1} - 1} = k' + x\sqrt{k-1}$$

in the particular case where $n = \frac{1}{2}$; in this case, the curve KCA solution of the problem is a straight line. The isochronic case, where $\psi a = c$ constant is another interesting case; here $n = 0$ and $s = C\sqrt{x}$ ($C = \frac{2c}{\pi}$), equation characterising the cycloid. This problem was initially solved by Huygens (1673).

Turning back to the general case Abel gives another interpretation of the solution as a derivative of ψ of non-integral order $-n$. Indeed, if $\psi x = \sum \alpha^{(m)} x^m$ and if k is a natural integer,

$$\frac{d^k \psi}{dx^k} = \sum \alpha^{(m)} \frac{\Gamma(m+1)}{\Gamma(m-k+1)} x^{m-k};$$

in which the right hand side is still meaningfull when k is not a natural integer, and then

$$\frac{\Gamma(m+1)}{\Gamma(m-k+1)} = \frac{1}{\Gamma(-k)} \int_0^1 \frac{t^m dt}{(1-t)^{1+k}},$$

so that the right hand side becomes $\frac{1}{x^k \Gamma(-k)} \int_0^1 \frac{\sum \alpha^{(m)} (xt)^m dt}{(1-t)^k} = \frac{1}{x^k \Gamma(-k)} \int_0^1 \frac{\psi(xt) dt}{(1-t)^k}$, whence the definition of $\frac{d^{-n}\psi}{dx^{-n}} = \frac{x^n}{\Gamma n} \int_0^1 \frac{\psi(xt)dt}{(1-t)^{1-n}}$ and the solution $s = \frac{1}{\Gamma(1-n)} \frac{d^{-n}\psi}{dx^{-n}}$ of the initial problem. The derivative of order n of $s = \varphi x$ is naturally $\frac{1}{\Gamma(1-n)} \psi x$, which means that

$$\frac{d^n \varphi}{da^n} = \frac{1}{\Gamma(1-n)} \int_0^a \frac{\varphi' x dx}{(a-x)^n} \qquad (n < 1);$$

for $n = \frac{1}{2}$, $\psi x = \sqrt{\pi} \frac{d^{\frac{1}{2}} s}{dx^{\frac{1}{2}}}$.

The idea of a derivative of non-integral order comes from Leibniz; it was based on the analogy, discovered by Leibniz, between the powers and the differentials in the celebrated formula for $d^n(xy)$, which has the same coefficient as $(x+y)^n = p^n(x+y)$ in Leibniz' notation. The general binomial formula, with exponent e non necessarily integral, suggests to Leibniz a formula for $d^e(xy)$ as an infinite series (letter to the Marquis de l'Hospital, 30 September 1695). Abel's procedure is an extension of a formula given by Euler in 1730: $\frac{d^n(z^e)}{dz^n} = z^{e-n} \frac{\int_0^1 dx(-lx)^e}{\int_0^1 dx(-lx)^{e-n}}$, where e and n are arbitrary numbers and l notes the logarithm. At Abel's time, some other authors also considered derivatives of arbitrary order, as Fourier and Cauchy, but the theory really began with Liouville in 1832 and Riemann in 1847.

At the end of this part, Abel reports that he has solved the more general integral equation $\psi a = \int \varphi(xa) fx \cdot dx$, where ψ and f are given functions and φ is unknown.

Abel published a German version of this study in *Crelle's Journal* (vol. I, 1826, *Œuvres*, t. I, p. 97–101). He finds the solution without any use of power series, starting from the Eulerian integral of the first kind $\int_0^1 \frac{y^{\alpha-1}dy}{(1-y)^n} = \frac{\Gamma\alpha\cdot\Gamma(1-n)}{\Gamma(\alpha+1-n)}$, which gives $\int_0^a \frac{z^{\alpha-1}dz}{(a-z)^n} = \frac{\Gamma\alpha\cdot\Gamma(1-n)}{\Gamma(\alpha+1-n)}a^{\alpha-n}$ and

$$\int_0^x \frac{da}{(x-a)^{1-n}}\int_0^a \frac{z^{\alpha-1}dz}{(a-z)^n} = \frac{\Gamma\alpha\cdot\Gamma(1-n)}{\Gamma(\alpha+1-n)}\int_0^x \frac{a^{\alpha-n}da}{(x-a)^{1-n}}$$

$$= \Gamma n\cdot\Gamma(1-n)\frac{\Gamma\alpha}{\Gamma(\alpha+1)}x^\alpha = \frac{x^\alpha}{\alpha}\Gamma n\cdot\Gamma(1-n).$$

Then, if $fx = \int\varphi\alpha\cdot x^\alpha d\alpha$, one has $\int_0^x \frac{da}{(x-a)^{1-n}}\int_0^a \frac{f'z.dz}{(a-z)^n} = \Gamma n\cdot\Gamma(1-n)\,fx$ and

$$fx = \frac{\sin n\pi}{\pi}\int_0^x \frac{da}{(x-a)^{1-n}}\int_0^a \frac{f'zdz}{(a-z)^n}.$$

Therefore, in the original problem $\varphi a = \int_{x=0}^{x=a} \frac{ds}{(a-x)^n}$, one has

$$\frac{\sin n\pi}{\pi}\int_0^x \frac{\varphi a da}{(x-a)^{1-n}} = \frac{\sin n\pi}{\pi}\int_0^x \frac{da}{(x-a)^{1-n}}\int_0^a \frac{ds}{(a-x)^n} = s.$$

In this paper, there is no mention of derivatives of non-integral order.

The second part of the Norwegian paper is devoted to the proof of the integral formula:

$$\varphi(x+y\sqrt{-1})+\varphi(x-y\sqrt{-1}) = \frac{2y}{\pi}\int_{-\infty}^{+\infty} e^{-v^2y^2}vdv\int_{-\infty}^{+\infty}\varphi(x+t)e^{-v^2t^2}dt,$$

giving as a particular case $\cos y = \frac{1}{\sqrt{\pi}}\int_{-\infty}^{+\infty} e^{-t^2+\frac{1}{4}\frac{y^2}{t^2}}dt$ when $\varphi t = e^t$, $x = 0$. Abel uses the developments

$$\varphi(x+y\sqrt{-1})+\varphi(x-y\sqrt{-1}) = 2\left(\varphi x - \frac{\varphi''x}{1\cdot 2}y^2 + \frac{\varphi''''x}{1\cdot 2\cdot 3\cdot 4}y^4 - \dots\right),$$

$$\varphi(x+t) = \varphi x + t\varphi'x + \frac{t^2}{1\cdot 2}\varphi''x + \frac{t^3}{1\cdot 2\cdot 3}\varphi'''x + \dots$$

and the definite integrals

$$\int_{-\infty}^{+\infty} e^{-v^2 t^2} t^{2n} dt = \frac{\Gamma\left(\frac{2n+1}{2}\right)}{v^{2n+1}}, \quad \int_{-\infty}^{+\infty} e^{-v^2 y^2} v^{-2n} dv = \Gamma\left(\frac{1-2n}{2}\right) y^{2n-1}.$$

The last two parts of the paper give summation formulae by means of definite integrals. From the development $\frac{1}{e^t-1} = e^{-t} + e^{-2t} + e^{-3t} + \dots$ and the value $\int_0^\infty e^{-kt} t^{2n-1} dt = \frac{\Gamma(2n)}{k^{2n}}$, Abel deduces $\int_0^\infty \frac{t^{2n-1}dt}{e^t-1} = \Gamma(2n)\zeta(2n)$; the Eulerian formula $\zeta(2n) = \frac{2^{2n-1}\pi^{2n}}{\Gamma(2n+1)} A_n$, where A_n is the n-th Bernoulli number, then gives $A_n = \frac{2n}{2^{2n-1}\pi^{2n}} \int_0^\infty \frac{t^{2n-1}dt}{e^t-1} = \frac{2n}{2^{2n-1}} \int_0^\infty \frac{t^{2n-1}dt}{e^{\pi t}-1}$. Using these values in the Euler–MacLaurin sum formula $\sum \varphi x = \int \varphi x.dx - \frac{1}{2}\varphi x + A_1 \frac{\varphi' x}{1\cdot 2} - A_2 \frac{\varphi''' x}{1\cdot 2\cdot 3\cdot 4} + \dots$ and Taylor series for $\varphi\left(x \pm \frac{t}{2}\sqrt{-1}\right)$, Abel finds

$$\sum \varphi x = \int \varphi x dx - \frac{1}{2}\varphi x + \int_0^\infty \frac{\varphi\left(x + \frac{t}{2}\sqrt{-1}\right) - \varphi\left(x - \frac{t}{2}\sqrt{-1}\right)}{2\sqrt{-1}} \frac{dt}{e^{\pi t} - 1}. \quad (24)$$

This formula was already published in 1820 by Plana in the Memoirs of the Turin Academy; Plana found it by the same type of formal manipulations as Abel. It was rigorously established by Schaar in 1848, using Cauchy's calculus of residues.

As particular applications of this formula, Abel gives the values of some definite integrals: for $\varphi x = e^{mx}$, $\int_0^\infty \frac{\sin mt dt}{e^{\pi t}-1} = \frac{1}{e^m-1} - m + \frac{1}{2}$, for $\varphi x = \frac{1}{x}$,

$$\int_0^\infty \frac{tdt}{\left(x^2 + \frac{1}{4}t^2\right)\left(e^{\pi t} - 1\right)} = 2\log x - \frac{1}{x} - 2\sum \frac{1}{x} + 3 + \int_0^\infty \frac{tdt}{\left(1 + \frac{1}{4}t^2\right)\left(e^{\pi t} - 1\right)}$$

and for $\varphi x = \sin ax$, $\int_0^\infty \frac{e^{at} - e^{-at}}{e^{\pi t}-1} dt = \frac{1}{a} - \cot a$.

The second Abel's summation formula is

$$\begin{aligned} &\varphi(x+1) - \varphi(x+2) + \varphi(x+3) - \varphi(x+4) + \dots \\ &= \frac{1}{2}\varphi x + 2\int_0^\infty \frac{dt}{e^{\pi t} - e^{-\pi t}} \frac{\varphi(x + t\sqrt{-1}) - \varphi(x - t\sqrt{-1})}{2\sqrt{-1}}. \end{aligned} \quad (25)$$

In order to obtain this, Abel puts *a priori* the first member equal to $\frac{1}{2}\varphi x + A_1 \varphi' x + A_2 \varphi'' x + \dots$ with unknown coefficients $A_1, A_2, \dots$; when $\varphi x = e^{cx\sqrt{-1}}$, one sees that $A_2 = A_4 = \dots = 0$ and $\frac{1}{2}\tan\frac{1}{2}c = A_1 c - A_3 c^3 + A_5 c^5 - \dots$. On the other hand, $\frac{1}{2}\tan\frac{1}{2}c = \int_0^\infty \frac{e^{ct} - e^{-ct}}{e^{\pi t} - e^{-\pi t}} dt$ after Legendre (*Exercices de Calcul Intégral*, t. II, p. 186), so the series for $e^{ct} - e^{-ct}$ gives the A_{2n+1} in the form of integrals and the reasoning ends as for the first formula. As an application, Abel takes $\varphi x = \frac{1}{x+1}$ and gets $\int\limits_0^\infty \frac{tdt}{(1+t^2)(e^{\pi t}-e^{-\pi t})} = \frac{1}{2}\log 2 - \frac{1}{4}$.

In the second volume of *Magasinet for Naturvidenskaberne* (1825), Abel published another derivation of the formula (24) and he extended it to the case of iterated sums (*Œuvres*, t. I, p. 34–39):

$$\begin{aligned}\sum^n \varphi x = {} & A_{n-1,n}\Gamma n \overset{n}{\int} \varphi x \cdot dx^n - A_{n-2,n}\Gamma(n-1) \overset{n-1}{\int} \varphi x \cdot dx^{n-1} \\ & + \ldots + (-1)^{n-1} \int \varphi x \cdot dx + (-1)^n \frac{1}{2}\varphi x \\ & +2(-1)^{n-1} \int\limits_0^\infty \frac{Pdt}{e^{2\pi t}-1} \frac{\varphi(x+t\sqrt{-1}) - \varphi(x-t\sqrt{-1})}{2\sqrt{-1}} \\ & +2(-1)^{n-1} \int\limits_0^\infty \frac{Qdt}{e^{2\pi t}-1} \frac{\varphi(x+t\sqrt{-1}) + \varphi(x-t\sqrt{-1})}{2}\end{aligned}$$

where the coefficients $A_{0,n}, A_{1,n}, \ldots, A_{n-1,n}$ are defined by the development of $p^n = \frac{1}{(e^v-1)^n}$ in the form $(-1)^{n-1}\left(A_{0,n}p + A_{1,n}\frac{dp}{dv} + A_{2,n}\frac{d^2p}{dv^2} + \ldots + A_{n-1,n}\frac{d^{n-1}p}{dv^{n-1}}\right)$ and $P = A_{0,n} - A_{2,n}t^2 + A_{4,n}t^4 - \ldots$, $Q = A_{1,n}t - A_{3,n}t^3 + A_{5,n}t^5 - \ldots$; by derivating p^n, Abel establishes recursive relations between the $A_{k,n}$: $A_{0,n+1} - A_{0,n} = 0$, $A_{1,n+1} - A_{1,n} = \frac{1}{n}A_{0,n}$, $A_{2,n+1} - A_{2,n} = \frac{1}{n}A_{1,n}, \ldots, A_{n-1,n+1} - A_{n-1,n} = \frac{1}{n}A_{n-2,n}$, $A_{n,n+1} = \frac{1}{n}A_{n-1,n}$. The proof of this formula is based on the expression of φ as a Laplace transform: $\varphi x = \int e^{vx} fv \cdot dv$, which naturally restricts the generality; it gives

$$\sum^n \varphi x = \int e^{vx} \frac{fv}{(e^v-1)^n} dv.$$

As an example, for $\varphi x = e^{ax}$ and $n = 2$, this formula gives:

$$\frac{1}{(e^a-1)^2} = \frac{1}{2} - \frac{1}{a} + \frac{1}{a^2} - 2\int\limits_0^\infty \frac{dt \cdot \sin at}{e^{2\pi t}-1} - 2\int\limits_0^\infty \frac{tdt \cdot \cos at}{e^{2\pi t}-1}.$$

Another example, with $\varphi x = \frac{1}{x^2}$ and $n = 1$, leads to

$$\frac{1}{a^2} + \frac{1}{(a+1)^2} + \frac{1}{(a+2)^2} + \ldots = \frac{1}{2a^2} + \frac{1}{a} + 4a\int\limits_0^\infty \frac{tdt}{(e^{2\pi t}-1)(a^2+t^2)^2};$$

in particular, for $a = 1$, $\frac{\pi^2}{6} = \frac{3}{2} + 4\int\limits_0^\infty \frac{tdt}{(e^{2\pi t}-1)(1+t^2)^2}$.

A posthumous paper of Abel is devoted to the study of the Laplace transform; its title is *Sur les fonctions génératrices et leurs déterminantes* (*Œuvres*, t. II, p. 67–81, mem. XI) and the study is purely formal. Abel writes an arbitrary function φ of several variables in the form:

$$\varphi(x, y, z, \dots) = \int e^{xu+yv+zp+\dots} f(u, v, p, \dots) du dv dp \dots ,$$

and he calls φ the *generating* function of f and f the *determinant* function of φ, in abbreviation $\varphi = \mathrm{fg} f$ and $f = \mathrm{D}\varphi$. The following properties of the transform are established in the case of one variable only: linearity, effect of a translation $\mathrm{D}\varphi(x+a) = e^{av}\mathrm{D}\varphi x$ and $\mathrm{fg}(e^{av}\mathrm{D}\varphi x) = \varphi(x+a)$, effect of derivations or integrations $\mathrm{D}\left(\frac{d^n\varphi x}{dx^n}\right) = v^n \cdot \mathrm{D}\varphi x$, $\mathrm{fg}(v^n\mathrm{D}\varphi x) = \frac{d^n\varphi x}{dx^n}$, $\mathrm{D}\left(\int^n \varphi x dx^n\right) = v^{-n}\mathrm{D}\varphi x$ and $\mathrm{fg}(v^{-n}\mathrm{D}\varphi x) = \int^n \varphi x dx^n$, effect of finite differences or iterated sums

$$\mathrm{D}\Delta_\alpha^n\varphi x = (e^{v\alpha}-1)^n fv,\ \mathrm{fg}((e^{v\alpha}-1)^n fv) = \Delta_\alpha^n\varphi x,\ \mathrm{D}\Sigma_\alpha^n(\varphi x) = (e^{va}-1)^{-n} fv$$
$$\text{and}\quad \mathrm{fg}((e^{v\alpha}-1)^{-n} fv) = \Sigma_\alpha^n(\varphi x).$$

Abel also states the effect of the composition of a translation, a derivation and a certain number of finite differences. More generally, if the operator δ is defined by

$$\delta(\varphi x) = A_{n,\alpha}\frac{d^n\varphi(x+\alpha)}{dx^n} + A_{n',\alpha'}\frac{d^{n'}\varphi(x+\alpha')}{dx^{n'}} + \dots , \tag{26}$$

where $A_{n,\alpha}$, $A_{n',\alpha'}$, ... are constant coefficients, one has $\mathrm{D}(\delta\varphi x) = \psi v \cdot \mathrm{D}\varphi x$ where

$$\psi v = A_{n,\alpha} v^n e^{v\alpha} + A_{n',\alpha'} v^{n'} e^{v\alpha'} + \dots ,$$

and Abel considers the composition of an arbitrary number of operators of the type of δ.

Abel clearly understood how the Laplace transform gives a symbolic calculus on the operators (26); he uses this calculus to obtain developments in series. For instance, he explains that the Taylor series for $\varphi(x+\alpha)$ amounts to the development $e^{v\alpha} = 1+v\alpha+\frac{v^2}{1\cdot 2}\alpha^2+\frac{v^3}{1\cdot 2\cdot 3}\alpha^3+\dots$ in the determinant function. A polynomial relation between the multiplicators ψ, ψ_1, ..., ψ_μ associated to operators δ, δ_1, ..., δ_μ gives an analogous relation between the operators themselves. Let us consider the operator $\delta\varphi x = \varphi(x+\alpha) + a\varphi x$; one has $\mathrm{D}\delta\varphi x = (e^{v\alpha}+a) fv$ where f is the determinant function of φ. Since

$$\begin{aligned}(a+e^{v\alpha})^n &= a^n + na^{n-1}e^{v\alpha} + \frac{n(n-1)}{2}a^{n-2}e^{2v\alpha} + \dots\\ &= e^{nv\alpha} + nae^{(n-1)v\alpha} + \frac{n(n-1)}{2}a^2e^{(n-2)v\alpha} + \dots ,\end{aligned}$$

$$\begin{aligned}\delta^n\varphi x &= a^n\varphi x + na^{n-1}\varphi(x+\alpha) + \frac{n(n-1)}{2}a^{n-2}\varphi(x+2\alpha) + \dots\\ &= \varphi(x+n\alpha)+na\varphi(x+(n-1)\alpha)+\frac{n(n-1)}{2}a^2\varphi(x+(n-2)\alpha)+\dots ;\end{aligned}$$

Abel writes down both forms, which are the same for n a natural integer, but which may be extended (under different conditions for the convergence of the series) to other values of n; he says nothing about that, but he may have envisaged this type of extension as we saw that he was interested by derivatives of non-integral order and we know that Cauchy defined pseudo-differential operators with constant coefficients using the Fourier transform (1827). For $a = -1$, $\Delta_\alpha^n \varphi x = \varphi(x+n\alpha) - n\varphi(x+(n-1)\alpha) + \frac{n(n-1)}{2}\varphi(x+(n-2)\alpha) - \dots$

Now let $\delta_1 \varphi x = \varphi(x+\alpha_1) + a_1\varphi x$, so that $\mathrm{D}\delta_1^n\varphi x = y^n fv$ with $y = e^{v\alpha_1} + a_1$; if $z = e^{v\alpha} + a$ then $y = a_1 + (z-a)^{\frac{\alpha_1}{\alpha}}$ and it is possible to get a development $y^n = \sum A_m z^m$. Therefore $\delta_1^n \varphi x = \sum A_m \delta^m \varphi x$. In the case where $\alpha_1 = \alpha$,

$$\begin{aligned}\delta_1^n \varphi x &= (a_1 - a)^n \varphi x + n(a_1-a)^{n-1}\delta\varphi x + \frac{n(n-1)}{2}(a_1-a)^{n-2}\delta^2\varphi x + \dots\\ &= \delta^n\varphi x + n(a_1 - a)\delta^{n-1}\varphi x + \frac{n(n-1)}{2}(a_1-a)^2\delta^{n-2}\varphi x + \dots\end{aligned}$$

For $a_1 = 0$, $\varphi(x+n\alpha) = \delta^n\varphi x - na\delta^{n-1}\varphi x + \frac{n(n-1)}{2}a^2\delta^{n-2}\varphi x + \dots$ and if moreover $a = -1$, $\varphi(x+n\alpha) = \Delta_\alpha^n \varphi x + n\Delta_\alpha^{n-1}\varphi x + \frac{n(n-1)}{2}\Delta_\alpha^{n-2}\varphi x + \dots$, a formula given by Euler (1755).

When $\delta\varphi x = \varphi(x+\alpha) - a\varphi x$ and $\delta_1\varphi x = c\varphi x + k\frac{d\varphi x}{dx}$, $\mathrm{D}\delta\varphi x = (e^{v\alpha} - a) fv = z fv$ and $\mathrm{D}\delta_1^n \varphi x = (c+kv)^n fv = y^n fv$; as $y = c + \frac{k}{\alpha}\log(z+a) = c + \frac{k}{\alpha}\log a + \frac{k}{\alpha}\left(\frac{z}{a} - \frac{1}{2}\frac{z^2}{a^2} + \frac{1}{3}\frac{z^3}{a^3} - \dots\right)$, one may write a development $y^n = \sum A_m z^m$, which gives $\delta_1^n\varphi x = \sum A_m \delta^m \varphi x$. For example, if $c = 0$, $a = k = 1$ and $n = 1$, $\frac{d\varphi x}{dx} = \frac{1}{\alpha}\left(\Delta\varphi x - \frac{1}{2}\Delta^2\varphi x + \frac{1}{3}\Delta^3\varphi x - \dots\right)$, a formula given by Lagrange (1772). Starting from a formula of Legendre:

$$b^v = 1 + lb\cdot vc^v + lb(lb - 2lc)\frac{(vc^v)^2}{2} + lb(lb-3lc)^2\frac{(vc^v)^3}{2\cdot 3} + \dots,$$

in which he makes $b = e^\alpha$ and $c = e^\beta$, Abel obtains in the same way

$$\begin{aligned}\varphi(x+\alpha) = \varphi x + \alpha\frac{d\varphi(x+\beta)}{dx} + \frac{\alpha(\alpha-2\beta)}{2}\cdot\frac{d^2\varphi(x+2\beta)}{dx^2}\\ + \frac{\alpha(\alpha-3\beta)^2}{2\cdot 3}\cdot\frac{d^3\varphi(x+3\beta)}{dx^3} + \dots\end{aligned} \tag{27}$$

and, in particular, $\varphi x = \varphi(0) + x\varphi'(\beta) + \frac{x(x-2\beta)}{2}\varphi''(2\beta) + \frac{x(x-3\beta)^2}{2\cdot 3}\varphi'''(3\beta) + \dots$ Abel published the special case of (27) in which $\varphi x = x^m$, m a natural integer, in the first issue of *Crelle's Journal* (*Œuvres*, t. I, p. 102–103); there, he proves the formula by induction on m and he observes that, when $\beta = 0$, the result reduces to the binomial formula. Another special case given in the posthumous memoir is that in which $\varphi x = \log x$; then

$$\log(x+\alpha) = \log x + \frac{\alpha}{x+\beta} + \frac{1}{2}\cdot\frac{\alpha}{x+2\beta}\cdot\frac{2\beta-\alpha}{x+2\beta} + \frac{1}{3}\cdot\frac{\alpha}{x+3\beta}\cdot\left(\frac{3\beta-\alpha}{x+3\beta}\right)^2 + \dots$$

and in particular $\log(1+\alpha) = \frac{\alpha}{1+\beta} + \frac{1}{2}\cdot\frac{\alpha}{1+2\beta}\cdot\left(1-\frac{1+\alpha}{1+3\beta}\right) + \frac{1}{3}\cdot\frac{\alpha}{1+3\beta}\cdot\left(1-\frac{1+\alpha}{1+3\beta}\right)^2 + \ldots$ When $\alpha = 3\beta$, this reduces to $\log(1+2\beta) = \frac{2\beta}{1+\beta} + \frac{2}{3}\cdot\frac{\beta^3}{(1+3\beta)^3} + \frac{1}{4}\cdot\frac{2\cdot 2^3\cdot\beta^4}{(1+4\beta)^4} + \ldots$; for example, $\log 3 = 1 + \sum_{n\geq 3} \frac{2}{n}\frac{1}{n+1}\left(\frac{n-2}{n+1}\right)^{n-1}$.

Abel also considers the developments of $\Delta_\alpha^n \varphi x$, $\frac{d^n \varphi x}{dx^n}$ and $\frac{d^n(e^x\varphi x)}{dx^n}$ in power series with respect to n; they are respectively obtained from the developments of

$$(e^{v\alpha}-1)^n = \exp(n\log(e^{v\alpha}-1)), \quad v^n = e^{n\log v}, \quad \text{and} \quad (1+v)^n = e^{n\log(1+v)}.$$

The coefficients respectively contain the powers of $\log(e^{v\alpha}-1)$, $\log v$ and $\log(1+v)$, so we must identify the operators δ respectively defined by

$$\delta\varphi x = \mathrm{fg}(\log(e^{v\alpha}-1)fv), \quad \log v \cdot fv \quad \text{and} \quad (1+v)fv;$$

these operators are respectively $\delta\varphi x = \alpha\varphi' x + \int d\alpha \sum_\alpha \varphi' x$, $\delta_1\varphi x - \frac{1}{2}\delta_1^2\varphi x + \frac{1}{3}\delta_1^3\varphi x - \ldots$, where $\delta_1\varphi x = \varphi' x - \varphi x$, and $\varphi' x - \frac{1}{2}\varphi'' x + \frac{1}{3}\varphi''' x - \ldots$

In the continuation of the paper, Abel expresses this last operator in the integral form

$$\delta\varphi x = \int_\infty^0 \frac{e^{-t}dt}{t}(\varphi(x-t)-\varphi x),$$

which is obtained in the following manner: the equality

$$\begin{aligned}\int_a^{a'} e^{(1-\alpha v)t}dt &= (e^a e^{-a\alpha v} - e^{a'}e^{-a'\alpha v})\frac{1}{1-\alpha v}\\ &= e^a(e^{-a\alpha v} + \alpha v e^{-a\alpha v} + \alpha^2 v^2 e^{-a\alpha v} + \ldots)\\ &\quad - e^{a'}(e^{-a'\alpha v} + \alpha v e^{-a'\alpha v} + \alpha^2 v^2 e^{-a'\alpha v} + \ldots)\end{aligned}$$

leads to

$$\begin{aligned}&\int_a^{a'} e^t\varphi(x-\alpha t)dt =\\ &\quad e^a(\varphi(x-\alpha a) + \alpha\varphi'(x-\alpha a) + \alpha^2\varphi''(x-\alpha a) + \alpha^3\varphi'''(x-\alpha a) + \ldots)\\ &\quad - e^{a'}(\varphi(x-\alpha a') + \alpha\varphi'(x-\alpha a') + \alpha^2\varphi''(x-\alpha a') + \alpha^3\varphi'''(x-\alpha a') + \ldots),\end{aligned}$$

from which Abel deduces $\varphi' x - \alpha\varphi'' x + \alpha^2\varphi''' x - \alpha^3\varphi'''' x + \ldots = \int_{-\infty}^0 e^t\varphi'(x+\alpha t)dt$ and, integrating with respect to α, $\alpha\varphi' x - \frac{1}{2}\alpha^2\varphi'' x + \frac{1}{3}\alpha^3\varphi''' x - \frac{1}{4}\alpha^4\varphi'''' x + \ldots = \int_{-\infty}^0 \frac{e^t dt}{t}(\varphi(x+\alpha t) - \varphi x)$.

Other classical relations between functions give Abel relations between operators. So the Fourier series $\frac{1}{2} = \cos\alpha v - \cos 2\alpha v + \cos 3\alpha v - \ldots$ leads to

$$\varphi x = \varphi(x+\alpha) + \varphi(x-\alpha) - \varphi(x+2\alpha) - \varphi(x-2\alpha) + \varphi(x+3\alpha) + \varphi(x-3\alpha) \\ -\varphi(x+4\alpha) - \varphi(x-4\alpha) + \ldots$$

and the formula $(e^{v\alpha}-1)^{-1} - (v\alpha)^{-1} + \frac{1}{2} = 2\int_0^\infty \frac{dt\cdot\sin(v\alpha t)}{e^{2\pi t}-1}$ leads to

$$\sum\nolimits_\alpha \varphi x - \frac{1}{\alpha}\int \varphi x dx + \frac{1}{2}\varphi x = 2\int_0^\infty \frac{dt}{e^{2\pi t}-1}\cdot\frac{\varphi(x+\alpha t\sqrt{-1}) - \varphi(x-\alpha t\sqrt{-1})}{2\sqrt{-1}},$$

which is formula (24). From the formula $\int_0^\infty \frac{dt\cdot\cos(\alpha v t)}{1+t^2} = \frac{\pi}{2}e^{-\alpha v}$ given by Legendre (*Exercices de Calcul intégral*, t. II, p. 176), Abel deduces

$$\int_0^\infty \frac{dt}{1+t^2}\cdot\frac{\varphi(x+\alpha t\sqrt{-1}) + \varphi(x-\alpha t\sqrt{-1})}{2} = \frac{\pi}{2}\varphi(x\pm\alpha), \tag{28}$$

for instance $\int_0^\infty \frac{dt}{(1+t^2)(\alpha^2t^2+x^2)} = \frac{\pi}{2}\cdot\frac{1}{x(x\pm\alpha)}$ for $\varphi x = \frac{1}{x}$ (where it is easy to verify that $\pm$ must be taken as $+$); when $\varphi x = \frac{1}{x^n}$, $\frac{\varphi(x+\alpha t\sqrt{-1})+\varphi(x-\alpha t\sqrt{-1})}{2} = z^{-n}\cos n\phi$, where $z = \sqrt{x^2+\alpha^2t^2}$ and $\phi = \arctan\frac{\alpha t}{x}$, so that $\int_0^\infty \frac{dt}{1+t^2}\cdot\frac{\cos\left(n\arctan\frac{\alpha t}{x}\right)}{(x^2+\alpha^2t^2)^{\frac{n}{2}}} = \frac{\pi}{2}\cdot\frac{1}{(x+\alpha)^n}$ or

$$\frac{\pi}{2}\cdot\frac{x^{n-1}}{\alpha(x+\alpha)^n} = \int_0^{\frac{\pi}{2}} \frac{(\cos\phi)^n\cos n\phi d\phi}{(x\sin\phi)^2 + (\alpha\cos\phi)^2}, \tag{29}$$

which reduces to

$$\frac{\pi}{2^{n+1}} = \int_0^{\frac{\pi}{2}} (\cos\phi)^n \cos n\phi d\phi \tag{30}$$

when $\alpha = x$.

From the integrals $\int_0^\infty \frac{dt\cdot\sin at}{t(1+t^2)} = \frac{\pi}{2}(1-e^{-a})$ and $\int_0^\infty \frac{tdt\cdot\sin at}{1+t^2} = \frac{\pi}{2}e^{-a}$ also given by Legendre, Abel deduces $\frac{\pi}{2}(\varphi x - \varphi(x\pm\alpha)) = \int_0^\infty \frac{dt}{t(1+t^2)}\cdot\frac{\varphi(x+\alpha t\sqrt{-1})-\varphi(x-\alpha t\sqrt{-1})}{2\sqrt{-1}}$, $\frac{\pi}{2}\varphi(x\pm\alpha) = \int_0^\infty \frac{tdt}{1+t^2}\cdot\frac{\varphi(x+\alpha t\sqrt{-1})-\varphi(x-\alpha t\sqrt{-1})}{2\sqrt{-1}}$ and $\frac{\pi}{2}\varphi x = \int_0^\infty \frac{dt}{t}\cdot\frac{\varphi(x+\alpha t\sqrt{-1})-\varphi(x-\alpha t\sqrt{-1})}{2\sqrt{-1}}$; for $\varphi x = \frac{1}{x^n}$ this gives $\int_0^{\frac{\pi}{2}} \frac{d\phi}{\sin\phi}(\cos\phi)^{n-1}\sin n\phi = \frac{\pi}{2}$ by putting $t = x\tan\phi$.

In an addition to this paper (*Sur quelques intégrales définies*, *Œuvres*, t. II, p. 82–86, mem. XII), Abel develops $(\cos\phi)^n\cos n\phi$ and $\frac{x^n}{(x+\alpha)^n}$ in power series with respect to n and, comparing the coefficients of the powers of n in (29), he gets the values of some definite integrals:

$$\frac{\pi}{2}\cdot\frac{1}{x\alpha}=\int_0^{\frac{\pi}{2}}\frac{d\phi}{x^2\sin^2\phi+\alpha^2\cos^2\phi},\ \frac{\pi}{2}\cdot\frac{1}{x\alpha}\log\frac{x}{x+\alpha}=\int_0^{\frac{\pi}{2}}\frac{\log\cos\phi d\phi}{x^2\sin^2\phi+\alpha^2\cos^2\phi},$$

$$\frac{\pi}{2}\cdot\frac{1}{x\alpha}\left(\log\frac{x}{x+\alpha}\right)^2=\int_0^{\frac{\pi}{2}}\frac{((\log\cos\phi)^2-\phi^2)d\phi}{x^2\sin^2\phi+\alpha^2\cos^2\phi}.$$

Putting $\varphi x=(\log x)^n$ and $\frac{\alpha t}{x}=\tan\phi$ in (28), he gets

$$\int_0^{\frac{\pi}{2}}\frac{d\phi}{x^2\sin^2\phi+\alpha^2\cos^2\phi}\cdot\frac{\left(\log\frac{x}{\cos\phi}+\phi\sqrt{-1}\right)^n+\left(\log\frac{x}{\cos\phi}-\phi\sqrt{-1}\right)^n}{2}$$
$$=\frac{\pi}{2x\alpha}(\log(x+\alpha))^n$$

and $\int_0^{\frac{\pi}{2}}d\phi\left(\left(\log\frac{x}{\cos\phi}+\phi\sqrt{-1}\right)^n+\left(\log\frac{x}{\cos\phi}-\phi\sqrt{-1}\right)^n\right)=\pi(\log 2)^n$ when $x=\alpha=1$.

More generally, putting $t=\tan u$ in (28) we get

$$\int_0^{\frac{\pi}{2}}du(\varphi(x+\alpha\sqrt{-1}\tan u)+\varphi(x-\alpha\sqrt{-1}\tan u))$$

$=\pi\varphi(x+\alpha)$ and $\int_0^{\frac{\pi}{2}}du(\varphi(1+\sqrt{-1}\tan u)+\varphi(1-\sqrt{-1}\tan u))=\pi\varphi(2)$ when $x=\alpha=1$; for $\varphi x=\frac{x^m}{1+\alpha x^n}$, this gives

$$\int_0^{\frac{\pi}{2}}\frac{(\cos u)^{n-m}(\cos mu(\cos u)^n+\alpha\cos(n-m)u)}{(\cos u)^{2n}+2\alpha\cos nu(\cos u)^n+\alpha^2}du=\frac{\pi}{2}\cdot\frac{2^m}{1+\alpha 2^n}.$$

In (30) Abel replaces n by a fraction $\frac{m}{n}$ and he puts $\frac{\phi}{n}=\theta$, so that

$$\frac{\pi}{2n}\cdot\frac{1}{2^{\frac{m}{n}}}=\int_0^{\frac{\pi}{2}}(\cos n\theta)^{\frac{m}{n}}\cos m\theta d\theta=-\int_1^{\cos\frac{\pi}{2n}}\sqrt[n]{(\psi y)^m}\,fy\frac{dy}{\sqrt{1-y^2}},$$

where $\psi y = y^n - \frac{n(n-1)}{2} y^{n-2}(1-y^2) + \frac{n(n-1)(n-2)(n-3)}{2\cdot 3\cdot 4} y^{n-4}(1-y^2)^2 - \dots$ and $fy = y^m - \frac{m(m-1)}{2} y^{m-2}(1-y^2) + \frac{m(m-1)(m-2)(m-3)}{2\cdot 3\cdot 4} y^{m-4}(1-y^2)^2 - \dots$; for instance $\frac{\pi}{8}\cdot\frac{1}{\sqrt[4]{2}} = -\int_1^{\cos\frac{\pi}{8}} \sqrt[4]{1-8y^2+8y^4}\frac{ydy}{\sqrt{1-y^2}} = \int_0^{\sin\frac{\pi}{8}} dz\sqrt[4]{1-8z^2+8z^4}$.

All this early work of Abel gives evidence of his carefull study of Legendre's *Exercices de Calcul intégral*, which also were his source of inspiration for the theory of elliptic integrals.

A paper on a related subject was published by Abel in the second volume of *Crelle's Journal* (1827, *Œuvres*, t. I, p. 251–262) under the title *Sur quelques intégrales définies*. It contains some applications of the relation discovered by Abel $y_2\frac{dy_1}{da} - y_1\frac{dy_2}{da} = e^{-\int pda}$, where y_1 and y_2 are two solutions of the linear differential equation $\frac{d^2u}{da^2} + p\frac{dy}{da} + qy = 0$. For instance $y_1 = \int_0^1 \frac{(x+a)^{\gamma+1}dx}{x^{1-\alpha}(1-x)^{1-\beta}}$ and $y_2 = \int_0^1 \frac{(x+a)^{\alpha+\beta+\gamma}dx}{x^{\beta}(1-x)^{\alpha}}$ are solutions of the hypergeometric equation

$$\frac{d^2y}{da^2} - \left(\frac{\alpha+\gamma}{a} + \frac{\beta+\gamma}{1+a}\right)\frac{dy}{da} + \frac{(\gamma+1)(\alpha+\beta+\gamma)}{a(a+1)}y = 0,$$

and this leads to the relation

$$\begin{aligned}(\alpha+\beta+\gamma)\int_0^1 \frac{dx(x+a)^{\gamma+1}}{x^{1-\alpha}(1-x)^{1-\beta}} \cdot \int_0^1 \frac{dx(x+a)^{\alpha+\beta+\gamma-1}}{x^{\beta}(1-x)^{\alpha}}\\ -(\gamma+1)\int_0^1 \frac{dx(x+a)^{\gamma}}{x^{1-\alpha}(1-x)^{1-\beta}} \cdot \int_0^1 \frac{dx(x+a)^{\alpha+\beta+\gamma}}{x^{\beta}(1-x)^{\alpha}} = Ca^{\alpha+\gamma}(1+a)^{\beta+\gamma},\end{aligned}$$

where the constant C is determined by making $a = \infty$:

$$\begin{aligned}C &= -(\alpha+\beta-1)\int_0^1 dx\cdot x^{\alpha-1}(1-x)^{\beta-1} \cdot \int_0^1 dx\cdot x^{-\beta}(1-x)^{-\alpha}\\ &= \pi(\cot\alpha\pi + \cot\beta\pi).\end{aligned}$$

In the same way $y_1 = \int_0^\infty \frac{x^{-\alpha}dx}{(1+x)^{\beta}(x+a)^{\gamma}}$ and $y_2 = \int_0^\infty \frac{x^{\beta-1}dx}{(1+x)^{1-\alpha}(x+a)^{\alpha+\beta+\gamma-1}}$ are solutions of the hypergeometric equation $\frac{d^2y}{da^2} + \left(\frac{\alpha+\gamma}{a} - \frac{\beta+\gamma}{1-a}\right)\frac{dy}{da} + \frac{\gamma(1-\alpha-\beta-\gamma)}{a(1-a)}y = 0$ and we get the relation $\int_0^\infty \frac{x^{-\alpha}dx}{(1+x)^{\beta}(x+a)^{\gamma}} = \frac{\Gamma(1-\alpha)\Gamma(\alpha+\beta+\gamma-1)}{\Gamma\beta\cdot\Gamma\gamma}\int_0^\infty \frac{x^{\beta-1}dx}{(1+x)^{1-\alpha}(x+a)^{\alpha+\beta+\gamma-1}}$. The function $y_3 = \int_0^\infty \frac{x^{-\beta}dx}{(1+x)^{\alpha}(x+1-a)^{\gamma}} = (1-a)^{-\beta-\gamma+1}\int_0^\infty \frac{x^{-\beta}dx}{(1+x)^{\gamma}(1+(1-a)x)^{\alpha}}$ is a third solution of the same hypergeometric equation, which, combined with $y_1 = a^{-\alpha-\gamma+1}\int_0^\infty \frac{x^{-\alpha}dx}{(1+x)^\gamma(1+ax)^\beta}$, gives

$$\begin{aligned} &a\int_0^\infty \frac{x^{-\alpha}dx}{(1+x)^\gamma(1+ax)^\beta}\cdot\int_0^\infty \frac{x^{-\beta}dx}{(1+x)^{\gamma+1}(1+(1-a)x)^\alpha} \\ &+(1-a)\int_0^\infty \frac{x^{-\beta}dx}{(1+x)^\gamma(1+(1-a)x)^\alpha}\cdot\int_0^\infty \frac{x^{-\alpha}dx}{(1+x)^{\gamma+1}(1+ax)^\beta} \\ &=\frac{\Gamma(1-\alpha)\Gamma(1-\beta)}{\Gamma(\gamma+1)}\Gamma(\alpha+\beta+\gamma-1). \end{aligned}$$

When $\beta = 1-\alpha$, this relation becomes $a\int_0^\infty \frac{x^{-\alpha}dx}{(1+x)^\gamma(1+ax)^{1-\alpha}}\cdot\int_0^\infty \frac{x^{\alpha-1}dx}{(1+x)^{\gamma+1}(1+(1-a)x)^\alpha} + (1-a)\int_0^\infty \frac{x^{-\alpha}dx}{(1+x)^{\gamma+1}(1+ax)^{1-\alpha}}\cdot\int_0^\infty \frac{x^{\alpha-1}dx}{(1+x)^\gamma(1+(1-a)x)^\alpha} = \frac{\pi}{\gamma\cdot\sin\alpha\pi}$; in particular, for $\alpha = \gamma = \frac{1}{2}$,

$$\begin{aligned} &a\int_0^\infty \frac{dx}{\sqrt{x(1+x)(1+ax)}}\cdot\int_0^\infty \frac{dx}{\sqrt{x(1+x)^3(1+(1-a)x)}} \\ &+(1-a)\int_0^\infty \frac{dx}{\sqrt{x(1+x)(1+(1-a)x)}}\cdot\int_0^\infty \frac{dx}{\sqrt{x(1+x)^3(1+ax)}} = 2\pi. \end{aligned}$$

As Abel observes, these integrals are elliptic and the change of variable $x = \tan^2\varphi$ transforms the preceding relation in

$$\begin{aligned} &a\int_0^{\frac{\pi}{2}} \frac{d\varphi}{\sqrt{1-(1-a)\sin^2\varphi}}\cdot\int_0^{\frac{\pi}{2}} \frac{d\varphi\cdot\cos^2\varphi}{\sqrt{1-a\sin^2\varphi}} \\ &+(1-a)\int_0^{\frac{\pi}{2}} \frac{d\varphi}{\sqrt{1-a\sin^2\varphi}}\cdot\int_0^{\frac{\pi}{2}} \frac{d\varphi\cdot\cos^2\varphi}{\sqrt{1-(1-a)\sin^2\varphi}} = \frac{\pi}{2}, \end{aligned}$$

which is equivalent to Legendre's famous relation between the complete integrals of the first two kinds (*Exercices de Calcul intégral*, t. I, p. 61). Legendre had proved this relation by a very similar method.

Starting from the integral $y = \int_0^x \frac{dx\cdot x^{\alpha-1}(1-x)^{\beta-1}}{(x+a)^{\alpha+\beta}}$, Abel finds that $\frac{dy}{da}+\left(\frac{\alpha}{1+a}+\frac{\beta}{a}\right)y = -\frac{x^\alpha(1-x)^\beta}{a(1+a)(x+a)^{\alpha+\beta}}$, so that $y\cdot a^\beta(1+a)^\alpha = C - x^\alpha(1-x)^\beta\int_0^a \frac{da\cdot a^{\beta-1}(1+a)^{\alpha-1}}{(a+x)^{\alpha+\beta}}$, where C is independent of a and is found to be $\frac{\Gamma\alpha\cdot\Gamma\beta}{\Gamma(\alpha+\beta)}$ by making $a=\infty$. Thus

$$\frac{\Gamma\alpha\cdot\Gamma\beta}{\Gamma(\alpha+\beta)}=a^{\beta}(1+a)^{\alpha}\int\limits_0^x\frac{dx\cdot x^{\alpha-1}(1-x)^{\beta-1}}{(x+a)^{\alpha+\beta}}$$
$$+x^{\alpha}(1-x)^{\beta}\int\limits_0^a\frac{da\cdot a^{\beta-1}(1+a)^{\alpha-1}}{(a+x)^{\alpha+\beta}};$$

when $\alpha+\beta=1$, this gives

$$\frac{(1+a)^{\alpha}}{a^{\alpha-1}}\int\limits_0^x\frac{dx\cdot x^{\alpha-1}(1-x)^{-\alpha}}{x+a}+\frac{x^{\alpha}}{(1-x)^{\alpha-1}}\int\limits_0^a\frac{da\cdot a^{-\alpha}(1+a)^{\alpha-1}}{a+x}=\frac{\pi}{\sin\pi\alpha}.$$

The integral $y=\int\limits_0^1 e^{-ax}x^{\alpha-1}(1-x)^{\beta-1}dx$ ($\alpha,\beta>0$) is a solution of the confluent hypergeometric equation $\frac{d^2y}{da^2}+\left(\frac{\alpha+\beta}{a}+1\right)\frac{dy}{da}+\frac{\alpha}{a}y=0$, and so is

$$y_1=\int\limits_1^{\infty}e^{-ax}x^{\alpha-1}(1-x)^{\beta-1}dx$$
$$=e^{-a}\int\limits_0^{\infty}e^{-ax}x^{\beta-1}(1+x)^{\alpha-1}dx=e^{-a}a^{-\alpha-\beta+1}\int\limits_0^{\infty}e^{-x}x^{\beta-1}(a+x)^{\alpha-1}dx$$

($a>0$). Abel derives from that the formula

$$\Gamma\alpha\cdot\Gamma\beta=\int\limits_0^1 e^{-ax}x^{\alpha-1}(1-x)^{\beta-1}dx\cdot\int\limits_0^{\infty}e^{-x}x^{\beta-1}(a+x)^{\alpha}dx$$
$$-a\int\limits_0^1 e^{-ax}x^{\alpha}(1-x)^{\beta-1}dx\cdot\int\limits_0^{\infty}e^{-x}x^{\beta-1}(a+x)^{\alpha-1}dx,$$

and, for $\beta=1-\alpha$,

$$\frac{\pi}{\sin\pi\alpha}=\int\limits_0^1\frac{dx}{x}e^{-ax}\left(\frac{x}{1-x}\right)^{\alpha}\cdot\int\limits_0^{\infty}e^{-x}dx\left(1+\frac{a}{x}\right)^{\alpha}$$
$$-a\int\limits_0^1 dx\cdot e^{-ax}\left(\frac{x}{1-x}\right)^{\alpha}\cdot\int\limits_0^{\infty}\frac{dx}{x+a}e^{-x}\left(1+\frac{a}{x}\right)^{\alpha}.$$

As a last example, Abel considers the integrals $y=\int\limits_0^{\infty}e^{ax-x^2}x^{\alpha-1}dx$ and $y_1=\int\limits_0^{\infty}e^{-ax-x^2}x^{\alpha-1}dx$ ($\alpha>0$), solutions of the differential equation $\frac{d^2y}{da^2}-\frac{1}{2}a\frac{dy}{da}-\frac{1}{2}\alpha y=0$, which is related to the so called Weber equation. The corresponding relation is

$$\frac{1}{2}\Gamma\left(\frac{\alpha+1}{2}\right)\Gamma\left(\frac{\alpha}{2}\right)e^{\frac{a^2}{4}} = \int_0^\infty e^{ax-x^2}x^{\alpha-1}dx \cdot \int_0^\infty e^{-ax-x^2}x^{\alpha}dx + \int_0^\infty e^{ax-x^2}x^{\alpha}dx \cdot \int_0^\infty e^{-ax-x^2}x^{\alpha-1}dx.$$

In a posthumous paper *Les fonctions transcendantes* $\sum\frac{1}{a^2}, \sum\frac{1}{a^3}, \sum\frac{1}{a^4}, \ldots, \sum\frac{1}{a^n}$ *exprimées par des intégrales définies* (*Œuvres*, t. II, p. 1–6), Abel gives integral formulae for these finite sums, extended from 1 to $a-1$. He also studies their continuation to non integral values of a and n. As $\frac{d^n\sum\frac{1}{a}}{da^n} = (-1)^n 2\cdot 3\cdots n\sum\frac{1}{a^{n+1}}$, one has

$$\sum\frac{1}{a^n} = (-1)^{n-1}\frac{d^{n-1}L(a)}{2\cdot 3\cdots(n-1)da^{n-1}}$$

where $L(a) = \sum\frac{1}{a} = \int_0^1 \frac{x^{a-1}-1}{x-1}dx$. From this Abel deduces

$$L(a,\alpha) = \sum\frac{1}{a^\alpha} = \frac{1}{\Gamma(\alpha)}\int_0^1 \frac{x^{a-1}-1}{x-1}\left(l\frac{1}{x}\right)^{\alpha-1}dx \tag{31}$$

for any value of α. Substituting $x^{a-1} = 1-(a-1)\left(l\frac{1}{x}\right)+\frac{(a-1)^2}{2}\left(l\frac{1}{x}\right)^2-\frac{(a-1)^3}{2\cdot 3}\left(l\frac{1}{x}\right)^3+\ldots$ and $\frac{1}{1-x} = 1+x+x^2+\ldots$, he obtains

$$\begin{aligned}L(a,\alpha) = {} & \frac{a-1}{1}\alpha\left(1+\frac{1}{2^{\alpha+1}}+\frac{1}{3^{\alpha+1}}+\frac{1}{4^{\alpha+1}}+\ldots\right)\\ & -\frac{(a-1)^2}{1\cdot 2}\alpha(\alpha+1)\left(1+\frac{1}{2^{\alpha+2}}+\frac{1}{3^{\alpha+2}}+\frac{1}{4^{\alpha+2}}+\ldots\right)\\ & +\frac{(a-1)^3}{1\cdot 2\cdot 3}\alpha(\alpha+1)(\alpha+2)\left(1+\frac{1}{2^{\alpha+3}}+\frac{1}{3^{\alpha+3}}+\frac{1}{4^{\alpha+3}}+\ldots\right)\\ = {} & \alpha(a-1)L'(\alpha+1)-\frac{\alpha(\alpha+1)}{2}(a-1)^2L'(\alpha+2)\\ & +\frac{\alpha(\alpha+1)(\alpha+2)}{2\cdot 3}(a-1)^3L'(\alpha+3)-\ldots,\end{aligned}$$

where $L'(\alpha) = L(\infty,\alpha) = 1+\frac{1}{2^\alpha}+\frac{1}{3^\alpha}+\frac{1}{4^\alpha}+\ldots$ is the zeta function. Putting $\frac{m}{a}$ instead of a in (31), Abel deduces $L\left(\frac{m}{a},\alpha\right) = -\frac{1}{\Gamma(\alpha)}\int_0^1 \frac{\left(l\frac{1}{y}\right)^{\alpha-1}}{y-1}dy + \frac{a^\alpha}{\Gamma(\alpha)}\int_0^1 \frac{y^{m-1}\left(l\frac{1}{y}\right)^{\alpha-1}}{y^a-1}dy$ or, writing $\frac{y^{m-1}}{y^a-1} = \frac{A}{1-cy}+\frac{A'}{1-c'y}+\ldots$ in the hypothesis $m-1<a$,

$$L\left(\frac{m}{a},\alpha\right) = a^\alpha(AL'(\alpha,c) + A'L'(\alpha,c') + A''L'(\alpha,c'') + \dots)$$

where $L'(\alpha,c) = 1 + \frac{c}{2^\alpha} + \frac{c^2}{3^\alpha} + \frac{c^3}{4^\alpha} + \dots$

The following paper *Sur l'intégrale définie* $\int_0^1 x^{a-1}(1-x)^{c-1}\left(l\frac{1}{x}\right)^{\alpha-1}dx$ (*Œuvres*, t. II, p. 7–13) is related to the same subject; it gives developments in series for La and associated functions. When $\alpha = 1$, the integral is equal to $\frac{\Gamma a\cdot\Gamma c}{\Gamma(a+c)}$. As the logarithmic derivative of Γa is equal to $La - C$, where C is the Euler constant, Abel deduces from this $\int_0^1 x^{a-1}(1-x)^{c-1}lx dx = (La - L(a+c))\frac{\Gamma a\cdot\Gamma c}{\Gamma(a+c)}$, $\int_0^1 x^{a-1}(1-x)^{c-1}l(1-x)dx = (Lc - L(a+c))\frac{\Gamma a\cdot\Gamma c}{\Gamma(a+c)}$. For $c = 1$, this gives $\int_0^1 x^{a-1}lx dx = -\frac{1}{a^2}$, $\int_0^1 x^{a-1}l(1-x)dx = -\frac{L(1+a)}{a}$. Developing $(1-x)^{c-1}$ in series, Abel obtains

$$(La - L(a+c))\frac{\Gamma a\cdot\Gamma c}{\Gamma(a+c)} = \frac{1}{a^2} - (c-1)\frac{1}{(a+1)^2} + \frac{(c-1)(c-2)}{2}\frac{1}{(a+2)^2} - \frac{(c-1)(c-2)(c-3)}{2\cdot 3}\frac{1}{(a+3)^2} + \dots \tag{32}$$

For example, if $c = 1-a$, $-La\cdot\frac{\pi}{\sin a\pi} = \frac{1}{a^2} + \frac{a}{(a+1)^2} + \frac{a(a+1)}{2(a+2)^2} + \frac{a(a+1)(a+2)}{2\cdot 3(a+3)^2} + \dots$, which becomes $2\pi\log 2 = 2^2 + \frac{2}{3^2} + \frac{3}{2\cdot 5^2} + \frac{3\cdot 5}{2^2\cdot 3\cdot 7^2} + \frac{3\cdot 5\cdot 7}{2^3\cdot 3\cdot 4\cdot 9^2} + \dots$ when $a = \frac{1}{2}$ for $L\frac{1}{2} = -2\log 2$. When $a = 1-x$ and $c = 2x-1$, $L(1-x) - Lx = \pi\cot\pi x$ and

$$-\pi\cot\pi x\frac{\Gamma(1-x)\Gamma(2x-1)}{\Gamma x} = \frac{1}{(1-x)^2} - \frac{2x-2}{(2-x)^2} + \frac{(2x-2)(2x-3)}{2(3-x)^2} - \frac{(2x-2)(2x-3)(2x-4)}{2\cdot 3(4-x)^2} + \dots$$

From (32) Abel deduces an expression of $\frac{L(a+c)-La}{L(a+c)-Lc}$ as a quotient of two series and, making $c = 1$, $L(1+a) = a - \frac{a(a-1)}{2^2} + \frac{a(a-1)(a-2)}{2\cdot 3^2} - \dots$. Thus

$$\begin{aligned}\pi\cot\pi a &= L(1-a) - La\\ &= -\left(2a - 1 + \frac{a(a+1)-(a-1)(a-2)}{2^2}\right.\\ &\quad\left.+ \frac{a(a+1)(a+2)-(a-1)(a-2)(a-3)}{2\cdot 3^2} + \dots\right).\end{aligned}$$

The integral of the title, with α an integer, is obtained by successive differentiations with respect to a:

$$\int_0^1 x^{a-1}(1-x)^{c-1}\left(l\frac{1}{x}\right)^{\alpha-1}dx = \Gamma\alpha\left(\frac{1}{a^\alpha} - \frac{c-1}{1}\frac{1}{(a+1)^\alpha} + \frac{(c-1)(c-2)}{1\cdot 2}\frac{1}{(a+2)^\alpha} - \dots\right).$$

Taking the successive logarithmic derivatives, Abel sees that this integral has an expression in terms of the sums La, $L'a = \sum \frac{1}{a^2}$, $L''a = \sum \frac{1}{a^3}, \ldots$; for example

$$\int_0^1 x^{a-1}(1-x)^{c-1}\left(l\frac{1}{x}\right)^3 dx = \Big(2(L''(a+c) - L''a) \\ +3(L'(a+c) - L'a)(L(a+c) - La) \\ +(L(a+c) - La)^3\Big)\frac{\Gamma a \cdot \Gamma c}{\Gamma(a+c)}.$$

The successive differentiations of the equality $\int_0^1 \left(l\frac{1}{x}\right)^{\alpha-1} dx = \Gamma\alpha$ with respect to α give the formula $\int_0^1 \left(l\frac{1}{x}\right)^{\alpha-1}\left(ll\frac{1}{x}\right)^n dx = \frac{d^n\Gamma\alpha}{d\alpha^n}$, whence $\int_0^\infty (lz)^n e^{-z^{\frac{1}{\alpha}}} dz = \alpha^{n+1}\frac{d^n\Gamma\alpha}{d\alpha^n}$ by a change of variable. Abel deduces from this the formulae $\int_0^\infty e^{-x^\alpha} dx = \frac{1}{\alpha}\Gamma\left(\frac{1}{\alpha}\right)$ $(n = 0)$ and $\int_0^\infty l\left(\frac{1}{x}\right)e^{-x^\alpha} dx = -\frac{1}{\alpha^2}\Gamma\left(\frac{1}{\alpha}\right)\left(L\left(\frac{1}{\alpha}\right) - C\right)$ ($n = 1$; C is the Euler constant), which leads to

$$\int_0^\infty e^{-nx}x^{\alpha-1}lx dx = \frac{\Gamma\alpha}{n^\alpha}(L\alpha - C - \log n).$$

A third posthumous paper is titled *Sommation de la série* $y = \varphi(0) + \varphi(1)x + \varphi(2)x^2 + \varphi(3)x^3 + \ldots + \varphi(n)x^n$, *n étant un entier positif fini ou infini, et* $\varphi(n)$ *une fonction algébrique rationnelle de n* (*Œuvres*, t. II, p. 14–18). Abel decomposes φ in terms of one of the forms An^α, $\frac{B}{(a+n)^\beta}$. He has first to sum $f(\alpha, x) = x + 2^\alpha x^2 + 3^\alpha x^3 + \ldots + n^\alpha x^n$; this is done using the identities $f(\alpha, x) = \frac{x d f(\alpha-1, x)}{dx}$ and $f(0, x) = \frac{x(1-x^n)}{1-x}$. Then Abel considers

$$F\alpha = \frac{1}{a^\alpha} + \frac{x}{(a+1)^\alpha} + \frac{x^2}{(a+2)^\alpha} + \ldots + \frac{x^n}{(a+n)^\alpha} = \frac{\int dx \cdot x^{\alpha-1}F(\alpha-1)}{x^\alpha},$$

for which $F(0) = \frac{1-x^{n+1}}{1-x}$. The formula (10) for the dilogarithm is thus obtained when $\alpha = 2$, $n = \infty$ and $a = 1$.

3 Algebraic Equations

We know that in 1821 Abel thought he had found a method to solve the general quintic equation by radicals; when he discovered his error and proved that such a solution was impossible, he wrote a booklet in french with a demonstration, *Mémoire sur les équations algébriques, où l'on démontre l'impossibilité de la résolution de l'équation générale du cinquième degré* (Christiania, 1824; *Œuvres*, t. I, p. 28–33).

The impossibility of an algebraic solution for the general quintic equation had already been published by P. Ruffini (1799, 1802, 1813), but his demonstration was incomplete for he supposed without proof that the radicals in a hypothetical solution were necessarily rational functions of the roots. Abel, who did not know of Ruffini's work, began with a proof of this fact.

Supposing the root of

$$y^5 - ay^4 + by^3 - cy^2 + dy - e = 0 \tag{33}$$

of the form

$$y = p + p_1 R^{\frac{1}{m}} + p_2 R^{\frac{2}{m}} + \ldots + p_{m-1} R^{\frac{m-1}{m}}, \tag{34}$$

with m a prime number and $p, p_1, \ldots, p_{m-1}, R$ of an analogous form ($R^{\frac{1}{m}}$ is a chosen exterior radical in a hypothetical solution by radicals and it is supposed that it is not a rational function of $a, b, \ldots, p, p_1, \ldots$), Abel first replaces R by $\frac{R}{p_1^m}$ in order to have an expression of the same form with $p_1 = 1$. Putting (34) in the equation, he gets a relation $P = q + q_1 R^{\frac{1}{m}} + q_2 R^{\frac{2}{m}} + \ldots + q_{m-1} R^{\frac{m-1}{m}} = 0$, with coefficients $q, q_1, \ldots$ polynomial in $a, b, c, d, e, p, p_2, \ldots, R$. These coefficients are necessarily 0 for otherwise the two equations $z^m - R = 0$ and $q + q_1 z + \ldots + q_{m-1} z^{m-1} = 0$ would have some common roots, given by the annulation of the greatest common divisor

$$r + r_1 z + \ldots + r_k z^k$$

of their first members. Since the roots of $z^m - R = 0$ are of the form $\alpha_\mu z$, where z is one of them and α_μ is an m-th root of 1, we get a system of k equations $r + \alpha_\mu r_1 z + \ldots + \alpha_\mu^k r_k z^k = 0$ ($0 \leq \mu \leq k-1$ and $\alpha_0 = 1$), from which it is possible to express z as a rational function of $r, r_1, \ldots$ (and the α_μ). Now the r_k are rational with respect to $a, b, \ldots, R, p, p_2, \ldots$ and we get a contradiction for, by hypothesis, z is not rational with respect to these quantities.

The relation $P = 0$ being identical, the expression (34) is still a root of (33) when $R^{\frac{1}{m}}$ is replaced by $\alpha R^{\frac{1}{m}}$, α an arbitrary m-root of 1, and it is easy to see that the m expressions so obtained are distinct; it results that $m \leq 5$. Then (34) gives us m roots y_k ($1 \leq k \leq m$) of (33), with $R^{\frac{1}{m}}, \alpha R^{\frac{1}{m}}, \ldots, \alpha^{m-1} R^{\frac{1}{m}}$ in place of $R^{\frac{1}{m}}$, and we have

$$\begin{aligned}
p &= \frac{1}{m}(y_1 + y_2 + \ldots + y_m), \\
R^{\frac{1}{m}} &= \frac{1}{m}(y_1 + \alpha^{m-1} y_2 + \ldots + \alpha y_m), \\
p_2 R^{\frac{2}{m}} &= \frac{1}{m}(y_1 + \alpha^{m-2} y_2 + \ldots + \alpha^2 y_m), \\
&\ldots, \\
p_{m-1} R^{\frac{m-1}{m}} &= \frac{1}{m}(y_1 + \alpha y_2 + \ldots + \alpha^{m-1} y_m);
\end{aligned}$$

this proves that $p, p_2, \ldots, p_{m-1}$ and $R^{\frac{1}{m}}$ are rational functions of the roots of (33) (and α). Now if, for instance $R = S + v^{\frac{1}{n}} + S_2 v^{\frac{2}{n}} + \ldots + S_{n-1} v^{\frac{n-1}{n}}$, the same reasoning shows that $v^{\frac{1}{n}}, S, S_2, \ldots$ are rational functions of the roots of (33) and continuing in this manner, we see that every irrational quantity in (34) is a rational function of the roots of (33) (and some roots of 1).

Abel next shows that the innermost radicals in (34) must be of index 2. Indeed if $R^{\frac{1}{m}} = r$ is such a radical, r is a rational function of the 5 roots $y_1, y_2, \ldots, y_5$ and R is a *symmetric* rational function of the same roots, which may be considered as independent variables for (33) is the general quintic equation. So we may arbitrarily permute the y_k in the relation $R^{\frac{1}{m}} = r$ and we see that r takes m different values; a result of Cauchy (1815) now says that $m = 5$ or 2 and the value 5 is easily excluded. We thus know that r takes 2 values and, following Cauchy, it has the form $v(y_1 - y_2)(y_1 - y_3) \ldots (y_2 - y_3) \ldots (y_4 - y_5) = vS^{\frac{1}{2}}$, where v is symmetric and S is the discriminant of (33).

The next radicals are of the form $r = \left(p + p_1 S^{\frac{1}{2}}\right)^{\frac{1}{m}}$, with p, p_1 symmetric. If $r_1 = \left(p - p_1 S^{\frac{1}{2}}\right)^{\frac{1}{m}}$ is the conjugate of r, then $rr_1 = \left(p^2 - p_1^2 S^{\frac{1}{2}}\right)^{\frac{1}{m}} = v$ must be symmetric (otherwise m would be equal to 2 and r would take on 4 values, which is not possible). Thus

$$r + r_1 = \left(p + p_1 S^{\frac{1}{2}}\right)^{\frac{1}{m}} + v\left(p + p_1 S^{\frac{1}{2}}\right)^{-\frac{1}{m}} = z$$

takes m values which implies that $m = 5$ and $z = q + q_1 y + q_2 y^2 + q_3 y^3 + q_4 y^4$, with $q, q_1, \ldots$ symmetric. Combining this relation with (33), we get y rationally in z, a, b, c, d and e, and so of the form

$$y = P + R^{\frac{1}{5}} + P_2 R^{\frac{2}{5}} + P_3 R^{\frac{3}{5}} + P_4 R^{\frac{4}{5}}, \tag{35}$$

with P, R, P_2, P_3 and P_4 of the form $p + p_1 S^{\frac{1}{2}}$, p, p_1 and S rational in a, b, c, d and e. From (35) Abel draws $R^{\frac{1}{5}} = \frac{1}{5}(y_1 + \alpha^4 y_2 + \alpha^3 y_3 + \alpha^2 y_4 + \alpha y_5) = \left(p + p_1 S^{\frac{1}{2}}\right)^{\frac{1}{5}}$, where α is an imaginary fifth root of 1; this is impossible for the first expression takes 120 values and the second only 10.

Euler (1764) had conjectured a form analogous to (35) for the solutions of the quintic equation, with R given by an equation of degree 4. In a letter to Holmboe (24 October 1826, *Œuvres*, t. II, p. 260), Abel states that if a quintic equation is algebraically solvable, its solution has the form $x = A + \sqrt[5]{R} + \sqrt[5]{R'} + \sqrt[5]{R''} + \sqrt[5]{R'''}$ where R, R', R'', R''' are roots of an equation of degree 4 solvable by quadratic radicals; this is explained in a letter to Crelle (14 March 1826, *Œuvres*, t. II, p. 266) for the case of a solvable quintic equation with *rational* coefficients, the solution being $x = c + A a^{\frac{1}{5}} a_1^{\frac{2}{5}} a_2^{\frac{4}{5}} a_3^{\frac{3}{5}} + A_1 a_1^{\frac{1}{5}} a_2^{\frac{2}{5}} a_3^{\frac{4}{5}} a^{\frac{3}{5}} + A_2 a_2^{\frac{1}{5}} a_3^{\frac{2}{5}} a^{\frac{4}{5}} a_1^{\frac{3}{5}} + A_3 a_3^{\frac{1}{5}} a^{\frac{2}{5}} a_1^{\frac{4}{5}} a_2^{\frac{3}{5}}$, where

$$a = m + n\sqrt{1+e^2} + \sqrt{h(1+e^2+\sqrt{1+e^2})},$$

$$a_1 = m - n\sqrt{1+e^2} + \sqrt{h(1+e^2-\sqrt{1+e^2})},$$

$$a_2 = m + n\sqrt{1+e^2} - \sqrt{h(1+e^2+\sqrt{1+e^2})},$$

$$a_3 = m - n\sqrt{1+e^2} - \sqrt{h(1+e^2-\sqrt{1+e^2})},$$

$A = K+K'a+K''a_2+K'''aa_2$, $A_1 = K+K'a_1+K''a_3+K'''a_1a_3$, $A_2 = K+K'a_2+K''a+K'''aa_2$ and $A_3 = K+K'a_3+K''a_1+K'''a_1a_3$, and $c, h, e, m, n, K, K', K''$ and K''' are rational numbers.

Abel published a new version of his theorem in the first volume of *Crelle's Journal* (1826, *Œuvres*, t. I, p. 66–87). In the first paragraph of this paper, Abel defines the *algebraic functions* of a set of variables $x', x'', x''', \ldots$ They are built from these variables and some constant quantities by the operations of addition, multiplication, division and extraction of roots of prime index. Such a function is *integral* when only addition and multiplication are used, and is then a sum of monomials $Ax'^{m_1}x''^{m_2}\ldots$ It is *rational* when division is also used, but not the extraction of roots, and is then a quotient of two integral functions. The general algebraic functions are classified in *orders*, according to the number of superposed radicals in their expression; a function $f(r', r'', \ldots, \sqrt[n']{p'}, \sqrt[n'']{p''}, \ldots)$ of order μ, with $r', r'', \ldots, p', p'', \ldots$ of order $< \mu$ and f rational, such that none of the $\sqrt[n_k]{p_k}$ is a rational function of the r and the other $\sqrt[n_\ell]{p_\ell}$, is said to be of *degree m* if it contains m radicals $\sqrt[n_k]{p_k}$. Such a function may be written $f(r', r'', \ldots, \sqrt[n]{p})$ with p of order $\mu - 1$, $r', r'', \ldots$ of order $\leq \mu$ and degree $\leq m-1$, and f rational; it is then easy to reduce it to the form $q_0 + q_1 p^{\frac{1}{n}} + q_2 p^{\frac{2}{n}} + \ldots + q_{n-1} p^{\frac{n-1}{n}}$, with coefficients $q_0, q_1, q_2, \ldots$ rational functions of $p, r', r'', \ldots$, so of order $\leq \mu$ and degree $\leq m-1$, $p^{\frac{1}{n}}$ not a rational function of these quantities. Abel carries out the supplementary reduction to the case $q_1 = 1$. In order to do this, he chooses an index μ such that $q_\mu \neq 0$ and puts $q_\mu^n p^\mu = p_1$, which will play the role of p. The starting point of his preceding paper has been completely justified.

In the second paragraph, Abel proves that if an equation is algebraically solvable, one may write its solution in a form in which all the constituent algebraic expressions are rational functions of the roots of the equation. The proof is more precise than that of the 1824 paper, but follows the same lines. The coefficients of the equation are supposed to be rational functions of certain independent variables $x', x'', x''', \ldots$

In the third paragraph, Abel reproduces the proof of Lagrange's theorem (1771) according to which the number of values that a rational function v of n letters may take under the $n!$ substitutions of these letters is necessarily a divisor of $n!$ and Cauchy's theorem (1815) which says that if p is the greatest prime number $\leq n$, and if v takes less than p values, then it takes 1 or 2 values. Indeed it must be invariant for any cycle of p letters, and it is possible to deduce from this that it is invariant for any cycle of 3 letters and from this by any even substitution. Thus, as Ruffini had proved, a rational function of 5 variable cannot take 3 or 4 values. Abel then gives, following Cauchy, the form of a function v of 5 letters $x_1, x_2, \ldots, x_5$ which takes 2 values: it may be written $p + q\rho$, where p and q are rational symmetric functions and

$$\rho \quad = (x_1 - x_2)(x_1 - x_3)\ldots(x_4 - x_5)$$

is the square root of the discriminant. Indeed if v_1 and v_2 are the two values of v, $v_1 + v_2 = t$ and $(v_1 v_2)\rho = t_1$ are symmetric and $v_1 = \frac{1}{2}t + \frac{t_1}{2\rho^2}\rho$. Finally, Abel gives the form of a function of 5 quantities which takes 5 values: it is $r_0 + r_1 x + r_2 x^2 + r_3 x^3 + r_4 x^4$, where r_0, r_1, r_2, r_3 and r_4 are symmetric functions of the five quantities and x is one of them. Indeed this is true for a rational function of x_1, x_2, x_3, x_4, x_5 which is symmetric with respect to x_2, x_3, x_4, x_5. Now if v is a function which takes 5 values v_1, v_2, v_3, v_4, v_5 under the substitutions of x_1, x_2, x_3, x_4, x_5, the number μ of values of $x_1^m v$ under the substitutions of x_2, x_3, x_4, x_5 is less than 5, otherwise it would give 25 values under the substitutions of x_1, x_2, x_3, x_4, x_5 and 25 does not divide 5!. If $\mu = 1$, v is symmetric with respect to x_2, x_3, x_4, x_5 and the result is true; it is also true if $\mu = 4$ for the sum $v_1 + v_2 + v_3 + v_4 + v_5$ is completely symmetric and $v_1 + v_2 + v_3 + v_4$ is symmetric with respect to x_2, x_3, x_4, x_5, so $v_5 = v_1 + v_2 + v_3 + v_4 + v_5 - (v_1 + v_2 + v_3 + v_4)$ is of the desired form. It is somewhat more work to prove that μ cannot be 2 or 3. Eliminating x between the equations

$$(x - x_1)(x - x_2)(x - x_3)(x - x_4)(x - x_5) = x^5 - ax^4 + bx^3 - cx^2 + dx - e = 0$$

and $r_0 + r_1 x + r_2 x^2 + r_3 x^3 + r_4 x^4 = v$ (a quantity taking 5 values), one obtains

$$x = s_0 + s_1 v + s_2 v^2 + s_3 v^3 + s_4 v^4$$

where s_0, s_1, s_2, s_3 and s_4 are symmetric functions. The paragraph ends with the following lemma: if a rational function v of the 5 roots takes m values under the substitutions of these roots, it is a root of an equation of degree m with coeffficients rational symmetric and it cannot be a root of such an equation of degree less than m.

The fourth paragraph finally gives the proof of the impossibility of a solution by radicals. As in the preceding paper, Abel proves that an innermost radical $R^{\frac{1}{m}} = v$ in a hypothetical solution has an index m (supposed prime) equal to 2 or 5; if $m = 5$, one may write

$$x = s_0 + s_1 R^{\frac{1}{5}} + s_2 R^{\frac{2}{5}} + s_3 R^{\frac{3}{5}} + s_4 R^{\frac{4}{5}}$$

$$\text{and} \quad s_1 R^{\frac{1}{5}} = \frac{1}{5}(x_1 + \alpha^4 x_2 + \alpha^3 x_3 + \alpha^2 x_4 + \alpha x_5)$$

where α is a fifth root of 1, and the second member takes 120 values, which is impossible for it is a root of the equation $z^5 - s_1^5 R = 0$. So $m = 2$ and $\sqrt{R} = p + qs$ with p, q symmetric and $s = (x_1 - x_2)\cdots(x_4 - x_5)$; the second value is $-\sqrt{R} = p - qs$, so $p = 0$. Then, at the second order appear radicals $\sqrt[5]{\alpha + \beta\sqrt{s^2}} = R^{\frac{1}{5}}$ with α, β symmetric as well as $\gamma = \sqrt[5]{\alpha^2 - \beta^2 s^2}$; $p = \sqrt[5]{R} + \frac{\gamma}{\sqrt[5]{R}}$ takes 5 values so that

$$x = s_0 + s_1 p + s_2 p^2 + s_3 p^3 + s_4 p^4 = t_0 + t_1 R^{\frac{1}{5}} + t_2 R^{\frac{2}{5}} + t_3 R^{\frac{3}{5}} + t_4 R^{\frac{4}{5}}$$

with t_0, t_1, t_2, t_3 and t_4 rational in a, b, c, d, e and R. From this relation, one deduces

$$t_1 R^{\frac{1}{5}} = \frac{1}{5}(x_1 + \alpha^4 x_2 + \alpha^3 x_3 + \alpha^2 x_4 + \alpha x_5) = p',$$

where α is a fifth root of 1; $p'^5 = t_1^5 R = u + u'\sqrt{s^2}$ and $(p'^5 - u)^2 = u'^2 s^2$, an equation of degree 10 in p', whereas p' takes 120 values, a contradiction.

Abel reproduced this demonstration in the *Bulletin de Férussac* (1826, t. 6, *Œuvres*, t. I, p. 8794).

In a short paper published in the *Annales de Gergonne* (1827, t. XVII, *Œuvres*, t. I, p. 212–218), Abel treated a problem of the theory of elimination: given two algebraic equations

$$\begin{aligned} \varphi y &= p_0 + p_1 y + p_2 y^2 + \ldots + p_{m-1} y^{m-1} + y^m = 0 \\ \text{and} \quad \psi y &= q_0 + q_1 y + q_2 y^2 + \ldots + q_{n-1} y^{n-1} + y^n \end{aligned}$$

with exactly one common solution y, compute any rational function fy of this solution rationally as a function of $p_0, p_1, \ldots, p_{m-1}, q_0, q_1, \ldots, q_{n-1}$. He denotes the roots of ψ by $y, y_1, \ldots, y_{n-1}$ and the product of the φy_j with $j \neq k$ by R_k ($y_0 = y$). As $\varphi y = 0$, $R_k = 0$ for $k \geqslant 1$ so that $fy = \frac{\sum fy_k \cdot \theta y_k \cdot R_k}{\sum \theta y_k \cdot R_k}$, where θ is any rational function. If $fy = \frac{Fy}{\chi y}$, with F and χ polynomial, one may take $\theta = \chi$ to get $fy = \frac{\sum Fy_k \cdot R_k}{\sum \chi y_k \cdot R_k}$.

Abel proposes a better solution, based on the observation that R, being a symmetric function of $y_1, y_2, \ldots, y_{n-1}$, may be expressed as $R = \rho_0 + \rho_1 y + \rho_2 y^2 + \ldots + \rho_{n-1} y^{n-1}$, with coefficients $\rho_0, \rho_1, \rho_2, \ldots, \rho_{n-1}$ polynomial in $p_0, p_1, \ldots, p_{m-1}, q_0, q_1, \ldots, q_{n-1}$, and the same is true for $Fy \cdot R = t_0 + t_1 y + t_2 y^2 + \ldots + t_{n-1} y^{n-1}$. Naturally, $R_k = \rho_0 + \rho_1 y_k + \rho_2 y_k^2 + \ldots + \rho_{n-1} y_k^{n-1}$ and $Fy_k \cdot R_k = t_0 + t_1 y_k + t_2 y_k^2 + \ldots + t_{n-1} y_k^{n-1}$. Now taking $\theta y = \frac{1}{\psi' y}$, we have

$$\begin{aligned} \sum \frac{R_k}{\psi' y_k} &= \rho_0 \sum \frac{1}{\psi' y_k} + \rho_1 \sum \frac{y_k}{\psi' y_k} + \rho_2 \sum \frac{y_k^2}{\psi' y_k} + \ldots + \rho_{n-1} \sum \frac{y_k^{n-1}}{\psi' y_k} \\ &= \rho_{n-1} \end{aligned}$$

and, in the same way, $\sum \frac{R_k \cdot Fy_k}{\psi' y_k} = t_{n-1}$, so that $Fy = \frac{t_{n-1}}{\rho_{n-1}}$. For a rational function $\frac{Fy}{F'y}$ (where F' is not the derivative of F!), the value is $\frac{t_{n-1}}{t'_{n-1}}$, where t'_{n-1} is the coefficient of y^{n-1} in $F'y \cdot R$. In the case of $fy = y$, let $R = \rho y^{n-1} + \rho' y^{n-2} + \ldots$; then

$$Ry = \rho y^n + \rho' y^{n-1} + \ldots = (\rho' - \rho q_{n-1}) y^{n-1} + \ldots$$

so that $y = -q_{n-1} + \frac{\rho'}{\rho}$.

In his researches about elliptic functions, published in the second and the third volume of *Crelle's Journal* (1827–28, *Œuvres*, t. I, p. 294–314 and 355–362) Abel met some new cases of algebraic equations solvable by radicals. Such equations were known from the time of A. de Moivre (1707), who showed that the equation of degree n (odd integer) giving $\sin \frac{a}{n}$ is solvable by radicals, when $\sin a$ is known, the formula involving $\sin \frac{2\pi}{n}$ in its coefficients. Then Gauss (1801) proved that the cyclotomic equation of degree $n-1$ (n a prime number) giving the n-th root $e^{\frac{2\pi i}{n}}$ is also solvable.

Abel developed an analogous theory for elliptic functions instead of circular functions. Let $x = \varphi u$ be defined by $u = \int\limits_0^x \frac{dt}{\sqrt{(1-c^2t^2)(1+e^2t^2)}}$ (e, c given real parameters); then φ is a uniform meromorphic function in the complex domain, with two independent periods

$$2\omega = 4\int\limits_0^{1/c} \frac{dx}{\sqrt{(1-c^2x^2)(1+e^2x^2)}},$$

$$2i\varpi = 4i\int\limits_0^{1/e} \frac{dx}{\sqrt{(1-e^2x^2)(1+c^2x^2)}}.$$

Abel discovered that the equation of degree n^2 giving $\varphi\left(\frac{a}{n}\right)$ when $\varphi(a)$ is known is solvable by radicals, the formula involving $\varphi\left(\frac{\Omega}{n}\right)$, Ω being a period, in its coefficients. The equation of degree $\frac{n^2-1}{2}$ giving the non-zero values of $\varphi^2\left(\frac{\Omega}{n}\right)$ may be decomposed in $n+1$ equations of degree $\frac{n-1}{2}$, all solvable, by means of an equation of degree $n+1$, which, in general, is not solvable by radicals. For certain singular *moduli* $\frac{e}{c}$, for instance when $\frac{e}{c} = 1, \sqrt{3}$ or $2 \pm \sqrt{3}$, the equation of degree $n+1$ is also solvable; Gauss already knew this lemniscatic case, where $c = e = 1$.

The base for these results is Euler's theorem of addition for elliptic integrals, which gives, in Abel's notation $\varphi(\alpha+\beta) = \frac{\varphi\alpha\, f\beta F\beta + \varphi\beta\, f\alpha F\alpha}{1+e^2c^2\varphi^2\alpha\varphi^2\beta}$, where $f\alpha = \sqrt{1-c^2\varphi^2\alpha}$, $F\alpha = \sqrt{1+e^2\varphi^2\alpha}$. So the roots of the equation for $\varphi\left(\frac{\alpha}{n}\right)$ are $\varphi\left((-1)^{m+\mu}\frac{\alpha}{n} + \frac{m\omega+\mu\varpi i}{n}\right)$, $|m|, |\mu| \le \frac{n-1}{2}$ and they are rational functions of $\varphi(\beta)$, $f\beta$, $F\beta$, where $\beta = \frac{\alpha}{n}$. Abel defines

$$\varphi_1\beta = \sum_{|m|\le\frac{n-1}{2}} \varphi\left(\beta + \frac{2m\omega}{n}\right), \quad \psi\beta = \sum_{|\mu|\le\frac{n-1}{2}} \theta^\mu \varphi_1\left(\beta + \frac{2\mu\varpi i}{n}\right)$$

$$\text{and } \psi_1\beta = \sum_{|\mu|\le\frac{n-1}{2}} \theta^\mu \varphi_1\left(\beta - \frac{2\mu\varpi i}{n}\right),$$

where θ is an n-th root of 1; he proves, by means of the addition theorem, that $\psi\beta\cdot\psi_1\beta$ and $(\psi\beta)^n + (\psi_1\beta)^n$ are rational functions of $\varphi\alpha$, so that $\psi\beta = \sqrt[n]{A+\sqrt{A^2-B^n}}$ with A and B rational in $\varphi\alpha$. Indeed, $\varphi_1\beta = \varphi\beta + \sum\limits_{m=1}^{\frac{n-1}{2}} \left(\varphi\left(\beta+\frac{2m\omega}{n}\right) + \varphi\left(\beta - \frac{2m\omega}{n}\right)\right)$ is rational with respect to $\varphi\beta = x$ and $\varphi_1\left(\beta \pm \frac{2\mu\varpi i}{n}\right) = R_\mu \pm R'_\mu\sqrt{(1-c^2x^2)(1+e^2x^2)}$, where R_μ and R'_μ are rational in x, so that $\psi\beta$ and $\psi_1\beta$ have the same form and $\psi_1\beta$ is deduced from $\psi\beta$ by changing the sign of the radical $\sqrt{(1-c^2x^2)(1+e^2x^2)}$. Now $\varphi_1\left(\beta+\frac{2k\omega}{n}\right) = \varphi_1\beta$, $\psi\left(\beta+\frac{2k'\varpi i}{n}+\frac{2k\omega}{n}\right) = \theta^{-k'}\psi\beta$ and $\psi_1\left(\beta+\frac{2k'\varpi i}{n}+\frac{2k\omega}{n}\right) = \theta^{k'}\psi_1\beta$ so that $\psi\beta\cdot\psi_1\beta$ and $(\psi\beta)^n$+$(\psi_1\beta)^n$, which are rational in $\varphi\beta$, take the same value when $\varphi\beta$ is replaced by any other root of the considered equation and are therefore rational in $\varphi\alpha$. The $n-1$ different values of θ give $n-1$ values A_j and B_j for A and B and one has

$$\varphi_1\left(\beta+\frac{2k\varpi i}{n}\right) = \varphi\alpha + \frac{1}{n}\sum_j \theta_j^{-k}\sqrt[n]{A_j+\sqrt{A_j^2-B_j^n}}. \tag{36}$$

Then Abel uses $\psi_2\beta = \sum_{|m|\le\frac{n-1}{2}} \theta^m\varphi_1\left(\beta+\frac{2m\omega}{n}\right)$, $\psi_3\beta = \sum_{|m|\le\frac{n-1}{2}} \theta^m\varphi_1\left(\beta-\frac{2m\omega}{n}\right)$, such that $\psi_2\beta\cdot\psi_3\beta$ and $(\psi_2\beta)^n+(\psi_3\beta)^n$ are rational functions of $\varphi_1\beta$. He gets $\psi_2\beta = \sqrt[n]{C+\sqrt{C^2-D^n}}$ with C, D rational in $\varphi_1\beta$ and

$$\varphi\beta = \frac{1}{n}\left(\varphi_1\beta + \sum_j \sqrt[n]{C_j+\sqrt{C_j^2-D_j^n}}\right). \tag{37}$$

The radicals in (36) and (37) are not independent (otherwise each formula should give n^{n-1} different values). Indeed, if $\psi^k\beta = \sum_{|\mu|\le\frac{n-1}{2}} \theta^{k\mu}\varphi_1\left(\beta+\frac{2\mu\varpi i}{n}\right)$, $\psi_1^k\beta = \sum_{|\mu|\le\frac{n-1}{2}} \theta^{k\mu}\varphi_1\left(\beta-\frac{2\mu\varpi i}{n}\right)$, where $\theta = \cos\frac{2\pi}{n}+i\sin\frac{2\pi}{n}$, $\psi^k\left(\beta+\frac{2\nu\varpi i}{n}\right) = \theta^{-k\nu}\psi^k\beta$ and $\psi_1^k\left(\beta+\frac{2\nu\varpi i}{n}\right) = \theta^{k\nu}\psi_1^k\beta$ so that $\frac{\psi^k\beta}{(\psi^1\beta)^k}+\frac{\psi_1^k\beta}{(\psi_1^1\beta)^k}$ and $\frac{\psi^k\beta}{(\psi^1\beta)^{k-n}}+\frac{\psi_1^k\beta}{(\psi_1^1\beta)^{k-n}}$ are rational functions of $\varphi\alpha$. As

$$(\psi^1\beta)^n = A_1+\sqrt{A_1^2-B_1^n} \quad\text{and}\quad (\psi_1^1\beta)^n = A_1-\sqrt{A_1^2-B_1^n},$$

it is easy to deduce $\psi^k\beta = (\psi^1\beta)^k\cdot(F_k+H_k\sqrt{A_1^2-B_1^n})$, with F_k and H_k rational in $\varphi\alpha$, that is $\sqrt[n]{A_k+\sqrt{A_k^2-B_k^n}} = (A_1+\sqrt{A_1^2-B_1^n})^{\frac{k}{n}}(F_k+H_k\sqrt{A_1^2-B_1^n})$. In the same way, $\sqrt[n]{C_k+\sqrt{C_k^2-D_k^n}} = (C_1+\sqrt{C_1^2-D_1^n})^{\frac{k}{n}}(K_k+L_k\sqrt{A_1^2-B_1^n})$, with K_k and L_k rational in $\varphi\alpha$.

For the other problem (division of the periods by an odd prime n), the roots are

$$\varphi^2\left(\frac{m\omega\pm\mu\varpi i}{n}\right), 0\le m,\mu\le\frac{n-1}{2}, (m,\mu)\ne(0,0);$$

Abel groups them according to the points $m{:}\mu$ of the projective line $\boldsymbol{P}_1(\boldsymbol{F}_n)$, which gives $n+1$ groups of $\frac{n-1}{2}$ roots each: $\nu(1,0)$, $\nu(m,1)$, $1\le\nu\le\frac{n-1}{2}$, $0\le m\le n-1$.

He puts $r_\nu = \varphi^2\left(\frac{\nu\omega}{n}\right), r_{\nu,m} = \varphi^2\left(\nu\frac{m\omega+\varpi i}{n}\right)$. If $p = \psi\left(\varphi^2\left(\frac{\omega'}{n}\right)\right)$ is any rational symmetric function of the $\varphi^2\left(\frac{k\omega'}{n}\right)$, $1 \leq k \leq \frac{n-1}{2}$, $\omega' = m\omega + \mu\varpi i$, then $\psi r_\nu = \psi r_1$ and $\psi r_{\nu,m} = \psi r_{1,m}$. Consider now the equation of degre $n+1$ of which the roots are ψr_1 and $\psi r_{1,m}$ $(0 \leq m \leq n-1)$. The sum of the k-th powers of these roots is $\frac{2}{n-1}$ times the sum of the $(\psi r_\nu)^k$ and of the $(\psi r_{\nu,m})^k$, so it is rational with respect to e and c, and the same is true for the coefficients of the considered equation of degree $n+1$. We thus see that the equations of degree $\frac{n-1}{2}$ respectively giving the r_ν and the $r_{\nu,m}$ for m fixed have coefficients given by an equation of degree $n+1$. In fact one such auxiliary equation of degree $n+1$ is sufficient, for when a function such as $p = \psi r_1$ is given, any other such function $q = \theta r_1$ is a rational function of p. Abel proves this by a method due to Lagrange, determining θr_1 and the $\theta r_{1,m}$ as solutions of the linear system

$$(\psi r_1)^k \theta r_1 + (\psi r_{1,0})^k \theta r_{1,0} + (\psi r_{1,1})^k \theta r_{1,1} + \ldots + (\psi r_{1,n-1})^k \theta r_{1,n-1} = s_k$$

where the s_k are easily seen to be symmetric with respect to $r_1, r_2, \ldots, r_{1,0}, \ldots$ and so rational functions of e and c.

In order to solve the $n+1$ equations of degree $\frac{n-1}{2}$, with roots $\varphi^2\left(\frac{k\omega'}{n}\right)$, Abel procedes as Gauss did for the cyclotomy: putting $\varepsilon = \frac{\omega'}{n}$ and α a primitive root *modulo* n, the roots may then be written $\varphi^2(\alpha^\ell \varepsilon)$, $0 \leq \ell \leq \frac{n-3}{2}$ and the Lagrange resolvant is

$$\psi(\varepsilon) = \varphi^2(\varepsilon) + \varphi^2(\alpha\varepsilon)\theta + \ldots + \varphi^2\left(\alpha^{\frac{n-3}{2}}\varepsilon\right)\theta^{\frac{n-3}{2}},$$

where θ is a $\frac{n-1}{2}$-th root of 1. It is a rational function of $\varphi^2(\varepsilon)$ and its $\frac{n-1}{2}$-th power v is symmetric with respect to the $\frac{n-1}{2}$ roots. Thus v is known when the equation of degree $n+1$ is solved. Varying θ, we get $\frac{n-1}{2}$ values v_k for v and

$$\varphi^2(\varepsilon) = \frac{2}{n-1}\left(-p_{\frac{n-3}{2}} + \sqrt[\frac{n-1}{2}]{v_1} + \ldots + \sqrt[\frac{n-1}{2}]{v_{\frac{n-3}{2}}}\right),$$

where $-p_{\frac{n-3}{2}} = \varphi^2(\varepsilon) + \varphi^2(\alpha\varepsilon) + \ldots + \varphi^2\left(\alpha^{\frac{n-3}{2}}\varepsilon\right)$ corresponds to $\theta = 1$ and is symmetric. The $n-1$ radicals are not independent: $s_k = \dfrac{\sqrt[\frac{n-1}{2}]{v_k}}{\left(\sqrt[\frac{n-1}{2}]{v_1}\right)^k}$ is a rational function of $\varphi^2(\varepsilon)$ which remains invariant when ε is replaced by $\alpha^m\varepsilon$, so it is a rational function of the root of the auxiliary equation of degree $n+1$.

When $e = c = 1, \omega = \varpi$ and there exists a *complex multiplication*, that is a formula

$$\varphi(m + \mu i)\delta = \varphi\delta \cdot T,$$

for m, μ integers and $m + \mu$ odd, with T a rational function if $\varphi^4(\delta)$. This permits the solution by radicals of the equation giving $\varphi\left(\frac{\omega}{n}\right)$ whenever n is a prime number of the form $4\nu+1$. Indeed, consider $n=\alpha^2+\beta^2$, where α and β are relatively prime integers such that $\alpha+\beta$ is odd, and there exists integers m, t such that $1=2m\alpha-nt$; then $\frac{m\omega}{\alpha+\beta i}+\frac{m\omega}{\alpha-\beta i}=\frac{2m\alpha}{n}$ and

$$\varphi\left(\frac{2m\alpha}{n}\right)=\varphi\left(\frac{\omega}{n}+t\omega\right)=(-1)^t\varphi\left(\frac{\omega}{n}\right),$$

so it is sufficient to consider the equation for $\varphi\left(\frac{\omega}{\alpha+\beta i}\right)=\varphi\delta$. This equation is equivalent to $\varphi(\alpha+\beta i)\delta=0$ and its roots are $x=\varphi\left(\frac{m+\mu i}{\alpha+\beta i}\omega\right)$, m, μ integers; it is easy to reduce these roots to the form $x=\varphi\left(\frac{\rho\omega}{\alpha+\beta i}\right)$, $|\rho|\leq\frac{n-1}{2}=2\nu$ and to see that, in this form, they are all distinct. Moreover, the equation has no multiple roots, and finally, since φ is an odd function, we have to consider an equation of degree 2ν, with roots $\varphi^2(\delta), \varphi^2(2\delta), \ldots, \varphi^2(2\nu\delta)$ or $\varphi^2(\delta), \varphi^2(\varepsilon\delta), \varphi^2(\varepsilon^2\delta), \ldots, \varphi^2(\varepsilon^{2\nu-1}\delta)$, where ε is a primitve root *modulo* n. Then, as for the preceding problem, one sees that $\varphi^2(\varepsilon^m\delta)=\frac{1}{2\nu}\left(A+\theta^{-m}v^{\frac{1}{2\nu}}+s_2\theta^{-2m}v^{\frac{2}{2\nu}}+s_{2\nu-1}\theta^{-(2\nu-1)m}v^{\frac{2\nu-1}{2\nu}}\right)$, where

$$\begin{aligned}
v &= (\varphi^2(\delta)+\theta\varphi^2(\varepsilon\delta)+\theta^2\varphi^2(\varepsilon^2\delta)+\ldots+\theta^{2\nu-1}\varphi^2(\varepsilon^{2\nu-1}\delta))^{2\nu},\\
s_k &= \frac{\varphi^2(\delta)+\theta^k\varphi^2(\varepsilon\delta)+\theta^{2k}\varphi^2(\varepsilon^2\delta)+\ldots+\theta^{(2\nu-1)k}\varphi^2(\varepsilon^{2\nu-1}\delta)}{(\varphi^2(\delta)+\theta\varphi^2(\varepsilon\delta)+\theta^2\varphi^2(\varepsilon^2\delta)+\ldots+\theta^{2\nu-1}\varphi^2(\varepsilon^{2\nu-1}\delta))^k},\\
A &= \varphi^2(\delta)+\varphi^2(\varepsilon\delta)+\varphi^2(\varepsilon^2\delta)+\ldots+\varphi^2(\varepsilon^{2\nu-1}\delta)
\end{aligned}$$

are rational with respect to the coefficients of the equation, so of the form $a+bi$ with a and b rational numbers. Abel oberves that, when n is a Fermat prime number 2^N+1, all the radicals in the solution are of index 2 for $2\nu=2^{N-1}$ and $\theta^{2^{N-1}}=1$. He applies these results to the division of the lemniscate of polar equation $x=\sqrt{\cos 2\theta}$ (x distance to the origin, θ polar angle), for which the elementary arc is $\frac{dx}{\sqrt{1-x^4}}$.

All these examples of solvable equations (Moivre, Gauss, elliptic functions) gave Abel models for a general class of solvable equations; following Kronecker, we call them Abelian equations and they are the object of a memoir published in the fourth volume of *Crelle's Journal* (1829, *Œuvres*, t. I, p. 478–507). To begin with, Abel defines (in a footnote) the notion of an irreducible equation with coefficients rational functions of some quantities $a, b, c, \ldots$ considered as known; his first theorem states that if a root of an irreducible equation $\varphi x=0$ annihilates a rational function fx of x and the same quantities $a, b, c, \ldots$, then the it is still true for any other root of $\varphi x=0$ (the proof is given in a footnote).

The second theorem states that if an irreducible equation $\varphi x=0$ of degree μ has two roots x' and x_1 related by a rational relation $x'=\theta x_1$ with known coefficients, then the given equation may be decomposed in m equations of degree n of which the coefficients are rational functions of a root of an auxiliary equation of degree m (naturally $\mu=mn$). First of all, the equation $\varphi(\theta x_1)=0$ with the theorem I shows that $\varphi(\theta x)=0$ for any root x of $\varphi x=0$; so $\theta x'=\theta^2x_1, \theta^3x_1, \ldots$ are all roots of $\varphi x=0$. If $\theta^m x_1=\theta^{m+n}x_1$ (the equation has only a finite number of roots), or $\theta^n(\theta^m x_1) - \theta^m x_1 = 0$, we have $\theta^n x - x = 0$ for any root of $\varphi x = 0$ by the theorem I and, in particular $\theta^n x_1 = x_1$; if n is minimal with this property, $x_1, \theta x, \dots, \theta^{n-1}x_1$ are distinct roots and the sequence $(\theta^m x_1)$ is periodic with period n. When $\mu > n$, there exists a root x_2 which does not belong to this sequence; then $(\theta^m x_2)$ is a new sequence of roots with exactly the same period for $\theta^n x_2 = x_2$ and if $\theta^k x_2 = x_2$ for a $k < n$, we should have $\theta^k x_1 = x_1$. When $\mu > 2n$, there exists a root x_3 different from the $\theta^m x_1$ and the $\theta^m x_2$, and the sequence $(\theta^m x_3)$ has a period n. Continuing in this way, we see that μ is necessarily a multiple mn of n and that the μ roots may be grouped in m sequences $(\theta^k x_j)_{0 \le k \le n-1}$ $(j = 1, 2, \dots, m)$. Note that this proof is analogous to that of Lagrange establishing that a rational function of n letters takes, under the substitutions of these letters, a number of values which divides $n!$. In order to prove his second theorem, Abel considers a rational symmetric function $y_1 = f(x_1, \theta x_1, \dots, \theta^{n-1}x_1) = Fx_1$ of the first n roots and the corresponding $y_j = Fx_j = F(\theta^k x_j)$ $(2 \le j \le m)$; for any natural integer ν, the sum $y_1^\nu + y_2^\nu + \dots + y_m^\nu$ is symmetric with respect to the mn roots of $\varphi x = 0$, so it is a known quantity and the same is true for the coefficients of the equation with the roots $y_1, y_2, \dots, y_m$. Since the equation with the roots $x_1, \theta x_1, \dots, \theta^{n-1}x_1$ has its coefficients rational symmetric functions of $x_1, \theta x_1, \dots, \theta^{n-1}x_1$, each of these coefficients is a root of an equation of degree m with known coefficients. In fact one auxiliary equation of degree m is sufficient: this is proved by the stratagem of Lagrange already used by Abel for the division of the periods of elliptic functions (Abel notes that it is necessary to choose the auxiliary equation without multiple roots, which is always possible).

When $m = 1$, $\mu = n$; the roots constitute only one sequence $x_1, \theta x_1, \dots, \theta^{\mu-1}x_1$ and the equation $\varphi x = 0$ is algebraically solvable as Abel states it in his theorem III. We now say that the equation $\varphi x = 0$ is *cyclic*. This result comes from the fact that the Lagrange resolvant

$$x + \alpha\theta x + \alpha^2\theta^2 x + \dots + \alpha^{\mu-1}\theta^{\mu-1}x$$

(x any root of the equation, α μ-th root of 1) has a μ-th power v symmetric with respect to the μ roots. We now get $x = \frac{1}{\mu}(-A + \sqrt[\mu]{v_1} + \sqrt[\mu]{v_2} + \dots + \sqrt[\mu]{v_{\mu-1}})$, where $v_0, v_1, \dots, v_{\mu-1}$ are the values of v corresponding to the diverse μ-th roots α of 1 ($\alpha = 1$ for v_0) and $-A = \sqrt[\mu]{v_0}$. The $\mu - 1$ radicals are not independent for if $\alpha = \cos\frac{2\pi}{\mu} + \sqrt{-1}\sin\frac{2\pi}{\mu}$ and if

$$\sqrt[\mu]{v_k} = x + \alpha^k\theta x + \alpha^{2k}\theta^2 x + \dots + \alpha^{(m-1)k}\theta^{\mu-1}x,$$

the quantity $\sqrt[\mu]{v_k}(\sqrt[\mu]{v_1})^{\mu-k} = a_k$ is a symmetric function of the roots of $\varphi x = 0$, so it is known. Abel recalls that this method was used by Gauss in order to solve the equations of the cyclotomy. The theorem IV is a corollary of the preceding one: when the degree μ is a prime number and two roots of $\varphi x = 0$ are such that one of them is a rational function of the other, then the equation is algebraically solvable.

As $a_{\mu-1} = \sqrt[\mu]{v_{\mu-1}} \cdot \sqrt[\mu]{v_1} = a$, does not change when α is replaced by its complex conjugate, it is real when the known quantities are supposed to be real. Thus v_1 and $v_{\mu-1}$ are complex conjugate and

$$v_1 = c + \sqrt{-1}\sqrt{a^\mu - c^2} = (\sqrt{\rho})^\mu(\cos\delta + \sqrt{-1}\cdot\sin\delta),$$

and so

$$\sqrt[\mu]{v_1} = \sqrt{\rho}\cdot\left(\cos\frac{\delta+2m\pi}{\mu} + \sqrt{-1}\cdot\sin\frac{\delta+2m\pi}{\mu}\right).$$

So in order to solve $\varphi x = 0$ it suffices to divide the circle in μ equal parts, to divide the angle δ (which is constructible) in μ equal parts and to extract the square root of ρ. Moreover, Abel notes that the roots of $\varphi x = 0$ are all real or all imaginary; if μ is odd they are all real.

The theorem VI is relative to a cyclic equation $\varphi x = 0$ of composite degree $\mu = m_1 \cdot m_2 \cdots m_\omega = m_1 \cdot p_1$. Abel groups the roots in m_1 sequences $(\theta^{km_1+j}x)_{0\le k\le p_1-1}$ $(0 \le j \le m_1 - 1)$ of p_1 roots each. This allows the decomposition of the equation in m_1 equations of degree p_1 with coefficients rational functions of a root of an auxiliary equation of degree m_1. In the same way, each equation of degree $p_1 = m_2 \cdot p_2$ is decomposed in m_2 equations of degree p_2 using an auxiliary equation of degree m_2, etc. Finally, the solution of $\varphi x = 0$ is reduced to that of ω equations of respective degrees $m_1, m_2, \ldots, m_\omega$. As Abel notes, this is precisely what Gauss did for the cyclotomy. The case in which $m_1, m_2, \ldots, m_\omega$ are relatively prime by pairs is particularly interesting. Here for $1 \le k \le \omega$ an auxiliary equation $fy_k = 0$ of degree m_k allows to decompose $\varphi x = 0$ in m_k equations $F_k(\theta^j x, y_k) = 0$ of degree $n_k = \frac{\mu}{m_k}$ $(0 \le j \le m_k - 1)$. Since x is the only common root of the ω equations $F_k(x, y_k) = 0$ (for $\theta^{km_p}x = \theta^{\ell m_q}x$ with $k \le n_p - 1$ and $\ell \le n_q - 1$ implies $km_p = \ell m_q$ and then $k = \ell = 0$ if $p \ne q$), it is rational with respect to $y_1, y_2, \ldots, y_\omega$. So, in this case, the resolution is reduced to that of the equations $f_1y_1 = 0$, $f_2y_2 = 0, \ldots, f_\omega y_\omega = 0$ of respective degrees $m_1, m_2, \ldots, m_\omega$ and with coefficients known quantities. One may take for the m_k the prime-powers which compose μ.

All the auxiliary equations are cyclic as is $\varphi x = 0$, so they may be solved by the same method. This follows from the fact that if

$$y = Fx = f(x, \theta^m x, \theta^{2m}x, \ldots, \theta^{(n-1)m}x)$$

is symmetric with respect to $x, \theta^m x, \theta^{2m}x, \ldots, \theta^{(n-1)m}x$, so is $F(\theta x)$. Then, by Lagrange's stratagem $F(\theta x)$ is a rational function λy of y.

Abel ends this part of the memoir with the theorem VII, relating to a cyclic equation of degree 2^ω: its solution amounts to the extraction of ω square roots. This is the case for Gauss' division of the circle by a Fermat prime.

The second part deals with algebraic equations of which all the roots are rational functions of one of them, say x. According to Abel's theorem VIII, if $\varphi x = 0$ is such an equation of degree μ and if, for any two roots θx and $\theta_1 x$ the relation $\theta\theta_1 x = \theta_1\theta x = 0$ is true, then the equation is algebraically solvable. Abel begins by observing that one may suppose that $\varphi x = 0$ is irreducible. So that if n is the period of $(\theta^k x)$, the roots are grouped in $m = \frac{\mu}{n}$ groups of n roots. Each group contains the roots of an equation of degree n with coefficients rational functions of a quantity $y = f(x, \theta x, \theta^2 x, \ldots, \theta^{n-1}x)$ given by an equation of degree m

with known coefficients, which is easily seen to be irreducible. The other roots of the equation in y are of the form $y_1 = f(\theta_1 x, \theta\theta_1 x, \theta^2\theta_1 x, \ldots, \theta^{n-1}\theta_1 x) = f(\theta_1 x, \theta_1\theta x, \theta_1\theta^2 x, \ldots, \theta_1\theta^{n-1}x)$ (by the hypothesis), so rational symmetric with respect to $x, \theta x, \theta^2 x, \ldots, \theta^{n-1}x$ and (again by Lagrange's stratagem) rational in y: $y_1 = \lambda y$. Now if $y_2 = \lambda_1 y = f(\theta_2 x, \theta\theta_2 x, \theta^2\theta_2 x, \ldots, \theta^{n-1}\theta_2 x)$,

$$\begin{aligned}\lambda\lambda_1 y = \lambda y_2 &= f(\theta_1\theta_2 x, \theta\theta_1\theta_2 x, \ldots, \theta^{n-1}\theta_1\theta_2 x)\\ &= f(\theta_2\theta_1 x, \theta\theta_2\theta_1 x, \ldots, \theta^{n-1}\theta_2\theta_1 x) = \lambda_1\lambda y\end{aligned}$$

so that the equation in y has the same property as the initial equation $\varphi x = 0$ and it is possible to deal with it in the same manner. Finally $\varphi x = 0$ is solvable through a certain number of cyclic equations of degrees $n, n_1, n_2, \ldots, n_\omega$ such that $\mu = nn_1n_2\cdots n_\omega$, this is Abel's theorem IX. In the theorem X, Abel states that when $\mu = \varepsilon_1^{\nu_1}\varepsilon_2^{\nu_2}\ldots\varepsilon_\alpha^{\nu_\alpha}$ with $\varepsilon_1, \varepsilon_2, \ldots, \varepsilon_\alpha$ prime, the solution amounts to that of ν_1 equations of degree ε_1, ν_2 equations of degree $\varepsilon_2, \ldots, \nu_\alpha$ equations of degree ε_α, all solvable by radicals.

As an example, Abel applies his general theorem to the division of the circle in $\mu = 2n + 1$ equal parts, where μ is a prime number; the equation with roots $\cos\frac{2\pi}{\mu}, \cos\frac{4\pi}{\mu}, \ldots, \cos\frac{2n\pi}{\mu}$ has rational coefficients and it is cyclic. If m is a primitive root *modulo* μ, the roots are $x, \theta x, \ldots, \theta^{n-1}x$ where $x = \cos\frac{2\pi}{\mu} = \cos a$ and $\theta x = \cos ma$, polynomial of degree m. As Gauss has proved, the division of the circle is reduced in μ parts is reduced to the division of the circle in n parts, the division of a certain (constructible) angle in n parts and the extraction of a square root of a quantity

$$\begin{aligned}\rho = |(x + \alpha\theta x + \alpha^2\theta^2 x + \ldots + \alpha^{n-1}\theta^{n-1}x)\\ \times (x + \alpha^{n-1}\theta x + \alpha^{n-2}\theta^2 x + \ldots + \alpha\theta^{n-1}x)|.\end{aligned}$$

It is not difficult to compute $\pm\rho = \frac{1}{2}n - \frac{1}{4} - \frac{1}{2}(\alpha + \alpha^2 + \ldots + \alpha^{n-1}) = \frac{1}{2}n + \frac{1}{4}$ so the square root is $\sqrt{\mu}$ conformally to Gauss' result. After his notebooks, we know that Abel also wanted to apply his theory to the division of periods of elliptic functions with a singular *modulus*, precisely in the case where $\omega = \varpi\sqrt{2n+1}$.

On the 8 of October 1828, Abel sent the statement of three theorems on algebraic equations to Crelle.

A. Given a prime number n and n unknown quantities $x_1, x_2, \ldots, x_n$ related by the relations $\varphi(x_1, x_2, \ldots, x_n) = \varphi(x_2, x_3, \ldots, x_1) = \ldots = \varphi(x_n, x_1, \ldots, x_{n-1}) = 0$, where φ is a polynomial of degree m, the equation of degree $m^n - m$ obtained by elimination of $n - 1$ of the quantities and division by the factor $\varphi(x, x, \ldots, x)$ is decomposable in $\frac{m^n - m}{n}$ equations of degree n, all algebraically solvable, with the help of an equation of degree $\frac{m^n - m}{n}$. Abel gives, as examples, the cases where $n = 2, m = 3$ and $n = 3, m = 2$; in these cases $m^n - m = 6$ and the equation of degree 6 is algebraically solvable.

B. If three roots of an irreducible equation of prime degree are so related that one of them is rationally expressed by the other two, then the equation is algebraically solvable.

This theorem is given, as a necessary and sufficient condition, by E. Galois as an application of his *Mémoire sur les conditions de résolubilité des équations par radicaux* (1831) and it was at first interpreted as the main result of this memoir.

C. If two roots of an irreducible equation of prime degree are so related that one of them is rationally expressed by the other, then the equation is algebraically solvable.

This statement is the same as that of the theorem IV in the 1829 memoir (which was composed in March 1828).

Abel left uncompleted an important paper *Sur la théorie algébrique des équations* (*Œuvres*, t. II, p. 217–243). In the introduction, he explains in a very lucid way his method in mathematics, saying that one must give a problem such a form that it is always possible to solve it. For the case of the solution by radicals of algebraic equations, Abel formulates certain problems:

(1) To find all the equations of a given degree which are algebraically solvable.
(2) To judge whether a given equation is algebraically solvable.
(3) To find all the equations that a given algebraic function may satisfy.

Here an algebraic function is defined, as in the 1826 paper, as built by the operations of addition, subtraction, multiplication, division and extraction of roots of prime index. There are two types of equations to consider: those for which the coefficients are rational functions of certain variables $x, z, z', z'', \ldots$ (with arbitrary numerical coefficients; for instance the general equation of a given degree, for which the coefficients are independent variables) and those for which the coefficients are constant; in the last case the coefficients are supposed to be rational expressions in given numerical quantities $\alpha, \beta, \gamma, \ldots$ with rational coefficients. An equation of the first type is said to be algebraically satisfied (resp. algebraically solvable) when it is verified when the unknown is replaced by an algebraic function of $x, z, z', z'', \ldots$ (resp. when all the roots are algebraic functions of $x, z, z', z'', \ldots$); there are analogous definitions for the second type, with "algebraic function of $x, z, z', z'', \ldots$" replaced by "algebraic expression of $\alpha, \beta, \gamma, \ldots$".

In order to attack his three problems, Abel is led to solve the following ones "To find the most general form of an algebraic expression" and "To find all the possible equations which an algebraic function may satisfy". These equations are infinite in number but, for a given algebraic function, there is one of minimal degree, and this one is irreducible.

Abel states some general results he has obtained about these problems:

(1) If an irreducible equation may be algebraically satisfied, it is algebraically solvable; the same expression represents all the roots, by giving the radicals in it all their values.
(2) If an algebraic expression satisfies an equation, it is possible to give it such a form that it still satisfies the equation when one gives to the radicals in it all their values.
(3) The degree of an irreducible algebraically solvable equation is the product of certain indexes of the radicals in the expression of the roots.

About the problem "To find the most general algebraic expression which may staisfy an equation of given degree", Abel states the following results:

(1) If an irreducible equation of prime degree μ is algebraically solvable, its roots are of the form $y = A + \sqrt[\mu]{R_1} + \sqrt[\mu]{R_2} + \ldots + \sqrt[\mu]{R_{\mu-1}}$, where A is rational and $R_1, R_2, \ldots, R_{\mu-1}$ are roots of an equation of degree $\mu - 1$.
This form was conjectured by Euler (1738) for the general equation of degree μ.

(2) If an irreducible equation of degree μ^α, with μ prime, is algebraically solvable, either it may be decomposed in $\mu^{\alpha-\beta}$ equations of degree μ^β of which the coefficients depend on an equation of degree $\mu^{\alpha-\beta}$, or each root has the form $y = A + \sqrt[\mu]{R_1} + \sqrt[\mu]{R_2} + \ldots + \sqrt[\mu]{R_\nu}$, with A rational and $R_1, R_2, \ldots, R_\nu$ roots of an equation of degree $\nu \le \mu^\alpha - 1$.

(3) If an irreducible equation of degree μ not a prime-power is algebraically solvable, it is possible to decompose μ in a product of two factors μ_1 and μ_2 and the equation in μ_1 equations of degree μ_2 of which the coefficients depend on equations of degree μ_1.

(4) If an irreducible equation of degree μ^α, with μ prime, is algebraically solvable, its roots may be expressed by the formula $y = f(\sqrt[\mu]{R_1}, \sqrt[\mu]{R_2}, \ldots, \sqrt[\mu]{R_\alpha})$ with f rational symmetric and $R_1, R_2, \ldots, R_\alpha$ roots of an equation of degree $\le \mu^\alpha - 1$.

A corollary of (3) is that when an irreducible equation of degree $\mu = \mu_1^{\alpha_1}\mu_2^{\alpha_2}\ldots\mu_\omega^{\alpha_\omega}$ ($\mu_1, \mu_2, \ldots, \mu_\omega$ prime) is algebraically solvable, only the radicals necessary to express the roots of equations of degrees $\mu_1^{\alpha_1}, \mu_2^{\alpha_2}, \ldots, \mu_\omega^{\alpha_\omega}$ appears in the expression of the roots. Abel adds that if an irreducible equation is algebraically solvable, its roots may be found by Lagrange's method. According to this method, an equation of degree μ is reduced to the solution of $\frac{(\mu-1)!}{\varphi(\mu)}$ equations of degree $\varphi(\mu)$ (φ the Euler function) with the help of an equation of degree $\frac{(\mu-1)!}{\varphi(\mu)}$ (Abel text leaves a blank at the place of these numbers). Abel announces that a necessary condition for the algebraic solvability is that the equation of degree $\frac{(\mu-1)!}{\varphi(\mu)}$ have a root rational with respect to the coefficients of the proposed equation; if μ is a prime number, this condition is also sufficient.

The first paragraph of the paper explains the structure of algebraic expressions, as was done in the published 1826 article; this time, the order of such an expression is defined as the minimum number of radicals necessary to write it. In the second paragraph, a polynomial

$$y^n + Ay^{n-1} + A'y^{n-2} + \ldots = \varphi(y, m)$$

is said to be of order m when the maximum order of its coefficients $A, A', \ldots$ is m. The first theorem states that an expression $t_0 + t_1 y_1^{\frac{1}{\mu_1}} + t_2 y_1^{\frac{2}{\mu_1}} + \ldots + t_{\mu_1-1} y_1^{\frac{\mu_1-1}{\mu_1}}$, with $t_0, t_1, \ldots, t_{\mu_1-1}$ rational with respect to a μ_1-th root ω of 1 and radicals different from $y_1^{\frac{1}{\mu_1}}$, is 0 only if $t_0 = t_1 = \ldots = t_{\mu_1-1} = 0$. The second theorem states that if an equation $\varphi(y, m) = 0$ of order m is satisfied by an algebraic expression

$y = p_0 + p_1 \sqrt[\mu_1]{y_1} + \dots$ of order $n > m$, it is still satisfied by the expression with $\omega \sqrt[\mu_1]{y_1}, \omega^2 \sqrt[\mu_1]{y_1}, \dots$ instead of $\sqrt[\mu_1]{y_1}$, where ω is a μ_1-th root of 1. After the third theorem when two equations $\varphi(y, m) = 0$ and $\varphi_1(y, n) = 0$ have a common root, the first one being irreducible and $n \le m$, then $\varphi_1(y, n) = f(y, m) \cdot \varphi(y, m)$. Then the fourth theorem says that $\varphi_1(y, n)$ is divisible by the product $\prod \varphi(y, m)$ of $\varphi(y, m)$ and the polynomial $\varphi'(y, m), \varphi''(y, m), \dots, \varphi^{(\mu-1)}(y, m)$ obtained by successively replacing in $\varphi(y, m)$ the outermost radical $\sqrt[\mu]{y_1}$ by $\omega \sqrt[\mu]{y_1}, \omega^2 \sqrt[\mu]{y_1}, \dots$ (ω μ-th root of 1); this comes from the fact that $\varphi'(y, m), \varphi''(y, m), \dots, \varphi^{(\mu-1)}(y, m)$ are relatively prime by pairs. In the fifth theorem, Abel states that if $\varphi(y, m) = 0$ is irreducible, so is $\prod \varphi(y, m) = \varphi_1(y, m') = 0$.

Now if $a_m = f\left(y_m^{\frac{1}{\mu_m}}, y_{m-1}^{\frac{1}{\mu_{m-1}}}, \dots\right)$, of order m, is a root of an irreducible equation $\psi(y) = 0$, ψ must be divisible by $y - a_m$, and so also by $\prod(y - a_m) = \varphi(y, m_1)$ (theorem IV), which is irreducible (theorem V). It now follows that ψ is divisible by $\prod \varphi(y, m_1) = \varphi_1(y, m_2)$ and by $\prod \varphi_1(y, m_2) = \varphi_2(y, m_3)$, etc., with $m > m_1 > m_2 > \dots$ Finally, we arrive at some $m_{\nu+1} = 0$ and $\varphi_\nu(y, 0)$ divides $\psi(y)$ and has rational coefficients, so that $\psi = \varphi_\nu$. This leads to the degree of ψ, for that of $\varphi(y, m_1)$ is μ_m, that of $\varphi_1(y, m_2)$ is $\mu_m \cdot \mu_{m_1}, \dots$ and that of φ_ν is $\mu_m \cdot \mu_{m_1} \dots \mu_{m_\nu} = \mu$. This is the third general result of the introduction, with the further explanation that the index of the outermost radical is always one of the factors of the degree μ. The first general result of the introduction is also a consequence of that fact, as the fact that an algebraic expression solution of an irreducible equation of degree μ takes exactly μ values.

In the third paragraph, Abel first deals with the case in which μ is a prime number; then $\mu_m = \mu$ and $a_m = p_0 + p_1 s^{\frac{1}{\mu}} + p_2 s^{\frac{2}{\mu}} + \dots + p_{\mu-1} s^{\frac{\mu-1}{\mu}}$, with $s = y_m$; giving to the radical $s^{\frac{1}{\mu}}$ its μ values $s^{\frac{1}{\mu}}, \omega s^{\frac{1}{\mu}}, \dots, \omega^{\mu-1} s^{\frac{1}{\mu}}$, where ω is a μ-th root of 1, we get μ values $z_1, z_2, \dots, z_\mu$ for a_m and, as the given equation has only μ roots, we cannot get new values by replacing the p_j or s by other values p'_j and s' obtained by changing the value of the radicals they contain. Now if

$$p'_0 + p'_1 \omega' s'^{\frac{1}{\mu}} + \dots + p'_{\mu-1} \omega'^{\mu-1} s'^{\frac{\mu-1}{\mu}} = p_0 + p_1 \omega s^{\frac{1}{\mu}} + \dots + p_{\mu-1} \omega^{\mu-1} s^{\frac{\mu-1}{\mu}},$$

we see that different values $\omega_0, \omega_1, \dots, \omega_{\mu-1}$ of the root ω correspond to different values of the root ω' of 1. Writing the corresponding equalities and adding, we obtain $\mu p'_0 = \mu p_0$, so $p'_0 = p_0$ and then

$$\mu p'_1 s'^{\frac{1}{\mu}} = p_1 s^{\frac{1}{\mu}} (\omega_0 + \omega_1 \omega^{-1} + \omega_2 \omega^{-2} + \dots + \omega_{\mu-1} \omega^{-\mu+1}) + \dots$$

So

$$s'^{\frac{1}{\mu}} = f\left(\omega, p_0, p'_0, p_1, p'_1, \dots, s', s^{\frac{1}{\mu}}\right) = q_0 + q_1 s^{\frac{1}{\mu}} + q_2 s^{\frac{2}{\mu}} + \dots + q_{\mu-1} s^{\frac{\mu-1}{\mu}}$$

and $s' = t_0 + t_1 s^{\frac{1}{\mu}} + t_2 s^{\frac{2}{\mu}} + \dots + t_{\mu-1} s^{\frac{\mu-1}{\mu}}$, although Abel's given proof of the fact that $t_1 = t_2 = \dots = t_{\mu-1} = 0$ is not quite complete. In the notes at the end of the second volume of Abel's *Works*, Sylow has explained how to complete the proof (p. 332–335) in order to finally obtain $p_1'^{\mu}s' = p_\nu^{\mu}s^\nu$ for some ν between 2 and $\mu - 1$; this shows that $p_1^{\mu}s$ is root of an equation of degree $\leq \mu - 1$.

Changing s, it is possible to get $p_1 = 1$ and then, we have as usual $p_0 = \frac{1}{\mu}(z_1 + z_2 + \ldots + z_\mu)$ a known quantity, $s^{\frac{1}{\mu}} = \frac{1}{\mu}(z_1 + \omega^{\mu-1}z_2 + \ldots + \omega z_\mu)$, $p_2 s^{\frac{2}{\mu}} = \frac{1}{\mu}(z_1 + \omega^{\mu-2}z_2 + \ldots + \omega^2 z_\mu), \ldots$; this gives

$$p_2 s = \left(\frac{1}{\mu}\right)^{\mu-1} (z_1+\omega^{-2}z_2+\ldots+\omega^{-2(\mu-1)}z_\mu)(z_1+\omega^{-1}z_2+\ldots+\omega^{-(\mu-1)}z_\mu)^{\mu-2},$$

$$p_3 s = \left(\frac{1}{\mu}\right)^{\mu-2} (z_1+\omega^{-3}z_2+\ldots+\omega^{-3(\mu-1)}z_\mu)(z_1+\omega^{-1}z_2+\ldots+\omega^{-(\mu-1)}z_\mu)^{\mu-3}, \text{ etc.}$$

By the usual Lagrangian stratagem, Abel proves that $q_1 = p_m s$ is a rational function of s and the known quantities for $2 \leq m \leq \mu - 1$. The ν distinct values of s are of the form $p_m^{\mu}s^m$ with $2 \leq m \leq \mu - 1$; Abel shows that the irreducible equation of which s is a root is *cyclic* of degree dividing $\mu - 1$, with roots $s, s_1 = \theta s, \ldots, s_{\nu-1} = \theta^{\nu-1}s$, where $\theta s = (fs)^\mu s^{m^\alpha}$, f rational, $2 \leq m \leq \mu - 1$ and α a divisor of $\mu - 1$. He finally arrives at the following form for the root z_1 of $\psi y = 0$:

$$\begin{aligned} z_1 = p_0 + s^{\frac{1}{\mu}} + s_1^{\frac{1}{\mu}} + \ldots + s_{\nu-1}^{\frac{1}{\mu}} + \varphi_1 s \cdot s^{\frac{m}{\mu}} + \varphi_1 s_1 \cdot s_1^{\frac{m}{\mu}} \\ + \ldots + \varphi_1 s_{\nu-1} \cdot s_{\nu-1}^{\frac{m}{\mu}} + \varphi_2 s \cdot s^{\frac{m^2}{\mu}} + \varphi_2 s_1 \cdot s_1^{\frac{m^2}{\mu}} + \ldots + \varphi_2 s_{\nu-1} \cdot s_{\nu-1}^{\frac{m^2}{\mu}} \\ + \ldots + \varphi_{\alpha-1} s \cdot s^{\frac{m^{\alpha-1}}{\mu}} + \varphi_{\alpha-1} s_1 \cdot s_1^{\frac{m^{\alpha-1}}{\mu}} + \ldots + \varphi_{\alpha-1} s_{\nu-1} \cdot s_{\nu-1}^{\frac{m^{\alpha-1}}{\mu}}, \end{aligned}$$

where the φ_j are rational functions and

$$\begin{aligned} s^{\frac{1}{\mu}} &= A a^{\frac{1}{\mu}} a_1^{\frac{m^\alpha}{\mu}} a_2^{\frac{m^{2\alpha}}{\mu}} \ldots a_{\nu-1}^{\frac{m^{(\nu-1)\alpha}}{\mu}}, \\ s_1^{\frac{1}{\mu}} &= A_1 a^{\frac{m^\alpha}{\mu}} a_1^{\frac{m^{2\alpha}}{\mu}} a_2^{\frac{m^{3\alpha}}{\mu}} \ldots a_{\nu-1}^{\frac{1}{\mu}}, \\ \ldots\ s_{\nu-1}^{\frac{1}{\mu}} &= A_{\nu-1} a^{\frac{m^{(\nu-1)\alpha}}{\mu}} a_1^{\frac{1}{\mu}} a_2^{\frac{m^\alpha}{\mu}} \ldots a_{\nu-1}^{\frac{m^{(\nu-2)\alpha}}{\mu}}, \end{aligned}$$

generalising the form communicated to Crelle in march 1826. Naturally $a, a_1, \ldots, a_{\nu-1}$ are roots of a cyclic equation of degree dividing $\mu - 1$, but Abel does not say anything about it, that part of the paper being almost reduced to computations. Kronecker (1853) rediscovered this result, and stated it more precisely, also studying the form of the roots of cyclic equations.

The last part of the paper contains computations to establish the second statement relative to the problem "To find the most general algebraic expression which may satisfy an equation of given degree"; Sylow gives an interpretation of these computations at the end of the volume (p. 336–337).

4 Hyperelliptic Integrals

Abel studied Legendre's *Exercices de Calcul intégral* at the fall of 1823 and this book inspired him a series of new important discoveries; we already saw some of them. A memoir presented in 1826 to the Royal Society of Sciences in Throndhjem (*Œuvres*, t. I, p. 40–60) is devoted to a generalisation of Legendre's formula for the exchange of the parameter and the argument in elliptic integrals of the third kind. Abel considers an integral $p = \int \frac{e^{fx}\varphi x dx}{x-a}$ taken from $x = c$, where f is a rational function and $\varphi x = k(x+\alpha)^{\beta}(x+\alpha')^{\beta'}\dots(x+\alpha^{(n)})^{\beta^{(n)}}$ with $\beta, \beta', \dots$ rational numbers; derivating with respect to the parameter a and comparing with the derivative of $\frac{e^{fx}\varphi x}{x-a}$ with respect to x, he obtains

$$\begin{aligned}\frac{dp}{da} - \left(f'a + \frac{\varphi' a}{\varphi a}\right)p = \\ -\frac{e^{fx}\varphi x}{x-a} + \frac{e^{fc}\varphi c}{c-a} + \sum\sum p\gamma^{(p)}a^{p'}\int e^{fx}\varphi x \cdot x^{p-p'-2}dx \\ -\sum \frac{\beta^{(p)}}{a+\alpha^{(p)}}\int \frac{e^{fx}\varphi x dx}{x+\alpha^{(p)}} + \sum\sum \frac{\mu^{(p)}\delta^{(p)}}{(a+\varepsilon^{(p)})^{\mu^{(p)}-p'+2}}\int \frac{e^{fx}\varphi x dx}{(x+\varepsilon^{(p)})^{p'}}\end{aligned} \tag{38}$$

if $fx = \sum \gamma^{(p)}x^p + \sum \frac{\delta^{(p)}}{(x+\varepsilon^{(p)})^{\mu^{(p)}}}$. When f is polynomial ($\delta^{(p)} = 0$) and

$$\psi x = (x+\alpha)(x+\alpha')\dots(x+\alpha^{(n)}),$$

there is another formula

$$\begin{aligned}\frac{dp}{da} - \left(f'a + \frac{\varphi' a}{\varphi a}\right)p = \frac{e^{fx}\varphi x \cdot \psi x}{\psi a(a-x)} - \frac{e^{fc}\varphi c \cdot \psi c}{\psi a(a-c)} \\ + \sum\sum \varphi(p, p')\frac{a^{p'}}{\psi a}\int e^{fx}\varphi x \cdot x^p dx,\end{aligned} \tag{39}$$

where $\varphi(p, p') = \frac{(p+1)\psi^{(p+p'+2)}}{2\cdot 3 \dots (p+p'+2)} + \frac{\left(\psi\frac{\varphi'}{\varphi}+f'\right)^{(p+p'+1)}}{2\cdot 3\dots(p+p'+1)}$ ($F, F', \dots$ denoting the values at $x = 0$ of the successive derivatives of a function Fx).

As $\int \left(dp - \left(\frac{\varphi' a}{\varphi a} + f'a\right)pda\right)\frac{e^{-fa}}{\varphi a} = \frac{pe^{-fa}}{\varphi a}$, taking c such that $e^{fc}\varphi c = 0$ in (38) or such that $e^{fc}\varphi c \cdot \psi c = 0$ in (39), Abel gets

$$\begin{aligned}\frac{e^{-fa}}{\varphi a}\int \frac{e^{fx}\varphi x dx}{x-a} - e^{fx}\varphi x \int \frac{e^{-fa}da}{(a-x)\varphi a} \\ = \sum\sum p\gamma^{(p)}\int \frac{e^{-fa}a^{p'}da}{\varphi a}\cdot\int e^{fx}\varphi x\cdot x^{p-p'-2}dx \\ -\sum \beta^{(p)}\int \frac{e^{-fa}da}{(a+\alpha^{(p)})\varphi a}\cdot\int \frac{e^{fx}\varphi x dx}{x+\alpha^{(p)}} \\ +\sum\sum \mu^{(p)}\delta^{(p)}\int \frac{e^{-fa}da}{(a+\varepsilon^{(p)})^{\mu^{(p)}-p'+2}\varphi a}\cdot\int \frac{e^{fx}\varphi x dx}{(x+\varepsilon^{(p)})^{p'}}\end{aligned} \tag{40}$$

and

$$\frac{e^{-fa}}{\varphi a}\int\frac{e^{fx}\varphi x.dx}{x-a}-e^{fx}\varphi x\cdot\psi x\int\frac{e^{-fa}da}{(a-x)\varphi a\cdot\psi a} \tag{41}$$
$$=\sum\sum\varphi(p,p')\int\frac{e^{-fa}a^{p'}da}{\varphi a\cdot\psi a}\cdot\int e^{fx}\varphi x\cdot x^p dx$$

when f is a polynomial; the integrals with respect to a must be taken from a value which annihilates $\frac{e^{-fa}}{\varphi a}$.

Abel gives special cases of these formulae, for instance when φ is the constant 1; if more-over $fx=x^n$, one has

$$e^{-a^n}\int\frac{e^{x^n}dx}{x-a}-e^{x^n}\int\frac{e^{-a^n}da}{a-x}$$
$$=n\left(\int e^{-a^n}a^{n-2}da\cdot\int e^{x^n}dx+\int e^{-a^n}a^{n-3}da\cdot\int e^{x^n}xdx+\dots\right.$$
$$\left.+\int e^{a^n}da\cdot\int e^{x^n}x^{n-2}dx\right).$$

When $fx=0$, (40) gives

$$\varphi x\int\frac{da}{(a-x)\varphi a}-\frac{1}{\varphi a}\int\frac{\varphi xdx}{x-a}$$
$$=\beta\int\frac{da}{(a+\alpha)\varphi a}\cdot\int\frac{\varphi xdx}{x+\alpha}+\beta'\int\frac{da}{(a+\alpha')\varphi a}\cdot\int\frac{\varphi xdx}{x+\alpha'}+\dots$$
$$+\beta^{(n)}\int\frac{da}{(a+\alpha^{(n)})\varphi a}\cdot\int\frac{\varphi xdx}{x+\alpha^{(n)}}$$

and (41) $\frac{1}{\varphi a}\int\frac{\varphi xdx}{x-a}-\varphi x\cdot\psi x\int\frac{da}{(a-x)\varphi a\cdot\psi a}=\sum\sum\varphi(p,p')\int\frac{a^{p'}da}{\varphi a\cdot\psi a}\cdot\int\varphi x\cdot x^p dx$. If, in this last formula, $\beta=\beta'=\dots=\beta^{(n)}=m$, as $\varphi x=(\psi x)^m$, $\varphi(p,p')=(p+1+m(p+p'+2))k^{(p+p'+2)}$, where $k^{(j)}$ is the coefficient of x^j in ψx, so

$$\frac{1}{(\psi a)^m}\int\frac{(\psi x)^m dx}{x-a}-(\psi x)^{m+1}\int\frac{da}{(a-x)(\psi a)^{m+1}}$$
$$=\sum\sum k^{(p+p'+2)}(p+1+m(p+p'+2))\int\frac{a^{p'}da}{(\psi a)^{m+1}}\cdot\int(\psi x)^m x^p dx,$$

equality which reduces to

$$\sqrt{\psi a}\int\frac{dx}{(x-a)\sqrt{\psi x}}-\sqrt{\psi x}\int\frac{da}{(a-x)\sqrt{\psi a}} \tag{42}$$
$$=\frac{1}{2}\sum\sum(p-p')k^{(p+p'+2)}\int\frac{a^{p'}da}{\sqrt{\psi a}}\cdot\int\frac{x^p dx}{\sqrt{\psi x}}$$

when $m=-\frac{1}{2}$ and this gives an extension to hyperelliptic integrals of Legendre's formula. If, for example $\psi x=1+\alpha x^n$, one has $\sqrt{1+\alpha a^n}\int\frac{dx}{(x-a)\sqrt{1+\alpha x^n}}$

$-\sqrt{1+\alpha x^n}\int\frac{da}{(a-x)\sqrt{1+\alpha a^n}} = \frac{\alpha}{2}\sum(n-2p'-2)\int\frac{a^{p'}da}{\sqrt{1+\alpha a^n}}\cdot\int\frac{x^{n-p'-2}dx}{\sqrt{1+\alpha x^n}}$. The elliptic case corresponds to $\psi x = (1-x^2)(1-\alpha x^2)$ and leads to

$$\begin{aligned}&\sqrt{(1-a^2)(1-\alpha a^2)}\int\frac{dx}{(x+a)\sqrt{(1-x^2)(1-\alpha x^2)}}\\&-\sqrt{(1-x^2)(1-\alpha x^2)}\int\frac{da}{(a+x)\sqrt{(1-a^2)(1-\alpha a^2)}}\\&=\alpha\int\frac{da}{\sqrt{(1-a^2)(1-\alpha a^2)}}\cdot\int\frac{x^2dx}{\sqrt{(1-x^2)(1-\alpha x^2)}}\\&-\alpha\int\frac{a^2da}{\sqrt{(1-a^2)(1-\alpha a^2)}}\cdot\int\frac{dx}{\sqrt{(1-x^2)(1-\alpha x^2)}}\end{aligned}$$

or, with $x=\sin\varphi$ and $a=\sin\psi$,

$$\begin{aligned}&\cos\psi\sqrt{1-\alpha\sin^2\psi}\int\frac{d\varphi}{(\sin\varphi+\sin\psi)\sqrt{1-\alpha\sin^2\varphi}}\\&-\cos\varphi\sqrt{1-\alpha\sin^2\varphi}\int\frac{d\psi}{(\sin\psi+\sin\varphi)\sqrt{1-\alpha\sin^2\varphi}}\\&=\alpha\int\frac{d\psi}{\sqrt{1-\alpha\sin^2\psi}}\cdot\int\frac{\sin^2\varphi d\varphi}{\sqrt{1-\alpha\sin^2\varphi}}\\&-\alpha\int\frac{\sin^2\psi d\psi}{\sqrt{1-\alpha\sin^2\psi}}\cdot\int\frac{d\varphi}{\sqrt{1-\alpha\sin^2\varphi}}.\end{aligned}$$

The formula (40) with $fx = x$ gives

$$\frac{e^{-a}}{\varphi a}\int\frac{e^x\varphi x dx}{x-a} - e^x\varphi x\int\frac{e^{-a}da}{(a-x)\varphi a} = -\sum\beta^{(p)}\int\frac{e^{-a}da}{(a+\alpha^{(p)})\varphi a}\cdot\int\frac{e^x\varphi x dx}{x+\alpha^{(p)}}$$

that is, for $\varphi x = \sqrt{x^2-1}$:

$$\begin{aligned}&e^x\sqrt{x^2-1}\int\frac{e^{-a}da}{(a-x)\sqrt{a^2-1}} - \frac{e^{-a}}{\sqrt{a^2-1}}\int\frac{e^x dx\sqrt{x^2-1}}{x-a}\\&=\frac{1}{2}\int\frac{e^{-a}da}{(a+1)\sqrt{a^2-1}}\cdot\int\frac{e^x dx\sqrt{x^2-1}}{x+1}\\&+\frac{1}{2}\int\frac{e^{-a}da}{(a-1)\sqrt{a^2-1}}\cdot\int\frac{e^x dx\sqrt{x^2-1}}{x-1}.\end{aligned}$$

Let us turn back to the formula (41) with $\beta=\beta'=\ldots=\beta^{(n)}=m$, but with f any polynomial:

$$\frac{e^{-fa}}{(\psi a)^m}\int\frac{e^{fx}(\psi x)^m dx}{x-a} - e^{fx}(\psi x)^{m+1}\int\frac{e^{-fa}da}{(a-x)(\psi a)^{m+1}}$$
$$=\sum\sum((p+p'+2)\gamma^{(p+p'+2)}+(p+1+m(p+p'+2))k^{(p+p'+2)})$$
$$\times\int\frac{e^{-fa}a^{p'}da}{(\psi a)^{m+1}}\cdot\int e^{fx}(\psi x)^m x^p dx,$$

that is

$$e^{-fa}\sqrt{\psi a}\int\frac{e^{fx}dx}{(x-a)\sqrt{\psi x}} - e^{fx}\sqrt{\psi x}\int\frac{e^{-fa}da}{(a-x)\sqrt{\psi a}}$$
$$=\sum\sum\left((p+p'+2)\gamma^{(p+p'+2)}+\frac{1}{2}(p-p')k^{(p+p'+2)}\right)$$
$$\times\int\frac{e^{-fa}a^{p'}da}{(\psi a)^{m+1}}\cdot\int e^{fx}(\psi x)^m x^p dx$$

when $m=-\frac{1}{2}$; if moreover $fx=x$ and $\psi x=1-x^2$, this gives $e^{-a}\sqrt{1-a^2} \times\int\frac{e^x dx}{(x-a)\sqrt{1-x^2}}=e^x\sqrt{1-x^2}\int\frac{e^{-a}da}{(a-x)\sqrt{1-a^2}}$ or $\cos\psi e^{\sin\psi}\int\frac{e^{\sin\varphi}d\varphi}{\sin\varphi+\sin\psi}=\cos\varphi e^{\sin\varphi} \times\int\frac{e^{\sin\psi}d\psi}{\sin\psi+\sin\varphi}$ (integrals from $\varphi,\psi=\frac{\pi}{2}$).

Abel also applies these formulae to definite integrals: the formula (40) with f polynomial gives

$$\int_{x'}^{x''}\frac{e^{fx}\varphi x dx}{x-a} \tag{43}$$
$$=e^{fa}\varphi a\sum\sum(p+p'+2)\gamma^{(p+p'+2)}\int_{a'}\frac{e^{-fa}a^{p'}da}{\varphi a}\cdot\int_{x'}^{x''}e^{fx}\varphi x dx$$
$$-e^{fa}\varphi a\sum\beta^{(p)}\int_{a'}\frac{e^{-fa}da}{(a+\alpha^{(p)})\varphi a}\cdot\int\frac{e^{fx}\varphi x dx}{x+\alpha^{(p)}}$$

when x', x'' annihilate $e^{fx}\varphi x$ and a' annihilates $\frac{e^{-fa}}{\varphi a}$. For $fx=0$, this gives

$$\int_{x'}^{x''}\frac{\varphi x dx}{x-a}=-\varphi a\sum\beta^{(p)}\int_{a'}\frac{da}{(a+\alpha^{(p)})\varphi a}\cdot\int_{x'}^{x''}\frac{\varphi x dx}{x+\alpha^{(p)}}$$

and for $\varphi x=1$, $\int_{x'}^{x''}\frac{e^{fx}dx}{x-a}=e^{fa}\sum\sum(p+p'+2)\gamma^{(p+p'+2)}\int_{a'}e^{-fa}a^{p'}da\cdot\int_{x'}^{x''}e^{fx}x^p dx.$

The formula (43), with $\frac{e^{-fa''}}{\varphi a''}=0$ gives

$$\sum \beta^{(p)} \int_{a'}^{a''} \frac{e^{-fa}da}{(a+\alpha^{(p)})\varphi a} \cdot \int_{x'}^{x''} \frac{e^{fx}\varphi x dx}{x+\alpha^{(p)}}$$

$$= \sum\sum (p+p'+2)\gamma^{(p+p'+2)} \int_{a'}^{a''} \frac{e^{-fa}a^{p'}da}{\varphi a} \cdot \int_{x'}^{x''} e^{fx}\varphi x dx,$$

for instance $\sum \beta^{(p)} \int_{a'}^{a''} \frac{e^{-ka}da}{(a+\alpha^{(p)})\varphi a} \cdot \int_{x'}^{x''} \frac{e^{kx}\varphi x dx}{x+\alpha^{(p)}} = 0$ for $fx = kx$ and $\varphi x = 1$,

$$\sum\sum (p+p'+2)\gamma^{(p+p'+2)} \int_{a'}^{a''} e^{-fa}a^{p'}da \cdot \int_{x'}^{x''} e^{fx}x^p dx = 0.$$

Using now (41) with x', x'' annihilating $e^{fx}\varphi x \cdot \psi x$, Abel gets

$$\int_{x'}^{x''} \frac{e^{fx}\varphi x dx}{x-a} = e^{fa}\varphi a \sum\sum \varphi(p,p') \int_{a'} \frac{e^{-fa}a^{p'}da}{\varphi a \cdot \psi a} \cdot \int_{x'}^{x''} e^{fx}\varphi x \cdot x^p dx$$

and, when $\beta = \beta' = \ldots = \beta^{(n)} = m$,

$$\int_{x'}^{x''} \frac{e^{fx}(\psi x)^m dx}{x-a} = e^{fa}(\psi a)^m \sum\sum \varphi(p,p') \int_{a'} \frac{e^{-fa}a^{p'}da}{(\psi a)^{m+1}} \cdot \int_{x'}^{x''} e^{fx}(\psi x)^m x^p dx. \tag{44}$$

For $fx = 0$ and $m = -\frac{1}{2}$, this gives

$$\int_{x'}^{x''} \frac{dx}{(x-a)\sqrt{\psi x}} = \frac{1}{2\sqrt{\psi a}} \sum\sum (p-p')k^{(p+p'+2)} \int_{a'} \frac{a^{p'}da}{\sqrt{\psi a}} \cdot \int_{x'}^{x''} \frac{x^p dx}{\sqrt{\psi x}}$$

expressing the periods of an hyperelliptic integral of the third kind by means of the periods of the integrals of the first two kinds; in the elliptic case,

$$\psi x = (1-x^2)(1-\alpha x^2),$$

one has

$$\sqrt{(1-a^2)(1-\alpha a^2)}\int\limits_{x'}^{x''}\frac{dx}{(x-a)\sqrt{(1-x^2)(1-\alpha x^2)}}$$

$$=\alpha\int\limits_{a'}\frac{da}{\sqrt{(1-a^2)(1-\alpha a^2)}}\cdot\int\limits_{x'}^{x''}\frac{x^2dx}{\sqrt{(1-x^2)(1-\alpha x^2)}}$$

$$-\alpha\int\limits_{a'}\frac{a^2da}{\sqrt{(1-a^2)(1-\alpha a^2)}}\cdot\int\limits_{x'}^{x''}\frac{dx}{\sqrt{(1-x^2)(1-\alpha x^2)}}$$

with x', x'' and $a' = \pm 1$ or $\pm\sqrt{\frac{1}{\alpha}}$.

From (44) with $\psi x = 1 - x^{2n}$, $x' = -1$, $x = 1$ and $a' = 1$, Abel deduces

$$\int\limits_{-1}^{1}\frac{dx}{(x-a)(1-x^{2n})^m}$$

$$=\frac{\Gamma(-m+1)}{n(1-a^{2n})^m}\sum(2p+1-2mn)\frac{\Gamma\left(\frac{1+2p}{2n}\right)}{\Gamma\left(-m+1+\frac{1+2p}{2n}\right)}\int\limits_{1}\frac{a^{2n-2p-2}da}{(1-a^{2n})^{1-m}}$$

for $m > 0$. If, for example $m = \frac{1}{2}$ and $n = 3$, this gives

$$\int\limits_{-1}^{1}\frac{dx}{(x-a)\sqrt{1-x^6}} = -\frac{2}{3}\frac{\sqrt{\pi}}{\sqrt{1-a^6}}\frac{\Gamma\left(\frac{1}{6}\right)}{\Gamma\left(\frac{2}{3}\right)}\int\limits_{1}\frac{a^4da}{\sqrt{1-a^6}}$$

$$+\frac{2}{3}\frac{\sqrt{\pi}}{\sqrt{1-a^6}}\frac{\Gamma\left(\frac{5}{6}\right)}{\Gamma\left(\frac{4}{3}\right)}\int\limits_{1}\frac{da}{\sqrt{1-a^6}}.$$

Now if $\frac{e^{-fa''}}{\varphi a''} = 0$, $\sum\sum\varphi(p,p')\int\limits_{a'}^{a''}\frac{e^{-fa}a^{p'}da}{\varphi a\cdot\psi a}\cdot\int\limits_{x'}^{x''}e^{fx}\varphi x\cdot x^p dx = 0$; for instance, when $\varphi x = (x+\alpha)^\beta(x+\alpha')^{\beta'}\dots(x+\alpha^{(n)})^{\beta^{(n)}}$ and $\psi a = (a+\alpha)^{1-\beta}(a+\alpha')^{1-\beta'}\dots(a+\alpha^{(n)})^{1-\beta^{(n)}}$, with $-1 < \beta, \beta', \dots, \beta^{(n)} < 0$, one has

$$\sum\sum\varphi(p,p')\int\limits_{a'}^{a''}\frac{a^{p'}da}{\psi a}\cdot\int\limits_{x'}^{x''}\frac{x^p dx}{\varphi x} = 0$$

if $x' = -\alpha^{(p)}$, $x'' = -\alpha^{(p')}$, $a' = -\alpha^{(q)}$ and $a'' = -\alpha^{(q')}$. When $\beta = \beta' = \dots = \beta^{(n)} = \frac{1}{2}$, denoting by φx a polynomial of degree n, with roots $\alpha, \alpha', \alpha'', \dots$ and by $F(p,x)$ the integral $\int\frac{x^p dx}{\sqrt{\varphi x}}$, this relation becomes

$$\begin{aligned}
&\sum\sum(p-p')k^{(p+p'+2)}F(p,\alpha)F(p,',\alpha')\\
&\quad+\sum\sum(p-p')k^{(p+p'+2)}F(p,\alpha'')F(p',\alpha''')\\
&=\sum\sum(p-p')k^{(p+p'+2)}F(p,\alpha)F(p,',\alpha''')\\
&\quad+\sum\sum(p-p')k^{(p+p'+2)}F(p,\alpha'')F(p',\alpha'),
\end{aligned}$$

and, $\varphi x = (1-x^2)(1-c^2x^2)$ (elliptic case), $\alpha = 1$, $\alpha' = -1$, $\alpha'' = \frac{1}{c}$ and $\alpha''' = -\frac{1}{c}$, we find

$$F(1)E\left(\frac{1}{c}\right) = E(1)F\left(\frac{1}{c}\right),$$

where $Fx = \int\limits_0 \frac{dx}{\sqrt{(1-x^2)(1-c^2x^2)}}$ and $Ex = \int\limits_0 \frac{x^2dx}{\sqrt{(1-x^2)(1-c^2x^2)}}$.

A short posthumous paper *Sur une propriété remarquable d'une classe très étendue de fonctions transcendantes* (*Œuvres*, t. II, p. 43–46, mem. VIII) contains some of the same formulae. Abel starts from a differential equation $0 = sy + t\frac{dy}{dx}$, where $t = \varphi x$ and $s = fx$ are polynomials; then, for $r = \frac{\varphi' x - fx}{x-a} - \frac{\varphi x}{(x-a)^2} = -\frac{\varphi a}{(x-a)^2} - \frac{fa}{x-a} + R$, with

$$\begin{aligned}
R &= \frac{1}{2}\varphi'' a - f' a + \left(\frac{1}{3}\varphi''' a - \frac{1}{2}f'' a\right)(x-a)\\
&\quad+\left(\frac{1}{2\cdot 4}\varphi'''' a - \frac{1}{2\cdot 3}f''' a\right)(x-a)^2 + \dots,
\end{aligned}$$

and $y = \psi x$ solution of the differential equation, one has

$$\int rydx = \frac{y\varphi x}{x-a} \quad\text{or}\quad \frac{y\varphi x}{x-a} = -\varphi a\frac{dz}{da} - fa\cdot z + \int Rydx,$$

where $z = \int \frac{ydx}{x-a}$. If $z = p\psi a$, this equality becomes $\int Rydx - \frac{\varphi x\cdot\psi x}{x-a} = \varphi a\cdot\psi a\frac{dp}{da}$ and we have

$$\begin{aligned}
p &= \frac{1}{\psi a}\int\frac{\psi x dx}{x-a} - \psi x\cdot\varphi x\int\frac{da}{(a-x)\psi a\cdot\varphi a} = \iint\frac{R\cdot\psi x}{\varphi a\cdot\psi a}dxda\\
&= \sum((n+1)\alpha_{m+n+2} - \beta_{m+n+1})\int\frac{a^m da}{\varphi a\cdot\psi a}\int x^n\psi x dx,
\end{aligned}$$

where α_k (resp. β_k) is the coefficient of x^k in φx (resp. fx); the origin of integration in x (resp. in a) must annihilate $\psi x\cdot\varphi x$ (resp. $\frac{1}{\psi a}$). Note that, up to a constant factor, $\psi x = e^{-\int\frac{fx}{\varphi x}dx}$ has the form $\frac{e^p}{(x-\delta)^m(x-\delta_1)^{m_1}\dots}$, where p is a rational function (not the preceding p !) and $0 < m, m_1, \dots < 1$. When $\psi x = \frac{1}{\sqrt{\varphi x}}$, $fx - \frac{1}{2}\varphi' x = 0$ and $\beta_m = \frac{1}{2}(m+1)\alpha_{m+1}$ and we find the formula (42).

The following paper in the second volume of Abel's *Works* (p. 47–54, mem. IX) extends this theory to the case of a linear differential equation of order m with polynomial coefficients:

$$0 = sy + s_1 \frac{dy}{dx} + \ldots + s_m \frac{d^m y}{dx^m}, \tag{45}$$

$s_k = \varphi_k(x)$ polynomial. Abel looks for a function r such that

$$\int rydx = vy + v_1 \frac{dy}{dx} + v_{m-2} \frac{d^{m-2}y}{dx^{m-2}} + ts_m \frac{d^{m-1}y}{dx^{m-1}},$$

with t a given function, taken equal to $\frac{1}{x-a}$. He finds that $v_{\mu-1} = s_\mu t - \frac{dv_\mu}{dx}$ for $0 \le \mu \le m-1$ (with $v_{-1} = -r$ and $v_{m-1} = ts_m$), so that $v_{\mu-1} = s_\mu t - \frac{d(s_{\mu+1}t)}{dx} + \frac{d^2(s_{\mu+2}t)}{dx^2} - \ldots$ and

$$-r = st - \frac{d(s_1 t)}{dx} + \frac{d^2(s_2 t)}{dx^2} - \ldots + (-1)^m \frac{d^m(s_m t)}{dx^m}.$$

Now $s_\mu t = \frac{\varphi_\mu(a)}{x-a} + R_\mu$, where R_μ is polynomial in x, so

$$-r = \frac{\varphi a}{x-a} + \frac{\varphi_1 a}{(x-a)^2} + \Gamma(3) \frac{\varphi_2 a}{(x-a)^3} + \ldots + \Gamma(m+1) \frac{\varphi_m a}{(x-a)^{m+1}} + \rho,$$

with $\rho = R - \frac{dR_1}{dx} + \ldots + (-1)^m \frac{d^m R}{dx^m}$. Thus the integral $z = \int \frac{ydx}{x-a}$ satisfies a differential equation $\varphi a \cdot z + \varphi_1 a \cdot \frac{dz}{da} + \ldots + \varphi_m a \cdot \frac{d^m z}{da^m} = -\chi' - \int \rho ydx$, where $\chi = vy + v_1 \frac{dy}{dx} + v_{m-2} \frac{d^{m-2}y}{dx^{m-2}} + ts_m \frac{d^{m-1}y}{dx^{m-1}}$ and $\chi' = \chi - \chi_0$ (where χ_0 is the value of χ at the origin of integration); Abel writes the solution of this equation in terms of a fundamental system of solutions of (45).

In the second part of the paper, Abel wishes to find coefficients $\alpha_1, \alpha_2, \ldots, \alpha_m$ depending on a such that $z = \int \left(\frac{\alpha_1}{x-a} + \frac{\alpha_2}{(x-a)^2} + \ldots + \frac{\Gamma m \cdot \alpha_m}{(x-a)^m} \right) ydx$ satisfy a differential equation of the form $\beta z + \gamma \frac{dz}{da} = \chi + \int \rho ydx$; he writes induction relations between the α_μ and, supposing $-\frac{\beta}{\gamma} = \varepsilon$ constant, a differential equation to determine γ.

The first article on Abelian integrals published by Abel in *Crelle's Journal* (1826, *Œuvres*, t. I, p. 104–144) is devoted to the search for differential forms $\frac{\rho dx}{\sqrt{R}}$, with ρ and R polynomials, such that their integrals have the form $\log \frac{p+q\sqrt{R}}{p-q\sqrt{R}} = z$, with p, q polynomials. As $dz = \frac{pqdR + 2(pdq - qdp)R}{(p^2 - q^2 R)\sqrt{R}}$, one must have $\rho = \frac{M}{N}$ with $M = pq\frac{dR}{dx} + 2\left(p\frac{dq}{dx} - q\frac{dp}{dx}\right) R$ and $N = p^2 - q^2 R$. It follows that $q\rho = 2\frac{dp}{dx} - p\frac{dN}{Ndx}$ and $p\frac{dN}{Ndx}$ must be polynomials. If

$$N = (x+a)^m (x+a_1)^{m_1} \ldots (x+a_n)^{m_n},$$

$\frac{dN}{Ndx} = \frac{m}{x+a} + \frac{m_1}{x+a_1} + \ldots + \frac{m_n}{x+a_n}$ and we see that p is divisible by

$$(x+a)(x+a_1) \cdots (x+a_n).$$

Hence $p = (x+a)(x+a_1)\cdots(x+a_n)p_1$ and

$$(x+a)^m(x+a_1)^{m_1}\cdots(x+a_n)^{m_n} = p_1^2(x+a)^2(x+a_1)^2\cdots(x+a_n)^2 - q^2R$$

which shows that $m = m_1 = \ldots = m_n = 1$ if p and q are relatively prime (R is supposed to be square free) and that $R = (x+a)(x+a_1)\cdots(x+a_n)R_1 = NR_1$. Now p_1, q, N and R_1 are determined by the equation $p_1^2N - q^2R_1 = 1$ and so $\rho = p_1q\frac{dR}{dx} + 2\left(p\frac{dq}{dx} - q\frac{dp}{dx}\right)R_1$.

Abel studies $p_1^2N - q^2R_1 = 1$, or more generally $p_1^2N - q^2R_1 = v$ (where v is polynomial of degree less than the mean of the degrees of N and R_1), as a Diophantine equation in the ring of polynomials. The first observation is that Np_1^2 and R_1q^2 must have the same degree and so $\delta R = \delta(NR_1) = 2(\delta q - \delta p_1 + \delta R_1)$ (where δ designates the degree) is an even number, so we can put $\delta N = n - m$ and $\delta R_1 = n + m$. By Euclidian division $R_1 = Nt + t'$, with $\delta t = 2m$ and $\delta t' < n - m$. Using the method of indeterminate coefficients, Abel shows that there exist t_1 and t_1' such that $\delta t_1' < m$ and $t = t_1^2 + t_1'$. The equation now becomes $(p_1^2 - q^2t_1^2)N - q^2s = v$, with $s = Nt_1' + t'$ and shows that $\left(\frac{p_1}{q}\right)^2$ and t_1^2 differ by a rational fraction of degree less than δt_1, and as a consequence the polynomial part of $\frac{p_1}{q}$ is $\pm t_1$, say t_1. Thus $p_1 = t_1q + \beta$ with $\delta\beta < \delta q$ and one can verify that $\left(\frac{q}{\beta} - \frac{t_1N}{s}\right)^2 = \frac{R_1N}{s^2} - \frac{v}{s\beta^2}$; if $R = R_1N = r^2 + r'$ with $\delta r' < \delta r$, we see as before that $\frac{q}{\beta} - \frac{t_1N}{s}$ and $\frac{r}{s}$ have the same polynomial part, so $\frac{q}{\beta}$ and $\frac{r+t_1N}{s}$ have the same polynomial part 2μ and $q = 2\mu\beta + \beta_1$ with $\delta\beta_1 < \delta\beta$. Now the Diophantine equation becomes

$$s_1\beta^2 - 2r_1\beta\beta_1 - s\beta_1^2 = v$$

where $s_1 = N + 4\mu t_1N - 4s\mu^2$ and $r_1 = 2\mu s - t_1N$; as $r_1^2 + ss_1 = R$, $\delta r_1 = n$ and $\delta s, \delta s_1 < n$.

Our equation may be written $\left(\frac{\beta}{\beta_1} - \frac{r_1}{s_1}\right)^2 = \left(\frac{r_1}{s_1}\right)^2 + \frac{s}{s_1} + \frac{v}{s_1\beta_1^2}$ which shows that $\frac{\beta}{\beta_1} - \frac{r_1}{s_1}$ and $\frac{r_1}{s_1}$ have the same polynomial part up to sign. Writing μ_1 for the polynomial part of $\frac{r_1}{s_1}$, the polynomial part of $\frac{\beta}{\beta_1}$ is $2\mu_1$ and $\beta = 2\beta_1\mu_1 + \beta_2$ with $\delta\beta_2 < \delta\beta_1$. The equation can now be rewritten $s_2\beta_1^2 - 2r_2\beta_1\beta_2 - s_1\beta_2^2 = -v$, where $r_2 = 2\mu_1s_1 - r_1$ and $s_2 = s + 4r_1\mu_1 - 4s_1\mu_1^2$, from which it is easy to see that $\delta r_2 = \delta r_1 > \delta s_2$. Continuing in this way, we obtain

$$s_n\beta_{n-1}^2 - 2r_n\beta_{n-1}\beta_n - s_{n-1}\beta_n^2 = (-1)^{n-1}v$$

with $\beta_{n-1} = 2\mu_n\beta_n + \beta_{n+1}$, μ_n integral part of $\frac{r_n}{s_n}$, $r_n = 2\mu_{n-1}s_{n-1} - r_{n-1}$ and $s_n = s_{n-2} + 4r_{n-1}\mu_{n-1} - 4s_{n-1}\mu_{n-1}^2$; as $\delta\beta > \delta\beta_1 > \ldots > \delta\beta_n > \ldots$, there is an m such that $\beta_m = 0$, giving the equation $s_m\beta_{m-1}^2 = (-1)^{m-1}v$. The sequence (β_k) is determined by the Euclidian algorithm and β_{m-1} is the g.c.d. of β and β_1, and so 1 if p and q are relatively prime and $v = (-1)^{m-1}s_m$.

In the initial problem, we had $v = 1$, so we must take s_m constant. As the s_k are of degree k, this gives $n - 1$ conditions on the $2n$ coefficients of R_1 and N, once the index m is chosen. Thus $\frac{p_1}{q} = t_1 + \cfrac{1}{2\mu + \cfrac{1}{2\mu_1 + \ddots + \cfrac{1}{2\mu_{m-1}}}}$ and $\rho = \left(p_1 \frac{dN}{dx} + 2N\frac{dp_1}{dx}\right) : q$, of degree $n-1$.

Abel adds some observations. The first is that one may put $N = 1$ in the problem, for

$$\log \frac{p + q\sqrt{R}}{p - q\sqrt{R}} = \frac{1}{2} \log \frac{p_1^2 N + q^2 R_1 + 2p_1 q\sqrt{R}}{p_1^2 N + q^2 R_1 - 2p_1 q\sqrt{R}} = \frac{1}{2} \log \frac{p' + q'\sqrt{R}}{p' - q'\sqrt{R}}$$

with $p' = p_1^2 N + q^2 R_1$ and $q' = 2p_1 q$. In that case, the equation takes the Pell-Fermat form $p'^2 - q'^2 R = 1$ and $\rho = \frac{2}{q'}\frac{dp'}{dx}$. The second is that, if $\frac{\alpha_k}{\beta_k}$ is the k-th convergent of the continued fraction for $\frac{p_1}{q}$, $\alpha_k^2 N - \beta_k^2 R_1 = (-1)^{k-1} s_k$ and so, putting

$$z_k = \alpha_k\sqrt{N} + \beta_k\sqrt{R_1} \quad \text{and} \quad z'_k = \alpha_k\sqrt{N} - \beta_k\sqrt{R_1},$$

we get $\frac{z_k}{z'_k} = \frac{r_k + \sqrt{R}}{r_k - \sqrt{R}} \frac{z_{k-1}}{z'_{k-1}}$. Lastly,

$$\log \frac{p_1\sqrt{N} + q\sqrt{R_1}}{p_1\sqrt{N} - q\sqrt{R_1}} = \log \frac{t_1\sqrt{N} + \sqrt{R_1}}{t_1\sqrt{N} - \sqrt{R_1}} + \log \frac{r_1\sqrt{N} + \sqrt{R}}{r_1\sqrt{N} - \sqrt{R}} + \ldots + \log \frac{r_m\sqrt{N} + \sqrt{R}}{r_m\sqrt{N} - \sqrt{R}}.$$

Abel also shows that

$$\rho dx = 2(\frac{1}{2} t_1 dN + N dt_1 + dr_1 + \ldots + dr_m - \mu ds - \ldots - \mu_{m-1} ds_{m-1}).$$

As $\left(\frac{p_1}{q}\right)^2 = \frac{R_1}{N} + \frac{v}{q^2 N}$, one sees that the continued fraction for $\frac{p_1}{q}$ is obtained from that for $\sqrt{\frac{R_1}{N}}$. Let us suppose that $N = 1$, so that $t_1 = r$ and $\sqrt{R} = r + \cfrac{1}{2\mu + \cfrac{1}{2\mu_1 + \ddots}}$. An easy computation shows that $r_m^2 + s_m s_{m-1} = r_{m-1}^2 + s_{m-1}s_{m-2} = \ldots = r_1^2 + s s_1 = R = r^2 + s$. If $s_m = a$ constant, we have $r_{m+1}^2 - r^2 = s - a s_{m+1}$ and, since $\delta r_{m+1} = \delta r > \delta s, \delta s_{m+1}$, this implies $r_{m+1} = r$ and $s_{m+1} = \frac{s}{a}$. This shows that the polynomial part μ_{m+1} of $\frac{r_{m+1}}{s_{m+1}}$ is equal to $a\mu$ and $s_{m+2} = a s_1$, $r_{m+2} = r_1$ and so on, in general we have $r_{m+n} = r_{n-1}, r_{m-n} = r_n, s_{m\pm n} = a^{(-1)^n} s_{n-1}$ and $\mu_{m\pm n} = a^{(-1)^{n-1}} \mu_{n-1}$. This shows that the continued fraction is periodic with partial quotients $r, 2\mu, 2\mu_1, \ldots, \frac{2\mu_1}{a}, 2a\mu, \frac{2r}{a}, 2a\mu, \frac{2\mu_1}{a}, \ldots, 2\mu, 2r, 2\mu, \ldots$; if $m = 2k-1$ is odd, $s_{k-1} = a^{(-1)^k} s_{k-1}$ and $a = 1$. Conversely, one sees that if $\sqrt{R}$ has a continued fraction of the preceding form, $s_m = a$. Abel may conclude by a criterion for the existence of a polynomial ρ such that $\int \frac{\rho dx}{\sqrt{R}} = \log \frac{y + \sqrt{R}}{y - \sqrt{R}}$ with y rational: $\sqrt{R}$ must be of the form $r + \cfrac{1}{2\mu + \cfrac{1}{2\mu_1 + \ddots + \cfrac{1}{2\mu_1 + \cfrac{1}{2\mu + \cfrac{1}{2r + \cfrac{1}{2\mu + \cfrac{1}{2\mu_1 + \ddots}}}}}}}$

$$\text{and then} \quad y = r + \cfrac{1}{2\mu + \cfrac{1}{2\mu_1 + \ddots + \cfrac{1}{2\mu_1 + \cfrac{1}{2\mu + \frac{1}{2r}}}}}.$$

Conversely, Abel proves that $s_k = s_{k-2}$ implies $r_k = r_{k-1}$, $\mu_k = \mu_{k-2}$, $r_{k+1} = r_{k-2}$, $s_{k+1} = s_{k-3}$ etc. and generally $r_{k+n} = r_{k-n-1}$, $\mu_{k+n} = \mu_{k-n-2}$, $s_{k+n} = s_{k-n-2}$; for $n = k-1$, this gives $s_{2k-1} = s_{-1} = 1$. If $s_k = cs_{k-1}$, he finds $\mu_k = \frac{1}{c}\mu_{k-1}$, $r_{k+1} = r_{k-1}$, $s_{k+1} = \frac{1}{c}s_{k-2}, \ldots, s_{2k} = c^{(-1)^k}$. The integral found is

$$\int \frac{2}{\sqrt{R}}\left(dr + dr_1 + \ldots + dr_{k-1} + \frac{1}{2}dr_k - \mu ds - \mu_1 ds_1 - \ldots - \mu_{k-1}ds_{k-1}\right)$$
$$= \log\frac{r+\sqrt{R}}{r-\sqrt{R}} + \log\frac{r_1+\sqrt{R}}{r_1-\sqrt{R}} + \ldots + \log\frac{r_{k-1}+\sqrt{R}}{r_{k-1}-\sqrt{R}} + \frac{1}{2}\log\frac{r_k+\sqrt{R}}{r_k-\sqrt{R}}$$

when $m = 2k$ and

$$\int \frac{2}{\sqrt{R}}\left(dr + dr_1 + \ldots + dr_{k-1} - \mu ds - \mu_1 ds_1 - \ldots - \mu_{k-2}ds_{k-2} - \frac{1}{2}\mu_{k-1}ds_{k-1}\right)$$
$$= \log\frac{r+\sqrt{R}}{r-\sqrt{R}} + \log\frac{r_1+\sqrt{R}}{r_1-\sqrt{R}} + \ldots + \log\frac{r_{k-1}+\sqrt{R}}{r_{k-1}-\sqrt{R}}$$

when $m = 2k - 1$.

In the elliptic case, where R is of degree $4 = 2n$, $n = 2$ and the s_m are of degree 1, so there is only one condition to write in order to have $s_m = \text{const}$. If $R = (x^2 + ax + b)^2 + ex$, we have $r = x^2 + ax + b$, $s = ex$, $\mu = \frac{x+a}{e}$ and then $r_1 = x^2 + ax - b$, $s_1 = \frac{4b}{e}x + \frac{4ab}{e} + 1$, $\mu_1 = \frac{e}{4b}x - \frac{e^2}{16b^2}$ and $s_2 = \left(\frac{ae^2}{4b^2} + \frac{e^3}{16b^3}\right)x - \frac{e^2}{4b^2}\left(\frac{ae}{4b} + \frac{e^2}{16b^2} - b\right)$. In order to make s_1 constant, we have only to put $b = 0$ and we find that $\int \frac{(3x+a)dx}{\sqrt{(x^2+ax)^2+ex}} = \log\frac{x^2+ax+\sqrt{R}}{x^2+ax-\sqrt{R}}$, with $R = (x^2+ax)^2+ex$. In order to make s_2 constant, we put $e = -4ab$, $R = (x^2+ax+b)^2 - 4abx$ and find

$$\int \frac{(4x+a)dx}{\sqrt{(x^2+ax+b)^2 - 4abx}} = \log\frac{x^2+ax+b+\sqrt{R}}{x^2+ax+b-\sqrt{R}}$$
$$+\frac{1}{2}\log\frac{x^2+ax-b+\sqrt{R}}{x^2+ax-b-\sqrt{R}}.$$

Abel also computes the case in which s_3 is constant, which is given by

$$e = -2b(a \pm \sqrt{a^2 + 4b})$$

and

$$\int \frac{(5x + \frac{3}{2} \mp \frac{1}{2}\sqrt{a^2 + 4b})dx}{\sqrt{(x^2 + ax + b)^2 - 2bx(a \pm \sqrt{a^2 + 4b})}} = \log \frac{x^2 + ax + b + \sqrt{R}}{x^2 + ax + b - \sqrt{R}} + \log \frac{x^2 + ax - b + \sqrt{R}}{x^2 + ax - b - \sqrt{R}} \quad ,$$

and the case where s_4 is constant, which leads to $e = -b(3a \pm \sqrt{a^2 + 8b})$ and

$$\int \frac{(6x + \frac{3}{2}a - \frac{1}{2}\sqrt{a^2 + 8b})dx}{\sqrt{(x^2 + ax + b)^2 - b(3a + \sqrt{a^2 + 8b})x}} = \log \frac{x^2 + ax + b + \sqrt{R}}{x^2 + ax + b - \sqrt{R}} + \log \frac{x^2 + ax - b + \sqrt{R}}{x^2 + ax - b - \sqrt{R}} + \frac{1}{2} \log \frac{x^2 + ax + \frac{1}{4}a(a - \sqrt{a^2 + 8b}) + \sqrt{R}}{x^2 + ax + \frac{1}{4}a(a - \sqrt{a^2 + 8b}) - \sqrt{R}}.$$

At the end of the memoir, Abel states a theorem according to which, whenever an integral $\int \frac{\rho dx}{\sqrt{R}}$, ρ and R polynomials, may be expressed by logarithms, it is always in the form $A \log \frac{p+q\sqrt{R}}{p-q\sqrt{R}}$, with A constant, p and q polynomials.

Chebyshev (1860) and Zolotarev (1872) studied the same problem in the elliptic case looking for arithmetical conditions on the coefficients of R, these latter supposed to be integers.

The first text written by Abel on elliptic functions (between 1823 and 1825), with the title *Théorie des transcendantes elliptiques* (*Œuvres*, t. II, p. 87–188), also deals with this problem but it was not published by Abel. In the first chapter, Abel studies the conditions under which an elliptic integral $\int \frac{Pdx}{\sqrt{R}}$, with P a rational function and $R = \alpha + \beta x + \gamma x^2 + \delta x^3 + \varepsilon x^4$, is an algebraic function. At first taking P polynomial, he observes that this algebraic function must be rational in x and $\sqrt{R}$, so of the form $Q' + Q\sqrt{R}$ with Q' and Q rational; since dQ' is rational, we may write $d(Q\sqrt{R}) = \frac{Pdx}{\sqrt{R}}$. The function Q is a polynomial otherwise its poles would remain as poles in the differential: $Q = f(0) + f(1)x + \ldots + f(n)x^n$ and $d(Q\sqrt{R}) = S\frac{dx}{\sqrt{R}}$, with

$$S = R\frac{dQ}{dx} + \frac{1}{2}Q\frac{dR}{dx} = \varphi(0) + \varphi(1)x + \ldots + \varphi(m)x^m.$$

This gives

$$\varphi(p) = (p+1)f(p+1)\ldots\alpha + \left(p + \frac{1}{2}\right) f(p) \cdot \beta + pf(p-1) \cdot \gamma + \left(p - \frac{1}{2}\right) f(p-2) \cdot \delta + (p-1)f(p-3) \cdot \varepsilon$$

and $m = n+3$. Abel draws the conclusion that the integrals $\int \frac{x^m dx}{\sqrt{R}}$ may be expressed as linear combinations of those with $0 \le m \le 2$ and an algebraic function; but $\int \frac{dx}{\sqrt{R}}$, $\int \frac{xdx}{\sqrt{R}}$ and $\int \frac{x^2dx}{\sqrt{R}}$ are independent when the reductions admitted involve only algebraic functions. The reduction of $\int \frac{x^m dx}{\sqrt{R}}$ is given by a system of $m-2$ linear equations $\varphi(p) = 0$ for $3 \le p \le m-1$, $\varphi(m) = -1$ to determine the $f(p)$ $(0 \le p \le m-3)$ and the formulae

$$\varphi(0) = f(1)\cdot\alpha + \frac{1}{2}f(0)\cdot\beta, \quad \varphi(1) = 2f(2)\cdot\alpha + \frac{3}{2}f(1)\cdot\beta + f(0)\cdot\gamma,$$
$$\varphi(2) = 3f(3)\cdot\alpha + \frac{5}{2}f(2)\cdot\beta + 2f(1)\cdot\gamma + \frac{3}{2}f(0)\cdot\delta.$$

For instance

$$\int \frac{x^4dx}{\sqrt{R}} = \left(\frac{5}{24}\frac{\beta\delta}{\varepsilon^2} - \frac{1}{3}\frac{\alpha}{\varepsilon}\right)\int \frac{dx}{\sqrt{R}} + \left(\frac{5}{12}\frac{\gamma\delta}{\varepsilon^2} - \frac{1}{2}\frac{\beta}{\varepsilon}\right)\int \frac{xdx}{\sqrt{R}}$$
$$+ \left(\frac{5}{8}\frac{\delta^2}{\varepsilon^2} - \frac{2}{3}\frac{\gamma}{\varepsilon}\right)\int \frac{x^2dx}{\sqrt{R}} - \left(\frac{5}{12}\frac{\delta}{\varepsilon^2} - \frac{1}{3}\frac{1}{\varepsilon}x\right)\sqrt{R}.$$

When the values found for $\varphi(0), \varphi(1)$ and $\varphi(2)$ are 0, the integral is algebraic; for instance, when $R = \frac{125}{256}\frac{\delta^4}{\varepsilon^3} + \frac{25}{32}\frac{\delta^3}{\varepsilon^2}x + \frac{15}{16}\frac{\delta^2}{\varepsilon}x^2 + \delta x^3 + \varepsilon x^4$, $\int \frac{x^4dx}{\sqrt{R}} = -\left(\frac{5}{12}\frac{\delta}{\varepsilon^2} - \frac{1}{3}\frac{1}{\varepsilon}x\right)\sqrt{R}$.

In a completely analogous manner, Abel reduces $\int \frac{dx}{(x-a)^m\sqrt{R}}$ to a linear combination of $\int \frac{dx}{\sqrt{R}}$, $\int \frac{xdx}{\sqrt{R}}$, $\int \frac{x^2dx}{\sqrt{R}}$, $\int \frac{dx}{(x-a)\sqrt{R}}$ and an algebraic function $Q\sqrt{R}$, Q having only one pole in a: $Q = \frac{\psi(1)}{x-a} + \frac{\psi(2)}{(x-a)^2} + \ldots + \frac{\psi(m-1)}{(x-a)^{m-1}}$. Indeed $d(Q\sqrt{R}) = S\frac{dx}{\sqrt{R}}$ with

$$S = \varphi(0) + \varphi(1)x + \varphi(2)x^2 + \frac{\chi(1)}{x-a} + \frac{\chi(2)}{(x-a)^2} + \ldots + \frac{\chi(m)}{(x-a)^m},$$
$$\varphi(0) = \left(\frac{1}{2}a\delta + \varepsilon a^2\right)\psi(1) - \frac{1}{2}(\delta + 4a\varepsilon)\psi(2) - \varepsilon\psi(3),$$
$$\varphi(1) = \frac{1}{2}\delta\psi(1), \quad \varphi(2) = \varepsilon\psi(1)$$

and $\chi(p) = -\alpha'(p-1)\psi(p-1) - \beta'(p-\frac{1}{2})\psi(p) - \gamma' p\psi(p+1) - \delta'(p+\frac{1}{2})\psi(p+2) - \varepsilon'(p+1)\psi(p+3)$; here $\alpha' = \alpha + \beta a + \gamma a^2 + \delta a^3 + \varepsilon a^4$, $\beta' = \beta + 2\gamma a + 3\delta a^2 + 4\varepsilon a^3$, $\gamma' = \gamma + 3\delta a + 6\varepsilon a^2$, $\delta' = \delta + 4\varepsilon a$ and $\varepsilon' = \varepsilon$, so that $R = \alpha' + \beta'(x-a) + \gamma'(x-a)^2 + \delta'(x-a)^3 + \varepsilon'(x-a)^4$. In order to get the announced reduction, we determine the $\psi(p)$ by a linear system $\chi(p) = 0$ for $2 \le p \le m-1$, $\chi(m) = -1$; then $\varphi(0), \varphi(1), \varphi(2)$ are given by the preceding formulae and

$$\chi(1) = -\frac{1}{2}\beta'\psi(1) - \gamma'\psi(2) - \frac{3}{2}\delta'\psi(3) - 2\varepsilon'\psi(4).$$

For instance,

$$\int \frac{dx}{(x-a)^2\sqrt{fx}} = -\frac{\varepsilon a^2 + \frac{1}{2}\delta a}{fa}\int \frac{dx}{\sqrt{fx}} + \frac{\delta}{2fa}\int \frac{xdx}{\sqrt{fx}} + \frac{\varepsilon}{fa}\int \frac{x^2dx}{\sqrt{fx}}$$
$$-\frac{1}{2}\frac{f'a}{fa}\int \frac{dx}{(x-a)\sqrt{fx}} - \frac{\sqrt{fx}}{(x-a)\,fa}, \tag{46}$$

where $R = fx$. This reduction does not work if $\alpha' = fa = 0$, which gives $\chi(m) = 0$ and in that case, we must take Q with a pole of order m in a and we see that $\int \frac{dx}{(x-a)^m\sqrt{R}}$ is reducible to $\int \frac{dx}{\sqrt{R}}$, $\int \frac{xdx}{\sqrt{R}}$ and $\int \frac{x^2dx}{\sqrt{R}}$ even for $m = 1$:

$$\int \frac{dx}{(x-a)\sqrt{R}} = -\frac{2\varepsilon a^2 + a\delta}{f'a}\int \frac{dx}{\sqrt{R}} + \frac{\delta}{f'a}\int \frac{xdx}{\sqrt{R}} + \frac{2\varepsilon}{f'a}\int \frac{x^2dx}{\sqrt{R}} - \frac{2}{f'a}\frac{\sqrt{R}}{x-a}. \tag{47}$$

The next task for Abel is to find the possible relations between integrals of the form

$$\int \frac{dx}{(x-b)\sqrt{R}}.$$

It is easy to see that the only possible relations have the form

$$\begin{aligned}
&\varphi(0)\int \frac{dx}{(x-a)\sqrt{R}} + \varphi(1)\int \frac{dx}{(x-a')\sqrt{R}} \\
&+\varphi(2)\int \frac{dx}{(x-a'')\sqrt{R}} + \varphi(3)\int \frac{dx}{(x-a''')\sqrt{R}} \\
&= \sqrt{R}\left(\frac{A}{x-a} + \frac{A'}{x-a'} + \frac{A''}{x-a''} + \frac{A'''}{x-a'''}\right),
\end{aligned}$$

where a, a', a'', a''' are the roots of R. Using the preceding reduction and the fact that $\int \frac{dx}{\sqrt{R}}$, $\int \frac{xdx}{\sqrt{R}}$ and $\int \frac{x^2dx}{\sqrt{R}}$ are independent, Abel finds $A = -\frac{2\varphi(0)}{f'a}$, $A' = -\frac{2\varphi(1)}{f'a'}$, $A'' = -\frac{2\varphi(2)}{f'a''}$, $A''' = -\frac{2\varphi(3)}{f'a'''}$, $A(2\varepsilon a^2 + a\delta) + A'(2\varepsilon a'^2 + a'\delta) + A''(2\varepsilon a''^2 + a''\delta) + A'''(2\varepsilon a'''^2 + a'''\delta) = 0$ and $A + A' + A'' + A''' = 0$; it is possible to choose $A''' = 0$ and this gives the relation

$$\begin{aligned}
&\varphi(0)\int \frac{dx}{(x-a)\sqrt{R}} + \varphi(1)\int \frac{dx}{(x-a')\sqrt{R}} + \varphi(2)\int \frac{dx}{(x-a'')\sqrt{R}} \\
&= \sqrt{R}\left(\frac{A}{x-a} + \frac{A'}{x-a'} + \frac{A''}{x-a''}\right)
\end{aligned}$$

with

$$\begin{aligned}
\varphi(0) &= \frac{1}{2}(a-a')(a-a'')(a-a''')(a'-a'')(a'+a''-a-a'''), \\
\varphi(1) &= \frac{1}{2}(a'-a)(a'-a'')(a'-a''')(a''-a)(a+a''-a'-a'''), \\
\varphi(2) &= \frac{1}{2}(a''-a)(a''-a')(a''-a''')(a-a')(a+a'-a''-a''').
\end{aligned}$$

Abel looks for linear relations between

$$\int \frac{dx}{\sqrt{R}}, \quad \int \frac{xdx}{\sqrt{R}}, \quad \int \frac{x^2dx}{\sqrt{R}} \quad \text{and} \quad \int \frac{dx}{(x-a)\sqrt{R}}, \quad \int \frac{dx}{(x-a')\sqrt{R}};$$

using (46), he finds $\int \frac{dx}{\sqrt{R}} = \frac{(a-a'')(a-a''')}{a''+a'''-a-a'} \int \frac{dx}{(x-a)\sqrt{R}} + \frac{(a'-a'')(a'-a''')}{a''+a'''-a-a'} \int \frac{dx}{(x-a')\sqrt{R}}$ and

$$\begin{aligned} &\int \frac{x^2dx}{\sqrt{R}} + \frac{\delta}{2}\int \frac{xdx}{\sqrt{R}} \\ &= \frac{a'(a'-a-a''-a''')\cdot f'a}{2(a'-a)(a+a'-a''-a''')}\int \frac{dx}{(x-a)\sqrt{R}} \\ &\quad + \frac{a(a'-a-a''-a''')\cdot f'a'}{2(a-a')(a+a'-a''-a''')}\int \frac{dx}{(x-a')\sqrt{R}} \\ &\quad + \frac{\sqrt{R}}{(a-a')(a+a'-a''-a''')}\left(\frac{a'(a'-a-a''-a''')}{x-a} - \frac{a(a-a'-a''-a''')}{x-a'}\right). \end{aligned}$$

When $a + a' = a'' + a'''$, these relations loose their sense and give $\int \frac{dx}{(x-a)\sqrt{R}} + \int \frac{dx}{(x-a')\sqrt{R}} = \frac{2\sqrt{R}}{(a''-a)(a''-a')(x-a)(x-a')}$.

In the second chapter of his memoir, Abel studies the integration of elliptic integrals by logarithmic functions. Such a function must be of the form

$$T = A\log(P + Q\sqrt{R}) + A'\log(P' + Q'\sqrt{R}) + \ldots + A^{(n)}\log(P^{(n)} + Q^{(n)}\sqrt{R})$$

with $P, Q, P', Q', \ldots$ polynomials and $A, A', \ldots$ constant, or, subtracting from dT a rational differential $T' = A\log\frac{P+Q\sqrt{R}}{P-Q\sqrt{R}} + A'\log\frac{P'+Q'\sqrt{R}}{P'-Q'\sqrt{R}} + \ldots$ Then dT' is a sum of terms of the form $\frac{M}{N}\cdot\frac{dx}{\sqrt{R}}$ with $M = A\frac{2N\frac{dP}{dx} - P\frac{dN}{dx}}{Q}$, $N = P^2 - Q^2R$; the fraction $\frac{M}{N}$ has only poles of order 1, these poles are not roots of R and it is easy to see that its polynomial part is of degree ≤ 1. Finally

$$T' = k\int \frac{dx}{\sqrt{R}} + k'\int \frac{xdx}{\sqrt{R}} + L\int \frac{dx}{(x-a)\sqrt{R}} + \ldots + L^{(\nu)}\int \frac{dx}{(x-a^{(\nu)})\sqrt{R}}, \tag{48}$$

and $\int \frac{x^2dx}{\sqrt{R}}$ cannot be reduced to other integrals by means of logarithms.

Let us suppose that T' contains $r+1$ logarithmic terms; looking at the degrees of $P, Q, P', Q', \ldots$ and at the corresponding number of indeterminate coefficients in T', Abel sees that the minimum value of ν is 2 and that r may be taken equal to 0. Moreover, one may take

$$P = f + f'x + f''x^2$$

of degree 2, $Q = 1$ and N of degree 2; then $f'' = \sqrt{\varepsilon}$, $f' = \frac{\delta}{2\sqrt{\varepsilon}}$ and $f = \frac{k(\delta^2-4\varepsilon\gamma)+2k'\varepsilon\beta}{2(\delta k'-4\varepsilon k)\sqrt{\varepsilon}}$. For $k = 0$ and $k' = 1$, one has

$$\int \frac{xdx}{\sqrt{R}} = (G + H\sqrt{K}) \int \frac{dx}{(x - \sqrt{K})\sqrt{R}} + (G - H\sqrt{K}) \int \frac{dx}{(x + \sqrt{K})\sqrt{R}}$$
$$+ \frac{1}{2\sqrt{\varepsilon}} \log \frac{\frac{\beta}{\delta}\sqrt{\varepsilon} + \frac{\delta}{2\sqrt{\varepsilon}}x + \sqrt{\varepsilon}x^2 + \sqrt{R}}{\frac{\beta}{\delta}\sqrt{\varepsilon} + \frac{\delta}{2\sqrt{\varepsilon}}x + \sqrt{\varepsilon}x^2 - \sqrt{R}},$$

where $G = -\frac{4\alpha\delta^2\varepsilon + \beta\delta^3 + 4\beta^2\varepsilon^2 - 4\beta\gamma\delta\varepsilon}{2(\delta^4 + 8\beta\delta\varepsilon^2 - 4\gamma\delta^2\varepsilon)}$, $H = -\frac{\delta}{4\varepsilon}$, $K = \frac{4\varepsilon}{\delta}\frac{\varepsilon\beta^2 - \alpha\delta^2}{4\gamma\delta\varepsilon - 8\beta\varepsilon^2 - \delta^3}$.

It is possible to have N of degree 1 when $(4\varepsilon\gamma - \delta^2)^2 + 4\delta^2(4\varepsilon\gamma - \delta^2) - 32\beta\delta\varepsilon^2 - 64\alpha\varepsilon^3 = 0$; then $f = \frac{4\varepsilon\gamma - \delta^2}{8\varepsilon\sqrt{\varepsilon}}$ and $\int \frac{xdx}{\sqrt{R}} = \frac{1}{3\varepsilon}(\mu' - \mu\delta) \int \frac{dx}{(x+\mu)\sqrt{R}} + \frac{1}{3\sqrt{\varepsilon}} \log \frac{\frac{\mu'}{\sqrt{\varepsilon}} + \frac{\delta}{2\sqrt{\varepsilon}}x + \sqrt{\varepsilon}x^2 + \sqrt{R}}{\frac{\mu'}{\sqrt{\varepsilon}} + \frac{\delta}{2\sqrt{\varepsilon}}x + \sqrt{\varepsilon}x^2 - \sqrt{R}}$, where $\mu' = \frac{4\varepsilon\gamma - \delta^2}{8\varepsilon}$ and $\mu = -\frac{\delta}{2\varepsilon}$.

Abel finds another reduction, writing $R = (p + qx + rx^2)(p' + q'x + x^2)$, $P = f(p' + q'x + x^2)$ and $Q = 1$ and choosing f such that

$$N = (f^2 - r)(p' + q'x + x^2)(x - a)^2;$$

then $\frac{M}{N} = 1 + L\frac{1}{x-a}$ with

$$L = \frac{pq' - qp' + (rq' - q)a^2}{(rq' - q)a}, \quad a = \frac{q - q'f^2}{2(f^2 - r)}.$$

This leads to the equation $f^4(q'^2 - 4p') - f^2(2qq' - 4p - 4p'r) + q^2 - 4pr = 0$ and to the relation $\int \frac{dx}{\sqrt{R}} = -L \int \frac{dx}{(x-a)\sqrt{R}} + A \log \frac{f(p' + q'x + x^2) + \sqrt{R}}{f(p' + q'x + x^2) - \sqrt{R}}$, where $A = \frac{f^2 - r}{f(rq' - q)}$.

Another formula is found by supposing $N = k(x - a)^4$; then

$$(p + p' - p'' - p''')a^2 - 2(pp' - p''p''')a + pp'(p'' + p''') - p''p'''(p + p') = 0,$$

where p, p', p'', p''' are the roots of R. In that case, $\frac{M}{N} = 1 - L\frac{1}{x-a}$, where

$$L = -\frac{2(f + af' + a^2f'')}{f' + 2af''}, \quad f = \sqrt{pp'p''p''' + ka^4},$$
$$f' = -\frac{p + p' + p'' + p''' + 4ka}{2\sqrt{1+k}}, \quad f'' = \sqrt{1+k}$$

and $k = \frac{(p + p' - p'' - p''')^2}{(2(p'' + p''') - 4a)(2(p + p') - 4a)}$; so

$$\int \frac{dx}{\sqrt{(x-p)(x-p')(x-p'')(x-p''')}}$$
$$= L \int \frac{dx}{(x-a)\sqrt{(x-p)(x-p')(x-p'')(x-p''')}}$$
$$+ A \log \frac{f + f'x + f''x^2 + \sqrt{(x-p)(x-p')(x-p'')(x-p''')}}{f + f'x + f''x^2 - \sqrt{(x-p)(x-p')(x-p'')(x-p''')}},$$

with $A = \frac{1}{2\sqrt{(p+p'-2a)(p''+p'''-2a)}}$. All this work is inspired by Legendre's reduction of elliptic integrals to canonical forms as it is presented in the *Exercices de Calcul intégral* but Abel's study is deeper and more general for he investigates all the possible relations between such integrals and proves the independance of the three canonical kinds.

Abel also studies the general case, where $\frac{M}{N} = \frac{x^m+k^{(m-1)}x^{m-1}+\ldots+k}{x^m+l^{(m-1)}x^{m-1}+\ldots+l}$; if Q is of degree n, P must be of degree $n+2$ and $m \leq 2n+4$ which is the degree of N. With the notations $R = \varphi x$, $P = Fx$, $Q = fx, a, a', \ldots, a^{(m-1)}$ roots of N (with multiplicities $\mu, \mu', \ldots, \mu^{(m-1)}$), one has

$$Fa^{(j)} = \pm fa^{(j)}\sqrt{\varphi a^{(j)}} \quad (0 \leq j \leq m-1),$$

whence, by successive derivations, a linear system to determine the coefficients of P and Q. Then $x^m + k^{(m-1)}x^{m-1} + \ldots + k$ takes in $a^{(j)}$ the value $\pm A\sqrt{\varphi a^{(j)}} \cdot \psi a^{(j)}$, where

$$\psi x = (x-a)(x-a')\ldots(x-a^{(m-1)})\frac{dN}{Ndx};$$

this gives a linear system to get $k, k', \ldots$ in function of $A, a, a', \ldots$ For instance, when $\mu = \mu' = \ldots = \mu^{(m-1)} = 1$, $m = 2n+4 = 4$ if $Q = 1$ and Abel finds, for the coefficients of P,

$$\begin{aligned}
-f &= i\frac{a'a''a'''}{(a-a')(a-a'')(a-a''')}\sqrt{\varphi a} + i'\frac{aa''a'''}{(a'-a)(a'-a'')(a'-a''')}\sqrt{\varphi a'} \\
&\quad + i''\frac{aa'a'''}{(a''-a)(a''-a')(a''-a''')}\sqrt{\varphi a''} + i'''\frac{aa'a''}{(a'''-a)(a'''-a')(a'''-a'')}\sqrt{\varphi a'''}, \\
f'' &= \frac{i\sqrt{\varphi a}}{(a-a')(a-a'')} + \frac{i'\sqrt{\varphi a'}}{(a'-a)(a'-a'')} + \frac{i''\sqrt{\varphi a''}}{(a''-a)(a''-a')}, \\
f' &= \frac{i\sqrt{\varphi a}}{a-a'} + \frac{i'\sqrt{\varphi a'}}{a'-a} - (a+a')f'',
\end{aligned}$$

where i, i', i'', i''' are equal to ± 1, and $A = -\frac{1}{(a+a'+a''+a''')f''+2f'}$. When $m = 2$, $Q = 1$ and $P^2 - R = C(x-a)(x-a')^3$, he finds

$$\begin{aligned}
f'' &= \frac{1}{8}\frac{2\varphi a' \cdot \varphi'' a' - (\varphi' a')^2}{\varphi a'\sqrt{\varphi a'}}, \\
f' &= \frac{1}{2}\frac{\varphi' a'}{\sqrt{\varphi a'}} - \frac{a'}{4}\frac{2\varphi a' \cdot \varphi'' a' - (\varphi' a')^2}{\varphi a'\sqrt{\varphi a'}}, \\
f &= \sqrt{\varphi a} - \frac{a'}{2}\frac{\varphi' a'}{\sqrt{\varphi a'}} + \frac{a'^2}{8}\frac{2\varphi a' \cdot \varphi'' a' - (\varphi' a')^2}{\varphi a'\sqrt{\varphi a'}}, \\
A &= -\frac{1}{(a+3a')f'' + 2f'}
\end{aligned}$$

and a, a' related by $\sqrt{\varphi a} \cdot \sqrt{\varphi a'} = \varphi a' + \frac{1}{2}(a-a')\varphi' a' + \frac{1}{8}(a-a')^2\frac{2\varphi a' \cdot \varphi'' a' - (\varphi' a')^2}{\varphi a'}$. When $P^2 - R = C(x-a)^2(x-a')^2$,

$$f'' = \frac{1}{4}\frac{\varphi' a}{(a-a')\sqrt{\varphi a}} + \frac{1}{4}\frac{\varphi' a'}{(a'-a)\sqrt{\varphi a'}},$$
$$f' = \frac{1}{2}\frac{a'\varphi' a}{(a-a')\sqrt{\varphi a}} + \frac{1}{2}\frac{a\varphi' a'}{(a'-a)\sqrt{\varphi a'}},$$
$$f = \frac{1}{4}\frac{aa'}{a-a'}\frac{\varphi' a}{\sqrt{\varphi a}} + \frac{1}{4}\frac{aa'}{a'-a}\frac{\varphi' a'}{\sqrt{\varphi a'}} - \frac{a'\sqrt{\varphi a} - a\sqrt{\varphi a'}}{a-a'},$$
$$A = -\frac{2}{\frac{\varphi' a}{\sqrt{\varphi a}} + \frac{\varphi' a'}{\sqrt{\varphi a'}}}$$

and a, a' related by $(p+p'+p''+p''')aa' - (pp' - p''p''')(a+a') + pp'(p''+p''') - p''p'''(p+p') = 0$, where p, p', p'', p''' are the roots of R. So $\int \frac{dx}{\sqrt{\varphi x}} = -\int \frac{2b+2b'x}{(x-a)(x-a')}\frac{dx}{\sqrt{\varphi x}} + A\log\frac{P+\sqrt{\varphi x}}{P-\sqrt{\varphi x}}$, with $b = -2\frac{a'\sqrt{\varphi a}+a\sqrt{\varphi a'}}{\frac{\varphi' a}{\sqrt{\varphi a}}+\frac{\varphi' a'}{\sqrt{\varphi a'}}}$, $b' = 2\frac{\sqrt{\varphi a}+\sqrt{\varphi a'}}{\frac{\varphi' a}{\sqrt{\varphi a}}+\frac{\varphi' a'}{\sqrt{\varphi a'}}}$. In a third case $P^2 - R = C(x-p)(x-a)(x-a')^2$ and $P = (x-p)(f+f'x)$ and a' is function of a.

The last case considered by Abel is that in which $m = 1$. Here $P^2 - Q^2R = C(x-a)^{2n+4}$ and $\frac{M}{N} = \frac{x+k}{x-a}$ with $k = -a - \mu A\sqrt{\varphi a}$. The coefficients of P and Q are determined by a linear system and then a is given by an algebraic equation; this leads to

$$\int \frac{dx}{(x-a)\sqrt{R}} = \frac{1}{\mu A\sqrt{\varphi a}}\int\frac{dx}{\sqrt{R}} - \frac{1}{\mu\sqrt{\varphi a}}\log\frac{P+Q\sqrt{R}}{P-Q\sqrt{R}}.$$

Abel observes that the equation $P^2 - Q^2R = C(x-a)^{2n+4}$ is equivalent to $P'^2 - Q'^2R' = C$, where $F(x-a) = (x-a)^{n+2}P'\left(\frac{1}{x-a}\right)$, $f(x-a) = (x-a)^nQ'\left(\frac{1}{x-a}\right)$ and $\varphi(x-a) = (x-a)^4R'\left(\frac{1}{x-a}\right)$.

As we know, the same equation is met in the problem to express $\int\frac{(k+x)dx}{\sqrt{R}}$ by a logarithm $A\log\frac{P+Q\sqrt{R}}{P-Q\sqrt{R}}$; here $\frac{M}{N} = x+k$, so N is constant and may be taken as 1. The conditions of the problem are $x+k = 2A\frac{dP}{Qdx}$, $1 = P^2 - Q^2R$; the first method proposed by Abel to determine $P = f + f'x + \ldots + f^{(n+2)}x^{n+2}$ and $Q = e + e'x + \ldots + e^{(n)}x^n$ is that of indeterminate coefficients. The first condition gives $A = \frac{e^{(n)}}{(2n+4)f^{(n+2)}}$, $k = \frac{f'e^{(n)}}{(n+2)ef^{(n+2)}}$ and the second gives a system of $2n+5$ equations between the $2n+4$ coefficients $e^{(p)}$, $f^{(p)}$: $f^2 - \alpha e^2 = 1, \ldots, f^{(n+2)2} - \varepsilon e^{(n)2} = 0$. The compatibility of this system imposes a relation between the coefficients α, β, γ, δ and ε of R; for instance, when $n = 0$, so that $Q = e$ and $P = f + f'x + f''x^2$, one has

$$2ff' - \beta e^2 = f'^2 + 2ff'' - \gamma e^2 = 2f'f'' - \delta e^2 = 0,$$

whence $f'' = \frac{\delta\sqrt{\varepsilon}}{\sqrt{\beta^2\varepsilon - \alpha\delta^2}}$, $f' = \frac{\delta^2}{2\sqrt{\beta^2\varepsilon^2 - \alpha\varepsilon\delta^2}}$, $f = \frac{\beta\sqrt{\varepsilon}}{\sqrt{\beta^2\varepsilon-\alpha\delta^2}}$, $e = \frac{\delta}{\sqrt{\beta^2\varepsilon-\alpha\delta^2}}$ and $\gamma = \frac{\delta^2}{4\varepsilon} + \frac{2\beta\varepsilon}{\delta}$, $A = \frac{1}{4\sqrt{\varepsilon}}$, $k = \frac{\delta}{4\varepsilon}$.

But it is possible to get a *linear* system for the coefficients $e^{(p)}$, $f^{(p)}$: if

$$\begin{aligned} Fy &= fy^{n+2} + f'y^{n+1} + \ldots + f^{(n+2)}, \\ fy &= ey^n + e'y^{n-1} + \ldots + e^{(n)} \\ \text{and} \quad \varphi y &= \alpha y^4 + \beta y^3 + \gamma y^2 + \delta y + \varepsilon, \end{aligned}$$

the second condition is $(Fy)^2 - (fy)^2\varphi y = y^{2n+4}$ and it gives $Fy = fy \cdot \sqrt{\varphi y}$ when $y = 0$. The system is obtained by differentiating $2n+3$ times this relation at $y = 0$. When $n = 0$, one finds $f'' = ce$, $f' = c'e$, $f = \frac{c''}{2}e$, $0 = c'''$, where $c^{(p)} = \frac{d^p\sqrt{\varphi y}}{dy^p}\big|_{y=0}$ and $\gamma = \frac{\delta^2}{4\varepsilon} + \frac{2\beta\varepsilon}{\delta}$ as above; when $n = 1$, the system is $0 = 2c' + c''\frac{e'}{e}$, $0 = 4c''' + c''''\frac{e'}{e}$, $0 = 5c'''' + c'''''\frac{e'}{e}$, whence

$$c'c'''' - 2c''c''' = 2c'c''''' - 5c''c'''' = 0.$$

Without restricting the generality, we may take $\varepsilon = 1$ and $\beta = -\alpha$; the preceding equations then give $\delta = 2$, $\gamma = -3$ and finally

$$\begin{aligned} &\int \frac{xdx}{\sqrt{x^4 + 2x^3 + 3x^2 - \alpha x + \alpha}} \\ &= \frac{1}{6}\log\frac{x^3 + 3x^2 - 2 - \frac{\alpha}{2} + (x+2)\sqrt{x^4 + 2x^3 - 3x^2 - \alpha x + \alpha}}{x^3 + 3x^2 - 2 - \frac{\alpha}{2} - (x+2)\sqrt{x^4 + 2x^3 - 3x^2 - \alpha x + \alpha}}. \end{aligned} \tag{49}$$

Abel proposes another way to study the equation $P^2 - Q^2R = 1$; he writes it $P + 1 = P'^2R'$, $P - 1 = Q'^2R''$, where $P'Q' = Q$ and $R'R'' = R$. Then $P = \frac{1}{2}(P'^2R' + Q'^2R'')$ and $2 = P'^2R' - Q'^2R''$; with $R' = x^2 + 2qx + p$, $R'' = x^2 + 2q'x + p'$ and P', Q' constant, one finds $q = q'$, $P' = Q' = \frac{\sqrt{2}}{\sqrt{p-p'}}$, $P = \frac{2x^2+4qx+p+p'}{p-p'}$, $Q = \frac{2}{p-p'}$, $k = q$ and $A = \frac{1}{4}$, so

$$\begin{aligned} &\int \frac{(x+q)dx}{\sqrt{(x^2 + 2qx + p)(x^2 + 2qx + p')}} \\ &= \frac{1}{4}\log\frac{2x^2 + 4qx + p + p' + 2\sqrt{R}}{2x^2 + 4qx + p + p' - 2\sqrt{R}}. \end{aligned}$$

With $P' = \frac{x+m}{c}$, $Q' = \frac{x+m'}{c}$, one finds $2q = r + m' - m$, $2q' = r + m - m'$,

$$\begin{aligned} p &= \frac{1}{2}r(3m' - m) + \frac{1}{2}m^2 - \frac{1}{2}m'^2 - mm', \\ p' &= \frac{1}{2}r(3m - m') + \frac{1}{2}m'^2 - \frac{1}{2}m^2 - mm', \\ 2c^2 &= \frac{1}{2}r(m' - m)^3 + \frac{1}{2}(m - m')(m^3 - m^2m' - m'^2m + m'^3), \end{aligned}$$

where $r = q + q'$, and then

$$P = \frac{(x^2+2mx+m^2)(x^2+2qx+p)-c^2}{c^2},$$
$$Q = \frac{x^2+(m+m')x+mm'}{c^2},$$
$$k = \frac{1}{4}(3r-m'-m).$$

If we impose $k=0, r=\frac{m+m'}{3}, m=2q'+q, m'=2q+q', p=-q^2-2qq'$ and $p'=-q'^2-2qq'$; we have

$$\int \frac{xdx}{\sqrt{(x^2+2qx-q^2-2qq')(x^2+2q'x-q'^2-2qq')}}$$
$$= \frac{1}{4}\log\frac{(x+q+2q')\sqrt{x^2+2qx-q^2-2qq'}+(x+q'+2q)\sqrt{x^2+2q'x-q'^2-2qq'}}{(x+q+2q')\sqrt{x^2+2qx-q^2-2qq'}-(x+q'+2q)\sqrt{x^2+2q'x-q'^2-2qq'}}.$$

The second method to study the equation $P^2-Q^2R=1$ is that used by Abel in his published memoir for the more general case of hyperelliptic integrals: putting $R=r^2+s$, with r of degree 2 and s of degree 1, the equation becomes $P^2-Q^2r^2-Q^2s=1$ and it shows that $P=Qr+Q_1$ with $\deg Q_1<\deg P$. Then $Q_1^2+2QQ_1r-Q^2s=1$ or, if $r=sv+u$, with v of degree 1 and u constant, $Q_1^2+2QQ_1u+Qs(2vQ_1-Q)=1$; thus $Q_2=Q-2vQ_1$ if of degree $<\deg Q=n$ and

$$s_1Q_1^2-2r_1Q_1Q_2-sQ_2^2=1,$$

with $s_1=1+4uv$, $r_1=r-2u$, $\deg Q_1=n-1$ and $\deg Q_2=n-2$. Iterating the process, one gets equations $s_{2\alpha-1}Q_{2\alpha+1}^2-2r_{2\alpha}Q_{2\alpha}Q_{2\alpha+1}-s_{2\alpha}Q_{2\alpha}^2=1$, $s_{2\alpha'+1}Q_{2\alpha'+1}^2-2r_{2\alpha'+1}Q_{2\alpha'+1}Q_{2\alpha'+2}-s_{2\alpha'}Q_{2\alpha'+2}^2=1$, with $\deg Q_p=n-p$; this gives $s_nQ_n^2=(-1)^{n+1}$, Q_n and s_n constant. The induction relations to determine the s_m are

$$s_m=s_{m-2}+4u_{m-1}v_{m-1},\ r_m=r_{m-1}-2u_{m-1}=s_mv_m+u_m. \tag{50}$$

A consequence of these relations is that $s_{m-1}s_m+r_m^2=s_{m-1}s_{m-2}+r_{m-1}^2$, so that this quantity does not depend on m and

$$s_{m-1}s_m+r_m^2=ss_1+r_1^2=r^2+s=R; \tag{51}$$

as $s_n=\mu$ is constant, it is easy to see that $r_{n-k}=r_k$, $s_{n-k}=s_{k-1}\mu^{(-1)^k}$, $v_{n-k}=v_{k-1}\mu^{(-1)^{k-1}}$ and $u_{n-k}=-u_{k-1}$. For $n=2\alpha+1$ and $k=\alpha+1$, this gives $\mu=1$ and $u_\alpha=0$; for $n=2\alpha$, $u_{\alpha-1}+u_\alpha=0$. The Q_m are determined from Q_n by the induction relations $Q_m=2v_mQ_{m+1}+Q_{m+2}$ and we see that $r, 2v, 2v_1, \ldots, 2v_{n-1}$ are the partial quotients of the continued fraction for $\frac{P}{Q}$, which is obtained by truncating that for $\sqrt{R}$. Putting $r_m=x^2+ax+b_m$, $s_m=c_m+p_mx$, $v_m=(g_m+x)\frac{1}{p_m}$ and $q_m=b-b_m$, Abel draws from (50) and (51) the relations

$$q_m = \frac{\frac{1}{2}p^2 + (ap - 2c)q_{m-1} - q_{m-2}q_{m-1}^2}{q_{m-1}^2}, \qquad \frac{c_{m-1}}{p_{m-1}} = \frac{c + q_{m-1}q_m}{p}$$

and $\quad p_m p_{m-1} = 2q_m$;

since $u_m = \frac{b_m - b_{m+1}}{2} = \frac{1}{2}(q_{m+1} - q_m)$ and $g_m = a - \frac{c_m}{p_m}$, these relations allow to determine r_m, s_m, u_m and v_m if we know the q_m, which are determined by an induction relation starting from $q = 0, q_1 = 2\frac{bp^2 - acp + c^2}{p^2}$ and are rational functions of a, b, c, p.

Abel applies this method to the elliptic integral $\int \frac{(x+k)dx}{\sqrt{(x^2+ax+p)^2+px+c}}$. The condition $s_n =$ constant is equivalent to $p_n = 0$ and it leads to $q_n = 0$ and $q_{n-k} = q_k$. The coefficient k is equal to $\frac{1}{n+2}a + \frac{1}{n+2}\left(\frac{c}{p} + \frac{c_1}{p_1} + \ldots + \frac{c_{n-1}}{p_{n-1}}\right)$ and the polynomials P and Q are determined by the continued fraction. When $c = 0$, Abel finds the results published in his 1826 paper, using

$$q_1 = 2b, \quad q_2 = \frac{p(p+4ab)}{8b^2},$$
$$q_3 = \frac{2b(16b^3 - p(p+4ab))}{(p+4ab)^2},$$
$$q_4 = \frac{4bp(p+4ab)(p^2+6abp+8a^2b^2-8b^3)}{(16b^3 - p(p+4ab))^2}.$$

From a relation $\int \frac{(y+k')dy}{\sqrt{R'}} = A' \log \frac{P'+Q'\sqrt{R'}}{P'-Q'\sqrt{R'}}$, Abel deduces

$$\int \frac{x+k}{x+l}\frac{dx}{\sqrt{R}} = A \log \frac{P+Q\sqrt{R}}{P-Q\sqrt{R}}$$

through the change of variable $y = \frac{1}{x+l}$; he finds $k = l + \frac{1}{k'}, A = -\frac{A'}{k'}$ and an algebraic equation to determine l in function of the coefficients of R. Indeed, when

$$R' = (y^2 + ay + b)^2 + c + py \quad \text{and} \quad R = (b^2 + c)(x^4 + \delta x^3 + \gamma x^2 + \beta x + \alpha),$$

$$2ab + p = (b^2 + c)(\delta - 4l), a^2 + 2b = (b^2 + c)(\gamma - 3\delta l + 6l^2),$$

$$2a = (b^2 + c)(\beta - 2\gamma l + 3\delta l^2 - 4l^3)$$

and $1 = (b^2+c)(\alpha - \beta l + \gamma l^2 - \delta l^3 + l^4)$. From this Abel deduces, with $-l$ instead of l, $\int \frac{dx}{(x-l)\sqrt{R}} = -\frac{1}{l+k}\int \frac{dx}{\sqrt{R}} - \frac{1}{(2n+4)\sqrt{\alpha+\beta l+\gamma l^2+\delta l^3+l^4}} \log \frac{P+Q\sqrt{R}}{P-Q\sqrt{R}}$, which gives a new proof of (49) when $l + k = \infty$.

In the third chapter of the *Théorie des transcendantes elliptiques*, Abel shows that the periods of an integral of the third kind $p = \int \frac{dx}{(x-a)\sqrt{R}}$ are combinations of the periods of the integrals $\int \frac{dx}{\sqrt{R}}, \int \frac{xdx}{\sqrt{R}}$ and $\int \frac{x^2dx}{\sqrt{R}}$. Taking the integral from a value $x = r$ which annihilates $R = fx$, differentiating with respect to a and using (46), he obtains

$$\frac{dp}{da}+\frac{1}{2}\frac{f'a}{fa}p=\frac{\sqrt{fx}}{(a-x)fa}+\frac{1}{fa}\int\frac{dx}{\sqrt{fx}}(A+Bx+Cx^2)$$

where $A=-\varepsilon a^2-\frac{1}{2}\delta a$, $B=\frac{1}{2}\delta$ and $C=\varepsilon$. From this he deduces

$$p\sqrt{fa}-\sqrt{fx}\int\frac{da}{(a-x)\sqrt{fa}}=\int\frac{da}{\sqrt{fa}}\int\frac{dx}{\sqrt{fx}}(A+Bx+Cx^2)+\text{constant}$$

and the constant is seen to be 0 by making $a=r$. Thus

$$\begin{aligned}&\sqrt{fa}\int\frac{dx}{(x-a)\sqrt{fx}}-\sqrt{fx}\int\frac{da}{(a-x)\sqrt{fa}}\\&=\int\frac{da}{\sqrt{fa}}\int\frac{(\frac{1}{2}\delta x+\varepsilon x^2)dx}{\sqrt{fx}}-\int\frac{dx}{\sqrt{fx}}\int\frac{(\frac{1}{2}\delta a+\varepsilon a^2)da}{\sqrt{fa}}\end{aligned}$$

which the formula (42) for the case of elliptic integrals. When r' is another root of fx, one obtains $\sqrt{fa}\int_r^{r'}\frac{dx}{(x-a)\sqrt{fx}}=\int_r\frac{da}{\sqrt{fa}}\int_r^{r'}\frac{(\frac{1}{2}\delta x+\varepsilon x^2)dx}{\sqrt{fx}}-\int_r^{r'}\frac{dx}{\sqrt{fx}}\int_r\frac{(\frac{1}{2}\delta a+\varepsilon a^2)da}{\sqrt{fa}}$. And if r'' is a third root of fx, $\int_r^{r''}\frac{da}{\sqrt{fa}}\int_r^{r'}\frac{(\frac{1}{2}\delta x+\varepsilon x^2)dx}{\sqrt{fx}}=\int_r^{r'}\frac{dx}{\sqrt{fx}}\int_r^{r''}\frac{(\frac{1}{2}\delta a+\varepsilon a^2)da}{\sqrt{fa}}$.

Abel finds new relations between periods starting from

$$\begin{aligned}s&=A\log\frac{P+Q\sqrt{R}}{P-Q\sqrt{R}}+A'\log\frac{P'+Q'\sqrt{R}}{P'-Q'\sqrt{R}}+\dots\\&=\int\frac{B+Cx}{\sqrt{R}}dx+L\int\frac{dx}{(x-a)\sqrt{R}}+L'\int\frac{dx}{(x-a')\sqrt{R}}+\dots\end{aligned}$$

(cf. (48)) which gives, by integrating from r to r':

$$\begin{aligned}s'-s=&\int_r^{r'}\frac{B+Cx}{\sqrt{fx}}dx\\&-\int_r^{r'}\frac{dx}{\sqrt{fx}}\left(\frac{L}{\sqrt{fa}}\int_r\frac{(\frac{1}{2}\delta a+\varepsilon a^2)da}{\sqrt{fa}}+\frac{L'}{\sqrt{fa'}}\int_r\frac{(\frac{1}{2}\delta a'+\varepsilon a'^2)da'}{\sqrt{fa'}}+\dots\right)\\&+\int_r^{r'}\frac{(\frac{1}{2}\delta x+\varepsilon x^2)dx}{\sqrt{fx}}\left(\frac{L}{\sqrt{fa}}\int_r\frac{da}{\sqrt{fa}}+\frac{L'}{\sqrt{fa'}}\int_r\frac{da'}{\sqrt{fa'}}+\dots\right)\end{aligned}$$

The end of the *Théorie des transcencantes elliptiques* (p. 173–188) is devoted to the proof that an integral of the third kind $\Pi(n)=\int\frac{dx}{(1+nx^2)\sqrt{(1-x^2)(1-c^2x^2)}}$ may be transformed in a linear combination of the integral of the first kind $F=\int\frac{dx}{\sqrt{(1-x^2)(1-c^2x^2)}}$, some logarithms (or arctangents) of algebraic functions and another integral of the third kind $\Pi(n')$ with a parameter n' arbitrarily large or arbitrarily close to a certain limit, and more generally to the relations between integrals of the third kind with different parameters. Let us consider $s = \arctan \frac{\sqrt{R}}{Q}$, with $R = (1-x^2)(1-c^2x^2)$ and $Q = x(a+bx^2)$; we have $ds = \frac{M}{N}\frac{dx}{\sqrt{R}}$ with $N = Q^2 + R$ and $M = \frac{1}{2}Q\frac{dR}{dx} - R\frac{dQ}{dx}$. If we impose that $N = k(1+nx^2)(1+n_1x^2)^2$, we find that $k = 1, b = \pm n_1\sqrt{n}, a = (1+n_1)\sqrt{1+n} \mp n_1\sqrt{n} = \chi(n)$ and $n_1 = \pm(\sqrt{1+n} \pm \sqrt{n})(\sqrt{c^2+n} \pm \sqrt{n}) = f(n)$. Then $\frac{M}{N} = A + \frac{L}{1+nx^2} + \frac{L'}{1+n_1x^2}$ with $A = 2a - \left(\frac{1}{n} + \frac{2}{n_1}\right)b$, $L = \frac{n_1}{\sqrt{n}} - a$ and $L' = 2\sqrt{n} - 2a$. Thus

$$\begin{aligned}\Pi(n) &= \frac{\sqrt{n}}{n_1 - a\sqrt{n}} \arctan \frac{\sqrt{R}}{ax+bx^3} - \frac{2a\sqrt{n} - (2n+n_1)}{n_1 a\sqrt{n}} F \\ &\quad + \frac{(2a - 2\sqrt{n})\sqrt{n}}{n_1 - a\sqrt{n}} \Pi(n_1) + C \\ &= \beta F + \gamma \Pi(n_1) + \alpha \arctan \frac{ax+bx^3}{\sqrt{R}}\end{aligned} \tag{52}$$

with $\alpha = \frac{\pm\sqrt{n}}{n_1 \mp a\sqrt{n}} = \varphi(n)$, $\beta = -\frac{\pm 2a\sqrt{n} - 2n - n_1}{n_1 \mp a\sqrt{n}} = \theta(n)$, $\gamma = \frac{\pm(2a \mp 2\sqrt{n})\sqrt{n}}{n_1 \mp a\sqrt{n}} = \psi(n)$. It is easy to see that $n_1 > 4n$ and that χ is an increasing function when both upper signs are chosen in $f(n)$. Thus, iterating the operation, we arrive at a parameter n_m as large as we wish, with α_m equivalent to $\frac{1}{\sqrt{n_m}}$, β_m remaining between 0 and 1 and $\lim \beta_m = 0$, $\lim \gamma_m = 4$. On the contrary, when both lower signs are chosen, n_m decreases and its limit is the root k of the equation $k = (\sqrt{k+1} - \sqrt{k})(\sqrt{k+c^2} - \sqrt{k})$. Applying (52) to $\Pi(k)$, we obtain

$$\Pi(k) = \frac{2a + 3\sqrt{k}}{3(a+\sqrt{k})} F - \frac{1}{3(a+\sqrt{k})} \arctan \frac{ax - k^{\frac{3}{2}}x^3}{\sqrt{R}}.$$

The formulae $n_1 = -(\sqrt{1+n} + \sqrt{n})(\sqrt{c^2+n} - \sqrt{n})$ and

$$n_1 = -(\sqrt{1+n} - \sqrt{n})(\sqrt{c^2+n} + \sqrt{n})$$

respectively lead to values of n_m between $-c^2$ and $-c$ and between -1 and $-c$. Abel also studies the case in which n is negative.

The transformed parameter n_1 is given by an equation of degree 4; inversely, one has

$$n = \frac{(n_1^2 - c^2)^2}{4n_1(n_1+1)(n_1+c^2)}.$$

When the sequence (n_m) is periodic, the integrals $\Pi(n_m)$ may be expressed as combinations of F and some arctangents.

Abel finds other relations as

$$\Pi(n) = -\frac{m'}{\mu}\frac{\psi(n_1)}{\psi(n)}\Pi(n_1) - \frac{A}{\mu n n_1 \psi(n)} F + \frac{1}{\mu\psi(n)} \arctan \frac{Q\sqrt{R}}{P},$$

with P, Q polynomials such that

$$P^2 + Q^2 R = (1 + nx^2)^{\mu}(1 + n_1 x^2)^{\mu'}, \quad \psi(n) = \frac{\sqrt{(1+n)(c^2+n)}}{\sqrt{n}},$$

A constant and $n_1 = \chi(n)$ a certain function. When for instance $P = 1 + bx^2$ and $Q = ex$, $\chi(n) = \frac{c(c-\sqrt{c^2+n})(1-\sqrt{1+n})}{n}$.

An essential discovery of Abel in the theory of elliptic functions is that these functions, obtained by inverting elliptic integrals, have 2 independent periods in the complex domain. In his posthumous memoir *Propriétés remarquables de la fonction* $y = \varphi x$ *déterminée par l'équation* $fydy - dx\sqrt{(a-y)(a_1-y)(a_2-y)\ldots(a_m-y)} = 0$, *fy étant une fonction quelconque de y qui ne devient pas nulle ou infinie lorsque* $y = a, a_1, a_2, \ldots, a_m$ (*Œuvres*, t. II, p. 40–42), he shows that the function φx, which is the inverse function of the hyperelliptic integral $x = \int \frac{fydy}{\sqrt{\psi y}}$ where $\psi y = (a-y)(a_1-y)(a_2-y)\ldots(a_m-y)$, must have each of the numbers $2(\alpha - \alpha_k)$ as period, where α_k is the values of the integral corresponding to $y = a_k$. Jacobi later proved (1834) that a regular uniform function of one complex variable cannot have more than 2 independent periods; thus the inverse function of a hyperelliptic integral cannot be uniform when $m > 4$. The inversion problem for hyperelliptic integrals or more generally for abelian integrals must involve functions of several complex variable, as Jacobi (1832) discovered through his intertretation of Abel theorem. Here Abel writes the Taylor series for the function φ:

$$\varphi(x+v) = y + v^2 Q_2 + v^4 Q_4 + v_6 Q_6 + \ldots + \sqrt{\psi y}(vQ_1 + v^3 Q_3 + v^5 q_5 + \ldots)$$

where the Q_j do not have poles at the a_k. Thus $\varphi(\alpha + v) = a + v^2 Q_2 + v^4 Q_4 + v_6 Q_6 + \ldots \varphi(\alpha + v)$ is an even function of v and $\varphi(2\alpha - v) = \varphi v$. In the same way $\varphi(2\alpha_1 - v) = \varphi v$ and $\varphi(2\alpha - 2\alpha_1 + v) = \varphi v$ and so on.

5 Abel Theorem

The most famous of Abel's results is a remarkable extension of Euler addition theorem for elliptic integrals. It is known as *Abel theorem* and gives the corresponding property for any integral of an algebraic function; such integrals are now called *abelian integrals*. This theorem, sent to the french Academy of Sciences by Abel in 1826 in a long memoir titled *Mémoire sur une propriété générale d'une classe très-étendue de fonctions transcendantes*, is rightly considered as the base of the following developments in algebraic geometry. Due to the negligence of the french Academicians, this fundamental memoir was published only in 1841, after the first edition of Abel's Work (1839).

In the introduction, Abel gives the following statement:

"When several functions are given of which the derivatives may be roots of the *same algebraic equation*, of which all the coefficients are *rational* functions of the same variable, one can always express the sum of any number of such functions by an *algebraic* and *logarithmic* function, provided that a certain number of *algebraic* relations be prescribed between the variables of the functions in question."

He adds that the number of relations does not depend on the number of the functions, but only on their nature. It is 1 for the elliptic integrals, 2 for the functions of which the derivatives contains only the square root of a polynmial of degree ≤ 6 as irrationality.

A second statement, which is properly Abel theorem says:

"One may always express the sum of a *given* number of functions, each of which is multiplied by a rational number, and of which the variable are arbitrary, by a similar sum of a *determined* number of functions, of which the variables are algebraic functions of the variables of the givent functions."

The proof of the first statement is short (§ 1–3, p. 146–150). Abel considers an algebraic equation $0 = p_0 + p_1 y + p_2 y^2 + \ldots + p_{n-1} y^{n-1} + y_n = \chi y$ with coefficients polynomials in x; this equation is supposed to be irreducible. He introduces another polynomial

$$\theta y = q_0 + q_1 y + q_2 y^2 + \ldots + q_{n-1} y^{n-1}$$

in x, y, certain coefficients $a, a', a'', \ldots$ of the polynomials $q_0, q_1, \ldots, q^{n-1}$ being indeterminates. The resultant $r = \theta y' \theta y'' \ldots \theta y^{(n)}$ of χ and θ, where $y', y'', \ldots, y^{(n)}$ are the roots of $\chi y = 0$, is a polynomial in $x, a, a', a'', \ldots$, which may be decomposed in $r = F_0 x F x$ where $F_0 x$ and Fx are polynomials in x and $F_0 x$ does not depend on $a, a', a'', \ldots$. Let $x_1, x_2, \ldots, x_\mu$ be the roots of $Fx = 0$ and $y_1, y_2, \ldots, y_\mu$ the corresponding common roots of the equations $\chi y = 0$, $\theta y = 0$. The y_k are rational functions of $x_k, a, a', a'', \ldots$ by the theory of elimination. Now let $f(x, y)dx$ be a differential form, with f a rational function of x, y. When $Fx = 0$, $dx = -\frac{\delta Fx}{F'x}$ where $F'x$ is the derivative with respect to x and δFx the differential with respect to $a, a', a'', \ldots$ Thus $f(x_k, y_k)dx_k = -\frac{f(x_k, y_k)}{F'x_k} \delta F x_k$ is rational with respect to $x_k, a, a', a'', \ldots$ and $dv = f(x_1, y_1)dx_1 + f(x_2, y_2)dx_2 + \ldots + f(x_\mu, y_\mu)dx_\mu$ is rational with respect to $a, a', a'', \ldots$ A consequence is that the function

$$\int f(x_1, y_1)dx_1 + \int f(x_2, y_2)dx_2 + \ldots + \int f(x_\mu, y_\mu)dx_\mu = v \tag{53}$$

is an algebraic and logarithmic function of $a, a', a'', \ldots$ Now if there are α indeterminate coefficients $a, a', a'', \ldots$ in θy, they may be determined by arbitrarily choosing α couples (x_j, y_j) of roots of $\chi y = 0$ and writing the equations $\theta y_j = 0$; the other y_k are then rational functions of x_k and the (x_j, y_j).

Abel gives a cleaver way to do the computation (§ 4, p. 150–159), first writing $\delta F x = \frac{\delta r}{F_0 x} = \frac{r \delta \theta y}{F_0 x \theta y}$ and

$$\begin{aligned} &f(x, y)dx \\ &= -\frac{1}{F_0 x F' x}\left(f(x, y')\frac{r}{\theta y'}\delta\theta y' + f(x, y'')\frac{r}{\theta y''}\delta\theta y'' + \ldots + f(x, y^{(n)})\frac{r}{\theta y^{(n)}}\delta\theta y^{(n)}\right) \\ &= -\frac{1}{F_0 x F' x}\sum f(x, y)\frac{r}{\theta y}\delta\theta y \end{aligned}$$

where the sum is extended to the n roots $y', y'', \ldots, y^{(n)}$. Then he writes $f(x, y) = \frac{f_1(x,y)}{f_2x\chi'y}$ where $f_1(x, y)$ is a polynomial in x, y, of degree $\leq n-1$ in y, f_2x a polynomial in x and $\chi'y$ the derivative of χy with respect to y. If $f_1(x, y)\frac{r}{\theta y}\delta\theta y = R^{(1)}y + Rx \cdot y^{n-1}$, with $R^{(1)}y$ polynomial in x, y of degree $\leq n-2$ in y and Rx a polynomial in x, it is easy to see that

$$\sum \frac{f_1(x, y)}{\chi'y}\frac{r}{\theta y}\delta\theta y = Rx. \tag{54}$$

Thus $dv = \sum f(x, y)dx = -\sum \frac{R_1x}{f_2x \cdot F_0x \cdot F'x}$; here the sum is extended to the μ roots x_k of $Fx = 0$. Grouping the roots of F_0x and of f_2x, we obtain $dv = -\sum \frac{R_1x}{\theta_1x \cdot F'x}$ where R_1x has no common root either with F_0x or with f_2x and the roots of θ_1x annihilate F_0x or f_2x. If R_2x is the quotient and R_3x the remainder of the Euclidian division of R_1x by θ_1x, one computes that $\sum \frac{R_1x}{F'x}$ is the coefficient of $\frac{1}{x}$ in the expansion of $\frac{R_1x}{\theta_1x \cdot Fx}$ in decreasing powers of x; this result comes from the development of $\frac{1}{F\alpha} = \sum \frac{1}{\alpha - x}\frac{1}{F'x}$ in decreasing powers of α. A rather more complicated computation, based on the decomposition of $\frac{R_3x}{\theta_1x}$ in simple elements, gives $\sum \frac{R_3x}{\theta_1x \cdot F'x} = -\sum{}' \nu \frac{d^{\nu-1}}{d\beta^{\nu-1}}\left(\frac{R_1\beta}{\theta_1^{(\nu)}\beta \cdot F\beta}\right)$ where the sum of the right hand side is extended to the roots β of θ_1x and, for each β, ν is the multiplicity of β . Unfortunately this result is incorrect and Sylow corrects it in the notes at the end of the second volume of Abel's Works (p. 295–296). The correct result is

$$\sum \frac{R_3x}{\theta_1x \cdot F'x} = \sum{}' \frac{1}{\Gamma\nu}\frac{d^{\nu-1}}{d\beta^{\nu-1}}\left(\frac{R_1\beta}{\vartheta\beta}\sum \frac{1}{(x-\beta)Fx}\right)$$

where $\vartheta x = \frac{\theta_1x}{(x-\beta)^\nu}$. From (54), we draw $R_1x = F_2x \cdot Fx \sum \frac{f_1(x,y)}{\chi'y}\frac{\delta\theta y}{\theta y}$ where $F_2x = \frac{\theta_1x}{f_2x}$ is a rational function of x independent of $a, a', a'', \ldots$ as are θ_1x, $f_1(x, y)$ and $\chi'y$. Thus $dv = -\prod \frac{F_2x}{\theta_1x}\sum \frac{f_1(x,y)}{\chi'y}\frac{\delta\theta y}{\theta y} + \sum{}' \frac{1}{\Gamma\nu}\frac{d^{\nu-1}}{dx^{\nu-1}}\left(\frac{F_2x}{\vartheta x}\sum \frac{f_1(x,y)}{\chi'y}\frac{\delta\theta y}{\theta y}\right)$ where the symbol $\prod$ denotes the coefficient of $\frac{1}{x}$ in the expansion of the following function in decreasing powers of x. Now the expression in the right hand side is integrable and gives

$$v = C - \prod \frac{F_2x}{\theta_1x}\sum \frac{f_1(x, y)}{\chi'y}\log\theta y + \sum{}' \frac{1}{\Gamma\nu}\frac{d^{\nu-1}}{dx^{\nu-1}}\left(\frac{F_2x}{\vartheta x}\sum \frac{f_1(x, y)}{\chi'y}\log\theta y\right). \tag{55}$$

In general $F_0x = 1$ and then $F_2x = 1, \theta_1x = f_2x$. If for example $f_2x = (x-\beta)^m$, the formula (55) takes the form

$$\sum \int \frac{f_1(x, y)dx}{(x-\beta)^m\chi'y} = C - \prod \sum \frac{f_1(x, y)}{(x-\beta)^m\chi'y}\log\theta y$$
$$+ \frac{1}{1 \cdot 2 \cdots (m-1)}\frac{d^{m-1}}{d\beta^{m-1}}\left(\sum \frac{f_1(\beta, B)}{\chi'B}\log\theta B\right)$$

where B is the value of y when $x = \beta$ (the second term disappears if $m = 0$).

In the fifth paragraph (p. 159–170), Abel studies under which conditions the right hand side of (54) is a constant independent of $a, a', a'', \ldots$ He supposes that $F_0x = 1$; then $\theta_1 x = f_2 x$ must be constant and $\sum \int \frac{f_1(x,y)dx}{(x-\beta)^m \chi' y} = C - \prod \sum \frac{f_1(x,y)}{\chi' y} \log \theta y$. In order that this expression be constant, he finds that the following condition must be realised: $\sup\limits_{1 \leqslant k \leqslant n} h \frac{f_1(x,y^{(k)})}{\chi' y^{(k)}} < -1$ where, for any function R of x, hR denotes the highest exponent of x in the expansion of R in decreasing powers of x. This condition is equivalent to $hf_1(x, y^{(k)}) < h\chi' y^{(k)} - 1$ for $1 \le k \le n$ and Abel deduces from it that $h(t_m y^{(k)} m) < h\chi' y^{(k)} - 1$ for $0 \le m \le n-1$ and $1 \le k \le n$ if

$$f_1(x, y) = t_0 + t_1 y + t_2 y^2 + \ldots + t_{n-1} y^{n-1}.$$

A proof of this deduction is given by Sylow in the notes (*Œuvres*, t. II, p. 296–297). The condition now takes the form $ht_m < \inf\limits_{1 \leqslant k \leqslant n} (h\chi' y^{(k)} - mhy^{(k)}) - 1$ for $0 \le m \le n-1$ and $1 \le k \le n$. Abel arranges the $y^{(k)}$ in a way such that $hy' \le hy'' \le \ldots \le hy^{(n)}$. Thus, in general $h(y^{(k)} - y^{(\ell)}) = hy^{(k)}$ for $\ell > k$ and $h\chi' y^{(k)} = hy' + hy'' + \ldots + hy^{(k-1)} + (n-k)hy^{(k)}$. Now one sees that

$$\inf_{1 \leqslant k \leqslant n} (h\chi' y^{(k)} - mhy^{(k)}) = hy' + hy'' + \ldots + hy^{(n-m-1)},$$

so that $ht_m = hy' + hy'' + \ldots + hy^{(n-m-1)} - 2 + \varepsilon_{n-m-1}$ with $0 \le \varepsilon_{n-m-1} < 1$. Let us suppose that

$$hy^{(j)} = \frac{m^{(\alpha)}}{\mu^{(\alpha)}}, \tag{56}$$

an irreducible fraction, for $k^{(\alpha-1)} + 1 \le j \le k^{(\alpha)}$, $1 \le \alpha \le \varepsilon$ (here $k^{(0)} = 0$ and $k^{(\varepsilon)} = n$). Since $k^{(\alpha)} - k^{(\alpha-1)}$ must be a multiple $n^{(\alpha)}\mu^{(\alpha)}$ of $\mu^{(\alpha)}$, we have $k^{(\alpha)} = n'\mu' + n''\mu'' + \ldots + n^{(\alpha)}\mu^{(\alpha)}$. If $k^{(\alpha)} \le n-m-1 < k^{(\alpha+1)}$ and $\beta = n-m-1-k^{(\alpha)}$,

$$ht_m = n'm' + n''m'' + \ldots + n^{(\alpha)}m^{(\alpha)} - 2 + \frac{\beta m^{(\alpha+1)} + A_\beta^{(\alpha+1)}}{\mu^{(\alpha+1)}}, \tag{57}$$

where $A_\beta^{(\alpha+1)} = \mu^{(\alpha+1)}\varepsilon_{k^{(\alpha)}+\beta}$ is the remainder of the division of $-\beta m^{(\alpha+1)}$ by $\mu^{(\alpha+1)}$. For $\alpha = 1$, this shows that $t_{n-\beta-1} = 0$ unless $\frac{\beta m' + A'_\beta}{\mu'} \ge 2$. This inequality signifies that the quotient of $-\beta m'$ by μ' is ≤ -2 or that $\frac{\mu'}{m'} < \beta \le 2\frac{\mu'}{m'}$, the least possible value of β being $\beta' = E\left(\frac{\mu'}{m'} + 1\right)$ (integral part of $\frac{\mu'}{m'} + 1$). In addition, one must impose $\beta \le n-1$ and $\beta < k' = n'\mu'$ (condition neglected by Abel). Now if $\beta' > n-1$, $\frac{\mu'}{m'} + 1 \ge n$ and, since $\mu' \le n$, $\frac{\mu'}{m'}$ is equal to $\frac{1}{n}$ or to $\frac{1}{n-1}$, which imposes to χy to be of degree 1 with respect to x; in this case, $\int f(x, y)dx = \int Rdy$ with R rational in y, is algebraic and logarithmic in y. Sylow (*Œuvres*, t. II, p. 298) observes that the least possible value of β is still β' in the case in which

$\frac{\mu'}{m'} + 1 \leq n'\mu'$, with the only exceptions of χy of degree 1 with respect to x or $\chi y = y^2 + (Ax + B)y + Cx^2 + Dx + E$; in these cases, $\int f(x, y)dx$ may be reduced to the integral of a rational function and is expressible by algebraic and logarithmic functions. Finally, the abelian integrals leading to a constant in the right hand side of (55) are of the form

$$\int \frac{(t_0 + t_1 y + \ldots + t_{n-\beta'-1} y^{n-\beta'-1})dx}{\chi' y}$$

where the degree ht_m of each coefficient t_m is given by (57). Such a function involves a number of arbitrary constant equal to $\gamma = ht_0 + ht_1 + \ldots + ht_{n-\beta'-1} + n - \beta' = ht_0 + ht_1 + \ldots + ht_{n-2} + n - 1$. Using (57), Abel transforms this expression into

$$\begin{aligned}\gamma = {} & \frac{A'_0}{\mu'} + \frac{m' + A'_1}{\mu'} + \frac{2m' + A'_2}{\mu'} + \ldots + \frac{(n'\mu' - 1)m' + A'_{n'\mu'-1}}{\mu'} \\ & + \frac{A''_0}{\mu''} + \frac{m'' + A''_1}{\mu''} + \frac{2m'' + A''_2}{\mu''} + \ldots + \frac{(n''\mu'' - 1)m'' + A''_{n''\mu''-1}}{\mu''} \\ & + n'm'n''\mu'' \\ & + \frac{A'''_0}{\mu'''} + \frac{m''' + A'''_1}{\mu'''} + \frac{2m''' + A'''_2}{\mu'''} + \ldots + \frac{(n'''\mu''' - 1)m''' + A'''_{n'''\mu'''-1}}{\mu'''} \\ & + (n'm' + n''m'')n'''\mu''' \\ & + \ldots\ldots\ldots\ldots\ldots\ldots\ldots - n + 1.\end{aligned}$$

Since $A_0^{(\alpha)} + A_1^{(\alpha)} + \ldots + A^{(\alpha)}_{n^{(\alpha)}\mu^{(\alpha)}-1} = n^{(\alpha)} \frac{\mu^{(\alpha)}(\mu^{(\alpha)}-1)}{2}$ and $n = n'\mu' + n''\mu'' + \ldots + n^{(\varepsilon)}\mu^{(\varepsilon)}$, this finally gives

$$\begin{aligned}\gamma = {} & n'\mu' \frac{m'n' - 1}{2} + n''\mu'' \left(m'n' + \frac{m''n'' - 1}{2}\right) \\ & + n'''\mu''' \left(m'n' + m''n'' + \frac{m'''n''' - 1}{2}\right) + \ldots \\ & + n^{(\varepsilon)}\mu^{(\varepsilon)} \left(m'n' + m''n'' + \ldots + m^{(\varepsilon-1)}n^{(\varepsilon-1)} + \frac{m^{(\varepsilon)}n^{(\varepsilon)} - 1}{2}\right) \\ & - \frac{n'(m' + 1)}{2} - \frac{n''(m'' + 1)}{2} - \ldots - \frac{n^{(\varepsilon)}(m^{(\varepsilon)} + 1)}{2} + 1. \end{aligned} \tag{58}$$

Abel indicates some particular cases, first the case in which $\varepsilon = 1$ and

$$\gamma = n'\mu' \frac{m'n' - 1}{2} - n' \frac{m' + 1}{2} + 1;$$

more particularly, if in addition $\mu' = n$, one has $n' = 1$ and $\gamma = (n - 1)\frac{m' - 1}{2}$. In the second particular case, $\mu' = \mu'' = \ldots = \mu^{(\varepsilon)} = 1 = n' = n'' = \ldots = n^{(\varepsilon)}$ and $\varepsilon = n$, thus

$$\gamma = (n-1)m' + (n-2)m'' + \ldots + 2m^{(n-2)} + m^{(n-1)} - n + 1$$

where $m^{(k)} = hy^{(k)}$ $(1 \leq k \leq n)$; when $hy^{(k)} = hy'$ for $1 \leq k \leq n-1$, this gives

$$\gamma = (n-1)\left(\frac{nhy'}{2} - 1\right).$$

Abel finally explains that the result remains true for the integrals of the form $\int \frac{f_1(x,y)dx}{\chi' y}$ even when $F_0 x$ is not constant, provided that $\frac{f_1(x,y)}{\chi' y}$ be finite whenever x is replaced by a root β of $F_0 x$ and y by the corresponding value B. In the final notes (*Œuvres*, t. II, p. 298), Sylow says under which precise conditions the number γ determined by Abel coincides with the *genus* p later defined by Riemann: the only multiple points of the curve defined by the equation $\chi y = 0$ must be at infinity in the directions of the axes and no two expansions of the $y^{(k)}$ in decreasing powers of x may begin by the same term.

As we have said, if there are α indeterminate coefficients $a, a', a'', \ldots$ in θy, one may choose arbitrarily α couple (x_j, y_j) of common roots to $\chi y = 0$ and $\theta y = 0$ and determine $a, a', a'', \ldots$ by the linear system $\theta y_j = 0$, $1 \leq j \leq \alpha$. If some couple (x_j, y_j) has a multiplicity k, one must replace the equation $\theta y_j = 0$ by $\theta y_j = \frac{d\theta y_j}{dx_j} = \ldots = \frac{d^{k-1}\theta y_j}{dx_j^{k-1}} = 0$. We get $a, a', a'', \ldots$ as rational functions of the (x_j, y_j) and we may substitute these functions in Fx. Abel (§ 6–7, p. 170–180) writes $Fx = B(x - x_1)(x - x_2)\ldots(x - x_\alpha)F^{(1)}x$ where $F^{(1)}x$ is a polynomial of degree $\mu - \alpha$ with coefficients rational in the (x_j, y_j) and $\psi x = \int f(x, y)dx$. According to (53),

$$\psi_1 x_1 + \psi_2 x_2 + \ldots + \psi_\alpha x_\alpha = v - (\psi_{\alpha+1} x_{\alpha+1} + \ldots + \psi_\mu x_\mu)$$

where $x_{\alpha+1}, \ldots, x_\mu$ are the roots of $F^{(1)}x = 0$, so algebraic functions of $x_1, \ldots, x_\alpha$, and v is an algebraic and logarithmic function.

Now α is of the form $hq_0 + hq_1 + \ldots + hq_{n-1} + n - 1 - hF_0 x + A$ with $0 \leq A \leq hF_0 x$ and $\mu = hr - hF_0 x = h\theta y' + h\theta y'' + \ldots + h\theta y^{(n)} - hF_0 x$. Thus

$$\mu - \alpha = h\theta y' + h\theta y'' + \ldots + h\theta y^{(n)} - (hq_0 + hq_1 + \ldots + hq_{n-1}) - n + 1 - A.$$

For any m, $h\theta y \geq h(q_m y^m) = hq_m + mhy$ and, according to (56),

$$h\theta y^{(j)} \geq hq_m + m\frac{m^{(\ell)}}{\mu^{(\ell)}} \quad \text{when } k^{(\ell-1)} + 1 \leq j \leq k^{(\ell)}. \tag{59}$$

Let us suppose that the maximum value of $h(q_m y^{(j)m})$ for $n - k^{(\ell)} \leq m \leq n - k^{(\ell-1)} - 1$ and $k^{(\ell-1)} + 1 \leq j \leq k^{(\ell)}$ is obtained for $m = \rho_\ell$:

$$hq_{\rho_\ell} + \rho_\ell \frac{m^{(\ell)}}{\mu^{(\ell)}} \geq hq_{n-\beta-1} + (n - \beta - 1)\frac{m^{(\ell)}}{\mu^{(\ell)}} \quad \text{or}$$

$$hq_{\rho_\ell} - hq_{n-\beta-1} \geq (n - \beta - 1 - \rho_\ell)\frac{m^{(\ell)}}{\mu^{(\ell)}}$$

for $k^{(\ell-1)} \le \beta \le k^{(\ell)} - 1$. Thus $hq_{\rho_\ell} - hq_{n-\beta-1} = (n-\beta-1-\rho_\ell)\frac{m^{(\ell)}}{\mu^{(\ell)}} + \varepsilon_\beta^{(\ell)} + A_\beta^{(\ell)}$ where $\varepsilon_\beta^{(\ell)}$ is a natural integer and $0 \le A_\beta^{(\ell)} < 1$. The sum of these equations for ℓ fixed and β variable gives

$$\begin{aligned}
n^{(\ell)}\mu^{(\ell)} & \left(hq_{\rho_\ell} + \rho_\ell \frac{m^{(\ell)}}{\mu^{(\ell)}}\right) \\
&= \frac{1}{2}(2n - k^{(\ell)} - k^{(\ell-1)} - 1)n^{(\ell)}\mu^{(\ell)} + A_0^{(\ell)} + A_1^{(\ell)} + \ldots + A^{(\ell)}_{n^{(\ell)}\mu^{(\ell)}-1} \\
&\quad + \varepsilon_0^{(\ell)} + \varepsilon_1^{(\ell)} + \ldots + \varepsilon^{(\ell)}_{n^{(\ell)}\mu^{(\ell)}-1} + hq_{n-1-k^{(\ell-1)}} + \ldots + hq_{n-k^{(\ell)}} \\
&= \frac{1}{2}(2n - k^{(\ell)} - k^{(\ell-1)} - 1)n^{(\ell)}\mu^{(\ell)} \\
&\quad + \frac{1}{2}n^{(\ell)}(\mu^{(\ell)} - 1) + C_\ell + hq_{n-1-k^{(\ell-1)}} + \ldots + hq_{n-k^{(\ell)}}
\end{aligned}$$

where $C_\ell = \varepsilon_0^{(\ell)} + \varepsilon_1^{(\ell)} + \ldots + \varepsilon^{(\ell)}_{n^{(\ell)}\mu^{(\ell)}-1}$. Let us write the inequalities (59) for $k^{(\ell-1)} + 1 \le j \le k^{(\ell)}$, $m = \rho_\ell$ and then sum up all these inequalities for ℓ variable. This gives

$$\begin{aligned}
h\theta y' + h\theta y'' + \ldots + h\theta y^{(n)} & \\
&\ge hq_{n-1} + hq_{n-2} + \ldots + hq_0 \\
&\quad + \sum_{\ell=1}^{\varepsilon}\left(\frac{1}{2}(2n - k^{(\ell)} - k^{(\ell-1)} - 1)n^{(\ell)}\mu^{(\ell)} + \frac{1}{2}n^{(\ell)}(\mu^{(\ell)} - 1) + C_\ell\right)
\end{aligned}$$

or $h\theta y' + h\theta y'' + \ldots + h\theta y^{(n)} - (hq_0 + hq_1 + \ldots + hq_{n-1}) \ge \gamma' + C_1 + C_2 + \ldots + C_\varepsilon$ where

$$\begin{aligned}
\gamma' &= n'm'\left(\frac{n'\mu' - 1}{2} + n''\mu'' + n'''\mu''' + \ldots + n^{(\varepsilon)}\mu^{(\varepsilon)}\right) + n'\frac{\mu' - 1}{2} \\
&\quad + n''m''\left(\frac{n''\mu'' - 1}{2} + n'''\mu''' + n''''\mu'''' + \ldots + n^{(\varepsilon)}\mu^{(\varepsilon)}\right) + n''\frac{\mu'' - 1}{2} \\
&\quad + \ldots \\
&\quad + n^{(\varepsilon-1)}m^{(\varepsilon-1)}\left(\frac{n^{(\varepsilon-1)}\mu^{(\varepsilon-1)} - 1}{2} + n^{(\varepsilon)}\mu^{(\varepsilon)}\right) + n^{(\varepsilon-1)}\frac{\mu^{(\varepsilon-1)} - 1}{2} \\
&\quad + n^{(\varepsilon)}m^{(\varepsilon)}\frac{n^{(\varepsilon)}\mu^{(\varepsilon)} - 1}{2} + n^{(\varepsilon)}\frac{\mu^{(\varepsilon)} - 1}{2}.
\end{aligned}$$

We finally obtain $\mu - \alpha \ge \gamma' - n + 1 - A + C_1 + C_2 + \ldots + C_\varepsilon$ and we remark that, according to (58),

$$\gamma' - n + 1 = \gamma,$$

so that $\mu - \alpha \ge \gamma - A + C_1 + C_2 + \ldots + C_\varepsilon$.

As Abel notes it, $\mu - \alpha = \gamma - A$ when $C_1 + C_2 + \ldots + C_\varepsilon = 0$ and, for each ℓ,

$$h\theta y^{k^{(\ell)}} = hq_{\rho_\ell} + \rho_\ell \frac{m^{(\ell)}}{\mu^{(\ell)}}. \tag{60}$$

He shows that, for a convenient choice of θy, these conditions are realised. The first one signifies that $\varepsilon_\beta^{(\ell)} = 0$ for $k^{(\ell-1)} \leq \beta \leq k^{(\ell)} - 1$ and $1 \leq \ell \leq \varepsilon$ or that

$$hq_{n-\beta-1} = hq_{\rho_\ell} - (n - \beta - 1 - \rho_\ell)\frac{m^{(\ell)}}{\mu^{(\ell)}} - A_\beta^{(\ell)} \tag{61}$$

for $k^{(\ell-1)} \leq \beta \leq k^{(\ell)} - 1$. The degrees hq_m will then be definite if we know the hq_{ρ_ℓ} and we have by (60),

$$hq_{\rho_\ell} + \rho_\ell \frac{m^{(\ell)}}{\mu^{(\ell)}} \geq hq_{\rho_\alpha} + \rho_\alpha \frac{m^{(\ell)}}{\mu^{(\ell)}} \tag{62}$$

for any ℓ and any α. Abel puts $\frac{m^{(\ell)}}{\mu^{(\ell)}} = \sigma_\ell$ and deduces from the preceding inequality

$$(\rho_{\ell-1} - \rho_\ell)\sigma_\ell \leq hq_{\rho_\ell} - hq_{\rho_{\ell-1}} \leq (\rho_{\ell-1} - \rho_\ell)\sigma_{\ell-1}.$$

Thus $hq_{\rho_\ell} - hq_{\rho_{\ell-1}} = (\rho_{\ell-1} - \rho_\ell)(\theta_{\ell-1}\sigma_{\ell-1} + (1 - \theta_{\ell-1})\sigma_\ell)$ where $0 \leq \theta_{\ell-1} \leq 1$, and $hq_{\rho_\ell} = hq_{\rho_1} + (\rho_1 - \rho_2)(\theta_1\sigma_1 + (1-\theta_1)\sigma_2) + (\rho_2 - \rho_3)(\theta_2\sigma_2 + (1-\theta_2)\sigma_3) + \ldots + (\rho_{\ell-1} - \rho_\ell)(\theta_{\ell-1}\sigma_{\ell-1} + (1-\theta_{\ell-1})\sigma_\ell)$. Inversely, for any choice of hq_{ρ_1} and of the θ_α (between 0 and 1), these values of hq_{ρ_ℓ} verify the inequalities (62). It is then possible, using (61) and some work, to prove that (60) is verified; a narrower limitation is imposed to the θ_α.

All this discussion was made in the hypothesis that the only condition limiting the indetermination of the coefficients of the q_m was that the polynomial F_0x divides the resultant r. When more conditions are imposed to limit the number α of the indeterminate coefficients $a, a', a'', \ldots$, the minimum value of $\mu - \alpha$ may be of the form $\gamma - A - B < \gamma - A$. In the final notes (*Œuvres*, t. II, p. 299–300), Sylow explains that A is the reduction due to the presence of singularities at a finite distance on the curve $\chi y = 0$ and that the additional reduction B is due to the eventual coincidence of the initial terms in some of the $y^{(k)}$. Moreover he explains how Abel's formula (58) may lead to a computation of A.

In the following paragraph 8 (p. 181–185), Abel explicitly deals with the case where χy is of degree $n = 13$ in y, the degrees in x of the coefficients p_m being 2 for $m = 0, 2, 8$; 3 for $m = 1, 3, 6, 9$; 4 for $m = 4, 7, 10$; 5 for $m = 5$ and 1 for $m = 11, 12$. He determines the exponents $hy^{(k)}$ by a method similar to that of the Newton polygon and finds $hy' = hy'' = hy''' = \frac{m'}{\mu'} = \frac{4}{3}$, $n' = 1$; $hy^{(4)} = hy^{(5)} = hy^{(6)} = hy^{(7)} = hy^{(8)} = \frac{m''}{\mu''} = \frac{1}{5}$, $n'' = 1$; $hy^{(9)} = hy^{(10)} = hy^{(11)} = hy^{(12)} = \frac{m'''}{\mu'''} = \frac{-1}{2}$, $n''' = 2$; $hy^{(13)} = \frac{m''''}{\mu''''} = -1$, $n'''' = 1$. These values give $\gamma = 38$ and the limitations $10 \leq \rho_1 \leq 12$, $5 \leq \rho_2 \leq 9$, $1 \leq \rho_3 \leq 4$ and $\rho_4 = 0$. Choosing for instance $\rho_1 = 11$, $\rho_2 = 6$, $\rho_3 = 4$, he finds $A_0' = \frac{2}{3}$, $A_2' = \frac{1}{3}$, $A_3'' = \frac{2}{5}$, $A_4'' = \frac{3}{5}$, $A_5'' = \frac{4}{5}$, $A_7'' = \frac{1}{5}$, $A_9''' = \frac{1}{2}$, $A_{10}''' = 0$, $A_{11}''' = \frac{1}{2}$ and then $\frac{12}{85} \leq \theta_1 \leq \frac{8}{17}$, $\frac{5}{14} \leq \theta_2 \leq 1$ and $\frac{1}{2} \leq \theta_3 \leq 1$. The values of the differences

$$hq_6 - hq_{11}, \quad hq_4 - hq_6, \quad hq_0 - hq_4$$

are correspondingly limited and they may be: $hq_6 - hq_{11} = 2, 3$; $hq_4 - hq_6 = 0$; $hq_0 - hq_4 = -3, -2$. It is now possible to determine all the degrees hq_m knowing $\theta = hq_{12}$. The possible values of α are $13\theta + 47$, $13\theta + 48$, $13\theta + 57$ or $13\theta + 58$. The corresponding values of μ are $13\theta + 85$, $13\theta + 86$, $13\theta + 95$ and $13\theta + 96$. Thus $\mu - \alpha = 38 = \gamma$ for every choice.

Then (§ 9, p. 185–188) Abel extends his relation (53) in the form

$$h_1\psi_1 x_1 + h_2\psi_2 x_2 + \ldots + h_\alpha \psi_\alpha x_\alpha = v$$

where the coefficients $h_1, h_2, \ldots, h_\alpha$ are rational numbers. In the paragraph 10 (p. 188–211), he deals with the case in which $\chi y = y^n + p_0$ where p_0 is a polynomial in x and the integral $\psi x = \int \frac{f_3 x \cdot dx}{y^m f_2 x}$ where $f_2 x$ and $f_3 x$ are polynomials in x. If $-p_0 = r_1^{\mu_1} r_2^{\mu_2} \ldots r_\varepsilon^{\mu_\varepsilon}$ where the polynomials $r_1, r_2, \ldots, r_\varepsilon$ are squarefree and relatively prime by pairs, let us put with Abel $R = r_1^{\frac{\mu_1}{n}} r_2^{\frac{\mu_2}{n}} \ldots r_\varepsilon^{\frac{\mu_\varepsilon}{n}}$, so that the determinations of y are $y^{(k)} = \omega^{k-1} R (1 \leq k \leq n)$, ω being a primitive n-th root of 1. The determinations of the integral ψx are of the form $\omega^{-em} \int \frac{f_3 x \cdot dx}{R^m f_2 x}$ where e is an integer and (53) takes the form

$$\begin{aligned} \omega^{-e_1 m}\psi x_1 + \omega^{-e_2 m}\psi x_2 + \ldots + \omega^{-e_\mu m}\psi x_\mu \\ = C - \prod \frac{\varphi_2 x}{f_2 x} + \sum{}' \frac{1}{\Gamma v} \frac{d^{v-1}}{dx^{v-1}} \left(\frac{F_2 x \cdot \varphi_2 x}{\vartheta x} \right) \end{aligned} \qquad (63)$$

where

$$\begin{aligned} \varphi_2 x = \frac{f_3 x}{R^m} (\log \theta R + \omega^{-m} \log \theta(\omega R) + \omega^{-2m} \log \theta(\omega^2 R) \\ + \ldots + \omega^{-(n-1)m} \log \theta(\omega^{n-1} R)). \end{aligned} \qquad (64)$$

Let us first suppose that all the coefficients in $q_0, q_1, \ldots, q_{n-1}$ are indeterminate, so that $\alpha = hq_0 + hq_1 + \ldots + hq_{n-1} + n - 1$. In our case, $hy' = hy'' = \ldots = hy^{(n)} = \frac{m'}{\mu'}$. We have $\varepsilon = 1$ and $n = n'\mu' = k'$. Let us determine the minimum value of $\mu - \alpha$. According to the relation (61),

$$hq_m = hq_{\rho_1} + (\rho_1 - m)\frac{m'}{\mu'} - A'_m \qquad (65)$$

with $0 \leq A'_m < 1$. Here the number μ is

$$hr = nhq_{\rho_1} + n'm'\rho_1 \qquad (66)$$

and, according to (58), $\mu - \alpha = \gamma = n'\mu' \frac{n'm'-1}{2} - n' \frac{m'+1}{2} + 1 = \frac{n-1}{2} nhR - \frac{n+n'}{2} + 1$. But this value can be lowered by a more convenient choice of θy. For $1 \leq m \leq \varepsilon$ and $0 \leq \pi \leq n-1$, Abel puts $\theta_m = E\frac{\mu_m}{n} + E\frac{2\mu_m}{n} + \ldots + E\frac{(n-1)\mu_m}{n}$ and $\delta_{m,\pi} = \theta_m - E\left(\frac{\pi\mu_m}{n} - \frac{\alpha_m}{n}\right)$ where E denotes the integral part of the following fraction and the α_m are natural integers. He takes the coefficients q_π of θy of the form $q_\pi = v_\pi r_1^{\delta_{1,\pi}} r_2^{\delta_{2,\pi}} \dots r_\varepsilon^{\delta_{\varepsilon,\pi}}$ where the v_π are polynomials in x. Then $q_\pi R^\pi = v_\pi r_1^{\theta_1+\frac{\alpha_1}{n}} r_2^{\theta_2+\frac{\alpha_2}{n}} \dots r_\varepsilon^{\theta_\varepsilon+\frac{\alpha_\varepsilon}{n}} R^{(\pi)}$ where $R^{(\pi)} = r_1^{k_{1,\pi}} r_2^{k_{2,\pi}} \dots r_\varepsilon^{k_{\varepsilon,\pi}}$, $k_{m,\pi} = \varepsilon\frac{\pi\mu_m - \alpha_m}{n}$ (ε denotes the excess of the following fraction over its integral part), and

$$\theta y^{(e+1)} = \theta'(x,e) r_1^{\theta_1+\frac{\alpha_1}{n}} r_2^{\theta_2+\frac{\alpha_2}{n}} \dots r_\varepsilon^{\theta_\varepsilon+\frac{\alpha_\varepsilon}{n}}$$

where $\theta'(x,e) = v_0 R^{(0)} + \omega^e v_1 R^{(1)} + \omega^{2e} v_2 R^{(2)} + \dots + \omega^{(n-1)e} v_{n-1} R^{(n-1)}$, $0 \le e \le n-1$. This gives $F_0 x = r_1^{n\theta_1+\alpha_1} r_2^{n\theta_2+\alpha_2} \dots r_\varepsilon^{n\theta_\varepsilon+\alpha_\varepsilon}$ and $Fx = \theta'(x,0)\theta'(x,1)\dots\theta'(x,n-1)$. Now (54) takes the form $Rx = \sum \frac{f_3 x}{y^m}\frac{r\cdot\delta\theta y}{\theta y} = F_0 x \sum \frac{f_3 x}{y^{(e+1)m}}\frac{Fx\cdot\delta\theta'(x,e)}{\theta'(x,e)}$ and $\frac{f_3 x}{y^m} = \frac{fx}{s_m}$ where

$$fx = f_3 x \cdot r_1^{-E\frac{m\mu_1}{n}} r_2^{-E\frac{m\mu_2}{n}} \dots r_\varepsilon^{-E\frac{m\mu_\varepsilon}{n}} \text{ and } s_m = r_1^{\varepsilon\frac{m\mu_1}{n}} r_2^{\varepsilon\frac{m\mu_2}{n}} \dots r_\varepsilon^{\varepsilon\frac{m\mu_\varepsilon}{n}}.$$

Thus

$$\begin{aligned} Rx = \frac{F_0 x \cdot fx}{s_m}\Bigg(&\frac{Fx}{\theta'(x,0)}\delta\theta'(x,0) + \omega^{-m}\frac{Fx}{\theta'(x,1)}\delta\theta'(x,1) \\ &+ \omega^{-2m}\frac{Fx}{\theta'(x,2)}\delta\theta(x,2) + \dots \\ &+ \omega^{-(n-1)m}\frac{Fx}{\theta'(x,n-1)}\delta\theta'(x,n-1)\Bigg). \end{aligned}$$

Since the $\frac{Fx}{\theta'(x,e)}\delta\theta'(x,e)$ are polynomial in x, $R^{(0)}, R^{(1)}, \dots, R^{(n-1)}$, so linear combinations of the s_m with coefficients polynomial in x, it results that $F_0 x$ divides Rx: $Rx = F_0 x \cdot R_1 x$. Now one sees that $F_2 x = 1$, $\theta_1 x = f_2 x$ and that (64) takes the form

$$\begin{aligned} \varphi_2 x = \frac{fx}{s_m}(&\log\theta'(x,0) + \omega^{-m}\log\theta'(x,1) + \omega^{-2m}\log\theta'(x,2) \\ &+ \dots + \omega^{-(n-1)m}\log\theta'(x,n-1)). \end{aligned}$$

Here

$$\begin{aligned} \mu &= hr - hF_0 x \\ &= nhq_{\rho_1} + n'm'\rho_1 - ((n\theta_1+\alpha_1)hr_1 + (n\theta_2+\alpha_2)hr_2 + \dots + (n\theta_\varepsilon+\alpha_\varepsilon)hr_\varepsilon) \end{aligned}$$

where $n'm' = nhR = \mu_1 hr_1 + \mu_2 hr_2 + \dots + \mu_\varepsilon hr_\varepsilon$. Thus, putting ρ for ρ_1,

$$\begin{aligned} \mu &= nhq_\rho + (\mu_1\rho - n\theta_1 - \alpha_1)hr_1 + (\mu_2\rho - n\theta_2 - \alpha_2)hr_2 \\ &\quad + \dots + (\mu_\varepsilon\rho - n\theta_\varepsilon - \alpha_\varepsilon)hr_\varepsilon \\ &= nhv_\rho + (n\delta_{1,\rho} - n\theta_1 + \rho\mu_1 - \alpha_1)hr_1 + (n\delta_{ys\rho} - n\theta_2 + \rho\mu_2 - \alpha_2)hr_2 \\ &\quad + \dots + (n\delta_{\varepsilon,\rho} - n\theta_\varepsilon + \rho\mu_\varepsilon - \alpha_\varepsilon)hr_\varepsilon \\ &\quad = nhv_\rho + n\varepsilon\frac{\rho\mu_1 - \alpha_1}{n}hr_1 + n\varepsilon\frac{\rho\mu_2 - \alpha_2}{n}hr_2 + \dots + n\varepsilon\frac{\rho\mu_\varepsilon - \alpha_\varepsilon}{n}hr_\varepsilon. \end{aligned}$$

On the other hand

$$\begin{aligned}\alpha =& hv_0 + hv_1 + \ldots + v_{n-1} + n - 1\\ =& hq_0 + hq_1 + \ldots + q_{n-1} + n - 1 - \sum_m (\delta_{m,0} + \delta_{m,1} + \ldots + \delta_{m,n-1}) hr_m\\ =& n\left(hv_\rho + \sum_m \delta_{m,\rho} hr_m\right) + \left(n\rho - \frac{n(n-1)}{2}\right)\frac{m'}{\mu'}\\ &- \frac{n'(\mu'-1)}{2} - \sum_m (\delta_{m,0} + \delta_{m,1} + \ldots + \delta_{m,n-1}) hr_m.\end{aligned}$$

Abel computes $\delta_{m,0} + \delta_{m,1} + \ldots + \delta_{m,n-1} = \alpha_m + (n-1)\theta_m$ and finally gets

$$\begin{aligned}\alpha = nhv_\rho &+ \left(n\varepsilon\frac{\rho\mu_1 - \alpha_1}{n} - \frac{n-k_1}{2}\right) hr_1 + \left(n\varepsilon\frac{\rho\mu_2 - \alpha_2}{n} - \frac{n-k_2}{2}\right) hr_2\\ &+ \ldots + \left(n\varepsilon\frac{\rho\mu_\varepsilon - \alpha_\varepsilon}{n} - \frac{n-k_\varepsilon}{2}\right) hr_\varepsilon - 1 + \frac{n+n'}{2}\end{aligned}$$

where k_m is the g.c.d. of μ_m and n . This gives

$$\mu - \alpha = \frac{n-k_1}{2} hr_1 + \frac{n-k_2}{2} hr_2 + \ldots + \frac{n-k_\varepsilon}{2} hr_\varepsilon + 1 - \frac{n+n'}{2} = \theta \tag{67}$$

independent of $\rho, \alpha_1, \alpha_2, \ldots, \alpha_\varepsilon$ and we have $\mu = nhv_\rho + nhR^{(\rho)}$ (cf. (66)). The degrees hv_m are determined from $hq_m = \delta_{1,m} hr_1 + \delta_{2,m} hr_2 + \ldots + \delta_{\varepsilon,m} hr_\varepsilon + v_m$ and (65) which give

$$\begin{aligned}hq_m =& hv_\rho + (\rho - m)\frac{m'}{\mu'} + (\delta_{1,\rho} - \delta_{1,m}) hr_1 + (\delta_{2,\rho} - \delta_{2,m}) hr_2\\ &+ \ldots + (\delta_{\varepsilon,\rho} - \delta_{\varepsilon,m}) hr_\varepsilon - A'_m\\ =& hv_\rho + E((k_{1,\rho} - k_{1,m}) hr_1 + (k_{2,\rho} - k_{2,m}) hr_2\\ &+ \ldots + (k_{\varepsilon,\rho} - k_{\varepsilon,m}) hr_\varepsilon)\\ =& hv_\rho + Eh\frac{R^{(\rho)}}{R^{(m)}}.\end{aligned}\tag{68}$$

Abel adopts new notations: $x_{\alpha+1} = z_1$, $x_{\alpha+2} = z_2, \ldots, x_\mu = z_\theta$; $e_{\alpha+1} = \varepsilon_1$, $e_{\alpha+2} = \varepsilon_2, \ldots, e_\mu = \varepsilon_\mu$; $\omega^{-e_\mu} = \omega_\mu$ and $\omega^{-\varepsilon_\mu} = \pi_\mu$ and he rewrites (63) in the form

$$\begin{aligned}&\omega_1^m \psi x_1 + \omega_2^m \psi x_2 + \ldots + \omega_\alpha^m \psi x_\alpha + \pi_1^m \psi z_1 + \pi_2^m \psi z_2 + \ldots + \pi_\theta^m \psi z_\theta\\ &= C - \prod \frac{fx\varphi x}{s_m(x) f_2 x} + \sum{}' \frac{1}{\Gamma \nu}\frac{d^{\nu-1}}{dx^{\nu-1}}\left(\frac{fx \cdot \varphi x}{s_m(x)\vartheta x}\right)\end{aligned}\tag{69}$$

where $\theta_1(x) = f_2 x = A(x-\beta_1)^{\nu_1}(x-\beta_2)^{\nu_2}\ldots$, fx is an arbitrary polynomial,

$$\begin{aligned}\varphi x = \log\theta'(x,0) &+ \omega^{-m}\log\theta'(x,1) + \omega^{-2m}\log\theta'(x,2)\\ &+ \ldots + \omega^{-(n-1)m}\log\theta'(x,n-1)\end{aligned}$$

and $\psi x = \int \frac{fxdx}{f_2x \cdot s_m(x)}$. Here, $x_1, x_2, \ldots, x_\alpha$ are considered as independent variables and $z_1, z_2, \ldots, z_\theta$ are the roots of the equation $\frac{\theta'(z,0)\theta'(z,1)\ldots\theta'(z,n-1)}{(z-x_1)(z-x_2)\ldots(z-x_\alpha)} = 0$. The coefficients $a, a', a'', \ldots$ are determined by the equations $\theta'(x_1, e_1) = \theta'(x_2, e_2) = \ldots = \theta'(x_\alpha, e_\alpha) = 0$ and the numbers $\varepsilon_1, \varepsilon_2, \ldots, \varepsilon_\theta$ by $\theta'(z_1, \varepsilon_1) = \theta'(z_2, \varepsilon_2) = \ldots = \theta'(z_\theta, \varepsilon_\theta) = 0$.

Some particular cases are explicited by Abel, first the case in which $f_2x = (x-\beta)^\nu$ with, for instance, $\nu = 1$ or 0. In this last case the right hand side of (69) reduces to

$$C - \prod \frac{fx \cdot \varphi x}{s_m(x) f_2 x}$$

which is constant when $h\,fx \le -E(-hs_m(x)) - 2$.

When $n = 1$, there is only one $s_m = s_0 = 1$ and $\psi x = \int \frac{fx \cdot dx}{f_2 x}$. Then $R^{(0)} = 1$, $\theta'(x, 0) = v_0$ and $\varphi x = \log v_0$. The relation (68) takes the form

$$\begin{aligned}\psi x_1 + \psi x_2 + \ldots + \psi x_\alpha + \psi z_1 + \psi z_2 + \ldots + \psi z_\theta \\ = C - \prod \frac{fx}{f_2 x} \log v_0 + \sum{}' \frac{1}{\Gamma \nu} \frac{d^{\nu-1}}{dx^{\nu-1}} \left(\frac{fx}{\vartheta x} \log v_0 \right)\end{aligned}$$

where $v_0(x) = a(x-x_1)(x-x_2)\ldots(x-x_\alpha)(x-z_1)(x-z_2)\ldots(x-z_\theta)$, but it is possible to make $\theta = 0$ in (67). For $\alpha = 1$, one finds the known integration of rational differential forms.

When $n = 2$ and $R = r_1^{\frac{1}{2}} r_2^{\frac{1}{2}}$, take $\alpha_1 = 1$ and $\alpha_2 = 0$. Then $s_0 = 1, s_1 = (r_1 r_2)^{\frac{1}{2}}$, $R^{(0)} = r_1^{\frac{1}{2}}$, $R^{(1)} = r_2^{\frac{1}{2}}$, $\theta'(x, 0) = v_0 r_1^{\frac{1}{2}} + v_1 r_2^{\frac{1}{2}}$, $\theta'(x, 1) = v_0 r_1^{\frac{1}{2}} - v_1 r_2^{\frac{1}{2}}$ and $\omega = -1$. For $m = 1$, we find $\varphi x = \log \frac{v_0 r_1^{\frac{1}{2}} + v_1 r_2^{\frac{1}{2}}}{v_0 r_1^{\frac{1}{2}} - v_1 r_2^{\frac{1}{2}}}$ and, writing $\varphi_0 x$ and $\varphi_1 x$ respectively for r_1 and r_2, (69) takes the form

$$\begin{aligned}\sum \omega \psi x + \sum \pi \psi z = C - \prod \frac{fx}{f_2 x \sqrt{\varphi_0 x \varphi_1 x}} \log \frac{v_0 \sqrt{\varphi_0 x} + v_1 \sqrt{\varphi_1 x}}{v_0 \sqrt{\varphi_0 x} - v_1 \sqrt{\varphi_1 x}} \\ + \sum{}' \frac{1}{\Gamma \nu} \frac{d^{\nu-1}}{dx^{\nu-1}} \frac{fx}{\vartheta x \sqrt{\varphi_0 x \varphi_1 x}} \log \frac{v_0 \sqrt{\varphi_0 x} + v_1 \sqrt{\varphi_1 x}}{v_0 \sqrt{\varphi_0 x} - v_1 \sqrt{\varphi_1 x}}\end{aligned}$$

where $\psi x = \int \frac{fx \cdot dx}{f_2 x \sqrt{\varphi_0 x \varphi_1 x}}$, v_0 and v_1 are determined by the equations $v_0 \sqrt{\varphi_0 x_1} + \omega_1 v_1 \sqrt{\varphi_1 x_1} = v_0 \sqrt{\varphi_0 x_2} + \omega_2 v_1 \sqrt{\varphi_1 x_2} = \ldots = 0$ and $z_1, z_2, \ldots, z_\theta$ by $\frac{(v_0(z))^2 \varphi_0 x - (v_1(z))^2 \varphi_1 x}{(z-x_1)(z-x_2)\ldots(z-x_\alpha)} = 0$. The signs π_k are given by $\pi_k = -\frac{v_0(z_k)\sqrt{\varphi_0 z_k}}{v_1(z_k)\sqrt{\varphi_1 z_k}}$. We have $k_1 = k_2 = 1$, $\theta = \frac{1}{2} h r_1 + \frac{1}{2} h r_2 - \frac{n'}{2} = \frac{1}{2}\left(h(r_1 r_2) - n'\right)$ where n' is the g.c.d. of 2 and $h(r_1 r_2)$. Thus $\theta = m - 1$ for $h(\varphi_0 x \varphi_1 x) = 2m - 1$ or $2m$. Taking $\rho = 1$, we have by (68)

$$hv_0 = v_1 + E\frac{1}{2}(h\varphi_1 x - h\varphi_0 x) = \begin{cases} hv_1 + \frac{1}{2}(h\varphi_1 x - h\varphi_0 x) - \frac{1}{2} \\ hv_1 + \frac{1}{2}(h\varphi_1 x - h\varphi_0 x) \end{cases}$$

depending on whether $h(\varphi_0 x \cdot \varphi_1 x)$ is odd or even. When $m = 1$, $\theta = 0$ and $\psi x = \int \frac{fx \cdot dx}{f_2 x \sqrt{R}}$ where R is a polynomial of degree 1 or 2. This integral is an algebraic and logarithmic function and Abel explicits the computation, taking $\varphi_0 x = \varepsilon_0 x + \delta_0$, $\varphi_1 x = \varepsilon_1 x + \delta_1$, $f_2 x = (x-\beta)^\nu$, $v_1 = 1$ and $v_0 = a$. When $m = 2$, $\theta = 1$ and $h(\varphi_0 x \cdot \varphi_1 x) = 3$ or 4 so that ψx is an elliptic integral. The relation (69) takes the form $\omega_1 \psi x_1 + \omega_2 \psi x_2 + \ldots + \omega_\alpha \psi x_\alpha = v - \pi_1 \psi z_1$ where v is algebraic and logarithmic. The product of the roots of the polynomial $(v_0 z)^2 \varphi_0 z - (v_1 z)^2 \varphi_1 z = A + \ldots + B z^{\alpha+1}$ is $x_1 x_2 \ldots x_\alpha z_1$, whence $z_1 = \frac{A}{B} \frac{(-1)^{\alpha+1}}{x_1 x_2 \ldots x_\alpha}$, where $\frac{A}{B}$ is a rational function of x_1, $x_2, \ldots, x_\alpha, \sqrt{\varphi_0 x_1}, \sqrt{\varphi_0 x_2}, \ldots, \sqrt{\varphi_0 x_\alpha}, \sqrt{\varphi_1 x_1}, \sqrt{\varphi_1 x_2}, \ldots, \sqrt{\varphi_1 x_\alpha}$. When

$$\varphi_0 x = 1, \varphi_1 x = \alpha_0 + \alpha_1 x + \alpha_2 x^2 + \alpha_3 x^3, v_1 = 1 \text{ and } v_0 = a_0 + a_1 x,$$

we must write $v_0 x_1 = -\omega_1 \sqrt{\varphi_1 x_1}$, $v_0 x_2 = -\omega_2 \sqrt{\varphi_1 x_2}$, whence

$$a_0 = \frac{\omega_1 x_2 \sqrt{\varphi_1 x_1} - \omega_2 x_1 \sqrt{\varphi_1 x_2}}{x_1 - x_2}, \quad a_1 = \frac{\omega_2 \sqrt{\varphi_1 x_2} - \omega_1 \sqrt{\varphi_1 x_1}}{x_1 - x_2}.$$

Then $A = a_0^2 - \alpha_0$, $B = -\alpha_3$ and $z_1 = \frac{1}{\alpha_3 x_1 x_2}\left(\frac{x_2^2 \varphi_1 x_1 + x_1^2 \varphi_1 x_2 - 2\omega_1 \omega_2 x_1 x_2 \sqrt{\varphi_1 x_1 \varphi_1 x_2}}{(x_1 - x_2)^2} - \alpha_0\right)$ which gives the addition theorem for elliptic integrals.

When $m = 3$, $\theta = 2$ and $h(\varphi_0 x \cdot \varphi_1 x) = 5$ or 6. Abel explains certain particular cases, for instance that in which $\psi x = \int \frac{(A_0 + A_1 x)dx}{\sqrt{\alpha_0 + \alpha_1 x + \ldots + \alpha_6 x^6}}$, which gives

$$\pm \psi x_1 \pm \psi x_2 \pm \ldots \pm \psi x_\alpha = \pm \psi z_1 \pm \psi z_2 + C,$$

where z_1, z_2 are the roots of a quadratic equation with coefficients rational in x_1, $x_2, \ldots, x_\alpha, \sqrt{R_1}, \sqrt{R_2}, \ldots, \sqrt{R_\alpha}$ (where R_k is the value of R corresponding to $x = x_k$). As we have said in our §1, Abel explained this result, with $\alpha = 3$, in a letter to Crelle.

The last example dealt with by Abel is not hyperelliptic for he takes $n = 3$, $R = r_1^{\frac{1}{3}} r_2^{\frac{2}{3}}$, $\alpha_1 = \alpha_2 = 0$. Then $s_0 = 1$, $s_1 = r_1^{\frac{1}{3}} r_2^{\frac{2}{3}}$, $s_2 = r_1^{\frac{2}{3}} r_2^{\frac{1}{3}}$, $R^{(0)} = s_0$, $R^{(1)} = s_1$, $R^{(2)} = s_2$ and

$$\theta'(x, 0) = v_0 + v_1 r_1^{\frac{1}{3}} r_2^{\frac{2}{3}} + v_2 r_1^{\frac{2}{3}} r_2^{\frac{1}{3}}, \theta'(x, 1) = v_0 + \omega v_1 r_1^{\frac{1}{3}} r_2^{\frac{2}{3}} + \omega^2 v_2 r_1^{\frac{2}{3}} r_2^{\frac{1}{3}},$$
$$\theta'(x, 2) = v_0 + \omega^2 v_1 r_1^{\frac{1}{3}} r_2^{\frac{2}{3}} + \omega v_2 r_1^{\frac{2}{3}} r_2^{\frac{1}{3}},$$

which give $Fx = \theta'(x,0)\theta'(x,1)\theta'(x,2) = v_0^3 + v_1^3 r_1 r_2^2 + v_2^3 r_1^2 r_2 - 3 v_0 v_1 v_2 r_1 r_2$. Here

$$\theta = hr_1 + hr_2 + 1 - \frac{3 + n'}{2}$$

where n' is the g.c.d. of 3 and $r_1 + 2hr_2$. Thus $\theta = h(\varphi_0 x \varphi_1 x) - 2$ if $hr_1 + 2hr_2$ is divisible by 3 and $\theta = h(\varphi_0 x \varphi_1 x) - 1$ in the contrary case.

Since the french Academy did not give any news of his memoir, Abel decided to published his theorem for the particular case of hyperelliptic integrals in Crelle's

Journal (vol. 3, 1828, *Œuvres*, t. I, p. 444–456). It is this publication which inspired Jacobi for the formulation of the inversion problem (1832). In a letter to Legendre (14 March 1829), Jacobi said of Abel theorem that it was "perhaps the most important discovery of what the century in which we live has made in mathematics ... though only a work to come, in a may be distant future, may throw light on its full importance". The statement is the following: "Let φx be a polynomial in x, decomposed in two factors $\varphi_1 x$, $\varphi_2 x$ and let fx be another polynomial and $\psi x = \int \frac{fx \cdot dx}{(x-\alpha)\sqrt{\varphi x}}$ where α is an any constant quantity. Let us designate by $a_0, a_1, a_2, \ldots, c_0, c_1, c_2, \ldots$ arbitrary quantities of which at least one is variable. Then if one puts

$$(a_0 + a_1 x + \ldots + a_n x^n)^2 \varphi_1 x - (c_0 + c_1 x + \ldots + c_m x^m)^2 \varphi_2 x = A(x - x_1)(x - x_2)\ldots(x - x_\mu)$$

where A does not depend on x, I say that

$$\varepsilon_1 \psi x_1 + \varepsilon_2 \psi x_2 + \ldots + \varepsilon_\mu \psi x_\mu = -\frac{f\alpha}{\sqrt{\varphi\alpha}} \log \frac{(a_0 + a_1\alpha + \ldots + a_n\alpha^n)\sqrt{\varphi_1\alpha} + (c_0 + c_1\alpha + \ldots + c_m\alpha^m)\sqrt{\varphi_2\alpha}}{(a_0 + a_1\alpha + \ldots + a_n\alpha^n)\sqrt{\varphi_1\alpha} - (c_0 + c_1\alpha + \ldots + c_m\alpha^m)\sqrt{\varphi_2\alpha}} + r + C$$

where C is a constant quantity and r the coefficient of $\frac{1}{x}$ in the expansion of

$$\frac{fx}{(x-\alpha)\sqrt{\varphi x}} \log \frac{(a_0 + a_1 x + \ldots + a_n x^n)\sqrt{\varphi_1 x} + (c_0 + c_1 x + \ldots + c_m x^m)\sqrt{\varphi_2 x}}{(a_0 + a_1 x + \ldots + a_n x^n)\sqrt{\varphi_1 x} - (c_0 + c_1 x + \ldots + c_m x^m)\sqrt{\varphi_2 x}}$$

in decreasing powers of x. The quantities $\varepsilon_1, \varepsilon_2, \ldots, \varepsilon_\mu$ are equal to $+1$ or to -1 and their values depend on those of the quantities $x_1, x_2, \ldots, x_\mu$."

Putting $\theta x = a_0 + a_1 x + \ldots + a_n x^n$, $\theta_1 x = c_0 + c_1 x + \ldots + c_m x^m$ and $Fx = (\theta x)^2 \varphi_1 x - (\theta_1 x)^2 \varphi_2 x$, the quantities $x_1, x_2, \ldots, x_\mu$ are the roots of $Fx = 0$. We have $F'xdx + \delta Fx = 0$ where

$$\delta Fx = 2\theta x \cdot \varphi_1 x \cdot \delta\theta x - 2\theta_1 x \cdot \varphi_2 x \cdot \delta\theta_1 x.$$

Now the equation $Fx = 0$ implies that $\theta x \cdot \varphi_1 x = \varepsilon\theta_1 x\sqrt{\varphi x}$ and $\theta_1 x \cdot \varphi_2 x = \varepsilon\theta x\sqrt{\varphi x}$ where $\varepsilon = \pm 1$. Thus $F'xdx = 2\varepsilon(\theta x \cdot \delta\theta_1 x - \theta_1 x \cdot \delta\theta x)\sqrt{\varphi x}$ and $\varepsilon\frac{fxdx}{(x-\alpha)\sqrt{\varphi x}} = \frac{2fx(\theta x \cdot \delta\theta_1 x - \theta_1 x \cdot \delta\theta x)}{(x-\alpha)F'x} = \frac{\lambda x}{(x-\alpha)F'x}$ where $\lambda x = (x-\alpha)\lambda_1 x + \lambda\alpha$ and $\lambda_1 x$ are polynomials. This leads to

$$\sum \varepsilon \frac{fx \cdot dx}{(x-\alpha)\sqrt{\varphi x}} = \sum \frac{\lambda_1 x}{F'x} + \lambda\alpha \sum \frac{1}{(x-\alpha)F'x} = -\frac{\lambda\alpha}{F\alpha} + \prod \frac{\lambda x}{(x-\alpha)Fx}$$

(the sums are extended to $x_1, x_2, \ldots, x_\mu$) and then to the relation of the statement. The values of the ε_k are determined by the equations $\theta x_k\sqrt{\varphi_1 x_k} = \varepsilon_k\theta_1 x_k\sqrt{\varphi_2 x_k}$.

In a second theorem, Abel explains that the same statement holds in the case in which some of the roots of Fx are multiple, provided that $\theta x \cdot \varphi_1 x$ and $\theta_1 x \cdot \varphi_2 x$

be relatively prime. The third theorem concerns the case in which $f\alpha = 0$, so that $\psi x = \int \frac{fx\cdot dx}{\sqrt{\varphi x}}$ where fx is a polynomial (written for $\frac{fx}{x-\alpha}$). In this case, the right hand side of the relation reduces to

$$C + \prod \frac{fx}{\sqrt{\varphi x}} \log \frac{\theta x\sqrt{\varphi_1 x} + \theta_1 x\sqrt{\varphi_2 x}}{\theta x\sqrt{\varphi_1 x} - \theta_1 x\sqrt{\varphi_2 x}}.$$

On the contrary (theorem IV), when the degree of $(fx)^2$ is less than the degree of φx, the right hand side reduces to $C - \frac{f\alpha}{\sqrt{\varphi\alpha}} \log \frac{\theta\alpha\sqrt{\varphi_1\alpha}+\theta_1\alpha\sqrt{\varphi_2\alpha}}{\theta\alpha\sqrt{\varphi_1\alpha}-\theta_1\alpha\sqrt{\varphi_2\alpha}}$. Abel deals with the case of the integrals $\psi x = \int \frac{dx}{(x-\alpha)^k\sqrt{\varphi x}}$ by successive differentiations starting from $k = 1$ (theorem V).

The sixth theorem concerns the case in which $\deg(fx)^2 < \deg \varphi x$, that is of integrals of the form $\psi x = \int \frac{(\delta_0+\delta_1 x+\ldots+\delta_{\nu'} x^{\nu'})dx}{\sqrt{\beta_0+\beta_1 x+\ldots+\beta_\nu x^\nu}}$ where $\nu' = \frac{\nu-1}{2} - 1$ when ν is odd and $\nu' = \frac{\nu}{2} - 2$ when ν is even or $\nu' = m - 2$ for $\nu = 2m - 1$ or $2m$. In this case, the right hand side of the relation is a constant.

The general case of $\psi x = \int \frac{rdx}{\sqrt{\varphi x}}$ where r is any rational function of x is reduced to the preceding ones by decomposing r in simple elements (theorem VII). As there are $m+n+2$ indeterminate coefficients $a_0, a_1, \ldots, c_0, c_1, \ldots$, Abel arbitrarily chooses $\mu' = m + n + 1$ quantities $x_1, x_2, \ldots, x_{\mu'}$ and determines $a_0, a_1, \ldots, c_0, c_1, \ldots$ as rational functions of $x_1, x_2, \ldots, x_{\mu'}, \sqrt{\varphi x_1}, \sqrt{\varphi x_2}, \ldots, \sqrt{\varphi x_{\mu'}}$ by the equations $\theta x_k\sqrt{\varphi_1 x_k} = \varepsilon_k\theta_1 x_k\sqrt{\varphi_2 x_k}$, $1 \leq k \leq \mu'$. Substituting these values in θx and $\theta_1 x$, Fx takes the form $(x-x_1)(x-x_2)\ldots(x-x_{\mu'})R$ where R is a polynomial of degree $\mu-\mu'$ with the roots $x_{\mu'+1}, x_{\mu'+2}, \ldots, x_\mu$. The coefficients of R are rational functions of $x_1, x_2, \ldots, x_{\mu'}, \sqrt{\varphi x_1}, \sqrt{\varphi x_2}, \ldots, \sqrt{\varphi x_{\mu'}}$. Putting $\varepsilon_1 = \varepsilon_2 = \ldots = \varepsilon_{\mu_1} = 1$, $\varepsilon_{\mu_1+1} = \varepsilon_{\mu_1+2} = \ldots = \varepsilon_{\mu'} = -1$, $x_{\mu_1+1} = x'_1, x_{\mu_1+2} = x'_2, \ldots, x_{\mu'} = x'_{\mu_2}$ and $x_{\mu'+1} = y_1, x_{\mu'+2} = y_2, \ldots, x_\mu = y_{\nu'}$, Abel rewrites the relation of the statement in the form $\psi x_1 + \psi x_2 + \ldots + \psi x_{\mu_1} - \psi x'_1 - \psi x'_2 - \ldots - \psi x'_{\mu_2} = v - \varepsilon_{\mu'+1}\psi y_1 - \varepsilon_{\mu'+2}\psi y_2 - \ldots - \varepsilon_\mu\psi y_{\nu'}$, where $x_1, x_2, \ldots, x_{\mu_1}, x'_1, x'_2, \ldots, x'_{\mu_2}$ are independent variables and $y_1, y_2, \ldots, y_{\nu'}$ algebraic functions of these variables. He determines the minimum value of $\nu' = \mu - \mu' = \mu - m - n - 1$, where $\mu = \sup(2n+\nu_1, 2m+\nu_2)$, ν_1 and ν_2 denoting the respective degrees of $\varphi_1 x$ and $\varphi_2 x$. The mean value of $2n + \nu_1$ and $2m + \nu_2$ is $m + n + \frac{\nu_1+\nu_2}{2}$. Thus $\nu' \geq \frac{\nu_1+\nu_2}{2} - 1 = \frac{\nu}{2} - 1$ where ν is the degree of φ, and this minimum value, which is the same as that of the theorem VI is attained (theorem VIII). The signs ε_j are determined by the equations $\theta y_j\sqrt{\varphi_1 y_j} = -\varepsilon_j\theta_1 y_j\sqrt{\varphi_2 y_j}$, $1 \leq j \leq \nu'$. Naturally, when some of the x_k or of the x'_k are equal, one must replace the corresponding equation $\theta x_k\sqrt{\varphi_1 x_k} = \varepsilon_k\theta_1 x_k\sqrt{\varphi_2 x_k}$ by a certain number of derivatives of this equation.

The memoir X (*Œuvres*, t. II, p. 55–66), unpublished by Abel, *Sur la comparaison des transcendantes*, gives a testimony of an early form of Abel theorem. It begins by the same demonstration as in the large memoir for the french Academy, to reach a relation of the form

$$\psi x_1 + \psi x_2 + \ldots + \psi x_\mu = C + \rho - (\psi z_1 + \psi z_2 + \ldots + \psi z_\nu) \qquad (70)$$

where $\psi x = \int f(x, y)dx$, y being an algebraic function of x and $f(x, y)$ a rational function of x and y. Here $z_1, z_2, \ldots, z_\nu$ are algebraic functions of $x_1, x_2, \ldots, x_\mu$, C is a constant and ρ is a function algebraic and logarithmic of $x_1, x_2, \ldots, x_\mu$. Abel proposes a method to compute C in the hypothesis in which $\mu > \nu$; we shall see examples of it later.

In the following, he applies his theorem to the particular case in which y is a rational function of x, defined by the equation $\alpha + \alpha_1 y = 0$ where α, α_1 are polynomials in x. Then the auxiliary equation $\theta y = 0$ is of degree 0 in y, of the form $0 = q = a + a_1 x + \ldots + a_{n-1}x^{n-1} + x^n = s$ and $ydx = \frac{da + xda_1 + x^{n-1}da_{n-1}}{\frac{ds}{dx}} \frac{\alpha}{\alpha_1}$. The n quantities $a, a_1, \ldots, a_{n-1}$ are determined in function of the n independent variables $x_1, x_2, \ldots, x_n$ by writing that these variables are roots of the equation $s = 0$, and $\nu = 0$. When $y = x^m$, $\psi x = \frac{x^{m+1}}{m+1}$ and the theorem states that

$$\frac{1}{m+1}(x_1^{m+1} + x_2^{m+1} + \ldots + x_n^{m+1}) = -\int (P_m da + P_{m+1}da_1 + \ldots + P_{m+n-1}da_{n-1})$$

where $P_k = \frac{x_1^k}{\frac{ds_1}{dx_1}} + \frac{x_2^k}{\frac{ds_2}{dx_2}} + \ldots + \frac{x_n^k}{\frac{ds_n}{dx_n}}$. Now the left hand side is a polynomial $\frac{1}{m+1}Q_{m+1}$ in $a, a_1, \ldots, a_{n-1}$ and we thus have $P_{m+k} = -\frac{1}{m+1}\frac{\partial Q_{m+1}}{\partial a_k}$. In particular $P_k = -\frac{\partial Q_1}{\partial a_k}$ where $Q_1 = -a_{n-1}$ and this gives

$$P_0 = P_1 = \ldots = P_{n-2} = 0,\ P_{n-1} = 1, \tag{71}$$

identities used several times by Abel. In the same manner, when $y = \frac{1}{(x-\alpha)^m}$,

$$\psi x = -\frac{1}{m-1}\frac{1}{(x-\alpha)^{m-1}}$$

and

$$\frac{1}{m-1}\left(\frac{1}{(x_1-\alpha)^{m-1}} + \frac{1}{(x_2-\alpha)^{m-1}} + \ldots + \frac{1}{(x_n-\alpha)^{m-1}}\right)$$
$$= \int (P_m^{(0)}da + P_m^{(1)}da_1 + \ldots + P_m^{(n-1)}da_{n-1})$$

where $P_m^{(k)} = \frac{x_1^k}{(x_1-\alpha)^m \frac{ds_1}{dx_1}} + \frac{x_2^k}{(x_2-\alpha)^m \frac{ds_2}{dx_2}} + \ldots + \frac{x_n^k}{(x_n-\alpha)^m \frac{ds_n}{dx_n}}$. Thus we have

$$P_m^{(k)} = \frac{1}{m-1}\frac{\partial Q'_{m-1}}{\partial a_k}$$

where

$$Q'_{m-1} = \frac{1}{(x_1-\alpha)^{m-1}} + \frac{1}{(x_2-\alpha)^{m-1}} + \ldots + \frac{1}{(x_n-\alpha)^{m-1}}.$$

When $m = 1$, $\psi x = \log(x - \alpha)$ and the left hand side of (70) is

$$\log(x_1-\alpha)(x_2-\alpha)\ldots(x_n-\alpha) = \log(-1)^n(a + a_1\alpha + \ldots + a_{n-1}\alpha^{n-1} + \alpha^n).$$

Thus $P_1^{(k)} = -\frac{\alpha^k}{a+a_1\alpha+\ldots+a_{n-1}\alpha^{n-1}+\alpha^n}$.

Now supposing that $s = (a + a_1x + \ldots + x_{\mu-1}x^{\mu-1} + x^\mu)\varphi x - fx$ where

$$\varphi x = \alpha_1 \text{ and } fx = -(\delta + \delta_1 x + \ldots + \delta_{n-1}x^{n-1}),$$

we have $ydx = \frac{\alpha(da+xda_1+\ldots+x^{\mu-1}da_{\mu-1})}{\frac{ds}{dx}}$ and we see that $\rho = 0$ in (70) if $\deg\alpha < \deg\alpha_1$. The quantities $x_1, x_2, \ldots, x_n$ are related by the equations $a + a_1x_k + \ldots + a_{\mu-1}x_k^{\mu-1} + x_k^\mu = \frac{fx_k}{\varphi x_k}$, $1 \leq k \leq n$. Let $x_1', x_2', \ldots, x_n'$ be another set of quantitites and suppose that $\deg\alpha < \deg\alpha_1$; we have $\psi x_1 + \psi x_2 + \ldots + \psi x_n = \psi x_1' + \psi x_2' + \ldots + \psi x_n'$. Now it is possible, by a convenient choice of $\delta, \delta_1, \ldots, \delta_{n-1}$, to impose $\psi x_n' = \psi x_{n-1}' = \ldots = \psi x_{\mu+1}' = 0$. Thus the theorem is written

$$\psi x_1 + \psi x_2 + \ldots + \psi x_n = \psi x_1' + \psi x_2' + \ldots + \psi x_\mu'.$$

For instance, if $\alpha = 1$ and $\alpha_1 = x$, $\psi x = -\log x$ and $s = \delta + ax + a_1x^2 + \ldots + a_{\mu-1}x^\mu + x^{\mu+1}$. Thus $\delta = (-1)^{\mu+1}x_1x_2\ldots x_{\mu+1} = (-1)^{\mu+1}x_1'x_2'\ldots x_{\mu+1}'$ and we may impose $x_2' = x_3' = \ldots = x_{\mu+1}' = 1$ to get $x_1' = x_1x_2\ldots x_{\mu+1}$. In this case, the theorem gives $\log x_1 + \log x_2 + \ldots + \log x_{\mu+1} = \log(x_1x_2\ldots x_{\mu+1})$. A second example is given by $\alpha = 1$, $\alpha_1 = 1+x^2$, $\psi x = -\arctan x$. Let x_1, x_2, x_3 be solutions of the equation $0 = \delta + \delta_1 x + (1+x^2)(a+x)$; we have $\arctan x_1 + \arctan x_2 + \arctan x_3 = C$ constant and $x_1x_2x_3 = -\delta - a$, $x_1+x_2+x_3 = -a$, $x_1x_2+x_1x_3+x_2x_3 = \delta_1 + 1$. Thus $x_1 + x_2 + x_3 - x_1x_2x_3 = \delta$ and $x_1x_2 + x_1x_3 + x_2x_3 - 1 = \delta_1$. Now putting $x_3 = x_2'$, $x_2 = -x_2'$ and $x_1 = x_1'$, we get $C = \arctan x_1'$ and $x_1' + x_1'(x_2')^2 = \delta$, $1 + (x_2')^2 = -\delta_1$, whence $x_1' = -\frac{\delta}{\delta_1} = \frac{x_1+x_2+x_3-x_1x_2x_3}{1-x_1x_2-x_1x_3-x_2x_3}$. Thus the theorem gives $\arctan x_1 + \arctan x_2 + \arctan x_3 = \arctan\frac{x_1+x_2+x_3-x_1x_2x_3}{1-x_1x_2-x_1x_3-x_2x_3}$.

At the end of this memoir, Abel generalises the relations (71). Considering the integral

$$\int fx \cdot dx = \psi x + \sum A\log(x-\delta)$$

where fx and ψx are rational functions and the auxiliary equation $\varphi x = a + a_1x + \ldots + a_nx^n = 0$, with the roots $x_1, x_2, \ldots, x_n$. By the theorem

$$\int fx_1 \cdot dx_1 + \int fx_2 \cdot dx_2 + \ldots + \int fx_n \cdot dx_n$$
$$= \psi x_1 + \psi x_2 + \ldots + \psi x_n + \sum A\log(x_1-\delta)(x_2-\delta)\ldots(x_n-\delta) = \rho$$

where $-d\rho = da\left(\frac{fx_1}{\varphi' x_1} + \frac{fx_2}{\varphi' x_2} + \ldots + \frac{fx_n}{\varphi' x_n}\right) + da_1\left(\frac{x_1\cdot fx_1}{\varphi' x_1} + \frac{x_2\cdot fx_2}{\varphi' x_2} + \ldots + \frac{x_n\cdot fx_n}{\varphi' x_n}\right) + \ldots + da_1\left(\frac{x_1^n fx_1}{\varphi' x_1} + \frac{x_2^n fx_2}{\varphi' x_2} + \ldots + \frac{x_n^n fx_n}{\varphi' x_n}\right)$.

Now $\psi x_1 + \psi x_2 + \ldots + \psi x_n$ is a rational function p of $a, a_1, \ldots, a_n$ and $(x_1-\delta)(x_2-\delta)\ldots(x_n-\delta) = (-1)^n\frac{\varphi\delta}{a_n}$ so that $\rho = p + \sum A(\log\varphi\delta - \log a_n)$ and

$$\frac{\partial \rho}{\partial a_m} = \frac{\partial p}{\partial a_m} + \sum A \left(\frac{1}{\varphi\delta} \frac{\partial \varphi\delta}{\partial a_m} - \frac{1}{a_n} \frac{\partial a_n}{\partial a_m} \right)$$
$$= - \left(\frac{x_1^m f x_1}{\varphi' x_1} + \frac{x_2^m f x_2}{\varphi' x_2} + \ldots + \frac{x_n^m f x_n}{\varphi' x_n} \right).$$

Abel deduces that $\frac{x_1^m fx_1}{\varphi' x_1} + \frac{x_2^m fx_2}{\varphi' x_2} + \ldots + \frac{x_n^m fx_n}{\varphi' x_n} = -\frac{\partial p}{\partial a_m} - \sum \frac{A\delta^m}{\varphi\delta} + \sum \frac{A}{a_n}\left(\frac{1}{2} \pm \frac{1}{2}\right)$ where the superior sign is taken when $m = n$ and the inferior sign when $m < n$. For $fx = 1$, $\psi x = x$, $p = x_1 + x_2 + \ldots + x_n = -\frac{a_{n-1}}{a_n}$ and $A = 0$. We find back (71) and the relation

$$\frac{x_1^n}{\varphi' x_1} + \frac{x_2^n}{\varphi' x_2} + \ldots + \frac{x_n^n}{\varphi' x_n} = -\frac{a_{n-1}}{a_n^2}.$$

For $fx = \frac{1}{x-\delta}$, $p = 0$ and $A = 1$; if $Fx = \beta + \beta_1 x + \ldots + \beta_n x^n$ we have

$$\frac{Fx_1}{(x_1 - \delta)\varphi' x_1} + \frac{Fx_2}{(x_2 - \delta)\varphi' x_2} + \ldots + \frac{Fx_n}{(x_n - \delta)\varphi' x_n} = \frac{\beta_n}{a_n} - \frac{F\delta}{\varphi\delta}$$

and other relations by differentiating this one.

6 Elliptic functions

Abel is the founder of the theory of elliptic functions. He partook this glory with Jacobi alone, for Gauss did not publish the important work he had done in this field; the 'grand prix' of the parisian Academy of sciences was awarded to Abel and Jacobi for their work on elliptic functions in 1830, after Abel's death. Abel's work on elliptic functions was published in the second and the third volumes of Crelle's *Journal* (1827–1828), in a large memoir titled *Recherches sur les finctions elliptiques* (*Œuvres*, t. I, p. 263–388).

Abel briefly recalls the main results of Euler, Lagrange and Legendre on elliptic integrals and defines his elliptic function $\varphi\alpha = x$ by the relation

$$\alpha = \int_0 \frac{dx}{\sqrt{(1 - c^2x^2)(1 + e^2x^2)}} \tag{72}$$

where c and e are real numbers. This definition is equivalent to the differential equation

$$\varphi'\alpha = \sqrt{(1 - c^2\varphi^2\alpha)(1 + e^2\varphi^2\alpha)}$$

with $\varphi(0) = 0$. Abel puts $f\alpha = \sqrt{1 - c^2\varphi^2\alpha}$ and $F\alpha = \sqrt{1 + e^2\varphi^2\alpha}$ and explains that the principal aim of his memoir is the resolution of the algebraic equation of degree m^2 which gives $\varphi\alpha$, $f\alpha$, $F\alpha$ when one knows $\varphi(m\alpha)$, $f(m\alpha)$, $F(m\alpha)$ (cf. our §3).

The first paragraph (p. 266–278) of Abel's memoir is devoted to the study of the functions $\varphi\alpha$, $f\alpha$ and $F\alpha$. According to (72), α is a positive increasing function of x for $0 \le x \le \frac{1}{c}$. Thus $\varphi\alpha$ is a positive increasing function of α for $0 \le \alpha \le \frac{\omega}{2}$ $= \int\limits_0^{1/c} \frac{dx}{\sqrt{(1-c^2x^2)(1+e^2x^2)}}$ and we have $\varphi\left(\frac{\omega}{2}\right) = \frac{1}{c}$. Since α is an odd function of x, $\varphi(-\alpha) = -\varphi(\alpha)$. Now Abel puts ix instead of x in (72) (where $i = \sqrt{-1}$) and gets a purely imaginary value $\alpha = i\beta$, so that $xi = \varphi(\beta i)$ where $\beta = \int\limits_0^x \frac{dx}{\sqrt{(1+c^2x^2)(1-e^2x^2)}}$. We see that β is a positive increasing function of x for $0 \le x \le \frac{1}{e}$ and that x is a positive increasing function of β for

$$0 \le \beta \le \frac{\varpi}{2} = \int\limits_0^{1/e} \frac{dx}{\sqrt{(1+c^2x^2)(1-e^2x^2)}}$$

and we have $\varphi\left(\frac{\varpi i}{2}\right) = i\frac{1}{e}$. Abel notes that the exchange of c and e transforms $\frac{\varphi(\alpha i)}{i}$ in $\varphi\alpha$, $f(\alpha i)$ in $F\alpha$, $F(\alpha i)$ in $f\alpha$ and exchanges ω and ϖ.

The function $\varphi\alpha$ is known for $-\frac{\omega}{2} \le \alpha \le \frac{\omega}{2}$ and for $\alpha = \beta i$ with $-\frac{\varpi}{2} \le \beta \le \frac{\varpi}{2}$. Abel extends its definition to the entire complex domain by the addition theorem:

$$\begin{aligned} \varphi(\alpha+\beta) &= \frac{\varphi\alpha \cdot f\beta \cdot F\beta + \varphi\beta \cdot f\alpha \cdot F\alpha}{1 + e^2c^2\varphi^2\alpha \cdot \varphi^2\beta}, \\ f(\alpha+\beta) &= \frac{f\alpha \cdot f\beta - c^2\varphi\alpha \cdot \varphi\beta \cdot F\alpha \cdot F\beta}{1 + e^2c^2\varphi^2\alpha \cdot \varphi^2\beta}, \\ F(\alpha+\beta) &= \frac{F\alpha \cdot F\beta + e^2\varphi\alpha \cdot \varphi\beta \cdot f\alpha \cdot f\beta}{1 + e^2c^2\varphi^2\alpha \cdot \varphi^2\beta}. \end{aligned} \tag{73}$$

This theorem is a consequence of Euler addition theorem for elliptic integrals, but Abel directly proves it by differentiating with respect to α and using $\varphi'\alpha = f\alpha \cdot F\alpha$, $f'\alpha = -c^2\varphi\alpha \cdot F\alpha$, $F'\alpha = e^2\varphi\alpha \cdot f\alpha$. Thus, denoting by r the right hand side of the first formula, he finds that $\frac{\partial r}{\partial\alpha} = \frac{\partial r}{\partial\beta}$ which shows that r is a function of $\alpha+\beta$. As $r = \varphi\alpha$ when $\beta = 0$, this gives $r = \varphi(\alpha+\beta)$. From (73), Abel deduces

$$\begin{aligned} \varphi(\alpha+\beta)+\varphi(\alpha-\beta) &= \frac{2\varphi\alpha\cdot f\beta\cdot F\beta}{R}, \quad \varphi(\alpha+\beta)-\varphi(\alpha-\beta) = \frac{2\varphi\beta\cdot f\alpha\cdot F\alpha}{R}, \\ f(\alpha+\beta)+f(\alpha-\beta) &= \frac{2f\alpha\cdot f\beta}{R}, \quad f(\alpha+\beta)-f(\alpha-\beta) = \frac{-2c^2\varphi\alpha\cdot\varphi\beta\cdot F\alpha\cdot F\beta}{R}, \\ F(\alpha+\beta)+F(\alpha-\beta) &= \frac{2F\alpha\cdot F\beta}{R}, \quad F(\alpha+\beta)-F(\alpha-\beta) = \frac{2e^2\varphi\alpha\cdot\varphi\beta\cdot f\alpha\cdot f\beta}{R} \end{aligned} \tag{74}$$

and

$$\begin{aligned} \varphi(\alpha+\beta)\varphi(\alpha-\beta) &= \frac{\varphi^2\alpha - \varphi^2\beta}{R}, \quad f(\alpha+\beta)f(\alpha-\beta) = \frac{f^2\beta - c^2\varphi^2\alpha \cdot F^2\beta}{R} \\ F(\alpha+\beta)F(\alpha-\beta) &= \frac{F^2\beta + e^2\varphi^2\alpha \cdot f^2\beta}{R} \end{aligned} \tag{75}$$

where $R = 1 + e^2c^2\varphi^2\alpha\,\varphi^2\beta$.

On the other hand $f\left(\pm\frac{\omega}{2}\right) = F\left(\pm\frac{\varpi}{2}i\right) = 0$ give

$$\varphi\left(\alpha \pm \frac{\omega}{2}\right) = \pm\frac{1}{c}\frac{f\alpha}{F\alpha},\ f\left(\alpha \pm \frac{\omega}{2}\right) = \mp\sqrt{e^2+c^2}\frac{\varphi\alpha}{F\alpha}, \tag{76}$$
$$F\left(\alpha \pm \frac{\omega}{2}\right) = \frac{\sqrt{e^2+c^2}}{c}\frac{1}{F\alpha},\ \varphi\left(\alpha \pm \frac{\varpi}{2}i\right) = \pm\frac{i}{e}\frac{F\alpha}{f\alpha},$$
$$F\left(\alpha \pm \frac{\varpi}{2}i\right) = \pm i\sqrt{e^2+c^2}\frac{\varphi\alpha}{f\alpha},\ f\left(\alpha \pm \frac{\varpi}{2}i\right) = \frac{\sqrt{e^2+c^2}}{e}\frac{1}{f\alpha}.$$

These relations imply that

$$\varphi\left(\frac{\omega}{2}+\alpha\right) = \varphi\left(\frac{\omega}{2}-\alpha\right),\ f\left(\frac{\omega}{2}+\alpha\right) = -f\left(\frac{\omega}{2}-\alpha\right), \tag{77}$$
$$F\left(\frac{\omega}{2}+\alpha\right) = F\left(\frac{\omega}{2}-\alpha\right),\ \varphi\left(\frac{\varpi}{2}i+\alpha\right) = \varphi\left(\frac{\varpi}{2}i-\alpha\right),$$
$$F\left(\frac{\varpi}{2}i+\alpha\right) = -F\left(\frac{\varpi}{2}i-\alpha\right),\ f\left(\frac{\varpi}{2}i+\alpha\right) = f\left(\frac{\varpi}{2}i-\alpha\right)$$

and $\varphi\left(\alpha \pm \frac{\omega}{2}\right)\varphi\left(\alpha + \frac{\varpi}{2}i\right) = \pm\frac{i}{ce}$, $F\left(\alpha \pm \frac{\omega}{2}\right) = F\alpha\frac{\sqrt{e^2+c^2}}{c} = f\left(\alpha \pm \frac{\varpi}{2}i\right) f\alpha$. We deduce that $\varphi\left(\frac{\omega}{2}+\frac{\varpi}{2}i\right) = f\left(\frac{\omega}{2}+\frac{\varpi}{2}i\right) = F\left(\frac{\omega}{2}+\frac{\varpi}{2}i\right) = \frac{1}{0}$ i.e. infinity. From (77) we have

$$\varphi(\alpha+\omega) = -\varphi\alpha = \varphi(\alpha+\varpi i),\ f(\alpha+\omega) = -f\alpha = -f(\alpha+\varpi i), \tag{78}$$
$$F(\alpha+\omega) = F\alpha = -F(\alpha+\varpi i)$$

and

$$\varphi(2\omega+\alpha) = \varphi\alpha = \varphi(2\varpi i+\alpha) = \varphi(\omega+\varpi i+\alpha), \tag{79}$$
$$f(2\omega+\alpha) = f\alpha = f(\varpi i+\alpha),\ F(\omega+\alpha) = F\alpha = F(2\varpi i+\alpha).$$

Thus the functions $\varphi\alpha$, $f\alpha$, $F\alpha$ are *periodic*:

$$\varphi(m\omega+n\varpi i\pm\alpha) = \pm(-1)^{m+n}\varphi\alpha,\ f(m\omega+n\varpi i\pm\alpha) = (-1)^m f\alpha, \tag{80}$$
$$F(m\omega+n\varpi i\pm\alpha) = (-1)^n F\alpha.$$

The equation $\varphi(\alpha+\beta i) = 0$ is equivalent to $\frac{\varphi\alpha\cdot f(\beta i)F(\beta i)+\varphi(\beta i)f\alpha\cdot F\alpha}{1+e^2c^2\varphi^2\alpha\cdot\varphi^2(\beta i)} = 0$ (cf. (73)) and, as $\varphi\alpha$, $f(\beta i)$, $F(\beta i)$ are real and $\varphi(\beta i)$ is purely imaginary, this signifies $\varphi\alpha\cdot f(\beta i)F(\beta i) = 0$ and $\varphi(\beta i)f\alpha\cdot F\alpha = 0$. These equations are satisfied by $\varphi\alpha = \varphi(\beta i) = 0$ or by $f(\beta i)F(\beta i) = f\alpha\cdot F\alpha = 0$. The first solution gives $\alpha = m\omega$, $\beta = n\varpi$ and it fits, for $\varphi(m\omega+n\varpi i) = 0$. The second solution gives $\alpha = \left(m+\frac{1}{2}\right)\omega$, $\beta = \left(n+\frac{1}{2}\right)\varpi$ and it does not fit, for $\varphi\left(\left(m+\frac{1}{2}\right)\omega+\left(n+\frac{1}{2}\right)\varpi i\right) = \frac{1}{0}$. In the same way, Abel determines the roots of the equation $fx = 0$, which are $x = \left(m+\frac{1}{2}\right)\omega + n\varpi i$ and those of the equation $Fx = 0$, which are $x = m\omega + \left(n+\frac{1}{2}\right)\varpi i$. From these results and the formulae

$$\varphi x = \frac{i}{ec}\frac{1}{\varphi\left(x-\frac{\omega}{2}-\frac{\varpi}{2}i\right)},\ fx = \frac{\sqrt{e^2+c^2}}{e}\frac{1}{f\left(x-\frac{\varpi}{2}i\right)},\ Fx = \frac{\sqrt{e^2+c^2}}{c}\frac{1}{F\left(x-\frac{\omega}{2}\right)}, \tag{81}$$

he deduces the poles of the functions φx, fx, Fx, which are $x = \left(m+\frac{1}{2}\right)\omega + \left(n+\frac{1}{2}\right)\varpi i$.

From (74) $\varphi x - \varphi a = \frac{2\varphi\left(\frac{x-a}{2}\right)f\left(\frac{x+a}{2}\right)F\left(\frac{x+a}{2}\right)}{1+e^2c^2\varphi^2\left(\frac{x+a}{2}\right)\varphi^2\left(\frac{x-a}{2}\right)}$. Thus the equation $\varphi x = \varphi a$ is equivalent to $\varphi\left(\frac{x-a}{2}\right) = 0$ or $f\left(\frac{x+a}{2}\right) = 0$ or $F\left(\frac{x+a}{2}\right) = 0$ or $\varphi\left(\frac{x-a}{2}\right) = \frac{1}{0}$ or $\varphi\left(\frac{x+a}{2}\right) = 0$. Thus the solutions are $x = (-1)^{m+n}a + m\omega + n\varpi i$. In the same way, the solutions of $fx = fa$ are given by $x = \pm a + 2m\omega + n\varpi i$ and those of $Fx = Fa$ by $x = \pm a + m\omega + 2n\varpi i$.

The second paragraph (p. 279–281) of Abel's memoir contains the proof by complete induction that $\varphi(n\beta)$, $f(n\beta)$ and $F(n\beta)$ are rational functions of $\varphi\beta$, $f\beta$ and $F\beta$ when n is an integer. Writing $\varphi(n\beta) = \frac{P_n}{Q_n}$, $f(n\beta) = \frac{P'_n}{Q_n}$ and $F(n\beta) = \frac{P''_n}{Q_n}$ where P_n, P'_n, P''_n and Q_n are polynomials in $\varphi\beta$, $f\beta$ and $F\beta$, we have, by (74)

$$\frac{P_{n+1}}{Q_{n+1}} = -\frac{P_{n-1}}{Q_{n-1}} + \frac{2f\beta\cdot F\beta\frac{P_n}{Q_n}}{1+e^2c^2\varphi^2\beta\frac{P_n^2}{Q_n^2}} = \frac{-P_{n-1}(Q_n^2+c^2e^2x^2P_n^2)+2P_nQ_nQ_{n-1}yz}{Q_{n-1}R_n}$$

where $x = \varphi\beta$, $y = f\beta$, $z = F\beta$ and $R_n = Q_n^2 + e^2c^2x^2P_n^2$, and we conclude that

$$Q_{n+1} = Q_{n-1}R_n,\ P_{n+1} = -P_{n-1}R_n + 2yzP_nq_nQ_{n-1}.$$

In the same way $P'_{n+1} = -P'_{n-1}R_n + 2yP'_nQ_nQ_{n-1}$ and $P''_{n+1} = -P''_{n-1}R_n + 2yP''_nQ_nQ_{n-1}$. These recursion formulae, together with $y^2 = 1 - c^2x^2$ and $z^2 = 1 + e^2x^2$, show that Q_n, $\frac{P_{2n}}{xyz}$, $\frac{P_{2n+1}}{x}$, P'_{2n}, $\frac{P'_{2n+1}}{y}$, P''_{2n} and $\frac{P''_{2n+1}}{z}$ are polynomials in x^2.

The equations $\varphi(n\beta) = \frac{P_n}{Q_n}$, $f(n\beta) = \frac{P'_n}{Q_n}$ and $F(n\beta) = \frac{P''_n}{Q_n}$ are studied in paragraph III (p. 282–291). When n is even, here noted $2n$, the first equation is written

$$\varphi(2n\beta) = xyz\psi(x^2) = x\psi(x^2)\sqrt{(1-c^2x^2)(1+e^2x^2)}$$

or $\varphi^2(2n\beta) = x^2(\psi x^2)^2(1-c^2x^2)(1+e^2x^2) = \theta(x^2)$, where $x = \varphi\beta$ is one of the roots. If $x = \varphi\alpha$ is another root, $\varphi(2n\alpha) = \pm\varphi(2n\beta)$ and, by the preceding properties,

$$\alpha = \pm((-1)^{m+\mu}2n\beta + m\omega + \mu\varpi i)).$$

Thus the roots of our equation are $\varphi\alpha = \pm\varphi\left((-1)^{m+\mu}\beta + \frac{m}{2n}\omega + \frac{\mu}{2n}\varpi i\right)$, formula in which we may replace m and μ by the remainders of their division by $2n$, because of (80). Abel remarks that, when $0 \le m, \mu < 2n$, all the values of $\varphi\alpha$ so obtained are different. It results that the total number of roots is equal to $8n^2$ and this is the degree of the equation, for it cannot have any multiple root. When $n = 1$, the equation is $(1+e^2c^2x^4)\varphi^2(2\beta) = 4x^2(1-c^2x^2)(1+e^2x^2)$ and its roots are $\pm\varphi\beta$, $\pm\varphi\left(-\beta+\frac{\omega}{2}\right)$, $\pm\varphi\left(-\beta+\frac{\varpi}{2}i\right)$ and $\pm\varphi\left(\beta+\frac{\omega}{2}+\frac{\varpi}{2}i\right)$.

When n is an odd number, here written $2n+1$, the equation is $\varphi(2n+1)\beta = \frac{P_{2n+1}}{Q_{2n+1}}$ and its roots $x = \varphi\left((-1)^{m+\mu}\beta + \frac{m}{2n+1}\omega + \frac{\mu}{2n+1}\varpi i\right)$ where $-n \leq m, \mu \leq n$. The number of these roots is $(2n+1)^2$ and it is the degree of the equation. For example $n = 1$ gives an equation of degree 9 with the roots $\varphi\beta$, $\varphi\left(-\beta - \frac{\omega}{3}\right)$, $\varphi\left(-\beta + \frac{\omega}{3}\right)$, $\varphi\left(-\beta - \frac{\varpi}{3}i\right)$, $\varphi\left(-\beta + \frac{\varpi}{3}i\right)$, $\varphi\left(\beta - \frac{\omega}{3} - \frac{\varpi}{3}i\right)$, $\varphi\left(\beta - \frac{\omega}{3} + \frac{\varpi}{3}i\right)$, $\varphi\left(\beta + \frac{\omega}{3} - \frac{\varpi}{3}i\right)$ and $\varphi\left(\beta + \frac{\omega}{3} + \frac{\varpi}{3}i\right)$.

Abel studies in the same way the equations $f(n\beta) = \frac{P'_n}{Q_n}$ and $F(n\beta) = \frac{P''_n}{Q_n}$ of which the roots are respectively $y = f\left(\beta + \frac{2m}{n}\omega + \frac{\mu}{n}\varpi i\right)$ and $z = F\left(\beta + \frac{m}{n}\omega + \frac{2\mu}{n}\varpi i\right)$, $(0 \leq m, \mu < n)$. Each of these equations is of degree n^2.

There are particular cases: $P_{2n}^2 = 0$, with the roots $x = \pm\varphi\left(\frac{m}{2n}\omega + \frac{\mu}{2n}\varpi i\right)$ $(0 \leq m, \mu \leq 2n-1)$, $P_{2n+1} = 0$, with the roots $x = \varphi\left(\frac{m}{2n+1}\omega + \frac{\mu}{2n+1}\varpi i\right)$ $(-n \leq m, \mu \leq n)$, $P'_n = 0$, with the roots $y = f\left(\left(2m + \frac{1}{2}\right)\frac{\omega}{n} + \frac{\mu}{n}\varpi i\right)$, $P''_n = 0$, with the roots $z = F\left(\frac{m}{n}\omega + \left(2\mu + \frac{1}{2}\right)\frac{\varpi i}{n}\right)$ $(0 \leq m, \mu \leq n-1)$ and $Q_{2n} = 0$, with the roots $x = \varphi\left(\left(m + \frac{1}{2}\right)\frac{\omega}{2n} + \left(\mu + \frac{1}{2}\right)\frac{\varpi i}{2n}\right)$ $(0 \leq m, \mu \leq 2n-1)$, $Q_{2n+1} = 0$ with the roots $x = (-1)^{m+\mu}\varphi\left(\left(m + \frac{1}{2}\right)\frac{\omega}{2n+1} + \left(\mu + \frac{1}{2}\right)\frac{\varpi i}{2n+1}\right)$ $(-n \leq m, \mu \leq n, (m, \mu) \leq (n, n))$.

The algebraic solution of the equations $\varphi(n\beta) = \frac{P_n}{Q_n}$, $f(n\beta) = \frac{P'_n}{Q_n}$ and $F(n\beta) = \frac{P''_n}{Q_n}$ is given in paragraph IV (p. 291–305). It is sufficient to deal with the case in which n is a prime number. The case $n = 2$ is easy for if $x = \varphi\frac{\alpha}{2}$, $y = f\frac{\alpha}{2}$ and $z = F\frac{\alpha}{2}$, we have

$$f\alpha = \frac{y^2 - c^2x^2z^2}{1 + e^2c^2x^4} = \frac{1 - 2c^2x^2 - c^2e^2x^4}{1 + e^2c^2x^4},$$
$$F\alpha = \frac{z^2 + e^2y^2x^2}{1 + e^2c^2x^4} = \frac{1 + 2e^2x^2 - e^2c^2x^4}{1 + e^2c^2x^4}.$$

Hence $\frac{F\alpha - 1}{1 + f\alpha} = e^2x^2$, $\frac{1 - f\alpha}{F\alpha + 1} = c^2x^2$ and $z^2 = \frac{F\alpha + f\alpha}{1 + f\alpha}$, $y^2 = \frac{F\alpha + f\alpha}{1 + F\alpha}$ and we draw $\varphi\frac{\alpha}{2} = \frac{1}{c}\sqrt{\frac{1 - f\alpha}{1 + F\alpha}} = \frac{1}{e}\sqrt{\frac{F\alpha - 1}{f\alpha + 1}}$, $f\frac{\alpha}{2} = \sqrt{\frac{F\alpha + f\alpha}{1 + F\alpha}}$, $F\frac{\alpha}{2} = \sqrt{\frac{F\alpha + f\alpha}{1 + f\alpha}}$. From these formulae, it is possible to express $\varphi\frac{\alpha}{2^n}$, $f\frac{\alpha}{2^n}$, $F\frac{\alpha}{2^n}$ with square roots in function of $\varphi\alpha$, $f\alpha$, $F\alpha$. Taking $\alpha = \frac{\omega}{2}$ as an example, Abel finds

$$\varphi\frac{\omega}{4} = \frac{1}{\sqrt{c^2 + c\sqrt{e^2 + c^2}}} = \frac{\sqrt{c\sqrt{e^2 + c^2} - c^2}}{ec},$$
$$f\frac{\omega}{4} = \frac{1}{e}\sqrt{e^2 + c^2 - c\sqrt{e^2 + c^2}},\ F\frac{\omega}{4} = \sqrt[4]{1 + \frac{e^2}{c^2}} = \sqrt{F\frac{\omega}{2}}.$$

The case n odd was explained in our §3. The essential point was that the auxiliary functions such as $\varphi_1\beta$ are *rational* functions of $\varphi\beta$ because of the addition theorem (73). At the same place, we have dealt with the equation $P_{2n+1} = 0$ (§V of Abel's memoir, p. 305–314)) which determines the quantities $x = \varphi\left(\frac{m\omega + \mu\varpi i}{2n+1}\right)$. We saw that the equation in $r = x^2$ is of degree $2n(n+1)$ and that it may be decomposed in $2n+2$ equations of degree n of which the coefficients are rational functions of the roots of an equation of degree $2n+2$. The equations of degree n are all solvable by radicals, but the equation of degree $2n+2$ is not solvable in general.

In paragraph VI (p. 315–323), Abel gives explicit formulae for

$$\varphi((2n+1)\beta),\ f((2n+1)\beta) \text{ and } F((2n+1)\beta)$$

in function of the quantities $\varphi\left(\beta + \frac{m\omega+\mu\varpi i}{2n+1}\right)$, $f\left(\beta + \frac{m\omega+\mu\varpi i}{2n+1}\right)$, $F\left(\beta + \frac{m\omega+\mu\varpi i}{2n+1}\right)$. Let $P_{2n+1} = Ax^{(2n+1)^2} + \ldots + Bx$, $P'_{2n+1} = A'y^{(2n+1)^2} + \ldots + B'y$, $P''_{2n+1} = A''z^{(2n+1)^2} + \ldots + B''z$ and $Q_{2n+1} = Cx^{(2n+1)^2-1} + \ldots + D = C'y^{(2n+1)^2-1} + \ldots + D' = C''z^{(2n+1)^2-1} + \ldots + D''$ (an even function). From the equations

$$\begin{aligned} Ax^{(2n+1)^2} + \ldots + Bx &= \varphi((2n+1)\beta \cdot (Cx^{(2n+1)^2-1} + \ldots + D), \\ A'y^{(2n+1)^2} + \ldots + B'y &= f((2n+1)\beta \cdot (C'y^{(2n+1)^2-1} + \ldots + D'), \\ A''z^{(2n+1)^2} + \ldots + B''z &= F((2n+1)\beta \cdot (C''z^{(2n+1)^2-1} + \ldots + D''), \end{aligned}$$

considering the sum and the product of the roots, Abel deduces that

$$\begin{aligned} \varphi((2n+1)\beta) &= \frac{A}{C} \sum_{m=-n}^{n} \sum_{\mu=-n}^{n} (-1)^{m+\mu} \varphi\left(\beta + \frac{m\omega+\mu\varpi i}{2n+1}\right) \\ &= \frac{A}{D} \prod_{m=-n}^{n} \prod_{\mu=-n}^{n} \varphi\left(\beta + \frac{m\omega+\mu\varpi i}{2n+1}\right). \end{aligned} \tag{82}$$

In the same way

$$\begin{aligned} f((2n+1)\beta &= \frac{A'}{C'} \sum_{m=-n}^{n} \sum_{\mu=-n}^{n} (-1)^{m} f\left(\beta + \frac{m\omega+\mu\varpi i}{2n+1}\right) \\ &= \frac{A'}{D'} \prod_{m=-n}^{n} \prod_{\mu=-n}^{n} f\left(\beta + \frac{m\omega+\mu\varpi i}{2n+1}\right) \end{aligned} \tag{82$'$}$$

and

$$\begin{aligned} F((2n+1)\beta) &= \frac{A''}{C''} \sum_{m=-n}^{n} \sum_{\mu=-n}^{n} (-1)^{\mu} F\left(\beta + \frac{m\omega+\mu\varpi i}{2n+1}\right) \\ &= \frac{A''}{D''} \prod_{m=-n}^{n} \prod_{\mu=-n}^{n} F\left(\beta + \frac{m\omega+\mu\varpi i}{2n+1}\right). \end{aligned} \tag{82$''$}$$

The coefficients $\frac{A}{C}, \frac{A'}{C'}, \frac{A''}{C''}$, which do not depend on β, are determined by letting β tend towards the pole $\frac{\omega}{2} + \frac{\varpi}{2}i$, for they are the respective limit values of

$$\frac{\varphi((2n+1)\beta)}{\varphi\beta},\ \frac{f((2n+1)\beta)}{f\beta},\ \frac{F((2n+1)\beta)}{F\beta}.$$

Putting $\beta = \frac{\omega}{2} + \frac{\varpi}{2}i + \alpha$, where α tends towards 0, and using (80) and (81), Abel determines $\frac{A}{C} = \frac{1}{2n+1}$, $\frac{A'}{C'} = \frac{A''}{C''} = \frac{(-1)^n}{2n+1}$. Since the limit of $\frac{\varphi((2n+1)\beta)}{\varphi\beta}$ when β tends towards 0 is $2n+1$ we find

$$2n+1 = \frac{A}{D}\prod_{m=1}^{n}\varphi^2\left(\frac{m\omega}{2n+1}\right)\prod_{\mu=1}^{n}\varphi^2\left(\frac{\mu\varpi i}{2n+1}\right) \times \prod_{m=1}^{n}\prod_{\mu=1}^{n}\varphi^2\left(\frac{m\omega+\mu\varpi i}{2n+1}\right)\varphi^2\left(\frac{m\omega-\mu\varpi i}{2n+1}\right).$$

In the same way, letting β tend respectively towards $\frac{\omega}{2}$ and $\frac{\varpi i}{2}$ we get

$$\begin{aligned}(-1)^n(2n+1) &= \frac{A'}{D'}\prod_{m=1}^{n}f^2\left(\frac{\omega}{2}+\frac{m\omega}{2n+1}\right)\prod_{\mu=1}^{n}f^2\left(\frac{\omega}{2}+\frac{\mu\varpi i}{2n+1}\right)\\ &\quad\times\prod_{m=1}^{n}\prod_{\mu=1}^{n}f^2\left(\frac{\omega}{2}+\frac{m\omega+\mu\varpi i}{2n+1}\right)f^2\left(\frac{\omega}{2}+\frac{m\omega-\mu\varpi i}{2n+1}\right)\\ &= \frac{A''}{D''}\prod_{m=1}^{n}F^2\left(\frac{\varpi}{2}i+\frac{m\omega}{2n+1}\right)\prod_{\mu=1}^{n}F^2\left(\frac{\varpi}{2}i+\frac{\mu\varpi i}{2n+1}\right)\\ &\quad\times\prod_{m=1}^{n}\prod_{\mu=1}^{n}F^2\left(\frac{\varpi}{2}i+\frac{m\omega+\mu\varpi i}{2n+1}\right)F^2\left(\frac{\varpi}{2}i+\frac{m\omega-\mu\varpi i}{2n+1}\right)\end{aligned}$$

from which it is possible to draw the values of $\frac{A}{D}$, $\frac{A'}{D'}$ and $\frac{A''}{D''}$. Abel further simplifies the expressions of $\varphi((2n+1)\beta)$, $f((2n+1)\beta)$ and $F((2n+1)\beta)$ as products by the formulae

$$\frac{\varphi(\beta+\alpha)\varphi(\beta-\alpha)}{\varphi^2\alpha} = -\frac{1-\frac{\varphi^2\beta}{\varphi^2\alpha}}{1-\frac{\varphi^2\beta}{\varphi^2(\alpha+\frac{\omega}{2}+\frac{\varpi}{2}i)}},$$

$$\frac{f(\beta+\alpha)f(\beta-\alpha)}{f^2\left(\frac{\omega}{2}+\alpha\right)} = -\frac{1-\frac{f^2\beta}{f^2(\frac{\omega}{2}+\alpha)}}{1-\frac{f^2\beta}{f^2(\alpha+\frac{\omega}{2}+\frac{\varpi}{2}i)}},$$

$$\frac{F(\beta+\alpha)F(\beta-\alpha)}{F^2\left(\frac{\varpi}{2}i+\alpha\right)} = -\frac{1-\frac{F^2\beta}{F^2(\frac{\varpi}{2}i+\alpha)}}{1-\frac{F^2\beta}{F^2(\alpha+\frac{\omega}{2}+\frac{\varpi}{2}i)}} \quad \text{(cf. (75) and (81))}$$

and he thus obtains

$$\begin{aligned}&\varphi((2n+1)\beta)\\ &= (2n+1)\varphi\beta\prod_{m=1}^{n}\frac{1-\frac{\varphi^2\beta}{\varphi^2\left(\frac{m\omega}{2n+1}\right)}}{1-\frac{\varphi^2\beta}{\varphi^2\left(\frac{\omega}{2}+\frac{\varpi}{2}i+\frac{m\omega}{2n+1}\right)}}\prod_{\mu=1}^{n}\frac{1-\frac{\varphi^2\beta}{\varphi^2\left(\frac{\mu\varpi i}{2n+1}\right)}}{1-\frac{\varphi^2\beta}{\varphi^2\left(\frac{\omega}{2}+\frac{\varpi}{2}i+\frac{\mu\varpi i}{2n+1}\right)}}\\ &\quad\times\prod_{m=1}^{n}\prod_{\mu=1}^{n}\frac{1-\frac{\varphi^2\beta}{\varphi^2\left(\frac{m\omega+\mu\varpi i}{2n+1}\right)}}{1-\frac{\varphi^2\beta}{\varphi^2\left(\frac{\omega}{2}+\frac{\varpi}{2}i+\frac{m\omega+\mu\varpi i}{2n+1}\right)}}\,\frac{1-\frac{\varphi^2\beta}{\varphi^2\left(\frac{m\omega-\mu\varpi i}{2n+1}\right)}}{1-\frac{\varphi^2\beta}{\varphi^2\left(\frac{\omega}{2}+\frac{\varpi}{2}i+\frac{m\omega-\mu\varpi i}{2n+1}\right)}},\end{aligned} \tag{83}$$

$$f((2n+1)\beta)$$

$$=(-1)^n(2n+1)f\beta\prod_{m=1}^{n}\frac{1-\frac{f^2\beta}{f^2\left(\frac{\omega}{2}+\frac{m\omega}{2n+1}\right)}}{1-\frac{f^2\beta}{f^2\left(\frac{\omega}{2}+\frac{\varpi}{2}i+\frac{m\omega}{2n+1}\right)}}\prod_{\mu=1}^{n}\frac{1-\frac{f^2\beta}{f^2\left(\frac{\omega}{2}+\frac{\mu\varpi i}{2n+1}\right)}}{1-\frac{f^2\beta}{f^2\left(\frac{\omega}{2}+\frac{\varpi}{2}i+\frac{\mu\varpi i}{2n+1}\right)}}$$

$$\times\prod_{m=1}^{n}\prod_{\mu=1}^{n}\frac{1-\frac{f^2\beta}{f^2\left(\frac{\omega}{2}+\frac{m\omega+\mu\varpi i}{2n+1}\right)}}{1-\frac{f^2\beta}{f^2\left(\frac{\omega}{2}+\frac{\varpi}{2}i+\frac{m\omega+\mu\varpi i}{2n+1}\right)}}\frac{1-\frac{f^2\beta}{f^2\left(\frac{\omega}{2}+\frac{m\omega-\mu\varpi i}{2n+1}\right)}}{1-\frac{f^2\beta}{f^2\left(\frac{\omega}{2}+\frac{\varpi}{2}i+\frac{m\omega-\mu\varpi i}{2n+1}\right)}} \tag{83'}$$

$$F((2n+1)\beta)$$

$$=(-1)^n(2n+1)F\beta\prod_{m=1}^{n}\frac{1-\frac{F^2\beta}{F^2\left(\frac{\varpi}{2}i+\frac{m\omega}{2n+1}\right)}}{1-\frac{F^2\beta}{F^2\left(\frac{\omega}{2}+\frac{\varpi}{2}i+\frac{m\omega}{2n+1}\right)}}\prod_{\mu=1}^{n}\frac{1-\frac{F^2\beta}{F^2\left(\frac{\varpi}{2}i+\frac{\mu\varpi i}{2n+1}\right)}}{1-\frac{F^2\beta}{F^2\left(\frac{\omega}{2}+\frac{\varpi}{2}i+\frac{\mu\varpi i}{2n+1}\right)}}$$

$$\times\prod_{m=1}^{n}\prod_{\mu=1}^{n}\frac{1-\frac{F^2\beta}{F^2\left(\frac{\varpi}{2}i+\frac{m\omega+\mu\varpi i}{2n+1}\right)}}{1-\frac{F^2\beta}{F^2\left(\frac{\omega}{2}+\frac{\varpi}{2}i+\frac{m\omega+\mu\varpi i}{2n+1}\right)}}\frac{1-\frac{F^2\beta}{F^2\left(\frac{\varpi}{2}i+\frac{m\omega-\mu\varpi i}{2n+1}\right)}}{1-\frac{F^2\beta}{F^2\left(\frac{\omega}{2}+\frac{\varpi}{2}i+\frac{m\omega-\mu\varpi i}{2n+1}\right)}} \tag{83''}$$

expressions of $\varphi((2n+1)\beta)$, $f((2n+1)\beta)$ and $F((2n+1)\beta)$ in rational functions of $\varphi\beta$, $f\beta$ and $F\beta$ respectively. Abel also transforms the last two to have $\frac{f((2n+1)\beta)}{f\beta}$ and $\frac{F((2n+1)\beta)}{F\beta}$ in rational functions of $\varphi\beta$.

In his paragraph VII (p. 323–351), Abel keeps $\alpha = (2n+1)\beta$ fixed in the formulae (82) and (83) and let n tend towards infinity in order to obtain expansions of his elliptic functions in infinite series and infinite products. From (82) with the help of (81), we have

$$\begin{aligned}\varphi\alpha = &\frac{1}{2n+1}\varphi\frac{\alpha}{2n+1}\\ &+\frac{1}{2n+1}\sum_{m=1}^{n}(-1)^m\left(\varphi\left(\frac{\alpha+m\omega}{2n+1}\right)+\varphi\left(\frac{\alpha-m\omega}{2n+1}\right)\right)\\ &+\frac{1}{2n+1}\sum_{\mu=1}^{n}(-1)^\mu\left(\varphi\left(\frac{\alpha+\mu\varpi i}{2n+1}\right)+\varphi\left(\frac{\alpha-\mu\varpi i}{2n+1}\right)\right)\\ &-\frac{i}{ec}\sum_{m=1}^{n}\sum_{\mu=1}^{n}(-1)^{m+\mu}\psi(n-m,n-\mu)\\ &+\frac{i}{ec}\sum_{m=1}^{n}\sum_{\mu=1}^{n}(-1)^{m+\mu}\psi_1(n-m,n-\mu)\end{aligned}$$

where

$$\psi(m,\mu)=\frac{1}{2n+1}\left(\frac{1}{\varphi\left(\frac{\alpha+\left(m+\frac{1}{2}\right)\omega+\left(\mu+\frac{1}{2}\right)\varpi i}{2n+1}\right)}+\frac{1}{\varphi\left(\frac{\alpha-\left(m+\frac{1}{2}\right)\omega-\left(\mu+\frac{1}{2}\right)\varpi i}{2n+1}\right)}\right)$$

and

$$\psi_1(m,\mu)=\frac{1}{2n+1}\left(\frac{1}{\varphi\left(\frac{\alpha+\left(m+\frac{1}{2}\right)\omega-\left(\mu+\frac{1}{2}\right)\varpi i}{2n+1}\right)}+\frac{1}{\varphi\left(\frac{\alpha-\left(m+\frac{1}{2}\right)\omega+\left(\mu+\frac{1}{2}\right)\varpi i}{2n+1}\right)}\right).$$

Now

$$\begin{aligned}A_m&=(2n+1)\left(\varphi\left(\frac{\alpha+m\omega}{2n+1}\right)+\varphi\left(\frac{\alpha-m\omega}{2n+1}\right)\right)\\&=(2n+1)\frac{2\varphi\left(\frac{\alpha}{2n+1}\right)f\left(\frac{m\omega}{2n+1}\right)F\left(\frac{m\omega}{2n+1}\right)}{1+e^2c^2\varphi^2\left(\frac{m\omega}{2n+1}\right)\varphi^2\left(\frac{\alpha}{2n+1}\right)}\end{aligned}$$

and

$$\begin{aligned}B_\mu&=(2n+1)\left(\varphi\left(\frac{\alpha+\mu\varpi i}{2n+1}\right)+\varphi\left(\frac{\alpha-\mu\varpi i}{2n+1}\right)\right)\\&=(2n+1)\frac{2\varphi\left(\frac{\alpha}{2n+1}\right)f\left(\frac{\mu\varpi i}{2n+1}\right)F\left(\frac{\mu\varpi i}{2n+1}\right)}{1+e^2c^2\varphi^2\left(\frac{\mu\varpi i}{2n+1}\right)\varphi^2\left(\frac{\alpha}{2n+1}\right)}\end{aligned}$$

remain bounded and the first part $\frac{1}{2n+1}\varphi\frac{\alpha}{2n+1}+\frac{1}{(2n+1)^2}\sum_{m=1}^{n}(-1)^m(A_m+B_m)$ of $\varphi\alpha$ has 0 for limit when n tends towards ∞. Thus

$$\begin{aligned}\varphi\alpha=&-\frac{i}{ec}\lim\sum_{m=1}^{n}\sum_{\mu=1}^{n}(-1)^{m+\mu}\psi(n-m,n-\mu)\\&+\frac{i}{ec}\lim\sum_{m=1}^{n}\sum_{\mu=1}^{n}(-1)^{m+\mu}\psi_1(n-m,n-\mu).\end{aligned}$$

It remains to compute the limit of $\sum_{m=0}^{n-1}\sum_{\mu=0}^{n-1}(-1)^{m+\mu}\psi(m,\mu)$ for the second part will be deduced from the first by changing the sign of i. We have $\psi(m,\mu)=\frac{1}{2n+1}\frac{2\varphi\left(\frac{\alpha}{2n+1}\right)\theta\left(\frac{\varepsilon_\mu}{2n+1}\right)}{\varphi^2\left(\frac{\alpha}{2n+1}\right)-\varphi^2\left(\frac{\varepsilon_\mu}{2n+1}\right)}$ where $\theta\varepsilon=f\varepsilon F\varepsilon$ and $\varepsilon_\mu=\left(m+\frac{1}{2}\right)\omega+\left(\mu+\frac{1}{2}\right)\varpi i$ (cf. (74) and (75)) and this has for limit $\theta(m,\mu)=\frac{2\alpha}{\alpha^2-\left(\left(m+\frac{1}{2}\right)\omega+\left(\mu+\frac{1}{2}\right)\varpi i\right)^2}$ when n

tends towards ∞. Abel tries to prove that $\sum_{\mu=1}^{n-1}(-1)^{\mu}\psi(m,\mu)-\sum_{\mu=1}^{n-1}(-1)^{\mu}\theta(m,\mu)$ is negligible with respect to $\frac{1}{2n+1}$ by estimating the difference $\psi(m,\mu)-\theta(m,\mu)$, but his reasoning is not clear. Then he replaces $\sum_{\mu=1}^{n-1}(-1)^{\mu}\theta(m,\mu)$ by the sum up to infinity using a sum formula to estimate $\sum_{\mu=n}^{\infty}(-1)^{\mu}\theta(m,\mu)$, again negligible with respect to $\frac{1}{2n+1}$. He finally obtains

$$\begin{aligned}\varphi\alpha &= \frac{i}{ec}\sum_{m=1}^{\infty}(-1)^m\sum_{\mu=1}^{\infty}(-1)^{\mu}\left(\frac{2\alpha}{\alpha^2-\left(\left(m+\frac{1}{2}\right)\omega-\left(\mu+\frac{1}{2}\right)\varpi i\right)^2}\right.\\ &\qquad\left.-\frac{2\alpha}{\alpha^2-\left(\left(m+\frac{1}{2}\right)\omega+\left(\mu+\frac{1}{2}\right)\varpi i\right)^2}\right)\\ &= \frac{1}{ec}\sum_{m=1}^{\infty}(-1)^m\sum_{\mu=1}^{\infty}(-1)^{\mu}\left(\frac{(2\mu+1)\varpi}{\left(\alpha-\left(m+\frac{1}{2}\right)\omega\right)^2+\left(\mu+\frac{1}{2}\right)^2\varpi^2}\right.\\ &\qquad\left.-\frac{(2\mu+1)\varpi}{\left(\alpha+\left(m+\frac{1}{2}\right)\omega\right)^2+\left(\mu+\frac{1}{2}\right)^2\varpi^2}\right). \end{aligned}\tag{84}$$

By the same method, Abel obtains

$$\begin{aligned}f\alpha &= \frac{1}{e}\sum_{\mu=0}^{\infty}\left(\sum_{m=0}^{\infty}(-1)^m\frac{2\left(\alpha+\left(m+\frac{1}{2}\right)\omega\right)}{\left(\alpha+\left(m+\frac{1}{2}\right)\omega\right)^2+\left(\mu+\frac{1}{2}\right)^2\varpi^2}\right.\\ &\qquad\left.-\sum_{m=0}^{\infty}(-1)^m\frac{2\left(\alpha-\left(m+\frac{1}{2}\right)\omega\right)}{\left(\alpha-\left(m+\frac{1}{2}\right)\omega\right)^2+\left(\mu+\frac{1}{2}\right)^2\varpi^2}\right),\end{aligned}\tag{84'}$$

$$\begin{aligned}F\alpha &= \frac{1}{c}\sum_{m=0}^{\infty}\left(\sum_{\mu=0}^{\infty}(-1)^{\mu}\frac{(2\mu+1)\varpi}{\left(\alpha-\left(m+\frac{1}{2}\right)\omega\right)^2+\left(\mu+\frac{1}{2}\right)^2\varpi^2}\right.\\ &\qquad\left.+\sum_{\mu=0}^{\infty}(-1)^{\mu}\frac{(2\mu+1)\varpi}{\left(\alpha+\left(m+\frac{1}{2}\right)\omega\right)^2+\left(\mu+\frac{1}{2}\right)^2\varpi^2}\right).\end{aligned}\tag{84''}$$

He deals with the formulae (83) in the same way by taking the logarithms. For any constants k and ℓ, $\frac{1-\frac{\varphi^2\left(\frac{\alpha}{2n+1}\right)}{\varphi^2\left(\frac{m\omega+\mu\varpi i+k}{2n+1}\right)}}{1-\frac{\varphi^2\left(\frac{\alpha}{2n+1}\right)}{\varphi^2\left(\frac{m\omega+\mu\varpi i+\ell}{2n+1}\right)}}$ has a limit equal to $\frac{1-\frac{\alpha^2}{(m\omega+\mu\varpi i+k)^2}}{1-\frac{\alpha^2}{(m\omega+\mu\varpi i+\ell)^2}}$. Abel tries to proves that the difference of the logarithms $\psi(m,\mu)$ and $\theta(m,\mu)$ of these expressions is dominated by $\frac{1}{(2n+1)^2}$, with the difficulty that m and μ vary in the sum to be computed. He deduces that the difference $\sum_{\mu=1}^{n}\psi(m,\mu)-\sum_{\mu=1}^{n}\theta(m,\mu)$ is negligible with respect to $\frac{1}{2n+1}$ and replaces $\sum_{\mu=1}^{n} \theta(m,\mu)$ by $\sum_{\mu=1}^{\infty} \theta(m,\mu)$. The proof that $\sum_{\mu=n+1}^{\infty} \theta(m,\mu) = \sum_{\mu=1}^{\infty} \theta(m,\mu+n)$ is negligible with respect to $\frac{1}{2n+1}$ is based on the expansion of $\theta(m,\mu+n)$ in powers of α but it is not sufficient. Abel finally gets $\lim \sum_{m=1}^{n} \sum_{\mu=1}^{n} \psi(m,\mu) = \sum_{m=1}^{\infty} \sum_{\mu=1}^{\infty} \theta(m,\mu)$. He deals in the same way with the simple products in (83) and obtains

$$\varphi\alpha = \alpha \prod_{m=1}^{\infty}\left(1-\frac{\alpha^2}{(m\omega)^2}\right)\prod_{\mu=1}^{\infty}\left(1+\frac{\alpha^2}{(\mu\varpi)^2}\right)$$

$$\times \prod_{m=1}^{\infty}\left(\prod_{\mu=1}^{\infty}\frac{1-\frac{\alpha^2}{(m\omega+\mu\varpi i)^2}}{1-\frac{\alpha^2}{\left(\left(m-\frac{1}{2}\right)\omega+\left(\mu-\frac{1}{2}\right)\varpi i\right)^2}}\prod_{\mu=1}^{\infty}\frac{1-\frac{\alpha^2}{(m\omega-\mu\varpi i)^2}}{1-\frac{\alpha^2}{\left(\left(m-\frac{1}{2}\right)\omega-\left(\mu-\frac{1}{2}\right)\varpi i\right)^2}}\right),$$

$$f\alpha = \prod_{m=1}^{\infty}\left(1-\frac{\alpha^2}{\left(m-\frac{1}{2}\right)^2\omega^2}\right)$$

$$\times \prod_{m=1}^{\infty}\prod_{\mu=1}^{\infty}\frac{1-\frac{\alpha^2}{\left(\left(m-\frac{1}{2}\right)\omega+\mu\varpi i\right)^2}}{1-\frac{\alpha^2}{\left(\left(m-\frac{1}{2}\right)\omega+\left(\mu-\frac{1}{2}\right)\varpi i\right)^2}}\,\frac{1-\frac{\alpha^2}{\left(\left(m-\frac{1}{2}\right)\omega-\mu\varpi i\right)^2}}{1-\frac{\alpha^2}{\left(\left(m-\frac{1}{2}\right)\omega-\left(\mu-\frac{1}{2}\right)\varpi i\right)^2}},$$

$$F\alpha = \prod_{\mu=1}^{\infty}\left(1+\frac{\alpha^2}{\left(\mu-\frac{1}{2}\right)^2\varpi^2}\right)$$

$$\times \prod_{m=1}^{\infty}\prod_{\mu=1}^{\infty}\frac{1-\frac{\alpha^2}{\left(m\omega+\left(\mu-\frac{1}{2}\right)\varpi i\right)^2}}{1-\frac{\alpha^2}{\left(\left(m-\frac{1}{2}\right)\omega+\left(\mu-\frac{1}{2}\right)\varpi i\right)^2}}\,\frac{1-\frac{\alpha^2}{\left(m\omega-\left(\mu-\frac{1}{2}\right)\varpi i\right)^2}}{1-\frac{\alpha^2}{\left(\left(m-\frac{1}{2}\right)\omega-\left(\mu-\frac{1}{2}\right)\varpi i\right)^2}}.$$

Abel also writes these formulae in a real form.

The Eulerian products for $\sin y$ and $\cos y$ lead to

$$\prod_{\mu=1}^{\infty}\frac{1-\frac{z^2}{\mu^2\pi^2}}{1-\frac{y^2}{\left(\mu-\frac{1}{2}\right)^2\pi^2}} = \frac{\sin z}{z\cos y} \quad\text{and}\quad \prod_{\mu=1}^{\infty}\frac{1-\frac{z^2}{\left(\mu-\frac{1}{2}\right)^2\pi^2}}{1-\frac{y^2}{\left(\mu-\frac{1}{2}\right)^2\pi^2}} = \frac{\cos z}{\cos y}$$

and this permits to transform the double products of Abel's formulae in simple products:

$$\varphi\alpha = \frac{\varpi}{\pi}\frac{\sin\left(\alpha\frac{\pi i}{\varpi}\right)}{i}\prod_{m=1}^{\infty}\left(1-\frac{\alpha^2}{m^2\omega^2}\right)$$

$$\times\prod_{m=1}^{\infty}\frac{\sin(\alpha+m\omega)\frac{\pi i}{\varpi}\sin(\alpha-m\omega)\frac{\pi i}{\varpi}\cos^2\left(m-\frac{1}{2}\right)\omega\frac{\pi i}{\varpi}}{\cos\left(\alpha+\left(m-\frac{1}{2}\right)\omega\right)\frac{\pi i}{\varpi}\cos\left(\alpha-\left(m-\frac{1}{2}\right)\omega\right)\frac{\pi i}{\varpi}\sin^2 m\omega\frac{\pi i}{\varpi}}$$

$$\times\frac{\left(m\omega\frac{\pi i}{\varpi}\right)^2}{(\alpha+m\omega)(\alpha-m\omega)\frac{\pi^2 i^2}{\varpi^2}}$$

$$=\frac{\varpi}{\pi}\frac{\sin\frac{\alpha}{\varpi}\pi i}{i}\prod_{m=1}^{\infty}\frac{1-\frac{\sin^2\alpha\frac{\pi}{\varpi}i}{\sin^2 m\omega\frac{\pi}{\varpi}i}}{1-\frac{\sin^2\frac{\pi}{\varpi}i}{\cos^2\left(m-\frac{1}{2}\right)\omega\frac{\pi}{\varpi}i}}$$

$$=\frac{1}{2}\frac{\varpi}{\pi}\left(h^{\frac{\alpha}{\varpi}\pi}-h^{-\frac{\alpha}{\varpi}\pi}\right)\prod_{m=1}^{\infty}\frac{1-\left(\frac{h^{\frac{\alpha}{\varpi}\pi}-h^{-\frac{\alpha}{\varpi}\pi}}{h^{m\frac{\omega}{\varpi}\pi}-h^{-m\frac{\omega}{\varpi}\pi}}\right)^2}{1+\left(\frac{h^{\frac{\alpha}{\varpi}\pi}-h^{-\frac{\alpha}{\varpi}\pi}}{h^{\left(m-\frac{1}{2}\right)\frac{\omega}{\varpi}\pi}-h^{-\left(m-\frac{1}{2}\right)\frac{\omega}{\varpi}\pi}}\right)}$$

$$=\frac{\omega}{\pi}\sin\frac{\alpha\pi}{\omega}\prod_{m=1}^{\infty}\frac{1+\frac{4\sin^2\frac{\alpha\pi}{\omega}}{\left(h^{\frac{m\varpi\pi}{\omega}}-h^{-\frac{m\varpi\pi}{\omega}}\right)^2}}{1-\frac{4\sin^2\frac{\alpha\pi}{\omega}}{\left(h^{\frac{(2m-1)\varpi\pi}{2\omega}}-h^{-\frac{(2m-1)\varpi\pi}{2\omega}}\right)^2}} \tag{85}$$

where $h = 2.712818\ldots$ is the basis of natural logarithms. In the same way, he obtains

$$F\alpha=\prod_{m=1}^{\infty}\frac{1+\frac{4\sin^2\frac{\alpha\pi}{\omega}}{\left(h^{\frac{(2m+1)\varpi\pi}{\omega}}-h^{-\frac{(2m+1)\varpi\pi}{\omega}}\right)^2}}{1-\frac{4\sin^2\frac{\alpha\pi}{\omega}}{\left(h^{\frac{(2m+1)\varpi\pi}{\omega}}-h^{-\frac{(2m+1)\varpi\pi}{\omega}}\right)^2}},$$

$$f\alpha=\cos\frac{\alpha\pi}{\omega}\prod_{m=1}^{\infty}\frac{1-\frac{4\sin^2\frac{\alpha\pi}{\omega}}{\left(h^{\frac{m\varpi\pi}{\omega}}+h^{-\frac{m\varpi\pi}{\omega}}\right)^2}}{1-\frac{4\sin^2\frac{\alpha\pi}{\omega}}{\left(h^{\frac{(2m-1)\varpi\pi}{2\omega}}+h^{-\frac{(2m-1)\varpi\pi}{2\omega}}\right)^2}}.$$

These expansions were known to Gauss and they were independently discovered by Jacobi, who used a passage to the limit in the formulae of transformation for the elliptic functions.

The expansion of $\frac{1}{\text{chy}}$ in simple fractions gives

$$\sum_{\mu=1}^{\infty}(-1)^{\mu}\frac{(2\mu+1)\varpi}{\left(\alpha\pm\left(m+\frac{1}{2}\right)\omega\right)^2+\left(\mu+\frac{1}{2}\right)^2\varpi^2}$$

$$=\frac{2\pi}{\varpi}\frac{1}{h^{\left(\alpha\pm\left(m+\frac{1}{2}\right)\omega\right)\frac{\pi}{\varpi}}+h^{-\left(\alpha\pm\left(m+\frac{1}{2}\right)\omega\right)\frac{\pi}{\varpi}}}$$

which permits to transform the formulae (84) in simple series. Thus

$$\varphi\alpha = \frac{2}{ec}\frac{\pi}{\varpi}\sum_{m=0}^{\infty}(-1)^m\left(\frac{1}{h^{\left(\alpha-\left(m+\frac{1}{2}\right)\omega\right)\frac{\pi}{\varpi}}+h^{-\left(\alpha-\left(m+\frac{1}{2}\right)\omega\right)\frac{\pi}{\varpi}}}\right.$$
$$\left.-\frac{1}{h^{\left(\alpha+\left(m+\frac{1}{2}\right)\omega\right)\frac{\pi}{\varpi}}+h^{-\left(\alpha+\left(m+\frac{1}{2}\right)\omega\right)\frac{\pi}{\varpi}}}\right)$$
$$= \frac{2}{ec}\frac{\pi}{\varpi}\sum_{m=0}^{\infty}(-1)^m\frac{\left(h^{\frac{\alpha\pi}{\varpi}}-h^{-\frac{\alpha\pi}{\varpi}}\right)\left(h^{\left(m+\frac{1}{2}\right)\frac{\omega\pi}{\varpi}}-h^{-\left(m+\frac{1}{2}\right)\frac{\omega\pi}{\varpi}}\right)}{h^{\frac{2\alpha\pi}{\varpi}}+h^{-\frac{2\alpha\pi}{\varpi}}+h^{(2m+1)\frac{\omega\pi}{\varpi}}+h^{-(2m+1)\frac{\omega\pi}{\varpi}}}$$
$$= \frac{4}{ec}\frac{\pi}{\omega}\sum_{m=0}^{\infty}(-1)^m\frac{\sin\frac{\alpha\pi}{\varpi}\cdot\left(h^{\left(m+\frac{1}{2}\right)\frac{\varpi\pi}{\omega}}-h^{-\left(m+\frac{1}{2}\right)\frac{\varpi\pi}{\omega}}\right)}{h^{(2m+1)\frac{\varpi\pi}{\omega}}+2\cos 2\alpha\frac{\pi}{\omega}+h^{-(2m+1)\frac{\varpi\pi}{\omega}}} \tag{86}$$

and

$$F\alpha = \frac{2}{c}\frac{\pi}{\varpi}\sum_{m=0}^{\infty}\frac{\left(h^{\frac{\alpha\pi}{\varpi}}+h^{-\frac{\alpha\pi}{\varpi}}\right)\left(h^{\left(m+\frac{1}{2}\right)\frac{\omega\pi}{\varpi}}+h^{-\left(m+\frac{1}{2}\right)\frac{\omega\pi}{\varpi}}\right)}{h^{\frac{2\alpha\pi}{\varpi}}+h^{-\frac{2\alpha\pi}{\varpi}}+h^{(2m+1)\frac{\omega\pi}{\varpi}}+h^{-(2m+1)\frac{\omega\pi}{\varpi}}},$$
$$f\alpha = \frac{4}{e}\frac{\pi}{\omega}\sum_{m=0}^{\infty}\frac{\cos\frac{\alpha\pi}{\omega}\cdot\left(h^{\left(m+\frac{1}{2}\right)\frac{\varpi\pi}{\omega}}+h^{-\left(m+\frac{1}{2}\right)\frac{\varpi\pi}{\omega}}\right)}{h^{(2m+1)\frac{\varpi\pi}{\omega}}+2\cos 2\alpha\frac{\pi}{\omega}+h^{-(2m+1)\frac{\varpi\pi}{\omega}}}.$$

In the lemniscatic case, where $e = c = 1$, one has $\omega = \varpi$ and these expansions take a simpler form

$$\varphi\left(\alpha\frac{\omega}{2}\right) = 2\frac{\pi}{\omega}\left(\frac{h^{\frac{\alpha\pi}{2}}-h^{-\frac{\alpha\pi}{2}}}{h^{\frac{\pi}{2}}+h^{-\frac{\pi}{2}}}-\frac{h^{\frac{3\alpha\pi}{2}}-h^{-\frac{3\alpha\pi}{2}}}{h^{\frac{3\pi}{2}}+h^{-\frac{3\pi}{2}}}+\frac{h^{\frac{5\alpha\pi}{2}}-h^{-\frac{5\alpha\pi}{2}}}{h^{\frac{5\pi}{2}}+h^{-\frac{5\pi}{2}}}-\dots\right)$$
$$= \frac{4\pi}{\omega}\left(\sin\left(\alpha\frac{\pi}{2}\right)\frac{h^{\frac{\pi}{2}}}{1+h^{\pi}}-\sin\left(3\alpha\frac{\pi}{2}\right)\frac{h^{\frac{3\pi}{2}}}{1+h^{3\pi}}+\sin\left(5\alpha\frac{\pi}{2}\right)\frac{h^{\frac{5\pi}{2}}}{1+h^{5\pi}}-\dots\right),$$
$$F\left(\alpha\frac{\omega}{2}\right) = 2\frac{\pi}{\omega}\left(\frac{h^{\frac{\alpha\pi}{2}}+h^{-\frac{\alpha\pi}{2}}}{h^{\frac{\pi}{2}}-h^{-\frac{\pi}{2}}}-\frac{h^{\frac{3\alpha\pi}{2}}+h^{-\frac{3\alpha\pi}{2}}}{h^{\frac{3\pi}{2}}-h^{-\frac{3\pi}{2}}}+\frac{h^{\frac{5\alpha\pi}{2}}+h^{-\frac{5\alpha\pi}{2}}}{h^{\frac{5\pi}{2}}-h^{-\frac{5\pi}{2}}}-\dots\right),$$
$$f\left(\alpha\frac{\omega}{2}\right) = \frac{4\pi}{\omega}\left(\cos\left(\alpha\frac{\pi}{2}\right)\frac{h^{\frac{\pi}{2}}}{h^{\pi}-1}-\cos\left(3\alpha\frac{\pi}{2}\right)\frac{h^{\frac{3\pi}{2}}}{h^{3\pi}-1}+\cos\left(5\alpha\frac{\pi}{2}\right)\frac{h^{\frac{5\pi}{2}}}{h^{5\pi}-1}-\dots\right)$$

and, taking $\alpha = 0$, $\frac{\omega}{2} = 2\pi\left(\frac{h^{\frac{\pi}{2}}}{h^{\pi}-1}-\frac{h^{\frac{3\pi}{2}}}{h^{3\pi}-1}+\frac{h^{\frac{5\pi}{2}}}{h^{5\pi}-1}-\dots\right) = \int_0^1\frac{dx}{\sqrt{1-x^4}}$,

$$\frac{\omega^2}{4} = \pi^2\left(\frac{h^{\frac{\pi}{2}}}{h^{\pi}+1}-3\frac{h^{\frac{3\pi}{2}}}{h^{3\pi}+1}+5\frac{h^{\frac{5\pi}{2}}}{h^{5\pi}+1}-\dots\right).$$

The second part of Abel's memoir, beginning with paragraph VIII (p. 352–362), was published in 1828. This paragraph is devoted to the algebraic solution of the equation $P_n = 0$ which gives $\varphi\left(\frac{\omega}{n}\right)$ in the lemniscatic case and for n a prime number of the form $4\nu + 1$. Abel announces that there is an infinity of other cases where the equation $P_n = 0$ is solvable by radicals.

Here, by the addition theorem (73), $\varphi(m+\mu i)\delta = \frac{\varphi(m\delta)f(\mu\delta)F(\mu\delta)+i\varphi(\mu\delta)f(m\delta)F(m\delta)}{1-\varphi^2(m\delta)\varphi^2(\mu\delta)}$ $= \varphi\delta T$ where T is a rational function of $(\varphi\delta)^2$ because of the formulae of multiplication. One says that there exists a *complex multiplication*. Putting $\varphi\delta = x$, we have $\varphi(m+\mu i)\delta = x\psi(x^2)$. Now $\varphi(\delta i) = i\varphi\delta = ix$ and $\varphi(m+\mu i)i\delta = i\varphi(m+\mu i)\delta = ix\psi(-x^2)$ and this shows that $\psi(-x^2) = \psi(x^2)$. In other words, ψ is an even function and T is a rational function of x^4. For instance

$$\varphi(2+i)\delta = \frac{\varphi(2\delta)f\delta \cdot F\delta + i\varphi\delta \cdot f(2\delta)F(2\delta)}{1-(\varphi 2\delta)^2\varphi^2\delta},$$

where $\varphi(2\delta) = \frac{2x\sqrt{1-x^4}}{1+x^4}$, $f\delta = \sqrt{1-x^2}$, $F\delta = \sqrt{1+x^2}$, $f(2\delta) = \frac{1-2x^2-x^4}{1+x^4}$ and $F(2\delta) = \frac{1+2x^2-x^4}{1+x^4}$. Thus $\varphi(2+i)\delta = xi\frac{1-2i-x^4}{1-(1-2i)x^4}$. Gauss had already discovered the complex multiplication of lemniscatic functions and the fact that it made possible the algebraic solution of the division of the periods. He made an allusion to this fact in the introduction to the seventh section of his *Disquisitiones arithmeticae*, but never publish anything on the subject. We have explained this algebraic solution in our §3.

The ninth paragraph of Abel's memoir (p. 363–377) deals with the transformation of elliptic functions. The transformation of order 2 was known since Landen (1775) and Lagrange (1784) and Legendre made an extensive suty of it in his *Exercices de calcul integral*. Later, in 1824, Legendre discovered another transformation, of order 3, which Jacobi rediscovered in 1827 together with a new transformation, of order 5. Then Jacobi announced the existence of transformations of any orders, but he was able to prove this existence only in 1828, using the idea of inversion of the elliptic integrals which came from Abel. Independently from Jacobi, Abel built the theory of transformations. Here is his statement:

"If one designates by α the quantity $\frac{(m+\mu)\omega+(m-\mu)\varpi i}{2n+1}$, where at least one of the two integers m and μ is relatively prime with $2n+1$, one has

$$\int \frac{dy}{\sqrt{(1-c_1^2y^2)(1+e_1^2y^2)}} = \pm a \int \frac{dx}{\sqrt{(1-c^2x^2)(1+e^2x^2)}} \qquad (87)$$

where $y = f\cdot x\frac{(\varphi^2\alpha-x^2)(\varphi^2 2\alpha-x^2)\cdots(\varphi^2 n\alpha-x^2)}{(1+e^2c^2\varphi^2\alpha\cdot x^2)(1+e^2c^2\varphi^2 2\alpha\cdot x^2)\cdots(1+e^2c^2\varphi^2 n\alpha\cdot x^2)}$,

$$\frac{1}{c_1} = \frac{f}{c}\left(\varphi\left(\frac{\omega}{2}+\alpha\right)\varphi\left(\frac{\omega}{2}+2\alpha\right)\cdots\varphi\left(\frac{\omega}{2}+n\alpha\right)\right)^2,$$

$$\frac{1}{e_1} = \frac{f}{e}\left(\varphi\left(\frac{\varpi i}{2}+\alpha\right)\varphi\left(\frac{\varpi i}{2}+2\alpha\right)\cdots\varphi\left(\frac{\varpi i}{2}+n\alpha\right)\right)^2,$$

$$a = f(\varphi\alpha\cdot\varphi 2\alpha\cdot\varphi 3\alpha\cdots\varphi n\alpha)^2.\text{"} \qquad (88)$$

Here f is indeterminate and e^2, c^2 may be positive or negative. By (80) (periodicity), we have $\varphi(\theta + (2n+1)\alpha) = \varphi\theta$ or $\varphi(\theta + (n+1)\alpha) = \varphi(\theta - n\alpha)$. Now if

$$\varphi_1\theta = \varphi\theta + \varphi(\theta+\alpha) + \ldots + \varphi(\theta + 2n\alpha),$$

we have $\varphi_1(\theta+\alpha) = \varphi_1\theta$ and $\varphi_1\theta$ admits the period α. This function may be written

$$\begin{aligned}\varphi_1\theta =& \varphi\theta + \varphi(\theta+\alpha) + \varphi(\theta-\alpha) + \varphi(\theta+2\alpha) + \varphi(\theta-2\alpha)\\ &+ \ldots + \varphi(\theta+n\alpha) + \varphi(\theta-n\alpha)\\ =& \varphi\theta + \frac{2\varphi\theta\cdot f\alpha\cdot F\alpha}{1+e^2c^2\varphi^2\alpha\cdot\varphi^2\theta} + \frac{2\varphi\theta\cdot f2\alpha\cdot F2\alpha}{1+e^2c^2\varphi^2 2\alpha\cdot\varphi^2\theta}\\ &+ \ldots + \frac{2\varphi\theta\cdot fn\alpha\cdot Fn\alpha}{1+e^2c^2\varphi^2 n\alpha\cdot\varphi^2\theta},\end{aligned} \tag{89}$$

a rational function ψx of $x = \varphi\theta$. Note that the auxiliary function $\varphi_1\theta$ used to solve the equation of division in the first part was precisely of this type (see §3).

For any ε, $R = \left(1 - \frac{\psi x}{\varphi_1\varepsilon}\right)(1+e^2c^2\varphi^2\alpha x^2)\ldots(1+e^2c^2\varphi^2 n\alpha x^2)$ is a polynomial of degree $2n+1$ in x. It is annihilated by $x = \varphi\varepsilon$ and so by $x = \varphi(\varepsilon + m\alpha)$, m any integer. Since $\varphi\varepsilon$, $\varphi(\varepsilon+\alpha)$, $\varphi(\varepsilon+2\alpha)$, ... , $\varphi(\varepsilon+2n\alpha)$ are all different, they are the roots of R and

$$R = A\left(1 - \frac{x}{\varphi\varepsilon}\right)\left(1 - \frac{x}{\varphi(\varepsilon+\alpha)}\right)\cdots\left(1 - \frac{x}{\varphi(\varepsilon+2n\alpha)}\right) \tag{90}$$

where A is found to be 1 by making $x = 0$. Multiplying by $\varphi\varepsilon$ and then making $\varepsilon = 0$, we obtain

$$\begin{aligned}\psi x &= gx\frac{\left(1-\frac{x}{\varphi\alpha}\right)\left(1-\frac{x}{\varphi 2\alpha}\right)\cdots\left(1-\frac{x}{\varphi 2n\alpha}\right)}{(1+e^2c^2\varphi^2\alpha\cdot x^2)\cdots(1+e^2c^2\varphi^2 n\alpha\cdot x^2)}\\ &= gx\frac{\left(1-\frac{x^2}{\varphi^2\alpha}\right)\left(1-\frac{x^2}{\varphi^2 2\alpha}\right)\cdots\left(1-\frac{x^2}{\varphi^2 2n\alpha}\right)}{(1+e^2c^2\varphi^2\alpha\cdot x^2)\cdots(1+e^2c^2\varphi^2 n\alpha\cdot x^2)}\end{aligned} \tag{91}$$

where $g = 1 + 2f\alpha\cdot F\alpha + 2f2\alpha\cdot F2\alpha + \ldots 2fn\alpha\cdot Fn\alpha$ is the value of $\frac{\varphi_1\varepsilon}{\varphi\varepsilon}$ for $\varepsilon = 0$.

Doing $\varepsilon = \frac{\omega}{2}$ in R, we have

$$\begin{aligned}&1 - \frac{\psi x}{\varphi_1\frac{\omega}{2}}\\ &= \left(1-\frac{x}{\varphi\frac{\omega}{2}}\right)\left(1-\frac{x}{\varphi\left(\frac{\omega}{2}+\alpha\right)}\right)\left(1-\frac{x}{\varphi\left(\frac{\omega}{2}+2\alpha\right)}\right)\cdots\left(1-\frac{x}{\varphi\left(\frac{\omega}{2}+2n\alpha\right)}\right)\frac{1}{\rho}\\ &= (1-cx)\left(1-\frac{x}{\varphi\left(\frac{\omega}{2}+\alpha\right)}\right)^2\left(1-\frac{x}{\varphi\left(\frac{\omega}{2}+2\alpha\right)}\right)^2\cdots\left(1-\frac{x}{\varphi\left(\frac{\omega}{2}+n\alpha\right)}\right)^2\frac{1}{\rho}\end{aligned}$$

where $\rho = (1+e^2c^2\varphi^2\alpha\cdot x^2)(1+e^2c^2\varphi^2 2\alpha\cdot x^2)\cdots(1+e^2c^2\varphi^2 n\alpha\cdot x^2)$. Changing x in $-x$, we have

$$1+\frac{\psi x}{\varphi_1 \frac{\omega}{2}} = (1+cx)\left(1+\frac{x}{\varphi\left(\frac{\omega}{2}+\alpha\right)}\right)^2\left(1+\frac{x}{\varphi\left(\frac{\omega}{2}+2\alpha\right)}\right)^2$$
$$\times\cdots\left(1+\frac{x}{\varphi\left(\frac{\omega}{2}+n\alpha\right)}\right)^2\frac{1}{\rho}.$$

Abel puts $y = k\psi x$, $c_1 = \frac{1}{k\varphi_1\frac{\omega}{2}}$ (k a constant),

$$t = \left(1-\frac{x}{\varphi\left(\frac{\omega}{2}+\alpha\right)}\right)\left(1-\frac{x}{\varphi\left(\frac{\omega}{2}+2\alpha\right)}\right)\cdots\left(1-\frac{x}{\varphi\left(\frac{\omega}{2}+n\alpha\right)}\right),$$
$$t_1 = \left(1+\frac{x}{\varphi\left(\frac{\omega}{2}+\alpha\right)}\right)\left(1+\frac{x}{\varphi\left(\frac{\omega}{2}+2\alpha\right)}\right)\cdots\left(1+\frac{x}{\varphi\left(\frac{\omega}{2}+n\alpha\right)}\right)$$

in order to have $1-c_1y = (1-cx)\frac{t^2}{\rho}$ and $1+c_1y = (1+cx)\frac{t_1^2}{\rho}$.

In the same way, $1 \mp e_1iy = (1-eix)\frac{s_1^2}{\rho}$, $1 \pm e_1iy = (1+eix)\frac{s^2}{\rho}$ where $e_1 = \pm\frac{i}{k\varphi_1\left(\frac{\varpi}{2}i\right)}$ and

$$s = \left(1-\frac{x}{\varphi\left(\frac{\varpi}{2}i+\alpha\right)}\right)\left(1-\frac{x}{\varphi\left(\frac{\varpi}{2}i+2\alpha\right)}\right)\cdots\left(1-\frac{x}{\varphi\left(\frac{\varpi}{2}i+n\alpha\right)}\right),$$
$$s_1 = \left(1+\frac{x}{\varphi\left(\frac{\varpi}{2}i+\alpha\right)}\right)\left(1+\frac{x}{\varphi\left(\frac{\varpi}{2}i+2\alpha\right)}\right)\cdots\left(1+\frac{x}{\varphi\left(\frac{\varpi}{2}i+n\alpha\right)}\right).$$

Thus $\sqrt{(1-c_1^2y^2)(1+e_1^2y^2)} = \pm\frac{tt_1ss_1}{\rho^2}\sqrt{(1-c^2x^2)(1+e^2x^2)}$. Now $dy = \frac{P}{\rho^2}dx$ where P is a polynomial of degree $4n$. Differentiating $1-c_1y = (1-cx)\frac{t^2}{\rho}$, we see that

$$P = \frac{t}{c_1}\left(ct\rho - (1-cx)\left(2\rho\frac{dt}{dx} - t\frac{d\rho}{dx}\right)\right)$$

is divisible by t and, in the same manner, it is divisible by t_1, s and s_1. Since these four polynomials of degree n cannot have any common factor, it results that $\frac{P}{tt_1ss_1}$ is a constant a and that $\frac{dy}{\sqrt{(1-c_1^2y^2)(1+e_1^2y^2)}} = \pm a\frac{dx}{\sqrt{(1-c^2x^2)(1+e^2x^2)}}$. When $x=0$, $t=t_1=s=s_1=1=\rho$ and $P = \frac{dy}{dx} = k\psi'(0) = kg$.

According to (90), the coefficient of x^{2n+1} in R is $-\frac{1}{\varphi\varepsilon\cdot\varphi(\varepsilon+\alpha)\cdots\varphi(\varepsilon+n\alpha)}$ and, comparing with (91) it is equal to $-\frac{(-1)^n}{\varphi_1\varepsilon}\frac{g}{(\varphi\alpha\cdot\varphi2\alpha\cdots\varphi n\alpha)^2}$. Thus

$$\varphi_1\varepsilon = \frac{(-1)^n g}{(\varphi\alpha\cdot\varphi2\alpha\cdots\varphi n\alpha)^2}\varphi\varepsilon\cdot\varphi(\varepsilon+\alpha)\cdots\varphi(\varepsilon+n\alpha).$$

Now (89) and (91) give two expressions for the limit of $\frac{\psi x}{x}$ for x infinite:

$$1 \text{ and } \frac{g(-1)^n}{(ce)^{2n}(\varphi\alpha\cdot\varphi 2\alpha\cdots\varphi n\alpha)^4}.$$

Thus, by comparison, $g=(-1)^n(ec)^{2n}(\varphi\alpha\varphi 2\alpha\cdots\varphi n\alpha)^4$ and

$$\varphi_1\varepsilon=(ec)^{2n}(\varphi\alpha\varphi 2\alpha\cdots\varphi n\alpha)^2\varphi\varepsilon\varphi(\varepsilon+\alpha)\cdots\varphi(\varepsilon+2n\alpha).$$

In particular

$$\varphi_1\left(\frac{\omega}{2}\right)=\frac{1}{kc_1}=(ec)^{2n}\delta^2\varphi\left(\frac{\omega}{2}\right)\varphi\left(\frac{\omega}{2}+\alpha\right)\cdots\varphi\left(\frac{\omega}{2}+2n\alpha\right),$$
$$\varphi_1\left(\frac{\varpi}{2}i\right)=\frac{\pm i}{ke_1}=(ec)^{2n}\delta^2\varphi\left(\frac{\varpi}{2}i\right)\varphi\left(\frac{\varpi}{2}i+\alpha\right)\cdots\varphi\left(\frac{\varpi}{2}i+2n\alpha\right)$$

where $\delta=\varphi\alpha\cdot\varphi 2\alpha\cdots\varphi n\alpha g$ The values (88) of the statement for $\frac{1}{c_1}$ and $\frac{1}{e_1}$ result if we put $f=k(e^2c^2)^n\delta^2$ and, as $\varphi\left(\frac{\omega}{2}+\alpha\right)\varphi\left(\frac{\varpi}{2}i+\alpha\right)=\frac{i}{ec}$ (cf. (76)), we obtain $c_1e_1=\pm\frac{(-1)^n(ec)^{2n+1}}{f^2}$. On the other hand

$$\pm\frac{e_1}{c_1}=(-1)^n\frac{e}{c}(ec)^{2n}\left(\varphi\left(\frac{\omega}{2}+\alpha\right)\varphi\left(\frac{\omega}{2}+2\alpha\right)\cdots\varphi\left(\frac{\omega}{2}+n\alpha\right)\right)^4,$$
$$\pm\frac{c_1}{e_1}=(-1)^n\frac{c}{e}(ec)^{2n}\left(\varphi\left(\frac{\varpi}{2}i+\alpha\right)\varphi\left(\frac{\varpi}{2}i+2\alpha\right)\cdots\varphi\left(\frac{\varpi}{2}i+n\alpha\right)\right)^4$$

and $a=kg=(-1)^nf\cdot\delta^2$.

Using $\left(\varphi\left(\frac{\omega}{2}+\alpha\right)\right)^2=\frac{1}{c^2}\frac{1-c^2\varphi^2\alpha}{1+c^2\varphi^2\alpha}$ and $\left(\varphi\left(\frac{\varpi}{2}i+\alpha\right)\right)^2=-\frac{1}{e^2}\frac{1+e^2\varphi^2\alpha}{1-e^2\varphi^2\alpha}$ (cf. (76)), one transforms the expressions of $\frac{1}{c_1}$ and $\frac{1}{e_1}$ in rational symmetric functions of $\varphi\alpha$, $\varphi 2\alpha,\ldots,\varphi n\alpha$. Reasoning as in his §V for the equation $P_{2n+1}=0$, Abel deduces that, when $2n+1$ is a prime number, c_1 and e_1 are determined by an equation of degree $2n+2$ (the 'modular equation' as it was called later). Now such an equation has roots not necessarily real and Abel says that the theory must be extended to the case of moduli c, e complex numbers.

When c and e are real, the only values of α giving c_1 and e_1 real are $\frac{2m\omega}{2n+1}$ and $\frac{2\mu\varpi i}{2n+1}$. The first value gives

$$\frac{1}{c_1}=\frac{f}{c}\left(\varphi\left(\frac{1}{2n+1}\frac{\omega}{2}\right)\varphi\left(\frac{3}{2n+1}\frac{\omega}{2}\right)\cdots\varphi\left(\frac{2n-1}{2n+1}\frac{\omega}{2}\right)\right)^2,$$

$$\frac{e_1}{c_1}=\pm(-1)^n\frac{e}{c}(ec)^{2n}\left(\varphi\left(\frac{1}{2n+1}\frac{\omega}{2}\right)\varphi\left(\frac{3}{2n+1}\frac{\omega}{2}\right)\cdots\varphi\left(\frac{2n-1}{2n+1}\frac{\omega}{2}\right)\right)^4.$$

Abel explains in particular the case in which $c=c_1=1, \pm(-1)^n=1$ and $0<e<1$. Then e_1 is very small when $2n+1$ is large. Abel carefully studies the sign in (87). Since ρ^2 is positive for x real, this sign is that of $tt_1ss_1\sqrt{\frac{1-x^2}{1-y^2}}$.

Now ss_1 is easily seen to be positive and the sign we are looking for is that of $tt_1 = \left(1 - \frac{x^2}{\varphi^2\left(\frac{1}{2n+1}\frac{\omega}{2}\right)}\right)\left(1 - \frac{x^2}{\varphi^2\left(\frac{3}{2n+1}\frac{\omega}{2}\right)}\right)\cdots\left(1 - \frac{x^2}{\varphi^2\left(\frac{2n-1}{2n+1}\frac{\omega}{2}\right)}\right)$ for the radical is positive. For instance, when $-\varphi\left(\frac{1}{2n+1}\frac{\omega}{2}\right) \leq x \leq \varphi\left(\frac{1}{2n+1}\frac{\omega}{2}\right)$ the sign is $+$ and we get $(-1)^n a = \frac{4n+2}{\omega}\int_0^1 \frac{dy}{\sqrt{(1-y^2)(1+e_1^2y^2)}}$ by doing $x = \varphi\left(\frac{1}{2n+1}\frac{\omega}{2}\right)$ to which corresponds $y = (-1)^n$. If we neglect e_1^2, this gives approximately $(-1)^n a = (2n+1)\frac{\pi}{\omega}$ and

$$\int_0^x \frac{dx}{\sqrt{(1-x^2)(1+e^2x^2)}} = \frac{(-1)^n\omega}{(2n+1)\pi} \arcsin y$$

for $y = (-1)^n(2n+1)\frac{\pi}{\omega}x\frac{\left(1-\frac{x^2}{\varphi^2\left(\frac{\omega}{2n+1}\right)}\right)\cdots\left(1-\frac{x^2}{\varphi^2\left(\frac{n\omega}{2n+1}\right)}\right)}{\left(1+e^2\varphi^2\left(\frac{\omega}{2n+1}\right)x^2\right)\cdots\left(1+e^2\varphi^2\left(\frac{n\omega}{2n+1}\right)x^2\right)}$. Abel also explains the effect of the other real transformation $\alpha = \frac{2\mu\varpi i}{2n+1}$. He states that every possible transformation is obtained by combining the transformations of order 2^k studied by Legendre with his new transformations. He will publish a proof of this statement in his *Précis d'une théorie des fonctions elliptiques* (1829).

The last paragraph of the *Recherches* (p. 377–388) is devoted to the study of the differential equation $\frac{dy}{\sqrt{(1-y^2)(1+\mu y^2)}} = a\frac{dx}{\sqrt{(1-x^2)(1+\mu x^2)}}$ and, in particular to the cases in which there is complex multiplication. Abel states two theorems: "I. Supposing a real and the equation algebraically integrable, it is necessary that a be a rational number."

"II. Supposing *a imaginary* and the equation *algebraically* integrable, it is necessary that a be of the form $m \pm \sqrt{-1}\cdot\sqrt{n}$ where m and n are rational numbers. In this case, the quantity μ is not arbitrary; it must satisfy an equation which has an infinity of roots, real and imaginary. Each value of μ satisfy to the question."

Here Abel only considers a particular case, that in which $e_1 = \frac{1}{e}$ for the first real transformation. Thus we have

$$\frac{dy}{\sqrt{(1-y^2)(1+e^2y^2)}} = a\sqrt{-1}\frac{dx}{\sqrt{(1-x^2)(1+e^2x^2)}} \text{ (changing } y \text{ in } \frac{ey}{i}\text{)}, \quad (92)$$

where $y = \pm\sqrt{-1}e^n x\frac{\left(\varphi^2\left(\frac{\omega}{2n+1}\right)-x^2\right)\cdots\left(\varphi^2\left(\frac{n\omega}{2n+1}\right)-x^2\right)}{\left(1+e^2\varphi^2\left(\frac{\omega}{2n+1}\right)x^2\right)\cdots\left(1+e^2\varphi^2\left(\frac{n\omega}{2n+1}\right)x^2\right)}$, e being determined by $1 = e^{n+1}\left(\varphi\left(\frac{1}{2n+1}\frac{\omega}{2}\right)\cdots\varphi\left(\frac{2n-1}{2n+1}\frac{\omega}{2}\right)\right)^2$ and a by $a = \pm\frac{1}{e}\left(\frac{\varphi\left(\frac{\omega}{2n+1}\right)\cdots\varphi\left(\frac{n\omega}{2n+1}\right)}{\varphi\left(\frac{1}{2n+1}\frac{\omega}{2}\right)\cdots\varphi\left(\frac{2n-1}{2n+1}\frac{\omega}{2}\right)}\right)^2$.

From (92), Abel deduces $\frac{\varpi}{2} = \int_0^{\frac{1}{e}} \frac{dz}{\sqrt{(1+z^2)(1-e^2z^2)}} = a\frac{\omega}{4n+2}$ (integration from $x = 0$ to $x = \varphi\left(\frac{\omega}{4n+2}\right)$, $y = z\sqrt{-1}$) and $\frac{\omega}{2} = a\frac{\varpi}{2}$. Thus $a = \sqrt{2n+1} = \frac{\omega}{\varpi}$. For instance, when $n = 1$, $\frac{dy}{\sqrt{(1-y^2)(1+e^2y^2)}} = \sqrt{-3}\frac{dx}{\sqrt{(1-x^2)(1+e^2x^2)}}$ where

$y = \sqrt{-1}ex\frac{\varphi^2(\frac{\omega}{3})-x^2}{1+e^2\varphi^2(\frac{\omega}{3})x^2}$,

$$1 = e^2\left(\varphi\left(\frac{1}{3}\frac{\omega}{2}\right)\right)^2 = \frac{e^2 - e^2\varphi^2\left(\frac{\omega}{3}\right)}{1 + e^2\varphi^2\left(\frac{\omega}{3}\right)} \text{ and } a = \frac{\varphi^2\left(\frac{\omega}{3}\right)}{\varphi^2\left(\frac{\omega}{6}\right)}\frac{1}{e} = \sqrt{3}.$$

This gives $\varphi\left(\frac{\omega}{3}\right) = \frac{\sqrt{3}}{e}$ and $e = \sqrt{3}+2$, $\varphi\left(\frac{\omega}{3}\right) = 2\sqrt{3}-3$. Changing x in $x\sqrt{2-\sqrt{3}}$ and y in $y\sqrt{2-\sqrt{3}}\sqrt{-1}$, Abel obtains $\frac{dy}{\sqrt{1-2\sqrt{3}y^2-y^4}} = \sqrt{3}\frac{dx}{\sqrt{1+2\sqrt{3}x^2-x^4}}$ where $y = x\frac{\sqrt{3}-x^2}{1+\sqrt{3}x^2}$.

For $n = 2$, $\frac{dy}{\sqrt{(1-y^2)(1+e^2y^2)}} = \sqrt{-5}\frac{dx}{\sqrt{(1-x^2)(1+e^2x^2)}}$ where $y = \sqrt{-1}e^2x\frac{\varphi^2(\frac{\omega}{5})-x^2}{1+e^2\varphi^2(\frac{\omega}{5})x^2}$ $\times\ \frac{\varphi^2\left(\frac{2\omega}{5}\right)-x^2}{1+e^2\varphi^2\left(\frac{2\omega}{5}\right).x^2}$, $1 = e^2\varphi^2\left(\frac{\omega}{10}\right)\varphi^2\left(\frac{3\omega}{10}\right)$, $\sqrt{5} = e^2\varphi^2\left(\frac{\omega}{5}\right)\varphi^2\left(\frac{2\omega}{5}\right)$. Using $\varphi^2\left(\frac{\omega}{10}\right) = \varphi^2\left(\frac{\omega}{2}-\frac{2\omega}{5}\right) = \frac{f^2\left(\frac{2\omega}{5}\right)}{F^2\left(\frac{2\omega}{5}\right)}$ and $\varphi^2\left(\frac{3\omega}{10}\right) = \varphi^2\left(\frac{\omega}{2}-\frac{\omega}{5}\right) = \frac{f^2\left(\frac{\omega}{5}\right)}{F^2\left(\frac{\omega}{5}\right)}$, Abel finally gets $-\frac{1}{e\sqrt{e}} = \frac{1}{e^2}\frac{1-e\sqrt{5}}{e-\sqrt{5}}$, which gives a cubic equation for e:

$$e^3 - 1 - (5+2\sqrt{5})e(e-1) = 0.$$

This equation has only one solution larger than 1, as e must be, $e = \left(\frac{\sqrt{5}+1}{2} + \sqrt{\frac{\sqrt{5}+1}{2}}\right)^2$. It is then easy to compute $\alpha = \varphi\left(\frac{\omega}{5}\right)$ and $\beta = \varphi\left(\frac{2\omega}{5}\right)$ for $\alpha^2\beta^2 = \frac{\sqrt{5}}{e^2}$ and

$$e^3 - 1 - e(e-1)\sqrt{5} = e^2(e+1)(\alpha^2+\beta^2).$$

Changing x in $\frac{x}{\sqrt{e}}$ and y in $\frac{y\sqrt{-1}}{\sqrt{e}}$, Abel obtains the equation

$$\frac{dy}{\sqrt{1-4\sqrt{2+\sqrt{5}}y^2-y^4}} = \sqrt{5}\frac{dx}{\sqrt{1+4\sqrt{2+\sqrt{5}}x^2-x^4}}$$

where $y = x\frac{\sqrt{5}-\sqrt{10+10\sqrt{5}x^2+x^4}}{1+\sqrt{10+10\sqrt{5}x^2+\sqrt{5}x^4}}$.

For higher orders n Abel says that the equation giving the singular modulus e is not necessarily algebraically solvable and he proposes an expansion of e in infinite series. He starts from (86) with $\alpha = \frac{\omega}{2}$, $\varphi\left(\frac{\omega}{2}\right) = \frac{1}{c} = 1$ and gets $e\omega = 4\pi\left(\frac{\rho}{\rho^2+1} + \frac{\rho^3}{\rho^6+1} + \frac{\rho^5}{\rho^{10}+1} + \dots\right)$ where $\rho = h^{\frac{\varpi}{\omega}\frac{\pi}{2}}$. With $\alpha = \frac{\varpi}{2}i$, $\varphi\left(\frac{\varpi}{2}i\right) = \frac{i}{e}$, Abel gets $\varpi = 4\pi\left(\frac{r}{r^2+1} + \frac{r^3}{r^6+1} + \frac{r^5}{r^{10}+1} + \dots\right)$ where $r = h^{\frac{\omega}{\varpi}\frac{\pi}{2}}$ and, since $\frac{\omega}{\varpi} = \sqrt{2n+1}$, $\omega = 4\pi\sqrt{2n+1}\left(\frac{h^{\frac{\pi}{2}\sqrt{2n+1}}}{h^{\pi\sqrt{2n+1}}+1} + \frac{h^{\frac{3\pi}{2}\sqrt{2n+1}}}{h^{3\pi\sqrt{2n+1}}+1} + \dots\right)$ and

$$e = \frac{4\pi}{\omega}\left(\frac{h^{\frac{\pi}{2}\frac{1}{\sqrt{2n+1}}}}{h^{\frac{\pi}{\sqrt{2n+1}}}+1} + \frac{h^{\frac{3\pi}{2}\frac{1}{\sqrt{2n+1}}}}{h^{\frac{3\pi}{\sqrt{2n+1}}}+1} + \dots\right).$$

At the end of this memoir, Abel explains how his theory of transformation gives the formulae published by Jacobi in 1827. Jacobi uses Legendre's notations, with a modulus k between 0 and 1 and the elliptic integral of the first kind $F(k,\theta) = \int_0^\theta \frac{d\theta}{\sqrt{1-k^2\sin^2\theta}}$, so that if $\alpha = F(k,\theta)$, $\varphi\alpha = \sin\theta$ where φ is Abel's elliptic function with $c = 1$ and $e^2 = -k^2$. Writing $c_1 = 1$, $e_1^2 = -\lambda^2$, $\mu = \frac{(-1)^n}{a}$, $x = (-1)^n \sin\theta$, $y = \sin\psi$ and $2n+1 = p$, Abel's formula for the first real transformation takes the form $\int \frac{d\theta}{\sqrt{1-k^2\sin^2\theta}} = \pm\mu \int \frac{d\psi}{\sqrt{1-\lambda^2\sin^2\psi}} + C$, where

$$\lambda = k^{2n+1}(\sin\theta' \cdot \sin\theta''' \ldots \sin\theta^{(2n-1)})^4, \qquad \mu = \left(\frac{\sin\theta' \cdot \sin\theta''' \cdots \sin\theta^{(2n-1)}}{\sin\theta'' \cdot \sin\theta'''' \cdots \sin\theta^{(2n)}}\right)^2$$

and $\sin\psi = \frac{k^{n+\frac{1}{2}}}{\sqrt{\lambda}} \sin\theta \frac{(\sin^2\theta''-\sin^2\theta)(\sin^2\theta''''-\sin^2\theta)\ldots(\sin^2\theta^{(2n)}-\sin^2\theta)}{(1-k^2\sin^2\theta''\cdot\sin^2\theta)(1-k^2\sin^2\theta''''\cdot\sin^2\theta)\ldots(1-k^2\sin^2\theta^{(2n)}\cdot\sin^2\theta)}$, the angles $\theta', \theta'', \ldots, \theta^{(2n)}$ being defined by $\sin\theta^{(m)} = \varphi\left(\frac{m}{p}\frac{\omega}{2}\right)$ or $F(k,\theta^{(m)}) = \frac{m}{p}\frac{\omega}{2}$. Since $c = c_1 = 1$, $\sqrt{\frac{1-y}{1+y}} = \frac{t}{t_1}\sqrt{\frac{1-x}{1+x}}$ or $\sqrt{\frac{1-\sin\psi}{1+\sin\psi}} = \sqrt{\frac{1-(-1)^n\sin\theta}{1+(-1)^n\sin\theta}} \frac{\sin\theta'-\sin\theta}{\sin\theta'+\sin\theta} \frac{\sin\theta'''+\sin\theta}{\sin\theta'''-\sin\theta} \cdots \frac{\sin\theta^{(2n-1)}+(-1)^n\sin\theta}{\sin\theta^{(2n-1)}-(-1)^n\sin\theta}$, relation which may be transformed in

$$\tan\left(45^\circ - \frac{1}{2}\psi\right) = \frac{\tan\frac{1}{2}(\theta'-\theta)}{\tan\frac{1}{2}(\theta'+\theta)} \cdot \frac{\tan\frac{1}{2}(\theta'''+\theta)}{\tan\frac{1}{2}(\theta'''-\theta)} \cdots$$
$$\times \frac{\tan\frac{1}{2}(\theta^{(2n-1)}+(-1)^n\theta)}{\tan\frac{1}{2}(\theta^{(2n-1)}-(-1)^n\theta)} \tan\left(45^\circ - (-1)^n \frac{1}{2}\theta\right).$$

In 1828, Abel had begun the redaction of a second memoir to continue the *Recherches sur les fonctions elliptiques* (*Œuvres*, t. II, p. 244–253). Putting $\alpha = \frac{(m+\mu)\omega+(m-\mu)\varpi i}{2n+1}$ where m, μ and n are integers such that $m+\mu$, $m-\mu$ and $2n+1$ have no common divisor, and

$$\begin{aligned}\varphi_1\theta &= \varphi\theta \cdot \varphi(\alpha+\theta)\varphi(\alpha-\theta)\varphi(2\alpha+\theta)\varphi(2\alpha-\theta)\cdots\varphi(n\alpha+\theta)\varphi(n\alpha-\theta)\\ &= \varphi\theta \frac{\varphi^2\alpha-\varphi^2\theta}{1+e^2c^2\varphi^2\alpha\cdot\varphi^2\theta} \frac{\varphi^2 2\alpha-\varphi^2\theta}{1+e^2c^2\varphi^2 2\alpha\cdot\varphi^2\theta} \cdots \frac{\varphi^2 n\alpha-\varphi^2\theta}{1+e^2c^2\varphi^2 n\alpha\cdot\varphi^2\theta},\end{aligned}$$

Abel remarks that this function is rational in $\varphi\theta$ and invariant by $\theta \mapsto \theta \pm \alpha$. It results that the roots of the equation

$$\begin{aligned}0 &= x(\varphi^2\alpha - x^2)(\varphi^2 2\alpha - x^2)\cdots(\varphi^2 n\alpha - x^2)\\ &\quad -\varphi_1\theta(1+e^2c^2\varphi^2\alpha x^2)(1+e^2c^2\varphi^2 2\alpha x^2)\cdots(1+e^2c^2\varphi^2 n\alpha x^2)\end{aligned}$$

are $\varphi\theta, \varphi(\theta+\alpha), \ldots, \varphi(\theta+2n\alpha)$. Now let $\psi\theta$ be a rational function of these roots and suppose that it is invariant by $\theta \to \theta+\alpha$. Using the addition theorem (73), one sees that

$$\psi\theta = \psi_1\theta + \psi_2\theta \cdot f\theta \cdot F\theta$$

where $\psi_1\theta$ and $\psi_2\theta$ are rational in $\varphi\theta$ and $\psi_1\theta = \frac{1}{2}(\psi\theta + \prime\psi\theta)$, $\psi_2\theta \cdot f\theta \cdot F\theta = \frac{1}{2}(\psi\theta - \prime\psi\theta)$ where $\prime\psi\theta$ is the function deduced from $\psi\theta$ by changing α into $-\alpha$. One has $\prime\psi\theta = \psi_1\theta - \psi_2\theta \cdot f\theta \cdot F\theta$ and $\prime\psi\theta$ is invariant by $\theta \mapsto \theta + \alpha$. Thus $\psi_1\theta$ is invariant by $\theta \mapsto \theta + \alpha$ and it is thus a rational symmetric function of $\varphi\theta$, $\varphi(\theta+\alpha), \dots, \varphi(\theta + 2n\alpha)$, that is a rational function of $\varphi_1\theta$. In the same way, one shows that the square of $\psi_2\theta f\theta F\theta$ is a rational function of $\varphi_1\theta$, so that $\psi\theta = p \pm \sqrt{q'}$ where p, q' are rational functions of $\varphi_1\theta$. Let

$$\begin{aligned}\chi\theta = (\varphi\theta)^2\varphi(\theta+\alpha) + (\varphi(\theta+\alpha))^2\varphi(\theta+2\alpha) + \dots \\ +(\varphi(\theta+(2n-1)\alpha))^2\varphi(\theta+2n\alpha) + (\varphi(\theta+2n\alpha))^2\varphi\theta\end{aligned}$$

and let $\prime\chi\theta$ be the function deduced from $\chi\theta$ by changing α into $-\alpha$. One has

$$\chi\theta = \chi_1\theta + \chi_2\theta \cdot f\theta \cdot F\theta, \quad \prime\chi\theta = \chi_1\theta - \chi_2\theta \cdot f\theta \cdot F\theta$$

where $\chi_1\theta$ and $\chi_2\theta$ are rational functions of $\varphi\theta$, and $\frac{1}{2}(\chi\theta - \prime\chi\theta) = \chi_2\theta \cdot f\theta \cdot F\theta = \pm\sqrt{r}$ where r is a rational function of $\varphi_1\theta$. Now $\frac{\psi_2^\theta}{\chi_2^\theta} = \frac{\psi\theta - \prime\psi\theta}{\chi\theta - \prime\chi\theta}$ is a rational function of $\varphi\theta$ invariant by $\theta \to \theta+\alpha$, so a rational function q of $\varphi_1\theta$ and $\frac{1}{2}(\psi\theta - \prime\psi\theta) = \pm q\sqrt{r}$, $\psi\theta = p \pm q\sqrt{r}$ where r does not depend of the function $\psi\theta$.

Abel proves that r is a polynomial in $\varphi_1\theta$, for if it had a pole $\varphi_1\delta$, we should have $\chi\delta - \prime\chi\delta = \frac{1}{0}$ which means that some $\varphi(\delta \pm \nu\alpha)$ would be infinite, but then $\varphi_1\delta$ would also be infinite, which is absurd. The expansions of our functions in decreasing powers of $x = \varphi\theta$ are $\varphi_1\theta = ax + \varepsilon$, $\chi\theta - \prime\chi\theta = Ax^2 + \varepsilon'$ where a, A are constant and ε and ε' contain powers of x respectively less than 1 and 2. If ν is the degree of r, the equation $r = \frac{1}{4}(\chi\theta - \prime\chi\theta)^2$ is rewritten $a'x^\nu + \dots = \frac{1}{4}A^2x^4 + \dots$ and it shows that $\nu = 4$. Since r must be annihilated by $\theta = \pm\frac{\omega}{2}, \pm\frac{\varpi}{2}i$, one has $r = C\left(1 - \left(\frac{\varphi_1\theta}{\varphi_1\frac{\omega}{2}}\right)^2\right)\left(1 - \left(\frac{\varphi_1\theta}{\varphi_1\frac{\varpi}{2}i}\right)^2\right)$ where C is a constant.

When $\psi\theta$ is a polynomial in $\varphi\theta, \varphi(\theta+\alpha), \dots, \varphi(\theta+2n\alpha)$, the same reasoning shows that p and q are polynomials in $\varphi_1\theta$ of respective degrees ν and $\nu - 2$ where ν is the degree of $\psi\theta$ with respect to any one of the quantities $\varphi\theta$, $\varphi(\theta+\alpha), \dots, \varphi(\theta+2n\alpha)$. If $\nu = 1$, one has $\psi\theta = A + B\varphi_1\theta$ where A and B are constants respectively determined by making $\theta = 0$ and $\theta = \frac{1}{0}$. For instance, let us put $\pi\theta = \varphi\theta \cdot \varphi(\theta+\nu_1\alpha)\varphi(\theta+\nu_2\alpha)\dots\varphi(\theta+\nu_\omega\alpha)$ and

$$P = \pi(\theta) + \pi(\theta+\alpha) + \pi(\theta+2\alpha) + \dots + \pi(\theta+2n\alpha)$$

where $\nu_1, \nu_2, \dots, \nu_\omega$ are distinct integers less than $2n+1$. One has $A = \pi(\alpha) + \pi(2\alpha) + \dots + \pi(2n\alpha)$ and B is the derivative of P for $\theta = 0$. When ω is odd (resp. even), B (resp. A) is equal to 0, for instance $\omega = 0$ gives

$$\varphi\theta + \varphi(\theta+\alpha) + \varphi(\theta+2\alpha) + \dots + \varphi(\theta+2n\alpha) = B\varphi_1\theta$$

and $\omega = 1$ gives

$$\begin{aligned}&\varphi\theta \cdot \varphi(\theta+\alpha) + \varphi(\theta+\alpha)\varphi(\theta+2\alpha) + \dots + \varphi(\theta+2n\alpha)\varphi\theta \\ &= \varphi\alpha \cdot \varphi 2\alpha + \varphi 2\alpha \cdot \varphi 3\alpha + \dots + \varphi(2n-1)\alpha \cdot \varphi 2n\alpha.\end{aligned}$$

The second paragraph is not very explicit; Abel considers the functions

$$\psi\theta = \sum_{k=0}^{2n} \delta^{k\mu}\varphi(\theta + k\alpha),\ \psi_1\theta = \sum_{k=0}^{2n} \delta^{-k\mu}\varphi(\theta + k\alpha)$$

where δ is a primitive $(2n+1)$-th root of 1. Since $\psi(\theta+\alpha) = \delta^{2n\mu}\psi\theta$ and $\psi_1(\theta+\alpha) = \delta^{-2n\mu}\psi_1\theta$, the product $\psi\theta\cdot\psi_1\theta$ is invariant by $\theta \mapsto \theta+\alpha$. It is an even polynomial in the transformed elliptic integral $y = \varphi_1(a\theta)$, of the form $A(y^2 - f^2)$ where $f = \varphi_1\left(a\frac{m\varpi i}{2n+1}\right)$. Thus this product is 0 when $\theta = \frac{m\varpi i}{2n+1}$ and this gives a remarkable identity

$$0 = \varphi\left(\frac{m\varpi i}{2n+1}\right) + \delta^{\mu}\varphi\left(\frac{m\varpi i}{2n+1} + \alpha\right) + \delta^{2\mu}\varphi\left(\frac{m\varpi i}{2n+1} + 2\alpha\right) + \dots$$
$$+\delta^{2n\mu}\varphi\left(\frac{m\varpi i}{2n+1} + 2n\alpha\right)$$

for a convenient m. Abel has announced this type of identity in the introduction of the *Précis d'une théorie des fonctions elliptiques*, published in 1829 (see our §8); Sylow and Kronecker have proposed proofs for them.

7 Development of the Theory of Transformation of Elliptic Functions

The theory of transformation and of complex multiplication was developed by Abel in the paper *Solution d'un problème général concernant la transformation des fonctions elliptiques* (*Astronomische Nachrichten* (6) 138 and (7) 147, 1828; *Œuvres*, t. I, p. 403–443), published in the Journal where Jacobi had announced the formulae for transformation. Abel deals with the following problem: "To find all the possible cases in which the differential equation

$$\frac{dy}{\sqrt{(1-c_1^2y^2)(1-e_1^2y^2)}} = \pm a\frac{dx}{\sqrt{(1-c^2x^2)(1-e^2x^2)}} \tag{93}$$

may be satisfied by putting for y an algebraic function of x, rational or irrational." He explains that the problem may be reduced to the case in which y is a rational function of x and he begins by solving this case. His notations are $x = \lambda\theta$ when $\theta = \int_0 \frac{dx}{\sqrt{(1-c^2x^2)(1-e^2x^2)}}$, $\Delta\theta = \sqrt{(1-c^2x^2)(1-e^2x^2)}$, $\frac{\omega}{2} = \int_0^{\frac{1}{c}} \frac{dx}{\sqrt{(1-c^2x^2)(1-e^2x^2)}}$, $\frac{\omega'}{2} = \int_0^{\frac{1}{e}} \frac{dx}{\sqrt{(1-c^2x^2)(1-e^2x^2)}}$ where e and c may be complex numbers. Abel recalls the addition theorem $\lambda(\theta\pm\theta') = \frac{\lambda\theta\cdot\Delta\theta'\pm\lambda\theta'\cdot\Delta\theta}{1-c^2e^2\lambda^2\theta\cdot\lambda^2\theta'}$ and the solution of the equation $\lambda\theta' = \lambda\theta$, which is $\theta' = (-1)^{m+m'}\theta+m\omega+m'\omega'$. Let $y = \psi(x)$ be the rational function we are looking for and $x = \lambda\theta$, $x_1 = \lambda\theta_1$ two solutions of the equation $y = \psi(x)$, y being given (it is supposed that this equation is not of the first degree). From the equation $\frac{dy}{\sqrt{R}} = \pm a d\theta = \pm a d\theta_1$, we deduce $d\theta_1 = \pm d\theta$. Thus $\theta_1 = \alpha \pm \theta$ where α is constant and $x_1 = \lambda(\alpha \pm \theta)$, where we may choose the sign $+$, for $\lambda(\alpha - \theta) = \lambda(\omega - \alpha + \theta)$. Now $y = \psi(\lambda\theta) = \psi(\lambda(\theta + \alpha)) = \psi(\lambda(\theta + 2\alpha)) = \ldots = \psi(\lambda(\theta + k\alpha))$ for any integer k. As the equation $y = \psi(x)$ has only a finite number of roots, there exist k and k' distinct such that $\lambda(\theta + k\alpha) = \lambda(\theta + k'\alpha)$ or $\lambda(\theta + n\alpha) = \lambda\theta$ where $n = k - k'$ (supposed to be positive). Then $\theta + n\alpha = (-1)^{m+m'}\theta + m\omega + m'\omega'$ and, necessarily, $(-1)^{m+m'} = 1$, $n\alpha = m\omega + m'\omega'$ or $\alpha = \mu\omega + \mu'\omega'$ where μ, μ' are rational numbers. If the equation $y = \psi(x)$ has roots other than $\lambda(\theta + k\alpha)$, any one of them has the form $\lambda(\theta + \alpha_1)$ where $\alpha_1 = \mu_1\omega + \mu_1'\omega'$ (μ_1, μ_1' rational) and all the $\lambda(\theta + k\alpha + k_1\alpha_1)$ are roots of the equation. Continuing in this way, Abel finds that the roots of $y = \psi(x)$ are of the form

$$x = \lambda(\theta + k_1\alpha_1 + k_2\alpha_2 + \ldots + k_\nu\alpha_\nu)$$

where $k_1, k_2, \ldots, k_\nu$ are integers and $\alpha_1, \alpha_2, \ldots, \alpha_\nu$ of the form $\mu\omega + \mu'\omega'$ (μ, μ' rational). The problem is to determine y in function of θ, the quantities $\alpha_1, \alpha_2, \ldots, \alpha_\nu$ being given.

Before the solution of this problem, Abel deals with the case in which $y = \frac{f'+fx}{g'+gx}$. In this case $1 \pm c_1 y = \frac{g' \pm c_1 f' + (g \pm c_1 f)x}{g'+gx}$, $1 \pm e_1 y = \frac{g' \pm e_1 f' + (g \pm e_1 f)x}{g'+gx}$ and $dy = \frac{fg' - f'g}{(g'+gx)^2} dx$ so that the differential equation (93) takes the form

$$\frac{fg' - f'g}{\sqrt{(g'^2 - c_1^2 f'^2)(g'^2 - e_1^2 f'^2)}} \times \frac{dx}{\sqrt{\left(1 + \frac{g+c_1 f}{g'+c_1 f'}x\right)\left(1 + \frac{g-c_1 f}{g'-c_1 f'}x\right)\left(1 + \frac{g+e_1 f}{g'+e_1 f'}x\right)\left(1 + \frac{g-e_1 f}{g'-e_1 f'}x\right)}}$$
$$= \pm a \frac{dx}{\sqrt{(1 - c^2x^2)(1 - e^2x^2)}}.$$

The solutions are $y = ax$, $c_1^2 = \frac{c^2}{a^2}$, $e_1^2 = \frac{e^2}{a^2}$; $y = \frac{a}{ec}\frac{1}{x}$, $c_1^2 = \frac{c^2}{a^2}$, $e_1^2 = \frac{e^2}{a^2}$; $y = m\frac{1 - x\sqrt{ec}}{1 + x\sqrt{ec}}$, $c_1 = \frac{1}{m}\frac{\sqrt{c} - \sqrt{e}}{\sqrt{c} + \sqrt{e}}$, $e_1 = \frac{1}{m}\frac{\sqrt{c} + \sqrt{e}}{\sqrt{c} - \sqrt{e}}$, $a = \frac{m\sqrt{-1}}{2}(c - e)$.

In order to deal with the general case, in which the solutions of $y = \psi(x)$ are

$$\lambda\theta, \lambda(\theta + \alpha_1), \ldots, \lambda(\theta + \alpha_{m-1}),$$

Abel writes $\psi(x) = \frac{p}{q}$ where p and q are polynomials of degree m in x, with respective dominant coefficients f and g. The equation $y = \psi(x)$ is rewritten

$$p - qy = (f - gy)(x - \lambda\theta)(x - \lambda(\theta + \alpha_1)) \ldots (x - \lambda(\theta + \alpha_{m-1})). \qquad (94)$$

If f' and g' are the respective coefficients of x^{m-1} in p and q, we see that

$$f' - g'y = -(f - gy)(\lambda\theta + \lambda(\theta + \alpha_1) + \ldots + \lambda(\theta + \alpha_{m-1}))$$

and $y = \frac{f' + f\cdot\varphi\theta}{g' + g\cdot\varphi\theta}$ where $\varphi\theta = \lambda\theta + \lambda(\theta + \alpha_1) + \ldots + \lambda(\theta + \alpha_{m-1})$. It remains to express $\varphi\theta$ rationally in function of x with the help of the addition theorem and to determine f, f', g, g', e_1, c_1 and a in order that (93) be satisfied. For some α_j, it is possible that $\lambda(\theta - \alpha_j) = \lambda(\theta + \alpha_j)$ or $\lambda(\theta + 2\alpha_j) = \lambda\theta$. Then $\alpha_j = \frac{m}{2}\omega + \frac{m'}{2}\omega'$ with $m + m'$ even and the distinct values of $\lambda(\theta + \alpha_j)$ are $\lambda\theta = x$, $\lambda(\theta + \omega) = -\lambda\theta = -x$, $\lambda\left(\theta + \frac{\omega}{2} + \frac{\omega'}{2}\right) = -\frac{1}{ec}\frac{1}{\lambda\theta} = -\frac{1}{ec}\frac{1}{x}$, $\lambda\left(\theta + \frac{3\omega}{2} + \frac{\omega'}{2}\right) = -\frac{1}{ec}\frac{1}{\lambda(\theta+\omega)} = \frac{1}{ec}\frac{1}{x}$. For the other $\lambda(\theta - \alpha_j) \neq \lambda(\theta + \alpha_j)$, so it is a root of the equation $y = \psi(x)$, of the form $\lambda(\theta + \alpha_{j'})$ and we have $\lambda(\theta + \alpha_j) + \lambda(\theta - \alpha_j) = \frac{2x\Delta\alpha_j}{1 - e^2c^2\lambda^2\alpha_j x^2}$. Thus

$$\begin{aligned}\varphi\theta &= \lambda\theta + k\lambda(\theta + \omega) + k'\lambda\left(\theta + \frac{\omega}{2} + \frac{\omega'}{2}\right) + k''\lambda\left(\theta + \frac{3\omega}{2} + \frac{\omega'}{2}\right)\\ &\quad + \lambda(\theta + \alpha_1) + \lambda(\theta - \alpha_1) + \ldots + \lambda(\theta + \alpha_n) + \lambda(\theta - \alpha_n)\\ &= (1 - k)x + \frac{k'' - k'}{ec}\frac{1}{x} + \sum \frac{2x\Delta\alpha_j}{1 - e^2c^2\lambda^2\alpha_j \cdot x^2}\end{aligned}$$

where k, k', k'' are equal to 0 or 1.

In the first case considered by Abel, $k = k' = k'' = 0$. Let $\delta, \delta', \varepsilon, \varepsilon'$ be the values of θ respectively corresponding to $y = \frac{1}{c_1}, -\frac{1}{c_1}, \frac{1}{e_1}, -\frac{1}{e_1}$. One has $1 - c_1y = \frac{g' - c_1f'}{r}\left(1 - \frac{\varphi\theta}{\varphi\delta}\right)$, $1 + c_1y = \frac{g' + c_1f'}{r}\left(1 - \frac{\varphi\theta}{\varphi\delta'}\right)$, $1 - e_1y = \frac{g' - e_1f'}{r}\left(1 - \frac{\varphi\theta}{\varphi\varepsilon}\right)$ and $1 + e_1y = \frac{g' + e_1f'}{r}\left(1 - \frac{\varphi\theta}{\varphi\varepsilon'}\right)$ where $r = g' + g\cdot\varphi\theta$. From the expression of $\varphi\theta$, one gets $1 - \frac{\varphi\theta}{\varphi\delta} = \frac{1 + A_1x + A_2x^2 + \ldots + A_{2n+1}x^{2n+1}}{(1 - e^2c^2\lambda^2\alpha_1\cdot x^2)(1 - e^2c^2\lambda^2\alpha_2\cdot x^2)\ldots(1 - e^2c^2\lambda^2\alpha_n\cdot x^2)}$, which must be annihilated by $\theta = \delta, \delta \pm \alpha_1, \ldots, \delta \pm \alpha_n$ (δ arbitrary). Thus

$$\begin{aligned}1 + A_1x + \ldots + A_{2n+1}x^{2n+1} &= \left(1 - \frac{x}{\lambda\delta}\right)\left(1 - \frac{x}{\lambda(\delta + \alpha_1)}\right)\left(1 - \frac{x}{\lambda(\delta - \alpha_1)}\right)\ldots\\ &\quad \times\left(1 - \frac{x}{\lambda(\delta + \alpha_n)}\right)\left(1 - \frac{x}{\lambda(\delta - \alpha_n)}\right).\end{aligned}$$

The differential equation (93) is written

$$\sqrt{(1 - c_1^2y^2)(1 - e_1^2y^2)} = \frac{1}{a}\frac{dy}{dx}\sqrt{(1 - c^2x^2)(1 - e^2x^2)}$$

and it shows that when $x = \pm\frac{1}{c}, \pm\frac{1}{e}$ or $\theta = \pm\frac{\omega}{2}, \pm\frac{\omega'}{2}$, the left hand side is 0. Thus, for instance, $\delta = \frac{\omega}{2}$, $\delta' = -\frac{\omega}{2}$, $\varepsilon = \frac{\omega'}{2}$, $\varepsilon' = -\frac{\omega'}{2}$ and $g' = c_1f\varphi\left(\frac{\omega}{2}\right) = e_1f\varphi\left(\frac{\omega'}{2}\right)$, $f' = \frac{g}{c_1}\varphi\left(\frac{\omega}{2}\right) = \frac{g}{e_1}\varphi\left(\frac{\omega'}{2}\right)$. A solution of this system is $g = f' = 0$, $\frac{f}{g'} = \frac{1}{k}$, $c_1 = \frac{k}{\varphi\left(\frac{\omega}{2}\right)}$, $e_1 = \frac{k}{\varphi\left(\frac{\omega'}{2}\right)}$ where k is arbitrary. Then $y = \frac{1}{k}\varphi\theta$ and $1 - \frac{\varphi\theta}{\varphi\left(\frac{\omega}{2}\right)} = \frac{1}{\rho}(1 - cx)\left(1 - \frac{x}{\lambda\left(\frac{\omega}{2} - \alpha_1\right)}\right)^2\left(1 - \frac{x}{\lambda\left(\frac{\omega}{2} - \alpha_2\right)}\right)^2\ldots\left(1 - \frac{x}{\lambda\left(\frac{\omega}{2} - \alpha_n\right)}\right)^2$ where

$$\rho = (1 - e^2c^2\lambda^2\alpha_1 \cdot x^2)(1 - e^2c^2\lambda^2\alpha_2 \cdot x^2)\dots(1 - e^2c^2\lambda^2\alpha_n \cdot x^2).$$

We obtain similar expressions for $1+\frac{\varphi\theta}{\varphi(\frac{\omega}{2})}$ and for $1\pm\frac{\varphi\theta}{\varphi\left(\frac{\omega'}{2}\right)}$ and, as in the *Recherches*,

$$1 - c_1^2y^2 = (1 - c^2x^2)\frac{t^2}{\rho^2}, \quad 1 - e_1^2y^2 = (1 - e^2x^2)\frac{t'^2}{\rho^2}$$

where

$$t = \left(1 - \frac{x^2}{\lambda^2\left(\frac{\omega}{2} - \alpha_1\right)}\right)\left(1 - \frac{x^2}{\lambda^2\left(\frac{\omega}{2} - \alpha_2\right)}\right)\dots\left(1 - \frac{x^2}{\lambda^2\left(\frac{\omega}{2} - \alpha_n\right)}\right),$$

$$t' = \left(1 - \frac{x^2}{\lambda^2\left(\frac{\omega'}{2} - \alpha_1\right)}\right)\left(1 - \frac{x^2}{\lambda^2\left(\frac{\omega'}{2} - \alpha_2\right)}\right)\dots\left(1 - \frac{x^2}{\lambda^2\left(\frac{\omega'}{2} - \alpha_n\right)}\right).$$

Thus $\sqrt{(1 - c_1^2y^2)(1 - e_1^2y^2)} = \pm\frac{tt'}{\rho^2}\sqrt{(1 - c^2x^2)(1 - e^2x^2)}$ and Abel shows, as in the *Recherches*, that $\frac{\rho^2\frac{dy}{dx}}{tt'}$ is a constant a, so that the desired result is obtained. The value of a is computed by comparing the limit values of $\frac{dy}{dx}$ for x infinite coming from $\frac{dy}{dx} = a\frac{tt'}{\rho^2}$ and from $y = \frac{1}{k}\left(x + 2x\sum\frac{\Delta(\alpha)}{1-e^2c^2\lambda^2\alpha\cdot x^2}\right)$. Abel finds $a = (e^2c^2)^n\frac{1}{k}\lambda^4\alpha_1 \cdot \lambda^4\alpha_2 \dots \lambda^4\alpha_n$. He gives some other forms for y, as

$$\begin{aligned} y &= a\frac{x\left(1 - \frac{x^2}{\lambda^2\alpha_1}\right)\left(1 - \frac{x^2}{\lambda^2\alpha_2}\right)\dots\left(1 - \frac{x^2}{\lambda^2\alpha_n}\right)}{(1 - e^2c^2\lambda^2\alpha_1 \cdot x^2)(1 - e^2c^2\lambda^2\alpha_2 \cdot x^2)\dots(1 - e^2c^2\lambda^2\alpha_n \cdot x^2)} \\ &= \frac{1}{k}(ec)^{2n}b\lambda\theta \cdot \lambda(\alpha_1 + \theta)\lambda(\alpha_1 - \theta)\dots\lambda(\alpha_n + \theta)\lambda(\alpha_n - \theta) \end{aligned}$$

where $b = \lambda^2\alpha_1 \cdot \lambda^2\alpha_2 \dots \lambda^2\alpha_n$. Doing $\theta = \frac{\omega}{2}$ and $\theta = \frac{\omega'}{2}$, he obtains

$$\frac{1}{c_1} = (-1)^n\frac{b}{k}e^{2n}c^{2n-1}\left(\lambda\left(\frac{\omega}{2} - \alpha_1\right)\lambda\left(\frac{\omega}{2} - \alpha_2\right)\dots\lambda\left(\frac{\omega}{2} - \alpha_n\right)\right)^2 \text{ and}$$

$$\frac{1}{e_1} = (-1)^n\frac{b}{k}e^{2n-1}c^{2n}\left(\lambda\left(\frac{\omega'}{2} - \alpha_1\right)\lambda\left(\frac{\omega'}{2} - \alpha_2\right)\dots\lambda\left(\frac{\omega'}{2} - \alpha_n\right)\right)^2.$$

As Abel remarks, the transformation defined by Jacobi corresponds to the case in which $\alpha_1 = \frac{2\omega}{2n+1}$, $c = c_1 = 1$ and the theory explained in the *Recherches* to the case in which

$$\alpha_1 = \frac{m\omega + m'\omega'}{2n + 1}$$

with $m+m'$ even, m, m' and $2n+1$ having no common factor. In both cases, $\alpha_2 = 2\alpha_1$, $\alpha_3 = 3\alpha_1, \dots, \alpha_n = n\alpha_n$. Jacobi independently found these transformations. The more general transformation $y = \frac{f'+f\cdot\varphi\theta}{g'+g\cdot\varphi\theta}$ is obtained by composing this particular one with a transformation of the type $y = \frac{f'+fx}{g'+gx}$.

A second case considered by Abel is that in which $k = 0$ and k' or k'' is equal to 1. It is impossible that $k' = k'' = 1$ for if $\lambda\left(\theta + \frac{\omega}{2} + \frac{\omega'}{2}\right)$ and $\lambda\left(\theta + \frac{3\omega}{2} + \frac{\omega'}{2}\right)$ are roots of $y = \psi(x)$, so is $\lambda\left(\theta + \frac{3\omega}{2} + \frac{\omega'}{2} - \frac{\omega+\omega'}{2}\right) = \lambda(\theta+\omega)$ and k is not 0. As in the first case, let $1 - c_1 y = 0$ for $x = \frac{1}{c}$. Then $1 \pm c_1 y = \frac{g' \pm c_1 f'}{r}\left(1 - \frac{\varphi\theta}{\varphi\left(\frac{\omega}{2}\right)}\right)$, $1 - c_1^2 y^2 = \frac{g'^2 - c_1^2 f'^2}{r^2}\left(1 - \left(\frac{\varphi\theta}{\varphi\left(\frac{\omega}{2}\right)}\right)^2\right)$ and

$$1 - \frac{\varphi\theta}{\varphi\delta} = -\frac{1}{\varphi\delta\cdot\rho}\left(1-\frac{x}{\lambda\delta}\right)\left(1-\frac{x}{\lambda(\delta+\beta)}\right)\left(1-\frac{x}{\lambda(\delta+\alpha_1)}\right)\left(1-\frac{x}{\lambda(\delta-\alpha_1)}\right) \times\cdots\left(1-\frac{x}{\lambda(\delta+\alpha_n)}\right)\left(1-\frac{x}{\lambda(\delta-\alpha_n)}\right) \tag{95}$$

where $\beta = \frac{\omega+\omega'}{2}$ (resp. $\frac{3\omega+\omega'}{2}$) if $k' = 1$ (resp. $k'' = 1$) and

$$\rho = \pm ecx(1 - e^2c^2\lambda^2\alpha_1\cdot x^2)(1 - e^2c^2\lambda^2\alpha_2\cdot x^2)\ldots(1 - e^2c^2\lambda^2\alpha_n\cdot x^2).$$

Abel takes $\delta = \pm\frac{\omega}{2}$ in order to compute $1 - c_1^2y^2$ and he finds

$$\sqrt{1 - c_1^2y^2} = \frac{\sqrt{c_1^2f'^2 - g'^2}}{\varphi\left(\frac{\omega}{2}\right)r\rho}t\sqrt{(1-c^2x^2)(1-e^2x^2)}.$$

Now it results from (93) that $\sqrt{1 - e_1^2y^2} = \frac{\varphi\left(\frac{\omega}{2}\right)}{a\sqrt{c_1^2f'^2 - g'^2}}\frac{r\rho}{t}\frac{dy}{dx}$ is a rational function of x. If we impose that $1 - e_1^2y^2$ be annihilated by $x = \pm\lambda\left(\frac{\omega-\beta}{2}\right)$, we effectively find

$$\sqrt{1 - e_1^2y^2} = \frac{\sqrt{e_1^2f'^2 - g'^2}}{\varphi\left(\frac{\omega-\beta}{2}\right)r\rho}\left(1 - \frac{x^2}{\lambda^2\left(\frac{\omega-\beta}{2}\right)}\right)\left(1 - \frac{x^2}{\lambda^2\left(\frac{\omega-\beta}{2} - \alpha_1\right)}\right)\cdots \times\left(1 - \frac{x^2}{\lambda^2\left(\frac{\omega-\beta}{2} - \alpha_n\right)}\right)$$

by doing $\delta = \pm\frac{\omega-\beta}{2}$ in the relation (95). Then we find that $g' = c_1 f\cdot\varphi\left(\frac{\omega}{2}\right) = e_1 f\cdot\varphi\left(\frac{\omega-\beta}{2}\right)$, $f' = \frac{g}{c_1}\varphi\left(\frac{\omega}{2}\right) = \frac{g}{e_1}\varphi\left(\frac{\omega-\beta}{2}\right)$. A solution is $f = g' = 0$, $\frac{f}{g'} = \frac{\varphi\left(\frac{\omega}{2}\right)}{c_1} = \frac{\varphi\left(\frac{\omega-\beta}{2}\right)}{e_1}$, which gives $c_1 = k\varphi\left(\frac{\omega}{2}\right)$, $e_1 = k\varphi\left(\frac{\omega-\beta}{2}\right)$, $y = \frac{1}{k\varphi\theta}$, $a = \pm\frac{ec}{k}$. Other solutions are obtained by composing with a transformation of the type $y = \frac{f'+fx}{g'+gx}$.

In the simplest case where $n=0$ and $c_1=c=1$, $\beta=\frac{3\omega}{2}+\frac{\omega'}{2}$, the formulae are $y=(1+e)\frac{x}{1+ex^2}$, $e_1=\frac{2\sqrt{e}}{1+e}$ and $a=1+e$.

In the third case, $k=1$ and one finds that $\varphi\theta=\varphi(\theta+\omega)=-\varphi\theta$ so that $\varphi\theta=0$. Let us return to (94), denoting by $\frac{1}{2}f'$ (resp. $\frac{1}{2}g'$) the coefficient of x^{m-2} in p (resp. q) and by $F\theta$ the function $\lambda^2\theta+\lambda^2(\theta+\alpha_1)+\ldots+\lambda^2(\theta+\alpha_{m-1})$. We have $f'-g'y=-(f-gy)F\theta$ and $y=\frac{f'+f\cdot F\theta}{g'+g\cdot F\theta}$ and we may proceed as in the preceding cases with $F\theta$ in place of $\varphi\theta$.

Abel states the general theorem for the first real transformation of arbitrary order n:

$$\frac{dy}{\sqrt{(1-y^2)(1-e_1^2y^2)}}=\pm\frac{adx}{\sqrt{(1-x^2)(1-e^2x^2)}} \tag{96}$$

where $a=k\lambda\frac{\omega}{n}\cdot\lambda\frac{2\omega}{n}\ldots\lambda\frac{(n-1)\omega}{n}$, $e_1=e^n\left(\lambda\frac{\omega}{2n}\cdot\lambda\frac{3\omega}{2n}\ldots\lambda\left(n-\frac{1}{2}\right)\frac{\omega}{n}\right)^2$, $1=k\lambda\frac{\omega}{2n}\cdot\lambda\frac{3\omega}{2n}$ $\times\ldots\lambda\left(n-\frac{1}{2}\right)\frac{\omega}{n}$ and $y=k\lambda\theta\cdot\lambda\left(\theta+\frac{\omega}{n}\right)\lambda\left(\theta+\frac{2\omega}{n}\right)\ldots\lambda\left(\theta+\frac{(n-1)\omega}{n}\right)$.

With Legendre's notation $x=\sin\varphi$, $y=\sin\psi$ and n very large, e_1 becomes negligible and Abel writes, with an approximation $\psi=a\int\limits_0^\varphi\frac{d\varphi}{\sqrt{1-e^2\sin^2\varphi}}=\sum\limits_{m=0}^{n-1}\arctan\left(\tan\varphi\sqrt{1-e^2\lambda^2\left(\frac{m\omega}{n}\right)}\right)$. For $\varphi=\frac{\pi}{2}$, $\psi=n\frac{\pi}{2}=a\frac{\omega}{2}$, so that $\frac{1}{a}=\frac{1}{\pi}\frac{\omega}{n}$ and passing to the limit for n infinite, Abel finds

$$\int\limits_0^\varphi\frac{d\varphi}{\sqrt{1-e^2\sin^2\varphi}}=\frac{1}{\pi}\int\limits_0^\omega\arctan\left(\tan\varphi\sqrt{1-e^2\lambda^2x}\right)dx.$$

The order of the transformation, that is the degree m of the equation $p-qy=0$ is the number of distinct values of $\lambda(\theta+k_1\alpha_1+\ldots+k_\nu\alpha_\nu)$ and Abel shows that $m=n_1n_2\ldots n_\nu$ where, for each j, n_j is the smallest strictly positive integer such that $\lambda(\theta+n_j\alpha_j)=\lambda(\theta+m_1\alpha_1+\ldots+m_{j-1}\alpha_{j-1})$ for some $m_1,\ldots,m_{j-1}$. Thus, when m is a prime number, $\nu=1$ and $m=n_1$. Abel states some theorems:

a) when the order of a transformation is a composite number mn, this transformation may be obtained by the composition of a transformation of order m and a transformation of order n;

b) the equation $y=\psi(x)$ is algebraically solvable and its roots x are rational functions of y and some radicals $r_1^{\frac{1}{n_1}}, r_2^{\frac{1}{n_2}},\ldots,r_\nu^{\frac{1}{n_\nu}}$ where $n_1,n_2,\ldots,n_\nu$ are prime numbers, $n_1n_2\cdots n_\nu$ is the degree of the equation and $r_1,r_2,\ldots,r_\nu$ have the form $\zeta+t\sqrt{(1-c_1^2y^2)(1-e_1^2y^2)}$ with ζ and t rational in y;

c) If the differential equation $\frac{dy}{\sqrt{(1-c^2y^2)(1-e^2y^2)}}=a\frac{dx}{\sqrt{(1-c^2x^2)(1-e^2x^2)}}$ has a solution algebraic in x and y, $a=\mu'+\sqrt{-\mu}$ where μ', μ are rational numbers and $\mu\geq0$. There is an infinity of values for the moduli e, c, given by algebraic equations solvable by radicals, for which $\mu>0$.

Recall that in the *Recherches* Abel doubted that these equations might be algebraically solvable. Kronecker (1857) gave a proof that they are solvable by radicals, as Abel states here.

d) If the differential equation $\frac{dy}{\sqrt{(1-y^2)(1-b^2y^2)}} = a\frac{dx}{\sqrt{(1-x^2)(1-c^2x^2)}}$ where $b^2 = 1-c^2$ has a solution algebraic in x and y, $a = \sqrt{\mu + \mu'\sqrt{-1}}$ where μ, μ' are rational numbers and $\mu \geq 0$. In particular, when a is real, it is the square root of a positive rational number. Thus, when $e_1^2 = 1 - e^2$ in (96), $a = \sqrt{n}$. Indeed the formula for k gives $y = 1$ when $\theta = \frac{\omega}{2n}$, thus

$$\int_0^1 \frac{dy}{\sqrt{(1-y^2)(1-e_1^2y^2)}} = \frac{a\omega}{2n}.$$

One may write $y = k\lambda\theta \cdot \lambda\left(\frac{\omega}{n} - \theta\right)\lambda\left(\frac{2\omega}{n} - \theta\right)\ldots\lambda\left(\frac{(n-1)\omega}{n} - \theta\right)$ because $\lambda\left(\theta + \frac{m\omega}{n}\right) = \lambda\left(\frac{(n-m)\omega}{n} - \theta\right)$, so that $y^2 = k^2x^2\frac{\lambda^2\frac{\omega}{n} - x^2}{1-e^2\lambda^2\frac{\omega}{n}x^2}\cdots\frac{\lambda^2\frac{(n-1)\omega}{n} - x^2}{1-e^2\lambda^2\frac{(n-1)\omega}{n}x^2}$. Now putting $x = p\sqrt{-1}$ and $y = z\sqrt{-1}$ and letting p and z tend towards infinity, Abel obtains $\frac{\omega}{2} = a\int_0^1 \frac{dy}{\sqrt{(1-y^2)(1-e_1^2y^2)}} = a\frac{a\omega}{2n}$, whence $a = \sqrt{n}$.

In the part of this memoir published in 1829, Abel gives another study of the same transformation in the case in which $0 < c, c_1 < 1$ and $e = e_1 = 1$. Then $\frac{\omega}{2} = \int_0^1 \frac{dx}{\sqrt{(1-x^2)(1-c^2x^2)}}$ is real but $\frac{\omega'}{2} = \int_0^{\frac{1}{c}} \frac{dx}{\sqrt{(1-x^2)(1-c^2x^2)}} = \frac{\omega}{2}\sqrt{-1}\int_1^{\frac{1}{c}} \frac{dx}{\sqrt{(1-x^2)(1-c^2x^2)}} = \frac{\omega}{2}\sqrt{-1}\frac{\varpi}{2}$, where $\frac{\varpi}{2} = \int_0^1 \frac{dx}{\sqrt{(1-x^2)(1-b^2x^2)}}$, $b = \sqrt{1-c^2}$, is complex. Let us suppose that the differential equation

$$\frac{dy}{\sqrt{(1-y^2)(1-c_1^2y^2)}} = a\frac{dx}{\sqrt{(1-x^2)(1-c^2x^2)}} \tag{97}$$

has a solution $f(y, x) = 0$ algebraic in x and y and define the function $y = \lambda_1\theta'$ by

$$\frac{dy}{\sqrt{(1-y^2)(1-c_1^2y^2)}} = d\theta' \quad \text{and} \quad \lambda_1(0) = 0.$$

The equation (97) takes the form $d\theta' = ad\theta$, so that $\theta' = \varepsilon + a\theta$ where ε is constant and $y = \lambda_1(\varepsilon + a\theta)$. Thus $f(\lambda_1(\varepsilon + a\theta), \lambda\theta) = 0$ identically in θ. This implies that $f(\lambda_1(\varepsilon + 2ma\omega + a\theta), \lambda\theta) = f(\lambda_1(\varepsilon + ma\varpi i + a\theta)\lambda\theta)$ for any integer m. Then there exists pairs of distinct integers (k, k') and (ν, ν') such that

$$\lambda_1(\varepsilon + 2k'a\omega + a\theta) = \lambda_1(\varepsilon + 2ka\omega + a\theta) \quad \text{and}$$
$$\lambda_1(\varepsilon + \nu'a\varpi i + a\theta) = \lambda_1(\varepsilon + \nu a\varpi i + a\theta)$$

or $2k'a\omega = 2ka\omega + 2m\omega_1 + m'\varpi\sqrt{-1}$, $\nu' a\varpi i = \nu a\varpi i + 2\mu\omega_1 + \mu'\varpi\sqrt{-1}$ where m, m', μ, μ' are integers, $\frac{\omega_1}{2} = \int\limits_0^1 \frac{dx}{\sqrt{(1-x^2)(1-c_1^2x^2)}}$, $\frac{\varpi_1}{2}\int\limits_0^1 \frac{dx}{\sqrt{(1-x^2)(1-b_1^2x^2)}}$, $b_1 = \sqrt{1-c_1^2}$. From these relations we draw $a = \frac{m}{\nu}\frac{\omega_1}{\omega} + \frac{m'}{2\nu}\frac{\varpi_1}{\omega}\sqrt{-1} = \frac{\mu'}{\nu'}\frac{\varpi_1}{\varpi} - \frac{2\mu}{\nu'}\frac{\omega_1}{\varpi}\sqrt{-1}$ and $\frac{m}{\nu}\frac{\omega_1}{\omega} = \frac{\mu'}{\nu'}\frac{\varpi_1}{\varpi}$, $\frac{m'}{2\nu}\frac{\varpi_1}{\omega} = -\frac{2\mu}{\nu'}\frac{\omega_1}{\varpi}$. Thus $\frac{\omega^2}{\varpi^2} = -\frac{1}{4}\frac{mm'}{\mu\mu'}\frac{\nu'^2}{\nu^2}$ and $\frac{\omega_1^2}{\varpi_1^2} = -\frac{1}{4}\frac{m'\mu'}{m\mu}$. As $\frac{\omega^2}{\varpi^2}$ is a continuous function of c, these equations can be satisfied for any c, c_1 only if $m' = \mu = 0$ or if $m = \mu' = 0$. In the first case we have $a = \frac{m}{\nu}\frac{\omega_1}{\omega} = \frac{\mu'}{\nu'}\frac{\varpi_1}{\varpi}$, $\frac{\omega_1}{\varpi_1} = \frac{\nu\mu'}{\nu' m}\frac{\omega}{\varpi}$ and in the second case $a = \frac{m'}{2\nu}\frac{\varpi_1}{\omega}\sqrt{-1} = -\frac{2\mu}{\nu'}\frac{\omega_1}{\varpi}\sqrt{-1}$, $\frac{\omega_1}{\varpi_1} = -\frac{1}{4}\frac{m'\nu'}{\mu\nu}\frac{\varpi}{\omega}$. Abel states that if (97) has a solution algebraic in x and y, then either $\frac{\omega_1}{\varpi_1}$ or $\frac{\varpi_1}{\omega_1}$ has a rational ratio to $\frac{\omega}{\varpi}$. In the first case $a = \delta\frac{\omega_1}{\omega}$ and in the second case $a = \delta\frac{\varpi_1}{\omega}\sqrt{-1}$, with δ rational. Both ratios k, k' are rational for certain particular values of c, c_1, determined by $\frac{\varpi}{\omega} = \sqrt{kk'}$, $\frac{\varpi_1}{\omega_1} = \sqrt{\frac{k'}{k}}$ and in these cases $a = \delta\frac{\omega_1}{\omega} + \delta'\frac{\varpi_1}{\omega}\sqrt{-1}$ with δ, δ' rational.

In order to prove that these conditions are sufficient for the existence of an algebraic solution to the equation (97), Abel observes that $\lambda\alpha = f\left(b\frac{\omega}{2} - b\alpha\right)$ where $f\alpha = \sqrt{1-x^2}$ is the function introduced in the *Recherches*. The expansion of $f\alpha$ in simple infinite product then gives

$$\lambda\alpha = A\frac{(1-t^2)(1-t^2r^2)(1-t^{-2}r^2)(1-t^2r^4)(1-t^{-2}r^4)\dots}{(1+t^2)(1+t^2r^2)(1+t^{-2}r^2)(1+t^2r^4)(1+t^{-2}r^4)\dots}$$
$$= A\psi\left(\alpha\frac{\pi}{\varpi}\right)\psi(\omega+\alpha)\frac{\pi}{\varpi}\psi(\omega-\alpha)\frac{\pi}{\varpi}\psi(2\omega+\alpha)\frac{\pi}{\varpi}\psi(2\omega-\alpha)\frac{\pi}{\varpi}\dots \quad (98)$$

where A is independent from α, $t = e^{-\frac{\alpha\pi}{\varpi}}$, $r = e^{-\frac{\omega}{\varpi}\pi}$ and $\psi(x) = \frac{1-e^{-2x}}{1+e^{-2x}}$. From this formula we draw

$$\lambda\theta\cdot\lambda\left(\theta+\frac{\omega}{n}\right)\lambda\left(\theta+\frac{2\omega}{n}\right)\dots\lambda\left(\theta+\frac{n-1}{n}\omega\right)$$
$$= A^n\psi\delta\frac{\pi}{\varpi_1}\cdot\psi(\omega_1+\delta)\frac{\pi}{\varpi_1}\psi(\omega_1-\delta)\frac{\pi}{\varpi_1}\psi(2\omega_1+\delta)\frac{\pi}{\varpi_1}\dots$$

where $\delta = \frac{\varpi_1}{\varpi}\theta$ and $\frac{\omega_1}{\varpi_1} = \frac{1}{n}\frac{\omega}{\varpi}$. On the other hand

$$\lambda_1\alpha = A_1\psi\left(\alpha\frac{\pi}{\varpi_1}\right)\psi(\omega_1+\alpha)\frac{\pi}{\varpi_1}\cdot\psi(\omega_1-\alpha)\frac{\pi}{\varpi_1}\dots$$

and, by comparison, we conclude that $\lambda_1\left(\frac{\varpi_1}{\varpi}\theta\right) = \frac{A_1}{A^n}\lambda\theta\cdot\lambda\left(\theta+\frac{\omega}{n}\right)\lambda\left(\theta+\frac{2\omega}{n}\right)$ $\times\dots\lambda\left(\theta+\frac{n-1}{n}\omega\right) = y$, algebraic function of $x = \lambda\theta$ such that $\frac{dy}{\sqrt{(1-y^2)(1-c_1^2y^2)}} = \frac{\varpi_1}{\varpi}d\theta = \frac{\varpi_1}{\varpi}\frac{dx}{\sqrt{(1-x^2)(1-c^2x^2)}}$.

There are three cases to consider: a real, a purely imaginary and a complex with $\frac{\varpi}{\omega} = \sqrt{kk'}$ and $\frac{\varpi_1}{\omega_1} = \sqrt{\frac{k'}{k}}$. In the first case, $a = \frac{\mu}{\nu}\frac{\varpi_1}{\varpi}$ and $\frac{\omega_1}{\varpi_1} = \frac{m}{n}\frac{\omega}{\varpi}$ where μ, ν, m, n

are integers. The equation (97) is satisfied by $x = \lambda(\nu\varpi\theta)$, $y = \lambda_1(\mu\varpi_1\theta)$. Let c' be a modulus such that $\frac{\omega'}{\varpi'} = \frac{1}{n}\frac{\omega}{\varpi}$, where ω' and ϖ' are the periods corresponding to c'. We have

$$\lambda'(\mu\nu\varpi'\theta) = \frac{A'}{A^n}\lambda(\mu\nu\varpi\theta)\lambda\left(\mu\nu\varpi\theta + \frac{\omega}{n}\right)\ldots\lambda\left(\mu\nu\varpi\theta + \frac{n-1}{n}\omega\right)$$

and since $\frac{\omega'}{\varpi'} = \frac{1}{m}\frac{\omega_1}{\varpi_1}$,

$$\lambda'(\mu\nu\varpi'\theta) = \frac{A'}{A_1^m}\lambda_1(\mu\nu\varpi_1\theta)\lambda_1\left(\mu\nu\varpi_1\theta + \frac{\omega_1}{m}\right)\cdots\lambda_1\left(\mu\nu\varpi_1\theta + \frac{m-1}{m}\omega_1\right).$$

Finally

$$\begin{aligned}&\frac{1}{A^n}\lambda(\mu\delta)\lambda\left(\mu\delta + \frac{\omega}{n}\right)\cdots\lambda\left(\mu\delta + \frac{n-1}{n}\omega\right)\\ &= \frac{1}{A_1^m}\lambda_1(\nu\delta_1)\lambda_1\left(\nu\delta_1 + \frac{\omega_1}{m}\right)\cdots\lambda_1\left(\nu\delta_1 + \frac{m-1}{m}\omega_1\right)\end{aligned} \tag{99}$$

where $\nu\varpi\theta = \delta$ and $\mu\varpi_1\theta = \delta_1$. The left hand side is an algebraic function of $\lambda(\mu\delta)$, so an algebraic function of $x = \lambda\delta$ and, in the same way, the right hand side is an algebraic function of $y = \lambda_1\delta_1$. Thus we have an algebraic integral of (97). One sees that $A = \frac{1}{\sqrt{c}}$ and $A_1 = \frac{1}{\sqrt{c_1}}$. As an example, Abel explains the case in which $a = \frac{\varpi_1}{\varpi}$ and $\frac{\omega_1}{\varpi_1} = \frac{2}{3}\frac{\omega}{\varpi}$; the equation (99) takes the form $c\sqrt{c}\cdot\lambda\left(\delta + \frac{\omega}{3}\right)\lambda\left(\delta + \frac{2\omega}{3}\right) = c_1\lambda_1\delta_1 \cdot \lambda_1\left(\delta_1 + \frac{\omega_1}{2}\right)$ or $y\frac{\sqrt{1-y^2}}{\sqrt{1-c_1^2y^2}} = \frac{c\sqrt{c}}{c_1}x\frac{\lambda^2\frac{\omega}{3}-x^2}{1-c^2\lambda^2\frac{\omega}{3}x^2}$.

In the second case $a = \frac{\mu}{\nu}\frac{\varpi_1}{\omega}\sqrt{-1}$ and $\frac{\omega_1}{\varpi_1} = \frac{m}{n}\frac{\varpi}{\omega}$ with μ, ν, m, n integers. Let us put $x = \frac{z\sqrt{-1}}{\sqrt{1-z^2}}$ so that $\frac{dx}{\sqrt{(1-x^2)(1-c^2x^2)}} = \sqrt{-1}\frac{dz}{\sqrt{(1-z^2)(1-b^2z^2)}}$ where $b = \sqrt{1-c^2}$. The equation (97) takes the form $\frac{dy}{\sqrt{(1-y^2)(1-c_1^2y^2)}} = \frac{\mu}{\nu}\frac{\varpi_1}{\omega}\frac{dz}{\sqrt{(1-z^2)(1-b^2z^2)}}$ and we are reduced to the preceding case, with the algebraic integral (99) where $z = \lambda\delta = \frac{x}{\sqrt{x^2-1}}$ and ω replaced by ϖ. For instance if $a = \frac{\varpi_1}{\omega}\sqrt{-1}$ and $\frac{\omega_1}{\varpi_1} = 2\frac{\varpi}{\omega}$, (99) is written $\sqrt{b}\cdot\lambda\delta = c_1\lambda_1(\delta_1)\lambda_1\left(\delta_1 + \frac{\omega_1}{2}\right)$ or $y\frac{\sqrt{1-y^2}}{\sqrt{1-c_1^2y^2}} = \frac{\sqrt{b}}{c_1}\frac{x}{\sqrt{x^2-1}}$.

In the third case $a = \frac{\mu}{\nu}\frac{\varpi_1}{\varpi} + \frac{\mu'}{\nu'}\frac{\varpi_1}{\omega}\sqrt{-1}$ where μ, ν, μ', ν' are integers and $\frac{\omega_1}{\varpi_1} = k\frac{\omega}{\varpi} = \frac{1}{k'}\frac{\varpi}{\omega}$ where k, k' are rational numbers. The two equations

$$\frac{dz}{\sqrt{(1-z^2)(1-c_1^2z^2)}} = \frac{\mu}{\nu}\frac{\varpi_1}{\omega}\frac{dx}{\sqrt{(1-x^2)(1-c^2x^2)}} \quad \text{and}$$

$$\frac{dv}{\sqrt{(1-v^2)(1-c_1^2v^2)}} = \frac{\mu'}{\nu'}\frac{\varpi_1}{\omega}\frac{dx}{\sqrt{(1-x^2)(1-c^2x^2)}}$$

have algebraic integrals and our equation (97) may be written

$$\frac{dy}{\sqrt{(1-y^2)(1-c_1^2y^2)}}=\frac{dz}{\sqrt{(1-z^2)(1-c_1^2z^2)}}+\frac{dv}{\sqrt{(1-v^2)(1-c_1^2v^2)}}$$

which is satisfied by $y=\frac{z\sqrt{(1-v^2)(1-c_1^2v^2)}+v\sqrt{(1-z^2)(1-c_1^2z^2)}}{1-c_1^2z^2v^2}$, algebraic function of x. Abel gives a translation of his general theorem in Legendre's notations (which were adopted by Jacobi).

The case where $c_1=c$ corresponds to complex multiplication. Then $a=\frac{m}{\nu}+\frac{m'}{2\nu}\frac{\varpi}{\omega}\sqrt{-1}=\frac{\mu'}{\nu'}-\frac{2\mu}{\nu'}\frac{\omega}{\varpi}\sqrt{-1}$ and $\frac{m}{\nu}=\frac{\mu'}{\nu'}$, $\frac{m'}{2\nu}\frac{\varpi}{\omega}=-\frac{2\mu}{\nu'}\frac{\omega}{\varpi}$. The multiplicator a is real if $m'=\mu=0$ but otherwise we must impose $\frac{\omega}{\varpi}=\frac{1}{2}\sqrt{-\frac{m'\nu'}{\mu\nu}}=\sqrt{k}$ where k is positive rational and we have $a=\delta+\delta'\sqrt{k}\sqrt{-1}$ where δ, δ' are rational numbers. Doing $\alpha=\frac{\omega}{2}$ in (98), Abel obtains

$$\sqrt[4]{c}=\frac{1-e^{-\pi\sqrt{k}}}{1+e^{-\pi\sqrt{k}}}\frac{1-e^{-3\pi\sqrt{k}}}{1+e^{-3\pi\sqrt{k}}}\frac{1-e^{-5\pi\sqrt{k}}}{1+e^{-5\pi\sqrt{k}}}\cdots\quad\text{and}$$

$$\sqrt[4]{b}=\frac{1-e^{-\frac{\pi}{\sqrt{k}}}}{1+e^{-\frac{\pi}{\sqrt{k}}}}\frac{1-e^{-\frac{3\pi}{\sqrt{k}}}}{1+e^{-\frac{3\pi}{\sqrt{k}}}}\frac{1-e^{-\frac{5\pi}{\sqrt{k}}}}{1+e^{-\frac{5\pi}{\sqrt{k}}}}\cdots$$

Abel continued the explanation of the theory of transformation in a memoir published in the third volume of *Crelle's Journal* (1828), *Sur le nombre des transformations différentes qu'on peut faire subir à une fonction elliptique par la substitution d'une fonction rationnelle dont le degré est un nombre premier donné* (*Œuvres*, t. I, p. 456–465). He puts

$$\Delta^2=(1-x^2)(1-c^2x^2),\quad \Delta'^2=(1-y^2)(1-c'^2x^2)$$

and supposes that the differential equation $\frac{dy}{\Delta'}=a\frac{dx}{\Delta}$ is satisfied by

$$y=\frac{A_0+A_1x+\ldots+A_{2n+1}x^{2n+1}}{B_0+B_1x+\ldots+B_{2n+1}x^{2n+1}}$$

where $2n+1$ is a prime number and one of the coefficients A_{2n+1}, B_{2n+1} is different from 0. He recalls that, according to the *Solution d'un problème général*, when $B_{2n+1}=0$, one has $y=\frac{\delta}{\varepsilon}\frac{p}{v}$, $c'=\varepsilon^2$ and $a=\frac{\delta}{\varepsilon}$ where $p=x\left(1-\frac{x^2}{\lambda^2\alpha}\right)\cdots\left(1-\frac{x^2}{\lambda^2(n\alpha)}\right)$, $v=(1-c^2\lambda^2\alpha x^2)\cdots(1-c^2\lambda^2(n\alpha)x^2)$, $\varepsilon=c^{n+\frac{1}{2}}\left(\lambda\left(\frac{\omega}{2}+\alpha\right)\ldots\lambda\left(\frac{\omega}{2}+n\alpha\right)\right)^2$, $\delta=c^{n+\frac{1}{2}}(\lambda\alpha\cdot\lambda(2\alpha)\cdots\lambda(n\alpha))^2$ and $\alpha=\frac{m\omega+m'\omega'}{2n+1}$ (m, m' integers). Other solutions are given by composing with $\frac{f'+fy}{g'+gy}$ where f', f, g, g' are constants such that $\left(1+\frac{g+f}{g'+f'}x\right)\left(1+\frac{g-f}{g'-f'}x\right)\left(1+\frac{g+c'f}{g'+c'f'}x\right)$ $\times\left(1+\frac{g-c'f}{g'-c'f'}x\right)=(1-x^2)(1-c'^2x^2)$. Thus, disregarding the signs, one finds 12 values for y and 6 values for c' for each choice of α:

	I	II	III	IV	V	VI
c'	ε^2	$\frac{1}{\varepsilon^2}$	$\left(\frac{1-\varepsilon}{1+\varepsilon}\right)^2$	$\left(\frac{1+\varepsilon}{1-\varepsilon}\right)^2$	$\left(\frac{1-\varepsilon i}{1+\varepsilon i}\right)^2$	$\left(\frac{1+\varepsilon i}{1-\varepsilon i}\right)^2$
a	$\pm\frac{\delta}{\varepsilon}$	$\mp\delta\varepsilon$	$\pm\frac{\delta}{2\varepsilon}(1+\varepsilon)^2 i$	$\mp\frac{\delta}{2\varepsilon}(1-\varepsilon)^2 i$	$\pm\frac{\delta}{2\varepsilon}(1+\varepsilon i)^2 i$	$\mp\frac{\delta}{2\varepsilon}(1-\varepsilon i)^2 i$
y	$\begin{cases}\frac{\delta}{\varepsilon}\frac{p}{v}\\ \frac{1}{\delta\varepsilon}\frac{v}{p}\end{cases}$	$\frac{\varepsilon}{\delta}\frac{v}{p}$ $\delta\varepsilon\frac{p}{v}$	$\frac{1+\varepsilon}{1-\varepsilon}\frac{v\pm\delta p}{v\mp\delta p}$	$\frac{1-\varepsilon}{1+\varepsilon}\frac{v\pm\delta p}{v\mp\delta p}$	$\frac{1+\varepsilon i}{1-\varepsilon i}\frac{v\pm\delta pi}{v\mp\delta pi}$	$\frac{1-\varepsilon i}{1+\varepsilon i}\frac{v\pm\delta pi}{v\mp\delta pi}.$

This comes from the fact that the modulus c^2 is a modular function of level 2 in Klein's sense: it is invariant by the group of 2×2 matrices congruent to the identity *modulo* 2 operating on the ratio of the periods, and this group is of index 6 in SL(2,**Z**).

It remains to count the number of α leading to different solutions. If α and α' lead to the same solution of type I, one finds that $p'=p$, $v'=v$ and $\frac{\delta'}{\varepsilon'}=\pm\frac{\delta}{\varepsilon}$. From $p'=p$ it results that $\lambda^2\alpha'=\lambda^2(\mu\alpha)$ for an integer μ between 1 and n, thus $\alpha'=k\omega+k'\omega'\pm\mu\alpha$ where k, k' are integers. For such a value of α', $p'=p$, $v'=v$, $\delta'=\delta$, and $\varepsilon'=\varepsilon$ and both solutions are effectively equal. Now when $\alpha=\frac{m\omega}{2n+1}$, there exist integers k, μ such that $k(2n+1)\pm\mu m=1$ and one has $k\omega\pm\mu\alpha=\frac{\omega}{2n+1}$. When $\alpha=\frac{m\omega+m'\omega'}{2n+1}$ with $m'\neq 0$, there exist integers k', μ such that $k'(2n+1)\pm m'\mu=1$ and one has $k\omega+k'\omega'\pm\mu\alpha=\frac{\omega'+\nu\omega}{2n+1}$ where $\nu=k(2n+1)\pm\mu m$. Thus the different choices for α are $\frac{\omega}{2n+1},\frac{\omega'}{2n+1},\frac{\omega'+\omega}{2n+1},\frac{\omega'+2\omega}{2n+1},\ldots,\frac{\omega'+2n\omega}{2n+1}$; their number is $2n+2$.

The values of y of types III, IV, V and VI may be written in another way with the help of the identities $v-\delta p=(1-x\sqrt{c})(1-2k_1x\sqrt{c}+cx^2)(1-2k_2x\sqrt{c}+cx^2)\cdots(1-2k_nx\sqrt{c}+cx^2)$,

$$v-\delta p\sqrt{-1}=(1-x\sqrt{-c})(1-2k_1'x\sqrt{-c}-cx^2)(1-2k_2'x\sqrt{-c}-cx^2)\cdots\times(1-2k_n'x\sqrt{-c}-cx^2)$$

and similar expressions for $v+\delta p$, $v+\delta p\sqrt{-1}$, where $k_\mu=\frac{\Delta(\mu\alpha)}{1-c\lambda^2(\mu a)}$, $k'_\mu=\frac{\Delta(\mu\alpha)}{1+c\lambda^2(\mu a)}$, $\Delta(\theta)=\pm\sqrt{(1-\lambda^2\theta)(1-c^2\lambda^2\theta)}$. When $0<c<1$, Abel explains that the only transformations for which c' is real correspond to $\alpha=\frac{\omega}{2n+1}$ or $\frac{\omega'-\omega}{2n+1}$ and that they are of type I, II, III or IV.

As we saw above, when $0<c<1$, ω is real and $\omega'=\omega+\varpi\sqrt{-1}$ where ϖ is real. Abel gives an expression of $\lambda\theta=f\left(b\left(\frac{\omega}{2}-\theta\right)\right)$ in infinite product

$$\lambda\theta=\frac{2}{\sqrt{c}}\sqrt[4]{q}\sin\left(\frac{\pi}{\omega}\theta\right)\frac{\left(1-2q^2\cos\left(\frac{2\pi}{\omega}\theta\right)+q^4\right)\left(1-2q^4\cos\left(\frac{2\pi}{\omega}\theta\right)+q^8\right)\cdots}{\left(1-2q\cos\left(\frac{2\pi}{\omega}\theta\right)+q^2\right)\left(1-2q^3\cos\left(\frac{2\pi}{\omega}\theta\right)+q^6\right)\cdots}$$

where $q=e^{-\frac{\varpi}{\omega}\pi}$ and computes ε with the help of this formula. If $\alpha=\frac{\omega}{2n+1}$, he finds

$$\varepsilon=2\sqrt[4]{q^{2n+1}}\left(\frac{1+q^{2(2n+1)}}{1+q^{2n+1}}\frac{1+q^{4(2n+1)}}{1+q^{3(2n+1)}}\cdots\right)^2.$$

The other values of α are of the form $\frac{\varpi i+2\mu\omega}{2n+1}$ $(0\le\mu\le 2n)$ and give

$$\varepsilon = 2\sqrt[4]{\delta_1^\mu q^{\frac{1}{2n+1}}}\left(\frac{1+\left(\delta_1^\mu q^{\frac{1}{2n+1}}\right)^2}{1+\delta_1^\mu q^{\frac{1}{2n+1}}}\,\frac{1+\left(\delta_1^\mu q^{\frac{1}{2n+1}}\right)^4}{1+\left(\delta_1^\mu q^{\frac{1}{2n+1}}\right)^3}\cdots\right)^2$$

where $\delta_1 = \cos\frac{2\pi}{2n+1} + \sqrt{-1}\sin\frac{2\pi}{2n+1}$ is a primitive $(2n+1)$-th root of 1. Thus the $2n+2$ values of ε are obtained by replacing q in the expression $2\sqrt[4]{q}\left(\frac{1+q^2}{1+q}\frac{1+q^4}{1+q^3}\cdots\frac{1+q^{2m}}{1+q^{2m-1}}\cdots\right)^2 = \sqrt{c}$ by

$$q^{2n+1},\, q^{\frac{1}{2n+1}},\, \delta_1 q^{\frac{1}{2n+1}},\, \delta_1^2 q^{\frac{1}{2n+1}},\,\ldots,\, \delta_1^{2n} q^{\frac{1}{2n+1}}.$$

The same substitutions in the expression $2\frac{\pi}{\omega}\sqrt[4]{q}\left(\frac{1-q^2}{1-q}\frac{1-q^4}{1-q^3}\cdots\right)^2$ give the $2n+1$ values of δ. Jacobi independently discovered similar rules for the transformations (1829).

In a very short paper published in *Crelle's Journal*(vol. 3, 1828; *Œuvres*, t. I, p. 466), Abel states the rule for the transformation of elliptic integrals of the third kind. Let $f(y,x)=0$ be an algebraic integral of the differential equation $\frac{dy}{\sqrt{(1-y^2)(1-c'y^2)}} = a\frac{dx}{\sqrt{(1-x^2)(1-c^2x^2)}}$. Then $\int\frac{A+Bx^2}{1-\frac{x^2}{n^2}}\frac{dx}{\sqrt{(1-x^2)(1-c^2x^2)}} = \int\frac{A'+B'y^2}{1-\frac{y^2}{m^2}}\frac{dy}{\sqrt{(1-y^2)(1-c'^2y^2)}} + k\log p$ where A', B', m and k are functions of A, B, n and p is an algebraic function of y and x. The transformed parameter m is determined by the equation $f(m,n)=0$. For n infinite, the integrals are of the second kind and the rule for the transformation is

$$\int(A+Bx^2)\frac{dx}{\sqrt{(1-x^2)(1-c^2x^2)}} = \int(A'+B'y^2)\frac{dy}{\sqrt{(1-y^2)(1-c'^2y^2)}} + v$$

where v is an algebraic function of x and y.

8 Further Development of the Theory of Elliptic Functions and Abelian Integrals

In the fourth volume of *Crelle's Journal* (1829; *Œuvres*, t. I, p. 467–477), Abel published a *Note sur quelques formules elliptiques*, devoted to the translation of the formulae given in the the *Recherches* into Legendre's notation in order to recover results published by Jacobi. Supposing that $c=1$, the problem is to pass from the case in which e is real to the case in which e^2 is negative. Abel puts $\lambda\alpha = f\left(\frac{\omega}{2}-b\alpha\right)$ where $b=\frac{1}{\sqrt{1+e^2}}$ so that $x=\lambda\alpha$ is equivalent to $\alpha=\int_0\frac{dx}{\sqrt{(1-x^2)(1-c^2x^2)}}$ where $c=\frac{e}{\sqrt{1+e^2}}$ and $b=\sqrt{1-c^2}$. One has

$$\frac{\omega}{2} = b\int_0^1 \frac{dx}{\sqrt{(1-x^2)(1-c^2x^2)}}, \quad \frac{\varpi}{2} = b\int_0^1 \frac{dx}{\sqrt{(1-x^2)(1-b^2x^2)}}.$$

Abel auxiliary functions $\lambda'\alpha = \sqrt{1-\lambda^2\alpha}$, $\lambda''\alpha = \sqrt{1-c^2\lambda^2\alpha}$ to play the roles of cosamα, Δamα in Jacobi's notation. One has $\lambda'\alpha = \varphi\left(\frac{\omega}{2} - b\alpha\right)$, $\lambda''\alpha = bF\left(\frac{\omega}{2} - b\alpha\right)$. As in the preceding papers, Abel uses the expansions of $f\alpha$, $\varphi\alpha$ and $F\alpha$ in simple infinite products to obtain expressions for $\lambda\theta$, $\lambda'\theta$ and $\lambda''\theta$ (cf. (98)):

$$\lambda\theta = A\frac{1-\rho^2}{1+\rho^2}\frac{1-\rho^2r^2}{1+\rho^2r^2}\frac{1-\rho^{-2}r^2}{1+\rho^{-2}r^2}\frac{1-\rho^2r^4}{1+\rho^2r^4}\frac{1-\rho^{-2}r^4}{1+\rho^{-2}r^4}\cdots,$$
$$\lambda'\theta = A'\frac{2\rho}{1+\rho^2}\frac{1-\rho^2r}{1+\rho^2r^2}\frac{1-\rho^{-2}r}{1+\rho^{-2}r^2}\frac{1-\rho^2r^3}{1+\rho^2r^4}\frac{1-\rho^{-2}r^3}{1+\rho^{-2}r^4}\cdots,$$
$$\lambda''\theta = A''\frac{2\rho}{1+\rho^2}\frac{1+\rho^2r}{1+\rho^2r^2}\frac{1+\rho^{-2}r}{1+\rho^{-2}r^2}\frac{1+\rho^2r^3}{1+\rho^2r^4}\cdot\frac{1+\rho^{-2}r^3}{1+\rho^{-2}r^4}\cdots \quad (100)$$

where $\rho = e^{-\frac{\theta\pi}{\varpi'}}$, $r = e^{-\frac{\omega'}{\varpi'}\pi}$, $\frac{\omega'}{2} = \int_0^{\frac{\pi}{2}} \frac{d\theta}{\sqrt{1-c^2\sin^2\theta}}$, $\frac{\varpi'}{2} = \int_0^{\frac{\pi}{2}} \frac{d\theta}{\sqrt{1-b^2\sin^2\theta}}$ and

$$\sqrt{A} = \frac{(1+r)(1+r^3)\cdots}{(1-r)(1-r^3)\cdots},$$
$$\sqrt{A'} = \frac{(1+r^2)(1+r^4)(1+r^6)\cdots}{(1-r)(1-r^3)(1-r^5)\cdots},$$
$$\sqrt{A''} = \frac{(1+r^2)(1+r^4)(1+r^6)\cdots}{(1+r)(1+r^3)(1+r^5)\cdots}. \quad (101)$$

Doing $\theta = \frac{\omega'}{2} + \frac{\varpi'}{2}i$ in (100), one obtains $\rho^2 = -r$ and

$$\lambda\theta = f\left(\frac{\varpi}{2}i\right) = \frac{\sqrt{1+e^2}}{e} = \frac{1}{c} = A\left(\frac{(1+r)(1+r^3)(1+r^5)\ldots}{(1-r)(1-r^3)(1-r^5)\ldots}\right)^2 = A^2,$$
$$\lambda'\theta = -\varphi\left(\frac{\varpi i}{2}\right) = -\frac{i}{e} = i\frac{b}{c} = 4A'i\sqrt{r}\left(\frac{1+r^2}{1-r}\frac{1+r^4}{1-r^3}\cdots\right)^2.$$

Thus $A = \frac{1}{\sqrt{c}}$, $A' = \frac{1}{2\sqrt[4]{r}}\sqrt{\frac{b}{c}}$. In a similar way, doing $\theta = \frac{\omega'}{2}$ one obtains $\rho^2 = r$ and $\lambda''\theta = b = 4A''\sqrt{r}\left(\frac{1+r^2}{1+r}\frac{1+r^4}{1+r^3}\cdots\right)^2 = 4A''\sqrt{r}\cdot A''$ so that $A'' = \frac{\sqrt{b}}{2\sqrt[4]{r}}$. These values compared with (101) give

$$\sqrt[4]{c} = \frac{1-r}{1+r}\frac{1-r^3}{1+r^3}\frac{1-r^5}{1+r^5}\cdots,$$
$$\sqrt[4]{\frac{b}{c}} = \sqrt{2}\sqrt[8]{r}\frac{1+r^2}{1-r}\frac{1+r^4}{1-r^3}\frac{1+r^6}{1-r^5}\cdots,$$
$$\sqrt[4]{b} = \sqrt{2}\sqrt[8]{r}\frac{1+r^2}{1+r}\frac{1+r^4}{1+r^3}\frac{1+r^6}{1+r^5}\cdots. \quad (102)$$

The limit value of $\frac{\lambda\theta}{1-\rho^2}$ for $\theta \to 0$ is $\frac{\varpi'}{2\pi}$ and, comparing with (100), this gives

$$\sqrt[4]{c}\sqrt{\frac{\varpi'}{\pi}} = \frac{(1-r^2)(1-r^4)(1-r^6)\cdots}{(1+r^2)(1+r^4)(1+r^6)\cdots} \text{ and}$$
$$\sqrt{\frac{\varpi'}{\pi}} = \frac{(1+r)(1-r^2)(1+r^3)(1-r^4)\cdots}{(1-r)(1+r^2)(1-r^3)(1+r^4)\cdots}$$
$$= ((1+r)(1+r^3)(1+r^5)\cdots)^2$$
$$\times(1+r)(1+r^2)(1+r^3)\cdots \times (1-r)(1-r^2)(1-r^3)\cdots .$$

Abel puts $P = (1+r)(1+r^3)(1+r^5)\cdots$ and $P' = (1+r^2)(1+r^4)(1+r^6)\cdots$ so that

$$PP' = (1+r)(1+r^2)(1+r^3)\cdots = \frac{1}{(1-r)(1-r^3)(1-r^5)\cdots}$$

and $\sqrt[4]{c} = \frac{1}{P^2P'}$, $\sqrt[4]{b} = \sqrt{2}\sqrt[8]{r}\frac{P'}{P}$. From these relations, he draws

$$P = \sqrt[6]{2}\sqrt[24]{\frac{r}{b^2c^2}}, \quad P' = \frac{\sqrt[6]{b}\sqrt[24]{r}}{\sqrt[3]{2}\sqrt[12]{c}}\frac{1}{\sqrt[8]{r}}, \tag{103}$$

$PP' = (1+r)(1+r^2)(1+r^3)(1+r^4)\cdots = \frac{\sqrt[12]{b}}{\sqrt[6]{2c}\sqrt[24]{r}}$, $(1-r)(1-r^2)(1-r^3)\cdots = \frac{\sqrt[12]{b}\sqrt[3]{c}}{\sqrt[6]{2}\sqrt[24]{r}}\sqrt{\frac{\varpi'}{\pi}}$, one of the formulae published by Jacobi.

Now putting $q = e^{-\frac{\varpi'}{\omega'}\pi}$ so that $\log r \log q = \pi^2$, $\theta = \frac{\varpi'}{2} + \frac{\omega'}{2}\sqrt{-1} + \frac{\omega'}{\pi}x\sqrt{-1}$ and exchanging b and c, Abel obtains from (100)

$$\lambda\left(\frac{\omega'}{\pi}x\right) = \frac{2}{\sqrt{c}}\sqrt[4]{q}\sin x\frac{1-2q^2\cos 2x+q^4}{1-2q\cos 2x+q^2}\frac{1-2q^4\cos 2x+q^8}{1-2q^3\cos 2x+q^6}\cdots,$$
$$\lambda'\left(\frac{\omega'}{\pi}x\right) = 2\sqrt{\frac{b}{c}}\sqrt[4]{q}\cos x\frac{1+2q^2\cos 2x+q^4}{1-2q\cos 2x+q^2}\frac{1+2q^4\cos 2x+q^8}{1-2q^3\cos 2x+q^6}\cdots,$$
$$\lambda''\left(\frac{\omega'}{\pi}x\right) = \sqrt{b}\frac{1+2q\cos 2x+q^2}{1-2q\cos 2x+q^2}\frac{1+2q^3\cos 2x+q^6}{1-2q^3\cos 2x+q^6}\cdots. \tag{104}$$

By comparison with Jacobi's formula for $\Delta\mathrm{am}\alpha$, Abel finds

$$\frac{1+2q\cos 2x+2q^4\cos 4x+2q^9\cos 6x+\dots}{1-2q\cos 2x+2q^4\cos 4x-2q^9\cos 6x+\dots}$$
$$= \frac{(1+2q\cos 2x+q^2)(1+2q^3\cos 2x+q^6)(1+2q^5\cos 2x+q^{10})\dots}{(1-2q\cos 2x+q^2)(1-2q^3\cos 2x+q^6)(1-2q^5\cos 2x+q^{10})\dots}.$$

The logarithms of (104) are written

$$\log\lambda\left(\frac{\omega'}{\pi}x\right) = \log 2 - \frac{1}{2}\log c - \frac{1}{4}\frac{\varpi'}{\omega'}\pi + \log\sin x$$
$$+2\left(\cos 2x\frac{q}{1+q} + \frac{1}{2}\cos 4x\frac{q^2}{1+q^2} + \frac{1}{3}\cos 6x\frac{q^3}{1+q^3} + \dots\right),$$
$$\log\lambda'\left(\frac{\omega'}{\pi}x\right) = \log 2 + \frac{1}{2}\log b - \frac{1}{2}\log c - \frac{1}{4}\frac{\varpi'}{\omega'}\pi + \log\cos x$$
$$+2\left(\cos 2x\frac{q}{1-q} + \frac{1}{2}\cos 4x\frac{q^2}{1+q^2} + \frac{1}{3}\cos 6x\frac{q^3}{1-q^3} + \dots\right),$$
$$\log\lambda''\left(\frac{\omega'}{\pi}x\right) = \frac{1}{2}\log b + 4\left(\cos 2x\frac{q}{1-q^2} + \frac{1}{3}\cos 6x\frac{q^3}{1-q^6} + \dots\right).$$

For $x = 0$, this last formula gives $\log\left(\frac{1}{b}\right) = 8\left(\frac{q}{1-q^2} + \frac{1}{3}\frac{q^3}{1-q^6} + \frac{1}{5}\frac{q^5}{1-q^{10}} + \dots\right)$ and the first one gives

$$\log\left(\frac{1}{c}\right) = \frac{1}{2}\frac{\varpi'}{\omega'}\pi - 2\log 2 + 4\left(\frac{q}{1+q} - \frac{1}{2}\frac{q^2}{1+q^2} + \frac{1}{3}\frac{q^3}{1+q^3} - \dots\right)$$

which is also equal to $8\left(\frac{r}{1-r^2} + \frac{1}{3}\frac{r^3}{1-r^6} + \frac{1}{5}\frac{r^5}{1-r^{10}} + \dots\right)$ according to (102). From the expansions of $\varphi\left(\alpha\frac{\omega}{2}\right)$, $f\left(\alpha\frac{\omega}{2}\right)$ and $F\left(\alpha\frac{\omega}{2}\right)$ in simple series (*Recherches*, formulae (86)), Abel deduces

$$\lambda\left(\frac{\omega'}{\pi}x\right) = \frac{4\pi}{c\omega'}\sqrt{q}\left(\sin x\frac{1}{1-q} + \sin 3x\frac{q}{1-q^3} + \sin 5x\frac{q^2}{1-q^5} + \dots\right)$$
$$\lambda'\left(\frac{\omega'}{\pi}x\right) = \frac{4\pi}{c\omega'}\sqrt{q}\left(\cos x\frac{1}{1+q} + \cos 3x\frac{q}{1+q^3} + \cos 5x\frac{q^2}{1+q^5} + \dots\right)$$
$$= \frac{2\pi}{c\varpi'}\left(\frac{r^x - r^{1-x}}{1+r} - \frac{r^{3x} - r^{3-3x}}{1+r^3} + \frac{r^{5x} - r^{5-5x}}{1+r^5} + \dots\right),$$
$$\lambda''\left(\frac{\omega'}{\pi}x\right) = \frac{2\pi}{\varpi'}\left(\frac{r^x + r^{1-x}}{1-r} - \frac{r^{3x} + r^{3-3x}}{1-r^3} + \frac{r^{5x} + r^{5-5x}}{1-r^5} + \dots\right).$$

Let c' be a modulus (between 0 and 1) such that there exists a transformation from the elliptic functions of modulus c to those of modulus c', and let ω'', ϖ'', r', q' be associated to c' as ω', ϖ', r, q are associated to c. The characterisation stated in *Solution d'un problème général* is $\frac{\omega''}{\varpi''} = \frac{n}{m}\frac{\omega'}{\varpi'}$ where n, m are integers, or $r' = r^{\frac{n}{m}}$, $q' = q^{\frac{m}{n}}$. For instance, let us take $c = \sqrt{\frac{1}{2}}$, so that $\varpi' = \omega'$ and $r = e^{-\pi}$. Any admissible value of c' is given by

$$\sqrt[4]{c'} = \frac{1-e^{-\mu\pi}}{1+e^{-\mu\pi}}\frac{1-e^{-3\mu\pi}}{1+e^{-3\mu\pi}}\frac{1-e^{-5\mu\pi}}{1+e^{-5\mu\pi}}\cdots$$
$$= \sqrt{2}e^{-\frac{\pi}{8\mu}}\frac{1+e^{-\frac{2\pi}{\mu}}}{1+e^{-\frac{\pi}{\mu}}}\frac{1+e^{-\frac{4\pi}{\mu}}}{1+e^{-\frac{3\pi}{\mu}}}\frac{1+e^{-\frac{6\pi}{\mu}}}{1+e^{-\frac{5\pi}{\mu}}}\cdots$$

where μ is a rational number and such a c' is expressible by radicals. Another example is that in which $b' = c$ or $c' = b$; then $\omega'' = \varpi'$, $\varpi'' = \omega'$. In this case $\frac{\varpi'}{\omega'} = \frac{\omega''}{\varpi''} = \frac{n}{m}\frac{\omega'}{\varpi'}$ and $\frac{\omega'}{\varpi'} = \sqrt{\frac{m}{n}} = \sqrt{\mu}$. Thus $\sqrt[4]{c} = \frac{1-e^{-\pi\sqrt{\mu}}}{1+e^{-\pi\sqrt{\mu}}}\frac{1-e^{-3\pi\sqrt{\mu}}}{1+e^{-3\pi\sqrt{\mu}}}\frac{1-e^{-5\pi\sqrt{\mu}}}{1+e^{-5\pi\sqrt{\mu}}}\cdots$ and $\sqrt[4]{b} = \frac{1-e^{-\frac{\pi}{\sqrt{\mu}}}}{1+e^{-\frac{\pi}{\sqrt{\mu}}}}\frac{1-e^{-\frac{3\pi}{\sqrt{\mu}}}}{1+e^{-\frac{3\pi}{\sqrt{\mu}}}}\frac{1-e^{-\frac{5\pi}{\sqrt{\mu}}}}{1+e^{-\frac{5\pi}{\sqrt{\mu}}}}\cdots$

At the end of this paper, Abel deduces the functional equation for a *theta*-function from (103). Exchanging c and b and r and q, he obtains $(1+q)(1+q^3)(1+q^5)\cdots = \sqrt[6]{2}\frac{\sqrt[24]{q}}{\sqrt[12]{bc}}$ and comparing with (103)

$$\frac{1}{\sqrt[24]{r}}(1+r)(1+r^3)(1+r^5)\cdots = \frac{1}{\sqrt[24]{q}}(1+q)(1+q^3)(1+q^5)\cdots$$

whenever r and q are between 0 and 1 and related by $\log r \cdot \log q = \pi^2$. He recalls some other results due to Cauchy (1818) and to Jacobi (1829).

In a second paper of the fourth volume of *Crelle's Journal* (1829), *Théorèmes sur les fonctions elliptiques* (*Œuvres*, t. I, p. 508–514), Abel considers the equation $\varphi(2n+1)\theta = R$ of which the roots are $x = \varphi(\theta + m\alpha + \mu\beta)$ where φ is the elliptic function of the *Recherches*, $\alpha = \frac{2\omega}{2n+1}$, $\beta = \frac{2\varpi i}{2n+1}$ and m, μ are integers. He proves that if $\psi\theta$ is a polynomial in these roots which is invariant when θ is changed into $\theta+\alpha$ or into $\theta+\beta$, one has

$$\psi\theta = p + qf(2n+1)\theta \cdot F(2n+1)\theta$$

where p and q are polynomials in $\varphi(2n+1)\theta$, of respective degrees ν and $\nu-2$, ν being the highest exponent of $\varphi\theta$ in $\psi\theta$. Indeed, by the addition theorem (73), $\varphi(\theta + m\alpha + \mu\beta)$ is a rational function of $\varphi\theta$ and $f\theta \cdot F\theta$. Since $(f\theta \cdot F\theta)^2 = (1 - c^2\varphi^2\theta)(1 + e^2\varphi^2\theta)$, one has

$$\psi\theta = \psi_1(\varphi\theta) + \psi_2(\varphi\theta)\, f\theta \cdot F\theta$$

where $\psi_1(\varphi\theta)$ and $\psi_2(\varphi\theta)$ are rational. They are respectively given by

$$\begin{aligned}\psi_1(\varphi\theta) &= \frac{1}{2}(\psi\theta + \psi(\omega-\theta)) \quad \text{and}\\ \psi_2(\varphi\theta)\, f\theta \cdot F\theta &= \frac{1}{2}(\psi\theta - \psi(\omega-\theta)).\end{aligned} \tag{105}$$

The invariance of $\psi\theta$ by $\theta \mapsto \theta+\alpha$ or $\theta+\beta$ implies that $\psi_1(\varphi(\theta + m\alpha + \mu\beta)) = \psi_1(\varphi\theta)$, so that $\psi_1(\varphi\theta)$ is a rational symmetric function of the roots of the considered equation. Thus $\psi_1(\varphi\theta) = p$ rational function of $\varphi(2n+1)\theta = y$. If $y = \varphi(2n+1)\delta$ is a pole of p, (105) shows that some $\delta + m\alpha + \mu\beta$ or some $\omega - \delta + m\alpha + \mu\beta$ is a pole of φ, but then $(2n+1)\delta$ is also a pole of φ, which is absurd. On the other hand, $f(2n+1)\theta = f\theta \cdot u$, $F(2n+1)\theta = F\theta \cdot v$ where u and v are rational functions of $\varphi\theta$. It results that $\frac{\psi_2(\varphi\theta)\, f\theta \cdot F\theta}{f(2n+1)\theta \cdot F(2n+1)\theta} = \chi(\varphi\theta)$ rational function of $\varphi\theta$ also equal to $\frac{1}{2}\frac{\psi\theta - \psi(\omega-\theta)}{f(2n+1)\theta \cdot F(2n+1)\theta}$ according to (105). Thus $\chi(\varphi\theta)$ is invariant by $\theta \mapsto \theta+\alpha$ or $\theta+\beta$

and one proves as above that is it a polynomial q in $\varphi(2n+1)\theta$. Abel computes the degrees of p and q by considering the behaviour of $\psi\theta$ and $\psi(\omega-\theta)$ when $\varphi\theta$ is infinite.

When $\nu = 1$, p is of degree 1 and $q = 0$, so that $\psi\theta = A + B\varphi(2n+1)\theta$ where A and B are constants. This is the case for $\psi\theta = \sum_{m=0}^{2n}\sum_{\mu=0}^{2n} \pi(\theta + m\alpha + \mu\beta)$ where $\pi\theta$ is the product of some roots of the equation and one finds that $A = 0$ when the number of factors of $\pi\theta$ is odd whereas $B = 0$ when this number is even.

In the same way, Abel obtains that if $\psi\theta$ is a polynomial in the quantities $f(\theta + m\alpha + \mu\beta)$ (resp. $F(\theta + m\alpha + \mu\beta)$) such that $\psi(\theta) = \psi(\theta+\alpha) = \psi(\theta+\beta)$, then $\psi\theta = p + q\varphi(2n+1)\theta \cdot F(2n+1)\theta$ (resp. $p + q\varphi(2n+1)\theta \cdot f(2n+1)\theta$) where p and q are polynomials in $f(2n+1)\theta$ (resp. $F(2n+1)\theta$) of respective degrees $\nu, \nu-2$, ν beeing the highest exponent of $f\theta$ (resp. $F\theta$) in $\psi\theta$.

As an application, Abel deduces a formula established by Jacobi (1828) for the division of elliptic integrals: $\varphi\left(\frac{\theta}{2n+1}\right) = \frac{1}{2n+1}\sum_{m=0}^{4n^2+4n} \sqrt[2n+1]{p_m + q_m f\theta \cdot F\theta}$ where p_m (resp. q_m) is an odd (resp. even) polynomial in $\varphi\theta$ of degree $2n+1$ (resp. $2n-2$) and $p_m^2 - q_m^2(f\theta)^2(F\theta)^2 = (\varphi^2\theta - a_m^2)^{2n+1}$ where a_m is a constant.

A third memoir of Abel in the volume 4 of *Crelle's Journal* (1829) is a small treatise on elliptic functions, titled *Précis d'une théorie des fonctions elliptiques* (*Œuvres*, t. I, p. 518–617). He uses the following notations: $\Delta(x,c) = \pm\sqrt{(1-x^2)(1-c^2x^2)}$, $\varpi(x,c) = \int \frac{dx}{\Delta(x,c)}$ (integral of the first kind), $\varpi_0(x,c) = \int \frac{x^2dx}{\Delta(x,c)}$ (integral of the second kind) and

$$\Pi(x,c,a) = \int \frac{dx}{\left(1-\frac{x^2}{a^2}\right)\Delta(x,c)} \qquad \text{(integral of the third kind).}$$

The general problem dealt with by Abel is the following: "To find all the possible cases in which one may satisfy an equation of the form

$$\begin{aligned} &\alpha_1\varpi(x_1,c_1) + \alpha_2\varpi(x_2,c_2) + \ldots + \alpha_n\varpi(x_n,c_n) \\ &+\alpha_1'\varpi_0(x_1',c_1') + \alpha_2'\varpi_0(x_2',c_2') + \ldots + \alpha_m'\varpi_0(x_m',c_m') \\ &+\alpha_1''\Pi(x_1'',c_1'',a_1) + \alpha_2''\Pi(x_2'',c_2'',a_2) + \ldots + \alpha_\mu''\Pi(x_\mu'',c_\mu'',a_\mu) \\ = {}& u + A_1\log v_1 + A_2\log v_2 + \ldots + A_\nu \log v_\nu \end{aligned} \tag{106}$$

where $\alpha_1, \alpha_2, \ldots, \alpha_n;\ \alpha_1', \alpha_2', \ldots, \alpha_m';\ \alpha_1'', \alpha_2'', \ldots, \alpha_\mu'';\ A_1, A_2, \ldots, A_\nu$ are constant quantities, $x_1, x_2, \ldots, x_n;\ x_1', x_2', \ldots, x_m';\ x_1'', x_2'', \ldots, x_\mu''$ variables related by *algebraic* equations and $u, v_1, v_2, \ldots, v_\nu$ *algebraic* functions of these variables." This problem is attacked by purely algebraic means, that is without the use of elliptic functions and their double periodicity.

Here are some results announced in the introduction: " If $\int \frac{rdx}{\Delta(x,c)}$, where r is an arbitrary rational function of x, is expressible by algebraic and logarithmic functions and by elliptic integrals $\psi, \psi_1, \psi_2, \ldots$, one may always suppose that

$$\int \frac{rdx}{\Delta(x,c)} = p\Delta(x,c) + \alpha\psi(y) + \alpha'\psi_1(y_1) + \alpha''\psi_2(y_2) + \dots$$
$$+ A_1 \log \frac{q_1 + q_1'\Delta(x,c)}{q_1 - q_1'\Delta(x,c)} + A_2 \log \frac{q_2 + q_2'\Delta(x,c)}{q_2 - q_2'\Delta(x,c)} + \dots$$

where all the quantities $p, q_1, q_2, \dots, q_1', q_2', \dots, y, y_1, y_2, \dots$ are *rational functions* of x." In this statement, $\Delta(x,c)$ may be the square root of a polynomial of any degree.

"If any equation of the form (106) takes place and one designates by c any one of the moduli which figure in it, among the other moduli there is at least one c' such that the differential equation $\frac{dy}{\Delta(y,c')} = \varepsilon \frac{dx}{\Delta(x,c)}$ may be satisfied by putting for y a *rational* function of x, and *vice versa*."

The second part of the memoir was not written by Abel and we have only the statement of its principal results in the introduction. Abel supposes that $0 < c < 1$ and introduces the elliptic function $\lambda\theta$ inverse of $\varpi(x,c)$, with its main properties: double periodicity, with the fundamental periods 2ϖ, ωi given by $\frac{\varpi}{2} = \int_0^1 \frac{dx}{\Delta(x,c)}$, $\frac{\omega}{2} = \int_0^1 \frac{dx}{\Delta(x,b)}$, determination of its zeros and poles, equation $\lambda(\theta' + \theta)\lambda(\theta' - \theta) = \frac{\lambda^2\theta' - \lambda^2\theta}{1 - c^2\lambda^2\theta \cdot \lambda^2\theta'}$, expansion in infinite product. He proves that if the equation $(\lambda\theta)^{2n} + a_{n-1}(\lambda\theta)^{2n-2} + \dots + a_1(\lambda\theta)^2 + a_0 = (b_0\lambda\theta + b_1(\lambda\theta)^3 + \dots + b_{n-2}(\lambda\theta)^{2n-3})\Delta(\lambda\theta, c)$ is satisfied by $\theta = \theta_1, \theta_2, \dots, \theta_{2n}$ such that $\lambda^2\theta_1, \lambda^2\theta_2, \dots, \lambda^2\theta_{2n}$ be different between them, then $\lambda(\theta_1 + \theta_2 + \dots + \theta_{2n}) = 0$ and $-\lambda(\theta_{2n}) = \lambda(\theta_1 + \theta_2 + \dots + \theta_{2n-1}) = \frac{-a_0}{\lambda\theta_1\lambda\theta_2 \dots \lambda\theta_{2n-1}}$. This statement gives a general theorem for the addition and its proof is given in the first part.

The roots of the equation of division of the periods are related by remarkable linear relations, where $\delta = \cos\frac{2\pi}{2\mu+1} + \sqrt{-1}\sin\frac{2\pi}{2\mu+1}$ is a primitive $(2\mu+1)$-th root of 1:

$$0 = \lambda\left(\frac{2m\varpi}{2\mu+1}\right) + \delta^k\lambda\left(\frac{2m\varpi + \omega i}{2\mu+1}\right) + \delta^{2k}\lambda\left(\frac{2m\varpi + 2\omega i}{2\mu+1}\right)$$
$$+ \delta^{3k}\lambda\left(\frac{2m\varpi + 3\omega i}{2\mu+1}\right) + \dots + \delta^{2\mu k}\lambda\left(\frac{2m\varpi + 2\mu\omega i}{2\mu+1}\right),$$
$$0 = \lambda\left(\frac{m\omega i}{2\mu+1}\right) + \delta^{k'}\lambda\left(\frac{2\varpi + m\omega i}{2\mu+1}\right) + \delta^{2k'}\lambda\left(\frac{4\varpi + m\omega i}{2\mu+1}\right)$$
$$+ \delta^{3k'}\lambda\left(\frac{6\varpi + m\omega i}{2\mu+1}\right) + \dots + \delta^{2\mu k'}\lambda\left(\frac{4\mu\varpi + m\omega i}{2\mu+1}\right).$$

Sylow gave a demonstration of these relations in 1864 and he explains how to deduce them from the theory of transformation in the final notes to Abel's Works. He also reproduces another proof communicated to him by Kronecker in a letter in 1876 (*Œuvres*, t. II, p. 314–316).

If there is a transformation of $\int \frac{dx}{\Delta(x,c)}$ (with $0 < c < 1$) into $\varepsilon \int \frac{dy}{\Delta(y,c')}$ (with c' arbitrary) by putting for y an algebraic function of x, then c' is given by one of the following formulae: $\sqrt[4]{c'} = \sqrt{2}\sqrt[8]{q_1}\frac{(1+q_1^2)(1+q_1^4)(1+q_1^6)\cdots}{(1+q_1)(1+q_1^3)(1+q_1^5)\cdots}$, $\sqrt[4]{c'} = \frac{1-q_1}{1+q_1}\frac{1-q_1^3}{1+q_1^3}\frac{1-q_1^5}{1+q_1^5}\cdots$ where $q_1 = q^\mu = e^{(-\mu\frac{\omega}{\varpi}+\mu' i)\pi}$, μ, μ' rational numbers.

As in the *Solution d'un problème général*, Abel obtains a statement concerning the rational transformation of a real elliptic integral of modulus c into another of modulus c', with $0 < c, c' < 1$. The periods $\varpi, \omega, \varpi', \omega'$ must be related by $\frac{\varpi'}{\omega'} = \frac{n'}{m}\frac{\varpi}{\omega}$ where n', m are integers and this condition is sufficient, the multiplicator being $\varepsilon = m\frac{\varpi'}{\varpi}$. Abel proposes to determine the rational function of x expressing y by means of its zeros and poles.

When c may be transformed into its complement $b = \sqrt{1-c^2}$ (singular modulus), $\frac{\varpi}{\omega} = \sqrt{\frac{m}{n}}$ and $\frac{dy}{\Delta(y,b)} = \sqrt{mn}\frac{dx}{\Delta(x,c)}$. Abel says that c is determined by an algebraic equation which "seems to be solvable by radicals"; he is thus doubtful about this fact, later proved by Kronecker. In the final notes (*Œuvres*, t. II, p. 316–318), Sylow gives a proof of this fact by reduction to the solvability of the equation of division of the periods. Abel gives an expression of $\sqrt[4]{c}$ by an infinite product. He also state that two moduli c and c' which may be transformed into one another are related by an algebraic relation and that, in general, it does not seem possible to draw the value of c' by radicals. But it is possible when c may be transformed into its complement. According to Abel, all the roots of a modular equation are *rationally* expressible by two of them, but this statement is mistaken; they are expressible with the help of radicals by one of them.

Abel gives an expression of $\lambda\theta$ as a quotient of two entire functions $\varphi\theta = \theta + a\theta^3 + a'\theta^5 + \ldots$ and $f\theta = 1 + b'\theta^4 + b''\theta^6 + \ldots$ related by the functional equations

$$\varphi(\theta'+\theta)\varphi(\theta'-\theta) = (\varphi\theta f\theta')^2 - (\varphi\theta f\theta')^2 - (\varphi\theta' f\theta)^2,$$
$$f(\theta'+\theta)f(\theta'-\theta) = (f\theta f\theta')^2 - c^2(\varphi\theta\varphi\theta')^2.$$

These functions are similar to the al-functions of Weierstrass, later replaced by σ. As we have said, Abel communicated the functional equations to Legendre, saying that they characterise the functions φ and f (see §1).

Abel adds that most of these properties are still valid when the modulus c is a complex number.

The first part of the memoir, the only one written and published, is divided in five chapters. In the first one (p. 528–545), Abel deals with the general properties of elliptic integrals, beginning by Euler addition theorem proved as a particular case of Abel theorem: "Let fx and φx be two arbitrary polynomials in x, one even and the other odd, with indeterminate coefficients. Let us put $(fx)^2 - (\varphi x)^2(\Delta x)^2 = A(x^2-x_1^2)(x^2-x_2^2)(x^2-x_3^2)\ldots(x^2-x_\mu^2)$ where A does not depend on x, I say that one will have

$$\Pi x_1 + \Pi x_2 + \Pi x_3 + \ldots + \Pi x_\mu = C - \frac{a}{2\Delta a}\log\frac{fa + \varphi a\cdot\Delta a}{fa - \varphi a\cdot\Delta a}, \qquad (107)$$

a denoting the parameter of the function Πx, such that $\Pi x = \int\frac{dx}{\left(1-\frac{x^2}{a^2}\right)\Delta x}$. The quantity C is the integration constant."

When all the coefficients of fx and φx are independent variables, $x_1, x_2, \ldots, x_\mu$ are all distinct and one has $\psi x = (fx)^2 - (\varphi x)^2(\Delta x)^2 = 0$ and $fx + \varphi x \Delta x = 0$ if x is any one of them. Let δ denote the differentiation with respect to the variable coefficients of fx and φx. One has

$$\psi' x \cdot dx + 2fx \cdot \delta fx - 2\varphi x \cdot \delta\varphi x \cdot (\Delta x)^2 = \psi' x \cdot dx - 2\Delta x(\varphi x \cdot \delta fx - fx \cdot \delta\varphi x)$$
$$= 0,$$

thus $\Pi x = \int \frac{2(\varphi x \cdot \delta fx - fx \cdot \delta\varphi x)}{\left(1-\frac{x^2}{a^2}\right)\psi' x}$ and

$$\Pi x_1 + \Pi x_2 + \Pi x_3 + \ldots + \Pi x_\mu$$
$$= \int \left(\frac{\theta x_1}{\left(1-\frac{x_1^2}{a^2}\right)\psi' x_1} + \frac{\theta x_2}{\left(1-\frac{x_2^2}{a^2}\right)\psi' x_2} + \ldots + \frac{\theta x_\mu}{\left(1-\frac{x_\mu^2}{a^2}\right)\psi' x_\mu} \right) \quad (108)$$

where $\theta x = 2(\varphi x \cdot \delta fx - fx \cdot \delta\varphi x)$ is a polynomial in x, of degree less than that of ψx. Therefore, the right hand side of (108) is equal to $\int \frac{a\theta a}{2\psi a} = a \int \frac{\varphi a \cdot \delta fa - fa \cdot \delta\varphi a}{(fa)^2 - (\varphi a)^2(\Delta a)^2} = C - \frac{a}{2\Delta a} \log \frac{fa + \varphi a \cdot \Delta a}{fa - \varphi a \cdot \Delta a}$. This proof is valid whenever Δx is the square root of an even polynomial in x, as was seen in the publication of Abel theorem for hyperelliptic functions in the third volume of *Crelle's Journal* (see our §5). It is naturally extended to the case in which the coefficients of fx and φx are no more independent and some of the x_j may be equal. Taking a infinite, Πx is reduced to the integral of the first kind ϖx and the logarithmic part vanishes, so that

$$\varpi x_1 + \varpi x_2 + \ldots + \varpi x_\mu = C.$$

An expansion of both members of (107) in ascending powers of $\frac{1}{a}$ gives, by comparison of the coefficients of $\frac{1}{a^2}$, $\varpi_0 x_1 + \varpi_0 x_2 + \ldots + \varpi_0 x_\mu = C - p$ where p is an *algebraic* function of the variables.

As in the general case of Abel theorem, one may choose $x_1, x_2, \ldots, x_m$ as independent variables and determine the coefficients of fx and φx in function of them. The $\mu - m$ quantities $x_{m+1}^2, x_{m+2}^2, \ldots, x_\mu^2$ are then the roots of an equation of degree $\mu - m$ and they are algebraic functions of $x_1, x_2, \ldots, x_m$. The minimum value of $\mu - m$ is 1. When $\mu = 2n$ is even, one may take

$$fx = a_0 + a_1 x^2 + a_2 x^4 + \ldots + a_{n-1} x^{2n-2} + x^{2n},$$
$$\varphi x = (b_0 + b_1 x^2 + b_2 x^4 + \ldots + b_{n-2} x^{2n-4})x,$$
$$(fx)^2 - (\varphi x)^2(1-x^2)(1-c^2x^2)$$
$$= (x^2 - x_1^2)(x^2 - x_2^2)\ldots(x^2 - x_{2n-1}^2)(x^2 - y^2), \quad (109)$$
$$fx_1 + \varphi x_1 \cdot \Delta x_1 = fx_2 + \varphi x_2 \cdot \Delta x_2 = \ldots$$
$$= fx_{2n-1} + \varphi x_{2n-1} \cdot \Delta x_{2n-1} = 0. \quad (110)$$

The linear equations of the last line determine the coefficients $a_0, a_1, \ldots, a_{n-1}$, $b_0, b_1, \ldots, b_{n-2}$ as *rational* functions of $x_1, x_2, \ldots, x_{2n-1}, \Delta x_1, \Delta x_2, \ldots, \Delta x_{2n-1}$. For $x = 0$, (109) gives

$$a_0^2 = x_1^2 x_2^2 \dots x_{2n-1}^2 y^2 \quad \text{whence } y = -\frac{a_0}{x_1 x_2 \dots x_{2n-1}}.$$

Since $fx_{2n} + \varphi x_{2n} \cdot \Delta x_{2n} = 0$, putting $\Delta x_{2n} = -\Delta y$ one has $\Delta y = \frac{fy}{\varphi y}$ a *rational* function of $x_1, x_2, \dots, \Delta x_1, \Delta x_2, \dots$ as is y. If, in (109), we put $x_1 = x_2 = \dots = x_{2n-1} = 0$, the right hand side becomes divisible by x^{4n-2} and we must have $a_0 = a_1 = \dots = a_{n-1} = b_0 = b_1 = \dots = b_{n-2} = 0$. Thus we obtain $x^{4n} = x^{4n-2}(x^2 - y^2)$ and $y = 0$. Abel shows that if $\Delta x_1 = \Delta x_2 = \dots = \Delta x_{2n-1} = 1$ for $x_1 = x_2 = \dots = x_{2n-1} = 0$, then $\Delta y = 1$. Indeed, for $x_1, x_2, \dots, x_{2n-1}$ infinitely small, the equations (110) reduce to

$$x^{2n} + a_{n-1}x^{2n-2} + b_{n-2}x^{2n-3} + \dots + b_0 x + a_0 = 0, \tag{111}$$

with $2n$ roots $x_1, x_2, \dots, x_{2n-1}$ and z such that $a_0 = z x_1 x_2 \cdots x_{2n-1}$. Thus $z = -y$ and consequently

$$y^{2n} + a_{n-1}y^{2n-2} + \dots + a_1 y^2 + a_0 = (b_0 + b_1 y^2 + \dots + b_{n-2}y^{2n-4})y,$$

relation equivalent to $\Delta y = 1$. Since the sum of the roots of (111) is 0, we have

$$y = x_1 + x_2 + \dots + x_{2n-1}$$

for $x_1, x_2, \dots, x_{2n-1}$ infinitely small.

It is also possible to take fx odd of degree $2n-1$ and φx even of degree $2n-2$. In this case, one finds that $\frac{1}{cy} = -\frac{a_0}{x_1 x_2 \dots x_{2n-1}}$.

When $\mu = 2n+1$ is odd, let us take

$$\begin{aligned}
&fx = (a_0 + a_1 x^2 + a_2 x^4 + \dots + a_{n-1}x^{2n-2} + x^{2n})x \quad \text{and} \\
&\varphi x = b_0 + b_1 x^2 + b_2 x^4 + \dots + b_{n-1}x^{2n-2}, \\
&(fx)^2 - (\varphi x)^2(1-x^2)(1-c^2x^2) = (x^2 - x_1^2)(x^2 - x_2^2)\dots(x^2 - x_{2n}^2)(x^2 - y^2), \\
&fx_1 + \varphi x_1 \cdot \Delta x_1 = fx_2 + \varphi x_2 \cdot \Delta x_2 = \dots = fx_{2n} + \varphi x_{2n} \cdot \Delta x_{2n} = 0.
\end{aligned}$$

As in the preceding case, one obtains $y = \frac{b_0}{x_1 x_2 \dots x_{2n}}$ and $\Delta y = \frac{fy}{\varphi y}$. For $x_1, x_2, \dots, x_{2n}$ infinitesimal, $\Delta x_1, \Delta x_2, \dots, \Delta x_{2n}$ being 1, one has $y = x_1 + x_2 + \dots + x_{2n}$ and $\Delta y = 1$. One may also suppose fx even and φx odd, and then $\frac{1}{cy} = \frac{b_0}{x_1 x_2 \dots x_{2n}}$.

When $n = 1$, $fx = a_0 x + x^3$, $\varphi x = b_0$ where a_0 and b_0 are determined by the equations

$$a_0 x_1 + x_1^3 + b_0 \Delta x_1 = a_0 x_2 + x_2^3 + b_0 \Delta x_2 = 0$$

which give $a_0 = \frac{x_2^3 \Delta x_1 - x_1^3 \Delta x_2}{x_1 \Delta x_2 - x_2 \Delta x_1}$, $b_0 = \frac{x_2 x_1^3 - x_1 x_2^3}{x_1 \Delta x_2 - x_2 \Delta x_1}$. Then

$$y = \frac{b_0}{x_1 x_2} = \frac{x_1^2 - x_2^2}{x_1 \Delta x_2 - x_2 \Delta x_1} = \frac{x_1 \Delta x_2 + x_2 \Delta x_1}{1 - c^2 x_1^2 x_2^2}.$$

One may verify that $(a_0x+x^3)^2-b_0^2(1-x^2)(1-c^2x^2) = (x^2-x_1^2)(x^2-x_2^2)(x^2-y^2)$. The addition theorem takes the form $\varpi x_1 + \varpi x_2 = \varpi y + C$, $\varpi_0 x_0 + \varpi_0 x_2 = \varpi_0 y - x_1x_2y + C$,

$$\Pi x_1 + \Pi x_2 = \Pi y - \frac{a}{2\Delta a}\log\frac{a_0a + a^3 + x_1x_2y\Delta a}{a_0a + a^3 - x_1x_2y\Delta a} + C,$$

and $\Delta y = \frac{a_0y+y^3}{b_0} = \frac{a_0+y^2}{x_1x_2}$.

When $x_1, x_2, \dots, x_\mu = x$ ($\mu = 2n-1$ or $2n$), the coefficients $a_0, a_1, \dots, b_0, b_1, \dots$ are determined by the equation $fx + \varphi x \cdot \Delta x = 0$ and its first $\mu - 1$ derivatives. Let $x_\mu = -\frac{a_0}{x^\mu}$ for $\mu = 2n-1$, $\frac{b_0}{x^\mu}$ for $\mu = 2n$ be the corresponding value of y, such that $\varpi x_\mu = C + \mu\varpi x$. One has $\varpi(x_{\mu+m}) = C + \varpi x_\mu + \varpi x_m = C + \varpi y$ if $y = \frac{x_m\Delta x_\mu + x_\mu\Delta x_m}{1-c^2x_m^2x_\mu^2}$ and this equation gives $x_{\mu+m} = \frac{y\Delta e+e\Delta y}{1-c^2e^2y^2}$ where e is a constant. Letting x tend towards 0, one sees that $x_{\mu+m}$ is equivalent to $(m+\mu)x$ as is y, so that $e = 0$, $\Delta e = 1$ and

$$x_{\mu+m} = \frac{x_m\Delta x_\mu + x_\mu\Delta x_m}{1-c^2x_m^2x_\mu^2}. \tag{112}$$

In the same way, $x_{\mu-m} = \frac{x_m\Delta x_\mu - x_\mu\Delta x_m}{1-c^2x_m^2x_\mu^2}$. For $m = 1$, this gives $x_{\mu+1} = -x_{\mu-1} + \frac{2x_\mu\Delta x}{1-c^2x^2x_\mu^2}$ and it is easy to deduce by induction that $x_{2\mu+1}$, $\frac{\Delta x_{2\mu+1}}{\Delta x}$, $\frac{x_{2\mu}}{\Delta x}$ and $\Delta x_{2\mu}$ are rational functions of x. One has $x_{\mu+m}x_{\mu-m} = \frac{x_\mu^2-x_m^2}{1-c^2x_\mu^2x_m^2}$; for $m = \mu-1$, this gives $x_{2\mu-1} = \frac{1}{x}\frac{x_\mu^2-x_{\mu-1}^2}{1-c^2x_\mu^2x_{\mu-1}^2}$. On the other hand, (112) with $m = \mu$ gives $x_{2\mu} = \frac{2x_\mu\Delta x_\mu}{1-c^2x_\mu^4}$.

Let us write $x_\mu = \frac{p_\mu}{q_\mu}$, $\Delta x_\mu = \frac{r_\mu}{q_\mu^2}$ where p_μ^2 and q_μ are polynomials in x without any common divisor. We have $\frac{p_{2\mu}}{q_{2\mu}} = \frac{2p_\mu q_\mu r_\mu}{q_\mu^4-c^2p_\mu^4}$ whence $p_{2\mu} = 2p_\mu q_\mu r_\mu$, $q_{2\mu} = q_\mu^4 - c^2p_\mu^4$, for these expressions are relatively prime. On the other hand, $\frac{xp_{2\mu-1}}{q_{2\mu-1}} = \frac{p_\mu^2q_{\mu-1}^2-q_\mu^2p_{\mu-1}^2}{q_\mu^2q_{\mu-1}^2-c^2p_\mu^2p_{\mu-1}^2}$ which is an irreducible fraction. Indeed the simultaneous equations $p_\mu^2q_{\mu-1}^2 - q_\mu^2p_{\mu-1}^2 = q_\mu^2q_{\mu-1}^2 - c^2p_\mu^2p_{\mu-1}^2 = 0$ would give $x_\mu^2 = x_{\mu-1}^2$ and $1 - c^2x_\mu^2x_{\mu-1}^2 = 0$. Since $x_{2\mu-1} = \frac{x_\mu\Delta x_{\mu-1}+x_{\mu-1}\Delta x_\mu}{1-c^2x_\mu^2x_{\mu-1}^2} = \frac{x_\mu^2-x_{\mu-1}^2}{x_\mu\Delta x_{\mu-1}-x_{\mu-1}\Delta x_\mu}$, we should have $x_\mu\Delta x_{\mu-1} = x_{\mu-1}\Delta x_\mu = 0$ and this is absurd for $x_\mu^2 = \frac{1}{c}$. Thus $p_{2\mu-1} = \frac{1}{x}(p_\mu^2q_{\mu-1}^2 - q_\mu^2p_{\mu-1}^2)$, $q_{2\mu-1} = q_\mu^2q_{\mu-1}^2 - c^2p_\mu^2p_{\mu-1}^2$ and, from these relations, Abel recursively deduces that $p_{2\mu-1}$ is an odd polynomial in x of degree $(2\mu-1)^2$, $p_{2\mu} = p'\Delta x$ where p' is an odd polynomial of degree $(2\mu)^2 - 3$, q_μ is an even polynomial of degree $\mu^2 - 1$ (resp. μ^2) when μ is odd (resp. even). More precisely,

$$p_{2\mu-1} = x(2\mu - 1 + A_2x^2 + \ldots + A_{(2\mu-1)^2-1}x^{(2\mu-1)^2-1}),$$
$$p_{2\mu} = x\Delta x(2\mu + B_2x^2 + \ldots + B_{4\mu^2-4}x^{(2\mu)^2-4}),$$
$$q_{2\mu-1} = 1 + A_4^1x^4 + \ldots + A^1_{(2\mu-1)^2-1}x^{(2\mu-1)^2-1},$$
$$q_{2\mu} = 1 + B_4^1x^4 + \ldots + B^1_{4\mu^2}x^{(2\mu)^2}.$$

In his second chapter (p. 545–557), Abel considers an exact differential form

$$y_1dx_1 + y_2dx_2 + \ldots + y_\mu dx_\mu \tag{113}$$

where the variables $x_1, x_2, \ldots, x_\mu$ are related by some algebraic relations in number less than μ and $y_1, y_2, \ldots, y_\mu$ are algebraic functions of them. He supposes that its primitive is of the form

$$\begin{aligned} u + A_1 \log v_1 + A_2 \log v_2 + \ldots + A_\nu \log v_\nu \\ + \alpha_1\psi t_1 + \alpha_2\psi t_2 + \ldots + \alpha_n\psi t_n \end{aligned} \tag{114}$$

where $A_1, A_2, \ldots, A_\nu, \alpha_1, \alpha_2, \ldots, \alpha_n$ are constants, $u, v_1, v_2, \ldots, v_\nu, t_1, t_2, \ldots, t_n$ algebraic functions of $x_1, x_2, \ldots, x_\mu$ and $\psi_m x = \int \frac{\theta' dx}{\Delta_m x}$ is an elliptic integral of modulus c_m ($1 \le m \le n$), with $\Delta_m x = \pm\sqrt{(1-x^2)(1-c_m^2x^2)}$ and $\theta' = 1, x^2$ or $\frac{1}{1-\frac{x^2}{a^2}}$. Let us suppose that $x_1, x_2, \ldots, x_m$ are independent variables and that $x_{m+1}, x_{m+2}, \ldots, x_\mu$ are algebraic functions of them. Abel introduces an algebraic function θ such that

$$u, v_1, v_2, \ldots, v_\nu, t_1, t_2, \ldots, t_n, \Delta_1(t_1), \Delta_2(t_2), \ldots, \Delta_n(t_n) \tag{115}$$

are rationally expressible in $\theta, x_1, x_2, \ldots, x_\mu, y_1, y_2, \ldots, y_\mu$. He says that a convenient linear combination of the functions (115) has this property. In other words, Abel uses what is now called a Galois resolvant, which is most remarkable. Let $V = 0$ the minimal algebraic equation satisfied by θ, with coefficients rational with respect to $x_1, x_2, \ldots, x_\mu, y_1, y_2, \ldots, y_\mu$, and let δ be its degree. Writing that (113) is the differential of (114), one obtains a relation

$$p_1dx_1 + p_2dx_2 + \ldots + p_mdx_m = 0$$

where $p_1, p_2, \ldots, p_m$ are rational functions of $\theta, x_1, x_2, \ldots, x_\mu, y_1, y_2, \ldots, y_\mu$ and this implies that $p_1 = p_2 = \ldots = p_m = 0$. These last relations are still verified when θ is replaced by any one of the δ roots $\theta_1, \theta_2, \ldots, \theta_\delta$ of $V = 0$ because it is an irreducible equation. It results that

$$\begin{aligned} \delta \textstyle\int(y_1dx_1 + y_2dx_2 + \ldots + y_\mu dx_\mu) = U + A_1 \log V_1 + \ldots + A_\nu \log V_\nu \\ + \alpha_1(\psi_1 t_1' + \psi_1 t_1'' + \ldots + \psi_1 t_1^{(\delta)}) + \ldots \\ + \alpha_n(\psi_n t_n' + \psi_n t_n'' + \ldots + \psi_n t_n(\delta)) \end{aligned}$$

where $U = u' + u'' + \ldots + u^{(\delta)}$ is the sum of the values taken by u when θ is successively replaced by $\theta_1, \theta_2, \ldots, \theta_\delta$, $\log V_m = \log v_m' + \log v_m'' + \ldots + \log v_m^{(\delta)}$ is the analogous sum associated to $\log v_m$ and $t_m', t_m'', \ldots, t_m^{(\delta)}$ are the values taken by t_m. Now, by the addition theorem for elliptic integrals, $\psi_m t'_m + \psi_m t''_m + \ldots + \psi_m t_m(\delta) = \psi_m T_m + p + B_1 \log q_1 + B_2 \log q_2 + \ldots + B_\nu \log q_\nu$ where $T_m, \Delta_m T_m, p, q_1, q_2, \ldots, q_\nu$ are rational functions of $t'_m, t''_m, \ldots, t_m^{(\delta)}, \Delta_m(t'_m), \Delta_m(t''_m), \ldots, \Delta_m(t_m^{(\delta)})$ and consequently of $\theta_1, \theta_2, \ldots, \theta_\delta, x_1, x_2, \ldots, x_\mu, y_1, y_2, \ldots, y_\mu$. But since they are symmetric with respect to $\theta_1, \theta_2, \ldots, \theta_\delta$, they are rational functions of $x_1, x_2, \ldots, x_\mu, y_1, y_2, \ldots, y_\mu$ as are U and the V_m. We finally obtain a relation of the form

$$\begin{aligned}&\delta \textstyle\int(y_1 dx_1 + y_2 dx_2 + \ldots + y_\mu dx_\mu \\ &= r + A' \log \rho' + A'' \log \rho'' + \ldots + A^{(k)} \log \rho^{(k)} \\ &\quad + \alpha_1 \psi_1 \theta_1 + \alpha_2 \psi_2 \theta_2 + \ldots + \alpha_n \psi_n \theta_n\end{aligned}$$

where δ is an integer, $\alpha_1, \alpha_2, \ldots, \alpha_n$ are the same as in (114), $A', A'', \ldots$ are constants and $\theta_1, \Delta_1(\theta_1), \theta_2, \Delta_2(\theta_2), \ldots, \theta_n, \Delta_n(\theta_n), r, \rho', \rho'', \ldots, \rho^{(k)}$ are *rational* functions of $x_1, x_2, \ldots, x_\mu, y_1, y_2, \ldots, y_\mu$.

A particular case concerns the differential forms (113) of which the primitive is of the form $u + A_1 \log v_1 + A_2 \log v_2 + \ldots + A_\nu \log v_\nu$ where $u, v_1, v_2, \ldots, v_\nu$ are *algebraic* functions of $x_1, x_2, \ldots, x_\mu$. Then one may suppose that $u, v_1, v_2, \ldots, v_\nu$ are *rational* functions of $x_1, x_2, \ldots, x_\mu, y_1, y_2, \ldots, y_\mu$. In a footnote, Abel announces a general theory, based on this result, for the reduction of integrals of algebraic differential forms by algebraic and logarithmic functions.

Applied to elliptic integrals, the preceding theorem takes the form: if

$$\begin{aligned}&\int\left(\frac{\alpha_1 r_1}{\Delta_1 x_1}dx_1 + \frac{\alpha_2 r_2}{\Delta_2 x_2}dx_2 + \ldots + \frac{\alpha_\mu r_\mu}{\Delta_\mu x_\mu}dx_\mu\right) \\ &= u + A_1 \log v_1 + A_2 \log v_2 + \ldots + A_\nu \log v_\nu\end{aligned} \tag{116}$$

where $r_1, r_2, \ldots, r_\mu$ are rational functions and $u, v_1, v_2, \ldots, v_\nu$ are *algebraic* functions of $x_1, x_2, \ldots, x_\mu$, one may suppose that $u, v_1, v_2, \ldots, v_\nu$ are rational functions of $x_1, x_2, \ldots, x_\mu, \Delta_1 x_1, \Delta_2 x_2, \ldots, \Delta_\mu x_\mu$. From (116), we may also conclude that there exists an integer δ such that

$$\begin{aligned}&\delta\alpha_1\psi_1 x_1 + \delta\alpha_2\psi_2 x_2 + \ldots + \delta\alpha_m\psi_m x_m + \alpha_{m+1}\psi_{m+1}\theta_1 + \ldots + \alpha_\mu\psi_\mu\theta_{\mu-m} \\ &= r + A' \log \rho' + A'' \log \rho'' + \ldots + A^{(k)} \log \rho^{(k)}\end{aligned}$$

where $\psi_j x = \int \frac{r_1}{\Delta_j x}dx$ and $\theta_1, \Delta_{m+1}\theta_1, \theta_2, \Delta_{m+2}\theta_2, \ldots, \theta_{\mu-m}, \Delta_\mu\theta_{\mu-m}, r, \rho', \rho'', \ldots, \rho^{(k)}$ are rational functions of $x_1, x_2, \ldots, x_m, \Delta_1 x_1, \Delta_2 x_2, \ldots, \Delta_m x_m$. When only one elliptic integral $\psi_m x$ is isolated, this gives

$$\begin{aligned}\delta\alpha_m\psi_m x = &-\alpha_1\psi_1\theta_1 - \alpha_2\psi_2\theta_2 - \ldots - \alpha_{m-1}\psi_{m-1}\theta_{m-1} - \alpha_{m+1}\psi_{m+1}\theta_{m+1} - \ldots \\ &-\alpha_\mu\psi_\mu\theta_\mu + r + A' \log \rho' + A'' \log \rho'' + \ldots + A^{(k)} \log \rho^{(k)}\end{aligned} \tag{117}$$

where $\theta_1, \Delta_{m+1}\theta_1, \theta_2, \Delta_{m+2}\theta_2, \ldots, r, \rho', \rho'', \ldots$ are rational functions of x and $\Delta_m x$, that is of the form $p + q\Delta_m x$ with p, q rational in x.

When $x_1 = x_2 = \ldots = x_\mu = x$ and $c_1 = c_2 = \ldots = c_\mu = c$, one obtains the following theorem: if there is a relation

$$\alpha\varpi x+\alpha_0\varpi_0 x+\alpha_1\Pi_1 x+\alpha_2\Pi_2 x+\ldots+\alpha_\mu\Pi_\mu x$$
$$=u+A_1\log v_1+A_2\log v_2+\ldots+A_\nu\log v_\nu$$

where $u, v_1, v_2, \ldots, v_\nu$ are algebraic functions of x, one may suppose that they are of the form $p+q\Delta x$ with p, q rational in x.

Differentiating (117) we obtain a relation of the form $P+Q\Delta_m x=0$ which implies $P=Q=0$ and therefore $P-Q\Delta_m x=0$. When the sign of $\Delta_m x$ is changed into the opposite, the θ_j take new values θ'_j and we have $-\delta\alpha_m\psi_m x=-\sum\alpha\psi\theta'+v'$ where v' designates the algebraic and logarithmic part. It results that $2\delta\alpha_m\psi_m x=\sum\alpha(\psi\theta'-\psi\theta)+v-v'$ where, by the addition theorem,

$$\psi\theta'-\psi\theta=\psi y-v''$$

if $y=\frac{\theta'\Delta\theta-\theta\Delta\theta'}{1-c^2\theta^2\theta'^2}$, v'' denoting an algebraic and logarithmic function. Now $\theta=p+q\Delta_m x$ and $\Delta\theta=r+\rho\Delta_m x$ where p, q, r, ρ are rational functions of x and it results that $\theta'=p-q\Delta_m x$, $\Delta\theta'=r-\rho\Delta_m x$ and that $y=t\Delta_m x$ where t is a rational function of x. Then it is easy to see that Δy is a rational function of x. One may replace y by $z=\frac{y\Delta e+e\Delta y}{1-c^2e^2y^2}$ where e is a constant because ψy and ψz differ by an algebraic and logarithmic function. For $e=1$, $z=\frac{\Delta y}{1-c^2y^2}$ is a rational function of x and $\Delta z=\frac{c^2-1}{1-c^2y^2}y$ has a rational ratio to $\Delta_m x$. We have $2\delta\alpha_m\psi_m x=\sum\alpha\psi z+V$ where V is an algebraic and logarithmic function. Then $V=u+A_1\log v_1+A_2\log v_2+\ldots$ where $u, v_1, v_2, \ldots$ are of the form $p+q\Delta_m x$ with p and q rational in x.

Taking $m=\mu$, we obtain $2\delta\alpha_\mu\psi_\mu x_\mu=\alpha_1\psi_1 z_1+\alpha_2\psi_2 z_2+\ldots+\alpha_{\mu-1}\psi_{\mu-1}z_{\mu-1}+V$ and we may eliminate $\psi_\mu x_\mu$ between this relation and (116), getting

$$\alpha_1(2\delta\psi_1 x_1-\psi_1 z_1)+\ldots+\alpha_{\mu-1}(2\delta\psi_{\mu-1}x_{\mu-1}-\psi_{\mu-1}z_{\mu-1})=V'.$$

Since δ is an integer, there exist $x'_1, x'_2, \ldots, x'_{\mu-1}$ such that $2\delta\psi_j x_j-\psi_j z_j=\psi_j x'_j+V_j$ $(1\le j\le\mu-1)$ and we have $\alpha_1\psi_1 x'_1+\alpha_2\psi_2 x'_2+\ldots+\alpha_{\mu-1}\psi_{\mu-1}x'_{\mu-1}=u'+A'_1\log v'_1+A'_2\log v'_2+\ldots+A'_{\nu'}\log v'_{\nu'}$ of the same form as (116) with one elliptic integral less. We may iterate until we arrive at a relation with only algebraic and logarithmic functions.

The general problem (106) has been reduced to the following one: "To satisfy in the most general manner the equation

$$\psi x=\beta_1\psi_1 y_1+\beta_2\psi_2 y_2+\ldots+\beta_n\psi_n y_n$$
$$+u+A_1\log v_1+A_2\log v_2+\ldots+A_\nu\log v_\nu \tag{118}$$

where $\psi, \psi_1, \psi_2, \ldots, \psi_n$ designate elliptic integrals of the three kinds, supposing that $y_1, y_2, \ldots, y_n$ are rational functions of x and that $\Delta_1 y_1, \Delta_2 y_2, \ldots, \Delta_n y_n$ are of the form $p\Delta x$ where p is rational in x and Δx designate the radical which appears in the function ψx." If $\Delta_m y_m=p_m\Delta x$ and $\psi_m x=\int\frac{\theta_m x\cdot dx}{\Delta_m x}$ where $\theta_m x$ is rational, we have $\psi_m y_m=\int\frac{\theta_m y_m}{p_m}\frac{dy_m}{dx}\frac{dx}{\Delta x}$ where $\frac{\theta_m y_m}{p_m}\frac{dy_m}{dx}$ is a rational function of x. Thus $\psi_m y_m=r+A\varpi x+A_0\varpi_0 x+A'\Pi(x,a')+A''\Pi(x,a'')+\ldots$ where r is an algebraic and logarithmic expression. Equation (118) finally takes the form

$$\alpha\varpi x + \alpha_0\varpi_0 x + \alpha_1\Pi(x,a_1) + \alpha_2\Pi(x,a_2) + \ldots + \alpha_\mu\Pi(x,a_\mu)$$
$$= u + A_1\log v_1 + A_2\log v_2 + \ldots \quad (119)$$

The problem (118) is thus reduced to the three following ones:

A) To find all the possible cases in which

$$(1-y^2)(1-c'^2y^2) = p^2(1-x^2)(1-c^2x^2)$$

with y and p rational functions of x (c, c' are constants).

B) To reduce $\varpi(y,c')$, $\varpi_0(y,c')$ and $\Pi(y,c',a)$, where y and c' are as in A), to the form

$$r + A\varpi x + A_0\varpi_0 x + A'\Pi(x,a') + A''\Pi(x,a'') + \ldots$$

C) To find the necessary and sufficient conditions for (119) to be satisfied.

The third chapter (p. 557–565) is devoted to the solution of problem C), where one may suppose that $u, v_1, v_2, \ldots, v_\nu$ are of the form $p + q\Delta x$, p and q rational in x. Abel takes the problems dealt with in the second chapter of his unpublished memoir *Théorie des transcendantes elliptiques* (see our §4) in a more general setting. Equation (119) is rewritten

$$\psi x = u + \sum A\log v,$$

where $\psi x = \beta\varpi x + \beta_0\varpi_0 x + \beta_1\Pi\alpha_1 + \beta_2\Pi\alpha_2 + \ldots + \beta_n\Pi\alpha_n$ and $\Pi\alpha_m = \int\frac{dx}{\left(1-\frac{x^2}{\alpha_m^2}\right)\Delta x}$; it is supposed that it is impossible to find any similar relation not containing all the $\Pi\alpha_m$ and that all the α_m are different from ± 1 and $\pm\frac{1}{c}$. Changing the sign of Δx, we obtain $-\psi x = u' + \sum A\log v'$ and then $2\psi x = u - u' + \sum A\log\frac{v}{v'}$. Changing the sign of x without changing that of Δx, we obtain $-2\psi x = u'' - u''' + \sum A\log\frac{v''}{v'''}$ and

$$\psi x = \frac{1}{4}(u - u' - u'' + u''') + \frac{1}{4}\sum A\log\frac{vv'''}{v'v''}.$$

If $v = p + qx + (p' + q'x)\Delta x$ where p, q, p', q' are even functions, $v' = (p+qx) - (p'+q'x)\Delta x$, $v'' = (p - qx) + (p' - q'x)\Delta x$ and $v''' = p - qx - (p' - q'x)\Delta x$. Thus $\frac{vv'''}{v'v''} = \frac{fx + \varphi x\cdot\Delta x}{fx - \varphi x\cdot\Delta x}$ where fx and φx are polynomials, one even and the other one odd. The algebraic part $\frac{1}{4}(u - u' - u'' + u''')$ is of the form $r\Delta x$ where r is an odd rational function of x and we may rewrite our equation in the form

$$\psi x = r\Delta x + \sum A\log\frac{fx + \varphi x\cdot\Delta x}{fx - \varphi x\cdot\Delta x} \quad (120)$$

with A in place of $\frac{1}{4}A$. We may suppose that there is no linear relation with integer coefficients between the A_m, otherwise it would be possible to reduce the number ν of the terms in the sum.

Differentiating one term $\rho = \log \frac{fx+\varphi x\cdot\Delta x}{fx-\varphi x\cdot\Delta x}$, we obtain $d\rho = \frac{v dx}{\theta x\cdot\Delta x}$ where

$$\theta x = (fx)^2 - (\varphi x)^2(\Delta x)^2 \quad \text{and} \quad v\varphi x = 2f'x\theta x - fx\theta' x,$$

so that v is an even polynomial. If the roots of θx are $\pm a_1, \pm a_2, \ldots, \pm a_\mu$, the decomposition of $\frac{v}{\theta x}$ in simple elements is of the form $k + \frac{\beta_1'}{a_1^2-x^2} + \frac{\beta_2'}{a_2^2-x^2} + \ldots + \frac{\beta_\mu'}{a_\mu^2-x^2}$ where k is a constant and $\beta_j' = 2m_j a_j \frac{fa_j}{\varphi a_j} = -2m_j a_j \Delta a_j$ where m_j is the multiplicity of a_j as a root of θx. Thus the differentiation of (120) gives, after multiplication by Δx:

$$\begin{aligned}
&\beta + \beta_0 x^2 + \frac{\alpha_1^2\beta_1}{a_1^2-x^2} + \frac{\alpha_2^2\beta_2}{a_2^2-x^2} + \ldots + \frac{\alpha_n^2\beta_n}{a_n^2-x^2} \\
&= \frac{dr}{dx}(\Delta x)^2 - r((1+c^2)x - 2c^2x^3) \\
&\quad + A_1\left(k_1 - \frac{2m_1a_1\Delta a_1}{a_1^2-x^2} - \frac{2m_2a_2\Delta a_2}{a_2^2-x^2} - \ldots\right) \\
&\quad + A_2\left(k_2 - \frac{2m_1'a_1'\Delta a_1'}{a_1'2-x^2} - \frac{2m_2'a_2'\Delta a_2'}{a_2'2-x^2} - \ldots\right) + \ldots.
\end{aligned}$$

From this relation, Abel deduces that $r = 0$ and that only one of the coefficients A_m may be different from 0. He takes $A_1 = 1$, $A_2 = A_3 = \ldots = A_\nu = 0$ and finds $\beta = k_1$, $\beta_0 = 0$, $\alpha_1 = a_1$, $\alpha_2 = a_2, \ldots, \beta_1 = -\frac{2m_1\Delta a_1}{a_1}$, $\beta_2 = -\frac{2m_2\Delta a_2}{a_2}, \ldots$. The most general relation between elliptic integrals with the same modulus is thus of the form

$$\begin{aligned}
&\beta\varpi x - \frac{2m_1\Delta\alpha_1}{\alpha_1}\Pi\alpha_1 - \ldots - \frac{2m_n\Delta\alpha_n}{\alpha_n}\Pi\alpha_n \\
&= \log\frac{fx+\varphi x\cdot\Delta x}{fx-\varphi x\cdot\Delta x} + C
\end{aligned} \tag{121}$$

where the parameters $\alpha_1, \alpha_2, \ldots, \alpha_n$ are related by the equation

$$(fx)^2 - (\varphi x)^2(1-x^2)(1-c^2x^2) = (x^2-\alpha_1^2)^{m_1}(x^2-\alpha_2^2)^{m_2}\ldots(x^2-\alpha_n^2)^{m_n}. \tag{122}$$

Abel remarks that this implies

$$\begin{aligned}
&m_1\varpi\alpha_1 + m_2\varpi\alpha_2 + \ldots + m_n\varpi\alpha_n = C \quad \text{and} \\
&m_1\Pi'\alpha_1 + m_2\Pi'\alpha_2 + \ldots + m_n\Pi'\alpha_n = C - \frac{a}{2\Delta a}\log\frac{fa+\varphi a\cdot\Delta a}{fa-\varphi a\cdot\Delta a}
\end{aligned}$$

if $\Pi'\alpha = \int \frac{d\alpha}{\left(1-\frac{\alpha^2}{a^2}\right)\Delta\alpha}$.

When $n = 1$, $\alpha_1 = \alpha$ and $m_1 = m$ we have $\Pi\alpha = \frac{\beta\alpha}{2m\Delta\alpha}\varpi x - \frac{\alpha}{2m\Delta\alpha}\log\frac{fx+\varphi x\cdot\Delta x}{fx-\varphi x\cdot\Delta x}$ if the parameter α verifies

$$(fx)^2 - (\varphi x)^2(1-x^2)(1-c^2x^2) = (x^2-\alpha^2)^m. \tag{123}$$

For $m = 2$, $fx = ax$ is of degree 1, $\varphi x = \frac{1}{c}\sqrt{-1}$ is constant and $(x^2-\alpha^2)^2 = x^4 - \left(\frac{1+c^2}{c^2} - a^2\right)x^2 + \frac{1}{c^2}$ and this gives $a = 1 \mp \frac{1}{c}$, $\alpha = \frac{1}{\sqrt{\pm c}}$. For $m = 3$ and fx odd, $fx = x^3 + ax$, $\varphi x = b$ and

$$(x^3+ax)^2 - b^2(1-x^2)(1-c^2x^2) = (x^2-\alpha^2)^3,$$

equation which leads to $\alpha^3 = b$, $\alpha^3 + a\alpha + b\Delta\alpha = 0$, $2a - c^2b^2 = -3\alpha^2$, $a^2 + (1+c^2)b^2 = 3\alpha^4$. This gives $a = \frac{1}{2}(c^2\alpha^6 - 3\alpha^2)$ and α determined by the equation $\Delta\alpha = \frac{1}{2}(1-c^2\alpha^4)$.

Generally, as Abel states, α must be a zero or a pole of the function x_m defined in the first chapter and such that $\frac{dx_m}{\Delta x_m} = m\frac{dx}{\Delta x}$ ($x_m = 0$ for $x = 0$). Indeed we have

$$p^2 - q^2(\Delta x)^2 = (x^2-\alpha^2)^m(x^2-\alpha_m)$$

where α_m is the value of x_m for $x = \alpha$ (chapter I) and, multiplying by (123), we obtain

$$\begin{aligned}(pfx \pm q\varphi x(\Delta x)^2)^2 - (p\varphi x \pm qfx)^2(\Delta x)^2 &= p^2(fx)^2 - q^2(\varphi x)^2(\Delta x)^4 \\ &= (x^2-\alpha^2)^{2m}(x^2-\alpha_m).\end{aligned}$$

It results that $pfx + q\varphi x(\Delta x)^2$ or $pfx - q\varphi x(\Delta x)^2$ is divisible by $(x^2-\alpha^2)^m$ and we have a relation $r^2 - \rho^2(\Delta x)^2 = x^2 - \alpha_m^2$ where r, ρ are polynomials, one even and the other odd. But this relation implies that $\rho = 0$ and that $\alpha_m = 0$ or $\frac{1}{0}$. Conversely, it is easy to see that such an α satisfies an equation (123). Abel remarks that, in these cases, the coefficient β in (121) is always different from 0, so that there is no elliptic differential of the third kind integrable by algebraic and logarithmic functions.

When $n = 3$ and $m_1 = m_2 = m_3 = 1$, (121) takes the form

$$\frac{\Delta\alpha_1}{\alpha_1}\Pi\alpha_1 + \frac{\Delta\alpha_2}{\alpha_2}\Pi\alpha_2 = \frac{\Delta\alpha}{\alpha}\Pi\alpha + \beta\varpi x - \frac{1}{2}\log\frac{fx + \varphi x\cdot\Delta x}{fx - \varphi x\cdot\Delta x}$$

where $fx = x^3 + ax$, $\varphi x = b$ and $(x^3+ax)^2 - b^2(1-x^2)(1-c^2x^2) = (x^2-\alpha^2) \times(x^2-\alpha_1^2)(x^2-\alpha_2^2)$. This gives $\alpha = \frac{\alpha_1\Delta\alpha_2 + \alpha_2\Delta\alpha_1}{1-c^2\alpha_1^2\alpha_2^2}$, $b = \alpha\alpha_1\alpha_2$,

$$a = \frac{1}{2}(c^2\alpha^2\alpha_1^2\alpha_2^2 - \alpha^2 - \alpha_1^2 - \alpha_2^2), \quad \frac{\Delta\alpha}{\alpha} = \frac{\alpha^2 + a}{\alpha\alpha_1\alpha_2}, \quad \beta = -c^2\alpha\alpha_1\alpha_2$$

(cf. chapter I). In particular, for α_2 infinite, $\alpha = \pm\frac{1}{c\alpha_1}$ and $\Pi\alpha + \Pi\left(\frac{1}{c\alpha}\right) = \varpi x + \frac{1}{2}\frac{\alpha}{\Delta\alpha}\log\frac{x\Delta\alpha + \alpha\Delta x}{x\Delta\alpha - \alpha\Delta x}$. Other relations between two elliptic integrals of the third kind are obtained by (121) with $n = 2$.

In the fourth chapter (p. 565–606), Abel solves the problem A) of the second chapter, that is to satisfy the equation $(1-y^2)(1-c'^2y^2) = r^2(1-x^2)(1-c^2x^2)$, y and r being rational functions of x. Since $1-y^2$ and $1-c'^2y^2$ have no common factor ($c' \neq 1$), this equation implies $1 - y^2 = r_1^2\rho$, $1 - c'^2y^2 = r_2^2\rho'$ where r_1, r_2 are rational functions of x, $r_1r_2 = r$ and $\rho\rho' = (1 - x^2)(1 - c^2x^2)$. Differentiating, we obtain $-2ydy = r_1(r_1d\rho + 2\rho dr_1)$, $-2c'^2ydy = r_2(r_2d\rho' + 2\rho' dr_2)$ which show that the numerator of $\frac{dy}{dx}$ is divisible by r_1 and r_2, and so by their product r: $\frac{dy}{dx} = rv$ where v is a rational function without any pole among the zeros of r. Let $y = \frac{p}{q}$, irreducible fraction where p, q are polynomials of respective degrees m, n. One has $r = \frac{\theta}{q^2}$ where θ is a polynomial and $\theta v = q^2\frac{dy}{dx} = \frac{qdp - pdq}{dx}$, whence v is a polynomial. If $m > n$, the equation

$$(q^2 - p^2)(q^2 - c'^2p^2) = \theta^2(1 - x^2)(1 - c^2x^2)$$

shows that $4m = 2\mu + 4$ where μ is the degree of θ. If ν is the degree of v, we then see that $\mu + \nu = m + n - 1$ and $\nu = m + n - 1 - \nu < 2m - \mu - 1 = 1$. Thus $\nu = 0$ and v is constant. In the same way, if $n > m$, we have $4n = 2\mu + 4$, $\nu < 2n - \mu - 1 = 1$ and $\nu = 0$. In the remaining case, in which $m = n$, it is for instance possible that $q - p = \varphi$ be of degree $m - k < m$. Then $4m - k = 2\mu + 4$ and $\mu + \nu = 2m - k - 1$ for $\theta v = \frac{pd\varphi - \varphi dp}{dx}$ and $\nu = 2m - k - 1 - \mu = 1 - \frac{1}{2}k$ is again 0. In any case v is a constant ε and

$$\frac{dy}{\sqrt{(1 - y^2)(1 - c'^2y_2)}} = \frac{\varepsilon dx}{\sqrt{(1 - x^2)(1 - c^2x^2)}}. \tag{124}$$

The second result announced in the introduction is thus demonstrated.

It remains to determine the rational function y and the transformed modulus c'. Abel begins by considering the case in which $y = \frac{\alpha + \beta x}{\alpha' + \beta' x}$ and he explains the 6 cases already met in *Sur le nombre de transformations différentes* ... (our §7). He then considers the case in which $y = \psi x = \frac{A_0 + A_1x + A_2x^2 + \ldots + A_\mu x^\mu}{B_0 + B_1x + B_2x^2 + \ldots + B_\mu x^\mu}$ (irreducible fraction, one of the coefficients A_μ, B_μ different from 0). The treatment uses only the addition theorem for elliptic integrals of chapter I and not the elliptic function λ and its double periodicity as in the preceding memoirs; but the lines are similar. If x, x' are two roots of the equation $y = \psi x$, one has $\frac{dx}{\Delta x} = \frac{1}{\varepsilon}\frac{dy}{\Delta' y} = \frac{dx\prime}{\Delta x\prime}$ and consequently $x' = \frac{x\Delta e + e\Delta x}{1 - c^2e^2x^2} = \theta x$ where e is a constant. Thus $\psi(\theta x) = \psi x$ and we see that the equation $y = \psi x$ has the roots $x, \theta x, \theta^2 x, \ldots, \theta^n x, \ldots$ where it is easy to see that

$$\theta^n x = \frac{x\Delta e_n + e_n\Delta x}{1 - c^2e_n^2x^2},$$

e_n being the rational function of e defined by $\frac{de_n}{\Delta e_n} = n\frac{de}{\Delta e}$ and $e_n = 0$ for $e = 0$ (see chapter I). Since the equation has only μ roots, there exists an n such that $\theta^n x = x$ that is $e_n = 0$ and $\Delta e_n = 1$. These equations are equivalent to $\frac{\Delta e_n}{1 - c^2e_n^2} = 1$, which is of degree n^2 in e. The number n must be minimal and we must eliminate the roots e which would lead to $e_\mu = 0$, $\Delta\varepsilon_\mu = 1$ for a $\mu < n$. If, for instance, n is a prime number, the root $e = 0$ is to be eliminated and it remains $n^2 - 1$ solutions e.

Let us suppose that two rational functions $\psi z = \frac{p}{q}$, $\psi' z = \frac{p'}{q'}$ where p, q, p', q' are polynomials of degree μ and the two fractions are irreducible. If the equations

$y = \psi(x)$ and $y' = \psi'(x)$ have the same roots $x, x', x'', \dots, x^{(\mu-1)}$ we have $\frac{p-qy}{p'-q'y'} = \frac{a-by}{a'-b'y'}$ where a, b, a', b' are the respective coefficients of z^μ in p, q, p', q' and z has any value. We draw $y' = \frac{\alpha+\beta y}{\alpha'+\beta' y}$; if moreover y and y' correspond to the same modulus c', we have $y' = \frac{1}{c'y}$.

When $n = \mu$, the roots of $y = \psi x$ are $x, \theta x, \dots, \theta^{n-1}x$ and

$$p - qy = (a - by)(z - x)(z - \theta x) \cdots (z - \theta^{n-1}x). \tag{125}$$

We can draw y from this equation, giving to z a particular value. If n is odd, noted $2\mu + 1$, putting $z = 0$, we obtain $y = \frac{a'+ax\cdot\theta x\cdot\theta^2 x\cdots\theta^{2\mu}x}{b'+bx\cdot\theta x\cdot\theta^2 x\cdots\theta^{2\mu}x}$ where a', b' are the respective constant terms of p, q. Since $e_{n-m} = -e_m$ and $\Delta e_{n-m} = \Delta e_m$, we see that $\theta^{n-m}x = \frac{x\Delta e_m - e_m\Delta x}{1-c^2e_m^2x^2}$ and $\theta^m x \cdot \theta^{n-m}x = \frac{x^2-e_m^2}{1-c^2e_m^2x^2}$. It results that the value found for y is rational in x. Moreover, it is invariant by the substitution $x \mapsto \theta x$ because $\theta^{2\mu+1}x = x$, and it results that (125) is verified for any value of z. For $x = \pm 1$ or $\pm\frac{1}{c}$, $\Delta x = 0$ and $\theta^m x = \theta^{2\mu+1-m}x$, so that

$$p - q\alpha = (a - b\alpha)(1 - z)\rho^2, \quad p - q\beta = (a - b\beta)(1 + z)\rho'^2,$$
$$p - q\gamma = (a - b\gamma)(1 - cz)\rho''^2, \quad p - q\delta = (a - b\delta)(1 + cz)\rho'''^2$$

where $\alpha, \beta, \gamma, \delta$ are the values of y corresponding to $x = 1, -1, \frac{1}{c}, -\frac{1}{c}$ and $\rho, \rho', \rho'', \rho'''$ are polynomials of degree μ in z. Now we want that

$$(q^2 - p^2)(q^2 - c'^2p^2) = r^2(1 - z^2)(1 - c^2z^2)$$

and this implies that $\{\alpha, \beta, \gamma, \delta\} = \left\{1, -1, \frac{1}{c'}, -\frac{1}{c'}\right\}$; conversely, this condition will be sufficient. Let us take $\alpha = 1$, $\beta = -1$, $\gamma = \frac{1}{c'}$, $\delta = -\frac{1}{c'}$. Since $y = \frac{a'+a\varphi x}{b'+b\varphi x}$ where

$$\varphi x = x \cdot \theta x \cdot \theta^2 x \cdots \theta^{2\mu}x = \frac{x(x^2 - e^2)(x^2 - e_2^2)\cdots(x^2 - e_\mu^2)}{(1 - c^2e^2x^2)(1 - c^2e_2^2x^2)\cdots(1 - c^2e_\mu^2x^2)}$$

is an odd function, we have $\alpha = \frac{a'+a\varphi(1)}{b'+b\varphi(1)}$, $\beta = \frac{a'-a\varphi(1)}{b'-b\varphi(1)}$, $\gamma = \frac{a'+a\varphi\left(\frac{1}{c}\right)}{b'+b\varphi\left(\frac{1}{c}\right)}$, $\delta = \frac{a'-a\varphi\left(\frac{1}{c}\right)}{b'-b\varphi\left(\frac{1}{c}\right)}$ or $a' \mp b' \pm (a \mp b)\varphi(1) = 0$, $a' \mp \frac{b'}{c'} \pm \left(a \mp \frac{b}{c'}\right)\varphi\left(\frac{1}{c}\right) = 0$. These equations are compatible only if a' or b' is 0 ($c' \neq 1$). Let us suppose that $a' = 0 = b$; we have $c' = \frac{\varphi(1)}{\varphi\left(\frac{1}{c}\right)}$, $y = \frac{a}{b'}\varphi x = \frac{\varphi x}{\varphi(1)}$ where $\varphi(1) = \frac{1-e^2}{1-c^2e^2}\frac{1-e_2^2}{1-c^2e_2^2}\cdots\frac{1-e_\mu^2}{1-c^2e_\mu^2}$, $\varphi\left(\frac{1}{c}\right) = \frac{1}{c^{2\mu+1}}\frac{1-c^2e^2}{1-e^2}\frac{1-c^2e_2^2}{1-e_2^2}\cdots\frac{1-c^2e_\mu^2}{1-e_\mu^2} = \frac{1}{c^{2\mu+1}\varphi(1)}$. Then $c' = c^{2\mu+1}(\varphi(1))^2$. In order to determine the multiplicator ε, Abel uses the value of $\frac{dy}{dx} = \varepsilon\frac{\Delta' y}{\Delta x}$ for $x = 0$, which is $\pm e^2e_2^2\cdots e_\mu^2\frac{1}{\varphi(1)}$; thus $\varepsilon = e^2e_2^2\cdots e_\mu^2\frac{c^{\mu+\frac{1}{2}}}{\sqrt{c'}}$. He has reconstituted the formulae for the transformations of odd order $2\mu + 1$: if e is a root of the equation $e_{2\mu+1} = 0$ which does not satisfy any other equation $e_{2m+1} = 0$ where $2m + 1$ is a divisor of $2\mu + 1$, let us put

$$y = \frac{c^{\mu+\frac{1}{2}}}{\sqrt{c'}} \frac{x(e^2-x^2)(e_2^2-x^2)\cdots(e_\mu^2-x^2)}{(1-c^2e^2x^2)(1-c^2e_2^2x^2)\cdots(1-c^2e_\mu^2x^2)},$$

$$c' = c^{2\mu+1}\left(\frac{(1-e^2)(1-e_2^2)\cdots(1-e_\mu^2)}{(1-c^2e^2)(1-c^2e_2^2)\cdots(1-c^2e_\mu^2)}\right)^2,$$

$$\varepsilon = \frac{c^{\mu+\frac{1}{2}}}{\sqrt{c'}} e^2 e_2^2 \ldots e_\mu^2. \tag{126}$$

Then we have $\frac{dy}{\sqrt{(1-y^2)(1-c'^2y^2)}} = \pm\varepsilon\frac{dx}{\sqrt{(1-x^2)(1-c^2x^2)}}$. Five other systems (y, c', ε) corresponding to the same value of e are obtained by composing with a transformation of order 1.

For instance, when $\mu = 1$, $2\mu+1 = 3$ is prime and we may take for e any root different from 0 of the equation $e_3 = 0$, that is $0 = 3 - 4(1+c^2)e^2 + 6c^2e^4 - c^4e^8$ of degree 4 in e^2 and the $c' = c^3\left(\frac{1-e^2}{1-c^2e^2}\right)^2$, $\varepsilon = c\sqrt{\frac{c}{c'}}e^2$, $y = \frac{c\sqrt{c}}{\sqrt{c'}}\frac{x(e^2-x^2)}{1-c^2e^2x^2}$. Eliminating e, we obtain the modular equation in the form

$$(c'-c)^2 = 4\sqrt{cc'}(1-\sqrt{cc'})^2.$$

The roots of the equation $0 = \frac{c^{\mu+\frac{1}{2}}}{\sqrt{c'}}z(z-e^2)(z-e_2^2)\cdots(z-e_\mu^2) + y(1-c^2e^2z^2)(1-c^2e_2^2z^2)\cdots(1-c^2e_\mu^2z^2)$ are $x, \theta x, \ldots, \theta^{2\mu}x$, thus $x + \theta x + \ldots + \theta^{2\mu}x = \frac{(-1)^{\mu+1}c^{2\mu}e^2e_2^2\cdots e_\mu^2}{c^{\mu+\frac{1}{2}}c'^{-\frac{1}{2}}}y$. Since $\theta^m x + \theta^{2\mu+1-m}x = \frac{2\Delta e_m x}{1-c^2e_m^2x^2}$, this gives $y = \left(x + \frac{2\Delta e\cdot x}{1-c^2e^2x^2} + \frac{2\Delta e_2\cdot x}{1-c^2e_2^2x^2} + \ldots + \frac{2\Delta e_\mu \cdot x}{1-c^2e_\mu^2x^2}\right)\frac{\sqrt{c}}{c^\mu\sqrt{c'}}\frac{(-1)^{\mu+1}}{e^2e_2^2\cdots e_\mu}$.

If n is even, noted 2μ, we have $\theta^\mu x = \frac{x\Delta e_\mu + e_\mu \Delta x}{1-c^2e_\mu^2x^2} = \frac{x\Delta e_\mu - e_\mu\Delta x}{1-c^2e_\mu^2x^2}$, which imposes $e_\mu = 0$ or $\frac{1}{0}$. In the last case, $\theta^\mu x = \pm\frac{1}{cx}$ and $\theta^{\mu+m}x = \pm\frac{1}{c\theta^m x}$. Thus the roots of $y = \psi x$ are $x, \pm\frac{1}{cx}, \theta x, \ldots, \theta^{\mu-1}x, \theta^{\mu+1}x, \ldots, \theta^{2\mu-1}x$ and we have

$$\begin{aligned} p - qy = (a-by)(z-x)\left(z \mp \frac{1}{cx}\right)(z-\theta x)(z-\theta^{2\mu-1}x)\cdots \\ \times(z-\theta^{\mu-1}x)(z-\theta^{\mu+1}x). \end{aligned} \tag{127}$$

We deduce from this equation that

$$a' - b'y = (by-a)\left(x \pm \frac{1}{cx} + \frac{2\Delta e\cdot x}{1-c^2e^2x^2} + \frac{2\Delta e_2\cdot x}{1-c^2e_2^2x^2} + \ldots + \frac{2\Delta e_{\mu-1}\cdot x}{1-c^2e_{\mu-1}^2x^2}\right)$$

where a' and b' are the coefficients of $z^{2\mu-1}$ in p and q. It results for y a rational expression in x, invariant by $x \mapsto \theta x$. Choosing $a = b' = 0$, we obtain

$$\begin{aligned} y &= \frac{a'}{b}\frac{1}{x \pm \frac{1}{cx} + \frac{2\Delta ex}{1-c^2e^2x^2} + \ldots + \frac{2\Delta e_{\mu-1}x}{1-c^2e_{\mu-1}^2x^2}} \\ &= A\frac{x(1-c^2e^2x^2)(1-c^2e_2^2x^2)\cdots(1-c^2e_{\mu-1}^2x^2)}{1+a_1x^2+a_2x^2+\ldots+a_\mu x^{2\mu}} = A\varphi x. \end{aligned}$$

If, for instance, $y = 1$ when $x = 1$, we have $A = \frac{1}{\varphi(1)}$ and, from (125), $q - p = (1-z)(1 \mp cz)\rho^2$ where ρ is a polynomial in z. Since q is even and p odd, $q + p = (1+z)(1 \pm cz)\rho'^2$ and

$$q^2 - p^2 = (1-z^2)(1-c^2z^2)(\rho\rho')^2.$$

It results that $q^2 - c'^2 p^2$ must be a square and $c' = \frac{1}{\alpha}$, where α is the value of y corresponding to $x = \frac{1}{\sqrt{\pm c}}$, satisfies to this condition. Indeed $\theta^{\mu+m}x = \theta\left(\pm\frac{1}{cx}\right) = \theta\left(\frac{1}{\sqrt{\pm c}}\right) = \theta x$ for $x = \frac{1}{\sqrt{\pm c}}$, so that $p - \alpha q$ is a square and the same may be said of $p + \alpha q$. Thus $p^2 - \alpha^2 q^2 = t^2$ where t is a polynomial in z and $(q^2-p^2)(q^2-c'^2p^2) = (1-z^2)(1-c^2z^2)r^2$ for $c' = \frac{1}{\alpha}$. Sylow observes that α is never 0 nor ∞, but it is equal to 1 when μ is even and this value does not work for c'. He explains how to find a correct value in this case (*Œuvres*, t. II, p. 520–521). Then $\frac{dy}{\Delta' y} = \varepsilon\frac{dx}{\Delta x}$ where ε is the value of $\frac{dy}{dx}$ for $x = 0$, that is $\varepsilon = A = \frac{1}{\varphi(1)}$. Abel gives an expression of the denominator q of φx as a product $b(z-\delta)(z-\theta\delta)\cdots(z-\theta^{2\mu-1}\delta)$ where δ is a pole of y. It is easy to see that $\delta = \frac{1}{\sqrt{\mp c}}$ is such a pole. Thus, if e is a pole of e_μ such the equations $e_m = 0$ and $\Delta e_m = 1$ cannot be satisfied for any divisor m of 2μ, the formulae

$$\pm\frac{\varepsilon}{c}\frac{1}{y} = x \pm \frac{1}{cx} + \frac{2\Delta ex}{1-c^2e^2x^2} + \frac{2\Delta e_2 x}{1-c^2e_2^2x^2} + \dots + \frac{2\Delta e_{\mu-1}x}{1-c^2e_{\mu-1}^2x^2},$$

$$\pm\varepsilon = c\left(1 \pm \frac{1}{c} + \frac{2\Delta e}{1-c^2e^2} + \frac{2\Delta e_2}{1-c^2e_2^2} + \dots + \frac{2\Delta e_{\mu-1}}{1-c^2e_{\mu-1}^2}\right)$$

lead to $\frac{dy}{\sqrt{(1-y^2)(1-c'2y^2)}} = \frac{\varepsilon dx}{\sqrt{(1-x^2)(1-c^2x^2)}}$. For instance, when $\mu = 1$, $\varepsilon = 1 \pm c$, $y = (1 \pm c)\frac{x}{1 \pm cx^2}$ and $c' = \frac{2\sqrt{\pm c}}{1 \pm c}$.

Another possible value for e is a root of $e_\mu = 0$ such that $\Delta e_\mu = -1$ (for $\Delta e_\mu = 1$ would lead to $\theta^\mu x = x$). Here $\theta^\mu x = -x$, $\theta^{\mu+m}x = -\theta^m x$ and equation (127) is replaced by

$$p - qy = (a - by)(z^2 - x^2)(z^2 - (\theta x)^2)\cdots(z^2 - (\theta^{\mu-1}x)^2)$$

which gives $a' - b'y = \pm(a - by)(x\theta x \cdots \theta^{\mu-1}x)^2$ for $z = 0$, a' and b' denoting the constant terms of p and q. Thus y is a rational function of degree 2μ of x and it remains to determine a, b, a', b' and c', ε. For instance, when $\mu = 1$, Abel finds $y = \frac{1+cx^2}{1-cx^2}$, $c' = \frac{1-c}{1+c}$, $\varepsilon = (1+c)\sqrt{-1}$ and he also gives the 5 other possible values for c'.

When the equation $y = \psi x$ has other roots than $x, \theta x, \dots, \theta^{n-1}x$, Abel shows that the degree μ of this equation is a multiple mn of n and that its roots may be distributed in m cycles $x^{(j)}, \theta x^{(j)}, \dots, \theta^{n-1}x^{(j)}, 0 \le j \le m-1$. The proof is identical with that used for the second theorem of the *Mémoire sur une classe particulière d'éqations ...* published in the same volume of *Crelle's Journal*. According to the precedind results, there exists a rational function $y_1 = \psi_1 x$ such that the roots of the equation $y_1 = \psi_1 x$ are $x, \theta x, \ldots, \theta^{n-1}x$ and that, for convenient c_1, ε_1

$$\frac{dy_1}{\sqrt{(1-y_1^2)(1-c_1^2y_1^2)}} = \varepsilon_1 \frac{dx}{\sqrt{(1-x^2)(1-c^2x^2)}}. \tag{128}$$

Let $\psi_1 z = \frac{p'}{q'}$, so that $p' - q'y = (a' - b'y)(z-x)(z-\theta x)\cdots(z-\theta^{n-1}x)$. If $y_{j+1} = \psi_1 x^{(j)}$ $(0 \le j \le m-1)$, we see that $\frac{p-qy}{a-by} = \frac{p'-q'y_1}{a'-b'y_1}\frac{p'-q'y_2}{a'-b'y_2}\cdots\frac{p'-q'y_m}{a'-b'y_m}$. Now let α be a zero and β a pole of ψz and let $\alpha_1, \alpha_2, \ldots, \alpha_m, \beta_1, \beta_2, \ldots, \beta_m$ be the corresponding values of $y_1, y_2, \ldots, y_m$; from the preceding relation we deduce that

$$p = A'(p' - \alpha_1 q')(p' - \alpha_2 q')\cdots(p' - \alpha_m q') \quad \text{and}$$
$$q = A''(p' - \beta_1 q')(p' - \beta_2 q')\cdots(p' - \beta_m q')$$

where A' and A'' are constants, and this gives $y = A\frac{(y_1-\alpha_1)(y_1-\alpha_2)\cdots(y_1-\alpha_m)}{(y_1-\beta_1)(y_1-\beta_2)\cdots(y_1-\beta_m)}$, rational function of degree m of y_1 where $A = \frac{A'}{A''}$. The combination of (124) and (128) gives the equation

$$\frac{dy}{\sqrt{(1-y^2)(1-c'^2y^2)}} = \frac{\varepsilon}{\varepsilon_1}\frac{dy_1}{\sqrt{(1-y_1^2)(1-c_1^2y_1^2)}}$$

and we see that the transformation of order $\mu = mn$ is obtained by composing a transformation ψ_1 of degree n and a transformation of order m. This result permits to reduce the theory of transformations to the case in which the order is a prime number.

In the general case, by the above reasoning $y = A\frac{(x-\alpha)(x-\alpha')\cdots(x-\alpha^{(\mu-1)})}{(x-\beta)(x-\beta')\cdots(x-\beta^{(\mu-1)})}$ where $\alpha, \alpha', \ldots, \alpha^{(\mu-1)}$ are the zeros and $\beta, \beta', \ldots, \beta^{(\mu-1)}$ the poles of ψx. Abel considers in particular the cases in which b or a is 0. When $b = 0$, the equation

$$p - qy = a(z-x)(z-x')\cdots(z-x^{(\mu-1)}) \tag{129}$$

implies that $a' - b'y = -a(x + x' + \ldots + x^{(\mu-1)})$ where a' and b' are the respective coefficients of $z^{\mu-1}$ in p and q. If $\frac{x\Delta e_m + e_m \Delta x}{1-c^2e_m^2x^2} \neq \frac{x\Delta e_m - e_m\Delta x}{1-c^2e_m^2x^2}$ for all m, $\mu = 2n+1$ is odd, $a' = 0$ and

$$y = Ax\left(1 + \frac{2\Delta e_1}{1-c^2e_1^2x^2} + \ldots + \frac{2\Delta e_n}{1-c^2e_n^2x^2}\right).$$

Therefore $q = (1-c^2e_1^2x^2)\cdots(1-c^2e_n^2x^2)$ and p is obtained by making $x = 0$ in (129):

$$p = az(z^2-e_1^2)\cdots(z^2-e_n^2) \quad \text{and}$$
$$y = a\frac{x(e_1^2-x^2)(e_2^2-x^2)\cdots(e_n^2-x^2)}{(1-c^2e_1^2x^2)(1-c^2e_2^2x^2)\cdots(1-c^2e_n^2x^2)}.$$

On the contrary, if $\frac{x\Delta e+e\Delta x}{1-c^2e^2x^2} = \frac{x\Delta e-e\Delta x}{1-c^2e^2x^2}$, $e = 0$ or $\frac{1}{0}$. When $e = \frac{1}{0}$, $x' = \pm\frac{1}{cx}$, $\mu = 2n$ is even, $a' = 0$ and $y = A\left(x \pm \frac{1}{cx} + \frac{2x\Delta e_1}{1-c^2e_1^2x^2} + \ldots + \frac{2x\Delta e_{n-1}}{1-c^2e_{n-1}^2x^2}\right) = \frac{a(1-\delta_1^2x^2)(1-\delta_2^2x^2)\cdots(1-\delta_n^2x^2)}{x(1-c^2e_1^2x^2)(1-c^2e_2^2x^2)\cdots(1-c^2e_{n-1}^2x^2)}$. When $e = 0$, $x' = -x$ and one finds that p and q have the same degree, contrary to the hypothesis.

When $a = 0$, $p - qy = by(z - x)(z - x')\cdots(z - x^{(\mu-1)})$ and it results that

$$y = a\frac{(1 - c^2e_1^2x^2)(1 - c^2e_2^2x^2)\cdots(1 - c^2e_n^2x^2)}{x(e_1^2 - x^2)(e_2^2 - x^2)\cdots(e_n^2 - x^2)} \quad \text{or}$$
$$a\frac{x(1 - c^2e_1^2x^2)(1 - c^2e_2^2x^2)\cdots(1 - c^2e_{n-1}^2x^2)}{(1 - \delta_1^2x^2)(1 - \delta_2^2x^2)\cdots(1 - \delta_n^2x^2)}$$

according to the parity of μ.

In particular

$$x_{2\mu+1} = a\frac{x(e_1^2 - x^2)(e_2^2 - x^2)\cdots(e_n^2 - x^2)}{(1 - c^2e_1^2x^2)(1 - c^2e_2^2x^2)\cdots(1 - c^2e_n^2x^2)}$$
$$= A\left(x + \frac{2\Delta e_1x}{1 - c^2e_1^2x^2} + \frac{2\Delta e_2x}{1 - c^2e_2^2x^2} + \ldots + \frac{2\Delta e_nx}{1 - c^2e_n^2x^2}\right)$$

where $2n = (2\mu + 1)^2 - 1$. Doing $x = \frac{1}{0}$ and 0, one finds $Ac^{2n}e_1^2e_2^2\cdots e_n^2 = a$, $A = \frac{1}{2\mu+1}$ and $ae_1^2e_2^2\cdots e_n^2 = 2\mu+1$. Thus $e_1^2e_2^2\cdots e_n^2 = \frac{2\mu+1}{c^n}$ and $a = c^n = c^{2\mu^2+2\mu}$. The roots of the equation $x_{2\mu+1} = y$ are x, $\frac{x\Delta e_1\pm e_1\Delta x}{1-c^2e_1^2x^2}$, $\frac{x\Delta e_2\pm e_2\Delta x}{1-c^2e_2^2x^2}$, $\ldots$, $\frac{x\Delta e_n\pm e_n\Delta x}{1-c^2e_n^2x^2}$. Let $\theta x = \frac{x\Delta e+e\Delta x}{1-c^2e^2x^2}$ and $\theta_1 x = \frac{x\Delta e'+e'\Delta x}{1-c^2e'^2x^2}$ be two of these roots such that neither e nor e' is a root of $x_{2m+1} = 0$ for a divisor $2m + 1$ of $2\mu + 1$ and such that $\theta_1 x$ is different from $x, \theta x, \ldots, \theta^{2\mu}x$. Then $x, \theta x, \ldots, \theta^{2\mu}x, \theta_1 x, \ldots, \theta_1^{2\mu}x$ are $4\mu + 1$ distinct roots of $x_{2\mu+1} = \psi x = y$. Thus, for any m and k, $\psi(\theta^m x) = \psi(\theta_1^k x)$ and it results that $\psi(\theta_1^k\theta^m x) = \psi(\theta^{2m}x) = x_{2\mu+1}$, so that $\theta_1^k\theta^m x$ is also a root. Now it is easy to prove that, for $0 \le m, k \le 2\mu$, all these roots are different when $2\mu + 1$ is a prime number. We have thus written the $(2\mu + 1)^2$ roots of our equation. Their expression is

$$\theta_1^k\theta^m x = \frac{x\Delta e_{m,k} + e_{m,k}\Delta x}{1 - c^2e_{m,k}^2x^2} \quad \text{where } e_{m,k} = \frac{e_m\Delta e'_k + e'_k\Delta e_m}{1 - c^2e_m^2e_k'^2}.$$

The roots of the equation $x_{2\mu+1} = 0$ are the $e_{m,k}$, where $e_{0,0} = 0$. The non-zero roots are given by an equation of degree $4\mu^2 + 4\mu$ which may be decomposed in $2\mu + 2$ equations of degree 2μ with the help equations of degree $2\mu + 2$. It is the result of the *Recherches* of 1827 (see our §3), demonstrated here by a purely algebraic way. Indeed, if p is a rational symmetric function of $e_1, e_2, \ldots, e_{2\mu}$, it may be expressed as a rational function φe_1 of e_1 such that $\varphi e_1 = \varphi e_2 = \ldots = \varphi e_{2\mu}$. Replacing e_1 by $e_{m,1}$, we see that $\varphi e_{m,1} = \varphi e_{m,2} = \ldots = \varphi e_{2\mu m,2\mu}$. It results that the sums $\rho_k = (\varphi e_1)^k + (\varphi e_{0,1})^k + \ldots + (\varphi e_{2\mu,1})^k$ are rational symmetric in the $4\mu^2 + 4\mu$

quantities $e_{m,k}$ different from 0 and so rational functions of c. Thus p is the root of an algebraic equation of degree $2\mu+2$ with coefficients rational in c. We may apply this result to the coefficients of the algebraic equation of which the roots are $e_1, e_2, \dots, e_{2\mu}$.

According to the formula (126), the modulus c' obtained from c by a transformation of order $2\mu+1$ is a rational symmetric function of $e_1, e_2, \dots, e_{2\mu}$. It is thus a root of an equation of degree $2\mu+2$ (the modular equation). Abel once more says that this equation seems not to be solvable by radicals. He adds that, since $\frac{dx_{2\mu+1}}{\Delta x_{2\mu+1}} = \frac{2\mu+1}{\varepsilon}\frac{dy}{\Delta' y}$, the multiplication by $2\mu+1$ (which is of degree $(2\mu+1)^2$) may be decomposed in the transformation of order $2\mu+1$ from x to y and another transformation of the same order from y to $x_{2\mu+1}$. Jacobi also used such a decomposition. The expressions of $x_{2\mu+1}$ and c in y and c' are given by (126) with a root e' determined from c' as e was from c. Thus the modular equation is symmetric in (c, c').

Abel recalls the total number of transformed moduli for a given order μ: 6 for $\mu = 1$, 18 for $\mu = 2$ and $6(\mu+1)$ for μ an odd prime number. Then he explains the algebraic solution of the equation $y = \psi x$ where ψx is a rational function defining a transformation. It is sufficient to consider the case in which the order is an odd prime number $2\mu+1$ and we know that, in this case, the roots are $x, \theta x, \dots, \theta^{2\mu} x$ where $\theta^m x = \frac{x\Delta e_m + e_m \Delta x}{1-c^2 e_m^2 x^2}$ and $\theta^{2\mu+1} x = x$. Let δ be a root of 1 and $v = x + \delta\theta x + \delta^2\theta^2 x + \dots + \delta^{2\mu}\theta^{2\mu} x$, $v' = x + \delta\theta^{2\mu} x + \delta^2\theta^{2\mu-1} x + \dots + \delta^{2\mu}\theta x$. They are of the form $v = p + q\Delta x$, $v' = p - q\Delta x$ where p and q are rational functions of x and $vv' = s$, $v^{2\mu+1} + v'^{2\mu+1} = t$ are rational functions of x. Since they are invariant by $x \to \theta x$, they are rational functions of y and we have $v = \sqrt[2\mu+1]{\frac{t}{2} + \sqrt{\frac{t^4}{4} - s^{2\mu+1}}}$. If $v_0, v_1, \dots, v_{2\mu}$ are the values of v corresponding to the $2\mu+1$ roots of 1, we obtain $x = \frac{1}{2\mu+1}(v_0 + v_1 + \dots + v_{2\mu})$, $\theta^m x = \frac{1}{2\mu+1}(v_0 + \delta^{-m} v_1 + \dots + \delta^{-1m\mu} v_{2\mu})$.

The last chapter of this first part deals with the following problem: "Given an elliptic integral of arbitrary modulus, to express this function by means of other elliptic integrals in the most general way." According to the results of the second chapter, this problem is expressed by the equation $\int \frac{rdx}{\Delta x} = k_1\psi_1 y_1 + k_2\psi_2 y_2 + \dots + k_m\psi_m y_m + V$ where $\varphi x = \int \frac{rdx}{\Delta x}$ is the given integral, $\psi_1, \psi_2, \dots, \psi_m$ are elliptic integrals of respective moduli $c_1, c_2, \dots, c_m$, $y_1, y_2, \dots, y_m$, $\frac{\Delta_1 y_1}{\Delta x}, \frac{\Delta_2 y_2}{\Delta x}, \dots, \frac{\Delta_m y_m}{\Delta x}$ are rational functions of x and V is an algebraic and logarithmic function. One may suppose that the number m is minimal and, according to a theorem of the fourth chapter, one has $\frac{dy_1}{\Delta_1 y_1} = \varepsilon_1 \frac{dx}{\Delta x}$, $\frac{dy_2}{\Delta_2 y_2} = \varepsilon_2 \frac{dx}{\Delta x}$, $\dots$, $\frac{dy_m}{\Delta_m y_m} = \varepsilon_m \frac{dx}{\Delta x}$ where $\varepsilon_1, \varepsilon_2, \dots, \varepsilon_m$ are constant. Now, for $1 \le j \le m$, there exists a rational function x_j of x such that $\varpi(x_j, c) = \varepsilon^{(j)}\varpi(x, c_j)$ and it results that there exits a rational function y of x such that φy be expressed as an elliptic integral of modulus c_j where x is the variable.

The part of the memoir published in Crelle's Journal stops here and Sylow completed it with a manuscript written by Abel and discovered in 1874. Here the transformation of elliptic integrals of the second and third kinds is explained. For the second kind, Abel proposes two methods. The first one is based on the differentiation with respect to the modulus c of the equation $\varpi(y, c') = \varepsilon\varpi(x, c)$, which gives

$$c'\frac{dc'}{dc}\int\frac{y^2dy}{(1-c'^2y^2)\Delta(y,c')}+\frac{dy}{dc}\frac{1}{\Delta(y,c')}$$
$$=\frac{d\varepsilon}{dc}\int\frac{dx}{\Delta(x,c)}+c\varepsilon\int\frac{x^2dx}{(1-c^2x^2)\Delta(x,c)}\,. \tag{130}$$

Now one can verify that $\int\frac{x^2dx}{(1-c^2x^2)\Delta(x,c)} = \frac{1}{c^2-1}\frac{x(1-x^2)}{\Delta(x,c)} + \frac{1}{1-c^2}\int\frac{(1-x^2)dx}{\Delta(x,c)}$ and there is a similar identity for $\int\frac{y^2dy}{(1-c'^2y^2)\Delta(y,c')}$. Thus (130) is rewritten

$$\frac{c'}{1-c'^2}\frac{dc'}{dc}\left(\varpi(y,c')-\varpi_0(y,c')-\frac{y(1-y^2)}{\Delta(y,c')}\right)+\frac{dy}{dc}\frac{1}{\Delta(y,c')}$$
$$=\frac{d\varepsilon}{dc}\varpi(x,c)+\frac{c\varepsilon}{1-c^2}\left(\varpi(x,c)-\varpi_0(x,c)-\frac{x(1-x^2)}{\Delta(x,c)}\right)$$

or $\varpi_0(y,c') = A\varpi(x,c)+B\varpi_0(x,c)+p$ where $A=\varepsilon\left(1-\frac{cdc(1-c'^2)}{c'dc'(1-c^2)}\right)-\frac{d\varepsilon(1-c'^2)}{c'dc'}$, $B=\frac{\varepsilon c(1-c'^2)dc}{c'(1-c^2)dc'}$ and $p=\frac{(1-c'^2)dc}{c'dc'}\frac{dy}{dc}\frac{1}{\Delta(y,c')}+B\frac{x(1-x^2)}{\Delta(x,c)}-\frac{y(1-y^2)}{\Delta(y,c')}$.

The second method is based on the decomposition of y^2 in partial fractions:

$$y^2=\frac{A}{(x-a)^2}+\frac{B}{x-a}+S$$

where a is a pole of y and A, B are constant. If $y=\frac{1}{\varphi x}$, $A=\frac{1}{(\varphi' a)^2}$ and $B=-\frac{\varphi'' a}{(\varphi' a)^3}$ and we have

$$(1-x^2)(1-c^2x^2)(\varphi' x)^2=\varepsilon^2\left((\varphi x)^2-1\right)\left((\varphi x)^2-c'^2\right). \tag{131}$$

For $x=a$, this gives $(1-a^2)(1-c^2a^2)(\varphi' a)^2=\varepsilon^2c'^2$. Let us differentiate (131) and make $x=a$; we obtain $2(1-a^2)(1-c^2a^2)\varphi' a\varphi'' a-\left(2(1+c^2)a-4c^2a^3\right)(\varphi' a)^2=0$ and we conclude that

$$A=\frac{1}{(\varphi' a)^2}=\frac{(1-a^2)(1-c^2a^2)}{\varepsilon^2c'^2},\qquad B=-\frac{\varphi'' a}{(\varphi' a)^3}=\frac{-(1+c^2)a+2c^2a^3}{\varepsilon^2c'^2}.$$

Thus

$$\int\frac{y^2dy}{\Delta(y,c')}=\frac{1}{\varepsilon c'^2}\int\left(\frac{(1-a^2)(1-c^2a^2)}{(x-a)^2}+\frac{2c^2a^3-(1+c^2)a}{x-a}\right)\frac{dx}{\Delta(x,c)}$$
$$+\varepsilon\int\frac{Sdx}{\Delta(x,c)}\,. \tag{132}$$

Now $d\frac{\Delta(x,c)}{x-a}=-\left(\frac{(1-a^2)(1-c^2a^2)}{(x-a)^2}+\frac{2c^2a^3-(1+c^2)a}{x-a}+c^2a^2-c^2x^2\right)\frac{dx}{\Delta(x,c)}$ and (132) takes the form:

$$\int \frac{y^2 dy}{\Delta(y, c')} = \frac{1}{\varepsilon c'^2}\left(\frac{\Delta(x, c)}{a - x} - c^2 a^2 \varpi(x, c) + c^2 \varpi_0(x, c)\right) + \varepsilon \int \frac{S dx}{\Delta(x, c)} \,.$$

If the poles of y are $a_1, a_2, \ldots, a_\mu$, we finally obtain

$$\begin{aligned}\varepsilon c'^2 \varpi_0(y, c') = \mu \varpi_0(x, c) - (c^2(a_1^2 + a_2^2 + \ldots + a_\mu^2) - \varepsilon^2 c'^2 k^2)\varpi(x, c) \\ + \Delta(x, c)\left(\frac{1}{a_1 - x} + \frac{1}{a_2 - x} + \ldots + \frac{1}{a_\mu - x}\right)\end{aligned}$$

where k is the value of y for x infinite. Abel separately considers the cases in which $k = 0$ or $k = \frac{1}{0}$. This last case is reduced to the first one by putting $x = \frac{1}{cz}$. For example, when

$$c' = \frac{2\sqrt{c}}{1 + c},\ y = (1 + c)\frac{x}{1 + cx^2} \text{ and } \varepsilon = 1 + c \,,$$

$\varpi_0(y, c') = \frac{c(1+c)}{2}\varpi_0(x, c) + \frac{1+c}{2}\varpi(x, c) - \frac{1+c}{2}\frac{x\Delta(x,c)}{1+cx^2}$.

For the integral of the third kind, Abel uses the equation

$$\int \frac{dy}{(a' - x)\Delta(y, c')} = \frac{1}{a'}\Pi(y, c', a') + \int \frac{y dy}{(a'^2 - y^2)\Delta(y, c')}$$

and the decomposition in partial fractions

$$\frac{1}{a' - y} = k' + \frac{1}{\varepsilon\Delta(a', c')}\left(\frac{\Delta(a_1, c)}{a_1 - x} + \frac{\Delta(a_2, c)}{a_2 - x} + \ldots + \frac{\Delta(a_\mu, c)}{a_\mu - x}\right)$$

which lead to

$$\begin{aligned}&\frac{\Delta(a', c')}{a'}\Pi(y, c', a') + \Delta(a', c') \int \frac{y dy}{(a'^2 - y^2)\Delta(y, c')} \\ &= k_1 \varpi(x, c) + \sum \frac{\Delta(a, c)}{a}\Pi(x, c, a) + v\end{aligned}$$

where k_1 is a constant and v is an algebraic and logarithmic function. Now the sum of μ integrals in the right hand side may be reduced to a single integral with the help of the result of the third chapter: if α is determined by

$$(fx)^2 - (\varphi x)^2\,(\Delta(x, c))^2 = (x^2 - a_1^2)(x^2 - a_2^2)\cdots(x^2 - a_\mu^2)(x^2 - \alpha^2)$$

where fx and φx are polynomials, one even and the other odd, according to (121) we have $\sum \frac{\Delta(a,c)}{a}\Pi(x, c, a) = k_2\varpi(x, c) + \frac{\Delta(\alpha,c)}{\alpha}\Pi(x, c, \alpha) - \frac{1}{2}\log\frac{fx+\varphi x\cdot\Delta(x,c)}{fx-\varphi x\cdot\Delta(x,c)}$. The coefficients of fx and φx are determined by the equations $fa_m + \varphi a_m \cdot \Delta(a_m, c) = 0 (1 \le m \le \mu)$ and the sign of $\Delta(\alpha, c)$ by $f\alpha + \varphi\alpha\Delta(\alpha, c) = 0$. Another way to do this reduction consists in observing that if a is any one of $a_1, a_2, \ldots, a_\mu$, that is a root of $a' = \psi(x)$, any other has the form $a_m = \frac{a\Delta(e_m,c)+e_m\Delta(a,c)}{1-c^2e_m^2a^2}$ where e_m does not depend of a. The same formula (121) with $n = 3$ and $m_1 = m_2 = m_3 = 1$ gives

$$\frac{\Delta(a_m, c)}{a_m}\Pi(x, c, a_m) = \frac{\Delta(a, c)}{a}\Pi(x, c, a) + \beta_m \varpi(x, c) + \frac{\Delta(e_m, c)}{e_m}\Pi(x, c, e_m) + \log S_m$$

and Abel shows that $\sum \frac{\Delta(e_m,c)}{e_m}\Pi(x, c, e) = 0$.

A posthumous paper, *Mémoire sur les fonctions transcendantes de la forme* $\int ydx$*, où* y *est une fonctions algébrique de* x (*Œuvres*, t. II, p. 206–216) contains extensions of the preceding results to more general Abelian integrals. Abel first considers μ integrals $r_j = \int y_j dx (1 \le j \le \mu)$ where y_j is an algebraic function of x and he supposes that they are related by an algebraic relation $R = \varphi(r_1, r_2, \dots, r_\mu) = 0$ where φ is a polynomial with coefficients algebraic with respect to x and μ is minimal. He proves that in that case there is a *linear* relation

$$c_1 r_1 + c_2 r_2 + \ldots + c_\mu r_\mu = P \tag{133}$$

where $c_1, c_2, \dots, c_\mu$ are constant and P is a rational function of $x, y_1, y_2, \dots, y_\mu$. Indeed, one may suppose that $R = r_\mu^k + P r_\mu^{k-1} + P_1 r_\mu^{k-2} + \ldots$ is irreducible with respect to r_μ (the coefficients $P, P_1, \ldots$ being rational with respect to $r_1, r_2, \dots, r_{\mu-1}$). By differentiation, one obtains

$$r_\mu^{k-1}(ky_\mu + P') + \left((k-1)Py_\mu + P_1'\right) r_\mu^{k-2} + \ldots = 0\,,$$

hence $ky_\mu + P' = 0$ and $kr_\mu + P =$ constant. This gives $k = 1$ and $R = r_\mu + P = 0$. Now the decomposition of P in partial fractions with respect to $r_{\mu-1}$ has the form

$$P = \sum \frac{S_k}{(r_{\mu-1} + t_k)^k} + \sum v_k r_{\mu-1}^k,$$

where t_k and v_k are rational with respect to $r_1, r_2, \dots, r_{\mu-2}$; by differentiation,

$$\sum \left(-\frac{kS_k(y_{\mu-1} + t_k')}{(r_{\mu-1} + t_k)^{k+1}} + \frac{S_k'}{(r_{\mu-1} + t_k)^k} \right) + \sum (v_k' r_{\mu-1}^k + k v_k r_{\mu-1}^{k-1} y_{\mu-1}) = -y_\mu$$

and this relation implies that $S_k = 0$ and $v_k' = 0$. Moreover, if k is not equal to 1, we must have $kv_k y_{\mu-1} + v_{k-1}' = 0$, but this would imply $kv_k r_{\mu-1} + v_{k-1} =$ constant, which is impossible. So $k = 1$ and $P = v_1 r_{\mu-1} + P_1$ where v_1 is a constant and P_1 is rational with respect to $r_1, r_2, \dots, r_{\mu-2}$. In the same way, we obtain, with a slight change of notation, $P_j = v_{\mu-1-j} r_{\mu-1-j} + P_{j+1} (0 \le j \le \mu - 2)$ where $v_1, v_2, \dots, v_{\mu-1}$ are constant and P_j is rational with respect to $r_1, r_2, \dots, r_{\mu-1-j}$. Finally, we have $r_\mu + v_{\mu-1} r_{\mu-1} + v_{\mu-2} r_{\mu-2} + \ldots + v_1 r_1 + v_0 = 0$ where v_0 is an algebraic function of x and this gives a relation of the form (133) where P is algebraic in x. Let $P^k + R_1 P^{k-1} + \ldots = 0$ be the minimal equation of P with coefficients rational in $x, y_1, y_2, \dots, y_\mu$. Differentiating, we get $(kdP + dR_1)P^{k-1} + ((k-1)R_1 dP + dR_2)\, P^{k-2} + \ldots = 0$ with $\frac{dP}{dx} = c_1 y_1 + c_2 y_2 + \ldots$, so that

$kdP + dR_1 = 0$ and $P = -\frac{R_1}{k} +$ constant. This gives $k = 1$ and $P = -R_1$, rational with respect to $x, y_1, y_2, \ldots, y_\mu$.

In his next theorem, Abel considers a relation

$$c_1 r_1 + c_2 r_2 + \ldots + c_\mu r_\mu = P + a_1 \log v_1 + a_2 \log v_2 + \ldots + a_m \log v_m \qquad (134)$$

where $v_1, v_2, \ldots, v_m$ are algebraic functions of x and P is a rational function of $x, y_1, y_2, \ldots, y_\mu, v_1, v_2, \ldots, v_m$. If v_m is root of an equation of degree n with coefficients rational in $x, y_1, y_2, \ldots, y_\mu, v_1, v_2, \ldots, v_{m-1}$, let $v'_m, v''_m, \ldots, v_m^{(n)}$ be its n values. One has

$$\begin{aligned} c_1 r_1 + c_2 r_2 + \ldots + c_\mu r_\mu = {} & \frac{1}{n}(P' + P'' + \ldots + P^{(n)}) \\ & + a_1 \log v_1 + a_2 \log v_2 + \ldots + a_{m-1} \log v_{m-1} \\ & + \frac{1}{n} a_m \log(v'_m v''_m \cdots v_m^{(n)}) \end{aligned}$$

where $P' + P'' + \ldots + P^{(n)}$ and $v'_m v''_m \cdots v_m^{(n)}$ are rational in $x, y_1, y_2, \ldots, y_\mu$, $v_1, v_2, \ldots, v_{m-1}$. Iterating we finally obtain $c_1 r_1 + c_2 r_2 + \ldots + c_\mu r_\mu = P + \alpha_1 \log t_1 + \alpha_2 \log t_2 + \ldots + \alpha_m \log t_m$ where $P, t_1, \ldots, t_m$ are rational functions of $x, y_1, y_2, \ldots, y_\mu$.

In particular, if y is an algebraic function of x and $\psi(x, y)$ a rational function such that the integral $\int \psi(x, y)dx$ is algebraic in $x, y, \log v_1, \log v_2, \ldots$, $\log v_m$, then this integral may be expressed in the form $P + \alpha_1 \log t_1 + \alpha_2 \log t_2 + \ldots + \alpha_m \log t_m$ where $P, t_1, \ldots, t_m$ are as above. If there is a relation $\int \psi(x, y)dx + \int \psi_1(x, y_1)dx = R$ where R is of the form of the right hand side of (134), and if the minimal equation for y_1 remains irreducible after adjunction of y, then one has separately $\int \psi(x, y)dx = R_1$ and $\int \psi_x(x, y_1)dx = R_2$. For if $y'_1, y''_1, \ldots, y_1^{(n)}$ are the values of the algebraic function x,

$$\begin{aligned} & n\psi(x, y)dx + (\psi_1(x, y'_1) + \psi_1(x, y''_1) + \ldots + \psi_1(x, y_1^{(n)}))dx \\ & = d(R' + R'' + \ldots + R^{(n)}) , \end{aligned}$$

hence a relation $\int \psi(x, y)dx = \frac{1}{n}(R' + R'' + \ldots + R^{(n)}) - \int f(x)dx = R_1$ and then $\int \psi_x(x, y_1)dx = R - R_1 = R_2$. If there is a relation $\int ydx = R$ where $y = p_0 + p_1 s^{-\frac{1}{n}} + p_2 s^{-\frac{2}{n}} + \ldots + p_{n-1} s^{-\frac{n-1}{n}}$, $p_0, p_1, \ldots, p_{n-1}, s$ algebraic functions such that $s^{\frac{1}{n}}$ is not rational in $p_0, p_1, \ldots, p_{n-1}, s$, then one has separately $\int \frac{p_m dx}{s^{\frac{m}{n}}} = R_j$ $(0 \leq m \leq n-1)$. Indeed $dR = df\left(s^{\frac{1}{n}}\right) = \psi\left(s^{\frac{1}{n}}\right) dx$ and the same relation is true for any value $\alpha^k s^{\frac{1}{n}}$ of $s^{\frac{1}{n}}$ (α primitive n-th root of 1). It is easy to deduce that

$$\int \frac{p_m dx}{s^{\frac{m}{n}}} = \frac{1}{n}(f(\sqrt[n]{s}) + \alpha^m f(\alpha\sqrt[n]{s}) + \ldots + \alpha^{(n-1)m} f(\alpha^{n-1}\sqrt[n]{s})) .$$

The rest of the paper is not finished. Abel studies the cases in which an integral

$$y = \int f\,(x, (x - a_1)^{\frac{1}{m_1}}, (x - a_2)^{\frac{1}{m_2}}, \ldots, (x - a_n)^{\frac{1}{m_n}})dx$$

(f is rational) is an algebraic function, and the corresponding reductions of Abelian integrals. According to the preceding results, he is reduced to

$$\int dx\cdot p\cdot(x-a_1)^{-\frac{k_1}{m_1}}(x-a_2)^{-\frac{k_2}{m_2}}\cdots(x-a_n)^{-\frac{k_n}{m_n}}$$
$$=P=v(x-a_1)^{1-\frac{k_1}{m_1}}(x-a_2)^{1-\frac{k_2}{m_2}}\cdots(x-a_n)^{1-\frac{k_n}{m_n}}$$

where p and v are rational and $\frac{k_1}{m_1},\frac{k_2}{m_2},\ldots,\frac{k_n}{m_n}$ are between 0 and 1. This gives

$$p=v(A_0+A_1x+\ldots+A_{n-1}x^{n-1})$$
$$+\frac{dv}{dx}(x-a_1)(x-a_2)\cdots(x-a_n)=v\varphi x+\frac{dv}{dx}fx$$

where

$$A_0+A_1x+\ldots+A_{n-1}x^{n-1}$$
$$=\left(1-\frac{k_1}{m_1}\right)(x-a_2)(x-a_3)\cdots(x-a_n)$$
$$+\left(1-\frac{k_2}{m_2}\right)(x-a_1)(x-a_3)\cdots(x-a_n)+\ldots$$
$$+\left(1-\frac{k_n}{m_n}\right)(x-a_1)(x-a_2)\cdots(x-a_{n-1})\,.$$

Abel explains the cases in which $v=x^m$ or $\frac{1}{(x-\alpha)^m}$. In the first case

$$p=x^m(A_0+A_1x+\ldots+A_{n-1}x^{n-1})$$
$$+\,mx^{m-1}(B_0+B_1x+\ldots+B_{n-1}x^{n-1}+x^n)$$
$$=mB_0x^{m-1}+(A_0+mB_1)x^m+(A_1+mB_2)x^{m+1}+\ldots$$
$$+\,(A_{n-1}+m)x^{n+m-1}.$$

Putting $\int x^\mu dx(x-a_1)^{-\frac{k_1}{m_1}}(x-a_2)^{-\frac{k_2}{m_2}}\cdots(x-a_n)^{-\frac{k_n}{m_n}}=R_\mu$, he gets

$$R_{m+n-1}=\frac{1}{m+A_{n-1}}x^m(x-a_1)^{1-\frac{k_1}{m_1}}(x-a_2)^{1-\frac{k_2}{m_2}}\cdots(x-a_n)^{1-\frac{k_n}{m_n}}$$
$$-\frac{mB_0}{m+A_{n-1}}R_{m-1}-\ldots-\frac{A_{n-2}+mB_{n-1}}{m+A_{n-1}}R_{m+n-2}$$

a recursion formula which permits to express R_{m+n-1} by $R_0,R_1,\ldots,R_{n-2}$.

In the second case

$$p=\frac{\varphi x}{(x-\alpha)^m}-\frac{mfx}{(x-\alpha)^{m+1}}$$
$$=-\frac{mf\alpha}{(x-\alpha)^{m+1}}+\frac{\varphi\alpha-mf'\alpha}{(x-\alpha)^m}+\frac{\varphi'\alpha-\frac{mf''\alpha}{2}}{(x-\alpha)^{m-1}}+\ldots$$
$$+\frac{\frac{\varphi^{(n-1)}\alpha}{1\cdot2\cdots(n-1)}-m\frac{f^{(n)}\alpha}{12\cdots n}}{(x-\alpha)^{m-n+1}}\,.$$

Putting $\int \frac{dx}{(x-\alpha)^\mu}(x-a_1)^{-\frac{k_1}{m_1}}(x-a_2)^{-\frac{k_2}{m_2}}\cdots(x-a_n)^{-\frac{k_n}{m_n}} = S_\mu$, he gets

$$\frac{(x-a_1)^{1-\frac{k_1}{m_1}}(x-a_2)^{1-\frac{k_2}{m_2}}\cdots(x-a_n)^{1-\frac{k_n}{m_n}}}{(x-\alpha)^m}$$
$$= -m\,f\alpha S_{m+1} + (\varphi\alpha - m\,f'\alpha)S_m + \ldots + \left(\frac{\varphi^{n-1}\alpha}{1\cdot 2\cdots(n-1)} - \frac{m\,f^{(n)}\alpha}{1\cdot 2\cdots n}\right)S_{m-n+1}.$$

If $f\alpha \neq 0$, this permits to express S_{m+1} in $S_1, R_0, R_1, \ldots, R_{n-2}$. If $f\alpha = 0$ but $\varphi\alpha - m\,f'\alpha \neq 0$, S_m is a linear combination of $R_0, R_1, \ldots, R_{n-2}$. Now

$$\varphi a_1 - m\,f'a_1 = \left(1 - \frac{k_1}{m_1} - m\right)(a_1-a_2)\cdots(a_1-a_n) \neq 0,$$

so that the S_m with 'parameter' a_1 are linear combinations of $R_0, R_1, \ldots, R_{n-2}$.

By the same method, Abel proves that a linear relation

$$c_0R_0 + c_1R_1 + \ldots + c_{n-2}R_{n-2} + \varepsilon_1 t_1 + \varepsilon_2 t_2 + \ldots + \varepsilon_\mu t_\mu$$
$$= v(x-a_1)^{1-\frac{k_1}{m_1}}(x-a_2)^{1-\frac{k_2}{m_2}}\cdots(x-a_n)^{1-\frac{k_n}{m_n}},$$

where $t_k = \int \frac{dx}{(x-\alpha_k)^\mu}(x-a_1)^{-\frac{k_1}{m_1}}(x-a_2)^{-\frac{k_2}{m_2}}\cdots(x-a_n)^{-\frac{k_n}{m_n}}$, is not possible. He finally proves that, in a relation $c_0R_0 + c_1R_1 + \ldots + c_{n-2}R_{n-2} + \varepsilon_1 t_1 + \varepsilon_2 t_2 + \ldots + \varepsilon_\mu t_\mu = P + \alpha_1 \log v_1 + \alpha_2 \log v_2 + \ldots + \alpha_m \log v_m$, the right hand side may be reduced to the form $\nu r_{\nu-1}\lambda_{\nu-1} + \sum \alpha \sum \omega^{k'} \log(\sum(s_k\lambda_k\omega^{k'k}))$ where ν is the g.c.d. of $m_1, m_2, \ldots, m_n$, for each $k \in [0, \nu-1]$, $\lambda_k = (x-a_1)^{\frac{\ell_1}{m_1}}(x-a_2)^{\frac{\ell_2}{m_2}}\cdots(x-a_n)^{\frac{\ell_n}{m_n}}$, λ_j being the remainder of the division of kk_j by m_j, ω is a primitive ν-th root of 1 and $r_{\nu-1}, s_0, s_1, \ldots, s_{\nu-1}$ are polynomials. First of all, the right hand side has the form

$$r_0 + r_1\lambda_1 + \ldots + r_{\nu-1}\lambda_{\nu-1} + \sum \alpha \log(s_0 + s_1\lambda_1 + \ldots + s_{\nu-1}\lambda_{\nu-1})$$

and when λ_1 is replaced by another value $\omega^{k'}\lambda_1$, λ_k becomes $\omega^{k'k}\lambda_k$. We thus get ν expressions for the considered integral $\int \frac{fx\cdot dx}{\lambda_1}$ and the terms $r_k\lambda_k$ with $k < \nu - 1$ disappear from the sum of these expressions. It is then possible to prove that $r_{\nu-1} = 0$ and that the relations of the considered type are combinations of those in which only one α is different from 0. In this case $\int \frac{fx\cdot dx}{\lambda_1} = \theta(x, \lambda_1) = \log\theta(\lambda_1) + \omega\log\theta(\omega\lambda_1) + \omega^2\log\theta(\omega^2\lambda_1) + \ldots + \omega^{\nu-1}\log\theta(\omega^{\nu-1}\lambda_1)$ where $\theta(\lambda_1) = s_0 + s_1\lambda_1 + \ldots + s_{\nu-1}\lambda_{\nu-1}$ and Abel attacks the determination of the possible forms for fx, but the paper is left incomplete (see Sylow's note, *Œuvres*, t. II, p. 327–329).

9 Series

We saw above (§3) that in his first papers, Abel did not hesitate to use infinite series in the 18th century manner, that is without any regard to questions of convergence. On the contrary, when dealing with expansions of elliptic functions (§6), he tried to treat the problem much more rigourously. In the meantime, he had read Cauchy's lectures at the *École Polytechnique* and he was impressed by this work. In a letter to Holmboe (16 January 1826), he writes "On the whole, divergent series are the work of the Devil and it is a shame that one dares base any demonstration on them. You can get whatever result you want when you use them, and they have given rise to so many disasters and so many paradoxes." Abel then explains that even the binomial formula and Taylor theorem are not well based, but that he has found a proof for the binomial formula and Cauchy's lectures contain a proof for Taylor theorem.

The memoir *Recherches sur la série* $1+\frac{m}{1}x+\frac{m(m-1)}{1\cdot 2}x^2+\frac{m(m-1)(m-2)}{1\cdot 2\cdot 3}x^3+\dots$, published in the first volume of Crelle's *Journal* (1826; *Œuvres*, t. I, p. 219–250) is devoted to a rigourous and general proof of the binomial formula. We have already explained the formal part of this memoir (§1) and we shall now analyse the part where Abel studies questions of convergence. Abel defines a convergent series as a series $v_0+v_1+v_2+\dots+v_m+\dots$ such that the partial sum $v_0+v_1+v_2+\dots+v_m$ gets indefinitely nearer to a certain limit, which is called the *sum* of the series, for increasing m, and he states Cauchy's criterium for convergence. The first theorem says that a series $\varepsilon_0\rho_0+\varepsilon_1\rho_1+\varepsilon_2\rho_2+\dots+\varepsilon_m\rho_m+\dots$ is divergent when $\rho_0,\rho_1,\rho_2,\dots$ are positive numbers such that $\frac{\rho_{m+1}}{\rho_m}$ has a limit $\alpha>1$ and the ε_m do not tend towards 0. On the contrary (theorem II), if the limit α is <1 and the ε_m remain ≤ 1, the series is convergent. The proof uses the comparison of $\rho_0+\rho_1+\dots+\rho_m+\dots$ with a convergent geometric series and Cauchy's criterium. In the third theorem, Abel considers a series

$$t_0+t_1+\dots+t_m+\dots$$

of which the partial sums $p_m=t_0+t_1+\dots+t_m$ remain bounded by some quantity δ and a decreasing sequence of positive numbers $\varepsilon_0,\varepsilon_1,\dots,\varepsilon_m,\dots$ The theorem states that

$$r=\varepsilon_0t_0+\varepsilon_1t_1+\varepsilon_2t_2+\dots+\varepsilon_mt_m$$

remains bounded by $\delta\varepsilon_0$. Abel uses what is now called 'Abel transformation', putting $t_0=p_0, t_1=p_1-p_0, t_2=p_2-p_1,\dots$ so that $r=p_0(\varepsilon_0-\varepsilon_1)+p_1(\varepsilon_1-\varepsilon_2)+\dots+p_{m-1}(\varepsilon_{m-1}-\varepsilon_m)+p_m\varepsilon_m\leq\delta\varepsilon_0$.

Theorem IV concerns a power series $f\alpha=v_0+v_1\alpha+v_2\alpha^2+\dots+v_m\alpha^m+\dots$ and it says that if the series is convergent for a (positive) value δ of α, it remains convergent for the (positive) values $\alpha\leq\delta$ and, for such an α, the limit of $f(\alpha-\beta)$ for $\beta\to 0$ is $f\alpha$. Abel puts $\varphi\alpha=v_0+v_1\alpha+v_2\alpha^2+\dots+v_{m-1}\alpha^{m-1}$ and $\psi\alpha=v_m\alpha^m+v_{m+1}\alpha^{m+1}+\dots=\left(\frac{\alpha}{\delta}\right)^m v_m\delta^m+\left(\frac{\alpha}{\delta}\right)^{m+1}v_{m+1}\delta^{m+1}+\dots\leq\left(\frac{\alpha}{\delta}\right)^m p$ where $p\geq v_m\delta^m, v_m\delta^m+v_{m+1}\delta^{m+1}, v_m\delta^m+v_{m+1}\delta^{m+1}+v_{m+2}\delta^{m+2},\dots$ (theorem III), and this bound is arbitrarily small for m sufficiently large. Now $f\alpha-f(\alpha-\beta)=\varphi\alpha-\varphi(\alpha-\beta)+\psi\alpha-\psi(\alpha-\beta)$ and, since $\varphi\alpha$ is a polynomial, it is sufficient to bound $\psi\alpha-\psi(\alpha-\beta)$ by $\left(\left(\frac{\alpha}{\delta}\right)^m+\left(\frac{\alpha-\beta}{\delta}\right)^m\right)p$, which is easy to do.

In the following theorem, the coefficients $v_0,v_1,\dots$ are continuous functions of x in an interval $[a,b]$ and Abel says that if the series is convergent for a value δ of α, its sum fx for $\alpha < \delta$ is a continuous function in $[a, b]$. Unfortunately, this theorem is not quite correct. Abel's proof consists in writing $fx = \varphi x + \psi x$ where φx is the sum of the terms up to $m-1$ and ψx is the corresponding remainder, which is bounded by $\left(\frac{\alpha}{\delta}\right)^m \theta x$ where $\theta x \geq v_m \delta^m$, $v_m \delta^m + v_{m+1}\delta^{m+1}$, $v_m \delta^m + v_{m+1}\delta^{m+1} + v_{m+2}\delta^{m+2}, \ldots$ (theorem III). For each x, this bound tends towards 0 as $m \to \infty$ but the convergence is not necessarily uniform in x and Abel's reasoning implicitly uses this uniformity. Recall that Cauchy stated more generally that the sum of a convergent series of continuous functions is continuous. In a footnote, Abel criticises this statement, giving the series $\sin x - \frac{1}{2}\sin 2x + \frac{1}{3}\sin 3x - \ldots$ as a counterexample: the series is everywhere convergent but its sum is discontinuous for $x = (2m+1)\pi$ (where it is 0).

Theorem VI correctly states the formula for the product of two *absolutely* convergent series $v_0 + v_1 + v_2 + \ldots = p$ and $v'_0 + v'_1 + v'_2 + \ldots = p'$. Let ρ (resp. ρ'_m) be the absolute value of v_m (resp. v'_m). The hypothesis is that $\rho_0 + \rho_1 + \rho_2 + \ldots = u$ and $\rho'_0 + \rho'_1 + \rho'_2 + \ldots = u'$ are convergent and the conclusion that the series of general term $r_m = v_0 v'_m + v_1 v'_{m-1} + v_2 v'_{m-2} + \ldots + v_m v'_0$ is convergent and that its sum is equal to pp'. Indeed $r_0 + r_1 + r_2 + \ldots + r_{2m} = p_m p'_m + t + t'$ where

$$\begin{aligned} p_m &= v_0 + v_1 + \ldots + v_m, \quad p'_m = v'_0 + v'_1 + \ldots + v'_m\,, \\ t &= p_0 v'_{2m} + p_1 v'_{2m-1} + \ldots + p_{m-1} v'_{m+1}\,, \\ t' &= p'_0 v_{2m} + p'_1 v_{2m-1} + \ldots + p'_{m-1} v_{m+1}\,. \end{aligned}$$

Now $|t| \leq u(\rho'_{2m} + \rho'_{2m-1} + \ldots + \rho'_{m+1})$, $|t'| \leq u'(\rho_{2m} + \rho_{2m-1} + \ldots + \rho_{m+1})$ so that t and t' tend towards 0. This result had been given by Cauchy in the sixth chapter of his *Analyse algébrique* (1821).

As an application, Abel considers two convergent series $t_0 + t_1 + t_2 + \ldots$, $t'_0 + t'_1 + t'_2 + \ldots$ with real terms and such that the series $t_0 t'_0 + (t_1 t'_0 + t_0 t'_1) + (t_2 t'_0 + t_1 t'_1 + t_0 t'_2) + \ldots$ is also convergent. Then the sum of this last series is equal to the product of the sums of the two given series. Indeed, by theorem IV, it is the limit of $t_0 t'_0 + (t_1 t'_0 + t_0 t'_1)\alpha + (t_2 t'_0 + t_1 t'_1 + t_0 t'_2)\alpha^2 + \ldots$ for $\alpha \to 1$ ($\alpha < 1$). Since both series $t_0 + t_1\alpha + t_2\alpha^2 + \ldots$ and $t'_0 + t'_1\alpha + t'_2\alpha^2 + \ldots$ are absolutely convergent for $\alpha < 1$ according to theorem II, the product of their sums is equal to

$$t_0 t'_0 + (t_1 t'_0 + t_0 t'_1)\alpha + (t_2 t'_0 + t_1 t'_1 + t_0 t'_2)\alpha^2 + \ldots$$

and the conclusion is clear.

In the third volume of Crelle's *Journal*, Abel published a *Note sur un mémmoire de M.L. Olivier, ayant pour titre "Remarques sur les séries infinies et leur convergence"* (1828; *Œuvres*, t. I, p. 399–402). In his memoir, Olivier stated a wrong criterium for the convergence of a series $\sum a_n$: that na_n must tend towards 0. As a counterexample, Abel gives the divergent series of general term $a_n = \frac{1}{n \log n}$ for which $na_n = \frac{1}{\log n}$ tends towards 0. He proves the divergence using the inequality $\log(1+x) < x$, which gives $\log\left(1 + \frac{1}{n}\right) < \frac{1}{n}$ or $\log\log(1+n) < \log\log n + \log\left(1 + \frac{1}{n\log n}\right) < \log\log n + \frac{1}{n \log n}$. It results that $\log\log(1+n) < \log\log 2 + \frac{1}{2\log 2} + \frac{1}{3\log 3} + \ldots + \frac{1}{n\log n}$ and the divergence follows from $\lim\,(\log\log(1+n)) = \infty$.

More generally, Abel proves that there is no function φn such that $\lim(\varphi n \cdot a_n) = 0$ be a criterium for the convergence of $\sum a_n$. Indeed, when $\sum a_n$ is divergent, the same is true for the series $\frac{a_1}{a_0} + \frac{a_2}{a_0+a_1} + \frac{a_3}{a_0+a_1+a_2} + \ldots + \frac{a_n}{a_0+a_1+\ldots+a_{n-1}} + \ldots$ for

$$\log(a_0 + a_1 + \ldots + a_n) - \log(a_0 + a_1 + \ldots + a_{n-1})$$
$$= \log\left(1 + \frac{a_n}{a_0 + a_1 + \ldots + a_{n-1}}\right) < \frac{a_n}{a_0 + a_1 + \ldots + a_{n-1}}$$

and $\log(a_0+a_1+\ldots+a_n) - \log a_0 < \frac{a_1}{a_0} + \frac{a_2}{a_0+a_1} + \frac{a_3}{a_0+a_1+a_2} + \ldots + \frac{a_n}{a_0+a_1+\ldots+a_{n-1}}$. Now if φn is a function such that $\varphi n \cdot a_n \to 0$ is a criterium of convergence, the series

$$\frac{1}{\varphi(1)} + \frac{1}{\varphi(2)} + \frac{1}{\varphi(3)} + \frac{1}{\varphi(4)} + \ldots + \frac{1}{\varphi n} + \ldots$$

is divergent but

$$\frac{1}{\varphi(2).\frac{1}{\varphi(1)}} + \frac{1}{\varphi(3)\left(\frac{1}{\varphi(1)} + \frac{1}{\varphi(2)}\right)} + \frac{1}{\varphi(4)\left(\frac{1}{\varphi(1)} + \frac{1}{\varphi(2)} + \frac{1}{\varphi(3)}\right)} + \ldots$$
$$+ \frac{1}{\varphi n\left(\frac{1}{\varphi(1)} + \frac{1}{\varphi(2)} + \frac{1}{\varphi(3)} + \ldots + \frac{1}{\varphi(n-1)}\right)} + \ldots$$

is convergent, which is contradictory.

Abel left unpublished a memoir *Sur les séries* (*Œuvres*, t. II, p. 197–205), probably written at the end of 1827. He begins by giving the definition of convergence and recalling Cauchy's criterium. Then the first part deals with series of positive terms and the second part with series of functions. The first theorem states that if a series $u_0 + u_1 + u_2 + \ldots + u_n + \ldots$ with $u_n \geq 0$ is divergent, then the same is true of $\frac{u_1}{s_0^\alpha} + \frac{u_2}{s_1^\alpha} + \frac{u_3}{s_2^\alpha} + \ldots + \frac{u_n}{s_{n-1}^\alpha} + \ldots$, where $s_n = u_0 + u_1 + u_2 + \ldots + u_n$ and $\alpha \leq 1$. It is an immediate extension of the preceding lemma, where α was taken equal to 1. The following theorem says that, under the same hypotheses, $\sum \frac{u_n}{s_n^{1+\alpha}}$ is convergent when $\alpha > 0$. Indeed $s_{n-1}^{-\alpha} - s_n^{-\alpha} = (s_n - u_n)^{-\alpha} - s_n^{-\alpha} > \alpha \frac{u_n}{s_n^{1+\alpha}}$. For example, if $u_n = 1$, the first theorem gives the divergence of the series $1 + \frac{1}{2} + \frac{1}{3} + \frac{1}{4} + \ldots + \frac{1}{n} + \ldots$ and the second theorem gives the convergence of the series $1 + \frac{1}{2^{\alpha+1}} + \frac{1}{3^{\alpha+1}} + \frac{1}{4^{\alpha+1}} + \ldots + \frac{1}{n^{\alpha+1}} + \ldots$ for $\alpha > 0$. When a series $\sum \varphi n$ is divergent, a necessary condition for the convergence of $\sum u_n$ is that

$$\liminf \frac{u_n}{\varphi n} = 0\,.$$

Indeed, if it is not the case, there exists $\alpha > 0$ such that $p_n = \frac{u_n}{\varphi n} \geq \alpha$ for n large enough and $\sum u_n \geq \sum \alpha \cdot \varphi n$ is divergent. Thus $\sum u_n$ is convergent only if $\liminf n u_n = 0$ but this condition is not sufficient and Abel recalls the final result of the preceding memoir. Abel next considers a function φn increasing without limit, implicitly supposed to be differentiable and concave, so that $\varphi(n+1) - \varphi n \leq \varphi' n$ and $\varphi'(0) + \varphi'(1) + \ldots + \varphi'(n) > \varphi(n+1) - \varphi(0)$ and this implies the divergence of $\varphi'(0) + \varphi'(1) + \ldots + \varphi'(n) + \ldots$ This applies to the iterated logarithm $\varphi_m n = \log^m(n+a)$: $\varphi_n' n = \frac{1}{(n+a)\log(n+a)\log^2(n+a)\cdots\log^{m-1}(n+a)}$ and the series

$$\sum \frac{1}{n \log n \log^2 n \cdots \log^{m-1} n}$$

is divergent. On the contrary, when $\varphi n = C - \frac{1}{\alpha-1}\frac{1}{(\log^m n)^{\alpha-1}}$ where $\alpha > 1$, $\varphi(n+1) - \varphi n > \varphi'(n+1)$ and $\varphi' n < \frac{1}{\alpha-1}\left(\frac{1}{(\log^m(n-1))^{\alpha-1}} - \frac{1}{(\log^m n)^{\alpha-1}}\right)$. It results that $\varphi'(a) + \varphi'(a+1) + \ldots + \varphi' n < \frac{1}{\alpha-1}\frac{1}{(\log^m(a-1))^{\alpha-1}}$ and the series $\sum \frac{1}{n \log n \log^2 n \cdots \log^{m-1} n (\log^m n)^{1+\alpha}}$ is convergent for $\alpha > 0$. Abel derives from this statement a rule for the convergence of a series $\sum u_n$: the series is convergent if $\lim \frac{\log\left(\frac{1}{u_n n \log n \cdots \log^{m-1} n}\right)}{\log^{m+1} n} > 1$ and it is divergent if this limit is < 1. For instance, in the first case, there exists an $\alpha > 0$ such that $u_n < \frac{1}{n \log n \cdots \log^{m-1} n (\log^m n)^{1+\alpha}}$ for n large enough.

The first result stated by Abel on the series of functions is that when a power series $\sum a_n x^n$ converges in $]-\alpha, \alpha[$, it may be differentiated term by term in this interval. Abel returns to theorem V of his memoir on the binomial formula, which shows that he was not satisfied with its proof. He considers $\varphi_0(y) + \varphi_1(y)x + \varphi_2(y)x^2 + \ldots + \varphi_n(y)x^n + \ldots = f(y)$ and he supposes that it is convergent for $0 \leq x < \alpha$ and y near a value β. Let A_n be the limit of $\varphi_n(y)$ when y tends towards β and suppose that $A_0 + A_1 x + \ldots + A_n x^n + \ldots$ is convergent. Then the sum R of this series is the limit of $f(y)$. Abel writes

$$\begin{aligned} f(\beta - \omega) - R = (\varphi_0(\beta - \omega) - A_0) + (x_1 \varphi_1(\beta - \omega) - A_1 x_1)\, x_2 + \ldots \\ + \left(\varphi_n(\beta - \omega) x_1^n - A_1 x_1^n\right) x_2^n + \ldots \end{aligned}$$

where $x = x_1 x_2$, $x_1 < \alpha$, $x_2 < 1$ and ω tends towards 0 and he chooses m such that

$$\varphi_m(\beta - \omega) x_1^m - A_1 x_1^m \geq \varphi_n(\beta - \omega) x_1^n - A_1 x_1^n$$

for all n, so that $f(\beta - \omega) = R + \frac{k}{1-x_2}(\varphi_m(\beta - \omega) x_1^m - A_1 x_1^m)$ where $-1 \leq k \leq 1$. Unfortunately, the value of m may depend on ω and the proof is still insufficient. As Lie remarks in the final notes (*Œuvres*, t. II, p. 326), it is sufficient to suppose that there exists M such that

$$(\varphi_n(\beta - \omega) - A_n)\, \alpha_0^n \leq M$$

for all n, for $x_1 < \alpha_0 < \alpha$ and for ω small enough in order to restaure a correct proof. Abel applies his theorem to the series

$$1^y x + 2^y x^2 + 3^y x^3 + \ldots + n^y x^n + \ldots ,$$
$$\sin y \cdot x + \frac{1}{2} \sin 2y \cdot x^2 + \frac{1}{3} \sin 3y \cdot x^3 + \ldots ,$$
$$\frac{y}{1+y^2} + \frac{y}{4+y^2} x + \frac{y}{9+y^2} x^2 + \ldots$$

continuous functions of $y \in \mathbb{R}$ when $0 \leq x < 1$; the second one is still convergent when $x = 1$, but its sum has discontinuities as a function of y. The third one has 0 for limit when y tends towards ∞, if $x < 1$, but the limit is $\frac{\pi}{2}$ if $x = 1$. Abel adds two remarks:

I. the series $\frac{\sin ay}{y} + \frac{\sin a^2 y}{y} x + \ldots + \frac{\sin a^{n+1} y}{y} x^n + \ldots$ is convergent for $0 \leq x < 1$ and $y > 0$, but when y tends towards 0, the limit A_n of $\frac{\sin a^{n+1} y}{y}$ is a^{n+1}, so that the series

$$A_0 + A_1 x + \ldots + A_n x^n + \ldots$$

is divergent when $ax > 1$.

II. the sum of

$$1+a+\ldots+a^y-(1+2a+\ldots+(y+1)a^y)x+(1+3a+\ldots+\frac{(y+1)(y+2)}{2}a^y)x^2-\ldots$$

is equal to $\frac{1}{1+x} + \frac{a}{(1+x)^2} + \ldots + \frac{a^y}{(1+x)^{y+1}} = fy$ for $0 \leq x < 1$ and y integer. When $y \to \infty$, this sum has for limit $\frac{1}{1+x-a}$ if $a < 1+x$, but, if $a \geq 1$, the limit of $\varphi_n(y) = 1 + (n+1)a + \ldots + \binom{y+n}{n} a^y$ is infinite and for $a < 1$, it is $\frac{1}{(1-a)^{n+1}} = A_n$. The series $A_0 + A_1 x + \ldots + A_n x^n + \ldots$ does not converge when $1 - x \leq a < 1$.

Abel gives an extension of his theorem IV of the memoir on the binomial series to the case in which $a_0 + a_1\alpha + a_2\alpha^2 + \ldots$ is divergent. In this case, if $a_n x^n$ is positive for n large, the limit of $a_0 + a_1 x + a_2 x^2 + \ldots$ for $x < \alpha$ tending towards α is infinite. The end of the paper contains a proof of Taylor theorem for a function $fx = a_0 + a_1 x + a_2 x^2 + \ldots$ defined by a power series convergent for $0 \leq x < 1$. A lemma states that if

$$fx = (a_0^{(0)} + a_1^{(0)} x + a_2^{(0)} x^2 + \ldots) + (a_0^{(1)} + a_1^{(1)} x + a_2^{(1)} x^2 + \ldots) + \ldots$$
$$+ (a_0^{(n)} + a_1^{(n)} x + a_2^{(n)} x^2 + \ldots) + \ldots$$

is convergent for $0 \leq x < 1$ and if $A_0 = a_0^{(0)} + a_0^{(1)} + \ldots + a_0^{(n)} + \ldots$, $A_1 = a_1^{(0)} + a_1^{(1)} + \ldots + a_1^{(n)} + \ldots, \ldots$ then $fx = A_0 + A_1 x + A_2 x^2 + \ldots + A_m x^m + \ldots$ whenever this series is convergent. Then Abel writes

$$\begin{aligned} f(x+\omega) &= a_0 + a_1(x+\omega) + a_2(x+\omega)^2 + \ldots \\ &= a_0 + a_1 x + a_2 x^2 + \ldots + (a_1 + 2a_2 x + \ldots)\omega + \ldots \\ &= fx + \frac{f'x}{1}\omega + \frac{f''x}{1 \cdot 2}\omega^2 + \ldots \end{aligned}$$

if this series is convergent. It remains to prove the convergence under the condition $x + \omega < 1$. Abel writes $x + \omega = x_1$ and $x = x_1 x_2$, so that $x_2 < 1$ and

$$x_1^n \frac{f^n x}{1 \cdot 2 \cdots n} = x_1^n a_n + (n+1) a_{n+1} x_1^{n+1} x_2 + \frac{(n+1)(n+2)}{1 \cdot 2} a_{n+2} x_1^{n+2} x_2^2 + \dots \leq v_n \frac{1}{(1-x_2)^{n+1}}$$

where v_n is the least upper bound of $a_{n+k} x_1^{n+k}$ for $k \geq 0$. This gives

$$\omega^n \frac{f^n x}{1 \cdot 2 \cdots n} \leq v_n \left(\frac{\omega}{x_1 - x_1 x_2} \right)^n \frac{1}{1 - x_2} = \frac{v_n}{1 - x_2}$$

where v_n tends towards 0.

10 Conclusion

Two main subjects constitute the core of Abel's work: algebraic equations and elliptic functions, with an extension to the most general abelian integrals. As we saw, they are intimately connected. Within our modern terminology, these subjects may be symbolised by the terms 'Abelian group', which refers to a class of solvable equations discovered by Abel, that is equations with a commutative Galois group, and by the theorem of Abel on Abelian integrals and the term 'Abelian variety'.

The theory of algebraic equations was one of the earliest fields of activity of Abel. He proved the impossibility to solve by radicals the general quintic equation. But later on he discovered that the so called Abelian equations are algebraically solvable and he attacked the general problem to characterise solvable equations. He obtained important results on the form of the solutions of solvable equations, and this part of the theory was the point of departure of Kronecker's work in algebra. Galois attacked the same problem from a different point of view, introducing the Galois group which measures the indiscernability between the roots.

Abel studied elliptic integrals in Legendre's *Exercices de Calcul Intégral*, following Degen's advice, and he immediately found fundamental new results. At the same time, Jacobi began to investigate this subject and Abel was stimulated by the competition with Jacobi. His theory contains all Jacobi's results up to the year 1829, but also some results of his own, as the study of the equation of division of an elliptic integral or of a period of such an integral. Particularly important is his discovery of complex multiplication which became a favourite subject for Kronecker and one of the sources of class field theory.

Abel's extension of the addition theorem for elliptic integrals to the general case of Abelian integrals is rightly considered as one of the most important discoveries in the first half of 19th century. It led Jacobi to formulate the inversion problem for hyperelliptic integrals. Through the works of Riemann and Clebsch, it became the base of a new method to study the geometry of algebraic curves. Abel's method to prove this theorem contains in germ the notions of divisors and of linear families of divisors on an algebraic curve and Riemann's interpretation of Abel's result leads to the notion of Jacobian of an algebraic curve.

With Gauss, Bolzano, Cauchy and Dirichlet, Abel is one of the reformators of rigour in the first half of 19th century. Abel's transformation of series gave him a way to prove the continuity of the sum of a power series up to the end of the interval of convergence in the case in which the series converges in this point. This theorem is the base of a method of summation for divergent series.

Abel always tried to attack problems in the most general way instead of studying particular cases and particular objects. In the theory of algebraic equations, he studied the structure of a general expression built with radicals and he asked under which conditions such an expression was the root of an algebraic equation of given degree. In the theory of Abelian integrals, he investigated the most general algebraic relation between given integrals and he proved that it is reducible to a linear relation. In the case of elliptic functions, a further reduction led to complex multiplication. This part of Abel's work announces Liouville's investigations on integration in finite terms and his classification of transcendental functions. We saw the same concern with generality in Abel's treatment of functional equations. This general method of Abel is well ahead of his time and close to the modern conception of axiomatic method.

References

[1] N.-H. Abel, *Œuvres complètes*, ed. by L. Sylow and S. Lie, Christiania, 1881, 2 vol. VIII+621+338 pages

[2] N.-H. Abel, *Il manoscritto parigino di Abel*, ed. A. Del Centina, Florence, 2002

[3] A.-L. Cauchy, Sur le nombre de valeurs qu'une fonction peut acquérir lorsqu'on y permute de toutes les manières possibles les quantités qu'elle renferme (1812), *Journ. Éc. Polytech.*, 17e cahier, t. X (1815), p. 1–28 = *Œuvres*, 2e série, t. 1, p. 64 sq.

[4] A.-L. Cauchy, Sur les fonctions qui ne peuvent obtenir que deux valeurs égales et de signes contraires par suite des transpositions opérées entre les variables qu'elles renferment (1812), *ibid.* = *Œuvres*, 2e série, t. 1, p. 90–169

[5] A.-L. Cauchy, Sur une loi de réciprocité qui existe entre certaines fonctions, *Bull. Soc. Philom.*, 1817, p. 121–124 = *Œuvres*, 2e série, t. 2, p. 223–227

[6] A.-L. Cauchy, Seconde note sur les fonctions réciproques, *ibid.*, 1818, p. 188–191 = *Œuvres*, 2e série, t. 2, p. 228–232

[7] A.-L. Cauchy, *Cours d'Analyse de l'École Royale Polytechnique; première partie: Analyse algébrique*, Paris, 1821 = *Œuvres*, 2e série, t. 3; modern ed. by U. Bottazzini, Bologne, 1992

[8] A.-L. Cauchy, *Résumé des Leçons données à l'École Royale Polytechnique sur le Calcul infinitésimal*, Paris, 1823

[9] A.-L. Cauchy, *Mémoire sur l'analogie des puissances et des différences, et sur l'intégration des équations linéaires*, Paris, 1825 = *Œuvres*, 2e série, t. 15, p. 23–40

[10] A.-L. Cauchy, Sur l'analogie des puissances et des différences, *Exerc. de Math.* 2, 1827 = *Œuvres*, 2e série, t. 7, p. 198–235

[11] A.-L. Cauchy, Sur les développements des fonctions en séries périodiques, *Mém. Ac. Sc. Paris*, t. 6, 1827, p. 603–612 = *Œuvres*, 1e série, t. 2, p. 12–19
[12] A. Clebsch, Ueber die Anwendung der Abelschen Functionen in der Geometrie, *Journ. für die reine und angew. Math.*, t. 63, p. 189–243
[13] A. Clebsch, Ueber diejenigen Curven, deren Coordinaten sich als elliptische Functionen eines Parameters darstellen lassen, *ibid.*, t. 64, p. 210–270
[14] L. Euler, De progressionibus transcendentibus, seu quarum termini generales algebraice dari nequeunt (1730), *Comm. Ac. Sc. Imper. Petrop.*, 1st ser., n° 5, 1738, p. 38–57 = *Opera omnia*, I_{14}, p. 1–24
[15] L. Euler, De formis radicum aequationum cuiusque ordinis coniectatio, *ibid.*, n° 6, 1738, p. 216–231 = *Opera omnia*, I_6, p. 1–19
[16] L. Euler, *Introductio in Analysin Infinitorum*, Lausanne, 1748, 2 vol. = *Opera omnia*, I_{8-9}
[17] L. Euler, *Institutiones Calculi differentialis*, Saint-Pétersbourg, 1755 = *Opera omnia*, I_{10}
[18] L. Euler, De resolutione aequationum cuiusvis gradus, *Nov. Comm. Petrop.*, 9, 1764, p. 70–98 = *Opera omnia*, I_6, p. 170–196
[19] L. Euler, *Institutiones Calculi integralis*, Saint-Pétersbourg, vol. II, 1769 = *Opera omnia*, I_{12}
[20] L. Euler, Demonstratio theorematis Newtoniani de evolutione potestatum binomii pro casibus, quibus exponentes non sunt numeri integri, 1773, *Opera omnia*, I_{16}, p. 207–216
[21] J.-B. Fourier, *Théorie analytique de la chaleur*, Paris, 1822 = *Œuvres*, t. I, p. 1–563
[22] J.-B. Fourier, Théorie du mouvement de la chaleur dans les corps solides (1811), *Mém. Acad. Sciences*, t. 4, 1824 = *Œuvres*, t. I
[23] É. Galois, Mémoire sur les conditions de résolubilité des équations par radicaux (1831), *Écrits et Mémoires mathématiques*, Paris, 1976, p. 43–101
[24] C.-F. Gauss, *Disquisitiones arithmeticæ*, Braunschweig, 1801 = *Werke*, t. I
[25] C.-F. Gauss, Varia imprimis de integrali $\int \frac{du}{\sqrt{1+\mu\mu\sin u^2}}$, *Werke*, t. III, p. 433–435
[26] C.-F. Gauss, Theoria sinus lemniscati universalissime accepti, *Werke*, t. X^1, p. 194–206
[27] C.-F. Gauss, Zur Theorie der Transcendenten Functionen gehörig, *Werke*, t. III, p. 436–445
[28] C.-F. Gauss, Zur Theorie der neuen Transcendenten, *Werke*, t. 3, p. 446–460
[29] C.-F. Gauss, Hundert Theoreme über die neuen Transcendenten, *Werke*, t. 3, p. 461–469
[30] A. Genocchi, Sulla formula sommatoria di Eulero e sulla teorica dei residui quadratici, *Ann. Sc. mat. fis.*, t. 3, 1852, p. 406–436
[31] A. Genocchi, Intorno ad alcune formule sommatorie, *ibid.*, t. 6, 1855, p. 70–114
[32] C. Huygens, *Horologium oscillatorium*, Paris, 1673
[33] C.G. Jacobi, Letter to Schumacher, *Astron. Nachr.*, 6, 1827 = *Ges. Werke*, t. 1, p. 31–33
[34] C.G. Jacobi, *Fundamenta nova theoriæ functionum ellipticarum*, Königsberg, 1829 = *Ges. Werke*, t. 1, p. 55–236
[35] C.G. Jacobi, Considerationes generales de Transcendentibus abelianis, *Journ. Für die reine und angew. Math.*, 9, 1832, p. 394–403 = *Ges. Werke*, t. 2, p. 7–16
[36] C.G. Jacobi, De functionibus duarum variabilium quadrupliciter periodicis, quibus theoria transcendentium abelianarum innititur, *ibid.*, 13, 1834, p. 55–78 = *Ges. Werke*, t. 2, p. 25–50
[37] L. Kronecker, Ueber die algebraisch auflösbaren Gleichungen I, *Monatsb. K. Preuss. Ak. Wissensch. Berlin*, 1853, p. 365–374 and 1856, p. 203–215 = *Math. Werke*, t. 4, p. 1–11 and 25–37

[38] L. Kronecker, Ueber die elliptischen Functionen, für welche complexe Multiplication stattfindet,, 1857, p. 455–460 = *Math. Werke*, t. 4, p. 179–183
[39] L. Kronecker, Ueber die complexe Multiplication der elliptischen Functionen, *ibid.*, 1862, p. 363–372
[40] L. Kronecker, Bemerkungen über die Darstellung von Reihen durch Integrale, *Journ. Für die reine und angew. Math.*, t. 105, p. 157–159 and 345–354 = *Mat. Werke*, t. 5, p. 329–342
[41] J.-L. Lagrange, Réflexions sur la résolution algébrique des équations, *Nouv. Mém. Ac. Roy. de Berlin*, (1771) 1773, p. 138–253 = *Œuvres*, t. 3, p. 305–421
[42] J.-L. Lagrange, Sur une nouvelle espèce de calcul relatif à la différentitation et à l'intégration des quantités variables, *ibid.*, 1772 = *Œuvres*, t. 3, p. 441–476
[43] J.-L. Lagrange, *Théorie des fonctions analytiques*, Paris, 1797
[44] J.-L. Lagrange, *Leçons sur le calcul des fonctions*, Paris, 1806 = *Œuvres*, t. 10
[45] P.S. Laplace, Mémoire sur les approximations des formules qui sont fonctions de très grands nombres (1782), *Mém. Ac. Royale des Sc.*, 1785 = *Œuvres*, t. 10, p. 209–291
[46] A.-M. Legendre, *Exercices de Calcul Intégral*, 2 vol., Paris 1811
[47] G.W. Leibniz, *Mathematische Schriften*, ed. C.I. Gerhardt, vol. 2, Berlin, 1849
[48] J. Liouville, Mémoire sur quelques questions de géométrie et de mécanique, et sur un nouveau genre de calcul pour résoudre ces questions, *Journal de l'Éc. Polytech.*, 13, sect. 21, 1832, p. 1–69
[49] J. Liouville, Mémoire sur le calcul des différentielles à indices quelconques, *ibid.*, p. 71–162
[50] J. Liouville, Mémoire sur l'intégration de l'équation $(mx^2+nx+p)\frac{d^2y}{dx^2}+(qx+r)\frac{dy}{dx}+sy=0$ à l'aide des différentielles à indice quelconque, *ibid.*, p. 163–186
[51] J. Liouville, Mémoire sur le théorème des fonctions complémentaires, *Journ. für die reine und angew. Math.*, t. 11, 1834, p. 1–19
[52] J. Liouville, Sur la classification des transcendantes et sur l'impossibilité d'exprimer les racines de certaines équations en fonction finie explicite des coefficients, *Journ. de Math. pures et appl.*, t. 2, 1837, p. 56–105, t. 3, 1838, p. 523–547 and t. 4, 1839, p. 423–456
[53] J. Liouville, Sur la détermination des intégrales dont la valeur est algébrique, *ibid.*, t. 3, 1838, p. 20–24
[54] A. de Moivre, Æquationum quarundam Potestatis tertiæ, quintæ, septimæ, nonæ, et superiorum, ad infinitum usque pergendo, in terminis finitis, ad instar Regularum pro cubicis, quæ vocantur Cardani, Resolutio Analytica, *Phil. Trans.*, 25, n° 209, 1707, p. 2368
[55] G.A.A. Plana, Sur une nouvelle expression analytique des nombres bernoulliens propre à exprimer en termes finis la formule générale pour la sommation des suites, *Mem. Ac. Torino*, t. 25, 1820, p. 403–418
[56] S.-D. Poisson, Note sur le développement des puissances des sinus et des cosinus en séries de sinus ou de cosinus d'arcs multiples, *Correspondance sur l'École Polytechnique*, vol. 2, n° 3 (1811), p. 212–217
[57] B. Riemann, Versuch einer allgemeinen Auffassung der Integration und Differentiation (1847), *Coll. Works*, New York, 1953, p. 353–366
[58] B. Riemann, Theorie der Abel'schen Functionen, *Journ. für die reine und angew. Math.*, 54, 1857 = *Coll. Works*, p. 88–144
[59] P. Ruffini, *Teoria generale delle equazioni in cui si dimostra impossibile la soluzione algebraica delle equazioni generali di grado superiore al quarto*, Bologne, 1799

[60] P. Ruffini, Della soluzione delle equazioni algebraiche determinate particolari di grado superiore al quarto, *Mem. Mat. Fis. Soc. Ital.*, 9, 1802 = *Op. Mat.*, t. 1, p. 343–406
[61] P. Ruffini, *Riflessioni intorno alla soluzione delle equazioni algebraiche generali*, Modène, 1813 = *Op. Mat.*, t. 2, p. 155–268
[62] M. Schaar, Mémoire sur les intégrales eulériennes et sur la convergence d'une certaine classe de séries, *Mém. Couronnés et Mém. des savants étrangers publ. Par l'Ac. De Belgique*, t. 22, 1848, p. 3–25
[63] A. Stubhaug, *Et foranskutt lyn. Niels Henrik Abel og hans tid*, Oslo, 1996, 598 pages; english transl. *Niels Henrik Abel and his times*, Springer-Verlag, Berlin, 2000, X+580 pages
[64] G. Zolotareff, Sur la méthode d'intégration de M. Tchébychef, *Math. Ann.*, 5, 1872, p. 560–580

The Legacy of Abel in Algebraic Geometry

Phillip Griffiths

This paper is based on a talk given at the bicentenary celebration of the birth of Niels Henrik Abel held in Oslo in June, 2002. The objectives of the talk were first to recall Abel's theorem in more or less its original form, secondly to discuss two of the perhaps less well known converses to the theorem, and thirdly to present two (from among the many) interesting issues in modern algebraic geometry that may at least in part be traced to the work of Abel. Finally, in the reprise I will suggest that the arithmetic aspects of Abel's theorem may be a central topic for the 21st century.

This talk was not intended to be a "documentary" but rather to tell the story – from my own perspective – of Abel's marvelous result and its legacy in algebraic geometry. Another talk at the conference by Christian Houzel gave a superb historical presentation and analysis of Abel's works.

In keeping with the informal expository style of this paper (the only proof given is one of Abel's original proofs of his theorem) at the end are appended a few general references that are intended to serve as a guide to the literature, and should not be thought of as a bibliography.

1 Origins of Abel's Theorem

During the period before and at the time of Abel there was great interest among mathematicians in *integrals of algebraic functions*, by which we mean expressions

$$\int y(x)dx \tag{1.1}$$

where $y(x)$ is a 'function' that satisfies an equation

$$f(x, y(x)) = 0 \tag{1.2}$$

where $f(x, y) \in \mathbb{C}[x, y]$ is an irreducible polynomial with complex coefficients. Although not formalized until later, it seems to have been understood that (1.1) becomes well-defined upon choosing a particular branch of the solutions to (1.2) along a path of integration in the x-plane that avoids the branch points where there are multiple roots. In more modern terms, one considers the algebraic curve F° in $\mathbb{C}^2$ defined by

$$f(x, y) = 0 ,$$

and on F° one considers the rational differential ω defined by the restriction to F° of

$$\omega = ydx .$$

On the closure F of F° in the compactification of $\mathbb{C}^2$ given either by the projective plane $\mathbb{P}^2$ or by $\mathbb{P}^1 \times \mathbb{P}^1$ one considers an arc γ avoiding the singularities of F and the poles of ω, and then (1.1) is defined to be

$$\int_\gamma \omega . \tag{1.3}$$

Actually, among mathematicians of the time the interest was in more general expressions

$$\int r(x, y(x))dx \tag{1.4}$$

where $r(x, y)$ is a rational function of x and y, and $y(x)$ is as above. The formal definition of (1.4) is as in (1.3) where now ω is the restriction to F of the rational differential 1-form $r(x, y)dx$.

Of special interest were the *hyperelliptic integrals*

$$\int \frac{p(x)dx}{\sqrt{q(x)}} \tag{1.5}$$

where $p(x)$ and $q(x)$ are polynomials with, say,

$$q(x) = x^n + q_1 x^{n-1} + \cdots + q_n$$

of degree n and having distinct roots. When $n = 1, 2$ it was well understood at the time of Abel that these integrals are expressible in terms of the "elementary" – i.e., trigonometric and logarithmic – functions. The geometric reason, which was also well understood, is that any plane conic curve may be rationally parametrized as expressed by the picture

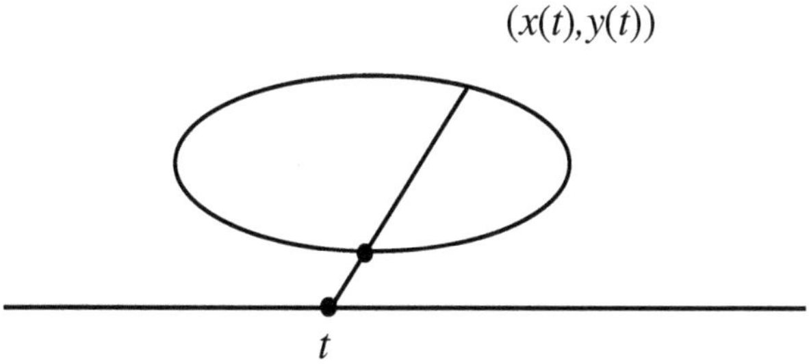

Plugging the rational functions $x(t)$ and $y(t)$ into (1.5) gives an integral

$$\int r(t)dt \ ,$$

where $r(t)$ is a rational function, and this expression may be evaluated by the partial fraction expansion of $r(t)$.

There was particular interest in the hyperelliptic integrals (1.5) when $n = 3, 4$ and important fragments were understood through the works of Euler, Legendre and others. They go under the general term of *elliptic integrals*, for the following reason: Just as the resolution of the arc length on a circle leads to the trigonometric functions as expressed by

$$\int \sqrt{dx^2 + dy^2} = \int \frac{dx}{\sqrt{1-x^2}} \, , \qquad x^2 + y^2 = 1 \, , \tag{1.6}$$

there was great interest in the functions that arise in the resolution of the arc length of an ellipse. Thus, through the substitution

$$t = \arcsin\left(\frac{x}{a}\right)$$

the arc length on the ellipse

$$\int \sqrt{dx^2 + dy^2} \, , \qquad \frac{x^2}{a^2} + \frac{y^2}{b^2} = 1 \qquad a > b$$

becomes the elliptic integral

$$a \int \frac{\left(1 - k^2 t^2\right) dt}{\sqrt{\left(1 - t^2\right)\left(1 - k^2 t^2\right)}} \, , \qquad k^2 = \left(a^2 - b^2\right)/a^2 \tag{1.7}$$

in Legendre form.

Of special interest were integrals (1.4) that possessed what was thought to be the very special property of having *functional equations* or *addition theorems*. For example, using the obvious synthetic geometric construction of doubling the length of an arc on the circle applied to the integral (1.6), one recovers the well known formulas for $\sin 2\theta$ and $\cos 2\theta$ expressed in terms of $\sin\theta$ and $\cos\theta$. More generally one may derive expressions for $\sin(\theta+\theta')$ etc. which are expressed as addition theorems for the integral (1.6). In the 18th century the Italian Count Fagnano discovered a synthetic construction for doubling the arc length on an ellipse, and when applied to (1.7) this construction leads to addition theorems for the "elliptic integral" (1.7). As alluded to above this was thought to be a very special feature, one that was the subject of intensive study in the late 18th and early 19th centuries.

2 Abel's Theorem and some Consequences

In Abel's work on integrals of algebraic functions there are two main general ideas

- abelian sums
- inversion

Together these led Abel to very general forms of

- functional equations

for the integrals. We will now explain these ideas.

Turning first to what are now called abelian sums, the integrals (1.1) and more generally (1.4) are highly transcendental functions of the upper limit of integration and consequently are generally difficult to study directly.[1] Abel's idea was to consider the sum of integrals to the variable points of intersection of $F = \{f(x, y) = 0\}$ with a family of curves $G_t = \{g(x, y, t) = 0\}$ depending rationally on a parameter t. Thus letting

$$F \cap G_t = \sum_i (x_i(t), y_i(t))$$

be the set of solutions to

$$\begin{cases} f(x, y) = 0 \\ g(x, y, t) = 0 \end{cases}$$

written additively using the notation of algebraic cycles, the *abelian sum* associated to (1.4) is defined to be

[1] The term "highly transcendental" needs care in interpretation – cf. the reprise below. Again Abel, in a paper published in 1826, showed the existence of polynomials R, F such that

$$\int \frac{F\,dx}{\sqrt{R}} = \ln\left(\frac{P+\sqrt{R}Q}{R-\sqrt{R}Q}\right)$$

has solutions for relatively prime polynomials P, Q. Here, R is a polynomial of degree $2n$ with distinct roots and F is a polynomial of degree $n-1$, so that the integrand is a differential of the 3rd kind. This is an "exceptional" case where the integral is transcendental but expressible in terms of elementary functions.

$$u(t) = \sum_i \int_{x_0}^{x_i(t)} r(x, y(x))dx \ . \qquad (2.1)$$

Below we will amplify on just how this expression is to be understood. A particularly important example is given by taking the G_t to be a family of lines as illustrated by the figures

(i)  $F = \{x^2 + y^2 = 1\}$

(ii) 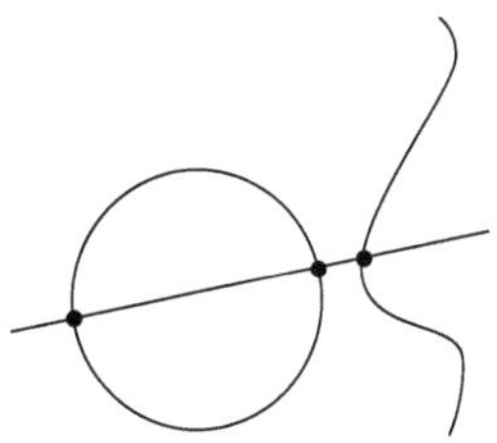 $F = \{y^2 = x^3 + ax + b\}$

In both cases we take $\omega = dx/y$ and the integrals (1.4) are respectively

$$\begin{cases} \text{(i)} \displaystyle\int \frac{dx}{\sqrt{1-x^2}} \\ \text{(ii)} \displaystyle\int \frac{dx}{\sqrt{x^3+ax+b}} \end{cases} \qquad (2.2)$$

Even though the individual terms in the abelian sum are in general highly transcendental functions, *Abel's theorem* expresses the abelian sum as an elementary function:

Theorem. *The abelian sum (2.1) is given by*

$$u(t) = r(t) + \sum_\lambda a_\lambda \log(t - t_\lambda) \qquad (2.3)$$

where $r(t)$ is a rational function of t.

One of the proofs given by Abel is as follows:

Proof. For reasons to appear shortly we define the rational function

$$q(x, y) = r(x, y) f_y(x, y) \ ,$$

so that the integrand in the integrals appearing in the abelian sum is the restriction to the curve F of

$$\frac{q(x, y)dx}{f_y(x, y)} .$$

Then by calculus

$$u'(t) = \sum_i \frac{q(x_i(t), y_i(t))x_i'(t)}{f_y(x_i(t), y_i(t))} .$$

From

$$\begin{cases} f(x_i(t), y_i(t)) & = 0 \\ g(x_i(t), y_i(t), t) & = 0 \end{cases}$$

we have

$$x_i'(t) = \left(\frac{g_t f_y}{f_x g_y - f_y g_x} \right) (x_i(t), y_i(t))$$

so that

$$u'(t) = \sum_i s(x_i(t), y_i(t)) \tag{2.4}$$

where $s(x, y)$ is the rational function given by

$$s(x, y) = \left(\frac{q g_t}{f_x g_y - f_y g_x} \right) (x, y) .$$

(The non-vanishing of the rational function in the denominator is a consequence of assuming that the curves F and G_t have no common component.) Abel now observes that the right hand side of (2.4) is a rational function of t – from a complex analysis perspective this is clear, since $u'(t)$ is a single-valued and meromorphic function of t for $t \in \mathbb{P}^1$. Integration of the partial fraction expansion of $u'(t)$ gives the result.

In his Paris memoir, and also in subsequent writings on the subject, in special cases Abel gave quite explicit expressions for the right hand side of (2.4), and therefore for the terms in the formula for $u(t)$ in his theorem. For example, when the curves G_t are lines the Lagrange interpolation formula gives the explicit expression for $u'(t)$.

We shall now give applications of Abel's theorem to the two integrals in (2.2). Both are based on the second of Abel's ideas mentioned above, namely to *invert* the integral (1.4) by defining the coordinates $x(u)$, $y(u)$ on the curve F as *single-valued* functions of the variable u by setting

$$u = \int_{(x_0, y_0)}^{(x(u), y(u))} \omega \tag{2.5}$$

where ω is the restriction to the curve F of $r(x, y)dx$. For example, for the integral (i) in (2.2) we obviously have

$$u = \int_{(0,1)}^{(\sin u, \cos u)} \omega \, .$$

The right hand side in (2.3) may be evaluated using the Lagrange interpolation formula and this leads to the relation

$$\int_0^{x_1} \frac{dx}{\sqrt{1-x^2}} + \int_0^{x_2} \frac{dx}{\sqrt{1-x^2}} = \int_0^{x_1 y_2 + x_2 y_1} \frac{dx}{\sqrt{1-x^2}}$$

which we recognize as the addition formula for the sin function.

Before turning to the second integral in (2.2), we remark that already in his Paris memoir Abel singled out a "remarkable" class of abelian integrals (1.4), *now called integrals of the* 1st *kind*, by the condition that the right hand side of (2.3) reduce to a constant – this is evidently equivalent to the abelian integral (1.4) being locally a bounded function of the upper limit of integration. Abel explicitly determined the integrals of the 1st kind for a large number of examples. For instance for the hyperelliptic curves

$$y^2 = p(x)$$

where $p(x)$ is a polynomial of degree $n+1$ with distinct roots, Abel showed that the integrals of the 1st kind are

$$\begin{cases} \omega = \dfrac{g(x)dx}{y} \\ \deg g(x) \leqq \left[\dfrac{n}{2}\right] \end{cases} .$$

In particular, assuming that the cubic $x^3 + ax + b$ has distinct roots, the expression (ii) in (2.2) is an integral of the 1st kind. Abel's theorem for the family of lines meeting the cubic may be expressed by the relation

$$u_1 + u_2 + u_3 = c \tag{2.6}$$

where c is a constant and

$$u = \int_{(x_0, y_0)}^{(x(u), y(u))} \frac{dx}{y} \tag{2.7}$$

with $u = u_i$ plugged into (2.7) for $i = 1, 2, 3$ in (2.6). Differentiation of (2.7) gives

$$1 = \frac{x'(u)}{y(u)}$$

so that

$$y(u) = x'(u) \, . \tag{2.8}$$

Choosing (x_0, y_0) appropriately (specifically the flex $(0,0,1)$ on the intersection of F with the line at infinity in $\mathbb{P}^2$) we will have

$$\begin{cases} c = 0 \\ x(-u) = x(u) \end{cases}$$

and (2.6) becomes the famous addition theorem for the elliptic integral

$$x(u_1 + u_2) = R(x(u_1), x'(u_1), x(u_2), x'(u_2)) \tag{2.9}$$

where R is a rational function that expresses the x-coordinates of the third point of intersection of a line with F as a rational function of the coordinates of the other two points.

Of course, $x(u)$ is the well-known Weierstrass p-function and the above discussion gives the functional equation (2.9) and differential equation

$$x'(u)^2 = x(u)^3 + ax(u) + b$$

satisfied by the p-function. We give two remarks amplifying this discussion.

The first is that in order to *define* the integral

$$\int \frac{dx}{\sqrt{x^3 + ax + b}} \tag{2.10}$$

one cuts the x-plane, including the point at infinity, along slits connecting two of the roots of $x^3 + ax + b$ and connecting the third root to $x = \infty$

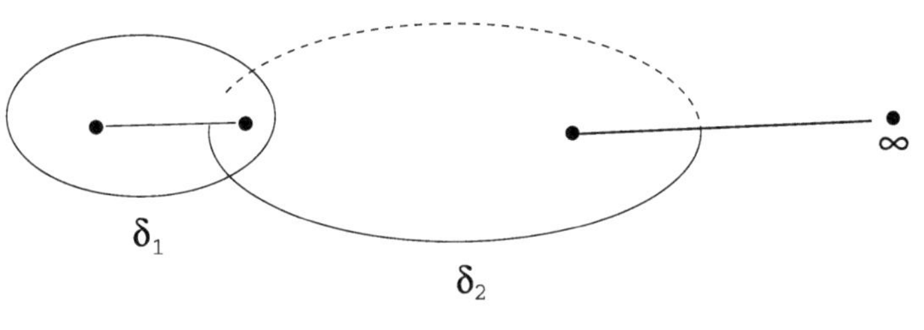

Then $\sqrt{x^3 + ax + b}$ is single-valued on the slit plane, and one may envision the algebraic curve F as a 2-sheeted covering of the x-plane where crossing a slit takes one to the "other sheet" – i.e., F is the Riemann surface associated to the algebraic function $\sqrt{x^3 + ax + b}$. The topological picture of F is the familiar torus

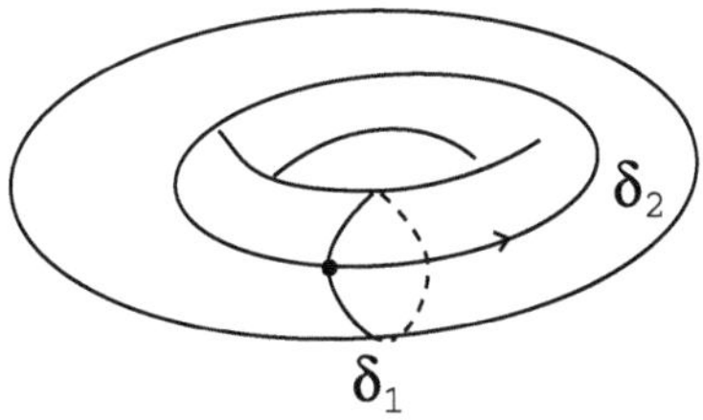

The integral (2.10) is then interpreted as an integral along a path on the Riemann surface. The choice of path is only well-defined up to linear combinations of δ_1 and δ_2. In particular, from (2.7) we infer that

$$\begin{cases} x(u+\lambda_i) = x(\lambda_i) \\ y(u+\lambda_i) = y(\lambda_i) \end{cases} \tag{2.11}$$

where

$$\lambda_i = \oint_{\delta_i} \frac{dx}{y}$$

are the *periods* of dx/y. Letting Λ be the lattice in the complex plane generated by λ_1 and λ_2, we have the familiar parametrizaton

$$\begin{array}{ccc} \mathbb{C}/\Lambda & \longrightarrow & F \\ \Cup & & \Cup \\ u & \longrightarrow & (x(u), x'(u)) \end{array}$$

of the cubic curve by the p-function and its derivative.

In his Paris memoir, Abel gave in general the essential analytic properties of elliptic functions, defined as those functions that arise by inversion of the integral of the first kind on curves having one such integral.

Remark that the dimension of the space of integrals of the first kind is one definition of the *genus* of the algebraic curve F. Following Abel's pioneering work, the extension of the above story to curves of arbitrary genus was carried out by Jacobi, Riemann and other 19th-century mathematicians.

A second remark is that the functions $x(u)$, $y(u)$ in (2.7) may be defined *locally* with (2.8) holding, and the *functional equation* (2.9) is valid where defined. But then this functional equation may be used to extend $x(u)$ and $y(u)$ to entire meromorphic functions – e.g., if $x(u)$ is defined for $|u| < \epsilon$, then from (2.9) we may define $x(2u)$ and continuing in this way proceed to define $x(u)$ for $|u| < 2\epsilon$, $|u| < 3\epsilon, \dots$. The principle that *a functional equation may be used to propagate a local object into a global one* is a central consequence of Abel's theorem, one that will be discussed further below.

In concluding this section we mention two direct consequences of Abel's theorem in algebraic geometry:

(i) the first beginnings of Hodge theory
(ii) the use of correspondences.

Under (i) we mean that Abel isolated what we now call the space of regular differentials $H^0(\Omega^1_F)$ as a basic invariant of an algebraic curve. He also computed $h^0(\Omega^1_F) = \dim H^0(\Omega^1_F)$ in a number of examples, which may be interpreted as taking the first steps toward identifying $h^0(\Omega^1_F)$ with the algebro-geometrically defined arithemetic genus. The further intepretation of $h^0(\Omega^1_F)$ as one-half the first Betti number – which marks the real beginning of Hodge theory – was to await Riemann.

Regarding (ii), the proof given above of Abel's theorem may be summarized by the diagram

$$\begin{array}{ccc} & I \subset F \times \mathbb{P}^1 & \\ {}^{\pi_1}\swarrow & & \searrow^{\pi_2} \\ F & & \mathbb{P}^1 \end{array}$$

where

$$I = \{(x, y, t) : f(x, y) = g(x, y, t) = 0\}$$

is the incidence correspondence, and where the map

$$\omega \to d\left(\sum_i \int_{x_0}^{x_i(t)} \omega\right)$$

in the proof is, in modern terms, the *trace*

$$\omega \to (\pi_2)_* \left(\pi_1^* \omega\right)$$

taking rational 1-forms on F to rational 1-forms on $\mathbb{P}^1$.

3 Converses to Abel's Theorem

In addition to recalling the usual global version of Abel's theorem and its converse, we will in this section give two less well known *local* converses to the result. These will illustrate the aforementioned principle that a local object having a functional equation may be propagated into a global one.

The usual version, found in textbooks, of Abel's theorem and its converse deals with the following question:

On a compact Riemann surface X we ask when a divisor

$$D = \sum_i n_i p_i \tag{3.1}$$

is the divisor of a meromorphic function; i.e., what is the *test* to determine if

$$D = \sum_{p \in X} \nu_p(f) p \tag{3.2}$$

for some function $f \in \mathbb{C}(X)^*$?

The answer is the following: For a regular 1-form $\omega \in H^0(\Omega^1_X)$ we recall that a period is defined to be the integral

$$\int_\delta \omega$$

where $\delta \in H_1(X, \mathbb{Z})$. Then there are two conditions that (3.2) hold. The first is that the degree of the divisor D

$$\deg D =: \sum_i n_i = 0 . \tag{3.3a}$$

If this is satisfied, then we may write $D = \partial\gamma$ for a 1-chain γ and then the second condition is that

$$\int_\gamma \omega \equiv 0 \quad \text{mod (periods)} \tag{3.3b}$$

for all $\omega \in H^0(\Omega^1_X)$.

The necessity of (3.3a) is a consequence of the residue theorem

$$\sum_{p \in X} \mathrm{Res}_p\left(\frac{df}{f}\right) = \sum_{p \in X} \nu_p(f) = 0 , \qquad f \in \mathbb{C}(X)^* .$$

The necessity of (3.3b) is essentially Abel's theorem as given above: Setting for $t \in \mathbb{P}^1$

$$f^{-1}(t) = \sum_i p_i(t) =: D_t$$

the configuration of points $\sum_i p_i(t)$ moves with a rational parameter and $D = D_0 - D_\infty$. Since ω is a regular differential the abelian sum

$$\sum_i \int_{p_i(0)}^{p_i(t)} \omega$$

is constant, and since $\int_\gamma \omega$ is only well-defined modulo periods the assertion (3.3b) follows. Alternatively, for

$$I \subset X \times \mathbb{P}^1$$

the incidence correspondence defined by

$$I = \{(p, t) : f(p) = t\}$$

we have as before that

$$d\left(\sum_i \int_{p_i(0)}^{p_i(t)} \omega\right) = (\pi_2)_* \left(\pi_1^* \omega\right)$$

is a regular 1-form on $\mathbb{P}^1$, hence equal to zero.

The usual global converse to Abel's theorem is that the conditions (3.3a) and (3.3b) are sufficient that (3.2) hold. This may be formulated by the statement that the map

$$\mathrm{Div}^0(X) \to J(X)$$

from the group of divisors of degree zero into the Jacobian variety

$$J(X) =: H^0\left(\Omega_X^1\right)^* / H_1(X, \mathbb{Z}) \, ,$$

given by the above construction

$$\langle D, \omega \rangle =: \int_\gamma \omega \qquad \text{mod periods}$$

where $D \in \mathrm{Div}^0(X)$, $\partial\gamma = D$ and $\omega \in H^0(\Omega_X^1)$, should be injective.

The first local converse deals with what we shall call *Abel's differential equations*. These simply state the conditions that a configuration of points $p_i \in X$ together with tangent vectors $\tau_i \in T_{p_i}X$ should satisfy the infinitesimal form

$$\sum_i \langle \omega(p_i), \tau_i \rangle = 0 \, , \qquad \omega \in H^0\left(\Omega_X^1\right) \tag{3.4}$$

of Abel's theorem. We may re-express (3.4) as follows: We consider $\sum_i p_i$ as a point in the symmetric product $X^{(d)}$. Each regular 1-form ω on X induces a 1-form $\mathrm{Tr}\,\omega$ on $X^{(d)}$ by

$$(\mathrm{Tr}\,\omega)(p_1 + \cdots + p_d) = \omega(p_1) + \cdots + \omega(p_d) \, .$$

Then (3.4) is equivalent to the differential system

$$\mathrm{Tr}\,\omega = 0 \, , \qquad \omega \in H^0\left(\Omega_X^1\right) \tag{3.5}$$

on $X^{(d)}$. From a differential equations perspective the remarkable fact is that *the maximal local integral manifolds of (3.5) are open sets in a global integral manifold* $\mathbb{P}^r \subset X^{(d)}$. Thus the DE's (3.5) truly do represent the condition for infinitesimal rational motion of divisors. The more precise statement is:

> *In each tangent space $T_z X^{(d)}$, $z = p_1 + \cdots + p_d$, the equations (3.5) define a subspace V with the properties (i) V is tangent to a local integral manifold of (3.5); and (ii) these local integral manifolds may be propagated to a global integral manifold isomorphic to $\mathbb{P}^r$.*

The property (i) is not automatic – it requires the *involutivity* of the exterior differential system (3.5), which imposes conditions beyond $d(\mathrm{Tr}\,\omega) = 0$ in neighborhoods where the rank of the equations (3.5) jumps. Property (ii) reflects the functional equation aspect of Abel's theorem discussed above.

The second converse to Abel's theorem was first formulated and proved by Abel's fellow countryman Sophus Lie. We shall state it in a special case based on the picture below. Here we are given local analytic arcs F_i in the plane and on each F_i a non-zero regular differential ω_i. For L in a neighborhood U of L_0 in the space of lines in the plane, we may define a mapping

$$L \to F_1 \times \cdots \times F_n \tag{3.6}$$

by

$$L \to (p_1(L), \ldots , p_n(L))$$

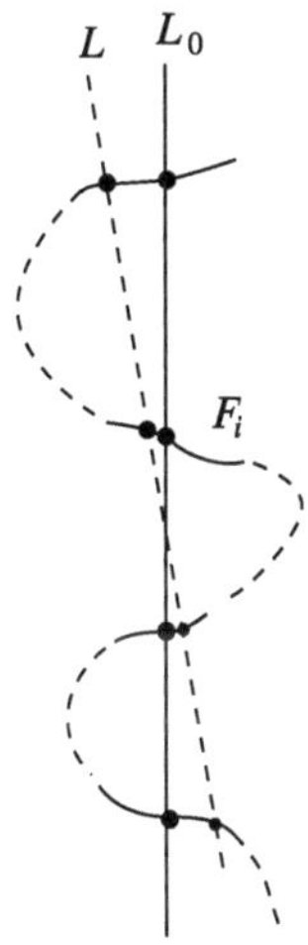

where $p_i(L) = L \cdot F_i$. In this situation Abel's relation is

$$\sum_i \omega_i(p_i(L)) = 0 \; ; \tag{3.7}$$

i.e., the pullback of $(\omega_1, \cdots \omega_n)$ on $F_1 \times \cdots \times F_n$ to U under the mapping (3.6) should be zero. Lie's result is:

Under the condition (3.7) there is a global algebraic curve F and regular differential ω on F such that

$$\begin{cases} F_i \subset F \\ \omega | F = \omega_i \; . \end{cases}$$

Again, using the functional equation (3.7) the local data (F_i, ω_i) may be propagated to give a global (F, ω).

4 Some Legacies of Abel's Theorem

Of course probably the main "legacy" of the theorem is the string of developments – by many mathematicians and continuing to modern times – leading to our understanding of the *Picard variety* or *divisor class group* of an algebraic variety. Abel's influence is reflected by the fact that the identity component of the Picard variety is an *abelian variety*, and at least in the complex case the functions on it are termed *abelian functions*.

Rather than recount these developments in any detail, however, I will discuss briefly two other legacies. One is the interesting but less well known subject of *webs* and the other is based on recent joint work with Mark Green.

4.1 Webs. We will restrict to plane webs – however, the subject is of interest in any dimension and codimension. Also, although the definition may be given globally on manifolds, thus far the main interest has been in the local geometry and so we shall work in an open set in $\mathbb{R}^2$.

Definition. *An n-web $\mathcal{W}(n)$ is given by n foliations in general position.*

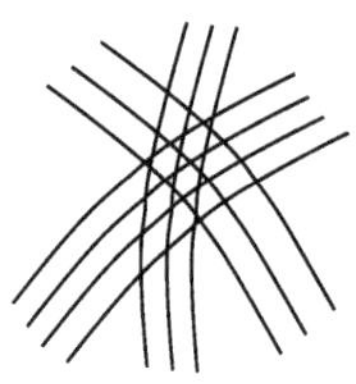

$F_i(x,y) = \text{constant}$

The leaves of the ith web are given by the level sets of a function $F_i(x, y)$; general position means that the tangent lines to the leaves through a point are distinct. It is sometimes convenient to give these tangent lines by a Pfaffian equation

$$\omega_i = 0$$

where

$$\omega_i = \lambda_i dF_i$$

for some non-zero function λ_i.

The subject of web geometry was initiated by Blaschke and his colleagues in Hamburg in the 1920's. A central problem was to find invariants of a web; in particular to find sufficient conditions that the web be *linearizeable*; i.e., after a diffeomorphism the leaves of the web should become lines in the plane.

From the beginning it was understood that webs were related to algebraic geometry. For example, the following picture gives a linear web

(i)

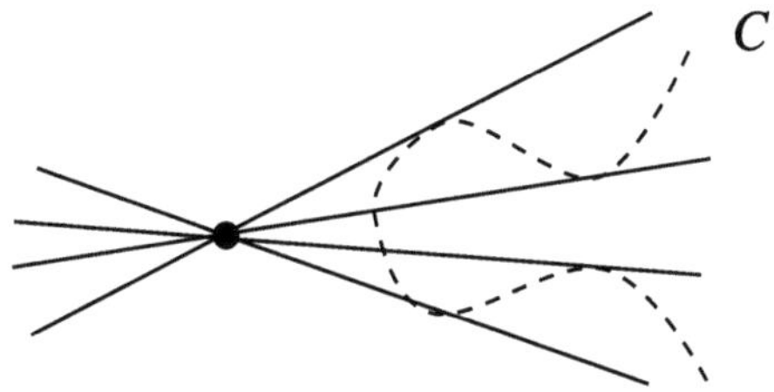

Here, C is an algebraic curve in the plane, and through a general outside point we draw the tangents to C (here, and similarly in what follows, for the purpose of illustration we assume that all the tangents are real). The *degree* n of the web is the usual algebro-geometric degree of the dual curve.

The projective dual of figure (i) associates to an algebraic curve C of degree n in the plane an n-web in an open set U in the dual projective space $\check{\mathbb{P}}^2$ of lines in the plane, as illustrated by the figure

(ii)

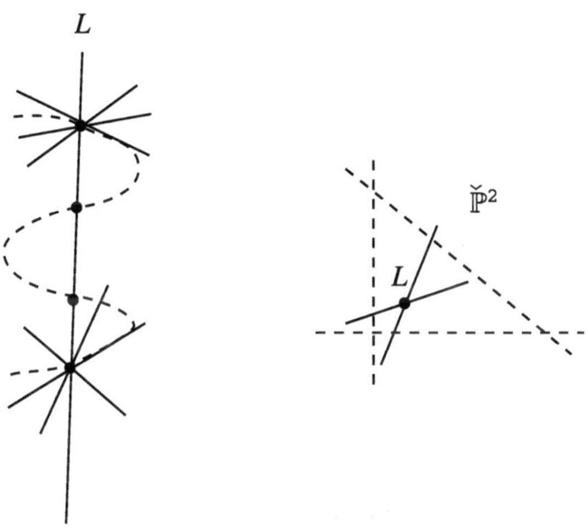

Here, a point in U is given by a line L in the plane. The lines through each of the n points of intersection of C with L give n pencils of lines, and by projective duality each such pencil gives a line in the dual space.

An additional source of examples of web is provided by the solution curves to an ODE

$$P(x, y, y') = (y')^n + P_1(x, y)(y')^{n-1} + \cdots + P_n(x, y) = 0$$

in the plane.

What has turned out to be thus far the most important invariant of a web was defined already by the school of Blaschke:

Definition. *An abelian relation is given by*

$$\sum_i g_i(F_i)dF_i = 0 \tag{4.1}$$

where $\mathbf{g}(\xi) = (g_1(\xi), \cdots g_n(\xi))$ *is a vector of functions of* 1*-variable. We denote by* $A(\mathcal{W})$ *the vector space of functions* $\mathbf{g}(\xi)$ *satisfying (4.1) and define the* **rank** $r(\mathcal{W})$ *of the web by*

$$r(\mathcal{W}) = \dim A(\mathcal{W}) .$$

As an example of an abelian relation we let ω be a non-zero differential of the first kind and consider first the local integral

$$I(L) = \int_{L_0 \cdot D}^{L \cdot D} \omega \tag{4.2}$$

as depicted by a local picture where we restrict attention to an arc D on C and on an

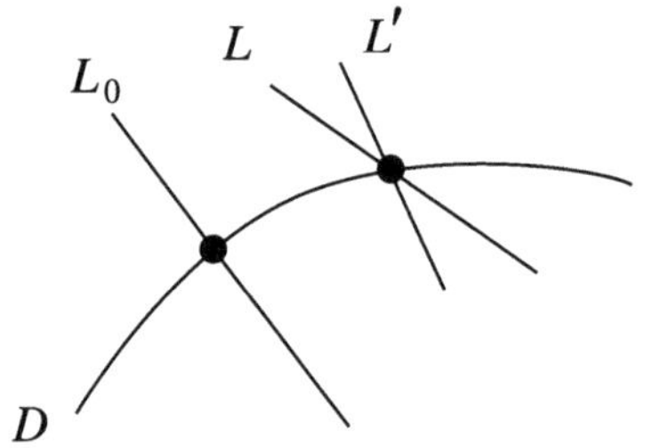

open neighborhood U of a line L_0 having one intersection point with the arc. Clearly in this picture

$$I(L) = I(L') ;$$

i.e., $I(L)$ is constant on the pencil of lines through a fixed point on C. Thus the level sets of I define lines in U, and the integral curves of the differential $dI(L)$ are exactly these lines. Referring to figure (ii) and adding up this discussion over the intersection points, if for example C is non-singular and we set $h^0(\Omega_C^1) = \dim H^0(\Omega_C^1)$, we then have for the web $\mathcal{W}_C$ associated to C as in figure (ii)

$$h^0\left(\Omega_C^1\right) \leqq r(\mathcal{W}_C) . \tag{4.3}$$

A result from the Blaschke school is that for any n-web

$$r(\mathcal{W}(n)) \leqq (n-1)(n-2)/2 . \tag{4.4}$$

For $\mathcal{W}(n) = \mathcal{W}_C$ as above it is well known that

$$h^0\left(\Omega_C^1\right) = (n-1)(n-2)/2$$

so that equality holds in (4.3). In general we say that a web $\mathcal{W}(n)$ has *maximum rank* if equality holds in (4.4). A central question in the subject is the

(4.5) **Problem.** *Determine all webs of maximum rank.*

Before discussing this problem we mention as another relation between web geometry and algebraic geometry, that Sophus Lie's converse to Abel's theorem discussed above has the following consequence

A **linear** *web with non-zero rank is algebraic* (4.6)

Here it should be understood that the abelian relation is *complete* in the sense that each g_i is not identically equal to zero. The assertion means that it is the web associated to an algebraic curve by the construction in figure (ii).

Another remark is that it is sometimes useful to give an abelian relation (4.1) in integrated form as

$$\sum_i G_i(F_i(x, y)) = \text{constant} \tag{4.7}$$

where $G_i(\xi)$ are functions of ξ with $G_i'(\xi) = g_i(\xi)$.

Turning to the problem (4.5), for $n = 3$ we have

$$r(\mathcal{W}(3)) \leqq 1 ,$$

and it was proved by Blaschke that if equality holds then the web is algebraic of type (ii) above. Moreover, the integrated form (4.7) of the abelian relation may be written as a functional equation for the logarithm

$$\varphi_1(x) - \varphi_1(y) + \varphi_1\left(\frac{y}{x}\right) = 0 \tag{4.8}$$

and up to a local diffeomorphism the web looks like

(iii)

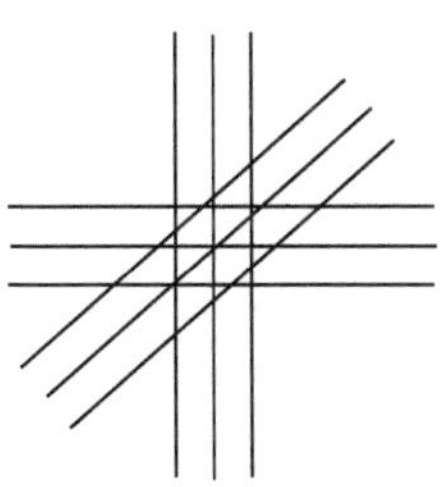

For $n = 4$ we have

$$r(\mathcal{W}(4)) \leqq 3 ,$$

and again if equality holds then the web is algebraic of type (ii) above. Using the result (4.6) this may be seen as follows: Writing a basis for the abelian relations as

$$\sum_j g_{ij}(F_j)dF_j = 0 \qquad i = 1, 2, 3 \tag{4.9}$$

we consider the matrix

$$\begin{vmatrix} g_{11} & g_{12} & g_{13} \\ g_{21} & g_{22} & g_{23} \\ g_{31} & g_{32} & g_{33} \end{vmatrix} .$$

The rows give a basis for the abelian relations and the columns give a map

$$U \to \mathbb{P}^2 \tag{4.10}$$

that, using (4.9), maps $\mathcal{W}(3)$ to a *linear* web in an open set in the projective plane. We may now apply (4.6) to conclude the result.

For $n = 5$ we have $r(\mathcal{W}(5)) \leqq 6$ and a non-linearizable (and therefore non-algebraic) web of maximum rank was found by Bol; it may be pictured as

(iv)

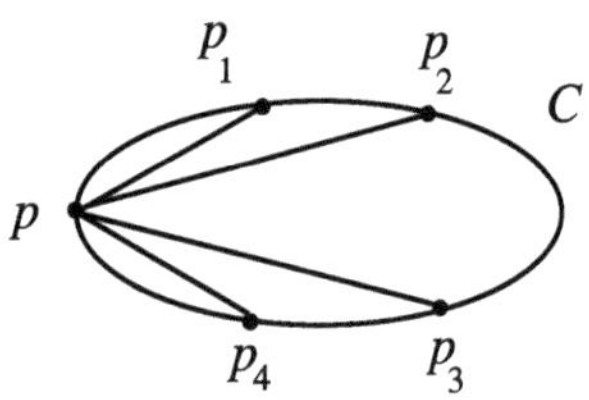

Here, the p_i are fixed points in general position. Through a general variable point p the leaves of the foliation are the lines $L_i = \overline{pp_i}$ and the conic C passing through $pp_1 \cdots p_5$. Five of the six independent abelian relations are derived from Abel's theorem applied to figure (iv). The integrated form of the sixth may be expressed by

$$\varphi_2(x) - \varphi_2(y) + \varphi_2\left(\frac{y}{x}\right) - \varphi_2\left(\frac{1-y}{1-x}\right) + \varphi_2\left(\frac{x}{y}\frac{1-y}{1-x}\right) = 0 , \tag{4.11}$$

which we recognize as Abel's form of the functional equation for the dilogarithm. So again Abel enters our story, only this time from a completely different perspective.

For $n = 6, 7, 8$ there exist exceptional planar webs of maximum rank – i.e., webs of maximum rank not of type (ii) above. All are based on the dilogarithm. For example, for $n = 6$ there are two exceptional webs; one uses the six term relation for the dilogarithm, and the other the usual five term relation (4.11). For $n = 9$ Hénaut has shown that the trilogarithm appears as an abelian relation in a 9-web of maximum rank not composed of one of type (ii) with the Bol web. This leads to the obvious

Question. Are all webs of maximum rank which are not algebraizable of this type?

We do not attempt to formulate this question precisely – intuitively, we are asking whether or not for each k there is an integer $n(k)$ such that there is a "new" $n(k)$-web of maximum rank one of whose abelian relations is a (the?) functional equation with $n(k)$ terms for the kth polylogarithm $\mathcal{L}i_k$? Here, "new" means the general extension of the phenomena above for the logarithm when $k = 1$ where $n(1) = 3$, for the Bol web when $k = 2$ and $n(2) = 5$, and for the Hénaut web when $k = 3$ and $n(3) = 9$.

4.2 Abel's DE's for points on a surface. The geometry of an algebraic variety is reflected by the configuration of its algebraic subvarieties. Stemming from Abel one has learned to study subvarieties modulo the relation of rational equivalence. That is, in a smooth complex algebraic variety X two subvarieties Z, Z' are *rationally equivalent* if there is a family $\{Z_t\}_{t\in\mathbb{P}^1}$ of subvarieties with $Z_0 = Z$, $Z_\infty = Z'$. Passing to the group $Z^p(X)$ of codimension-p algebraic cycles modulo the relation generated by rational equivalence one obtains the *Chow groups* $CH^p(X) = Z^p(X)/Z^p_{\text{rat}}(X)$.

For X an algebraic curve, Abel's theorem and its converse give a complete set of Hodge-theoretic invariants for the identity component $CH^1(X)_0$ (which is of course the *Jacobian variety* of X). In general, $CH^1(X)$ is the Picard variety whose identity component is an abelian variety – the story has much the same general flavor as in the case of algebraic curves.

However, already for configurations of points on an algebraic surface the story is much different – since Mumford's result in the 1960's we know that $CH^2(X)$ may be infinite dimensional. A few years ago, motivated by Spencer Bloch's formula for the *formal* tangent space $T_f CH^2(X)$, Mark Green and I wanted to understand what geometric content might lie behind Spencer's formula. This led us to propose a geometric definition for the tangent space $TZ^2(X)$ (cf. the example below) and to then define the *geometric* tangent space

$$T_g CH^2(X) = TZ^2(X)/TZ^2_{\text{rat}}(X) \tag{4.12}$$

where $TZ^2_{\text{rat}}(X)$ is the tangent space to the subgroup of 0-cycles rationally equivalent to zero. It is then a theorem that

$$T_g CH^2(X) \cong T_f CH^2(X)\,. \tag{4.13}$$

We shall denote either of these simply by $TCH^2(X)$ and refer to this vector space as *the* tangent space to $CH^2(X)$.

Implicit in (4.12) and (4.13) are the infinitesimal geometric conditions that a configuration of points move to 1st order in a rational equivalence class. Recall that the condition a 0-cycle Z on X be rationally equivalent to zero is that

$$Z = \sum_\nu (f_\nu)$$

where f_ν is a rational function on an irreducible curve Y_ν and (f_ν) is its divisor. A 1st order variation of the data (Y_ν, f_ν) gives a 1st order variation of $\sum_\nu (f_\nu)$, and we seek the geometric conditions on a configuration of points $p_1 + \cdots + p_d$ (assumed for simplicity to be distinct) and tangent vectors $\tau_i \in T_{p_i}X$ to be a 1st order variation of $\sum_\nu (f_\nu)$.

The answer to the corresponding question for configurations of points on an algebraic curve is given by Abel's DE's (3.5). We shall now explain the answer in the case of an algebraic surface. For this we first observe that Abel's construction of the trace extends to differential forms of any degree on a smooth algebraic variety X; the formula

$$\omega(p_1 + \cdots + p_d) = \omega(p_1) + \cdots + \omega(p_d)$$

defines a map

$$H^0\left(\Omega_X^q\right) \xrightarrow{\text{Tr}} H^0\left(\Omega_{X^{(d)}}^q\right) .$$

If $\dim X \geqq 2$ the symmetric products are singular along the diagonal; and regular differential forms are then defined to be rational forms that are regular on any desingularization.

We now let p_i, τ_i be as above and set

$$\tau = \sum_i (p_i, \tau_i) \in TX^{(d)} .$$

The first set of conditions that

$$\tau \in TZ^2_{\text{rat}}(X) \tag{4.14}$$

are, as in the curve case, that

$$\langle \text{Tr}\,\varphi, \tau \rangle = \sum_i \langle \varphi(p_i), \tau_i \rangle = 0 \tag{4.15}$$

for all regular 1-forms $\varphi \in H^0(\Omega_X^1)$. Equation (4.15) simply says that τ should be in the kernel of the differential of the Albanese map.

The new ingredient comes from the 2-forms on X. Already from the works of Mumford and Bloch one knew that the 2-forms are relevant; the following is a geometric explanation. First remark that if one considers any n-dimensional complex manifold X; e.g. an open set in $\mathbb{C}^n$, and if one then considers collections of forms $\varphi_d \in H^0(\Omega^q_{X^{(d)}})$ that have the *hereditary property*

$$\varphi_{d+1} \mid_{X^{(d)}} = \varphi_d$$

where the inclusion $X^{(d)} \hookrightarrow X^{(d+1)}$ is given by

$$p_1 + \cdots + p_d \to p + p_1 + \cdots + p_d$$

for some fixed point $p \in X$, then it is a theorem that

(4.16) *The hereditary forms are generated as an exterior algebra by the traces of the q-forms on X where* $0 \leqq q \leqq n = \dim X$. *All of these forms are needed to generate.*

The geometric point is this: Taking X to be a germ of a neighborhood of a point in $\mathbb{C}^n$, if $X^{(d)}$ were smooth, then of course the 1-forms would generate the forms of all degrees. Exactly along the diagonals – which reflect the infinitesimal structure of X – to generate we need forms of all degrees up to dim X.

The other new ingredient is that

(4.17) *The field of definition of the* $p_i \in X$ *enters into the condition (4.14).*

To explain this we assume for simplicity of exposition that the algebraic surface X is defined over $\mathbb{Q}$ (or over a number field); e.g., we may think of $X \subset \mathbb{P}^N$ as being projected to $X^0 \subset \mathbb{P}^3$ where X^0 has an affine equation

$$f(x, y, z) = 0$$

where $f \in \mathbb{Q}[x, y, z]$. We may assume that x, y give local uniformizing parameters around $p_i = (x_i, y_i, z_i)$ and write

$$\tau_i = \lambda_i \frac{\partial}{\partial x} + \mu_i \frac{\partial}{\partial y} .$$

The regular 2-forms on X are given by the pullbacks to X of

$$\omega = \left. \frac{g(x, y, z) dx \wedge dy}{f_z(x, y, z)} \right|_{X^0} \tag{4.18}$$

where $\deg g \leqq \deg f - 4$ and g vanishes on the double curve of X^0. Since X is defined over $\mathbb{Q}$ we may take a basis for $H^0(\Omega_X^2)$ to be given by 2-forms (4.18) where $g \in \mathbb{Q}[x, y, z]$. Recalling that the *Kähler differentials*

$$\Omega^1_{\mathbb{C}/\mathbb{Q}}$$

is the complex vector space generated by expressions δa, $a \in \mathbb{C}$, modulo the relations

$$\begin{cases} \delta(a+b) = \delta a + \delta b \\ \delta(ab) = a\delta b + b\delta a \\ \delta a = 0 \text{ if } a \in \mathbb{Q} , \end{cases}$$

we now *define*

$$\langle \omega(p_i), \tau_i \rangle = \frac{g(x_i, y_i, z_i)}{f_z(x_i, y_i, z_i)} (\mu_i \delta y_i - \lambda_i \delta x_i) \in \Omega^1_{\mathbb{C}/\mathbb{Q}} \tag{4.19}$$

and

$$\langle \mathrm{Tr}\,\omega, \tau \rangle = \sum_i \langle \omega(p_i), \tau_i \rangle \ . \tag{4.20}$$

Abel's DE's for the 2-forms are then defined to be the $\Omega^1_{\mathbb{C}/\mathbb{Q}}$-valued equations

$$\mathrm{Tr}\,\omega = 0 \tag{4.21}$$

where ω is as above. It is then a theorem that *the equations (4.15) and (4.21) define infinitesimal rational motion as explained above.*

(4.22) **Corollary.** *If $\omega(p_i) \neq 0$ and if the x_i, y_i are independent transcendentals, then (4.20) has no non-zero solutions.*

In other words, no matter how large d is the 0-cycle $z = p_1 + \cdots + p_d$ is rigid in its rational equivalence class. (This includes allowing rational motions of $(z+z')-z'$ for any $z' \in X^{(d')}$.) This result gives a proof of Mumford's theorem and provides rather precise meaning to the use of "generic" in Mumford's argument and the subsequent developments by Roitman, Voisin, and others.

At the other extreme we have the

(4.23) **Corollary.** *If $x_i, y_i \in \bar{\mathbb{Q}}$, then (4.20) is zero for any choice of the τ_i.*

This is an infinitesimal version of a well known conjecture of Beilinson–Bloch – it gives a geometric existence result, albeit only to 1st order. Understanding the "integration" of Abel's DE's (4.15) and (4.21) is a deep and fundamental question.[2]

One may quite reasonably ask how the essentially arithmetic object $\Omega^1_{\mathbb{C}/\mathbb{Q}}$ gets into the purely geometric question of tangents to arcs in the space of 0-cycles on an algebraic surface. The following example illustrates how this comes about.

Example. The issue already appears locally, so we considert he space of arcs $z(t)$ in $Z^2(\mathbb{C}^2)$. We may define an arc to be a finite linear combination with integer coefficients of analytic maps of the t-disc into the symmetric products $(\mathbb{C}^2)^{(d)}$. One may then define an equivalence relation $\sim$ on the space of arcs and the tangent space is the complex vector space defined by

$$TZ^2(\mathbb{C}^2) = \{\text{arcs in } Z^2(\mathbb{C}^2)\}/\sim \ .$$

The equivalence relation $\sim$ is characterized by the properties:

(i) $z_i(t) \sim \tilde{z}_i(t)$ for $i = 1, 2 \quad \Rightarrow \quad z_1(t) \pm z_2(t) \sim \tilde{z}_1(t) \pm \tilde{z}_2(t)$;

[2] Integrating a DE means finding a solution by an iterative process. Since there are no derivations of $\mathbb{Q}$ the methods of calculus break down – one must break the problem into "increments" by some other means, perhaps either by an iteration process that at each stage decreases the "arithmetic complexity" of the 0-cycle, or by analyzing the DE's (4.15) and (4.20) in the completions of $\mathbb{Q}$ under all valuations.

(ii) $z(\alpha t) \sim \alpha z(t)\,, \qquad \alpha \in \mathbb{Z}$
(iii) $\alpha z(t) \sim \alpha \tilde{z} \quad \Rightarrow \quad z(t) \sim \tilde{z}(t)$ for $\alpha \in \mathbb{Z}^*$; and
(iv) if $z(t), \tilde{z}(t)$ are arcs in $\mathrm{Hilb}_0(X)$ with the same tangent vector in $T\,\mathrm{Hilb}_0(X)$, then

$$z(t) \sim \tilde{z}(t)\,.$$

Now let

$$z_{\alpha\beta}(t) = \mathrm{Var}\left(x^2 - \alpha y^2, xy - \beta t\right)\,, \qquad \alpha \neq 0$$

and F be the free group generated by the 0-cycles

$$w_{\alpha\beta}(t) = z_{\alpha\beta}(t) - z_{1\beta}(t).$$

Then we have the result:

(4.24) *The map*

$$F/\sim \,\to \Omega^1_{\mathbb{C}/\mathbb{Q}}$$

given by

$$w_{\alpha\beta}(t) \to \beta \frac{\delta\alpha}{\alpha}$$

is a well-defined isomorpism.

As a non-obvious geometric corollary, we see that if α is a root of unity then

$$z_{\alpha\beta}(t) \sim z_{1\beta}(t)\,.$$

This result illustrates the very interesting and subtle interplay between geometry and arithmetic in higher codimension.

In summary, Abel's DE's (4.21) for the rational motion of configurations of points on a surface have an arithmetic/geometric character – the integration of these equations presents a major challenge (cf. footnote (2)).

5 Reprise

In the beginning we considered the integral (1.1)

$$\int y(x)dx$$

of an algebraic function; there we observed that at the time of Abel such integrals were seen as "highly transcendental" functions of the upper limit of integration, and Abel's great insight was to find a general recipe for generating simple relations among them. In recent years there has been renewed interest in the integrals (1.1), *exactly because they are generally transcendental.* Whether or not relations of the Abel type generate all such relations then leads into one of the deepest questions in arithmetic algebraic geometry. We shall now briefly discuss this.

For this we assume that the algebraic equation

$$f(x, y(x)) = 0$$

satisfied by $y(x)$ is defined over $\mathbb{Q}$ (or over a number field); i.e., $f(x, y) \in \mathbb{Q}[x, y]$. For $\xi_0, \xi \in \mathbb{Q}$ we set

$$u(\xi) = \int_{\xi_0}^{\xi} y(x)dx \,. \tag{5.1}$$

The transcendence properties of the numbers $u(\xi)$ have been studied over many years by many mathematicians, including Enrico Bombieri who in 1981 proved a result first enunciated by Siegel in 1929 which may informally be stated as follows:

(5.2) *Assume that $u(\xi)$ is not an algebraic function of ξ, and for convenience take $\xi_0 = 0$, assumed to be a regular value of $y(x)$. Then for each integer l there is a constant $C(l)$ such that if*

$$|\xi| < C(l)$$

then $u(\xi)$ does not satisfy an algebraic equation over $\mathbb{Q}$ of degree l.

This result also applies to the more general integrals (1.4), in particular to the integral of a differential of the first kind. Taking F to be a cubic curve and ξ_0 the x-coordinate of a flex, small perturbations of the flex tangent

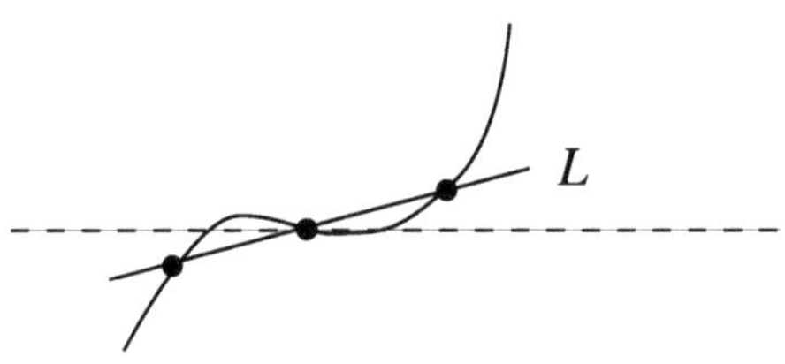

lead by Abel's theorem to linear relations

$$u(\xi_1) + u(\xi_2) + u(\xi_3) = 0 \tag{5.3}$$

for arbitrarily small $|\xi_i|$. Thus, thinking of the $u(\xi_i)$ as nearly transcendental numbers in the sense of (5.2), for geometric reasons stemming from Abel's theorem they satisfy algebraic relations defined over $\mathbb{Q}$ (actually, linear relations).

Integrals of the form

$$\int_{(\xi_0,\eta_0)}^{(\xi,\eta)} r(x, y(x))dx \tag{5.4}$$

where $(\xi_0, \eta_0), (\xi, \eta) \in F(\bar{\mathbb{Q}})$, together with their higher dimensional analytic analogues, are termed *periods* by Kontsevich and Zagier. Periods include the case when $(\xi, \eta) = (\xi_0, \eta_0)$; i.e., the integral around a closed loop $\gamma \in H_1(F, \mathbb{Z})$. In fact, by identifying (ξ, η) with (ξ_0, η_0) the integral (5.4) becomes an integral over a closed loop on a singular curve

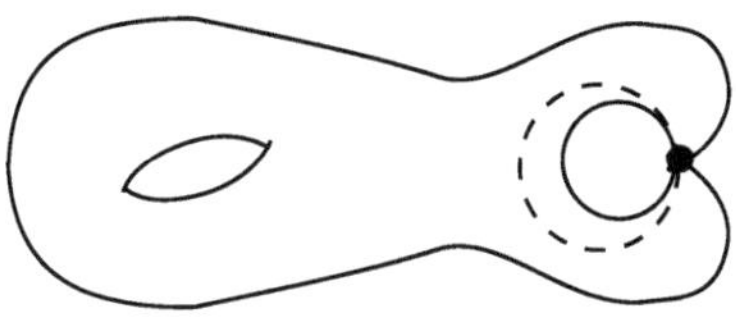

The periods generate a countable field $\mathbb{P}i$ with

$$\bar{\mathbb{Q}} \subset \mathbb{P}i \subset \mathbb{C}\,.$$

Kontsevich and Zagier point out that there is no known explicit example of a transcendental number that is not a period.

A general philosophy is

(5.5) *The relations of* $\mathbb{P}i$ *over* $\bar{\mathbb{Q}}$ *should be defined by geometric conditions.*

One example of this was just given. For another example, if we choose differentials $\omega_1, \dots, \omega_g$ of the 1st kind defined over $\bar{\mathbb{Q}}$ and which give a basis for $H^0(\Omega^1_F)$, and if $\gamma_1, \dots, \gamma_{2g} \in H_1(F, \mathbb{Z})$ is a canonical basis for the integral 1st homology, then the periods

$$\pi_{\alpha j} = \int_{\gamma_j} \omega_\alpha$$

satisfy the 1st Riemann bilinear relations

$$\sum \pi_{\alpha i} Q_{ij} \pi_{\beta j} = 0 \tag{5.6}$$

where $Q = \|Q_{ij}\|$ is the inverse of the intersection matrix. Geometrically, this relation arises from the class of the diagonal $\Delta \subset F \times F$. More generally, any generalized correspondence $T \subset \underbrace{F \times \cdots \times F}_{n}$ gives a polynomial relation of degree n over $\mathbb{Q}$. A beautiful and deep conjecture of Grothendieck is that *all* relations of the $\pi_{\alpha j}$ over $\mathbb{Q}$ arise in this way. In fact, Grothendieck conjectures the analogous statement for smooth varieties of any dimension defined over $\bar{\mathbb{Q}}$ and for all of the algebraic de Rham cohomology defined over $\bar{\mathbb{Q}}$.

I do not know a precise formulation of (5.5) which includes Grothendieck's conjecture – which is in some sense global – and relations of the Abel type – which are in some sense local although they arise from the global constraint $h^0(\Omega^1_{\mathbb{P}^1}) = 0$. In any case, taking into account the arithmetic aspect of Abel's DE's discussed in the preceeding section and the arithmetic questions concerning periods discussed above, I believe that one may with some confidence expect that *the arithmetic aspects of Abel's theorem and its legacies will be a central and deep topic for mathematicians in the third century after the time of Abel.*

6 Guide to the Literature

Abel's famous "Paris memoir" entitled *Mémoire sur une propriété générale d'une classe très étendue des fonctions transcendantes* was presented to l'Académie des sciences à Paris in 1826 and published in t. VII in 1841. It appears in Oeuvres complètes de Niels Henrik Abel, pages 145–211.

The paper referred to in footnote (1) is *Sur l'intégration de la formule différentielle $\rho dx/\sqrt{R}$, R and ρ étant des fonctions entières*, Oeuvres complètes, pages 104–144. The recent paper *Abel equations*, St. Petersburg Math. J., vol. 13 (2002), pages 1–45, by V. A. Malyshev gives an extension of Abel's result and a further guide to the literature.

The usual version of Abel's theorem and its (global) converse appears in standard books on Riemann surfaces, e.g., the famous *Die Idee der Riemannscher Fläche* by Herman Weyl.

The local converses, including the theorem of Sophus Lie and its extensions by Darboux and others, are discussed in the paper by the author *Variations on a theorem of Abel*, Invent. Math., vol. 35 (1976), pages 321–390. The recent paper by G. Henkin, *Abelian differentials on singular varieties and variations on a theorem of Lie–Griffiths*, Invent. Math., vol. 135 (1991), pages 297–328, presents new results and references that have appeared after the paper mentioned above.

The theory of webs was first presented in the book *Geometrie der Gewebe. Topologische Fragen der Differentialgeometrie*, Springer, Berlin (1938) by W. Blaschke and G. Bol. Two recent works *Analytic web geometry*, Toulouse (1996), 6–47, World Sci. Publishing by A. Hénaut, and *Differential geometry of webs* in Handbook of differential geometry, vol. I, pages 1–152, North-Holland (2000) by M. Akivis and V. Goldberg, give surveys of recent works and a further guide to the literature.

Abel's form of the functional equation for the dilogarithm is given in his paper *Note sur la fonction* $\psi x = x + \frac{x^2}{2^2} + \frac{x^3}{3^2} + \cdots + \frac{x^n}{n^2} + \cdots$, Oeuvres complètes, pages 189–193.

Bloch's formula for $T_f CH^2(X)$ may be found in his *Lectures on algebraic cycles*, Duke Univ. Math. Ser. IV (1980).

A discussion of the extension of Abel's theorem to configurations of points on an algebraic surface may be found in *Abel's differential equations*, Houston J. of

Math. (volume in honor of S. S. Chern), vol. 28 (2002), pages 329–351, by Mark Green and the author.

A general survey of the arithmetic properties of algebraic integrals is given in the paper *Periods*, Mathematics unlimited – 2001 and beyond, Springer, Berlin (2001), pages 771–808, by M. Kontsevich and D. Zagier. The Bombieri–Siegel result appears in the paper by Bombieri *On G-functions*, Recent progress in analytic number theory, Vol. 2, Durham (1979), pages 1–67. The book *G-functions and geometry*, Aspects of mathematics, E/3, Friedr. Vieweg and Sohn, Brandenberg (1989) by I. André contains a "geometrization" of these issues, including a discussion of Grothendieck's conjecture.

Solving Quintics by Radicals

Daniel Lazard

Abstract. A formula is given for solving by radicals any polynomial of degree 5 which is solvable by radicals. This formula is valid over any field of characteristic different from 2 and 5. The field extension which is generated by the radicals which appear in the result is always minimal, when only one root is produced, as well as when all roots are given. This formula has been implemented in Maple.

1 Introduction

It is well known that Abel has shown that not every equation is solvable by radicals. Slightly later, Galois has provided criteria for such a solvability. It consists in looking if some auxiliary equation (resolvent) has a linear factor with multiplicity one. Galois was aware that the computations needed by these criteria were far outside the human capability.

As a consequence of such intractable computations, most of the work of the mathematicians of twentieth century, were oriented toward abstract mathematics, in order to replace, as far as possible, huge computations by qualitative understanding of the structures. Nevertheless, the power of modern computers allows computations which are far beyond human capabilities. This opens the way to mathematical works in the spirit of the mathematics of Abel's time. This paper is an attempt in this direction.

In the case of equations of degree 5, Cayley [1] has computed such a resolvent (called R_θ in this paper) which never has a multiple factor. Thus, for testing if a fifth degree equation is solvable by radicals, it suffices to write Cayley's resolvent in a computer and to call the factorization package of any Computer Algebra system.

If the equation is solvable by radicals, it remains to explicitly write down the solution. When we began to write this paper we did not know any previous work on this subject for degree greater than 4. François Morain brought [4] to our attention, where a method is given for solving quintics.

The aim of this paper is to provide a formula for the degree 5. It differs from the method of [4] in several ways. A first but minor difference lies in the fact that he expresses the root as a sum of four fifth roots. Without specifying which determinations of the roots have to be chosen, this gives 625 solutions instead of 5. This is a minor difference because it may be easily corrected using Paxton Young's paper.

The second difference lies in genericity. In [4] several cases are considered, depending on the values of the coefficients. In our formula the only branching instructions consist in choosing (twice) the sign of a square root in order to avoid that some denominators become zero. This means that our implementation in Maple is really what is usually called a formula, namely an expression involving only basic operations of arithmetics ($+$, $-$, $*$, $/$ and root extraction). This means also that our formula applies also with symbolic coefficients. However testing equality with 0 for an algebraic expression depending on symbols is a difficult task; this has the consequence that our Maple program may provide a false answer in the symbolic case, because of a bad detection of zero.

A third difference is that our method may apply to any field of coefficients of characteristic different from 2 and 5. We have not verified if this is also true for the method of Paxton-Young.

A fourth difference, the most important in our opinion, is that our formula is presented in a systematic way, without *ad hoc* computational tricks. Thus, it would be possible, in theory, to extend it to degree 7 for example. However this would lead to awful computations (the Cayley resolvent of degree 6 should be replaced by a resolvent of degree 120).

Even for degree 5, the size of our formula shows that solving by radicals is the worst method, even when possible. In our opinion this is the main conclusion of our paper.

Finally, we prove that our formula is optimal in following sense: The field extension defined by the radicals appearing in our formula is always the smallest radical extension containing one root (resp. all the roots).

2 Notations and Basic Facts of Galois Theory

In all this paper, we consider a base field $\boldsymbol{Q}$ of characteristic different from 2 and 5, which may be thought of as the field of the rationals, but may be any other field. We consider also a univariate *irreducible* polynomial f of degree d over $\boldsymbol{Q}$ which one wants to solve by radicals (in most of the paper, we will have $d = 5$). The *Galois group* $\mathcal{G}$ of f is the Galois group over $\boldsymbol{Q}$ of the field generated by all the roots of f.

The main result of Galois may be rephrased as : *"A polynomial is solvable by radicals iff its Galois group is solvable,"* a group $\mathcal{G}$ being *solvable* iff it contains a tower of subgroups $\{e\} = \mathcal{G}_0 \subset \mathcal{G}_1 \subset \cdots \subset \mathcal{G}_k = \mathcal{G}$ such each $\mathcal{G}_i$ is a normal subgroup of $\mathcal{G}_{i+1}$ and that each $\mathcal{G}_{i+1}/\mathcal{G}_i$ is an Abelian group.

To be more precise, we introduce d variables $x_0, \ldots, x_{d-1}$, corresponding to the d roots of f. The symmetric group $\mathcal{S}_d$ acts on the x_i and, by this action, the Galois group $\mathcal{G}$ of f is a subgroup of $\mathcal{S}_d$.

For simplification of notations, most indices will be assumed to be integers modulo d; this will be always the case when, otherwise, the expression in which the index appears would not be defined. For example, $x_d = x_0$ and $x_{d+1} = x_1$.

For solving by radicals, we will consider some polynomials in the x_i and we will use their invariance under some subgroup of the Galois group for expressing their values in terms of the elementary symmetric functions, i.e., the coefficients of f.

The basic such polynomials are

$$s_0 = x_0 + x_1 + \cdots + x_{d-1} = \sum_{i=0}^{d-1} x_i$$

$$s_1 = x_0 + \omega x_1 + \cdots + \omega^{d-1} x_{d-1} = \sum_{i=0}^{d-1} \omega^i x_i$$

$$s_k = x_0 + \omega^k x_1 + \cdots + \omega^{k(d-1)} x_{d-1} = \sum_{i=0}^{d-1} \omega^{ik} x_i \qquad \text{for } k = 2, \ldots, d-1\,,$$

where ω is a primitive d-th root of unity.

It is important to note that the cyclic permutation $x_i \to x_{i-1}$ has the same action on the s_k as the multiplication by ω^k. Similarly, for a prime to d, the permutation $x_i \to x_{i/a}$ and the substitution $\omega \to \omega^a$ induce on the s_k the permutation $s_k \to s_{ak}$ (let us recall that indices are integers modulo d).

It follows that $S_1 := s_1^d$ and $S_k := s_k s_1^{d-k}$ are invariant under cyclic permutations of the indices. Solving the equation f is now reduced to compute the values of the S_k when the x_i are substituted by the roots of f. In fact, s_0 is easily expressed in terms of the coefficients of f, and we have

$$\begin{aligned} s_1 &= \sqrt[d]{S_1} \\ s_k &= S_k / s_1^{d-k} \qquad \text{for } k > 1 \\ x_0 &= \frac{1}{d} \sum_{k=0}^{d-1} s_k \\ x_i &= \frac{1}{d} \sum_{k=0}^{d-1} \omega^{-ik} s_k \qquad \text{for } i > 0\,. \end{aligned}$$

Notation. We denote by $\mathcal{S}_d$ the symmetric group of order d and by $\mathcal{C}_d$ the cyclic group of order d generated by the permutation $i \to i+1 \bmod d$. When d is fixed we will simply write $\mathcal{S}$ and $\mathcal{C}$. As we shall encounter many polynomials which, like the S_i, are invariant under the action of some group, we will need the following abbreviations. For any polynomial p in the x_i and any subgroup $\mathcal{G}$ of the symmetric group $\mathcal{S}_d$, we will denote by $\sum_{\mathcal{G}} p$ the sum of the polynomials of the orbit of p under $\mathcal{G}$. Thus for $d = 3$, we have

$$\sum_{\mathcal{C}} x_0 x_1^2 = x_0 x_1^2 + x_1 x_2^2 + x_2 x_0^2$$

$$\sum_{\mathcal{S}} x_0 x_1^2 = \sum_{\mathcal{C}} x_0 x_1 (x_0 + x_1) = x_0 x_1^2 + x_1 x_2^2 + x_2 x_0^2 + x_0^2 x_1 + x_1^2 x_2 + x_2^2 x_0\,.$$

3 Invariants and Resolvents

It follows from the last section that solving by radicals reduces to computing the value of some polynomials which are invariant under some subgroup of the symmetric group (here the cyclic group). This computation will be done by expressing them in terms of elementary symmetric polynomials, i.e., in terms of the coefficients of f. The main tool for this computation is the notion of resolvent.

As we are working with polynomials in d variables over $\boldsymbol{Q}$, let us set $\boldsymbol{R}_d := \boldsymbol{Q}[x_1, \ldots x_d]$, the ring of polynomials, and $\boldsymbol{K}_d := \boldsymbol{Q}(x_1, \ldots x_d)$, the field of rational functions. For a subgroup $\mathcal{G}$ of $\mathcal{S}_d$, we denote by $\boldsymbol{R}_d^{\mathcal{G}}$ the subring of $\boldsymbol{R}_d$ consisting of the polynomials which are invariant under the action of $\mathcal{G}$, acting by permutation of the indices. Similarly $\boldsymbol{K}_d^{\mathcal{G}}$ is the subfield of $\boldsymbol{K}_d$ of invariant rational functions.

It is well known that $\boldsymbol{R}_d^{\mathcal{S}} = \boldsymbol{Q}[\sigma_1, \ldots, \sigma_d]$ and $\boldsymbol{K}_d^{\mathcal{S}} = \boldsymbol{Q}(\sigma_1, \ldots, \sigma_d)$, where $\sigma_i = \sum_{\mathcal{S}} x_0 x_1 \cdots x_{i-1}$ is the i-th elementary symmetric function. This allows to express any symmetrical function of the roots of f in terms of the coefficients of f.

Given two subgroups $\mathcal{G} \subset \mathcal{H}$ of $\mathcal{S}_d$, we call *resolvent invariant of* $\mathcal{G}$ *relatively to* $\mathcal{H}$ any element of $\boldsymbol{R}_d^{\mathcal{G}}$ which generates the field extension $\boldsymbol{K}_d^{\mathcal{G}} / \boldsymbol{K}_d^{\mathcal{H}}$. We call *resolvent equation* or simply *resolvent* its minimal polynomial over $\boldsymbol{K}_d^{\mathcal{H}}$. When $\mathcal{H}$ is not specified, it is assumed to be $\mathcal{S}_d$. If f is an univariate irreducible polynomial with Galois group included in $\mathcal{H}$, a resolvent equation R becomes a polynomial in $\boldsymbol{Q}[X]$, by substituting the roots of f to the x_i. We denote by R_f the result of this substitution. A resolvent R is said to be *separable* for f if R_f is square free.

A resolvent is *always separable* if it is separable for all irreducible polynomials with Galois group included in $\mathcal{H}$.

For example, $\prod_{i<j}(x_i - x_j)$ is a resolvent invariant for the alternate group $\mathcal{A}_d$, and the corresponding resolvent is $x^2 - \Delta$ where Δ is the discriminant of the polynomial having the x_i as roots. This resolvent is always separable because f is never irreducible if its discriminant is zero.

The interest of the resolvents is that they allow to test if a Galois group is contained in $\mathcal{G}$, and to express any polynomial invariant under $\mathcal{G}$ in terms of a root of a polynomial whose coefficients are invariant under $\mathcal{H}$.

Theorem 1. *Let R be a resolvent of $\mathcal{G} \subset \mathcal{H}$ and let f be a univariate polynomial of degree d.*

- *If the Galois group of f is included in $\mathcal{G}$, then R_f has a root in $\boldsymbol{Q}$.*
- *If R is separable for f and R_f has a root in $\boldsymbol{Q}$, then, if the Galois group of f is included in $\mathcal{H}$, it is also included in $\mathcal{G}$.*

We will need another theorem on invariants. Despite it is not new, we give a proof of it, because this proof will be used for our computations. This proof is based on Gröbner bases, for which we refer to [2].

Theorem 2. *Given a subgroup $\mathcal{G}$ of $\mathcal{S}_d$, the ring $\boldsymbol{R}_d^{\mathcal{G}}$ is a free $\boldsymbol{R}_d^{\mathcal{S}}$-module with a basis consisting of invariant homogeneous polynomials of degree at most $d(d-1)/2$*

Let us consider the elementary symmetric functions σ_i, and let $e_1, \ldots, e_d$ be new indeterminates which will be viewed as "names" for the σ_i. We consider the ideal I in $\boldsymbol{Q}[x_0, \ldots, x_{d-1}, e_1, \ldots, e_d]$, which is generated by the $\sigma_i - e_i$. We will compute a Gröbner basis of I for any admissible ordering such that $x_0 < x_1 < \cdots < x_{d-1}$ and $m_1 < m_2$ for any pair of monomials satisfying the same inequality for their total degrees in the x_i.

Lemma 1. *For such an ordering, a reduced Gröbner basis of I is*

$$\begin{aligned} J = \{ & x_0^d - e_1 x_0^{d-1} + \cdots + (-1)^{d-1} e_{d-1} x_0 + (-1)^d e_d, \\ & C_{d-1}^{(2)} - e_1 C_{d-2}^{(2)} + \cdots + (-1)^{d-2} e_{d-2} C_1^{(2)} + (-1)^{d-1} e_{d-1}, \\ & \ldots, \\ & C_2^{(d-1)} - e_1 C_1^{(d-1)} + e_2, \\ & C_1^{(d)} - e_1 = \sigma_1 - e_1 \}, \end{aligned}$$

where $C_k^{(i)}$ is the sum of all monomials of degree k in $x_0, \ldots, x_{i-1}$.

Proof (of the Lemma 1). The leading monomials of the polynomials in J are $x_0^d, x_1^{d-1}, \ldots, x_{d-1}$, which are pairwise coprime. It follows immediately that J is a reduced Gröbner basis.

For showing that the $J \subset I$, it suffices to show that the elements of J become zero when the e_i are replaced by the σ_i. Let J_i be the i-th element of J and $J_{i,k}$ be the result of the substitution of x_{i-1} by x_k in J_i. We prove that the $J_{i,k}$ for $k \geq i-1$ becomes zero by the replacement of the e_j by the σ_j. This is clearly true for $i=1$ and results by recursion from the formula $J_{i,k} - J_i = (x_k - x_{i-1})J_{i+1,k}$ for $k \geq i$.

Finally, we have to prove that J generates I, i.e. that J is the complete Gröbner basis of I. There are $n!$ monomials in the x_i which are irreducible by J. If J were not a Gröbner basis of I, there would be a polynomial in I, irreducible by J, and these $n!$ monomials would be linearly dependent over $\boldsymbol{Q}(\sigma_1, \ldots \sigma_d)$. This would be a contradiction to the fact that the Galois group of $\boldsymbol{Q}(x_1, \ldots x_d)$ over $\boldsymbol{Q}(\sigma_1, \ldots \sigma_d)$ is the symmetric group. □

Lemma 2. *Let $\mathcal{G}$ be a sub-group of the symmetric group. The $\boldsymbol{Q}[e_1, \ldots e_d]$-module of the polynomials in the x_i which are invariant under $\mathcal{G}$ is generated by the elements for which the leading monomial, after reduction by J, is independent of the e_i.*

Proof. In Lemma 1, we have just specified the ordering on the monomials to be sharper than the partial ordering of the total degree in the x_i. We now make this ordering precise by letting monomials of the same total degree in the x_i be first compared by comparing their e_i part.

Let f be a polynomial which is invariant under $\mathcal{G}$, and let g be its normal form after reduction by J. Let E be the monomial in the e_i appearing in the leading term $\mathrm{lt}(g)$ of g, and let h be the homogeneous polynomial in the x_i, such that hE is the part of g consisting in all the monomials which are products of E by a monomial in the x_i, of the same degree as $\mathrm{lt}(g)$. The above choice of the ordering implies that the monomials of g, which are not in hE, are of lower order than any monomial of hE, i.e., that hE is the beginning of g.

As g becomes invariant under $\mathcal{G}$ after substitution of the e_i by the σ_i, the polynomial $\sum_{\gamma\in\mathcal{G}} g^\gamma$, which is the sum of the orbit of g under $\mathcal{G}$ is reduced by J to $\mathrm{card}(\mathcal{G})g$. Similarly, $H := \sum_{\gamma\in\mathcal{G}} h^\gamma$ is a polynomial invariant under $\mathcal{G}$. The reduction by J of both sums may be done in parallel, showing that H reduces to a polynomial with $\mathrm{card}(\mathcal{G})h$ as leading part. Thus $f - EH/\mathrm{card}(\mathcal{G})$ reduces to a polynomial with a lower leading term than g.

Iterating this process shows immediately the Lemma. □

End of proof of Theorem 2: The leading terms of J being $\{x_0^d, x_1^{d-1}, \ldots, x_d\}$, any polynomial irreducible by J has a degree in the x_i at most $d(d-1)/2$, and the same is true for the polynomials H of Lemma 2. Thus, it remains to show that one may extract a $\boldsymbol{Q}[e_1, \ldots, e_d]$-basis from the set of these polynomials H.

This may be done effectively by the following process. For each monomial m of degree at most $d(d-1)/2$ in the x_i, compute $M := \sum_{\gamma\in\mathcal{G}} m^\gamma$, and let R be the result of the reduction of M by J. If the leading term of R is independent of the e_i and is different from the leading terms of any previous R, add M to the basis.

The proof that the set thus defined is a basis is straightforward, by looking at the leading terms. □

This proof induces an algorithm for computing a basis of the invariants and the decomposition of any invariant on this basis: Compute the normal form under J of any orbit of a monomial. The theorem tells us if it appears in a basis. When enough such invariants with linearly independent first terms are obtained, one has got a basis. The decomposition of another invariant on this basis may easily be done by normal form computation.

As presented, it seems that the search for a basis implies to consider a lot of orbits of monomials. In fact, only few orbits need to be tested: The theorem of Molien [3] allows to compute easily the Hilbert series $\sum a_i t^i$ of $\boldsymbol{R}_d^{\mathcal{G}}$, such that its coefficient a_i is the dimension over $\boldsymbol{Q}$ of the homogeneous part of degree i of $\boldsymbol{R}_d^{\mathcal{G}}$. This series may be expressed as $P(t)/\prod_{i=1}^{d}(1-t^i)$, where $P(t)$ is a polynomial in t. The coefficient of t^i in P is the number of basis elements that are to be found in degree i. As an example, for the group $\mathcal{M}_5$, this gives $P(t) = 1 + t^4 + t^5 + t^6 + t^7 + t^8$, showing that there is exactly one basis element in each of the degrees 4 to 8, and none in the other positive degrees.

This algorithm, which we have described, is not very efficient, but it is very short to implement it in Maple and it is efficient enough for our purpose.

4 Equations of Small Degree

In this section, we will show how the polynomials s_i of Sect. 2 can be used to recreate the classical formulas for solving the equations of small degree, showing that these formulas may be deduced without tricks from a general method, which we will apply again for degree 5.

4.1 Degree 2. Let $f = ax^2 + bx + c$; we have

$$\begin{aligned}
\omega &= -1, \quad s_0 = x_0 + x_1 = -b/a\ , \\
S_1 &= (x_0 - x_1)^2 = (x_0 + x_1)^2 - 4x_0x_1 = (b/a)^2 - 4c/a\ , \\
x_0 &= \left(s_0 + \sqrt{S_1}\right)/2\ , \qquad x_1 = \left(s_0 - \sqrt{S_1}\right)/2\ ,
\end{aligned}$$

which is exactly the standard formula.

4.2 Degree 3. Let us suppose that $f = x^3 + px + q$. The primitive third root of unity is as usual denoted by j.

The group $\mathcal{S}_3$ has only one subgroup $\mathcal{A}_3 = \mathcal{C}_3$. It has $(x_0 - x_1)(x_1 - x_2)(x_2 - x_0) = \sum_{\mathcal{C}}(x_0x_1^2 - x_0^2x_1)$ as a resolvent invariant with resolvent $X^2 + \Delta$, where Δ is a symmetric polynomial which becomes the discriminant $4p^3 + 27q^2$ after substituting the x_i by the roots of f.

However, to have a better formula, in which j does not appear, we will choose $(j - j^2)(x_0 - x_1)(x_1 - x_2)(x_2 - x_0)$ as a resolvent invariant; it has $X^2 - 3\Delta$ as resolvent. This choice is relevant because the conjugation of j, exchanging j and j^2, has the same action as the transposition of x_1 and x_2.

This allows us to express the S_i in terms of this resolvent invariant and symmetric polynomials:

$$\begin{aligned} S_1 &= \left(x_0 + jx_1 + j^2x_2\right)^3 \\ &= s_0^3 - \frac{3}{2}\left(j - j^2\right)\sum_{\mathcal{C}}\left(x_0x_1^2 - x_0^2x_1\right) - \frac{9}{2}\sum_{\mathcal{S}} x_0^2x_1 \\ S_2 &= \left(x_0 + jx_1 + j^2x_2\right)\left(x_0 + j^2x_1 + jx_2\right) \\ &= s_0^2 - 3\sum_{\mathcal{S}} x_0x_1 \end{aligned}$$

Thus, after substituting the x_i by the roots of f, all we need may be expressed in terms of p, q, and $\sqrt{3\Delta} = 18\sqrt{\frac{p^3}{27} + \frac{q^2}{4}}$. This gives the resulting expressions

$$s_1 = 3\sqrt[3]{-\frac{q}{2} + \sqrt{\frac{p^3}{27} + \frac{q^2}{4}}}\,, \qquad S_2 = -3p\,, \qquad s_2 = -3p/s_1$$

and the solutions

$$x_0 = \frac{s_1}{3} - \frac{p}{s_1}\,, \qquad x_1 = \frac{j^2 s_1}{3} - \frac{jp}{s_1}\,, \qquad x_2 = \frac{js_1}{3} - \frac{j^2 p}{s_1}\,.$$

This formula is only valid if $s_1 \neq 0$, because it appears as denominator. Thus, if $p = 0$, the determination of the square root has to be chosen in order that $s_1 \neq 0$. This is always possible when f is irreducible, because this implies $q \neq 0$.

5 Degree Five

The transitive subgroups of $\mathcal{S} = \mathcal{S}_5$ are well known. There are:

- The alternate group $\mathcal{A}$ of order 60 and of index 2 in $\mathcal{S}$
- The meta-cyclic group $\mathcal{M}$ of order 20 and of index 6 in $\mathcal{S}$. It is the maximal solvable subgroup of $\mathcal{S}$. Its elements act as $x_i \to x_{ai+b}$ (indices modulo 5 and $a \neq 0 \bmod 5$). It is bi-transitive and has 6 conjugates, corresponding to the 6 permutations of the x_i fixing x_0 and x_1.
- The dyadic group $\mathcal{D}$ of order 10, of index 2 in $\mathcal{M}$ and of index 6 in $\mathcal{A}$. Its elements are those of $\mathcal{M}$ for which $a = \pm 1$.
- The cyclic group $\mathcal{C}$ of order 5 and of index 2 in $\mathcal{D}$, consisting of the elements $\mathcal{M}$ such that $a = 1$.

The meta-cyclic group $\mathcal{M}$ being the maximal solvable subgroup of $\mathcal{S}$, an equation is solvable by radicals iff its Galois group is contained in $\mathcal{M}$. This implies that a separable resolvent of $\mathcal{M}$ has a root in $\boldsymbol{Q}$. Thus our problem consists in expressing the roots of f in terms of the coefficients of f and of this root.

As we have already expressed the roots of f in terms of polynomials which are invariant under $\mathcal{C}$, we have to do the following in order to solve by radicals:

- Find relative resolvent invariants and the corresponding resolvents equations for each inclusion in the chain $\mathcal{C} \subset \mathcal{D} \subset \mathcal{M} \subset \mathcal{S}$. These resolvents will be of the form $X^2 - a$, except the last one which will be of degree 6.
- Express any polynomial which is invariant under one of these groups in terms of the resolvent invariant and of polynomials which are invariants under the bigger group.
- This will finally express (through two square roots) our invariants under $\mathcal{C}$ in terms of symmetric polynomials and of the root of the resolvent of degree 6, which will be in $\boldsymbol{Q}$ iff the polynomial f is solvable by radicals.

As said above, this process starts with the fact that the root x_0 of $f = x^5 + px^3 + qx^2 + rx + s$ that we want to compute is $(s_0 + s_1 + s_2 + s_3 + s_4)/5$, whith

$$
\begin{aligned}
s_0 &= x_0 + x_1 + x_2 + x_3 + x_4 = 0\\
s_1 &= x_0 + \omega x_1 + \omega^2 x_2 + \omega^3 x_3 + \omega^4 x_4\\
s_2 &= x_0 + \omega^2 x_1 + \omega^4 x_2 + \omega x_3 + \omega^3 x_4\\
s_3 &= x_0 + \omega^3 x_1 + \omega x_2 + \omega^4 x_3 + \omega^2 x_4\\
s_4 &= x_0 + \omega^4 x_1 + \omega^3 x_2 + \omega^2 x_3 + \omega x_4 ,
\end{aligned}
$$

and where ω is a primitive 5-th root of unity.

The circular permutation $x_i \to x_{i-1}$ acts on the s_i as the multiplication by ω^i. It follows that the polynomials $S_1 := s_1^5$, $S_2 := s_2 s_1^3$, $S_3 := s_3 s_1^2$ and $S_4 := s_4 s_1$ are invariant under the cyclic permutations of the x_i. Thus, if $S_1 \neq 0$, the root x_0 is a rational function of the S_i and $\sqrt[5]{S_1}$.

We shall need to consider two actions on the invariants: The permutation φ such $\varphi(x_i) = x_{2i}$ (let us recall that indices are computed modulo 5) and the conjugation ψ of ω acting as $\psi(\omega) = \omega^2$. *All the invariants we shall compute, including the s_i and the S_i, will be invariant under $\varphi \circ \psi = \psi \circ \varphi$.* This will imply that the invariants under $\mathcal{M}$, which appear in the final resolution will be independent of ω.

Thus the main difficulty of the solving process is to choose the best resolvent invariants. The resolvent invariants for $\mathcal{C}$ and $\mathcal{D}$ of lowest degrees are of degrees 3 and 2 respectively. However they are not always separable because any invariant of $\mathcal{C}$ of degree less than 5 is zero, together with its conjugates by $\mathcal{M}$, when specialized for the equation $x^5 - a$.

Thus a resolvent invariant for $\mathcal{C}$ or $\mathcal{D}$ needs to be of degree at least 5 for being always separable. We explicit now such invariants.

Let $T' = (x_0 - x_1)(x_1 - x_2)(x_2 - x_3)(x_3 - x_4)(x_4 - x_0)$ and $U' = (x_0 - x_2)(x_1 - x_3)(x_2 - x_4)(x_3 - x_0)(x_4 - x_1)$. They have resolvent polynomials for $\mathcal{C}$ which are clearly always separable (if they were not, f would have a multiple root). For managing the relation between φ and ψ we take $T = (\omega - \omega^4)T' + (\omega^2 - \omega^3)U'$ as resolvent invariant for $\mathcal{C}$ relatively to $\mathcal{D}$, and we consider also $U = \psi(T) = (\omega^2 - \omega^3)T' - (\omega - \omega^4)U'$. They satisfy the following properties:

- $\varphi(T) = -U$ and $\varphi(U) = -T$
- T and U are never both zero (if they were, T' would be 0 and f not irreducible).

- If $\boldsymbol{Q}$ does not contain $\sqrt{-1}$, then $T^2 + U^2$ is never zero (if it were, $\sqrt{-1} = T/U$ would be in the field generated by the roots of f, and the complex conjugation would be in the Galois group; as φ^2 is, up to a circular permutation of the roots, the only involution of $\mathcal{M}$, it would be equal to complex conjugation, which is impossible since $\varphi^2(T/U) = T/U$).

We do not know if this last property is true without hypothesis on $\boldsymbol{Q}$.

The most evident resolvent polynomial for $\mathcal{D}$ relatively to $\mathcal{M}$, is $T'U'$, the square root of the discriminant of f. It is always separable, but of rather high degree (10) and a resolvent polynomial of lower degree would lead to a simpler formula. On the other hand, $\sum_{\mathcal{C}} x_i x_{i+1}$ is the resolvent polynomial of least degree, but not always separable.

The best choice seems to be $\varepsilon = (\omega - \omega^2 - \omega^3 + \omega^4) \prod_{\mathcal{C}} (x_1 - x_2 - x_3 + x_4)$, which is of degree 5. It satisfies $\varphi(\varepsilon) = \psi(\varepsilon) = -\varepsilon$. It is always separable; in fact, its resolvent equation is of the shape $x^2 - D$, and if it is not separable, some factor of ε, say $x_1 - x_2 - x_3 + x_4$, is zero. It is now easy to extract an equation of degree 3 in x_0 from

$$\begin{aligned}
0 &= x_0 + x_1 + x_2 + x_3 + x_4 \\
0 &= x_1 + x_4 - (x_2 + x_3) \\
p &= x_0(x_1 + x_2 + x_3 + x_4) + (x_1 + x_4)(x_2 + x_3) + x_1x_4 + x_2x_3 \\
-q &= x_0((x_1 + x_4)(x_2 + x_3) + x_1x_4 + x_2x_3) \\
&\quad +(x_1x_4 + x_2x_3)(x_2 + x_3) + x_2x_3(x_1 + x_4 - x_2 - x_3)
\end{aligned}$$

which shows that f is not irreducible.

Now, if we know how to compute polynomials in the x_i which are invariant under $\mathcal{M}$ (this will be the object of next section), we are able to solve f:

$\varepsilon = \sqrt{D}$, where D is such an invariant under $\mathcal{M}$.

$E = T^2 + U^2$ and $F = \varepsilon(T^2 - U^2)$ are invariant under $\mathcal{M}$. Thus $T = \sqrt{E + \frac{F}{\varepsilon}}$.

If $T = 0$, which implies that $\varepsilon \in \boldsymbol{Q}$, one may change the determination of $\sqrt{D}$, i.e., the sign of ε in order that $T \neq 0$.

$G = \varepsilon TU$ is invariant under $\mathcal{M}$. Thus $U = \frac{G}{\varepsilon T}$.

Thus, the computation of resolvent invariants is reduced to the computation of 4 invariants under $\mathcal{M}$, and it remains to compute the S_i, which are invariant under $\mathcal{C}$. For each such polynomial S, we consider the four following invariants under $\mathcal{M}$:

$$\begin{aligned}
I_1(S) &:= S + \varphi(S) + \varphi^2(S) + \varphi^3(S) \\
I_2(S) &:= \varepsilon(S - \varphi(S) + \varphi^2(S) - \varphi^3(S)) \\
I_3(S) &:= TS - U\varphi(S) - T\varphi^2(S) + U\varphi^3(S) \\
I_4(S) &:= US + T\varphi(S) - U\varphi^2(S) - T\varphi^3(S)
\end{aligned}$$

from which one may easily deduce S by solving this linear system of determinant $-8\varepsilon(T^2 + U^2)$.

If $T^2 + U^2 = 0$ (which is not possible over the field of the rationals), one may replace the last invariant by

$$I_4'(S) := \varepsilon((T+2U)S + (U-2T)\varphi(S) - (T+2U)\varphi^2(S) - (U-2T)\varphi^3(S))$$

The determinant of the new system is $-16\varepsilon^2(TU - T^2 + U^2) = -16\varepsilon^2(G - F)$. As $TU - T^2 + U^2 = T'U'(\omega + \omega^4 - \omega^2 - \omega^3)$, this determinant is never zero.

The computation of $I_1(S)$, $I_2(S)$, $I_3(S)$, $I_4'(S)$ always allows to compute S over any field $\boldsymbol{Q}$ of characteristic different from 2 and 5. However, the degree of $I_4'(S_1)$ in the x_i is 15 instead of 10 for $I_4(S_1)$. This leads to a much more complicate formula. For this reason, we have only done the computation with $I_1(S)$, $I_2(S)$, $I_3(S)$, $I_4(S)$, which gives a formula which is only proved for fields $\boldsymbol{Q}$ in which -1 is not a square.

6 Degree 5 – Invariants of the Meta-cyclic Group

In the preceding section, we have reduced our problem to the computation of a number of invariants of the meta-cyclic group $\mathcal{M}$. We explain now how to compute them.

The proof of Theorem 2 contains an algorithm for computing a basis of the ring of invariants of $\mathcal{M}$ and for expressing any such invariant on this basis. This has been implemented in Maple as follows.

```
with(grobner);
vars:=[x4,x3,x2,x1,x0,p,q,r,s]:
J:=[x0+x1+x2+x3+x4,
x0^2+x0*x1+x0*x2+x0*x3+x1^2+x1*x2+x1*x3+x2^2+x2*x3+x3^2+p,
x0^3+x0^2*x1+x0^2*x2+x0*x1^2+x0*x1*x2+x0*x2^2+x1^3+x1^2*x2+x1*x2^2
                                                +x2^3+x0*p+x1*p+x2*p+q,
x0^4+x0^3*x1+x0^2*x1^2+x0*x1^3+x1^4+x0^2*p+x0*x1*p+x1^2*p+x0*q+x1*q+r,
x0^5+x0^3*p+x0^2*q+x0*r+s]:

morbit:=proc(exp) # sum of an orbit and reduction by J
         sum(sum(subs(
             '[x0=x[b],x1=x['modp(a+b,5)'],x2=x['modp(2*a+b,5)'],
             x3=x['modp(3*a+b,5)'],x4=x['modp(4*a+b, 5)']]'
                         ,exp),a=1..4),b=0..4) ;
       subs([x[0]=x0,x[1]=x1,x[2]=x2,x[3]=x3,x[4]=x4],%);
       sort(normalf(%,J,vars)); primpart(%)
end:

i4:= morbit(x_0^2*x_1*x_4); i5:= morbit(x_0^3*x_1*x_4);
i6:= morbit(x_0^4*x_1*x_4); i7:= morbit(x_0^3*x_1^2*x_4^2);
i8:= morbit(x_0^4*x_1^2*x_4^2);

reduc:=proc(exp) # expression of an invariant on the basis
         normalf(exp,J,vars);
         normalf(%,[i8-i_8,i7-i_7,i6-i_6,i5-i_5,i4-i_4],vars);
         sort(%)
end:
```

Fig. 1. Maple computation

The function `morbit`[1] of Fig. 1 computes the sum of the orbit under $\mathcal{M}$ of any monomial and reduces it by the Gröbner basis J of Lemma 1. When applied to the

[1] This program is written in Maple, version V.5

monomials $x_0^2 x_1 x_4$, $x_0^3 x_1 x_4$, $x_0^4 x_1 x_4$, $x_0^3 x_1^2 x_4^2$, $x_0^4 x_1^2 x_4^2$, this shows that the sums of the orbit of these monomials satisfy the condition of Lemma 2. Thus, the $\boldsymbol{R}_d^{\mathcal{S}}$-module $\boldsymbol{R}_d^{\mathcal{M}}$ of rank 6 has a basis consisting of 1 and these sums, namely

$$i_4 = \sum_{\mathcal{C}} x_0^2 (x_1 x_4 + x_2 x_3)$$

$$i_5 = \sum_{\mathcal{C}} x_0^3 (x_1 x_4 + x_2 x_3)$$

$$i_6 = \sum_{\mathcal{C}} x_0^4 (x_1 x_4 + x_2 x_3)$$

$$i_7 = \sum_{\mathcal{C}} x_0^3 \left(x_1^2 x_4^2 + x_2^2 x_3^2\right)$$

$$i_8 = \sum_{\mathcal{C}} x_0^4 \left(x_1^2 x_4^2 + x_2^2 x_3^2\right)$$

The function `reduc` expresses any polynomial invariant under $\mathcal{M}$ as a linear combination of i_4, i_5, i_6, i_7 and i_8. In this function, `i4, ...` are the normal forms (reduced by J) of these invariants, and `i_4, ...` are names for them.

When we have got these invariants, we have to express them as a rational function of a resolvent invariant. We chose i_4 for such a resolvent invariant, because it has both advantages to be the simplest one and to be always separable (see below).

For expressing the i_α in terms of i_4, we express first the $i_4 i_\alpha$ as linear combinations of the i_α; this is done by applying the function `reduc` to the corresponding products of normal forms, which are polynomials in the x_i of degree at most 12. Then, by the successive substitutions of Fig. 2, we get the expression of the powers of i_4 as linear expressions of $i_4, \ldots, i_8$. Thus solving the linear system {`g2, g3, g4, g5`} in {`i_5, i_6, i_7, i_8`} gives the expression of these invariants as a rational function of i_4.

```
g2:=    i_4^2 = reduc(i4^2);
f2:=    i_4*i_5 = reduc(i4*i5);
f3:=    i_4*i_6 = reduc(i4*i6);
f4:=    i_4*i_7 = reduc(i4*i7);
f5:=    i_4*i_8 = reduc(i4*i8);
su:=[i_5=rhs(f2)/i_4,i_6=rhs(f3)/i_4,i_7=rhs(f4)/i_4,i_8=rhs(f5)/i_4]:

g3:=    i_4^3  = sort(expand(subs(su,g2,expand(i_4*rhs(g2)))));
g4:=    i_4^4  = sort(expand(subs(su,g2,expand(i_4*rhs(g3)))));
g5:=    i_4^5  = sort(expand(subs(su,g2,expand(i_4*rhs(g4)))));
```

Fig. 2. Maple computation, continued

It remains to computes the resolvent equation of i_4 and to show that it is always separable, i.e. to show that the determinant of the above linear system never vanishes. For computing the resolvent equation, it would suffice to find a linear relation between the powers of i_4, i.e., to eliminate `i_5, i_6 ,i_7, i_8` between `g2, g3, g4, g5`, and also `g6`, which may be computed in the same way. But for proving that i_4 is always separable, we proceed in another way.

Let $V := \sum_{\mathcal{C}} x_0x_1$ and $W := \sum_{\mathcal{C}} x_0x_2$.

It is clear that V and W are invariant under $\mathcal{D}$ and exchanged by φ. Thus $\theta := (V - W)^2$, sometimes called Cayley resolvent [1], is invariant under the action of $\mathcal{M}$. In fact $\theta = 4i_4 + p^2 + 12r$, and the resolvent equation of i_4 and its separability are easily deduced by translation from those of θ.

With the notations of the last section, $(W - V)T'U'$ is invariant under $\mathcal{M}$ and thus may be expressed as a polynomial in θ whose coefficients are rational fractions of p, q, r, s. In fact, this is a polynomial in θ, p, q, r, s:

$$P = \frac{1}{32}\left(\theta^3 - \left(20r + 3p^2\right)\theta^2 - \left(8p^2r - 16pq^2 - 240r^2 + 400sq - 3p^4\right)\theta - p^6 + 28p^4r \right.$$
$$\left. - 16p^3q^2 - 176p^2r^2 - 80p^2sq + 224prq^2 - 64q^4 + 4000ps^2 + 320r^3 - 1600rsq\right)$$

It follows that the resolvent equation for θ is $R_\theta = P^2 - \theta\Delta$, where Δ, the square of $T'U'$, is the discriminant of f.

If this resolvent R_θ would have a multiple root for some values of the coefficients of f, this root would have a multiplicity at least 5: In fact, if the polynomial f is irreducible, its Galois group contains a cyclic permutation, and this permutation acts as a cycle of order 5 on the conjugates of θ. As the multiple root cannot have 5 distinct conjugates, this cycle fixes it, and its multiplicity is at least 5.

As we have $R_\theta = P^2 - \theta\Delta$, $R'_\theta = 2PP' - \Delta$ and $\Delta \neq 0$, such a multiple root is not 0 nor a root of PP'. From $R_\theta = R'_\theta = 0$ and $P \neq 0$, we deduce $P = 2\theta P'$. Substituting in the derivatives of order $2, 3, 4$ of R_θ which involve only P and its derivatives, we easily get the contradiction $P = 0$, proving that R_θ and, by translation, the resolvent of i_4 are always separable.

7 The Formula

In this section, we describe and make explicit the formula deduced from the preceding sections. There are no proofs here, for the reason that they are all given in the preceeding sections, as well as the the way of getting all the invariants which are made explicit here.

One wants to solve the *irreducible* polynomial

$$f := x^5 + px^3 + qx^2 + rx + s\ .$$

As seen before, it is solvable by radicals iff the following polynomial R has a root i_4 in $\boldsymbol{Q}$, which may be tested by any factorization algorithm. This polynomial R is a monic polynomial of degree 6 in θ, which is the sum of

$$-\left(\theta + 3r + \frac{p^2}{4}\right)\left(108p^5s^2 - 72p^4qrs + 16p^4r^3 + 16p^3q^3s - 4p^3q^2r^2 - 900p^3rs^2 \right.$$
$$+ 825p^2q^2s^2 + 560p^2qr^2s - 128p^2r^4 - 630pq^3rs + 144pq^2r^3 - 3750pqs^3$$
$$\left. + 2000pr^2s^2 + 108q^5s - 27q^4r^2 + 2250q^2rs^2 - 1600qr^3s + 256r^5 + 3125s^4\right)$$

(note that the second factor is the discriminant of f) and of the square of

$$\frac{1}{2}\left(2\theta^3 + 8\theta^2 r + \left(-6p^2r + 2pq^2 - 50qs + 24r^2\right)\theta \right.$$
$$\left. -15p^2qs - 16p^2r^2 + 13pq^2r + 125ps^2 - 2q^4 - 200qrs + 64r^3\right).$$

From now on, i_4 is a root of R in $\boldsymbol{Q}$. We will express the roots of f as functions of i_4, p, q, r, s. Let us define i_5, i_6, i_7, i_8, related to i_4 by equations of Fig. 3. These equations are linear in i_5, i_6, i_7, i_8. By solving this linear system, these invariants are expressed as polynomials in i_4 whose coefficients are rational fractions in p, q, r, s with the determinant of the system as common denominator. As i_4 is always separable, this determinant is never 0.

We do not give here the expression of i_5, i_6, i_7, i_8 as polynomials in i_4 because of their size: i_8 needs 330 lines in our Maple program.

With i_4, i_5, i_6, i_7, i_8 one computes

$$\begin{aligned}
i_4^2 &= 5i_8 - 2pi_6 + 4qi_5 - 2p^2i_4 - 6p^2r + 2pq^2 + 10qs + 4r^2\\
i_4^3 &= \frac{1}{2}\big((3p^2 - 20r)i_8 + (-pq - 50s)i_7 + (-3p^3 + 28pr - 12q^2)i_6\\
&\quad + (3p^2q - 45ps - 6qr)i_5 + (-3p^4 + 36p^2r - 15pq^2 + 60qs - 32r^2)i_4\\
&\quad - 6p^4r + 3p^3q^2 + 41p^2qs + 52p^2r^2 - 54pq^2r - 250ps^2 + 14q^4 + 140qrs - 80r^3\big)\\
i_4^4 &= (19p^2r - 9pq^2 + 225qs - 60r^2)i_8 + (15p^2s - 8pqr + 3q^3 + 100rs)i_7\\
&\quad + (-4p^3r + 4p^2q^2 - 105pqs - 16pr^2 + 29q^2r + 125s^2)i_6\\
&\quad + (-9p^3s + 17p^2qr - 8pq^3 + 140prs + 155q^2s - 68qr^2)i_5\\
&\quad + (-4p^4r + 4p^3q^2 - 79p^2qs - 16p^2r^2 + 15pq^2r - 25ps^2 + 4q^4 + 80qrs)i_4\\
&\quad + 6p^4qs - 22p^4r^2 + 16p^3q^2r - 4p^2q^4 - 404p^2qrs + 68p^2r^3 +\\
&\qquad 132pq^3s + 42pq^2r^2 + 550prs^2 - 30q^4r - 50q^2s^2 + 20qr^2s + 16r^4\\
i_4^5 &= \frac{1}{2}\big((15p^4r - 5p^3q^2 + 290p^2qs - 152p^2r^2 - 27pq^2r - 1375ps^2 + 22q^4 - 700qrs + 240r^3)i_8\\
&\quad + (18p^4s - 11p^3qr + 3p^2q^3 - 530p^2rs + 110pq^2s + 124pqr^2 - 41q^3r - 2375qs^2 + 200r^2s)i_7\\
&\quad + (-15p^5r + 5p^4q^2 - 212p^3qs + 168p^3r^2 - 83p^2q^2r + 325p^2s^2\\
&\qquad + 10pq^4 + 1560pqrs - 176pr^3 - 620q^3s - 12q^2r^2 - 1500rs^2)i_6\\
&\quad + (15p^4qr - 5p^3q^3 - 147p^3rs + 351p^2q^2s - 90p^2qr^2 - 43pq^3r\\
&\qquad - 3175pqs^2 - 420pr^2s + 20q^5 + 215q^2rs + 152qr^3 + 625s^3)i_5\\
&\quad + (-15p^6r + 5p^5q^2 - 200p^4qs + 200p^4r^2 - 110p^3q^2r + 355p^3s^2 + 15p^2q^4 + 1728p^2qrs\\
&\qquad - 432p^2r^3 - 752pq^3s + 220pq^2r^2 - 200prs^2 - 43q^4r + 1825q^2s^2 - 2640qr^2s + 512r^4)i_4\\
&\quad - 30p^6r^2 + 25p^5q^2r + 198p^5s^2 - 5p^4q^4 - 491p^4qrs + 364p^4r^3 + 181p^3q^3s - 286p^3q^2r^2\\
&\qquad - 810p^3rs^2 + 95p^2q^4r + 3005p^2q^2s^2 + 4120p^2qr^2s - 1088p^2r^4 - 12pq^6 - 4095pq^3rs\\
&\qquad + 612pq^2r^3 - 15875pqs^3 + 900pr^2s^2 + 858q^5s - 34q^4r^2 + 10700q^2rs^2 - 6240qr^3s\\
&\qquad + 960r^5 + 6250s^4\big)
\end{aligned}$$

Fig. 3.

$$
\begin{aligned}
D = & \left(-24p^3 + 100pr + 24q^2\right) i_4 + (88pq - 300s) i_5 + \left(-24p^2 + 100r\right) i_6 - 120qi_7 \\
& + 40pi_8 - 80p^3r + 40p^2q^2 - 480pqs + 160pr^2 + 332q^2r + 125s^2 \\
E = & \left(3p^3 + 12pr + 3q^2\right) i_4 + (-pq - 50s) i_5 + \left(3p^2 + 20r\right) i_6 + \\
& + 4p^3r - 3p^2q^2 + 40pqs + 16pr^2 - 21q^2r + 125s^2 \\
F = & \left(85p^4q - 418p^3s - 440p^2qr + 419pq^3 + 1590prs - 1040q^2s + 524qr^2\right) i_4 \\
& + \left(4p^3r - 73p^2q^2 + 2095pqs - 56pr^2 - 748q^2r - 4875s^2\right) i_5 \\
& + \left(85p^3q - 520p^2s - 298pqr + 366q^3 + 2100rs\right) i_6 \\
& + \left(-65p^2q + 875ps - 550qr\right) i_8 + \left(-58p^2r + 41pq^2 - 275qs + 440r^2\right) i_7 \\
& - 12p^5s + 158p^4qr - 85p^3q^3 - 1462p^3rs - 159p^2q^2s + 142p^2qr^2 + 896pq^3r \\
& + 175pqs^2 + 2900pr^2s - 402q^5 - 1925q^2rs - 448qr^3 - 1875s^3 \\
G = & \left(15p^4q + 213p^3s + 50p^2qr + pq^3 - 940prs + 515q^2s - 184qr^2\right) i_4 \\
& + \left(-4p^3r - 27p^2q^2 - 270pqs + 96pr^2 - 182q^2r + 3000s^2\right) i_5 \\
& + \left(15p^3q + 195p^2s + 68pqr - 6q^3 - 1100rs\right) i_6 \\
& + \left(-35p^2q - 250ps - 200qr\right) i_8 + \left(-22p^2r + 19pq^2 + 650qs - 40r^2\right) i_7 \\
& + 12p^5s + 42p^4qr - 15p^3q^3 + 492p^3rs - 156p^2q^2s + 358p^2qr^2 - 246pq^3r \\
& + 2825pqs^2 - 1400pr^2s + 42q^5 + 550q^2rs - 232qr^3 - 1250s^3
\end{aligned}
$$

and then

$$
\varepsilon = \sqrt{5D}\,, \qquad T = \sqrt{\frac{5}{2}\left(E + \frac{F}{\varepsilon}\right)}\,, \qquad U = \frac{5G}{T\varepsilon}\,.
$$

If $T = 0$, which implies that ε is in $\boldsymbol{Q}$, we change the sign of ε. In fact, if both values of T were 0, then the same would be true for the values of E, F, and of the invariants T and U of the preceding sections, which is impossible.

Consider now

$$
\begin{aligned}
H = & \; 25\,(2i_5 - pq - 5s) \\
I = & \; 25\left(\left(-24p^3 + 100pr - 46q^2\right) i_4 + (68pq - 300s) i_5 + \left(-24p^2 + 100r\right) i_6 - 70qi_7\right. \\
& \left. +40pi_8 - 80p^3r + 20p^2q^2 - 255pqs + 160pr^2 - 28q^2r + 125s^2\right) \\
J = & \left(-p^3 - 96pr + 11q^2\right) i_4 + (-7pq + 525s) i_5 + \left(-9p^2 - 60r\right) i_6 - 25qi_7 - 25pi_8 \\
& + 50p^3r - 7p^2q^2 - 145pqs - 308pr^2 + 128q^2r - 1000s^2 \\
K = & \left(63p^3 - 412pr + 27q^2\right) i_4 + (-109pq + 1175s) i_5 + \left(67p^2 - 420r\right) i_6 + 75qi_7 \\
& - 125pi_8 + 210p^3r - 79p^2q^2 - 415pqs - 676pr^2 + 496q^2r - 750s^2
\end{aligned}
$$

which allows us to define

$$Q_1 = \frac{5}{4}\left(H + \frac{I}{\varepsilon} + \frac{TJ + UK}{E}\right), \qquad P_1 = \sqrt[5]{Q_1}\,.$$

From now on we need $P_1 \neq 0$. If this is not the case, we arrange this by changing the sign of one or two of the above square roots, which is equivalent to one of the following substitutions (or their composition) applied to the above formulas for Q_1 and P_2, P_3, P_4.

$$(\varepsilon \to -\varepsilon, \quad T \to U, \quad U \to -T)$$
$$(T \to -T, \quad U \to -U)$$

With P_1, one computes

$$P_{41} = -5p$$
$$P_{42} = 5\left(10i_7 - 4pi_5 - 14qi_4 - 4p^2q + 45ps - 72qr\right)$$
$$P_{31} = -25q$$
$$P_{32} = 25\left(-10i_8 + 2pi_6 - 22qi_5 + 2p^2i_4 + 20p^2r + 2pq^2 - 35qs - 40r^2\right)$$
$$P_{33} = 5\left(35i_8 - 4pi_6 + 23qi_5 + (-6p^2 + 12r)i_4 - 58p^2r + 14pq^2 - 105qs + 76r^2\right)$$
$$P_{34} = 5\left(5i_8 - 22pi_6 + 14qi_5 + \left(-18p^2 + 16r\right)i_4 - 34p^2r + 22pq^2 - 140qs + 68r^2\right)$$
$$P_{21} = 5\left(3i_4 + 2p^2 - 16r\right)$$
$$P_{22} = 25\left(-10qi_6 + \left(8p^2 - 50r\right)i_5 + (-2pq - 25s)i_4 + 8p^3 + 70q^3q - 20p^2s - 26pqr \right.$$
$$\left. +50rs\right)$$
$$P_{23} = 25\left(-4pi_7 - qi_6 + 4ri_5 + (-3pq + 15s)i_4 + 26p^2s - 26pqr + 7q^3 - 40rs\right)$$
$$P_{24} = 25\left(3pi_7 - 18qi_6 + 22ri_5 + (-14pq + 20s)i_4 + 18p^2s - 33pqr + 21q^3 + 30rs\right)$$

Then

$$P_4 = \frac{P_{41}}{2P_1} + \frac{P_{42}}{2\varepsilon P_1}$$
$$P_3 = \frac{P_{31}}{4P_1^2} + \frac{P_{32}}{4\varepsilon P_1^2} + \frac{P_{33}T + P_{34}U}{10EP_1^2}$$
$$P_2 = \frac{P_{21}}{4P_1^3} + \frac{P_{22}}{4\varepsilon P_1^3} + \frac{P_{23}T + P_{24}U}{10EP_1^3}$$

A root of f is then $x_0 = \frac{P_1+P_2+P_3+P_4}{5}$.

The other roots may be obtained by changing the determination of the fifth root. They may also be obtained as

$$x = \frac{\omega P_1 + \omega^2 P_2 + \omega^3 P_3 + \omega^4 P_4}{5}\,.$$

where ω is any primitive fifth root of unity, i.e., a root of $x^4 + x^3 + x^2 + x + 1$, easily solvable by radicals: ω is a rational function of $\sqrt{\frac{-5+\sqrt{5}}{2}}$.

8 Optimality

In this section, we shall prove that our formula is optimal with respect to the number of roots to extract, and, moreover, that the field extension defined by the roots in our formula is the (unique up to an isomorphism) smallest field in which one of the roots of f is expressible by radicals. We show also that the same optimality propreties are true for the expression of all roots together (using above expression of ω).

Definition 1. *A* simple radical extension *is a simple field extension $k(x)$ of k such x has a prime power in k. A* radical extension *is an extension obtained by a finite number of simple radical extensions.*

The main result of this section is the following.

Theorem 3. *Let f be an irreducible quintic polynomial over the field $\boldsymbol{Q}$ which is solvable by radicals. Let $\boldsymbol{L}$ be a radical extension of $\boldsymbol{Q}$ which contains a root x of f. Then $\boldsymbol{L}$ contains a radical extension containing x, and isomorphic to the extension defined by our formula.*

We may suppose that $\boldsymbol{L} = \boldsymbol{K}(z)$ is a simple radical extension of some subfield $\boldsymbol{K}$ which is a radical extension of $\boldsymbol{Q}$ and does not contain any root of f: Otherwise, we could replace $\boldsymbol{L}$ by $\boldsymbol{K}$.

We prove now that $z = \sqrt[5]{y}$ for some $y \in \boldsymbol{K}$: The Galois group over $\boldsymbol{K}$ of the field generated by the five roots of f is a subgroup of the Galois group of f, itself included in $\mathcal{M}$. As $\boldsymbol{K}$ does not contain any root of f, this subgroup has no fixed point and thus contain $\mathcal{C}$ and is transitive on the roots. Thus f is irreducible over $\boldsymbol{K}$ and $\boldsymbol{K}(z)/\boldsymbol{K}$ has a degree multiple of 5. As z has a prime power in $\boldsymbol{K}$, we have

$$\boldsymbol{L} = \boldsymbol{K}(z) = \boldsymbol{K}\left(\sqrt[5]{y}\right) = \boldsymbol{K}(x)$$

Let ω be a primitive fifth root of unit. The conjugates of z over $\boldsymbol{K}$ are the $\omega^i z$ and ω is in the field generated by the conjugates of z. Thus, the splitting field of $\boldsymbol{L}$ over $\boldsymbol{K}$ is

$$\boldsymbol{K}(\omega, z) = \boldsymbol{K}(\omega, x) = \boldsymbol{K}(x, x_1, x_2, x_3, x_4)$$

where $x, x_1, \ldots$ are all the roots of f.

It follows immediately that all radicals which appear in our formula are contained in $\boldsymbol{K}(\omega, x)$, being polynomials in $\omega, x, x_1, \ldots$; thus it remains to prove that they are in $\boldsymbol{K}(x)$. For this we have three cases to consider.

If $\omega \in \boldsymbol{K}$ there is nothing to prove.

If $\omega \notin \boldsymbol{K}$ but $\sqrt{5} \in \boldsymbol{K}$, we have $\boldsymbol{K}(\omega) : \boldsymbol{K} = 2$, $\boldsymbol{K}(\omega, x) : \boldsymbol{K} = 10$, and the Galois group of f over $\boldsymbol{K}$ is $\mathcal{D}$. Let ϑ be its unique element of order 2 which fixes x. It maps ω to ω^4 and x_i to x_{4-i} (for some numbering of the x_i). Thus it changes the signs of $\omega - \omega^4, \omega^2 - \omega^3, T', U'$ and fixes T, U and ε. As these invariants are also fixed by the circular permutation of the roots of f, they are in $\boldsymbol{K}$.

Finally, if $\sqrt{5} \notin \boldsymbol{K}$, we have $\boldsymbol{K}(\omega) : \boldsymbol{K} = 4$, $\boldsymbol{K}(\omega, x) : \boldsymbol{K} = 20$, and the Galois group of f over $\boldsymbol{K}$ is $\mathcal{M}$. Let ϑ be its element which fixes x and maps ω on ω^2. As it is of order 4, it acts on the roots of f as a circular permutation. If we chose the numbering of the x_i such that it maps x_i to x_{2i}, it follows from the definition of the invariants T,U, and ε, that they are fixed by ϑ, and thus that, in all cases, $\boldsymbol{K}$ contains all square roots of our formula. Thus, the theorem is proved.

Moreover, the following characterization of the smallest radical extension containing all roots of f, is immediate from what precedes.

Theorem 4. *The smallest radical extension containing all roots of a solvable quintic f which is irreducible over $\boldsymbol{Q}$ is the extension generated by all roots of f and a primitive fifth root of unit ω. Our formula expresses all the roots of f in this extension.*

If ω is of degree 2^d over $\boldsymbol{Q}$ and if the Galois group of f over $\boldsymbol{Q}(\omega)$ is of order $5 \cdot 2^e$, this smallest radical extension is of degree $5 \cdot 2^{d+e}$ and is defined by $d + e$ square roots and one fifth root.

9 Implementation and Conclusion

The formula of the preceding section has been implemented in Maple almost as it is described. The main difference is that the linear system in $i_4, \ldots, i_8$, which has to be solved, has been solved once and for all.

The file contains also all examples of quintics solvable by radicals that we know of.

This implementation is available from http://www-calfor.lip6.fr/~dl. As this implementation contains mainly polynomial expressions and root extractions, its translation to any computer algebra system is very easy.

However, such an implementation leads to a side problem which is not so easy: *To verify the correctness of the solution which is provided.* This seems easy by substituting the solution in the quintic polynomial and simplifying to 0. But such a task needs three ingredients.

The first one is a precise semantic for the square and quintic roots. The one that we need must be such that, given two occurrences of the same root, the same determination is chosen, whichever it is.

The second ingredient is that the number of root extractions which appear in the solution is kept minimal. If not, the determination of the extra roots may be wrongly chosen. We have already mentioned this in relation to the Paxton Young formula which involves four fifth roots implying 625 solutions. For this reason, we could not use the Maple value for the primitive fifth roots of unit. In fact the answer given by Maple to `solve(1+x+x^2+x^3+x^4)` is equivalent with the following (in fact, in the output of Maple, $\sqrt{2}$ and $\sqrt{-1}$ are factored out from the nested radicals).

$$-\frac{1}{4}+\frac{1}{4}\sqrt{5}+\frac{1}{4}\sqrt{-10-2\sqrt{5}}\,,\quad -\frac{1}{4}+\frac{1}{4}\sqrt{5}-\frac{1}{4}\sqrt{-10-2\sqrt{5}}\,,$$
$$-\frac{1}{4}-\frac{1}{4}\sqrt{5}+\frac{1}{4}\sqrt{-10+2\sqrt{5}}\,,\quad -\frac{1}{4}-\frac{1}{4}\sqrt{5}-\frac{1}{4}\sqrt{-10+2\sqrt{5}}$$

This contains two different iterated square roots, and, consequently, Maple is unable to simplify to 0 the difference between the square of the first solution and any other solution. Thus, in our program, we have replaced these expressions by the powers of the first one.

The third ingredient is a good simplifier for expressions involving root extractions. From version V.3 on, Maple is able to verify the solution for the equations with integer coefficients. Nevertheless this remains a long computation, especially when one want to verify all solutions and not only the one which does not involves fifth roots of unit: for the 27 examples included in the program, this verification takes more than 1400s., with a 600Mherz PC running Maple 8.

For the case of equations with symbolic coefficients, there was already a problem for the simple case $x^5 - a = 0$. Maple was unable to test that some expressions involved in our formula were zero, and provided a false result in this case. The problem comes from the fact that, for Maple V.5, $\sqrt{a^2} = \mathrm{csgn}(a)a$; this has been easily solved by replacing `sqrt(exp)` by `sqrt(exp,symbolic)` which returns a when applied to a^2. Nevertheless, even with this patch, we were unable to verify the solution of the three non trivial examples of solvable quintics with symbolic coefficients, which are given in our program.

All of this enforces our opinion that solving by radicals, when possible, leads to expressions which are too huge to be useful.

Thus, in our opinion, the only usefulness for our program is to be a very good test for simplifiers for radical expressions.

References

[1] A. Cayley. On a new auxiliary equation in the theory of equation of the fifth order, *Phil. Trans. Royal Soc. of London* (1861).

[2] D. Cox, J. Little and D. O'Shea. *Ideals, Varieties and Algorithms*. Undergraduate Texts in Mathematics (Springer-Verlag, 1992).

[3] T. Molien. Über die Invarianten der linearen Substitutionsgruppe. *Sitzungber. Königl. Preuss. Akad. Wiss.*, 1897, 1152–1156.

[4] G. Paxton Young. Solvable Quintics Equations with Commensurable Coefficients. *Amer. J. Math.*, **10** (1888), 99–130.

From Abel to Kronecker: Episodes from 19th Century Algebra

Birgit Petri and Norbert Schappacher

This paper is about Leopold Kronecker reading Niels Henrik Abel's results and ideas on the resolution of algebraic equations. When the young Kronecker began to work on algebraic equations, he went back and forth between Abel's works and his own ideas. And throughout his career he continued to position himself much nearer to Abel than to Galois. At the same time, his own creativity transformed Abel's results and questions into something more arithmetic and fairly different. For instance, already in his very first publication on algebraic equations [41], when unfolding Abel's problems on solvable equations, Kronecker essentially claimed both what is known today as the 'Theorem of Kronecker and Weber,' to the effect that every abelian extension of $\mathbf{Q}$ is cyclotomic, and its analogue for abelian extensions of $\mathbf{Q}(\sqrt{-1})$, and even indicated further generalizations.

Our article was written with two different kinds of readers in mind: interested mathematicians as well as historians of mathematics. The mathematician, especially the number theorist, may appreciate for example our attempt at partially reconstructing Kronecker's reasoning for the prime-degree-case of the 'Kronecker-Weber-Theorem'. The historian will notice the newly used documents. For instance, we draw on, and publish here for the first time, a few letters from Kronecker to Dirichlet, of which Harold M. Edwards has made us aware and shared his personal transcription with us. We also make use of some of the unpublished, handwritten notes of Kronecker's lecture courses which are preserved in the library of the Strasbourg mathematical institute IRMA.

Both the mathematical and the historical reader, however, will not fail to realize the peculiar position that the mathematical events discussed here occupy with respect to what can be considered (at least with hindsight) as the mainstream development of Galois Theory in the nineteenth century. Comparing for instance Kronecker's first paper in the subject [41] to Enrico Betti's big memoir [9], which had appeared only a year before, and which Kronecker knew, the two texts almost seem to belong to different mathematical cultures: Betti's treatise is justly regarded to be a milestone in the development of Galois Theory in that it treats permutations first – if in a way which is still quite far from our group theory, and actually quite hard to penetrate for mathematician and historian alike –, and their applications to the theory of algebraic equations in a separate part thereafter. Kronecker's Berlin Academy Note on the other hand barely sketches proofs, and takes a dramatic number theoretic turn at the end, which may well have been the very starting point of his work.

The final Sect. 4 of our paper concentrates on elements of the history of Kronecker's 1861 sharpening of Abel's theorem on the non-resolubility of the general quintic equation: Kronecker saw that such an equation does not admit one-parameter resolvents. This was preceded by several proposals – due to Hermite, Kronecker, and Brioschi – to use elliptic functions in the resolution of the quintic equation; we will describe them briefly in Sect. 3. On the other hand, Kronecker's theorem of 1861 gave rise to a lasting difference of opinions between Kronecker and Felix Klein about the limits of algebraic resolutions of the quintic equation. It was Klein who supplied the first published proof of Kronecker's result; but we will try to explain the difference of his point of view from Kronecker's.

The present paper thus highlights a few seminal episodes from nineteenth century algebra, which are directly linked to the name of Abel, and which, even though partly forgotten, are part of our mathematical heritage.

1 Recalling Niels Henrik Abel on the Resolution of Algebraic Equations

1.1. Since early mathematics in several different cultures had some knowledge of solving what we consider today as special cases of quadratic polynomial equations, it is probably impossible to pin down a precise first occurence of the general formula for the solutions of all quadratic equations $x^2 + px + q = 0$, i.e., of the formula

$$x_\kappa = -\frac{p}{2} + \frac{1}{2}(-1)^\kappa \sqrt{\Delta} \qquad \text{(for } \kappa = 0, 1\text{)},$$

where $\Delta = p^2 - 4q$, and where the square root is fixed in some way.

1.2. The resolution of cubic equations was largely accomplished by Tartaglia and Cardano around 1540. In today's notation, if $x^3 + px + q = 0$ (a case to which one is easily reduced by what was later to be called a *Tschirnhausen transformation*), and if we set $\Delta = -4p^3 - 27q^2$, then, for a suitable choice of the two cubic roots, one obtains the solutions:

$$x_\kappa = \varrho^\kappa \sqrt[3]{\left(-\frac{q}{2} + \frac{1}{2}\sqrt{\frac{-\Delta}{27}}\right)} + \varrho^{2\kappa} \sqrt[3]{\left(-\frac{q}{2} - \frac{1}{2}\sqrt{\frac{-\Delta}{27}}\right)} \qquad (\text{for } \kappa = 0, 1, 2),$$

where ϱ is a primitive third root of 1.[1]

1.3. Skipping the case of degree four (in fact, we will only consider equations of prime degree further on), the next bit of general mathematical culture in the theory of algebraic equations is of course the *impossibility of resolving the general quintic equation by radicals*, which was proved completely for the first time by Abel – see [2]. This fact is usually linked today to the simplicity of the alternating group on 5 letters, even though Abel's original proof was *not* group-theoretical in nature, but proceeded, in a fashion very typical of Abel, via an analysis of the most general form of an algebraic expression involving radicals which could possibly be a solution to the quintic equation.

In order to prove the impossibility rigorously, Abel carefully spelled out, what 'resolution by radicals' was supposed to mean. As a consequence he proved in particular:[2]

> If an equation is algebraically solvable, then the root can always be written in such a way that all algebraic functions of which it is composed can be expressed by rational functions of the roots of the given equation.

In modern words, this means that the constructions of the resolution by radicals are realized within the splitting field of the given polynomial. This statement was turned by Kronecker in 1861 into a criterion that every notion of algebraic solvability of an algebraic equation should meet. He called it **Abel's postulate**, and we shall see in Sect. 4 below that it played an important role in his strengthening of Abel's impossibility theorem.[3]

---

[1] Let us note in passing that Eisenstein published a short note [16] where he tried to write down the formulae for equations of degree 1,2,3, and 4 in a coherent way, and where he added a cryptic footnote indicating an analytic resolution of the quintic equation – see [61].

[2] See the indented passsage at the end of §II of [2]. Kronecker quoted this in [46, p. 55–56] from the German translation of Abel's paper as it appeared in Crelle's journal: *Journal für die reine und angewandte Mathematik* **1** (1826), p. 73, tacitly adjusting the spelling in two places: *Wenn eine Gleichung algebraisch auflösbar ist, so kann man der Wurzel allezeit ein[e] solche Form geben, daß sich alle algebraische[n] Functionen, aus welchen sie zusammengesetzt ist, durch rationale Functionen der Wurzeln der gegebenen Gleichung ausdrücken lassen.*

[3] In his lecture course [52, p. 164], Kronecker introduced this postulate with these words: *Eine Gleichung auflösen heißt sie ersetzen durch eine Kette von Gleichungen vorgeschriebener Beschaffenheit; das Wesen der "Kette" besteht darin, dass immer aus einer Gleichung eine Größe für die folgenden bestimmt wird. Bei der Auflösung durch reine Gleichungen*

1.4. It is less well-known that Abel also investigated, around the same time, the general shape of the roots of an equation of degree 5 (or higher) which does happen to be solvable by radicals. On 14 March 1826, he explained in a letter to Crelle[4] that, if a quintic equation with rational coefficients is solvable by radicals, then

$$x = c + A \cdot a^{\frac{1}{5}} \cdot a_1^{\frac{2}{5}} \cdot a_2^{\frac{4}{5}} \cdot a_3^{\frac{3}{5}} + A_1 \cdot a_1^{\frac{1}{5}} \cdot a_2^{\frac{2}{5}} \cdot a_3^{\frac{4}{5}} \cdot a^{\frac{3}{5}} +$$
$$+ A_2 \cdot a_2^{\frac{1}{5}} \cdot a_3^{\frac{2}{5}} \cdot a^{\frac{4}{5}} \cdot a_1^{\frac{3}{5}} + A_3 \cdot a_3^{\frac{1}{5}} \cdot a^{\frac{2}{5}} \cdot a_1^{\frac{4}{5}} \cdot a_2^{\frac{3}{5}},$$

where $a = m + n\sqrt{1+e^2} + \sqrt{h\left(1+e^2+\sqrt{1+e^2}\right)}$,

$$a_1 = m - n\sqrt{1+e^2} + \sqrt{h\left(1+e^2-\sqrt{1+e^2}\right)},$$
$$a_2 = m + n\sqrt{1+e^2} - \sqrt{h\left(1+e^2+\sqrt{1+e^2}\right)},$$
$$a_3 = m - n\sqrt{1+e^2} - \sqrt{h\left(1+e^2-\sqrt{1+e^2}\right)},$$

and $A = K + K'a + K''a_2 + K'''aa_2$, $A_1 = K + K'a_1 + K''a_3 + K'''a_1a_3$, $A_2 = K + K'a_2 + K''a + K'''aa_2$, $A_3 = K + K'a_3 + K''a_1 + K'''a_1a_3$, with $c, h, e, m, n, K, K', K'', K'''$ rational numbers. Abel also claimed he had analogous formulæ in degree "7, 11, 13, etc."

1.5. Abel came back to this problem of writing down the general form of the roots of solvable equations shortly before his death in a manuscript [7] of which only the first pages are fully worked out, and which ends in a sequence of unconnected formulas. One of the claims in this manuscript is that the root x of any solvable equation of prime degree μ with rational coefficients takes the form:

$$x = A + \sqrt[\mu]{R_1} + \sqrt[\mu]{R_2} + \cdots + \sqrt[\mu]{R_{\mu-1}},$$

with rational A and where $R_1, \ldots, R_{\mu-1}$ are roots of an equation of degree $\mu - 1$.

---

gilt nun der von Abel entdeckte Satz, – auf dem Abels Unmöglichkeitsbeweis wesentlich beruht –, dass alle successive eingeführten Größen rationale Funktionen der Wurzeln der aufzulösenden Gleichung sind. Die Forderung, dass bei jeder Auflösung die successive auftretenden Größen rationale Funktionen der Wurzeln der aufzulösenden Gleichung seien, soll als das "Abel'sche Postulat" bezeichnet werden. Die Berechtigung desselben liegt darin, daß, sowie man das Postulat aufgiebt, Gleichung derselben Gattung verschiedene Auflösungsmethoden nötig haben, dass also bei der Vermittlung durch irrationale Funktionen die Einteilung der algebraischen Funktionen in Gattungen durchbrochen wird. Das Abelsche Postulat ist daher identisch mit der Forderung, dass man keine Methoden der Auflösung algebraischer Gleichungen einführen darf, welche Gleichungen die derselben Gattung angehören auseinander reißen. – The term *Gattung* here corresponds more or less to the splitting field of a polynomial.

4 See *Journal für die reine und angewandte Mathematik* **5** (1830), p. 336; the French version is reproduced in [1, vol. 2, p. 266].

2 1853–1856: The General Form of Solvable and Cyclic Equations

It was in particular Abel's result recalled in 1.5 which the twenty-nine year old Leopold Kronecker (1823–1891) had apparently discovered himself before he received, on the evening of 30 January 1853, his copy of Abel's Collected Papers where he could read [7] himself. Kronecker received Abel's Collected Works at Liegnitz (at the time Lower Silesia, today Legnica, Poland), where he worked overseeing the family estate and his uncle's banking business, trying at the same time – as much as recurrent health problems would permit – to get actively back into mathematics.[5]

In order to appreciate Abel's and Kronecker's points of view, one should compare them to today's standard practice of teaching the Galois Theory of solvable equations by showing that an equation can be solved by radicals if and only if its Galois group is solvable. Such a presentation – if not its modern formulation in terms of field extensions, automorphisms, etc. – is relatively close in spirit to Galois' original point of view. That point of view actually took quite some time in the nineteenth century to be understood and further developed. For instance, the criterion for solvability which was especially popular in the middle of the nineteenth century was not group theoretical, but rather the statement that "all roots of the equation can be expressed rationally in terms of two of them."[6]

2.1 Principles. Abel's and Kronecker's approach to solvable equations was of a different nature. As for Abel, one may notice throughout his works a certain tendency to completely classify classes of mathematical objects by a fitting formalism. His systematic treatment of elliptic integrals [4] can be seen as an example of this, which the famous memoir on "abelian functions" sent to the Paris Academy [3] carries even further.[7] And in the long opening of the unfinished manuscript that Kronecker was finally able to read in 1853, Abel presented his method of asking only questions which specify the formal type of acceptable answers as "the only scientific" one, and proudly recalled[8]:

> I have treated several branches of analysis in this way, and even if I often took up problems that were beyond my forces, I did obtain a lot of general results

[5] See Kronecker's letter to Dirichlet of 31 January 1853 reproduced in Appendix I below. – The edition of Abel's works that Kronecker read was of course the first edition, edited by Holmboe in 1839. In the 1870's, Kronecker, together with Clebsch and Weierstrass, would be among the first mathematicians to plead for a new edition of Abel's Works, which was then realized by Sylow and Lie in 1881 – see the preface of [1].

[6] Cf. [19], in particular p. 69.

[7] Abel even tried this approach (unsuccessfully) on Fermat's Last Theorem – see [Abel *Œuvres*, vol. II, p. 254–255]

[8] [7, p. 218]: *J'ai traité plusieurs branches de l'analyse de cette manière, et quoique je me sois souvent proposé des problémes qui ont dépassé mes forces, je suis néansmoins parvenu à un grand nombre de résultats généraux qui jettent un grand jour sur la nature des quantités dont la connaissance est l'objet des mathématiques.*

> which elucidate the nature of the quantities the knowledge of which is the object of mathematics.

Abel had already applied his general principles to solvable algebraic equations once, by singling out those which (in nowaday's parlance) have an abelian Galois group – see [6]. In the fragment [7], however, he was led to the following two questions concerning *all* solvable algebraic equations[9]:

> 1. Find all equations of a given degree which are algebraically solvable.
> 2. Determine whether or not a given equation is algebraically solvable.

And he announced that his investigation would produce[10]

> several general propositions about the solvability of equations and about the nature of their roots. It is these general properties that really make up the theory of algebraically solvable equations, because it is of little importance to know whether a particular equation is solvable or not. One of these general properties is for example that it is impossible to solve algebraically the general equations of degree higher than four.

This plea for a theory yielding the general form of the roots of all solvable equations (with rational coefficients) apparently squared perfectly with Kronecker's own approach.[11] He expressed his overall goals at many places, appealing in a very German way to 'essence' (*das Wesen*) and the 'true nature' of things. For instance, in the beginning of the letter to Dirichlet of 31 January 1853 reproduced in Appendix I, we read:

> Having looked a bit more closely at the theory of solvable equations for a few weeks, I have noticed a few things which seem to me to cast a great deal of light on the essence of this matter.

And a little later in the same letter:

---

[9] [7, p. 219]: *1. Trouver toutes les équations d'un degré déterminé quelconque qui soient résolubles algébriquement. 2. Juger si une équation donnée est résoluble algébriquement, ou non.*

[10] [7, p. 219]: *Dans le cours des recherches on parviendra à plusieurs propositions générales sur les équations par rapport à leur résolubilité et à la forme des racines. C'est en ces propriétés générales que consiste véritablement la théorie des équatiosn quant à leur résolution algébrique, car il importe peu si l'on sait qu'une équation d'une forme particulière est résoluble ou non. Une des ces propriétés générales est par exemple qu'il est impossible de résoudre algébriquement les équations générales passé le quatrième degré.* – This announces Abel's second proof of this famous impossiblility result, of which Malmsten obtained a variant in [59].

[11] According to Kronecker, the topic of solvable equations had been mentioned to him by his formal thesis advisor Dirichlet already around 1843, as an easily accessible domain of research – see the letter from Kronecker to Dirichlet dated 6 May 1853, reproduced in Appendix I below.

> From this one sees the true nature of the solvable equations, which it is impossible to glean from Galois's investigations. For Galois only takes up the task to find "the condition of solvability", whereas Abel also takes into account the other one, "to find all solvable equations."

At another occasion he speaks of the "clear insight into the nature of the roots of solvable equations with integer coefficients" which Abel's and his own results afford.[12] And in order to forestall any possible confusion about what, for Kronecker, constituted true insight into the nature and essence of mathematical objects, let us add here the quintessential quote from a letter of the older Kronecker to Cantor, dated 21 August 1884[13]:

> True scientific value I can recognize – in the domain of mathematics – only in concrete mathematical truths, or to put it more pointedly: 'only in mathematical formulae'. Only these are eternal, as the history of mathematics shows. The various theories about the foundations of mathematics (like Lagrange's) have been swept away by time, but Lagrange's resolvent has remained.

2.2 Statements. Let us now look at the actual results that Kronecker held in such high esteem. To Abel's statement recalled in 1.5. above – which had been proved by Malmsten in 1847[14] – Kronecker had two immediate points to make.

First, on a purely foundational note, Kronecker grew increasingly conscious of rationality questions. In 1853, his standard way of dealing with them was to refer to what we would call elements of the ground-field, as "rational functions of $A, B, C, \ldots$," where $A, B, C, \ldots$ indicate generators of the "ground-field" over the rationals. Later on in his life, he would use the terminology of "rationality domain" to describe the groundfield that all the coefficients of the original equations have to belong to, and over which the equation has to be irreducible. This notion resembles in many ways our notion of field. It is adumbrated as of 1853 in Kronecker's letters to Dirichlet.[15]

---

12 [42, p. 31]: *Die vorstehenden Bemerkungen gewähren allerdings schon eine klare Einsicht in die Natur der Wurzeln ganzzahliger auflösbarer Gleichungen μten Grades* ... – Since such remarks, with reference to explicit formulae rather than a conceptual theory of algebraic equations, may strike us as unusual today, let us stress that the young Kronecker was in good company – see for instance [30, p. 258]: *Die bisher noch nirgends angegebene wahre Form der Wurzeln der Gleichung $x^p = 1$ ist die folgende* ...

13 [12, p. 196]: *Einen wahren wissenschaftlichen Werth erkenne ich – auf dem Felde der Mathematik – nur in concreten mathematischen Wahrheiten, oder schärfer ausgedrückt, 'nur in mathematischen Formeln'. Diese allein sind, wie die Geschichte der Mathematik zeigt, das Unvergängliche. Die verschiedenen Theorien für die Grundlagen der Mathematik (so die von Lagrange) sind von der Zeit weggeweht, aber die Lagrangesche Resolvente ist geblieben!*

14 See [59]. Malmsten's proofs apparently left something to be desired for Kronecker: see his letter to Dirchlet of 30 January 1853 in Appendix I below, and [41, p. 5.] We do not speculate what the gaps were that he would have liked to see filled.

15 Already in the first letter reproduced in Appendix I below, there is an allusion to the notion of irreducibility in this context. This is taken up explicitly in Kronecker's letter to Dirichlet

Secondly, concerning Abel's result itself, Kronecker first added the "simple remark"[16] that the equation of degree $\mu - 1$ satisfied by the R_i's had to be cyclic over the 'ground field', i.e., "not only the symmetric functions of $R_1, R_2, \ldots$, but – if they are taken in a suitable order – also their cyclic functions must be rational functions of the $A, B, C, \ldots$."[17] Furthermore, Kronecker gave a more precise form that all R_i occurring as above must have:[18]

$$\text{(III)} \qquad R_\kappa = F(r_\kappa)^\mu \cdot r_\kappa^{\gamma_{-1}} \cdot r_{\kappa+1}^{\gamma_{-2}} \cdot r_{\kappa+2}^{\gamma_{-3}} \cdot \cdots \cdot r_{\kappa+\mu-2}$$

where $r_1, \ldots, r_{\mu-1}$ are the roots of a cyclic equation of degree $\mu - 1$ over the same rationality domain (cyclically permuted in this order), F is a rational function with coefficients in the same rationality domain, and for an integer m, γ_m is the smallest non-negative remainder of g^m modulo μ, for a fixed primitive root g modulo μ.

Finally, Kronecker claimed that if the roots of an equation of degree μ arise in the way just shown from a cyclic equation of degree $\mu - 1$, then the first equation is solvable. He corrected this last claim slightly in 1854, excluding trivial exceptions – not in print, but in a letter to Dirichlet.[19]

2.3 The Argument. There is only one proof-like passage in Kronecker's 1853 paper (and the letters do not provide much more in terms of explicit arguments).[20]

---

of 31 March 1854 reproduced in Appendix I, and the connection to the notion of rationality domain is made in [64, p. 412], note *). – But as far as we know, neither in the 1850's nor later in his life was Kronecker prepared to accept fields of finite characteristic as rationality domains when treating algebraic equations. Notwithstanding the use of 'finite fields' via higher congruences from Galois onwards (even for indexing permutations on p^n letters), the first to have explicitly discussed finite fields as a special case of the axiomatic notion of field seems to have been Heinrich Weber, as late as 1893 – see [68].

16 See the letter to Dirichlet of 30 January 1853 in Appendix I.

17 Kronecker in 1853 still used the term *Abelian* here instead of *cyclic*. When Kronecker adopted the term *Abelian* in the sense which is still current today, in the 1870's, he switched to calling cyclic equations *simply Abelian*.

18 This and the following results were all announced to Dirichlet in the letters reproduced in Appendix I below (where the first letter only deals with equations with rational coefficients), and published in [41] after Dirichlet's answer: [64, p. 409–410]. For ease of reference, we reproduce here and in the sequel, the labeling of the formulae used in [41].

19 See Kronecker's letter to Dirichlet of 31 March 1854 reproduced in Appendix I, point 1 towards the end.

20 See [41, p. 9–10]. – A sad, recurring theme of Kronecker's letters to Dirichlet from this time are his health problems. He often referred to them to explain why he was unable to write up a more elaborate treatment of his results. One may, of course, also suspect that his algebraic language and notation was simply not malleable enough to be adequate for his ideas. In his letter to Dirichlet dated 31 March 1854 reproduced below, Kronecker writes: "For the purpose of publication, I will still have to improve on my methods quite a lot – and I am actually very much intimidated by the difficulty of exposition of these matters." Kronecker himself recalled in 1886 that the algebraic terminology customary in 1861 was insufficient for an adequate treatment of the material presented in [46], which we will discuss in Sects. 3 and 4 – see the introductory remarks of [49].

It concerns the following situation: Let n be a prime number, α a primitive n-th root of unity, and write the Lagrange resolvent

$$(\alpha, z) = z_0 + z_1\alpha + z_2\alpha^2 + \cdots + z_{n-1}\alpha^{n-1},$$

with respect to the roots z_κ, for $\kappa \in \mathbf{Z}/n\mathbf{Z}$, of a cyclic equation of degree n with coefficients in (and irreducible over) a given rationality domain $\mathfrak{R}$.[21] Kronecker first notes the following identity (today's 'character orthogonality relations')

$$\text{(V)} \qquad n \cdot z_\kappa = (1, z) + \alpha^{-\kappa}(\alpha, z) + \alpha^{-2\kappa}(\alpha^2, z) + \cdots + \alpha^{-(n-1)\kappa}(\alpha^{n-1}, z).$$

Furthermore, let g be a primitive root modulo n. Since both (α^g, z) and $(\alpha, z)^g$ get multiplied by α^g if one applies the generating cyclic substitution $z_k \mapsto z_{k+1}$, the quotient $(\alpha, z)^g \cdot (\alpha^g, z)^{-1}$ belongs to $\mathfrak{R}(\alpha)$; Kronecker calls it $f(\alpha)$. Thus, for all $j = 0, \ldots, n-2$:

$$(\alpha^{g^j}, z)^g = (\alpha^{g^{j+1}}, z) \cdot f(\alpha^{g^j}).$$

Taking the j-th equation to the power g^{n-2-j}, and multiplying all of them together makes the product telescope, and yields:

$$\text{(VII)} \qquad (\alpha, z)^{g^{n-1}-1} = f(\alpha)^{g^{n-2}} \cdot f(\alpha^g)^{g^{n-3}} \cdot \cdots \cdot f(\alpha^{g^{n-2}}).$$

This already resembles formula (III) above. But Kronecker preferred to play with it some more in order to come close to a formula which he knew from Kummer, and which is familiar nowadays in the arithmetic of cyclotomic fields under the name of Stickelberger's theorem. – Indeed, in (VII) we can substitute any other n-th root of unity instead of α. On the other hand, we can also choose g carefully: Kronecker picked g in such a way that $g^{n-1} - 1 = m \cdot n$, with *m prime to n*. Heinrich Weber, in his digest of Kronecker's theory at the end of volume I of his *Lehrbuch der Algebra*, went a little further and chose g such that $m \equiv 1 \pmod{n}$.[22]

Anyway, taking n-th roots, Kronecker wound up with the following identity

$$\text{(VIII)} \quad (\alpha^m, z) = F(\alpha^m) \left\{ f(\alpha^m) \cdot f(\alpha^{2m})^{\frac{1}{2}} \cdot f(\alpha^{3m})^{\frac{1}{3}} \cdot \cdots \cdot f(\alpha^{(n-1)m})^{\frac{1}{n-1}} \right\}^{\frac{1}{n}}$$

for suitable $f(\alpha), F(\alpha) \in \mathfrak{R}(\alpha)$, where one has to interpret an exponent "$\frac{1}{j}$" inside the big curly bracket as the inverse of j in $(\mathbf{Z}/n\mathbf{Z})^*$.

This means in today's terminology, that the Lagrange resolvent (α, z), which belongs to the compositum $\mathfrak{R}(\alpha; z_k)$ of two cyclic extensions, is essentially the n-th root of an element of the form β^Θ, where $\beta \in \mathfrak{R}(\alpha)$ and

$$\Theta = \sum_{j \in (\mathbf{Z}/n\mathbf{Z})^*} j^{-1} \cdot \sigma_j = \sum_{j \in (\mathbf{Z}/n\mathbf{Z})^*} j \cdot \sigma_j^{-1} \in \mathbf{Z}[\mathrm{Gal}(\mathfrak{R}(\alpha)/\mathfrak{R})].$$

Here, σ_j denotes the automorphism $\alpha \mapsto \alpha^j$.

---

21 As mentioned above, this last notation does not occur in Kronecker's writings of the 1850's. We add it here for our convenience.

22 See [69, pp. 686–691]. The corresponding chapter is called *Wurzeln metacyclischer Gleichungen*; Weber (generalizing a usage of the later Kronecker) had tried in vain to introduce the term "metacyclic group" for 'solvable group' – see [69, p. 647, footnote].

Kummer had published such an identity in 1847 in the following, slightly different setting (with different notations): Let n, α be as before, and let $p = \ell \cdot n + 1$ be a prime, and z a primitive p-th root of unity. Choose a primitive root h modulo p, and consider the Gaussian period of length ℓ

$$\eta = z + z^{h^n} + z^{h^{2n}} + \cdot + z^{h^{(\ell-1)n}},$$

i.e., the trace of z down to the unique degree n subextension $K/\mathbf{Q}$ of $\mathbf{Q}(z)/\mathbf{Q}$. Then, for a suitable identification $k\colon \operatorname{Gal}(K/\mathbf{Q}) \longrightarrow \mathbf{Z}/n\mathbf{Z}$, the element that Kummer looked at may be written:

$$(\alpha, z) = \sum_{k=0}^{p-2} z^{h^k} \cdot \alpha^k = \sum_{\sigma \in \operatorname{Gal}(K/\mathbf{Q})} \alpha^{k(\sigma)} \cdot \eta^\sigma.$$

Then again $(\alpha, z)^n \in \mathbf{Q}(\alpha)$, and Kummer factored this element into ideal prime factors of the field of n-th roots of unity like this :

(2.3.1) $$(\alpha, z)^n = \pm\alpha^s \cdot f(\alpha)^\Theta,$$

for a suitable (ideal) factor $f(\alpha)$ of p. – Reading Kummer's argument closely, one realizes that his identity (2.3.1) between a product of 'ideal prime numbers' with roots of unity on the one hand, and an 'actual complex number', i.e., an element of $\mathbf{Q}(\alpha)$, on the other, splits up into two different statements in our modern terminology: (i) For a suitable prime ideal $\mathfrak{p} \subset \mathbf{Z}[\alpha]$ above p, the principal ideal $(\alpha, z)^n \cdot \mathbf{Z}[\alpha]$ equals $\mathfrak{p}^\Theta$; (ii) $(\alpha, z)^n$ is a generator of $\mathfrak{p}^\Theta$ which is distinguished by the fact that the product of it with its complex conjugate is the n-th power of a rational number, in fact $(\alpha, z)^n \cdot (\alpha^{-1}, z)^n = p^n$. This normalizes it up to a unit in the (-1)-part of the unit group under complex conjugation, i.e., up to a root of unity in $\mathbf{Z}[\alpha]$.[23]

That it was this result of Kummer's which inspired Kronecker in his study of solvable equations is also clear from the fact that, when he first announced his findings to Dirichlet, he still thought he had discovered the necessity to actually use Kummer's *ideal* prime factors in the theory of algebraic equations; but he soon corrected himself. Kummer apparently never believed in such a necessity.[24]

2.4 The "Theorem of Kronecker and Weber." In his 1853 paper, Kronecker refers to Kummer's result by comparing it to his formula (VIII):[25]

---

[23] See [54, §11], in particular p. 362–363. Kummer also records in a footnote a remark made by Eisenstein which in principle fixes the unit factor modulo n. – For a recent treatment and generalization, see for instance [56, Chap. 1, §2].

[24] See the letter to Dirichlet of 31 January 1953 in Appendix I. The claim is taken back in the subsequent letter of 2 April 1953. The same must have happened in Kronecker's letters to Kummer (which are not extant today), as can be gathered from Kummer's letter to Kronecker (12 March 1853) [53, p. 92]: *Daß die idealen Primzahlen für eine algebraische Untersuchung nicht wesentlich, sondern allenfalls Erleichterungs- und Hilfsmittel sein können, war mir eigentlich von vornherein klar.*

[25] See [41, p. 10]: *Das bei Weitem größte Interesse aber gewährt die Vergleichung des Ausdrucks (VIII) (unter der Annahme, daß A, B, C, etc. ganze Zahlen seien) mit seinem*

> The greatest interest by far lies in the comparison of the expression (VIII) (under the assumption that A, B, C, etc. are rational integers) with the corresponding expression for certain special Abelian equations which occur in the theory of the division of the circle; to wit, with the very important expression for (α, x) given by Kummer .... For this comparison yields the remarkable result, which is not only valid in the case of prime degree: "that the root of any abelian equation with rational integer coefficients can be written as a rational function in roots of unity."

Thus, remembering that 'abelian' meant 'cyclic' for him at the time, Kronecker seems to claim the famous 'Kronecker–Weber-Theorem' for cyclic extensions of **Q**. In fact, this is not quite what he claimed to have proved then, because he acknowledged having trouble with the powers of 2 divisible by 8.[26] But even for cyclic equations of odd prime power degree, no complete argument appears in the letters or the paper. Let us try to reconstruct as far as possible Kronecker's 'comparison of (VIII) to Kummer's decomposition' in the case for which (VIII) was established in [41], i.e., for cyclic equations of prime degree n.

Thus, starting from a cyclic equation over **Q** of prime degree n, and having applied to it the above reasoning all the way to formula (VIII), we would now like to understand how Kronecker may have gone on to deduce the

Proposition. *Let n be a prime and α a primitive n-th root of unity. If $\beta \in \mathbf{Q}(\alpha)$, then $(\beta^{\Theta})^{\frac{1}{n}}$ lies in a field which can be generated by roots of unity.*

For the proof, decompose β into ideal prime factors:

$$(\beta) = (1-\alpha)^a \cdot \mathfrak{b} \cdot \mathfrak{c}$$

where $\mathfrak{b}$ groups together the ideal prime factors of degree 1 occurring in β, except $(1-\alpha)$, and $\mathfrak{c}$ all those of degree > 1.[27] Now, $(1-\alpha)^{\Theta} = (1-\alpha)^{n\frac{(n-1)}{2}}$ up to units. Furthermore, every ideal prime $\mathfrak{q}$ of degree $f > 1$ is invariant (up to units) under $\alpha \mapsto \alpha^{g^{ei}}$ for $i = 1, \dots, f$, where $n = e \cdot f$ (and g as before a primitive root modulo n). Since $1 + g^e + g^{2e} \dots + g^{(f-1)e} \equiv 0 \pmod{n}$, it follows that $\mathfrak{q}^{\Theta}$ is the n-th power of an ideal complex number. This was already explicitly deduced by Kummer in his

---

entsprechenden Ausdrucke für gewisse spezielle in der Theorie der Kreistheilung vorkommende Abelsche Gleichungen; nämlich mit der überaus wichtigen, von Kummer ... angegebenen Form für (α, x). Diese Vergleichung ergiebt nämlich das bemerkenswerthe und nicht bloß für den Fall eines Primzahlgrades sondern ganz allgemein geltende Resultat: "daß die Wurzel jeder Abelschen Gleichung mit ganzzahligen Coëfficienten als rationale Function von Wurzeln der Einheit dargestellt werden kann."

[26] See [41, p. 8]; cf. the letters to Dirichlet reproduced in Appendix I. This problem is discussed at length in a letter of Kronecker to Dirichlet dated 5 January 1954 which deserves a separate study and is not reproduced in the Appendix below.

[27] The modern reader trained in Dedekind's way may think here primarily of decomposing the principal ideal $\beta \cdot \mathbf{Z}[\alpha]$ into prime ideals of that ring, as our notation suggests. (We do not speculate on the notation which Kronecker would have used at this point.) But we have to keep track of the points where Kummer's setting gives more information than an identity of ideals, like in 2.3.1 above – see further on in the proof.

1847 paper,[28] but he did not stop there. In fact, what Kummer pointed out was that Θ kills the whole class group of $\mathbf{Q}(\alpha)$, because every ideal class can be represented by a product of ideal prime numbers of degree 1. Thus, for $\mathfrak{q}$ as above, $\mathfrak{q}^{\Theta}$ is an "actual complex number" in Kummer's terminology (i.e., in modern language, the ideal of elements of $\mathbf{Z}[\alpha]$ divisible by $\mathfrak{q}^{\Theta}$ is principal). But even more holds for these primes of degree > 1: $\mathfrak{q}^{\Theta}$, which Kummer writes $\mathbf{P}(\alpha)$, is in fact (up to units) the n-th power of an "actual complex number."[29]

Using this, it follows from the above decomposition of (β) that there exists $\gamma \in \mathbf{Q}(\alpha)$ such that $(\beta^{\Theta} \cdot \gamma^{-n}) = \mathfrak{b}^{\Theta}$ is only divisible by ideal prime factors of degree one, i.e., by $\mathfrak{p}$'s dividing rational primes $p \equiv 1 \pmod{n}$. And Kummer's decomposition (2.3.1) discussed above provides a generator $\xi = \prod_j (\alpha, z_j)^n$ of the ideal $\mathfrak{b}^{\Theta} = (\beta^{\Theta} \cdot \gamma^{-n})$, such that $\xi \cdot \overline{\xi}$ is the n-th power of a rational number. But $\beta^{\Theta} \cdot \overline{\beta^{\Theta}} = N_{\mathbf{Q}(\alpha)/\mathbf{Q}}(\beta)^n$ is also the n-th power of a rational number. Now write the unit

$$\frac{\beta^{\Theta}}{\prod_j (\alpha, z_j)^n \cdot \gamma^n} = \pm\alpha^{?} \cdot u$$

as a product of a root of unity with a real unit $u \in \mathbf{Z}[\alpha+\alpha^{-1}]$. Then u^2, and therefore u itself, must be an n-th power in $\mathbf{Z}[\alpha]$, and the proposition follows.

It seems plausible to assume that Kronecker knew already in 1853 that every unit of $\mathbf{Z}[\alpha]$ is a product of a real unit by a root of 1. This statement is equivalent to the fact that, for every unit ε of $\mathbf{Z}[\alpha]$, the quotient $\varepsilon/\overline{\varepsilon}$ is a root of unity (because all its absolute values are 1). This latter statement was used in Kummer's 1847 proof of Fermat's Last Theorem for regular prime exponents.[30] Kronecker generalized the first statement to arbitrary cyclotomic fields in [44], acknowledging without precise reference that Kummer had been the first to establish this for roots of unity of prime order. In his general proof, Kronecker used the theorem that a totally imaginary algebraic integer all of whose absolute values are 1 is a root of unity, which he established in [43]. Let us note in passing that these two short papers of 1857 are among the not so frequent publications of Kronecker's that read like ordinary mathematical papers, with statements *and* explicit proofs.

In conclusion, even though we cannot be sure of all the details of Kronecker's way to derive the prime order cyclic case of the "Kronecker–Weber Theorem" from what he could find in Kummer's work, it seems certain that what we would consider

[28] See [54, §12, pp. 364–367]. Note that Kronecker, in [41] and in his letters to Dirichlet, refers to the immediately preceding §11 of this same paper of Kummer's, and there can be no doubt that he was familiar with the last pages of the paper as well.

[29] See [54, p. 365–366]. – Again, Kummer deduced this from his decomposition of the Lagrange resolvents into Jacobi sums $\psi(\alpha)$.

[30] According to H. M. Edwards [13, p. 169], the fact that all $\varepsilon/\overline{\varepsilon}$ are roots of unity "would have been common knowledge in 1847 to Kummer, Kronecker, Dirichlet, and others interested in the structure of the units of cyclotomic integers". – Cf. [60] for a mostly mathematical, but very much historically inspired discussion of the general case of the "Kronecker–Weber Theorem."

as one of the beginnings of class field theory was for Kronecker an outgrowth, with a specific number theoretic twist, of an investigation of the general form of solvable algebraic equations in the spirit of Abel. When Kronecker told Liouville in Paris that he had proven all abelian equations over the rationals to be essentially cyclotomic, Liouville replied he had always thought so, but as far as he knew it had not been proved previously.[31]

However, Kronecker did not stop with the 'Kronecker–Weber Theorem'. In [41], and then again in [42], this statement is immediately followed by its analogue for abelian extensions of $\mathbf{Q}(i)$, and even further generalizations are announced. This stirs above all our amazement at this visionary mathematician. We do not see any evidence that would permit even a beginning of a reconstruction of Kronecker's precise thoughts on his *liebsten Jugendtraum* in the 1850s.[32]

3 1858–1861: Elliptic Functions and the Quintic Equation

Kronecker's 1861 paper on the quintic equation [46] was at the crossroads of several developments of the preceding decades. In this section we briefly review various solutions of the quintic equation via elliptic functions proposed between 1858 and 1861, leaving the new turn that Kronecker took in 1861 to the next section.

3.1. In 1858, Charles Hermite succeeded in solving the general quintic equation via elliptic functions. More precisely, he treated the equation

$$x^5 - x - a = 0 \tag{3.1.1}$$

to which the general case can be reduced by extracting a suitable square and a cube root.[33] Hermite started from an elliptic integral in Legendre's normal form $\int \frac{d\phi}{\sqrt{1-\kappa^2 sin^2\phi}}$, with modulus κ^2, and considered the eighth root of the modulus: $\sqrt[4]{\kappa} = \phi(\omega) = u$, as a function of the period ratio ω of the elliptic function which is the inverse function of the integral.[34] Then κ, and hence u, are invariant under the operation of $\mathrm{SL}_2(\mathbf{Z})$ on ω. Hermite's principal tool was the modular equation for transformations of order five.

The theory of *transformations of elliptic integrals and functions* was the historic predecessor, in the nineteenth century, of today's theory of isogenies between elliptic

[31] See letter Kronecker to Dirichlet of 15 June 1853 in Appendix I below.

[32] See also [63].

[33] This had been published in the 1786 paper *Meletemata quaedam mathematica circa transformationem aequationum algebraicarum* of the Swedish amateur mathematician Erland Samuel Bring. The paper was successfully presented as a dissertation at the University of Lund by the son of a friend of Bring's. See [20, p. 5–7].

[34] See [24]. A textbook treatment of what we are about to describe, which is more readable for modern mathematicians, but does point out some historical origins of the ideas involved is [17, Chapter 6]. Note that u can be effectively computed from ω by a fast converging power series in $q = e^{i\pi\omega}$.

curves, and of modular curves.[35] It had been Jacobi who discovered transformations in all odd orders n, after having read [4]. In Jacobi's notation, they have the general form

$$(3.1.2)\qquad \frac{dy}{\sqrt{(1-y^2)(1-\lambda^2y^2)}} = \frac{dx}{M\sqrt{(1-x^2)(1-k^2x^2)}}\,, \qquad y = \frac{U(x)}{V(x)}\,,$$

where M is called the multiplier of the transformation, and $n = \deg(U) = \deg(V)+1$ its order. He presented a first systematic treatment of transformations in his treatise *Fundamenta nova* [29]. There he also introduced the modular equations, whose roots are precisely the moduli λ of all elliptic integrals obtained from a given one by transformations of a given order n.[36] Coming back to Hermite's setting, we are talking about the orbit of $v := \phi(n\omega)$ under the operation of $\mathrm{SL}_2(\mathbf{Z})$. In the case of transformations of order 5, the roots of the corresponding modular equation are $v_\infty = \phi(5\omega)$, and $v_i = \phi(\frac{\omega+16i}{5})$ for $i = 0, \ldots, 4$, and the equation reads:

$$u^6 - v^6 + 5u^2v^2\left(u^2 - v^2\right) + 4uv\left(1 - u^4v^4\right) = 0\,.$$

Évariste Galois, in his letter to Auguste Chevalier the night before his death in 1832, had recorded that the group of the modular equation of prime order n is (in modern notation) $\mathrm{PGL}_2(\mathbf{F}_n)$. In the special case $n = 5$, the Galois group over $k(u)$ is therefore isomorphic to the alternating group $\mathcal{A}_5$. Taking an element of $k(v_\infty, v_0, \ldots, v_4)$ which is invariant under a subgroup of index 5, we find a minimal polynomial of degree 5. In the terminology of the time, this is a *resolvent of degree* 5 of the modular equation.

Hermite chose the element $\Phi(\omega) = (v_\infty - v_0)(v_1 - v_4)(v_2 - v_3)$, and found the resolvent $x^5 - x - \psi(u) = 0$. Furthermore, he succeeded in determining u such that $\psi(u) = a$, for the given constant term a of (3.1.1). In this way, he could compute the roots of the given quintic equation via the formula for Φ, resp. for its conjugates – see [24, 10].

3.2. As early as 1828, Jacobi [28, p. 261] had considered the so-called *multiplier equation (Multiplikatorgleichung)* for transformations of order n. Let us take n to be prime. Then expressing M in (3.1.2) by k and λ and taking into account that λ is a zero of the modular equation for transformations of order n, one obtains an equation for M of degree $n+1$ (the same as for the modular equation) whose

[35] The detailed history of this theory in the nineteenth century remains to be written. The source literature is enormous. Houzel's report [26] is best as a guide to the most important sources.

[36] In [29, §24 and §29], Jacobi uses the verb *nuncupare* = to solemnly give name to (or appoint), instead of the simpler *vocare* = to call, when introducing the terminology 'modular equation'. The same verb *nuncupare* occurs before, at the end of §19, when he introduces his *principle of double periodicity* for the elliptic functions. On the other hand, when the *multiplier M* of a transformation receives its name, it is simply "called" (... *quem vocabimus*) this at the end of §21.

coefficients are rational functions in k^2, i.e., in the modulus.[37] Jacobi realized that the $n+1$ roots $z_\infty, z_0, \ldots, z_{n-1}$ of this multiplier equation for transformations of order n, with n prime, can be obtained from only $\frac{n+1}{2}$ other quantities.

The corresponding formulæ in the case of a transformation of order 5 are: $\sqrt{z_\infty} = \sqrt{5}A_0$, $\sqrt{z_i} = A_0 + \epsilon^i A_1 + \epsilon^{4i} A_2$, with $i = 0, \ldots, 4$, and ϵ a primitive fifth root of unity. In 1858, Francesco Brioschi first proved Jacobi's claim, and then studied the more general equations whose roots are given by the above formulæ, relative to arbitrary coefficients A_0, A_1, A_2. He called these equations *generalized Jacobian equations of degree six*. In particular he computed the coefficients of the equations that the z_i satisfy, and found:

$$(3.2.1)\quad (z-A)^6 - 4A(z-A)^5 + 10B(z-A)^3 - C(z-A) + (5B^2 - AC) = 0,$$

where $A = A_0^2 + A_1 A_2$, but B and C are homogeneous polynomials of degree 6, resp. 10 in the A_i. The Galois group of this generalized Jacobian equation of degree 6 is again isomorphic to $\mathcal{A}_5$, just as for the multiplier equation. Imitating Hermite, Brioschi first computed a resolvent of degree 5 for $f = (z_\infty - z_0)(z_1 - z_2)(z_3 - z_4)$, but then realized that even the resolvent for $\sqrt{f}$ had coefficients rational in the A_i, and that it had the special form $x^5 + ax^3 + bx + c = 0$. The original case of the multiplier equation itself is the one where $B = 0$, and in this case one finds $a = 0$.[38]

3.3. In a letter to Hermite dated 6 June 1858 [45], Kronecker showed that it is also possible to resolve the quintic equation without first reducing it to the form (3.1.1). For this he wrote down a resolvent of the general quintic equation which he could solve via elliptic functions. He only gave the relevant formulæ and explained why his procedure did yield the roots of the quintic equation.

Kronecker considered the following function of those roots:

$$f(\nu, x_0, \ldots, x_4) = \sum_{m=0}^{m=4}\sum_{n=1}^{n=4} \left(x_m x_{m+n}^2 x_{m+2n}^2 + \nu \cdot x_m^3 x_{m+n} x_{m+2n}\right) \cdot \sin\frac{2n\pi}{5},$$

along with the following conjugates under permutations from $\mathcal{A}_5$:

$$f_r = f(\nu, x_r, x_{r+3}, x_{r+4}, x_{r+1}, x_{r+2})$$

for $r = 0, \ldots, 4$. Since f is invariant under the permutation $x_i \mapsto x_{i+1}$, one can determine the x_i by radicals once f has been obtained. Kronecker now chooses the parameter ν in such a way that one has $f^2 + f_0^2 + \cdots + f_4^2 = 0$. *This involves solving a quadratic equation with coefficients that are rationally known, i.e. in modern parlance, which belong to the fixed field* $\mathfrak{R}(x_0, \ldots, x_4)^{\mathcal{A}_5}$. In particular, it will happen in general that, in this solution of our quintic equation, a square root is

[37] Fricke remarks [17, p. 508] that the multiplier equation is the first historical example of a transformation equation for a modular form, as opposed to a modular function.

[38] From this one sees that one can also use the multiplier equation directly to resolve a quintic equation (3.1.1).

employed which does not belong to the splitting field $\mathfrak{R}(x_0, \ldots, x_4)$. Such a square-root was later called an *accessory irrationality* by Klein. Its introduction violates what Kronecker would later call *Abel's postulate* (see 1.3 above), and would thus be at the heart of the difference of opinions between Kronecker and Klein – see Sect. 4 below.

Then Kronecker finds that the f_r are roots of an equation

$$f^{12} - 10\phi f^6 + 5\psi^2 = \psi f^2,$$

for ϕ and ψ which are rationally known. And he furthermore gave the cubic equation from which one has to determine $\kappa^2(1-\kappa^2)$, as well as the function of the period ratio ω of the elliptic integral with modulus κ^2 which provide the f_i.

In [46], Kronecker commented that his point of departure had been Jacobi's relations between the roots of the multiplier equation (see 3.2.), that the other conjugate values of f differ from the given ones only by signs, and that the $f, f_0, \ldots, f_4$ satisfy three more linear relations. This explains why the f_i correspond to the $\sqrt{z_i}$ of the generalized Jacobian equation of degree 6 (3.2.1), and why the normalization $\sum f_i^2 = 0$ leads precisely to the simplest case $A = 0$.

Since Kronecker himself dates his investigations to the year 1856, and since Brioschi published his formula for the generalized Jacobian equation of degree 6 only in June 1858, one may assume that Kronecker had the idea to generalize Jacobi's relations independently of Brioschi.[39]

4 1861–1921: Leopold Kronecker vs. Felix Klein

In the last section, we have described the positive contributions to the solution of the general quintic equation via elliptic functions. We now turn to the, so to say, negative result that Kronecker announced in 1861, and which apparently gave him a new vantage point when looking at the solutions we have seen in Sect. 3.

4.1 Kronecker's Strengthening of Abel's Impossibility Theorem. Abel had proved in [2] that the general quintic equation cannot be solved by radicals. In 1861, Kronecker claimed a strengthening of Abel's result:

Theorem 1 (Kronecker 1861). The general quintic equation does not admit one-parameter resolvents.

Since the terminology may not be self-explanatory, let us first explain in modern language what this means: Given

$$f(x) = \prod_{i=0}^{4}(x - x_i) = \sum_{k=0}^{5} a_k x^k \in \mathfrak{R}(a_0, \ldots, a_5)[x]$$

[39] The genesis of the above function f and proofs for the claims he had made in [46] were extensively explained in Kronecker's lecture course [52], starting on page 114. – For more or other details cf. [37, p. 154f] and [62].

of discriminant Δ, with a_k algebraically independent over $\mathfrak{R}$, the fixed field K of $\mathcal{A}_5$ inside the splitting field $L = \mathfrak{R}(x_0, \ldots, x_4)$ is $K = \mathfrak{R}(a_0, \ldots, a_5, \sqrt{\Delta})$. Given any $y \in L \setminus K$, we call its minimal polynomial $P(X) = \sum_{j=0}^{N} b_j X^j$ over K a *resolvent* for f. Such a resolvent is called a *one-parameter resolvent*, if there exist rational functions $g_j(t) \in \mathfrak{R}(t)$, for $j = 0, \ldots, N$, and $z \in K$, such that $b_j = g_j(z)$ for all j.

Comparing with the solutions described in 3.3, the resolvents used there are all special cases of the generalized Jacobi equation (3.2.1). This equation apparently depends on three parameters A, B, C. But these can actually be reduced to the two parameters $a = \frac{B}{A^3}$, $b = \frac{C}{A^5}$. For details we refer to Felix Klein's lucid exposition of these matters in [37, §II.1, esp. pp. 156ff].

Realizing this, Kronecker – so he told his students later – tried to push further to show that there was also a one-parameter resolvent, until he finally envisaged the possibility that such a resolvent could not exist, which led him –so he told his students later – to a proof of the theorem.[40] It is in this proof that Kronecker needed what he would later call "Abel's postulate" – see 1.3 above. In fact, it is not hard to show that the resolution 3.3 of the quintic proposed by Kronecker himself does yield a one-parameter resolvent. Indeed, his determination of the f_r such that $f^2 + f_0^2 + \cdots + f_4^2 = 0$, has the effect that the parameter A of (3.2.1) is zero; and the remaining B and C can then easily be reduced to a single parameter. But Kronecker's one-parameter resolvent of 3.3 was of course obtained by paying the price of adjoining a square-root to the ground field which will not in general belong to the splitting field, in other words, the price paid was to violate "Abel's postulate."

4.2 Kronecker's Proof of his Theorem. No proof by Kronecker of his Theorem of 4.1 ever appeared in print. As mentioned earlier, Kronecker later blamed the lack of proof in his publications of 1861 on the inadequacy, for his "purely arithmetical treatment of algebraic quantities", of the "algebraic terminology used at the time, which had developed from an analytic-geometric point of view".[41]

---

[40] See the notes of Kronecker's lecture course [52, p. 158]: … *Professor Kronecker hat die Mühe nicht gescheut, diese Gleichung wirklich aufzustellen, da er unbedingt annahm, eine solche Auflösung müsse existieren. Aber er konnte sie nie finden. Es dauerte lange, ehe er die Möglichkeit ins Auge fasste, dass es keine solche Auflösung gebe. Es giebt nun in der That keine, sogar, wenn man noch spezielle Beziehungen zwischen a, b, c zulässt. Der Beweis, dass es nicht geht, lässt sich mit den jetzigen Mitteln wohl direkt führen. Professor Kronecker hat ihn indirekt geführt, indem er davon ausging, dass, wenn es eine rationale Funktion von* x_0, x_1, x_2, x_3, x_4 *giebt, welche einer Gleichung 5.ten Grades mit einem Parameter genügt, dies gar nichts anderes heißt, als dass es rationale Funktionen von 5 Größen giebt, welche so beschaffen sind, dass zwischen je zwei conjugierten eine Gleichung besteht. Er fand dann, dass dies nicht der Fall ist.*

[41] See [49, p. 277]: *An einer Veröffentlichung durch den Druck hat mich namentlich die Schwierigkeit gehindert, meine bezüglichen Entwickelungen, welche von einer rein arithmetischen Behandlung der algebraischen Größen ausgingen, in der damals gebräuchlichen, aus analytisch-geometrischer Anschauungsweise hervorgegangenen algebraischen Terminologie auseinanderzusetzen.* – And he added that the situation was now different, in particular after the publication of his *Grundzüge*, i.e., [48].

Felix Klein [31, Vol. II, p. 503] reported after Kronecker's death that Kronecker showed him a personal manuscript from 1861 containing his original proof of the theorem when they met in Berlin on 23 March 1881. Klein copied parts of it by hand and found a gap towards the end of Kronecker's proof, although the basic idea was the same as Klein had used to prove the theorem in 1877.[42] A proof of Kronecker's Theorem was also placed *coronidis loco* at the very end of Klein's *Ikosaederbuch* [37]. Then, during the Winter term of 1885–86, Kronecker lectured on the subject in Berlin. Staeckel's notes [52] from this lecture course were in Kurt Hensel's possession before they were sold, with many other books from Hensel's library, to the Nazi *Reichsuniversität Straßburg* during World War II, which explains their present whereabouts.

In his 1885–86 lectures, Kronecker did not, as Klein had hoped: [38, p. 491], comment on his *Ikosaederbuch* [37], but only casually mentioned Klein's terminology to his students. He also claims that he had studied the question, for which equations all roots are linear functions of one of them, as early as 1861. Such equations he called "singular equations" in 1885–86, quoting several sources in the theory of second order linear homogeneous differential equations, but no reference to Klein. Contrary to Klein, he did not attach great importance to these singular equations. In one algebraic theorem, he almost forgot to exclude them.[43]

When Felix Klein prepared the commentaries to his Collected Papers in 1921, he asked Hensel to send him Staeckel's notes and Kronecker's papers. Hensel complied, but Klein failed to find the papers he had copied in 1881. We reproduce in Appendix II below the letter from Klein to Hensel with which Klein returned the material. It clearly expresses Klein's anger that Kronecker refuses to acknowledge any algebraic merit of his theory, and it contains the personal comment:

> But there is still the personal side: Kronecker has not brought himself to admit in his lecture course of 85/86 that it was me who first completed his proof of 1861. Rather, he creates the impression with his listeners that he has done everything himself, and

[42] See [38, p. 491], where Klein tells Kurt Hensel, and thereby the reader, that Kronecker at the time of his first proof *did not know the icosahedral substitutions, even though he had come so close to them.* (Klein's contribution [38] to the special issue of Crelle's Journal for the centenary of Dirichlet is stylistically a mixture between an ordinary scientific paper and a personal letter to Hensel. This unusual form corresponds well to the complexity of the issue between Klein and Kronecker.) In his 1921 comment on his 1905 paper, Klein gives a little more detail: [31, Vol. II, p. 503].

[43] See [52, p. 195]: [The problem of the one-parameter resolvents] *ist damit zurückgeführt auf die Frage nach den Gleichungen, für welche alle Wurzeln lineare Funktionen von einer von ihnen sind. Auf diese Frage ist Prof. Kronecker schon vor 25 Jahren gekommen. Sie ist später in anderer Beziehung vielfach untersucht worden (Schwarz, Serret).* And [52, p. 221]: *Dann, sagt Prof. Kronecker, gilt der Satz, dass, wenn es in einer Gattung eine Gleichung vom Range 1 giebt, es auch stets eine reine Gleichung in dieser Gattung giebt, das heißt aber, dass diese Gleichung vom Range 1 eine Abel'sche Gleichung ist.* This sentence carries a footnote where one reads in particular: ... *ausgenommen sind die dort erwähnten "singulären Gleichungen".*

> that only the terminology 'icosahedral equation' is due to me. Since the lectures are unpublished I avoid any mention of this in my commentaries; I do not want to stir up mud unnecessarily.[44]

Even though it is true that the lecture notes criticize any method of solving the quintic equation which violates "Abel's postulate", and therefore in particular Klein's whole approach via the icosahedral equation, one may also point out that Kronecker does not forbid analytic resolutions as such, he just tries to regulate the domain of algebra the way he sees it. For instance, when describing the method which we discussed in 3.3 above, we read in [52, p. 152]:[45]

> ... The discovery that this works has given a lot of pleasure at the time. But Professor Kronecker could not share this pleasure. It is sort of lottery. And the most essential point is that the degree 6 equation can only be reduced to one parameter via a square-root.

And after the passage about Abel's postulate which we quoted in footnote 2 above, Kronecker calls the resolution of the quintic via elliptic functions "algebraically worthless."[46]

The debate about how to continue after Kronecker's Theorem: whether to admit a mild but inevitable violation of "Abel's postulate" in order to develop fruitful methods of solving the quintic equation, or whether to abandon these approaches because they cannot be called "solving the equation" in the strictest sense, leaves us fairly indifferent today. At the same time, a modern algebraic geometer will naturally keep track of the 'accessory irrationalities' necessary for whatever constructions he

---

[44] The allusion to Kronecker only attributing the terminology 'icosahedral equation' to Klein refers to [52, p. 228–229]: *Außer den reinen Gleichungen gibt es also noch einige singuläre Gleichungen, deren Wurzeln alle lineare Funktionen einer sind. Das sind die Gleichungen der regelmäßigen Körper, wie Herr Klein sie nennt, die Beziehung zu diesen hat schon Herr Schwarz erkannt. Die singulären Gleichungen sind: 1) Die Ikosaedergleichung von Herrn Klein, das ist eine spezielle Galoissche Gleichung der Gleichungen 5. Grades bei Adjunktion der alternierenden Funktionen, bei der jede Wurzel eine lineare gebrochene Funktion jeder anderen ist. Der Grad dieser Gleichung ist 60. [...] Das sind die Ausnahmen. Für die Algebra liegt die Sache nun ungemein einfach. Diese Ausnahmefälle haben wir schon abgemacht. Es ist bereits gezeigt worden, dass diese Galoisschen Gleichungen als Gattungs-gleichungen nicht vom Range 1 sein können.*

[45] *Nun giebt uns die Natur eine Gleichung 6.ten Grades, welche denselben Affekt hat, und deren Koeffizienten auch nur von einem Parameter abhängen, nämlich die Modulargleichung 6-ten Grades. Die Frage ist daher: Lassen sich diese beiden Parameter identifizieren? Die Entdeckung, dass dies geht, hat seiner Zeit ungeheuer viel Freude gemacht. Prof. Kronecker konnte diese Freude nicht teilen. Es ist eine Art Zufallsspiel. Und dann ist das Wesentliche, dass die Gleichung 6.ten Grades nur durch ein Quadratwurzelzeichen auf einen Parameter gebracht werden kann.*

[46] [52, p. 164]: ... *Da sich nun zeigen läßt, daß die* λ, μ, ν *notwendig mit Quadratwurzeln in* x_0, x_1, x_2, x_3, x_4 *behaftet sind, ist die Auflösung durch elliptische Funktionen algebraisch wertlos.*

performs.[47] But between Kronecker and Klein, this divergence of appreciation is likely to have contributed to, if not caused Kronecker's very negative opinion of Felix Klein, and especially of his *Ikosaederbuch*. This negative opinion can not only be gathered from the lecture notes, but it shows clearly through the crushing remarks about Klein made in Kronecker's name by Helmholtz, about a month after Kronecker's death, when the succession to Kronecker and Weierstrass was discussed in Berlin.[48]

4.3 Klein's Perspective. In 1905, Felix Klein regretted the fact that the fundamental importance of his approach to the theory of equations had still not been fathomed by many:[49]

> These are not considerations which find their place next to other investigations about the resolution of the quintic equation, but they claim to represent the proper substance of those earlier investigations.

It is characteristic for Klein's theory of equations that the computations are guided by geometric interpretations. In other words, there is an interplay of geometry, invariant theory, and the theory of equations. We will try to highlight a few typical features of Klein's approach. For a re-interpretation in a modern algebro-geometric setting, we refer to [67].

Already in 1871, Klein had the idea (which was important to him) to view the permutations of the Galois group as "linear transformations of a continuous space" [32, p. 263]. This developed into the programme to approach algebraic equations as *Formenprobleme*, thus integrating the theory of algebraic equations into invariant theory – see [36, p. 396]. More precisely, suppose that, for a finite group $G \subset \mathrm{GL}_n(k)$, one has generators F_i of those homogeneous forms in $k[x_1, \ldots, x_n]$ which are invariant under G. And suppose also that the relations between them are known. Then the *Formenproblem* for G is to determine all possible values x_i from given values of the F_i which are compatible with the relations. For every isomorphism $G \longrightarrow H$, the solution of the *Formenproblem* for G can be reduced rationally to the *Formenproblem* for H, i.e., the solutions for G can be rationally expressed in terms of those for H.

The 'correct' treatment of an equation with Galois group G according to Klein therefore consists in finding an isomorphism $G \longrightarrow H \subset \mathrm{GL}_m(k)$, with minimal

[47] Cf. for instance J.-P. Serre's account of the 'accessory irrationalities' in [67].

[48] See [10, p. 306]: *von Helmholtz: Kronecker sprach sehr ab über Klein. Er betrachtete ihn als Faiseur.*

[49] See [38, p. 482]: *Es handelt sich nicht um Überlegungen, welche sich* neben *die früheren Untersuchungen über die Auflösung der Gleichungen fünften Grades stellen, sondern um solche, die den Anspruch erheben, den eigentlichen Kern dieser früheren Untersuchungen auszumachen.*

m, which reduces the *Formenproblem* for G to that of H, and reduces the number of parameters of the equation as far as possible.[50]

The coefficients of the Jacobian equation of degree six are precisely the invariant forms F_i for a group $\mathcal{A}_5 \cong I_3 \subset GL_3(k)$. (The isomorphism follows from Jacobi's relations.) In this sense, Kronecker and Brioschi "really" proposed this *Formenproblem*.[51]

Working on the Riemann sphere, Klein passed from the group of rotations of the icosahedron to a group $I \subset PGL_2(k)$ of fractional linear transformations $x \mapsto \frac{ax+b}{cx+d}$, which is again isomorphic to $\mathcal{A}_5$. The orbit equation of this operation is Klein's *icosahedral equation* $I(z) = X$. Klein viewed it as the natural generalization of pure equations $z^p = a$ and thus asked for a resolution of the quintic equation via an icosahedral equation.[52]

The following observation is essential for Klein: If $\lambda \in k(x_1, \dots, x_5)$ is transformed by the permutations of $\mathcal{A}_5$ via fractional linear transformations, and not invariant under any permutation, then a suitable quantity $\lambda_1 = \frac{c_1\lambda+c_2}{c_3\lambda+c_4}$ satisfies an icosahedral equation. – Klein solved the *principal quintic equation* $x^5+5ax^2+5bx+c = 0$ via an icosahedral equation, determining geometrically a λ satisfying an icosahedral equation. In fact, the quintuples of solutions of the principal equations lie in the projective space $\sum x_i = 0$, on the quadric $\sum x_i^2 = 0$. This quadric carries two generating families of lines which are transformed into themselves by the $\mathcal{A}_5$-operation. The parameters of these families therefore transform via fractional linear transformations, as wanted.[53]

4.4 Klein's Proof of Kronecker's Theorem. As mentioned in 4.1, Klein proved Kronecker's Theorem to the effect that the general quintic equation does not admit a one-parameter resolvent. His proof is by contradiction and proceeds in three steps which we shall sketch here very briefly.[54]

First, from the existence of a one-parameter resolvent Klein deduced that there is an element of the splitting field which transforms under $\mathcal{A}_5$ via fractional linear transformations, the group of these transformations being isomorphic to $\mathcal{A}_5$. This element is obtained from Lüroth's theorem as a primitive element θ of $k(y_1, \dots, y_k)$, if $\prod(X - y_i) \in k(z)[X]$ is the resolvent for y_1, with parameter $z \in K$.[55] θ transforms

[50] This requirement is Klein's reply to Kronecker who had found it remarkable in 1861 that a reduction of the number of parameters of an equation was possible according to its 'affect', i.e., its Galois group. See [46, p. 61].

[51] In order to check this match, Klein had to modify geometrically the forms initially computed via invariant theory – see [34, p. 353].

[52] For the determination of such orbit equations, see [33, p. 286f]. For the idea that this naturally generalizes the classical resolution via pure equations, see [38, p. 483].

[53] The chosen, normalized λ then generates the splitting field of the principal equation. This finally yields explicit formulæ for the roots of the principal equation.

[54] A detailed comparison with the proof that Kronecker presented in his 1885–86 lectures will be left to a later occasion.

[55] This is a precise translation of Klein's argument into modern notation – see [34, p. 379f]. Cf. also [70, pp. 470ff].

in a fractional linear way since $\sigma(\theta)$, for $\sigma \in \mathcal{A}_5$, also generates $k(y_1, \dots, y_k)$. The transformation group is isomorphic to $\mathcal{A}_5$ because $k(\theta)|K$ is Galois and therefore $k(\theta) = L$. It follows after suitable normalization that there is an element $\lambda \in L$ which satisfies an icosahedral equation.

In the second step, one deduces from unique factorization in $k[x_1, \dots, x_5]$ that numerator and denominator of the element θ found in the first step, would have to be transformed linearly and independently of each other: If $\theta = \frac{\phi}{\psi}$ is in reduced form, then the identity

$$\frac{\phi_1}{\psi_1} = \theta_1 = \sigma(\theta) = \frac{a\theta + b}{c\theta + d} = \frac{a\phi + b\psi}{c\phi + d\psi}$$

(up to a possible constant factor which can incorporated into the coefficients) implies $\phi_1 = a\phi + b\psi$, $\psi_1 = c\phi + d\psi$, and therefore the existence of a subgroup $I_2 \subset \mathrm{GL}_2(k)$ isomorphic to $\mathcal{A}_5$.

In the third step, one has to show that such a subgroup cannot exist. Klein's proof of this uses the fact that $\mathcal{A}_5$ contains a Klein's four group V_4, which already cannot be embedded into $\mathrm{GL}_2(k)$ – see [37, p. 46f]. Indeed, $\mathcal{A}_5$ can be generated by elements S and T of order 5, resp. 2, such that ST has order 3. Therefore, the embedding would have to be into $\mathrm{SL}_2(k)$. But the only element of order 2 in $\mathrm{SL}_2(k)$ is $A(z_1, z_2) = (-z_1, -z_2)$. But combined with an element of order 5, this will never produce an element of order 3.

4.5 Klein's "Accessory Irrationalities" and Modular Functions. As Klein insisted on using the icosahedral equation, he had to live with what he called "accessory irrationalities" *(akzessorische Irrationalitäten)*. On the other hand, after adjoining a suitable square-root, the general quintic equation can be transformed into a principal equation, so that Klein had a resolution of the general quintic. In the same vein, Klein could solve by means of an icosahedral equation, the *Formenproblem* for I_3, and thus the generalized Jacobian equation of degree six, after adjoining an accessory square-root. This settled the problem of the general quintic equation in the framework of his research programme.

There was another reason for Klein not to mind accessory irrationalities: his interest in complex function theory (of one variable) and coverings of the complex plane. It was in this context that he formulated his principle to the effect that "the resolution of quintic equations has to be reduced to the icosahedral equation":[56]

> One will certainly try, if at all possible, to reduce the resolution of general algebraic equations to the resolution of special equations with only one parameter. For only the algebraic functions of one variable are currently somewhat understood. Among these special equations, I find that those are always the most important whose Galois

[56] See [35, p. 75]: *Sicher wird man suchen, solange es angeht, die Auflösung der allgemeinen algebraischen Gleichungen auf diejenige spezieller Gleichungen zurückzuführen, welche nur einen Parameter enthalten. Denn nur die algebraischen Funktionen einer Variablen beherrscht man zur Zeit einigermaßen. Unter diesen speziellen Gleichungen scheinen nun immer diejenigen die wichtigsten zu sein, deren Galoissche Resolvente das kleinstmögliche Geschlecht besitzt ... dem eben entspricht, daß man die Ikosaedergleichung einführt.*

> resolvent has the lowest possible genus ... which means introducing the icosahedral equation.

This idea meshes perfectly with the theory of elliptic modular functions. It was in fact through his work on elliptic modular functions that Klein obtained a more precise understanding of the non-algebraic part of the resolution of the quintic equation by Hermite and Kronecker.

Klein worked with the J-invariant of an elliptic integral, interpreting it as a function of the period ratio ω (normalized to have positive imaginary part), and with the operation of $SL_2(\mathbf{Z})$ on the upper half-plane. For the Riemann surface (with infinitely many sheets) of ω over the J-plane, Klein found a geometric picture which allows to read off the ramification. Klein's *Elementarviereck* $E = \{\omega = x + iy \in \mathbf{C} : |x| \leq \frac{1}{2}, |\omega| \geq 1\}$ (with suitably identified edges) is the fundamental domain still routinely used today, in which every J-value occurs exactly once and whose translates by $SL_2(\mathbf{Z})$ cover the whole upper half-plane.

In this setting, Klein called a one-parameter equation "solvable by elliptic modular functions", if the roots of this equation, with J substituted for the parameter, can be written as functions of ω.

Relevant modular functions were not only J (which is invariant under all of $SL_2(\mathbf{Z})$), but also more general modular functions ϕ with respect to congruence subgroups G of $SL_2(\mathbf{Z})$, in particular the icosahedral irrationality η which Klein defined by the identity $I(\eta) = J$. It is invariant under the congruence subgroup $G = \{A \in SL_2(\mathbf{Z}) \mid A \equiv I \pmod 5\} \cong SL_2(\mathbf{Z})/SL_2(\mathbf{F}_5)$. In this language, transformations of order n are induced by $\omega \mapsto \frac{\omega}{n}$, and the modular equation is satisfied by the different values $\phi(\frac{T\omega}{n})$; if the T_i are representatives of the cosets of $\frac{1}{n}G/G$, then the modular equation is $\prod(X - \phi(\frac{T_i\omega}{n})) = 0$, and its coefficients lie in the function field $k(\phi)$.

Given such a modular equation and putting $\phi_i' = \phi(\frac{T_i\omega}{n})$, Klein would look for Galois resolvents, i.e., for elements that generate $k(\phi_1', \ldots, \phi_n')|k(\phi)$, resp. for their minimal polynomials. In our original setting, let us look at the transformation of order five of the modular function J. The corresponding modular equation has degree 6. A Galois resolvent of this modular equation turns out to be precisely the icosahedral equation (of the lowest possible genus 0) and the multiplier of this transformation satisfies a generalized Jacobian equation of degree 6 with $A = 0$[57].

So, starting from J simplifies the situation considerably (Jacobi's transformations yielded such an equation with $B = 0$ instead). In fact, given a generalized Jacobian equation of degree 6, $z^6 - 10z^3 - Cz + 5 = 0$, the corresponding invariant J can simply be computed as $J = -C^3/12^3$, whereas Kronecker (see 3.3) wound up with a cubic equation for $\kappa^2(1 - \kappa^2)$ which had to be solved in order to get the correct modulus κ^2.

---

[57] See [36, p. 46–55].

In conclusion, Klein could claim to have thoroughly *understood* the solutions of the quintic equation by elliptic functions, which to Kronecker and Brioschi would have appeared like lucky coincidences, in terms of modular functions:[58]

> Substituting the absolute invariant of the elliptic modular functions J instead of X in the icosahedral equation, turns the icosahedral quantity x into the "principal modulus of the principal congruence subgroup of level 5." All methods relating the resolution of quintic equations to elliptic functions rest on this fundamental theorem.

As to Hermite's method (3.1), Klein could account for the additional accessory cube root which it employs to reduce a principal equation of degree 5 to the form $x^5 - x - a = 0$: this corresponds precisely to the cube root needed to determine κ^2 from J [36, p. 71f]. According to Klein, the equation $x^5 - x - a = 0$ is not suitable as normal form for the general quintic equation because its Galois resolvent has genus 4.

5 Glimpses Beyond

The tension between Felix Klein and Leopold Kronecker about one-parameter resolvents and accessory irrationalities in the resolution of the quintic equation tells a very interesting little parable about the potential for interpretation of mathematical theorems: Both Klein and Kronecker always agreed on the mathematical correctness of Kronecker's theorem, and they even hit upon essentially the same proof. There was, it is true, a difference of appreciation of Kronecker's first proof from 1861: Klein found it incomplete; we have no direct comment by Kronecker on this, also not in his 1885–86 lectures on the subject. But even assuming that Kronecker took Klein's criticism of his proof very badly, this element of the story does not seem to account for the overall almost ideological confrontation of these two mathematicians. This is why it seems to us that this parable is an especially valuable historical example against which historiographical or sociological theories about the (development of) mathematics, or the sciences in general, ought to be checked. We do not go into this here, all the more so as a thorough analysis would have to look at many other factors besides the discussions about the quintic equation; it would have to take into account the respective roles of both mathematicians in the German and European mathematical communities, and even in German society on the whole. A kind of parallel, comparative biography of these two strong personalities in science would have to be drawn up, in order to try and understand in detail what the factors were that made their mathematical practice differ so profoundly, in spite of their essentially

[58] [38, p. 484]: *Setzt man in* [der Ikosaedergleichung $\frac{H^3(x)}{1728 f^5(x)} = X$] *für X die absolute Invariante J der elliptischen Modulfunktionen, so bekommt die Ikosaedergröße x die einfache Bedeutung des "Hauptmoduls der Hauptkongruenzgruppe fünfter Stufe." Alle Arten, die Auflösung der Gleichungen fünften Grades mit den elliptischen Funktionen in Zusammenhang zu bringen, beruhen auf diesem Fundamentalsatz.*

coherent and consensual attitude about the validity of the mathematical results under discussion – cf. [22].

From Klein's point of view, the story provided a contradictory image of Kronecker's scientific personality:[59]

> This is the peculiar double nature of Kronecker: on the one hand the profound researcher who finds, or at least divines, the most beautiful results, on the other hand the dogmatist who believes he can erect barriers for the method of science in its thrust towards progress. Cf. his declaration against the irrational numbers.

But viewed from today's vantage point, not only Kronecker's Theorem, also Klein's celebration of the icosahedral equation have long ceased to attract modern mathematicians. Klein's broad research programme does live on in various contexts – see Slodowy's introduction to [37]. Kronecker's strengthening of Abel's theorem must have played a role in the formulation of Hilbert's 13th problem, about the non-existence of 2-parameter resolvents for the general equation of degree 7 – even though Hilbert unwittingly opened the door too wide (allowing continuous, not necessarily algebraic functions as resolvents), and wound up with a wrong conjecture.

In this way, the disproof of Hilbert's 13th problem by Arnold and Kolmogorov has become an internal affair of topology and analysis, with no palpable connection to Kronecker's 1861 discovery. Likewise, the theory of modular functions and forms is so productive and strong in our days that there seems to be very little time for reminiscences of Klein's favourite icosahedral equation.

It is Abel's proof of the impossibility of resolving the quintic equations by radicals which has firmly entered the general mathematical culture. And Kronecker's 1853 comparison between abelian and cyclotomic extensions of the rationals and its generalizations, motivated as they were by Abel's algebra and Kummer's arithmetic, have found a home in class field theory, i.e., in one of the pillars of XXth century Number Theory.

---

[59] Quoted from Klein's letter to Hensel in Appendix II below. The last sentence alludes of course to Kronecker's very far-reaching programme of the arithmetization of analysis – see for instance [11].

Appendix I. Letters from Kronecker to Dirichlet

The following five letters come from a folder of altogether eight letters from Kronecker to Dirichlet dating from the 1850's, which are preserved in the "Nachlaß Dirichlet" *at the Sammlung Preußischer Kulturbesitz, Staatsbibliothek zu Berlin. For each letter, we indicate the corresponding sheet numbers of the folder "Kronecker, Leopold, an D.G." in the above mentioned Nachlaß, and within the letters we indicate pagebreaks by vertical strokes.*

Correspondence between Peter Gustav Lejeune-Dirichlet and Leopold Kronecker from other sources has been published by Schering while Kronecker was still alive – see [64]. This contains in particular Dirichlet's answer to the first two letters published here. Schering had no access to these.

Our hearty thanks go to Harold M. Edwards who not only mentioned the existence of these letters in our discussions at Oslo, but actually sent us a copy of his personal transcriptions from the Berlin file. We have subsequently checked his notes against the originals in Berlin, and take complete responsibility for the deciphering given here.[60]

Kronecker to Dirichlet, 31 January 1853.– Sheet 9 and 10.

Verehrtester Herr Professor!

Nur Ihre mir so oft bewiesene freundliche Gesinnung ist es, welche mich zu den folgenden Zeilen und zu der darin enthaltenen Bitte ermuntert. Indem ich mich nämlich seit wenigen Wochen genauer mit der Theorie der auflösbaren Gleichungen beschäftigte, habe ich einige Bemerkungen gemacht, welche mir über das Wesen der Sache viel Licht zu verbreiten schienen. Ich habe diese Bemerkungen vorgestern unserm Freunde Kummer, den ich zu seinem Geburtstage in Breslau besuchte, mitgeteilt und dieser empfiehlt es mir dringend, Ihnen eine kleine Mittheilung darüber zu machen. Freilich wußte Kummer da so wenig wie ich, daß das Wesentliche davon schon in dem bezüglichen Aufsatze in Abels nachgelassenen Werken enthalten ist. Ich habe diese Werke erst gestern abend erhalten und nun zum ersten Male die erwähnte fragmentarische Abhandlung gelesen, - wenn man hier überhaupt lesen kann. Nun schien mir aber wiederum grade die Hauptsache in seiner Abhandlung so "vergraben," daß es für Jemanden, der jene Eigenschaft der auflösbaren Gleichungen nicht schon kennt, fast unmöglich sein dürfte, dieselbe herauszufinden. Daß es vielen, die sich doch viel mit der Sache beschäftigt haben, <u>nicht</u> aufgefallen ist, beweisen die Malmstenschen[61], Lutherschen[62], Bettischen[63] Abhandlungen. Ich kenne übrigens außer den erwähnten und der Galoisschen Arbeit keinerlei Schriften über den vorliegenden Gegenstand und habe von Heine

[60] As was customary at the time, Kronecker wrote many 's' or 'ss' as: long-h-tied-to-an-s. We have rendered these as 's', 'ss', or 'ß', according to German spelling before the current, messy reform. To give an example, in the word *verhältnismäßig*, Kronecker would have written long-h-tied-to-an-s *twice*. – Otherwise, we carefully adhere to Kronecker's spelling.

[61] Kronecker alludes here, and again below, to the paper [59] by Carl Johann Malmsten (1814–1886), professor at Uppsala. Cf. [20, pp. 18–19, 27–29].

[62] Eduard Luther (1816–1887) was at the time *Privatdozent*, and as of 1859 professor, of astronomy at Königsberg, today Kaliningrad, Russia. Kronecker probably alludes to his two Crelle papers, on the solvable equations of degree 5, resp. 6 – see [57] and [58].

[63] Surely a reference, here as well as later on in the letter, to Betti's seminal memoir [9], one of the milestones in the development of Galois theory.

nur gehört, daß Schoenemann darüber geschrieben haben soll, eine Schrift deren Hauptresultate wenigstens Ihnen wohl bekannt sein dürften.[64] Nach diesen einleitenden Worten kommt schon die große Bitte, mir wenn auch mit ganz wenigen Zeilen baldgütigst zu sagen, ob Ihnen die folgenden Bemerkungen beziehungsweise neu und ob sie Ihnen etwas werth erscheinen.

Ich spreche nun nur von Primzahl Graden, da ich nach meinen Untersuchungen in den anderen Graden keine prinzipiellen Schwierigkeiten sehe. Abel stellt den – abgesehen von einigen Lücken durch Malmsten im 34sten Bande des Crelle[65] bewiesenen – Satz auf: daß für jede irreductible Gleichung des μ-ten Grades (wo μ Primzahl) die Wurzel die Form haben müsse Const. $+ \sqrt[\mu]{R_1} + \sqrt[\mu]{R_2} + \cdots + \sqrt[\mu]{R_{\mu-1}}$, wo die $\mu - 1$ $R's$ Wurzeln einer rationalen
1|2 Gleichung $\mu - 1$ sten Grades sind. Hierzu füge ich die einfache Bemerkung, | daß diese letztere Gleichung eine Abelsche (von der einfachsten im 4ten Bande des Crelle behandelten Art) sein muß. Mit andern Worten: Nicht nur die symmetrischen, sondern auch alle cyklischen Funktionen der $R's$ müssen rational sein. Der Beweis dieser Bemerkung ist leicht, und die Bemerkung läßt sich sogar aus den Galoisschen Betrachtungen fast herauslesen. Dennoch habe ich sie in der Bettischen Schrift nicht gefunden und es zeigt sich auch bald, daß nur bei der Abelschen Betrachtung der Werth dieser Bemerkung zu schätzen ist. Unter den zahlreichen unmittelbaren Folgerungen will ich nur einige hier erwähnen:

1. Es geht daraus hervor, daß der Malmstensche Aufsatz im 34sten Bande schon beweist, – daß die Bedingung, daß die Lagrangesche Resolvente einen rationalen Factor habe nicht bloß nothwendig zur Auflösbarkeit sondern auch hinreichend ist.

2. Die Wurzel jeder auflösbaren Gleichung μ-ten Grades ist ausdrückbar durch μ-te Wurzelzeichen und solche deren Exponent ein Divisor von $\mu - 1$ ist.

Überhaupt sieht man daraus die wahre Beschaffenheit der auflösbaren Gleichungen, die man aus den Galoisschen Untersuchungen durchaus nicht erkennen kann: denn Galois nimmt sich nur die eine Aufgabe vor "die Bedingung der Auflösbarkeit["] zu finden, während Abel auch die andere berücksichtigt "alle auflösbaren Gleichungen zu finden." Durch die obige Bemerkung kommt diese Frage darauf hinaus "alle Abelschen Gleichungen $\mu - 1$sten Grades zu finden." Und hierbei zeigt sich erst wieder nicht nur die Wichtigkeit dieser speziellen Art von Gleichungen, sondern auch, wie wenig dieselben seither beachtet wurden. Bevor ich aber hierauf weiter eingehe, muß ich Ihnen noch erwähnen, daß wie sich von selbst versteht, der obige Ausdruck $R_1^{1/\mu} + R_2^{1/\mu} + \cdots + R_{\mu-1}^{1/\mu}$ nicht jedesmal – wenn auch die $R's$ Wurzeln einer Abelschen Gleichung sind – die Wurzel einer rationalen Gleichung μ-ten Grades ist, sondern, wenn die Wurzeln irgend einer Abelschen Gleichung $\mu - 1$sten Grades nach der richtigen Ordnung geschrieben $\rho_1, \rho_2, \ldots, \rho_{\mu-1}$ sind, so ist folgender Ausdruck[66]

$$C + f(\rho_1)\Big(\rho_1 \cdot \rho_2^{g-1} \cdot \rho_2^{g-2} \cdot \rho_3^{g-3} \cdot \cdots \cdot \rho_{\mu-1}^{g-\mu+2}\Big)^{1/\mu} + f(\rho_2)\Big(\rho_2 \cdot \rho_3^{g-1} \cdots\Big)^{1/\mu} + \cdots$$

die Wurzel einer rationalen Gleichung μ-ten Grades. Nach bekannter Weise ist hier $g_{-k} \equiv g^{-k} \bmod \mu$. Dieser Ausdruck ist aber nicht nur immer die Wurzel einer solchen Gleichung, sondern es ist derselbe auch die allgemeinste Form einer solchen Wurzel, wenn man
2|3 nur für die ρ's nicht bloß alle wirklichen sondern auch | alle idealen Wurzeln aller Abelschen Gleichungen $\mu - 1$sten Grades nimmt. Wie der Begriff des Idealen hier mit Notwendigkeit

[64] Heine may have heard of the forthcoming paper [65] by Theodor Schoenemann (1812–1868), *Professor am Gymnasium zu Brandenburg*.

[65] See [59].

[66] In the first bracket, Kronecker garbles the factors – it should read as the μ-th root of $(\rho_1 \cdot \rho_2^{g-1} \cdot \rho_3^{g-2} \cdot \rho_4^{g-3} \cdot \cdots \cdot \rho_{\mu-1}^{g-\mu+2})$.

und Sicherheit in der Algebra erscheint – so zeigt sich recht, daß er gewissermaßen dem Imaginären gleicht. Sie werden in dem Produkt der $\rho's$, welches oben erscheint, sofort die Form wiedererkennen, welche Kummer für die Zerlegung der Jacobischen $(\alpha, x)^\lambda$ in ideale Primfactoren gegeben hat.[67] Wie dies auch wirklich zusammengehört, werden Sie aus folgenden Bemerkungen ersehen, denen ich nur vorausschicken will, daß obige Form mit der von Abel im 5ten Bande des Crelle für die Wurzel einer Gleichung 5ten Grades aufgestellten Form [überein]stimmt.[68]

Um die allgemeine Form Abelscher Gleichungen oder vielmehr deren Wurzeln zu untersuchen, will ich wieder zuerst voraussetzen, daß deren Grad eine Primzahl sei. Im anderen Falle erfordert die Untersuchung nämlich die genaue Kenntnis der complexen Zahlen $f(\alpha)$ wo $\alpha^\lambda = 1$ und λ zusammengesetzt. Diese Ihnen gewiß geläufige Theorie hat nur praktische, – keine prinzipiellen Schwierigkeiten. Es sei also λ Primzahl und Grad einer Abelschen Gleichung, deren Wurzeln $x, \theta(x)$, $\theta\theta(x), \ldots.$ Die Gleichung sei ganzzahlig. Spezielle Fälle dieser Gleichung sind die Gleichungen für die λ Perioden aus k-Gliedern von $k\lambda + 1$sten Wurzeln der Einheit (wenn $k\lambda + 1$ Primzahl). Schon für diese giebt es keine allgemeine Form der Wurzeln als die:

$$C + (\alpha, z) + \left(\alpha^2, z\right) + \left(\alpha^3, z\right) + \cdots + \left(\alpha^{\lambda-1}, z\right),$$

wo das Jacobische Zeichen gewählt[,] $\alpha^\lambda = 1$ und $z^{k\lambda+1} = 1$ genommen ist. In dieser allgemeinen Form soll aber z nicht erscheinen und diese Form ergiebt daher nach der Kummerschen Factorenzerlegung folgende andere Form:

$$C + \left(f(\alpha) f\left(\alpha^2\right)^{1/2} f\left(\alpha^3\right)^{1/3} \ldots\right)^{1/\lambda} + \left(f\left(\alpha^2\right) f\left(\alpha^4\right)^{1/2} \ldots\right)^{1/\lambda} + \ldots$$

Diese der umstehenden genau entsprechende Form giebt in der That die Wurzel aller Periodengleichungen wenn man für $f(\alpha)$ alle idealen und wirklichen complexen Primzahlen nimmt. Ich habe nun weiter bemerkt, daß für jede Abelsche ganzzahlige Gleichung des λten Grades, die Wurzel eine ganz ebensolche Form hat, wenn man nur jedem Theile irgend eine ganze complexe Zahl $\varphi(\alpha)$ und resp. $\varphi(\alpha^2)$ etc. außerhalb des λten Wurzelzeichens als Factor hinzufügt und wenn man sich unter $f(\alpha)$ nicht bloß alle complexen Primzahlen deren Norm
3|4 $k\lambda + 1$ ist denkt, sondern auch jedes Produkt solcher Primzahlen. Hieraus folgt aber, | daß die Wurzel jeder ganzzahligen Abelschen Gleichung vom λten Grade "eine ganze complexe Zahl von $k\lambda + 1$ sten Wurzeln der Einheit oder sogar nur deren λ Perioden gebildet ist."

Das erscheint mir in der That recht bemerkenswerth 1. weil es zeigt, daß man ja auch darauf gehen kann, die Wurzeln einer Gleichung λten Grades in einer Form darzustellen, die durch Einführung höherer als λter Wurzelzeichen sich vereinfacht[.] 2. weil es zeigt, daß Abelsche Gleichungen auch die allgemeinsten mit ganzzahligen Coeffizienten zu der einzigen Reihe derjenigen, die bei der Kreisteilung vorkommen, in gewisser Weise gehören. Ich füge noch hinzu, daß ganz ebenso alle Abelsche Gleichungen mit in Ihrem Sinne complexen Coeffizienten von der Form $a + bi$ nur zu der Reihe der aus der Lemniscaten Theilung entstehenden Gleichungen gehören etc. D.h. jedenfalls geht dies weiter auch für die Abelschen Transcendenten. Ich bemerke hierbei, daß ich nicht übersehen habe, wie aus dem Produkt z.B. zweier Abelscher Gleichungen λten Grades mit complexen Coeff. $(a + bi)$, eine (nicht

[67] See Kummer, Werke, vol. I, p. 245–248.

[68] See the (German translation of an) excerpt of a letter from Abel to Crelle, dated Freyberg 14 March 1826, printed after Abel's death in *Journal für die reine und angewandte Mathematik* **5** (1830), p. 336; the French version of this is reproduced in [1, Vol. 2, p. 266].

Abelsche) aber eben auflösbare Gleichung 2λten (also zusammengesetzten Grades) entsteht, welche reelle ganze Coeff. hat. Dies, d.h. das Aehnliche giebt auch die wahre Schwierigkeit der Gleichungen zusammengesetzter Grade. Es ist das übrigens kein Mangel, daß ich immer von ganzen Zahlen hier spreche. Ich halte es im Gegenteil für naturgemäß; denn ich kann (nicht bloß aus obigen Resultaten) beweisen, daß das zahlentheoretische Element bei diesen algebraischen Untersuchungen ein Wesentliches ist. Schon der überall vorausgesetzte Begriff der Irreductibilität ist ein Unding, wenn man nicht Rationale oder nur algebraisch irrationale Coefficienten voraussetzt. Das aber und vieles Andere hier auseinanderzusetzen ist nicht bloß wegen des "kurzen Raumes" sondern auch wegen der "Langeweile" unstatthaft. Also nur nochmals die unverschämte Bitte um möglichst umgehende Antwort, wenn auch nur in zwei Zeilen. Sagen Sie mir doch gütigst (für den Fall daß Ihr Urtheil über das Ihnen Mitgetheilte günstig ausfällt) darin, ob Sie es (wie Kummer) für angemessen halten, daß ich Ihnen die Hauptpunkte und Gesichtspunkte in ganz kurzer und geordneter Form noch einmal für eine geeignete Art öffentlicher Mitteilung aufsetzen soll; – denn mit der großen Ausarbeitung würde es doch noch eine Weile dauern. Mit meiner Gesundheit gehts noch immerfort schlecht genug. Meine Frau läßt sich gleich mir Ihnen und Ihrer Frau Gemahlin bestens empfehlen.

Mit alter aufrichtiger Verehrung

Ihr dankbarer Schüler

Liegnitz den 31. Januar 1853 Leopold Kronecker

Kronecker to Dirichlet, 2 April 1853. – Sheet 11 and 12.

Geehrtester Herr Professor,

Beiliegende 50 [Taler][69] als jetztfällige Rate für den jungen Jacobi sind die Veranlassung dieser Zeilen. Ich hätte auch ohne eine solche Veranlassung es nicht gewagt, Sie nochmals mit einem Schreiben zu behelligen, nachdem das Ausbleiben der erbetenen Antwort auf meinen Brief vom Anfang Februar mich eben von einem abermaligen Schreiben abmahnen mußte. Trotzdem will ich jetzt meine damaligen etwas unreifen Mitteilungen vervollständigen und hoffe dabei wegen dieser erneuten Zudringlichkeit auf Ihre mir sonst so oft bewiesene Nachsicht.

Ich habe nun jetzt folgendes Problem gelöst, welches meines Erachtens das Wesentliche der Untersuchungen über auflösbare Gleichungen enthält: "Die allgemeinste algebraische Funktion von beliebigen Größen $x', x'', \ldots$ zu finden, welche einer irreductiblen Gleichung vom p ten Grade genügt, deren Coefficienten rationale Functionen der x's sind." Hier ist p Primzahl und ich habe nur wegen des Wortes "irreductibel" einiges hinzuzufügen. Dieses Wort ist nämlich in allen mir bekannten algebraischen Abhandlungen etwas leichtsinnig gebraucht worden. Wie ich Ihnen schon damals schrieb, ist "Irreductibilität" ein wesentlich zahlentheoretischer Begriff. Die Irreductibilität einer ganzzahligen oder rationalen Function einer Unbekannten ist leicht zu erklären. Ganz ebenso kann man die Irreductibiliät einer Function von $x', x'', \ldots$ mit rationalen Coeffizienten definiren. Es versteht sich, daß eine solche irreductible Function von $x', x'', \ldots$ reductibel werden kann, wenn man einem oder mehreren x'en gewisse rationale oder algebraisch-irrationale Werthe beilegt. In dem obigen Problem hat "irreductibel" nur den Sinn, daß die Gleichung als rationale Function der allgemeinen x'e irreductibel ist. So ist z.B. die Gleichung $z^3 + x'z^2 + x''z + x''' = 0$ natürlich irreductibel. Obiges Problem kann auch so ausgesprochen werden: "Die allgemeinste algebr. Function von $x', x'', \ldots$ zu finden, welche nur p Werthe hat und so beschaffen ist, daß die symmetrischen

[69] Kronecker uses the standard sign for this currency.

Functionen jener p Ausdrücke rationale Functionen der x'e ergeben." Hierbei ist nur zu bemerken, daß sich alle durch die verschiedenen Wurzelwerthe ergebenden Werthe der algebraischen Function identisch auf p Werthe reducieren müssen. Ein zweites Problem, welches ich zur Lösung obigen Problems brauche, ist: "Die allgemeinste algebraische Function von $x', x'', \ldots$ zu finden, welche n Werthe hat und so beschaffen ist, daß die symmetrischen und cyklischen Functionen jener n-Werthe rationale Functionen der x'e ergeben." Hierbei ist n eine beliebige ganze Zahl. Nur weil ich mich schon seit 14 Tagen mit der einfachsten Form der Auflösung dieses letzteren Problems für den Fall quäle, wenn n eine Potenz von 2 enthält – nur deshalb habe ich Ihnen heut noch nicht, wie ich es mir vornahm, ein vollständiges zur eventuellen Veröffentlichung reifes Resumé gesandt. Die Auflösung der Probleme habe ich Ihnen fast ganz schon voriges Mal mitgetheilt; nur die Probleme selbst hatte ich mir noch nicht genügend klar gemacht. Ich füge nur bei, daß der Begriff des Idealen zur Auflösung jener Probleme nicht nöthig ist. Es ist leicht zu ersehen, daß sich aus der Auflösung obiger Probleme die Erledigung der engeren von Galois aufgestellten Aufgabe über das Criterium der Auflösbarkeit einer gegebenen Gleichung ergiebt. Und zwar ergiebt sich eben nicht bloß
1|2 ein oder | das andere Criterium, sondern es ergeben sich alle Criterien. Nach der in den nachgelassenen Werken Abels enthaltenen Abhandlung scheint übrigens selbst Abel sich die Probleme noch nicht so klar gestellt zu haben. – Wie Sie sofort sehen werden, enthält das zweite oben erwähnte Problem alle "Abelschen" Gleichungen. Aus dessen Auflösung ergiebt sich leicht das interessante Resultat: "daß zu jeder Abelschen Gleichung beliebigen Grades mit ganzzahligen Coefficienten die Wurzel als rationale Function von Wurzeln der Einheit dargestellt werden kann." Aehnliche Resultate erhält man für Abelsche Gleichungen, deren Coeffizienten ganze complexe Zahlen $a + bi$, in Bezug auf die Lemniscatentheilung, und man erhält hieraus ein Mittel zur Auffindung der "einfachsten Lemniscatentheilungsgleichungen." Die Ausführung dieses interessanten Punktes mußte ich vorläufig verschieben, um nicht von der Hauptsache abzukommen. Es ergiebt sich ferner aus der Auflösung jenes zweiten Problems das Criterium für die Irreductibilität Abelscher Gleichungen. Bei dem großen Interesse, das gerade Abelsche Gleichungen für Zahlentheorie und Analysis haben, ist dies sehr wichtig. Denn die aus Wurzeln Abelscher Gleichungen gebildeten complexen Zahlen haben die Eigenschaft, daß die conjugierten Zahlen zu derselben Sorte gehören ... Das erste Hauptproblem vom p ten auf jeden beliebigen Grad auszudehnen, wird mir mit Hülfe der Bettischen Abhandlung keine Schwierigkeiten machen; alsdann ist aber auch wohl die Hauptaufgabe dieses Theils der Algebra erledigt.

Damit sei es genug um Sie nicht zu sehr zu ennuyiren, denn sonst könnte ich Ihnen noch Massen von Folgerungen mittheilen. Nun, geehrter Herr Professor, nur die eine Bitte, daß Sie bei der Empfangsbestätigung der 50 [Taler][70] nur in einer Zeile hinzufügen möchten, ob Ihnen obige Mittheilungen neu erscheinen oder nicht. – Bei der Fortdauer meines Uebels habe ich die Absicht Mitte Mai nach Paris zu gehen, um einen dortigen berühmten Arzt zu consultieren. Ich werde diese Reise über Berlin machen und darf vielleicht durch Ihre Güte einige Empfehlungsschreiben erhalten, die ich mir persönlich bei Ihnen abzuholen erlauben würde. – Vielleicht habe ich schon in kurzer Zeit die Freude, aus ein paar Zeilen von Ihnen zu ersehen, daß an Ihrem bisherign Schweigen nicht das Aufhören des Interesses Schuld ist, das Sie sonst immer an mir zu nehmen die Güte hatten.

Mit dankbarer Verehrung

Liegnitz den 2. April 1853. Ihr ergebener Schüler

Leopold Kronecker

[70] Kronecker uses the standard sign for this currency.

Kronecker to Dirichlet, 6 May 1853.– Sheet 5[71]and 6.

Verehrtester Herr Professor,

Ihr ausführlicher Brief vom 3ten d. M.[72] hat mir so außerordentlich viel Freude gemacht, daß ich nicht umhin kann, Ihnen schon jetzt meinen herzlichsten Dank dafür zu sagen. Sie wissen es ja längst, wie unendlich wohlthuend mir Ihr Wohlwollen, und also auch ein Zeichen desselben ist, und kann ich Ihnen nur versichern, daß ich stets bemüht sein werde, durch sorgsame Arbeiten dieses ehrende Wohlwollen zu verdienen, das Sie mir von vornherein entgegen getragen haben ... [73] Meine jetzigen Arbeiten selbst anlangend; so sind dieselben theils durch äußere, theils durch innere Hindernisse ein wenig aufgehalten worden. Zu den äußeren gehören familiäre Abhaltungen und Ereignisse, namentlich eines, von dem ich Sie hiermit in Kenntniss setze: Die vor drei Wochen erfolgte Entbindung meiner geliebten Frau von einem Töchterchen – inneres Hinderniss war nur das schon in meinem vorigen Brief berührte. Für diejenigen Abelschen Gleichungen nämlich, deren Grad eine Potenz von 2 ist, fehlt mir eine Kleinigkeit, um die von mir vermuthete einfachste Form der Wurzeln zu erweisen. Dies soll mich indessen nicht hindern, ein recht sorgsam ausgearbeitetes Resumé meiner Arbeiten (soweit ich eben gekommen bin) Ihnen Ihrer gütigen Aufforderung gemäß mitzubringen. Da ich mich ungemein sehne, Sie wieder persönlich zu sehen, so werde ich meinen Aufenthalt in Berlin nach Ihren Angaben einrichten, d.h. ich werde den 23sten früh dort ankommen und zwei bis drei Tage dort bleiben. Dann muß ich freilich nach Paris eilen, da ich außerdem für diesen Sommer noch eine Kur in Kreuznach und Helgoland vorhabe. Ich hätte nie geglaubt, daß ich so zeitig würde anfangen müssen, die Unannehmlichkeiten des Körpers zu merken. Doch glaube ich eben bei meiner Jugend noch an eine radikale Heilung glauben zu dürfen. Doch ich bin gedankenloser Weise zu weit in mein Lieblingsthema, mein Unwohlsein, hineingekommen, und wende mich nun aber gleich wieder zu Ihrem Briefe. Wenn Sie nämlich darin sagen, daß Sie das über meine Arbeiten Berichtete nicht gleich verstanden haben, so ist die Schuld einzig und allein auf meine schlechten Berichte zu schieben. Denn meine mündliche Auseinandersetzung wird Ihnen zeigen, daß die Sachen wirklich so erstaunlich einfach sind, daß ich eben deshalb an deren Neuheit zweifeln mußte. Ich erinner mich noch sehr gut, daß Sie mich vor ungefähr zehn Jahren einmal an dieses Thema der auflösbaren Gleichungen mit dem Bemerken gewiesen haben, daß die bezüglichen Untersuchungen verhältnismäßig leicht sein müßten. Und Sie hatten ganz Recht. Denn abgesehen von Weitläufigkeiten, die sich bei so fundamentalen Dingen kaum vermeiden lassen, wird sogar auch die Ausführung einfach sein. Nun will ich aber Ihre Geduld nicht länger mißbrauchen, sondern lieber alles Uebrige unserer mündlichen Unterhaltung aufsparen. Ich bitte meine Frau und mich selbst Ihrer Frau Gemahlin angelegentlich zu empfehlen und unsere besten Wünsche für den Erfolg der Karlsbader Cur auszurichten. Indem ich uns auch Ihnen bestens empfehle bleibe ich

in treuer Anhänglichkeit und dankbarer Verehrung Ihr ergebener

Liegnitz den 6. Mai 1853 Leopold Kronecker

[71] On the back of sheet 5, there are a few lines of standard identities for the Legendre symbol $(\frac{\pm 2}{p})$, not in Kronecker's handwriting.

[72] See [64, p. 409–410].

[73] Suspension dots in Kronecker's text.

Kronecker to Dirichlet, 15 June 1853. – Sheet 7 and 8.

Paris 15. Juni 1853
Hôtel des étrangers, rue Vivienne.

Verehrtester Herr Professor,

meinem Versprechen gemäß sende ich Ihnen einliegend die nach Ihrer gütigen Anleitung verbesserten Mitteilungen zu und hoffe, daß dieselben noch rechtzeitig vor der Sitzung der Akademie in Ihren Besitz gelangen werden.[74] Es ist freilich eine arge Zumutung meinerseits, daß ich Ihnen ein durch vielfache Veränderungen so unsauber gewordenes Schriftstück zukommen lasse. Aber einerseits wollte ich eben von allen Ihren errata profitiren und andrerseits erlaubte es mir mein körperlicher Zustand durchaus nicht eine neue Reinschrift anzufertigen. Ich hoffe indessen, daß die Unsauberkeit der Leserlichkeit nicht geschadet haben wird, und daß die Verbesserungen zur größeren Deutlichkeit wesentlich beigetragen haben werden. So habe ich namentlich – um gleich von vorne herein den etwaigen (?)[75] Leser in medias res zu versetzen – den Schluß mit einigen Abänderungen zum Anfange umgestaltet, und hoffe mit dieser Umgestaltung Ihren Beifall zu erhalten. Das Ganze fängt also jetzt auf der Seite 1.a an und endet auf Seite 6. Zur Verhütung jeglichen Mißverständnisses habe ich sogar närrischer Weise den Anfang mit "Anfang" bezeichnet. Sollten Sie nun noch bei Empfang dieses der Sache für Werth halten der Academie mitzuteilen und den Abdruck in den Monatsberichten zu beantragen, so würde ich Sie nur bitten eine Anzahl Exemplare gütigst für mich zu asserviren. Ich werde Joachimstal bitten dieselben zur Zeit bei Ihnen in Empfang zu nehmen und mir zuzuschicken. – Was mein hiesiges Leben anlangt, so wird dasselbe fast vollständig von meinem Arzte Civiale in Anspruch genommen. Da ich nämlich alltäglich von 11 bis 12 Uhr ungemein schmerzhafte Curen bei ihm auszustehen habe, da ich nachher durch Bäder und Ruhe diese Schmerzen abzustumpfen suchen muß, so bleiben mir nur etwa die Stunden von 6 Uhr ab, die ich mir einigermaßen zu Nutze machen kann. Sie werden mir glauben, daß diese Lebensweise für mathematische Speculationen nicht eben günstig ist. Denn die wenigen Stunden, dir mir von schmerzhafter Erregung frei bleiben, absorbieren mir die mannigfaltigen neuen Eindrücke von Paris. – Liouville, bei dem Sie mich durch Ihre Carte eingeführt haben, habe ich bis jetzt einmal besucht. Er ist auch sehr beschäftigt, war übrigens sehr zuvorkom-
1|2 mend und wir haben über eine Stunde verplaudert. Ich habe ihm einige wenige | Resultate meiner algebraischen Arbeiten mitgetheilt, namentlich "daß alle ganzzahligen Abelschen Gleichungen wesentlich Kreistheilungsgleichungen seien." Er meinte, daß er sich dies schon immer gedacht habe, bewiesen sei es aber auch seines Wissens bisher noch nicht. Er erzählte mir auch, daß zum großen concours viele Arbeiten über den Fermatschen Satz eingegangen seien, was mich sehr wunderte, daß aber keine einzige wohl reüssiert habe, was mich gar nicht erstaunte. In Hinsicht der algebraischen Dinge ist Liouville ganz in den Galoisschen Anschauungen befangen und ist in Folge derselben in eine Art von Extrem namentlich in eine zu große Verallgemeinerung der algebraischen Probleme verfallen. Es versteht sich, daß dieser Fehler andererseits ganz ungeheure Vortheile darbietet; und dabei ist Liouville's Gewandtheit in der Behandlung der Galoisschen Theorie ganz erstaunlich. – Hermite ist auf Monate verreist, was mir sehr leidthut. Cauchy habe ich nur in der Sitzung des Instituts gesehen. – Mein Aufenthalt hier dehnt sich vorläufig ins Unbestimmte namentlich jedenfalls bis auf noch 14 Tage aus; denn ich will nun schon aushalten, da mir Civiale Hoffnung giebt, daß ich die vielen schrecklichen Schmerzen nicht umsonst erlitten haben werde.

[74] The session of the Academy at which Dirichlet presented the paper [41] was on 20 June 1853.

[75] Question mark in parentheses in the text.

Nun, geehrter Herr Professor, habe ich Sie nur nochmals um Entschuldigung zu bitten, daß ich Ihnen so viele Angelegenheiten verursache. Aber daran ist nur die freundliche Gesinnung Schuld, die Sie mir so oft beweisen. Denn diese ermuntert mich dazu, Ihre Güte so vielfach in Anspruch zu nehmen. Uebrigens würde ich mich glücklich schätzen, wenn ich Ihnen etwa hier oder späterhin irgend einen Dienst erweisen könnte und bitte ich Sie über mich zu disponiren.

Mit vorzüglicher Hochachtung
Ihr dankbarer Schüler
Leopold Kronecker

Kronecker to Dirichlet, 31 March 1854. – Sheet 15 and 16.

Breslau den 31. Maerz 1854
Herrnstraße 26

Geehrtester Herr Professor,

Entschuldigen Sie gütigst, daß die Abwesenheit von meinem eigentlichen Domizil die Zusendung der vierten Rate jener 50 [Taler][76] um einige Tage verzögert hat – aber wenn die häusliche Ordnung fehlt, so vergißt man gar leicht die ordnungsgemäße Erfüllung seiner Verpflichtungen. Indessen wird die Verzögerung wohl keine wesentlichen Knappheit herbeigeführt haben, da Sie ja äußersten Falls gewiß die vorschußweise Berichtigung an Ihr Mündel veranlaßt haben. – Meinen viel zu ausgedehnten Brief vom 6. Januar d.J. haben Sie hoffentlich erhalten – seitdem (wenn ich Sie noch einen Augenblick von mir unterhalten darf) hatte sich mein Leiden wieder so viel verschlimmert, daß eine Erneuerung meiner Pariser Cur unausweichlich wurde. Ich reiste deshalb Mitte Februar hierher, wo ich unter den Händen eines geschickten Arztes meine ebenso angreifende als schmerzhafte Cur wieder aufnahm. Erst in den letzten Tagen aber fange ich an einigen Erfolg zu sehen und hoffe nun durch Ausdauer im Dulden wenigstens eine erträgliche Existenz zu erlangen. Daß eine solche Cur, die mich etwa die Hälfte jedes Tages durch Wehe und Nach-Wehe in Anspruch nimmt, mir ein wesentliches Hindernis für meine Arbeiten ist, können Sie sich wohl denken; und Sie werden dadurch gewiß eine Lässigkeit entschuldigen, die sonst unverantwortlich wäre. Uebrigens hat mein hiesiger Aufenthalt auch seine Lichtseiten. Namentlich bewährt mir Kummer seine Freundschaft im außerordentlichsten Maße, und es vergeht buchstäblich fast kein Tag, an dem er mir nicht einen großen Theil seiner Zeit widmet und auch – wenn nöthig – den peinlichen Gedanken entreißt, die so oft das Gefolge meiner Schmerzen sind. Bei alle dem habe ich indessen nach Möglichkeit meine mathematischen Arbeiten gefördert, und bin jetzt mit der Ausarbeitung einer kleinen Abhandlung für das Journal von Liouville fast fertig. Dieser Mann, der mir überhaupt in Folge Ihrer gütigen Empfehlung viel Freundlichkeit erwiesen hat, versprach mir nämlich in Paris ein kleines mémoire über Irreductibilität von Kreisteilungsgleichungen, von dem ich ihm die Grundzüge mittheilte, in sein Journal aufzunehmen.[77]

1|2 |

Die Hauptabsicht dieser kleinen Abhandlung ist nur der Beweis der Irreductibilität jener Gleichung, die die primitiven n-ten Wurzeln der Einheit enthält, ein Beweis, den ich bereits in meiner Dissertation versprochen hatte. Dieser Beweis selbst ist deshalb so interssant, weil er eine wesentlich andere Methode enthält, als die, welche für den Fall, wo n Primzahl, ausreichen. Außerdem ergiebt sich noch eine sehr intereßante Verallgemeinerung nämlich, daß

[76] Kronecker uses the standard sign for this currency.

[77] Cf. [39, Vol. p. 413], as well as p. 510, note 32.

> diejenige Gleichung welche die primitiven n-ten Wurzeln der Einheit enthält auch in der Weise irreductibel ist, daß sie nicht in Factoren zerlegt werden kann, deren Coefficienten rationale Functionen von $\alpha, \beta, \gamma, \ldots$ sind, wenn jede dieser Größen $\alpha, \beta, \gamma, \ldots$ eine Wurzel einer Gleichung mit ganzzahligen Coefficienten ist, deren Determinante mit "n" keinen gemeinsamen Factor hat. Diese Gleichungen müssen aber dabei so eingerichtet sein, daß der Coeff. des höchsten Grades 1 ist.

Diese Verallgemeinerung wird in der kleinen Abhandlung mit bewiesen. Uebrigens habe ich noch mancherlei nette Sachen dabei über Irreductibilität von Gleichungen gemacht. Hauptsächlich habe ich mich aber mit Fundierung derjenigen Sachen von Abel beschäftigt, die ich bei meinen Arbeiten über Gleichungen voraussetzen mußte. Diese scheinbar undankbare Arbeit hat aber mich und auch Kummer (dem ich natürlich alles mittheile) sehr interessiert. Ich bin vollständig damit ins Reine gekommen und habe nun wenigstens die Beruhigung, alles, was ich Ihnen für die Academie mitgetheilt hatte, richtig zu wissen. Für den Zweck einer Veröffentlichung werde ich freilich noch Vieles an meinen Methoden zu feilen haben – und ich fürchte mich überhaupt sehr vor der Schwierigkeit der Darstellung dieser Dinge.[78] Aber die Hauptsache – die Bewahrheitung der wirklich unbewiesenen Abelschen Resultate – ist doch wenigstens erreicht. Uebrigens darf ich Ihnen dabei nicht unerwähnt laßen, daß die Galoisschen Methoden, die ich aber nicht so gründlich wie die Abelschen mit Critik studiert habe, mir ganz ebenso schwer auszufüllende Lücken zu haben scheinen. Uebrigens ist mir bei dieser ganzen mehr kritischen Arbeit doch so manches aufgefallen, d.h. nicht sowohl Resultate als einfach Beweismethoden für bekannte Dingen und Beweise für bekannte uner-
2|3 wiesene Thatsachen. Doch eignen | sich derlei Dinge überhaupt nicht für eine briefliche Mittheilung, am allerwenigsten sind sie werth Ihnen einige Zeit des Lesens zu rauben. – Aber Eines habe ich noch auf dem Herzen, dessen Erwähnung Ihre Freundlichkeit entschuldigen muß. Die unglückselige Eile nämlich, womit ich kurz vor meiner Abreise nach Paris vorm Jahre meine Notiz für die Academie schreiben mußte, hat nämlich – gelind gesagt – einige Ungenauigkeiten hinein gebracht, die mir einmal getadelt werden könnten. Ich schreibe sie Ihnen deshalb, um Sie gewissermaaßen bei Ihnen zu deponieren und dadurch meine Kenntnis derselben beweisen zu können.

1) Wenn ich nämlich pag. 5 des Abdrucks meiner Notiz in den Monatsberichten[79] sage "daß die aufgestellte Form nur Ausdrücke enthält, die dem Probleme genügen," so ist das insofern ungenau, als man für die Form III erst noch die Beschränkung hinzufügen muß, daß R_k sich nicht als eine vollständige μ-te Potenz der Größen $r, A, B, C, \ldots$ darstellen lasse, weil nämlich in diesem Falle die resultierende Gleichung keine irreductible wäre, wie es das Problem verlangt.

2) Wenn ich auf derselben pag. 5[80] die Gleichungen $z_2 = f(z_1, r_1)$ etc. aufstelle, sind diese insofern unrichtig, als dies r_1 nicht dasselbe wie in III ist, sondern die Wurzel einer anderen Abelschen Gleichung ist, die mit jener deren Wurzel die r's sind, sehr nahe und merkwürdig zusammenhängt.

Das erste dieser errata habe ich schon Heine mitgetheilt; das zweite ist nicht so wesentlich, und ist mir durch die Nachlässigkeit entstanden, daß ich schon auf meinen Arbeitsbogen für zwei verschiedene aber ähnliche Größen dieselbe Bezeichnung eingeführt hatte.

---

78 For the further development of Kronecker's work, see his letter to Dirichlet dated 3 March 1856, in [64, 411–414].

79 See [41, p. 6/7]: "Substituiert man ..., so erhält man eine Form, welche nicht nur jeder dem Problem genügende Ausdruck haben *muß*, sondern welche auch,(und das ist die Hauptsache) *nur* solche Ausdrücke enthält, die dem Probleme genügen."

80 [41, p. 7].

Ich habe Ihnen nur noch herzliche Grüße von Kummer auszurichten und Sie zu bitten mir Heine bestens zu grüßen und ihm gelegentlich zu sagen, daß, weil meine Person jetzt noch Wochen lang meinen Aerzten gehören wird, er statt meiner selbst nur einen sehr ausführlichen Brief in der nächsten Zeit noch in Berlin von mir zu erwarten habe. Daß Sie mir selbst die
3|4 langweilige Ausdehnung dieses Briefes | verzeihen werden, der doch nur ein Geleitschein der einliegenden 50 [Taler] sein sollte – hoffe ich von Ihrer Freundlichkeit, die Sie mir so oft bewiesen haben. Indem ich mich schließlich noch nebst meiner Frau Ihnen und Ihrer Frau Gemahlin angelegentlichst empfehle, bleibe ich

mit alter Anhänglichkeit
und Verehrung

Ihr dankbarer Schüler
Leopold Kronecker

Appendix II. A Letter from Felix Klein to Kurt Hensel

The following letter turned up as a lose sheet in the handwritten notes of Leopold Kronecker's lecture course: "Theorie der algebraischen Gleichungen" from the Winter term 1885–86. The notes were taken "by Dr. G. Stäckel," belonged to Kurt Hensel's library and ended up in the Strasbourg IRMA mathematical library when most of Hensel's library was sold to the Nazi "Reichuniversität Straßburg" in 1943. It is evident from this letter and from [31, Vol. III, p. 503–504], that Hensel had sent Klein this volume of lecture notes when the latter was preparing the commentaries for the last edition of his Collected Works. The manuscript from Kronecker's Nachlaß that Klein refers to at the beginning of the letter and which Hensel had sent along as well, has apparently been destroyed, with most of Kronecker's papers, shortly after World War II in the explosion of a mineshaft in the Harz mountains, not far from Göttingen, where books and documents had been moved during the war to protect them from possible air raids. On the fate of Kronecker's Nachlass, see [14]. (On p. 421 of this report, Edwards mentions a letter from Hensel to Klein dated 25 November 1925 which belongs to the same exchange between Hensel and Klein as the letter published here.)

Göttingen 19–XII.21.

Verehrter Hr. Kollege!

Ihr Material habe ich nun glücklich vor dem Feste noch durchgearbeitet und lasse es Ihnen durch meinen Assistenten Dr. Vermeil, da dieser in die Ferien geht, mit bestem Dank noch zuschicken.

Die Manuskriptblätter 1–23 bieten in der Tat nichts Neues: p. 1–10 beziehen sich auf die Arbeit von 1858, p. 11–23 auf die von 1861. Auf der Rückseite von p. 17–18 finden sich Rechnungen mit fünften Einheitswurzeln, aus denen hervorgeht, dass sich Kronecker hinterher von der Existenz der G_{60} der nicht homogenen Ikosaedersubstitutionen überzeugt hat. Ob Kr[onecker] die Blätter, die ich 1881 abschrieb, hinterher vernichtet hat?

In der Vorlesung von 1885/86 wurden die ursprünglichen Ueberlegungen von 1861 im wesentlichen genau so vervollständigt, wie ich es getan habe. Nun ist der Unmöglichkeitsbeweis betr. eine holoedrisch isomorphe Gruppe binärer Substitutionen etwas weniger einfach als bei mir. Während ich ihn an der einfachsten Gruppe, der Vierergruppe, führe, zieht Kr.

die G_{12} aller geraden Vertauschungen von 4 Dingen in Betracht und kombiniert eine Substitution der bez. Vierergruppe mit der zyklischen Vertauschung von 3 Elementen. Wichtiger ist mir, dass er auch nun, nicht ohne Pathos, den Gebrauch einer "akzessorischen" Irrationalität als "algebraisch wertlos" verbietet.[81] Das ist die merkwürdige Doppelnatur von Kronecker: auf der einen Seite der tiefeindringende Forscher, der die schönsten Resultate findet oder doch erschaut, auf der anderen Seite der Dogmatiker, der glaubt, für die Methode der vorwärtsdringenden Wissenschaft Schranken aufrichten zu können. Vergl[eiche] seine Stellungsnahme gegen die Irrationalzahlen.

Nun kommt noch eine persönliche Seite. Kr[onecker] hat es in der Vorlesung von 85/86 nicht über sich gebracht, anzugeben, dass ich erst seinen Beweis von 1861 fertig gestellt habe, sondern erreicht beim Zuhörer die Ueberzeugung entstehen zu lassen, dass er das alles selbst gemacht habe und von mir nur die "Benennung" Ikosaedergleichung herrühre. Da die Vorlesung von 85/86 nicht publiziert ist, unterlasse ich es in den Bemerkungen, die ich nun meinem Wiederabdruck zufüge, hierauf einzugehen; ich habe keine Lust, unnötig Schlamm aufzurühren. Dagegen erörtere ich den math. Sachverhalt, wie ich ihn oben andeutete, um so genauer. Ich hoffe, dass Springer meine bez. Ausführungen in kürzester Zeit setzt und lass Ihnen bez. Fahnenkorrektur von ihm direkt zugehen, mit der Bitte, Sie wollen mir alle Bedenken, die Sie diesbezüglich haben werden, möglichst ungehend mitteilen.

Ueber die komplexen Elemente habe ich nichts Sonderliches zu sagen; für mich ist immer die analytische Geometrie der Ausgangspunkt gewesen.

Mit angelegentlichen Empfehlungen

Ihr ergebener
Klein

References

[1] Niels Henrik Abel: Œuvres complètes (Sylow, Lie, eds.), 2 volumes, Christiania 1881

[2] Niels Henrik Abel: Démonstration de l'impossibilité de la résolution algébrique des équations générales qui passent le quatrième degré, (German version: Journal für die reine und angewandte Mathematik **1** (1826), 65–84), *Œuvres*, vol. I, pp. 66–87

[3] Niels Henrik Abel: Mémoire sur une propriété générale d'une classe trés-étendue de fonctions transcendantes, *Œuvres*, vol. I, pp. 145–211; Abel's manuscript of this paper is wonderfully reproduced in: Il manoscritto parigino di Abel / The Abel's Parisian manuscript (a cura di Andrea del Centina), Provinicia di Firenze, Collana Cultura e memoria **25**, Firenze (L.S. Olschki) 2002.

[4] Niels Henrik Abel: Recherches sur les fonctions elliptiques, Journal für die reine und angewandte Mathematik **2** (1827), 101–181; **3** (1828), 160–190 = *Œuvres*, vol. I, pp. 263–388

[5] Niels Henrik Abel: Solution d'un probléme général concernant la transformation des fonctions elliptiques, Astronomische Nachrichten, Altona (Schumacher) **6** (1828) = *Œuvres*, vol. I, pp. 403–428

[6] Niels Henrik Abel: Mémoire sur une classse particulière d'équations résolubles algébriquement, Journal für die reine und angewandte Mathematik **4** (1829) = *Œuvres*, vol. I, pp. 478–507

[81] This alludes to [52, p. 164].

[7] Niels Henrik Abel: Sur la résolution algébrique des équations, *Œuvres*, vol. II, pp. 217–243

[8] Enrico Betti: Opere matematiche, 2 volumes, Milano (Hoepli)

[9] Enrico Betti: Sulla resoluzione delle equazioni algebraiche, Annali di Scienze matematiche e fisiche **III** (1852), 49–115 = *Opere*, vol. I, pp. 31–98

[10] Kurt-R. Biermann: Die Mathematik und ihre Dozenten an der Berliner Universität 1810–1933, Stationen auf dem Wege eines mathematischen Zentrums von Weltgeltung, Berlin (Akademie-Verlag) 1988

[11] Jacqueline Boniface & Norbert Schappacher: 'Sur le concept de nombre en mathématique' – cours inédit de Leopold Kronecker à Berlin (1891), Revue d'histoire des mathématiques **7** (2001), 207–275

[12] Georg Cantor: Georg Cantor, Briefe, edited by Herbert Meschkowski & Winfried Nilson, Berlin · Heidelberg · New York (Springer) 1991

[13] Harold M. Edwards: Fermat's Last Theorem, A Genetic Introduction to Algebraic Number Theory, GTM **50**, New York · Berlin · Heidelberg (Springer) 1977

[14] Harold M. Edwards: On the Kronecker *Nachlass*, Historia Mathematica **5** (1978), 419–426

[15] Gotthold Eisenstein: Mathematische Werke, 2 vol.s, New York (Chelsea) 1975

[16] Gotthold Eisenstein: Allgemeine Auflösung der Gleichungen von den ersten vier Graden, Journal für die reine und angewandte Mathematik **27** (1844), 81–83 = *Werke*, vol. I, 7–9

[17] Robert Fricke: Die elliptischen Funktionen und ihre Anwendungen, Zweiter Teil: Die algebraischen Ausführungen, Leipzig & Berlin (Teubner) 1922

[18] Évariste Galois: Ecrits et mémoires mathématiques d'Évariste Galois, edited by R. Bourgne, J.-P. Azra, Paris (Gauthier-Villars) 1997

[19] Évariste Galois: Mémoires sur les conditions de résolubilité des équations par radicaux, *in:* [*EM*], 43–71

[20] Lars Gårding: Mathematics and Mathematicians. Mathematics in Sweden before 1950, Providence, RI (American Mathematical Society) 1997

[21] Lars Gårding & C. Skau: Niels Henrik Abel and solvable equations, Archive for the History of Exact Sciences **48** (1994), 81–103

[22] Bettina Heintz: Die Innenwelt der Mathematik, zur Kultur und Praxis einer beweisenden Disziplin, Wien · New York (Springer) 2000

[23] Charles Hermite: Œuvres de Charles Hermite, edited by Emile Picard, four volumes, Paris, Gauthier-Villars, 1905–1917

[24] Charles Hermite: Sur la résolution de l'équation du cinquième degré, Comptes rendus de l'Académie des Sciences, XLVI (1858), 508–515 = *Œuvres*, vol. II, 5–12

[25] Otto Hesse: Algebraische Auflösung derjenigen Gleichungen 9ten Grades, deren Wurzeln die Eigenschaft haben, daß eine gegebene rationale und symmetrische Function $\theta(x_1, x_2)$ je zweier Wurzeln x_1, x_2 eine dritte Wurzel x_k giebt, so daß gleichzeitig $x_\kappa = \theta(x_\lambda, x_\mu), x_\lambda = \theta(x_\mu, x_\kappa), x_\mu = \theta(x_\kappa, x_\lambda)$, Journal reine angew. Math. **34** (1847), 193–208

[26] Christian Houzel: Fonctions elliptiques et intégrales abéliennes, *in:* J. Dieudonné (ed.), Abrégé d'histoire des mathématiques 1700–1900, vol. II, Paris (Hermann), 1–113

[27] Carl Gustav Jacob Jacobi: Gesammelte Werke, edited by C.W. Borchardt, eight volumes, 1881–1891

[28] Carl Gustav Jacob Jacobi: Suite des notices sur les fonctions elliptiques, aus einem Brief an Crelle vom 21.07.1828, *Werke*, vol. I, 255–275

[29] Carl Gustav Jacob Jacobi: Fundamenta nova theoriæ functionum ellipticarum, Königsberg (Bornträger) 1829 = *Werke*, vol. I, 49–239

[30] Carl Gustav Jacob Jacobi: Über die Kreisteilung und ihre Anwendung auf die Zahlentheorie. Auszug eines Schreibens an die Akademie der Wissenschaften zu Berlin; Monatsbericht Akademie Berlin October 1837 (also contained in Journal reine angew. Math. **30** (1837), 166–182) = *Werke*, vol. VI, 254–264
[31] Felix Klein: Gesammelte mathematische Abhandlungen, three volumes, edited by R. Fricke, A. Ostrowski, H. Vermeil and E. Bessel-Hagen, Berlin (Springer) 1921–1923
[32] Felix Klein: Über eine geometrische Repräsentation der Resolventen algebraischer Gleichungen, Math. Ann. **4** (1871), 346–358 = *GMA*, vol. II, 262–274
[33] Felix Klein: Über binäre Formen mit linearen Transformationen in sich selbst, Math. Ann. **9** (1875/76), 183–208 = *GMA*, vol. II, p. 275–301
[34] Felix Klein: Weitere Untersuchungen über das Ikosaeder, Math. Ann. **12** (1877), 505–560 = *GMA*, vol. II, 321–380
[35] Felix Klein: Über die Transformation der elliptischen Funktionen und die Auflösung der Gleichungen fünften Grades, Math. Ann. **14** (1878/79), 111–172 = *GMA*, vol. III, 13–75
[36] Felix Klein: Über die Auflösung gewisser Gleichungen vom siebenten und achten Grade, Math. Ann. **15** (1879), 251–282 = *GMA*, vol. II, 390–425
[37] Felix Klein: Vorlesungen über das Ikosaeder und die Auflösung der Gleichungen vom fünften Grade, (Teubner) 1884; reprinted with an introduction and commentaries by Peter Slodowy, (Birkhäuser & Teubner) 1993
[38] Felix Klein: Über die Auflösung der allgemeinen Gleichungen fünften und sechsten Grades, Journal reine angew. Math. **129**, 1905, p. 151–174 (also contained in Math. Ann. **61** (1905), 50–71) = *GMA*, vol. II, 481–502
[39] Leopold Kronecker: Leopold Kroneckers Werke, five volumes, edited by Kurt Hensel, Leipzig, 1895–1930, reprinted New-York, 1968
[40] Leopold Kronecker: De unitatibus complexis, Dissertatio inauguralis arithmeticæ, Berlin 1845 [§§1–16]; complete version [§§1–20] published in Journal reine angew. Math. **93** (1882), 1–52 = *Werke*, vol. I, 5–73
[41] Leopold Kronecker: Über die algebraisch auflösbaren Gleichungen, Monatsber. Kgl. Preuss. Akad. Wiss. Berlin (1853), 365–374 = *Werke*, vol. IV, 3–11
[42] Leopold Kronecker: Über die algebraisch auflösbaren Gleichungen, Monatsber. Kgl. Preuss. Akad. Wiss. Berlin (1856), 203–215 = *Werke*, vol. IV, 27–37
[43] Leopold Kronecker: Zwei Sätze über Gleichungen mit mit ganzzahligen Koeffizienten, Journal reine angew. Math. **53** (1857), 173–175 = *Werke*, vol. I, 105–108
[44] Leopold Kronecker: Über complexe Einheiten, Journal reine angew. Math. **53** (1857), 176–181 = *Werke*, vol. I, 111–118
[45] Leopold Kronecker: Sur la résolution de l'équation du cinquième degré. Extrait d'une lettre adressée à M. Hermite, Berlin, le 6 juin 1858, Comptes rendus des séances de l'Académie des Sciences de Paris, **XLVI**, Sem. I., séance du 14 juin, 1150–1152 = *Werke*, vol. IV, 45–47
[46] Leopold Kronecker: Mitteilung über algebraische Arbeiten (Über Gleichungen fünften Grades), Monatsber. Kgl. Preuss. Akad. Wiss. Berlin (1861), 609–617 = *Werke*, vol. IV, 55–62
[47] Leopold Kronecker: Auseinandersetzung einiger Eigenschaften der Klassenzahl idealer complexer Zahlen, Monatsber. Kgl. Preuss. Akad. Wiss. Berlin (1870), 881–889 = *Werke*, vol. I, 273–282
[48] Leopold Kronecker: Grundzüge einer arithmetischen Theorie der algebraischen Grössen, Festschrift zu Herrn Ernst Kummers fünfzigjährigem Doctor-Jubiläum, 10. September 1881, Berlin, Reimer; published again in Journal reine angew. Math. **92** (1882), 1–122 = Werke, vol. II, 239-387

[49] Leopold Kronecker: Zur Theorie der Gattungen rationaler Functionen von mehreren Variabeln, Monatsber. Kgl. Preuss. Akad. Wiss. Berlin (1886), 251–253 = *Werke*, vol. III–1, 277–280

[50] Leopold Kronecker: Zur Theorie der allgemeinen complexen Zahlen und der Modulsysteme, Sitzungsber. Kgl. Preuss. Akad. Wiss. Berlin (1888), 429–438, 447–465, 557–578, 595–612, 983–1016 = *Werke*, vol. III–2, 3–114.

[51] Leopold Kronecker: Vorlesungen über Zahlentheorie, Erster Band, 1. bis 31. Vorlesung, bearbeitet und herausgegeben von Kurt Hensel, Leipzig (Teubner) 1901 ; reprint (Springer) 1978

[52] Leopold Kronecker & G. Staeckel: Über die Affecte der Gleichungen, welche in der Theorie der elliptischen Funktionen vorkommen; unpublished handwritten lecture notes of Kronecker's Berlin lecture course, winter 1885–86; Bibliothèque mathématique de l'IRMA, Strasbourg

[53] Ernst Eduard Kummer: Collected Papers, vol. I, Contributions to number theory, edited by A. Weil, Berlin · Heidelberg · New York, (Springer) 1975

[54] Ernst Eduard Kummer: Über die Zerlegung der aus Wurzeln der Einheit gebildeten complexen Zahlen in ihre Primfactoren, Journal reine angew. Math. **35** (1847), 327–367 = *CP*, 211–251

[55] Ernst Eduard Kummer: Mémoire sur la théorie des nombres complexes composés de racines de l'unité et de nombres entiers, Journal de mathématiques pures et appliquées **XVI** (1851), 377–498

[56] Serge Lang: Cyclotomic Fields I and II, combined second edition, Graduate Texts in Math. **121**, New York · Berlin · Heidelberg (Springer) 1990

[57] Eduard Luther: De criteriis quibus cognoscatur an æquatio quinti gradus irreductibilis algebraice resolvi possit, Journal reine angew. Math. **34** (1847), 244–254

[58] Eduard Luther: Über die Factoren der algebraisch-lösbaren irreductibeln Gleichungen vom sechsten Grad und ihrer Resolventen, Journal reine angew. Math. **37** (1848), 193–220

[59] Carl Johann Malmsten: In solutionem Æquationum Algebraicarum Disquisitio, Journal reine angew. Math. **34** (1847), 46–74

[60] Olaf Neumann: Two proofs of the Kronecker–Weber theorem "according to Kronecker, and Weber," Journal reine angew. Math. **323** (1981), 105–126

[61] Samuel J. Patterson: Eisenstein and the quintic equation, Historia Mathematica **17** (1990), 132–140

[62] James Pierpont: Zur Geschichte der Gleichungen des fünften Grades (bis 1858), Monatshefte für Mathematik und Physik, Wien 1895, 15–68

[63] Norbert Schappacher: On the History of Hilbert's Twelfth Problem, A Comedy of Errors, *in:* "Matériaux pour l'histoire des mathématiques au XXe siècle," Actes du colloque à la mémoire de Jean Dieudonné (Nice, 1996), Séminaires et Congrès (Société Mathématique de France) **3** (1998), 243–273

[64] Ernst Schering: Briefwechsel zwischen Gustav Lejeune-Dirichlet und Herrn Leopold Kronecker, herausgegeben von Ernst Schering, Nachrichten von der Gesellschaft der Wissenschaften zu Göttingen 1885, S. 361–382 = [39, vol. V, 409–431]

[65] Theodor Schoenemann: Über die Beziehungen, welche zwischen den Wurzeln irreductibler Gleichungen stattfinden, besonders wenn der Grad derselben eine Primzahl ist, Denkschrift Wiener Akademie mathematisch-physikalische Classe V

[66] Jean-Pierre Serre: Œuvres – Collected Papers, vol. III: 1972–1984, Berlin · Heidelberg (Springer) 1986

[67] Jean-Pierre Serre: Extensions icosaédriques, *CP*, 550–554

[68] Heinrich Weber: Die allgemeinen Grundlagen der Galois'schen Gleichungstheorie, Mathematische Annalen **43** (1893), 521–549
[69] Heinrich Weber: Lehrbuch der Algebra, Second edition: volume I, Braunschweig (Viehweg) 1898
[70] Heinrich Weber: Lehrbuch der Algebra, Second edition: volume II, Braunschweig (Viehweg) 1899

On the History of the Artin Reciprocity Law in Abelian Extensions of Algebraic Number Fields: How Artin was Led to his Reciprocity Law

Günther Frei

1 Introduction: Abelian Extensions and Abelian Equations

Artin's Reciprocity Law (1923/26) for Abelian extensions of algebraic number fields is the central theorem of *Class Field Theory* which, by the Theory of Takagi (1920) [64], is the Theory of Abelian Extensions of Algebraic Number Fields. Abelian extensions of algebraic number fields have been studied extensively in the second half of the 19-th century, in particular by Kronecker (1853, 1856, 1877, 1882), Weber (1886/87, 1897/98) and Hilbert (1896, 1897, 1898) who laid the foundations and discovered many fundamental properties of the Class Fields. These discoveries were made possible by the thorough study of a particular kind of Abelian extensions, namely the study of cyclotomic fields initiated by Gauss (1801) [24, Sect. 7] and carried further by Kummer (1847–1874). The term *Abelian* in connection with algebraic extensions (or, at that time, algebraic equations) was coined by Kronecker in 1853, first related to algebraic equations as '*Abelian equation*.' [1] By this Kronecker referred to polynomials with cyclic Galois group. However, Kronecker was already

[1] see [45], Werke, Bd. 4, p. 1 and in particular p. 6.

then aware of the more general polynomials with Abelian (in the modern sense) Galois group, since he is referring on page 6 (in [45], Werke, Bd. 4) to the fundamental treatise of Abel of 1829 which appeared in Volume 4 of Crelle's Journal [1], just two months before Abel's death, and where Abel states explicitly the condition for commutativity $\theta\theta_1 x = \theta_1\theta x$ in his assertion that a polynomial with Abelian Galois group is solvable (with radicals). Abel says there on page 479 of [1]: "En général j'ai démontré le théorème suivant: 'Si les racines d'une équation d'un degré quelconque sont liées entre elles de telle sorte, que *toutes* ces racines puissent être exprimées rationnellement au moyen de l'une d'elles, que nous désignerons par x; si de plus, en désignant par θx, $\theta_1 x$ deux autres racines quelconques, on a $\theta\theta_1 x = \theta_1\theta x$, l'équation dont il s'agit sera toujours résoluble algébriquement.' " (In general I have demonstrated the following theorem: If the roots of an equation of any degree are related to each other in such a way that *all* these roots can be expressed rationally by means of one of them, which will be denoted by x; if, in addition, we have $\theta\theta_1 x = \theta_1\theta x$, where we denote by θx, $\theta_1 x$ any two other roots, then the equation in question is algebraically solvable [that is, solvable by radicals].) Later in 1877 [48] Kronecker used the term '*Abelian equation*' in the larger sense to mean a polynomial with Abelian Galois group in our modern sense. He says there on page 66 (in [48], Werke, Bd. 4) that he now calls '*Abel*ian equation' an equation having the property that all its roots x are rational functions of any one of them and if θ_1 and θ_2 are two of these functions then $\theta_1\theta_2 x = \theta_2\theta_1 x$. He then calls '*simple Abel*ian equation' the equation he treated in 1853, namely the equation with cyclic Galois group of prime order. That he had already used the term '*Abelian equation*' in the paper of 1853 in the special case of cyclic equations is justified by Kronecker by the fact that the '*Abelian equation*' can be reduced to the '*simple Abelian equation*,' or as Kronecker says on page 69 (in [48], Werke, Bd. 4), that every root of any *Abel*ian equation is a rational function of roots of *simple Abel*ian equations.[2] This justification is repeated in his paper of 1882 on the composition of Abelian equations.[3] In the paper of 1853 Kronecker also states the famous theorem that any root of a (simple) Abelian equation [that is a cyclic equation] with *integral rational* coefficients can be represented by roots of unity, that is, is contained in a cyclotomic field over the rational number field.[4] The same theorem is stated by Kronecker in his paper of 1877 in the case of the general Abelian equation with *integral rational* coefficients.[5] It was later proved by Weber (1886) [67] and more simply by Hilbert (1896) [41]. In both papers Kronecker also stated his Jugendtraum for an analogous theorem for Abelian equations with coefficients in a quadratic imaginary number field. This theorem was proved partially by Weber (1908) [69] and Fueter (1914) [23] and completely by Takagi (1920) [64]. Both theorems have plaid a crucial rôle in the history of class field theory.[6]

---

[2] That Abelian equations can be reduced to cyclic equations was already shown by Abel in his paper of 1829, see [1].

[3] see [50], Werke, Bd. 4, pp.118–119.

[4] see [45], Werke, Bd 4, pp. 10–11.

[5] see [48], Werke, Bd 4, p. 69.

[6] For the history of class field theory see [17], [30] and [57].

This paper must be viewed as a continuation of the paper *How Hasse was led to the Theory of Quadratic Forms, the Local–Global Principle, the Theory of the Norm Residue Symbol, the Reciprocity Laws, and to Class Field Theory* [19] and as a preparation for a forthcoming paper on the correspondence between Artin and Hasse related to the Reciprocity Laws and on the history of the Reciprocity Laws [21]. It is part of a project, jointly with Peter Roquette, on the Artin–Hasse correspondence and on the work of Helmut Hasse.

2 Artin's Thesis Advisor: Gustav Herglotz. Artin's Thesis

2.1 Artin's Thesis Advisor: Gustav Herglotz. 1. *Emil Artin* (1898–1962) completed his doctoral dissertation in June 1921 under the direction of *Gustav Herglotz* (1881–1953) in Leipzig. That he was working quite independently appears from his letters to Herglotz.[7] However, from the same letters, we can also infer that Herglotz must have had some influence on Artin which went far beyond of Herglotz just being a listener to Artin's progress reports. For, Artin writes from Göttingen where he spent a year, after having obtained his doctorate, in a letter dated 13 November 1921:

> "Nach meiner Ankunft [in Göttingen] machte ich die Besuche bei Courant, Hilbert, Klein und Landau und wurde von Hilbert in die Mathematische Gesellschaft eingeladen. Dienstag in acht Tagen soll ich dort über meine Dissertation berichten. Leider habe ich hier sehr wenig Fühlung mit den Dozenten so dass mir die persönliche Anregung fehlt, die ich in Leipzig in so weitgehendem Masse durch Sie Herr Professor hatte. Dafür werde ich Ihnen immer Dank schulden."

> "After my arrival [in Göttingen] I paid the [obligatory] visits to Courant, Hilbert, Klein and Landau, and was invited to the Mathematische Gesellschaft by Hilbert. Tuesday in 8 days, I am to report on my thesis. Unfortunately, I have here very little contact with the Dozenten [faculty]. For this reason, I am missing the personal inspiration I got in Leipzig, to such a large extent, from you, Herr Professor. For this, I will forever be grateful to you."[8]

Artin was not lucky with his lecture in the Mathematische Gesellschaft, for he reports on the 30th of November 1921 to Herglotz:

> "Meinen Vortag habe ich gehalten, doch habe ich bei Hilbert kein Glück damit gehabt. Landau und den Zahlentheoretikern hat er ja sehr gut gefallen wie sie auch während des Vortrags, als Hilbert mich öfters unterbrach,

[7] These letters are kept, together with the Herglotz-Nachlass, in the Handschriftenabteilung of the Niedersächsische Staats- und Universitätsbibliothek Göttingen.

[8] This is also in agreement with what the editors of *The Collected Papers of Emil Artin* say in the *Preface*: "Herglotz was the only person whom Artin recognized as having been his 'teacher'." (see [7]).

sagten. Aber Hilbert unterbrach mich häufig, zum Schluss konnte ich gar nicht mehr reden[,] und sagte[,] er habe von Anfang an überhaupt nicht zugehört da er Alles für Trivialitäten gehalten habe. Von dieser Meinung ist er nun aber abgekommen als ich (ich musste dies ganz ausser dem Zusammenhang tun da ich nicht reden konnte und die letzten Resultate meiner Dissertation und meiner letzten Untersuchungen nicht vorbringen konnte) die erwähnten Primzahlzerlegungen angab. Ich bin aber doch damit reingefallen und Hilbert hat mir die ganze Lust am Arbeiten verdorben durch seine Kritik die ich übrigens (und die anderen auch) für nicht gerechtfertigt halte. Ich weiss ja nicht wie Sie darüber denken aber das verdirbt die ganze Freude an den Ergebnissen."

(I have now given my lecture, but, as far as Hilbert is concerned, I was not lucky. Landau and the number theorists liked it very much, as they also expressed, while Hilbert was interrupting me frequently. But Hilbert was interrupting me so often that, at the end, I could not talk [finish] any more. And Hilbert said that, from the beginning, he did not listen at all, since he took everything for trivial. Now, he has changed his mind, when I indicated the [before mentioned] prime-factorization (I had to do this completely out of context, since I could not finish my talk and present my last results of my dissertation and my latest investigations). So it was nevertheless a flop after all, and Hilbert has spoiled my joy for work completely by his critic which, by the way, I (and also the others) find not justified. I do not know how you think about this, but this event has spoiled all my joy for the results.)

Then Artin finishes this letter by saying:

"Verzeihen Sie nun Herr Professor, dass ich Sie schon wieder mit einem so langen Brief belästigt habe aber es wird wohl mit diesem Thema nicht mehr vorkommen da ich es wohl an den Nagel hängen werde."

(Please, excuse me for bothering you again with such a long letter, but it will hardly happen again with this subject, since I will probably give it up.)

2. Indeed, Artin's thesis and his first papers start right in the middle of where Herglotz' main interest were in the years from 1920 to 1923. Namely, following up Hecke's investigations between 1917 and 1920 on zeta-functions and L-functions, Herglotz published 6 papers on number theory in the years from 1921 to 1923, mainly on applying elliptic functions and the theory of the ζ-functions to important problems in algebraic number theory. These publications are closely related to Artin's first publications, and thus must have influenced the beginning of Artin's scientific career decisively. In particular, Artin's thesis must go back to a suggestion of Herglotz, possibly following a publication by Kornblum.[9] But also Artin's first interest in ζ-functions, L-functions and the connected class number relations is visibly a direct result of Herglotz' interests at that time. This can be inferred from the publications by Herglotz on number theory in the period 1921–1923. Let us therefore review them briefly.

[9] see section 6, art. 10.

3. In a paper entitled *Über das quadratische Reziprozitätsgesetz in imaginären quadratischen Zahlkörpern,*[10] Herglotz gives a proof of the quadratic reciprocity law in imaginary quadratic number fields by means of complex multiplication, that is, by means of the derivative $\wp'$ of the Weierstrass $\wp$-function, in analogy to Eisenstein's classical proof of the quadratic reciprocity law for the rational numbers by means of the trigonometric sinus-function. In *Zur letzten Eintragung im Gaußschen Tagebuch,*[11] Herglotz takes up the statement of the last entry in Gauss' Diary where Gauss states that the number of points, including the points at infinity, on the elliptic curve $x^2y^2+x^2+y^2 \equiv 1$ modulo a complex primary Gaussian prime $\pi \in \mathbb{Z}[i]$, is $N(\pi-1)$ where N denotes the norm, and that this "most important observation connects the theory of biquadratic residues most elegantly with the lemniscatic functions" [29, p. 127]. Herglotz proves this conjecture of Gauss by showing that these points are exactly the roots modulo π of a certain division equation of the Gaussian lemniscatic functions *sin lemn* and *cos lemn*, namely the division equation of the multiplicator $\pi - 1$ viewed as an element of the endomorphism ring of the lemniscate. In *Über einen Dirichletschen Satz,*[12] on a theorem by Dirichlet, Herglotz extends a result by Dirichlet on the class number of – what Hilbert called – the Dirichlet biquadratic number field[13], from a biquadratic field to a number field composed of n quadratic fields, namely, that the class number of a field K of degree 2^n over $\mathbb{Q}$, composed of n disjoint quadratic extensions k_i over $\mathbb{Q}$, is the product of the class numbers of all the $2^n - 1$ quadratic sub-fields, up to a well determined factor which depends on the units. To get this result, Herglotz uses Hilbert's theory of the Galois extension, results by Dedekind on the sub-fields of a Galois extension, and by Hecke on the zeta-functions of sub-fields, viewed as sub-fields of cyclotomic fields, and on the functional equation of the zeta-function. In *Über die Entwicklungskoeffizienten der Weierstraßschen $\wp$-Funktion*[14] on the coefficients in the expansion of the Weierstrass $\wp$-function, Herglotz gives another approach to Hurwitz' analogue of the v. Staudt–Clausen Theorem and of v. Staudt's partial fraction decomposition of the Bernoulli numbers in the case of the lemniscate. Via the theory of complex multiplication, Herglotz then derives, for imaginary quadratic number fields, arithmetic properties and explicit arithmetic formulae for singular values of functions arising from Weierstrass' $\wp$-function. Finally, in *Über die Kroneckersche Grenzformel für reelle quadratische Körper, I, II*[15] on Kronecker's limit formula for real quadratic number fields, Herglotz builds on a paper by Hecke of 1917 with almost the same title, where Hecke determines the constant term in the development of the Dedekind ζ-function at $s = 1$ for a real quadratic number field k. In each of the two papers I

[10] Leipziger Ber. 73 (1921), pp. 303–310; Ges. Schriften, pp. 396–403.

[11] Leipziger Ber. 73 (1921), pp. 271–276; Ges. Schriften, pp. 415–420.

[12] Math. Zeitschrift 12 (1922), pp. 255–261; Ges. Schriften, pp. 429–435. Received on the 17th of September 1921.

[13] Hilbert's paper on Dirichlet's biquadratic number field (Math. Ann. 45, 1894, pp. 309–340) was a preparation to his theory of the general relativ-quadratic field (Math. Ann. 51, 1899, pp. 1–127) where Hilbert outlined in detail his concept of class field theory.

[14] Leipziger Ber. 74 (1922), pp. 269–289; Ges. Schriften, pp. 436–456.

[15] Leipziger Ber. 75 (1923), pp. 3–14 and pp. 31–37; Ges. Schriften, pp. 466–484.

and II, Herglotz computes this number in a new way. This constant term contains, as an essential part, the logarithm of the Dedekind η-function at a specific value ω belonging to the field k. By extending the study of this constant term to any number field k, Hecke was hoping to find transcendental analytic functions that would allow to construct class fields over k, in a way similar to the construction of the Kronecker class field over an imaginary quadratic number field by means of the Klein modular j-function [17] or to the construction of all abelian extensions over $\mathbb{Q}$ by means of the exponential function. This hope of Hecke's was based on the connection that exists between the theory of the linear transformation of the η-function and Kronecker's limit formula, and between the number field k and algebraic extensions of k generated by singular values $\eta(\omega)$ of η with ω in k.

4. So we see that Herglotz was studying difficult and important problems in number theory lying right at the intersection where number theory, complex analysis and algebraic geometry meet, problems which are still at the focus of today's research and which are, for a large part, still unsolved. They are essentially revolving around the analytic class number formula and around fundamental properties of class field theory and its generalizations, and the construction of class fields by means of transcendental functions. They are closely connected with the ones investigated by Artin in his dissertation, and in his papers on zeta- and L-functions immediately following the dissertation, in the sense that these also focus on the analytic class number formula and on the relation between the zeta-function of a field K and the zeta-function of a sub-field k of K. Herglotz' influence on the paper by Artin on zeta-functions and L-functions will be examined more closely in Sect. 6, art. 3.

2.2 Function Fields over Finite Fields. 1. In his dissertation, Artin initiates the theory of quadratic extensions of function fields whose constants lie in a finite field of characteristic $p \neq 0$. More precisely, and put in other words, Artin studies the theory of hyper-elliptic curves over finite prime fields [4]. In doing this he was following up investigations by Dedekind in 1856[16] on what was called by Gauss *"higher congruences with respect to a prime number"* or *"double congruences"*, and by Kronecker *"modular systems of level 2"*[17].

2. The theory of higher congruences was very crucial for the elaboration of the analogy between function fields and number fields. This analogy was, from the time of Gauss on up to our days, one of the most driving forces for the creation of new concepts and ideas and is responsible for many important discoveries in algebra and number theory. It was Gauss who introduced the theory of higher congruences, that is, the theory of function fields over a finite field of constants, and it was also Gauss who first stressed and elaborated on the close analogy between the arithmetic of these function fields and the arithmetic of the field of rational numbers. We will study in a separate paper the development of the theory of function fields over finite fields from Gauss to Dedekind and Artin [20]. Here, we will only briefly indicate

[16] see [10]; J. Reine Angew. Math. 54 (1857), 1–26. The paper was written up by Dedekind in October 1856 in Göttingen, but was published only in 1857.

[17] see [49], § 21; Werke, Bd. 2, pp. 336–339.

the main lines of this development as much as they are relevant for the discussion of Artin's thesis. For the details we refer to [20] and [22].

3 Gauss and the Beginning of the Theory of Function Fields

1. It is well known that Gauss laid the foundation of modern number theory in his fundamental treatise *Disquisitiones arithmeticae* which is divided up into seven sections.[18] What is less well known is that Gauss had intended to include an eighth section on function fields and on the cyclotomy of these function fields, running parallel to his theory of cyclotomy for rational numbers as developed in Section Seven. A draft of this Section Eight was already written up in 1797. But, by lack of time, and also because Gauss intended to develop it further, it was not included in the *Disquisitiones arithmeticae* in order to give space to a more detailed treatment of sections Five and Seven on quadratic forms and on cyclotomy. These notes by Gauss on Section Eight, entitled *Disquisitiones generales de congruentiis* (General treatise on congruences) were only found after his death among his papers as a *Chapter Eight* of a manuscript, entitled *Analysis residuorum.* For various reasons, they must have belonged to an earlier version of a part of the *Disquisitiones arithmeticae* [22]. They were edited by Dedekind in 1863, second printing in 1876, in the second volume of Gauss' collected works under the title *Disquisitiones generales de congruentiis.*[19]

2. Already at the very beginning of his *Disquisitiones generales de congruentiis,* Gauss insists strongly on the analogy between number fields and function fields (art. 330), or more precisely, between properties in the ring of integers $\mathbb{Z}$ and the rings of polynomials $\mathbb{Q}[x]$ and $\mathbb{F}_p[x]$, where $\mathbb{F}_p$ is the prime field of characteristic $p \neq 0$ [56, p. 602].

3. Then Gauss establishes the fundamental theorem of arithmetic for the ring of polynomials – Gauss calls them 'functions' – $\mathcal{R} = \mathbb{F}_p[x]$ of the function field $\mathcal{K} = \mathbb{F}_p(x)$ with respect to a fixed prime number p (art. 340). It is based on the existence of a Euclidean algorithm in $\mathcal{R}$, proved in art. 334.

Next, Gauss starts out to solve what he calls the main problem of his investigations, namely to determine the number of irreducible polynomials – Gauss calls them 'prime functions' – $P(x)$ in $\mathcal{R}$ of a given degree m (artt. 341–347). This number is first obtained by recursion (artt. 343–346), and then by an explicit formula (art. 347).

Gauss then goes on to study in more detail the irreducible polynomials $P(x)$ in $\mathbb{F}_p[x]$ of a given degree m. He obtains many properties belonging, in fact, to the theory of finite fields. Several remarks made by Gauss seem to indicate that he was fully aware of this fact, although the notion of a finite field does not appear explicitly in his papers.[20]

[18] see [24], or [56] for a German translation, or [27] for an English translation.

[19] see [26, pp. 212–240], [56, pp. 602–629] and [22].

[20] That Gauss knew much more on general congruences than is presented in his notes is also apparent from his *Preface* to the *Disquisitiones arithmeticae* and from letters to Bolyai and Dirichlet. See [20] and [22].

4. In the sequel of his presentation, Gauss systematically builds on what we now call the *Frobenius automorphism* in order to deduce properties of a finite field and its sub-fields.

Then Gauss sketches a theory of cyclotomy for the polynomial $F(x) = x^\nu - 1 \in \mathbb{F}_p[x]$ over the field $\mathbb{F}_p$ with $\nu \in \mathbb{N}$ (artt. 362–364), by referring to his theory of cyclotomy for the polynomial $\phi(x) = x^\nu - 1 \in \mathbb{Z}[x]$ over the field of rational numbers $\mathbb{Q}$.[21] This amounts to establishing Galois theory over the finite field $\mathbb{F}_p$ by means of Gaussian periods.

As a consequence, Gauss obtains a new proof for the Quadratic Reciprocity Law in $\mathbb{Q}$ (art. 365), his third proof (Proof VII in the official counting). For the details see [20] and [22].

4 Dedekind's Paper of 1857 on Function Fields over a Finite Field

1. *Dedekind* in his paper of 1857, *Abriß einer Theorie der höheren Kongruenzen in bezug auf einen reellen Primzahl-Modulus*,[22] first refers to *Gauss* who initiated the subject and to *E. Galois, I. A. Serret* and *Th. Schönemann* who had taken it up afterwards. In his commentary in the Collected Mathematical Works of Dedekind, Ore remarks that most of Dedekind's results can already be found in the above mentioned papers by Galois, Serret and Schönemann, but that Dedekind "reduces the theory to its simplest, purely arithmetic form".[23] By means of this presentation, Dedekind also laid the foundation for his later fundamental paper on ideal theory, *Über den Zusammenhang zwischen der Theorie der Ideale und der Theorie der höheren Kongruenzen*.[24] From what he says in that later paper, it appears that it was Kummer's discovery of ideal numbers (defined in terms of congruences) which motivated Dedekind's investigations on higher congruences, that is, on function fields over a finite field of constants. What Dedekind was doing in his paper of 1857 was to extract the common structure of function fields and number fields. At the same time he developed the structure theory of finite fields, as it turned out. This follows from the result by *E. H. Moore*, stating that every finite field is isomorphic to some residue field of a ring of polynomials $\mathbb{F}_p[x]$ over a finite field of constants $\mathbb{F}_p$ with respect to an irreducible polynomial $P(x)$ in $\mathbb{F}_p[x]$ for some prime p [58].

2. Dedekind in his exposition is following the presentation Dirichlet had given in his lectures on number theory [14, 15]. Thus, Dedekind is also following essentially the exemplary presentation of elementary number theory given by Gauss in his *Disquisitiones arithmeticae* [24, 56, 27]. This can be seen from the following description of the content of Dedekind's paper.

---

21 see *Disquisitiones arithmeticae*, artt. 339–356, in particular artt. 342–343. Note that ν is a prime number in the case $\phi(x) = x^\nu - 1 \in \mathbb{Z}[x]$ of the *Disquisitiones arithmeticae*.

22 see [10]; also [13, Bd. I, V, pp. 40–67]. The paper was presented to *Journal für die reine und angewandte Mathematik* in October 1856.

23 see [13, Bd. I, p. 66]

24 see [11]; also [13, Bd. I, XV, pp. 202–232].

3. After having introduced the elementary notions of degree, divisor and unit in the ring of polynomials $\mathcal{R} = \mathbb{F}_p[x]$ (artt. 2–3),[25] Dedekind shows that $\mathcal{R}$ is Euclidean, that is, admits a Euclidean algorithm (art. 4). From there he deduces the fundamental theorem of arithmetic for $\mathcal{R}$ in the way we consider today as "standard" or "classical" (art. 6).

4. Next, Dedekind introduces the notions of congruence in $\mathcal{R}$ and of a complete system of representatives of in-congruent polynomials in $\mathcal{R}$ (art. 7), solves the linear congruence in $\mathcal{R}$ with respect to a polynomial M in $\mathcal{R}$, and derives the Chinese Remainder Theorem for $\mathcal{R}$ (art. 9). He then introduces the analogue of Euler's φ-function for a polynomial M in $\mathcal{R}$ and derives the fundamental properties of φ (art. 10). From there, Dedekind deduces the analogue of Euler's theorem and, as a special case, Fermat's theorem (art. 11), furthermore Lagrange's theorem on the maximal number of roots of a polynomial $F(y)$ of degree n with coefficients in $\mathcal{R}$ modulo an irreducible polynomial P in $\mathcal{R}$, from where he gets the analogue of Wilson's theorem (art. 12). Now he is ready to introduce the notions of a power residue modulo a polynomial M in $\mathcal{R}$, of a primitive root modulo an irreducible polynomial P in $\mathcal{R}$, the notion of the index modulo P (art. 13), and to deduce Euler's criterion for power residues (art. 14). Then he specializes the theory of power residues to quadratic residues (art. 15), defines the analogue of the Legendre symbol, derives its fundamental properties, including the analogue of Gauss' Lemma (art. 16). Finally, he gets to the Quadratic Reciprocity, dealt with in article 17.

5. There, Dedekind says[26] that the analogy between the arithmetic in $\mathbb{F}_p[x]$ and the arithmetic in $\mathbb{Z}$, which was valid up to here for all theorems and proofs, does not leave any doubt that there exists in $\mathbb{F}_p[x]$ a theorem analogue to the quadratic reciprocity law. He continues by saying that, indeed, the theorem has the form

$$\left[\frac{P}{R}\right]\left[\frac{R}{P}\right] = \left(\frac{-1}{p}\right)^{\pi\cdot\varrho},$$

where P and R are *monic*[27] irreducible polynomials[28] of degrees π and ϱ respectively, and where $\left(\frac{-1}{p}\right) = (-1)^{\frac{1}{2}(p-1)}$ denotes the *Legendre symbol* in $\mathbb{Q}$.[29]

Then Dedekind says that the case where P and Q are not monic can be easily reduced to the case where both are monic. And he adds that the proof of the theorem

[25] Instead of the modern term *polynomial* Dedekind uses the term *integer rational function* or simply *function*.

[26] see [13, Bd. I, pp. 58–59]; or [20] for the original citation.

[27] P is called *monic* if the leading coefficient of P is 1. Instead of the modern term *monic* Dedekind uses the term *primary (primär)*.

[28] For the term *irreducible polynomial* Dedekind uses the term *prime function (Primfunktion)*.

[29] Dedekind writes

$$\left(\frac{P}{R}\right)\left(\frac{R}{P}\right) = \left(\frac{-1}{p}\right)^{\pi\cdot\varrho},$$

but we have replaced, following Artin, ( )( ) by [ ][ ], in order to distinguish the Legendre symbol ( ) in $\mathbb{Z}$ from the Legendre symbol [ ] in $\mathcal{R} = \mathbb{F}_p[x]$.

can be given quite analogously to Gauss' fifth proof of the reciprocity law in $\mathbb{Z}$, by applying the lemma proved in the preceding article, namely Gauss' Lemma. He sketches this proof for $\mathbb{F}_p[x]$ and finishes by saying:

> "... die Schlußfolgerungen [...] bis zu dem letzten Resultat hin, in welchem der Beweis des Theorems enthalten ist, sind denen der zitierten Abhandlung von *Gauss* so ähnlich, daß die vollständige Durchführung Niemandem entgehen kann. Und hiermit wollen wir diesen Teil unserer Theorie verlassen, da seine weitere Entwicklung sich von selbst ergibt."

> "... the deductions [conclusions] [...] up to the last result which contains the proof of the theorem [Quadratic Reciprocity Law] are so similar to the ones in the cited treatise by *Gauss* that no one can fail to find the complete proof. Herewith, we will leave our theory, since its further development follows automatically."

6. Gauss' fifth proof of the quadratic reciprocity is Gauss' most elementary proof of the quadratic reciprocity law for rational integers for which Gauss has given altogether eight proofs, six of which were published, and two more were found in his papers after his death. The fifth proof was published in 1818, together with the sixth proof, under the title *Theorematis fundamentalis in doctrina de residuis quadraticis demonstrationes et amplificationes novae.*[30] But Gauss found it probably already between 1807 and 1808 [18, 59]. As is the third proof, so is the fifth proof based on *Gauss' Lemma*. It takes on the following form in Dedekind's paper:[31]

If P is an irreducible polynomial of degree π in $\mathcal{R} = \mathbb{F}_p[x]$ and A any polynomial in $\mathcal{R}$ prime to P, then

$$\left[\frac{A}{P}\right] = (-1)^{\mu},$$

where μ is obtained in the following way. Divide up all $p^{\pi} - 1$ in-congruent polynomials modulo P into two groups, $\mathcal{G}$ and $\mathcal{H}$, in such a way that, if $F \in \mathcal{G}$ then $-F \in \mathcal{H}$. Then μ denotes the number of elements in the set $A \cdot \mathcal{G} = \{A \cdot F : F \in \mathcal{G}\}$ lying in $\mathcal{H}$, that is, the cardinality of $A \cdot \mathcal{G} \cap \mathcal{H}$.

7. In the remaining artt. 18–21 on irreducible polynomials, Dedekind determines the structure of finite fields. For more details see [20].

5 Artin's Dissertation: Arithmetic Part

1. Here is where Artin set in in his doctoral dissertation, *Quadratische Körper im Gebiete der höheren Kongruenzen I, II,* [4] with a systematic study of the arithmetic of quadratic extensions of function fields over finite fields of characteristic $p \neq 0, 2$. The center of his study was the analytic class number formula which led him to

[30] see [26, pp. 47–64]; or the German translation in [56, pp. 497–501].

[31] see [10, art. 16]; or [13, Bd. I, p. 58].

introduce the ζ-function of function fields and to formulate and conjecture the Riemann Hypothesis for these fields.

In the Einleitung (Introduction) Artin says: [4, p. 153]

> "Die *Dedekind*schen Untersuchungen über höhere Kongruenzen[32] legen folgende Erweiterung der Theorie nahe.
> Es werde dem Körper K der rationalen Funktionen modulo p die Funktion $\sqrt{D(t)}$ adjungiert, wo $D(t)$ eine ganze im Sinne *Dedekind*s quadratfreie Funktion des Parameters t ist. Der entstehende quadratische Körper $K(\sqrt{D(t)})$ weist dann ähnliche Eigenschaften auf wie ein quadratischer Zahlkörper.
> So gilt zum Beispiel der Satz über eindeutige Zerlegbarkeit der Ideale in Primideale, der Satz von der Endlichkeit der Klassenzahl, die Sätze über die Einheiten.
> Zur Klassenzahlformel gelangt man durch Einführung der Zetafunktionen. Hier läßt sich die Frage nach der Richtigkeit der *Riemann*schen Vermutung in jedem speziellen Fall entscheiden. Eine Durchrechnung der ersten Fälle – es handelt sich um zirka vierzig Körper – ergab stets die Richtigkeit der *Riemann*schen Vermutung. Einem allgemeinen Beweis ihrer Richtigkeit scheinen sich aber noch Schwierigkeiten ähnlicher Art wie beim *Riemann*schen $\zeta(s)$ entgegenzustellen, doch liegen die Verhältnisse hier insofern klarer und durchsichtiger, als es sich (im wesentlichen) um ganze rationale Funktionen handelt. Auf Fragen, die damit im Zusammenhang stehen, werde ich noch zurückkommen.
> Von den sonstigen Eigenschaften unserer Zetafunktionen sei noch hervorgehoben: Sie besitzen eine einfache Funktionalgleichung, welche als Folge merkwürdige Reziprozitätsbeziehungen gewisser Charaktersummen nach sich zieht. Ihre Nullstellen stehen in einfachem Zusammenhang mit den Wurzeln einer algebraischen Gleichung, wodurch eben die Entscheidung über die *Riemann*sche Vermutung gefällt werden kann."

> "*Dedekind*'s investigations on higher congruences[33] suggest the following generalization of the theory.
> Let's adjoin to the field K of rational functions modulo p the function $\sqrt{D(t)}$, where $D(t)$ is an integer square-free function of the parameter t in the sense of *Dedekind*.[34] The generated quadratic field $K(\sqrt{D(t)})$ has then similar properties as a quadratic number field.
> For instance, one has the theorem on unique factorization of ideals into prime ideals, the theorem on the finiteness of the class number, the theorems on the units.
> The class number formula is obtained by introducing the zeta functions. Here the question whether the *Riemann* Hypothesis is true can be decided in each particular

[32] Artin refers to [10].

[33] Artin refers to [10].

[34] We will use the terms *integer function* or *integer rational function* used by Artin synonymously with the modern term *polynomial*, but we will mostly keep Artin's terminology.

> case. The computation of the first cases – it refers to about forty fields – always confirmed the *Riemann* Hypothesis. But yet for a general proof of its validity one still has to overcome difficulties of a similar kind as with *Riemann*'s $\zeta(s)$, however, the situation here is clearer and more transparent since it refers (essentially) to integral rational functions. I will come back yet to questions related to it.
> Among the other properties of our zeta functions the following shall be pointed out: They satisfy a simple functional equation which implies remarkable reciprocity relations of certain character sums. Their zeros are related in a simple manner to the roots of an algebraic equation, by means of which for that reason the *Riemann* Hypothesis can be decided."

The thesis was sent for publication to the *Mathematische Zeitschrift* on the 14th of October 1921, but appeared only in 1924. It comprises 94 pages and is divided into two parts.

2. In the first part, Artin develops the arithmetic of hyper-elliptic function fields with finite constant field, $\mathbb{F}_p(t)(\sqrt{D(t)})$, p an odd prime number, following Dedekind's presentation of the arithmetic of $\mathbb{F}_p(t)$: ring of integers, integral basis, discriminant, units, ideal theory,[35] prime ideals, the fundamental theorem of arithmetic for ideals, ideal classes, the finiteness of the class number and the theory of units. The finiteness of the class number is obtained, following the classical path traced by Gauss, and before him by Lagrange, for binary quadratic forms, that is, by reduction theory: Every class contains only a finite number of reduced ideals. The whole arithmetic theory is completely analogous to the corresponding theory of quadratic number fields.

Dedekind had only considered the ring of integer functions $\mathbb{F}_p[t]$ (see [10]), whereas Artin takes the whole field of rational functions $\mathbb{F}_p(t)$ as a base field of $\mathbb{F}_p(t)(\sqrt{D(t)})$, where $\mathbb{F}_p$ is the finite field with p elements, p a prime $\neq 2$, and $D(t)$ is a non-square element in the ground field of rational functions $\mathbb{F}_p(t)$. It is interesting that Artin does not mention the fundamental treatise by Dedekind and Weber, *Theorie der algebraischen Funktionen einer Veränderlichen* (J. Reine Angew. Math. 92 (1882), pp. 181–290; [13], Bd. 1, XVIII, pp. 238–350), where Dedekind and Weber systematically develop Riemann's (geometric) theory of functions on a Riemann surface in a purely algebraic way, and in close analogy to Dedekind's theory of algebraic numbers, as an arithmetic theory of algebraic functions (over the field of constants $\mathbb{C}$, but still true essentially over an arbitrary perfect field of constants). From the way Artin approached the subject, we can conclude that he, then, did not yet know this paper. Artin was concentrating on the analytic side, and his starting point were surely papers by Herglotz and Hecke, and very likely also the thesis by Kornblum (see the section 6, art. 10). Note that Artin's thesis was published in the same journal where Kornblum's paper had appeared two and a half years before Artin submitted his thesis, and that Herglotz was a member of the scientific board (wissenschaftlicher Beirat) of this journal. The theory of Dedekind and Weber was applied to function fields of characteristic p only later, namely by F. K. Schmidt in his paper [63] (see also [61], Sect. 4.3).

[35] All ideals in Artin's thesis are integral ideals, that is, consisting entirely of integral functions.

3. At the end of this first part, Artin proves the *Quadratic Reciprocity Law* in $K = \mathbb{F}_p(t)$. We have seen in the preceding section that this law was already stated by Dedekind in 1857, and that Dedekind sketched the proof in a way that anybody familiar with Gauss' fifth proof of the quadratic reciprocity law in $\mathbb{Q}$ is able to complete the analogous proof for function fields, by filling in the details. The statement of the quadratic reciprocity law in the function field K is completely analogous to the corresponding law in the rational number field $\mathbb{Q}$, except that there is no second complementary law.

The First Complementary Law takes on the following form:[36]

$$\left[\frac{a}{P}\right] = \left(\frac{a}{p}\right)^{\nu},$$

where [ ] stands for the Legendre symbol in $\mathbb{F}_p(t)$, and ( ) for the Legendre symbol in the rational field $\mathbb{Q}$. P is a monic irreducible polynomial in $\mathbb{F}_p[t]$, a a unit in $\mathbb{F}_p[t]$ and hence a rational integer not congruent to 0 modulo p, and ν is the degree of P.

This law is derived by Artin from the following two properties:

(1) If ε_0 is the fundamental unit[37] in the (real) quadratic function field $R = \mathbb{F}_p(t)(\sqrt{P}) = K(\sqrt{P})$, where P is a monic irreducible polynomial in $\mathbb{F}_p[t]$ of even degree, then the norm of ε_0 is equal to the fixed generator g of the unit group $\mathbb{F}_p^*$ of $\mathbb{F}_p : N(\varepsilon_0) = g$. That is to say, the Pellian equation $x^2 - Py^2 = g$ has a solution (x, y) with x and y in $\mathbb{F}_p(t)$.

(2) The imaginary quadratic function field $J = \mathbb{F}_p(t)(\sqrt{g}) = K(\sqrt{g})$ has class number one.

In addition, Artin uses the decomposition law for quadratic function fields $K(\sqrt{D})$.

(1) is obtained by means of *Hilbert's Theorem 90* for quadratic function fields [4, p. 189 (Hilfssatz)], that is, from Galois cohomology with coefficients in M^*: $H^1(G, M^*) = 1$, where $G = \mathrm{Gal}(M/K)$ is the Galois group of M/K, and $M = K(\sqrt{D})$, with a non-square integral function $D \in K$.

4. The General Quadratic Reciprocity Law for the function field $K = \mathbb{F}_p(t)$ has the following form:

$$\left[\frac{P}{Q}\right] \cdot \left[\frac{Q}{P}\right] = \left(\frac{-1}{p}\right)^{\nu \cdot \mu},$$

[36] see [4, § 15, I, p. 202]. The First Complementary Law is not given by Dedekind.

[37] $\varepsilon_0 = U + V\sqrt{D}$ is called *a fundamental unit* in the real quadratic function field $\mathbb{F}_p(t)(\sqrt{D})$ if ε_0 is a unit of smallest degree $n > 0$ such that $N(\varepsilon_0) = 1$ or $N(\varepsilon_0) = g$. ε_0 is uniquely determined up to a sign. ε_0 with a fixed choice of the sign is called *the fundamental unit* of $\mathbb{F}_p(t)(\sqrt{D})$ (see [4], p. 197). If $D = P$ is prime, then $N(\varepsilon_0) \neq 1$, hence $N(\varepsilon_0) = g$ (see [4], p. 198).

where P and Q are monic irreducible polynomials in $\mathbb{F}_p[t]$ of degrees ν and μ respectively.[38]

Dedekind in his proof was referring to Gauss' fifth proof of the quadratic reciprocity law for rational numbers.[39] His aim was to show that the quadratic reciprocity law in $\mathbb{F}_p[t]$, like the other arithmetical properties in $\mathbb{F}_p[t]$, can be obtained in a completely analogous way as those in $\mathbb{Z}$, and in addition, that the quadratic reciprocity law in $\mathbb{F}_p[t]$ can be proved in an elementary way, that is, by using only elementary properties of the ring $\mathbb{F}_p[t]$. Artin, however, is following a completely different path by making use of properties of the quadratic algebraic extensions $\mathbb{F}_p(t)(\sqrt{D})$ of $\mathbb{F}_p(t)$ he was studying; namely the same path he had already followed for the proof of the first complementary law [4, § 15, pp. 202–204], a path very likely inspired by a paper of Kummer from 1861.[40]

Namely, he obtains the general quadratic reciprocity law in $K = \mathbb{F}_p(t)$ for two monic irreducible polynomials P and Q in $\mathbb{F}_p[t]$, via the decomposition law for Q in the (real) quadratic extension $R = K(\sqrt{P}) = \mathbb{F}_p(t)(\sqrt{P})$, again from the property (1) and from the finiteness of the class number for quadratic function fields, and from criteria for this class number to be odd [4, § 14, p. 199]. These criteria are, in their turn, simply consequences of *Hilbert's Theorem 90* for quadratic function fields, via the theory of ambiguous ideals and ambiguous classes.

In the second part of his thesis, Artin will need the quadratic reciprocity law in $\mathbb{F}_p(t)$ in order to show that the zeta-function $Z(s)$ he defined for the function field $\mathbb{F}_p(t)(\sqrt{D})$ is a rational function of p^{-s}.

5. That the quadratic reciprocity for the rational number field $\mathbb{Q}$, for two odd prime numbers p and q, follows solely from the existence of a solution of the Pellian equation $x^2 - pqy^2 = 1$ and from the finiteness of the class number of quadratic number fields $\mathbb{Q}(\sqrt{d})$, was already stressed by *Kronecker* in a commentary to *Kummer*'s paper, *Zwei neue Beweise der allgemeinen Reciprocitätsgesetze unter den Resten und Nichtresten der Potenzen, deren Grad eine Primzahl ist*.[41] *Kronecker* promised to give, in a separate paper, the details for his assertion and for the relation between the reciprocity law on the one hand and the existence of a solution of the Pellian equation and the finiteness of the class number for algebraic number fields on the other. But it seems that this paper was never published.[42] It is very likely that Artin knew of this remark by Kronecker and of Kummer's new proof of the quadratic reciprocity based on the existence of a solution of the Pellian equation and the finiteness of the class number given in [54], pp. 14–16.[43] For, Artin is

---

[38] see [10, art. 17] (or [13, Bd. I, p. 58]) and [4, §15, p. 204].

[39] see section 4, art. 5; and [10, art. 17].

[40] see below, art. 5.

[41] see J. Reine Angew. Math. 100 (1887), pp. 10–50; Reprint of *Abhandl. der königl. Akademie der Wiss. zu Berlin* 1861, pp. 81–122. See also Kummer, Collected Papers, Vol. I, p. 848.

[42] see also Kronecker's paper *Über den Gebrauch der Dirichlet'schen Methoden in der Theorie der quadratischen Formen*, Monatsberichte der Königl. Preuss. Akademie der Wiss. zu Berlin, 1864, pp. 294–295; also Kronecker, Math. Werke, Bd. IV, p. 237.

[43] see also Kummer, Collected Papers, Vol. I, pp. 846–848.

following exactly Kummer's steps in his own proof of the quadratic reciprocity law for function fields.

6. It is quite remarkable that *Hecke* in his proof that $(s-1)\zeta_k(s)$ is an entire function, where ζ_k is the Dedekind zeta-function of an algebraic number field k, also uses, among the arithmetical properties of k, only the existence of a system of fundamental units and the finiteness of the class number of k.[44] Hecke's fundamental idea was to express $(s-1)\zeta_k(s)$ in terms of certain theta-functions of n variables attached to k, if k is of degree n over $\mathbb{Q}$, and then use the transformation formula for these theta-functions, in a way similar to what Riemann had done in 1859 for the ordinary Riemann zeta-function of $\mathbb{Q}$. It is quite likely that Artin knew this paper by Hecke, since Herglotz who himself was building mostly on Hecke's papers must have drawn Artin's attention to these papers by Hecke, right from the beginning. See also the citation from Artin's letter to Herglotz of the 3rd of December 1921 given in the footnote in section 7, art. 3.

A simple proof for the more general case where M is an arbitrary field extension of $F(t)$ of finite degree, with F a field algebraic over $\mathbb{F}_p$, and for the nth power reciprocity law, was given by F. K. Schmidt, in 1926, in his paper *Zur Zahlentheorie in Körpern der Charakteristik p* (see [62] and also [63]). For a detailed analysis see [61], Sect 3.2.2. A proof of the general nth power reciprocity law for function fields of characteristic p was already given by H. Kühne in 1902 in his paper *Eine Wechselbeziehung zwischen Functionen mehrerer Unbestimmten, die zu Reciprocitätsgesetzen führt*, J. Reine Angew. Math. 124 (1902), 121–133 (I am indebted to Franz Lemmermeyer for bringing this paper to my attention). Kühne also has proved the analogue of Dirichlet's unit theorem for algebraic function fields of characteristic p in the paper *Angenäherte Auflösung von Congruenzen nach Primmodulsystemen in Zusammenhang mit den Einheiten gewisser Körper*, J. Reine Angew. Math. 126 (1903), 102–115. It seems that these papers by Kühne have gone mostly unnoticed. For more information on Kühne see [22].

6 Artin's Dissertation: Analytic Part

7. In the second part of his dissertation, Artin develops the analytic theory of these quadratic extensions, that is, the analogue for quadratic function fields of the theory of the zeta-function, of the resulting analytic class number formula, of the L-functions and, as another consequence, the analogue of Dirichlet's Theorem on irreducible polynomials in an arithmetic progression.

In § 17 Artin introduces, for the first time, the zeta-function $Z(s)$ for a function field. Namely, he writes for the field $M = K(\sqrt{D}) = \mathbb{F}_p(t)(\sqrt{D})$, with a function $D \in \mathbb{F}_p[t]$ which is not a square:

$$Z(s) = \sum_{\mathfrak{a}} \frac{1}{|N(\mathfrak{a})|^s}$$

[44] see Hecke, *Über die Zetafunktion beliebiger algebraischer Zahlkörper*, Nachr. K. Ges. Wiss. zu Göttingen, 1917, pp. 77–89; Werke, pp. 159–171.

for $s \in \mathbb{C}$, and the sum taken over all ideals[45] $\mathfrak{a}$ in $M = \mathbb{F}_p(t)(\sqrt{D})$, where $|N(\mathfrak{a})| = p^\nu$, if the norm of $\mathfrak{a}$, $N(\mathfrak{a}) \in \mathbb{F}_p[t]$, is of degree ν [4, § 5, p. 167].

Next, Artin examines the fundamental properties of $Z(s)$: convergence and the Euler product. Then he defines the important numbers

$$\sigma_\nu = \sum_{|F|=p^\nu} \left[\frac{D}{F}\right],$$

the sum taken over all monic polynomials $F \in \mathbb{F}_p[t]$ of degree ν and with $(D, F) = 1$, where $\left[\frac{D}{F}\right]$ denotes the *Jacobi symbol* for the function field $K = \mathbb{F}_p(t)$.

These numbers were already introduced by Artin at the end of the first part where he showed, by means of the quadratic reciprocity law (for the Jacobi symbol), that $\sigma_\nu = 0$ if $\nu \geq n = deg\, D$ and $D \neq g$ [4, § 16, p. 206]. They are crucial for the determination of the class number h.

From the *Euler Product* of $Z(s)$:

$$Z(s) = \prod_{\mathfrak{p}} \frac{1}{1 - \frac{1}{|N(\mathfrak{p})|^s}},$$

the product taken over all prime ideals $\mathfrak{p}$ in $K(\sqrt{D})$,[46] and from the decomposition law for irreducible polynomials in $K(\sqrt{D})$, Artin obtains

$$Z(s) = \frac{1}{1 - p^{-(s-1)}} \sum_{\nu=0}^{\infty} \frac{\sigma_\nu}{p^{\nu s}}.$$

Since $\sigma_\nu = 0$ if $\nu \geq n = deg\, D$ and $D \neq g$, it follows that $Z(s)$ is a rational function of p^{-s}, and that the sum on the right hand side is a polynomial in p^{-s}.

8. In the sequel, Artin carefully examines the poles and zeroes of $Z(s)$ and derives the *class number formula* for $M = \mathbb{F}_p(t)(\sqrt{D})$. Namely, if M is imaginary and $D \neq g$,[47] then the class number h is given by

$$h = \sigma_0 + \sigma_1 + \ldots + \sigma_{n-1},$$

where n is the degree of D;
and

$$h = -\frac{1}{R} \sum_{\nu=1}^{n-1} \nu \sigma_\nu,$$

[45] The ideals are always integral ideals, consisting of integral functions only.

[46] Later, in 1926, F. K. Schmidt introduced the zeta-function for arbitrary algebraic function fields of characteristic p, whereby the contribution given by the infinite primes (archimedean places) $\mathfrak{p}$ are also taken into account (see [62] and [63] and also [61]).

[47] For $D = g$, the class number h is equal to 1. Remember that g is a fixed generator of the unit group $\mathbb{F}_p^*$ of $\mathbb{F}_p$. See [4], p. 216.

if M is real, where $R = \frac{\log|\varepsilon_0|}{\log p}$ is the *regulator* and ε_0 the *fundamental unit* of M [4, § 21, p. 225].

Or else, by evaluating the residue of $Z(s)$ at $s = 1$, he obtains

$$h = \kappa \sum_{\nu=0}^{n-1} \frac{\sigma_\nu}{p^\nu},$$

with κ depending only on $\sqrt{|D|}$ and p if M is imaginary, and on $\sqrt{|D|}$, p and the regulator R if M is real.[48]

From the functional equation of $Z(s)$,[49] Artin derives reciprocity relations for the σ_ν which allow him to compute the class number in a new and simpler way [4, § 21].

9. Since

$$Z(s) = \frac{1}{1 - p^{-(s-1)}} \sum_{\nu=0}^{n-1} \frac{\sigma_\nu}{p^{\nu s}}$$

is a rational function in p^{-s} and $\sigma_0 = 1$, $Z(s)$ can be written as

$$Z(s) = \frac{1}{1 - p^{1-s}} \prod_{\nu=1}^{n-1} (1 - \beta_\nu p^{-s}),$$

where $\beta_1, \dots, \beta_{n-1}$ are the roots of the algebraic equation with rational coefficients

$$z^{n-1} + \sigma_1 z^{n-2} + \sigma_2 z^{n-3} + \dots + \sigma_{n-1} = 0,$$

and $z = p^s$.

Hence the zeroes ρ of $Z(s)$ are given by $p^\rho = \beta_\nu$, for $\nu = 1, \dots, n-1$, or

$$\rho = \frac{\log|\beta_\nu|}{\log p} + i \frac{\arg \beta_\nu}{\log p}, \qquad \nu = 1, \dots, n-1.$$

So the *Riemann Hypothesis* for $K(\sqrt{D})$, $\Re(\rho) = \frac{1}{2}$, is equivalent to the relation $\log|\beta_\nu| = \frac{1}{2} \log p$, or else, $|\beta_\nu| = \sqrt{p}$, for $\nu = 1, \dots, n-1$.

Artin then computes the class number and verifies the *Riemann Hypothesis* for the field $\mathbb{F}_p(t)(\sqrt{D})$ with functions $D \in \mathbb{F}_p[t]$ of degree $n = 3$ and primes $p = 3, 5, 7$, with functions D of degree $n = 4$ and primes $p = 3, 5$, and with functions D of degree $n = 5$ and the prime $p = 3$ [4, § 23].

10. Finally, Artin introduces, following Dirichlet and Dedekind, the character group X of the group $\mathcal{C}$ of residue classes modulo a monic polynomial $F \in \mathbb{F}_p[t]$ and relatively prime to F, and then for each character $\chi \in X$ an L-series

---

[48] see [4, §§ 18–19]. '§ 13' on page 217 should be corrected to '§ 19'.

[49] There is no doubt that Artin had studied, already then, Hecke's papers of 1917 on the functional equation of $\zeta(s)$, although there is no reference to Hecke in Artin's thesis. See also the citation from Artin's letter to Herglotz of the 3rd of December 1921 given in the footnote in section 7, art. 3.

$L(s, \chi)$ [4, § 25]. From his proof that $L(s, \chi)$ has no zeroes on the vertical line $\mathcal{R}(s) = 1$ [4, § 26], Artin deduces, following the classical proof of *Landau* in his *Handbuch der Lehre von der Verteilung der Primzahlen*,[50] the analogue for function fields of the *Theorem of Dirichlet* on prime numbers in an arithmetic progression. It states that an arithmetical progression, modulo a monic polynomial $F \in \mathbb{F}_p[t]$, contains irreducible polynomials P for any degree n, from a certain degree on, and the monic irreducible polynomials are asymptotically equally distributed among the congruence classes modulo F and prime to F. Thereby, Artin makes use of results by *Kornblum* on $L(s, \chi)$; e. g. that $L(s, \chi)$ is entire if χ is not the principal character χ_0.[51] Kornblum's doctoral thesis on irreducible polynomials in arithmetical progressions was published posthumously after the war by Landau.[52]

Kornblum's starting point was also Dedekind's paper of 1857 (see [10]). Kornblum discovered that the classical proof for the non-vanishing of $L(s, \chi)$, for $s = 1$ and $\chi \neq \chi_0$, can be adapted to function fields $\mathbb{F}_p(t)$. However, Kornblum does not introduce algebraic extensions of $\mathbb{F}_p(t)$. It is possible that Herglotz suggested to Artin to study, in his thesis, quadratic algebraic extensions of a function field, after having seen Kornblum's thesis. Remember that Herglotz was a wissenschaftlicher Beirat of the Mathematische Zeitschrift.

7 Artin's Paper on Zeta-functions and L-series in Algebraic Number Fields

1. In his next paper, Artin takes up a new topic,[53] namely, the question whether the zeta-function $\zeta_K(s)$ of a finite algebraic extension K of an algebraic number field k divides the zeta-function $\zeta_k(s)$ of the ground field k; that is, whether $\zeta_K(s)/\zeta_k(s)$ is an entire function.[54] Artin refers to *Dedekind* who proved this to be true if k is the field of rational numbers, $k = \mathbb{Q}$, and K is a pure cubic extension of $\mathbb{Q}$.[55] In the case of an abelian extension K/k, Artin refers to *Takagi*'s fundamental paper on class field theory[56] and to *Heinrich Weber*'s *Lehrbuch der Algebra*.[57] Namely, from Takagi's main theorem on class field theory, stating that any abelian algebraic

[50] Artin refers to p. 460.

[51] *Heinrich Kornblum* (1890–1914) was a student of Landau in Göttingen. He fell in the First World War in October 1914 at the age of 24 near Poël-Capelle.

[52] see Kornblum, *Über die Primfunktionen in einer arithmetischen Progression.* Math. Zeitschrift 5 (1919), pp. 100–111.

[53] For the reasons why Artin probably changed the subject see art. 3.

[54] see [2] Artin, *Über die Zetafunktionen gewisser algebraischer Zahlkörper*, Math. Ann. 89 (1923), pp. 147–156. The manuscript was sent for publication to the *Mathematische Annalen* on the 4th of September 1922.

[55] see Dedekind, *Über die Anzahl der Idealklassen in reinen kubischen Körpern*, J. Reine Angew. Math. 121 (1900), 40–123; Werke, XXIX.

[56] see Takagi, *Ueber eine Theorie des relativ-Abel'schen Zahlkörpers*, J. Coll. Science, Univ. of Tokyo, Vol. 41, Art. 9, 1920, pp. 1–133; in particular §§ 23–24.

[57] Bd. 3, § 163 ff.

extension K of an algebraic number field k is a class field over k, corresponding to a certain congruence class group $\mathfrak{C}_\mathfrak{m}$ in k with respect to an integral ideal $\mathfrak{m}$ of k (which can be taken to be the conductor of K/k), and from Weber's theory of L-series [68], attached to a congruence class group $\mathfrak{C}_\mathfrak{m}$ in k, it follows that

$$\zeta_K(s) = \zeta_k(s) \prod_\chi L(s, \chi),$$

the product taken over all – what we will call – *Weber characters* χ [68], different from the principal character χ_0, of the congruence class group $\mathfrak{C}_\mathfrak{m}$ corresponding to K/k, according to Takagi's class field theory. This fundamental relation for abelian extensions K/k was conjectured by Weber in 1897 [68] and proved by Takagi in 1920 [64]. It was viewed by *Hecke*, already in 1917, as the quintessential form of the decomposition law and the ramification law (theorem on the prime ideals dividing the discriminant) of K/k.[58] This relation goes back to *Dedekind* in the case where K is a cyclotomic field $K = \mathbb{Q}(\zeta_p)$, $\zeta_p = e^{\frac{2\pi i}{p}}$, p an odd prime number (see Dedekind, Ges. math. Werke, Bd. 3, XLVI, § 185, formula (47), p. 190) and to *Dirichlet* in the case where K is a quadratic number field $K = \mathbb{Q}(\sqrt{d})$. Since *Hecke* proved, in 1917,[59] that these *Weber L-functions* $L(s, \chi)$ are entire if $\chi \neq \chi_0$, it follows that $\zeta_K(s)/\zeta_k(s)$ is entire if K/k is abelian.

2. The L-functions $L(s, \chi)$ of a so called *congruence class group* $\mathfrak{C}_\mathfrak{m}$ with respect to an integral ideal $\mathfrak{m}$ belonging to an algebraic number field k were introduced by *Heinrich Weber* in 1897 in a series of three papers.[60] Weber introduced them, together with certain fields attached to $\mathfrak{C}_\mathfrak{m}$ which he later named *class fields*. Among these fields were the fields constructed by Kronecker over an imaginary quadratic number field $\mathbb{Q}(\sqrt{-d})$ by means of what Kronecker called *singular moduli*. Weber's aim was to study the conjecture made by Kronecker, that – what we have called – the *Kronecker class field K* over $\mathbb{Q}(\sqrt{-d})$, i. e. the absolute class field of $\mathbb{Q}(\sqrt{-d})$, is not ramified over $\mathbb{Q}(\sqrt{-d})$. Weber observed that these fields, which are class fields in the modern sense, must have interesting properties. Among other things, they make

[58] see Hecke, *Über eine neue Anwendung der Zetafunktionen auf die Arithmetik der Zahlkörper*, Nachr. K. Ges. Wiss. Göttingen, 1917, pp. 90–95, in particular p. 95; Werke, 8, p. 177.

[59] see Hecke, *Über die L-Funktionen und den Dirichletschen Primzahlsatz für einen beliebigen Zahlkörper*, Nachr. K. Ges. Wiss. Göttingen, 1917, pp. 299–318; Werke, 9, pp. 178–197, in particular § 4. The proof is again based on Hecke's idea to express $L(s, \chi)$ by means of theta-functions in n variables, if n is the degree of k, and to apply the transformation formula for these functions. See also the footnote in section 5, art. 6. Following a suggestion of Artin, *Tate* in his doctoral thesis has given a new and more conceptual proof for the functional equation and the analytic continuation of the ζ- and L-functions, by means of the theory of harmonic analysis on the group of idèles, that is, by introducing the Haar measure on locally compact abelian groups (see Cassels and Fröhlich, *Algebraic Number Theory*, Academic Press, London, 1967, Chapter XV, pp. 305–347).

[60] see Weber, *Über Zahlengruppen in algebraischen Zahlköpern I, II, III*, Math. Ann. 48 (1897), pp. 433–473; 49 (1897), pp. 83–100; 50 (1898), pp. 1–26; i. e. [68]. See also [17].

it possible to apply to k Dirichlet's analytic methods for primes in an arithmetic progression. Namely, the existence of the class field over k attached to $\mathfrak{C}_{\mathfrak{m}}$ implies the non-vanishing of $L(s, \chi)$ for $s = 1$ and $\chi \neq \chi_0$. This result allowed Weber to deduce, following the method of Dirichlet, the analogue for an algebraic number field k of Dirichlet's Theorem. That is, any arithmetic progression in k modulo an integral ideal $\mathfrak{m}$ and prime to $\mathfrak{m}$ contains infinitely many prime ideals $\mathfrak{p}$ in k, even infinitely many prime ideals $\mathfrak{p}$ of absolute (residue class) degree $f = 1$ [17].

3. Artin's study of the quotient $\zeta_K(s)/\zeta_k(s)$ was undoubtedly motivated by Herglotz' paper *Über einen Dirichletschen Satz* [37], where Herglotz established the relation between the class number of a number field K, composed of n disjoint quadratic fields k_i, and the class number of the $2^n - 1$ quadratic sub-fields. For this relation, the zeta-functions of K and of k_i play a crucial rôle. Herglotz had submitted this paper on the 17th of September 1921 to the Mathematische Zeitschrift. At that time, Artin was still occupied with generalizing his thesis from a prime field $\mathbb{F}_p$, as a field of constants, to a finite algebraic extension $\mathbb{F}_{p^n}$ of degree n over $\mathbb{F}_p$. Originally, Artin had planned to publish this generalization[61] as a "little note."[62] But since this generalization was relatively easy, and the few more examples he obtained, confirming the Riemann Hypothesis, did not shed much more light on the Riemann Hypothesis,[63] it may well be that the unfortunate reception by Hilbert of Artin's talk on his dissertation in the Mathematische Gesellschaft in Göttingen, on the 22th of November 1921, made Artin finally decide to abandon the subject altogether.[64] These might also have been the reasons why Artin changed to a different subject, a subject certainly inspired by Herglotz' paper [37], published just two months ago, and probably also by Herglotz' ongoing work, published in [39], on Kronecker's limit formula. One is also led to conjecture that Herglotz encouraged Artin to continue the study of the papers by Hecke, particularly the one on Hecke's generalization of Kronecker's limit formula which produces relations between the class number h_K of K and the class numbers h_k of sub-fields k of K.[65] For, it is quite

[61] A recent paper by Peter Ullrich discusses Artin's notes on this generalization (see Ullrich, Peter: *Emil Artin's unveröffentlichte Verallgemeinerung seiner Dissertation.* Mitt. Math. Ges. Hamburg XIX (2000), 173–194).

[62] see Artin's letter to Herglotz, dated 13 November 1921.

[63] see Artin's letter to Herglotz, dated 13 November 1921.

[64] see section 2.1, art. 1, in particular the passage in Artin's letter to Herglotz of the 30th of November 1921 "... aber es wird wohl mit diesem Thema nicht mehr vorkommen da ich es wohl an den Nagel hängen werde" (... but it will hardly happen again with this subject, since I will probably give it up). The subject was taken up again, in 1925, only one year after the publication of Artin's thesis, independently by Rauter, a Ph. D. student of Hasse, Sengenhorst and F. K. Schmidt, for arbitrary field extensions of $F(t)$ of finite degree, with F a field algebraic over $\mathbb{F}_p$. In the sequel, the latter developed the theory up to class field theory of characteristic p, and discovered the connection, in fact equivalence, between the functional equation of the zeta-function and the Theorem of Riemann–Roch. For a detailed discussion see [61].

[65] see Hecke, *Über die Kroneckersche Grenzformel für reelle quadratische Körper und die Klassenzahl relativ-Abelscher Körper*, Verhandl. Naturf. Ges. Basel, Bd. 28, 1917, pp. 363–

clear that Artin must have well studied Hecke's papers on the ζ- and L-functions, as they leave a noticeable trace in Artin's own papers. This is also confirmed by a letter of Artin, from Göttingen, to his teacher Herglotz, dated 3 December 1921, where Artin writes: "Die Heckeschen Arbeiten habe ich mir jetzt ordentlich zu Gemüte geführt. Es steckt kolossal viel drinnen. Der junge Siegel hat ganz prächtige Resultate auf dieser Grundlage erhalten und darüber in der Math. Ges. berichtet." (I have now seriously studied Hecke's papers. They contain an enormous amount of material. The young Siegel has obtained very beautiful results on that basis, and he reported on them in the Mathematische Gesellschaft.)

The relation between the zeta-function and the class number is as follows. If $\zeta_K(s)/\zeta_k(s)$ can be expressed in terms of Weber's L-functions $L(s, \chi)$, then the quotient h_K/h_k of the corresponding class numbers can be expressed by the values of $L(s, \chi)$ at $s = 1$, since the residue of $\zeta_M(s)$ at $s = 1$ contains the class number h_M of a number field M (Analytic Class Number Formula). Already in his doctoral thesis, Artin had computed extensively tables of class numbers for function fields $M = \mathbb{F}_p(t)(\sqrt{D})$ by means of his zeta-functions introduced for these fields. Also, if $\zeta_K(s)/\zeta_k(s)$ were entire, information would be gained on the decomposition law for K/k, and it would follow that the zeroes of $\zeta_k(s)$ are among those of $\zeta_K(s)$.

4. In the paper [2], Artin was able to compute explicitly the quotient $\zeta_K(s)/\zeta_k(s)$ in terms of Weber's L-functions in many new cases, e. g. if K/k is metacyclic (non-galoisian) with degree $n = [K/k]$ square free,[66] or if K/k is a Galois extension whose Galois group G is the group of the icosahedron.[67] In the last case, Artin also gave an explicit example of two sub-fields M_1 and M_2 of K, $K \supset M_1, M_2 \supset k$, for which $\zeta_{M_2}(s)/\zeta_{M_1}(s)$ is entire, but M_1 is not a sub-field of M_2, moreover, m_1 does not even divide m_2, where $m_i = [M_i : k]$, $i = 1, 2$, are the degrees of M_1 and M_2 over k.

8 Artin's Paper on a New Kind of L-series and Artin's Reciprocity Law in Abelian Extensions of Algebraic Number Fields

1. A completely new turn to the study of $\zeta_K(s)/\zeta_k(s)$ for a Galois extension K/k was taken in the fundamental paper *Über eine neue Art von L-Reihen*, where Artin introduced, in July 1923, non-abelian L-functions, now called *Artin L-functions*, by means of a representation Γ of the Galois group G of K/k by matrices with non-vanishing determinant and with elements in the field of complex numbers (or more generally, in a field of characteristic zero). These Artin L-functions were defined by Artin as follows [3].

372; Werke, 10, pp. 198–207. See also Herglotz, *Über die Kroneckersche Grenzformel für reelle quadratische Körper, I, II*, Leipziger Ber. 75 (1923), pp. 3–14 and pp. 31–37; or Herglotz, Ges. Schriften, Göttingen, 1979, art. 28, pp. 466–484.

[66] In this case, the quotient is a product of radicals of L-functions of certain sub-fields of the normal closure Ω of K over k. See [2], formula (4) on p. 152.

[67] In this case, the quotient is a product of L-series and square radicals of L-series of certain sub-fields of K. See [2], formula (11) on p. 155.

For each prime ideal $\mathfrak{p}$ in k not dividing the relative discriminant $\mathfrak{d}(K/k)$ of K/k, we denote by $\sigma_\mathfrak{p}$ the *Frobenius automorphism* of $\mathfrak{p}$, defined by the property

$$\sigma_\mathfrak{p}(\alpha) \equiv \alpha^{N(\mathfrak{p})} \pmod{\mathfrak{P}}.$$

for all $\alpha \in K$, where $\mathfrak{P}$ is a prime ideal in K dividing $\mathfrak{p}$, and $N(\mathfrak{p})$ is the (absolute) norm of $\mathfrak{p}$. $\sigma_\mathfrak{p}$ depends on the prime ideal $\mathfrak{P}$ chosen in K, but the conjugacy class $[\sigma_\mathfrak{p}]$ of $\sigma_\mathfrak{p}$ in G does not. Denote by $M_\mathfrak{p} = \Gamma(\sigma)$ the matrix associated by Γ to an arbitrary element $\sigma \in [\sigma_\mathfrak{p}]$. Then the characteristic polynomial $\mid E - tM_\mathfrak{p} \mid$, where E is the unit matrix and $\mid \mid$ denotes the determinant, is independent of the particular element σ chosen in $[\sigma_\mathfrak{p}]$. That is, the determinant $\mid E - tM_\mathfrak{p} \mid$ depends only on the class $[\sigma_\mathfrak{p}]$ and hence only on $\mathfrak{p}$. For, the determinant is an invariant of the equivalence class of the matrix $M_\mathfrak{p}$ as well as of the representation Γ. Let us further denote by χ_Γ what we will call the *Γ-character* associated with the representation Γ, that is, $\chi_\Gamma(\sigma) = tr\, \Gamma(\sigma)$ for $\sigma \in G$, where $\Gamma(\sigma)$ is the matrix associated to σ by Γ, and *tr* denotes the trace.

Then Artin defines the new *(Artin) L-function*, belonging to the *Γ-character* χ_Γ, by

$$L(s, \chi_\Gamma; K/k) = \prod_\mathfrak{p} \frac{1}{\left| E - \frac{M_\mathfrak{p}}{N(\mathfrak{p})^s} \right|},$$

the product taken over all prime ideals $\mathfrak{p}$ not dividing the discriminant $\mathfrak{d}(K/k)$.[68] Then $\zeta_K(s)$ can again be expressed as a product of (Artin) L-functions.[69] However, it is still unknown whether these Artin L-functions are entire if χ_Γ is not the principal character. On the other hand, Hideo Aramata in 1933 and Richard Brauer in 1947[70] proved, based on important preparatory results derived by Artin, that the quotient $\zeta_K(s)/\zeta_k(s)$ is, in fact, entire if K/k is a Galois extension. Brauer also showed in 1947 that the Artin L-functions are meromorphic.

2. If K/k is abelian, then the class $[\sigma_\mathfrak{p}]$ contains only the one element $\sigma_\mathfrak{p}$, and $\sigma_\mathfrak{p}$ becomes independent of the prime ideal $\mathfrak{P}$ in K, taken above $\mathfrak{p}$. Furthermore, Γ becomes one-dimensional if Γ is irreducible. Hence $M_\mathfrak{p} = \Gamma(\sigma_\mathfrak{p}) = tr\, \Gamma(\sigma_\mathfrak{p}) = \chi_\Gamma(\sigma_\mathfrak{p})$, and therefore the Artin L-function for χ_Γ becomes

$$L\big(s, \chi_\Gamma; K/k\big) = \prod_\mathfrak{p} \frac{1}{1 - \frac{\chi_\Gamma(\sigma_\mathfrak{p})}{N(\mathfrak{p})^s}},$$

the product taken over all prime ideals $\mathfrak{p}$ not dividing the discriminant $\mathfrak{d} = \mathfrak{d}(K/k)$. Since K/k is abelian, K is class field to some Weber congruence class group $\mathfrak{C}_\mathfrak{f}$ in

[68] Later, in 1930, Artin has given a more complete definition of his L-series by including also the contribution coming from the prime ideals $\mathfrak{p}$ dividing the discriminant $\mathfrak{d}$. See [6], *Zur Theorie der L-Reihen mit allgemeinen Gruppencharakteren*, Abh. Math. Sem. Hamburg 8 (1930), pp. 292–306; Coll. Papers, pp. 165–179. See also [60].

[69] see [3], 4, formulae (19) and (20), p. 96.

[70] see R. Brauer, *On Artin's L-series with general group characters*, Ann. Math. (2) 48 (1947), pp. 502–514.

k with conductor $\mathfrak{f}$, according to Takagi's class field theory, and $\mathfrak{C}_\mathfrak{f}$ is isomorphic to the Galois group G of K/k. To each *Weber character* χ of this congruence class group $\mathfrak{C}_\mathfrak{f}$ belongs a *Weber L-function* in k

$$L(s, \chi; k) = \prod_{\mathfrak{p}} \frac{1}{1 - \frac{\chi(\mathfrak{p})}{N(\mathfrak{p})^s}},$$

the product taken over all prime ideals $\mathfrak{p}$ not dividing $\mathfrak{f}$, where the Weber character χ of $\mathfrak{p}$ is defined by $\chi(\mathfrak{p}) = \chi(C_\mathfrak{p})$, if $C_\mathfrak{p}$ is the class in $\mathfrak{C}_\mathfrak{f}$ containing the prime ideal $\mathfrak{p}$.[71] According to another theorem by Takagi, the prime ideals dividing the relative discriminant $\mathfrak{d}$ of K/k are the same as those dividing the conductor $\mathfrak{f}$ of $\mathfrak{C}_\mathfrak{f}$.[72] Hence the question arises, whether the Artin L-series $L(s, \chi_\Gamma; K/k)$, defined for the Galois group $G = \mathrm{Gal}(K/k)$ and for the Γ-character χ_Γ on G, coincides with the Weber L-series $L(s, \chi; k)$, defined for the congruence class group $\mathfrak{C}_\mathfrak{f}$ and for the Weber character χ on $\mathfrak{C}_\mathfrak{f}$, whereby G is isomorphic to $\mathfrak{C}_\mathfrak{f}$ by Takagi's class field theory.

Hence the Artin L-series are identical with the Weber L-series if the following two properties are satisfied [3, 5, Satz 2, p. 98]:

(1) $\sigma_\mathfrak{p} \in G$ depends only on the class $C_\mathfrak{p} \in \mathfrak{C}_\mathfrak{f}$.

(2) There is an *explicit* isomorphism α from G to $\mathfrak{C}_\mathfrak{f}$, $\alpha : G \to \mathfrak{C}_\mathfrak{f}$, such that $\alpha(\sigma_\mathfrak{p}) = C_\mathfrak{p}$.

Artin called (1) together with (2) the *Reciprocity Law* of the abelian extension K/k, since in the case of a relative cyclic extension K/k it is identical with the classical reciprocity laws, e. g. the Reciprocity Law of Hilbert in k, if k contains the nth roots of unity, $n = [K : k]$.[73]

3. Artin could prove his Reciprocity Law only in the cases where an explicit reciprocity law was known, that is, in the cases where the appropriate roots of unity lie in the ground field k. In the other cases it was only a conjecture. It was only in July 1927 that he succeeded in giving a general proof of his conjecture [5], after a paper by Tchebotarev on the density of primes, published in 1926 in the Mathematische Annalen 95 (pp. 191–228), had provided him with a key idea [65]. The history of the Tchebotarev density theorem will be studied in another paper.

---

[71] see [68]; or [17], Sect. 5.3, in particular Theorem 25.

[72] see Takagi, *Ueber eine Theorie des relativ-Abel'schen Zahlkörpers*, J. Coll. Science, Univ. of Tokyo, Vol. 41, Art. 9, 1920, pp. 1–133; in particular § 15, Satz 23, (3).

[73] see Hilbert, *Die Theorie der algebraischen Zahlen*, J. bericht DMV 4 (1897), pp. 175–546; in particular Chapter 33.

Chronology

1797 Gauss, *Disquisitiones generales de congruentiis.*

1829 Abel, *Mémoire sur une classe particulière d'équations résolubles algébriquement.*

1853 Kronecker, *Über die algebraisch auflösbaren Gleichungen.*

1856 (October) **Dedekind**, *Abriß einer Theorie der höheren Kongruenzen in bezug auf einen reellen Primzahl-Modulus.*

1863 Gauss, *Disquisitiones generales de congruentiis*, edited by Dedekind.

1886/87 Weber, *Theorie der Abelschen Zahlkörper I, II.*

1887 Kummer, *Zwei neue Beweise der allgemeinen Reciprocitätsgesetze unter den Resten und Nichtresten der Potenzen, deren Grad eine Primzahl ist.*

1897/98 Weber, *Über Zahlengruppen in algebraischen Zahlköpern I, II, III.*
1898 (March 3) **Artin** born in Vienna.

1900 Dedekind, *Über die Anzahl der Idealklassen in reinen kubischen Körpern.*

1917 Hecke, *Über die Zetafunktion beliebiger algebraischer Zahlkörper.*

1917 Hecke, *Über die Kroneckersche Grenzformel für reelle quadratische Körper und die Klassenzahl relativ-Abelscher Körper.*

1919 (January) **Artin** begins his studies in Leipzig.

1919 (April 10) **Kornblum**, *Über die Primfunktionen in einer arithmetischen Progression.*

1920 Takagi, *Ueber eine Theorie des relativ-Abel'schen Zahlkörpers.*

1921 (June 20) **Artin** finished his thesis, *Quadratische Körper im Gebiete der höheren Kongruenzen I, II.*

1921 (September 17) **Herglotz**, *Über einen Dirichletschen Satz.*

1921 (October 14) **Artin's thesis** received for publication.

1921 (October) **Artin** spends a year in Göttingen.

1921 (November 22): **Artin's lecture** in the Mathematische Gesellschaft in Göttingen.

1922 (September 4) **Artin**, *Über die Zetafunktionen gewisser algebraischer Zahlkörper.*

1922 (October) **Artin** moves to Hamburg as an assistant to Blaschke.

1923 Herglotz, *Über die Kroneckersche Grenzformel für reelle quadratische Körper, I, II.*

1923 (July) **Artin's habilitation**, *Über eine neue Art von L-Reihen* (Artin's Reciprocity Law as a conjecture).

1923 (July) **Artin** becomes a Privatdozent (Lecturer) in Hamburg.

1924 Artin's thesis published.

1925 (April) **Artin** becomes an Ausserordentlicher (Associate) Professor.

1926 (October) **Artin** becomes an Ordentlicher (Full) Professor.

1926 Tchebotarev, *Die Bestimmung der Dichtigkeit einer Menge von Primzahlen, welche zu einer gegebenen Substitutionsklasse gehören.*

1927 Artin, *Beweis des allgemeinen Reziprozitätsgesetzes*. (Proof of the General Reciprocity Law).

Acknowledgements. It is a great pleasure to thank *Peter Roquette* for his hospitality, his support and for the many interesting discussions. My thanks go also to the University of Heidelberg and to the Deutsche Forschungsgemeinschaft for their generous practical and financial support and to the editor of this volume, *Arnfinn Laudal.*

References

[1] Abel, Niels Henrik: *Mémoire sur une classe particulière d'équations résolubles algébriquement.* J. Reine Angew. Math. 4 (1829). Oeuvres complètes, Nouvelle édition par MM. L. Sylow et S. Lie, tome I, XXV. Christiania, 1881, pp. 478–507.

[2] Artin, Emil: *Über die Zetafunktionen gewisser algebraischer Zahlkörper.* Math. Ann. 89 (1923), 147–156.

[3] Artin, Emil: *Über eine neue Art von L-Reihen.* Abh. Math. Sem. Hamburg 3 (1923), 89–108.

[4] Artin, Emil: *Quadratische Körper im Gebiete der höheren Kongruenzen I, II.* Math. Zeitschrift 19 (1924), 153–246.

[5] Artin, Emil: *Beweis des allgemeinen Reziprozitätsgesetzes.* Abh. Math. Sem. Hamburg 5 (1927), 353–363.

[6] Artin, Emil: *Zur Theorie der L-Reihen mit allgemeinen Gruppencharakteren*, Abh. Math. Sem. Hamburg 8 (1930), pp. 292–306; Coll. Papers, pp. 165–179.

[7] Artin, Emil: *The Collected Papers of Emil Artin.* Edited by Serge Lang and John T. Tate. Addison-Wesley, 1965.

[8] Brauer, Richard: *On Artin's L-series with general group characters.* Ann. Math. (2) 48 (1947), pp. 502–514.

[9] Cassels, J. W. S. and Fröhlich, A. : *Algebraic Number Theory.* Academic Press, London, 1967.

[10] Dedekind, Richard: *Abriß einer Theorie der höheren Kongruenzen in bezug auf einen reellen Primzahl-Modulus.* J. Reine Angew. Math. 54 (1857), 1–26.

[11] Dedekind, Richard: *Über den Zusammenhang zwischen der Theorie der Ideale und der Theorie der höheren Kongruenzen.* Abh. Königl. Ges. Wiss. Göttingen 23 (1878), 1–23; Ges. math. Werke, Bd. 1, XV, pp. 202–232.

[12] Dedekind, Richard: *Über die Anzahl der Idealklassen in reinen kubischen Körpern.* J. Reine Angew. Math. 121 (1900), 40–123; Werke, Bd. 2, XXIX, pp. 148–235.

[13] Dedekind, Richard: *Gesammelte mathematische Werke.* Herausgegeben von Robert Fricke, Emmy Noether, Öystein Ore. Erster Band, Braunschweig 1930; Zweiter Band, Braunschweig 1931; Dritter Band, Braunschweig 1932.

[14] Dirichlet, Lejeune; Dedekind, Richard: *Vorlesungen über Zahlentheorie.* Erste Auflage. Vieweg, Braunschweig, 1869.

[15] Dirichlet, Lejeune; Dedekind, Richard: *Vorlesungen über Zahlentheorie.* Vierte Auflage. Vieweg, Braunschweig, 1894.

[16] Dedekind, Richard and Weber, Heinrich: *Theorie der algebraischen Funktionen einer Veränderlichen.* J. Reine Angew. Math. 92 (1882), pp. 181–290; Dedekind, Ges. math. Werke, Bd. 1, XVIII, pp. 238–350.

[17] Frei, Günther: *Heinrich Weber and the Emergence of Class Field Theory.* In: The History of Modern Mathematics, edited by David Rowe and John McCleary (2 Vol.), Academic Press, Boston (1989), Vol. 1, pp. 425–450.

[18] Frei, Günther: *The Reciprocity Law from Euler to Eisenstein.* In: The Intersection of History and Mathematics (Editors: Sasaki Ch., Sugiura M., Dauben J.W.). Birkhäuser, Basel, 1994, pp. 67–88.

[19] Frei, Günther: *How Hasse was led to the Theory of Quadratic Forms, the Local–Global Principle, the Theory of the Norm Residue Symbol, the Reciprocity Laws, and to Class Field Theory.* In: Class Field Theory – Its Centenary and Prospect, Edited by Katsuya Miyake. Advanced Studies in Pure Mathematics 30. Mathematical Society of Japan. Tokyo, 2001, pp. 31–62.

[20] Frei, Günther: *On the Development of the Theory of Function Fields over a Finite Field from Gauss to Dedekind and Artin.* Preprint 2001. 62 pages.

[21] Frei, Günther: *On the History of the Reciprocity Laws. Eleven letters from Artin to Hasse in the years 1923–1927.* Preprint 2002.

[22] Frei, Günther: *Gauss' unpublished Section Eight of the Disquisitiones arithmeticae: The Beginning of the Theory of Function Fields over a Finite Field.* To appear in: The shaping of arithmetic two hundred years of number theory after C.F. Gauß's Disquisitiones Arithmeticae (edited by C. Goldstein, N. Schappacher, J. Schwermer). Springer-Verlag, Heidelberg, 2003.

[23] Fueter, Rudolf: *Abelsche Gleichungen in quadratisch-imaginären Zahlkörpern.* Math. Ann. 75 (1914), 177–255.

[24] Gauss, Carl Friedrich: *Disquisitiones arithmeticae.* Werke, Erster Band, Göttingen, 1863. (German translation in [56]; English translation in [27], Second corrected printing, Springer, Heidelberg, 1986.)

[25] Gauss, Carl Friedrich: *Werke.* Zweiter Band. Göttingen, 1863.

[26] Gauss, Carl Friedrich: *Werke.* Zweiter Band, Zweiter Abdruck. Göttingen, 1876.

[27] Carl Friedrich Gauss: *Disquisitiones arithmeticae.* Translated by Arthur A. Clarke. Yale University Press, New Haven, 1966.

[28] Gauss, Carl Friedrich: *Mathematisches Tagebuch 1796–1814.* Mit einer historischen Einführung von Kurt-R. Biermann. Ins Deutsche übertragen von Elisabeth Schuhmann. Durchgesehen und mit Anmerkungen versehen von Hans Wußing und Olaf Neumann. 4. Auflage. Ostwalds Klassiker der exakten Wissenschaften 256. Akademische Verlagsgesellschaft Geest & Portig, Leipzig, 1985.

[29] Gray, J. J.: *A commentary on Gauss's mathematical diary, 1796–1814, with an English translation.* Expo. Math. 2 (1986), pp. 97–130.

[30] Hasse, Helmut: *History of Class Field Theory.* In: Algebraic Number Theory, Edited by J. W. S. Cassels and A. Fröhlich. Academic Press, London, 1967, pp. 266–279.

[31] Hecke, Erich: *Über die Zetafunktion beliebiger algebraischer Zahlkörper.* Nachr. K. Ges. Wiss. Göttingen, 1917, pp. 77–89; Werke, 7, pp. 159–171.

[32] Hecke, Erich: *Über eine neue Anwendung der Zetafunktionen auf die Arithmetik der Zahlkörper.* Nachr. K. Ges. Wiss. Göttingen, 1917, pp. 90–95; Werke, 8, pp. 172–177.

[33] Hecke, Erich: *Über die L-Funktionen und den Dirichletschen Primzahlsatz für einen beliebigen Zahlkörper.* Nachr. K. Ges. Wiss. Göttingen, 1917, pp. 299–318; Werke, 9, pp. 178–197.

[34] Hecke, Erich: *Über die Kroneckersche Grenzformel für reelle, quadratische Körper und die Klassenzahl relativ-Abelscher Körper.* Verh. Naturf. Ges. Basel 28 (1917), pp. 363–372; Werke, 10, pp. 198–207.

[35] Herglotz, Gustav: *Über das quadratische Reziprozitätsgesetz in imaginären quadratischen Zahlkörpern.* Leipziger Ber. 73 (1921), pp. 303–310; Ges. Schriften, pp. 396–403.

[36] Herglotz, Gustav: *Zur letzten Eintragung im Gaußschen Tagebuch.* Leipziger Ber. 73 (1921), pp. 271–276; Ges. Schriften, pp. 415–420.

[37] Herglotz, Gustav: *Über einen Dirichletschen Satz.* Math. Zeitschrift 12 (1922), pp. 255–261; Ges. Schriften, pp. 429–435.

[38] Herglotz, Gustav: *Über die Entwicklungskoeffizienten der Weierstraßschen $\wp$-Funktion.* Leipziger Ber. 74 (1922), pp. 269–289; Ges. Schriften, pp. 436–456.

[39] Herglotz, Gustav: *Über die Kroneckersche Grenzformel für reelle quadratische Körper, I, II.* Leipziger Ber. 75 (1923), pp. 3–14 and pp. 31–37; Ges. Schriften, pp. 466–484.

[40] Hilbert, David: *Über den Dirichletschen biquadratischen Zahlkörper.* Math. Ann. 45 (1894), 309–340; Ges. Abh., Bd. I, 5, pp. 24–52.

[41] Hilbert, David: *Ein neuer Beweis des Kroneckerschen Fundamentalsatzes über Abelsche Zahlkörper.* Nachr. Ges. Wiss. Göttingen, 1896, pp. 29–39 ; Ges. Abh., Bd. I, 6, pp. 53–62.

[42] Hilbert, David: *Die Theorie der algebraischen Zahlkörper*, J.-bericht DMV 4 (1897), pp. 175–546; Ges. Abh., Bd. I, 7, pp. 63–363.

[43] Hilbert, David: *Über die Theorie des relativquadratischen Zahlkörpers.* Math. Ann. 51 (1899), 1–127; Ges. Abh., Bd. I, 9, pp. 370–482.

[44] Kornblum, Heinrich: *Über die Primfunktionen in einer arithmetischen Progression.* Math. Zeitschrift 5 (1919), 100–111.

[45] Kronecker, Leopold: *Über die algebraisch auflösbaren Gleichungen (I. Abhandlung).* Monatsberichte der Königl. Preuss. Akademie der Wiss. zu Berlin, 1853, pp. 365–374; Werke, Bd. 4, I, pp. 1–11.

[46] Kronecker, Leopold: *Über die algebraisch auflösbaren Gleichungen (II. Abhandlung).* Monatsberichte der Königl. Preuss. Akademie der Wiss. zu Berlin, 1856, pp. 203–215; Werke, Bd. 4, IV, pp. 25–37.

[47] Kronecker, Leopold: *Über den Gebrauch der Dirichlet'schen Methoden in der Theorie der quadratischen Formen.* Monatsberichte der Königl. Preuss. Akademie der Wiss. zu Berlin, 1864, pp. 285–303; Werke, Bd. 4, XXII, pp. 227–244.

[48] Kronecker, Leopold: *Über Abelsche Gleichungen.* Monatsberichte der Königl. Preuss. Akademie der Wiss. zu Berlin, 1877, pp. 845–851; Werke, Bd. 4, IX, pp. 63–71.

[49] Kronecker, Leopold: *Grundzüge einer arithmetischen Theorie der algebraischen Grössen.* J. Reine Angew. Math. 92 (1881), 1–122; Werke, Bd. 2, XI, pp. 237–387.

[50] Kronecker, Leopold: *Die Composition Abelscher Gleichungen.* Sitzungsberichte der Königl. Preuss. Akademie der Wiss. zu Berlin, 1882, pp. 1059–1064; Werke, Bd. 4, XII, pp. 113–121.

[51] Kronecker, Leopold: *Werke.* Herausgegeben von K. Hensel. Bd. 1 (1895), Bd. 2 (1897), Bd. 3.1 (1899), Bd. 3.2 (1931), Bd. 4 (1929), Bd. 5 (1930).

[52] Kühne, H.: *Eine Wechselbeziehung zwischen Functionen mehrerer Unbestimmten, die zu Reciprocitätsgesetzen führt.* J. Reine Angew. Math. 124 (1902), 121–133.

[53] Kühne, H.: *Angenäherte Auflösung von Congruenzen nach Primmodulsystemen in Zusammenhang mit den Einheiten gewisser Körper*, J. Reine Angew. Math. 126 (1903), 102–115.

[54] Kummer, Ernst Eduard: *Zwei neue Beweise der allgemeinen Reciprocitätsgesetze unter den Resten und Nichtresten der Potenzen, deren Grad eine Primzahl ist.* J. Reine Angew. Math. 100 (1887), 10–50; Reprint of Abhandl. der königl. Akademie der Wiss. zu Berlin 1861, pp. 81–122; Collected Papers, Vol. I, pp. 842–882.

[55] Landau, Edmund: *Handbuch der Lehre von der Verteilung der Primzahlen.* Göttingen, 1909.

[56] Maser, H.: *Carl Friedrich Gauss' Untersuchungen über höhere Arithmetik.* Julius Springer, Berlin, 1889.

[57] Miyake, Katsuya: *The Establishment of the Takagi–Artin Class Field Theory.* In: The Intersection of History and Mathematics, Edited by S. Sasaki, M. Sugiura, J. W. Dauben. Birkhäuser, Basel, 1994, pp. 109–128.

[58] Moore, E. H.: *Mathematical Papers Read at the International Mathematics Congress.* Chicago, 1893; Macmillan, New York, 1896, pp. 208–226.

[59] Pieper, Herbert: *Variationen über ein zahlentheoretisches Thema von Carl Friedrich Gauss.* Birkhäuser Verlag, Basel, 1978.

[60] Roquette, Peter: *On the history of Artin's L-functions and conductors.* Mitt. Math. Ges. Hamburg 19 (2000), 5–50.

[61] Roquette, Peter: *Class Field Theory in Characteristic p, its Origin and Development.* In: Class Field Theory – Its Centenary and Prospect, edited by K. Miyake, Advanced Studies in Pure Mathematics 30, Tokyo (2001), pp. 549–631.

[62] Schmidt, F. K.: *Zur Zahlentheorie in Körpern der Charakteristik p. Vorläufige Mitteilung.* Sitz.-Ber. phys. med. Soz. Erlangen 58/59 (1926/27), 159–172.

[63] Schmidt, F. K.: *Analytische Zahlentheorie in Körpern der Charakteristik p.* Math. Zeitschr. 33 (1931), 1–32.

[64] Takagi, Teiji: *Über eine Theorie des relativ Abel'schen Zahlkörpers.* Journ. of Coll. of Science, Univ. of Tokyo 41, Art 9, 1920, 1–133.

[65] Tschebotareff (Tchebotarev), N.: *Die Bestimmung der Dichtigkeit einer Menge von Primzahlen, welche zu einer gegebenen Substitutionsklasse gehören.* Math. Ann. 95 (1926), 191–228.

[66] Ullrich, Peter: *Emil Artin's unveröffentlichte Verallgemeinerung seiner Dissertation.* Mitt. Math. Ges. Hamburg XIX (2000), 173–194.

[67] Weber, Heinrich: *Theorie der Abelschen Zahlkörper I, II.* Acta Math. 8 (1886), pp. 193–263; 9 (1887), pp. 105–130.

[68] Weber, Heinrich: *Über Zahlengruppen in algebraischen Zahlkörpern I, II, III.* Math. Ann. 48 (1897), pp. 433–473; 49 (1897), pp. 83–100; 50 (1898), pp. 1–26.

[69] Weber, Heinrich: *Lehrbuch der Algebra.* Dritter Band. Vieweg, Braunschweig, 1908.

The Italian School of Algebraic Geometry and Abel's Legacy

Aldo Brigaglia, Ciro Ciliberto, and Claudio Pedrini

Introduction

Exactly one century ago, in the spring of 1902, the prestigious mathematical journal *Acta Mathematica*, decided to mark the centenary of Abel's birth by dedicating three issues to the memory of this Norwegian mathematician. Some of the most influential Italian mathematicians of the day were invited to contribute to these issues. Among those invited were Guido Castelnuovo, at the time the most well-known representative of the Italian school of algebraic geometry, and Federigo Enriques. Enriques, at first, accepted the invitation and, as he communicated to Castelnuovo, decided to write

> *... una specie di Bericht, breve, intorno ai nostri lavori, in cui mi propongo di esporre la teoria sotto l'aspetto analitico.*
>
> *(... a kind of Bericht, a short one, concerning our papers, in which I would like to expose the theory using the analytic viewpoint.)*
>
> (letter of 13/3/1902; for all citations from the correspondence between Enriques and Castelnuovo, see [15]).

The project was never completed. However this short citation is important for us: it tells us, in fact, that the Italian algebraic geometers, at the beginning of last century, used to look at their connections with Abel via the analytic and transcendental methods. Although these techniques never really became central inside the Italian school, they were always considered as a natural counterpart of the projective-geometric aspects typical of the Italians. Furthermore, as we will see, at the time Enriques was writing, there was in Italy an important revival of the analytic ideas.

In this paper we will try to make clear an interesting set of contributions which, inside the Italian school, were, in Enriques' sense, inspired by Abel's analytic viewpoint. For this we will go back to the real beginning of the school with Luigi Cremona, who is considered to be its founder. Then we will follow the evolution of these ideas up until the 1930's and we will indicate how they are thought of today.

Abel, apparently, never had any contact with the Italian mathematical environment, in his short life, nor even specifically during his travels to Italy (see [132]). Therefore, his influence on Italian mathematicians was late and indirect. However, we believe that, despite this, it has been deep and longstanding.

In the first chapter we will give an overview of the influence of Abel's ideas on the beginnings of the Italian school, filtered via Riemann's viewpoint and the geometric interpretation of it by Clebsch and Gordan, and later by Brill and Noether. This chapter mainly centers around the character of Luigi Cremona, the founder of the Italian school.

In the second chapter we will see how Abel's influence was still active at the beginning of the XXth century and played a basic role in understanding the notion of irregularity of surfaces. The main characters, in this period, are the young Severi and the well established Castelnuovo, whose contributions we will review, indicating also their subsequent far reaching influence inside and outside the Italian school, until A. Weil's proof of Riemann's hypothesis for the zeta function of an algebraic curve over a finite field.

The third chapter is devoted to Severi's ideas on rational equivalence of 0-cycles on a surface. We will indicate how some of these ideas were related, in Severi's mind, to Abel's viewpoint. As is well known, Severi's contributions on the subject have been very controversial. We will briefly report on the main criticisms but we will also try to elucidate which of them have a present validity. In particular we will direct our, and we hope the reader's attention, to some of his ideas which are very closely related to Bloch's conjecture. We dedicate a few, more technical, sections at the end of Chap. 3 to open a window on the present developments of this last subject, specifically to some motivic interpretations which we think are rather close to Severi's original viewpoint.

Due to the different tastes and attitudes of the authors, which we deliberately did not make too much effort to hide, the paper is rather uneven. The first chapter is expository and more historiographical in nature. The other two, though sharing with the first a historiographical perspective, have a different flavour: in the second some technical aspects start to appear, and they become even more relevant in Chap. 3, especially, as we said, in its last sections. We hope that the uneveness of the paper will attract, rather than repel, readers with different interests.

Acknowledgments. The authors wish to warmly thank A. Geramita for the careful reading of a first version of the paper and several suggestions which considerably improved its English. Useful remarks have been made also by A. del Centina, D. Portelli, and R. Piene. Thanks to N. Schappacher who pointed out to the second author the existence of the correspondence between F. Severi and A. Weil concerning the debate at the ICM of Amsterdam in 1954. Finally the authors also thank C. Safina for her good job on the TeX version of the paper.

1 Abel's Influence at the Beginnings of the Italian School

This chapter is intended to indicate how Abel's ideas, via Riemann and Clebsch, influenced Italian mathematicians around the first half of the XIX century. In particular we will follow this influence on L. Cremona, the master of the Italian school of algebraic geometry. Specifically we will discuss his relations with Clebsch, how Cremona's geometric viewpoint was affected by it, and how he tried to communicate this to his students. Finally we will comment upon a certain decline of the analytic viewpoint until the beginning of the new century.

1.1 Abel according to Volterra and Enriques. It seems appropriate to us to start with two citations which show, in our opinion, how some influential Italian mathematicians of about one century ago used to look at Abel:

> *È appena un secolo – osservava acutamente il Mittag–Leffler, scrivendo le belle pagine dedicate alla memoria di Abel – che questo grande analista proclamò apertamente essere la matematica fine sufficiente a sè medesima e portare il suo ideale in sè stessa. ... La matematica, ripiegandosi in sè medesima, come pensava Abel, onde costituire prima e consolidare poi quella teoria delle funzioni e quella geometria che furon il fondamento delle ricerche degli ultimi anni, condusse a tal perfezione l'analisi del pensiero con l'esame assiduo e profondo dei propri concetti e dei mezzi di cui dispone, che questi acquistarono tanta acutezza, flessibilità e potenza da penetrare e commuovere tutta la speculazione scientifica e filosofica.*
>
> *(It is just one century ago – as Mittag–Leffler observed when he wrote his beautiful pages in memory of Abel – that this great analyst openly claimed that mathematics is self–sufficient and carries its ideal inside itself. ... As Abel thought, when he solidly developed the theory of functions and the geometry which were the fundamental researches of his last years, mathematics, looking inside itself, brought to a remarkable level of perfection its analysis of thinking, through a deep and constant examination of its own concepts and tools. In such a way that it greatly influenced the entire scientific and philosophical research.)*

This citation from Volterra (see [137]) shows the idea that, we think, several Italian mathematicians had of Abel: he was considered to be the precursor of a new way of doing mathematics, abstract, rigorous and general and, above all, a way capable of facing newer and newer subjects, with greater and greater generality.

Volterra's voice in fact is echoed, some years later, by Enriques (see the Preface of [60]):

> *Il criterio di ricerca così splendidamente fatto valere da Abel "porre i problemi nell'aspetto più generale per scoprirne la vera natura," designava l'indirizzo dell'Analisi che vuol liberare la conoscenza dei rapporti qualitativi dalle complicazioni accidentali dei calcoli, cioè appunto quell'indirizzo di cui è massima attuazione la teoria geometrica delle equazioni e delle funzioni algebriche.*
>
> *(The research criterion so splendidly taken by Abel – to pose problems in their most general aspects in order to discover their true nature – set the course of Analysis, which wanted to liberate the knowledge of qualitative relationships from the accidental complications of calculations; this is precisely that course of maximum realization for the geometric theory of equations and algebraic functions.)*

Enriques' citation shows another important feature that the Italian school attributed to, and inherited from, Abel: the prevalence of the *qualitative relationships* in contrast with the *accidental complications of calculations*. Volterra's and Enriques' opinion is that, in some sense, the modern way of doing mathematics has one of its bench-marks in Abel's work. We believe that this viewpoint was shared by most of the Italian mathematicians of their time.

1.2 Abel's and Riemann's influence on L. Cremona and the Italian mathematicians of his time. Volterra's and Enriques' viewpoint has old roots. It goes back, in fact, more than fifty years, when the most active Italian mathematicians tried to understand Riemann's new ideas and to publicize them in Italy. In doing so, they first had to study Abel's contributions, on which Riemann's contruction was based.

In this respect one of the main characters is L. Cremona, the founder of the Italian school of algebraic geometry (for more information than the little we will be able to give on Cremona in this notes, see [89] and [16], Sect. 1.1 and references therein). Indeed there are strong lines which connect Cremona to Abel, which we will try to indicate.

To start, one of the first papers (actually the second) of Cremona in 1856 is a note [42] entitled *Intorno ad un teorema di Abel* (*On a theorem of Abel*). Here Cremona gives the proof of a particular case of Abel's famous theorem. It is apparent that, at that time, Abel's original proof was probably not yet fully understood and there was therefore the need for simpler and clearer proofs, even though they could be applied only to some particular situations. Cremona's approach in [42] is not geometrical. However his systematic use of resultants indicates the algebraic attitude of the young mathematician. Furthermore, besides its limited scientific interest, the paper has, for us, a symbolic meaning: it shows the interest with which, in the second half of the 1850's, Italian mathematicians were considering Abel's work. Indeed, as Cremona says [89], about his master, Francesco Brioschi:

> *... nei successivi anni che passai in Pavia, cioè sino al 1857, mi iniziò allo studio delle funzioni ellittiche e abeliane e alle opere magistrali di Abel e Jacobi.*
>
> *(... in the subsequent years that I spent in Pavia, namely till 1857, he pointed me to the study of elliptic and abelian functions and to the masterly works of Abel and Jacobi.)*

As we said already, the diffusion of Abel's works in Italy developed in the same years in which the same started, mainly due to Betti's teaching in Pisa, for the work of Riemann. It is worth recalling that in 1857 the fundamental article [107] of Riemann on abelian functions appeared and Betti, Brioschi, and Casorati visited the German mathematician in Göttingen in 1858. However Riemann's paper was not easily understood at that time. For instance Betti in 1860-61, in the prolusion to his course in Pisa on Riemann's ideas applied to elliptic functions, says (see [14, p. 129]):

> *Ha sopra gli altri il vantaggio della sua immensa generalità e di soddisfare compiutamente alle principali tendenze dell'analisi moderna, perché il meccanismo del calcolo non ci entra quasi per niente, ed è quasi tutto un magnifico lavoro di pensiero. Ma quanto è la forza della mente altrettanta è la concisione e l'oscurità dello stile di questo eminente geometra. In modo che i suoi lavori per ora è quasi come non esistessero nel mondo scientifico.*
>
> *(It has over the others the advantage of its immense generality. Moreover it completely satisfies the main tendencies of the modern analysis, since computations, in their mechanical forms, are almost absent and one sees there a beautiful construction of the mind. However, so great is the strenght of the mind, so concise and obscure is the style of this eminent geometer. To the extent that, so far, the impact of his papers on the scientific world has been almost absent.)*

This is important evidence of two facts. On one side one sees in Riemann the real follower of the methods, i.e., generality and absence of calculations, which will be noted by Volterra and Enriques, as typical of Abel. On the other side one notes the great difficulty with which these new methods were absorbed. Perhaps the algebro-geometric translation of Riemann's ideas, which will come up in a while, has been crucial for the Italians since it contributed to the understanding and diffusion in the country of the work of the German mathematician.

In 1862 Betti published a paper [11] with the purpose of making *più facile e più chiara l'esposizione* (*easier and clearer the exposition*) of the work of Riemann, as he says in a letter to Placido Tardy (see [14]). Betti's opinion was confirmed by Riemann himself, who in 1862 visited Italy and wrote to Betti from Florence (see [14, p. 133]):

Ho potuto leggere solo quello "Sopra le funzioni algebriche di una variabile complessa." Mi è piaciuto molto e desidererei proprio che l'avessimo in tedesco, poiché contribuirebbe in maniera essenziale a rendere accessibile quella teoria.

(I could only read the paper "On the algebraic functions of one complex variable." I liked it a lot and I would like that we could have it in German, since it would contribute in an essential manner to make that theory accessible.)

As is well known, Riemann lived in Pisa in the years 1863–1865, where he had daily discussions with Betti and, from 1864, also with Eugenio Beltrami.

1.3 Clebsch and Cremona. This is the lively mathematical atmosphere in which, in 1864, another epochal contribution, the paper [35] of Clebsch, appeared. Clebsch, a former student of Riemann, in the aforementioned paper made a first successfull attempt at giving a geometric interpretation, in terms of algebraic plane curves, of Riemann's ideas and of Abel's theorem on the integrals of algebraic functions. Clebsch's geometric interpretation, which was somehow implicit in Riemann's analytic treatment, appears very natural and its impact on the mathematical ambient was extremely positive, in the short, as well as in the long, run.

A few more words on Clebsch's ideas, which have been at the base of the developments of algebraic geometry, will be said in Sect. 2.1 (see also [74]). We will limit ourselves here to recall that Clebsch interprets the abelian integrals on a Riemann surface of genus (*Geschlecht* according to Clebsch) g in terms of *adjoint curves* to a plane model of the Riemann surface as a curve of degree d with $\delta = \binom{d-1}{2} - g$ nodes (see Clebsch formula 2.1). Then Clebsch's geometric interpretation of Abel's theorem, which relies on a famous identity of Jacobi (see [67, p. 671]), gives the necessary and sufficient conditions for dn points on a curve C of degree d to be cut out on C by a curve of degree n. Clebsch applies this to study curves having multiple contacts with C, e.g., he studies the configuration of the 28 bitangents to a smooth plane curve of degree 4. Then he makes further applications to a systematic study of curves of genus 1, called *elliptic curves*, since, according to Abel and Jacobi, they can be parameterized by elliptic functions.

These results, springing in a natural way from Abel's and Riemann's work on abelian integrals, were, at the same time, elegantly suggested by geometric intuition and, in turn, ended up further stimulating the geometric fantasy.

Clebsch's approach was immediately noticed by the Italians, especially by Cremona, who, as we saw, was at the same time busy with the same kind of questions, i.e., understanding Abel and Riemann from a more algebro–geometric viewpoint. Cremona was then at the top of his activity, refining, or rather, creating new geometric techniques for the understanding of natural objects like space curves, surfaces etc. For instance, between 1862 and 1865, he published his famous memoirs on birational transformations, called, after him, *Cremona transformations*. One therefore understands why Clebsch's ideas attracted the Italian mathematician, even though the methodological differences between the two were relevant: e.g. Riemann's analytic methods are still very present to Clebsch, whereas Cremona's viewpoint is certainly more intuitive and geometric, a difference which will mark, from then on, the difference between the Italian and the German school, despite the great affinities and exchanges of ideas (see [14, §1.2]).

The collaboration between Clebsch and Cremona, followed after a while by a real friendship, started in August 1863, when Cremona sent some of his papers to the German colleague. The exchanges of ideas soon became frequent and intense and went on until Clebsch's premature death in 1872. In December 1863, Clebsch communicates to his friend that he is about to publish his aforementioned paper [35], and, in doing so, he points out at the same time how their methods and viewpoints are somehow complementary (Letter of December 23, 1863; for the correspondence of Cremona we refer to [18, 64, 92, 92, 93, 102]):

> *Wir leben glücklicher Weise in einer Zeit wo analytische und synthetische Geometrie aufgehört haben einander zu befehden, und zufrieden sind von einander zu lernen. Ich glaube in meine Vorlesungen über analytische Geometrie wesentlich dadurch gefördert zu sein, dass ich synthetische Prinzipien mit Vorliebe in dieselben eingeführt habe; und dass ich so wenigstens die Elemente derjenigen Theorien bringen kann, welche Sie in Ihrer* Introduzione *behandeln; ... In einiger Zeit hoffe ich Ihnen eine Arbeit zu übersenden, welche wie ich glaube von einigem Interesse sein wird, und auch die Synthese mit einer Reihe von Fragen versicht. Ich habe nämlich Riemanns Theorie der Abelschen Integrale auf die Untersuchung ebener und räumlicher Curven angewandt, und daraus die Lösung sehr allgemeiner und schwieriger Berhrungsaufgaben gezogen.*
>
> *(Luckily enough we live at a time in which analytic and synthetic geometry are no longer hostile to each other, rather they are eager to learn from each other. I think in my lessons on analytic geometry I succeeded in promoting a preferential use of synthetic principles and in presenting the bases of the theory that you develop in your* Introduzione ... *I hope I'll be able to send to you in a short time a paper that I think will be of some interest, and I'll also add a series of questions. Indeed, I applied Riemann's theory of abelian integrals to questions about plane and skew curves, thus getting the solution of several general, difficult problems.)*

On March 18, 1864, the reprint of Clebsch's paper is sent to Cremona and it starts an interesting discussion, through which Cremona perceives, in a clearer and clearer way, the power of the transcendental methods of Abel, Jacobi and Riemann and, at the same time, he strengthens his ability to *translate* into geometric terms the analytic results. This has been, after all, one of the characteristic features and main qualities of the algebro–geometric viewpoint typical of the Italian school.

Cremona and Clebsch used to keep each other informed, on a very frequent basis, about the developments of their research. In particular Clebsch informs Cremona about the work he is doing in collaboration with Gordan in order to write a treatise on the subject. Indeed, in the summer 1865 Clebsch writes to Cremona:

> *Mein Hauptbeschäftigung in dem Sommer war eine Vorlesung über die Theorie der Abelschen Funktionen, welche mich genötigt hat, in diese Theorie selbst tiefer einzudringen. Ich habe daraus die Überzeugung geschöpft, dass es möglich ist, diese Theorie ohne die eigentümlichen und schwer verständlichen Grundlagen vorzutragen, welche Riemann derselben gegeben hat.*
>
> *(My main occupation during the summer has been to work on a treatise on the theory of abelian functions. This forced me to go deeper into this theory. I got convinced of the fact that it is possible to expose this theory without relying on the foundations Riemann gave to it, which are so difficult to understand.)*

The book [39] is published one year later, in 1866: it will become the basis for all the studies concerning the interplay between abelian integrals and algebraic geometry. Notice that Cremona's viewpoint is at least partially present in the book: certainly Clebsch perceived the interest in birational maps and he was able to prove the invariance of the genus of a curve under these transformations.

Probably the main common area of interest for Clebsch and Cremona was the study of space curves and of surfaces, with their birational transformations. The two however proceeded, as we indicated above, with a substantially different approach. Cremona was more systematical and purely algebro–geometric, Clebsch did not use this approach: together with geometric tools, he used also analytic ones. This fact finds its resonance in the following citation from Enriques [60], Preface, p. VIII, where, strangely enough, Cremona's *puristic* viewpoint is criticized:

> *Non è più il tempo in cui possa presumersi di dominare l'intera materia con una sola veduta, come il trattato di Salmon volle fare ordinando lo studio delle curve algebriche sulla base dello studio delle forme invariantive e il Cremona movendo da pochi principi sintetici sulle polari. La nostra epoca ha superato decisamente il purismo delle scuole analitiche e geometriche, traendo da ciascuna gl'istrumenti della ricerca; il ravvicinamento dei metodi che risponde al programma eclettico di Clebsch, ha segnato un reale e fecondo progresso.*
>
> *(It is no longer the time in which one can pretend to master the whole matter by taking one single point of view, as Salmon wanted to do in his treatise by founding the theory of algebraic curves on the study of invariant forms, and as Cremona did by moving from a few synthetic principles about polars. Today we boldly overcame the purisistic approach of the analytic and geometric schools, rather we took from each of them the right tools for our research; the mixture of various techniques, according to Clebsch's eclectic programme, marked a real, fruitful progress with respect to the past.)*

On the other hand it is known that Enriques, in his young days in Rome in 1891, was *somewhat disappointed by the teaching of the, by now, very old, Cremona* (see [16, p. 26]). Hence it is likely that the above opinion shows the signs of this experience.

The previous citation shows that, according to Enriques, the route of modernity in algebraic geometry passes through Abel, Riemann and Clebsch. It is very possible that Cremona himself would have agreed with Enriques on this, since around the end of the 1860's he tried, as we will see, to follow exactly this route. Actually he even expressed opinions very close to the ones later formulated by Enriques. Rather the question is, why this viewpoint of Cremona's was not shared by the rest of the Italian school, even if Cremona was considered, with no discussion, its master? We will come back to this later.

Going back to the relationship between Clebsch and Cremona and to their useful and sound competition, we see that it particularly develops after the publication of Clebsch's memoir of 1866 and of the book with Gordan in 1866. As typical examples of it, and of the diversity, or rather the complementarity, of tools used by the two authors, we should mention the two papers [36] and [43] about the so–called *Steiner's Roman surface*, which we know today to be the projection in $\mathbb{P}^3$ of a Veronese surface in $\mathbb{P}^5$. These papers come from a collaboration between Clebsch and Cremona which is witnessed by the following letter, written by the former to the latter on July 21, 1866:

> *Es macht mich sehr glücklich, mit Ihnen in diesen Untersuchungen zu begegnen, und ich werde mich sehr freuen, Ihre Methoden kennenzulernen: ein solches Zusammentreffen zeigt jedenfalls, dass man an der richtigen Stelle arbeitet. Ich habe mich neuerdings auch, Ihrem Beispiele folgend, mit der Steinerschen Fläche beschäftigt. Diese hat für mich namentlich das Interesse, dass sie durch die Gleichungen*
>
> $$rx_1 = f_1(c, l, m)$$
>
> $$rx_2 = f_2(c, l, m)$$
>
> $$rx_3 = f_3(c, l, m)$$
>
> $$rx_4 = f_4(c, l, m)$$
>
> *dargestellt wird, wo die f allgemeine Funktionen zweiten Grades sind. Die durch diese Formeln begründete Abbildung hat Kummer erwähnt, aber nicht ausgeführt. Sie liefert einiges Interessante, in geometrische Beziehung aber namentlich folgendes. Wenn man in jedem Puncte der Fläche Linienelemente zieht, welche die Fläche dreipunctig berühren, so sind sie Bogenelemente von Curven, welche ich Curven der Haupttangenten nennen will (wie ich glaube hat Hesse diesen Ausdruck). Diese Curven nun, welche durch Integration gefunden werden, sind hier algebraisch; es sind Raumcurven vierter Ordnung zweiter species...*

(I like very much to work on these problems at the same time you do and I will be happy to learn your methods: this coincidence shows that we are on the right track. Recently, following your example, I dealt with Steiner's surface. This is particularly interesting for me, because it is represented by the equations

$$rx_1 = f_1(c, l, m)$$

$$rx_2 = f_2(c, l, m)$$

$$rx_3 = f_3(c, l, m)$$

$$rx_4 = f_4(c, l, m)$$

where the f's are general functions of degree two. Kummer was aware of this representation, but he never really treated it in detail. Concerning geometric aspects, it leads to something interesting, and precisely to this. If we draw, at any point of the surface the lines having with the surface a triple contact, they are the arc elements of curves which I (following Hesse) call the asymptotic curves. These curves, that I found via integration, are algebraic; they are skew curves of the fourth order and of the second kind...)

Notice that Clebsch's article [36] is dated July 24, 1866, only three days after the above letter. A few months later, on September 25 of the same year, Cremona presented to Clebsch his geometric viewpoint. Cremona points out that what Clebsch obtained *durch Integration* (*via integration*), i.e., the fact that the asymptotic curves of Steiner's surface are algebraic rational quartic curves, he has been able to obtain by geometric means by looking at the plane representation of the surface:

Un risultato così elegante m'invogliò a cercarne la dimostrazione per via geometrica; e la trovai nella rappresentazione della superficie sopra un piano, la quale forma l'argomento della presente lettura, e che già ho comunicata per intero al signor Clebsch con lettera del 25 settembre.

(Such an elegant result stimulated me to try to find a geometric proof of it; which I was able to do by using the plane representation of the surface, that is the topic of the present note, which I already communicated to Mr. Clebsch with a letter on the 25th of September.)

Indeed Cremona's paper is very geometric and already in the full spirit of the Italian school.

The plane representation of a rational surface, a typically cremonian theme, was certainly the subject of close collaboration between the two, to the point that they separately wrote two memoirs [37, 49], each published in the journal of mathematical societies of which the colleague was a member, even exchanging the languages: Clebsch wrote in Italian, Cremona in German. In these researches the full importance of the concept of the genus of a curve occurs to Cremona, who uses it in the paper [44] and for the first time he translates the German term *Geschlecht* with the Italian *genere*, a word which gives the idea of a *botanical* invariant, suitable indeed for a classification.

It is apparent that the exchanges of information and opinions with Clebsch induced Cremona to realize the importance of fully absorbing the analytic tools into the main frame of the Italian school. As we indicated, this viewpoint was certainly shared by Betti, Brioschi, and Casorati who, at that time, were, with Cremona, the élite of the Italian mathematical community. As we said, a lot of work had been made by Betti to spread out Riemann's ideas in the Italian ambient. Brioschi, Casorati, and Cremona tried to do the same a few years later.

1.4 A course on elliptic and abelian functions in Milan in the academic year 1868–1869. The opportunity was given to the three colleagues (Brioschi, Casorati, Cremona) by a course that they taught together in Milan in the academic year 1868–69 (see [7]). The title of the course was *Lezioni sulla teorica delle funzioni ellittiche e abeliane* (*Lectures on the theory of elliptic and abelian functions*), and it was attended by several students of the three mathematicians, like Armenante, Ascoli, Bertini, Jung, Misani, Paladini, and the Austrian mathematician Weyr. The course was dedicated to the results of Abel, Jacobi, Riemann and Clebsch. Cremona's lectures were specifically devoted to Clebsch's algebro-geometric approach to the theory of abelian integrals.

Cremona's course was divided in three parts. The first part was dedicated to the preliminaries of the classical theory of plane curves as it had been elaborated by Cremona in the previous years. It is interesting to notice that the presentation is given with a more extended use of the analytic methods than Cremona used in his previous courses of *Geometria superiore* (*Higher geometry*). The first part ends with the introduction of the concept of the genus of a curve, according to Clebsch, as it is presented in [44].

The second part is devoted to the first three chapters of Clebsch and Gordan's treatise [39] on abelian functions. Several results and proofs are presented, with respect to the original, in more geometrical terms, certainly closer to Cremona's viewpoint. In particular Clebsch's proof of the invariance of the genus by Cremona transformations is proposed as well as the transformation of any curve of genus g into a curve of degree $g+1$ with, in general, $\frac{g(g-3)}{2}$ nodes.

The third part of the course was entirely dedicated to Abel's theorem and its algebro-geometric applications.

Also part of Casorati's course was dedicated to some chapters of the Clebsch-Gordan book. In the same circle of ideas Casorati published, in 1868, his famous treatise [19] in which Riemann's theory of abelian functions is presented from the analytic viewpoint, with the declared purpose of helping the Italian students to become familiar with these new ideas. This is clearly remarked by Cremona in his review of the book (see [45]):

> *Scritto qual'è con linguaggio limpido, preciso, rigoroso, ci apre le porte di un misterioso recinto, sinora vietato ai moltissimi: ci spiana e allarga vie impraticabili: ci dà il sugo essenziale di vaste teorie elaborate da molti intelletti in tempi e per vie affatto differenti: insomma ci abilita a fare, con tanto minor fatica e in un tempo incomparabilmente più breve quegli studj che prima erano giudicati fra i più ardui delle scienze matematiche.*

(Written as it is with a clear, precise and rigourous language, it opens to us the doors of a beautiful room, so far forbidden to the majority of mathematicians: it makes for us easy and large roads which were so far impassable: it gives to us the true essence of wide theories which have been formulated in different times by several mathematicians with completely diverse approaches: in conclusion, it enables us to study, with considerably less effort and in incomparably less time, subjects which have been considered so far among the hardest in the world of mathematical sciences.)

It is perhaps interesting to open a parenthesis here. As U. Bottazzini points out in [14, p. 161], this is:

Un obbiettivo che mobilitò molti dei matematici della generazione "risorgimentale" sia con la redazione di manuali ispirati alle proprie lezioni, sia con la traduzione di testi stranieri.

(An objective which involved most of the mathematicians of the generation of the "revival," who wrote books inspired by their lectures and translated foreign texts as well.)

(See also [14, pp. 6–7].)

One has to look at the Milan course from this viewpoint too. In fact, updating mathematical studies in Italy has to be seen also in the light of the recently acquired national independence and of the necessity of creating a new generation of experts, able to promote the scientific progress of the newly founded nation. In this perspective the full acquisition of the most advanced techniques of the time, in particular of Abel's and Riemann's ideas, becomes a *political* issue.

The Brioschi–Casorati–Cremona course played a pivotal role in the development of the Italian school and its importance for the education of the new generation of mathematicians can hardly be underestimated. It is interesting to record the position of the well established mathematician A. Genocchi, from Turin, who perceives that the Italian mathematical community is coming to a turning point in its history and writes to Cremona (letter of January 10, 1867):

... beati i giovani che possono prendere parte alla presente trasformazione delle matematiche!

(... God bless the young people who can take part in the present transformation of mathematics!)

It is also remarkable that the careful reading and understanding of the Clebsch–Gordan book, which was necessary for the teachers of the course, seems to have been a task *collectively* taken by the Italian mathematicians. For instance, in 1868 Cremona meets Genocchi, who shows him a letter of Placido Tardy, pointing out some mistakes in the chapter *Eindeutige Transformationen* of that book. In August Cremona writes about this to Clebsch, who, on August 31 answers and confirms the validity of Tardy's observations:

Ich komme endlich auf Ihre Bemerkungen über s. 77–79 der Abelschen Funktionen. Sie haben leider Recht, diese Seiten, an denen Hr. Gordan gar keinen Theil hat, sind mit einer mir jetzt unbegreiflichen Nachlässigkeit geschrieben. Zunächst sind auf p. 77 die 2n Puncte nicht ganz beliebig, sondern natürlich der durch das Zustandekommen des Büschels gegebenen Bedingung unterworfen. Ich bitte Sie also das Wort beliebig auszustreichen. Ebenso dass im Weitern vorkommenden "einen ausgenommenen" Wahrscheinlich rührt dies von einem Irrtum her, der sich beim Abschreiben der Resultate für $p = 1$ *eingeschlichen hat. Auch liegt die Correction schon, wie Sie sagen, in den letzten Zeilen des §.*

(I come now to your remarks about pages 77–79 of Abelian Functions. Unfortunately you are right: these pages, to which Mr. Gordan did not contribute, have been written with a superficiality that I cannot understand. First of all at p. 77 the 2n points are not arbitrary, but are clearly subject to the conditions imposed by the pencil. Therefore I beg you to delete the word "any." Similarly, in what follows, the words "except one." It probably depends on an accidental mistake made in copying the result for $p = 1$*. The correction takes place already, as you say, in the last lines of the paragraph.)*

And the discussion about themes raised by the reading of the book apparently went on for the rest of the year (see letter from Cremona to Clebsch on October 10).

1.5 Cremona and the analytic methods. It is important to remark that Cremona himself, in the same period in which he was teaching the course, confesses that he is undergoing a process of revision of his methods and ideas. Indeed, he writes to Genocchi expressing concepts very close to the ones stated by Enriques several years later (letter of March 19, 1868):

Io credo l'avvenire riservato non già all'analisi pura né alla geometria pura, ma piuttosto alla combinazione dei due metodi. Chi trascura o diniega uno dei due metodi parmi faccia come chi si troncasse un braccio. Sia pure l'analisi il braccio destro, ma la geometria è il braccio sinistro. Cerchiamo adunque di perfezionare entrambi i metodi, se vogliamo che il progresso delle matematiche sia armonico e completo.

(I believe that the future is neither reserved to pure analysis nor to pure geometry, but rather to the combinations of the two methods. The one who overlooks or denies one of the two acts, in my opinion, as one who would cut one of his arms. Let us take for granted that the analysis is the right arm, but then geometry is the left arm. Therefore, let us try to improve both methods if we really want the progress of mathematics to be harmonious and complete.)

On the other hand the Italian scientific community interprets Cremona's interest in the analytic methods as an important modification of his views. Indeed Genocchi writes to him in a letter of July 15, 1869:

Ho visto che in un'adunanza dell'Istituto di Bologna avete comunicata una vostra costruzione del teorema generale di Abel più semplice di quella di Clebsch e Gordan. Desidero vivamente ch'essa sia pubblicata e intanto vedo con piacere che vi scostate sempre più dalla geometria pura.

(I saw that at a meeting of the Institute of Bologna you communicated your proof of the general theorem of Abel which is simpler than the one of Clebsch and Gordan. I strongly wish that it be published, while I see with pleasure that you are more and more diverting from pure geometry.)

This is an important citation for us, since it confirms the constant interest of Cremona in Abel's theorem. An interest which is stressed by the collaboration of Cremona, in the same period, with S. Lie, during the preparatory work for the publication of Abel's collected papers [1] (see [92]). Lie, in fact, of whom, according to Loria [89], Cremona was *un des premiers mathématiciens qui surent mesurer la valeur des idées* (*one of the first mathematicians who were able to measure the value of the ideas*), consulted the Italian colleague about Abel's manuscripts which were among the documents of Guglielmo Libri.

Even in his correspondence with foreign mathematicians Cremona becomes an enthusiastic supporter of the new analytic methods, writing, among others, to Cayley, Zeuthen and Hoüel. To the latter he proposed a translation of the Clebsch–Gordan book, as a sequel to the one of Riemann's papers he was at that time doing (letter of January 29, 1869):

Peut-être qu'un jour ce sera vous même qui ferez connaître aussi le livre magnifique de MM. Clebsch et Gordan sur les fonctions abéliennes. Cependant, même en Italie, quoiqu'on étudie un peu l'allemand, on a plus de facilité à lire le français que toute autre langue étrangère.

(Maybe we will have to be grateful to you if the beautiful book of Clebsch and Gordan will be better known. Even in Italy, though we study some German, we understand French better than any other language.)

Hoüel, in a first moment, answered negatively. But some years later he reconsidered Cremona's proposal, though the project was never been begun. It is also interesting to notice Beltrami's reaction to Cremona's papers in this field, recorded in this letter to Hoüel (letter of October 8, 1873):

Vous avez en une très–bonne idée en entreprenant de traduire les fonctions abéliennes de MM. Clebsch et Gordan. Mais je préférerais de beaucoup que la vôtre ne fût pas une simple traduction. Cet ouvrage important a été, à ce que j'ai entendu dire par des juges très–versés, redigé un peu trop à la hâte par ses auteurs. Il paraît que ses différentes parties présentent des inégalités de fond et de forme très saisissantes. Lorsque MM. Cremona et Casorati entreprirent de faire deux cours parallèles sur ce sujet, à Milan, ils trouvèrent tous les deux, du côté analytique non moins que du côté géométrique, beaucoup de choses à simplifier, à modifier, à corriger même. Les fruits d'une

> *partie de cette révision ont été consignés dans deux mémoires insérés aux* Annali *et dans les volumes de l'Académie de Bologne.*
>
> *(You had a very good idea to translate the abelian functions of Clebsch and Gordan. But I would by far have preferred that this would not be just a mere translation. This important book, as I heard from competent judges, was written by the authors in a hurry. It seems that they presented different viewpoints with surprising differences of style and form. When Cremona and Casorati started teaching these topics in two parallel courses in Milan, both from the analytical side and from the geometric one, they found several things that could be simplified, modified and even corrected. The outcome of a part of this revision appeared in two papers inserted in the* Annali *and in the proceedings of the Academy of Bologna.)*

The intention of Cremona to spread the analytic methods inside the Italian school, is witnessed by papers written by some of his students and by Cremona himself. Actually, the papers to which Genocchi and Beltrami refer in the above citations certainly were motivated, as often occured to Cremona, by the need to clarify the topics he was teaching in his courses. It is the case of mentioning here the note [8], which seems to be a sort of homework, a dictation of the master, and the papers [46, 47, 48]. Particular importance, in this historical reconstruction, seems to have the short joint note [21] with Casorati, followed by Casorati's paper [20].

It is probably useful to dwell briefly on some of these papers. For instance [46] is devoted to showing that any hyperelliptic curve, i.e., a double cover of the projective line, of genus g, can be birationally transformed into a curve of degree $g+2$ with a unique singular point of multiplicity g, with some extra tangential conditions which it is not important to discuss here. Note that in the Clebsch–Gordan book only the cases $g = 1, 2$ were treated. It should be noted that these are the first instances of the idea, which will soon become standard and fruitful, of determining a particularly interesting projective birational model of a curve to be studied.

This approach is also taken in [47], which studies the abelian integrals on a curve and Abel's theorem. This paper directly stems from the Milan lectures. It contains no new results, but it simplifies parts of the Clebsch–Gordan treatise. Cremona defines here the abelian integrals of three kinds: where being of the first kind means regular, being of the second kind means with a single double pole and being of the third kind means having two poles with residues 1 and -1. Then, using geometric arguments, he shows how to use these integrals to express other more complicated ones. As an outcome he finds the fact, well known to Clebsch, that the abelian integrals of the first kind are in one–to–one correspondence with the *adjoint polynomials* to the curve (see Sect. 2.1 for details). Finally a proof of Abel's theorem is presented. The new technique, and the *simplification* with respect to Clebsch, is in the use of geometry, and in particular of linear systems of curves and related counts of parameters, in order to manipulate algebraic functions and show reductions of integrals which were, since then, proved only by means of heavy calculations. This was very fascinating at that time, but these synthetic ideas also started raising some objections, in as much as not every step was carefully motivated: in particular, some arguments often used implicit, unverified genericity assumptions.

Finally the paper [21] concerns a well know problem: the one of counting the parameters, i.e., the *moduli*, on which the curves of genus g depend. Riemann in [107] gave an answer by stating that the number in question is $3g - 3$ as soon as $g \geq 2$. Riemann's plausibility argument is today known as *Riemann's count* (see [67]) and can be made perfectly rigorous. At the time however, it did not seem too clear. In fact, Cayley was not convinced by it and he published the paper [29] in which he proposes the number $4g - 6$. Notice that for $g = 3$, the only case which was known for sure at the time, the two formulas coincide. In the same year 1869 Brill came up with a paper in which he proved the validity of Riemann's count for $g = 4$. In [21] Casorati and Cremona also contribute. They support Riemann by producing other plausibility arguments based now on birational transformations of plane curves. The arguments they are able to make are complete for $g \leq 6$. This result was immediately communicated to Cayley who, on June 28, 1869, wrote to Cremona: he acknowledges the mistake, saying that he already mentioned it to Clebsch six months before and he also indicates the possible origin of his mistake in a wrong way of computing some parameters. On the same topic Cayley will later write another paper [31].

1.6 The analytic spring dries up for a while. In conclusion, certainly Cremona's idea at the end of the 1860's was to arrive at a substantial enrichment of Italian projective techniques with Abel's and Riemann's analytic methods as presented by Clebsch and Gordan. Cremona's programme however, although enthusiastically started, has never been completed. It is perhaps appropriate to dwell for a while on the reasons for this failure and on its consequences.

Certainly one reason resides in the fact that the years between 1870 and 1872 are the last ones in which Cremona was able to produce high level contributions in algebraic geometry. Indeed, already in the period he spent in Milan until 1873, he was partly absorbed by the extremely successful studies on statics and descriptive geometry. In 1873 he moved to Rome (recently annexed to Italy and declared the capital of the newborn state), with the task of founding there a new polytechnical school, like the one he was teaching at in Milan. This project absorbed most of his energies. But even more he was overloaded by teaching and bureaucratic duties after his election to the Senate in 1877. He was a little more than forty then and his research activity almost stopped at that time. We believe that Clebsch's death in November 1872 also played a crucial role, since it took away from Cremona one of his main scientific references.

More important however are, in our opinion, the scientific reasons. In the 1870's new characters come to the stage. We allude specifically to Brill and Noether, who, starting from Clebsch and Gordan but using also several geometric ideas introduced by Cremona, made a substantial and extremely successful effort towards a further algebrization and geometrization of the theory of Riemann surfaces. It is, perhaps, appropriate to refer to [16, §1.2], from which we cite:

The natural outcome of these geometric applications is the development of an algebro–geometric approach to Riemann's theory. This complex and ambitious program was undertaken in the 1870's by M. Noether (1844–1922) and A. Brill (1842–1935). Their ideas can be considered as the beginning of the systematic use in geometry of algebraic techniques, in particular of those techniques which today belong to the realm of commutative algebra. Noether (1869) in fact places at the foundation of the algebraic viewpoint the so-called $Af+Bg$ theorem that gives necessary and sufficient conditions for a polynomial in two variables to belong to the ideal generated by the two polynomials. Noether (1875) shows as well how Cremona transformations can be used to reduce every plane curve to another having simpler singularities, called ordinary *singularities (multiple points with distinct tangents) and how a curve possesses smooth birational models in every projective space of dimension $r \geq 3$ and plane models with only nodal singularities. In the fundamental memoir [17], making great use of Cremona's lessons, the analytic theory of Riemann is translated into the revolutionary projective–geometric language of* linear series *on an algebraic curve.*

As explained in [16, Sect. 1.3], Brill and Noether's approach was soon incorporated by the Italian school and actually developed in even more geometric terms by Bertini, Veronese, and C. Segre, the former two being students of Cremona. This gave an enormous impulse towards projective geometry rather than the analytic methods. Later the validity of these methods was confirmed by the spectacular contributions of Castelnuovo and Enriques to the classification of surfaces. We guess that the very success of these ideas constituted a cage in which the Italian school was relegated for about thirty years.

In contrast with this, there was an initial lack of success in extending the analytic approach to higher dimensional varieties. This probably played a role in driving the interests of the Italians away from the trascendental techniques. Indeed, starting with Clebsch and, a bit later, with Noether, there was a first attempt to extend the analytic approach to the theory of surfaces. This revealed itself to be complicated and more difficult than the algebro–geometric approach, and, in addition, gave rise to substantial difficulties, like the appearence of the mysterious concept of *irregularity* (see Sect. 2.1 and [10]). As we will recall in Chap. 2, it will take three or four decades before the mist around this subject would start becoming less thick, without really totally dissolving till much more recent times.

From the 1870's until the first years of the new century the main success of the Italian school was Castelnuovo and Enriques' classification of surfaces. On the other hand the French school with Humbert, Picard, Poincaré and others started reconsidering and refining the analytic methods. These two schools somehow faced each other in the 1880's, without, we think, a real effort by either one of the two to fully understand and incorporate the methods of the other. A deeper interest arises only a bit later. For example in 1894 Picard asks information about the results of the Italian school (see [15], letter of July 7, 1894), whereas Enriques reads the papers of Picard and Poincaré in the fall of the same year (see letter of October 16, 1894). At this point the contacts of the Italians with the heritage of Abel and Riemann were reestablished. They will become more stable and intense after the important papers of Castelnuovo, Enriques and Severi at the beginning of 1900 (see Sect. 2.2 below). And this is also witnessed by the invitation of Mittag–Leffler, director of the *Acta Mathematica*, which we mentioned in the Introduction.

In this scenario it is not by chance, in our opinion, that C. Segre, in the academic year 1903–1904, when his young student Severi was still with him in Turin, tried to repeat Cremona's experience by teaching, in his course of Higher Geometry, some *Applicazioni degli integrali abeliani alla geometria* (*Applications of abelian integrals to geometry*) (see [65]). Abel's theorem is contained in the last chapter and it takes the form which we know today, i.e., a criterion of linear equivalence of divisors on a curve. As a consequence Segre gives an analytic proof of the theorem of Riemann–Roch. Essentially the same line was followed by Enriques and Chisini in their famous book [60], vol. IV. The analytic techniques, at least in one variable, i.e., for the curve case, are well established by that time. Thus Segre's course contains no news. However it is witness to a renewed interest in the analytic methods.

2 Abel's Theorem for Curves on a Surface and Variations

As indicated at the end of the previous chapter, the beginning of the XXth century sees a new interest of the Italian school for the analytic techniques. In this framework, in 1905 and 1906 two basic papers of Severi [117] and [119] appear. As their titles show, both are dedicated to extensions of Abel's theorem to the case of surfaces. They are very interesting for our historical discussion, and we want to illustrate their motivations, the results contained in them and the interesting consequences they had, not only inside the Italian school, but also outside it.

It should be noticed that this is not the first time one tries to extend Abel's theorem to higher dimensional varieties. As Severi points out in [117], an attempt was made by Humbert in the interesting paper [75], which however did not have followers at the time. There, among other things, one extends Jacobi's identity (see [67, p. 671]) to more than two variables and one proves a result which is a first instance of Proposition 3.2 below.

2.1 The irregularity of a surface. In 1905 G. Castelnuovo started the introduction of the important paper [25] with the following observation:

> *Quando si tenta di estendere alle superficie algebriche i risultati ottenuti, nella teoria delle curve, da Riemann, Clebsch, Brill e Noether, si incontrano, accanto ad affinità previste, singolari anomalie.*
>
> *(When one tries to extend to algebraic surfaces the results obtained, in the theory of curves, by Riemann, Clebsch, Brill and Noether, one encounters at the same time some similarities which were easy to foresee, but also some singular anomalies.)*

The main anomaly, or difficulty, which Castelnuovo was referring to was the one, concerning *irregular surfaces*, already discovered by Cayley in 1871 [30]. Indeed, according to Clebsch [38] and Noether [101], if one wants to consider the *canonical system* of a smooth, algebraic surface S, one is led, in analogy to the case of curves, to look at the vector space of holomorphic 2-forms on S, which today we denote by $H^0(S, \Omega_S^2)$. For each non zero holomorphic 2-form ω, one can consider its zero locus (ω), which is a curve K on S. As ω varies in $H^0(S, \Omega_S^2) - \{0\}$, K varies describing the canonical linear system $|K|$ of S. The dimension $p := h^0(S, \Omega_S^2)$ plays therefore, in the surface case, the same role as the genus in the curve case, and in fact Clebsch and Noether called this number the *genus* of the surface.

As we said already, the meaning of the genus g of a curve C in terms of projective geometry had been pointed out, following Riemann's ideas, by Clebsch and Gordan [39] and Brill and Noether [17]. Indeed one can take a birational morphism $\phi : C \to C' \subset \mathbb{P}^2$, of C to a plane curve C' of a certain degree d with singular points $p_1, ..., p_\delta$ which are nodes of C'. Then *Clebsch formula* says that:

$$g = \binom{d-1}{2} - \delta \tag{2.1}$$

The functional meaning of this is the following. If $f(x, y) = 0$ is the affine equation of C', then the pull-back to C via ϕ of a meromorphic 1-form on the plane of the type $\frac{g(x,y)}{\frac{\partial f}{\partial y}(x,y)}dx$, with $g(x, y)$ a polynomial, is holomorphic if and only if $g(x, y) = 0$ is a curve of degree $d' \leq d - 3$ *adjoint* to C', i.e., passing through the nodes $p_1, ..., p_\delta$ of C'.

The situation in the case of surfaces is, at a first glance, similar. One may take a birational morphism $\phi : S \to S' \subset \mathbb{P}^3$, in such a way that the image surface S' has degree d and *ordinary singularities*, i.e., a curve Γ of nodal double points with some little extra singularities which we do not care about now (see [67, p. 611]). Then the dimension of the canonical system of S equals the dimension of the linear system of surfaces of degree $d - 4$ in $\mathbb{P}^3$ *adjoint* to S', i.e., containing the double curve Γ. In other words $p = h^0(S, \Omega_S^2)$ is the number of independent homogeneous polynomials of degree $d - 4$ in four variables vanishing on Γ. This was noticed by Clebsch in [38].

The next step is to look for a formula for p similar to 2.1. This was in fact the goal of Cayley [30]. In order to achieve it he took a more general viewpoint: take any curve Γ in $\mathbb{P}^3$ and find the *postulation formula* for Γ, i.e., compute for each n the dimension of the vector space $I_n(\Gamma)$ of all homogeneuos polynomials of degree n in four variables vanishing on Γ, or rather its codimension $h_\Gamma(n)$ in the vector space S_n of all homogeneous polynomials of degree n in four variables. Today we call $h_\Gamma(n)$ the *Hilbert function* of Γ, whereas, in the old times, it was called, as we said, the *postulation* of Γ.

Now, computing h_Γ itself is quite hard, but one notices that for $n >> 0$, $h_\Gamma(n)$ is a linear polynomial in n which is of the form $p_\Gamma(n) = \delta n - \gamma + 1$, where δ is the degree and γ is the genus of Γ. This is the so-called *Hilbert polynomial* of Γ (for the basics about Hilbert functions and Hilbert polynomials, see [71, Chap. I, §7]).

The question of what is the minimum integer N such that for all $n \geq N$ one has $h_\Gamma(n) = p_\Gamma(n)$ is a difficult one. If we knew, in this specific case, that $N \leq d - 4$, we would get *Cayley's formula*, the analogue of Clebsch formula, by simply saying that:

$$p = \binom{d-1}{3} - p_\Gamma(d-4) = \binom{d-1}{3} - \delta(d-4) + \gamma - 1 \qquad (2.2)$$

As Cayley noticed however, this is not always the case. Indeed, if S' is a scroll with plane sections of genus g, formula 2.2 gives $p = -g$, whereas $p = 0$ for a scroll. Thus, if $g > 0$, formula 2.2 does not give the correct value of p. This remark naturally leads to the following questions:

(i) What is the meaning of the number p provided by formula 2.2?

(ii) How can we compute the genus p?

As for question (i) Noether [100, 101] and Zeuthen [143] were able to prove that the quantity given by formula 2.2 is, as well as p, a birational invariant, in particular it does not depend on the map $f : S \to S' \subset \mathbb{P}^3$. This led Noether to call this new invariant the *arithmetic genus* or *numerical genus* of the surface, denoting it by p_a or p_n, in opposition to p, which was since then denoted by p_g and called the *geometric genus* of the surface. They also noted that the difference $q = p_g - p_a$ is non–negative, and called it the *irregularity* of the surface. For our convenience, we will denote it, for the time being, by q_a, calling it the *arithmetic irregularity* of the surface (see also [59, p. 326]). In today's cohomological language, one has $q_a = h^1(S, \Omega_S^2)$. As we will see in a moment, a few other notions of irregularity naturally arise and the problem of proving that they are all the same invariant has been crucial in the development of the theory of surfaces around the end of XIXth century.

For a smooth, algebraic curve C, the genus g is the dimension of the vector space $H^0(C, \Omega_C^1)$ of holomorphic 1-forms on C. Hence, if one wants to extend to a surface S the notion of genus, one is also tempted to consider the dimension of the space $H^0(S, \Omega_S^1)$ of holomorphic 1-forms on S, which is again a birational invariant. This viewpoint was taken by the French school, starting with Picard [103] in 1885. The new invariant, which we will denote by q_{an} and temporarily call the *analytic irregularity*, was noticed to be equal to the irregularity q_a in several cases, like the plane and the scrolls. Are the two invariants equal? This was the next question to be solved.

It should be noticed that, according to Castelnuovo [25] (see footnote $(***)$), it turns out, from some unpublished paper by Riemann, collected by Noether and Weierstrass [106, p. 99], that the idea of associating to an algebraic surface the invariants that we denoted by p_g and q_{an} actually goes back to Riemann himself, who noticed that $q_{an} = 2$ for the jacobian surface of a curve of genus 2.

A third way of extending the notion of genus to surfaces, was again considered by the French school, more or less in the same years, and it is quite natural. Indeed, in the curve case, one has $b_1 = 2g$, where b_1 is the *first Betti number* of the curve, considered as a Riemann surface. Thus, given a surface S, considered as a 4-dimensional manifold, one can consider its first Betti number $b_1(S)$ and ask: is $b_1(S)$ even? if yes, is $\frac{b_1(S)}{2}$ equal to q_{an} as in the curve case? In any event, $b_1(S)$ is a birational invariant and we denote the invariant $\frac{b_1(S)}{2}$ by q_t and call it the *topological irregularity* of the surface.

Finally there is a subtler, perhaps more typically algebro–geometric, way of extending the concept of genus, which is in a sense close to the viewpoint implicit in the classical Abel's theorem. In fact the genus g of a curve C is, according to Abel's theorem, the maximum dimension of a family of effective divisors on C which are not linearly equivalent to each other. Indeed there is an abelian variety $J(C)$ of dimension g, the *Jacobian variety* of C, whose points parameterize the equivalence classes of divisors on C (see [67, p. 235] for details).

Now, on a surface S, one can consider analogous concepts. Namely one can look at the maximum dimension of a continuous system of curves on S which are not linearly equivalent to each other. Of course this is a new birational invariant, which, for our convenience, we will denote by q_g and call the *geometric irregularity* of the surface (see again [59, p. 326]). This invariant, which a priori is not even clear to be a finite number, soon attracted the attention of algebraic geometers, starting from Noether, who noticed that q_g is zero for rational surfaces and is g for a scroll with plane sections of genus g. In cohomological terms one sees that $q_g = h^1(S, \mathcal{O}_S)$.

Though the foundations of the classification of surfaces via the behaviour of their plurigenera had been set during the last decade of the XIXth century by Castelnuovo [23] and Enriques [54, 55], the relations between the various concepts of irregularity and the nature itself of *irregular surfaces* were still very unclear at that time. It is interesting to notice that irregular surfaces seemed to Noether, who discovered them, mysterious objects more to be exorcised than to be studied, seeing that he conjectured, without any evidence, that except from the irrational ruled surfaces, no other irregular surfaces existed (see [100]). This totally ungrounded conjecture, later partially withdrawn by Noether in [101], was disproved by Castelnuovo [22] only several years after it had been formulated and also Humbert (see [77]) discovered other surfaces for which q_g is non-zero, e.g., the abelian surfaces like the jacobians of curves of genus 2.

However the embarassment of the algebraic geometers of the time in dealing with irregular surfaces is fully shown by this quotation from Castelnuovo [28]:

> *Avevamo costruito, in senso astratto s'intende, un gran numero di modelli di superficie del nostro spazio o di spazi superiori; e questi modelli avevamo distribuito, per dir così in due vetrine. Una conteneva le superficie regolari per le quali tutto procedeva come nel migliore dei mondi possibili; l'analogia permetteva di trasportare ad esse le proprietà più salienti delle curve piane. Ma quando cercavamo di verificare queste propretà sulle superficie dell'altra vetrina, le irregolari, cominciavano i guai, e si presentavano eccezioni d'ogni specie.*

(We had constructed, in the abstract sense of course, a large number of models of surfaces in our space or in higher spaces; and we had distributed these models in two display cases. One contained the regular surfaces, for which everything proceeded as in the best of all possible worlds; the most salient properties of plane curves were transported to these by analogy. But when we tried to verify these properties for surfaces in the other display case, the one with the irregular surfaces in it, then the trouble began, and exceptions of every sort turned up.)

An embarassment which, as we will see, the Italian school was never really able to overcome (see also [16, pp. 46–47]).

2.2 The fundamental theorem of irregular surfaces. The last decade of the XIXth century and the first of the XXth century have been crucial for the understanding of the questions we indicated above. Here we recall some of the main contributions in this direction (for more informations on this subject we refer to [142, Chap. VII] and related appendix of Mumford):

(i) (Humbert 1893, [76]) If $q_g > 0$ then $q_{an} > 0$.
(ii) (Enriques 1899, [56]) If $q_g > 0$ then $q_a > 0$.
(iii) (Enriques 1901, [57]) If $q_t > 0$ then $q_g > 0$,
(iv) (Severi 1905, [116, 118]) $q_{an} \leq q_a$.
(v) (Enriques 1904, [58]) $q_a = q_g$.

The last paper contained a mistake, pointed out by Severi only in 1921 [123]. This was the beginning of a long and harsh fight between Enriques and Severi, which we will not dwell on now (see [59, Chap. IX, §6] and [16], l.c.). Based on Enriques' result, in 1905 Castelnuovo [25] and, independently, Severi [117], proved that $q_a = q_g = q_{an} = q_t$, a result that was referred to as the *fundamental theorem of irregular surfaces*. However in 1910 Poincaré [105] independently gave the first correct proof of the same crucial theorem. Today we know the proof is an immediate consequence of Serre duality, De Rham, Dolbeaut and Hodge theorems (see [67]). Indeed $q_a = h^1(S, \Omega_S^2)$ which, by Serre duality, equals $h^1(S, \mathcal{O}_S) = q_g$. This in turn, by Dolbeaut theorem, is equal to $h^0(S, \Omega_S^1) = q_{an}$. Finally by De Rham and Hodge theorems, one has $2q_{an} = 2h^0(S, \Omega_S^1) = h^1(S, \mathbb{C}) = b_1(S) = 2q_t$.

Humbert's proof of (i) is a nice application of Abel's theorem. It is worth explaining the idea in modern language. Suppose that the surface S has a flat family $\mathcal{F}$ of curves parameterized by a smooth, irreducible curve Γ. The family in question can be seen as a correspondence between Γ and S, i.e., as a divisor F in $\Gamma \times S$. As such, it determines a morphism $\phi : \mathrm{Alb}(S) \to J(\Gamma)$ (see Sect. 2.5 and Sect. 3 below for similar concepts). Now an application of Abel's theorem plus some easy geometric considerations tell us that ϕ is the zero map if and only if the curves in $\mathcal{F}$ are linearly equivalent to each other. Therefore if $\mathcal{F}$ is formed by curves which are not linearly equivalent to each other, then ϕ is non-zero, hence $\mathrm{Alb}(S)$ is non–zero. But $\dim \mathrm{Alb}(S) = q_{an}$, whence the assertion.

As for (ii), another basic tool comes into play. Indeed Enriques looks at the *adjoint linear system* $|K + C|$ to a curve C on a surface S. The *adjunction formula* tells us that adjoint curves cut out on C canonical divisors. Enriques notices that if $q_a = 0$ then $|K + C|$ cuts out on C the complete canonical series. We see this by looking at the cohomology sequence of

$$0 \to \mathcal{O}_S(K) \to \mathcal{O}_S(K + C) \to \mathcal{O}_C(K + C) \simeq \Omega^1_C \to 0$$

By contrast he shows that if $q_g > 0$ and if C varies in a continuous, non linear system, then $|K + C|$ cannot cut on C the complete canonical series.

The paper (iv) is based on a similar idea as (ii), or, to be more precise, on *dual* considerations. In fact Severi takes a very ample curve C on the surface S and considers the *characteristic series* cut out on C by the complete linear system $|C|$. It is a result of Castelnuovo [24], that the characteristic series has codimension at most q_a in its complete linear series. We understand this by looking at the cohomology sequence of

$$0 \to \mathcal{O}_S \to \mathcal{O}_S(C) \to \mathcal{O}_C(C) \to 0$$

and using Serre duality. Castelnuovo instead has a very ingenuous geometric argument for this.

In particular, if $q_a = 0$ then the characteristic series is complete. On the other hand Severi proves, using residues of meromorphic 1-forms on S, that the characteristic series has codimension at least q_{an} in its complete linear series (see [142, p. 201], for a modern interpretation).

We skip the comments on (iii), which, though important, did not play in our opinion an essential role in the development of the subject, and we turn to (v). Enriques' idea was based again on Castelnuovo's theorem on the *defect of completeness* of the characteristic series. Enriques' intuition was that this defect of completeness should be accounted, under mild conditions, by the fact that C is contained in a *complete continuous system* $\mathfrak{C}$, so that the curves *infinitely near* to C in $\mathfrak{C}$ should cut out on C the complete characteristic series. In modern terms, Enriques looks at the infinitesimal deformations of C on S, which are given by the sections of the normal bundle $N_{C|S} \simeq \mathcal{O}_S(C)_{|C}$. Since one has the restriction map $r : H^0(S, \mathcal{O}_S(C)) \to H^0(C, \mathcal{O}_S(C)_{|C})$, some infinitesimal deformations, namely those in $Im(r)$, are accounted for by moving C inside the linear system $|C|$. The others, the ones which contribute to $Coker(r)$, should be accounted for, in Enriques' mind, by deforming the curve off $|C|$ in some continuous, non linear, way. In order to prove this, Enriques uses the so-called *principio di spezzamento* (*splitting principle*), which essentially says that when an irreducible, may be singular, curve, varying in a continuous family on a smooth surface, becomes reducible in, say, two irreducible components, then, roughly speaking, it acquires at least *one more singularity* which *connects* the two components. It would be too long, and perhaps a desperate task, to try to explain Enriques' application of this principle to the proof of what he had in mind. The interested reader may consult Enriques' book [59, Chap. IX], which contains the author's last attempt to write down a proper version of his argument: nonetheless the presentation there is extremely longwinded and obscure! The point is that, though the splitting principle is correct (see [2, 114, 142] for extensions and comments), Enriques argument fails since nobody ensures us that the *infinitesimal* deformations contributing to *Coker*(r) are not *obstructed*, i.e., that they correspond to some *true* deformation. The Italian school never put on a solid ground these considerations and it is interesting to notice that counterexamples to Enriques' arguments have been first given by Italian algebraic geometers like Albanese [2] and more specifically by Zappa [141].

As we mentioned above, the theorem showing that $q_a = q_g = q_{an} = q_t$ was proved by Poincaré in 1910. This proof, somewhat simplified by Severi [123], is interesting from our viewpoint of Abel's followers. Indeed it is based on an ingenuous application of Abel's theorem and Jacobi inversion theorem. What Poincaré does is to consider a pencil $\mathcal{P}$ of hyperplane sections of genus g of the surface S in a suitable projective embedding. Then one can *rationally determine* on the curves C of $\mathcal{P}$ a basis $\omega_1, ..., \omega_g$ for the holomorphic differentials, i.e., one may take a basis of $H^0(C, \Omega^1_C)$ which is defined over the base field of $\mathcal{P}$. This enables Poincaré to also rationally determine the Abel–Jacobi map $\alpha : C(g) \to J(C)$, which, by Jacobi's inversion theorem, is a birational map from the g-fold symmetric product $C(g)$ of the curve C to the jacobian variety $J(C)$. Now, if one takes a sufficiently general point $\xi \in J(C)$ which is also rational on the base field of $\mathcal{P}$, then $\alpha^{-1}(\xi)$ is an effective divisor of degree g on C which, as C varies in $\mathcal{P}$, describes a curve Ξ on S. One may hope that Ξ is an algebraic curve which varies in a continuous, non linear, system on S. As a matter of fact Poincaré proves that this is not the case if ξ is general in $J(C)$ but it is so if and only if ξ is general in an appropriate abelian subvariety of $J(C)$. It is interesting to see what this abelian subvariety is: indeed, in our notation, there is a surjection $J(C) \to \mathrm{Alb}(S)$ and dually a map $\mathrm{Pic}^0(S) \to J(C)$ with finite kernel, and the image of $\mathrm{Pic}^0(S)$ under this map is the variety in question.

The Italians considered Poincaré's proof as obtained by *trascendental methods*, i.e., using complex analysis and topology, basically what we call today *Hodge theory*. As mentioned, Enriques, and from time to time, other Italians, like Severi and B. Segre, went on for years fighting with the difficulty of the subject and between them in order to find a *geometric proof* of it (see, for a, rather partisan, account [59, Chap. IX, §6]). Clearly Enriques, who somehow postulated the primacy of the algebro-geometric, projective methods over the topological and analytic ones, could not stand the sense of defeat caused by the incapacity of proving a basic result in the theory of surfaces, which he felt to be the main creation of the Italian school. The subtle problems related to infinitesimal deformations of curves on a surface have been put in the right algebro–geometric setting only in more recent times by Mumford [97].

2.3 Abel's theorem for curves on a surface. It is time now to go back to the papers [25] by Castelnuovo and [117] by Severi. In them the authors prove that $q_a = q_g = q_{an} = q_t$ using Enriques' result on the implication : $q_a > 0 \Rightarrow q_g > 0$.

Both papers are, in our opinion, important for various reasons. First they certainly influenced Poincaré's paper of 1910. Secondly they introduce, and partially formalize, ideas and techniques which became since then basic in algebraic geometry. Thirdly they contain results which have been applied to other important problems in algebraic geometry. Finally they motivated further research and opened new fields in which there is still today room for further explorations.

In [25] Castelnuovo starts from Enriques' statement that a *sufficiently general curve* C on a surface moves in a continuous system of dimension $\dim |C| + q_a$, which is formed by a q_a-dimensional family of linear systems of dimension equal to $\dim |C|$. Then Castelnuovo introduces the *Picard variety* of S, which we denote by $\mathrm{Pic}^0(S)$, whose points correspond to the linear equivalence classes of curves on S. Castelnuovo then proves that $\mathrm{Pic}^0(S)$ is an abelian variety of dimension q_a. He proves that S can be mapped to $\mathrm{Pic}^0(S)$ and he deduces from this that $q_{an} \geq q_a$. The converse inequality $q_{an} \leq q_a$ had been established, as we saw, by Severi. Thus $q_a = q_{an}$. Notice that Castelnuovo's map $S \to \mathrm{Pic}^0(S)$ is, in modern terminology, the dual of the Albanese map.

Severi's paper [117] is inspired by Abel's theorem. Indeed, what Severi does is to prove an interesting equivalence criterion for curves on a surface, i.e., an *Abel's theorem* for curves on a surface, apparently conjectured by Castelnuovo and Enriques (see [15], letter 643). We state it in modern language. The most expressive form of it is as follows. Let $\mathcal{C}$ be a flat family of curves on a surface S parameterized by an irreducible variety of positive dimension. Let A be a smooth, irreducible ample curve on S. Let $Z^0(S)$ be the group of 0-cycles of degree 0 on S and let $a : Z^0(S) \to \mathrm{Alb}(S)$ be the *Albanese map*.

Theorem 2.1. *The system $\mathcal{C}$ of curves on S is contained in a linear system if and only if for any two curves C, C' in $\mathcal{C}$, one has $a(C \cdot A - C' \cdot A) = 0$.*

In modern terms, the proof is easy. One implication is trivial. As for the other, as we said, we have a surjection $\rho : J(A) \to \mathrm{Alb}(S)$ and therefore a map $\rho^* : \mathrm{Pic}^0(S) \to J(A)$ with finite kernel. By considering $C \cdot A - C' \cdot A$ as a point in $J(A)$, the hypothesis yields $\rho(C \cdot A - C' \cdot A) = 0$ for any two curves C, C' in $\mathcal{C}$. This implies that $\rho^*(C - C') = 0$ for any two curves C, C' in $\mathcal{C}$, hence $C - C'$ is a torsion element for any two curves C, C' in $\mathcal{C}$, which yields the assertion.

Notice however that Severi does not formally introduce the Albanese variety here: this he will do in the later paper [121]. This makes his arguments rather involved. He also does not seem to recognise at first glance the surjectivity of ρ, which is *Lefschetz's theorem* (again Severi will partially formalize this only in [121, §2, thm. 1]). Therefore he is forced to use ad hoc arguments to overcome these difficulties, and this makes the paper heavy. It is on the other hand clear how from Enriques' and Abel–Severi's theorem one deduces $q_a \leq q_{an}$. Indeed, according to Enriques, one can consider a flat family $\mathcal{C}$ of curves on S, parameterized by a q_a-dimensional variety V, such that the general curve of $\mathcal{C}$ is not linearly equivalent to infinitely many curves of $\mathcal{C}$. Then by Abel–Severi one has a generically finite map $V \to \mathrm{Alb}(S)$, which implies that $q_a = \dim(V) \leq \dim \mathrm{Alb}(S) = q_{an}$.

In view of the above considerations we can say that Castelnuovo's and Severi's papers [25] and [117] basically share the same ideas, though, $\mathrm{Pic}^0(S)$ and $\mathrm{Alb}(S)$ being dual abelian varieties (see [5, 111, 88]), they take somehow *dual* viewpoints.

In the same paper [117], Severi proves what he calls the *second Abel's theorem* for surfaces:

Theorem 2.2. *If $f : S \to S'$ is a morphism of surfaces, then S' is regular if and only if $a(f^{-1}(x) - f^{-1}(x')) = 0$ for two general points $x, x' \in S$.*

This is clear from our viewpoint. Indeed from f one deduces the existence of a map $a_f : \mathrm{Alb}(S') \to \mathrm{Alb}(S)$ which has finite kernel. By composing with the Albanese map $\alpha : S' \to \mathrm{Alb}(S')$, one has the map $S' \to \mathrm{Alb}(S)$ and it is clear that this is constant if and only if $\mathrm{Alb}(S') = 0$, i.e., S' is regular. This implies Severi's assertion. As Severi himself recognizes, the importance of this theorem is smaller than the one of Theorem 2.1 (see [117, end of §1]).

Finally, in [121] Severi improves on his first Abel's theorem, proving that:

Theorem 2.3. *Two curves C, C' on a surface S, which are numerically equivalent, have integral multiples which are linerly equivalent if and only if $a(C \cdot A - C' \cdot A) = 0$ for some smooth ample curve A on S.*

Severi's proof here is somewhat more transparent than the one in [117]. He introduces what we call now the Albanese variety, makes the crucial observation about the surjectivity of $\rho : J(A) \to \mathrm{Alb}(S)$, which we now deduce by Lefschetz's theorem (see [67, p. 156]), and essentially makes the same argument we outlined above. It is perhaps interesting to notice that, commenting in [131, p. 54], on the denomination of Albanese variety (see [79]), Severi attributes to himself the introduction of this concept, quoting however the later paper [121]. Actually in [121] not only the Albanese variety is considered, but also the Albanese map as well as the isogeny $\mathrm{Pic}^0(S) \to \mathrm{Alb}(S)$. It is remarkable that also the title of [121] has, so to speak, an *Abelian reminiscence*. Indeed it is devoted to *an inversion theorem* for the holomorphic 1-forms on a surface. The motivation for this paper was a beautiful result of Picard [104, p. 469], proving that given a surface S, there is no positive integer d such that the d-fold symmetric product $S(d)$ is birational to an abelian variety, unless $d = 1$ and S itself is birational to an abelian surface. As Picard himself pointed out, this result shows the difference between the surface and the curve case, in which instead $C(g)$ is birational to $J(g)$ if C is a curve of genus g. Picard suggests at the end of his paper the analogous question as an open problem for higher dimensional varieties: this has been recently solved by F. Catanese (see this volume). Severi's approach via the Albanese variety clarifies Picard's question and shows that even the injectivity of the Abel–Jacobi map $C \to J(C)$ does not hold in general in the surface case, though the image of $S \to \mathrm{Alb}(S)$ has the same irregularity as S.

2.4 Algebraic series of divisors on curves. For the purpose of a historical analysis of Abel's legacy within the Italian school, it is worth remarking that the influence of Severi's paper [117] shows up also in other directions.

A basic step in the proof of Abel–Severi's Theorem 2.1, is a proposition, which we are now going to state. Let C be a smooth, irreducible curve and let $\gamma^1_{\nu,n}$ be an irreducible 1–dimensional series of effective divisors of degree ν of C, of *index* n, i.e., such that n is the number of divisors in the series containing the general point x of C. Hence, given x, one has a divisor Γ_x of degree $n\nu$ containing nx, i.e., the sum of all divisors of the series containing x. Severi's proposition says that:

Proposition 2.4. *With the above notation, the linear equivalence class of* Γ_x *does not depend on* x *if and only if all the divisors of the series* $\gamma^1_{\nu,n}$ *are contained in one and the same linear series.*

Severi's original proof makes use of Abel's theorem, and actually Proposition 2.4 can be seen, after all, as a geometric formulation of Abel's theorem, i.e., as an equivalence criterion for divisors varying in a continuous system on a curve. Translated in modern terms, the proof of Proposition 2.4 goes as follows. Let Γ be the irreducible curve which parameterizes the series $\gamma^1_{\nu,n}$, which we can consider as a curve inside $C(\nu)$. Restricting the Abel–Jacobi mapping $a : C(\nu) \to J(C)$ to Γ, one has a map $a : \Gamma \to J(C)$, and therefore a map $f : J(\Gamma) \to J(C)$. On the other hand, dually, we can also consider C as mapping to $\Gamma(\nu)$, and therefore we have a map $g : J(C) \to J(\Gamma)$, which is easily seen to be the dual of the map f. The hypothesis implies that g is the zero map. Thus f is also zero, which yields that $a : \Gamma \to J(C)$ is constant, whence the assertion.

It is interesting to notice that Severi's lemma plays an important role also in Castelnuovo's paper [25] at the crucial point where he proves that $q_{an} \geq q_a$. As a matter of fact, as Castelnuovo says in a footnote at §11 of his paper:

> *Nella redazione preliminare di questo mio lavoro ... io mi appoggiavo ... sopra considerazioni di Analysis situs ... Nel preparare questa redazione definitiva mi sono accorto che quelle considerazioni potevano prestarsi a qualche obbiezione, per superare la quale si andava incontro a serie complicazioni. Intanto il signor SEVERI, imbattutosi in difficoltà analoghe, era riuscito a vincerle mediante il lemma ed il teorema sopra enunciati.*
>
> *(In the preliminary version of the present paper ... I relied ... on considerations of Analysis situs .. In preparing this final version I became aware that those considerations could give rise to some objections, and, in order to overcome them, one would meet serious complications. Meanwhile Mr. SEVERI, who met similar difficulties, has been able to overcome them by means of the aforementioned lemma and theorem.)*

As we presented it above, Severi's proof has already been put in algebro–geometric terms, once one accepts Abel's theorem. Severi's original argument however was analytic, and this stimulated Castelnuovo to try and find a geometric proof of it in order to have a coherent algebro–geometric treatment of the subject. Castelnuovo succeeded in doing this one year later, in 1906, in the paper [26], where he proves a stronger result, which implies Severi's one. Castelnuovo's theorem is again an equivalence criterion for divisors varying in a series $\gamma^1_{\nu,n}$ on a curve C of genus g. Therefore it can be seen as an *Abel's type theorem*. Castelnuovo looks at the number d of divisors of the $\gamma^1_{\nu,n}$ counted with appropriate multiplicities, having a *double point*. Looking at the $\gamma^1_{\nu,n}$ as a curve Γ in $C(\nu)$ as we did above, d is the intersection number of Γ with the big diagonal $\Delta \subset C(\nu)$. Then Castelnuovo's theorem says that:

Theorem 2.5. *One has $d \leq 2n(\nu + g - 1)$ and the equality holds if and only if the divisors of the series $\gamma^1_{\nu,n}$ are all contained in one and the same linear series on C.*

The proof of Castelnuovo is based on considerations of enumerative geometry, i.e., on the so–called *principle of correspondence* on a curve. Translated in modern terms, this amounts to computing the class of Δ in $C(\nu)$ (see [6, p. 358]).

According to Theorem 2.5, the number $z = n(\nu + g - 1) - \frac{d}{2}$ is a non negative invariant of the series which has been later called by Torelli [135] the *equivalence defect* of the series. Castelnuovo's theorem can be formulated by saying that a series $\gamma^1_{\nu,n}$ is contained in a linear series if and only if its equivalence defect vanishes (for a modern treatment, see, for instance, [33] and [83]).

Castelnuovo's paper was a seed which fell on a fertilized ground. Indeed, as we indicated, the study of the series $\gamma^1_{\nu,n}$ on a curve C can be considered as equivalent to the study of correspondences between C and other curves. This had been started by Hurwitz in [78] from an analytic viewpoint. In fact a correspondence between two curves C and C' induces a morphism between the two jacobians $J(C)$ and $J(C')$, i.e., recalling that $J(C) = \frac{H^0(C,\Omega^1_C)^*}{H_1(C,\mathbb{Z})}$ (see [67, p. 327]), there is an induced linear transformation $H^0(C, \Omega^1_C)^* \to H^0(C', \Omega^1_{C'})^*$ which maps $H_1(C, \mathbb{Z})$ to $H_1(C', \mathbb{Z})$. Thus the subject can be studied in terms of linear algebra. This viewpoint has been extensively developed inside and outside the Italian school (see [16], chapt. 6; for modern presentations, see [99, 88] and references therein).

The algebro–geometric treatment of the subject started inside the Italian school with two papers by de Franchis [50] and Severi [115] on the same subject, i.e., surfaces which are product of two curves. It is interesting to remark that these have been among the first publications of these two authors, and somehow directed their subsequent research towards several open problems concerning irregular surfaces (see [34]). It would be too long to recall here all the contributions of the Italian school to this theory. We will limit ourselves to referring only to those more closely related to Castelnuovo's equivalence criterion, also in connection with important developments outside the Italian school.

In this respect, probably the most relevant is a consequence of Castelnuovo's criterion which is known as *Castelnuovo–Severi's inequality* (see [26, §3] and [117]). The result refers to a curve D of *type* (a, b) in the product $C \times C'$ of two curves. The curve D is said to be of type (a, b) if it intersects the fibres F and F' of the projections $C \times C' \to C', C \times C' \to C$ in a and b points respectively. A typical example of curve of type (a, b) is a curve algebraically equivalent to a curve of the form $bF + aF'$. If D is linearly equivalent to a curve of the form $bF + aF'$, then D gives rise to a correspondence such that when x varies in C then the corresponding divisor $D(x)$ varies in a linear series on C'. Looking at this from Hurwitz viewpoint, D determines the zero map between $J(C)$ and $J(C')$. This is the reason why they have been called *correspondences with zero valence*. The result is as follows:

Theorem 2.6. *If D is a curve of type (a, b) in the product $C \times C'$ of two curves, then $D^2 \leq 2ab$ and equality holds if and only if D is a correspondence with zero valence.*

It is not difficult to deduce Theorem 2.6 from Castelnuovo's criterion 2.5. Indeed, D gives rise to two series $\gamma^1_{a,b}$ on C and $\gamma^1_{b,a}$ on C'. As remarked by Rosati [110], these two series have the same equivalence defect z and one has $2ab - D^2 = z$, which implies the assertion (for a modern treatment see [33]).

From another viewpoint Castelnuovo–Severi's inequality is an exercise in surface theory, once one takes into account the *Hodge index theorem* (see [71, p. 368]). It is remarkable indeed that Severi's proof in [122] uses in fact an ad hoc version of the Hodge index theorem, a basic tool which has been used by the Italians in several forms, but apparently formalized by B. Segre [113] only in 1937.

As is well known, Castelnuovo–Severi's inequality is a fundamental step in Weil's proof of the analogue of the *Riemann hypothesis* for curves over finite fields (see [71], loc.cit. and Appendix C for information about Weil's conjectures, on which we cannot dwell here). In the same perspective, it is worth mentioning Hasse's [72, 73] and Deuring's [51, 52, 53] transpositions of Castelnuovo's and Severi's theory to the arithmetic setting (see also the recent interesting historical paper by S. Lang [87]).

For a few years after Castelnuovo's discovery, his equivalence criterion became the object of intensive study by a group of young algebraic geometers. Their results are interesting and we want to briefly mention them. The first extension of Castelnuovo's criterion to higher dimensional series of divisors on a curve is due to Torelli [133]. Torelli's viewpoint was made more precise by Allen [9], who introduced for an irreducible series γ^r_ν of dimension r of divisors of degree ν a sequence of equivalence defects $z_i, i = 0, ..., r$, which account for *partial lacks of linear equivalence* of the divisors of the series. If $r = 1$, then z_0 is the index and z_1 is Castelnuovo's equivalence defect. By interpreting the series γ^r_ν as an irreducible r–dimensional variety Z in $C(\nu)$, one has the Abel–Jacobi map $a : C(\nu) \to J(C)$, and we can consider two divisor classes ξ, θ in $C(\nu)$: ξ is the class of $C(\nu - 1)$ inside $C(\nu)$, θ is the pull–back via a of the class of the *Riemann theta divisor* Θ on $J(C)$. Then z_i is the intersection number $Z \cdot \theta^i \cdot \xi^{r-i}$. Using these invariants Allen proves a beautiful formula which extends the famous *formula of de Jonquières* concerning the number of divisors in a linear series having suitable multiple points. Allen-de Jonquières' formula implies Casteluovo-Torelli's criterion. Allen's results basically amount to compute the classes of the diagonals inside a symmetric product (see [6, chapt. 8, §5], see also [33]).

A new direction of research in this area was indicated by Torelli in [133], who tried to extend Castelnuovo's theorem to systems of curves on a surface. Let $\mathcal{C}$ be a flat family of curves on a surface S parameterized by a smooth, irreducible curve, and assume that the general curve of $\mathcal{C}$ is smooth whereas only finitely many curves in C are singular, with only a node. One can consider the following invariants: δ is the number of singular curves in $\mathcal{C}$, $n := C^2$ with C a curve of $\mathcal{C}$, ν is the *index* of $\mathcal{C}$, i.e., the number of curves in $\mathcal{C}$ passing through the general point of S, g is the genus of the general curve of $\mathcal{C}$ and e is the Euler-Poincaré characteristic of S. Torelli's results is the following:

Theorem 2.7. *In the above setting one has $\delta \leq \nu(n+4g-4+e)$ and the equality holds if and only if $\mathcal{C}$ is contained in a linear system.*

For a modern proof, see again [33]. Needless to say, Torelli's theorem can be considered as another Abel's type theorem for curves on a surface. It would be interesting to extend it to higher dimensional series of curves on a surface, or of divisors on a variety (see again [33, Oss. (5.17)] and [134]).

A different extension of Castelnuovo's equivalence defects has been provided by Comessatti [41], who defined, for a $\gamma^1_{\nu,n}$ on a curve C of genus g, a sequence of integers $Z_0, Z_1, \ldots, Z_{g-1}$, where Z_0 is Castelnuovo's equivalence defect. By interpreting $\gamma^1_{\nu,n}$ as a curve Γ in $C(\nu)$, Comessatti's defects essentially measure the *non degeneracy* of the image of Γ in $J(C)$ via the Abel–Jacobi map. In his beautiful paper [27] of 1921, his last one in algebraic geometry before a silence of more than twenty years, Castelnuovo gave, among other things, an interpretation of Comessatti's defects in terms of the characteristic polynomial of the endomorphism of $H^0(C, \Omega^1_C)$ induced by the correspondence of C determined by $\gamma^1_{\nu,n}$. It would be interesting to understand deeper Comessatti's geometric viewpoint and see if suitable extensions of his defects could be defined for higher dimensional series.

2.5 The topological Abel's theorem. In conclusion of this chapter, we want to go back to the original theorem of Abel and mention a topological interpretation of it started by Rosati in [110] and completed by Chisini in [32]. We will follow the nice exposition in [60, vol. III, chapt. IV, §44]. Chisini's *topological Abel's theorem* has certainly been motivated by the developments of the theory of series of divisors on a curve and of correspondences mentioned above. Furthemore it probably played a role in forming Severi's viewpoint on possible extensions of Abel's theorem to 0-cycles on a surface, which we will address in the next chapter.

In order to state Chisini's result, we need a definition. In giving it we start using some notions which are *motivic* in nature, as we will see in Sects. 3.4 and 3.6. Consider a series γ^r_ν of dimension $r \geq 1$ of divisors of degree ν on a curve C. Thus γ^r_ν is parameterized by an irreducible variety V of dimension r and, as we indicated before, one can consider the incidence correspondence $\Gamma \subset V \times C$, which is a correspondence of V to C. As such it defines, in a natural way, a map $g : H^1(C, \mathbb{C}) \to H^1(V, \mathbb{C})$, which can be easily described once one knows the expression of the class of Γ as a linear combination of its Künnet's components inside $H^2(V \times C, \mathbb{C})$. If we use this language, then Chisini defines the series γ^r_ν to have *null–circulation* if the map g is the zero map. Equivalently, this means that Γ has zero component in $H^1(V, \mathbb{C}) \otimes H^1(C, \mathbb{C})$. Roughly speaking, this means that when a point describes a cycle on V, the corresponding points on C describe a cycle which is homologous to zero. Notice that if γ^r_ν is a linear series, i.e., V is a $\mathbb{P}^r$, then of course it has null-circulation. Furthermore if γ^r_ν is contained in γ^s_ν and the latter has null-circulation, then clearly also γ^r_ν has null-circulation. Hence if γ^r_ν is contained in a linear series g^s_ν, then it has null-circulation. With all this in mind, Chisini's *topological Abel's theorem* says that:

Theorem 2.8. *A series γ_ν^r on the curve C of genus g has null-circulation if and only if γ_ν^r is contained in a linear series.*

This is, of course, a trivial consequence of the classical Abel's theorem, but Chisini wants to give an independent proof, based only on topological and algebro–geometric considerations. He gives a very nice argument, which we briefly reproduce. Thus, assuming that γ_ν^r has null–circulation, we want to prove that it is contained in a linear series. The first step is to make a trivial reduction to the case $r = 1$ which we skip. Then we have a series $\gamma_{\nu,n}^1$, and the second step is to look at the correspondence $\Gamma \subset C \times C$ where $(x, y) \in \Gamma$ if and only if $x + y$ is contained in some divisor of $\gamma_{\nu,n}^1$. Notice that Γ is symmetric and of type $n(\nu - 1)$, i.e., it meets the *coordinate curves* $C \times \{y\}$ and $\{x\} \times C$ at $n(\nu - 1)$ points. Then Γ determines an endomorphism of $H^1(C, \mathbb{C})$. Now, one moment of reflection shows that, since $\gamma_{\nu,n}^1$ has null–circulation, the endomorphism in question is the multiplication by $-n$. By applying the *Lefschetz fixed point formula*, which Chisini proves in this specific case, one has that Γ intersects the diagonal $\Delta \subset C \times C$ at $2n(\nu + g - 2)$ points. The third and final step is now to remark that the intersection points of Γ and Δ correspond to the double points of $\gamma_{\nu,n}^1$. Hence one finishes the proof by applying Castelnuovo's equivalence criterion.

Though interesting, Chisini's viewpoint is no news: the action of a correspondence on C on $H^1(C, \mathbb{C})$ is reflected by its action on $H^0(C, \Omega_C^1)$ and ultimately by its action on the jacobian *J*(*C*), as Hurwitz pointed out in [78]. Actually Hurwitz proved in this way that the correspondences between two curves form a finitely generated abelian group and Chisini is able to deduce Hurwitz's theorem from Theorem 2.8. As is well known, Severi [120] was convinced by the consideration of this specific case to formulate his general *theory of the base*, proving that divisors on a variety modulo numerical equivalence always form a finitely generated abelian group. These ideas were transposed in a purely algebraic setting by Weil [138, 139], in view of applications to curves and varieties over finite fields (see also [63]).

3 Rational Equivalence of 0–cycles on a Surface

The last chapter of our story starts in 1932 and the main character is again Severi. In that year Severi published the paper [124] in which he set the foundations for a theory of rational equivalence of 0-cycles on a surface. In our opinion the basic papers are [124, 125, 126, 127] and especially [128], to which we will mainly refer. A summary of Severi's ideas can be found in the treatise [131]. We will illustrate in a critical way Severi's contributions, indicating how, in his view, these were related to Abel's theorem. Then we will briefly open a window on the more recent *motivic* developments of the subject, trying to show how some of them have their roots in Severi's work.

3.1 Series of equivalence. To put it as D. Mumford does in one of his beautiful appendices to Zariski's book [142] (see p. 238):

In the period 1935 to 1950, SEVERI published many papers on series of equivalence and its generalizations to higher dimensions. It is hard to untangle everywhere what he conjectured and what he proved and, unfortunately, some of his conclusions are incorrect.

However we want to indicate that, despite this, Severi put his hands on ideas that still form a very important and widely open research area.

Certainly Severi even failed to give a rigorous and solid definition of rational equivalence. To the extent that, in 1954, some twenty years after his first paper on the subject, at the International Congress of Mathematics in Amsterdam, *the definition itself of series of equivalence was debated sharply* (see Mumford, loc.cit.; see also the abstract of Severi's talk at that congress [129] and the unpublished correspondence between Severi and Weil [130, 140], kindly pointed out to us by prof. N. Schappacher). What we intend to do, however, is to consider his failures without dwelling too long on them, but rather trying to understand the positive features of his work. To put it as Fulton does in [61]:

It would be unfortunate if Severi's pioneering works in this area were forgotten; and if incompleteness or the presence of errors are grounds for ignoring Severi's work, few of the subsequent papers on rational equivalence would survive.

So we will go on *as if* Severi and we would have agreed on the definition of rational equivalence for algebraic cycles that currently we accept, which is the one given by Chevalley and Samuel around 1956 (see [71, p. 426]). Thus, for every projective variety X of pure dimension n, we have the graded Chow ring $CH(X) = \oplus_{i=0}^{n} CH_i(X)$ where $CH_i(X)$ is the group of cycles of dimension i modulo rational equivalence. One also denotes by $CH^j(X)$ the group of codimension j cycles of X, so that $CH_i(X) = CH^{n-i}(X)$. We will often specialize to the case $n = 2$ where the ambient variety is a surface, which is the one where some of the problems of the theory already show up in a remarkable complexity. In this case the only part of the Chow ring which is difficult to understand is $CH_0(X)$, inasmuch as $CH_1(X) = \mathrm{Pic}(X)$ and $CH_2(X) = \mathbb{Z} < X >$. In order to understand the group $CH_0(X)$ it clearly suffices to do so for its *degree zero part* $CH_0^0(X)$, i.e., the kernel of the natural homomorphism $\deg : Z \in CH_0(X) \to \deg(Z) \in \mathbb{Z}$.

Notice that if X is a curve, then $CH_0^0(X) = \mathrm{Pic}^0(X) = J(C)$ is a variety and, if one fixes a base point $x \in X$ one has, for each positive integer n, the Abel–Jacobi map $\alpha_n : X(n) \to CH_0^0(X)$ defined as $\alpha_n(Z) = [Z - nx]$. This map is surjective as soon as $n \geq g$.

Also in the surface case, if we fix a base point $x \in X$, one has, for each positive integer n, an *Abel–Jacobi map* $\alpha_n : X(n) \to CH_0^0(X)$ defined as $\alpha_n(Z) = [Z - nx]$. A related map is $\beta_n : (Z, Z') \in X(n) \times X(n) \to [Z - Z'] \in CH_0^0(X)$.

Two basic questions now are in order and Severi, more or less explicitely, considered both of them:

(i) Can we describe $CH_0^0(X)$ in *finite terms*, e.g. is there an n such that the map $\alpha_n : X(n) \to CH_0^0(X)$ is surjective? This is equivalent to ask whether there is an n such that the map $\beta_n : X(n) \times X(n) \to CH_0^0(X)$ is surjective.

(ii) How can we describe the *series of equivalence*, i.e., the fibres of the Abel–Jacobi maps $\alpha_n : X(n) \to CH_0^0(X)$? This is what Severi used to refer to as an *Abel's theorem for series of equivalence*.

As for question (i), this is the object of Mumford's epochal paper [98], in which he proves that, if $p_g(X) = h^0(X, \Omega_X^2) > 0$, then, contrary to what happens in the case of curves, the answer is negative (see Sect. 3.7). As Mumford points out, Severi definitely missed this point, and, even worse than this, he considered as *evident* a general statement which implies a positive answer to question (i) (see [127, §13], and Definition 3.32, (3) below):

> *... ammettiamo che per ogni varietà k dimensionale (virtuale) A di W_r, esista un numero finito di caratteri numerativi $c_1, c_2, \ldots$ tali che le relazioni $c_1 > 0, c_2 > 0, \ldots$ sieno sufficienti per affermare che A è una varietà effettiva. Con questo si vuol dire che nel sistema di equivalenza $|A|$ esistono varietà totali effettive.*
>
> *(... let us suppose that for every (virtual) k-dimensional variety A of W_r, there is a finite number of enumerative characters $c_1, c_2, \ldots$ such that the relations $c_1 > 0, c_2 > 0, \ldots$ are sufficient to claim that A is an effective variety. With this we mean that in the equivalence system $|A|$ there are total varieties which are effective.)*

As Mumford indicates, it seems difficult, not to say hopeless, to *untangle* what Severi afterwards really proved and what he deduced from this wrong *postulate*. Despite the difficulty of the task, we would like nevertheless to try to, at least partially, shed some light on this point.

To start with, we remark that, at the same point [127], loc.cit., Severi says:

> *Ammesso il postulato, ne segue che le equivalenze aritmetiche, algebriche e topologiche (con divisione) non sono che aspetti diversi di un medesimo concetto.*
>
> *(Given the postulate, it follows that arithmetic, algebraic and topological equivalences (with division) are all different aspects of one and the same concept.)*

As we will see, Severi does not use the *wrong postulate*, but rather he appeals to this *consequence*. Which is, by the way, wrong as well. In fact, as we know today, it is still a crucial open problem whether numerical and homological equivalence are the same (see Sect. 3.4). However we do know that in general homological equivalence is different from algebraic equivalence: this is the non triviality of the *Griffiths group*, which is actually even not finitely generated for some varieties like quintic hypersurfaces in $\mathbb{P}^4$ (see [66, 40], and [136, §7] for a nice account and related questions). Nonetheless, as we will try to indicate in a few moments, Severi never really used the aforementioned *consequence* of the *wrong postulate* in its full strenght, but rather in a specific situation, which will be useful to examine and discuss in some detail. We will do this in Sect. 3.3.

Going back to Mumford's theorem, as this author indicates (see [98], pp. 196–7):

> *Now, after criticizing Severi like this, I have to admit the following: the method of disproof of (1)–(3) is due entirely to Severi: Severi created, in fact, a very excellent tool for analyzing the influence of regular 2-forms on F on his systems of equivalence. One must admit that in this case the technique of the Italians was superior to their vaunted intuition.*

As for Propositions (1)–(3), they are equivalent formulations of the surjectivity of α_n or β_n for large n (see Sect. 3.7). What Mumford refers to, is an idea which actually enters in Severi's analysis of question (ii). It is not clear whether this idea has to be attributed to Severi or to somebody else. For example it can be retraced in Albanese's beautiful paper [3] where one proves that a surface is rational if and only if its twofold symmetric product is rational (see [91] for a modern version). Earlier and particular versions of the same tool can be found in [75] and in Picard's note [104, p. 469], mentioned above.

Let us explain what this idea is. To do this, it is convenient to put things in a general setting, as Mumford does. Let V be a smooth variety, let G be a finite group acting on V, let $W = V/G$ and $\pi : V \to W$ be the canonical morphism, let $\omega \in H^0(V, \Omega_V^p)$ be a G-invariant p-form on V. If G acts freely on V, then there is a unique form $\eta \in H^0(W, \Omega_W^p)$ such that $\omega = \pi^*(\eta)$. in other words $H^0(W, \Omega_W^p) \simeq H^0(V, \Omega_V^p)^G$. This is no longer true if G is not acting freely, since in this case W turns out to be singular. However, what one proves is the following fact. Let S be any smooth variety and $f : S \to W$ a morphism. One can consider $\tilde{S} := (S \times_W V)_{red}$, which is equipped with the morphisms $\tilde{f} : \tilde{S} \to V$ and $p : \tilde{S} \to S$. Then (see [98]):

Lemma 3.1. *In the above setting, given a G–invariant form $\omega \in H^0(V, \Omega_V^p)$, for all non singular varieties S and morphisms $f : S \to W$, there is a unique p-form $\eta_f \in H^0(S, \Omega_S^p)$ such that $p^*(\eta_f) - \tilde{f}^*(\omega)$ is torsion in $\Omega_{\tilde{S}}^p$.*

The situation we will have in mind is the following: X is a surface, $V = X^n$, G is the symmetric group S_n and $W = X(n)$ is the n-fold symmetric product of X.

If $\omega \in H^0(X, \Omega_X^p)$ is a holomorphic p-form on X, we can define the form $\omega^{(n)} \in H^0(X^n, \Omega_{X^n}^p)$ as $\omega^{(n)} = \sum_{i=1}^n p_i^*(\omega)$, where $p_i : X^n \to X$ is the projection to the i-th factor. Of course $\omega^{(n)}$ is S_n-invariant. Let $\Delta \subset X(n)$ be the main diagonal, let U be the open set $X(n) - \Delta$ and let $\iota : U \to X(n)$ be the inclusion. Since U is smooth, for any form $\omega \in H^0(X, \Omega_X^p)$ we can consider the form $\omega_\iota^{(n)}$ on U which we denote by $\omega^{[n]}$.

A series of dimension r of effective 0-cycles of degree n on X can be looked at as given by a map $f : S \to X(n)$ where S is a smooth, irreducible variety of dimension r and f is birational to its image. One has a series of equivalence if and only if $\alpha_n \circ f$ is constant, where $\alpha_n : X(n) \to CH_0^0(X)$ is an Abel–Jacobi map.

What Severi proves [128, §14], is that:

Proposition 3.2. *If $f : S \to X(n)$ is a series of equivalence, then for any p–form $\omega \in H^0(X, \Omega_X^p)$, $p = 1, 2$, the form $\omega_f^{(n)}$ vanishes on S.*

Mumford applies this result to the case $p = 2$ and he obtains that:

Theorem 3.3. *Given a surface X with $p_g > 0$, there is no n such that the maps α_n or β_n are surjective.*

Proof. Indeed, Let $Z \in X(n)$ be a general 0–cycle. Thus $Z = x_1 + \ldots + x_n$, where $x_1, \ldots, x_n \in X$ are distinct point, i.e., Z lies in U. Notice that the tangent space to $X(n)$ at Z can be identified with the tangent space to X^n at $(x_1, \ldots, x_n)$. Consider a series of equivalence containing Z. This is, locally around Z in U, given by a subvariety S of U containing Z. Suppose $p_g(X) > 0$. Take any non–zero holomorphic 2–form ω on S. This gives rise to the holomorphic 2–form $\omega^{[n]}$ on U, which is non-degenerate at Z. Furthermore, by Severi's result 3.2, S has to be an integral variety for $\omega^{[n]}$, which clearly implies $\dim(S) \leq n$.

The above argument shows that the general fibre of α_n has dimension at most n. Hence the image of $X(n)$ in $C_0^0(X)$ via α_n contains a constructible set of dimension at least n and at most $2n$. This implies that no map α_n can be surjective and the maps β_n cannot be surjective either. □

The situation described in Theorem 3.3 can be expressed by saying that $CH_0^0(X)$ is not *finite dimensional* (see Sect. 3.7).

It is worth pointing out a striking consequence of Mumford's analysis (see [136, p. 186]), which explains how strangely rational equivalence of 0–cycles behaves on surfaces with $p_g > 0$:

Proposition 3.4. *If X is a surface with $p_g > 0$, then the general 0–cycle $Z \in X(n)$ is rationally isolated, i.e., there is no positive dimensional series of equivalence $f : S \to X(n)$ containing Z.*

In conclusion, as Mumford noticed in [98, §§7–9], Severi found in [128] the right way of handling problem (i), i.e., considering pull–backs of 2–holomorphic form. However, as we will see in the next section, he makes a crucial mistake which puts him off the right track and he seems unable to deduce, as Mumford does, the correct upper bound for the dimension of a general series of equivalence.

3.2 Abel's theorems for 0–cycles. Turning to question (ii), Severi takes a point of view which is very modern in essence (see Sect. 3.4 below) and he introduces two basic concepts which play an important role in the matter, i.e., the concepts of a *series with null-circulation* and of a *series with algebraic circulation*. Let us explain what they are.

The first concept we already met in the case of curves (see Sect. 2.5), and it is basically the same. Let S be a smooth, irreducible variety of dimension r and let $f : S \to X(n)$ be a series of effective 0–cycles on the surface X parameterized by an irreducible variety S. One can then consider the related correspondence $\Sigma \subset S \times X$, i.e., the cycle $\Sigma \in CH_r(S\times X) = CH^2(S\times X)$ defined by $\Sigma = \{(s, x), x \in f(s)\}$. Of course considering $f : S \to X(n)$ is equivalent to considering the correspondence Σ.

The correspondence $\Sigma \subset S \times X$ has real codimension 4 in $S \times X$. As such, it defines a map $\sigma_i : H^i(X, \mathbb{C}) \to H^i(S, \mathbb{C})$, for every $i = 0, \ldots, 4$. The relevant maps are σ_i, $i = 1, 2, 3$, where σ_i is determined by the $(4-i, i)$- *Künneth's component* $\Sigma_{4-i,i}$ of Σ in $H^{4-i}(S, \mathbb{C}) \otimes H^i(X, \mathbb{C})$, $i = 1, 2, 3$.

Severi defines the series to have *null–circulation* if σ_1 and σ_3 are the zero maps, i.e., if $\Sigma_{3,1}$ and $\Sigma_{1,3}$ are zero. Notice that, fixing a point $x \in X$ we can define for every positive integer n, the Albanese map $a_n : X(n) \to \mathrm{Alb}(X)$, with x as the base point for the integration paths. Now a series $f : S \to X(n)$ has *null–circulation* if and only if $a_n \circ f$ is constant. This is what Severi calls *Abel's theorem* for the series with null–circulation (see [128, §4]). Of course a series of equivalence has null–circulation.

Severi notes that, if $f : S \to X(n)$ is a complete series with null–circulation, then:

$$2n - q \leq \dim(S) \leq 2n - q + i(Z), \tag{3.1}$$

where q is the irregularity of X and $i(Z)$ is the *first index of speciality* of a cycle of the series, i.e., $i(Z) := h^0(X, \Omega^1_X(-Z))$ is the number of independent 1–forms on X vanishing at Z. Severi calls equation 3.1 the *Riemann–Roch theorem for series with null–circulation* (see [128, §6]).

Turning to σ_2, one remarks that there is an orthogonal decomposition with respect to the intersection product $H^2(X, \mathbb{C}) = NS_{\mathbb{C}}(X) \oplus \mathrm{Tr}(X)$, where $NS(X)$ is the Néron-Severi group of X, i.e., the image of the map $\mathrm{Pic}(X) \to H^2(X, \mathbb{Z})$, $NS_{\mathbb{C}}(X) := NS(X) \otimes_{\mathbb{Z}} \mathbb{C}$ and $\mathrm{Tr}(X)$ is the group of *transcendental cycles*. Severi defines the series $f : S \to X(n)$ to have *algebraic circulation* if σ_2 vanishes on $\mathrm{Tr}(X)$, which is the same as asking that $\Sigma_{2,2}$ sits in $NS(S) \otimes NS(X)$. *Abel's theorem* in this case is the following (see [128, §24]):

Proposition 3.5. *The series $f : S \to X(n)$ has algebraic circulation if and only if for every holomorphic 2–form ω on X, the form $\omega_f^{(n)}$ vanishes on S.*

A series of equivalence has also algebraic circulation by Proposition 3.2.

Proceeding further in his theory, Severi seeks for a *Riemann–Roch theorem for series with algebraic circulation* (see [128, §9]). This is essentially the probem considered by Mumford in [98]. Actually Mumford's argument in the proof of Theorem 3.3 shows that:

Proposition 3.6. *If X is a surface with $p_g > 0$, $Z \in X(n)$ is a general 0–cycle and $f : S \to X(n)$ is a series with algebraic circulation containing Z, then* $\dim(S) \leq n$.

Severi makes the following assertion: *if the series $f : S \to X(n)$ has algebraic circulation and is complete, then*

$$2n - p_g \leq \dim(S) \leq 2n - \frac{p_g - j(Z)}{2}, \tag{3.2}$$

where $j(Z)$ is the second index of speciality of a 0–cycle Z of the series, i.e., $j(Z) := h^0(X, \Omega^2_X(-Z))$ is the number of independent 2–forms on X vanishing at Z.

Now the left hand side inequality in 3.2 is false! Suppose in fact that $p_g > 0$ and consider an arbitrary n. If Severi's inequality $\dim(S) \geq 2n - p_g$ were true, by Proposition 3.6 one would have $p_g \geq n$ for all n, a contradiction. Severi's argument is indeed wrong: his idea is that, since p_g independent holomorphic 2–forms ω of X give rise to independent forms $\omega^{[n]}$ vanishing on S, then S has to verify at most p_g conditions, hence one should have $\dim(S) \geq 2n - p_g$. The mistake is that the vanishing of $\omega^{[n]}$ on S does not impose one single condition on S!

On the other hand, by comparison with Mumford's argument, we see that the right hand side of 3.2 cannot be expected to be sharp and therefore cannot be too useful. It is interesting to remark that, in proving the right hand side of 3.2 (see [128, Sect. 28]), Severi makes an argument which is close to the one by Mumford that we recalled above. Therefore he should have been able to come to the same conclusions as Mumford. However, in our opinion, he was deceived by his first wrong inequality, at the point of being unable to make any serious deduction even from a right argument.

3.3 Severi's claims. The most interesting part of Severi's paper [128] is perhaps section no. 30. Here Severi asks the following question: *suppose we have a series $f : S \to X(n)$ which has null–circulation and algebraic circulation: is it a series of equivalence?* It is interesting to analyse his approach to the problem.

Let us give one more definition. Let $f : S \to X(n)$ be a series. Severi says that it is a *series of pseudo–equivalence* if there is a positive integer m such that the map $f^m : x \in S \to mf(x) \in X(mn)$ is a series of equivalence. Notice that a series of pseudo–equivalence, as well as a series of equivalence, has null–circulation and algebraic circulation.

Then Severi makes the following statement:

Claim 3.7. A series $f : S \to X(n)$ of dimension $r \geq 2$ is a series of pseudo–equivalence if and only if it has null–circulation and algebraic circulation.

As an immediate consequence of this, Severi makes a claim which can be stated by us in the following way:

Claim 3.8. Let X be a smooth, projective surface. $CH^0_0(X) \otimes \mathbb{Q} = 0$ if and only if $p_g = q = 0$ for X.

This is essentially Bloch's conjecture in its form 3.30, to which we shall soon come back (see Sect. 3.6). However Claim 3.7 is false as it is: we will see this in a moment (see Example 3.9 below).

Before going further, one word about the difference between series of equivalence and series of pseudo–equivalence is in order. Severi claims more than once that the two concepts are basically different. For example he says that the series of points of an Enriques surface is not a series of equivalence, whereas its double is. This is wrong, as noticed in [13]: the series in question is a series of equivalence. More precisely it has been proved by Roitman [108, 109], that the Albanese morphism determines a map $a : CH_0^0(X) \to \mathrm{Alb}(X)$ which induces an isomorphism $\mathrm{Tors}(CH_0^0(X)) \simeq \mathrm{Tors}(\mathrm{Alb}(X))$ (see Theorem 3.20 below). Thus the torsion of $CH_0^0(X)$ can be considered to be well known. In particular, for a regular surface like an Enriques surface, $CH_0^0(X)$ has no torsion at all. However, though Severi was wrong about the difference between rational equivalence and pseudo–equivalence, by Roitman's theorem his intuition that in order to understand equivalence series it suffices to understand pseudo–equivalence series is quite valid.

Example 3.9. Let X be a smooth, projective surface and let C be a smooth curve on it. The obvious map $f : C(2) \to X(2)$ determines a series of 0–cycles of degree 2 on X. We leave to the reader to see that this series has algebraic circulation.

Suppose that X is regular. Then every series on X has null–circulation. If Severi's Claim 3.7 were true, then $f : C(2) \to X(2)$ would be a series of pesudo–equivalence and, according to Roitmann, a series of equivalence.

Assume also that $p_g > 0$. Let x, y be a general pair of points of X. By Proposition 3.4, the 0–cycle $x + y$ is rationally isolated in $X(2)$. By taking C passing through x and y we find a contradiction. This shows that Severi's Claim 3.7 is false.

Though Severi's *argument* for the proof of Claim 3.7 is wrong, we want to analyse it and explain where the gap is. The argument goes as follows. First Severi reduces, in a standard, harmless way, the problem to proving the assertion for the series of dimension 2. Then we have a correspondence $\Sigma \subset S \times X$ between two surfaces. The assumption that the series has null–circulation and algebraic circulation means that $\Sigma_{3,1}$ and $\Sigma_{1,3}$ are zero, whereas $\Sigma_{2,2}$ sits in $NS(S) \times NS(X)$. At this point Severi invokes his *postulate* to make the following:

Claim 3.10. If X and S are surfaces and $\Sigma \subset S \times X$ is a 2–dimensional correspondence between them, then the class of Σ in $CH^2(S \times X)$ depends only on the cohomology class of Σ in $H^4(S \times X, \mathbb{C})$.

If this were true, then one would easily conclude that the series $f : S \to X(n)$ is, up to torsion, a series of equivalence, i.e., that it is a series of pseudo–equivalence. We do not dwell on this implication here, since it will be essentially explained in Sect. 3.6 (see in particular the proof of Proposition 3.28). The problem is that Claim 3.10, as well as the *postulate*, is not valid. The only way of remedying this is to try and ask less than Claim 3.10. As we will see, this is the content of Bloch's conjecture stated below as Conjecture 3.23 (see [12, conjecture (1.8)], and [136, p. 188]).

The above considerations indicate that Severi was often wrong and certainly too bold in making conjectures. However he was the first who stated 3.8, i.e., Bloch's conjecture in its popular form 3.30, and he was able to somehow perceive the *motivic* content of the matter, by considering correspondences and their action both on Chow groups and cohomology groups. We will explain this in the next sections.

3.4 Motives. A useful way of reformulating Severi's idea in a modern framework is to use the notion of *motives*, created by Grothendieck in the sixties [68]. As Murre puts it in [96, p. 139]:

> *One of Grothendieck's aims was to understand and explain the striking similarities between the various cohomology theories of algebraic varieties. Going beyond that Grothendieck states: "the theory of motives is a systematic theory of arithmetic properties of algebraic varieties, as embodied in their groups of classes of cycles for numerical equivalence."*

As a general reference for motives, we refer to [85] and also to [96, Chaps. VII–VIII] and the references therein. We will briefly recall the basics about *pure motives*.

Let $\mathcal{V}_{\mathbb{C}}$ be the category of smooth projective complex varieties and let $\sim$ be an *adequate* equivalence relation for cycles (see [81]). For every $X \in \mathcal{V}_{\mathbb{C}}$ let $A^i_{\sim}(X) = (Z^i(X)/\sim)_{\mathbb{Q}}$ be the *Chow group* of codimension i cycles on X modulo the choosen equivalence relation $\sim$ with $\mathbb{Q}$-coefficients. Let $\mathcal{M}_{\sim}$ be the category whose objects are triples $M = (X, p, m)$ where $X \in \mathcal{V}_{\mathbb{C}}$, and $p \in A^{\dim(X)}_{\sim}(X \times X)$ is a projector in the ring of correspondences on X and $m \in \mathbb{Z}$ (see [112, 1.4]). Morphisms from $M = (X, p, m)$ to $N = (Y, q, n)$ in the category $\mathcal{M}_{\sim}$ are given by correspondences $f \in A^{n-m}_{\sim}(X, Y)$ of *degree* $n - m$ such that $f \circ p = q \circ f = f$, where $A^r_{\sim}(X, Y) := A^{d+r}_{\sim}(X \times Y)$ if X is purely d-dimensional.

$\mathcal{M}_{\sim}$ is a pseudoabelian and a $\mathbb{Q}$–linear tensor category. $\mathcal{M}_{\sim}$is also rigid in the sense that there exist internal Hom's and dual objects M^* for all $M \in \mathcal{M}_{\sim}$ (see [112]).

Though we will not need this, we recall that the *unit motive* is $1 = (Spec(k), id, 0)$. One defines the *Lefschetz motive* $\mathbb{L}$ to be $(Spec(k), id, -1)$. For every motive $M = (X, p, m)$ and every integer r, one defines the *Tate twist* $M(r)$ to be the motive $M \otimes \mathbb{L}^{-r} = (X, p, m + r)$, where $\mathbb{L}^r = \mathbb{L}^{\otimes^r}$ if r is positive, $\mathbb{L}^0 = 1$ and $\mathbb{L}^r = \mathbb{L}^{\otimes^{-r}}$ if r is negative (see [112, 1.9]).

If $\alpha \in Hom_{\mathcal{M}_{\sim}}(X, X)$ then α acts on $A^d_{\sim}(X)$, where $d = \dim(X)$ as follows:

$$\alpha_*(x) = p_{2*}(p_1^*(x) \cdot \alpha)$$

where $x \in A^d_{\sim}(X)$ and $p_i : X \times X \to X$ are the projections.

$\mathcal{M}_{\sim}$ is called the category of *pure motives* with respect to the given equivalence relation $\sim$.

We will denote by $M_{\sim} : \mathcal{V}_{\mathbb{C}} \to \mathcal{M}_{\sim}$ the functor which associates to every smooth projective variety X its *motive* $M_{\sim}(X) = (X, id, 0)$, where $id = [\Delta_X]$, i.e., the class of the diagonal Δ_X in $A^d_{\sim}(X \times X)$ and d is the dimension of X.

We will consider the following adequate equivalence relations $\sim$ on cycles:

(1) rational equivalence;
(2) algebraic equivalence;

(3) homological equivalence;
(4) numerical equivalence.

It is known that: (1) $\Rightarrow$ (2) $\Rightarrow$ (3) $\Rightarrow$ (4). Rational equivalence is strictly finer than algebraic equivalence already for divisors on curves. As we noticed above, in Sect. 3.1, a famous example by Griffiths [66] shows that algebraic equivalence is strictly finer than homological equivalence, even modulo torsion, for codimension 2 cycles on a 3-fold. According to Grothendieck's Standard Conjectures on algebraic cycles, homological equivalence and numerical equivalence should coincide (see [96, 136]).

By a result of Jannsen [80], the category $\mathcal{M}_\sim$ is abelian semisimple if and only if $\sim$ is the numerical equivalence relation.

Let now $\mathcal{M}_{rat}$ be the category of motives modulo rational equivalence and let A^i be the Chow group of codimension i cycles modulo rational equivalence with $\mathbb{Q}$–coefficients. Notice the change of notation for the Chow group with respect to Sect. 3.3. We will however use, at our convenience, either the present notation A^* or the previous one $CH_{*\mathbb{Q}}$ to denote the Chow groups with $\mathbb{Q}$–coefficients.

We will often simply denote by M the functor $M_{rat} : \mathcal{V}_\mathbb{C} \to \mathcal{M}_{rat}$.

If H^* denotes singular cohomology with $\mathbb{Q}$–coefficients for smooth projective varieties over $\mathbb{C}$, then one defines a functor H^i on $\mathcal{M}_{rat}$, for every $i \in \mathbb{Z}$, by setting $H^i((X, p, m)) = p_* H^{i+2m}(X)$. We will denote by $cl : A^i(X) \to H^{2i}(X, \mathbb{Q})$ the cycle map. Then $A^i(X)_{hom}$ denotes the kernel of cl, i.e., the subgroup of codimension i cycles modulo rational equivalence which are homologically trivial. We will similarly denote by $A^i(X)_{num}$ the subgroup of codimension i cycles modulo rational equivalence which are numerically trivial.

If $\alpha \in A^d(X \times X)$ then, using Künneth decomposition, its class in $H^{2d}(X \times X, \mathbb{Q})$ decomposes as the direct sum of classes in $H^{2d-i}(X, \mathbb{Q}) \otimes H^i(X, \mathbb{Q})$. Therefore if we denote by $h = M_{hom}$ the functor which associates to every $X \in \mathcal{V}_\mathbb{C}$ its motive in $\mathcal{M}_{hom}$ we have a natural decomposition:

$$h(X) \simeq \bigoplus_{1 \le i \le 2d} h^i(X),$$

where $h^i(X) = (X, \sigma_i, 0)$ and σ_i is the $(i, 2d - i)$–Künneth component of the diagonal Δ_X in $H^*(X \times X)$. If we assume that the Künneth components σ_i of the diagonal are algebraic (which is known to be the case for curves, surfaces and abelian varieties see [85]), then the idempotent $H^*(X) \to H^i(X) \to H^*(X)$ is given by an algebraic correspondence π_i which is an idempotent in $A^d(X \times X)_{hom}$.

Following Murre [94] we will say that $X \in \mathcal{V}_\mathbb{C}$ has a *Chow–Künneth decomposition* if there exist orthogonal idempotents π_i, $0 \le i \le 2d$, in $A^d(X \times X)$ such that $cl(\pi_i)$ is the $(2d - i, i)$–component of Δ_X in $H^d(X \times X)$ and

$$[\Delta_X] = \sum_{0 \le i \le 2d} \pi_i \in A^d(X \times X). \tag{3.3}$$

This implies that in $\mathcal{M}_{rat}$ the motive $M(X)$ decomposes as follows:

$$M(X) = \bigoplus_{0 \le i \le 2d} M^i(X) \tag{3.4}$$

where $M^i(X) = (X, \pi_i, 0)$.

A natural question that arises in this context, and, as we saw in Sect. 3.3, was somehow touched by Severi himself, is the following: *how much of the structure of the Chow ring $A^*(X)$ can be recovered from the motive $h(X)$ in $\mathcal{M}_{hom}$?*

A conjectural answer to the above question is given by the following [81, 2.1]:

Conjecture 3.11. (Bloch–Beilinson's conjectural filtration of Chow groups) For every smooth projective variety X over $\mathbb{C}$ there exists a descending filtration $F^\bullet$ on $A^j(X)$, for all $j \ge 0$, such that:

(a) $F^0A^j(X) = A^j(X)$, $F^1A^j(X) = A^j(X)_{hom}$;
(b) $F^\bullet$ is compatible with the intersection product of cycles;
(c) $F^\bullet$ is compatible with f^* and f_* if $f : X \to Y$ is a morphism;
(d) The associated graded group $Gr_F^*A^j(X)$, where

$$Gr_F^\nu A^j(X) := F^\nu A^j(X)/F^{\nu+1}A^j(X)$$

depends only on the motive $h^{2j-\nu}(X)$ of X in $\mathcal{M}_{hom}$;
(e) $F^\nu A^j(X) = 0$ for $\nu >> 0$.

Remark 3.12. If we assume that the Künneth components σ_i of the diagonal are algebraic, then condition (d) is equivalent to the following (see [82, p. 236]):

(d′) Let Y be any smooth, projective variety and let $\Gamma \in A^{\dim(Y)}(Y \times X)$. If the induced map Γ_* between $H^{2j-\nu}(Y)$ and $H^{2i-\nu}(X)$ is zero then so is the map:

$$Gr_F^\nu\Gamma : Gr_F^\nu A^i(Y) \to Gr_F^\nu A^j(X).$$

Bloch–Beilinson's conjecture on the existence of such a filtration $F^\bullet$ is equivalent to the following conjectural complete formulation of the existence of a Chow–Künneth decomposition (see [94] and [81, 5.1]):

Conjecture 3.13. (Murre's conjectural filtration of Chow groups) For every smooth projective variety X over $\mathbb{C}$, one has:

(1) There exist orthogonal idempotents π_i, $0 \le i \le 2d$, in $A^d(X \times X)$ such that $cl(\pi_i)$ is the $(2d-i, i)$-component of Δ_X in $H^d(X \times X)$ and 3.3 holds;
(2) the correspondences $\pi_{2j+1}, \cdots, \pi_{2d}$ act as 0 on $A^j(X)$;
(3) Let $F^\nu A^j(X) = \mathrm{Ker}(\pi_{2j}) \cap \mathrm{Ker}(\pi_{2j-1}) \cdots \cap \mathrm{Ker}(\pi_{2j-\nu+1})$: then $F^\bullet$ is independent of the choice of the π_i's;
(4) $F^1A^j(X) = A^j(X)_{hom}$.

For a curve the existence of such a filtration is trivial. Murre [94, 95] has shown that it exists if X is either surface or the product of a surface and a curve. For abelian varieties the existence of such a decomposition follows from results of Shermenev, Denninger–Murre and Künnemann [86]. More generally it exists for any product of curves (see Corollary 3.18 below).

3.5 Finite dimensional motives. In this section we will briefly compare the above definitions and conjectures with the notion of *finite dimensionality* of a motive, as defined by S.I. Kimura in [84] (see also [69, §1]). In order to do so, we will need the following result proved by Jannsen [81, 5.3]:

Lemma 3.14. *Assume X is a smooth projective variety of dimension d such that $A^d(X\times X)_{hom}$ is a nilpotent ideal of $A^d(X\times X)$. Then condition (1) in Conjecture 3.13 holds for X.*

Let $\mathcal{C}$ be a pseudoabelian, $\mathbb{Q}$–linear, tensor category and let X be an object in $\mathcal{C}$. Let Σ_n be the symmetric group of order n. Any $\sigma \in \Sigma_n$ defines a map $\sigma : (x_1, ..., x_n) \to (x_{\sigma(1)}, ..., x_{\sigma(n)})$ on the n–fold tensor product X^n of X by itself. There is a one–to–one correspondence between all irreducible representations of the group Σ_n over $\mathbb{Q}$ and all partitions of the integer n (see [62]). Let V_λ be the irreducible representation corresponding to a partition λ of n and let χ_λ be the character of the representation V_λ. Let

$$d_\lambda = \frac{\dim(V_\lambda)}{n!} \sum_{\sigma\in\Sigma_n} \chi_\lambda(\sigma)\cdot\Gamma_\sigma .$$

where Γ_σ is the correspondence associated to σ. Then $\{d_\lambda\}$ is a set of pairwise orthogonal idempotents in $Hom_{\mathcal{C}}(X^n, X^n)$ such that $\sum_\lambda d_\lambda = \Delta_{X^n}$. The category $\mathcal{C}$ being pseudoabelian, they give a decomposition of X^n. The n–th symmetric product $S^n X$ of X is then defined to be the image $Im(d_\lambda)$ where λ corresponds to the partition (n), and the n–th exterior power $\wedge^n X$ is $Im(d_\lambda)$ where λ corresponds to the partition $(1,\ldots,1)$.

The object X in $\mathcal{C}$ is said to be *evenly (resp. oddly) finite dimensional* if $\wedge^n X = 0$ (resp. $S^n X = 0$) for some n. An object X is *finite dimensional* if it can be decomposed into a direct sum $X_+ \oplus X_-$ where X_+ is evenly finite dimensional and X_- is oddly finite dimensional (see [84, 69]).

Remark 3.15. (a) If two motives are finite dimensional so is their direct sum and their tensor product. If N is a subobject of a finite dimensional object M then N is finite dimensional. Similarly if N is a quotient object of a finite dimensional object M it is finite dimensional. In particular if $f : X \to Y$ is a proper surjective morphism of smooth projective variety, and $M(X)$ is finite dimensional then $M(Y)$ is also finite dimensional.

(b) The dual motive M^* of M is finite dimensional if and only if M is finite dimensional.

(c) The motive of a smooth projective curve is finite dimensional, hence the motive of a variety X which is the quotient of a product $C_1 \times \cdots \times C_n$ of curves under the action of a finite group G acting freely on $C_1\times\cdots\times C_n$ is finite dimensional (see [69]).

(d) Finite dimensionality is a birational invariant for surfaces [69, Theorem 2.8].

One has the following result (see [84, 7.2]):

Theorem 3.16. *Let $M \in \mathcal{M}_{rat}$ be finite dimensional and let $f \in \mathrm{End}_{\mathcal{M}_{rat}}(M)$ be homologically trivial, i.e f induces the 0 map on $H^*(M)$. Then f is nilpotent in* $\mathrm{End}_{\mathcal{M}_{rat}}(M)$.

As a consequence one has the following corollaries:

Corollary 3.17. *Let X be a smooth projective variety and let $M(X)$ be a finite dimensional motive. Then condition (1) in Conjecture 3.13 holds for X.*

Corollary 3.18. *Let X be a smooth projective variety of dimension d which is the quotient of a product $C_1 \times \cdots \times C_n$ of curves under the action of a finite group G acting freely on $C_1 \times \cdots \times C_n$. Then condition (1) in Conjecture 3.13 holds for X.*

The proof of Corollary 3.17 follows by Theorem 3.16 and Lemma 3.14, whereas the proof of Corollary 3.18 follows by Remark 3.15, Theorem 3.16 and Lemma 3.14.

The following recent result, which appears in [4, 9.2.2], relates Conjecture 3.11 with the finite dimensionality of the motive $M(X)$, where X is any smooth, projective variety:

Theorem 3.19. *If Conjecture 3.11 holds for smooth projective varieties over $\mathbb{C}$ and numerical equivalence equals homological equivalence, then, for any smooth projective variety X, the motive $M(X)$ is finite dimensional.*

3.6 Bloch's conjecture. We want to relate now the definitions and conjectures of Sect. 3.4 with the so–called Bloch's conjecture for surfaces, which, as we said in Sect. 3.3, is related with some of Severi's considerations on series of equivalence.

Let X be any smooth, complex variety. Notice the short exact sequence of abelian groups

$$0 \to T(X) \to CH_0^0(X) \to \mathrm{Alb}(X) \to 0 \tag{3.5}$$

where $a : CH_0^0(X) \to \mathrm{Alb}(X)$ is the Albanese map and $T(X) := \mathrm{Ker}(a)$.

One can define a filtration of the Chow group of 0–dimensional cycles of X in the following way:

$$\begin{aligned} F^1 CH_0(X) &= CH_0^0(X) = \mathrm{Ker}\{\deg : CH_0(X) \to \mathbb{Z}\} \\ F^2 CH_0(X) &:= T(X) = \mathrm{Ker}\{a : F^1 CH_0(X) \to \mathrm{Alb}(X)\} \end{aligned}$$

and the associated graded group is:

$$Gr_F^* CH_0(X) = \mathbb{Z} \oplus \mathrm{Alb}(X) \oplus T(X).$$

We notice the following result of Roitmann [108, 109] we quoted already in Sect. 3.3:

Theorem 3.20. *If X is a smooth, projective surface, then $T(X)$ is a uniquely divisible group and moreover the Albanese map induces an isomorphism*

$$\mathrm{Tors}(CH_0^0(X)) \simeq \mathrm{Tors}(\mathrm{Alb}(X)).$$

If X is a smooth projective surface, by results of Murre [94], X has a Chow–Künneth decomposition whose associated filtration is the one we introduced above, with $\mathbb{Q}$ coefficients. Indeed by the properties of the Chow–Künneth decomposition one has:

$$F^0A^i(X) = A^i(X), \quad F^{i+1}A^i(X) = 0,$$

$$F^1A^i(X) = A^i(X)_{hom} = \begin{cases} \mathrm{Ker}(\pi_2) = A^1(X)_{num}, \text{ if } & i=1 \\ \mathrm{Ker}(\pi_4) = A^2(X)_{num} := A^2_0(X), \text{ if } & i=2 \end{cases}$$

Furthermore, by tensoring the sequence 3.5 with $\mathbb{Q}$ and by taking into account Theorem 3.20, we have the exact sequence

$$0 \to T(X) \to A^2_0(X) \to \mathrm{Alb}(X)_{\mathbb{Q}} \to 0 \tag{3.6}$$

where, of course $\mathrm{Alb}(X)_{\mathbb{Q}} := \mathrm{Alb}(X) \otimes \mathbb{Q}$. Hence

$$F^2A^2(X) = \mathrm{Ker}(\pi_3 | F^1A^2(X)) = T(X).$$

and the graded group $Gr^*_F A^2(X)$ associated to the filtration above is:

$$Gr^*_F A^2(X) = \mathbb{Q} \oplus \mathrm{Alb}(X)_{\mathbb{Q}} \oplus T(X). \tag{3.7}$$

Remark 3.21. (a) The filtration $F^*CH_0(X)$ we defined in general for any smooth, complex variety X can be considered as a *truncated* Chow–Künneth filtration.

(b) It is worth noticing that, if X is a smooth surface, then $A^2(X) \simeq Gr^*_F A^2(X)$. Indeed of course $A^2(X) \simeq \mathbb{Q} \oplus A^2_0(X)$. On the other hand, the exact sequence 3.6 splits (see [90], Cor. 4.3, p. 16, Cor. 7.3, p. 93 and Thm. 7.1, p. 92).

Now let Y be any smooth projective variety and let Γ be a cycle of codimension 2 on $Y \times X$, i.e $[\Gamma] \in A^2(Y \times X) = Hom_{\mathcal{M}_{rat}}(M(Y), M(X))$. Then Γ, being a correspondence, determines a map:

$$\Gamma_* : CH_0(Y) \to CH_0(X)$$

which, of course preserves the above described filtrations (see 3.7). Hence it defines a map:

$$Gr\Gamma_* : Gr^*_F CH_0(Y) \to Gr^*_F CH_0(X)$$

(see Remark 3.12), which reads

$$Gr\Gamma_* : \mathbb{Z} \oplus \mathrm{Alb}(Y) \oplus T(Y) \to \mathbb{Z} \oplus \mathrm{Alb}(X) \oplus T(X).$$

Remark 3.22. Suppose both X and Y are surfaces. Then, by working with $\mathbb{Q}$ coefficients, as we saw in Remark 3.21, we have two maps:

$$\Gamma_* : A^2(Y) \to A^2(X)$$

and

$$Gr\Gamma_* : A^2(Y) \simeq Gr_F^* A^2(Y) \to A^2(X) \simeq Gr_F^* A^2(X)$$

It is important to notice that they are by no means the same map. Indeed the former is not graded, whereas the latter, defined as in Remark 3.12, is. The reader may verify that the action of $Gr\Gamma_*$ on the graded pieces of $A^2(Y) \simeq Gr_F^* A^2(Y)$ consists in first applying Γ_* and then projecting on the appropriate graded pieces of $A^2(X) \simeq Gr_F^* A^2(X)$. We will later explain this with an example (see 3.29).

In [12] Bloch formulates the following:

Conjecture 3.23. (Bloch's Conjecture for Surfaces) In the above setting, the map $Gr\Gamma_*$ depends only upon the cohomology class of Γ in $H^4(Y \times X, \mathbb{C})$.

Remark 3.24. (a) As observed in Sect. 3.3, Bloch's Conjecture 3.23 is related with Severi's Claim 3.10. Actually 3.10 can be expressed, in the above setting, by asserting that *the map Γ_* depends only upon the cohomology class of Γ in $H^4(Y \times X, \mathbb{C})$.* The subtle replacement of Γ_* with $Gr\Gamma_*$ makes the difference between Severi's false claim and Bloch's conjecture.

(b) By the results of [94], Conjecture 3.11 is equivalent, for a surface, to condition (d') (see Remark 3.12), which, in turn, is clearly equivalent to Conjecture 3.23.

The motivic content of Bloch's conjecture is indicated by the following:

Theorem 3.25. *Let X be a smooth projective surface over $\mathbb{C}$. If the motive $M(X)$ is finite dimensional, then Conjecture 3.23 holds for X.*

Proof. Suppose that $M(X)$ is finite dimensional. In order to show that X satisfies Conjecture 3.23 it suffices to show that the filtration $F^\bullet$ defined above satisfies Conjecture 3.11 with condition (d) replaced by the equivalent condition (d') of Remark 3.12 (see Remark 3.24, (b)). Now Conjecture 3.11 is equivalent to Murre's Conjecture 3.13, so we are left to show that the filtration $F^\bullet$ satisfies condition (3) in Conjecture 3.13, i.e., that it is independent of the choices of the projectors π_i. Let $\tilde{\pi}_i$ be another set of orthogonal idempotents lifting the Künneth components of the diagonal Δ_X. Then the endomorphism $f = \pi_i - \tilde{\pi}_i$ of $M(X)$, being homologically trivial, is nilpotent by Theorem 3.16, i.e., there exists an integer n such that $f^n = 0$. This implies $f^n(x) = 0$ for all $x \in A^i(X)$. On the other hand, π_i and $\tilde{\pi}_i$ being idempotents, $\pi_i^n(x) = 0$ implies $\pi_i(x) = 0$, and similarly for $\tilde{\pi}_i$. This shows that $\mathrm{Ker}(\pi_i) = \mathrm{Ker}(\tilde{\pi}_i)$ for all i. From the definition of F^ν it follows that it is independent of the choice of the π_i's. □

The following version of Bloch's conjecture can be found in [136], 1.7:

Conjecture 3.26. Let X, Y be smooth, projective surfaces over $\mathbb{C}$ and let Γ be a codimension 2 cycle on $Y \times X$ whose (2, 2)–Künneth component in $H^2(Y, \mathbb{C}) \otimes H^2(X, \mathbb{C})$ lies in $NS(Y)_{\mathbb{C}} \otimes NS(X)_{\mathbb{C}}$. Then the action of $Gr\Gamma_*$ factors through the Albanese variety of Y, i.e., it vanishes on $T(Y)$.

Remark 3.27. Conjecture 3.26 can be considered as the *right formulation* of Severi's Claim 3.7. Notice indeed that, in Severi's terminology, the correspondence Γ in the statement of Conjecture 3.26 has algebraic circulation. If, in addition, it has also null–circulation then the induced map $\mathrm{Alb}(Y) \to \mathrm{Alb}(X)$ is zero. Therefore, if Conjecture 3.26 holds, then $Gr\Gamma_*$ vanishes on $CH_0^0(Y)$. By contrast, Severi's Claim 3.7 would have required Γ_* to act as zero on $CH_0^0(Y)$ (see Remark 3.24, (a)).

The proof of the following proposition is essentially a transposition of Severi's plausibility argument presented in Sect. 3.3 to show that Claim 3.10 implies Claim 3.7.

Proposition 3.28. *Conjecture 3.23 implies Conjecture 3.26.*

Proof. As usual, let $\Gamma_{2,2}$ be the $(2,2)$–Künneth component of Γ in $H^4(X \times Y, \mathbb{C})$. Let e_j for $j = 1, \ldots, \rho_1$ and f_i for $i = 1, \ldots, \rho_2$ be bases of $NS(X)$ and $NS(Y)$ respectively. Of course we may assume the e_j's and f_i's are represented by classes of divisors, whose classes in homology we will denote by $[e_j]$, $[f_i]$. Then $\Gamma_{2,2} = \sum_{i,j} q_{i,j}[f_i] \otimes [e_j]$ acts as zero on $CH_0(Y)$ because any 0–cycle can be moved away from a finite number of divisors. On the other hand it is easy to see that, if Conjecture 3.23 holds, then the other Künneth components of Γ act trivially on $T(Y)$ (see [12, Lemma (1.9)]). Then, again by Conjecture 3.23, Conjecture 3.26 holds.

Example 3.29. It is perhaps worth at this point to go back to Example 3.9 and explain the difference between Severi's claims and Bloch's conjecture.

In Example 3.9, we have $Y = C(2)$ and we suppose that X has $q = 0$ and $p_g > 0$. Notice that $CH_0^0(Y) = \mathrm{Alb}(Y) \oplus T(Y) \simeq J(C) \oplus T(Y)$ and $CH_0^0(X) = T(X)$. Let $\Gamma \subset Y \times X$ be the related correspondence. Notice that $\Gamma_{1,3}$ and $\Gamma_{3,1}$ vanish, and $\Gamma_{2,2}$ sits in $NS(Y) \otimes NS(X)$.

Assume now Bloch's conjecture. By altering Γ with constant correspondences, we may assume that $\Gamma_{0,4}$ and $\Gamma_{4,0}$ are zero. By the argument in the proof of Proposition 3.28, the action of $\Gamma_{2,2*}$ on $CH_0^0(Y)$ is zero. Hence Bloch's conjecture predicts that $Gr\Gamma_*$ is the zero map. This is indeed the case. In fact $\Gamma \subset Y \times C \subset Y \times X$. Thus Γ can be also seen as a correspondence between Y and C. As such it determines a map $\Gamma'_* : CH_0^0(Y) \simeq J(C) \oplus T(Y) \to CH_0^0(C) \simeq J(C)$, which is clearly the projection onto the first factor. Since Γ_* factors through Γ'_*, we see that Γ_* is zero on $T(Y)$, and therefore $Gr\Gamma_*$ is the zero map.

On the other hand, Severi's claims would imply that $\Gamma_* : CH_0^0(Y) \to CH_0^0(X)$ is the zero map. This would in turn imply that $f : C(2) \to X(2)$ is a series of equivalence, which leads to a contradiction as we saw in Example 3.9. Indeed, if C is a general curve on X, according to Proposition 3.4, the induced map $\Gamma_* : J(C) \to T(X)$ is non trivial.

The following sub–conjecture of Conjecture 3.23 is also usually referred to as Bloch's Conjecture:

Conjecture 3.30. Let X be a smooth, projective surface over $\mathbb{C}$ with $p_g = 0$. Then $T(X) = 0$. In particular, if X is of general type, then $CH_0^0(X) = 0$.

One has the following implication:

Proposition 3.31. *Conjecture 3.26 implies Conjecture 3.30.*

Proof. From $p_g = 0$ we get $H^2(X, \mathbb{C}) = NS(X)_{\mathbb{C}}$, therefore every correspondence $\Gamma \in A^2(X \times X)$ satisfies the hypotheses of Conjecture 3.26. In particular this applies to the identity $[\Delta_X]$. Thus, if Conjecture 3.26 holds, then $[\Delta_X]_*$ acts as 0 on $T(X)$, hence $T(X) = 0$. □

Conjecture 3.30 has been proved for all surfaces with Kodaira dimension less than 2 in [13] (see also [69], Corollary 2.12, for a motivic proof). Conjecture 3.30 is still open for surfaces of general type, though it has been verified for several known classes of these.

3.7 Mumford's theorem and finite dimensionality of motives. As indicated in Sect. 3.1, for a smooth projective variety X of dimension d over $\mathbb{C}$, Mumford introduced the following notion of *finite dimensionality* of the Chow group $CH_0^0(X)$ of 0–cycles of degree 0, modulo rational equivalence (see [98, 80, 1.6], 1.6):

Definition 3.32. *$CH_0^0(X)$ is finite dimensional if one of the following equivalent conditions are satisfied:*

(1) There exists an integer n such that the natural map $\beta_n : X(n) \times X(n) \to CH_0^0(X)$, which sends (A, B) to the class of $A - B$, is surjective;
(2) There exist a smooth projective curve C and a morphism $f : C \to X$ such that the induced map:

$$f_* : \mathrm{Pic}^0(C) \to CH_0^0(X)$$

is surjective;
(3) there exists an integer n such that any 0–cycle Z on X of degree $\geq n$ is rationally equivalent to an effective cycle $\tilde{Z}$;
(4) The Abel–Jacobi map $a : CH_0^0(X) \to \mathrm{Alb}(X)$ is an isomorphism, i.e the Albanese kernel $T(X) = \mathrm{Ker}(a)$ vanishes.

As we recalled already, in 1969 Mumford [98] showed that if X is a surface with $p_g > 0$ then $CH_0^0(X)$ is not finite dimensional.

Therefore if $X = C \times D$ where C and D are curves of genus ≥ 1, then the Chow group $CH_0^0(X)$ is not finite dimensional while, as we saw in Remark 3.15, (c), the motive $M(X)$ is finite dimensional.

However the following result shows that, for surfaces X with $p_g = 0$, the two different notions of finite dimensionality coincide.

Proposition 3.33. *Let X be smooth projective surface with $p_g = 0$. Then the following conditions are equivalent:*

(i) $M(X)$ is finite dimensional;
(ii) $CH_0^0(X)$ is finite dimensional;
(iii) Conjecture 3.30 holds for X.

Proof. Parts (ii) and (iii) are equivalent by (4) of Definition 3.32. Part (i) implies part (ii) by Theorem 3.19 and Proposition 3.31. Part (iii) implies part (i) by proposition 10 in [70]. □

References

[1] N.H. Abel, *Oeuvres complètes de Niels Henrik Abel, Nouvelle édition publiée aux frais de l'état norvégien par MM. L. Sylow et S. Lie, Volumes I and II*, Christiania (1881).

[2] G. Albanese, *Intorno ad alcuni concetti e teoremi fondamentali sui sistemi agebrici di curve di una superficie algebrica*, Ann. Mat. Pura e Appl., (III) **24**, 159–233 (1915).

[3] G. Albanese, *Condizione per la razionalità della varietà delle coppie di punti di due superficie algebriche distinte o coincidenti*, Rend. R. Accad. Lincei, (V) **33**–II, 73–76 (1924).

[4] Y. André, B. Kahn, *Nilpotence, radicaux et structures monoidales*, preprint, posted on arXiv:math.CT/0203273 v2, (2002).

[5] A. Andreotti, *Sopra le varietà di Picard d'una superficie algebrica e sulla classificazione delle superficie irregolari*, Rend. R. Accad. Lincei, (VIII) **10**–V, 380–85 (1951).

[6] E. Arbarello, M. Cornalba, Ph. Griffiths, J. Harris, *Geometry of algebraic curves, vol. I*, Grundlehren der Math. Wiss., **267**, Springer–Verlag, 1985.

[7] A. Armenante, G. Jung, *Relazione sopra tre corsi paralleli dei professori Brioschi, Cremona e Casorati sopra le funzioni ellittiche ed abeliane*, Giornale di Matematiche, **7**, 224–234, (1869).

[8] A. Armenante, G. Jung, *Sulle trasformazioni birazionali o univoche e sulle curve normali e subnormali del genere p*, Giornale di Matematiche, **7**, 235–245, (1869).

[9] E.S. Allen, *Su alcuni caratteri di una serie algebrica, e la formula di De Jonquières per serie qualsiasi*, Rend. Circ. Mat. Palermo, **37**, 345–370, (1919).

[10] F. Bardelli, *On the origins of the concept of irregularity of an algebraic surface*, Proceedings of the Conference *1860–1940, Algebra e Geometria: il contributo italiano*, Cortona 1992, Rend. Circ. Mat. Palermo, II **36**, 11–26 (1994).

[11] E. Betti, *Sopra le funzioni algebriche di una variabile complessa*, Annali delle Università Toscane, (1862).

[12] S. Bloch, *Lectures on Algebraic Cycles*, Duke Univ. Math. Ser. Vol IV, Durham, NC, 1980.

[13] S. Bloch, A. Kas, D. Lieberman, *Zero cycles on surfaces with $p_g = 0$*, Compositio Math. 33, 135–145, (1976)

[14] U. Bottazzini, *Và pensiero*, Il Mulino, Bologna, 1994.

[15] U. Bottazzini, A. Conte, P. Gario, *Riposte Armonie, Lettere di Federigo Enriques a Guido Castelnuovo*, Bollati Boringhieri, Torino, 1996.

[16] A. Brigaglia, C. Ciliberto, *Italian algebraic geometry between the two world wars*, Queen's Papers in Pure and Applied Mathematics, Kingston, 1995.

[17] A. Brill, M. Noether, *Über die Algebraische Functionen und ihre Anwendungen in der Geometrie*, Math. Ann. **7**, 269–310 (1874).

[18] L. Carbone, R. Gatto, F. Palladino, *L'epistolario Cremona – Genocchi*, Olschki, Firenze, 2001.

[19] F. Casorati, *Teorica delle funzioni di variabile complessa*, Pavia, 1868.

[20] F. Casorati, *Le relazioni fondamentali tra i moduli di periodicità degli integrali abeliani di prima specie*, Ann. Mat. Pura ed Applicata, (2), **3**, 1–27 (1870).

[21] F. Casorati, L. Cremona, *Intorno al numero de' moduli dell'equazioni e delle curve algebriche di dato genere. Osservazioni.*, Rend. R. Ist. Lombardo Sci. Lett. e Arti, (2), **2**, 620–625 (1869).
[22] G. Castelnuovo, *Osservazioni intorno alla geometria sopra una superficie algebrica, Note I e II*, Rend. R. Ist. Lombardo Sci. Lett. e Arti, Ser. II **24**, 127–137, 307–318 (1891).
[23] G. Castelnuovo, *Sulla razionalità delle involuzioni piane*, Math. Ann., **44**, 125–155 (1894).
[24] G. Castelnuovo, *Alcune proprietà fondamentali dei sistemi lineari di curve tracciati sopra una superficie algebrica*, Ann. Mat. Pura e Appl., (2) **25**, 235–318 (1897).
[25] G. Castelnuovo, *Sugli integrali semplici appartenenti ad una superficie irregolare*, Rend. R. Accad. Lincei, V **14**, 545–556, 593–598, 655–663 (1905).
[26] G. Castelnuovo, *Sulle serie algebriche di gruppi di punti appartenenti ad una curva algebrica*, Rend. R. Accad. Lincei, V **15**, 337–344 (1906).
[27] G. Castelnuovo, *Sulle funzioni abeliane*, Rend. R. Accad. Lincei, **30**, 50–55, 99–103, 195–200, 355–361 (1921).
[28] G. Castelnuovo, *La geometria algebrica e la scuola italiana*, Atti Congresso Internazionale dei Matematici, Bologna, I, 191–201, 1928.
[29] A. Cayley, *On the transformation of plane curves*, Proc. London Math. Soc., **1**, III, 3–11 (1866).
[30] A. Cayley, *On the deficiency of certain surfaces*, Math. Ann., **3**, 526–529 (1871).
[31] A. Cayley, *Note on the theory of invariants*, Math. Ann., **3**, 268–271 (1871).
[32] O. Chisini, *Il teorema d'Abel e il principio di corrispondenza nel loro aspetto topologico*, Rend. R. Ist. Lombardo Sci. Lett. e Arti, 552–569 (1921).
[33] C. Ciliberto, F. Ghione, *Serie algebriche di divisori su una curva e su una superfcie algebrica*, Ann. Mat. Pura e Appl., IV **136**, 329–353 (1984).
[34] C. Ciliberto, E. Sernesi, *Some aspects of the scientific activity of Michele de Franchis*, Rend. Circ. Mat. Palermo, Suppl. II **27**, 3–33 (1991).
[35] A. Clebsch, *Über die Anwendung der Abel'schen Functionen in Geometrie*, J. für reine und angew. Mathematik, **63**, 189–243 (1864).
[36] A. Clebsch, *Über die Steinersche Fläche*, J. für die reine und angew. Mathematik, **67**, 1–22 (1866).
[37] A. Clebsch, *Intorno alla rappresentazione di superficie algebriche sopra un piano*, Rend. R. Ist. Lombardo Sci. Lett. e Arti, **5**, 794–806 (1868).
[38] A. Clebsch, *Sur les surfaces algébriques*, C. R. Acad. Sc. Paris, **67**, 1238–1239 (1868).
[39] A. Clebsch, P. Gordan, *Theorie des Abelschen Funktionen*, Teubner, Leipzig, 1866.
[40] H. Clemens, *Homological equivalence, modulo algebraic equivalence, is not finitely generated*, Publ. Math. IHES, **58**, 19–38 (1983).
[41] A. Comessatti, *Sulle serie algebriche semplicemente infinite di gruppi di punti appartenenti ad una curva algebrica*, Rend. Circ. Mat. Palermo, **36**, 35–57 (1913).
[42] L. Cremona, *Intorno a un teorema di Abel*, Ann. Sci. Mat. e Fisiche, **7**, 99–105 (1856).
[43] L. Cremona, *Rappresentazione della superficie di Steiner e delle superficie gobbe di 3° grado sopra un piano*, Rend. R. Ist. Lombardo Sci. Lett. e Arti, **4**, 15–23 (1867).
[44] L. Cremona, *Preliminari di una teoria geometrica delle superficie*, Mem. dell'Acc. delle Scienze di Bologna, **6**, 91–136; **7**, 29–78 (1867).
[45] L. Cremona, *Sull'opera del prof. Casorati "Teoria delle funzioni di variabili complesse" (vol. 1 Relazione)*, Rend. R. Ist. Lombardo Sci. Lett. e Arti, (2),**1**, 420–424 (1868).
[46] L. Cremona, *Sulla trasformazione delle curve iperellittiche*, Rend. R. Ist. Lombardo Sci. Lett. e Arti, (2), **2**, 566–571 (1869).

[47] L. Cremona, *Sugli integrali a differenziale algebrico*, Mem. dell'Acc. delle Scienze di Bologna, (2), **10**, 3–33 (1870).
[48] L. Cremona, *Observations gèomètriques a propos de la note di Mr. Brioschi "Sur les tangents doubles d'une curbe du 4 ordre avec un point double"*, Math. Ann., 4, 99–102 (1870).
[49] L. Cremona, *Über Abbildung algebraischer Flächen*, Math. Ann., **4**, 213–220 (1870).
[50] M. de Franchis, *Sulla varietà ∞^2 delle coppie di punti di due curve o di una curva algebrica*, Rend. Circ. Mat. Palermo, **17**, 104–121 (1903).
[51] M. Deuring, *Arithmetische Theorie Multiplikationenringe elliptischer Funktionenkorper, I*, J. reine und angew. Math., **177**, 161–191 (1937).
[52] M. Deuring, *Arithmetische Theorie Multiplikationenringe elliptischer Funktionenkorper, II*, J. reine und angew. Math., **183**, 25–36 (1941).
[53] M. Deuring, *Die Typen der Multiplikationenringe elliptischer Funktionenkorper*, Abh. Math. Sem. Hamburg, **14**, 191–272 (1941).
[54] F. Enriques, *Ricerche di geometria sulle superficie algebriche*, Mem. Accad. Sci. Torino, **44**, 171–252 (1893).
[55] F. Enriques, *Introduzione alla geometria sopra le superficie algebriche*, Mem. Soc. It. Sci., **10** (3), 1–81 (1896).
[56] F. Enriques, *Una proprietà delle serie continue di curve appartenenti ad una superficie algebrica*, Rend. Circ. Mat. Palermo, **13**, 95–98 (1899).
[57] F. Enriques, *Sur les surfaces algèbriques admettant des intègrales de diffèrentielles totale de première espèce*, Ann. de Toulouse, **3** (2), 57–64 (1901).
[58] F. Enriques, *Sulla proprietà caratteristica delle superficie algebriche irregolari*, Rend. Accad. Sci. Bologna, 5–13 (1905).
[59] F. Enriques, *Le superficie algebriche*, Zanichelli, Bologna, 1949.
[60] F. Enriques, O. Chisini, *Lezioni sulla teoria geometrica delle equazioni e delle funzioni algebriche*, 4 vol., Zanichelli, Bologna, 1915–1934.
[61] W. Fulton, *Intersection Theory*, Springer–Verlag, Ergebnisse der Math., 3 **2**, 1984.
[62] W. Fulton, J. Harris, *Representation Theory*, Graduate Texts in Math., Springer–Verlag, 1991.
[63] F. Ghione, *La disuguaglianza di Castelnuovo–Severi in teoria dei numeri*, Atti del Convegno *La teoria dei numeri nella Cina antica e di oggi*, Ferrara, 1980.
[64] L. Giacardi, *La corrispondenza tra Jules Hoüel e Luigi Cremona*, in A. Milan Gasca (ed.), *La corrispondenza di Luigi Cremona*, v. I, Roma, 77–88, 1992.
[65] L. Giacardi, *I quaderni di Corrado Segre*, CDROM, Università di Torino, Dipartimento di Matematica, 2002.
[66] P. Griffiths, *On the periods of certain rational integrals I, II*, Ann. of Math., **90**, 460–541 (1969).
[67] P. Griffiths, J. Harris, *Principles of Algebraic Geometry*, Wiley Intersc. (1978).
[68] A. Grothendieck, *Standard conjectures on algebraic cycles*, In Agebraic Geometry, Bombay, 1968, Oxford Univ. Press, (1969), 193–199.
[69] V. Guletskii, C. Pedrini, *The Chow motive of the Godeaux Surface*, Proceedings of the Conference on Algebraic Geometry in memory of P. Francia, W. De Gruyter, Berlin-New–York, 179–196 (2002).
[70] V. Guletskii, C. Pedrini, *Finite dimensionality of motives and the Conjectures of Bloch and Beilinson*, preprint 2002.
[71] R. Hartshorne, *Algebraic Geometry*, Graduate Texts in Math., Springer Verlag, 1977.
[72] H. Hasse, *Abstracte Begründung der komplexen Multiplikation und Riemannsche Vermutung in Funktionenkörpern*, Abh. Math. Sem. Hamburg, **10**, 325–348 (1934).

[73] H. Hasse, *Zur Theorie der abstrakten elliptischen Funktionenkörper, I, II, III*, J. reine und angew. Math., **175**, 55–62, 69–88, 193–208 (1936).
[74] C. Houzel, *Fonctions elliptiques et intégrales abéliennes*, in J. Dieudonné (ed.), *Abrégé d'histoire des mathématiques*, vol. II, 1–113 (1978).
[75] G. Humbert, *Sur le théorème d'Abel et quelques-unes de ses applications à la géométrie*, J. de Math. Pures et Appl., Ser. IV **5**, 81–134 (1889).
[76] G. Humbert, *Sur une propriété d'une classe de surfaces algébriques*, C. R. Acad. Sc. Paris, 361–363 (1893).
[77] G. Humbert, *Sur le fonctions abéliennes singulières, I, II, III*, J. de Math. Pures et Appl., Ser. IV **5**, 233–350 (1899); **6**, 279–3860 (1900); **7**, 97–123 (1901).
[78] A. Hurwitz, *Über algebraische Correspondenzen und das verallgemeinerte Correspondenzprinzip*, Math. Ann., **28**, 561–585 (1886).
[79] J. Igusa, *On some problems in abstract algebraic geometry*, Proc. Nat. Acad. Sci. USA, bf 41, 964–967(1955).
[80] U. Jannsen, *Motives, numerical equivalence and semi–simplicity*, Inventiones Math., **107**, 447–452 (1992).
[81] U. Jannsen, *Motivic Sheaves and Filtrations on Chow groups*, Proceedings of Symposia in Pure Mathematics, **55**, Part. I, 245–302 (1994).
[82] U. Jannsen, *Equivalence relations on algebraic cycles*, The Arithmetic and Geometry of Algebraic Cycles, NATO Sc. Series, Kluwer Acad. Press., **548**, 225–260 (2000).
[83] E. Kani, *On Castelnuovo's equivalence defect*, J. reine und angew. Math., 24–70 (1984).
[84] S.I. Kimura, *Chow groups can be finite dimensional, in some sense* (preprint), to appear in Journal of Algebraic Geometry.
[85] S. Kleiman, *Algebraic Cycles and the Weyl conjectures*, Dix Exposés sur la cohomologie des schémas, North–Holland Amsterdam, 1359–386, 1968.
[86] K. Künnemann, *On the Chow motive of an abelian Scheme*, Proceedings of Symposia in Pure Mathematics, **55**, Part. 1, 189–205 (1994).
[87] S. Lang, Comments on non–references in Weil's work, DMV–Mitteilungen, 49–56 (2002)
[88] H. Lange, Ch. Birkenhake, *Complex Abelian Varieties*, Grundlehren der Math., Springer–Verlag, 1992.
[89] G. Loria, *Luigi Cremona et son oeuvre mathématique*, Bibliotheca Mathematica, 125–195 (1904).
[90] S. MacLane, *Homology*, Springer–Verlag, New York, 1963.
[91] A. Mattuck, *On the symmetric product of a rational surface*, Proc. AMS, **21**, 683–688 (1969).
[92] M. Menghini, *Lettere di Sophus Lie a Luigi Cremona*, in M. Menghini (ed.), *La corrispondenza di Luigi Cremona*, vol. II, 89–105, 1994.
[93] M. Menghini, *Lettere di Alfred Clebsch a Luigi Cremona (1863 – 1871)*, in M. Menghini (ed.), *La corrispondenza di Luigi Cremona*, vol. III, Palermo, 1996.
[94] J. Murre, *On the motive of an algebraic surface*, J. Reine und Angw. Math., **409**, 190–204 (1990).
[95] J. Murre, *On a conjectural filtration on the Chow groups of an algebraic variety, Part I and II*, Indagationes Mathem. N. S., **4**, 177–201 (1993).
[96] J. Murre, *Algebaric cycles, and algebraic aspects of cohomology and K–theory*, Algebraic Cycles and Hodge Theory, L. N. M. **1594**, Springer–Verlag, 93–152 (1993).
[97] D. Mumford, *Lectures on Curves on an Algebraic Surface*, Ann. of Math. Studies **59**, Princeton Univ. Press, Princeton, 1966.
[98] D. Mumford, *Rational equivalence of 0–cycles on surfaces*, J. Math. Kyoto Univ. **9**, 195–204 (1969).

[99] D. Mumford, *Abelian Varieties*, Oxford Univ. Press, 1970.

[100] M. Noether, *Sulle curve multiple di superficie algebriche*, Ann. Mat. Pura e Appl., **5**, ser. II, 163–177 (1871).

[101] M. Noether, *Zur Theorie des eindeutigen Entsprechens algebraischer Gebilde von beliebig vielen Dimensionen, I–II*, Math. Ann., **2**, 293–316 (1870), Math. Ann., **8**, 495–533 (1875).

[102] L. Nurzia, *Lettere di Arthur Cayley a Luigi Cremona*, A. Milan Gasca (ed.), *La corrispondenza di Luigi Cremona*, vol. I, Roma, 107–128, 1992.

[103] É. Picard *Sur les intégrales de différentielles totales algébriques de première espèce*, J. Math. Pures et Appl., IV **1**, 281–346 (1885).

[104] É. Picard, G. Simart, *Théorie des fonctions algébriques de deux variables indépendantes*, vol. I–II, Paris, 1897–1906.

[105] H. Poincaré, *Sur les courbes tracées sur les surfaces algébriques*, Ann. École. Norm. Sup. **27**, 3, 55–108 (1910).

[106] B. Riemann, *Gesammelte Werke, Nachträge*, 1902.

[107] B. Riemann, *Theorie der Abel'schen functionen*, in Math. Werke, Teubener, Leipzig, 1876.

[108] A. Roitmann, *Rational equivalence of 0–cycles*, Math. USSR, Sbornik **18**, 571–688 (1972).

[109] A. Roitmann, *The torsion in the group of 0–cycles modulo rational equivalence*, Ann. of Math., (2) **111**, 553–569 (1980).

[110] C. Rosati, *Sulle corrispondenze algebriche fra i punti di una curva algebrica* (2 notes), Rend. R. Accad. Lincei, Ser. V **22**, 385–390, 431–436 (1913).

[111] M. Rosati, *Sull'equivalenza birazionale delle due varietà di Picard associate ad una varietà superficialmente irregolare*, Rend. Accad. Lincei, 708–715 (1954).

[112] A.J. Scholl, *Classical motives*, Proc. Symposia in Pure Math., **55**, Part 1, 163–187 (1994).

[113] B. Segre, *Intorno ad un teorema di Hodge sulla teoria della base per le curve di una superficie algebrica*, Ann. Mat. Pura e Appl., (4) **16**, 157–163 (1937).

[114] B. Segre, *Un teorema fondamentale della geometria sulle superficie algebriche ed il principio di spezzamento*, Ann. Mat. Pura e Appl., (4) **17**, 107–126 (1938).

[115] F. Severi, *Sulle superficie che rappresentano le coppie di punti di una curva algebrica*, Atti R. Accad. Sci. Torino, **38**, 119–134 (1903).

[116] F. Severi, *Sulle superficie che posseggono integrali di Picard della* 2^a *specie*, Rend. R. Accad. Lincei, 5 **13**, 20–49 (1905).

[117] F. Severi, *Il teorema d'Abel sulle superficie algebriche*, Annali di Mat. 3 **12**, 55–79 (1905).

[118] F. Severi, *Sulla differenza fra i numeri degli integrali di Picard della prima e della seconda specie appartenenti ad una superficie algebrica*, Atti Accad. Sci. Torino, **40**, 254–262 (1905).

[119] F. Severi, *Intorno al teorema d'Abel sulle superficie algebriche ed alla riduzione a forma normale degli integrali di Picard*, Rend. Circ. Mat. Palermo, **21**, 257–282 (1906).

[120] F. Severi, *La base minimale pour la totalité des courbes tracées sur une surface algébrique*, Ann. de l'École Norm. Sup. de Paris, 3, **25**, 449–468 (1908).

[121] F. Severi, *Un teorema di inversione per gli integrali semplici di* I^a *specie appartenenti ad una superficie algebrica*, Atti R. Istituto Veneto Sci. Lettere ed Arti, **72**, 765–772 (1913).

[122] F. Severi, *Sopra alcune proprietà aritmetiche delle corrispondenze fra i punti di una curva algebrica*, Atti R. Accad. Sci. Torino, **48**, 660–674 (1913).

[123] F. Severi, *Sulla teoria degli integrali semplici di* 1^a *specie appartenenti ad una superficie algebrica* (7 notes), Rend. R. Accad. Lincei, 5 **30**, 163–167, 204–208, 231–235, 296–301, 276–280, 328–332, 365–367 (1921).

[124] F. Severi, *La serie canonica e la teoria delle serie principali di gruppi di punti sopra una superficie algebrica*, Comm. Math. Helv., **4**, 268–326 (1932).

[125] F. Severi, *Un nuovo campo di ricerche nella geometria sopra una superficie e sopra una varietà algebrica*, Mem. R. Accad. Italia, (3) **5**, 268–326 (1932).

[126] F. Severi, *Nuovi contributi alla teoria delle serie di equivalenza sulle varietà algebrica*, Mem. R. Accad. Italia, (4) **6**, 71–129 (1933).

[127] F. Severi, *La base per le varietà algebriche di dimensione qualunque contenute in una data e la teoria generale delle corrispondenze fra i punti di due superficie algebriche*, Mem. R. Accad. Italia, (5) **6**, 239–283 (1934).

[128] F. Severi, *Ulteriori sviluppi della teoria delle serie di equivalenza sulle superficie algebriche*, Pontificia Accad. Sci. (6) **6**, 977–1029 (1942).

[129] F. Severi, *Problèmes résolus et problèmes nouveaux dans la théorie des systèmes d'équivalence*, Proceedings of the Intern. Congress of Math., 1954, Amsterdam, **3**, 529–541 (1956).

[130] F. Severi, *Letter to A. Weil of March 29, 1956 (in Italian)*, ETH Archives, no. HS 652:8402.

[131] F. Severi, *Il teorema di Riemann–Roch per curve, superficie e varietà, questioni collegate*, Erg. der Math., Springer–Verlag, Berlin, 1958.

[132] A. Stubhaug, *Niels Herik Abel and his times: Called too soon by flames afar*, Springer–Verlag, Berlin, 2000.

[133] R. Torelli, *Sui sistemi algebrici di curve appartenenti ad una superfcie algebrica*, Atti Accad. Sci. Torino, **42**, 86–99 (1906–07).

[134] R. Torelli, *Sulla linearità delle involuzioni più volte infinite appartenenti ad una curva algebrica*, Atti R. Ist. Veneto, **67**, 831–835 (1908).

[135] R. Torelli, *Sulle serie algebriche semplicemente infinite di gruppi di punti appartenenti a una curva algebrica*, Rend. Circ. Mat. Palermo, **37**, 25–46 (1914).

[136] C. Voisin, *Transcendental methods in the study of algebraic cycles*, in Algebraic Cycles and Hodge theory, L. N. M. **1594**, Springer–Verlag, 153–222 (1993).

[137] V. Volterra, *Il movimento scientifico presente e la nuova società italiana per il progresso delle scienze*, Scientia, 1, II, 225–237 (1907).

[138] A. Weil, *Sur les courbes algébriques et les variétés qui s'en déduisent*, Act. Sci. Ind., Publ. Inst. Math. Univ. Strasbourg, VII, Hermann et C., Paris, 1948.

[139] A. Weil, *Variétés abéliennes et courbes algébriques*, Act. Sci. Ind., Publ. Inst. Math. Univ. Strasbourg, VIII, Hermann et C., Paris, 1948.

[140] A. Weil, Letter to F. Severi of May 9, 1956 (in French), ETH Archives, no. HS 652:8403a.

[141] G. Zappa, *Sull'esistenza di curve algebricamente non isolate, a serie caratteristica non completa, sopra una rigata algebrica*, Acta Pont. Accad. Sci. **7**, 4–8 (1943).

[142] O. Zariski, *Algebraic surfaces*, Second suplemented edition, Springer–Verlag, Ergebnisse der Math., **61**, 1971.

[143] H.G. Zeuthen, *Études géometriques de quelques–unes des propriétés de deux surfaces dont les points se correspondent un–à–un*, Math. Ann., **4**, (1871).

From Abel's Heritage: Transcendental Objects in Algebraic Geometry and Their Algebraization

Fabrizio Catanese

Welcome the Abel Prize and long live the memory of Abel, long live mathematics! [1]

1 Introduction

The Encyclopedic Dictionary of Mathematics (edited in Japan) is not a book devoted to the history of mathematics, but tries instead to briefly introduce the reader to the current topics of mathematical research. By a not only lexicographical coincidence it starts with "Abel, Niels Henrik" as topic 1.

It contains a succinct biography of Abel:

"Niels Henrik Abel (August 5, 1802 – April 6, 1829) ... In 1822, he entered the University of Christiania [today's Oslo] ... died at twenty-six of tuberculosis. His best known works are: the result that algebraic equations of order five or above cannot generally be solved algebraically; the result that [3]*Abelian

The present research, an attempt to treat history and sociology of mathematics and mathematics all at the same time, took place in the framework of the Schwerpunkt "Globale Methode in der komplexen Geometrie", and of the EAGER EEC Project.

[1] Was the greatest enemy of mathematics Alexander the Great, who cut the Gordian knot instead of peacefully writing a book about it?

[2] This topic has in fact been treated quite extensively in the contribution by Ciliberto.

[3] A star next to a theme denotes that a section of the dictionary is devoted to the discussion of the topic.

equations [i.e., with Abelian Galois group] can be solved algebraically; the theory of *binomial series and of *elliptic functions; and the introduction of *Abelian functions. His work in both algebra and analysis, written in a style conducive to easy comprehension, reached the highest level of attainment of his time."

Talking about Abel's heritage entails thus talking about a great part of modern mathematics, as it is shown by the ubiquity of concepts such as Abelian groups, Abelian integrals and functions, Abelian varieties, and of relatives of theirs such as anabelian geometry, nonabelian Hodge theory ...

Writing is certainly a more difficult task than talking, when the time limits force us to plan our way on secure and direct tracks: for this reason we decided that the present text, with the exception of a couple of protracted mathematical discussions, should essentially be the text of our oral exposition at the Abel Bicentennial Conference. Thus its aim is just to lead the reader along quite personal views on history and development of mathematics, and on certain topics in the still very alive subject of transcendental algebraic geometry.

One could declare its Leitmotiv to be G. B. Vico's theory of cycles in the history of mankind, adapted to the analysis of mathematical evolutions and revolutions:

Geometry in ancient Greece, Algebra by the Arabs and in early Renaissance, Geometry again by B. Cavalieri and his indivisibles, Analysis and Physics by the Bernoulli's, ... and later on an intricate succession of points of view and methods, often alternative to each other, or striving in directions opposite one to the other, which all together enriched our knowledge and understanding of the mathematical reality.

Therefore, if we conceive algebra, geometry ... more as methodologies than as domains of knowledge, the difficult question naturally comes out: which way of doing mathematics is the one we are considering?

This question, probably a sterile question when considering the history of mathematics, is however a very important one when we are making choices for future directions of mathematical research: to purport this assertion it will suffice only to cite the (for me, even exaggerated) enthusiasm of nowadays algebraic geometers for the new insights coming to their field by physical theories, concepts and problems.

In any case, in our formerly bourgeois world, idle questions with provocative answers to be defended at tea time at home or in a Café, have often motivated interesting discussions, and it is just my hope to be able to do the same thing here.

2 Abel, the Algebraist?

So I will start, as due, by citing Hermann Weyl's point of view (cf. [165, 166] and also [169, pp. 26 and 151], for comments), expressed in an address directed towards mathematics teachers, and later published in the Journal : Unterrichtsblätter für Mathematik und Naturwissenschaften, Band 38 (1933), S. 177–188.

There are two Classes of mathematicians:

- ALGEBRAISTS: as Leibniz, Weierstrass
- GEOMETERS–PHYSICISTS: as Newton, Riemann, Klein

and people belonging to different classes may tend to be in conflict with each other.

The tools of the algebraists are: logical argumentation, formulae and their clever manipulation, algorithms. The other class relies more on intuition, and graphical and visual impressions. For them it is more important to find a new truth than an elegant new proof. The concept of rigour is the battlefield where the opposite parties confront themselves, and the conflicts which hence derived were sometimes harsh and longlasting.

The first well known example is the priority conflict between Leibniz and Newton concerning the invention of the Calculus (which however was invented independently by the two scientists, as it is currently agreed upon). The inputs which the two scientists provided did indeed integrate themselves perfectly. On one hand the pure algebraic differential quotient dy/dx would be a very dry concept (algebraists nevertheless are still nowadays very keen on inflicting on us the abstract theory of derivations!) without the intuition of velocities and curve tangents; on the other hand, in the analysis of several phenomena, a physical interpretation of Leibniz's rule can turn out to be amazingly complicated.

More closely related with Abelian integrals and their periods was Weierstrass' constructive criticism of the "Principles" by Riemann and Dirichlet.

As also pointed out in the contribution[4] by Schappacher, this conflict soon became the Berlin–Göttingen conflict, and almost deflagrated between Weierstrass and Felix Klein (who continued on the way he started as a student of Clebsch).[5]

Klein's antipathy for Weierstrass was more intellectual than personal: Klein put a special enphasys on geometrical and physical intuition, which he managed to develop in the students by letting them construct solid (plaster, or metal) models of curves and surfaces in ordinary 3-space, or letting them draw very broad (1×2 meters) paper tables of cubic plane curves with an explicit plotting of their $\mathbb{Q}$-rational points.

A concrete witness to this tradition is the exhibition of plaster models of surfaces which are to be found still nowadays in the Halls of the mathematical Institute in Göttingen. These models were then produced by the publishing company L. Brill in Darmstadt, later by the Schilling publishing company, and sold around the world: I have personally seen many of those in most of the older Deparments I have visited (cf. the 2 Volumes edited by G. Fischer on "Mathematical Models" [85][6]).

[4] Here and after, I will refer to the oral contributions given at the Abel Bicentennial Conference, and not to the articles published in this Volume.

[5] We have noticed that another article devoted to this topic has appeared after we gave the talk, namely [26] by U. Bottazzini.

[6] Writes G. Fischer: "There were certainly other reasons than economic for the waning interest in models.... More and more general and abstract viewpoints came to the forefront of mathematics ... Finally Nicolas Bourbaki totally banned pictures from his books."

A similar trick, with computer experiments replacing the construction of models, is still very much applied nowadays in the case where professors have to supervise many more students' theses than they can really handle.

Weierstrass had a victory, in the sense that not only the theory of calculus, but also the theory of elliptic functions is still nowadays taught almost in the same way as it was done in his Berlin lectures.

But Klein's "defeat" (made harder by the long term competition with Poincaré about the proof of the uniformization theorem, see later) was however a very fertile humus for the later big growth of the Göttingen influence, and certainly Weyl's meditations which we mentioned above were reflecting this historically important controversy.

Where does then Abel stay in this classification? I already took position, with my choice of the title of this section: Abel is for me an algebraist and I was glad[7] to hear Christian Houzel stressing in his contribution the role of Abel's high sense of rigour. Abel's articles on the binomial coefficients, on the summation of series, and on the solution of algebraic equations testify his deep concern for the need of clear and satisfactory proofs.[8]

Of course, like many colours are really a mixture of pure colours, the same occurs for mathematicians, and by saying that he was deep down an "algebraist" I do not mean to deny that he possessed a solid geometrical intuition, as we shall later point out.

In fact, Abel himself was proud to introduce himself during his travels as 'Professor of Geometry.'[9]

The best illustration of his synthetical point of view is shown by the words (here translated from French) with which he begins his Memoir XII-1 ("Mémoire sur une propriété générale d'une classe très étendue de fonctions transcendantes."):[10]

"The transcendental functions considered until nowadays by the geometers are a very small number. Almost all the theory of transcendental functions is reduced to the one of logarithmic, exponential and circular functions, which are essentially only one kind. Only in recent times one has begun to consider some other functions. Among those, the elliptic transcendentals, of which M. Legendre developed so many elegant and remarkable properties, stay in the first rank."

The statement "are essentially only one kind" is the one we want now to comment upon.

---

[7] For many years, until after the first world war, there used to be a course in Italian Universities entitled "Lezioni di analisi algebrica ed infinitesimale". This shows that the birth of Analysis as a new separated branch, trying to appear more on the side of applied mathematics, is a relative novelty which ends the reconciliation made for the dualism Leibniz–Newton.

[8] cf. the article by J. J. O'Connor and E. F. Robertson in http://www-history.mcs.st-andrews.ac.uk/References/Abel.html, citing his letter to Holmboe from Berlin. Here we can read: "In other words, the most important parts of mathematics stand without foundation."

[9] To be perfectly honest, all we know is that in 1826 he signed himself in at the "Goldenes Schiff" in Predazzo as 'Abel, professore della geometria.'

[10] This important article was lying at the centre of the contribution by Griffiths, and is also amply commented upon in Kleiman's contribution.

Algebraists like indeed short formulae, and these are in this case available. It suffices to consider the single formula : $\exp(x) := \Sigma_{n=0}^{\infty} \frac{x^n}{n!}$, and then, by considering $\exp(ix)$ and the inverse functions of the ones we can construct by easy algebraic manipulations, we get easily hold of the wild proliferation of functions which for instance occupy the stage of the U.S.A. calculus courses (sin, cos, sec, cosec, tg, cotg, and their hyperbolic analogues).

No doubt, synthesis is a peculiarity of pure mathematics, and applied or taught mathematics may perhaps need so many different names and functions: but, for sure, Abel stood by the side of synthesis and conciseness.

His well known saying, that he was able to learn so rapidly because he had been "studying the Masters and not their pupils", is actually quite timely. Today there is certainly an inflation of books and divulgations, many are second, third hand or even further. Abel's point of view should be seriously considered by some pedagogists who want to strictly regulate children's learning, forcing them to study n-th hand knowledge. Perhaps this is a strictly democratic principle, by which one wants to prevent some children from becoming precociously wise (as Abel did), and possibly try to stop their intellectual growth (this might be part of a more general ambitious program, sponsored by Television Networks owners).

In any case, Abel had read the masters, and he knew many functions: he still belongs to the mathematical era where functions are just concretely defined objects and not subsets of a Cartesian product satisfying a geometrical condition. One of Abel's main contributions was to consider not only ample classes of functions defined by integrals of algebraic functions, but to study their inverse functions and their periodicity (the so called elliptic functions being among the latter).

At his time did not yet exist the concept of "Willkürliche Funktionen" (= "arbitrary functions") quite central e.g. for Weierstrass, Dini, Peano and Hilbert (cf. [65, 158, 86]) and which motivated much of the developments in the theory of sets leading to the construction of several pathological situations (as Lebesgue's non constant function with derivative almost everywhere zero, [111]).

Most of the functions he considered were in fact written as $\int_{x_0}^{x} y(t)dt$, where $y(x)$ is an **algebraic function** of x, which simply means that the function is defined on some interval in $\mathbf{R}$ and that there exists a polynomial $P(x, y) \in \mathbf{C}[x, y]$ [11] such that $P(x, y(x)) \equiv 0$. The functions given by these integrals, or by sums of several of these, are nowadays called **Abelian functions**.

The above statement is by and large true, with however a single important exception, concerning Abel's treatment of functional equations: there he considers quite generally the functions which occur as solution of certain **functional equations**.

As example, we take the content of an article also considered in Houzel's talk, VI-1 (published in Crelle, Bd. 1 (1826)). This article is a real gem: it anticipates

[11] Observe that we wrote $\mathbf{C}[x, y]$ instead of $\mathbf{R}[x, y]$. It is commonly agreed that, if we want to summarize in two words which the greatest contribution of Abel and Jacobi was, then it was to consider the elliptic integrals not just as functions of a real variable, but also as functions of a complex variable. So, we owe to them the birth of the theory of holomorphic functions.

S. Lie's treatment of Lie group germs, and yields actually a stronger result (although, under the assumption of commutativity):

Theorem 2.1. *Let $f : U \to \mathbb{R}$ be a germ of function defined on a neighbourhood of the origin, $\mathbb{R}^2 \supset U \ni 0$, such that*

$$f(z, f(x, y))$$

is symmetric in x, z, y (i.e., in today's terminology, we have an Abelian Lie group germ in 1 variable). Then there exists a germ of change of variable $\psi : (\mathbb{R}, 0) \to (\mathbb{R}, 0)$ such that our Lie group becomes $(\mathbb{R}, +)$, or, more concretely, such that

$$\psi(f(x, y)) = \psi(x) + \psi(y) .$$

Among the Masters' work which Abel studied was certainly, as already mentioned, M. Legendre and his theory of elliptic integrals. For these, already considered by Euler and Lagrange, Legendre devised a **normal form** (here $R(x)$ is a rational function of x):

$$\int_{x_0}^{t} \frac{R(x)}{\sqrt{(1-x^2)(1-k^2x^2)}} dx .$$

In the remarkable paper XVI (published on Crelle, Bd. 2, 3 (1827, 1828)), entitled "Recherches sur les fonctions elliptiques," Abel, as we already mentioned, writes clearly, after observing that the study of these elliptic integrals can be reduced to the study of integrals of the first, second, and third species[12]

$$\int \frac{d\theta}{\sqrt{1-c^2\sin^2\theta}} \, ; \int d\theta\sqrt{1-c^2\sin^2\theta} \, , \int \frac{d\theta}{(1+n\sin^2\theta)\sqrt{1-c^2\sin^2\theta}} \, ,$$

"These three functions are the ones that M. Legendre has considered, especially the first, which enjoys the most remarkable and the simplest properties. I am proposing myself, in this memoir, to consider the inverse function, i.e., the function $\phi(a)$, determined by the equations

$$a = \int \frac{d\theta}{\sqrt{1-c^2\sin^2\theta}}$$

$$\sin\theta = \phi(a) = x ."$$

In the following pages, where he manages to give a simple proof of the double periodicity of the given function ϕ, Abel shows the clarity of his geometric intuition.

He simply observes that in Legendre's normal form one should overcome the natural resistance to consider non real roots, and actually it is much better to consider the case where $k^2 < 0$, and thus he considers (his notation) the integral

[12] Since an elliptic integral is the one where we consider a square root $\sqrt{P(x)}$ where P is a polynomial of degree 3 or 4, one can reduce it, after applying a projective transformation of the line $\mathbb{P}^1$, to a square root $\sqrt{Q(y^2)}$, where Q is quadratic, and then we can view it as an integral on the unit circle, whose projection to the x-line $\mathbb{P}^1$ yields a double cover.

$$\int_{x_0}^{t} \frac{1}{\sqrt{\left(1-c^2x^2\right)\left(1+e^2x^2\right)}}dx\ ,$$

where c, e are strictly positive real numbers.

The roots of the radical are the points $\pm 1/c$, $\pm(1/e)\sqrt{-1}$, we have a rectangular symmetry around the origin and we have two periods ω, $\tilde{\omega}$ obtained by integrating on the two closed paths lying over the segments joining pairs of opposite roots:

$$\omega = 4\int_0^{1/c} \frac{1}{\sqrt{\left(1-c^2x^2\right)\left(1+e^2x^2\right)}}dx\ ,\ \tilde{\omega} = 4\int_0^{i/e} \frac{1}{\sqrt{\left(1-c^2x^2\right)\left(1+e^2x^2\right)}}dx\ .$$

It is straightforward to observe that the two periods (of the **real curve**) are such that $\omega \in \mathbb{R}$, respectively $\tilde{\omega} \in i\mathbb{R}$, thus they are linearly independent over the real numbers.

The conclusion is that the inverse function is **doubly periodic** with one real and one imaginary period, so that its fundamental domain is a rectangle with sides parallel to the real, resp. imaginary axis. Since every elliptic integral of the first kind can be reduced to this form, their inverse functions are all doubly periodic (unlike the circular functions, which possess only one period).

Abel does not bother to highlight the geometry underlining his argument, it is clear that he has developed a good geometrical intuition, but his style is extremely terse and concise. This conciseness becomes almost abrupt in the other article XIII-2[13], "Théorie des transcendantes elliptiques."

This long memoir starts nulla interposita more (it was probably unfinished): "For more simplicity I denote the radical by $\sqrt{R}$, whence we have to consider the integral

$$\int \frac{Pdx}{\sqrt{R}},$$

P denoting a rational function of x."

It is divided into three chapters, the first devoted to the reduction of elliptic integrals by means of algebraic functions, the second to the reduction of elliptic integrals by means of logarithmic functions, and finally the third is entitled "A remarkable relation which exists among several integrals of the form

$$\int \frac{dx}{\sqrt{R}}, \int \frac{xdx}{\sqrt{R}}, \int \frac{x^2dx}{\sqrt{R}}, \int \frac{dx}{(x-a)\sqrt{R}}\ ."$$

[13] The second volume of the edition [1] by Sylow and Lie, Christiania 1881, contains the unpublished papers of Abel, with a few exceptions. As the editors remark, this edition, posterior by more than 30 years to the edition of 1839 edited by the friend and colleague of Abel, Holmboe, was financed by the Norwegian Parliament after the great demand for Abel's works (Holmboe's edition went rapidly out of print) of which especially the French Mathematical Society made itself interpreter. The two editors decided to omit in the second volume three articles which were partly based on an erroneous memoir of his youth, written in Norwegian, where Abel thought he could prove that the general equation of degree n can be solved by radicals. This is the only published article which is not appearing in the Holmboe edition, nor in Volume I of the edition by Sylow and Lie.

The memoir contains the explicit discussions of several concrete problems concerning such reductions, and contains for many of those problems explicit references to Legendre. It looks to me a rather early work, because we directly see the influence of the study of Legendre, but rather important for two reasons.

The first reason, already clear from the title of Chapter III, is that Abel here for the first time considers the question of the relations holding among sums of elliptic integrals. This problem will be considered more generally for all algebraic integrals of arbitrary genus g in his fundamental Memoir XII-1, entitled "Mémoire sur une propriété générale d'une classe très étendue de fonctions transcendantes", presented on October 10, 1826 to the Académie des Sciences de Paris, and published only in 1841.

The main theorem of the latter was so formulated: "If we have several functions whose derivatives can be roots of the same algebraic equation [if $y(x)$ is an algebraic function of x, i.e., there is a polynomial P such that $P(x, y(x)) \equiv 0$, then for each rational function $f(x, y)$ there is a polynomial $F(x, y)$ such that $F(x, f(x, y(x)) \equiv 0]$[14], with coefficients rational functions of one variable [x], one can always express the sum of an arbitrary number of such functions by means of an algebraic and a logarithmic function, provided that one can establish among the variables of these functions a certain number of algebraic relations".

The first theorem is then given through formula (12):

$$\int f(x_1, y_1)dx_1 + \int f(x_2, y_2)dx_2 + \cdots + \int f(x_\mu, y_\mu)dx_\mu = v[(t_1, \dots, t_k)] \ :$$

here $f(x, y)$ is a rational function, we take the μ points which form the complete intersection of $[P(x, y) = 0]$ and $[G_t(x, y) = 0]$ where G_t depends rationally upon the parameter $t = (t_1, \dots, t_k)$, and the conclusion is, as we said, that v is the sum of a rational and of a logarithmic function.

Abel also explains clearly in the latter memoir that the number of these relations is a number, which later on was called the genus of the curve C birational to the plane curve of equation $P(x, y) = 0$. The way we understand the hypothesis of the theorem nowadays is through the geometric condition: if the Abel sum of these points is constant in the Jacobian variety of C. I will come back to the geometric interpretations in the next section, let me now return to the second reason of importance of the cited memoir XIII-2.

For instance, in Chapter I, Abel gives very explicit formulae, e.g. for the reduction of integrals of the form

$$\int \frac{x^m dx}{\sqrt{R}},$$

where $R(x)$ is a polynomial of degree 3, 4, to the integrals

$$\int \frac{dx}{\sqrt{R}}, \int \frac{xdx}{\sqrt{R}}, \int \frac{x^2 dx}{\sqrt{R}}.$$

[14] Here and elsewhere, [..] stands for an addition of the present author

The above integral is calculated by recursions, starting from the equation

$$d\left(Q\sqrt{R}\right) = S\frac{dx}{\sqrt{R}},$$

and writing explicitly

$$S = \phi(0) + \phi(1)x + \cdots + \phi(m)x^m$$

$$Q = f(0) + f(1)x + \cdots + f(m-3)x^{m-3}.$$

This is an example of Abel's mastery in the field of **Differential Algebra**. Although the modern reader, as well as Sylow and Lie, may underscore the impact of these very direct calculations, it seems to me that there has been a resurgence of this area of mathematics, especially in connection with the development of computer algorithms and programs which either provide an explicit integration of a given function by elementary functions[15], or decide that the given function does not admit an integration by elementary functions (this problem was solved by Risch, and later concrete decision procedures and algorithms were given by J. Davenport and Trager [137, 62, 63, 64, 155]).

Differential algebra is also the main tool in the article (XVII-2) "Mémoire sur les fonctions transcendantes de la forme $\int y\, dx$ ou y est une fonction algébrique de x."

This paper looks very interesting and somehow gave me the impression (or at least I liked to see it in this way) of being a forerunner of the applications of Abelian integrals to questions of transcendence theory.

This time I will state the main theorem by slightly altering Abel's original notation

Theorem 2.2. *Assume that ϕ is a [non trivial] polynomial*

$$\phi(r_1, \ldots, r_\mu, w_1, \ldots w_p)$$

and that $\phi \equiv 0$ if we set

$$r_i = \int y_i(x)\, dx, \;\; w_j = u_j(x),$$

where y_i, u_j are algebraic functions. Then there is a [non trivial] linear relation $\Sigma_i c_i \int y_i(x)\, dx = P(x)$, with constant coefficients c_i and with P an algebraic function.

Corollary 2.3. *Let $y_i(x)\, dx$ be linearly independent Differentials of the I Kind on a Riemann surface. Then the respective integrals are algebraically independent.*

---

15 i.e., by rational functions or by logarithms (more generally, one can consider algebraic functions and logarithms of these).

Nowadays, we would use the periodicity of these functions on the universal cover of the algebraic curve, or the order of growth of the volume of a periodic hypersurface (this argument was the one used later by Cousin, cf. [61]) to infer that the polynomial must be linear (vanishing on the complex linear span of the periods). Abel's argument is instead completely algebraic in nature, he chooses in fact the polynomial ϕ to be of minimal degree with respect to r_μ, and applies d/dx to the relation in order to obtain (Abel's notation)[16]

$$\Sigma_j \phi'(r_j)\, y_j + \phi'(x) = 0 \, .$$

Abel shows then that ϕ has degree 1 in r_μ, since writing

$$R = r_\mu^k + P_0 r_\mu^{k-1} + P_1 r_\mu^{k-2} + \cdots = 0$$

he gets:

$$0 = \frac{d}{dx}(R) = r_\mu^{k-1}(k y_\mu + P_0') + \{\ldots\} r_\mu^{k-2} + \cdots = 0$$

whence, by the minimality of the degree k, this polynomial in r_μ has all coefficients identically zero, in particular $(k y_\mu + P_0') \equiv 0$, therefore

$$r_\mu = \int y_\mu dx = -(1/k) P_0$$

is the desired degree one relation. By induction Abel derives the full statement that ϕ is linear in $r_1, \ldots, r_\mu$.

3 The Geometrization of Abel's Methods

The process of geometrization and of a deeper understanding of Abel's discoveries went a long way, with alternate phases, for over 150 years. We believe that[17] a fundamental role for the geometrization was played by the Italian school of algebraic geometry, which then paved the way for some of the more abstract developments in algebraic geometry.

Although it was very depressing for Abel that his fundamental Memoir XII-1 was not read by A. Cauchy (this is the reason why it took more than 15 years before it was published), still in Berlin Abel found the enthusiastic support of L. Crelle, who launched his new Journal by publishing the articles of Abel and Jacobi. Recall that finally fame and recognition were reaching Abel through the offer of a professorship in Berlin, which, crowning the joint efforts of Crelle and Jacobi, arrived however a few days after Abel had died.

Especially inspired by the papers of Abel was Jacobi, in Berlin, who also wrote a revolutionary article on elliptic function theory, entitled "Fundamenta nova theoriae functionum ellipticarum."

[16] Why was not Abel using the notation $\partial/\partial r_j$?

[17] As amply illustrated by Ciliberto in his contribution.

It was Jacobi who introduced the words 'Abelian integrals,' 'Abelian functions': Jacobi's competing point of view (few competitions however were so positive and constructive in the history of mathematics) started soon to prevail.

Jacobi introduced the socalled **elliptic theta functions**, denoted $\theta_{00}, \theta_{01}, \theta_{10}, \theta_{11}$, (cf. e.g. [156] or [125]) and expressed the elliptic functions, like Abel's ϕ, inverse of the elliptic integral of the first kind, as a ratio of theta functions.

A much more general definition of **theta-series** (the expression was later coined by Rosenhain and Göpel, followers of Riemann) was given by B. Riemann, who defined his **Riemann Theta function** as the following series of exponentials

$$\theta(z, \tau) := \Sigma_{n \in \mathbb{Z}^g} \exp\left(2\pi i \left[\frac{1}{2}\, {}^t n \tau n + {}^t n z\right]\right)$$

where $z \in \mathbb{C}^g$, $\tau \in \mathcal{H}_g = \{\tau \in \mathrm{Mat}(g, g, \mathbb{C}) | \tau = {}^t\tau, \mathrm{Im}(\tau) \text{ is positive definite}\}$.

The theta function converges because of the condition that $\mathrm{Im}(\tau)$ is positive definite, it admits $\mathbb{Z}^g$ as group of periods, being a Fourier series, and it has moreover a $\tau\mathbb{Z}^g$-quasi-periodicity which turns out to be the clue for constructing $2g$-periodic meromorphic functions as quotients of theta series.

With a small variation (cf. [125]) one defines the theta-functions with characteristics

$$\theta[a, b](z, \tau) := \Sigma_{n \in \mathbb{Z}^g} \exp\left(2\pi i \left[\frac{1}{2}\, {}^t(n + a)\tau(n + a) + {}^t (n + a)(z + b)\right]\right),$$

and the Jacobi functions $\theta_{a,b}(z, \tau)$ are essentially the functions $\theta[a/2, b/2](2z, \tau)$.

Beyond the very explicit and beautiful formulae, what lies beyond this apparently very analytic approach is the **pioneeristic principle** that any **meromorphic function f** on a complex manifold X can be written as

$$f = \frac{\sigma_1}{\sigma_2}$$

of two relatively prime sections of a unique **line bundle** L on X.

This formulation came quite long after Jacobi, but Jacobi's work had soon a very profound impact. For instance, one of the main contributions of Jacobi was the solution of the **inversion problem** explicitly for genus $g = 2$.

Concretely, Jacobi considered a polynomial $R(x)$ of degree 6, and then, given the two Abelian integrals

$$u_1(x_1, x_2) := \int_{x_0}^{x_1} \frac{dx}{\sqrt{R}} + \int_{x_0}^{x_2} \frac{dx}{\sqrt{R}},$$
$$u_2(x_1, x_2) := \int_{x_0}^{x_1} \frac{xdx}{\sqrt{R}} + \int_{x_0}^{x_2} \frac{xdx}{\sqrt{R}},$$

he found that the two symmetric functions

$$s_1 := x_1 + x_2, s_2 := x_1 x_2,$$

are 4-tuple periodic functions of u_1, u_2.

Under the name **Jacobi inversion problem** went the generalization of this result for all genera g, and the solution to the Jacobi inversion problem was one of the celebrated successes of Riemann.

Nowadays the result is formulated as follows: given a compact Riemann surface C of genus g, let $\omega_1, \dots \omega_g$ be a basis of the space $H^0(\Omega^1_C)$ of holomorphic differentials on C **adapted** to a **symplectic** basis $\alpha_1, \dots \alpha_g, \beta_1, \dots \beta_g$ for the Abelian group of closed paths $H_1(C, \mathbb{Z})$: this means that

$$\left(\int_{\alpha_i} \omega_j\right) = \left(\delta_{i,j}\right) = I_g, \left(\int_{\beta_i} \omega_j\right) = \left(\tau_{i,j}\right),$$

where I_g is the Identity $(g \times g)$ matrix and $\tau \in \mathcal{H}_g$, and that the intersection matrices satisfy $(\alpha_i, \alpha_j) = 0$, $(\beta_i, \beta_j) = 0$, $(\alpha_i, \beta_j) = I_g$.

Then the **Abel–Jacobi** map $C^g \to \mathbb{C}^g/(\mathbb{Z}^g + \tau\mathbb{Z}^g) := \mathrm{Jac}(C)$, associating to the g-tuple $P_1, \dots P_g$ of points of C the sum of integrals (taken modulo $(\mathbb{Z}^g + \tau\mathbb{Z}^g)$)

$$a_g(P_1, \dots P_g) := \int_{P_0}^{P_1} (\omega) + \cdots \int_{P_0}^{P_g} (\omega),$$

(ω being the vector with i-th component ω_i), is **surjective** and yields a **birational map** of the **symmetric product**

$$C^{(g)} := Sym^g(C) := C^g/\mathcal{S}_g,$$

$\mathcal{S}_g$ being the symmetric group of permutation of g elements, onto the **Jacobian variety of C**

$$\mathrm{Jac}(C) := \mathbb{C}^g/(\mathbb{Z}^g + \tau\mathbb{Z}^g).$$

We see the occurrence of a matrix τ in the so called Siegel upper half space $\mathcal{H}_g$ of symmetric matrices with positive definite imaginary part: this positive definiteness ensures the convergence of Riemann's theta series, and indeed Riemann used explicitly his theta function to express explicitly the symmetric functions of the coordinates of a g-tuple $P_1, \dots, P_g$ of points of C as rational functions of theta factors.

Today, we tend to forget about these explicit formulae, and we focus our attention to the geometric description of the Abel–Jacobi maps for any n-tuple of points of C in order to grasp the power of the discoveries of Abel, Jacobi, and Riemann.

The **Abel–Jacobi** maps $C^n \to \mathbb{C}^g/(\mathbb{Z}^g + \tau\mathbb{Z}^g) =: \mathrm{Jac}(C)$, given by

$$a_n(P_1, \dots, P_n) := \int_{P_0}^{P_1} (\omega) + \cdots + \int_{P_0}^{P_n} (\omega),$$

and which naturally factor through the symmetric product $C^{(n)}$, enjoy the following properties (cf. the nice and concise lecture notes by D. Mumford [124] for more on this topic, and also [60] for a clear modern presentation)

(1) **Abel's Theorem:** The fibres are projective spaces corresponding to **linearly equivalent** divisors, i.e. $a_n(P_1, \dots P_n) = a_n(Q_1, \dots Q_n)$ if and only if there is a rational function f on C with polar divisor $P_1 + \dots + P_n$ and divisor of zeros $Q_1 + \dots + Q_n$.
(2) For $n = g - 1$ we have that the image of a_{g-1} equals, up to translation, the hypersurface $\Theta \subset \mathrm{Jac}(C)$ whose inverse image in $\mathbb{C}^g$ is given by the vanishing of the Riemann theta function $\Theta = \{z | \theta(z, \tau) = 0\}$.
(3) For $n = g$ we have that a_g is onto and birational.
(4) For $n \geq 2g - 1$ we have that a_n is a fibre bundle with fibres projective spaces $\mathbb{P}^{n-g}$.
(5) For $n = 1$ we have an embedding of the curve C inside $\mathrm{Jac}(C)$, and in such a way that C meets a general translate of the theta divisor Θ in exactly g points. And in this way one gets an explicit geometrical description of the inverse to the Abel–Jacobi map a_g. Indeed, to a point $y \in \mathrm{Jac}(C)$ we associate the g-tuple of points of C given by the intersection $C \cap (\Theta + y)$ (here we think of $C \subset \mathrm{Jac}(C)$ under the embedding a_1).

Many of the above properties are very special and lead often to a characterization of curves among algebraic varieties.

The above formulations are moreover the fruit of a very long process of maturation whose evolution is not easy to trace. For instance, when did the concept of the **Jacobian variety** of a curve (Riemann surface) C make its first appearance?

This notion certainly appears in the title of the papers by R. Torelli in 1913 [154] but apparently[18] the name was first used by F. Klein and became very soon extremely popular. Observe however that in the classical treatise by Appell and Goursat [9], dedicated to analytical functions on Riemann surfaces, and appeared first in 1895, although Jacobi's inversion theorem is amply discussed, no Jacobian or whatsoever variety is mentioned.

The French school of Humbert, Picard, Appell, and Poincaré was very interested about the study of the so called "hyperelliptic varieties", generalization of the elliptic curves in the sense that they were defined as algebraic varieties X of dimension n admitting a surjective entire holomorphic map[19] $u : \mathbb{C}^n \to X$.

Among those are the so called **Abelian varieties,**[20] which are the projective varieties which have a structure of an algebraic group.

In particular, Picard proved a very nice result in dimension $d = 2$, which was observed by Ciliberto (cf. his article in the present volume, also for related historical references) to hold quite generally. We want to give here a simple proof of this result, which we found during the Conference, and which makes clear one basic aspect in which the higher dimensional geometry has a different flavour than the theory of

[18] We heard this claim from S. J. Patterson in Göttingen, soon after he had written the article [133].

[19] The letter u clearly stands for "uniformization map".

[20] Which, however, at the time of Torelli's paper, 1913, and also afterwards, were called Picard Varieties. With the Prize winning Memoir by Lefschetz, [112], the terminology Abelian varieties became the only one in use.

curves[21]: namely there is no ramification in passing from Cartesian to symmetric products.

Theorem 3.1. *Let X be an algebraic variety of dimension $d \geq 2$ and assume that there is a natural number n such that the n-th symmetric product $X^{(n)}$ is birational to an Abelian variety A. Then $n = 1$ (whence, X is birational to an Abelian variety).*

Moreover, this result ilustrates another main difference between the geometry of curves and the one of higher dimensional varieties X: the latter has a quite different flavour, because only seldom one can resort to the help of subsidiary Abelian varieties for the investigation of a higher dimensional variety X.

Proof. Without loss of generality we may assume that X be smooth. Let us consider the projection π of the Cartesian product onto the symmetric product, and observe that $\pi : X^n \to X^{(n)}$ is unramified in codimension 1.

We get a rational map $f : X^n \to A$ by composing π with the given birational isomorphism, and then we observe that every rational map to an Abelian variety is a morphism.

It is moreover clear from the construction that f is not branched in codimension 1: in particular, it follows that X^n is birational to an unramified covering of A, whence X^n is birational to an Abelian variety.

Let us introduce now the following notation: for a smooth projective variety Y we consider the algebra of global holomorphic forms $H^0(\Omega^*_Y) := \oplus_{i=0,\dots,\dim(Y)} H^0(\Omega^i_Y)$ (**holomorphic algebra**, for short).

This graded algebra is a birational invariant, and for an Abelian variety A it is the free exterior algebra over $H^0(\Omega^1_A)$.

Now, the holomorphic algebra $H^0(\Omega^*_{X^n})$ of a Cartesian product X^n is the tensor product of n copies of the holomorphic algebra $H^0(\Omega^*_X)$ of X.

Denote $H^0(\Omega^*_X)$ by B: then we reached the conclusion that $B^{\otimes n}$ is a free exterior algebra over its part of degree 1, $B^1_1 \oplus B^1_2 \cdots \oplus B^1_n$.

It follows that also B is a free exterior algebra (i.e., X enjoys the property that its holomorphic algebra $H^0(\Omega^*_X)$ is a free exterior algebra over $H^0(\Omega^1_X)$).

Moreover, the holomorphic algebra of the symmetric product is the invariant part of this tensor product (for the natural action of the symmetric group in n-letters $\mathcal{S}_n$), and by our assumption $C := (B^{\otimes n})^{\mathcal{S}_n}$ is also a free exterior algebra. However, $C^1 = B^1$, and $r := \dim(B^1)$ is also the highest degree i such that $B^i \neq 0$. But then $C^{nr} \neq 0$, contradicting the property that C is a free exterior algebra with $\dim(C^1) = r$, if $n \neq 1$.

Thus $n = 1$, and X is birational to an Abelian variety. □

[21] One could argue whether the beauty of 1-dimensional geometry bears similarities to the surprising isomorphisms of classical groups of small order. These also, by the way, are related to Abel's heritage. For instance, the isomorphism $\mathcal{S}_4 \cong A(2, (\mathbb{Z}/2))$, whence $\mathcal{S}_3 \cong A(2, (\mathbb{Z}/2))/(\mathbb{Z}/2)^2 \cong PGL(2, (\mathbb{Z}/2)) \cong \mathrm{Aut}(\mathbb{H})$. Here, $\mathbb{H}$ is the group of order 8 of unit integral quaternions, and the last isomorphism is related to some later development in the theory of algebraic curves, namely to Recilla's tetragonal construction, see [136, 71].

Remark 3.2. In the case of dimension $d = 1$, the same algebraic arguments easily yield that the n-th symmetric product of a curve C is not birational to an Abelian variety if $n \neq g$, $g := \dim\ H^0(\Omega^1_C)$.

Once more, the algebra of differential forms, as in Abel's work, has played the pivotal role.

The importance of this algebra was observed also by Mumford [122] who used it to show that on an algebraic surface X with $H^0(\Omega^2_X) \neq 0$, the group of 0-cycles (sums $\Sigma_i m_i P_i$ of points $P_i \in X$ with integer multiplicities $m_i \in \mathbb{Z}$) modulo rational equivalence is not finite dimensional, contrary to the hope of Severi, (Mumford sarcastically wrote: "One must admit that in this case the *technique* of the italians was superior to their vaunted intuition"[22]) who unfortunately was basing his proposed theory on a wrong article ([146], where not by chance the error was an error of ramification).

It must be again said that the italian school, and especially Castelnuovo, gave a remarkable impetus to the geometrization of the theory of Abelian varieties.

This approach, especially through the work of Severi, influenced André Weil who understood the fundamental role of Abelian varieties for many questions of algebraic number theory. Weil used these ideas to construct [161] the Jacobian variety of a curve as a quotient of the symmetric product $C^{(g)}$, and then, for a d-dimensional variety with $d \geq 2$, the Albanese variety $\mathrm{Alb}(X)$ as a quotient (in the category of Abelian varieties) of the Jacobian $\mathrm{Jac}(C)$ of a sufficiently general linear section $C = X \cap H_1 \cap H_2 \cdots \cap H_{d-1}$.

It must be however said that also the later geometric constructions were deeply influenced by the bilinear relations which Riemann, through a convenient dissection of his Riemann surface C, showed to hold for the periods of the Abelian integrals of the first kind of C.

Nowadays, the usual formulation is (according to Auslander and Tolimieri, [12] pages 267 and 274, the first formulation is essentially due to Gaetano Scorza in [143], while the second is essentially due to Hermann Weyl in [167, 168], with refinements from A. Weil's book [164])

Definition 3.3. *Let Γ be a discrete subgroup of a complex vector space V, such that the quotient V/Γ is compact (equivalently, $\Gamma \otimes \mathbb{R} \cong V$): then we say that the complex torus V/Γ satisfies the two Riemann bilinear relations if*

I) There exists an alternating form $A : \Gamma \times \Gamma \to \mathbb{Z}$ such that A is the imaginary part of an Hermitian form H on V

II) H is positive definite.

Remark 3.4. Or, alternatively, a complex structure on $\Gamma \otimes \mathbb{C}$, i.e., a decomposition $\Gamma \otimes \mathbb{C} = V \oplus \bar{V}$ and an element $A \in \Lambda^2(\Gamma)^\vee$ yield a polarized Abelian variety if

[22] However, as well known, there are Italians with techniques and ideas, and others who are not perfect. In particular, while it is not difficult to find errors or wrong assertions in Enriques and Severi, it is rather hard to do this with Castelnuovo.

the component of A in $\Lambda^2(V)^\vee \subset \Lambda^2(\Gamma \otimes \mathbb{C})^\vee$ is zero and then its component in $(V)^\vee \otimes (\bar{V})^\vee$ is a positive definite Hermitian form.[23]

The basic theorem characterizing complex Abelian varieties is however due to Henri Poincaré [130, 131] who proved the linearization of the system of exponents, i.e., the more difficult necessary condition in the theorem, by an averaging procedure (integrating the ambient Hermitian metric of $X \to \mathbb{P}^N$ with respect to the translation invariant measure of $X = V/\Gamma$, he obtained a translation invariant Hermitian metric).

Theorem 3.5. *A complex torus $X = V/\Gamma$ is an algebraic variety if and only if the two Riemann bilinear relations hold true for V/Γ.*

Both conditions are equivalent to the existence of a meromorphic function f on the complex vector space V whose group of periods is exactly Γ (i.e., $\Gamma = \{v \in V \mid f(z+v) \equiv f(z)\}$).

Poincaré had an extensive letter exchange with Klein (cf. Klein's Collected Works, where pages 587 to 621 of Vol. III are devoted to the "Briefwechsel" between the two, concerning the problem of uniformization, and their early attempts, which were based on a 'principle of continuity' which was not so easy to justify[24]), especially related to the study of discontinuous groups, acting not only on $\mathbb{C}^n$ as in the case of tori, but also on the hyperbolic upperhalf plane $\mathbb{H}$. The main result, whose complete proof was obtained in 1907 by a student of Klein, Koebe, and by Poincaré independently, was the famous uniformization theorem that again we state in its modern formulation for the sake of brevity.

Theorem 3.6. *If a Riemann surface is not the projective line $\mathbb{P}^1_{\mathbb{C}}$, the complex plane $\mathbb{C}$, nor $\mathbb{C}^*$ or an elliptic curve, then its universal covering is the (Poincaré) upper half plane $\mathbb{H}$.*

The reason why the upper half plane "belongs" to Poincaré is that Klein prefered to work with the biholomorphically equivalent model given by the unit disk $\mathbb{D} := \{z \in \mathbb{C} \mid |z| < 1\}$. In this way Klein was capable of making us the gift of beautiful symmetries given by tesselations of the disk by fundamental domains for the action of very explicit Fuchsian groups (discrete subgroups Γ of $PSU(1,1,\mathbb{C})$) with compact quotient $\mathbb{D}/\Gamma$.

To summarize the highlights of the turn of the century, when geometry was a very central topic, one should say that several new geometries came to birth at that time: but the new developments were based on new powerful analytic tools, which were the bricks of the new building.

However, although the birth of differential geometry lead to new geometrical theories based on infinite processes where metric notions played a fundamental role, algebraic geometry went on with alternating balance between geometrical versus algebraic methods.

---

[23] The second characterization is very useful for the study of fibre bundles of Abelian varieties, as we had opportunity to experience ourselves, cf. [43]

[24] In his unpublished Fermi Lectures held in Pisa in 1976, D. Mumford explained how this approach was working, using clarifications due to Chabauty, [50].

4 Algebraization of the Geometry

At this moment, a witty reader, tired of the distinction "algebra"–"non algebra," might also remind us that the popular expression "This is algebra for me" simply means: 'I do not understand a single word of this.'

There is a serious point to it: the concept algebra is slightly ambiguous, and a very short formula could be not very inspiring without a thoroughful explanation of its meaning(s), and of all the possible consequences and applications.

One of the best ways to understand a formula is for instance to relate it to a picture, to see it thus related to a geometrical or dynamical process.

Needless to say, the best example of such an association is the Weierstrass equation of a plane cubic curve

$$C_3^1 = \left\{(x, y, z) \in \mathbb{P}^2_{\mathbb{C}} | y^2 z = 4x^3 - g_2 x z^2 - g_3 z^3\right\}.$$

To this equation we immediately associate the picture yielding the group law of $C = C_3^1$, i.e., a line L intersecting C in the three points P, Q and $T = -(P + Q)$.

Where have we seen this picture first?[25] Well, in my case, I (almost) saw it first in the book by Walker on Algebraic curves, exactly in the last paragraph, in the section 9.1 entitled "Additions of points on a cubic". The book was written in 1949, and if we look at books on algebraic curves written long before, the group law is not mentioned there. For instance, Coolidge's book "A treatise on algebraic plane curves" has a paragraph entitled "elliptic curves", pages 302–304, and the main theorems are first that an elliptic curve is birational to a plane cubic, and then the **Cross-Ratio Theorem** asserting that if $P \in C_3^1 \subset \mathbb{P}^2$ is any point, through P pass exactly 4 tangent lines, and their cross ratio is independent of the choice of $P \in C$. The Weierstrass equation, and the explanation that the cubic curve is uniformized through the triple $(1, \mathcal{P}, \mathcal{P}')$, where $\mathcal{P}$ is the Weierstrass function, comes later, as due after the Riemann Roch theorem[26], on pages 363–367 in the paragraph "Curves of genus 1."

Going to important textbooks of the Italian tradition, like Enriques and Chisini's 4 Volumes on the "Lezioni sulla teoria geometrica delle equazioni e delle funzioni algebriche"[27] we see that Volume IV contains the Book 6, devoted to "Funzioni ellittiche ed Abeliane." Here Abel's theorem is fully explained, and on page 77 we see Abel's theorem for elliptic curves, on page 81 the addition theorem for the $\mathcal{P}$ function of Weierstrass: the geometry of the situation is fully explained, i.e., that three collinear points sum to zero in the group law given by the sum of Abelian integrals of the first kind

[25] Not always mathematical concepts have a birthday. But sometimes it happens, as for another Legacy of Abel: the concept of an abstract group was born in 1878, with a "Desiderata and suggestions" by A. Cayley [48]. In this note appears first the multiplication table of a group. Immediately afterwards, however, in [49], Cayley realized that it is much better to work with what is now called the "Cayley graph" of a group endowed with a set of generators.

[26] "With proof communicated to the author by prof. Osgood verbally in Nov. 1927".

[27] Edited by Zanichelli in the respective years 1915, 1918, 1924, 1934.

$$u_1 + u_2 + u_3 = \int_{x_0}^{x_1} \frac{dx}{y} + \int_{x_0}^{x_2} \frac{dx}{y} + \int_{x_0}^{x_3} \frac{dx}{y} = 0 \, .$$

It is also observed that the inverse of the point $P = (1, x, y)$ is the point $P = (1, x, -y)$, as a consequence of the fact that $\mathcal{P}$ is an even function.

This Book 6 is clearly influenced by Bianchi's Lecture Notes "Lezioni sulla teoria delle Funzioni di variabile complessa"[28] whose Part 2 is entirely devoted to "Teoria delle funzioni ellittiche", and in the pages 315–322 the addition theorem for $\mathcal{P}$ is clearly explained, moreover "Alcune applicazioni geometriche" are given in the later pages 415–418.

So, the picture is there, missing however the wording: a plane cubic is an Abelian group through the sum obtained via linear equivalence of divisors, namely, the sum of three points P, Q, T is zero if and only if the divisor $P + Q + T$ is linearly equivalent to a fixed divisor D of degree 3 (in the Weierstrass model, D is the divisor $3O$, O being the flex point at infinity).

As it was explained to me by Norbert Schappacher (cf. [139]), the works of Mordell and Weil in the 1920's are responsible for this new wording and perspective. In fact, these authors considered a cubic curve C whose equations has coefficients in a field K, and noticed that the set $C(K)$ of K-rational points, i.e., the points whose coordinates are in K, do indeed form a subgroup. For this they did not need that one flex point should be K-rational, since essentially, once we have a K-rational point O, we can reembed the elliptic curve C by the linear system $|3O|$, and then obtain a new cubic C' whose K rational points are exactly those of C.

Through these works started an exciting new development, namely the geometrization of arithmetic, which was one of the central developments in the 20-th century mathematics.

For instance, the theory of elliptic curves over fields of finite characteristic was (cf. [159]) built by Weil and then Tate, who (cf. the quite late appearing in print [153]), starting from the Weierstrass equation, slightly modified into

$$y^2 z = x^3 - p_2 x z^2 - p_3 z^3$$

started to construct analogues of the theory of periods.[29]

Going back to Bianchi, it is Klein's and Bianchi's merit to have popularized the geometric picture of elliptic curves, and actually Bianchi went all the way through in some of his papers to describe the beautiful geometry related to the embeddings of elliptic curves as non degenerate curves of degree n in $\mathbb{P}^{n-1}$ ($\forall n \geq 3$).

It took however quite long till a purely algebraic interpretation of Weierstrass' equation made its way through.

Nowadays we would associate to an elliptic curve C and to a point $O \in C$ the $\mathbb{N}$-graded ring

---

28 Spoerri, Pisa, 1916.

29 De hoc satis, because the talks by Faltings and Wiles were exactly dealing with these aspects of Abel's legacy.

$$\bigoplus_{m=0}^{\infty} H^0(C, \mathcal{O}_C(mO)) := \mathcal{R}(C, \mathcal{L})$$

where C is defined over a (non algebraically closed) field K of characteristic $\neq 2$, O is a K-rational point and $\mathcal{L} := \mathcal{O}_C(O)$.

As a consequence of the Riemann Roch theorem we obtain the following

Theorem 4.1.

$$\mathcal{R}(C, \mathcal{L}) \cong K[u, \xi, \eta]/\left(\eta^2 - \xi^3 + p_2\xi u^4 + p_3 u^6\right),$$

where $\deg(u) = 1$, $\deg(\xi) = 2$,$\deg(\eta) = 3$, $\operatorname{div}(u) = O$.

To go back to the original Weierstrass equations it suffices to observe that

$$\mathcal{P} = \xi/u^2, \mathcal{P}' = \eta/u^3,$$

and that $x := \xi u$, $y := \eta$, $z := u^3$ are a basis of the vector space $H^0(C, \mathcal{O}_C(3O))$.

One sees also clearly how the Laurent expansion at O of $\mathcal{P}$ is determined by and determines p_2, p_3.

Surprisingly, the following general problem is still almost completely open, in spite of a lot of research in this or similar directions

Problem 4.2. Describe the graded ring $\mathcal{R}(A, \mathcal{L})$ for $\mathcal{L}$ an ample divisor on an Abelian variety, for instance in the case where $\mathcal{L}$ yields a principal polarization.

Before explaining the status of the question, I would first like to explain its importance.

Take for instance the case of an elliptic curve C whose ring is completely described (the ring does not depend upon the choice of O because we have a transitive group of automorphisms provided by translations for the group law of C).

We want for instance to describe the geometry of the embedding of C as a curve of degree 4 in $\mathbb{P}^3$. We observe that, at least in the case where K is algebraically closed, any such embedding is given by the linear system $|4O|$, for a suitable choice of O.

The coordinates of the map are given by a basis of the vector space $H^0(C, \mathcal{O}_C(4O))$, i.e., by 4 independent homogeneous elements of degree 4 in our graded ring $\mathcal{R}(C, \mathcal{L})$. These are easily found to be equal to $s_0 := u^4$, $s_1 := u^2\xi$, $s_2 := \xi^2$, $s_3 := u\eta$. Then we obviously have the two equations

$$s_0 s_2 = s_1^2\ , \quad s_3^2 = s_1 s_2 + p_2 s_0 s_1 + p_3 s_0^2\ ,$$

holding for the image of C (the second is obtained by the "Weierstrass" equation once we multiply by u^2).

These are all the equations, essentially by Bezout's theorem, since C maps to a curve of degree 4 and $4 = 2 \times 2$.

From an algebraic point of view, what we have shown is the process of determining a subring of a given ring, and the nowadays computer algebra programs like "Macaulay" have standard commands for this operation (even if sometimes the computational complexity of the process may become too large if one does not use appropriate tricks).

Classically, a lot of attention was devoted to the geometric study of the maps associated to the linear systems $|m\Theta|$, where Θ is the divisor yielding a principal polarization of the given Abelian variety.

For instance, the quartic Kummer surface is the image in $\mathbb{P}^3$ of a principally polarized Abelian surface under the linear system $|2\Theta|$, yielding a $2:1$ morphism which identifies a point v with $-v$, and blows up the 16 2-torsion points to the 16 nodal singularities of the image surface.

There is a wealth of similar results, which can be found for instance in the books of Krazer,[30] Krazer–Wirtinger, and Coble [104, 105, 59], written in the period 1890–1926.

The first books are directly influenced by the Riemann quadratic relations, i.e., linear relations between degree two monomials in theta functions with characteristics (and their coefficients being also products of "Thetanullwerte", i.e., values in 0 of such thetas with characteristics), and show an attempt to use geometrical methods starting from analytic identities. Coble's book is entitled "Algebraic geometry and theta functions", and is already influenced by the breakthrough made by Lefschetz in his important Memoir [112].

Lefschetz used systematically the group law to show that if $s_1(v), \dots, s_r(v) \in H^0(m\Theta)$ and we choose points $a_1, \dots, a_r$ such that $\Sigma_i a_i = 0$, then the product $s_1(v+a_1) \cdots s_r(v+a_r) \in H^0(rm\Theta)$. Then he chooses sufficiently many and sufficiently general points $a_1, \dots, a_r$ so that these sections separate points and tangent vectors. Thus Lefschetz proves in particular

Theorem 4.3. *$|m\Theta|$ yields a morphism for $m \geq 2$ and an embedding for $m \geq 3$.*

The direction started by Lefschetz was continued by many authors, notably Igusa, Mumford, Koizumi, Kempf, who proved several results concerning the equations of the image of an Abelian variety (e.g., that the image of $|m\Theta|$ is an intersection of quadrics for $m \geq 5$).

I will later return to some new ideas related to these developments, I would like now to focus on the status of the problem I mentioned.

The case where A has dimension $g = 1$ being essentially solved by Weierstrass, the next question is whether the answer is known for $g = 2$. This is the case, since the description of the graded ring was obtained by A. Canonaco in 2001 [28]; an abridged version of his result is as follows

Theorem 4.4. *Let A be an Abelian surface and Θ be an effective divisor yielding a principal polarization: then the graded ring $\mathcal{R}(A, \mathcal{O}_A(\Theta))$ has a presentation with 11 generators, in degrees $(1, 2^3, 3^5, 4^2)$, and 37 relations, in degrees $(4, 5^6, 6^{17}, 7^{10}, 8^3)$.*

[30] "Theorie der Thetafunktionen," Teubner, 1894.

Remark 4.5. Canonaco gives indeed explicitly the 37 relations, whose shape however is not always the same. One obtains in this way an interesting stratification of the moduli space of p. p. Abelian surfaces. Does this stratification have a simple geometrical meaning in terms of invariant theory?

The proof uses at a certain point some computer algebra aid, since the equations are rather complicated.

Nevertheless, we would like to sketch the simple geometric ideas underlying the algebraic calculations, since, as one can easily surmise, they are related to the aforementioned Riemann's developments of Abel's investigations.

Proof. The key point is thus that, in the case where Θ is an irreducible divisor, Θ is isomorphic to a smooth curve C of genus 2 (the other case where Θ is reducible is easier, since then A is a product of elliptic curves, and Θ is the union of a vertical and of a horizontal curve).

One uses first of all the exact sequence (for $n \geq 2$, since for $n = 1$ the right arrow is no longer surjective)

$$0 \to H^0(A, \mathcal{O}_A((n-1)\Theta)) \to H^0(A, \mathcal{O}_A(n\Theta)) \to H^0(\Theta, \mathcal{O}_\Theta(n\Theta)) \to 0$$

and of the isomorphism

$$H^0(\Theta, \mathcal{O}_\Theta(n\Theta)) \cong H^0(C, \mathcal{O}_C(nK_C)) .$$

One relates thus our graded ring to the canonical ring of the curve C, which is well known, the canonical map of C yielding a double covering of $\mathbb{P}^1$ branched on 6 points.

In more algebraic terms, there is a homogeneous polynomial $R(y_0, y_1)$ of degree 6 such that

$$\mathcal{R}(C, \mathcal{O}_C(K_C)) \cong K[y_0, y_1, z]/(z^2 - R(y_0, y_1)) .$$

One can summarize the situation by observing that, if $\mathcal{R}_A := \mathcal{R}(A, \mathcal{O}_A(\Theta))$, $\mathcal{R}_C := \mathcal{R}(C, \mathcal{O}_C(K_C))$, then $\mathcal{R}_A$ surjects onto the subring $\mathcal{R}'$ of $\mathcal{R}_C$ defined by $\mathcal{R}' := \bigoplus_{n \geq 2} H^0(C, \mathcal{O}_C(nK_C))$.

To lift the ring structure of $\mathcal{R}'$, which is not difficult to obtain, to the ring structure of $\mathcal{R}_A$ we use again Abel's theorem, i.e., the sequence of maps

$$C^2 \to C^{(2)} \to A$$

where the last is a birational morphism contracting to a point the divisor E consisting of the set of pairs $\{(P, i(P)) | P \in C\}$ in the canonical system of C (here i is the canonical involution of the curve C).

Letting D_i be the pull back of a fixed Weierstrass point of C under the i-th projection of $C \times C$ onto C, we obtain that

$$H^0(A, \mathcal{O}_A(n\Theta)) \cong H^0(C^2, \mathcal{O}_{C^2}(n(D_1 + D_2 + E)))^+,$$

where the superscript $^+$ denotes the $+1$-eigenspace for the involution of C^2 given by the permutation exchange of coordinates. It also helps to consider that the Θ divisor of A is the image of a vertical divisor $\{P\} \times C$. Instead, the diagonal Δ_C of C^2 enters also in the picture because the pull back (under $h = \phi_K \times \phi_K : C \times C \to \mathbb{P}^1 \times \mathbb{P}^1$, ϕ_K being the canonical map of C) of the diagonal of $\mathbb{P}^1 \times \mathbb{P}^1$ is exactly the divisor $\Delta_C + E$, whence $\Delta_C + E$ is linearly equivalent to $2D_1 + 2D_2$.

We omit the more delicate parts of the proof, which are however based on the above linear equivalences and on the action of the dihedral group $D_{4\times 2}$ on $C \times C$ (this is a lift, via the $(\mathbb{Z}/2)^2$ Galois cover $h : C \times C \to \mathbb{P}^1 \times \mathbb{P}^1$ of the permutation exchange of coordinates on $\mathbb{P}^1 \times \mathbb{P}^1$). □

Remark 4.6. 1) This approach should work in principle for the more general case of the Jacobian variety of a hyperelliptic curve. In fact, for each Jacobian variety we have the sequence of maps

$$C^g \to C^{(g)} \to A$$

and the Θ divisor is the image of a big vertical divisor $\{P\} \times C^{g-1}$. Again we have a $(\mathbb{Z}/2)^g$ Galois cover $h : C^g \to (\mathbb{P}^1)^g$ and a semidirect product of the Galois group with the symmetric group in g letters $\mathcal{S}_g$ (the group of the g-dimensional cube).

2) Another question is whether there exists a more elegant, or just shorter presentation for the ring.

The algebra of theta functions and the algebra of representation theory. A more conceptual understanding of the several identities of general theta functions came through the work of Mumford ([121], cf. also [95] and [163]).

In Mumford's articles and in Igusa's treatise one finds a clear path set by choosing representation theory as a guide line, especially as developed by Weyl, Heisenberg, and von Neumann.

The basic idea is shortly said: let G be a compact topological group, endowed henceforth with the (translation invariant) Haar measure $d\mu_G$.

Consider then the vector space $V = L^2(G, \mathbb{C})$: then we have an action τ of the group G on V by translations, $\tau_\gamma(f)(g) := f(g\gamma^{-1})$.

Defining the group of characters G^* as $G^* := Hom(G, \mathbb{C}^*)$ we have an action of G^* on V given by multiplication $\chi f(g) := \chi(g) f(g)$.

The two actions fail to commute, but by very little, since

$$\chi[\tau_\gamma(f)](g) = \chi(g) f(g\gamma^{-1})$$

$$\tau_\gamma(\chi f)(g) = \chi(g\gamma^{-1}) f(g\gamma^{-1}) = \chi(\gamma)^{-1}\chi(g) f(g\gamma^{-1})$$

thus commutation fails just up to multiplication with the constant function $\chi(\gamma)^{-1}$.

Together, the action of G and of G^* generate a subgroup of the Heisenberg group, a central extension

$$1 \to \mathbb{C}^* \to \text{Heis}(G) \to G \times G^* \to 1.$$

It turns out that the algebra of theta functions is deeply related to the representation theory of the Heisenberg group of the Abelian variety A (we see A as the given group G).

But, as Mumford pointed out, we have a more precise relation which takes into account a given line bundle $\mathcal{L}$.

In the case of an Abelian variety A, the group of characters is endowed with a complex structure viewing it as the **Picard variety** $\mathrm{Pic}^0(A)$, the connected component of 0 in $H^1(A, \mathcal{O}_A^*)$. $\mathrm{Pic}^0(A)$ is also called the dual Abelian variety, and a non degenerate line bundle $\mathcal{L}$ is one for which the homomorphism $\phi_{\mathcal{L}} : A \to \mathrm{Pic}^0(A)$, defined by $\phi_{\mathcal{L}}(x) = T_x^*(\mathcal{L}) \otimes \mathcal{L}^{-1}$ (T_x denoting translation by x), is surjective (hence with finite kernel $K(\mathcal{L})$).

Mumford introduces the finite Heisenberg group associated to $\mathcal{L}$ via the so called Thetagroup of $\mathcal{L}$, defined as $\mathrm{Theta}(\mathcal{L}) := \{(x, \psi) | \psi : \mathcal{L} \cong T_x^*(\mathcal{L})\}$.

$\mathrm{Theta}(\mathcal{L})$ is a central extension of $K(\mathcal{L})$ by $\mathbb{C}^*$, but since $K(\mathcal{L})$ is finite, if n is the exponent of $K(\mathcal{L})$, the central extension is induced (through extension of scalars) by another central extension

$$1 \to \mu_n \to \Theta(\mathcal{L}) \to K(\mathcal{L}) \to 1,$$

where μ_n is the group of n-th roots of unity.

Moreover, the alternating form $\alpha : \Gamma \times \Gamma \to \mathbb{Z}$ given by the Chern class of $\mathcal{L}$ gives a non degenerate symplectic form on $K(\mathcal{L})$ with values in μ_n, thus allowing to easily obtain from $\Theta(\mathcal{L})$ the Heisenberg group of a finite group G.

The geometry of the situation is that the group $K(\mathcal{L})$ acts on the projective space associated to the vector space $H^0(A, \mathcal{L})$: but if we want a linear representation on the vector space $H^0(A, \mathcal{L})$ we must see this vector space as a representation of the finite Heisenberg group $\Theta(\mathcal{L})$ (thus we have a link with Schur's theory of multipliers of a projective representation).

In the case of $\mathcal{L} = \mathcal{O}_A(n\Theta)$, $K(\mathcal{L})$ consists of the subgroup A_n of n-torsion points, and another central idea, when we have to deal with a field of positive characteristic, is to replace the vector space $\Gamma \otimes \mathbb{R}$ with the inverse limit of the subgroups A_n.

The story is too long and too recent to be further told here: Mumford used this idea in order to study the moduli space of Abelian varieties over fields of positive characteristics, and in turn this was used to take the reduction modulo primes of Abelian varieties defined over number fields.

These results were crucial for arithmetic applications, especially Faltings' solution [83] of the

Mordell conjecture. *Let a curve C of genus $g \geq 2$ be defined over a number field K: then the set $C(K)$ of its rational points is finite.*

It must be furthermore said that the algebraic calculations allowed by the study of the characters of representations of the finite Heisenberg groups has lead also to a better concrete understanding of equations and geometry of Abelian varieties.

Surprisingly enough, even in the case of elliptic curves this has led, together with Atiyah's study of vector bundles on elliptic curves (cf. [11]) to a deeper understanding of the geometry of symmetric products of elliptic curves and their maps to projective spaces (cf. [45]).

The recent literature is so vast that we have chosen to mention just a single but quite beautiful example, due to Manolache and Schreyer (cf. [114]).

The authors give several equivalent descriptions of the moduli space $X(1,7)$ of Abelian surfaces S with a polarization L whose elementary divisors are $(1,7)$.

Their main result is that this moduli space is birational to the Fano 3-fold V_{22} of polar hexagons to the Klein plane quartic curve C (of equation $x^3y+y^3z+z^3x=0$) which is a compactification of the moduli space $X(7)$ of elliptic curves E with a level 7 structure, i.e., elliptic curves given with an additional isomorphism of the group E_7 of torsion points with $(\mathbb{Z}/7)^2$.

The Klein quartic is rightly famous because it admits then as group of automorphisms the group $SL(2,\mathbb{Z}/7)/\{\pm I\}'$, a group of cardinality 168, and as it is well known this makes the Klein quartic the curve of genus 3 with the maximal number of automorphisms (cf. [2]).

Now, it is easy to suspect some connection between the pairs (S, L) and the pairs $(E, \mathcal{L} := \mathcal{O}_E(7O))$ once we have learnt of the finite Heisenberg group: in fact, the respective groups for L and for $\mathcal{L}$ are isomorphic, and the respective complete linear systems $|L|$ and $|\mathcal{L}|$ yield embeddings for E, respectively for the general S, into $\mathbb{P}^6$.

That is, we view both E and S as Heisenberg invariant subvarieties of the same $\mathbb{P}^6$ with an action of $(\mathbb{Z}/7)^2$ provided by the projectivization of the standard representation of the Heisenberg group on $\mathbb{C}^{\mathbb{Z}/7}$.

The geometry of the situation tells us that E is an intersection of $28-14=14$ independent quadrics, while we expect S to be contained in $28-28=0$ quadrics, so we seem to be stuck without a new idea.

The central idea of the authors is to think completely in algebraic terms, looking at a self dual locally free Hilbert resolution of the ideal of S, which has length 5 instead of 4 (because of $H^1(S,\mathcal{O}_S)\neq 0$).

It turns out that the middle matrix, because of Heisenberg symmetry, boils down to a 3×2 matrix of linear forms on a certain $\mathbb{P}^3$. Then, the 3 determinants of the 2×2-minors yield three quadric surfaces whose intersection is a twisted cubic curve Γ_S which is shown to completely determine S. In this way one realizes the moduli space as a certain subvariety of the Grassmann variety of 3-dimensional vector subspaces of the vector space $H^0(\mathbb{P}^3, \mathcal{O}_{\mathbb{P}^3}(2))$ of quadrics in the given $\mathbb{P}^3$ (one takes the locus of subspaces where a certain antisymmetric bilinear map of vector bundles restricts to zero).

This subvariety is the Fano 3-fold V_{22} mentioned above, which is a rational variety.

At this point we don't want to deprive the reader of the pleasure of learning the intricate details from the original sources [114, 141]: but we need at least to explain what is a polar hexagon of a plane quartic curve C with equation $f(x)=0$.

Observe that the polynomial $f(x)$ depends upon exactly 15 coefficients, whence, if we take 6 linear forms $l_i(x)$, they depend on 18 affine parameters, and we expect therefore to have a 3-dimensional variety Hex(C) parametrizing the 6-tuples of such linear forms such that $f(x) = \Sigma_i l_i(x)^4$.

Hex(C) is called the variety of polar hexagons, and it is indeed a 3-fold in the case of the Klein quartic. Finally, these constructions allow to find the Klein quartic as the discriminant of the net of quadrics in $\mathbb{P}^3$ associated to the Hilbert resolution of S.

Also the identification of the Fano 3-fold with Hex(C) is based on the study of higher syzygies, but beyond this many other beautiful classical results are used, which are due to Klein, Scorza, and Mukai (cf. [99, 142, 119]).

Especially nice is the old theorem of Scorza [142], proved in 1899, that the variety of plane quartic curves is birational to the variety of pairs of a plane quartic curve D given together with an even theta-characteristic (this amounts to writing the equation of D as the discriminant of a net of quadrics in $\mathbb{P}^3$). This theorem is a clear example of the geometrization of ideas coming from the theory of theta functions (which, as we saw, are certain Fourier series, and therefore, seemingly, purely analytical objects).

Abelian varieties and multilinear algebra. The roots of these developments, which historically go under the name "The problem of Riemann matrices", and occupied an important role for the birth of the theory of rings, modules and algebras, are readily explained by the following basic

Remark 4.7. Given two tori $T = V/\Gamma$, $T' = V'/\Gamma'$, any holomorphic map between $f : T \to T'$ between them is induced by a complex linear map $F : V \to V'$ such that $F(\Gamma) \subset \Gamma'$.

Whence, for a g-dimensional torus, the ring

$$\mathrm{End}(T) : \{f : T \to T | f \text{ is holomorphic}, f(0) = 0\}$$

is the subring of the ring of matrices Mat$(2g, 2g, \mathbb{Z})$ given by

$$\mathrm{End}(T) : \{B \in \mathrm{End}(\Gamma) | B \in (V^\vee \otimes V) \oplus (\bar{V}^\vee \otimes \bar{V}) \subset (\Gamma \otimes \mathbb{C})^\vee \otimes (\Gamma \otimes \mathbb{C})\}$$

since then the restriction of B to $V \subset (\Gamma \otimes \mathbb{C})$ is complex linear.

In general, the study of endomorphism rings of complex tori is not completely achieved.

The main tool which makes the case of Abelian Varieties easier is the famous **Poincaré's complete reducibility theorem** (cf. [130])

Theorem 4.8. *Let A' be a subabelian variety of an Abelian variety A: then there exists another Abelian variety A'' and an* **isogeny** *(a surjective homomorphism with finite kernel) $A' \times A'' \to A$.*

Proof. The datum of A' amounts to the datum of a sublattice $\Gamma' \subset \Gamma$ which is saturated (Γ/Γ' is torsion free) and complex (i.e., there is a complex subspace $W \subset V$ with $W \oplus \overline{W} = \Gamma' \otimes \mathbb{R} \subset \Gamma \otimes \mathbb{R} = V \oplus \overline{V}$).

Now, given the alternating form A, its orthogonal in Γ yields a sub-lattice Γ'' spanning the complex subspace U orthogonal to W for the Hermitian bilinear product associated to H: since H is positive definite, we obtain an orthogonal direct sum $V = W \oplus U$, and we define $A'' := U/\Gamma''$. □

We sketched the above proof just with the purpose of showing how the language of modern multilinear algebra is indeed very appropriate for these types of questions.

The meaning of the reducibility theorem is that, while for general tori a subtorus $T' \subset T$ only yields a quotient torus T/T', here we get a direct sum if we consider an equivalence relation which identifies two isogenous Abelian vareties.

Algebraically, the winning trick was thus to classify first Endomorphism Rings tensored with the rational integers, because

Remark 4.9. If $T = V/\Gamma, T' = V/\Gamma'$ are isogenous tori, then $\mathrm{End}(T) \otimes \mathbb{Q} \equiv \mathrm{End}(T') \otimes \mathbb{Q}$.

And then the study is restricted to the one of **Simple Abelian Varieties**, i.e., of the ones which do not admit any Abelian subvariety whatsoever (naturally, this concept was very much inspired by the analogous concept of curves which do not admit a surjective and not bijective mapping onto a curve of positive genus).

The classification of endomorphism rings of Abelian varieties was achieved through a long series of works by Scorza, Rosati, Lefschetz, and Albert (cf. e.g. [143, 112, 5, 138]) and today one can find an exposition in Chapters 5 and 9 of the book by Lange and Birkenhake [107], cf. also, for an historical account, the article by Auslander and Tolimieri [12].

Although the methods of Scorza and Rosati were more geometrrical, certainly more than the later ones by Albert, who essentially worked in the new direction set up by Emmy Noether, i.e., of the abstract algebra, a central role is played by a notion due to Rosati, the socalled **Rosati involution**.

Given an endomorphism with integral matrix B, the Rosati involution associates to it (A being a Riemann integral matrix as in (3.3), (3.4)) the matrix $B' : A^{-1}\,{}^tBA$. The Rosati involution is positive in the sense that the symmetric bilinear form $(B_1, B_2) := \mathrm{Tr}(B_1' B_2 + B_1 B_2')$ yields a positive definite scalar product.

It turns out that the classification of Riemann matrices is very close to the study of rational Algebras with a positive involution, and abstract arguments imply that these simple algebras are skew fields $\mathbb{F}$ of finite dimension over $\mathbb{Q}$ of two types

(I) The centre K of $\mathbb{F}$ is a totally real number field and, if $K \neq \mathbb{F}$, then $\mathbb{F}$ is a quaternion algebra over K. Moreover, for every embedding $\sigma : K \to \mathbb{R}$, $\mathbb{F} \otimes \mathbb{R}$ is always definite ($\mathbb{F} \otimes \mathbb{R} \cong \mathbb{H}$), or always indefinite ($\mathbb{F} \otimes \mathbb{R} \cong M(2, \mathbb{R})$).
(II) The centre K of $\mathbb{F}$ is a totally complex quadratic extension of a totally real number field K_0, and then $\mathbb{F} \otimes \mathbb{C}$ is a matrix algebra $M(r, \mathbb{C})$ such that the positive involution extends to the standard involution $C \to \overline{{}^tC}$.

The analytic moduli theory of Abelian varieties owes much to the work of Siegel and to his ‘Symplectic geometry’ [150]: today the space of matrices

$$\mathcal{H}_g := \{\tau \in \mathrm{Mat}(g, g, \mathbb{C}) | \tau =^t \tau, \ \mathrm{Im}(\tau) > 0\}$$

is called the Siegel upper half space, and it is a natural parameter space for Abelian varieties, since, depending on the polarization, there is a subroup Γ of $Sp(2g, \mathbb{Q})$ such that the moduli space is, analytically, the quotient $\mathcal{H}_g/\Gamma$.

The moduli theory of Abelian varieties with a certain polarization and endomorphism structure was pursued relatively recently by Shimura (cf. [149]), and it is a currently very active field of research for the arithmetic applications of the theory of such Shimura varieties.

I do not need to cite for instance the (recently proved) so called Shimura–Taniyama–Weil conjecture about the modularity of elliptic curves defined over $\mathbb{Q}$: I can simply refer to the talk by Wiles.

In this direction, however, the current tendency is to develop also much the geometry, since one has to look at the reduction of these modular varieties modulo primes. The hope is that this study will play a primary role for the pursuing of the so called Langlands program, which is a vast generalization of the previously cited conjecture, proposing to relate modular forms arising in different contexts (cf. [108], and [109, 110, 70] for early accounts of the story).

I hope that some more competent author than me will report about this development in the present volume.

I want instead to end this section by pointing out (cf. [170]) how important the role of Scorza was for the development of the field of abstract algebra in Italy: his path started with correspondences between curves, but, as we contended here, his researches centered about Riemann matrices made him realize about the relevance of the powerful new algebraic concepts.

5 Further Links to the Italian School

We mentioned in the previous section how the research of Rosati and Scorza was very much influenced by the new geometric methods of the Italian school of algebraic geometry.

As we said, a crucial role was played by Castelnuovo: concerning Abel's theorem, in the article [31], entitled "The 1–1 correspondences between groups of p points on a curve of genus p", he explained how one could e.g. formulate the fundamental theorem about the inversion of Abelian integrals as a consequence of the theory of linear series on a curve (a development starting with the geometric interpretation of the Riemann Roch theorem).

It is interesting to observe that, when he wrote a final note in the edition appearing in his collected works, he points out that the results can be formulated in a simpler way if one introduces the concept of the Jacobian variety of the curve.

These notes added around 1935 are rather interesting: for instance, in the note to the paper entitled "On simple integrals belonging to an irregular surface" [32] he pointed out that exactly in this memoir he introduced the concept of the socalled **Picard variety**, applying this concept to the study of algebraic surfaces. In fact, the theorem of Picard to which Castelnuovo refers, proved by Picard in [128], and with precisions by Painlevé in [127], is indeed the characterization of the Abelian varieties (we add to it a slight rewording in modern language).

Theorem 5.1. *Let V^p be a p-dimensional algebraic variety admitting a transitive p-dimensional abelian group of birational transformations: then the points $\xi \in V$ are uniformized by entire 2p-periodic functions on $\mathbb{C}^p$,*

$$\xi_k = \phi_k(u_1, \dots, u_p)$$

(i.e., V is birational to a complex torus of dimension p).

The first main result of Castelnuovo in [32] is the construction of the so called Albanese variety and Albanese map of an algebraic surface X. Recall that, in modern language, the Albanese variety of a projective variety X is the Abelian variety $(H^0(\Omega^1_X))^\vee / H_1(X)$ where $H_1(X)$ is the lattice, in the dual vector space of $H^0(\Omega^1_X)$, given by integration along closed paths.

The Albanese map, defined up to translation, as a result of the choice of a base point x_0, associates to a point x the linear functional $\alpha(x) := \int_{x_0}^{x} (\text{mod } H_1(X))$.[31]

The second result, obtained independently by F. Severi in [145], concerns the equality of the irregularity of an algebraic surface and the dimension of the space of holomorphic one forms. Both proofs were relying on a shaky proof given by Enriques one year before, in [76], claiming the existence a continuous system Σ of dimension $q := p_g - p_a$ of "inequivalent curves" (i.e., such that for a generic curve $C \in \Sigma$, the set of curves in Σ which are linearly equivalent to C has dimension zero).

Fortunately, a corrrect analytical proof was later found by Poincaré in [132]. Enriques and Severi tried for a long time to repair the flaw in Enriques' geometrical arguments, although in the end it started to become clear the need for higher order differential elements (i.e., higher order terms in the Taylor expansion of the curve variation). The fruit of the researches carried on much later in the 50's was to show that indeed, for varieties defined over algebraically closed fields of positive characteristic, the **arithmetic irregularity** $q := p_g - p_a$ was in general larger than the **geometric irregularity** defined as the maximal dimension of a continuous

[31] Giacomo Albanese emigrated from Italy in 1936 to Sao Paulo, Brasil, where, soon after the war, he became closely acquainted with André Weil, who taught there, as well as Zariski. Weil is responsible for the name "Albanese variety," but Ciliberto and Sernesi in [4] write: "Whilst the attribution of this concept to Albanese is dubious, ... ". Indeed the basis for this is an article, [3], where Albanese studies correspondences between algebraic surfaces through the consideration of the induced action on the space of holomorphic 1-forms. The coupled names "Albanese" and "Picard" appear in the title of the article by W. L. Chow, "Abstract theory of the Picard and Albanese varieties," [52].

Indeed, in the 50's, one main purpose was to distinguish among the two dual varieties, which are only isogenous, and not in general isomorphic. According to the historical note of Lang on page 52 of [106], Matsusaka was the first to give a construction of the Albanese variety using the generic curve.

system of "inequivalent curves". We refer the reader to Mumford's classical book [120], relating this question to the non reducedness of the **Picard scheme** $H^1(\mathcal{O}_X^*)$.

Castelnuovo was instead more interested in the applications of the previous theorem, the most important one being the theorem [33] that an algebraic surface with arithmetic genus p_a smaller than -1 is birationally ruled.

This theorem is indeed one of the key theorems of the classification of algebraic surfaces, since it also implies the well known

Castelnuovo rationality criterion: *A surface is rational if and only if the bigenus* P_2 *and the irregularity* q *vanish.*

Without opening a new story, I would like to observe that the socalled "Enriques classification" of algebraic surfaces, done by Castelnuovo and Enriques, was one of the most interesting cooperations in the history of mathematics; it took place in the years from 1892 to 1914 (and was especially intense in the period 1892 to 1906).

Besides the published papers, one can consult today the book entitled "Riposte armonie" [135] ("Hidden harmonies," as are the ones governing algebraic surfaces), which contains around 670 letters (or postcards) written from Federigo Enriques to Guido Castelnuovo.

Naturally, also Castelnuovo wrote quite many letters, but apparently Enriques did not bother to keep them. This already shows where Castelnuovo and Enriques respectively belong, in the rough distinction made by H. Weyl which we quoted above, although both of them were obviously geometers.

In fact, Enriques used to discuss mathematics with his assistants during long walks in gardens or parks, and would only sometime stop to write something with his stock on the gravel. Moreover, as Guido Castelnuovo wrote of him[32] with affection, he was a "mediocre reader, who saw in a page not what was written, but what he wanted to see"; certainly his brain was always active like a vulcano. After the first world war the collaboration of the two broke up, more on the side of Castelnuovo. As his daughter Emma Castelnuovo writes, Enriques would regularly visit his sister (Castelnuovo's wife) at their house, and after dinner the two mathematicians went to a separate sitting room, where Enriques wanted to discuss his many new ideas, while Castelnuovo had prudently instructed his wife to come after some time and interrupt their conversation with some excuse.

Castelnuovo was 6 years older than Enriques, was always calm and mature, and, after the appearance of his ground breaking two notes over algebraic surfaces [30] he was a natural reference for the brilliant student Enriques, who graduated in Pisa in 1891 (just at the time when Guido became a full professor). Enriques wanted first to perfection his studies under the guidance of Segre in Torino, but instead got a fellowship in Rome by Cremona, and there, in 1892, started the intense mathematical interchange with Castelnuovo.

In his first Memoir [75] Enriques, after an interesting historical introduction, sketches the main tools to be used for the birational study of algebraic surfaces, namely: the theory of linear systems of curves, the canonical divisor and the operation

[32] In the commemoration opening vol. I of his selected papers in geometry, [78], IX–XXII.

of adjunction. Some results, as the claim that two birational quartic surfaces are necessarily projectively equivalent, are today known not to hold. It took a long time to make things work properly, and it is commonly agreed that the joint paper [35] marks the achievement of the classification theory.

For the later steps (main ones, as well known to anyone who understands the structure of the classification theorem), a very important role played the

(IP) De Franchis' theorem on irrational pencils [66, 33] and the
(HS) Classification of hyperelliptic surfaces.

(IP) De Franchis' theorem, obtained independently also by Castelnuovo and Enriques, asserts that if on a surface S there are several linearly independent one forms $\omega_i, i = 1, \dots r$ which are pointwise proportional, then there is a mapping $f : S \to C$ to an algebraic curve C such that these forms are pull backs of holomorphic one forms on C. It was used by Castelnuovo to show that under the inequality $p_g \leq 2q - 4$ there is a mapping to a curve of positive genus.

This theorem leads to a typical example of the algebraization of the geometry: although the Hodge theory of Kähler varieties was established with the use of hard analytic tools which underlie Hodge's theorem on harmonic integrals [94, 101], what turned out to be very fertile was the simple algebraic formulation in terms of the cohomology algebra of a projective variety.

Using this, Z. Ran, M. Green, and myself [134, 38] were independently able to extend the result of de Franchis to the case of higher dimensional varieties and higher dimensional targets. In this way the ideas of the italian school came back to intense life, and became an important tool for the investigation of the fundamental groups of algebraic varieties (for instance N. Mok [117] tried to extend this result to infinite dimensional representations, with the hope of using such a result for the solution of the so called Shafarevich conjecture about the universal coverings of algebraic varieties).

(HS) The second main work of de Franchis,[33] together with Bagnera, was the classification of hyperelliptic surfaces, i.e. of surfaces whose universal covering space is biholomorphic to $\mathbb{C}^2$ (cf. [14, 13, 14, 79, 80, 81, 82]). This classification was also obtained by Enriques and Severi, and the Bordin Prize was awarded to Enriques and Severi in 1907, and to the Sicilian couple in 1909[34]. Strange as it may seem that two couples get two prices for the same theorem, instead of sharing one, this story is even more complicated, since the first version of the paper by Enriques and Severi was withdrawn after a conversation of Severi with de Franchis, and soon corrected. Bagnera and de Franchis were only a little later, since they had to admit a restriction (a posteriori useless, since no curve on an Abelian surface is contractible, cf. [67]): their proof was however simpler, and further simplified by de Franchis much later [68].

[33] Michele de Franchis, born in Palermo, 1875, was very much influenced by the teaching of Scorza. He was also the Director of the Rendiconti Circolo Matematico di Palermo in the years 1914–1946, as the successor of the founder G. B. Guccia.

[34] Lefschetz was instead recipient of the Bordin Prize in 1919 for the cited memoir [112].

Another beautiful result, and this one even more related to Abelian integrals, is the famous

Torelli's Theorem. *Let C, C' be two algebraic curves whose Jacobian varieties are isomorphic as polarized Abelian varieties (equivalently, admitting the same matrix of periods for Abelian integrals of the first kind): then C and C' are birationally isomorphic, [154].*

Torelli was born in 1884 and was a student of Bertini, in Pisa, where he attended also Bianchi's lectures; he was for short time assistant of Severi, and died prematurely in the first world war, in 1915.

These years at the beginning of the 20-th century in Italy were thus quite exciting. In the book by J. Dieudonné, [69] vol.1, Chapter VI, entitled "Developpement et chaos", contains a paragraph dedicated to "L'école italienne et la théorie des systemes linéaires", namely, devoted to the second generation of the Italian school. As third generation of the Italian school, we became very interested, also because of this criticism, to become fully acquainted with the results of these precursors.

What I found as a very interesting peculiarity reading the book by Enriques on "Algebraic surfaces" [77], was the mixture of theorems, proofs, speculations, and history of the genesis of the mathematical ideas.

For instance, Chapter IX, entitled "Irregular surfaces and continuous systems of disequivalent curves," has a section 6 on "History of the theory of continuous systems" (pages 339–347). One can read there that in 1902 Francesco Severi, who had just graduated in Torino, following the advice of his master C. Segre, accepted a position in Bologna as assistant of Enriques. Under the influence of Enriques, Severi started the investigation of the particular surfaces which occur as the symmetric square of a curve (cf. [144]). According to Enriques, this research lead him to consider the **Problem of the base** for the class of divisors modulo numerical equivalence.

We may also observe that these surfaces have a very special geometry, and, although a general characterization has not yet been described in general, we have some quite recent (cf. [46, 93, 129]) results.

Theorem 5.2. *The symmetric squares $C^{(2)}$ of a curve of genus 3 are the only irregular surfaces S of general type with $p_g \geq 3$ presenting the* **non-standard case** *for the bicanonical map, i.e., such that*

- *their bicanonical map ϕ_{2K} is not birational onto its image*
- *S does not contain any continuous system of curves of genus 2.*

Moreover, any algebraic surface with $p_g = q = 3$ is either such a symmetric product (iff $K^2 = 6$ for its minimal model), or has $K^2 = 8$ and is the quotient of a product $C_1 \times C_2$ of two curves of respective genera 2 and 3 by an involution $i = i_1 \times i_2$ where C_1/i_1 has genus 1, while i_2 operates freely.

The symmetric square $C^{(2)}$ of a curve of genus 2 occurs in another characterization of the **non-standard case** given by Ciliberto and Mendes Lopes [56, 57].

Theorem 5.3. *The double covers S of a principal polarized Abelian surface A, branched on a divisor algebraically equivalent to 2Θ, are the only irregular surfaces of general type with $p_g = 2$ presenting the* **non-standard case** *for the bicanonical map.*

In these theorems an essential role is played by the continuous system of **paracanonical curves**, i.e., those curves which are algebraically equivalent to a canonical divisor. To this system is devoted section 8 of the cited Chapter IX of [77], and there Enriques, after mentioning false attempts by Severi and himself to determine the dimension of the paracanonical system $\{K\}$, analyses the concrete case of a surface $C^{(2)}$, with C of genus 3, in order to conjecture that $\dim\{K\} = p_g$. The assumptions conjectured by Enriques were not yet the correct ones, but, under the assumptions that the surface does not contain any irrational pencil of genus ≥ 2, the conjecture of Enriques was proved by Green and Lazarsfeld, via the so called "Generic vanishing theorems" (cf. [87, 88]).

I will not dwell further on this very interesting topic, referring the reader to the survey, resp. historical, articles [39] and [54].

I should however point out that further developments are taking place in this direction, following a seminal paper by Mukai [118] who extended the concept of Fourier transforms (don't forget that theta functions are particular Fourier series!) to obtain an isomorphism between the derived category of coherent sheaves on an Abelian variety X and the one of its dual Abelian variety $\hat{X} := \mathrm{Pic}^0(X)$.

One specimen is the combination of Mukai's technique with the theory of generic vanishing theorems by Green and Lazarsfeld to obtain limitations on the singularities of divisors on an Abelian variety (cf. [92], where one can also find references to previous work by Kollár, Ein, and Lazarsfeld).

Speaking about links with the Italian school I should not forget the beautiful lectures I heard in Pisa from Aldo Andreotti on complex manifolds and on complex tori. Through his work with F. Gherardelli [8], I got in touch with a problem of transcendental nature which occupies a central place in Severi's treatise on Quasi Abelian Varieties.

Quasi Abelian varieties, in Severi's terminology, are the Abelian complex Lie groups which sit as Zariski open sets in a projective variety.

Whence, they are quotients $\mathbb{C}^n/\Gamma$ where Γ is a discrete subgroup of $\mathbb{C}^n$, thus of rank $r \leq 2n$, and the above algebraicity property leads again to the two Riemann bilinear relations:

I) There exists an alternating form $A : \Gamma \times \Gamma \to \mathbb{Z}$ such that A is the imaginary part of an Hermitian form H on V.
II) There exists such an H which is positive definite (in this case H is not uniquely determined by A).

Andreotti and Gherardelli conjectured that **the Riemann bilinear relations hold if and only if there is a meromorphic function of $\mathbb{C}^n$ with group of periods equal**

to Γ. This conjecture was the first Ph.D. problem I gave, and after some joint efforts, Capocasa and I were able to prove it in [29].

Homological algebra for Abelian and irregular varieties. As already mentioned, David Mumford's ground breaking articles [121] set up the scope of laying out a completely algebraic theory of theta functions. His attempt was not the only one, for instance Barsotti (cf. [16, 17, 18, 19], and [106] for references about his early work) had another approach to Abelian varieties, based on power series, Witt vectors and generalizations of them (Witt covectors, cf. [18]), and the socalled "Prostapheresis formula" (cf. [19]).

Discussing here the respective merits of both approaches would be difficult, but at least I can say that, while Barsotti's work is mainly devoted to Abelian varieties in positive characteristic, the theory of theta groups, as already mentioned, is also a very useful tool in characteristic zero.

The title of Mumford's series of articles is "On the equations defining abelian varieties," which has a different meaning than "The equations defining abelian varieties."[35] Thus, he set up a program which has been successfully carried out in the case of several types of Abelian varieties. More generally, one can set as a general target the one of studying the equations of irregular varieties, i.e., of those which admit a non trivial Albanese map.

Since otherwise the problem is set in too high a generality, let me give a concrete example (for many topics I will treat now, consult also the survey paper [40], which covers developments until 1996).

Chapter VIII of the book by F. Enriques (finished by his assistants Pompilj and Franchetta after the death of Enriques) is devoted to the attempt to find explicitly the canonical surfaces S of low canonical degree in $\mathbb{P}^3$: i.e., one considers surfaces with $p_g = 4$ and with birational canonical map $\phi := \phi_K : S \to \mathbb{P}^3$. If the canonical system has no base points, then we will have a surface of degree $d = K^2 \geq 5$. The cases $K^2 = 5, 6$ are easy to describe, and for $K^2 \geq 7$ Enriques made some proposals to construct some regular surfaces (with a different method, Ciliberto [53] was able to construct these for $7 \leq K^2 \leq 10$, and to sketch a classification program, later developed in [37], based essentially on Hilbert's syzygy theorem).

It was possible to treat the irregular case (cf. [47]) using a new approach based on Beilinson's theorem [23] for coherent sheaves $\mathcal{F}$ on $\mathbb{P}^n$, which allows to write every such sheaf as the cohomology of a **monad** (a complex with cohomology concentrated at only one point) functorially associated with $\mathcal{F}$.

The natural environment for rings not necessarily generated in degree 1 is however the **weighted projective space**, which is the projective spectrum $Proj(\mathbf{A})$ of a polynomial ring $\mathbf{A} := \mathbb{C}[x_0, \dots, x_n]$ graded in a non standard way, so that the indeterminates x_i have respective degrees m_i which are positive integers, possibly distinct.

35 Computers however have not only helped us to do some explicit calculations, but, according to F. O. Schreyer, they have also made us wiser: by showing us explicit equations which need several pages to be written down, they make us wonder whether having explicit equations means any better understanding.

Canonaco [27] was first able to extend Beilinson's theorem to the weighted case under some restriction on the characteristic of the base field, and later [28] not only removed this restriction, but succeeded to construct a unique functorial Beilinson type complex, making use of a new theory, of so called **graded schemes**. A concrete application given was to determine the canonical ring of surfaces with $K^2 = 4$, $p_g = q = 2$ (for these surfaces there is no good canonical map to the ordinary 3-space $\mathbb{P}^3$).

Although a general theory appears to be very complicated, it thus turns out that Abelian varieties (and for instance related symmetric products of curves) offer crucial examples (admitting sometime a geometrical characterization) for the study of irregular varieties. Some of them were already discussed before, we want to give a new one which is particularly interesting, and yields (cf. [47] for more details) an easy counterexample to an old "conjecture" by Babbage (cf. [36] and [22] for references and the first counterexample).

Example 5.4. Let $J = \mathrm{Jac}(C)$ be the Jacobian of a curve C of genus 3, and let $A \to J$ be an isogeny of degree 2. The inverse image S of the theta divisor $\Theta \cong_{bir} C^{(2)}$ provides A with a polarization of type $(1, 1, 2)$.

The canonical map of S factors through an involution ι with 32 isolated fixed points, and the canonical map of the quotient surface $\Sigma := S/\iota$, whose image is a surface Y of degree 6 in $\mathbb{P}^3$ having 32 nodal isolated singularities, and a plane cubic as double curve. Moreover, Σ is the normalization of Y.

It is now difficult for me to explain and to foresee exactly what principles these examples should illustrate, let me however try.

In Theorem 3.1 I tried to give an explicit example of the "algebraization" of the geometry, showing how the question of the birationality to Abelian varieties of symmetric products of varieties can be reduced to pure exterior algebra arguments. Abelian varieties are just, so to say, the complex incarnation of exterior algebras.

On the other hand, a companion article by Bernstein–Gelfand–Gelfand [24] appeared next to the cited article [23] by Beilinson. In abstract setting, it shows the equivalence of the derived category of coherent sheaves on $\mathbb{P}^n$ and the category of finite modules over the exterior algebras. While it is not yet completely clear how to extend this result to the weighted case, quite recently in [72] it was shown how the BGG method allows to write functorially not only the sheaves, but also the homomorphisms in the Beilinson monad.

Note that the exterior algebra of BGG is the exterior algebra over the indeterminates of a polynomial ring, and is therefore apparently geometrically unrelated for the moment to the exterior algebra of an Abelian variety. Given however a morphism $f : X \to A$, we attach to it the induced homomorphism between the respective holomorphic algebras of A and X, $f^* : H^0(\Omega^*_A) \to H^0(\Omega^*_X)$. Whence, we obtain a module over the exterior algebra, and we associate to it a Beilinson monad. This procedure shows that to f we associate some geometric objects related to the Gauss maps corresponding to f.

It is just the converse which looks more problematic: is it possible to associate, to a map $f : X \to \mathbb{P}^n$ to a projective space, a geometric map to an Abelian variety giving a realization of the module $f_* \mathcal{O}_X$?

In general, progress on the question of canonical rings or equations of irregular varieties might require at least further combinations of the several existing techniques which we have mentioned.

6 More New Results and Open Problems

6.1 The Torelli problem. The Torelli theorem, mentioned in the previous section, was again at the centre of attention in the 50's, when several new proofs were found, by Weil, Matsusaka, Andreotti (and many others afterwards).

Particularly geometrical was the proof given by A. Andreotti [6], who showed that, given a Jacobian variety Jac(C) with a theta divisor $\Theta \cong_{bir} C^{(g-1)}$, the Gauss map of Θ is ramified on the dual variety $C^{\vee} \subset \mathbb{P}^{g-1}$ of the canonical image of the curve C. Using projective duality, one sees therefore that (Jac(C), Θ) determines C (Andreotti had then to treat the hyperelliptic case separately).

Andreotti and Mayer [7] pushed the study of the geometry of canonical curves, especially of the quadrics of rank 4 containing them, to obtain some equations valid for the period matrices of curves inside the Siegel upper half plane. This paper, written at a time when the fashion was oriented in quite different directions, had a great impact on the revival of classical researches about algebraic curves and Abelian varieties (cf. [10] and [107] for references).[36]

In the same years Phillip Griffiths (cf. [89, 90, 91] and references therein) greatly extended the theory of the periods of Abelian integrals, proposing to use the Hodge structures of varieties, i.e., the isomorphism class of the datum

$$H^*(X, \mathbb{Z}) \to H^*(X, \mathbb{Z}) \otimes \mathbb{C} = \oplus_{p,q} H^{p,q}(X)$$

to study their moduli (note that a modern formulation of Torelli's theorem is that the birational isomorphism class of an algebraic curve is determined by the Hodge structure on its cohomology algebra).

A prominent role played in his program the attempt to find a reasonable generalization of Torelli's theorem, and indeed (cf. [91]) a lot of Torelli type theorems were proved for very many classes of special varieties.

Since then, a basic question has been the one of finding sufficient conditions for the validity of an infinitesimal Torelli theorem for the period map of holomorphic

[36] I came to read this beautiful paper under the instigation of Francesco Gherardelli. He explained to me that the citation from "Il teatro alla moda" by Benedetto Marcello: "As a first duty shall the modern poet ignore all about the ancient Roman and Greek poets, because these last too ignored everything about modern poetry" was motivated by the extreme difficulties that Aldo had when he wanted to lecture in Paris about this result without first explaining the excellence of the rings of coefficients he was using, or use the notation g^1_r without explaining the representability of such functors.

n-forms (i.e., for the Hodge structure on $H^n(X, \mathbb{Z}) \otimes \mathbb{C}$) for a variety X of dimension n. The question is then, roughly speaking, whether the period map is a local embedding of the local moduli space. Observe moreover that the n-forms are the only forms which surely exist on simple cyclic ramified coverings Y_D of a variety X of general type, branched on pluricanonical divisors D (divisors $D \in |mK_X|$), and in this context the Torelli problem is a quantitative question about how large m has to be in order that the variation of Hodge structure distinguishes the Y_D's (cf. the article [116] which gave a very interesting application of these ideas to families of higher dimensional varieties, opening a new direction of research).

The validity of such an infinitesimal Torelli theorem can be formulated in purely algebraic terms as follows: is the cup product

$$H^1(X, T_X) \times H^0(X, \Omega^n_X) \to H^1(X, \Omega^{n-1}_X)$$

non degenerate in the first factor?

Classical examples by Godeaux and Campedelli and modern ones (cf. [20] for references) produce surfaces of general type with $p_g = q = 0$ (thus with a trivial Hodge structure), yet with moduli.

In view of Andreotti's interpretation of Torelli's theorem, one suspects then that a good condition might be the very ampleness of the canonical divisor, i.e., the condition that the canonical map be an embedding.

This might unfortunately not be the case, as shown in a joint paper with I. Bauer [21]:

Theorem 6.1. *There are surfaces of general type with injective canonical morphism and such that the infinitesimal Torelli theorem for holomorphic 2-forms does not hold for each surface in the moduli space. Examples of such behaviour are quotients* $(C_1 \times C_2)/(\mathbb{Z}/3)$*, where* $C_1 \to C_1/(\mathbb{Z}/3) \cong \mathbb{P}^1$ *is branched on* $3k+2$ *points, and* $(\mathbb{Z}/3)$ *acts freely on* C_2 *with a genus 3 quotient.*

It would be interesting to establish stronger geometrical properties of the canonical map which guarantee the validity of the infinitesimal Torelli theorem for the holomorphic n-forms.

6.2 The Q.E.D. problem. If higher dimensional varieties were products of curves, life would be much simpler. It obviously cannot be so, since there are plenty of varieties which are simply connected (e.g., smooth hypersurfaces in $\mathbb{P}^n$ with $n \geq 3$), without being rational.

Can life be simpler?

It is a general fact of life that, in order to make the study of algebraic varieties possible, one must introduce some equivalence relation.

The most classical one is the so called birational equivalence, which allows in particular not to distinguish between the different projective embeddings of the same variety.

Definition 6.2. *Let X and Y be projective varieties defined over the field K: then they are said to be birational if their fields of rational functions are isomorphic:* $K(X) \cong K(Y)$.

Moreover, one must allow algebraic varieties to depend on parameters, for instance the complex hypersurfaces of degree d in $\mathbb{P}^n$ depend on the coefficients of their equations: but if these are complex numbers, we can have uncountably many birational classes of algebraic varieties.

To overcome this difficulty, Kodaira and Spencer introduced the notion of **deformation equivalence** for complex manifolds: they [103] defined two complex manifolds X', X to be **directly deformation equivalent** if there is a proper holomorphic submersion $\pi : \Xi \to \Delta$ of a complex manifold Ξ to the unit disk in the complex plane, such that X, X' occur as fibres of π. If we take the equivalence relation generated by direct deformation equivalence, we obtain the relation of deformation equivalence, and we say that X is a deformation of X' in the large if X, X' are deformation equivalent.

These two notions extend to the case of compact complex manifolds the classical notions of irreducible, resp. connected, components of moduli spaces.

It was recently shown however [113, 98, 42] that it is not possible to give effective conditions in order to guarantee the deformation equivalence of algebraic varieties, as soon as the complex dimension becomes ≥ 2.

Thus in [44] I introduce the following relation

Definition 6.3. *Let n be a positive integer, and consider, for complex algebraic varieties X, Y of dimension n, the equivalence relation generated by*

(1) Birational equivalence
(2) Flat deformations with connected base and with fibres having only at most canonical singularities
(3) Quasi-étale maps, i.e., morphisms which are unramified in codimension 1.

This equivalence will be denoted by $X \cong_{QED} Y$ (QED standing for: quasi-étale-deformation).

Remark 6.4. 1) Singularities play here an essential role. Note first of all that, without the restriction on these given in (2), we obtain the trivial equivalence relation (since every variety is birational to a hypersurface).

2) Assume that a variety X is rigid, smooth, with trivial algebraic fundamental group: then X has no deformations, and there is no non trivial quasi-étale map $Y \to X$.

In this case the only possibility, to avoid that X be isolated in its QED-equivalence class, is that there exists a quasi-étale map $f : X \to Y$.

If f is not birational, however, the Galois closure of f yields another quasi-étale map $\phi : Z \to X$, thus it follows that f is Galois and we have a contradiction if $\mathrm{Aut}(X) = \{1\}$.

It does not look so easy to construct such a variety X.

Are there invariants for this equivalence?

A recent theorem of Siu [152] shows that the Kodaira dimension is invariant by QED equivalence.

It is an interesting question to determine the QED equivalence classes inside the class of varieties with fixed dimension n, and with Kodaira dimension k. For curves and special surfaces, there turns out to be only one class [44]:

Theorem 6.5. *In the case $n \leq 2, k \leq 1$ the following conditions are equivalent:*

(i) $X \cong_{QED} Y$
(ii) $dim X = dim Y = n$, $Kod(X) = Kod(Y) = k$.

The previous result uses heavily the Enriques classification of algebraic surfaces. We can paraphrase the problem, for the open case of surfaces of general type, using Enriques' words [77]: "We used to say in the beginning that, while curves have been created by God, surfaces are the work of the devil. It appears instead that God wanted to create for surfaces a finer order of hidden harmonies ... "

Are here then new hidden harmonies to be found?

Acknowledgements. I wish to thank Arnfinn Laudal for convincing me to try to learn about Abel's life and to look at his papers, and for allowing me to do it comfortably at home.

I would also like to thank Ragni Piene for some useful comments.

Bibliographical remark. We are not in the position to even mention the most important references. However, some of the references we cite here contain a vast bibliography, for instance [151, pp. 193–240], and [171, pp. 248–268].

References

[1] N. H. Abel, "*Oeuvres complètes, Tomes I et II* ", L. Sylow and S. Lie ed., Grøndahl, Christiania (1881), 621 pp., 341 pp.
[2] R. D. M. Accola, "*Topics in the theory of Riemann surfaces*," **Springer Lecture Notes in Mathematics. 1595** (1994), ix, 105 p.
[3] G. Albanese, "*Corrispondenze algebriche fra i punti di due superficie algebriche.*" **Ann. R. Scuola Normale Sup. Pisa, (II), 3** (1934), 1–26, 149–182.
[4] G. Albanese, "*Collected papers*," with an historical note by C. Ciliberto and E. Sernesi, **Queen's papers in pure and applied mathematics, vol. 103** (1996), 408 pp.
[5] A. A. Albert, "*Structure of algebras.*" **Amer. Math. Soc. Coll. Publ. 24, New York** (1939).
[6] A. Andreotti, "*On a theorem of Torelli*," **Am. J. Math. 80** (1958), 801–828.
[7] A. Andreotti, A. Mayer, "*On period relations for abelian integrals on algebraic curves*," **Ann. Scuola Normale Sup. Pisa 21** (1967), 189–238.
[8] A. Andreotti, F. Gherardelli, "*Some remarks on quasi-abelian manifolds*," in 'Global analysis and its applications', **I.A.E.A. Vienna** (1976), 203–206.
[9] P. Appell, E. Goursat "*Théorie des fonctions algébriques et de leurs intégrales.*" **Gauthier-Villars, Paris**(1895), 530 pp.
[10] E. Arbarello, M. Cornalba, P. Griffiths, J. Harris, "*Geometry of algebraic curves I.*" **Grundlehren der math. Wiss. 267, Springer** (1984), 386 pp.

[11] M. F. Atiyah, "*Vector bundles over an elliptic curve.*" **Proc. Lond. Math. Soc., III. Ser. 7**, 414–452 (1957).

[12] L. Auslander, R. Tolimieri "*A matrix free treatment of the problem of Riemann matrices.*" **Bull. Amer. Math. Soc. 5, 3** (1981), 263–312.

[13] L. Bagnera, M. de Franchis "*Sur les surfaces hyperelliptiques.*" **Comptes Rendus Acad. Sci. Paris, 145** (1907), 747–749.

[14] L. Bagnera, M. de Franchis "*Sopra le superficie algebriche che hanno le coordinate del punto generico esprimibili con funzioni meromorfe quadruplamente periodiche di due parametri, I, II.*" **Rend. Accad. Lincei, 16** (1907), 492–498, 596–603.

[15] L. Bagnera, M. de Franchis "*Le superficie algebriche le quali ammettono una rappresentazione parametrica mediante funzioni iperellittiche di due argomenti.*" **Memorie Soc. Italiana delle Scienze, detta Accad. dei XL, 15** (1908), 251–343.

[16] I. Barsotti, "*Medoti analitici per varietá abeliane in caratteristica positiva. Cap. 1–2, Cap. 3–4, Cap. 5, 6, 7.*" **Ann. Sc. Norm. Super. Pisa, Sci. Fis. Mat., III.** Ser. 18 (1964), 1–25, Ser. 19 (1965), 277–330, 481–512; Ser. 20 (1966), 101–137, 331–365.

[17] I. Barsotti, "*Considerazioni sulle funzioni theta,*" **Sympos. Math., Roma 3, Probl. Evolut. Sist. solare, Nov. 1968 e Geometria, Febb. 1969** (1970), 247–277.

[18] I. Barsotti, "*Bivettori,*" in 'Algebraic geometry, int. Symp. Centen. Birth F. Severi', Roma 1979, **Symp. Math. 24** (1981), 23–63.

[19] I. Barsotti, "*Differential equations of theta functions, Continuation.*" **Rend. Accad. Naz. Sci. Detta XL, V. Ser., Mem. Mat. 7, No.1.** (1983), 227–276 and **Mem. Mat. 9** (1985), 215–236.

[20] W. Barth, C. Peters, A. Van de Ven, "*Compact Complex Surfaces*", **Ergebnisse der Mathematik und ihrer Grenzgebiete, Folge 3, B.4, Springer-Verlag** (1984), 304 pp.

[21] I. Bauer, F. Catanese, "*Symmetry and variation of Hodge structure,*" preprint 2002.

[22] A. Beauville, "*L'application canonique pour les surfaces de type général,*" **Inv. Math. 55** (1979), 121–140.

[23] A. Beilinson, "*Coherent sheaves on* $\mathbb{P}^n$ *and problems of linear algebra,*" **Funkt. Anal. i Pril. 12 (3)** (1978), 68–69, translated in **Funct. Anal. Appl. 12**, 214–216.

[24] I. N. Bernstein, I. M. Gelfand, S. I. Gelfand, "*Algebraic bundles on* $\mathbb{P}^n$ *and problems of linear algebra,*" **Funkt. Anal. i Pril. 12 (3)** (1978), 66–68, translated in **Funct. Anal. Appl. 12**, 212–214.

[25] L. Bianchi, "*Lezioni sulla teoria delle funzioni di variabile complessa,*" **Spoerri, Pisa** (1916).

[26] U. Bottazzini, "*'Algebraic truths' vs 'geometric fantasies': Weierstrass' response to Riemann.*" Li, Ta Tsien (ed.) et al., Proceedings of the International Congress of Mathematicians, ICM 2002, Beijing, China, August 20–28, 2002. Vol. III: Invited lectures. **Beijing: Higher Education Press** (2002), 923–934.

[27] A. Canonaco, "*A Beilinson-type theorem for coherent sheaves on weighted projective spaces,*" **Jour. of Algebra 225** (2000), 28–46.

[28] A. Canonaco, "*The Beilinson complex and canonical rings of irregular surfaces,*" **Tesi di Perfezionamento, Scuola Normale Superiore, Pisa** (2002).

[29] F. Capocasa, F. Catanese, "*Periodic meromorphic functions.*" **Acta Math. 166** (1991) 27–68.

[30] G. Castelnuovo, "*Osservazioni intorno alla geometria sopra una superficie. Note I, II,*" **Rendiconti del R. Ist. Lombardo, s.2, vol. 24** (1891), pages 246–254 and 255–265 of the Memorie Scelte.

[31] G. Castelnuovo, "*Le correspondenze univoche tra gruppi di p punti sopra una curva di genere p.*" **Rendiconti del R. Ist. Lombardo, s.2, vol. 25** (1893), pages 79–94 of the Memorie Scelte.

[32] G. Castelnuovo, "*Gli integrali semplici appartenenti ad una superficie irregolare.*" **Rendiconti della R. Accad. dei Lincei, s.V, vol. XIV** (1905), pages 473–500 of the Memorie Scelte.

[33] G. Castelnuovo, "*Sulle superficie aventi il genere aritmetico negativo.*" **Rendiconti del Circ. Mat. di Palermo, t. XX** (1905), pages 501–506 of the Memorie Scelte.

[34] G. Castelnuovo, "*Memorie scelte.*" **Zanichelli, Bologna** (1937).

[35] G. Castelnuovo, F. Enriques "*Die algebraischen Flächen vom Gesichtspunkte der birationalen Transformationen aus.*" **Enzycl. der math. Wissensch. III, 2, 6-6b** (1914), 674–768.

[36] F. Catanese, "*Babbage's conjecture, contact of surfaces, symmetric determinantal varieties and applications.*" **Inv. Math. 63** (1981), 433–465.

[37] F. Catanese, "*Commutative algebra methods and the equations of regular surfaces.*" in 'Algebraic Geometry Bucharest 1982', **Springer Lect. Notes in Math. 1056** (1981), 68–111.

[38] F. Catanese, "*Moduli and classification of irregular Kähler manifolds (and algebraic varieties) with Albanese general type fibrations. Appendix by Arnaud Beauville.*" **Inv. Math. 104** (1991) 263–289; Appendix 289.

[39] F. Catanese, "*Recent results on irregular surfaces and irregular Kähler manifolds.*" in 'Geometry and complex variables,' **Lecture Notes in pure appl. Math. 132, Marcel Dekker, New York** (1991), 59–88.

[40] F. Catanese, "*Homological algebra and algebraic surfaces*," in 'Algebraic Geometry', J.Kollár. et al. ed. **Proc. Symp. Pure Math. 62, Amer. Math. Soc., Providence R.I.** (1997), 3–56.

[41] F. Catanese, "*Deformation types of real and complex manifolds*," in the Proc. of the Chen–Chow Memorial Conference, Advanced Topics in Algebraic Geometry and Algebraic Topology, **Nankai Tracts in Math. 5, World Scientific** (2002), 193–236.

[42] F. Catanese, "*Moduli Spaces of Surfaces and Real Structures*," **Ann. Math. 158, No.2** (2003), 539–554.

[43] F. Catanese, "*Deformation in the large of some complex manifolds, I*," preprint 2002, to appear in a Volume in memory of Fabio Bardelli, Ann. Mat. pura e appl.

[44] F. Catanese, *Q.E.D. for algebraic varieties*, Preprint (2002).

[45] F. Catanese, C. Ciliberto, "*Symmetric products of elliptic curves and surfaces of general type with* $p_g = q = 1$." **J. Algebr. Geom. 2, No.3** (1993), 389–411.

[46] F. Catanese, C. Ciliberto, M. Mendes Lopes, "*On the classification of irregular surfaces of general type with nonbirational bicanonical map*," **Trans. Am. Math. Soc. 350, No.1** (1998), 275–308.

[47] F. Catanese, F.O. Schreyer, *Canonical projections of irregular algebraic surfaces*, in 'Algebraic Geometry. A volume in memory of Paolo Francia,' **De Gruyter, Berlin, New York** (2002), 79–116.

[48] A. Cayley, "*Desiderata and suggestions n.1. The theory of groups*," **Amer. Jour. Math, Vol. 1, 1** (1878), 50–52.

[49] A. Cayley, "*Desiderata and suggestions n. 2. The theory of groups: graphical representation*," **Amer. Jour. Math, Vol. 1, 2** (1878), 174–176.

[50] C. Chabauty, "*Limite d'ensembles et géométrie des nombres*," **Bull. Soc. Math. France 78** (1950), 143–151.

[51] S. S. Chern, "*Complex manifolds*," **Publ. Mat. Univ. Recife** (1958).

[52] W. L. Chow, "*Abstract theory of the Picard and Albanese varieties*," unpublished manuscript (1955).

[53] C. Ciliberto, "*Canonical surfaces with* $p_g = p_a = 4$ *and* $K^2 = 5, \ldots, 10$," **Duke Math. J. 48** (1981), 121–157.

[54] C. Ciliberto, "*A few comments on some aspects of the mathematical work of F. Enriques.*" in 'Geometry and complex variables' **Lecture Notes in pure appl. Math. 132, Marcel Dekker, New York** (1991), 89–109.

[55] C. Ciliberto, "*M. de Franchis and the theory of hyperelliptic surfaces,*" in 'Studies in the history of modern mathematics', **Supplemento Rend. Circ. Mat. di Palermo, S.II 55** (1998), 45–73.

[56] C. Ciliberto, M. Mendes Lopes, "*On surfaces with $p_g = q = 2$ and non birational bicanonical map,*" **Adv. in Geom. 2, No.3** (2002), 281–300.

[57] C. Ciliberto, M. Mendes Lopes, "*On surfaces with $p_g = 2, q = 1$ and non-birational bicanonical map,*" in 'Algebraic Geometry. A volume in memory of Paolo Francia, **De Gruyter, Berlin, New York** (2002), 117–126.

[58] C. Ciliberto, E. Sernesi, "*Some aspects of the scientific activity of Michele de Franchis,*" in 'Opere di de Franchis', **Supplemento Rend. Circ. Mat. di Palermo, S.II 27** (1991), 3–36.

[59] Coble, Arthur B. "*Algebraic geometry and theta functions.*" Revised printing. **American Mathematical Society Colloquium Publication, vol. X American Mathematical Society, Providence, R.I** (1929), (II edition 1961) vii+282 pp.

[60] M. Cornalba, "*Complex tori and Jacobians.*" **Complex Anal. Appl., int. Summer Course Trieste 1975, Vol. II** (1976) 39–100.

[61] P. Cousin, "*Sur les fonctions periodiques,*" **Ann. Sci. École Norm. Sup. 19** (1902), 9–61.

[62] J. H. Davenport, "*Algorithms for the integration of algebraic functions.*" Symbolic and algebraic computation, EUROSAM '79, int. Symp., Marseille 1979,**Lect. Notes Comput. Sci. 72** (1979), 415–425.

[63] J. H. Davenport, "*The computerisation of algebraic geometry.*" Symbolic and algebraic computation, EUROSAM '79, int. Symp., Marseille 1979, **Lect. Notes Comput. Sci. 72** (1979), 119–133.

[64] J. H. Davenport, "*On the integration of algebraic functions.*" **Lecture Notes in Computer Science, 102. Berlin Heidelberg New York: Springer-Verlag** (1981), 197 p.

[65] U. Dini, "*Fondamenti per la teorica delle funzioni di variabile reale.*" **Spoerri, Pisa** (1878).

[66] M. de Franchis, "*Sulle superficie algebriche le quali contengono un fascio irrazionale di curve,*" **Rendiconti Circ. Mat. di Palermo, XX** (1905), 49–54.

[67] M. de Franchis, "*Dimostrazione del teorema fondamentale sulle superficie iperellittiche,*" **Rendiconti Accad. Lincei, 24** (1936), 3–6.

[68] M. de Franchis, "*Sulla classificazione delle superficie iperellittiche,*" **Scritti matematice in onore di Berzolari** (1936), 613–615.

[69] J. Dieudonné, "*Cours de géométrie algébrique, vol. 1,2,*" **Collection SUP 10,11, Presses Universitaires de France** (1974), 234 pp., 222 pp.

[70] P. Deligne, "*Variétés de Shimura : interprétation modulaire, et techniques de construction de modèles canoniques,*" in 'Automorphic forms, representations, and L-functions,' **Proc.Symp. Pure Math. 33, vol. 2, A.M.S.** (1979), 247–290.

[71] R. Donagi, "*The tetragonal construction,*" **Bull. Am. Math. Soc., New Ser. 4** (1981), 181–185 .

[72] D. Eisenbud, G. Fløystad, F.-O. Schreyer, "*Sheaf cohomology and free resolutions over exterior algebras,*" preprint, math.AG/0104203.

[73] "*Encyclopedic dictionary of mathematics. Vol. I. Abel to multivariate analysis.*" Translated from the second Japanese edition. Edited by Shokichi Iyanaga and Yukiyoshi Kawada. **MIT Press, Cambridge, Mass.-London** (1977). xv+883 pp.

[74] *"Encyclopedic dictionary of mathematics. Vol. II. Networks to zeta functions."* Translated from the second Japanese edition. Edited by Shokichi Iyanaga and Yukiyosi Kawada. **MIT Press, Cambridge, Mass.-London** 1977, 885–1750.
[75] F. Enriques, *"Ricerche di geometria sulle superficie algebriche."* **Memorie Accad. Torino, s.2 vol. 44** (1893), 171–232.
[76] F. Enriques, *"Sulla proprietá caratteristica delle superficie irregolari."* **Rend. Accad. Sci. Bologna, nuova Serie vol. 9** (1905), 5–13.
[77] F. Enriques, *"Le superficie algebriche."* **Zanichelli, Bologna** (1949), 464 pp. (Translation by F. Catanese, C. Ciliberto, R. Pardini, to be published by Cambridge University Press).
[78] F. Enriques, *"Memorie scelte di geometria, vol. I, II, III."* **Zanichelli, Bologna**, (1956), 541 pp., (1959), 527 pp., (1966), 456 pp.
[79] F. Enriques, F. Severi, *"Intorno alle superficie iperellittiche."* **Rend. Accad. Lincei, s. V vol. 16** (1907), 443–453.
[80] F. Enriques, F. Severi, *"Intorno alle superficie iperellittiche irregolari."* **Rend. Accad. Lincei, s. V vol. 17** (1908), 4–9.
[81] F. Enriques, F. Severi, *"Mémoire sur les surfaces hyperelliptiques, I."* **Acta Math. vol. 32** (1909), 283–392.
[82] F. Enriques, F. Severi, *"Mémoire sur les surfaces hyperelliptiques, II."* **Acta Math. vol. 33** (1910), 223–403.
[83] G. Faltings, *"Endlichkeitssätze für abelsche Varietäten über Zahlkörpern."* **Invent. Math. 73, no. 3** (1983), 349–366. *"Erratum: Finiteness theorems for abelian varieties over number fields."* **Invent. Math. 75, no. 2** (1984), 381.
[84] G. Faltings, C. L. Chai, *"Degeneration of abelian varieties."* **Ergebnisse der Mathematik und ihrer Grenzgebiete (3), 22. Springer-Verlag, Berlin** (1990). xii+316 pp.
[85] G. Fischer (ed.), *"Mathematische Modelle."* Aus den Sammlungen von Universitäten und Museen. Mit 132 Fotografien. (Mathematical models. From the collections of universities and museums. With 132 photographs). Bildband und Kommentarband. **Braunschweig/Wiesbaden: Friedr. Vieweg & Sohn** (1986), XII, 129 pp.; VIII, 89 pp.
[86] A. Genocchi, *"Calcolo differenziale,"* con aggiunte del prof. G. Peano **Torino** (1884).
[87] M. Green, R. Lazarsfeld, *"Deformation theory, generic vanishing theorems, and some conjectures of Enriques, Catanese and Beauville."* **Inv. Math. 90** (1987), 416–440.
[88] M. Green, R. Lazarsfeld *"Higher obstructions to deforming cohomology groups of line bundles."* **Jour. Amer. Math. Soc. 4** (1991), 87–103.
[89] P. Griffiths, *"Periods of integrals on algebraic manifolds I,II."* **Amer. Jour. Math. 90** (1968), 568–626, 805–865.
[90] P. Griffiths, *"Periods of integrals on algebraic manifolds III."* **Publ. Math. I.H.E.S. 38** (1970), 125–180.
[91] P. Griffiths (ed.), *"Topics in transcendental algebraic geometry."* **Annals of Math. Studies 106, Princeton Univ. Press** (1984), 316 pp.
[92] C. D. Hacon, *"Fourier transforms, generic vanishing theorems and polarizations of abelian varieties."* **Math. Zeit. 235** (2000), 717–726.
[93] C. D. Hacon, R. Pardini, *"Surfaces with $p_g = q = 3$,"* **Trans. Amer. Math. Soc. 354, No.7** (2002), 2631–2638.
[94] W. V. D. Hodge, *"Theory and application of the harmonic integrals,"* **Cambridge Univ. Press** (1952).
[95] J. I. Igusa, *"Theta functions."* **Die Grundlehren der mathematischen Wissenschaften, Band 194. Springer-Verlag, New York-Heidelberg** (1972) x+232 pp.
[96] G. Kempf, *"On the geometry of a theorem of Riemann,"* **Ann. of Math. (2) 98** (1973), 178–185.

[97] G. Kempf, "*Projective coordinate rings of abelian varieties.*" **Algebraic analysis, geometry, and number theory (Baltimore, MD, 1988) Johns Hopkins Univ. Press, Baltimore** (1989), 225–235.

[98] V. Kharlamov, V. Kulikov, "*On real structures of real surfaces,*" **math.AG/0101098**.

[99] F. Klein, "*Über die Transformation siebenter Ordnung der elliptischen Functionen,*" **Math. Ann. 14** (1879), 428–471.

[100] F. Klein, "*Gesammelte mathematische Abhandlungen, Bände I, II, III.*"
Band I: Liniengeometrie, Grundlegung der Geometrie, zum Erlanger Programm. Edited by R. Fricke und A. Ostrowski (with additions by Klein), ii+xii+612 pp. (1 plate).
Band II: Anschauliche Geometrie, Substitutionsgruppen und Gleichungstheorie, zur mathematischen Physik. Edited by R. Fricke und H. Vermeil (with additions by Klein), ii+vii+713 pp.
Band III: Elliptische Funktionen, insbesondere Modulfunktionen, hyperelliptische und Abelsche Funktionen, Riemannsche Funktionentheorie und automorphe Funktionen, Anhang verschiedene Verzeichnisse. Edited by R. Fricke, H. Vermeil und E. Bessel-Hagen (with additions by F. Klein), ii+ix+774+36 pp.
Reprint of the first edition **Julius Springer Verlag, Berlin**, (1923), **Springer-Verlag, Berlin-New York**, 1973.

[101] K. Kodaira, "*The theory of harmonic integrals and their applications to algebraic geometry*" (1952), in 'Kunihiko Kodaira: Collected Works,' **Princeton Univ. Press and Iwanami Shoten** (1975), 488–582.

[102] K. Kodaira, J. Morrow, "*Complex manifolds.*" **Holt, Rinehart and Winston**, New York–Montreal, Que.-London (1971).

[103] K. Kodaira, D. Spencer, "*On deformations of complex analytic structures I-II,*" **Ann. of Math. 67** (1958), 328–466.

[104] A. Krazer, "*Lehrbuch der Thetafunktionen.*" **Teubner, Leipzig** (1903).

[105] A. Krazer, W. Wirtinger "*Abelsche Funktionen und allgemeine Thetafunktionen,*" **Enzykl. Math. Wiss. II B 7** (1920), 604–873.

[106] S. Lang, "*Abelian varieties.*" **Interscience Publ. Inc.** (1959), reprinted by Springer Verlag (1983).

[107] H. Lange, C. Birkenhake, "*Complex Abelian varieties.*" **Grundlehren der mathematischen Wissenschaften 302, Springer Verlag** (1992).

[108] R. P. Langlands, "*Problems in the theory of automorphic forms,*" in 'Lectures in modern analysis and applications', **Springer Lecture Notes in Math. 170** (1970), 18–86.

[109] R. P. Langlands, "*Some contemporary problems with origins in the Jugendtraum,*" in 'Mathematical developments arising from Hilbert problems,' **Proc. Symp. Pure Math. 28, vol. 2, A.M.S.** (1976), 401–418.

[110] R. P. Langlands, "*Automorphic representations, Shimura varieties, and motives, ein Märchen,*" in 'Automorphic forms, representations, and L-functions,' **Proc. Symp. Pure Math. 33, vol. 2, A.M.S.** (1979), 205–246.

[111] H. Lebesgue, "*Intégrale, longuer, aire,*" **Annali di Mat. pura ed appl. 7** (1902), 231–358.

[112] S. Lefschetz, "*On certain numerical invariants of algebraic varieties with application to abelian varieties,*" **Trans. Amer. Math. Soc. 22** no. 3 (1921), 327–406 and no. 4, (1921), 407–482.

[113] M. Manetti, "*On the Moduli Space of diffeomorphic algebraic surfaces,*" **Inv. Math. 143** (2001), 29–76.

[114] N. Manolache, F. O. Schreyer, "*Moduli of* $(1,7)$*-polarized abelian surfaces via syzygies,*" **Math. Nachr. 226** (2001), 177–203.

[115] T. Matsusaka, "*On a theorem of Torelli*," **Am. J. Math. 80** (1958), 784–800.
[116] L. Migliorini, "*A smooth family of minimal surfaces of general type over a curve of genus at most one is trivial*," **Jour. Alg. Geom. 4** (1995), 668–684.
[117] N. Mok, "*The generalized theorem of Castelnuovo-de Franchis for unitary representations*," in 'Geometry from the Pacific Rim. Proceedings of the Pacific Rim geometry conference, National University of Singapore, Republic of Singapore, December 12–17, 1994,' Berrick, et al.,ed., **Walter de Gruyter, New York** (1997), 261–284.
[118] S. Mukai, "*Duality between $D(X)$ and $D(\hat{X})$ with its applications to Picard sheaves*," **Nagoya Math. Jour. 81** (1981), 153–175.
[119] S. Mukai, "*Fano 3-folds*," Complex projective geometry (Trieste, 1989/Bergen, 1989), **London Math. Soc. Lecture Note Ser., 179**, Cambridge Univ. Press, Cambridge (1992), 255–263.
[120] D. Mumford, "*Lectures on curves on an algebraic surface*." **Annals of Math. Studies, n. 59, Princeton Univ. Press** (1966).
[121] D. Mumford, "*On the equations defining abelian varieties.I, II, III*", **Invent. Math. 1** (1966), 287–354, **3** (1967), 75–135, **3** (1967), 215–244.
[122] D. Mumford, "*Rational equivalence of* 0*-cycles on surfaces*," **J. Math. Kyoto Univ. 9** (1968), 195–204.
[123] D. Mumford, "*Abelian varieties*." **Tata Institute of Fundamental Research Studies in Mathematics, Oxford University Press. VIII** (1970).
[124] D. Mumford, "*Curves and their Jacobians*." **The University of Michigan Press, Ann. Arbor** (1975), 104 p.
[125] D. Mumford, "*Tata lectures on theta. I*." With the assistance of C. Musili, M. Nori, E. Previato and M. Stillman. **Progress in Mathematics, 28. Birkhäuser Boston, Inc., Boston, MA** (1983). xiii+235 pp.
"*Tata lectures on theta. II. Jacobian theta functions and differential equations*." With the collaboration of C. Musili, M. Nori, E. Previato, M. Stillman and H. Umemura. **Progress in Mathematics, 43. Birkhäuser Boston, Inc., Boston, MA** (1984), xiv+272 pp.
"*Tata lectures on theta. III*." With the collaboration of Madhav Nori and Peter Norman. **Progress in Mathematics, 97. Birkhäuser Boston, Inc., Boston, MA** (1991), viii+202 pp.
[126] A. C. Norman, J. H. Davenport, "*Symbolic integration – the dust settles?*," Symbolic and algebraic computation, EUROSAM '79, Int. Symp., Marseille 1979, **Lect. Notes Comput. Sci. 72**, 398–407 (1979).
[127] P. Painlevé, "*Sur les fonctions qui admettent un théorème d'addition*," **Acta Math. 27** (1903), 1–54.
[128] É. Picard, "*Sur la théorie des groupes et des surfaces algébriques*." **Rendiconti del Circ. Mat. di Palermo, IX** (1895).
[129] P. Pirola, "*Algebraic surfaces with $p_g = q = 3$ and no irrational pencils*," **Manuscripta Math. 108, No.2** (2002), 163–170.
[130] H. Poincaré, "*Sur la réduction des intégrales abéliennes*," **Bull. Soc. Math. France 12** (1884), 124–143.
[131] H. Poincaré, "*Sur les fonctions abéliennes*," **Acta Math. 26** (1902), 43–98.
[132] H. Poincaré, "*Sur les courbes tracées sur les surfaces algebriques*," **Annales de l'École normale Sup.,III, 27** (1910), 43–98.
[133] S. J. Patterson, "*Abel's theorem*." Bottazzini, U. (ed.), Studies in the history of modern mathematics. IV. Palermo: **Circolo Matem^tico di Palermo, Suppl. Rend. Circ. Mat. Palermo, II. Ser. 61**, 9–48 (1999)

[134] Z. Ran, "*On subvarieties of abelian varieties*," **Inv. Math. 62** (1981), 459–479.

[135] U. Bottazzini, A. Conte, P. Gario editors, F. Enriques, author, "*Riposte Armonie*." **Bollati Boringhieri, Torino** (1996), 722 pp.

[136] S. Recillas, "*Jacobians of curves with g^1_4's are the Prym's of trigonal curves*," **Bol. Soc. Mat. Mex., II. Ser. 19** (1974) 9–13.

[137] R. H. Risch, "*The solution of the problem of integration in finite terms*," **Bull. Amer. Math. Soc. 76** (1970), 605–608.

[138] C. Rosati, "*Sulle matrici di Riemann*," **Rend. Circ. Mat. Palermo, 53** (1929), 79–134.

[139] N. Schappacher, "*Développement de la loi de groupe sur une cubique. [Development of the law of composition on a cubic curve]*." **Séminaire de Théorie des Nombres, Paris 1988–1989, Progr. Math., 91, Birkhäuser Boston, Boston, MA** (1990), 159–184.

[140] N. Schappacher, R. Schoof, "*Beppo Levi and the arithmetic of elliptic curves*," **Math. Intelligencer 18** (1996), no. 1, 57–69.

[141] F. O. Schreyer, "*Geometry and algebra of prime Fano 3-folds of genus* 12," **Comp. Math. 127** (2001), 297–319.

[142] G. Scorza, "*Sopra la teoria delle figure polari delle curve piane del* 4° *ordine*," **Ann. di Mat.(3) 2** (1899), 155–202.

[143] G. Scorza, "*Intorno alla teoria generale delle matrici di Riemann*," **Rend. Circ. Mat. Palermo 41** (1916), 263–379.

[144] F. Severi, "*Sulle superficie che rappresentano le coppie di punti di una curva algebrica*." **Atti della R. Accad. delle Scienze di Torino, 38** (1903), 185–200.

[145] F. Severi, "*Il teorema d'Abel sulle superficie algebriche*," **Annali di Mat. s.III, t.XII** (1905), 55–79.

[146] F. Severi, "*La serie canonica e la teoria delle serie principali di gruppi di punti sopra una superficie algebrica*," **Comment. Math. Helv. 4** (1932), 268-326.

[147] F. Severi, "*Funzioni quasi-abeliane*." **Pontificiae Academiae Scientiarum Scripta Varia 4. Roma**, 327 p. (1947).

[148] F. Severi, "*Memorie scelte. Vol. I*." **Cesare Zuffi Editore. XX, Bologna**, 458 p. (1950).

[149] G. Shimura, "*On analytic families of polarized abelian varieties and automorphic functions*," **Ann. of Math. 78** (1963), 149–193.

[150] C. L. Siegel, "*Symplectic geometry*," **Amer. Jour. Math., n. 65** (1943), 1–86, reprinted by Academic Press, New York and London, (1964).

[151] C. L. Siegel, "*Topics in complex function theory, Vol. III*." **Interscience Tracts in pure and appl. Math., n. 25, Wiley, N.Y.** (1973).

[152] Y.-T. Siu, "*Extension of twisted pluricanonical sections with plurisubharmonic weight and invariance of semipositively twisted plurigenera for manifolds not necessarily of general type*." in 'Complex Geometry, Collections of papers dedicated to Hans Grauert,' I. Bauer et al ed. **Springer Verlag** 223–277 (2002).

[153] J. T. Tate, "*The arithmetic of elliptic curves*," **Invent. Math. 23** 179-206 (1974).

[154] R. Torelli, "*Sulle varietá di Jacobi, I, II*," **Rendiconti R. Accad. dei Lincei 22-2** (1913), 98-103, 437–441.

[155] B. M. Trager, "*Integration of simple radical extensions*," Symbolic and algebraic computation (EUROSAM '79, Internat. Sympos., Marseille, 1979), pp. 408–414, **Lecture Notes in Comput. Sci. 72**, Springer, Berlin-New York (1979).

[156] F. Tricomi, "*Funzioni ellittiche*." Second edition, **Nicola Zanichelli Editore, Bologna** (1951), ix+343 pp.

[157] K. Ueno, "*Classification theory of algebraic varieties and compact complex spaces*." **Lecture Notes in Mathematics 439**, Springer-Verlag, XIX, 278 p. (1975).

[158] K. Weierstraß, "*Einleitung in die Theorie der analytischen Funktionen.*" Vorlesung Berlin 1878, Lecture notes taken and with supplementary material by Adolf Hurwitz. Preface by R. Remmert, ed. Peter Ullrich. **Dokumente zur Geschichte der Mathematik 4. Deutsche Mathematiker Vereinigung, Freiburg; Friedr. Vieweg Sohn, Braunschweig** (1988) xxx+184 pp.

[159] A. Weil, "*L'arithmétique sur les courbes algébriques,*" **Acta Math. 52** (1929), 281–315.

[160] A. Weil, "*Sur les courbes algébriques et les variétés qui s'en deduisent.*" **Actualités scientifiques et industrielles. 1041. Paris: Hermann & Cie. IV**, 85 p. (1948); Publ. Inst. Math. Univ. Strasbourg 7 (1945).

[161] A. Weil, "*Variétés abeliennes et courbes algébriques.*" **Paris: Hermann & Cie.** 163 p. (1948).

[162] A. Weil, "*Zum Beweis des Torellischen Satzes,*" **Nachr. Akad. Wiss. Göttingen, Math.-Phys. Kl., IIa 1957**, 32–53 (1957).

[163] A. Weil, "*Sur certains groupes d'opérateurs unitaires,*" **Acta Math. 111** (1964), 143–211.

[164] A. Weil, "*Introduction a l'étude des variétés kählériennes.*" **Paris: Hermann & Cie.** (1958).

[165] H. Weyl, "*Topologie und abstrakte Algebra als zwei Wege mathematischen Verständnisses,*" **Unterrichtsblätter für Mathematik und Naturwissenschaften, band 38** (1933), pp. 177–188, also in

[166] H. Weyl, "*Gesammelte Abhandlungen. Bände I, II, III, IV.*" Edited by K. Chandrasekharan, **Springer-Verlag, Berlin–New York** (1968), Band I: vi+698 pp., Band II: iv+647 pp., Band III: iv+791 pp., Band IV: ix+694 pp.

[167] H. Weyl, "*On generalized Riemann matrices,*" **Ann. Math. 35** (1934), 714–729.

[168] H. Weyl, "*Generalized Riemann matrices and factor sets,*" **Ann. Math. 37** (1936), 709–745.

[169] I. M. Yaglom, "*Felix Klein and Sophus Lie. Evolution of the idea of symmetry in the nineteenth century.*" Translated from the Russian by Sergei Sossinsky. Translation edited by Hardy Grant and Abe Shenitzer. **Birkhäuser Boston, Inc., Boston, MA** (1988). xii+237 pp.

[170] G. Zappa, "*The papers of Gaetano Scorza on group theory,*" **Atti Accad. Naz. Lincei, Cl. Sci. Fis. Mat. Nat., IX. Ser., Rend. Lincei, Mat. Appl. 2, No. 2** (1991), 95–101.

[171] O. Zariski, "*Algebraic surfaces.*" **Ergebnisse der Math. und ihrer Grenzgebiete, Band III Heft 5, Springer Verlag** (1935), second supplemented edition in **Ergebnisse 61** (1970), 270 pp.

What is Abel's Theorem Anyway?

Steven L. Kleiman

Abstract. Supplementing other treatments, this article discusses the history and meaning of four theorems that have been accepted as Abel's Theorem. The discussion explains Abel's own proofs, and pays due attention to the hyperelliptic case. Section 1 explains how each theorem came to be called Abel's Theorem. Section 2 treats Abelian integrals, Clebsch's geometric reformulation, and Abel's Elementary Function Theorem. Section 3 treats Abel's Equivalence Theorem, the genus, and adjoints. Section 4 treats the Riemann–Roch Theorem and Abel's Relations Theorem. Section 5 explains the various forms of Abel's Addition Theorem and Abel's proofs of them. Section 6 discusses the Abel map, and uses it to prove the Addition Theorem in its most elaborate form. Section 7 discusses the Picard and Albanese varieties, and explains Abel's version of the genus. Section 8 sums it all up, and concludes that only the Addition Theorem can rightfully be called *Abel's Theorem.*

1 Why Four Theorems?

Did Abel prove just *one* theorem? The name *Abel's Theorem* has always referred to a theorem Abel proved about sums of integrals of algebraic functions – never mind, he was the first to prove the insolvability of the quintic by radicals, and the first to investigate the convergence of Newton's binomial formula with an arbitrary complex exponent (see [68], pp. 754, 964).

However, four different theorems have been touted as the one and only Abel's Theorem. Why? Because Abel published three papers on the subject.

To make it easier to refer to the four, let us call them the Elementary Function Theorem, the Equivalence Theorem, the Relations Theorem, and the Addition Theorem. Their statements are explained informally later in this section, and more formally in Sects. 2, 3, 4, and 5.

Abel's first paper [1] treats the Addition Theorem in some detail, and proves versions of the other three theorems as preliminary results along the way. Abel submitted it in 1826, but it did not appear until 1841.

Meanwhile, virtually two years after Abel submitted the first paper, he feared it was lost forever, and so he wrote the second paper [2] (see [109], pp. 453, 469). In it, he summarized the Addition Theorem informally, then he treated it in a major special case, the case of hyperelliptic integrals.

The following January, Abel wrote the third paper [3]. In it, he proved only the Elementary Function Theorem, but in full generality. It is just a preliminary result, but Abel didn't say so. At the time, he was slowly dying of tuberculosis, and writing the paper was a strenuous effort (see [109], p. 482).

Abel's first paper is fairly long: sixty-one handwritten pages. Abel was proud of its quality, and hopeful it would secure his career. He finished it in Paris, and showed it to Cauchy. Alas, Cauchy merely glanced at it. Shortly afterwords, on October 30, Abel submitted it in person to the Royal Academy of Sciences, a part of the French Institute. Fourier, the secretary, read the introduction aloud, and the academy appointed Cauchy and Legendre as referees. (See [87], p. 56; see [12], p. 309; see [82], p. 150; see [109], p. 411.)

The treatment itself is actually elementary – "really only a wonderful exercise in the integral calculus," as Forsyth [47], p. 580, put it. However, the treatment is poorly organized and hard to read. Moreover, Gray [53], p. 365, observed: "it is occasionally wrong, which is more serious, and it operates at such a level of generality that it was and is not clear that matters are always as Abel described." And Kline [68], p. 655, explained that Abel's "paper was very difficult to understand, partly because he tried to prove what we would today call an existence theorem by actually computing the result." Also, Cooke [37], p. 411, felt that "the paper owes its great length to Abel's insecurity about the foundations of analysis, based on the bitter experience of errors he had fallen into by using insufficiently grounded methods."

So the paper was reworked, "in a shortened and simplified form," in 1881 by Rowe [97], in part on the basis of Boole's 1857 work in the same direction. However, in 1894, Brill and Noether [17], Fn., pp. 213–214, said that Rowe "did not cover all the difficult parts." Then, in 1900, Forsyth [47], pp. 579–601, gave an exposition, which was "adopted" from Rowe's, and further shortened and simplified by using the theory of Riemann surfaces.

At any rate, the details in Abel's paper have been less important than the concepts when others have developed the paper's mathematical content.

The first paper was not published until 1841. So it did not appear in the 1839 edition of Abel's collected works, which was "prepared at the request" of King Oscar I of Sweden and Norway, and paid for by the state through the Church Department, which handled education. The editor, Holmboe, had to ask the department for help in obtaining a copy of the paper, and a formal diplomatic request was sent to the French authorities. There was no reply, but the paper was published at last. (See [82], p. 258, and [109], pp. 548–549.)

Abel's original manuscript, in his handwriting, is preserved in Italy, except for pp. 21–24, which remain lost. The manuscript has had an interesting, but long and sad history. It was related by Ore [82], pp. 246–261, summarized by Stubhaug [109], pp. 549–551, and recently updated by Del Centina [41].

Del Centina edited the book [42] containing his update. The book also contains facsimiles of the manuscript, of the four missing pages in Libri's handwriting, and of Legendre's cover letter. (G. Libri was a professor of mathematics and member of the academy, which charged him with overseeing the printing in 1841; see [41], p. 91.) In addition, the book contains some other articles about Abel's life and work and about Florence's Moreniana Library, which houses fifty-three pages of the manuscript and Legendre's letter.

Legendre's letter is undated; likely, it was written in June 1829. Indeed, on June 4, Legendre learned of Abel's death. On June 29, Cauchy submitted the referees' report, which Legendre cosigned. The report was weak: it said most of the results had already appeared in print, but owing to Abel's great achievements and tragic death, publication was being recommended. The recommendation was accepted. (See [82], pp. 247 and 233.)

In the letter, Legendre wrote (see the transcription in [41] on p. 96) that he was the first to look at the manuscript, but found the script illegible and the symbols poorly formed. So he asked Cauchy to referee it. However, Cauchy was distracted by other matters, and forgot about the paper until March 1829 when a German scientist expressed his astonishment that it had not yet been refereed. Cauchy then found the manuscript, but the two of them were concerned since Abel had already published part of his work in Crelle's journal and would probably continue with the rest. They then learned of Abel's unfortunate death, which changed the situation.

But the facsimile shows perfectly legible script and well-formed symbols!

The German scientist was Jacobi. He was astonished by a footnote in Abel's second paper [2], p. 445. In the introduction, Abel said he would prove the Addition Theorem in full generality in a subsequent issue of the journal, but here he would treat a special case, which includes elliptic functions. (At the time, the term "elliptic function" referred to the integral, not, as later and now, its inverse.) In the footnote, Abel added sadly that he had presented a memoir on the general case to the Royal Academy toward the end of 1826. Legendre wrote to Jacobi about the paper in February 1929, and Jacobi replied, asking how come Abel's great discovery could have been ignored for so long. (See [41], p. 89, and [82], p. 246; also see [111], p. 402, about Steiner's minor role.)

In [2], Abel treated the hyperelliptic case. It is considerably simpler; indeed, the paper is one-fifth as long as the Paris memoir [1]. Yet, this case has the flavor of the general case. Moreover, "since the longer paper didn't appear until 1841," as Cooke [37], p. 411, pointed out, "it was primarily the hyperelliptic case that introduced Abel's new methods to the world."

Abel's treatment in [2] is computationally involved too. Indeed, eight pages of its twelve are devoted to obtaining, in the hyperelliptic case, a relatively explicit form of the key Elementary Function Theorem. Nevertheless, the degree of abstraction was strikingly novel. Indeed, in a letter to Gauss on 2 January 1829, Bessel wrote (as quoted in [82], p. 213): "I have been very surprised about Abel's theorem, which makes it possible to discover the properties of integrals without executing them. I believe an entirely new phase of the integral calculus has been brought to light. I wish that Abel would pursue it further in many applications. It appears unintelligible to me that such a theorem has been presented to the Academy in Paris, and yet remains unknown."

Abel's treatment was analyzed carefully both historically and mathematically by Brill and Noether [17], pp. 209–210, by Houzel [58], pp. 73–74, and by Cooke [37], pp. 413–418. In particular, Brill and Noether felt they found the "bridge that leads from Euler to Abel in a certain transformation that Abel applied." However, Cooke [37], p. 405, argued that a somewhat different route is more plausible.

In the introduction to [2], Abel put the **Addition Theorem** roughly this way: *the integrals of arbitrary algebraic differentials form a very large class of transcendental functions with an addition formula; if an arbitrary sum of functions arising from the same differential cannot be given by a single function of the same sort, as is the case for elliptic functions, at least the sum can be given by a sum of a specific number of them, plus certain algebraic and transcendental terms.*

The preceding statement is basically what Cooke [37], p. 400, called the "commonly-accepted informal statement of Abel's theorem." In fact, the Addition Theorem was accepted as Abel's Theorem by Gray [52], p. 142, and [53], p. 366, by Shafarevich [107], pp. 416–417, by Kline [68], p. 654, by Ore [83], p. 15, by Klein [67], p. 104, by Baker [8], p. 209, and by Weierstrass [111], pp. 403–404. (The half-page statement of the theorem in [68] is strikingly similar to that in [8], although [8] is not cited.)

However, contrary to what Abel announced in [2], in his third paper [3], he proved only the key, but preliminary, **Elementary Function Theorem.** It provides a sufficient condition for this sum of integrals to be given by the algebraic and transcendental terms alone, so by an elementary function. The condition is not imposed on the common integrand, but on the variables of the integrals, their upper limits; they must form the full set of roots of a certain resultant equation with variable coefficients.

The paper is very short: two pages. The statement fills the first page; the proof, the second. The mathematics is more elegant and more conceptual than in [2]. There is no introduction, but in the very last sentence, Abel said he would develop a number of illuminating applications later.

The Elementary Function Theorem too has been commonly accepted as Abel's Theorem by historians and mathematicians. Some have also explained that Abel went on to derive his Addition Theorem in full generality. Notably, Dieudonné [43], pp. 19–20, and Wirtinger [120], p. 159, did so.

Other authors have simply gone on to derive the Addition Theorem, without clearly crediting it to Abel. For example, see the treatments by Hensel and Landsberg [57], by Goursat [51], pp. 244–250 and 665–685, by Bliss [14], pp. 119–132, and by Griffiths [55], pp. 327–345. Goursat, however, treated the Addition Theorem only in the hyperelliptic case.

Still other authors have also explained, at most, how to recover the older addition theorems for circular functions and elliptic functions. For example, see the treatments by Clebsch and Gordan [33], Ch. 2, by Clebsch and Lindemann [34], Ch. 6, by Cremona [40], p. 161, by Enriques and Chisini [44], p. 141, by Markusevich [76], pp. 37–40, and by Laudal [72], p. 68.

Some treatments require more discussion here. First, Rowe [97], p. 713 and p. 721, accepted the Elementary Function Theorem as Abel's Theorem, and described it as "the principal theorem of Abel's memoir [1]." Correspondingly, Rowe cited Equation (37) in [1], but nothing in [3]. Nevertheless, Rowe devoted twelve pages, a third of [97], to Abel's derivation of the Addition Theorem. Later, Forsyth [47], pp. 579–601, did about the same thing.

Brill and Noether [17], p. 212, referred to the Elementary Function Theorem as the *celebrated* (berühmt) Abel's Theorem. However, they then called the Addition Theorem the *main result* (Hauptergebnis) in [1]. Thus they felt that Abel's main result was not commonly accepted as Abel's Theorem.

By contrast, Ore [82], p. 219, and [83], p. 17, felt that Abel had, in [3], reproved the "main theorem from his Paris memoir [1]." Thus he felt that the Elementary Function Theorem is Abel's main result. Yet Ore [83], p. 15, accepted the Addition Theorem as Abel's Theorem.

Under the heading *Abel's Addition Theorem*, Birkhoff and Merzbach [11], pp. 188–190, gave a translation of the introduction to Abel's first paper [1], which includes informal statements of the Relations Theorem and the Addition Theorem. Then they said: "We reproduce below Picard's [*sic*] simplified proof of Abel's Addition [*sic*] Theorem." Thus they ended with a statement labeled *Abel's Theorem*, but which is, in fact, a statement of Abel's Elementary Functions Theorem.

Birkhoff and Merzbach's version of Picard's statement and proof is nearly identical to Coolidge's statement and proof in [38], p. 213. They cited both Picard and Coolidge, whereas Coolidge just cited Abel's third paper [3]. But Coolidge gave a valid explanation: "We can do no better than to reproduce his [Abel's] original form, with slight changes of wording and notation."

Coolidge did cite Picard on a related matter. Both he, p. 213, and Birkhoff and Merzbach, pp. 176–177, quoted Picard's famous pronouncement on Abel's Theorem: "Perhaps never in the history of science has so important a proposition been obtained using such simple considerations."

Birkhoff and Merzbach cited the second edition (1905) of Picard's book [90], pp. 437–439, whereas Coolidge cited the third edition (1926, p. 464). In the former, Picard cited only Abel's Paris paper [1], and then proved a version of Abel's Elementary Function Theorem. Picard's treatment is rather different from Abel's, but also rather different from the one "reproduced" by Birkhoff and Merzbach. Later, on p. 426, Picard arrived at Abel's Addition Theorem, although only for integrals of the first kind; he did not credit it to Abel, but did say that it follows from Abel's Theorem.

For an integral of the first kind (one finite everywhere), the elementary function reduces to a constant. So two sums of these integrals have the same value (modulo periods) if the variables of each form the set of roots of the same resultant equation, but with different choices of coefficients. Two such sets are called (linearly, or rationally) *equivalent*. And Abel's **Equivalence Theorem** says this: *if two sets of variables are equivalent, then the corresponding sums have the same value for every integral of the first kind.*

Abel, however, did not deal with integrals of the first kind as such, although he certainly did seek conditions guaranteeing the constancy of the sum of integrals. According to Brill and Noether [17], p. 275, Riemann was the first to formulate and prove the Equivalence Theorem in full generality essentially as above. Riemann [92], pp. 123–124, called it Abel's Addition Theorem, and referred to Jacobi's proof of it in the hyperelliptic case [64], § 8. Also, he said that, for an arbitrary Abelian integral, a similar argument yields the Elementary Function Theorem, although he did not use the name.

The converse of the Equivalence Theorem holds too, although Abel did not recognize it. It and the direct assertion were combined and proved under the heading of *Abel's Theorem* by Weyl on p. 149 of his celebrated book [119] on Riemann surfaces, although Weyl did mention that Abel proved only the direct assertion. In a footnote, Weyl explained further that Abel's theorem was "developed in splendid simplicity in the short note [3]," and "is more general, in that it concerns not only integrals of the first kind ... The converse of Abel's theorem for integrals of the first kind may be read between the lines in Riemann; it was stated explicitly (without completely adequate proof) by Clebsch [29], p. 198." But Weyl failed to mention Weierstrass.

After the appearance in 1913 of Weyl's book, it has become common in the theory of Riemann surfaces – as Patterson [86], p. 10, pointed out – to do as Weyl did, and accept the full necessary and sufficient criterion as Abel's Theorem, even though Weyl noted doing so is historically misleading.

However, Severi called the full criterion the *Abel–Riemann Theorem* on p. 271 of his 1921 book [104], which was based on his 1907–8 lectures in Padua, but the name "Abel–Riemann Theorem" never caught on. Severi also said the result can be found somewhere unspecified in Weierstrass's work, and indeed Weierstrass [111], pp. 407–419, did prove it all.

The Relations Theorem is the fourth theorem accepted as Abel's Theorem. Notably, it was accepted by Houzel [58], p. 72, and by Bell [9], p. 322. Later Bell [10], pp. 501–502, accepted the Addition Theorem, citing Baker [8], p. 209, as his authority.

Abel stated the Relations Theorem informally in the introduction to his Paris paper [1], p. 146. Then he said that this theorem yields the Addition Theorem, which he also stated informally. Unfortunately, he did not make comprehensive formal statements of his results, let alone use the heading of "Theorem," in [1], as he did in his later two papers. So there is some uncertainty about precisely how to formulate these two theorems.

Moreover, Abel included two assertions in this statement of the Addition Theorem, which he omitted from his later statement in the introduction to [2]. The first assertion concerns the variables of the given sum of integrals and those of the reduced sum: the latter must be algebraic functions of the former. The second assertion is that, in the given sum, the integrals can be multiplied by arbitrary rational numbers.

The **Relations Theorem** lies a step beyond the Elementary Function Theorem, but they sound similar. Both assert that the same sum of integrals is given by an elementary function. However, in the Elementary Function Theorem, the integration variables are assumed to form the set of roots of a single resultant equation whose coefficients are polynomials in certain variables; moreover, these variables are the arguments of the elementary function. In the Relations Theorem, the integration variables are assumed to satisfy a specific number p of algebraic relations. In fact, these relations make the last p variables into algebraic functions of the first variables; moreover, the first variables are also the arguments of the elementary function.

The situation is a bit more tangled. Abel began the technical discussion in [1] with a short conceptual proof of the Elementary Function Theorem; it is similar to his proofs in [3] and [4]. Unfortunately, on p. 149 of [1], he said this theorem is what he had announced on p. 146. It isn't! There is no mention in the Elementary Function Theorem of a specific number of relations among the integration variables; Abel stressed the matter in the introduction, and devoted pp. 170–180 to it. Fortunately, neither Houzel nor Bell were mislead; both indicated the importance of the number.

Legendre, it seems, never appreciated the full extent of Abel's Addition Theorem. But, he did come to appreciate its importance for hyperelliptic integrals, to which he devoted the third supplement to his *Traité des fonctions elliptiques* in two volumes, 1825–26. (See [82], pp. 204, 208, 213, 219–220, 233–234, and [68], p. 421.) In the supplement, Legendre introduced the term "ultra-elliptic" to refer to the more general integrals.

Legendre sent a copy of the supplement to Crelle for review on March 24, 1832, and at Crelle's request, Jacobi reviewed it. He [63], p. 415, translated "ultra-elliptiques" by *hyperelliptischen*, and the prefix "hyper" has stuck, even in French. Moreover, since Abel had introduced the integrals, Jacobi suggested calling them *Abelian transcendents* (Abelschen Transcendenten). However, by 1847, he [65], p. 151, had begun calling them *Abelian integrals* (Abelschen Integralen) as well. Furthermore, Jacobi was inspired to give in [64], dated 12 July 1832, the first of several treatments of Abel's Addition Theorem in the hyperelliptic case, which he called simply *Abel's Theorem.*

In his cover letter, Legendre [75] praised Abel's Addition Theorem, calling it, in the immortal words of Horace's Ode 3, XXX.1 (see [9], p. 307), a monument more lasting than bronze (monumentum aere perennius). In turn, Jacobi [63], p. 415, said, "Surely, as the noblest monument to this extraordinary genius, it is preferable that this theorem acquire the name of Abel's Theorem." And ever after, it has been called **Abel's Theorem!**

2 Abel's Elementary Function Theorem

An *Abelian integral* is simply an integral of the form

$$\psi x := \int R(x, y)\, dx$$

where x is an independent variable, R is a *rational function*, and $y = y(x)$ is an (integral) *algebraic function*. That is, R is the quotient of two polynomials. And y is the implicit multivalued function defined by an irreducible equation of the form

$$f(x, y) := y^n + f_1(x)y^{n-1} + \cdots + f_n(x) = 0 \tag{2.1}$$

where the $f_i(x)$ are polynomials in x.

The notation here and below is not always the same as Abel's, but is essentially self-consistent, whereas Abel's changed from paper to paper.

For example, $y := \sqrt{x}$ is defined by $f(x, y) := y^2 - x$. More generally, let $\varphi(x)$ be a squarefree polynomial. Then $y := \sqrt{\varphi(x)}$ is defined by the irreducible polynomial $f(x, y) := y^2 - \varphi(x)$. Now, any polynomial in x and y can be reduced to one of the form $\varphi_1(x)y + \varphi_2(x)$; just replace each even power y^{2r} by $\varphi(x)^r$, and each odd power y^{2r+1} by $\varphi(x)^r y$.

Hence ψx can be written in the form

$$\psi x = \int \frac{\pi_1(x)\, dx}{\pi_2(x)} + \int \frac{\rho_1(x)\, dx}{\rho_2(x)\sqrt{\varphi(x)}} \tag{2.2}$$

where the $\pi_i(x)$ and $\rho_i(x)$ are polynomials. The first summand can be integrated by the method of partial fractions, and the integral expressed as a sum of rational functions and logarithms, so as an elementary function.

Let d be the degree of $\varphi(x)$. If d is 1 or 2, then, via a rational change of variables, the second summand can be transformed into one like the first, and then integrated. Indeed, if $\varphi(x) = ax + b$, then

$$\int \frac{\rho_1(x)\, dx}{\rho_2(x)\sqrt{\varphi(x)}} = \int \frac{2\rho_1\big((y^2 - b)/a\big)\, dy}{a\rho_2\big((y^2 - b)/a\big)}\,.$$

Suppose $\varphi(x) = ax^2 + bx + c$. Then by completing the square and replacing x by $u := \alpha x + \beta$ and y by $v := \gamma y$ for suitable constants $\alpha,\ \beta,\ \gamma$, the second summand can be transformed into one with $\varphi(x) = 1 - x^2$. Finally, setting

$$x = \frac{2t}{1 + t^2} \quad \text{and} \quad y = \frac{1 - t^2}{1 + t^2}\,,$$

transforms the second summand in (2.2) into one like the first.

If $d \geq 3$, then the second summand in (2.2) usually cannot be expressed as an elementary function. It is called an *elliptic* integral when d is 3 or 4, as such integrals arise in rectifying the ellipse. The second summand was termed a *hyperelliptic* integral when $d \geq 5$ in 1832 by Legendre and Jacobi, since such integrals share many formal properties with the elliptic integrals.

Usually, the Abelian integral ψx is interpreted, following the dictates of rigor, as a contour integral with fixed lower limit on the Riemann surface of f. Furthermore, Dieudonné [43], p. 19, felt Abel himself meant just such an integral. And, indeed, Abel [1], p. 149, spoke of "integrating between certain limits." But it seems more likely, as Cooke [37], pp. 400, 412–413, suggested, that Abel really meant an indefinite complex integral of the sort used by Euler. Indeed, when Abel [1], p. 145, summarized his work, he spoke of a function with a given sort of derivative.

Abel began his study of advanced mathematics by reading Euler's three standard texts (*Introductio, Institutiones calculi differentialis,* and *Institutiones calculi integralis*). He read them together with his mathematics teacher and mentor, Holmboe, at the Christiania Cathedral School, a preparatory school. Furthermore, Abel discovered his theorem before leaving for Paris; indeed, it is proved in full generality in the posthumous fragment [4]. (See [12], p. 2; see [111], pp. 401–402; see [17], pp. 213, 274; and see [37], p. 407.)

When Abel was in Paris, Cauchy had just begun to develop his theory of integration, and he published a preliminary version in his own journal. Abel wrote to Holmboe that he read the issues diligently, but they seem not to have had much influence on his work on his theorem. (See [109], p. 409; see [82], p. 147; and see [37], p. 412.)

Furthermore, as Cooke [37], p. 412, observed, "absolutely no knowledge of contour integrals or the residue theorem is even remotely hinted at in the present paper [2], even though their implicit presence is obvious to a modern mathematician." On the contrary, as Cooke noted on p. 414, Abel "used a principle that we recognize nowadays as a special case of the residue theorem." In addition, Cooke noted on p. 416 that Abel used a formula "that seems to foreshadow Cauchy's formulas for the derivatives of an analytic function."

Abel used this principle and this formula in [1] too, but not in [3], nor in [4]. Indeed, Abel used them to obtain fairly explicit, though intricate, forms of the Elementary Function Theorem in [1] and [2]. However, he did not use them in his strikingly short proofs of the more conceptual forms of the theorem in [1] again, and in [3] and [4].

There is no geometry in Abel's work, even though it might seem he considered himself a geometer (see [72], p. 68). Indeed, on his roundabout trip to Paris, Abel registered at an inn in Predazzo as "professore della geometria" (see [109], p. 391). And he began his Paris paper [1] with these words: "The transcendental functions considered up to now by geometers [*sic*] are very few in number." Thus, under geometry, Abel included, as did others, the analysis of functions arising as arc lengths, areas, and the like.

Yet, Abel's work might have benefited if he had known more about the singular points and the points at infinity of an algebraic curve in the projective plane. Moreover, projective geometry was a subject of considerable interest in France and

Germany when Abel visited them (see [68], pp. 834–860). Also, he was a good friend of the geometer Steiner (see [82], p. 96).

However, in 1864 Clebsch [29] reformulated Abel's theorem geometrically, and derived results about contacts between various curves and surfaces. In the introduction, Clebsch said that such applications had not been sought earlier, because Riemann's 1857 theory of Abelian integrals was difficult to understand despite the more recent efforts of younger mathematicians. Moreover, according to Brill and Noether [17], p. 320, Clebsch wrote Roch in August 1864, saying that, even after great effort, he understood little of Riemann's paper [92].

Clebsch studied in Königsberg from 1850 to 1854, learning function theory from Richlot and geometry from Hesse, both of whom had studied under Jacobi. Hesse encouraged Clebsch to pursue the algebraic geometry of Cayley, Salmon, and Sylvester. Clebsch's later melding of their work with that of Abel and Riemann led to a sea change in algebraic geometry, which turned toward the study of birational invariants. Clebsch learned about Riemann's great conceptual theory in 1863 from Gordan. However, Riemann's approach was analytic and topological, not really algebra-geometric.

Gordan studied in Berlin, where he followed Kummer's lectures on number theory in 1855, in Königsberg, where he was influenced by Jacobi's school, and in Breslau, where he completed his dissertation in 1862. He then visited Göttingen to pursue his interest in Riemann's theory. Unfortunately, Riemann's health was failing, and he spent that winter and spring in Italy. However, he left behind several students who understood his work, including Neumann, Prym, and Roch.

At the time, Clebsch was a professor in Karlsruhe, about to move to Giessen, and he invited Gordan to go there with him. In 1868, Clebsch succeeded Riemann in Göttingen. Alas, Clebsch died of diphtheria at 39 in November 1872, and Gordan left Giessen the next year to become a professor in Erlangen. (See [18]; see [46]; see [48], p. 448; see [52], p. 141; see [53], pp. 367–368; and see [68], pp. 934–937.)

Clebsch [29] reformulated Abel's setup much as follows. View the equation $f(x, y) = 0$ as defining a (projective) plane curve C. Let $g(x, y) = 0$ define a second curve D. Say C and D intersect in μ points, coalescing allowed, with abscissas $x_1, \dots, x_\mu$. View $x_1, \dots, x_\mu$ as (a complete set of conjugate algebraic) functions of the coefficients $a_1, \dots, a_r$ of $g(x, y)$. See Fig. 1.

In Abel's two-page third paper [3], the Elementary Function Theorem is formulated in its conceptual form basically as follows.

Theorem 2.1 (Elementary Function). *There exist constants $k_1, \dots, k_m$ and rational functions $u, v_1, \dots, v_m$ of $a_1, \dots, a_r$ such that*

$$\psi x_1 + \cdots + \psi x_\mu = u + k_1 \log v_1 + \cdots + k_m \log v_m \ . \tag{2.3}$$

For example, let us work out the case $y := \varphi(x)$ where $\varphi(x)$ is a polynomial (compare with Abel [4], pp. 60–64); here $f(x, y) := y - \varphi(x)$. Expand $R(x, \varphi(x))$ in partial fractions in x. Since the assertion is linear in R, we may assume $R = (x - b)^q$.

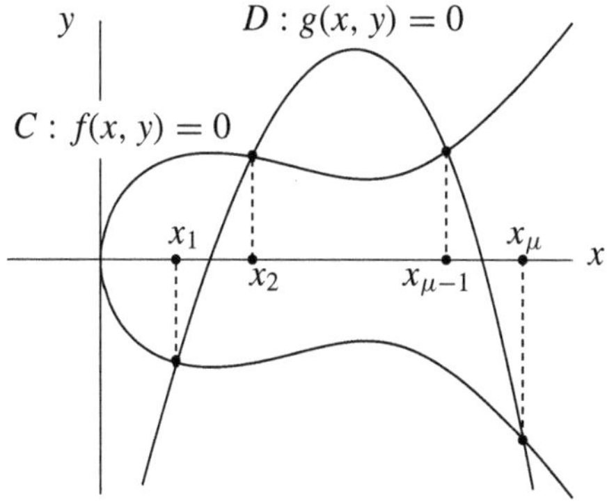

Fig. 1. Clebsch's geometric reformulation (courtesy of J.-M. Økland)

If $q \neq -1$, then $\psi x_1 + \cdots + \psi x_\mu$ is equal to

$$u := \frac{(x_1 - b)^{q+1}}{(q+1)} + \cdots + \frac{(x_\mu - b)^{q+1}}{(q+1)} .$$

However, u is a symmetric rational function in $x_1, \ldots, x_\mu$. And they are all the roots of $g(x, \varphi(x))$. Therefore, u is a rational function in the coefficients of $g(x, \varphi(x))$. But they are linear combinations of the coefficients $a_1, \ldots, a_r$ of $g(x, y)$. So the sum is a rational function in $a_1, \ldots, a_r$, as asserted. If $q = -1$, then $\psi x_1 + \cdots + \psi x_\mu$ is equal to $\log(x_1 - b) + \cdots + \log(x_\mu - b)$, so to $\log v$ where

$$v := (x_1 - b) \cdots (x_\mu - b) .$$

But v is, similarly, a rational function in $a_1, \ldots, a_r$, as asserted.

The key here is a basic theorem about a polynomial equation: *any symmetric rational function in its roots can be expressed as a rational function in its coefficients.* This theorem is also the key to the general case (see [107], p. 417, for example). Indeed, Dieudonné [43], p. 20, described Abel's proof as "remarkable" since it yields such a "general" result, yet is "hardly more than an exercise in the theory of the symmetric functions of the roots of a polynomial." When Abel used this theory in [3], p. 516, he just said it was "known," and gave no reference.

In fact, this theorem on symmetric functions was proved, for the first time, in 1771 by Vandermonde, although, a hundred years earlier, Newton had proved it for various sums and products of the roots (see [68], p. 600). Moreover, Brill and Noether [17], pp. 209–211, suggested it was just such advances in the theory of equations that allowed Abel to generalize Euler's addition formula of 1756–57 for elliptic integrals. Cooke [37], p. 400, added: "Although Brill and Noether do not go into details, it seems clear that a still-primitive understanding of complex numbers had retarded the full understanding of the significance of symmetric polynomials in Euler's day."

Similarly, Kline [68], p. 645, suggested Abel may have gotten some ideas from Gauss, especially from his *Disquisitiones Arithmeticae* of 1801. Also, as Cox, Little, and O'Shea [39], p. 314, noted, in 1816 Gauss published another proof of this theorem on symmetric functions (which introduced the lex ordering) as part of his second proof of the fundamental theorem of algebra.

On the other hand, on p. 56 in [59], Houzel gives a rather different impression of the history of this key theorem on symmetric functions. He says that Newton discovered the general case, but published only a particular case; the general case was published, for the first time, by Waring in 1762, at a period when it waws "common knowledge." If so, then Euler must have known it when he worked on his addition formula. Why then was euler unable to generalize it? It would be very nice to know the full story.

Of course, the method of partial fractions also plays a key role here. But, the method was well known, already when Euler worked, having been introduced in 1702 independently by Leibniz and James Bernoulli. And both used logarithms of complex numbers then too. (See [68], p. 407.)

3 Abel's Equivalence Theorem

Abel proved only the Elementary Function Theorem, Theorem 1, in his third paper [3]. But, in his other three papers, [1], [2], and [4], he addressed two further questions. First, when is the sum $\psi x_1 + \cdots + \psi x_\mu$ in (2.3) constant? Second, how many of the x_i can be varied independently?

Remarkably, both questions lead to the same number p, the (geometric, or true) genus of $C : f(x, y) = 0$. Furthermore, the first question leads to the Equivalence Theorem, and the second, to the Relations Theorem and from there to the Addition Theorem. The first question is discussed in this section; the second, in the next, Sect. 4.

Suppose the integral ψx is of the *first kind,* that is, everywhere finite, even at infinity – in other words, bounded. (The division of Abelian integrals into three kinds was introduced in 1793 by Legendre for elliptic integrals, and in 1857 by Riemann in general; see [68], p. 421 and p. 663.) It follows, as explained next, that, when we vary the coefficients $a_1, \ldots, a_r$ of the auxiliary equation $g(x, y) = 0$, the function

$$\sigma(a_1, \ldots, a_r) := \psi x_1 + \cdots + \psi x_\mu .$$

remains constant.

Indeed (compare with [76], p. 38), since ψx_i remains bounded as x_i varies, $\sigma(a_1, \ldots, a_r)$ remains bounded as the a_j vary. Now, owing to Theorem 1, each partial derivative $\partial\sigma/\partial a_j$ is a rational function. So $\partial\sigma/\partial a_j$ too is bounded. Hence it is constant. Therefore, σ is linear and bounded, so constant.

Fix $a_1, \ldots, a_r$. Say $D : g(x, y) = 0$ meets $C : f(x, y) = 0$ in the points

$$z_1 = (x_1, y_1), \ldots, z_\mu = (x_\mu, y_\mu) .$$

Make a second choice of the coefficients, say $a'_1, \ldots, a'_r$. Let $g'(x, y) = 0$ and

$$z'_1 = \left(x'_1, y'_1\right), \ldots, z'_\mu = \left(x'_\mu, y'_\mu\right)$$

be the corresponding items. Then the constancy simply means the equation

$$\psi x_1 + \cdots + \psi x_\mu = \psi x'_1 + \cdots + \psi x'_\mu \tag{3.1}$$

holds (modulo periods) for all ψ of the first kind.

Possibly, some of the z's are equal to some of the z''s. If so, then some of the terms in (3.1) cancel. Also, the other z's and z''s are the zeros and poles of the restriction of g/g' to C. So the z's and z''s form sets (or point-groups) that are equivalent in the following sense.

In general, two sets of points of C, repetitions allowed, are called (linearly, or rationally) *equivalent* if, after all common points are removed, the remaining points are the zeros and poles of a meromorphic (or rational) function on C. Extra care must be taken if C is singular; one way to proceed is to work on the desingularization of C, the Riemann surface of f.

Consider two equivalent sets, say consisting of $z_1, \ldots, z_\nu$ and $z'_1, \ldots, z'_\nu$. Pass a curve D of suitably high degree through $z_1, \ldots, z_\nu$. Say D intersects C in $z_{\nu+1}, \ldots, z_\mu$ as well. As the two sets are equivalent, there is a second curve D' of the same degree so that D' intersects C in $z'_1, \ldots, z'_\mu, z_{\mu+1}, \ldots, z_\nu$. (In basically this context, equivalent sets are said to be "co-residual"; see [8], p. 136 and [38], p. 206.) Hence, the preceding discussion yields Abel's half of the full Equivalence Theorem.

Theorem 3.2 (Equivalence). *Consider two sets of points of C with abscissas $x_1, \ldots, x_\nu$ and $x'_1, \ldots, x'_\nu$. The sets are equivalent if and only if the equation*

$$\psi x_1 + \cdots + \psi x_\nu = \psi x'_1 + \cdots + \psi x'_\nu ,$$

holds for all ψ of the first kind.

As mentioned in Sect. 1, the other half of the theorem was stated explicitly for the first time by Clebsch [29], p. 198, in 1864. In fact, there Clebsch assumed C is nonsingular. However, as Brill and Noether [17], pp. 326–327, pointed out, Clebsch shortly afterward published two papers in which he allowed C to have nodes and cusps.

A necessary and sufficient condition that the integral $\psi x := \int R(x, y)\, dx$ be of the first kind is that the integrand be a differential of the form

$$R(x, y)\, dx = \frac{h(x, y)}{\partial f/\partial y}\, dx \tag{3.2}$$

where h satisfies two conditions. First, h must vanish suitably wherever $\partial f/\partial y$ vanishes on C, and h must behave similarly at infinity; such an h is called an *adjoint* of C, and more is said about adjoints below. Second, h must be a *special* adjoint;

that is, the bound $\deg h \leq \deg f - 3$ must hold in addition. (See [8], pp. 121–122; see [38], p. 206; and see [68], p. 936.) Such differentials too are said to be *of the first kind.*

For example, suppose $f(x, y) := y^2 - \varphi(x)$ where $\varphi(x)$ is a squarefree polynomial of degree d. Set

$$p := \begin{cases} (d-1)/2, & \text{if } d \text{ is odd ;} \\ (d-2)/2, & \text{if } d \text{ is even .} \end{cases} \tag{3.3}$$

Then every differential of the first kind is a linear combination of the following p basic ones (see [76], p. 33, for example):

$$\frac{dx}{\sqrt{\varphi(x)}}, \frac{x\,dx}{\sqrt{\varphi(x)}}, \ldots, \frac{x^{p-1}\,dx}{\sqrt{\varphi(x)}} . \tag{3.4}$$

Abel, though, was not interested in integrals of the first kind, although some commentators, such as Sylow [110], p. 298, felt he should have been. Rather, Abel was interested in the constancy of the sum $\psi x_1 + \cdots + \psi x_\mu$. In the case $f := y^2 - \varphi(x)$, his Theorem III on p. 450 in [2] provides an explicit expression for the right side of (2.3). Then, on p. 451, he said it is easy to see that this expression is constant when ψx is any linear combination of the integrals of the differentials listed in (3.4). Abel did not claim to have found every integral for which the sum is constant, but likely he felt he had.

For any $C : f = 0$, Abel investigated this constancy in [1], pp. 159–170, but his results are not definitive. "At the cost of rather complicated computations," as Dieudonné [43], p. 20, put it, Abel concluded constancy obtains when the integrand is of the form (3.2) and the coefficients of $h(x, y)$ satisfy certain linear relations (see also [17], p. 216). Dieudonné added that the form (3.2) is "central in Riemann's theory."

Abel denoted the number of linearly independent h by γ, and gave several formulas for γ. They are complicated, but notably depend only on the degree d of C and on certain numbers associated to the singularities. He worked with the singularities by expressing y as a fractional power series in x, a method introduced by Newton in 1671 and treated by De Gua in 1740, Euler in 1748–1749, and Cramer in 1750 (see [68], pp. 552–554, and [17], pp. 133, 135, 139, 141, 216).

Notably, Abel found that

$$\gamma \leq \pi \text{ where } \pi := (d-1)(d-2)/2 . \tag{3.5}$$

At one time, π was called the "virtual" genus of C, but now π is normally called its "arithmetic" genus.

Abel found that equality holds in (3.5) when the coefficients of f satisfy no special relation. In fact, equality holds whenever C is nonsingular everywhere, including at infinity. Notably, however, if C is hyperelliptic, then C is nonsingular at finite distance, but singular at infinity.

For any $C : f = 0$, the number of linearly independent integrals of the first kind is equal to the genus p, where by the *genus* is meant half the first Betti number of the associated Riemann surface. This result was proved in essence in 1857 by Riemann himself [92], p. 137, (or [11], p. 198, in English translation). In fact, he used p to denote the genus; this use of the letter p soon became common, and has remained so. Notably, however, Weierstrass [111] used ϱ instead to denote this quantity, and used no term to refer to it.

Whether p is viewed topologically, as half the Betti number, or analytically, as the number of integrals, p is plainly a *birational invariant.* In other words, p is invariant under a map from C to another plane curve if the map is given by rational functions in x and y and has an inverse given similarly. Such a transformation is just a kind of change of variables in the integrals, and it induces an analytic isomorphism between the associated Riemann surfaces.

The term "genus" is not due to Riemann. Rather, it was introduced in 1865 by Clebsch [30], p. 43, in order to indicate his aim of using this invariant p as a means for classifying curves. And indeed, he proved in § 1 of [30] that every curve of genus 0 is birationally equivalent to a line and in [32] that every curve of genus 1 is birationally equivalent to a nonsingular plane cubic.

Earlier, plane curves $C : f(x, y) = 0$ were classified according to their *degree* (or order), that is, the degree of f. This method was introduced in 1637 by Descartes, and used by Wallis, Newton, James Bernoulli and others. (See [68], pp. 308–334 and 547–554.)

Also in 1865, Clebsch [31], p. 98, gave an important algebra-geometric formula for p, namely,

$$p = (d-1)(d-2)/2 - \delta - \kappa \tag{3.6}$$

where d is the degree of C and where δ and κ are the numbers of nodes and cusps, provided C has no higher singularities. Clebsch then related this formula to the Plücker formulas.

When C is nonsingular, $p = \gamma$ by (3.6) and the statement after (3.5). When C is singular, p and γ can differ. Indeed, Clebsch and Gordan [33], p. 49, observed that the sum $\psi x_1 + \cdots + \psi x_\mu$ can remain constant for certain integrals ψx of the third kind when C has nodes and cusps. (See also [17], pp. 216–217.) The full story involves the restrictions placed on the variables a_i, the coefficients of $g(x, y)$, on which the sum depends. The situation was not understood fully until after 1950; see Sect. 7.

The term "genus" was accepted right away by most all, except writers in English; they used the term "deficiency" instead. The notion of deficiency had already been introduced in 1729 by Maclaurin: he proved an irreducible plane curve of degree d has at most $(d-1)(d-2)/2$ nodes and cusps, and he considered the difference between this maximum number and the actual number, that is, the right side of (3.6) (see [68], p. 552). This difference was termed the deficiency and denoted D, not p, by Cayley [27], p. 1, also in 1865, although Cayley cited Cramer's book of 1750, not Maclaurin's of 1720. The term deficiency was used for about fifty years.

Cayley [27], p. 1, also cited Clebsch's paper [30], not because the term genus was introduced there, but because there "it was first explicitly stated" that, if a curve of degree d has $(d-1)(d-2)/2$ nodes and cusps, then it is birational to the line, and so can be traced globally by a single algebraic parameter. Cayley noted that this fact follows from Riemann's general theory, and then gave it a simple direct algebra-geometric proof. Cayley termed such curves *unicursal.*

For example, let $C : f(x, y) = 0$ be the *lemniscate*. It was introduced by James Bernoulli in 1694 in connection with his study of bending rods. Here

$$f(x, y) := \left(x^2 + y^2\right)^2 - a^2\left(x^2 - y^2\right)$$

where a is a constant, and C is a figure eight. (See [68], p. 412.) Hence C has a node at the origin, but it also has two more imaginary nodes at infinity. So (3.6) yields $p = 0$; whence, C is unicursal. Therefore, any Abelian integral $\psi x := \int R(x, y)\, dx$ is given by an elementary function in x.

However, Bernoulli tried to find the arc length s of the lemniscate C in terms of the polar radius r, and was led to this formula:

$$s = \int_0^r \frac{a^2\, dr}{\sqrt{a^4 - r^4}} .$$

He "surmised that this integral ... could not be integrated in terms of the elementary functions," as Kline [68], p. 412, put it, and continued, "Seventeenth-century attempts to rectify the ellipse, whose arc length is important for astronomy, ... [and] the problem of finding the period of a simple pendulum led" to similar elliptic integrals.

The birational invariance of the genus was proved algebra-geometrically in 1866 by Clebsch and Gordan [33], pp. 54–60. They begin with (3.6), and proceeded to determine the degree and the numbers of nodes and cusps on the transformed curve. Klein, who [66], p. 2, called Clebsch "one of my principal teachers," nevertheless, on p. 5, criticized the proof for involving "a long elimination, without affording the true geometrical insight into [the] meaning" of the genus.

On the other hand, as Kline [68], pp. 939–940, explained, Clebsch was no longer "satisfied merely to show the significance of Riemann's work for curves. He sought now to establish the theory of Abelian integrals on the basis of the algebraic theory of curves ... One must appreciate that at this time Weierstrass's more rigorous theory of Abelian integrals was not known and Riemann's foundation – his proof of existence based on Dirichlet's principle – was not only strange but not well established."

Clebsch and Gordan's proof was not the first algebra-geometric proof of the invariance of the genus, as is sometimes said (for example, on p. 939 in [68]). Indeed, Cayley [27], p. 3, found one the year before, in 1865. However, Clebsch and Gordan's work inspired further advances; Cayley's didn't really.

The term *adjoint* (adjungirte) was introduced in 1874 by Brill and Noether on p. 272 of their great paper [16]. (See [52], p. 147; see [17], p. 335; and see [34], p. 429. In [58], p. 110, the term is attributed to Clebsch, but not documented.) The notion is implicit in Abel's work, in Riemann's, and in Clebsch's, but it is explicit and central in Brill and Noether's; so they had to give it a name.

Brill and Noether's paper [16] is the next milestone in algebraic geometry after Clebsch and Gordan's book [33]. Indeed, as Gray [52], p. 148, wrote, "It should be stressed that with this work the algebraic geometers not only caught up with Riemann, they surpassed him in generality and rigour. It is from this point that they began to make new discoveries ... "

Brill studied under Clebsch in Karlsruhe and Giessen, graduating in 1864 and passing his *Habilitation* in 1867. He taught as *Dozent* in Giessen until 1869, then as professor in Darmstadt until 1875, in Munich until 1884, and in Tübingen until 1918. (See [91].)

Noether (Emmy's father, Max) earned his doctorate in 1868 under the physicist Kirchhoff in Heidelberg without a dissertation. He learned about Riemann's theory of algebraic functions from Kirchhoff and by reading Riemann and Clebsch–Gordan. Lüroth was in Heidelberg; he had studied with Clebsch in Giessen, and encouraged Noether to go there. There Noether joined the circle of young mathematicians around Clebsch and Gordan in lectures, in seminars, on strolls, and over coffee. Noether passed his *Habilitation* in 1870 back in Heidelberg; he taught there until 1874, when he moved to Erlangen permanently. (See [15], p. 213; see [25], p. 161; and see [69], p. 125.)

A curve $C : f(x, y) = 0$ is said to have *multiplicity* m at a point z, and z is said to be an m*-fold* point of C, if the Taylor expansion of f at z begins with a form of degree m. This form has m linear factors, and they define the tangent lines to C at z. If $m > 1$ and if the lines are distinct, then z is an *ordinary* multiple point. For example, a *node* is an ordinary double point.

When Brill and Noether [16], p. 272, defined adjoints, they assumed every singularity (multiple point) is ordinary. In § 7, they explained how to make do in general by reducing the singularities to ordinary ones through a sequence of quadratic transformations of the ambient plane, a method Noether had developed in 1871. "Actually he merely indicated a proof which was perfected and modified by many writers," as Kline [68], p. 941, pointed out.

It is also possible to transform the curve into one only with nodes by using birational transformations that do not extend to the ambient plane. Notably, Kronecker developed a method of doing so, and, as Kline [68], p. 941, observed, "he communicated this method verbally to Riemann and Weierstrass in 1858, lectured on it from 1870 on, and published it in 1881." Many others worked on the matter as well, and Bliss devoted his 1923 AMS Presidential Address [13] to an account of this work.

There are occasions, however, when it is neither customary nor desirable to reduce the singularities of the given curve. For example, a hyperelliptic curve C is given by an equation of the form $y^2 = \varphi(x)$ where $\varphi(x)$ is a squarefree polynomial, and an equation of any other form would mask the hyperellipticity. Yet, if say $d := \deg \varphi$, then C has a point z of multiplicity $d - 2$ at the end of the y-axis, and if $d \geq 4$, then this singularity is not ordinary as the line at infinity is the only tangent to C at z.

Let $C : f = 0$ and $\Gamma : h = 0$ be curves. Then Γ and h are called *adjoints* of C under the following conditions. First, Γ must be allowed to be reducible and to have multiple components, including the line at infinity, so that the various h of given degree or less, plus 0, form a vector space. Now, suppose every multiple point z of C is ordinary. If z is an m-fold point of C, then z must be an $(m-1)$-fold point of Γ. This definition is basically that given by Brill and Noether in [16], p. 272.

The definition of adjoint was extended in 1884 by Noether [79], p. 337, to curves C with arbitrary singularities. He proceeded by induction on the number of quadratic transformations required to reduce the singularities of C to ordinary multiple points. Namely, Γ is an *adjoint* of C if the first transform Γ' of Γ is an adjoint of the first transform of C. However, care must be taken in forming Γ'. Let the quadratic transform be centered at the point z of multiplicity m on C and μ on Γ. Then Γ' must include the exceptional line corresponding to z with multiplicity $\mu - m + 1$.

The first rigorous general theory of adjoints was developed by Gorenstein in his 1950 Harvard thesis (see [94], p. 191), which was supervised by Zariski (see [85], p. 119) and published in 1952 in [50]. In effect, on p. 431, Gorenstein termed Γ an *adjoint* of C if the restriction $h|C$ belongs to the conductor, in the field of meromorphic (or rational) functions on C, of the subring generated by x and y, and if a corresponding condition holds at infinity. He proved, on pp. 434–435, that a differential is of the first kind if and only if it is of the form (3.2) where h is an adjoint in his sense. His sense and Noether's are shown to be equivalent, for instance, in Casas's text [20], p. 152.

4 Abel's Relations Theorem

In Abel's Paris paper [1], he first studied these two questions. Given an Abelian integral ψx, when is the sum $\psi x_1 + \cdots + \psi x_\mu$ equal to an elementary function? When does the sum remain constant? His answers are discussed above in Sects. 2 and 3.

On p. 170, Abel began his study of a third question. How many of the x_i in (2.3) can be varied independently? That is, in the intersection of the curves $C : f(x, y) = 0$ and $D : g(x, y) = 0$, how many of the common points can be assigned at will on C by making suitable choices of the coefficients of $g(x, y)$? The answer is, as Abel put it on p. 172, "rather remarkable."

The answer is also rather subtle. Indeed, in 1898, Scott [99], p. 261, wrote: "The theory of the intersections of curves has probably led its investigators into more errors than any other modern theory. Even the history of the central question, the so-called Cramer paradox, is usually given incorrectly, with the omission of all reference to Maclaurin. This [omission] is all the more surprising, inasmuch as Cramer himself ascribes [the paradox] to Maclaurin."

The Cramer paradox is this. A curve $C : f(x, y) = 0$ of degree d is determined by $(d^2 + 3d)/2$ points. Indeed, each point places one linear relation on the coefficients of f; moreover, f has $(d+2)(d+1)/2$ coefficients, but one can be normalized to be 1 since C is also defined by the vanishing of cf for any nonzero scalar c. Now, a second curve $D : g(x, y) = 0$ of degree d meets C in d^2 points, in general. So these d^2 points do not determine a unique curve of degree d through them, yet

$$d^2 \geq \left(d^2 + 3d\right)/2 \text{ if } d \geq 3 \, .$$

How can this be?

According to Scott [99], pp. 262–263, Maclaurin, in 1720, made "no attempt to explain" the paradox. But in 1750, independently, both Cramer and Euler explained it. Namely, the d^2 relations are not independent. In fact, according to Brill and Noether [17], pp. 289–290, Euler recognized that precisely $(d^2 + 3d)/2 - 1$ are independent, and in 1818, Lamé gave the reason why: the d^2 points also lie on the curve defined by the vanishing of any linear combination $af + bg$; so the relations determine all but one of the available coefficients. For example, through eight arbitrary (general) points on a cubic, there pass infinitely many other cubics, but they all pass through the same ninth point; it is not arbitrary, but depends on the eight.

What did Abel do? Once again, his work is simpler in [2], where he treated the hyperelliptic case. So let $f(x, y) := y^2 - \varphi(x)$ where $\varphi(x)$ is a nonzero squarefree polynomial of degree d, but any $d \geq 1$ works here. Now, as noted before, $g(x, y)$ can be reduced to the form

$$g(x, y) = \theta_1(x)y - \theta_2(x)$$

by replacing each even power y^{2r} by $\varphi(x)^r$, and each odd power y^{2r+1} by $\varphi(x)^r y$. Observe that this reduction preserves the common zeros of f and g, which are the common points of $C : f = 0$ and $D : g = 0$ at finite distance. Assume θ_1 and θ_2 are not identically zero.

Abel allowed for the possibility that, as the coefficients of $g(x, y)$ are varied, $\theta_2(x)$ and $\varphi(x)$ retain a fixed common factor $\varphi_1(x)$. Say

$$\theta_2(x) = \theta(x)\varphi_1(x) \text{ and } \varphi(x) = \varphi_1(x)\varphi_2(x) \, .$$

Mathematically, $\varphi_1(x)$ is needed at times, see below. Abel's recognition of this need was one milestone on the route to his work from Euler's; for an interesting historical discussion of $\varphi_1(x)$, see Cooke's [37], pp. 408–409, 413.

Eliminating y between $f(x, y)$ and $g(x, y)$ gives the "resultant" polynomial

$$F(x) := \theta(x)^2\varphi_1(x) - \theta_1(x)^2\varphi_2(x) \, .$$

Let its roots be $x_1, \ldots, x_\mu$. They are the abscissas of the variable points of intersection of C and D. Let $y_1, \ldots, y_\mu$ be the corresponding ordinates. So y_i is determined by the equation $g(x_i, y_i) = 0$, or

$$\theta_1(x_i)y_i - \theta(x_i)\varphi_1(x_i) = 0 \, . \tag{4.1}$$

Note that, if x' is a root of $\varphi_1(x)$, then $(x', 0)$ is a fixed point of intersection. Conversely, every fixed point at finite distance is of this form. Say $x_1, \ldots, x_\alpha$ can be assigned at will by choosing the coefficients of $g(x, y)$, and say $x_{\alpha+1}, \dots, x_\mu$ are determined by $(x_1, y_1), \dots, (x_\alpha, y_\alpha)$.

Equation (4.1) may be viewed as linear relation among the coefficients of $\theta_1(x)$ and $\theta(x)$. However, Abel did not consider (4.1) explicitly in [2]. Rather, he formed the equivalent equation

$$\theta(x_i)\sqrt{\varphi_1(x_i)} = \varepsilon_i\theta_1(x_i)\sqrt{\varphi_2(x_i)} \tag{4.2}$$

where $\varepsilon_i = \pm 1$. Since $F(x_i) = 0$, (4.2) holds; it is equivalent to (4.1) since $f(x_i, y_i) = 0$. The value of ε_i depends on the choice of signs in the square roots, but remains constant as x_i varies, as Abel noted on p. 449.

Normalize $g(x, y)$ by assuming $\theta_1(x)$ is monic. Let m and n be the degrees of θ_1 and θ. Then g has $m + n + 1$ free coefficients. They yield the coefficients of F. Hence $\alpha \le m + n + 1$. In fact,

$$\alpha = m + n + 1\ , \tag{4.3}$$

and this equality is "easy to see from the shape of (4.2)," as Abel said on p. 453, but without further explanation.

Here's an explanation, which can be rephrased in terms of (4.2). Suppose $\alpha < m+n+1$. Then $(x_1, y_1), \dots, (x_\alpha, y_\alpha)$ fail to determine the values of the coefficients of g. Make two different choices of these values, obtaining g_1 and g_2 say. Then g_1 and g_2 vanish at (x_i, y_i) for $i = 1, \dots, \alpha$, so for $i = 1, \dots, \mu$ since the first α points determine them all. Let (x_0, y_0) be another point of C. Set $c_i := g_i(x_0, y_0)$ and $g' := c_1g_2 - c_2g_1$. Then g' vanishes at (x_i, y_i) for $i = 0, 1, \dots, \mu$. But then eliminating y between $f(x, y)$ and $g'(x, y)$ gives a resultant polynomial of degree μ with $\mu + 1$ roots, a contradiction.

Let d_1 and d_2 be the degrees of φ_1 and φ_2. Note $\mu = \deg F$. Hence

$$\mu = \max\{\, 2n + d_1,\ 2m + d_2 \,\} \text{ and } d_1 + d_2 = d\ . \tag{4.4}$$

Therefore, (4.3) yields

$$2\mu \ge 2n + d_1 + 2m + d_2 = 2\alpha - 2 + d\ ;$$

furthermore, the inequality is strict if d is odd. Hence (3.3) yields

$$\mu - \alpha \ge p\ ; \tag{4.5}$$

furthermore, $\mu - \alpha = p$ if and only if computing $\Delta := (2m + d_2) - (2n + d_1)$ gives $\Delta = \pm 1$ when $d = 2p + 1$, and gives $\Delta = 0$ when $d = 2p + 2$.

For example, given $\alpha \ge 0$, write $\alpha + p = q(2p+2) + d_2$ with $0 \le d_2 \le 2p+1$. Set $m := q(p + 1)$. Define n by (4.3). Define d_1 and μ by (4.4). Then $\mu - \alpha = p$.

On p. 454, Abel arrived at Inequality (4.5). He asserted that equality can be achieved, but did not say how. Of course, equality does not hold for every m and n, despite what is said in [58], p. 74. Moreover, when equality holds, if μ is odd and d

is even, then d_1 must be odd; in particular, then φ_1 cannot be constant, despite what is said in [37], p. 409.

Abel noted that the minimum value p of $\mu - \alpha$ is equal to the number p of differentials in (3.4), which he found to have constant sum $\psi x_1 + \cdots + \psi x_\mu$. This equality appears to be a coincidence in Abel's work, but likely he felt something deeper is involved since he called attention to it.

Earlier, Abel [1], pp. 170–180, studied a similar situation with an arbitrary curve $C : f(x, y) = 0$. His goal was the same, "to find the minimum ... of $\mu - \alpha$," as he said on p. 172. And he tried to relate this minimum to the number γ of independent integrals with constant sum $\psi x_1 + \cdots + \psi x_\mu$; this γ was discussed above in Sect. 3.

In [1], Abel's computations are more involved, and his results are less conclusive. On p. 180, he arrived at the equation

$$\mu - \alpha = \gamma - \iota$$

where $\iota \geq 0$. Moreover, he observed that ι vanishes in general. Brill and Noether [17], p. 219, gave a concrete example where $\iota > 0$.

In fact, Abel did not write ι, but $A + B$. The nature of A and B was clarified a bit by Sylow [110], p. 298, and a bit more by Brill and Noether [17], pp. 218–219; notably, A reflects the behavior at the singularities of C at finite distance, and B does so at infinity. However, the situation is subtle; in particular, care is required in the matters of which D to take and of how to let D vary; see Sect. 7.

The situation has been completely clarified over the course of time through the efforts of many. First, in 1857, Riemann [92], §5, studied the number, $\alpha + 1$ say, of independent meromorphic functions with poles only at μ given points, repetition allowed, on a Riemann surface of genus p; he found $\mu - \alpha \leq p$. Then, in 1864, Roch [93] refined Riemann's inequality into the equation

$$\mu - \alpha = p - i \tag{4.6}$$

where i is the number of independent differentials of the first kind vanishing at the μ points. The number i is called the *index of specialty*.

In 1874, Brill and Noether [16] gave the first algebra-geometric treatment of (4.6), whose assertion they named the *Riemann–Roch Theorem*. "They took from Clebsch the idea that it was to be studied geometrically, that is, in terms of a linear family of adjoint curves," as Gray [54], p. 814, observed. Moreover, later they themselves [17], p. 358, pointed out that their theory served to replace Abel's Theorem as a tool in algebraic geometry.

The idea is essentially this (but see [43], p. 39; see [38], pp. 205–207; see [25], p. 165; and see [49], Ch. 8). Under the relation of (linear) equivalence, which was discussed in Sect. 3, the sets of points of C form equivalence classes, called *complete linear systems*, or series. Each has a *degree*, or order, μ and a *dimension* α. By definition, μ is the number of points that vary from set to set. And α is the maximum number of points that can be assigned at will on C; that is, α general points lie in one and only one set in the system.

Plainly $\alpha + 1$ is equal to the number of independent functions with poles only at the μ variable points in any given set in the system. Furthermore, $i - 1$ is equal to the dimension of the system cut out by the special adjoint curves through those μ points. And $p - 1$ is equal to the dimension of the system cut out by all the special adjoint curves; it is called the *canonical system*. Thus (4.6) has a purely algebra-geometric interpretation directly on C, and this interpretation is compatible with birational transformation.

For instance, consider the canonical system. As just observed, for it, $\alpha = p - 1$. Also, plainly, $i - 1 = 0$. Hence, $\mu = 2p - 2$. And, indeed, $2p - 2$ is the number of zeros of a differential of the first kind.

Hence, if $\mu \geq 2p - 1$, then $i = 0$. Indeed, as just observed, every special adjoint curve cuts C in $2p - 2$ variable points; so no such curve can pass through the μ variable points in a set of the given system. So then $\mu - \alpha = p$.

Remarkably, up to a suitable fixed set of points on C, every complete linear system is cut by the adjoint curves of suitably high degree passing through that fixed set. Conversely, consider all the adjoint curves of given degree through a fixed set of points on C; up to a possibly larger fixed set, these curves cut out a complete linear system on C.

Already Abel recognized the need for fixed points. Of course, he worked with abscissas, roughly as follows. Eliminating y between $f(x, y)$ and $g(x, y)$, he [1], p. 147, obtained their resultant polynomial $r(x)$, and then factored it,

$$r(x) = F_0(x) \cdot F(x)\,,$$

so that, as the coefficients of $g(x, y)$ vary, the roots of $F_0(x)$ remain fixed and those of $F(x)$ vary. (See also [17], p. 214.)

Let the roots of $F(x)$ be $x_1, \ldots, x_\mu$. Let $y_1, \ldots, y_\mu$ be the ordinates of the corresponding points of intersection of C and D. Say $(x_1, y_1), \ldots, (x_\alpha, y_\alpha)$ can be assigned at will, and $(x_{\alpha+1}, y_{\alpha+1}), \ldots, (x_\mu, y_\mu)$ are determined by them. Abel [1], pp. 170–171, considered the quotient

$$F^{(1)}(x) := \frac{F(x)}{(x - x_1) \cdots (x - x_\alpha)}\,,$$

whose roots are precisely $x_{\alpha+1}, \ldots, x_\mu$.

Abel observed that the coefficients of $F^{(1)}(x)$ are rational functions in

$$x_1, y_1; \ldots ; x_\alpha, y_\alpha\,. \tag{4.7}$$

Indeed, these quantities provide α linear relations among the coefficients of $g(x, y)$. These relations can be solved owing to the choice of α. Therefore, the coefficients of $g(x, y)$ are rational functions in the quantities in (4.7). Hence so are the coefficients of $r(x)$, since they are polynomials in those of $f(x, y)$ and $g(x, y)$. Finally, the coefficients of $F^{(1)}(x)$ are polynomials in those of $r(x)$ and $F_0(x)$ and in $x_1, \ldots, x_\alpha$.

Since y is an algebraic function of x, the $\mu - \alpha$ relations

$$F^{(1)}(x_{\alpha+1}) = 0, \ldots, F^{(1)}(x_\mu) = 0 \tag{4.8}$$

make $x_{\alpha+1}, \ldots, x_\mu$ into algebraic functions of $x_1, \ldots, x_\alpha$. Hence $y_{\alpha+1}, \ldots, y_\mu$ are algebraic functions of them as well. Abel viewed the relations in (4.8) as distinct conditions on the individual variables $x_{\alpha+1}, \ldots, x_\mu$.

On p. 170, Abel also combined these considerations with the more conceptual form of the Elementary Function Theorem, which he stated on p. 149. Specifically, he observed that the right-hand side of (2.3) becomes a linear combination of an algebraic function and of logarithms of algebraic functions of $x_1, \ldots, x_\alpha$.

Strictly speaking, the roots of $F_0(x)$ yield terms on the left-hand side of (2.3). However, since these terms remain constant, they can be moved to the right-hand side, and incorporated in u. Abel did not include these terms in his statement on p. 149.

The preceding considerations yield the following theorem, which appears to be what Abel was aiming for.

Theorem 4.3 (Relations). *Let ψx be an Abelian integral, and let $\mu \geq 2p - 1$ where p is the genus. Set $\alpha := \mu - p$. Then, for some m,*

$$\psi x_1 + \cdots + \psi x_\mu = u + k_1 \log v_1 + \cdots + k_m \log v_m \tag{4.9}$$

where $k_1, \ldots, k_m$ are constants and where $u, v_1, \ldots, v_m$ are algebraic functions of $x_1, \ldots, x_\alpha$, provided that $x_{\alpha+1}, \ldots, x_\mu$ satisfy precisely p relations of the form (4.8) *that arise from a system of all adjoint curves of the same degree through some fixed set of points.*

5 Abel's Addition Theorem

The final step is to prove the Addition Theorem. However, its statement appears in different forms in different places. The simplest form is an immediate consequence of the Relations Theorem. The most elaborate form is explained in this section, but its proof calls for the development of an additional tool, the Abel map, which is the subject of Sects. 6 and 7.

On pp. 170 and 185 in [1], Abel gave the Addition Theorem in its simplest form. It asserts that the sum of an arbitrary number α of Abelian integrals ψx_i can be reduced (modulo periods) to an elementary function v of the x_i diminished by a sum of p such integrals, where, as always, p is the genus:

$$\psi x_1 + \cdots + \psi x_\alpha = v - (\psi x_{\alpha+1} + \cdots + \psi x_\mu) \text{ where } \mu := \alpha + p\,. \tag{5.1}$$

Furthermore, $x_{\alpha+1}, \ldots, x_\mu$ are algebraic functions of $x_1, \ldots, x_\alpha$; these same functions work for any integral ψx associated to the same algebraic function $y(x)$, but v depends on the choice of ψx.

To derive this assertion, apply the Relations Theorem as follows. As usual, denote the curve associated to ψx by C. Form the system of adjoint curves of high enough degree, so that it cuts out on C a complete linear system of degree μ' with $\mu' \geq \mu$ and $\mu' \geq 2p - 1$. The Relations Theorem now provides (4.9), but with μ' in place of μ.

Fix $\mu' - \mu$ general points on C, say with abscissas $x_{\mu+1}, \ldots, x_{\mu'}$, and require the adjoint curves to pass through them. In effect, we have reordered the variables $x_1, \ldots, x_{\mu'}$ so that $x_{\alpha+1}, \ldots, x_\mu$ are algebraic functions of $x_1, \ldots, x_\alpha$ and $x_{\mu+1}, \ldots, x_{\mu'}$. Then we fix the values of the latter so that, of the p relations of algebraic dependence, none become trivial. Thus $x_{\alpha+1}, \ldots, x_\mu$ become algebraic functions of $x_1, \ldots, x_\alpha$. Hence, so do the functions $u, v_1, \ldots, v_m$ in (4.9); thus they form the desired v. Finally, the sum $\psi x_{\mu+1} + \cdots + \psi x_{\mu'}$ is constant; so it may be incorporated in v.

Let us now see how to tie in the hyperelliptic case as discussed in Sect. 4. Using its notation, assume $\mu = \alpha + p$. The Addition Theorem is trivial when $\alpha \leq p$: just fix any $p - \alpha$ points on C, and use their abscissas as the needed additional x's on the right. So assume $\alpha \geq p + 1$. Finally, assume $d \geq 2$, so that $d = \deg C$.

Let E be the line at infinity. Then $(d-3)E$ is a special adjoint curve since, as noted in Sect. 3, the constant 1 is an adjoint polynomial. Hence the sum $D + (d-3)E$ is an adjoint curve. Consider all the adjoint curves that have the same degree as $D + (d-3)E$ and that pass through the fixed points of the system of sets cut out on C by $D + (d-3)E$ as D varies. These fixed points serve as the $\mu' - \mu$ points considered above.

These adjoints cut on C a complete linear system. It includes the system cut by the various D, up to the fixed set cut by $(d-3)E$. So both systems have the same number μ of variable points. Hence both have the same number α of points that can be assigned at will; indeed, the smaller one has α by assumption, and the larger has α by the Riemann–Roch (4.6), since $\mu \geq 2p+1$, so $i = 0$. Therefore, the two systems have the same sets of variable points. Thus Abel's procedure in the hyperelliptic case in [2] is, indeed, a special case of his general procedure in [1], at least when the latter is interpreted in terms of the theory of adjoints.

Aesthetically, (5.1) would be more pleasing without the minus sign on the right-hand side, and Abel [1], p. 186, showed how to replace it easily with a plus sign. At the same time, he showed how to allow minus signs on the left. Namely, given any $x'_1, \ldots, x'_{\alpha'}$, (5.1) yields

$$\psi x'_1 + \cdots + \psi x'_{\alpha'} + \psi x_{\alpha+1} + \cdots + \psi x_\mu = v' - \left(\psi x''_1 + \cdots + \psi x''_p\right) .$$

Subtract this formula from (5.1), and set $V := v - v'$. The result is

$$\psi x_1 + \cdots + \psi x_\alpha - \psi x'_1 - \cdots - \psi x'_{\alpha'} = V + \psi x''_1 + \cdots + \psi x''_p , \tag{5.2}$$

which is essentially Abel's Formula (112), p. 186.

In the hyperelliptic case, Abel's main result is Théorème VIII on p. 454 in [2]. Its Formula (39) is a lot like (5.2), but is essentially this:

$$\psi x_1 + \cdots + \psi x_\alpha - \psi x'_1 - \cdots - \psi x'_{\alpha'} = V + \varepsilon_1 \psi x''_1 + \cdots + \varepsilon_p \psi x''_p \tag{5.3}$$

where $\varepsilon_k = \pm 1$ for $k = 1, \ldots, p$. Abel's use of ε_k is interesting. Cooke [37], p. 401, pointed out that Euler had ignored signs in the elliptic case, $p = 1$.

On p. 455, Abel noted that the value of ε_k is determined by the equation

$$\theta\left(x_k''\right)\sqrt{\varphi_1\left(x_k''\right)} = -\varepsilon_k\theta_1\left(x_k''\right)\sqrt{\varphi_2\left(x_k''\right)}\,.$$

This equation is similar to (4.2), but has a minus sign because, when (5.3) is derived, $\psi x_k''$ arises on the left-hand side, and is moved to the right.

On p. 455, Abel also noted that the coefficients of the polynomials $\theta(x)$ and $\theta_1(x)$ are determined by the system of $\alpha + \alpha'$ equations

$$\theta(x_i)\sqrt{\varphi_1(x_i)} = \theta_1(x_i)\sqrt{\varphi_2(x_i)} \text{ and } \theta\left(x_j'\right)\sqrt{\varphi_1\left(x_j'\right)} = -\theta_1\left(x_j'\right)\sqrt{\varphi_2\left(x_j'\right)}$$

where $i = 1, \ldots, \alpha$ and $j = 1, \ldots, \alpha'$. Thus the curve D is determined.

So, in the intersection $C \cap D$, the point with abscissa x_k'' has ordinate y_k'' given by (4.1) with x_k'', y_k'' in place of x_i, y_i. But Abel did not use this equation. Rather, it would seem he set

$$y_k'' := \sqrt{\varphi_1\left(x_k''\right)}\sqrt{\varphi_2\left(x_k''\right)}\,.$$

And choosing the wrong signs for the square roots leads to the ordinate $-y_k''$. In fact, Abel did not use ordinates in [2]. (And he wrote $y_1, \ldots,$ not $x_1'', \ldots$.)

But why does the ordinate $-y_k''$ yield as integral $-\psi x_k''$ up to an elementary function? Abel didn't say. However, $-y_k''$ does so because of the Elementary Function Theorem; just apply it to the system of vertical lines $D : x - a = 0$.

With (5.1) in mind and the algebraic function $y(x)$ fixed, Abel stressed two fundamental properties of the number p: first, p depends on $y(x)$, but is independent of ψx and α; second, p is minimal.

The first property, independence, means that, given any integrand, the sum of α integrals, for any α, can be reduced to a sum of p integrals, plus an elementary function. This property plays a key role in Abel's derivation above of (5.2) from (5.1), as he himself noted on p. 186. Of course, if the sum of $p + 1$ integrals can be reduced to a sum of p, then the sum of any number α can be reduced to p too by a simple induction on α.

The second property, minimality, means there exists some integrand such that the sum of p integrals cannot be reduced to a sum of $p - 1$ integrals plus an elementary function. In fact, there exists one such of the first kind.

This minimality shows "clearly why Euler's original statement could not be generalized," as Dieudonné [43], p. 20, put it. Indeed, Cooke [37], pp. 392–393, explained:"In 1751 Euler was given the duty of reading the collected works of Fagnano, which led him to discover an addition theorem for elliptic integrals." Let us see why Euler's Addition Theorem cannot be generalized.

To be sure, from 1714 to 1720, Fagnano found ad hoc algebraic relations among the lengths of cords and arcs of lemniscates, ellipses, and hyperbolas. Earlier, in 1698, John Bernoulli (James's brother) noted there are algebraic relations among the arguments of sums and differences of logarithms and inverse trigonometric functions;

he asked if this property is shared by other functions that arise as integrals. (See [68], pp. 413–418, and [58], pp. 6–8.)

Fagnano's work led Euler in 1757 to discover the addition formula

$$\int_0^{x_1} \frac{dx}{\sqrt{1-x^4}} \pm \int_0^{x_2} \frac{dx}{\sqrt{1-x^4}} = \int_0^{x_3} \frac{dx}{\sqrt{1-x^4}}$$

where the variables x_1, x_2, x_3 must satisfy the symmetric relation

$$x_1^4x_2^4x_3^4 + 2x_1^4x_2^2x_3^2 + 2x_1^2x_2^4x_3^2 + 2x_1^2x_2^2x_3^4 + x_1^4 + x_2^4 + x_3^4 - 2x_1^2x_2^2 - 2x_1^2x_3^2 - 2x_2^2x_3^2 = 0 .$$

(Also see [43], pp. 18–19, [107], pp. 412–413, [58], p. 8, [76], pp. 1–4, 28–40.)

In 1759, Euler generalized this addition formula to some other elliptic integrals; he expressed the sum or difference of two integrals as a third plus an elementary function. Moreover, as Kline [68], p. 420, observed, "He expressed regret that his methods were not extensible to higher roots than the square root or to radicands of higher than fourth degree." In fact, no methods work for these integrals, as Abel showed: for them, $p \geq 2$, and p is the minimal number of integrals in the reduced sum. Moreover, as an example, Abel devoted the last part of his paper [1], pp. 188–211, to the case $y^n = \varphi(x)$ where $\varphi(x)$ is a polynomial.

The importance of minimality was recognized soon after the appearance of Abel's second and third papers, [2] and [3], in 1828 and 1829. Indeed, between 1838 and 1841, when Abel's [1] first paper finally appeared, Jürgensen, Broch, and Minding treated the minimality in various cases; their work was compared with Abel's by Brill and Noether [17], pp. 225–231. On p. 217, they noted that "Abel's competitors (Mitbewerber)" had more difficulty than he did because they did not know his normal form (3.2) for the integrands.

The minimality of p is related to another significant property of (5.1): the uniqueness of $x_{\alpha+1}, \ldots, x_\mu$ when $\alpha \geq p$; that is, if $x'_{\alpha+1}, \ldots, x'_\mu$ work too, then they are equal to $x_{\alpha+1}, \ldots, x_\mu$ up to order, given the v. Abel did not notice this uniqueness. However, it and the minimality are implicit in Riemann's paper [92], and are explicit in Weierstrass's paper [111], pp. 419, 421; a version of their treatments is given in the next section.

In its most elaborate form, Abel's Addition Theorem asserts that any linear combination of Abelian integrals is equal to the sum of p integrals, plus an elementary function; the combining coefficients can be any rational numbers – positive or negative, integers or fractions. Abel [1], pp. 187–188, derived this assertion in short order, but without adequate justification.

Basically, Abel suggested that, if the combining coefficients are integers, then it suffices, in (5.2), to take additional x_i and x'_i, and let them coalesce; however, he did not explain why the x''_i remain determinate. Furthermore, when he introduced a common denominator n, he claimed, unconvincingly, that it is also possible to introduce arbitrary positive integers as combining coefficients on the right. Then he took all these coefficients equal to n, and finally he divided both sides by n. This argument was repeated, with little change, by Rowe [97], p. 743. Its spirit reappears in Sect. 6.

Nevertheless, the division of Abelian integrals is a subject with a substantial history (see [68], pp. 417–419, 756; see [58], pp. 8, 23–26, 84; and see [43], p. 21). In 1718, Fagnano showed how to find the midpoint of any arc of the lemniscate. In 1825, Legendre treated the division of any elliptic integral of the first kind by any odd number n; he claimed there is an equation of degree n^2 for the new upper limit, but did not prove it. In 1827 and 1829, Abel made a systematic study of this matter, among others, in two one-hundred-page papers on elliptic functions; of course, he wrote both after his Paris paper.

In 1832, the night before his fatal duel, Galois wrote an account of his research, which has been preserved. In particular, he wrote that, for any genus p, the division by any n of the sum of p Abelian integrals leads to an equation of degree n^{2p}. This statement was proved by Clebsch [29], § 15, in 1864. Starting in 1834 with Jacobi, there has been a lot of work done over the years on two related matters: on the behavior of theta functions when their arguments are divided by n, and on arbitrary isogenies of Abelian varieties.

Here is a formal statement of the Addition Theorem in its most elaborate form. Its (5.4) is essentially Abel's Formula (119) in [1], p. 188. A proof is discussed at the end of Sect. 6.

Theorem 5.4 (Addition). *Let $y(x)$ be an algebraic function, p its genus, and ψx an associated Abelian integral. Let α be a positive integer, and $h_1, \ldots, h_\alpha$ rational numbers. Then*

$$h_1\psi x_1 + \cdots + h_\alpha \psi x_\alpha = v + \psi x'_1 + \cdots + \psi x'_p \tag{5.4}$$

where v is an elementary function of $x_1, \ldots, x_\alpha$, and where $x'_1, \ldots, x'_p$ are algebraic functions of them.

More precisely, v is a $\mathbb{C}$-linear combination of an algebraic function of $x_1, \ldots, x_\alpha$ and of logarithms of algebraic functions; if ψx is of the first kind, then v is constant. Moreover, $x'_1, \ldots, x'_p$ work for every choice of ψx; also, if $h_1 = 1, \ldots, h_\alpha = 1$ and $\alpha \geq p$, then $x'_1, \ldots, x'_p$ are unique, given the v.

Lastly, p is minimal: given algebraic functions $x'_1, \ldots, x'_{p-1}$ of $x_1, \ldots, x_p$, there exists an integral of the first kind ψx such that, for any constant v,

$$\psi x_1 + \cdots + \psi x_p \neq v + \psi x'_1 + \cdots + \psi x'_{p-1}\,.$$

6 The Abel Map

Theorem 4 is proved at the end of this section through an elaboration of the ideas of Riemann [92] and Weierstrass [111]. The proof involves a new tool, the Abel map, and so it is discussed first. The map combines the Equivalence Theorem and the Riemann–Roch Theorem in a powerful sophisticated geometric form.

To begin, note that the definition of ψx involves an inherent ambiguity. It is a fancy sort of constant of integration. It arises in part from the choice of the starting point of integration, but in part from the choice of the path of integration. For example, notice that

$$\int_1^x \frac{1}{x}\,dx = \log x + 2m\pi\sqrt{-1} \tag{6.1}$$

if the path is chosen to wind m times around the origin in the complex plane.

Euler "drew attention to the fact that the function $\int_0^y dx/\sqrt{1-x^4}$ has in the real domain a 'modulus of multi-valuedness' similar to the inverse trigonometric functions," as Shafarevich [107], p. 414, said crediting Slavutin. Abel [1], p. 149, noted the presence of the ambiguity in connection with the sum of integrals in (2.3), but he simply dismissed it as unimportant so long as the variables are kept within suitable limits. In particular, he did not observe that the ambiguity is related to the genus p.

The ambiguity was discussed by Cauchy in 1846 (see [68], pp. 640–641). However, it was treated definitively by Riemann in 1857 and 1865 (see [68], pp. 662–665; see [107], pp. 417–421; see [58], p. 97; and see [43], pp. 85–87). Riemann proved every integral of the first kind has $2p$ *periods*, which are numbers like the number $2\pi\sqrt{-1}$ in (6.1). These periods generate, by $\mathbb{Z}$-linear combination, all possible changes in the value of the integral arising from changes in the path of integration, with fixed end points. (Note, however, that $\int_1^x \frac{1}{x}\,dx$ is of the third kind and genus 0.)

In effect, Riemann did the following. He fixed a basis $\psi_1 x, \ldots, \psi_p x$ of the integrals of the first kind. Then, inside the vector space $\mathbb{C}^p$, he formed the discrete subgroup, or lattice, generated by the corresponding $2p$ p-vectors of periods. He proved the lattice has rank $2p$, and he formed the quotient

$$J := \mathbb{C}^p/(\text{period lattice}). \tag{6.2}$$

By 1920, for example in Severi's book [104], p. 272, J was called the *Jacobian* (or Jacobian variety, or Jacobi variety), and still is. (In 1907, Enriques and Severi [45], spoke of the "Jacobi surface" of a curve of genus 2.)

Let C' be the Riemann surface of $y(x)$; so C' is the desingularization of the plane curve C associated to $y(x)$. Let $C^{(\alpha)}$ be the α-fold *symmetric product* of C'; by definition, $C^{(\alpha)}$ is the quotient of the direct product C^α of α copies of C' under permutation of the factors. So $C^{(\alpha)}$ parameterizes the (unordered) sets of α points of C', repetitions allowed. Let us continue to abuse notation, in the spirit of Abel's work, by using $x_1, \ldots, x_\alpha$ to refer both to points of C' and to the abscissas of their images in C.

In effect, Riemann introduced and studied the following map:

$$\Psi_\alpha\colon C^{(\alpha)} \to J \text{ given by } \Psi_\alpha\{x_1, \ldots, x_\alpha\} = \left(\sum_{i=1}^{\alpha} \psi_1 x_i, \ldots, \sum_{i=1}^{\alpha} \psi_p x_i\right).$$

This map is rather important, but for a long time, it had no name.

In 1948, Weil [113, (c)], p. 150, gave Ψ_α a nondescript name, the "canonical map," and this name has often been used since then. Since about 1970 (see for example [35], p. 284), Griffiths's school has used the name "Abel–Jacobi map" for a major generalization of Ψ_α, introduced in 1952 by Weil [115], p. 94. Maybe the best name for Ψ_α is just the *Abel map*; it is historically correct and conveniently short.

The fibers of Ψ_α are the complete linear systems owing to the Equivalence Theorem. It now follows from the Riemann–Roch Theorem that Ψ_α is surjective for $\alpha \geq p$ (since $\dim \operatorname{Im}(\Psi_\alpha) = p$ and $C^{(\alpha)}$ is compact) and it also follows that Ψ_α is injective on a saturated dense open subset U_α for $\alpha \leq p$. In fact, U_α is *Zariski* open; that is, its complement is defined by polynomial equations. In particular, Ψ_p restricts to an isomorphism between dense open subsets of $C^{(p)}$ and J.

A priori, J is a complex analytic manifold, and Ψ_α is a complex analytic map; indeed, they are constructed by means of integrals. A posteriori, J and Ψ_α are algebraic; they are given by polynomials! This result was established through the efforts of many mathematicians over the course of a century (see [58], pp. 110–111, and [43], pp. 85–87). This result is remarkable! And it is essential in the proof below of Theorem 4, as it guarantees that the functions $x_1', \ldots, x_p'$ are, in fact, algebraic.

Furthermore, J is an *Abelian variety,* or compact algebraic group variety: not only do its points form a group, but the operations of adding and of inverting are given by polynomial maps. This abstract definition was given in 1948 by Weil [113], (c) p. 110, and it is now the standard definition in algebraic geometry. In fact, an Abelian variety is not simply compact, but is projective; this result was proved in 1957 by Weil [116].

Of course, J is an Abelian (commutative) group by construction. However, every Abelian variety is an Abelian group. This result was discovered in 1889 by Picard [88], p. 222, in the case where the variety is a surface (see [74], p. 357). The general case was proved in 1948 by Weil [113], (c) p. 107. However, Weil said Chevalley had proved the result earlier in an unpublished work using differential forms and the adjoint group. Weil gave a different proof, which is rather elementary and purely algebraic.

It would be reasonable to guess this result gave rise to the name "Abelian variety," but it did not. The name stems rather from another important result: an Abelian variety is parameterized globally by Abelian functions with common periods. This result was proved in 1889 for surfaces and in 1895 in general by Picard [88], p. 222, and [89], p. 250; his proof was completed at certain points of analysis in 1903 by Painlevé [84]. In the general case, Picard had to assume to begin with that the Abelian variety is an Abelian group. (See [74], pp. 355, 369; see [121], p. 104; and see [60], p. 256.)

An *Abelian function* is, by definition, a meromorphic function in p complex variables with $2p$ independent periods (see also [76], pp. 41–44). The general theory of Abelian functions was begun by Riemann [92] in 1857, and developed by many others, particularly by Weierstrass in 1869, 1876, and 1879, by Frobenius in 1884, and by Poincaré in 1897 and 1902.

Riemann [92] was initially interested in inverting Abelian integrals of the first kind, that is, in finding an expression for the inverse of the Abel map $\Psi_p\colon C^{(p)} \to J$ on the dense open subset $\Psi_p U_p$ of J. The inverse can be expressed using the coordinate functions on $C^{(p)}$, so in terms of functions on $\Psi_p U_p$. Since $J := \mathbb{C}^p/\text{periods}$, these functions can be lifted to an open subset of $\mathbb{C}^p$, and then continued to meromorphic functions on $\mathbb{C}^p$ with $2p$ periods. These special meromorphic functions, Riemann named "Abelian functions."

Later, in 1859, Riemann studied arbitrary meromorphic functions in p variables. He proved, according to Hermite, that they have at most $2p$ independent periods, which satisfy bilinear relations, known today as "Riemann's bilinear relations," like those satisfied by the Abelian functions coming from a curve. In 1869, Weierstrass observed that not every Abelian function arises from a curve; in other words, not every Abelian variety is a Jacobian.

Consider the set of all the Abelian functions whose periods include a given lattice of rank p. Plainly, this set is a field K. In fact, K has transcendence degree p over $\mathbb{C}$; in other words, any $p+1$ Abelian functions satisfy a polynomial equation, but some p satisfy none. This statement is implicit in Riemann's work. Weierstrass formulated it explicitly, but did not offer a proof. The first complete proof was published in 1902 by Poincaré.

It follows that K is the field of rational functions on a p-dimensional projective algebraic variety, which is parameterized on a Zariski open set by p of them. There are many such varieties! Among them, there is a unique Abelian variety. The latter result was proved in 1919 by Lefschetz [73], p. 82. At the time, all the varieties with field K were called "Abelian varieties." However, Lefschetz [73], p. 83, said he would use the name only for the distinguished variety.

Some authors – including Severi, Siegel [108], p. 98, and Markusevich [76], p. 148 – never accepted the more restrictive definition of Abelian variety. Rather, they called the distinguished variety a "Picard variety." In this usage, for example see Severi's paper [102], p. 381, the pth symmetric product $C^{(p)}$ is an Abelian variety since the Abel map $\Psi_p\colon C^{(p)} \to J$ is birational; that is, pullback identifies the function fields. So beware!

Originally, the notion of Abelian variety was even broader: it included any variety whose functions form a subfield of finite index in a given field of Abelian functions, in other words, a variety that is parameterized on a classical open set by Abelian functions (see [98], p. 213). For example, an Abelian surface is just a hyperelliptic surface (see [101], p. 436, and [98], p. 213). The index was termed the "rank" by Enriques and Severi in 1909 according to Scorza [98]; however, on p. 134, he said he would use the term "Abelian variety" only when the rank is 1.

In 1832, Jacobi [64], p. 10, posed the famous problem that inspired Riemann; it is known today as the *Jacobi Inversion Problem.* In Jacobi's own words (as translated in [11], p. 210), he asked: "what, in the general case, are those functions whose inverses are Abelian transcendents, and what does Abel's theorem show about them?" On p. 11, Jacobi answered the question for hyperelliptic curves of genus 2. He did so "after long, fruitless efforts, to discover the natural generalization of 'inversion' of elliptic integrals," as Dieudonné [43], p. 20, put it. Did Abel consider inversion in general? Bjerknes [12], p. 216, and Shafarevich [107], p. 419, suggested that Abel did at least have it in mind. At any rate, it is ultimately due to this work of Jacobi's that J is called the Jacobian.

The inversion of elliptic integrals was introduced in 1827 first by Abel and then, a couple of months later, by Jacobi, but Jacobi did not credit Abel. (Thirty years earlier, Gauss inverted them, but his work appeared only much later, after his death in 1855.) Nevertheless, here too Abel inspired Jacobi; this conclusion is universally accepted today; indeed, it is inescapable given all the evidence (see [12], pp. 183–216; see [82], pp. 180–190; see [58], pp. 17–18; see [43], p. 20; and see [109], pp. 452–454, 538).

The properties of the Abel map $\Psi_\alpha\colon C^{(\alpha)} \to J$ yield Theorem 4 as follows. First, suppose p is not minimal. Then there exist algebraic functions $x'_1, \ldots, x'_{p-1}$ of $x_1, \ldots, x_p$ such that, for each $\psi_i x$ in the basis of integrals of the first kind, there exists a constant v_i such that, modulo periods,

$$\psi_i x_1 + \cdots + \psi_i x_p = v_i + \psi_i x'_1 + \cdots + \psi_i x'_{p-1}\,.$$

Hence $\Psi_p C^{(p)} \subseteq (v_1, \ldots, v_p) + \Psi_{p-1} C^{(p-1)} \subseteq J$. But $\Psi_p C^{(p)} = J$ as Ψ_p is surjective. Hence $\Psi_{p-1} C^{(p-1)} = J$. But this equation is false, since $C^{(p-1)}$ is of complex dimension $p-1$ and J is of complex dimension p.

Next, to prove the asserted uniqueness, suppose that there exist constants v_i and algebraic functions $x'_1, \ldots, x'_p$ and $x''_1, \ldots, x''_p$ of $x_1, \ldots, x_\alpha$ such that

$$\psi_i x_1 + \cdots + \psi_i x_\alpha = v_i + \psi_i x'_1 + \cdots + \psi_i x'_p$$
$$\psi_i x_1 + \cdots + \psi_i x_\alpha = v_i + \psi_i x''_1 + \cdots + \psi_i x''_p$$

modulo periods for $i = 1, \ldots, p$. Then

$$\Psi_p\{x'_1, \ldots, x'_p\} = \Psi_p\{x''_1, \ldots, x''_p\} \text{ in } J\,.$$

Form $U := \Psi_\alpha^{-1}\big((v_1, \ldots, v_p) + \Psi_p U_p\big)$. Then U is open and dense in $C^{(\alpha)}$ since Ψ_α is surjective as $\alpha \geq p$. Let $\{x_1, \ldots, x_\alpha\}$ vary in U. Then $\{x'_1, \ldots, x'_p\}$ and $\{x''_1, \ldots, x''_p\}$ belong to U_p. Hence these two points of U_p are equal since Ψ_p is injective on U_p. Thus the asserted uniqueness is proved.

Finally, to prove (5.4), let n be the least common denominator of $h_1, \ldots, h_\alpha$, and say $h_1 = n_1/n, \ldots, h_\alpha = n_\alpha/n$. Form the map Φ_α from the direct product C^α to J,

$$\Phi_\alpha\colon C^\alpha \to J, \text{ given by } \Phi_\alpha(x_1, \ldots, x_\alpha) = n_1\Psi_1 x_1 + \cdots + n_\alpha \Psi_1 x_\alpha\,. \tag{6.3}$$

Fix $(v_1, \ldots, v_p) \in J$ so that the open set $(v_1, \ldots, v_p) + \Psi_p U_p$ meets $\Phi_\alpha C^\alpha$. Then, since Ψ_p is an isomorphism on U_p, the equation

$$\Phi_\alpha(x_1, \ldots, x_\alpha) = (v_1, \ldots, v_p) + \Psi_p\left\{x'_1, \ldots, x'_p\right\} \tag{6.4}$$

defines $x'_1, \ldots, x'_p$ as a full set of conjugate algebraic functions of $x_1, \ldots, x_\alpha$ as the latter vary as a tuple in the open set $\Phi_\alpha^{-1}((v_1, \ldots, v_p) + \Psi_p U_p) \subset C^\alpha$.

If $n > 1$, then the sum $h_1\psi x_1 + \cdots + h_\alpha \psi x_\alpha$ is ambiguous, even modulo periods. For example, $\frac{1}{2}\int_1^x \frac{1}{x}\,dx$ might mean either $\frac{1}{2}\log x$ or $\frac{1}{2}\log x + \pi\sqrt{-1}$; the two are distinct modulo the period $2\pi\sqrt{-1}$, but their doubles are equal. Similarly, $\frac{1}{n}\int_1^x \frac{1}{x}\,dx$ has n distinct interpretations. Likewise, since an integral of the first kind ψx has $2p$ periods, $\frac{1}{n}\psi x$ has n^{2p} distinct interpretations. Thus multiplication by n defines a covering map $\nu_n\colon J \to J$ of degree n^{2p}; it is a group homomorphism, and its kernel is $(\mathbb{Z}/n)^{2p}$.

Hence $\nu_n(\Psi_p U_p)$ is open. Therefore, after $x_1, \ldots, x_\alpha$ are further restricted, there exist algebraic functions $x_1'', \ldots, x_p''$ of them (in fact, there are n^{2p} different choices, and they form a full set of conjugates) such that

$$n\Psi_p\left\{x_1'', \ldots, x_p''\right\} = \Psi_p\left\{x_1', \ldots, x_p'\right\}\,.$$

Therefore, (6.3) and (6.4) yield

$$n_1\Psi_1 x_1 + \cdots + n_\alpha \Psi_1 x_\alpha = (v_1, \ldots, v_p) + n\Psi_p\left\{x_1'', \ldots, x_p''\right\}\,. \tag{6.5}$$

Notice, (6.5) yields (5.4) for all integrals ψx of the first kind; however, this case is recovered below.

To derive (5.4) in full generality, restrict $x_1, \ldots, x_\alpha$ further so that there exist algebraic functions $x_1''', \ldots, x_p'''$ of them such that

$$\Psi_p\left\{x_1''', \ldots, x_p'''\right\} = -\Psi_p\left\{x_1'', \ldots, x_p''\right\}\,. \tag{6.6}$$

Put $q := n_1 + \cdots + n_\alpha + np$. Form the two sets

$$\left\{x_1''', \ldots, x_p''', x_1'', \ldots, x_p''\right\} \in C^{(2p)} \text{ and}$$
$$\left\{n_1x_1, \ldots, n_\alpha x_\alpha, nx_1''', \ldots, nx_p'''\right\} \in C^{(q)}$$

where n_1x_1 stands for x_1 repeated n_1 times, and so forth.

Then, as $x_1, \ldots, x_\alpha$ vary, the first set above varies in the fiber of Ψ_{2p} over 0. And owing to (6.5) and (6.6), the second set varies in the fiber of Ψ_q over $(v_1, \ldots, v_p)$. So each set varies in a linear system. It now follows from the Elementary Function Theorem that, for any integral ψx,

$$\psi x_1''' + \cdots + \psi x_p''' + \psi x_1'' + \cdots + \psi x_p'' = v'$$
$$n_1\psi x_1 + \cdots + n_\alpha\psi x_\alpha + n\psi x_1''' + \cdots + n\psi x_p''' = v''$$

where v' and v'' are linear combinations of an algebraic function of $x_1, \ldots, x_\alpha$ and of logarithms of algebraic functions. Set $v := v''/n - v'$. Then (5.4) follows. The proof of Theorem 4 is now complete.

In this proof, the integrals of the first kind predominate. Those of the second and third kind only enter briefly at the end, and these integrals are handled by the Elementary Function Theorem. Nevertheless, they had a rich history in the nineteenth and twentieth centuries, and it would be a worthwhile endeavor to write an account of it.

7 Abel's Version of the Genus

In the preceding section, Abel's Addition Theorem was proved in its most elaborate form, Theorem 4. In the proof, the Jacobian J and the Abel map Ψ_α played a major role, and their algebraicity was crucial. Historically, they also played a substantial role in the development of the theory of algebraic surfaces from about 1870 to 1920; often implicit, this role involved both their direct application to suitable curves and families of curves and their indirect application through their generalization to the Picard and Albanese varieties. However, it was the needs of arithmetic in the 1920s and 1930s that provided the impetus to develop in the 1940s and 1950s a purely algebra-geometric theory. And this theory has led finally to a full understanding of Abel's version of the genus.

Mathematically, J and Ψ_α arise from the symmetric product $C^{(\alpha)}$, which is algebraic by construction. The fibers of Ψ_α are the complete linear systems of degree α; so they are algebraic subvarieties. Why not construct J algebra-geometrically as the quotient of $C^{(\alpha)}$ divided by the relation of linear equivalence? Why not define Ψ_α as the quotient map?

Of course, algebraic geometers have constructed J and Ψ_α in just this way for over a century. Furthermore, they have defined the group structure on J essentially as described in the next paragraph. This definition was attributed by Severi in [104], p. 272, and again in [105], p. 12, to Castelnuovo [21], although Castelnuovo did not mention J explicitly in this 1893 paper.

Fix a base point x_0; it corresponds to the fixed lower limit of integration. Given two linear systems of degree α, take a set from each. Their union determines a linear system of degree 2α. So it is of dimension at least $2\alpha - p$ by the Riemann–Roch Theorem. Assume $\alpha \geq p$. Then this system contains a set of the form $\{\alpha x_0, x_1, \ldots, x_\alpha\}$. Define the sum to be the linear system determined by $\{x_1, \ldots, x_\alpha\}$. It is easy to see that this sum is well defined, and compatible with the sum in J as defined by (6.2).

In 1905, Castelnuovo [22], pp. 478–481, showed how this algebra-geometric construction of J and Ψ_α generalizes naturally to yield an Abelian variety parameterizing the complete linear systems fibering an algebraic system of curves on an algebraic surface. Castelnuovo did not use the term "Abelian variety"; it was not in use at the time. Rather, he spoke of the *Picard variety* of the surface to honor Picard's discovery (see the preceding section) that a compact commutative group variety is parameterized by Abelian functions. Castelnuovo needed these functions to complete research of his own, of Enriques's, and of Severi's into the fundamental nature of irregular surfaces. (See [23], pp. 492–493; see [24], pp. 709–710; see [74], p. 369; and see [121], pp. 104, 162–163.)

In 1950, Weil [114], p. 439, subtly refined the use of the term "Picard variety," restricting it to mean the Abelian variety parameterizing complete linear systems of subvarieties of codimension 1 of a given variety of any dimension; the parameterization is part of the package. In his commentary [117], I, p. 572, Weil explained: "Historically speaking, it would have been justified to give it Castelnuovo's name, but it was a question of tampering as little as possible with common usage rather than rendering unto this master his due honor." Weil's refined usage quickly became standard.

Also, Weil [114], p. 438, discussed a related Abelian variety, and named it the *Albanese variety*, which is the name that has stuck. Weil also suggested the name "dual Picard variety," and used it in his paper [115] in 1952.

The Albanese variety is the target of a map Ψ whose source is the given variety X. Like the Abel map, Ψ is given by simple integrals of the first kind. More abstractly, Ψ is characterized by a universal property: any map from X to an Abelian variety factors through Ψ. For a curve, the Albanese variety and the Picard variety coincide; both are equal to the Jacobian. This statement is an abstract algebra-geometric version of the Equivalence Theorem.

However, the Albanese variety of a surface had already been introduced in 1913 by Severi [103]. He noted that it is distinct from the Picard variety, but each admits a homomorphism onto the other (the two are "isogenous," to use another common term Weil introduced in [113]; see [117], I, p. 570). For the next forty years, Severi and his students spoke of there being two Picard varieties associated to a surface (see [7], p. 1, for example). So "historically speaking, it would have been justified to give it" Severi's name.

Also, Weil offered rather weak mathematical justification for having given the variety Albanese's name: in [114], p. 438, he said only "it had been considered mainly by Albanese"; in [117], I, p. 571, he said roughly that Albanese had obtained a bit better understanding of the fibers of Ψ than Severi had.

Perhaps, Weil acted instead out of a sense of poetic justice; this idea comes from reading between the lines in [117], I, on pp. 553 and 562 as well. Furthermore, Lang [71] documents other cases where Weil did not simply ignore social conventions for giving credit, but took "license for obscuring and misrepresenting works and original ideas of others who opened up the field, and for poking fun at them."

On the one hand, perhaps Weil felt Severi had not given Castelnuovo enough credit, and on the other, perhaps Weil felt he owed Albanese a debt of gratitude for having established in São Paulo, Brazil, a mathematical library where Weil "profited greatly" from reading for the first time the works of Castelnuovo, Torelli, and others. Both Albanese and Weil were in São Paulo for about a year, the period of time after Albanese returned to Brazil from four years back in Italy and before he died on 8 June 1947, but there is no evidence the two men ever met (see [26], p. 2, and [118], pp. 188, 192).

Severi was not pleased by the name! (See [117], I, p. 571, and [106], p. 80.)

In 1946, Weil [112] carefully rebuilt the foundations of algebraic geometry, and in 1948, he [113] developed an algebra-geometric theory of the Riemann–Roch Theorem, Abelian varieties, and the Jacobian. This theory is the first valid in arbitrary characteristic, and it served as the basis for Weil's two great proofs in [113] of E. Artin's celebrated 1921 conjectures about the arithmetic of a curve defined by polynomial equations with coefficients in a finite field.

In fact, in 1882, Dedekind and Weber gave an abstract algebraic proof of the Riemann–Roch Theorem, and in 1929, F. K. Schmidt observed that their proof required little change to make it valid in arbitrary characteristic. In the 1930s, Hasse proved Artin's "Riemann Hypothesis" in genus 1 using an analogue of the theory of elliptic functions, and he and Deuring suggested a geometric way to generalize the proof to higher genus. This work inspired Weil. (See [43], pp. 65 and 82–83, and [71], pp. 50–52.)

Weil constructed the Jacobian J by patching together copies of the set U_p parameterizing linear systems of degree p and dimension 0 on the curve C. But Weil left open two questions: Is J defined over the same coefficient field as C? Is J embeddable in projective space?

These questions were answered in the affirmative in the 1950s by Chow, Matsusaka, and Weil himself. All three constructed both the Albanese and the Picard varieties of a variety of any dimension and in any characteristic. Matsusaka gave the first construction of the Picard variety. It is a lot like Castelnuovo's, but Matsusaka used the theory of Chow coordinates (developed by Chow and van der Waerden in 1938) to form the quotient, which, from the start, has the right coefficient field and is projective.

Chow gave a similar construction of the Jacobian, and then obtained the Albanese and Picard varieties by using his theory of the "image" and "trace" of the Jacobian of a "general curve," which is a 1-dimensional section of the ambient variety by a general linear space. Chow's work appeared after Matsusaka's, but his construction of the Jacobian was announced earlier by Weil [114], p. 438, in 1950, and the year before by Chow himself, according to a footnote in [28], p 453.

Weil gave a more complete and elegant treatment, based on the "see-saw principle," taken directly from Severi, and the theorems of the square and the cube, in a course at the University of Chicago. He also developed a suitable theory of descent of coefficient fields. His work became the core of Lang's book [70] of 1959. (See [116], p. 248, and [117], II, pp. 540–541.) This book also contains many historical notes describing the work of Chow, Matsusaka, and many others, along with references to the original papers.

In the 1950s, Rosenlicht published a remarkable series of papers, which grew out of his 1950 Harvard thesis. Like Gorenstein's thesis on adjoints, Rosenlicht's thesis was supervised by Zariski, who had studied Abelian functions and algebraic geometry with Castelnuovo, Enriques, and Severi in Rome from 1921 to 1927 (see [85], Ch. 3). Rosenlicht generalized the theory of the Abel map to a curve C with arbitrary singularities in arbitrary characteristic. His work leads to a full explanation of the final mystery in Abel's Paris paper [1]: the significance of the number γ, Abel's version of the genus.

In [94], Rosenlicht generalized the two basic notions: (linear) equivalence of sets of points, and differentials of the first kind. To discuss them, let $\mathcal{O}$ be the (semi-local) ring of rational functions on C that can be expressed as a fraction whose numerator and denominator are polynomials in the coordinate functions on C and whose denominator vanishes at no singularity of C.

If two point sets contain no singularity, then they are called *equivalent* if, after all common points are removed, the remaining points are the zeros and poles of a function in $\mathcal{O}$ whose reciprocal is also in $\mathcal{O}$.

Care must be taken with a point set that contains a singularity. It is best to interpret the set as a subscheme defined locally by a principal ideal. This idea was introduced by Cartier in his 1958 Paris thesis [19], p. 222. The theory was developed further by Grothendieck in his February 1962 Bourbaki talk [56], No. 232, § 4, and he called these subschemes *Cartier divisors*. This name is now common, but for the sake of simplicity, not used below.

A generalized differential of the first kind, or *Rosenlicht differential*, is a meromorphic differential ω on the desingularization C' of C possessing two properties: first, ω has no pole at any $z \in C'$ lying over a simple point of C; second, for each $h \in \mathcal{O}$, the sum $\sum_z \operatorname{res}_z h\omega$ vanishes as $z \in C'$ ranges over the points lying over the singularities of C.

Let π be the number of independent Rosenlicht differentials; it is known today as the *arithmetic genus* of C. For example, if C is plane a curve of degree d, then $\pi = (d-1)(d-2)/2$ by Rosenlicht's lemma on p. 186. For any C, his Theorem 7 asserts a generalized Riemann–Roch Formula: $\mu - \alpha = \pi - i$ where i is the number of independent Rosenlicht differentials that vanish at a given set of μ points and where α is the dimension of the system of all sets that are equivalent, in the generalized sense, to the given set.

Let $\overline{\mathcal{O}}$ be the integral closure of $\mathcal{O}$ in the function field of C, and put $\delta := \dim \overline{\mathcal{O}}/\mathcal{O}$. Rosenlicht's Theorem 8 asserts $\pi = p + \delta$. For example, if C has only nodes and cusps, then δ is the number of them; if also C is plane, then Rosenlicht's formula recovers Clebsch's formula, (3.6).

In [95], Rosenlicht generalized "Abel's theorem," that is, the Equivalence Theorem, Theorem 2. Namely, his Theorem 1 asserts that, over $\mathbb{C}$, two sets of points are equivalent if and only if, for every Rosenlicht differential, the two corresponding sums of integrals (from a fixed point) have the same value.

Rosenlicht then fixed a basis $\omega_1, \ldots, \omega_\pi$ of the Rosenlicht differentials. His Theorem 2 asserts that, inside the vector space $\mathbb{C}^\pi$, the corresponding period vectors form a discrete subgroup. Its rank is strictly less than 2π if C is singular. So the quotient P is a complex Lie group, but P is not compact, so is not an Abelian variety. Rosenlicht termed P the *generalized Jacobian,* and observed that it is the target of a generalized Abel map Φ_α.

Let ω_0 be a meromorphic differential. Suppose that, for any two equivalent sets of points, the two corresponding sums of integrals of ω_0 have the same value. Then ω_0 is a Rosenlicht differential. Indeed, suppose not. Then the following argument (inspired by [95], p. 507) yields a contradiction.

Given $\alpha \geq \pi + 1$, let $z_1, \ldots, z_\alpha$ vary independently in an open subset U of the smooth locus of C containing no pole of ω_0. Form the $\pi + 1$ by α matrix $M := \left[\omega_i/dz_j\right]$ of functions. Since $\omega_0, \ldots, \omega_\pi$ are independent, M is of rank $\pi + 1$ on an open subset of the self-product U^α. But M is equal to the Jacobian matrix of the following generalized Abel map:

$$\Lambda\colon U^\alpha \to \mathbb{C}^{\pi+1} \text{ given by}$$
$$\Lambda(z_0,\ldots,z_\alpha) = \left(\sum_{j=0}^{\alpha}\int^{z_j}\omega_0,\ldots,\sum_{j=0}^{\alpha}\int^{z_j}\omega_\pi\right)$$

Therefore, by the Implicit Function Theorem, the image ΛU^α contains a nonempty open subset V of $\mathbb{C}^{\pi+1}$.

By hypothesis, Λ remains constant as the set $\{z_1,\ldots,z_\alpha\}$ varies in a linear system. Hence, by the Riemann–Roch Theorem, the fibers of Λ are all of dimension at least $\alpha-\pi$. Consequently, no such V can exist, a contradiction.

For example, suppose C is plane. Then the curves D of given degree cut on C a complete linear system of equivalent sets by Rosenlicht's Corollary on p. 175 in [94]; whence, up to a fixed set of points, so do all the D that pass through that fixed set. Conversely, every complete linear system on C is of the latter form, because any set of points lies in some D of suitable degree.

By the observations above, the Rosenlicht differentials ω are characterized by the constancy of the sum

$$\int^{z_1}\omega+\cdots+\int^{z_\mu}\omega \tag{7.1}$$

where $\{z_1,\ldots,z_\mu\}=C\cap D$ as D ranges over the curves of given degree passing through a given set, for all sets and degrees.

For instance, if C has only nodes and cusps, then the Rosenlicht differentials ω of the third kind are just the differentials flagged by Clebsch and Gordan [33], p. 49, because, for these ω, the sum (7.1) remains constant.

Is the constancy of (7.1) what Abel considered? Did he define γ as the number of independent such ω? If so, then $\gamma=\pi$. But, as Brill and Noether [17], pp. 216 and 222, pointed out, Abel found instead that $\gamma=p$ if C is smooth at finite distance and has mild singularities at infinity, for instance, if C is hyperelliptic. Rowe [97] carried Abel's computations further; in more cases, he arrived at an expression for γ (actually, for $\mu-\alpha$), and in a supplement, Cayley proved that this expression is equal to p.

How can the constancy of (7.1) lead to different values of γ? Through different restrictions on D. Clebsch and Gordan, in effect, placed no restrictions on D. Abel [1], p. 147, however, required D, to begin with, to be defined by the vanishing of a polynomial $g(x,y)$ of the form:

$$g(x,y) := g_1(x)y^{n-1}+g_2(x)y^{n-2}+\cdots+g_n(x)$$

where n is the highest power of y in the polynomial $f(x,y)$ defining C as in (2.1). See the discussion of the hyperelliptic case in Sects. 4 and 5.

More precisely, $\gamma=p$ if the D are required to satisfy the adjunction conditions, which are discussed in Sect. 3, and $\gamma=\pi$ if not.

There is, however, a middle ground: require the D to satisfy only some of the adjunction conditions. This idea was introduced in an 1879 monograph by Lindemann, who aimed to prove a generalized Riemann–Roch theorem. However, the same year in [78], Noether detected a gap in Lindemann's transcendental proof, and replaced it by an algebraic proof, which also yields additional results. (See the footnote in [78], p. 507, and also [106], p. 9.)

In fact, there are a number of partial desingularizations $C^{\dagger}$ that lie between C and its full desingularization. Indeed, it follows from Theorems 4, 5, and 8 in Rosenlicht's paper [94] that there exists a $C^{\dagger}$ whose arithmetic genus γ is any given number between p and π.

Given a $C^{\dagger}$, both it and C have canonical systems associated to their respective Rosenlicht differentials. The system on $C^{\dagger}$ is obtained from the system on C by imposing the appropriate adjunction conditions, and the latter are defined by the conductor ideal associated to the two curves. This statement can be seen as part of Grothendieck's extensive duality theory, first described in his May 1957 Bourbaki talk [56], No. 149.

If $C^{\dagger}$ has bad (non-Gorenstein) singularities, then additional care must be taken with the notion of canonical system, and for that matter, of any linear system. From a technical point of view, it is best to work with "ω-pseudo-divisors" as introduced in [6]; they are defined by the invertible subsheaves of the torsion-free sheaf of Rosenlicht differentials, the dualizing sheaf ω. From a conceptual point of view, the theory of ω-pseudo-divisors can been seen to be close in spirit to Brill and Noether's theory in [16] and Noether's in [78].

For any curve C, plane or not, Rosenlicht proved the generalized Jacobian P and the Abel map Φ_{α} are algebraic. To do so, he [95], Thm. 7, gave a second construction of them, a purely algebraic one. It is modeled on Weil's construction for smooth curves via patching, and works in any characteristic.

In 1962, Oort [81] gave another algebraic construction by a "method of group extensions," which works in any characteristic. The generalized Jacobian P is an extension of the Jacobian J of the desingularized curve by a linear group. Oort assumed J exists. Then he constructed the extension in a way that shows how its nature depends on the nature of the singularities.

In his February 1962 Bourbaki talk [56], No. 232, Grothendieck outlined a sophisticated new construction, yielding a Picard variety associated to any given variety. When the variety is a curve, Grothendieck's Picard variety is just Rosenlicht's generalized Jacobian. (See also Grothendieck's letter [36], p. 197, of 9 August 1960 to Serre.) In spirit, the construction is a lot like Castelnuovo's and Matsusaka's, but it is a lot more refined. To replace Matsusaka's use of Chow coordinates, Grothendieck introduced a new parameter space, the "Hilbert scheme."

Grothendieck, however, had to appeal to the theory of Chow coordinates for a key finiteness result: in projective space, the subvarieties of given degree form a bounded family (see [56], No. 221, p. 7). Grothendieck's action was "scandalous," since, on p. 1, he proclaimed that Hilbert schemes were "destined to replace" Chow coordinates as a tool – a prophetic statement.

In 1966, Mumford [77], Lect. 14, saved the day. He introduced an important new concept, now known as "Castelnuovo–Mumford" regularity; it yields a simple direct proof of the needed finiteness. In addition, Mumford gave the details of Grothendieck's construction of the Picard variety, worked out in the case of surfaces.

Furthermore, Mumford showed how Grothendieck's work yields the first algebraic proof of a basic theorem, which asserts the completeness of the characteristic linear system of a suitably positive curve on a surface; in fact, this proof is the goal of [77]. In 1904, first Enriques and then Severi gave algebraic proofs, which were later found to be deficient. Meanwhile, in 1905, Castelnuovo used the theorem to find the dimension of the Picard variety. In 1910, Poincaré gave a rigorous analytic proof (see [121], pp. 98–104).

For any curve C, the algebraicity and other properties of P and Φ_α yield a new version of Theorem 4, the elaborate form of the Addition Theorem. In it, p is replaced by π, and integrals of the first kind are replaced by integrals of Rosenlicht differentials. The proof at the end of Sect. 6 carries over with almost no change; there is no need for P to be compact, nor for multiplication by n to define a map $\nu_n\colon P \to P$ of any particular (finite) degree.

It is interesting to compare the two versions of Theorem 4. The new one asserts that an arbitrary sum of integrals of the same Rosenlicht differential can be reduced to a sum of π such integrals, plus a constant. The old version asserts that the initial sum can be reduced to a sum of p such integrals, plus an elementary function v. If $p < \pi$, then v need not be constant.

In 1957, Rosenlicht [96] proved a sort of universal mapping property for generalized Jacobians. However, he did not begin with a singular curve and its Jacobian. Rather, he began with a map $\Upsilon\colon U \to G$ where U is a smooth open curve and G is a commutative algebraic group. Then he embedded U in a suitable compact singular curve C so that Υ extends to a map $C \to G$ that factors through the generalized Jacobian of C. Consequently, this universal mapping property does not amount to an abstract algebra-geometric version of the Equivalence Theorem for a given singular curve.

Nevertheless, Rosenlicht [96], p. 81, observed that his universal mapping property "can be used to prove theorems about algebraic groups themselves." Also, he noted that Serre too had proved this property independently and that Serre and Lang had applied it to class field theory. Then in 1959 Serre published a monograph [100], where he developed all this theory from scratch.

Generalized Jacobians appear in another major part of the theory of Abelian integrals, which concerns the way they vary and degenerate, or specialize, as the underlying curve does. The subject was traced by Severi [105], p. 14, back to some of Klein's work in 1874, in which he studied the limits of integrals of the first kind on a smooth plane quartic as the quartic varies, acquiring a double point. From 1890 to 1910, the theory was developed by Picard, Poincaré, and Lefschetz, and by Castelnuovo, Enriques, and Severi, and applied to the theory of algebraic surfaces. (See [24], pp. 725–727; see [121], Chaps. 6–7; see [43], pp. 35, 53; and see [60], pp. 253–255.)

In 1947, Severi published a monograph [105], about generalized Jacobians and their behavior under specialization for curves with at most nodes and cusps. In fact, according to Serre [100], p. 114, Severi was the first author to discuss the generalized Jacobian explicitly: he studied its analytic structure and its algebraic structure, but "without always properly distinguishing the one from the other."

In 1956, Igusa [62], Thm. 3, established the compatibility of specializing a curve with specializing its generalized Jacobian when the general curve is smooth and the special curve has at most one node. Moreover, in Sect. 1, he proved that such a specialization obtains in the important case now known as a "Lefschetz pencil": the curves are cut on a smooth surface by a general 1-parameter family of hyperplanes. Igusa worked in arbitrary characteristic, and was the first to do so; he explained on p. 171, that Néron had, in 1952, studied the total space of such a family of Jacobians, but Néron had not explicitly considered the special fiber.

Igusa's approach is like Castelnuovo's, Chow's, and Matsusaka's before him and Grothendieck's after him. However, Grothendieck went considerably further: not only did he prove compatibility with specialization for an arbitrary family of (irreducible) curves, in equicharacteristic or mixed, but he worked with a family of varieties of arbitrary dimension.

In particular, when a curve varies and degenerates, its arithmetic genus π, which is the dimension of its generalized Jacobian, remains constant, whereas its geometric genus p drops. So, for the purpose of studying the degeneration of Abelian integrals, Abel's version of the genus, π or γ, is better suited to the job than Riemann's version p.

Thus, though it once seemed odd that Abel had focused on the constancy of the sum of integrals, rather than on the boundedness of the summands, in fact the constancy turned out to be the deeper, subtler condition. Indeed, after the significance of the boundedness was explained by Riemann, it took a hundred years more before the significance of the constancy was fully understood. Once again, Abel's intuition has proven to be sound!

8 Conclusion

For various reasons, four different theorems have been commonly accepted by respected scholars as the celebrated theorem known as *Abel's Theorem*. In fact, some scholars have implicitly accepted two of the four as the same. But the four theorems are not simply versions of each other. Rather, they are mathematically distinct, and are the results of successive steps forward.

The Elementary Function Theorem is the result of the first step. It is the key to the other theorems. Abel proved it in relatively explicit forms in [1] and [2]; he also proved it, with far less work, in more elegant and more conceptual forms in [1], [3], and [4]. In [1], p. 150, he explained why he proved it twice: the conceptual method is not computationally effective.

However, in [3], Abel proved only the Elementary Function Theorem, and he did not identify it as a preliminary result. Moreover, it is the one theorem common to Abel's four papers. Doubtless, these reasons are ultimately why this theorem is so widely accepted as Abel's Theorem.

The Equivalence Theorem is the result of the second step. Combined with its converse, which Abel did not recognize, it has become a fundamental result in an important part of modern mathematics, the theory of Riemann surfaces. So the theorem needs a name for ready reference, and it is common and fitting to use Abel's name since he made the seminal contribution. Historical correctness being a secondary consideration, mathematicians will doubtless always refer to the full Equivalence Theorem as Abel's Theorem.

The Relations Theorem is the result of the third step. In the introduction to [1], Abel stated it informally in a way that makes it sound deceptively like a more refined form of the Elementary Function Theorem. Furthermore, he reinforced this false idea a few pages later. There appears to be no further reason for interest in this theorem; so there appears to be no strong reason to accept it as Abel's Theorem.

The Addition Theorem is the result of the fourth and last step. In the introductions to both [1] and [2], Abel makes it clear that this theorem is his goal. However, he stated the theorem in different forms.

In the introduction to [2], Abel stated the Addition Theorem in its simplest form, which is an immediate consequence of the Relations Theorem. However, he gave proofs only for hyperelliptic integrals.

In the introduction to [1], Abel stated the Addition Theorem in a more elaborate form, and in the course of the paper, he refined the statement even further. Although Abel's proof is, apparently, not fully valid, nevertheless, in its most elaborate form, the Addition Theorem can be proved using the Jacobian variety and the Abel map, which combine the full Equivalence Theorem and the Riemann–Roch Theorem in a powerful algebra-geometric form.

The Addition Theorem involves a numerical invariant, the genus. But the genus comes in two versions, Abel's and Riemann's. In fact, Abel's definition is nebulous, but it can be made clear and precise. Then the two genera may differ in value; if so, then Riemann's has the smaller value, ostensibly an advantage. With either genus, the Addition Theorem sounds the same, mutatis mutandis. However, Abel's genus is better suited for the study of the degeneration of Abelian integrals.

In 1832, Jacobi successfully advocated that, as a monument to Abel's genius, the Addition Theorem be called Abel's Theorem. Yet, Jacobi knew only the simplest form of the theorem; the more elaborate forms appear only in [1], which was, at the time, temporarily misplaced.

Thus, of the four theorems, only the Addition Theorem can rightfully be called **Abel's Theorem!**

Acknowledgments. The author is indebted to Beverly, his wife, and to Ragni Piene, the editor, for finding numerous obscure passages and typographical errors in previous drafts.

References

[1] N. H. Abel, *Mémoire sur une propriété générale d'une classe très étendue de fonctions transcendantes.* Mémoires présentés par divers savants. **VII** (1841), 176–264. = [5] **I**, XII, pp. 145–211. ▶ Introduction trans. in [11], pp. 188–189.

[2] N. H. Abel, *Remarques sur quelques propriétés générales d'une certaine sorte de fonctions transcendantes.* J. für Math. **3** (1828), 313–323. = [5] **I,** XXI, pp. 444–456. ▶ Partly trans. in [11], pp. 190–195.

[3] N. H. Abel, *Démonstration d'une propriété générale d'une certaine classe de fonctions transcendantes.* J. für Math. **4** (1829), 200–201. = [5] **I,** XXVII, pp. 515–517.

[4] N. H. Abel, *Sur la comparaison des fonctions transcendantes.* Posthumous fragment. = [5] **II,** X, pp. 55–66.

[5] N. H. Abel, *Œvres complètes,* **I, II.** Sylow L. and Lie S. (eds.), Grøndahl, Christiania (Oslo), 1881.

[6] A. B. Altman and S. L. Kleiman, *Compactifying the Jacobian.* Bull. Amer. Math. Soc. **82** (1976), 947–949.

[7] A. Andreotti, *Sopra le varietà di Picard di una superficie algebrica.* Rend. Accad. Nazionalle dei XL (IV) **2** (1951), 1–9.

[8] H. F. Baker, *Abelian functions. Abel's theorem and the allied theory of theta functions.* Cambridge U. Press, 1897, reissued 1995.

[9] E. T. Bell, "Men of Mathematics." Simon and Schuster, 1937, reprinted 1965.

[10] E. T. Bell, "The development of mathematics." Second ed., McGraw-Hill, 1945.

[11] G. Birkhoff and U. Merzbach, "A source book in classical analysis." Harvard U. Press, 1973.

[12] C. A. Bjerknes, "Niels Henrik Abel." Mémoire Soc. Sci. physiques et naturelles de Bordeaux, 3*e* série, I, Gauthier-Villars, 1884.

[13] G. A. Bliss, *The reduction of singularities of plane curves by birational transformation.* Bull. AMS **29** (1923), 161–183.

[14] G. A. Bliss, "Algebraic functions." AMS Colloquium Publ. **XVI**, 1933, republished by Dover, 1966.

[15] A. Brill, *Max Noether.* Jahresbericht DMV **32** (1923), 211–233.

[16] A. Brill and M. Noether, *Ueber die algebraischen Functionen und ihre Anwendung in der Geometrie.* Math Ann. **7** (1874), 269–310.

[17] A. Brill and M. Noether, *Die Entwickelung der Theorie der algebraischen Functionen in älterer und neuerer Zeit.* Jahresbericht DMV **3** (1894), 107–566.

[18] W. Burau, *Clebsch, Rudolf Friedrich Alfred.* In "Dictionary of Scientific Biography." Gillispie (ed. in chief) Vol. III. Scribner's 1971, pp. 313–315.

[19] P. Cartier, "Dérivationes et diviseurs en géométrie algébrique." Gauthier-Villars, 1959.

[20] E. Casas, "Singularities of plane curves." LMS Lecture Notes Series **276**, 2000.

[21] G. Castelnuovo, *Le corrispondenze univoche tra gruppi di p punti sopra una curva di genere p.* Rend. Ist. Lomb. (2) **25** (1893), 1189–1205. = "Memorie scelte." Zanichelli, 1937, pp. 79–94.

[22] G. Castelnuovo, *Sulle integrali semplici appartenenti ad una superficie irregolare* (3 notes). Rend. R. Accad. Lincei **V 14** (1905), 545–546, 593–598, 655–663. = "Memorie scelte." Zanichelli, 1937, pp. 473–507.

[23] G. Castelnuovo and F. Enriques, *Sur quelques résultats nouveaux dans la théorie des surfaces algébriques.* Note V. in "Théorie des fonctions algébriques de deus variables indépendantes" by E. Picard and G. Simart, t. II, Gauthier-Villars, 1906, republished by Chelsea, 1971, pp. 485–522.

[24] G. Castelnuovo and F. Enriques, *Die algebraischen Flächen vom Gesichtspunkte der birationalen Transformationen aus.* Dated Dec. 1914. In "Encyklopädie math. Wissenschaften." Vol. **III C 6b**. Teubner 1927, pp. 674–768.

[25] G. Castelnuovo, F. Enriques, and F. Severi, *Max Noether.* Math. Ann. **93** (1925), 161–181.

[26] B. Castrucci, *Prof. Giacomo Albanese.* Bull. Soc. Saõ Paulo **2** (1947), 1–5.

[27] A. Cayley, *On the transformation of plane curves.* Proc. LMS **I** (1865–1866), 1–11. = Papers VI, Cambridge U. Press, 1893, pp. 1–8.

[28] W-L. Chow, *The Jacobian variety of an algebraic curve.* Amer. J. Math. **76** (1954), 453–476.

[29] A. Clebsch, *Ueber die Anwendung der Abelschen Functionen in der Geometrie.* J. für Math. **63** (1864), 189–243.

[30] A. Clebsch, *Ueber diejenigen ebenen Curven, deren Coordinaten rationale Functionen eines Parameters sind.* J. für Math. **64** (1865), 43–65.

[31] A. Clebsch, *Ueber die Singularitäten algebraischer Curven.* J. für Math. **64** (1865), 98–100.

[32] A. Clebsch, *Ueber diejenigen Curven, deren Coordinaten sich als elliptische Functionen eines Parameters darstellen lassen.* J. für Math. **64** (1865), 210–271.

[33] A. Clebsch and P. Gordan, "Theorie der Abelschen Funktionen." Teubner, 1866.

[34] A. Clebsch and F. Lindemann, "Vorlesungen über geometrie." Vol. 1, Teubner, 1876.

[35] C. H. Clemens and P. A. Griffiths, *The intermediate Jacobian of the cubic threefold.* Ann. Math. (2) **95** (1972), 281–356.

[36] P. Colmez and J.-P. Serre, "Correspondance Grothendieck–Serre." SMF, 2001.

[37] R. Cooke, *Abel's Theorem.* In "The History of Modern Math. Vol. 1: Ideas and their reception." Rowe D. E. and McCleary J. (Eds.) Academic Press, 1989, pp. 389–421.

[38] J. L. Coolidge, "A history of geometrical methods." Oxford U. Press, 1940.

[39] D. Cox, J. Little, and D. O'Shea, "Ideals, Varieties, and Algorithms." Springer-Verlag, Second edition, 1996.

[40] L. Cremona, *Sugli integrali a differenziale algebrico.* Mem. dell'Acc. delle Scienze di Bologna, (2) **10** (1870), 3–33.

[41] A. Del Centina, *Abel's manuscripts in the Libri collection: their history and their fate.* In [42], pp. 87–103.

[42] A. Del Centina, "Abel's Parisian manuscript, preserved in the Moreniana Library of Florence." Olschki, Firenze, 2002.

[43] J. A. Dieudonné, "History of algebraic geometry." J. D. Sally (trans.) Wadsworth, 1985.

[44] F. Enriques and O. Chisini, "Lezioni sulla teoria geometrica delle equazioni e delle funzioni algebriche 2." Zanichelli. Vol. 4, 1st ed., 1934. Reprinted 1989.

[45] F. Enriques and F. Severi *Intorno alle superficie iperellittiche.* Rend. Acc. Lincei **XVI** (1907), 443–453. = "Memorie Scelte." Vol. **2**, Zanichelli 1959, pp. 283–294.

[46] C. S. Fisher, *Gordan, Paul Albert.* In "Dictionary of Scientific Biography." Gillispie (ed. in chief) Vol. V. Scribner's 1972, pp. 472–473.

[47] A. R. Forsyth, "Theory of functions of a complex variable." Third edition, Cambridge 1918. Reprinted in two volumes by Dover, 1965.

[48] H. Freudenthal, *Riemann, Georg Friedrich Bernhard.* In "Dictionary of Scientific Biography." Gillispie (ed. in chief) Vol. XI. Scribner's 1975, pp. 447–456.

[49] W. Fulton, "Algebraic curves." Benjamin Inc., 1968.

[50] D. Gorenstein, *An arithmetic theory of adjoint plane curves.* Trans. AMS **72** (1952), 414–436.

[51] E. Goursat, "Functions of a complex variable. Vol. II, Part One." E. R. Hedrick and O. Dunkel. (trans.) Ginn and Co., 1916. Reprinted by Dover, 1959.

[52] J. J. Gray, *The Riemann–Roch theorem: acceptance and rejection of geometric ideas.* Cahiers d'histoire et de philosophie des sciences **20** (1987), 139–151.

[53] J. J. Gray, *Algebraic geometry in the late nineteenth century.* In "The history of modern mathematics, Vol. I: Ideas and their reception." Rowe D. E. and McCleary J. (Eds.) Academic Press, 1989, pp. 361–385.

[54] J. J. Gray, *The Riemann–Roch theorem and geometry, 1854–1914.* Doc. Math. J. DMV – Extra Volume ICM 1998. Vol. III, pp. 811–822.

[55] P. A. Griffiths, *Variations on a theorem of Abel.* Inv. Math. **35** (1976), 321–390.

[56] A. Grothendieck, "Fondements de la géométrie algébrique." Séminaire Bourbaki 1957–1962. Secrétariat Math., Paris, 1962.

[57] K. Hensel and G. Landsberg, "Theorie der algebraischen Functionen einer Variabeln und ihre Anwendung auf algebraische Kurven und abelsche Integrale." Teubner, 1902. Reprinted by Chelsea, 1965.

[58] C. Houzel, *Fonctions elliptiques et intégrales abéliennes.* In "Abrégé d'histoire des mathématiques 1700–1900, Vol. II." J. Dieudonné (ed.) Hermann, 1978 = [61], Ch. VIII, pp. 81–190.

[59] C. Houzel, *Histoire de la théorie des équations algébriques de Lagrange à Galois.* Journées X-UPS, **4** (1987), 55–66 = [61], Ch. V, pp. 53–61.

[60] C. Houzel, *Aux origines de la géométrie algébrique: les travaux de Picard sur les surfaces (1884–1905).* Cahiers d'histoire et de philosophie des sciences. Vol. **34** Société Mathématique de France. (1991), 243–276. = [61], Ch. X, pp. 203–234.

[61] C. Houzel, "La géométrie algébrique, Recherches Historiques." Albert Blanchard, 2002.

[62] J. Igusa, *Fibre systems of Jacobian varieties.* Amer. J. Math. **78** (1956), 171–199.

[63] C. G. J. Jacobi, Review of Legendre's "Théorie des fonctions elliptiques, troisième supplément." J. für Math. **8** (1832), 413–417.

[64] C. G. J. Jacobi, *Considerationes generales de transcendentibus Abelianis.* J. für Math. **9** (1832), 394–403. = "Werke II." Reimer, 1882, pp. 5–16. ► Partly trans. in [11] pp. 207–212.

[65] C. G. J. Jacobi, *Notiz über A. Göpel.* J. für Math. **35** (1847), 277–282. = "Werke II." Reimer, 1882, pp. 145–152.

[66] F. Klein, "Lectures on Mathematics." Reported by A. Ziwet. Macmillan, 1894.

[67] F. Klein, "Vorlesungen über die Entwicklung der Mathematik im 19. Jarhundert." Prepared by R. Courant and O. Neugebauer. Part I. Springer, 1926.

[68] M. Kline, "Mathematical thought from ancient to modern times." Oxford U. Press, 1972.

[69] E. Kramer, *Noether, Max.* In "Dictionary of Scientific Biography." Gillispie (ed. in chief) Vol. X. Scribner's 1974, pp. 193–141.

[70] S. Lang, "Abelian varieties." Interscience Tract **7**. Interscience, 1959.

[71] S. Lang, *Comments on non-references in Weil's works.* Mitteilungen DMV. **1** (2002), 49–56.

[72] O. A. Laudal, *The mathematical significance of Abel's Paris memoir.* In [42], pp. 65–68.

[73] S. Lefschetz, *On certain numerical invariants of algebraic varieties with applications to Abelian varieties.* (Bordin prize, Paris, 1919, and Bôcher prize, AMS, 1924.) Trans. AMS **22** (1921), 327–482. = "Selected Papers, Chelsea, 1971, pp. 41–196.

[74] S. Lefschetz, *Hyperelliptic surfaces and Abelian varieties.* In "Selected topics in algebraic geometry." Bull. National Res. Council, No. 96 (Washington, 1934). Chelsea reprint, 1970, pp. 349–395.

[75] A. M. Legendre, *Cover letter.* J. für Math. **8** (1832), 413.

[76] A. I. Markusevich, "Introduction to the classical theory of Abelian functions." Transl. Math. Monographs **96**, AMS, 1992.

[77] D. Mumford, "Lectures on curves on an algebraic surface." Annals of Math. Studies, No. 59. Princeton U. Press, 1966.

[78] M. Noether, *Ueber Schnittpunktsysteme einer algebraischen Curve mit nicht-adjungirten Curven.* Math. Ann. **15** (1879), 507–528.

[79] M. Noether, *Rationale Ausführung der Operationen in der Theorie der algebraischen Functionen.* Math. Ann. **23** (1884), 311–358.

[80] J. J. O'Connor and E. F. Robertson, *Rudolf Friedrich Alfred Clebsch.* http://www-history.mcs.st-andrews.ac.uk/history/References/Clebsch.html

[81] F. Oort, *A construction of generalized Jacobian varieties by group extensions.* Math. Ann. **147** (1962), 277–286.

[82] O. Ore, "Niels Henrik Abel, mathematician extraordinary." U. Minn. Press, 1957, reprinted Chelsea Publishing, 1974.

[83] O. Ore, *Abel, Niels Henrik.* In "Dictionary of Scientific Biography." Gillispie (ed. in chief) Vol. I. Scribner's 1971, pp. 12–17.

[84] P. Painlevé, *Sur les fonctions qui admettent un théorème d'addition.* Acta Math. **27** (1903), 1–54.

[85] C. Parikh, "The unreal life of Oscar Zariski." Academic Press 1991.

[86] S. J. Patterson, "Abel's Theorem." R. Circ. Mat. Palermo II, Supplemento N **61** (1999), 9–48.

[87] L. Pepe, "Genius and orderliness, Abel in Berlin and Paris." In [42], pp. 87–103.

[88] E. Picard, *Mémoire sur la théorie des fonctions algébriques de deux variables.* J. Math. Pures et Appl. **5** (1889), 135–319. = "Oeuvres" Tome III, Cen. nat. rech. sc., 1980, pp. 201–385.

[89] E. Picard, *Sur la théorie des groupes et des surfaces algébriques.* R. Circ. Mat. Palermo **9** (1895), 244–255. = "Oeuvres" Tome III, Cen. nat. rech. sc., 1980, pp. 417–428.

[90] E. Picard, "Traité d'analyse." Vol. 2, 2nd ed., Gauthier-Villars, 1905.

[91] J. B. Pogrebyssky, *Brill, Alexander Wilhelm von.* In "Dictionary of Scientific Biography." Gillispie (ed. in chief) Vol. II. Scribner's 1970, p. 465.

[92] B. Riemann, *Theorie der Abelschen Functionen.* J. für Math. **54** (1857), 115–155. = "Werke." Springer, 1990, pp. 120–174. ▶ Partly trans. in [11], pp. 196–201.

[93] G. Roch, "Ueber die Anzahl willkürlichen Constanten in algebraischen Functionen." J. für Math. **64** (1864), 372–376. ▶ partly trans. in [11], pp. 201–203.

[94] M. Rosenlicht, *Equivalence relations on algebraic curves.* Annals Math. **56** (1952), 169–191.

[95] M. Rosenlicht, *Generalized Jacobian varieties* Annals Math. **59** (1954), 505–530.

[96] M. Rosenlicht, *A universal mapping property of generalized Jacobian varieties.* Annals Math. **66** (1957), 80–88.

[97] R. C. Rowe, *Memoir on Abel's theorem.* Phil. Trans. Royal Acad. **172** (1881), 713–750.

[98] G. Scorza, *Intorno alla teoria generale delle matrici di Riemann ed alcune se applicazione* R. Circ. Mat. Palermo **41** (1916), 263–380. = "Opere scelte." Cremonese, Vol. 2, 1961, pp. 126–275.

[99] C. A. Scott, *On the intersections of plane curves.* Bull. AMS **2** (1897/98), 260–273.

[100] J.-P. Serre, "Groupes algébriques et corps de classes." Hermann, 1959.

[101] F. Severi, *Sulle superficie algebriche che ammettonon un gruppo continuo permutabile a due parametri, di trasformazioni birazionali.* Atti del R. Ist. Veneto di Scienze, Lettere ed Arti, **67** (1907), 409–419.

[102] F. Severi, *Relazioni fra gl'integrali semplici e multipli di I^a specie d'una varietà algebrica.* Annali Mat. **20** (1913), 201–216.
[103] F. Severi, *Un teorema d'inversione per gli integrali semplici di 1^a specie appartenentu ad una superficie algebrica.* Atti del R. Ist. Veneto di Scienze, Lettere ed Arti, **72** (1913), 765–772.
[104] F. Severi, "Vorlesungen über algebraische Geometrie." E. Löffler (trans.) Teubner, 1921. Reprinted by Johnson 1968.
[105] F. Severi, "Funzioni Quasi Abeliane." Pontificia Accademia delle Scienze. Scipta Varia, **4**, 1947.
[106] F. Severi, "Geometria dei sistemi algebrici sopra una superficie e sopra una varietà algebrica. Vol. 3. sviluppo delle teorie degli integrali semplici e multipli sopra una superficie o varietà e delle teorie collegate." Cremonese, 1959.
[107] I. R. Shafarevich, "Basic algebraic geometry." K. A. Hirsch (trans.) Springer, 1974.
[108] C. L. Siegel, "Topics in complex function theory." Vol. 3. E. Gottschling and M. Tretkoff (trans.) Interscience Tracts, No. 25. Wiley, 1973.
[109] A. Stubhaug, "Niels Henrik Abel and his times: called too soon by flames afar." Translated from the second Norwegian edition of 1996 by R. H. Daly. Springer 2000.
[110] L. Sylow, *Remarques sur les nombres γ et $\mu - \alpha$.* In [5], **II**, pp. 298–300.
[111] K. Weierstrass, "Theorie der Abelschen Transcendenten." = Werke 4. Mayer & Müller, 1902.
[112] A. Weil, "Foundations of algebraic geometry." AMS Coll., Vol. **XXIX**, 1946. Second ed. 1962.
[113] A. Weil, (a) "Sur les courbes algébriques et les variétés qui s'en déduisent," Hermann, Paris, 1948; (b) Variétés abéliennes et courbes algébriques," ibid., 1948; (c) 2nd edition of (a) and (b), under the collective title "Courbes algébriques et variétés abéliennes," ibid., 1971.
[114] A. Weil, *Variétés abéliennes.* In "Colloque d'Algèbre et Théorie des Nombres." CNRS, Paris, pp. 125–127. = "Papers" I, Springer, 1979, pp. 437–440.
[115] A. Weil, *On Picard varieties.* Amer. J. Math. **90** (1952), 865–893. = "Papers" II, Springer, 1979, pp. 73–102.
[116] A. Weil, *On the projective embedding of Abelian varieties.* In "Algebraic geometry and topology, A symposium in honor of S. Lefschetz," Princeton U. Press, 1957. = "Papers" II, Springer, 1979, pp. 177–181.
[117] A. Weil, *Commentaire.* In "Collected Papers." Springer, 1979, Vol. I, pp. 518–574, I,I pp. 526–553, and III, pp. 443–465.
[118] A. Weil, "The Apprenticeship of a mathematician." J. Gage (trans.) Birkhäuser 1992.
[119] H. Weyl, *The concept of a Riemann surface.* 3rd ed., Teubner, 1955, G. R. Maclane (trans.) Addison-Wessley, 1964.
[120] W. Wirtinger, *Algebraische Funktionen und ihre Integrale.* Dated Oct. 1901. In "Encyklopädie math. Wissenschaften." Vol. **II B 2.** Teubner 1921, pp. 115–175.
[121] O. Zariski, "Algebraic Surfaces." Springer. First edition, Ergebnisse **III, 5,** 1935. Second supplemented edition, Ergebnisse **61**, 1971.

On Abel's Hyperelliptic Curves

Torsten Ekedahl

Summary. In this note we discuss a class of hyperelliptic curves introduced by Abel in an 1826 paper. After some indications of the context in which he introduced them and a description of his main result we give some results on the moduli space of such curves.

In particular we compute the dimension of it at each of its points as well as giving a combinatorial formula for the number of components.

In his paper [1] Abel takes up a very special case of the problem of deciding when a rational differential form is the logarithmic differential of a rational function. Even though it is easy to imagine that the problem later led Abel to his famous Paris dissertation, the solution that Abel proposed is very special and quite different from the approach he chose later. Nevertheless, the present article is concerned with investigating the problem of [1].

A characteristic, and for the time quite unusual, feature of all of Abel's work is his insistence on treating general cases rather than special examples. It thus seems entirely fitting to study the moduli problem of all the solutions to his proposed problem and we shall indeed consider a number of aspects of the moduli spaces (or more precisely stacks) that classify his solutions.

We start by introducing a (small) number of variants of an attempt to formulate Abel's condition as a moduli problem. In particular Abel's solution of his problem in terms of a polynomial type Pell's equation appears not as the moduli problem that has been chosen as the central moduli problem of this article but as a chart for it. We then study the relations between these variants, the end result being that they are indeed closely related.

After a discussion of the case of genus 0 and 1 we give a reformulation of the moduli problem in terms of families of maps between genus zero curves. We then proceed to make an infinitesimal study of the moduli problem which allows us to conclude that it is smooth in characteristic 0. We then go on to study a Lyashko–Looijenga type map and show that it is a covering map. This allows us to give a topological covering space type description of the moduli stack which in particular gives us a combinatorial description of the set of components of the moduli space.

We shall, except for the last section, adopt a purely algebraic approach. Apart from reasons of taste there are some arguments in favour of such a choice. The reader's attention should be particularly directed to Theorem 6.2, where we shall discover that some naturally defined "equi-ramification strata" turn out to be non-reduced. It seems likely that the multiplicity with which those strata appear is significant.

We shall also use the language of algebraic stacks. This may seem unnecessary particularly as our stacks are very close to being spaces (cf. Proposition 8). However, I claim that it is the technically most convenient as well as most intuitive way of doing things. In particular when defining maps between solutions to moduli problems, representing these solutions as stacks means that in order to define maps between them one may often follow the path of first deciding what the map should do on points and then verify that this pointwise construction is natural enough so that it makes sense for families of objects. This is in fact what we shall do most of the time. Sometimes, however, we shall discover that some choices that were made in the point case can not be made in the case of a family and we shall then have to incorporate those choices in the definition of the moduli problem. This will lead for instance to the three slightly different versions of Abel's hyperelliptic curve.

Conventions. By a *monic polynomial* we shall mean a polynomial in one variable whose highest degree coefficient is equal to 1. Such a polynomial will be said to be *normalised* if its next to highest degree coefficient is equal to 0.

As we shall deal extensively with stacks it seems natural to use the term 'scheme' to denote an algebraic space and hence by 'locally' mean 'locally in the étale topology'.[1] Though we shall do so, this is not strictly necessary, however, and then 'locally' may at times be interpreted as 'locally in the Zariski topology,' though consistently using the étale topology will always work.

We have made only a token attempt at formulating our results in arbitrary characteristics.[2] Starting with Sect. 3 all our schemes and stacks will be over Spec $\mathbb{Z}[1/2]$ and starting with Sect. 6 we shall work exclusively in characteristic zero, this will

[1] Note that in practice the only difference between ordinary schemes and algebraic spaces is that for the latter the Zariski topology is not available.

[2] It will be clear that if the characteristic is large enough with respect to the degree n, the situation will be similar to that of characteristic 0.

also be true at the end of the preliminary Sect. 1 and at points in Sect. 3 (which will be explicitly spelled out).

As usual a *multiset* is a set whose members are counted with certain multiplicities, formally it is a set provided with a multiplicity function from it to the integers ≥ 1. If S is a multiset, we shall use $\underline{S}$ to denote the domain of the multiplicity function and μ_S for the multiplicity function itself. We shall use set-theoretic notation when dealing with multisets:

- $S := \{1, 1, 1, 2, 2, 3\}$ will denote the multiset for which $\underline{S} = \{1, 2, 3\}$ and $\mu_S(n) = 4 - n$.
- $\sum_{s \in S} s^2$ should be interpreted as $\sum_{s \in \underline{S}} \mu_S(s)s^2$, i.e., $3 \cdot 1^2 + 2 \cdot 2^2 + 3^2$.
- Similarly, $\{\lfloor s/2 \rfloor \mid s \in S\}$ should be interpreted as $\{0, 0, 0, 1, 1, 1\}$.

A multiset S is finite if $\underline{S}$ is and then its *cardinality*, $|S|$, equals $\sum_{s \in S} 1$. A multiset S is said to be a submultiset of the multiset T if $\underline{S} \subseteq \underline{T}$ and $\mu_S(s) \leq \mu_T(s)$ for all $s \in \underline{S}$.

1 Preliminaries

We shall sometimes speak about the universal object over a stack which classifies some type of geometric object. Note that, contrary to the case when the moduli problem of classifying such objects is representable by a scheme, this is somewhat ambiguous and is not quite as strong. Firstly, for a family of objects over S, the family may not be the pullback of the universal family but is so only locally on S. Secondly, the universal object is not unique; two such objects are only locally isomorphic. Thirdly, a universal object may in fact not even exist over the stack itself, but only locally. It would be more proper to speak about the stack of universal objects, but we shall allow ourselves the luxury of not doing that. The first phenomena are shown quite clearly in the case of the classifying stack, BG, of a finite group G. A universal object is given by the trivial G-torsor, and a non-trivial G-torsor over S is of course not the pullback of the trivial one. In fact, any G-torsor over the base is universal and there may very well be non-trivial G-torsors over the base.

Assume that $X \to S$ is a scheme and $X \to \mathbb{P}^1 \times S$ an S-morphism. Let C and D be the schematic inverse images of $0 \times S$ and $\infty \times S$ and assume C and D are Cartier divisors. We shall repeatedly use the (obvious) fact that such a morphism is the same thing as an isomorphism $\mathcal{O}_X(C) \xrightarrow{\sim} \mathcal{O}_X(D)$.

Let S be a scheme. A line bundle $\mathcal{L}$ and a trivialisation φ of $\mathcal{L}^2$ will be called an *involutive line bundle*. Consider further $\mathbb{P}(\mathcal{O}_S \oplus \mathcal{L}) \to S$, the projective bundle on the vector bundle $\mathcal{O}_S \oplus \mathcal{L}$, the two sections ∞ and $\mathbf{0}$ associated to the two projections of $\mathcal{O}_S \oplus \mathcal{L}$ and the involution σ of $\mathbb{P}(\mathcal{O}_S \oplus \mathcal{L}) \to S$ defined as the composite of the map $\mathbb{P}(\mathcal{O}_S \oplus \mathcal{L}) \to \mathbb{P}(\mathcal{L} \oplus \mathcal{O}_S)$ that switches the two factors, the standard identification and distributivity $\mathbb{P}(\mathcal{L} \oplus \mathcal{O}_S) = \mathbb{P}((\mathcal{L} \oplus \mathcal{O}_S) \bigotimes \mathcal{L}) = \mathbb{P}(\mathcal{L}^2 \oplus \mathcal{L})$ and φ applied to the first factor $\mathbb{P}(\mathcal{L}^2 \oplus \mathcal{L}) \to \mathbb{P}(\mathcal{O}_S \oplus \mathcal{L})$. We shall call the data $(\mathbb{P}(\mathcal{O}_S \oplus \mathcal{L}), \mathbf{0}, \infty, \sigma)$ the *involutive projective bundle* associated to the involutive line bundle $(\mathcal{L}, \varphi)$ and denote it $\mathbb{P}_{\mathcal{L}}$.

Example 1. If $\mathcal{L} = \mathcal{O}_S$ and $\varphi = \lambda \in \mathcal{O}_S^\times$, then the involutive bundle is $(\mathbb{P}^1 \times S, 0 \times S, \infty \times S, x \mapsto \lambda/x)$. Locally this is the general situation.

Note also that the fixed point locus of σ is a double covering of the base that is isomorphic to the double cover associated to $\mathcal{L}$ and φ; something which is seen for instance by using the local description just given. We shall call it the *involutive locus*.

Similarly to the remark above, an S-morphism $X \to \mathbb{P}_{\mathcal{L}}$ such that the inverse images of ∞ and $\mathbf{0}$ are Cartier divisors C and D is the same thing as an isomorphism $\mathcal{O}_X(C) \xrightarrow{\sim} \mathcal{O}_X(D) \otimes \mathcal{L}$.

The involutive bundle will be said to be *split* if one is given a trivialisation of $\mathcal{L}$ for which φ becomes the identity. Then the involutive projective bundle is identified with $\mathbb{P}^1 \times S$ in such a way that 0 corresponds to the zero section, ∞ to the section at infinity and the involutive locus is given by $\{(s\colon t) \mid s^2 = t^2\}$ which when 2 is invertible is $\{(\pm 1 : 1)\}$.

We may explicitly construct the quotient of $\tau\colon \mathbb{P}(\mathcal{O}_S \oplus \mathcal{L}) \to S$ by the action of σ in the following way. We define an S-map $\mathbb{P}(\mathcal{O}_S \oplus \mathcal{L}) \to \mathbb{P}(\mathcal{O}_S \oplus \mathcal{L})$ by giving $\mathcal{O}_{\mathbb{P}(\mathcal{O}_S \oplus \mathcal{L})}(2)$ as a quotient of $\tau^*(\mathcal{O}_S \oplus \mathcal{L})$. By adjunction giving such a map is the same as giving a map $\mathcal{O}_S \oplus \mathcal{L} \to \tau_*\mathcal{O}(1) = S^2(\mathcal{O}_S \oplus \mathcal{L})$. We do this by mapping 1 of the $\mathcal{O}$-factor to $1 \otimes 1 \oplus \varphi(1)$ in $S^2\mathcal{O}_S \oplus \mathcal{L}^{\otimes 2} \subset S^2(\mathcal{O}_S \oplus \mathcal{L})$ and the $\mathcal{L}$-factor to $\mathcal{O}_S \otimes \mathcal{L} \subset S^2(\mathcal{O}_S \oplus \mathcal{L})$ through $1 \otimes id$. In the local normal form above – homogenised – this map is given by $(x : y) \mapsto (x^2 + \lambda y^2 : xy)$ which evidently has no base points, i.e., it is surjective and hence gives a map $\pi\colon \mathbb{P}(\mathcal{O}_S \oplus \mathcal{L}) \to \mathbb{P}(\mathcal{O}_S \oplus \mathcal{L})$. As σ locally has the form $(x\colon y) \mapsto (\lambda y\colon x)$ it is clear that π is equivariant with trivial action on the target. Using again the local form it is easily verified that it is the quotient map. We shall speak of it as the *involutive quotient map* associated to $\mathcal{L}$ and φ. Note that the involutive locus maps to the image under $x \mapsto 2x$ of itself. For this reason, starting with Sect. 3, we shall instead use $(x\colon y) \mapsto (x^2 + \lambda y^2\colon 2xy)$ as quotient map so that the involutive locus is mapped to itself.

Seen from the point of view of its target the involutive quotient map π is a double covering. Restricting ourselves to the case when 2 is invertible we may describe this covering as follows. We get a map $\mathcal{O}_{\mathbb{P}(\mathcal{O}_S \oplus \mathcal{L})}(-1) \to \pi_*\mathcal{O}_{\mathbb{P}(\mathcal{O}_S \oplus \mathcal{L})}$: By adjunction it corresponds to a map $\pi^*\mathcal{O}_{\mathbb{P}(\mathcal{O}_S \oplus \mathcal{L})}(-1) \to \mathcal{O}_{\mathbb{P}(\mathcal{O}_S \oplus \mathcal{L})}$ and by construction we have $\pi^*\mathcal{O}_{\mathbb{P}(\mathcal{O}_S \oplus \mathcal{L})}(-1) = \mathcal{O}_{\mathbb{P}(\mathcal{O}_S \oplus \mathcal{L})}(-2)$ so that such a map corresponds to a section of $\mathcal{O}_{\mathbb{P}(\mathcal{O}_S \oplus \mathcal{L})}(2)$, i.e., a section of $S^2(\mathcal{O}_S \oplus \mathcal{L})$ and we choose $\otimes 1 \oplus -\varphi(1)$. In the local form above the section $\otimes 1 \oplus -2\varphi(1)$ corresponds to $1/2(x^2 - \lambda y^2)$. From that it is easily verified that the map $\mathcal{O}_{\mathbb{P}(\mathcal{O}_S \oplus \mathcal{L})}(-1) \to \pi_*\mathcal{O}_{\mathbb{P}(\mathcal{O}_S \oplus \mathcal{L})}$ is injective and has as image the -1-eigenspace of σ. The double cover π is now determined by the square map $\mathcal{O}_{\mathbb{P}(\mathcal{O}_S \oplus \mathcal{L})}(-1) \otimes \mathcal{O}_{\mathbb{P}(\mathcal{O}_S \oplus \mathcal{L})}(-1) \to \mathcal{O}_{\mathbb{P}(\mathcal{O}_S \oplus \mathcal{L})}$, i.e., a section of $S^2(\mathcal{O}_S \oplus \mathcal{L})$. A local calculation shows that that section is $1 \otimes 1 \oplus -\varphi(1)$.

We shall have need of the following technical result on Cartier divisors.

Proposition 1. *Let $\pi\colon X \to S$ be a smooth, proper map of schemes with connected fibres and $D \subset X$ a relative (wrt to π) effective Cartier divisor, and let n be an integer such that $n\mathcal{O}_S = \mathcal{O}_S$ (i.e., n is invertible in $\mathcal{O}_S$).*

The functor that to a $T \to S$ associates the set of relative effective Cartier divisors $D \subset X \times_S T$ such that $nD = C \times_S T$, is representable by a closed subscheme of S.

In particular, if S is reduced and if for every geometric fibre of $X \to S$, C is n times an effective Cartier divisor, C itself is n times an effective Cartier divisor.

Proof. The functor is clearly representable by some S-scheme $V \to S$ locally of finite type, and we may, by a standard limit argument, assume that S is noetherian. What needs to be proven is that $V \to S$ is proper, injective on geometric points and unramified. For the properness we may use the valuative criterion so that S is the spectrum of a discrete valuation ring and we assume given a D_η over the generic point with $nD_\eta = nC_\eta$. We then let D be the schematic closure of D_η, which is a Cartier divisor, as X is regular and is relative as it does not have any horisontal components.

As for injectivity on geometric points we may assume that S is the spectrum of an algebraically closed field and then the uniqueness of D is clear as X is regular and thus the group of Cartier divisors is torsion free.

Finally, to prove that $V \to S$ is unramified, it is enough to show that it is formally unramified, so we may assume that $S = \operatorname{Spec} R$, where $(R, \mathfrak{m}_R)$ is a local Artinian ring, $0 \neq \delta \in R$ annihilates $\mathfrak{m}$ and we assume that a D exists over the closed subscheme defined by δ. We then are to prove that there is at most one lifting of D to $X \to S$. Now, a Cartier divisor E is given by specifying a line bundle $\mathcal{L}$ and an injective $\mathcal{O}$-homomorphism $\mathcal{O} \to \mathcal{L}$. As $X \to S$ is flat, the injectivity follows from injectivity over the special fibre and is hence automatic in our situation. Furthermore, if D is given by $s\colon \mathcal{O} \to \mathcal{L}$, then nD is given by $s^{\otimes n}\colon \mathcal{O} \to \mathcal{L}^{\otimes n}$. In our situation we assume a pair $(\mathcal{M}, t)$ over X representing C and two pairs $(\mathcal{L}, s)$ and $(\mathcal{L}', s')$ whose n'th powers are isomorphic to $(\mathcal{M}, t)$ and whose reductions modulo δ are isomorphic. Now the kernel and cokernel of the reduction $\operatorname{Pic}(X) \to \operatorname{Pic}(X/\delta)$ are $\mathcal{O}_S$-modules, so that multiplication by n is by assumption bijective on them, which shows that $\mathcal{L}'$ and $\mathcal{L}$ are isomorphic, and we may assume them to be equal. Hence s' is of the form $s + \delta w$ with w a section of $\overline{\mathcal{L}} = \mathcal{L}/\mathfrak{m}$ by the flatness of $X \to S$. By assumption their n'th powers are isomorphic so that $(1+\delta\lambda)s^n = (s+w)^n = s^n + ns^{n-1}\delta w$ for some $\lambda \in R$. This gives $\overline{s}^{n-1}(\lambda\overline{s} + nw) = 0$, where $\overline{s}$ is the reduction of s modulo $\mathfrak{m}$. As $\overline{s}$ is a non-zero divisor this gives $w = -\lambda/n\overline{s}$, i.e., $\delta w = -\lambda/n\delta s$, which gives $s' = (1 - \lambda/n\delta)s$, so that the pairs $(\mathcal{L}, s)$ and $(\mathcal{L}', s')$ are isomorphic.

The last statement follows immediately from the previous ones as under its assumptions V has the same topological space as S. □

The following result is no doubt well known but I do not know of a reference.

Proposition 2. *i) Let $f\colon C \to D$ be a separable non-constant map of smooth proper curves over a field* **k** *and consider the deformation functor whose values on a nil-thickening of* Spec **k** *are isomorphism classes of deformations of C and the map f. The map that maps such deformations to similar deformations of the formal completions of C resp. D along the ramification resp. branch locus is an isomorphism.*

ii) Let **k** *be a field, n an integer invertible in* **k** *and $f\colon \operatorname{Specf} \mathbf{k}[[x]] \to \operatorname{Specf} \mathbf{k}[[t]]$ be the map $t \mapsto x^n$. Then $t \mapsto x^n + \sum_{0 \leq i < n-1} a_i x^i$, where the a_i are power series variables, is a miniversal deformation of f.*

Proof. The first part can be proved by noticing that outside of the ramification/branch loci the map is unramified and hence extends uniquely along any nil-thickening. This shows that the deformation problem is the same as that for the localisation along the ramification/branch loci. The comparison between the deformation for the localisations and the completions is also clear as when one inverts generators for the ramification/branch loci then the map is étale. This means that the map is specified by choosing a lattice in the ring of functions in the source over the ring of functions of the target. Specifying such a lattice is the same in the localisation as in the completion.

Alternatively one can use deformation theory. If $R \to S$ is a small extension of local Artinian algebras with residue field $\mathbf{k}$, small meaning that the kernel K is killed by the maximal ideal of R, then the liftings of a deformation over S to one over R is in bijection with $H^0(C, f^*T_D/T_C) \otimes_{\mathbf{k}} K$. Indeed, if the deformation of C is kept fixed, then liftings of deformations of f, given one, are in bijection with $H^0(C, f^*T_D) \otimes K$. Taking into account the possibility of varying also deformations of C, we have to divide out by the action of liftings of automorphisms of the deformation of C, i.e., sections of $T_C \otimes K$. This action is given by addition composed with the map $T_C \to f^*T_D$ and hence the full problem is in bijection with $H^0(C, f^*T_D/T_C) \otimes_{\mathbf{k}} K$. As we never used the properness, the same is true for the local or complete problem as f^*T_D/T_C is supported on the ramification locus.

As for the last part, the formula $t \mapsto x^n + \sum_{0 \le i < n-1} a_i x^i$ gives a deformation over $\mathbf{k}[[a_0, \ldots, a_{n-2}]]$ and hence a map to the miniversal deformation. As $\mathbf{k}[[a_0, \ldots, a_{n-2}]]$ is (formally) smooth, to show that this map is an isomorphism it is enough to show that it induces an isomorphism on tangent spaces, and for that we can use the description of deformations over $\mathbf{k}[\delta]$, $\delta^2 = 0$, just given to show that. Indeed, the action of the sections of T_C on such deformations is by interpreting a derivation of $\mathbf{k}[[x]]$ as an automorphism of the scalar extension to $\mathbf{k}[\delta]$, $\mathbf{k}[[x]][\delta]$, that is the identity modulo δ, and then composing the given map $\mathbf{k}[[t]][\delta] \to \mathbf{k}[[x]][\delta]$ with that automorphism. If the vector field is $h(x)d/dx$ and the map has the form $f(x) + g(x)\delta$ with $f, g, h \in \mathbf{k}[[x]]$, then this composite is $f(x) + (h(x)f'(x) + g(x))\delta$. This shows that the tangent vector of the map is given by the residue of g modulo $f'(x)$, which makes it clear that the tangent map is an isomorphism. □

When we make an infinitesimal study of the moduli stack we shall not just deal with the stack as such but also with the natural stratification of it given by the ramification exponents of a map between curves. We recall its definition and first properties given in [3, App.]. We begin by noting that for technical reasons we shall need to assume that we deal with schemes over Spec $\mathbb{Q}$ for the rest of this section.

Remark 1. Note that this restriction is not just due to the fact that one would need some slight modifications to get similar results in positive characteristic. In fact there are some truly new phenomena in positive characteristic. Consider for instance the case of Proposition 3. Condensed it says that for a finite flat map there is a stratification of the base such that on each stratum there is a closed subscheme of the total space which is étale over the base and whose defining ideal is nilpotent.

A similar result is not possible in positive characteristic. Consider for instance an inseparable field extension $k \subset K$ of degree p say. For the corresponding map of schemes $\operatorname{Spec} K \to \operatorname{Spec} k$ if it had a similar stratification then there could only be one stratum but $\operatorname{Spec} K$ does not have a closed subscheme which is étale over $\operatorname{Spec} k$ and whose ideal is nilpotent.

Recall that if $f\colon Y \to X$ is a finite flat map then we define its *trace form* to be the symmetric bilinear form $(r, s) \mapsto \operatorname{Tr}(rs)$. We then define, for each natural number n, the closed subscheme of X given by the condition that the corank of the map $f_*\mathcal{O}_Y \to \operatorname{Hom}_{\mathcal{O}_X}(f_*\mathcal{O}_Y, \mathcal{O}_X)$ induced by the trace form is $\geq n$. We shall call the stratification thus obtained the *trace stratification* wrt f. In an open stratum we get the following primary decomposition result.

Proposition 3. *Let $f\colon Y \to X$ be a finite flat map for which X equals a single open trace stratum. Then the radical of the trace form (i.e., the kernel of the map $f_*\mathcal{O}_Y \to \operatorname{Hom}_{\mathcal{O}_X}(f_*\mathcal{O}_Y, \mathcal{O}_X)$) is a subbundle and an ideal. The closed subscheme Y' defined by it is an étale covering of X. Furthermore, there is a unique X-retraction $Y \to Y'$ which makes Y a flat Y'-scheme.*

Finally, if $Y \to Y'$ has rank n then the n'th power of the radical of the trace map is zero.

Proof. The fact that the radical is a subbundle follows directly from the fact that by assumption $f_*\mathcal{O}_Y \to \operatorname{Hom}_{\mathcal{O}_X}(f_*\mathcal{O}_Y, \mathcal{O}_X)$ has constant rank (in the schematic sense defined by the vanishing and non-vanishing of subdeterminants when X is not reduced) and then its kernel is a subbundle. That it is an ideal follows directly from the definition of the trace form. Then $Y' \to X$ is flat so to prove that it is étale it is enough to do it when X is the spectrum of a field in which case it is well known and easy to see that Y' is étale. Assume that we know that the existence and unicity of a retraction locally. Then the unicity forces it to exist globally by (étale) descent. By descent again, the flatness of the retraction needs to be checked only in the case when Y' is the disjoint union of copies of X in which case it is clear.

For the last statement, replacing X by Y' we may assume that f has rank n. If $\mathcal{I}$ is the radical and $x \in \mathcal{I}$ (i.e., is a local section of that sheaf) then we have that $x^i \in \mathcal{I}$ for all $i > 0$ and hence $\operatorname{Tr}(x^i) = 0$. As we are in characteristic zero this implies that the characteristic polynomial of multiplication by x is t^n and by the Cayley–Hamilton theorem $x^n = 0$. Again, as we are in characteristic zero, we get by polarisation that any product of n local sections of $\mathcal{I}$ is zero.

It remains to prove local existence and unicity of a retraction. Locally we may assume that Y' is a disjoint union of copies of X which makes the existence of a retraction obvious and the unicity clear. □

When the corank of the trace form is constant, the proposition shows that the function on the points of Y defined by the rank at a point of the radical is a locally constant function and hence the function which to a point of X associates the multisets of those ranks is locally constant. The stratification obtained in this way will be referred to as the *stratification by multiplicity*. Thus while the trace stratification is a decreasing sequence of closed subschemes, the stratification by multiplicity is a further decomposition of the open strata. We shall also in this case denote the subscheme defined by the radical by X^{fred} and call it the *fibrewise reduced subscheme*.

Using the primary decomposition we get an extension of the pointwise result that expresses an effective Cartier divisor on a smooth curve as the sum of points.

Proposition 4. *Let $f\colon X \to S$ be a smooth map of* Spec $\mathbb{Q}$*-schemes of relative dimension* 1 *and suppose $D \subset X$ is a relative effective Cartier divisor and assume that the corank of the trace form of $D \to S$ is of locally constant rank.*

i) The fibrewise reduced subscheme D^{fred} of Spec $\mathcal{O}_X/\mathcal{I}_D$ *is a relative effective Cartier divisor. It can be written as the disjoint of subschemes D_i having the property that at a point $d \in D_i$ the defining ideal of D_i in D has rank e_i as $\mathcal{O}_{D_i}$-module. If that is done then we have that $D = \sum_i e_i D_i$ as divisors.*

ii) Conversely if D can be written as $D = \sum_i e_i D_i$ with D_i étale disjoint Cartier divisors then the corank of the trace form of $D \to S$ is locally constant and the union of the D_i is the fibrewise reduced subscheme of D.

Proof. That D^{fred} is a Cartier divisor is clear and we get from Proposition 3 that its defining ideal $\mathcal{I}$ in D is a locally free $\mathcal{O}_{D^{\mathrm{fred}}}$-module and hence its rank is locally constant. This gives the components D_i and to prove the equality of Cartier divisors we may work locally around one of the D_i, i.e., assume that the rank of $\mathcal{I}$ as $\mathcal{O}_{D^{\mathrm{fred}}}$-module is everywhere equal to some n. By Proposition 3 the n'th power of $\mathcal{I}_{C^{\mathrm{fred}}}$ is contained in $\mathcal{I}_D$. To check that it is an equality it is enough to check on fibres over closed points of S and then it is true as they have the same degree at all the points of D^{fred}.

Finally, if $D = \sum_i e_i D_i$ and D' is the union of the D_i then $\mathcal{I}_{D'}/\mathcal{I}_D$ is an S-flat nilpotent ideal of $\mathcal{O}_X/\mathcal{I}_D$ such that the quotient by it is étale. This shows that $\mathcal{I}_{D'}/\mathcal{I}_D$ is the radical of the trace form and thus that D' is the fibrewise reduced subscheme of D. □

The multiplicities e_i can be considered as locally constant functions on D^{fred} and we may choose the D_i such that the e_i are all distinct. Having done that the decomposition is unique and we shall call it the *primitive decomposition* of D. We shall also consider the locally constant function on the base S which to a point s associates the multiset of the multiplicities of the points of D in the fibre over s. This will be called the *multiplicity multiset* associated to D.

If S is a scheme and $f\colon Y \to X$ is a finite S-map between smooth (possibly formal) S-schemes of relative dimension 1, then we get two finite S-schemes, the ramification locus which is a relative Cartier divisor of Y and the branch locus which is a relative Cartier divisor of X (by definition the branch locus is the norm wrt to f of the ramification locus considered as a Cartier divisor which is defined as f is finite flat). We shall call the stratifications by multiplicity on S induced by them the *ramification stratification* and *branch stratification* respectively.

2 The Original Problem

We shall begin by formulating in modern terms Abel's question and the answer he gave to it. The initial setup is that of a square free monic polynomial $R(x)$ of even degree over the complex numbers and the rational differential form $\omega := \rho dx/\sqrt{R}$, $\rho \in \mathbb{C}(x)$, on the compact Riemann surface C with field of rational functions $\mathbb{C}(x, \sqrt{R}) := \mathbb{C}(x)[y]/(y^2 - R(x))$. The general question Abel poses is when this form is the logarithmic differential $\operatorname{dlog} f := df/f$ for a non-zero rational function f. If ι is the hyperelliptic involution of C which takes x to x and $\sqrt{R}$ to $-\sqrt{R}$, then $\iota^*\omega = -\omega$ and as f is determined up to a constant we get $f \circ \iota \cdot f = \lambda \in \mathbb{C}^*$. Modifying f by multiplying by a square root of λ^{-1} allows us to assume that $f \circ \iota \cdot f = 1$. By a somewhat anachronistic appeal to Hilbert's Theorem 90 we get that f has the form $g/g \circ \iota$ and by clearing denominators we may assume that $g = P + \sqrt{R}Q$ with $P, Q \in \mathbb{C}(x)$, which is indeed the form that Abel assumes the solution to have. We note that g is uniquely determined up to a rational function in x. Abel then almost immediately restricts himself to the case where ρ is a polynomial. This implies that ω is regular over $C^\circ := \operatorname{Spec} \mathbb{C}[x, \sqrt{R}]$ and hence in particular that f does not have poles or zeroes in C°, or otherwise put, if ∞_1 and ∞_2 are the two points of C in the complement of C° in C, then $(f) = m\infty_1 - m\infty_2$ for some integer m. For a divisor D on C we denote by D° the part of D that has support on C° and then we have $0 = (f)^\circ = ((g) - \iota(g))^\circ$. Now by assumption $(g)^\circ \geq 0$ and hence for $r \in C^\circ$, r and ιr appear with the same multiplicity in g. Now, for any $r \in C^\circ$ $r + \iota r - (\infty_1 + \infty_2)$ is the divisor of a rational function in x so that if r is a non-Weierstrass point (i.e., $r \neq \iota r$) we may modify g by a rational function in x so that r does not appear in (g). Similarly, if r is a Weierstrass point we may assume that it appears with at most multiplicity 1 in (g). In particular, $(g)^\circ$ has support at the Weierstrass points of C.

Even though Abel treats the general case, we shall only be interested in the case when $(g)^\circ = 0$. The reason for this is that we shall be mainly interested in the existence of a g as a condition on the curve C and we have that $2\omega = \operatorname{dlog} f^2 = \operatorname{dlog} g^2/g^2 \circ \iota$ and all the Weierstrass points appear with even multiplicity in (g^2), and they can therefore be removed completely. Hence at the price of possibly replacing ρ with 2ρ we see that Abel's problem has been reduced to the problem of finding $g \in \mathbb{C}(x, \sqrt{R})$ with $(g)^\circ = 0$, i.e., $(g) = n(\infty_1 - \infty_2)$ for some integer n and excluding the trivial case of $\rho = 0$ we may assume that n is non-zero. In any case $(g)^\circ = 0$ and hence g is a unit in $\mathbb{C}[x, \sqrt{R}]$. This implies that $g = P + \sqrt{R}Q$, with $P, Q \in \mathbb{C}[x]$ and furthermore that the norm $N(g) = P^2 - RQ^2$ of g with respect to the finite flat extension $\mathbb{C}[x, \sqrt{R}]/\mathbb{C}[x]$ is a unit in $\mathbb{C}[x]$, i.e., a non-zero constant. After changing g by a constant we may assume that $P^2 - RQ^2 = 1$. Abel then notes that this is analogous to Pell's equation and proceeds to use continued fractions in analogy with the case of Pell's equation. There is a difference however in that the number theoretic case gives a method for solving Pell's equation while the geometric case gives a criterion for the existence of a solution (as well as a method for constructing it when it does exist).

Remark 2. Abel's approach gives a condition on R for a solution to the problem with n arbitrary to exist. This is not appropriate for our purposes as when n varies we get a countable union of closed subvarities in the space of R's which is unnatural from a geometric point of view.

3 Abel Curves

We shall now give the formal definition of an Abel curve. To simplify the presentation (and make it closer to Abel's original results) we shall *from now on assume that* 2 *is invertible in all our schemes*. If we want to consider the moduli space of Abel curves we want to make as few choices as possible as any choice leads to a larger space which is the reason for the somewhat lengthy definition. To avoid ambiguities in the case of genus 1 to us a *hyperelliptic curve* will be a smooth proper curve C together with a *choice* of an involution ι such that the quotient C/ι is of genus zero.[3]

Definition 1. *A* (smooth) Abel curve of genus g and order n *over a scheme S consists of*

- *a smooth and proper S-curve $\pi: C \to S$,*
- *an S-involution ι of C making each fibre a hyperelliptic curve of genus g,*
- *two disjoint sections ∞_1 and ∞_2 of π such that $\infty_2 = \iota\infty_1$,*
- *a line bundle $\mathcal{L}$ on S together with a trivialisation $\varphi: \mathcal{O}_S \xrightarrow{\sim} \mathcal{L}^2$, and*
- *a finite flat S-map $f: C \to P$, where (P, σ) is the involutive bundle associated to $\mathcal{L}$ and the trivialisation φ, of degree n such that the sections ∞_1 and ∞_2 map to the sections $\mathbf{0}$ and ∞ of the involutive bundle $\mathbb{P}$ and $\sigma \circ f = f \circ \iota$.*

A split Abel curve *is an Abel curve together with a splitting of the involutive bundle.*

An isomorphism between Abel curves consists of isomorphisms between the C and P parts of the curves transporting all the structures of the first curve to those of the second.

Associating to each S the groupoid of Abel curves and isomorphisms between them gives a stack (in say the flat topology) that we shall denote $\mathcal{H}_g^n$ and similarly we get the stack of split Abel curves $\mathcal{H}_g^{s,n}$.

When the base is an algebraically closed field we get exactly the description that came out of Abel's problem. Note that in that case it follows from the equation $P^2 - RQ^2 = 1$ that $2\deg P \geq \deg R$, i.e., $2n \geq 2g+2$ which means $n \geq g+1$.

It is not immediately clear that this is the right definition for families as one could worry that we have made an unnecessary choice in choosing two sections ∞_1 and ∞_2 instead of a divisor of degree 2 that only after a base change splits up into two disjoint sections. The following definition expresses that concern.

Definition 2. *A* twisted (smooth) Abel curve of order n *over a scheme S consists of*

- *a smooth and proper S-curve $\pi: C \to S$,*
- *an S-involution ι of C making each fibre a hyperelliptic curve,*
- *a ι-invariant relative effective divisor D of degree 2 of C which is étale over S and on which ι acts freely,*
- *a smooth and proper S-curve $\rho: P \to D$ all of whose fibres have genus zero,*
- *an S-involution σ of P,*

[3] With this definition we can have hyperelliptic curves of genus zero which for our purposes is quite acceptable though rather trivial.

- *a σ-invariant relative effective divisor D' of P which is étale over S and on which σ acts freely,*
- *a finite flat S-map $f\colon C \to P$ of degree n such that the inverse image (as effective divisors or equivalently as subschemes) of D' is nD and for which $f \circ \iota = \sigma \circ f$.*

Isomorphisms between Abel curves consist of an isomorphism g between the C-parts preserving the ι's, ∞_1's, and ∞_2's and an automorphism h of $\mathbb{P}^1 \times S$ such that $f \circ g = h \circ f$.

Associating to each S the groupoid of twisted Abel curves and isomorphisms between them gives a stack (in say the flat topology) that we shall denote $\mathcal{H}_g^{t,n}$.

The relation between these definitions is expressed in the following result.

Proposition 5. *i) The stack of twisted Abel curves of genus g and order n, $\mathcal{H}_g^{t,n}$, is equivalent to $B\Sigma_2 \times \mathcal{H}_g^n$, where $B\Sigma_2$ is the stack of Σ_2-torsors, i.e., the stack of étale double covers. The projection on the first factor associates to a twisted Abel curve, using the notation of Definition 2, the étale double cover $D \to S$.*

ii) The forgetful map $\mathcal{H}_g^{s,n} \to \mathcal{H}_g^n$ is an étale double cover.

Proof. Using the notation of Definition 2 we get from a twisted Abel curve over S an étale double cover $D \to S$ which gives a map from the stack of twisted Abel curves to $B\Sigma_2$. On the other hand, (ι, σ) gives an involution of the Abel curve and we may use it and the double cover $D \to S$ to twist the Abel curve, in particular the twist, $\tilde{C}$, of C is obtained by taking the quotient of $D \times_S C$ by the action of (ι, ι). The section given by the graph of the inclusion of D in C is invariant under this map and hence descends to a section of $\tilde{C} \to S$ and the same is true of the group of the map $D \to C$ composed with ι. In other words, the divisor D twists to give a divisor that is the disjoint union of two sections. Now, the map $f\colon C \to P$ maps D isomorphically to D' so that also D' is the disjoint union of two sections. The existence of these two disjoint sections makes $P \to S$ isomorphic to $\mathbb{P}(\mathcal{L} \oplus \mathcal{M})$ for some line bundles $\mathcal{L}$ and $\mathcal{M}$ on S, where the two sections correspond to the two summands. Now, σ permutes the two sections, which forces $\mathcal{L}$ and $\mathcal{M}$ to be isomorphic, so that $P \to S$ is isomorphic to $\mathbb{P}^1 \times S \to S$ with the two sections given by $0 \times S$ and $\infty \times S$. As the inverse images of $0 \times S$ and $\infty \times S$ are n times the two sections of D, we get an Abel curve, and consequently a map $\mathcal{H}_g^{t,n} \to \mathcal{H}_g^n$ and combining the two constructed maps we get a map $\mathcal{H}_g^{t,n} \to B\Sigma_2 \times \mathcal{H}_g^n$. Conversely, given an Abel curve over S we can consider the map f as an isomorphism $\mathcal{O}_C(n\infty_1) \xrightarrow{\sim} \mathcal{O}_C(n\infty_2)$. Letting ι act on that isomorphism gives another isomorphism $\mathcal{O}_C(n\infty_2) \xrightarrow{\sim} \mathcal{O}_C(n\infty_1)$. Their composites are then multiplication by an invertible function λ on S. That means that if we define σ on $\mathbb{P}^1 \times S$ by $(x : y) \mapsto (\lambda y : x)$ then $f \circ \iota = \sigma \circ f$ so that we have a twisted Abel curve over S. Now, (ι, σ) is an involution of that object and so that if we have an étale double cover $\tilde{D} \to S$ we can use it to twist our twisted Abel curve and we obtain thus a map $B\Sigma_2 \times \mathcal{H}_g^n \to \mathcal{H}_g^{t,n}$ which is clearly an inverse to the map just constructed.

As for the second part it is clear. $\square$

The proposition shows that it is no real loss in generality to restrict ourselves to Abel curves which we shall do from now on with the exception of the following result which confirms the representability of the two stacks.

Proposition 6. *The stacks* $\mathcal{H}_g^n$, $\mathcal{H}_g^{s,n}$, *and* $\mathcal{H}_g^{t,n}$ *are Deligne–Mumford stacks of finite type over* $\operatorname{Spec}\mathbb{Z}[1/2]$.

Proof. This is quite standard as soon as we have verified that the automorphism group scheme of an (twisted) Abel curve over an algebraically closed field is finite étale. For $g \geq 2$ this is clear as it is true for all curves of genus g. For $g = 1$ we have to use the fact that the hyperelliptic involution is part of the structure so that an automorphism has to commute with it. For a hyperelliptic involution ι we may choose a fixed point as origin and in the thus obtained group structure on the curve, the involution is multiplication by -1; then it is clear that the automorphism group scheme centralizer of the curve is finite étale. Finally, for genus 0 we have to look at the automorphism group scheme of automorphisms of $\mathbb{P}^1$ fixing two points and commuting with an involution that permutes the two points. It is clear that the points and the involution is conjugate to 0, ∞, and $x \mapsto 1/x$ and then the automorphism group scheme that fixes these is clearly finite étale. □

Our definition of an Abel curve is chosen to be closely modeled on Abel's original condition. On the other hand – at least punctually – the relevant condition is that the divisor class $\infty_1 - \infty_2$ is killed by n as there is then a map to $\mathbb{P}^1$ whose zero and pole divisor is $n(\infty_1 - \infty_2)$. This turns out to be true for families.

Proposition 7. *Let* $\mathcal{H}_{g,2}$ *be the stack of hyperelliptic curves with two distinct points* (C, ι, a, b) *of genus* g *and let* $s\colon \mathcal{H}_{g,2} \to J_g$ *be the section of the Jacobian of the universal curve given by* $a - b$. *Let* $\mathcal{H}$ *be the closed substack of* $\mathcal{H}_{g,2}$ *defined by the conditions* $\iota a = b$ *and* $ns = 0$. *Let* ρ *be the involution of* $\mathcal{H}_g^n$ *which takes an object* $(C \to S, \iota, \infty_1, \infty_2, f)$ *to* $(C \to S, \iota, \infty_1, \infty_2, -f)$. *Then the map given by*

$$\begin{array}{ccc} \mathcal{H}_g^n & \to & \mathcal{H} \\ (C, \infty_1, \infty_2, f) & \mapsto & (C, \infty_1, \infty_2) \end{array}$$

is an isomorphism of stacks.

Proof. As has been noted above, f may be thought of as an isomorphism $\phi\colon \mathcal{O}_C(n\infty_1) \xrightarrow{\sim} \mathcal{O}_C(n\infty_2)$ and then ρ takes it to $-\phi$. On the other hand, an S-object of $\mathcal{H}$ has the property that $\mathcal{O}(n\infty_1 - n\infty_2)$ is a pullback of a (unique) line bundle $\mathcal{L}$ on S. Now, applying ι to $\mathcal{O}(n\infty_1 - n\infty_2)$ gives its inverse which translates into an isomorphism $\mathcal{L} \xrightarrow{\sim} \mathcal{L}^{-1}$, i.e., a trivialisation of $\mathcal{L}^2$. This gives an object of $\mathcal{H}_g^n$ over S. □

As the zero-section in an abelian scheme is a local complete intersection subscheme we get one immediate consequence.

Corollary 1. $\mathcal{H}_g^n \to \operatorname{Spec}\mathbb{Z}[1/2]$ *is of relative dimension at least* g *at each of its points and at a point where the relative dimension is* g *it is a local complete intersection.*

Proof. The substack $\mathcal{H}$ of $\mathcal{H}_{g,2}$ fulfilling $\iota a = b$ is an open substack of the stack $\mathcal{H}_{g,1}$ of hyperelliptic curves with one chosen point, namely the complement of the locus of fixed points of the hyperelliptic involution, where the isomorphism maps (C, ι, a) to $(C, \iota, a, \iota a)$. Hence that substack is smooth of relative dimension $2g - 1 + 1 = 2g$. Now, by the proposition $\mathcal{H}_g^n$ is the inverse image in $\mathcal{H}$ of the zero section of $J_g \to \mathcal{H}_{g,2}$ under the map ns and $J_g \to \mathcal{H}_{g,2}$ being smooth, the zero section is a local complete intersection map of codimension g. □

Remark 3. In characteristic 0 we shall show that the codimension is in fact g and that $\mathcal{H}_g^n$ is in fact smooth.

Fix n and g with $n \geq g+1$ and consider $\mathbf{A} := \mathbf{A}^{2n+g+3}_{\mathbb{Z}[1/2]}$ that we shall regard as the parameter space for triples (P, Q, R) of polynomials of degrees n, $n-g-1$, and $2g+2$ respectively with R monic. We let $\mathcal{V}_g^n$ be the subscheme of triples that fulfill $P^2 - RQ^2 = 1$ and for which R is square free (i.e., its discriminant is invertible) and P and Q have invertible top coefficients. We let $\mathcal{U}_g^n$ be the subscheme of $\mathcal{V}_g^n$ defined by the condition that R is normalised and P and Q are monic. Over $\mathcal{V}_g^n$ we have an Abel curve given by

$$C := \operatorname{Proj} \mathcal{O}_{\mathcal{U}_g^n}[s, t, y] / \left(y^2 - t^{2g+2} R(s/t)\right) ,$$

where $\deg s = \deg t = 1$ and $\deg y = g+1$, ι is given by $(s\colon t\colon y) \mapsto (s\colon t\colon -y)$, ∞_1 and ∞_2 are given by $(0\colon 1\colon 1)$ resp. $(0\colon 1\colon -1)$, and f is given by $(s\colon t\colon y) \mapsto (t^n(P(s/t) + yQ(s/t))\colon t^n)$. This therefore gives a map $\mathcal{V}_g^n \to \mathcal{H}_g^n$. We shall call any Abel curve that is a pullback of this family by a map to $\mathcal{V}_g^n$ a *Pell family* and if it is given as a pullback by a map to the closed subscheme $\mathcal{U}_g^n$ we shall call it a *normalised Pell family*.

Theorem 3.1. *i)* $\mathcal{V}_g^n \to \mathcal{H}_g^n$ *factors through the map* $\mathcal{H}_g^{s,n} \to \mathcal{H}_g^n$.

ii) Over $\operatorname{Spec} \mathbb{Q}$ *the map* $\mathcal{V}_g^n \to \mathcal{H}_g^{s,n}$ *is a torsor under the subgroup of* $\mathbf{G}_m \times \mathrm{Aff}$, *where* Aff *is the group of affine transformations of the affine line, of pairs* $(\lambda, z \mapsto az + b)$ *for which* $\lambda^2 = a^{2g+2}$.

iii) Over $\operatorname{Spec} \mathbb{Q}$ *the map* $\mathcal{U}_g^n \to \mathcal{H}_g^{s,n}$ *is a torsor under the subgroup of* $\mathbf{G}_m \times \mathrm{Aff}$ *of pairs* $(\lambda, z \mapsto az+b)$ *for which* $\lambda = a^{g+1}$, $a^n = 1$ *and* $b = 0$, *a group isomorphic to the group* μ_n *of n'th roots of unity.*

iv) In particular the map $\mathcal{V}_g^n \to \mathcal{H}_g^n$ *is a chart. i.e., smooth and surjective, and* $\mathcal{U}_g^n \to \mathcal{H}_g^n$ *is even an étale chart.*

Proof. To prove the first part we note that for a Pell family the involution on $\mathbb{P}^1 \times S$ compatible with f and ι is $x \mapsto 1/x$ whose fixed point scheme is ± 1 and by ordering it as $\{1, -1\}$ we get a family in $\mathcal{H}_g^{s,n}$.

Assume now that $(C \to S, \iota, \infty_1, \infty_2, f)$ is a family in $\mathcal{H}_g^{s,n}$. By assumption, using the notation of Definition 1, P is isomorphic to $\mathbb{P}^1 \times S$ in a way such that $\mathbf{0}$ on P is $0 \times S$ and ∞ is $\infty \times S$ and the involution σ is $x \mapsto 1/x$. Consider now the quotient D of C by ι. As 2 is invertible, taking the quotient by ι commutes with base change so that in particular $\pi\colon D \to S$ is a smooth proper map with genus 0 fibres. Furthermore, either of the sections ∞_1 or ∞_2 give a section ∞ of π. Now, again as 2 is invertible, the double cover $C \to D$ is given by a line bundle $\mathcal{M}$ on D and a section of $\mathcal{M}^2$. As $\mathcal{M}$ has degree $g+1$ on each fibre $\mathcal{M}(-(g+1)\infty)$ is the pullback from S of a line bundle $\mathcal{L}$.

We shall now show that giving an isomorphism of D with $\mathbb{P}^1 \times S$ taking ∞ to $\infty \times S$ and trivialising $\mathcal{L}$ is the same thing as giving a Pell family over S and an isomorphism with it and our split Abel curve. This will prove the second part and the third follows as the group of affine transformations is smooth.

In one direction it is clear as a Pell family gives by construction a trivialisation of D as well as $\mathcal{L}$.

For the converse we shall need to use (cf. Corollary 2 which assumes that we are over $\mathbb{Q}$) that $\mathcal{U}_g^n$ is smooth so that we may assume that S is smooth.[4]

Assume now that an isomorphism $D \xrightarrow{\sim} \mathbb{P}^1 \times S$ and a trivialisation of $\mathcal{L}$ has been given. This means that $\mathcal{M}$ is isomorphic to $\mathcal{O}(g+1)$ so that the section of $\mathcal{M}^2$ is a homogeneous form $R(s,t)$ of degree $2g+2$ with coefficients in $\Gamma(S, \mathcal{O}_S)$. The existence of the sections ∞_1 and ∞_2 show that $R(1,0)$ is a non-zero square and hence after scaling R we can assume that $R(s,1)$ is monic. Now, as the Abel curve is split we may regard f as an isomorphism $f\colon \mathcal{O}_C(\infty_1) \to \mathcal{O}_C(\infty_2)$ and then, again by the fact that the curve is split, $f \circ \iota^*(f)$ is scalar multiplication by a square and hence by scaling f we may assume that $f \circ \iota^*(f) = 1$. On $C^\circ := C \setminus \{\infty_1\} \cup \{\infty_2\}$ f maps into $\mathbf{G}_{m\,S}$ so that f is a unit in $\Gamma(C^\circ, \mathcal{O})$. This ring is equal to $\Gamma(S, \mathcal{O}_S)[s,y]/(y^2 - R(s,1))$ so that f has the form $P(s) + yQ(s)$ and the condition $f \circ \iota^*(f) = 1$ translates into $P^2 - RQ^2 = 1$. Now, if the base is a field it is easy to see that the degree of P is equal to the degree n and hence, as S is reduced P is of degree n and its top coefficient is a unit. The equation $P^2 - RQ^2 = 1$ and the fact that R is monic shows that Q has degree $n-g-1$ with invertible top coefficient, i.e., we have a map to $\mathcal{V}_g^n$. The possible changes in choices is given by a scaling factor λ, which is a unit in $\mathcal{O}_S$, in the choice of trivialisation of $\mathcal{L}$ and an affine transformation $s \mapsto as + b$ where $s \in \Gamma(S, \mathcal{O}_S^\times)$ and $b \in \Gamma(S, \mathcal{O}_S)$. This change takes y to λy and then $(P(s), Q(s), R(s))$ to $(P(as+b), \lambda^{-2}R(as+b), \lambda Q(as+b))$ so that if we want to keep R monic we need $\lambda^2 = a^{2g+2}$ which shows ii).

Turning to iii) we may after an étale extension which extracts an n'th root of the top coefficient compose with a change of trivialisation and affine transformation such that P is monic. As R is also monic this forces the top coefficient of Q to be ± 1 and if -1 we may change the trivialisation by -1 to get that Q is also monic. We may then by an appropriate affine transformation of the form $s \mapsto a + s$ assume that R is normalised, i.e., we have obtained an S-point of $\mathcal{U}_g^n$. The ambiguities in our choices are then reduced to a pair $(\lambda, s \mapsto as)$ with $a^n = 1$, $\lambda^2 = a^{2g+2}$ and $1 = \lambda a^{n-g-1}$ conditions which are equivalent to $a^n = 1$ and $\lambda = a^{g+1}$.

The last statement is now clear. □

Remark 4. Despite the very explicit form of these charts it seems difficult to use them. I have for instance not been able to show the smoothness of the moduli space

[4] In fact we only use that it is reduced.

using the Pell equation directly (in the generic case when R and Q have no common zeros it can be done).

We may use this result to show that $\mathcal{H}_g^{s,n}$ is almost a scheme by computing the fixed point sets for the action of subgroups of μ_n on $\mathcal{U}_g^n$. For this we introduce $\mathcal{W}_g^n$ as the closed subscheme of $\mathcal{V}_g^n$ consisting of tuples (P, R, Q) if $\mathcal{V}_g^n$ for which P and Q are monic.

Proposition 8. *Let $m > 1$ be an integer that divides n so that $\mu_m \subseteq \mu_n$. Then the fixed point locus for μ_m acting on $\mathcal{U}_g^n$ is empty unless $2g+2 \equiv 0, 1 \bmod m$.*

i) If $m|g+1$ then the fixed point scheme is of the form $(p(s^m), r(s^m), q(s^m))$, where $(p(t), r(t), q(t))$ is the universal family of $\mathcal{W}_{(g+1)/m-1}^{n/m}$.

ii) If $m|2g+2$ but $m\nmid g+1$ then the fixed point scheme is of the form $(p(s^m), r(s^m), s^{m/2}q(s^m))$, with $(p(t), r(t)t, q(t))$ the universal family (P, R, Q) of $\mathcal{W}_{(g+1)/m-1/2}^{n/m}$ restricted to the closed subscheme given by $R(0) = 0$.

iii) Assuming that $m|2g+1$ then the fixed point scheme is of the form $(p(s^m), sr(s^m), s^{(m-1)/2}q(s^m))$, where $(p(t), r(t)t, q(t))$ is the universal family (P, R, Q) of $\mathcal{W}_{(2g+1)/(2m)-1/2}^{n/m}$ restricted to the closed subscheme given by $R(0) = 0$.

Proof. If a tuple $(P(s), R(s), Q(s)$ is a point of $\mathcal{U}_g^n$ and ζ an m'th root of unity, then ζ takes the tuple to $(P(\zeta s), \zeta^{-2g-2}R(\zeta s), \zeta^{g+1}Q(\zeta s)$. Hence, that the tuple is fixed under μ_m, is equivalent to P, R, resp. Q being homogeneous of degrees 0, $2g+2$, resp. $-g-1$, where the grading takes values in $\mathbb{Z}/m\mathbb{Z}$ and s has degree 1. This means that for a tuple that is a fixed point, $R(s)$ is of the form $r(s^m)$. Furthermore, if k is the residue modulo m of $2g+2$, then s^k will be the lowest order non-zero monomial of R and as R does not have any multiple roots this implies that k is 0 or 1. Assume that $m|g+1$. Then $R(s)$ has the form $r(s^m)$ and $Q(s)$ has the form $q(s^m)$. Clearly, p, q, and r are all monic and as $p^2(s^m) - r(s^m)q^2(s^m) = 1$ we get $p^2(t) - r(t)q^2(t) = 1$ so that (p, r, q) gives a family in $\mathcal{W}_{(g+1)/m-1}^{n/m}$ and conversely such a family gives a fixed point $(p(s^m), r(s^m), q(s^m))$ (note that as $m > 1$ $p(s^m)$ is automatically normalised and that $r(t)$ is multiplicity free precisely when $r(s^m)$ is). Assume that $m|2g+2$ but $m\nmid g+1$. Then we still have $P(s) = p(s^m)$ and $R(s) = r(s^m)$, but $Q(s) = s^{m/2}q(s)$ and $P^2(s) - R(s)Q^2(s) = 1$ gives $p^2(t) - r(t)tq^2(t) = 1$ so that $(p(t), r(t)t, q(t))$ gives a family in $\mathcal{W}_{(g+1)/m-1/2}^{n/m}$ for which the R-component is 0 at 0. Finally if $2g+2 \equiv 1 \bmod m$ we get $P(s) = p(s^m)$, $R(s) = r(s^m)s$, and $Q(s) = s^{(m-1)/2}q(s)$, which gives $p^2(t) - r(t)tq^2(t) = 1$. □

Remark 5. By the arguments of the proof of Theorem 3.1 (and assuming we are in characteristic zero) $\mathcal{W}_g^n$ is isomorphic to $\mathbf{G}_a \times \mathcal{U}_g^n$ through affine translations $s \mapsto s + a$ in the polynomial variable. The subscheme defined by $R(0) = 0$ is by the same argument isomorphic to the finite étale cover of $\mathcal{U}_g^n$ whose S-object are (P, Q, R), an S-object of $\mathcal{U}_g^n$, together with a choice of zero of R.

4 Low Genera

It should come as no surprise that the cases of Abel curves of genus 0 and 1 are special and we start by treating them.

Proposition 9. *i)* $\mathcal{H}_0^n$ *is isomorphic to* $B\Sigma_2$ *with universal family having* $\mathbb{P}^1$ *as curve with hyperelliptic involution* $x \mapsto 1/x$*, function* $f\colon \mathbb{P}^1 \to \mathbb{P}^1$ *given by* $x \mapsto x^n$ *and involution* $\sigma(x) = 1/x$*. The mapping to* $B\Sigma_2$ *giving the isomorphism is given by associating to an Abel curve the fixed point locus of its hyperelliptic involution.*

ii) Let $\mathcal{A}_1 \to \mathcal{M}_1$ *be the universal elliptic curve. Let* $\mathcal{U}$ *be the open substack of the fibre square of* $\mathcal{A}_1 \to \mathcal{M}_1$ *which is the complement of the diagonal and let* $\varphi\colon \mathcal{U} \to \mathcal{A}_1$ *be the map* $(x, y) \mapsto x - y$*. Then* $\mathcal{H}_1^n$ *is isomorphic to the inverse image of the kernel of multiplication by n by* φ

Proof. Starting with the genus zero case suppose we have a family of Abel curves of genus zero and degree n $(C \to S, f, \infty_1, \infty_2, \mathcal{L}, \varphi)$. Then $\mathcal{O}(\infty_1 - \infty_2)$ is the pullback of a (unique) line bundle $\mathcal{M}$ on S and the involution ι induces an isomorphism $\mathcal{M} \xrightarrow{\sim} \mathcal{M}^{-1}$ (which identifies C and ι with the involutive bundle and involution associated to the obtained trivialisation of $\mathcal{M}^2$). Now, f corresponds to an isomorphism $\mathcal{O}(n\infty_1) \xrightarrow{\sim} \mathcal{O}(n\infty_2) \otimes \mathcal{L}$, i.e., an isomorphism $\mathcal{M}^{\otimes n} \xrightarrow{\sim} \mathcal{L}$ and the fact that $f \circ \iota = \sigma \circ f$, where σ is involutive involution, implies that φ equals the n'th power of the given trivialisation $\mathcal{M} \xrightarrow{\sim} \mathcal{M}^{-1}$. This shows the whole Abel curve is determined by the involutive line bundle $\mathcal{M}$.

As for the genus 1 case we start by identifying the closed substack of $\mathcal{H}_{1,2}$ of triples (ι, a, b) with $\iota a = b$. In fact for any two disjoint sections a and b of a family of genus 1 curves there is a unique hyperelliptic involution that takes a to b, namely $x \mapsto -x + a + b$. This implies is isomorphic to $\mathcal{U}$ and the rest follows from Proposition 7. □

5 Hurwitz Type Description

If $f\colon C \to \mathbb{P}^1$ is a split Abel curve with hyperelliptic involution ι then $f \circ \iota = f^{-1}$. The map $\tau\colon \mathbb{P}^1 \to \mathbb{P}^1$ given by $\tau(x) = 1/2(x + x^{-1})$ is a quotient map for the action of the involution $x \mapsto x^{-1}$. We therefore get a commutative diagram

$$\begin{array}{ccc} C & \longrightarrow & D := C/\iota \\ {\scriptstyle f}\big\downarrow & & {\scriptstyle g}\big\downarrow \\ \mathbb{P}^1 & \xrightarrow[\tau]{} & \mathbb{P}^1 \end{array}$$

and we see that we may recover C from the map g by taking the normalisation of its pullback along τ. The map f is then also determined. This gives the possibility of describing Abel curves in terms of maps of the form g. This is precisely what we are going to do in this section.

Definition 3. *An* Abel map *of genus g and degree n over a scheme S consists of*

- *a smooth proper map $\pi\colon P \to S$, the fibres of which are genus 0 curves,*
- *a section ∞ of π and an effective Cartier divisor C of P that is étale over S,*
- *an involutive line bundle $(\mathcal{L}, \varphi)$ over S with $\pi\colon Q \to S$ the associated projective bundle,*
- *an S-morphism $g\colon P \to Q$ fibrewise of degree n such that, $g^*\infty = n\infty$ as Cartier divisors, and*
- *a relative effective Cartier divisor $D \subset P$ such that $g^*\mathcal{F} = C + 2D$, where $\mathcal{F}$ is the fixed point scheme of φ which is an effective Cartier divisor.*

A split Abel map *is an Abel map together with a splitting of the involutive bundle.*

Remark 6. Note that g is flat so that g^* of Cartier divisors is well defined.

Given an Abel family $(f\colon X \to P, (\mathcal{L}, \varphi), \iota, \sigma)$ over a scheme S, where P is involutive bundle associated to the involutive line bundle $(\mathcal{L}, \varphi)$, we may consider the induced map $g\colon C/\iota \to P/\sigma$. As 2 is invertible, taking the quotient by an involution commutes with base change, so that $X/\iota \to S$ is a smooth genus 0 fibration, whereas P/σ is the involutive quotient and hence is isomorphic to $\mathbb{P}(\mathcal{O} \oplus \mathcal{L})$. Consider now the induced map $X \to P \times_{P/\sigma} X/\iota$. The composite with it and the projection $P \times_{P/\sigma} C/\iota \to X/\iota$ is the quotient map and both $X \to X/\iota$ and $P \times_{P/\sigma} X/\iota \to X/\iota$ are double covers. As such they are specified by line bundles $\mathcal{M}$ and $\mathcal{N}$ and sections s and t of $\mathcal{M}^{-2}$ resp. $\mathcal{N}^{-2}$. The map $C \to P \times_{P/\sigma} C/\iota$ corresponds to a map $\mathcal{N} \to \mathcal{M}$ compatible with the sections of $\mathcal{M}^{-2}$ and $\mathcal{N}^{-2}$. The map $\mathcal{N} \to \mathcal{M}$ defines a relative Cartier divisor D as it defines a Cartier divisor on each fibre (over S). Let C be the divisor of s and note that the divisor of t is the pullback by g of the divisor of the involutive quotient map, i.e., $\mathcal{F}$, where $\mathcal{F}$ is the involutive locus. The compatibility between the coverings then gives that $g^*\mathcal{F} = C + 2D$, and as $g^*\infty = n\infty$ as $f^*\infty = n\infty_1$, we have an Abel map. Finally, as X is smooth, C is étale over S. This construction can be reversed.

Proposition 10. *The stack of Abel curves is isomorphic to the stack of Abel maps.*

Proof. We have just defined a map in one direction. Conversely, assume given an Abel curve and using the notations of Definition 3 we recall that the involutive quotient map is given by $\mathcal{O}_{\mathbb{P}(\mathcal{O}\oplus\mathcal{L})}(-1)$ and the section $\otimes 1 \oplus \varphi(1)$ of $\mathcal{O}_{\mathbb{P}(\mathcal{O}\oplus\mathcal{L})}(2)$ whose Cartier divisor is the involutive locus $\mathcal{F}$. The pullback of it by g is then given by $\mathcal{N} := g^*\mathcal{O}_{\mathbb{P}(\mathcal{O}\oplus\mathcal{L})}(-1)$ and the Cartier divisor $g^*\mathcal{F}$. If we put $\mathcal{M} := \mathcal{N}(D)$ then by assumption the section of $\mathcal{N}^{-2}$ comes from one of $\mathcal{M}^{-2}$ and hence gives a double covering $X \to P$ that maps to the g-pullback of the involutive double cover and X is smooth as C is étale and 2 is invertible. This gives an inverse map. □

In the future we shall pass freely back and forth between Abel maps and Abel curves.

6 Infinitesimal Calculations

In this section we shall study the deformation theory of Abel maps (and hence of Abel curves). To avoid problems with wild ramification (and worse still, inseparable maps) we shall from now on assume that all our schemes and stacks are over Spec $\mathbb{Q}$.

If $(p_1(x), p_2(x), \dots, p_n(x))$ is a sequence of monic polynomials over S (i.e., with coefficients in $\Gamma(X, \mathcal{O}_X)$) then we may put X and Y equal to Spec $(\oplus_i \mathcal{O}_S[[x]])$ and let f be given by $x \mapsto p_i$ on the i'th component and we shall refer to the ramification and branch stratifications associated to f as the ramification resp. branch stratifications of the sequence $(p_1(x), p_2(x), \dots, p_n(x))$.

Definition 4. *Let $T = (S, S_1, S_2)$ be a sequence of disjoint finite sets and r a function from S', the disjoint union of the components of T to the positive integers; for $i = 1, 2$ let S_i^e and S_i^o be the subsets of S_i where r takes even resp. odd values and set*

$$n := \sum_{s \in S} (r(s) - 1) + \sum_{s \in S_1^e \dot{\cup} S_2^e} (r(s)/2 - 1) + \sum_{s \in S_1^o \dot{\cup} S_2^o} (r(s) - 1)/2.$$

We define $\mathcal{P}(T)$ to be the affine space $\mathbf{A}^n$ seen as the parameter space of tuples $(p_s)_{s \in S'}$ where p_s is a normalised polynomial of degree $r(s)$ if $s \in S$, a normalised polynomial of degree $r(s)/2$ if $s \in S_1^e \dot{\cup} S_2^e$ and a monic polynomial of degree $(r(s) - 1)/2$ if $s \in S_1^o \dot{\cup} S_2^o$. Despite this interpretation we shall continue to refer to the origin as the origin.

To a point (p_s) of $\mathcal{P}(T)$ we associate the tuple $(q_s)_{s \in S'}$, where $q_s = p_s$ if $s \in S$, $q_s = p_s^2$ if $s \in S_1^e \dot{\cup} S_2^e$ and $q_s(x) = (x - a_s)p_s^2(x)$ with a_s being twice the next to highest coefficient of p_s. (Thus q_s is always a normalised polynomial of degree $r(s)$.) The ramification and branch stratifications of the sequence (q_s) associated to the tautological sequence will be referred to as simply the ramification resp. branch stratification of $\mathcal{P}(T)$.

We have the following characterisation of the points of a stratum.

Proposition 11. *Let $f\colon Y \to X$ be a finite map of (possibly formal) smooth 1-dimensional schemes over a field $\mathbf{k}$. Then the corank of the trace map of a closed point s of the ramification locus of f is equal to the ramification index at s minus 1 and the corank of the trace map of a closed point s of the branch locus of f is equal to the sum of the ramification indices of points of the fibres over s of f minus the number of points of the fibre.*

Proof. This is clear for the ramification locus. For the branch locus it follows from the fact that locally at a closed point branch divisor is the sum of the norms of the ramification divisors at the points of the fibres and that the norm of a closed point considered as a divisor equals to the image point which is seen by looking at valuations of a defining element. □

We are now ready to give a description of the deformation theory of Abel maps (and equivalently Abel curves). To simplify descriptions, for an Abel map $(f\colon P \to Q, C, D)$ over a base S by its *assigned branch points* we shall mean the divisor of Q which is the sum of the involutive locus and the ∞-divisor.

Theorem 6.2. *Let* $(f: P \to Q, C, D)$ *be a split Abel map over an algebraically closed field* $\mathbf{k}$*. Let* $S \subset P(\mathbf{k})$ *be the ramification points that do not map to the assigned branch points, let* S_1^e *and* S_2^e *resp.* S_1^o *and* S_2^o *be the ramification points over* $(1:1)$ *and* $(-1:1)$ *with even resp. odd ramification index (wrt to* f*) and let* $T := (S, S_1^o, S_1^e, S_2^o, S_2^e)$*. Finally, let* r *associate to a point its ramification index wrt the map* f*. Then the completion of the local ring of the stack of Abel maps at the Abel map is isomorphic to the completion of the local ring of* $\mathcal{P}(T)$ *at the origin and the isomorphism may be assumed to be stratification preserving.*

Proof. We shall give an isomorphism of deformation functors so we consider a deformation of the given Abel map over a local Artinian ring R with residue field $\mathbf{k}$. Note that as 2 is invertible, the involutive bundle has just the trivial deformation so we may restrict ourselves to split Abel maps. If we just consider deformations of the map f, then Proposition 2 shows that such deformations are in bijection with tuples $(q_s)_{s\in S'}$, S' being as in Definition 4, where q_s is a normalised polynomial over R of degree $r(s)$. It remains to understand the influence the choice of relative Cartier divisors has. Now, the Weierstrass preparation theorem is equivalent to saying that the ideal of a relative Cartier divisor of Spec $R[[x]] \to$ Spec R is generated by a unique Weierstrass polynomial (i.e., of the form $x^n + a_1x^{n-1} + \cdots + a_0$ with $a_i \in \mathfrak{m}_R$) and it is clear from the uniqueness that inclusion of divisors corresponds to divisibility of polynomials and addition of divisors corresponds to product of divisors.

Hence for $s \in S_1^o \dot\cup S_1^e \dot\cup S_2^o \dot\cup S_2^e$ the inverse image of the assigned branch points at s is defined by q_s, C_s being étale is defined by a polynomial of degree 0 or 1 depending on whether the degree of q_s is odd or even (as the difference is even). In the even case, if p_s is the polynomial of D_s we have $q_s = p_s^2$ and p_s is normalised as q_s is. In the odd case, if C_s is given by $x - a_s$ and D_s by p_s we have that $q_s = (x - a_s)p_s^2$ and as q_s is normalised we have that a_s is twice the next to highest coefficient of p_s. This shows that the q_s for $s \in S$ and the p_s for s in the complement gives an R-point of $\mathcal{P}(T)$ and the converse is also clear. The definition of the stratification of $\mathcal{P}(T)$ has been set up so that the constructed isomorphism preserves the strata. □

Remark 7. The assumption of an algebraically closed field as base is just for notational convenience as is the existence of a splitting.

We put the most important consequences of this theorem in the following corollary. Note that we have identified the stack of Abel maps with that of Abel curves.

Corollary 2. *i)* $\mathcal{H}_g^n$ *is a smooth stack everywhere of dimension* g*.*

ii) The open substack of $\mathcal{H}_g^n$ *consisting of Abel maps with only simple ramification (i.e., all ramification indices are* ≤ 2*) and for which for all branch points outside of the assigned branch points there is only one ramification point above it, is dense.*

Proof. The map $\mathcal{H}_g^{s,n} \to \mathcal{H}_g^{s,n}$ is an étale cover so we may deal with the split case instead. The smoothness follows immediately from the theorem and we postpone the calculation of the dimension. For the second part we may complete the local ring at a point and then transfer the problem to the complete local ring at the origin of $\mathcal{P}(T)$. We shall now show that the set of $\mathcal{P}(T)$ where the corank of the trace form is 0 is non-empty and by Proposition 11 it is enough to show that generically on $\mathcal{P}(T)$ the derivative of each q_s has no multiple roots. If $s \in S$ this is clear as then q_s is a generic monic polynomial and then so is $1/nq'_s$, where n is the degree of q_s. If $s \in S_i^e$, $i = 1, 2$, then $q_s = p_s^2$ where p_s is a generic monic polynomial and thus $q'_s = 2p_s p'_s$. Generically p_s and p'_s have no roots in common, p_s has no double roots and neither has p'_s by the argument just given. If $s \in S_i^e$ then $q_s = (x - a_s)p_s^2$ where p_s is a generic monic polynomial with next to highest coefficient a_s. Then we have $q'_s = p_s(p_s + 2(x - a_s)p'_s)$. Again, generically p_s has no double roots. Roots that are common to p_s and $p_s + 2(x - a_s)p'_s$ are also roots of either $x - a_s$ but generically a_s is not a root of p_s or of p'_s which again is not the case generically. We are left with showing that generically $p_s + 2(x - a_s)p'_s$ has no double roots. Now, $p_s + 2(x - a_s)p'_s$ divided by $2n + 1$, n being the degree of p_s, is a generic monic polynomial. Indeed, it is easily seen that the coefficients of p_s can be expressed as polynomials in those of $p_s + 2(x - a_s)p'_s$.

We have thus shown that generically all ramification points are simple and it remains to show that away from the assigned branch points there is generically only one ramification point above one branch point. For this we note that for a given $s \in S$ the contribution from that ramification point to the branch locus is defined by the norm of q'_s and hence what needs to be shown is that for two $s, s' \in S$ that map to the same point under f, the two norms do not have a common component. Now q_s and $q_{s'}$ are generic polynomials with independent coefficients. Hence the locus defined by the common components would have to be independent of both the coefficients of q_s and $q_{s'}$ (and of course only depend on their union) and would hence have to be constant. At the origin the full ramification loci consist just of 0 and so the common locus would have to be 0 everywhere. However, q_s has generically no factor in common with q'_s.

Finally, to compute the dimension we may by what has just been proved, look only at the case where all the ramification is simple and outside of the assigned branch points there is only one ramification point over a given branch point. We may also assume that the Abel curve is split. Now, if $s \in S$ is a ramification point, then the local deformation at that point depends on one parameter as q_s is a normalised second degree polynomial, whereas for a ramification point over the assigned branch points the local deformation at that point depends on zero parameters, as p_s is a normalised first degree polynomial. Hence the dimension is equal to the cardinality of S. Let now e_i, $i = 1, 2$, be the number of ramification points over ± 1, let $e' := e_1 + e_2$, and let e be the number of ramification points not above ± 1. By the Hurwitz formula applied to f we have

$$-2 = -2n + n - 1 + e' + e$$

and by the Hurwitz formula applied to double covering ramified at the non-ramification points over ± 1 we have

$$2g - 2 = -4 + (2n - 2e')$$

and elimination gives $e = g$. □

7 The Lyashko–Looijenga Map

By the Lyashko–Looijenga map is generally meant the map that to a family of finite maps between smooth curves associates the branch locus of each member. Sometimes one restricts oneself to families where the trace corank of the branch locus is constant and then it is natural to consider the fibrewise reduced subscheme of the branch locus. Furthermore, sometimes some of the branch points are by assumption fixed and then of course it is natural to exclude them from consideration. Our situation is of this type as the involutive fixed points are essentially fixed (i.e., they can not move non-trivially in a continuous fashion) and actually fixed in the split case.

We shall see that the situation is not completely straightforward; our strata on which the LL-map is defined will generally turn out to be non-reduced which certainly kills all hope of the LL-map being étale. All is not lost, however, as a stratum is locally the product of a smooth stack and a zero-dimensional one and the LL-map turns out to be étale on the reduced substack. The most obvious reason for the stratum being non-reduced is our definition of the branch locus. This definition is, however, more or less forced upon us if one wants the branch divisor to vary continuously (i.e., be a relative Cartier divisor) as generically the branch divisor is étale and hence determined by the condition that its support be the branch locus.

Definition 5. *i) A* ramification specification of degree *n consists of a finite multiset S of multisets of (strictly) positive integers such that for each multiset s in S $\sum_{e\in s} e = n$. The* multiplicity multiset *associated to S is the multiset $\{\rho(s) \mid s \in S\}$, where $\rho(s) = \{e-1 \mid e \in S, e > 1\}$. The* total ramification *of S is*

$$\sum_{e\in s\in S} (e-1).$$

ii) An Abel ramification specification of order *n is a ramification specification S of order n whose total ramification equals $n-1$ together with the choice of a sub-multiset T of S of cardinality 2. If t is the number of odd integers, counted with multiplicity, of the members of T, then the* genus *of S is equal to $(t-4)/2$.*

Remark 8. A ramification specification is determined by its associated multiplicity multiset and the degree n. A multiplicity multiset is the same as a *passport* of [4].

We shall now consider stratifications of $\mathcal{H}_g^n$. First we consider the ramification stratification of the universal map of $\mathcal{H}_g^n$ giving a multiset of multiplicities associated to each stratum. Then we consider its intersection with the trace stratification associated to the branch locus which gives a further division of the multiplicities according to which branch point they are mapped to. This gives exactly an Abel ramification specification S of order n associated to each such stratum. Conversely, for each Abel ramification specification S of order n we denote by $\mathcal{H}_{g,S}^n$ the corresponding stratum.

If $X \to S$ is a map of algebraic stacks and m a positive integer then $\mathrm{Conf}^m(X/S)$ (or just $\mathrm{Conf}^m(X)$ if S is understood) is the *m-point configuration space*, i.e., the stack quotient by the permutation action of the symmetric group Σ_m on the open substack of the m'th fibre power of $X \to S$ consisting of distinct points.

Our main use of this construction is to the universal involutive projective bundle; namely the projective bundle $\mathbb{P} \to B\Sigma_2$ that to an involutive line bundle over S (i.e., a map $S \to B\Sigma_2$) associates the involutive projective bundle. We then let $\mathbb{P}' \to B\Sigma_2$ be the open substack of $\mathbb{P}$ obtained by removing the section of infinity and the involutive fixed point set. Note that the universal involutive line bundle of $\mathcal{H}_g^n$ gives a map $\mathcal{H}_g^n \to B\Sigma_2$ and the base of the universal Abel map is just the pullback of $\mathbb{P}$ under this map.

Definition 6. *Let S be an Abel ramification specification* (S, T) *of order n and let m be the cardinality of S minus* 2. *We define the Lyashko–Looijenga map* $LL\colon \mathcal{H}_{g,S}^n \to \mathrm{Conf}^m(\mathbb{P}'/B\Sigma_2)$ *by associating to an Abel map over S its reduced branch locus minus assigned base points.*

We can now prove the major result on the LL-map after we have proven the following lemma.

Lemma 1. *Let* R *be a commutative ring which contains* $\mathbb{Q}$ *and* $a \in R$. *Then for a strictly positive integer n the polynomial* $(t^{2n+1} - a^{2n+1})/(t-a)$ *is a square of a polynomial precisely when* $a^{n+1} = 0$.

Proof. In the ring of Laurent power series in t^{-1}, $R((t^{-1}))$ the polynomial has the unique square root

$$t^n\sqrt{1 + at^{-1} + \cdots + a^{2n}t^{-2n}}$$

and hence the polynomial has a polynomial square root precisely when all powers beyond t^{-n} have zero coefficients in $\sqrt{1 + at^{-1} + \cdots + a^n t^{-2n}}$. This series is obtained by substituting $s \mapsto at^{-1}$ in $\sqrt{(1 - s^{2n+1})/(1-s)}$ which makes it clear that if $a^{n+1} = 0$ then the square root is a polynomial. It is equally clear that the converse is true if the coefficient of s^{n+1} in $\sqrt{(1 - s^{2n+1})/(1-s)}$ is non-zero. However, as $n > 0$, $2n + 1 \geq n + 2$, and thus modulo s^{n+2} $\sqrt{(1 - s^{2n+1})/(1-s)}$ is congruent to $(1-t)^{-1/2}$ which clearly has all of its coefficients non-zero. □

Theorem 7.3. *Let S be an Abel ramification specification* (S, T) *of order n and let m be the cardinality of S minus* 2.

i) The completion of $\mathcal{H}_{g,S}^n$ *at any geometric point* $s = \mathrm{Spec}\,\mathbf{k}$ *is isomorphic to*

$$\prod_{\substack{2n+1 \in t \in T \\ n>0}} \mathrm{Specf}\,\mathbf{k}[[a]]/\left(a^{n+1}\right) \times \prod_{s \in S \setminus T} \mathrm{Specf}\,\mathbf{k}[[\sigma, a_1, \ldots, a_{m(s)}]]/(\sigma_1, \ldots, \sigma_e)$$

where $m(s) := |\{e \mid e \in s;\, e \geq 2\}|$, $e(s) := \sum_{e \in s}(e-1)$ *and*

$$\prod_i (s - a_i)^{e_i - 1} = s^e + \sum_{1 \leq j \leq e} (-1)^j \sigma_j s^{e-j}$$

as polynomials in s. In particular, $\mathcal{H}^n_{g,S}$ is smooth precisely when there is exactly one ramification point over each unassigned branch point and no ramification point of odd ramification index above involutive fixed points. It is always the case that the reduced substack $(\mathcal{H}^n_{g,S})^{\text{red}}$ is smooth.

ii) The Lyashko–Looijenga map $LL\colon (\mathcal{H}^n_{g,S})^{\text{red}} \to \text{Conf}^m(\mathbb{P}'/B\Sigma_2)$ restricted to the reduced subscheme is an étale covering map.

Proof. We start by making a local calculation. It is clear from Theorem 6.2 that we get a product over the elements of S. Let us first consider an unassigned branch point. Let $\{e_1,\dots,e_k\}$ be an element of $S\setminus T$ with the members equal to 1 removed. Hence a deformation over a local Artinian ring R is given by a collection $(p_i)_{1\le i\le k}$ of normalised polynomials with $\deg p_i = e_i$. Now the condition, that the deformation stay inside the stratum given by $\{e_1,\dots,e_k\}$ means, according to Proposition 4 and the identification of Cartier divisors with Weierstrass polynomials, that each p_i' has the form $e_i(x-\alpha_i)^{e_i-1}$ and as p_i is normalised we get that $\alpha_i = 0$ and hence that $p_i(x) = x^{e_i}+b_i$. Furthermore the ramification divisor is defined by x^{e-1}. To compute the branch divisor we have to compute the norm of x^{e-1}, and using the multiplicativity of the norm it is enough to compute the norm of x. Now, it is clear that under the map $R[[t]] \to R[[x]]$ given by p_i we have that $R[[x]]$ is isomorphic to $R[[t,x]]/(x^{e_i}+b_i-t)$ which gives that the norm of x is $\pm(t-b_i)$. Hence the branch divisor is given by $\prod_i(t-b_i)^{e_i-1}$. Now, we are working in the stratum where the fibrewise reduced branch divisor exists, which means that there is a $\sigma\in\mathfrak{m}_R$ such that $\prod_i(t-b_i)^{e_i-1} = (t-\sigma)^e$, where $e=\sum_i(e_i-1)$. Comparing next to highest coefficients gives $\sigma = \sum_i(e_i-1)b_i$ and changing variable $s=t-\sigma$ and putting $a_i = b_i-\sigma$ gives us $\prod_i(s-b_i)^{e_i-1} = s^e$. This shows that the universal R is $\mathbf{k}[[\sigma,a_1,\dots,a_k]]/(\sigma_1,\dots,\sigma_e)$. Now, for degree reasons, as soon as $k>1$, $\mathbf{k}[[\sigma,a_1,\dots,a_k]]/(\sigma_1,\dots,\sigma_e)$ is strictly larger than $\mathbf{k}[[\sigma]]$. On the other hand putting s equal to a_i gives $a_i^e=0$, which shows that when dividing out by the nilradical of $\mathbf{k}[[\sigma,a_1,\dots,a_k]]/(\sigma_1,\dots,\sigma_e)$, this ring equals $\mathbf{k}[[\sigma]]$.

Considering now instead one of the involutive fixed points again as we are in a fixed ramification stratum, we get the form $p_i = x^{e_i}+b_i$. This time, however, we have that when e_i is even, p_i is a square and when it is odd, p_i is a square times a linear polynomial. In the first case it is easy to see that if $x^{e_i}+b_i$ is a square, then $a_i=0$. In the second case, if $x^{e_i}+b_i = (x-a_i)q^2(x)$, then setting $x=a_i$ we get $b_i = -a_i^{e_i}$, so that $(x^{e_i}-a_i^{e_i})/(a-a_i) = q^2(x)$ and we conclude from Lemma 1 that this is possible precisely when $a_i^{(e_i+1)/2}=0$. This concludes the proof of i).

Turning to ii), that the map is étale is clear from the local calculation, as the fibrewise reduced branch divisor is defined by $t-\sigma$, using the notations of the first part. It remains to prove that it is proper and for that we shall use the valuative criterion, and as everything is of finite type over $\operatorname{Spec}\mathbb{Q}$ we may restrict ourselves to discrete valuations, which we may assume to be strictly Henselian. Hence we may assume that the map is split and by Theorem 3.1 we may assume that it is given by a Pell family (P,Q,R) such that $P^2-RQ^2=1$, and P is then the Abel map in question. By for instance [4, Lemma 3.1] (and the fact also noted in [loc. cit.] that the inverse image of the origin under the LL-map is the origin) P has coefficients in R.

By Gauss' lemma so does R and Q. The next step is to show that the discriminant of R is a unit. For this one may reduce modulo the maximal ideal of R and apply the Hurwitz formula to the map given by P. Indeed, by assumption the number of branch points of P is fixed and hence by Hurwitz' formula the number of ramification points is also fixed. This makes it impossible for zeros of R to come together. □

Remark 9. i) The local description of the stratum contradicts [3, Prop. A.3] which claims that the LL-map always is étale. In view of the theorem (very slightly modified to fit into the context of [loc. cit.]) this is now seen to be false when there is more than one ramification point over a branch point. It thus has to be modified to saying that the restriction to the reduced subscheme of an equisingular stratum is étale. Luckily, this is what is used in the main text and it is also given a topological proof in [3, Thm. 4.2].

ii) It is possible to get a natural interpretation of the reduced structure on the strata. This will be treated elsewhere.

8 Topological Construction

In this section we shall study the covering given by the LL-map. Even though we can easily get a description for all strata, using the same methods, we shall only deal with the open stratum as that gives a combinatorial algorithm for computing the number of connected components of the stack of Abel curves. Note however that other strata are also interesting. For instance the lowest stratum where all the branch points are assigned has been considered in connection with Grothendieck's "dessin d'enfants" (cf., [5]).

As usual the fibres of the LL-map are in bijection with conjugacy classes of certain sequences of the symmetric group. In order to give a procedure for computing the number of components of a stratum we need generators for the fundamental group of the appropriate configuration space. The following result can most certainly be extracted from the literature but for the convenience of the reader as well as the author we give a proof. We start by giving some notation. If C is a simple non-closed oriented curve in $\mathbb{C}$ and $S \subset C$ is a finite set then the orientation of C induces a total order on S. If i is a positive integer strictly smaller than the number s of elements of S then we define, as usual, the elements σ_i of the braid group on s strands given by letting the i'th point move along C to the position of the $i+1$'st point to the right of C and letting the $i+1$'st point move along C to the position of the i'th point to the left of C ("right" and "left" being from the point of view of the orientation of C). If $s \geq 3$ and we define τ_1 resp. τ_2 to be the braids that takes the second resp. $s-1$'st point and moves along C on the left resp. right hand side till just before the first resp. last point, then circles that point once counter-clockwise, and returns back to its original position along the right resp. left hand side of C. (They are equal to σ_1^2 resp. σ_{s-1}^{-2}.)

Proposition 12. *Let $C := [-1, 1]$ oriented in any direction and $S \subset C$ a finite subset with s elements containing ± 1, and let $A := \mathbb{C} \setminus \{\pm 1\}$. Then the map induced by the inclusion $\{x_1, \dots, x_{s-2}\} \mapsto \{-1, 1, x_1, \dots, x_{s-2}\}$*

$$\pi_1\left(\mathrm{Conf}^{s-2}(A), S\setminus\{\pm 1\}\right) \to \pi_1(\mathrm{Conf}^s(\mathbb{C}), S)$$

is an injection whose image is generated by σ_i, $2 \le i \le s-2$, τ_1 and τ_2.

Proof. By possibly applying $z \to -z$ we may assume that the orientation of $[-1, 1]$ is such that -1 becomes its first element.

To begin with it is clear the σ_i, $2 \le i \le s-2$, τ_1 and τ_2 lie in the image. Recall that we have a surjection $\pi_1(\mathrm{Conf}^s(\mathbb{C}), S) \to \Sigma_s$ taking σ_i to the transposition $(i-1, i)$. The image of $\pi_1(\mathrm{Conf}^A(s-2), S\setminus\{\pm 1\})$ maps into Σ_{s-2}, considered as the subgroup that fixes the first and last elements, and as σ_i maps to $(i-1, i)$, the subgroup generated by them maps surjectively onto Σ_{s-2}. Hence both for injectivity and generation it suffices to consider $\pi_1(\widetilde{\mathrm{Conf}}^{s-2}(A), S\setminus\{\pm 1\}) \to \pi_1(\widetilde{\mathrm{Conf}}^{s}(\mathbb{C}), S)$, where $\widetilde{\mathrm{Conf}}^{t}(X)$ is the space of ordered t-subsets of X, and to show that the map is injective and the image is generated by the conjugates of σ_i^2, $1 \le s \le s-1$ in the group generated by $2 \le i \le s-2$, τ_1 and τ_2.

Now, by conjugating by the σ_i, $2 \le i \le s-1$, we can get from τ_1 and τ_2 all braids A_i^1 and A_i^2 defined like τ_1 resp. τ_2 only starting at the i'th point for $2 \le i \le s-1$, as well as the A_{ij}, $2 \le i < j \le s-2$, defined like τ_1 only starting at the j'th point and encircling the i'th point. (The A_{ij} are the A_{ij} of [2, 1-11], A_i^1 is A_{1i} and A_i^2 is a mirror image of $A_{s-i,s}$.) Our aim is to show the injectivity and that these elements generate the image. In this we shall follow the proof of [2, Lemma 1.8.2] and we start following [2] in using the notation $F_{m,n}(X)$ for $\widetilde{\mathrm{Conf}}^{n}(X\setminus Q_m)$ where Q_m is a fixed subset of X of cardinality m and will use of the theorem of Fadell and Neuwirth (cf. [2, Thm. 1.2]) which says that when X is a manifold, then the projection on the first r factors $F_{m,n} \to F_{m,r}$ is a fibration with fibre $F_{m+r,n-r}$. Applied to $r = n-1$ and $X = \mathbb{C}$ and $X = A$ this will allow us to prove the statement by induction. As the involved spaces are acyclic, the fibrations give short exact sequences, and by induction we are reduced to showing that for $1 \le i < s$

$$\pi_1\left(A \setminus S_i, x_{i+1}\right) \to \pi_1\left(\mathbb{C}\setminus S_i, x_{i+1}\right),$$

where S_i consists of the i first elements of S and x_{i+1} is the $i+1$'st element, is an injection and that the image is contained in the subgroup generated by $A_{k,i+1}$ and A_{i+1}^1 (and when $i = s-1$ also the A_k^2) for $1 \le i \le k$. This however is clear. □

This result, combined with Theorem 7.3 and Corollary 2 ii), allows us to give a combinatorial description of the number of components of $\mathcal{H}_g^n$ and $\mathcal{H}_g^{s,n}$. For this we first introduce the following definition.

Definition 7. *Let $N_{g,n}$ be the set of tuples $(\sigma, \sigma_1, \ldots, \sigma_g, \tau) \in (\Sigma_n)^{g+2}$ fulfilling the conditions*

- *$\sigma\sigma_1 \cdots \sigma_g\tau$ is an n-cycle and*
- *the σ_i are transpositions and σ and τ are products of disjoint transpositions and the sum of the number of fixed points of σ and of τ equals $2g+2$.*

Let $M_{g,n}$ be the set of orbits of the action of Σ_n on $N_{g,n}$ given by

$$(\rho, (\sigma, \sigma_1, \ldots, \sigma_g, \tau)) \mapsto \left(\rho\sigma\rho^{-1}, \rho\sigma_1\rho^{-1}, \ldots, \rho\sigma_g\rho^{-1}, \rho\tau\rho^{-1}\right).$$

Thus armed we can give a combinatorial description of the set of connected components of the stacks of (split) Abel curves.

Theorem 8.4. *The set of connected components of $\mathcal{H}_g^{s,n}$ is in bijection with equivalence classes of $M_{g,n}$ under the equivalence relation generated by the relations.*

- $(\sigma, \sigma_1, \ldots, \sigma_i, \sigma_{i+1}, \ldots, \sigma_g, \tau) \sim (\sigma, \sigma_1, \ldots, \sigma_i\sigma_{i+1}\sigma_i^{-1}, \sigma_i, \ldots, \sigma_g, \tau)$ *for all* $1 \leq i < g$.
- $(\sigma, \sigma_1, \ldots, \sigma_g, \tau) \sim (\sigma[\sigma_1, \sigma], \sigma\sigma_1\sigma^{-1}, \ldots, \sigma_g, \tau)$, *with* $[\sigma_1, \sigma] = \sigma_1\sigma\sigma_1^{-1}\sigma^{-1}$.
- $(\sigma, \sigma_1, \ldots, \sigma_g, \tau) \sim (\sigma, \sigma_1, \ldots, \tau^{-1}\sigma_g\tau, [\tau^{-1}, \sigma_g^{-1}]\tau)$.

The set of connected components of $\mathcal{H}_g^n$ is in bijection with equivalence classes of $M_{g,n}$ under the equivalence relation generated by the above relations together with the relation

$$(a_1, a_2, \ldots, a_{g+2}) \sim (b_{g+2}a_{g+2}b_{g+2}^{-1}, \ldots, b_2a_2b_2^{-1}, a_1),$$

where $b_i = a_1 \ldots a_{i-1}$ for $i \geq 2$.

Proof. The part on $\mathcal{H}_g^{s,n}$ follows directly from the fact that the LL-map is an étale covering (Theorem 7.3), that the fibres of the LL-mapping are in bijection with $M_{g,n}$, the description of the generators for the fundamental group for the target of the LL-map (Proposition 12) and the formula for the action of the σ_i on $M_{g,n}$.

As for the $\mathcal{H}_g^n$-part the Lyashko–Looijenga map has as target the quotient of $\mathrm{Conf}^g(\mathbf{A}^1 \setminus \{\pm 1\})$ divided by the map induced by $z \mapsto -z$. Hence we have to add the relation that identifies an equivalence class of maps from the fundamental group to Σ_n with the one obtained by composing with the action of the (outer) automorphism induced by $z \mapsto -z$. For that we choose as basepoint of $\mathrm{Conf}^g(\mathbf{A}^1 \setminus \{\pm 1\})$ the set $\{-1/2, -1/3, \ldots, 1/3, 1/2\}$ (with 0 included if g is odd) and as basepoint for $\mathbf{A}^1 \setminus \{-1, -1/2, -1/3, \ldots, 1/3, 1/2, 1\}$ i. Acting by $z \mapsto -z$ gives us $-i$ as new basepoint and we identify fundamental groups by choosing a curve from $-i$ to i going to the left of $\{-1, -1/2, -1/3, \ldots, 1/3, 1/2, 1\}$. □

References

[1] N. H. Abel, Sur l'intégration de la formule différentielle $\frac{\varrho\, dx}{\sqrt{r}}$, r et ϱ étant des fonctions entiéres. *J. Reine Angew. Math.* 1:105–144, 1826.

[2] J. S. Birman, *Braids, links, and mapping class groups.* Princeton University Press, Princeton, N.J., 1974. Annals of Mathematics Sudies, No. 82.

[3] T. Ekedahl, S. Lando, M. Shapiro, and A. Vainshtein, Hurwitz numbers and intersections on moduli spaces of curves. *Invent. Math.* 146(2):297–327, 2001.

[4] S. K. Lando and D. Zvonkin, On multiplicities of the Lyashko–Looijenga mapping on strata of the discriminant. *Funktsional. Anal. i Prilozhen.* 33(3):21–34, 1999.

[5] G. Shabat and A. Zvonkin, Plane trees and algebraic numbers. In *Jerusalem combinatorics '93*, volume 178 of *Contemp. Math.*, pages 233–275. Amer. Math. Soc., Providence, RI, 1994.

Formal Deformation of Chow Groups

Mark Green* **and Phillip Griffiths**

Contents

1 Introduction

1.1 Summary of results. The problem of constructing algebraic cycles with given properties – e.g., in a given homology class – is one of the longest standing and deepest questions in algebraic geometry. Over the years there have been several variational approaches centered around the question

(1.1) *Given a family $\{X_t\}_{t\in B}$ of smooth algebraic varieties with $X_{t_0} = X$, and given a codimension-p algebraic subvariety $Z \subset X$ whose fundamental class $\psi_0(Z)$ remains of Hodge type (p, p) in $H^{2p}(X_t, \mathbb{C})$ for t in an analytic neighborhood of t_0, is there a family of subvarieties $\{Z_t\}_{t\in B}$ with $Z_{t_0} = Z$?*

Especially notable here is Spencer Bloch's semi-regularity paper [2]. The advantage of this approach is that one can use the full strength of the deformation theory of

* Research partially supported by NSF grant DMS 9970307.

subvarieties. One might similarly ask whether there is a family of algebraic cycles $\{Z_t\}_{t\in B}$ deforming Z; some first steps toward an infinitesimal theory for this problem were the subject of [11]. A third possible context, which we will consider here, is to work in the Chow group. Thus, letting $\xi \in CH^p(X)$ be the rational equivalence class represented by $[Z]$, the question (1.1.1) should be first modified to

(1.2) *Under the assumptions in (1.1.1), is there a family $\xi_t \in CH^p(X_t)$ with $\xi_{t_0} = \xi$?*

There is a basic obstruction to solving this problem, namely that the fundamental class $\psi_0(\xi)$ must remain of Hodge type (p, p) in $H^{2p}(X_t, \mathbb{C})$ for t in an analytic neighborhood of t_0. For $p = 1$, this is the only obstruction. We will see below, for $p \geq 2$ that even modulo torsion there are other obstructions to deforming ξ even when $\psi_0(\xi) = 0$. The full question (1.1.2) is interesting, as is the question

(1.3) *Under the assumptions of (1.1.1), is there a family $\xi_t \in CH^p(X_t)$ with*

$$\psi_0(\xi_0) = \psi_0(\xi) \; ?$$

This question has a formal analogue where, so to speak, we try to construct ξ_t as a formal power series in t, and the purpose of this paper is to develop an obstruction calculus that will help to understand the formal analogue of (1.1.3). Two applications of this calculus are the following, where again everything is modulo torsion

(1.4) *Suppose that $\xi \in CH^p(X)$ and as in (1.1.1) assume that the fundamental class $\psi_0(\xi)$ remains of Hodge type (p, p) for t in an analytic neighborhood of t_0. Suppose further that*

$$H^{p+1}\left(\Omega^{p-2}_{X/\mathbb{C}}\right) = \cdots = H^{p+1}(\mathcal{O}_X) = 0 \,. \tag{1.5}$$

Then (1.1.2) is true formally.

For $p = 2$, we are hypothesizing that

$$H^3(\mathcal{O}_X) = 0$$

which is equivalent to saying

$$J^2(X) = J^2(X)_{ab} \,,$$

i.e., $J^2(X)$ is an abelian variety. By upper-semicontinuity the same should be true for nearby X_t. By the Generalized Hodge Conjecture, all of the $J(X_t)$ should lie in the image of the Abel–Jacobi maps, and we expect no further obstructions beyond $\psi_0(\xi)$ remaining of Hodge type. However, if

$$H^3(\mathcal{O}_X) \neq 0$$

then there exist cycles ξ with $\psi_0(\xi) = 0$ and $AJ^2_X(\xi)$ non-torsion, yet where

$$\mathrm{im}(AJ^2_{X_t}) \subseteq \text{torsion for } 0 < |t| < \epsilon$$

as in [9] and [V]. So (1.1.2) cannot hold in this situation. The precise explanation of what "being true formally" means will be given below. We will also see that the assumption (1.1.5) implies that, at least formally, questions (1.1.2) and (1.1.3) coincide; i.e., there are no obstructions to formally lifting $F^1CH^p(X)$.

In general we will see that:

(1.6) *If*

$$H^{p+1}\left(\Omega^{p-m-1}_{X/\mathbb{C}}\right) = \cdots = H^{p+1}(\mathcal{O}_X) = 0\,, \qquad m \geqq 1\,,$$

then $F^mCH^p(X)$ is formally unobstructed to all orders in t. Taking $m = p$ we have that

$$F^pCH^p(X) \text{ is formally unobstructed}\,. \tag{1.7}$$

Here we use the version of the conjectural filtration on Chow groups of [10].

Since the fundamental works of Bloch and others (cf. [B], [Q] and [13]) leading to the identification, again modulo torsion,

$$CH^p(X) \otimes \mathbb{Q} \cong H^p\left(\mathcal{K}^M_p(\mathcal{O}_X)\right) \otimes \mathbb{Q} \tag{1.8}$$

where $\mathcal{K}^M_p(\mathcal{O}_X)$ is the sheaf associated to the pth-Milnor K-groups of the local rings $\mathcal{O}_{X,x}$ for some time it has been understood that for $p \geqq 2$ the groups $CH^p(X)$ have an arithmetic aspect. Below we shall show that:

(1.9) *Suppose that X is defined over $\mathbb{Q}$ and that $\xi \in CH^p(X(\mathbb{Q}))$ and, as in (1.1.4), $\psi_0(\xi)$ remains of Hodge type (p, p) for t in an analytic neighborhood of t_0. Then formally (1.1.2) has an affirmative answer.*

We shall also see that, in general, the lifting of $\xi \in CH^p(X(\mathbb{Q}))$ can *only* be formal. Of course, if the Hodge Conjecture is true then question (1.1.3) *will* have an affirmative answer. Thus, there is some arithmetic/geometric subtlety in the difference between (1.1.2) and (1.1.3).

Before turning to a more detailed description of the contents of this paper we might summarize by saying that *this work contains good news and bad news*: The good news is that, under only Hodge-theoretic assumptions, we are able to construct *something*; the bad news is that there are subtle obstructions to the convergence of this iterative construction. There has been a significant recent progress in tackling such issues, cf. the excellent survey of Chambert-Loir [6].

1.2 Discussion of techniques. We will work with an algebraic family of algebraic varieties, given by

$$\mathcal{X} \xrightarrow{p} B\,. \tag{2.1}$$

Here, $\mathcal{X}$ and B are smooth complex varieties and p is a smooth projective map. We shall sometimes write (1.2.1) as

$$\{X_t\}_{t\in B} \tag{2.2}$$

where $X_t = p^{-1}(t)$ is a smooth, projective variety. It will cause no essential loss in generality if we assume that B is one-dimensional. We let $t_0 \in B$ be a reference point and set $X = X_{t_0}$; we then may think of (1.2.1) as a deformation of X. We denote by $\mathcal{I}_X \subset \mathcal{O}_{\mathcal{X}}$ the ideal sheaf of X_{t_0} and set

$$\begin{cases} \mathcal{O}_{\mathcal{X}_m} = \mathcal{O}_{\mathcal{X}/\mathcal{I}_X^{m+1}} \\ \mathcal{X}_m = \text{scheme}\left(X_{t_0}, \mathcal{O}_{\mathcal{X}_m}\right) . \end{cases}$$

Geometrically the scheme $\mathcal{X}_m$ is the mth order neighborhood of X in $\mathcal{X}$, or equivalently the deformation to order m of X given by (1.2.2). We have natural maps

$$\mathcal{X}_{m+1} \longrightarrow \mathcal{X}_m$$

and we let

$$\hat{\mathcal{X}} = \lim_m \mathcal{X}_m$$

be the scheme $(X, \mathcal{O}_{\hat{\mathcal{X}}})$ where $\mathcal{O}_{\hat{\mathcal{X}},x} = \lim_m \mathcal{O}_{\mathcal{X}_m}$; we shall refer to $\hat{\mathcal{X}}$ as the *formal neighborhood* of X in $\mathcal{X}$.

For a codimension-p algebraic cycle Z of X we denote by $\xi = [Z] \in CH^p(X)$ the corresponding rational equivalence class. Throughout this paper *we will work modulo torsion*. We denote by

$$\psi_0(\xi) = \psi_0(Z) \in Hg^p(X)$$

the fundamental class of ξ or Z. Here

$$Hg^p(X) = H^{2p}(X, \mathbb{Z}) \cap H^{p,p}(X)$$

is the usual Hodge group of integral cohomology classes of Hodge type (p, p), considered modulo torsion.

Fundamental to this work is the Soulé variant of the Bloch–Quillen identification

$$CH^p(X) \cong H^p\left(\mathcal{K}_p^M(\mathcal{O}_X)\right) \quad \text{mod torsion} . \tag{1.1.8}$$

Relative to a Zariski covering U_α of X, we may represent elements on the right as cocycles formed from symbols

$$\left\{f_1, \dots, f_p\right\}, \qquad f_i \in \mathcal{O}_X^*\left(U_{\alpha_0} \cap \cdots \cap U_{\alpha_p}\right) . \tag{2.3}$$

If Z is a codimension-p subvariety whose ideal sheaf has a resolution by locally free sheaves, then from this data we may construct the corresponding cocycle (1.2.3).

Thus, in a sense we may think of the RHS as giving the "equations" of the rational equivalence class of an algebraic cycle.

The central thrust of this work is to develop a calculus for successively lifting a class $\xi \in H^p\left(\mathcal{K}_p^M(\mathcal{O}_X)\right)$ to classes $\xi_m, \in H^p\left(\mathcal{K}_p^M(\mathcal{O}_{\mathcal{X}_m})\right)$ over the infinitesimal neighborhoods of X in $\mathcal{X}$. We are *not* saying that the identification (1.1.8) extends to the schemes $\mathcal{X}_m$ (this may be an interesting question). Rather, if we can lift $\xi = [Z]$ to a class in $H^p(U, \mathcal{K}_p^m(\mathcal{O}_U))$ where U is a Zariski neighborhood of X in $\mathcal{X}$, then from

$$CH^p(U) \cong H^p\left(U, \mathcal{K}_p^M(\mathcal{O}_U)\right)$$

we will have lifted $[Z]$ to $[\mathcal{Z}] \in CH^p(U)$ where $\mathcal{Z}$ is a codimension-p algebraic cycle in U. Our work will enable us to analyze the corresponding formal question of lifting ξ to $\hat{\xi} \in H^p\left(\mathcal{K}_p^M(\mathcal{O}_{\hat{\mathcal{X}}})\right)$.

The general mechanisms for lifting geometric objects given by "equations" over successive infinitesimal neighborhoods is well known and classical, dating from the earliest works of Kodaira and Spencer. The fundamental observation in the present study is the following: Recall that for any ring R there is a natural map

$$K_p^M(R) \longrightarrow \Omega^p_{R/\mathbb{Q}} \tag{2.4}$$

where the RHS are the absolute Kähler differentials. The map (1.2.4) is induced by

$$\{r_1, \ldots, r_p\} \longrightarrow \frac{dr_1}{r_1} \wedge \cdots \wedge \frac{dr_p}{r_p}, \qquad r_i \in R^*\ .$$

By abuse of notation we denote this $d\log r_1 \wedge \cdots \wedge d\log r_p$.

For any scheme Y the maps (1.2.4) give maps

$$\mathcal{K}_p^M(\mathcal{O}_Y) \longrightarrow \Omega^p_{Y/\mathbb{Q}}$$

inducing on cohomology what is called the *arithmetic cycle class mapping* (cf. [14])

$$\eta : H^p\left(\mathcal{K}_p^M(\mathcal{O}_Y)\right) \longrightarrow H^p\left(\Omega^p_{Y/\mathbb{Q}}\right)\ . \tag{2.5}$$

The basic observation in this work is:

(2.6) *The obstruction to lifting $\xi_m \in H^p\left(\mathcal{K}_p^M(\mathcal{O}_{\mathcal{X}_m})\right)$ to $H^p\left(\mathcal{K}_p^M(\mathcal{O}_{\mathcal{X}_{m+1}})\right)$ depends only on the arithmetic cycle class $\eta(\xi_m) \in H^p\left(\Omega^p_{\mathcal{X}_m/\mathbb{Q}}\right)$.*

To interpret (1.2.6) we recall from [10] that, assuming the important conjecture of Bloch–Beilinson, the arithmetic cycle class $\eta(Z)$ captures some – but not all – of the information in $[Z] \in CH^p(X)$. For $p = 1$ this "lost" information is not crucial to the obstruction to lifting $\xi = [Z]$, this being essentially due to the $p = 1$ case of (1.1.7). However, for $p \geqq 2$ this lost information is non-trivial for the lifting problem and may be related to the issue of non-convergence of the formal iterative construction leading to the results (1.1.4) and (1.1.9).

In Sect. 2.1 we will review the classical 1st order Kodaira–Spencer theory, especially the extension class and subsequent coboundary maps associated to the exact sheaf sequence

$$0 \longrightarrow \Omega^{p-1}_{X/\mathbb{Q}} \oplus \Omega^{p}_{X/\mathbb{Q}} \longrightarrow \Omega^{p}_{\mathfrak{X}_1/\mathbb{Q}} \longrightarrow \Omega^{p}_{X/\mathbb{Q}} \longrightarrow 0 \tag{2.7}$$

where the first map is

$$\varphi \oplus \omega \mapsto dt \wedge \varphi + t\omega , \qquad t^2 = 0 .$$

The central fact, deriving ultimately from Kähler geometry, is that the coboundary map in the exact cohomology sequence of (1.2.7)

$$H^q\left(\Omega^{p}_{X/\mathbb{Q}}\right) \xrightarrow{\delta} H^{q+1}\left(\Omega^{p-1}_{X/\mathbb{Q}}\right) , \tag{2.8}$$

is zero. We also establish another result (Proposition 2.1.13) that follows from the degeneracy at E_2 of the Leray spectral sequence associated to a smooth, projective morphism.

In Sect. 2.2 we analyze the 1st order obstruction theory for the groups $H^p(\mathcal{K}^M_p(\mathcal{O}_X))$. The basic observation is the commutative diagram

$$\begin{array}{ccccccccc}
0 & \longrightarrow & \Omega^{p-1}_{X/\mathbb{Q}} & \longrightarrow & \mathcal{K}^M_p\left(\mathcal{O}_{\mathfrak{X}_1}\right) & \longrightarrow & \mathcal{K}^M_p(\mathcal{O}_X) & \longrightarrow & 0 \\
 & & \downarrow & & \downarrow \eta & & \downarrow \eta & & \\
0 & \longrightarrow & \Omega^{p-1}_{X/\mathbb{Q}} \oplus \Omega^{p}_{X/\mathbb{Q}} & \longrightarrow & \Omega^{p}_{\mathfrak{X}_1/\mathbb{Q}} & \longrightarrow & \Omega^{p}_{X/\mathbb{Q}} & \longrightarrow & 0
\end{array} \tag{2.9}$$

where the left hand vertical arrow is

$$\varphi \longrightarrow \varphi \oplus d\varphi$$

and the top left hand horizontal arrow is induced by

$$f\frac{dg_1}{g_1} \wedge \cdots \wedge \frac{dg_{p-1}}{g_{p-1}} \mapsto \left\{1 + tf, g_1, \ldots , g_{p-1}\right\}, \qquad t^2 = 0 \tag{2.10}$$

where $f \in \mathcal{O}_{X,x}, g_i \in \mathcal{O}^*_{X,x}$. The $m = 1$ result (1.2.6) follows from (1.2.8) and (1.2.9). The geometric understanding of the relationship between first order obstruction theory and the filtration on $H^p\left(\mathcal{K}^M_p\left(\mathcal{O}_X\right)\right) \cong CH^p(X)$ is also given.

In Sect. 3.1 we give the rudiments of higher order Kodaira–Spencer theory in a form that is amenable to the calculations we need to make. (We make no attempt to provide a systematic general theory as is done in [12] and [8].) In particular the higher order analogues of (1.2.7) and (1.2.8) are given.

Finally in Sect. 3.2 we turn to the higher order deformation for the groups $H^p\left(\mathcal{K}^M_p(\mathcal{O}_X)\right)$. The analogue of (1.2.9) is the diagram

$$\begin{array}{ccccccccc} 0 & \longrightarrow & \Omega^{p-1}_{X/\mathbb{Q}} & \longrightarrow & \mathcal{K}^M_p(\mathcal{O}_{\mathcal{X}_m}) & \longrightarrow & \mathcal{K}^M_p(\mathcal{O}_{\mathcal{X}_{m-1}}) & \longrightarrow & 0 \\ & & \downarrow \eta & & \downarrow \eta & & \downarrow \eta & & \\ 0 & \longrightarrow & \Omega^{p-1}_{X/\mathbb{Q}} \oplus \Omega^p_{X/\mathbb{Q}} & \longrightarrow & \Omega^p_{\mathcal{X}_m/\mathbb{Q}} & \longrightarrow & \Omega^p_{\mathcal{X}_{m-1}/\mathbb{Q}} & \longrightarrow & 0 \end{array}$$

$$m \geqq 1 \tag{2.11}$$

where the left hand map in the top row is the same as (1.2.10) with t^m replacing t; the exactness here requires a somewhat intricate lemma on ϵ-Steinberg symbols (cf. (3.2.3)). With this formalism in place we give the proofs of the main applications (1.1.5)–(1.1.7) and (1.1.9).

In Sect. 4 we discuss failures of the formal theory; i.e., examples where all the obstructions to formally lifting vanish but where *no* choice of successive extensions will lead to a convergent, or geometric, result. This phenomenon was encountered in an earlier work [10], and it may have to do with theory of G-functions (cf. [1]). We feel that its understanding presents a significant issue.

2 First Order Obstruction Theory

2.1 Variants of Kodaira–Spencer theory. In this section, we collect together some classical results of deformation theory. Given a family $\mathcal{X} \xrightarrow{p} B$, the basic object in classical Kodaira–Spencer theory is the exact sequence

$$0 \longrightarrow \Theta_{\mathcal{X}/B} \longrightarrow \Theta_{\mathcal{X}} \longrightarrow p^*\Theta_B \longrightarrow 0 .$$

Letting t denote a local uniformizing parameter on B with $p^{-1}(t_0) = X$, we may rewrite this sequence as

$$0 \longrightarrow \Theta_{\mathcal{X}/B} \longrightarrow \Theta_{\mathcal{X}} \longrightarrow \mathcal{O}_{\mathcal{X}} \otimes \partial/\partial t \longrightarrow 0 .$$

Reducing this sequence modulo t^{m+1} and setting

$$\Theta_m = \Theta_{\mathcal{X}/B}/\left(t^{m+1}\right)$$

we obtain

$$0 \longrightarrow \Theta_m \longrightarrow \Theta_{\mathcal{X}_m} \longrightarrow \mathcal{O}_{\mathcal{X}_m} \otimes \partial/\partial t \longrightarrow 0 . \tag{1.1}$$

The 1st order classical Kodaira–Spencer theory is derived from the $m = 0$ case of this sequence, which setting $\Theta_0 = \Theta_X$ we write as

$$0 \longrightarrow \Theta_X \longrightarrow \Theta_{\mathcal{X}_1} \longrightarrow \mathcal{O}_X \otimes \partial/\partial t \longrightarrow 0 . \tag{1.2}$$

Denoting by δ the coboundary map in exact cohomology sequences we have the classical

Definition. *The (1st order) Kodaira–Spencer class associated to $\mathcal{X} \longrightarrow B$ is given by*

$$\theta_1 = \delta\,(\partial/\partial t) \in H^1(\Theta_X)\;.$$

Geometrically, we think of θ_1 as the obstruction to lifting $\partial/\partial t$ to the fibre over $t = 0$ in the family $\mathcal{X} \longrightarrow B$. This is clearly the obstruction to trivializing $\mathcal{X} \longrightarrow B$ to 1st order, i.e.,

$$\theta_1 = 0 \Leftrightarrow \mathcal{X}_1 \cong X \times \operatorname{Spec}\left(\mathbb{C}[t]/t^2\right)\;.$$

Equivalently, θ_1 is the obstruction to splitting the sequence given by (2.1.2), this obstruction being the same as the extension class in $H^1(\operatorname{Hom}(\mathcal{O}_X, \Theta)) \otimes \partial/\partial t \cong H^1(\Theta_X)$ of (2.1.2).

When we study the geometry of $\mathcal{X} \longrightarrow B$ inductively over the successive infinitesimal neighborhoods $\mathcal{X}_k$, especially in the study of algebraic cycles and cohomology, there are a number of related sequences to (2.1.1). We shall now discuss their 1st order versions.

On the 1st order neighborhood $\mathcal{X}_1$, we have $t^2 = 0$ and the three basic sequences for our study are

$$\begin{cases} \text{(i)}\;\; 0 \longrightarrow \mathcal{O}_X \longrightarrow \mathcal{O}_{\mathcal{X}_1} \longrightarrow \mathcal{O}_X \longrightarrow 0 \\ \text{(ii)}_1\; 0 \longrightarrow \mathcal{O}_X \oplus \Omega^1_{X/\mathbb{C}} \longrightarrow \Omega^1_{\mathcal{X}_1/\mathbb{C}} \longrightarrow \Omega^1_{X/\mathbb{C}} \longrightarrow 0 \\ \text{(iii)}\; 0 \longrightarrow \Theta_X \longrightarrow \Theta_{\mathcal{X}_1} \longrightarrow p^*\Theta_B \longrightarrow 0\;. \end{cases} \tag{1.3}$$

Here, the right arrows are restriction mappings and the left arrows are given respectively by

$$\begin{cases} \text{(i)} \qquad f \longrightarrow tf \\ \text{(ii)}\;\; f \oplus \varphi \longrightarrow fdt + t\varphi \\ \text{(iii)} \qquad \theta \longrightarrow i_*\theta, \qquad \text{where } i : X \to \mathcal{X}\;. \end{cases}$$

More precisely, under (i) we have to give for each $x \in X$ a map

$$\mathcal{O}_{X,x} \xrightarrow{\;t\;} \mathcal{O}_{\mathcal{X}_1,x}\;.$$

For $f \in \mathcal{O}_{X,x}$ we choose any $\tilde{f} \in \mathcal{O}_{\mathcal{X}_1,x}$ with $\tilde{f}|_X = f$ and then the map is

$$f \longrightarrow t\tilde{f}\;.$$

Since $t^2 = 0$, the choice of lifting $\tilde{f}$ does not matter. Similar remarks apply to (ii) and (iii). We may write

$$\begin{cases} \text{(i)}\;\; = t \\ \text{(ii)}\;\; = dt \oplus t \\ \text{(iii)} = t\;. \end{cases}$$

The extension classes of the sequences (2.1.3) are of interest. By the remarks above we expect that all of these can be derived from the 1st order Kodaira–Spencer class $\theta_1 \in H^1(\Theta_X)$, and we shall now explain how this goes. For this we note that there are pairings

$$\begin{cases} \Theta_X \otimes_{\mathbb{C}} \mathcal{O}_X & \longrightarrow \mathcal{O}_X \\ \Theta_X \otimes_{\mathbb{C}} \Omega^1_{X/\mathbb{C}} & \longrightarrow \mathcal{O}_X \oplus \Omega^1_{X/\mathbb{C}} \\ \Theta_X \otimes_{\mathbb{C}} \Theta_X & \longrightarrow \Theta_X \end{cases} \tag{1.4}$$

given respectively by

$$\begin{cases} \theta \otimes f \longrightarrow \mathcal{L}_\theta f & (= \theta \cdot f) \\ \theta \otimes \omega \longrightarrow \theta \rfloor \omega \oplus \mathcal{L}_\theta \omega & \\ \theta \otimes \sigma \longrightarrow \mathcal{L}_\theta \sigma & (= [\theta, \sigma]) \end{cases}$$

where $\mathcal{L}$ denotes Lie derivative. Of these maps only $\theta \otimes \omega \longrightarrow \theta \rfloor \omega$ is $\mathcal{O}_X$-linear; the remainder are $\mathbb{C}$-linear derivations.

Proposition 1.5. *The extension classes associated to the exact sequences (2.1.3) are respectively*

$$\begin{cases} \text{(i)} & \mathcal{L}_{\theta_1} \in H^1\left(\mathrm{Hom}_{\mathbb{C}}\left(\mathcal{O}_X, \mathcal{O}_X\right)\right) \\ \text{(ii)} & \theta_1 \rfloor \oplus \mathcal{L}_{\theta_1} \in H^1\left(\mathrm{Hom}_{\mathcal{O}_X}\left(\Omega^1_{X/\mathbb{C}}, \mathcal{O}_X\right)\right) \oplus H^1\left(\mathrm{Hom}_{\mathbb{C}}\left(\Omega^1_{X/\mathbb{C}}, \Omega^1_{X/\mathbb{C}}\right)\right) \\ \text{(iii)} & \mathcal{L}_{\theta_1} \in H^1\left(\mathrm{Hom}_{\mathbb{C}}\left(\Theta_{X_1/B}, \Theta_X\right)\right) . \end{cases}$$

Proof. We shall prove the most interesting case (ii); a similar argument will apply to the other cases. We shall also work in the analytic topology; the algebraic case may be established by a similar argument, or alternatively the result may be deduced from GAGA. Finally, we shall carry out the computations when $\dim X = 1$; the extension to the general case is only notationally more complicated.

We may cover $\mathcal{X}$ by open sets U_α in which there are box coordinates (z_α, t). In the overlaps $U_{\alpha\beta} =: U_\alpha \cap U_\beta$ we will have

$$z_\alpha = f_{\alpha\beta}(z_\beta, t) = f_{\alpha\beta}(z_\beta) + f_{\alpha\beta 1}(z_\beta)t + f_{\alpha\beta 2}(t_\beta)\frac{t^2}{2} + \cdots . \tag{1.6}$$

We first claim that the Kodaira–Spencer class is represented by the cocycle

$$\theta_{1\alpha\beta} = f_{\alpha\beta 1}(z_\beta)\partial/\partial z_\alpha \in H^1\left(\{U_\alpha\}, \Theta_X\right) . \tag{1.7}$$

To see this, we let $(\partial/\partial t)_\alpha$ be the lift of $\partial/\partial t$ in the (z_α, t) coordinate system. Then in $U_{\alpha\beta}$

$$\begin{cases} \left((\partial/\partial t)_\beta - (\partial/\partial t)_\alpha\right) \cdot t = 0 \\ \left((\partial/\partial t)_\beta - (\partial/\partial t)_\alpha\right) \cdot z_\alpha = \frac{\partial f_{\alpha\beta}(z_\beta, t)}{\partial t}\Big|_{t=0} = f_{\alpha\beta 1}(z_\beta) , \end{cases}$$

which is what is wanted.

We now let e_α be the product extension of a function or a differential form from $U_\alpha \cap X$ to U_α. Then by (2.1.6) we have

$$(e_\beta - e_\alpha)dz_\alpha = \partial_{z_\beta} f_{\alpha\beta}(z_\beta)dz_\beta + f_{\alpha\beta 1}(z_\beta)dt + \partial_{z_\beta} f_{\alpha\beta 1}(z_\beta)t \mod t^2, tdt\ .$$

The first term is dz_α expressed in the z_β-coordinate system. The second term is, by (2.1.7),

$$\theta_{\alpha\beta 1} \rfloor dz_\alpha\ ,$$

and using the Lie derivative formula the third term is the coefficient of dz_β in

$$\mathcal{L}_{\theta_{\alpha\beta 1}} dz_\alpha$$

expressed in the z_β-coordinate system. □

From (2.1.3) (ii)$_1$ we infer the exact sequence

$$\text{(ii)}_p \qquad 0 \longrightarrow \Omega^{p-1}_{X/\mathbb{C}} \oplus \Omega^p_{X/\mathbb{C}} \longrightarrow \Omega^p_{\mathcal{X}_1/\mathbb{C}} \longrightarrow \Omega^p_{X/\mathbb{C}} \longrightarrow 0 \qquad (2.1.3)$$

whose extension class is

$$\theta_1 \rfloor \oplus \mathcal{L}_{\theta_1}\ ,$$

where the first term is induced from the pairing

$$\Theta_X \otimes \Omega^p_{X/\mathbb{C}} \longrightarrow \Omega^{p-1}_{X/\mathbb{C}}\ .$$

It is well-known and elementary that the coboundary maps in the long exact cohomology sequences associated to short exact sheaf sequences are given by cup-products with the extension classes.

Proposition 1.8. *The coboundary maps in the long exact cohomology sequences arising from (2.1.3)(i) and (ii)$_p$ corresponding to $\mathcal{L}_{\theta_1}$ are zero. Thus the only non-zero coboundary maps are*

$$H^q\left(\Omega^p_{X/\mathbb{C}}\right) \xrightarrow{\theta_1\rfloor} H^{q+1}\left(\Omega^{p-1}_{X/\mathbb{C}}\right)\ .$$

Proof. Using the Lie derivative formula

$$\mathcal{L}_\theta \omega = d(\theta \rfloor \omega) + \theta \rfloor d\omega\ ,$$

the proposition follows from the Kähler manifold fact that the mapping

$$H^q\left(\Omega^p_{X/\mathbb{C}}\right) \xrightarrow{d} H^q\left(\Omega^{p+1}_{X/\mathbb{C}}\right)$$

induced by the exterior derivative is zero. Indeed, there is a commutative diagram

$$\begin{array}{ccc} H^q\left(\Omega^p_{X/\mathbb{C}}\right) & \xrightarrow{d} & H^q\left(\Omega^{p+1}_{X/\mathbb{C}}\right) \\ \wr\| & & \wr\| \\ H^{p,q}_{\bar\partial}(X) & \xrightarrow{\partial} & H^{p+1,q}_{\bar\partial}(X) \end{array}$$

(since $d = \partial$ on $\Omega^p_{X/\mathbb{C}}$), and it is a Kähler fact that every class in $H^{p,q}_{\bar\partial}(X)$ is represented by a ∂-closed form (e.g., take the harmonic representative for a given Kähler metric). □

Remark. This result has the following elementary geometric meaning: Given $\omega \in H^q\left(\Omega^p_{X/\mathbb{C}}\right)$,

$$\mathcal{L}_{\theta_1}\omega \in H^{q+1}\left(\Omega^p_{X/\mathbb{C}}\right)$$

is the 1st obstruction to finding a family $\omega_t \in H^q\left(\Omega^p_{X_t/\mathbb{C}}\right)$ with $\omega_0 = \omega$. But from

$$H^r(X_t, \mathbb{C}) \cong \bigoplus_{p+q=r} H^q\left(\Omega^p_{X_t/\mathbb{C}}\right)$$

it follows that $\dim H^q\left(\Omega^p_{X_t/\mathbb{C}}\right)$ is constant and hence ω extends to a family ω_t.

The 1st obstruction to extending $\theta_1 \in H^1(\Theta_X)$ to $\theta_{1,t} \in H^1(\Theta_{X_t})$ is given by

$$\mathcal{L}_{\theta_1}(\theta_1) = [\theta_1, \theta_1]$$

which is also zero, although the map

$$H^q(\Theta_X) \xrightarrow{\mathcal{L}_{\theta_1}} H^{q+1}(\Theta_X)$$

is in general non-zero.

Remark. A result of Herb Clemens (cf. [5]) states that for $\theta \in H^1(\Theta_X)$ the map

$$[\theta, \theta] : H^q\left(\Omega^p_{X/\mathbb{C}}\right) \xrightarrow{\rfloor} H^{q+2}\left(\Omega^{p-1}_{X/\mathbb{C}}\right)$$

is zero. More generally, for $\theta_i \in H^{r_i}(\Theta_X)$, $i = 1, 2$, the map

$$[\theta_1, \theta_2] : H^q\left(\Omega^p_{X/\mathbb{C}}\right) \longrightarrow H^{q+r_1+r_2}\left(\Omega^{p-1}_{X/\mathbb{C}}\right)$$

is zero. This follows from the standard differential-geometric formula

$$\left(\theta_1 \rfloor \mathcal{L}_{\theta_2} - \mathcal{L}_{\theta_2}(\theta_1 \rfloor)\right)\omega = [\theta_1, \theta_2] \rfloor \omega$$

together with

$$\mathcal{L}_{\theta_i} : H^q\left(\Omega^p_{X/\mathbb{C}}\right) \longrightarrow H^q\left(\Omega^p_{X/\mathbb{C}}\right)$$

being the zero map as a consequence of

$$\mathcal{L}_{\theta_i}\omega = \theta_1 \rfloor d\omega + d(\theta_i \rfloor \omega)$$

as above.

Corollary. *If X is a Calabi–Yau manifold then for* $\theta_i \in H^{r_1}(\Theta_X)$,

$$[\theta_1, \theta_2] = 0\,.$$

Proof. This follows from

$$H^{r_1+r_2}(\Theta_X) \cong H^{r_1+r_2}\left(\Omega^{n-1}_{X/\mathbb{C}}\right)$$

and the commutativity of the diagram

$$\begin{array}{ccc} H^r(\Theta_X) \otimes H^{n-r}\left(\Omega^1_{X/\mathbb{C}}\right) & \longrightarrow & H^n(\mathcal{O}_X) \\ \wr\| & & \wr\| \\ H^r\left(\Omega^{n-1}_{X/\mathbb{C}}\right) \otimes H^{n-r}\left(\Omega^1_{X/\mathbb{C}}\right) & \longrightarrow & H^n\left(\Omega^n_{X/\mathbb{C}}\right)\,, \end{array}$$

using that the pairing in the bottom row is non-degenerate. □

Corollary (Bogomolov). *Calabi–Yau varieties are unobstructed.*

Proof. This is a well-known consequence of the fact that

$$H^1(\Theta_X) \otimes H^1(\Theta)_X \xrightarrow{[\ ,\ ]} H^2(\Theta_X)$$

is zero. □

Finally we want to extend the above discussion to absolute differentials. Here we assume that $\mathcal{X} \longrightarrow B$ is algebraic and we work in the algebraic setting – Zariski topology, sheaf of regular (in the algebraic sense) functions, etc. We will discuss

$$(\text{iv})_1 \qquad 0 \longrightarrow \mathcal{O}_X \oplus \Omega^1_{X/\mathbb{Q}} \longrightarrow \Omega^1_{\mathcal{X}_1/\mathbb{Q}} \longrightarrow \Omega^1_{X/\mathbb{Q}} \longrightarrow 0 \qquad (2.1.3)$$

and the sequences derived from it by exterior algebra. Here, for any scheme Y we denote by $\Omega_{Y/\mathbb{Q}}$ the sheaf of absolute Kähler differentials. From (2.1.3)(iv)$_1$ we deduce a diagram

$$
\begin{array}{ccccccccc}
 & & 0 & & 0 & & 0 & & \\
 & & \downarrow & & \downarrow & & \downarrow & & \\
0 & \longrightarrow & \Omega^1_{\mathbb{C}/\mathbb{Q}} \otimes \mathcal{O}_X & \longrightarrow & \Omega^1_{\mathbb{C}/\mathbb{Q}} \otimes \mathcal{O}_{\mathcal{X}_1} & \longrightarrow & \Omega^1_{\mathbb{C}/\mathbb{Q}} \otimes \mathcal{O}_X & \longrightarrow & 0 \\
 & & \downarrow & & \downarrow & & \downarrow & & \\
0 & \longrightarrow & \mathcal{O}_X \oplus \Omega^1_{X/\mathbb{Q}} & \longrightarrow & \Omega^1_{\mathcal{X}_1/\mathbb{Q}} & \longrightarrow & \Omega^1_{X/\mathbb{Q}} & \longrightarrow & 0 \\
 & & \downarrow & & \downarrow & & \downarrow & & \\
0 & \longrightarrow & \mathcal{O}_X \oplus \Omega^1_{X/\mathbb{C}} & \longrightarrow & \Omega^1_{\mathcal{X}_1/\mathbb{C}} & \longrightarrow & \Omega^1_{X/\mathbb{C}} & \longrightarrow & 0 \\
 & & \downarrow & & \downarrow & & \downarrow & & \\
 & & 0 & & 0 & & 0 & &
\end{array}
\tag{1.9}
$$

The left hand column is the same as the right hand column direct summed with $\mathcal{O}_X$ in the $\Omega^1_{X/\mathbb{Q}}$ and $\Omega^1_{X/\mathbb{C}}$ terms and with $\mathrm{id}_{\mathcal{O}_X}$ being the corresponding map.

The commutativity of the diagram (2.1.9) results from the following considerations. We let k be a field of definition of X and choose a smooth variety S defined over $\mathbb{Q}$ and with

$$k \cong \mathbb{Q}(S) .$$

We then denote by

$$\mathcal{S}(X) \xrightarrow{\pi} S$$

the *k-spread* of X.

If $s_0 \in S$ is a $\mathbb{Q}$-generic point of S, i.e., s_0 does not lie on any proper subvariety of S defined over $\mathbb{Q}$, then

$$\pi^{-1}(s_0) \cong X$$

and under this identification,

$$\Omega^1_{X(k)/\mathbb{Q}} \cong \Omega^1_{\mathcal{S}(X)(\mathbb{Q})/\mathbb{Q}}\big|_{\pi^{-1}(s_0)} .$$

We also note that

$$\Omega^1_{k/\mathbb{Q}} \cong \Omega^1_{S(\mathbb{Q})/\mathbb{Q}}\big|_{s_0} .$$

If $\mathcal{X} \to B$ is defined over k, then for $t \in B(k)$ we have the k-spread

$$\mathcal{S}(X_t) \to S$$

of X_t. Passing to complex points, the family $\{\mathcal{S}(X_t)\}_{t\in B}$ gives

$$
\begin{array}{ccc}
\mathcal{S}(\mathcal{X}) & \longrightarrow & B \times S \\
\cup & & \Psi \\
X & \longrightarrow & (t_0, s_0) .
\end{array}
$$

Replacing $\Omega^1_{\mathbb{C}/\mathbb{Q}}$ by $\Omega^1_{k/\mathbb{Q}}$ in the diagram (2.1.9) loses no geometric information. If

$$\mathcal{X} \xrightarrow{p} B$$

is a family defined over k, then we get

$$\begin{array}{ccc} \mathcal{S}(\mathcal{X}) & \longrightarrow & \mathcal{S}(B) \\ & \searrow \;\; \swarrow & \\ & S\,. & \end{array}$$

We may identify

$$\Omega^1_{\mathcal{X}_1(k)/\mathbb{Q}} \cong \Omega^1_{\mathcal{S}(\mathcal{X})_1(\mathbb{Q})/\mathbb{Q}}\big|_{\pi^{-1}(s_0)}\,.$$

The natural exact sequence

$$0 \to \Omega^1_{k/\mathbb{Q}} \otimes \mathcal{O}_{\mathcal{X}_1(k)} \to \Omega^1_{\mathcal{X}_1(k)/\mathbb{Q}} \to \Omega^1_{\mathcal{X}_1(k)/k} \to 0$$

becomes

$$0 \to \Omega^1_{S(\mathbb{Q})/\mathbb{Q}} \otimes \mathcal{O}_{\mathcal{S}(\mathcal{X})_1(\mathbb{Q})/\mathbb{Q}}\big|_{s_0} \to \Omega^1_{\mathcal{S}(\mathcal{X})_1(\mathbb{Q})/\mathbb{Q}}\big|_{s_0} \to \Omega^1_{\mathcal{S}(\mathcal{X})_1/S}\big|_{s_0} \to 0\,.$$

so that by base change the term in the middle in (2.1.9) may be thought of as the complexification of the 1-forms in a 1st order neighborhood of X in $\mathcal{S}(\mathcal{X})$ and the horizontal and vertical arrows reflect the sequences of the type (2.1.3)(ii)$_1$ in the B and S directions respectively.

Let $\sigma \in H^1(\Theta_{X(k)}) \otimes \Omega^1_{k/\mathbb{Q}}$ denote the 1st order Kodaira–Spencer class of $\mathcal{S}(X) \to S$, where we identify $T^*_{s_0}S \cong \Omega^1_{k/\mathbb{Q}}$. Then σ gives the extension class of

$$0 \to \Omega^1_{k/\mathbb{Q}} \otimes \mathcal{O}_{X(k)} \to \Omega^1_{\mathcal{S}(X)(k)_1/\mathbb{Q}} \to \Omega^1_{X(k)/k} \to 0\,,$$

and by the process described in Proposition 2.1.5 it induces the extension classes of the vertical sequences in (2.1.9). Moreover we have

$$[\theta_1, \sigma] = 0\,,$$

and this implies the commutativity of (2.1.9).

For $\theta \in \Theta_{X(k)}$ the Lie derivative

$$\mathcal{L}_\theta : \Omega^1_{X/\mathbb{Q}} \to \Omega^1_{X/\mathbb{Q}}$$

may be defined by

$$\mathcal{L}_\theta(\omega) = \theta \rfloor d\omega + d(\theta \rfloor \omega)$$

where $d = d_{X/\mathbb{Q}}$. We will see that:

The induced map

$$\mathcal{L}_{\theta_1} : H^q\left(\Omega^1_{X/\mathbb{Q}}\right) \to H^{q+1}\left(\Omega^1_{X/\mathbb{Q}}\right)$$

is zero.

For general p, the sheaves $\Omega^p_{Y/\mathbb{Q}}$ are filtered by

$$F^m\Omega^p_{Y/\mathbb{Q}} = \text{image}\left\{\Omega^m_{\mathbb{C}/\mathbb{Q}} \otimes \Omega^{p-m}_{Y/\mathbb{Q}} \to \Omega^p_{Y/\mathbb{Q}}\right\}. \tag{1.10}$$

There is a spectral sequence which abuts to $H^q(\Omega^p_{Y/\mathbb{Q}})$ and whose E_1-term is $\Omega^m_{\mathbb{C}/\mathbb{Q}} \otimes H^q(\Omega^{p-m}_{Y/\mathbb{C}})$. For $Y = X$ the maps $\mathcal{L}_{\theta_1}$ are zero on these groups and hence vanish on $H^q(\Omega^p_{X/\mathbb{Q}})$. Thus we have the

Proposition 1.11. *The coboundary maps in the exact cohomology sequence of*

$$0 \to \Omega^{p-1}_{X/\mathbb{Q}} \oplus \Omega^p_{X/\mathbb{Q}} \to \Omega^p_{X_1/\mathbb{Q}} \to \Omega^p_{X/\mathbb{Q}} \to 0$$

are $\delta = \theta_1 \rfloor \oplus \mathcal{L}_{\theta_1}$.

Theorem. $H^q(\Omega^p_{X/\mathbb{Q}}) \xrightarrow{\mathcal{L}_{\theta_1}} H^{q+1}\left(\Omega^p_{X/\mathbb{Q}}\right)$ *is* 0.

The proof proceeds by a number of steps.

(a) $\Omega^p_{X(k)/\mathbb{Q}} \cong \Omega^p_{\mathcal{S}(X)/\mathbb{Q}}\big|_{s_0}$ where s_0 is a $\mathbb{Q}$-generic point of S.
(b) Thus we need to show

$$H^q\left(\Omega^p_{\mathcal{S}(X)(\mathbb{Q})/\mathbb{Q}}\big|_{s_0}\right) \xrightarrow{\mathcal{L}_{\theta_1}} H^{q+1}\left(\Omega^p_{\mathcal{S}(X)(\mathbb{Q})/\mathbb{Q}}\big|_{s_0}\right)$$

is zero.
(c)

$$H^q\left(\Omega^p_{\mathcal{S}(X)(\mathbb{Q})/\mathbb{Q}}\big|_{s_0}\right) \cong \lim_{\substack{U \text{ Zar open} \\ \text{def}/\mathbb{Q}\subseteq S}} H^q\left(U, \Omega^p_{\mathcal{S}(X)(\mathbb{Q})/\mathbb{Q}}\right)(\log D)$$

where for U Zariski open, def$/\mathbb{Q}$ in S, write $U = X - D$, D divisor with normal crossings.
(d) The map

$$H^q\left(U, \Omega^p_{\mathcal{S}(X)(\mathbb{Q})/\mathbb{Q}}(\log D)\right) \xrightarrow{\mathcal{L}_{\theta_1}} H^{q+1}\left(U, \Omega^p_{\mathcal{S}(X)(\mathbb{Q})/\mathbb{Q}}(\log D)\right)$$

is zero.

Proof (of (d)). Since $\mathcal{L}_{\theta_1} = (\rfloor\theta_1) \circ d - d \circ (\rfloor\theta_1)$, we just need that

$$H^q\left(U, \Omega^p_{\mathcal{S}(X)(\mathbb{Q})/\mathbb{Q}}(\log D)\right) \xrightarrow{d} H^q\left(0, \Omega^{p+1}_{\mathcal{S}(X)(\mathbb{Q})/\mathbb{Q}}(\log D)\right)$$

is zero. But this follows by base change from showing

$$H^q\left(U, \Omega^p_{\mathcal{S}(X)(\mathbb{C})/\mathbb{C}}(\log D)\right) \xrightarrow{d} H^1\left(U, \Omega^{p+1}_{\mathcal{S}(X)(\mathbb{C})/\mathbb{C}}(\log D)\right)$$

is zero. This follows from degeneracy of the Hodge–De Rham spectral sequence for logarithmic differentials.

To conclude this section we shall give for later use a technical result which, however, has interesting geometric content. For this we consider the diagram

$$\begin{array}{ccccccccc} & & 0 & & 0 & & 0 & & \\ & & \downarrow & & \downarrow & & \downarrow & & \\ 0 & \longrightarrow & F^1\Omega^{p-1}_{X/\mathbb{Q}} \oplus F^1\Omega^p_{X/\mathbb{Q}} & \longrightarrow & F^1\Omega^p_{\mathcal{X}_1/\mathbb{Q}} & \longrightarrow & F^1\Omega^p_{X/\mathbb{Q}} & \longrightarrow & 0 \\ & & \downarrow & & \downarrow & & \downarrow & & \\ 0 & \longrightarrow & \Omega^{p-1}_{X/\mathbb{Q}} \oplus \Omega^p_{X/\mathbb{Q}} & \longrightarrow & \Omega^p_{\mathcal{X}_1/\mathbb{Q}} & \longrightarrow & \Omega^p_{X/\mathbb{Q}} & \longrightarrow & 0 \\ & & \downarrow & & \downarrow & & \downarrow & & \\ 0 & \longrightarrow & \Omega^{p-1}_{X/\mathbb{C}} \oplus \Omega^p_{X/\mathbb{C}} & \longrightarrow & \Omega^p_{\mathcal{X}_1/\mathbb{C}} & \longrightarrow & \Omega^p_{X/\mathbb{C}} & \longrightarrow & 0 \\ & & \downarrow & & \downarrow & & \downarrow & & \\ & & 0 & & 0 & & 0 & & \end{array}$$

A piece of the cohomology diagram is, using Proposition 2.1.11, given by

$$\begin{array}{ccccc} & & 0 & & \\ & & \downarrow & & \\ & & F^1H^q\left(\Omega^p_{X/\mathbb{Q}}\right) & & \\ & & \downarrow & & \\ H^q\left(\Omega^p_{\mathcal{X}_1/\mathbb{Q}}\right) & \longrightarrow & H^q\left(\Omega^p_{X/\mathbb{Q}}\right) & \xrightarrow{\theta_1\rfloor} & H^{q+1}\left(\Omega^{p-1}_{X/\mathbb{Q}}\right) \\ & & \downarrow & & \downarrow \\ & & H^q\left(\Omega^p_{X/\mathbb{C}}\right) & \xrightarrow{\theta_1\rfloor} & H^{q+1}\left(\Omega^{p-1}_{X/\mathbb{C}}\right) \end{array} \quad . \tag{1.12}$$

The replacement of $H^q(F^1\Omega^p_{X/\mathbb{Q}})$ by $F^1H^q(\Omega^p_{X/\mathbb{Q}})$ will be explained below.

Proposition 1.13. *Let $\eta \in H^q(\Omega^p_{X/\mathbb{Q}})$ have image denoted by $\eta_{\mathbb{C}} \in H^q(\Omega^p_{X/\mathbb{C}})$. Suppose that*

$$\theta_1\rfloor\eta_{\mathbb{C}} = 0$$

in $H^{q+1}(\Omega^{p-1}_{X/\mathbb{C}})$. Then we can find $\eta' \in F^1H^q(\Omega^p_{X/\mathbb{Q}})$ such that

$$\theta_1\rfloor(\eta + \eta') = 0$$

in $H^{q+1}(\Omega^{p-1}_{X/\mathbb{C}})$.

Corollary 1.14. *Under the assumptions of the proposition, we may modify η by $\eta' \in F^1H^q(\Omega^p_{X/\mathbb{C}})$ so that $\eta + \eta'$ lifts to $H^q(\Omega^p_{\mathcal{X}_1/\mathbb{Q}})$.*

Geometrically, $\eta_{\mathbb{C}} \in H^q\left(\Omega^p_{X/\mathbb{C}}\right) = H^{p,q}(X) \subset H^{p+q}(X,\mathbb{C})$. The assumption in the proposition is that, to 1st order, $\eta_{\mathbb{C}}$ remains of Hodge type (p,q) when we identify all the $H^{p+q}(X_t,\mathbb{C})$ with $H^{p,q}(X,\mathbb{C})$ for $|t-t_0|<\epsilon$ (Gauss–Manin connection). This implies that $\eta_{\mathbb{C}}$ lifts to $H^q(\Omega^p_{\mathfrak{X}_1/\mathbb{C}})$. What the proposition states is that we may modify the absolute class $\eta \in H^q(\Omega^p_{X/\mathbb{Q}})$ by η' keeping the same $(\eta+\eta')_{\mathbb{C}} = \eta_{\mathbb{C}}$ so that $\eta+\eta'$ lifts to $H^q(\Omega^p_{\mathfrak{X}_1/\mathbb{Q}})$.

Proof (of (2.1.13)). We first recall the relative de Rham formulation of the degeneracy at E_2 of the Leray spectral sequence of a proper smooth mapping

$$W \xrightarrow{f} S$$

of quasi-projective varieties, where for our application we may take S to be affine. Now $\Omega^\bullet_{W/\mathbb{C}}$ is filtered by

$$F^m\Omega^\bullet_{W/\mathbb{C}} = \text{image}\left\{f^*\left(\Omega^m_{S/\mathbb{C}}\right)\otimes\Omega^{\bullet-m}_{W/\mathbb{C}} \to \Omega^\bullet_{W/\mathbb{C}}\right\}$$

with

$$Gr^m\Omega^\bullet_{W/\mathbb{C}} \cong \Omega^m_{S/\mathbb{C}}\otimes\Omega^{\bullet-m}_{W/S}$$

(we drop the f^*). Since S is assumed to be affine the Leray spectral sequence has

$$E_1^{p,q} = H^0\left(\Omega^p_{S/\mathbb{C}}\right)\otimes\mathbb{H}^q\left(\Omega^\bullet_{W/S}\right)$$

and

$$d_1 = \nabla_{W/S}$$

is the Gauss–Manin connection. By the theorem of Blanchard–Deligne this spectral degenerates at E_2. The cohomology group

$$H^q(W) \cong \mathbb{H}^q\left(\Omega^\bullet_{W/\mathbb{C}}\right)$$

has a natural filtration with

$$Gr^pH^q(W) \cong \left(H^0\left(\Omega^p_{S/\mathbb{C}}\right)\otimes\mathbb{H}^{p+q}\left(\Omega^\bullet_{W/S}\right)\right)_{\nabla_{W/S}}$$

where $(\quad)_{\nabla_{W/S}}$ denotes the cohomology of the sequence

$$\cdots \to H^0\left(\Omega^p_{S/\mathbb{C}}\right)\otimes\mathbb{H}^q\left(\Omega^\bullet_{W/S}\right) \xrightarrow{\nabla_{W/S}} H^0\left(\Omega^{p+1}_{S/\mathbb{C}}\right)\otimes\mathbb{H}^q\left(\Omega^\bullet_{W/S}\right) \to \cdots .$$

Suppose now that we have

$$\mathcal{W} \to S\times T$$

which we think of as a family of families

$$W_t \to S, \quad t \in T \text{ and } W_0 = W$$

to each of which the above considerations apply. We then have

$$\cdots \to \Omega^p_{T/\mathbb{C}} \otimes \mathbb{H}^q\left(\Omega^\bullet_{\mathcal{W}/T}\right) \xrightarrow{\nabla_{\mathcal{W}/T}} \Omega^{p+1}_{T/\mathbb{C}} \otimes \mathbb{H}^q\left(\Omega^\bullet_{\mathcal{W}/T}\right) \to \cdots .$$

The issue we want to consider is whether $\nabla_{\mathcal{W}/T}$ induces non-trivial maps

$$\begin{aligned} &\ker\left(Gr^p H^q(W) \xrightarrow{\nabla_{\mathcal{W}/T}} \Omega^q_{T/\mathbb{C}} \otimes Gr^p H^q(W)\right) \\ &\qquad \longrightarrow \operatorname{coker}\left(Gr^{p+1} H^q(W) \xrightarrow{\nabla_{\mathcal{W}/T}} \Omega^q_{T/\mathbb{C}} \otimes Gr^{p+1} H^q(W)\right) \\ &\qquad\qquad \vdots \end{aligned}$$

We may rewrite this as

$$\begin{aligned} &\ker\left\{ \left(H^0\left(\Omega^p_{S/\mathbb{C}}\right) \otimes \mathbb{H}^{q-p}\left(\Omega^\bullet_{W/S}\right)\right)_{\nabla_{W/S}} \right. \\ &\qquad \left. \longrightarrow \Omega^1_{T/\mathbb{C}} \otimes \left(H^0\left(\Omega^p_{S/\mathbb{C}} \otimes \mathbb{H}^{q-p}\left(\Omega^\bullet_{W/S}\right)\right)\right) \right\}_{\nabla_{W/S}} \\ &\qquad \longrightarrow \operatorname{coker}\{\text{similar groups}\} . \end{aligned} \tag{1.15}$$

Now, and this is the point, assuming that T is contractible

$$\mathcal{W} \longrightarrow S \times T$$

has its own Leray spectral sequence which degenerates at E_2, and we claim that *the maps in (2.1.15) are part of d_2 for this spectral sequence.* First, denoting by X a typical fibre of $W \to S$ we have that

$$\mathbb{H}^q\left(\Omega^\bullet_{\mathcal{W}/S\times T}\right) \cong \mathbb{H}^q\left(\Omega^\bullet_{W/S}\right)$$

localizes over $s_0 \in S$ to $H^q(X)$. Now

$$\left(H^0\left(\Omega^p_{S\times T/\mathbb{C}}\right) \otimes H^{q-p}(X)\right)_{\nabla_{\mathcal{W}/S\times T}} \xrightarrow{d_2} \left(H^0\left(\Omega^{p+2}_{S\times T/\mathbb{C}}\right) \otimes H^{q-p+1}(X)\right)_{\nabla_{\mathcal{W}/S\times T}} .$$

The LHS contains

$$\ker\left\{\left(H^0\left(\Omega^p_{S/\mathbb{C}}\right) \otimes H^{q-p}(X)\right)_{\nabla_{\mathcal{W}/S}} \xrightarrow{\nabla_{\mathcal{W}/T}} \Omega^1_{T/\mathbb{C}} \otimes H^0\left(\Omega^p_{S/\mathbb{C}}\right) \otimes H^{q-p}(X)\right\} ,$$

the RHS contains

$$\frac{\Omega^1_{T/\mathbb{C}} \otimes \left(H^0\left(\Omega^{p+2}_{S/\mathbb{C}}\right) \otimes H^{q-p+1}(X)\right)_{\nabla_{W/S}}}{\nabla_{W/T}\left(H^0\left(\Omega^{p+1}_{S/\mathbb{C}}\right) \otimes H^{q-p+1}(X)\right)_{\nabla_{W/S}}}$$

and the d_2 mapping contains (2.1.15).

Finally, we apply these considerations to the situation where

$$\mathcal{W} \to T \times S$$

is given by

$$\begin{array}{ccc} \mathcal{W}_t & = & \mathcal{S}(X_t) \\ \downarrow & & \downarrow \\ S & = & S \end{array}, \qquad t \in T(k)$$

and

$$\begin{array}{ccc} \mathcal{W} \mid T \times \{s_0\} & = & \mathcal{X} \\ \downarrow & & \downarrow \\ T & = & T \end{array}$$

to conclude (2.1.13). □

2.2 First order obstructions for lifting Chow groups

Basic assumption: We are working *modulo torsion*, and for simplicity of notation we use

$$CH^p(X)$$

to denote the usual Chow group modulo torsion. By basic results of Bloch–Quillen and Soulé we have

$$CH^p(X) \cong H^p\left(\mathcal{K}^M_p(\mathcal{O}_X)\right) \tag{2.1}$$

where $\mathcal{K}^M_p(\mathcal{O}_X)$ is the sheaf (in the Zariski topology) given by the pth Milnor K-groups constructed from the local rings $\mathcal{O}_{X,x}$.

The basic idea is to make the identification (2.2.1) and then calculate the obstruction to lifting a class in $H^p(\mathcal{K}^M_p(\mathcal{O}_X))$ over the successive infinitesimal neighborhoods $\mathcal{X}_k$ of X in $\mathcal{X}$. For this the basic sequence following Van der Kallen and Bloch is

$$0 \longrightarrow \Omega^{p-1}_{X/\mathbb{Q}} \longrightarrow \mathcal{K}^M_p\left(\mathcal{O}_{\mathcal{X}_1}\right) \longrightarrow \mathcal{K}^M_p(\mathcal{O}_X) \longrightarrow 0\,. \tag{2.2}$$

The right hand map is given by restriction of Steinberg symbols

$$\{f_1, \dots, f_p\} \to \{f_1\,|_X, \dots, f_p\,|_X\}, \qquad f_i \in \mathcal{O}^*_{\mathcal{X}_1,x}$$

and the left hand map is induced by

$$f\frac{dg_1}{g_1}\wedge\cdots\wedge\frac{dg_{p-1}}{g_{p-1}}\to\{1+t\tilde{f},\tilde{g}_1,\dots\tilde{g}_p\}\,,\quad t^2=0 \tag{2.3}$$

where $f\in\mathcal{O}_{X,x}$, $g_i\in\mathcal{O}^*_{X,x}$ and $\tilde{f},\tilde{g}_1,\dots,\tilde{g}_p$ denote lifts of these functions to $\mathcal{O}_{\mathcal{X}_{1,x}}$. We shall prove later that this mapping is well-defined and that (2.2.2) is exact. Also, it will simplify the notation to write (2.2.3) as

$$f\frac{dg_1}{g_1}\wedge\cdots\wedge\frac{dg_{p-1}}{g_{p-1}}\to\{1+tf,g_1,\dots g_{p-1}\}$$

with the extensions being understood.

To explain the basic observation behind this work, we recall the *arithmetic cycle class* mapping

$$H^p\left(\mathcal{K}^M_p(\mathcal{O}_X)\right)\xrightarrow{\eta}H^p\left(\Omega^p_{X/\mathbb{Q}}\right)$$

induced by

$$\{f_1,\dots,f_p\}\to\frac{df_1}{f_1}\wedge\cdots\wedge\frac{df_p}{f_p},\qquad d=d_{X/\mathbb{Q}} \tag{2.4}$$

(we omit the usual $(2\pi\sqrt{-1})^{-p}$ factor, as it will play no role in this paper and would only serve as a notational distraction).

Proposition 2.5. *The coboundary map δ in the exact cohomology sequence of (2.2.2) is given by*

$$\delta(\xi)=\theta_1\rfloor\eta(\xi)\,,\qquad\xi\in H^p\left(\mathcal{K}^M_p(\mathcal{O}_X)\right)\,.$$

Since $\delta(\xi)$ is the 1st order obstruction to lifting ξ to $\xi_1\in H^p\left(\mathcal{K}^M_p(\mathcal{O}_{\mathcal{X}_1})\right)$, we have the to us remarkable conclusion:

(2.6) *The obstruction to lifting ξ to ξ_1 depends only on the arithmetic cycle class $\eta(\xi)$.*

In our paper [10] we have proposed a set of Hodge-theoretic invariants $\varphi_0(Z)$, $\dots,\varphi_{2p-2}(Z)$ of an algebraic cycle $Z\in Z^p(X)$ which, assuming a conjecture of Bloch–Beilinson, would be a complete set of invariants of the rational equivalence class $[Z]\in CH^p(X)$ modulo torsion. The invariants $\varphi_0(Z),\varphi_1(Z),\varphi_2(Z),\dots,$ $\varphi_{2p-3}(Z)$ determine the arithmetic cycle class $\eta(Z)$, so at first glance it appears that the obstruction to lifting $\xi=[Z]$ only "sees" slightly more than one half of the invariants of ξ. The actual situation is somewhat subtle and will be discussed later.

For the proof of (2.2.5) we consider the diagram

$$\begin{array}{ccccccccc} 0 & \longrightarrow & \Omega^{p-1}_{X/\mathbb{Q}} & \longrightarrow & \mathcal{K}^M_p(\mathcal{O}_{\mathcal{X}_1}) & \longrightarrow & \mathcal{K}^M_p(\mathcal{O}_X) & \longrightarrow & 0 \\ & & \downarrow & & \downarrow & & \downarrow & & \\ 0 & \longrightarrow & \Omega^{p-1}_{X/\mathbb{Q}} \oplus \Omega^p_{X/\mathbb{Q}} & \longrightarrow & \Omega^p_{\mathcal{X}_1/\mathbb{Q}} & \longrightarrow & \Omega^p_{X/\mathbb{Q}} & \longrightarrow & 0\,. \end{array} \tag{2.7}$$

Here, the bottom row is the exact sheaf sequence in Proposition 2.1.11, the first map being

$$\varphi \oplus \omega \mapsto dt \wedge \varphi + t\omega\,. \tag{2.8}$$

The two right hand vertical arrows are $\wedge^p d\log$ as in (2.2.4). The left hand vertical arrow is

$$\varphi \mapsto \varphi \oplus d\varphi\,. \tag{2.9}$$

The commutativity of (2.2.7) follows from (2.2.8) and

$$\begin{aligned} &\wedge^p d\log\{1+tf, g_1, \dots g_{p-2}\} = \\ &dt \wedge f\frac{dg_1}{g_1} \wedge \dots \wedge \frac{dg_{p-1}}{g_{p-1}} + tdf \wedge \frac{dg_1}{g_1} \wedge \dots \wedge \frac{dg_{p-1}}{g_{p-1}}\,. \end{aligned}$$

Passing to cohomology we have

$$\begin{array}{ccccc} H^p\left(\mathcal{K}^M_p(\mathcal{O}_{\mathcal{X}_1})\right) & \longrightarrow & H^p\left(\mathcal{K}^M_p(\mathcal{O}_X)\right) & \xrightarrow{\ \delta\ } & H^{p+1}\left(\Omega^{p-1}_{X/\mathbb{Q}}\right) \\ \downarrow \eta & & \downarrow \eta & & \downarrow \\ H^p\left(\Omega^p_{\mathcal{X}_1/\mathbb{Q}}\right) & \longrightarrow & H^p\left(\Omega^p_{X/\mathbb{Q}}\right) & \xrightarrow{\ \hat{\delta}\ } & H^{p+1}\left(\Omega^{p-1}_{X/\mathbb{Q}}\right) \oplus H^{p+1}\left(\Omega^p_{X/\mathbb{Q}}\right). \end{array} \tag{2.10}$$

Here, the right hand vertical arrow is, by (2.2.10)

$$\varphi \mapsto \varphi \oplus d\varphi\,, \qquad \varphi \in H^{p+1}\left(\Omega^{p-1}_{X/\mathbb{Q}}\right).$$

Since by Proposition 2.1.11

$$\begin{aligned} \hat{\delta} &= \theta_1 \rfloor \oplus \mathcal{L}_{\theta_1} \\ &= \theta_1 \rfloor \end{aligned}$$

since $\mathcal{L}_{\theta_1} = 0$ on cohomology, we have

$$\delta(\xi) = \hat{\delta}\eta(\xi)\,.$$

This proves the proposition. □

Remark. It is to us quite interesting that the $\Omega^p_{X/\mathbb{Q}}$ summand in the bottom row of (2.2.7), which plays no role in the obstruction to lifting a class in $H^p(\Omega^p_{X/\mathbb{Q}})$ to $H^p(\Omega^p_{\mathcal{X}_1/\mathbb{Q}})$, is absent from the top row in (2.2.10). We may say that *the geometric question of lifting a class in $H^p(\mathcal{K}^M_p(\mathcal{O}_X))$ to $H^p(\mathcal{K}^m_p(\mathcal{O}_{\mathcal{X}_1}))$ does not see irrelevant information.*

We will now give the geometric interpretation of the obstruction in Proposition 2.2.5 and of the conclusion (2.2.6) in the first two cases $p = 1$, $p = 2$; the interpretation in the case $p = 2$ will extend to the cases $p \geqq 2$.
$p = 1$. In this case – which is classical – the sequence (2.2.2) reduces to

$$0 \to \mathcal{O}_X \to \mathcal{O}^*_{\mathcal{X}_1} \to \mathcal{O}^*_X \to 0$$

with the first map being

$$f \to 1 + tf\ , \qquad t^2 = 0\ .$$

The basic obstruction diagram (2.2.10) is

$$\begin{array}{ccccc} H^1\left(\mathcal{O}^*_{\mathcal{X}_1}\right) & \longrightarrow & H^1\left(\mathcal{O}^*_X\right) & \longrightarrow & H^2\left(\mathcal{O}_X\right) \\ \downarrow & & \downarrow & & \Vert \\ H^1\left(\Omega^1_{\mathcal{X}_1/\mathbb{Q}}\right) & \longrightarrow & H^1\left(\Omega^1_{X/\mathbb{Q}}\right) & \longrightarrow & H^2\left(\mathcal{O}_X\right)\ . \end{array}$$

To interpret $\theta_1 \rfloor \eta(\xi)$ where $\xi = H^1\left(\mathcal{O}^*_X\right)$, we have

$$\begin{array}{ccc} H^1\left(\mathcal{O}^*_X\right) & \xrightarrow{\ \eta\ } & H^1\left(\Omega^1_{X/\mathbb{Q}}\right) \\ & \overset{\eta_{\mathbb{C}}}{\searrow} & \downarrow \\ & & H^1\left(\Omega^1_{X/\mathbb{C}}\right) \end{array}$$

where the dotted arrow has the interpretation

$$\eta_{\mathbb{C}}(\xi)\ \textit{is the fundamental class}\ \psi_0(\xi) \in H^{1,1}(X).$$

Moreover,

> $\theta_1 \rfloor \eta(\xi) = \theta_1 \rfloor \eta_{\mathbb{C}}(\xi)$ *is the* 1st *order deviation of* $\psi_0(\xi)$ *from remaining of Hodge type* $(1,1)$ *in the family* $\{X_t\}_{t \in B}$.

Geometrically, to lift $\xi \in \operatorname{Pic}(X)$ to $\operatorname{Pic}(\mathcal{X})$, the Chern class of ξ must remain in $Hg^1(X_t)$ for all t, and once this happens ξ lifts since the family $\operatorname{Pic}^0(X_t)$ is unobstructed. The above gives the 1st order calculation that this is the case. Nothing new happens here.

$p = 2$. In this case the obstruction diagram (2.2.10) together with the exact cohomology sequence of

$$0 \to \Omega^1_{\mathbb{C}/\mathbb{Q}} \otimes \mathcal{O}_X \to \Omega^1_{X/\mathbb{Q}} \to \Omega^1_{X/\mathbb{C}} \to 0$$

combine to give a diagram ($\mathcal{K}_2^M = \mathcal{K}_2$ in this case)

$$\begin{array}{ccccc}
 & & & & H^1(\Omega^1_{X/\mathbb{C}}) \\
 & & & & \downarrow \nabla_a \\
 & & & & \Omega^1_{\mathbb{C}/\mathbb{Q}} \otimes H^3(\mathcal{O}_X) \\
 & & & \overset{\mathcal{O}_2}{\nearrow} & \downarrow \\
H^2(\mathcal{K}_2(\mathcal{O}_{X_1})) \to & H^2(\mathcal{K}_2(\mathcal{O}_X)) & \longrightarrow & & H^3(\Omega^1_{X/\mathbb{Q}}) \\
 & & \underset{\mathcal{O}_1}{\searrow} & & \downarrow \\
 & & & & H^3(\Omega^1_{X/\mathbb{C}}).
\end{array} \qquad (2.11)$$

Here, $\mathcal{O}_1(\xi)$ is what we shall call the *primary* 1st *order obstruction* to lifting $\xi \in H^2(\mathcal{K}_2(\mathcal{O}_X))$ to $H^2(\mathcal{K}_2(\mathcal{O}_{X_1}))$. From the discussion above we may infer the interpretation

(2.12) *The primary* 1st *order obstruction is given by*

$$\mathcal{O}_1(\xi) = \theta_1 \rfloor \psi_0(\xi);$$

it represents the 1st *order deviation of the fundamental class* $\psi_0(\xi) \in H^{2,2}(X)$ *from remaining of Hodge type* $(2, 2)$ *in the family* $\{X_t\}_{t \in B}$.

Proof. Let $\eta_{\mathbb{C}} : H^2(\mathcal{K}_2(\mathcal{O}_X)) \to H^2(\Omega^2_{X/\mathbb{C}})$ be the mapping induced on cohomology by

$$\{f_1, f_2\} \to d_{\mathbb{C}} \log f_1 \wedge d_{\mathbb{C}} \log f_2$$

where $d_{\mathbb{C}} = d_{X/\mathbb{C}}$. It is well known that under the isomorphism

$$H^2\left(\Omega^2_{X/\mathbb{C}}\right) \cong H^{2,2}(X)$$

we have (up to the factor $(2\pi\sqrt{-1})^{-2}$)

$$\eta_{\mathbb{C}}(\xi) = \psi_0(\xi)$$

corresponds to the fundamental class of ξ. Since

$$\mathcal{O}_1(\xi) = \theta_1 \rfloor \eta_{\mathbb{C}}(\xi)$$

the result (2.2.12) follows. □

In (2.2.11) the dotted arrow means that $\mathcal{O}_2(\xi)$ is defined only if $\mathcal{O}_1(\xi) = 0$. Suppose this is the case. We shall refer to $\mathcal{O}_2(\xi)$ as the *secondary* 1st *order obstruction* to lifting $\xi = H^2(\mathcal{K}_2(\mathcal{O}_X))$ to $H^2(\mathcal{K}_2(\mathcal{O}_{\mathcal{X}_1}))$, and we shall now give its geometric interpretation.

For this we need a short discussion recalling some of the material from [10]. Suppose that $\xi = [Z] \in CH^p(X(k))$ where both X and Z are defined over field k. Taking k-spreads we have first the spread

$$\mathcal{S}(X) \xrightarrow{\tilde{\omega}} S \tag{2.13}$$

of X and then the spread

$$\mathcal{S}(Z) \in Z^p(\mathcal{S}(X)(\mathbb{Q})) \tag{2.14}$$

of Z. Both $\mathcal{S}(X)$ and $\mathcal{S}(Z)$ are only defined up to ambiguities as discussed in (loc. cit). The arithmetic cycle class

$$\eta(\mathcal{S}(Z)) \in H^p\left(\Omega^p_{\mathcal{S}(X)(\mathbb{Q})/\mathbb{Q}}\right) .$$

By base change

$$H^p\left(\Omega^p_{\mathcal{S}(X)(\mathbb{Q})/\mathbb{Q}}\right) \otimes \mathbb{C} = H^p\left(\Omega^p_{\mathcal{S}(X)/\mathbb{C}}\right) \subset H^{2p}(\mathcal{S}(X), \mathbb{C})$$

and under this inclusion

$$\eta(\mathcal{S}(Z)) \in Hg^p(\mathcal{S}(X))$$

is the fundamental class of $\mathcal{S}(Z)$. We shall denote by

$$\psi_0(\mathcal{S}(Z)) \in H^{2p}(\mathcal{S}(X), \mathbb{C})/\text{ambiguities}$$

the image of $\eta(\mathcal{S}(Z))$ when we factor out the ambiguities in $H^{2p}(\mathcal{S}(X), \mathbb{C})$ induced by the ambiguities in (2.2.13) and (2.2.14). From (loc. cit) we have

$$\psi_0(\mathcal{S}(Z)) \textit{ may be identified with } \eta(\xi) . \tag{2.15}$$

This has the following precise meaning. We consider the Leray spectral sequence of (2.2.13) for the sheaf $\Omega^r_{\mathcal{S}(X)/\mathbb{C}}$ and localized over the generic point $s_0 \in S$. This spectral sequence degenerates at E_2 (cf. [7]), and making the identification

$$\Omega^1_{S/\mathbb{C},s_0} = \Omega^1_{k/\mathbb{Q}} \otimes \mathbb{C} \subset \Omega^1_{\mathbb{C}/\mathbb{Q}} \tag{2.16}$$

we have (loc. cit)

$$E_2^{p,q} = \left(\Omega^q_{\mathbb{C}/\mathbb{Q}} \otimes H^{p+q}\left(\Omega^r_{X/\mathbb{C}}\right)\right)_{\nabla_a} . \tag{2.17}$$

Here the right hand side is the cohomology of the complex

$$\Omega^{q-1}_{\mathbb{C}/\mathbb{Q}} \otimes H^{p+q-1}\left(\Omega^{r+1}_{X/\mathbb{C}}\right) \xrightarrow{\nabla_a} \Omega^{q}_{\mathbb{C}/\mathbb{Q}} \otimes H^{p+q}\left(\Omega^{r}_{X/\mathbb{C}}\right) \xrightarrow{\nabla_a} \Omega^{q+1}_{\mathbb{C}/\mathbb{Q}} \otimes H^{p+q+1}\left(\Omega^{r-1}_{X/\mathbb{C}}\right)$$

where ∇_a is the arithmetic Gauss–Manin connection. This spectral sequence abuts to

$$H^{p+q}\left(\Omega^{r}_{\mathcal{S}(X)/\mathbb{C}}\right) \text{ localized at } s_0 \cong H^{p+q}\left(\Omega^{r}_{X/\mathbb{Q}}\right) \tag{2.18}$$

where the isomorphism results from (2.2.16). Taking $q = 0$ and $r = p$, $\psi_0(\mathcal{S}(Z))$ belongs to the LHS in (2.2.18), $\eta(\xi)$ belongs to the RHS and under this identification

$$\psi_0(\mathcal{S}(Z)) \to \eta(\xi) \,. \tag{2.19}$$

Finally, from [10]

$$\left\{H^{p}\left(\Omega^{p}_{\mathcal{S}(X)/\mathbb{C}}\right) \text{ localized at } s_0\right\} \cong H^{p}\left(\Omega^{p}_{\mathcal{S}(X)/\mathbb{C}}\right)/\text{ambiguities} \,. \tag{2.20}$$

Combining (2.2.18) and (2.2.20) we see that the map (2.2.19) is *injective*, and this is the meaning of (2.2.15).

After these preliminaries we may finally give the geometric interpretation of the secondary 1st order obstruction. Namely, if $\xi \in H^2(\mathcal{K}_2(\mathcal{O}_X))$ deforms to $\xi_t \in H^2(\mathcal{K}_2(\mathcal{O}_{X_t}))$ then we will have $\xi_t = [Z_t]$ for a family of cycles $Z_t \in Z^2(X_t)$. Assuming everything is defined over k, we will have the family of spreads

$$\begin{cases} \mathcal{S}(X_t) \longrightarrow S \,, & t \in B(k) \\ \mathcal{S}(Z_t) \in Z^2(\mathcal{X}(X_t)(\mathbb{Q})) \,. & \end{cases} \tag{2.21}$$

In particular,

$$\psi_0(\mathcal{S}(Z)) \textit{ remains of Hodge type } (2,2) \textit{ in the family (2.2.21)} \,. \tag{2.22}$$

The vanishing of the primary 1st order obstruction $\mathcal{O}_1(\xi)$ is the 1st order implication of this in the top piece of the Leray filtration on $H^4(\mathcal{S}(X), \mathbb{C})$ mod ambiguities. Assuming $\mathcal{O}_1(\xi) = 0$, the vanishing of $\mathcal{O}_2(\xi)$ is well-defined and geometrically gives the condition that in (2.2.22) the second piece in the Leray filtration of $H^4(\mathcal{S}(X), \mathbb{C})$ mod ambiguities be true to 1st order.

This interpretation extends in the evident way for all p. We may summarize the above discussion as follows:

(2.23) *The 1st order obstruction to lifting a rational equivalence class $[Z] \in CH^p(X)$ is expressed by the diagram*

There are filtrations on $CH^p(X) \cong H^p(\mathcal{K}^M_p(\mathcal{O}_X))$, $H^p(\Omega^p_{X/\mathbb{Q}})$ and $H^{p+1}(\Omega^{p-1}_{X/\mathbb{Q}})$, and the maps in the above diagram all preserve these filtrations. The filtration on the obstruction space $H^{p+1}(\Omega^{p-1}_{X/\mathbb{Q}})$ reflects the geometric condition that, to 1st order, the pieces of the fundamental class of

$$\begin{array}{ccccc} H^p(\mathcal{K}_p^M(\mathcal{O}_{X_1})) & \longrightarrow & H^p(\mathcal{K}_p^M(\mathcal{O}_X)) & \xrightarrow{\delta} & H^{p+1}(\Omega^{p-1}_{X/\mathbb{Q}}) \\ & & \downarrow \eta & \nearrow \theta_1 \rfloor & \\ & & H^p(\Omega^p_{X/\mathbb{Q}}) . & & \end{array}$$

the spread $\mathcal{S}(Z)$ *in the Leray filtration on* $H^{2p}(\mathcal{S}(X), \mathbb{C})$ *mod ambiguities remain of Hodge type* (p, p).

Since the filtrations on $CH^p(X)$ and $H^p(\Omega^p_{X/\mathbb{Q}})$ have p steps and the filtration on the obstruction space $H^{p+1}(\Omega^{p-1}_{X/\mathbb{Q}})$ has only $p-1$ steps, and since the obstruction map preserves the filtrations, we may infer the following

Corollary 2.24. *To* 1st *order,* $F^p CH^p(X)$ *is unobstructed.*

For $p = 1$ this is the well-known fact that $\mathrm{Pic}^0(X)$ moves smoothly with X in a family. For $p \geqq 2$ this appears to be a new observation whose geometric interpretation we will discuss later.

3 Higher Order Obstruction Theory

3.1 Higher order Kodaira–Spencer theory. There are two well-developed systematic higher order Kodaira–Spencer theories [8], [12]. The rudimentary theory given here is amenable to the computations we need to make.

Recalling the exact sequence (2.1.1) we give the

Definition. *The* m^{th} *order Kodaira–Spencer class is*

$$\theta_m = \delta(\partial/\partial t) \in H^1(\Theta_{m-1}) .$$

For $m = 1$ we have the usual 1st order Kodaira–Spencer class $\theta_1 \in H^1(\Theta_X)$. The class θ_m depends on the choice of local coordinate t on B; however we observe that it's vanishing does not depend on this choice

$$\theta_m = 0 \text{ if, and only if, } \mathcal{X}_m \cong X \times \mathrm{Spec}\left(\mathbb{C}[t]/t^{m+1}\right) .$$

The above definition is convenient for calculations, and the geometric conclusions that will be drawn from these calculations will not depend on the choice of parameter t. One may of course give an intrinsic definition, but for our purposes this is not necessary. Remark that a change of parameter

$$t \to at + bt^2$$

induces

$$\begin{aligned}\theta_1 &\longrightarrow \left(\frac{1}{a}\right)\theta_1\\ \theta_2 &\longrightarrow \left(\frac{1}{a}\right)\left[\theta_2+\left(\frac{b}{a}\right)\theta_1\right],\end{aligned}$$

from which the general pattern of how the θ_m transform is clear.

Using the notation (2.1.6) and setting

$$\tilde{\theta}_{m\alpha\beta} = f_{\alpha\beta m}(z_\beta)\partial/\partial z_\alpha \tag{1.1}$$

we have

$$e_\beta(\partial/\partial t) - e_\alpha(\partial/\partial t) = \tilde{\theta}_{1\alpha\beta} + t\tilde{\theta}_{2\alpha\beta} + \cdots .$$

Thus the Čech representative cocycles for $\theta_1, \theta_2, \ldots$ are given by

$$\begin{cases}\theta_{1\alpha\beta} = \tilde{\theta}_{1\alpha\beta}\\ \theta_{2\alpha\beta} = \tilde{\theta}_{1\alpha\beta} + t\tilde{\theta}_{2\alpha\beta}\\ \quad\vdots\end{cases} \tag{1.2}$$

Under a change of t as above

$$\begin{aligned}\tilde{\theta}_{1\alpha\beta} &\longrightarrow \left(\frac{1}{a}\right)\tilde{\theta}_{1\alpha\beta}\\ \tilde{\theta}_{2\alpha\beta} &\longrightarrow \left(\frac{1}{a}\right)\tilde{\theta}_{2\alpha\beta} + \left(\frac{b}{a^2}\right)\tilde{\theta}_{1\alpha\beta} .\end{aligned}$$

Since $\Theta_{\mathcal{X}/B}$ is a sheaf of Lie algebras and since vertical vector fields for $\mathcal{X} \to B$ vanish on the function t, the Θ_m are sheaves of Lie algebras. Moreover, there is an exact sequence

$$0 \longrightarrow \Theta_X \xrightarrow{t^m} \Theta_m \longrightarrow \Theta_{m-1} \longrightarrow 0, \qquad m \geqq 1 . \tag{1.3}$$

We claim that:

(1.4) *The extension class of (3.1.3) is given by an element*

$$e_m \in H^1(\mathrm{Der}_{\mathbb{C}}(\Theta_{m-1}, \Theta_X)) ,$$

where $\mathrm{Der}_{\mathbb{C}}(\Theta_{m-1}, \Theta_X)$ *are the* $\mathbb{C}$*-linear Lie algebra homomorphisms.*

Proof. The cases $m = 1, 2$ will illustrate the general pattern. Using again the notation (2.1.6) and setting $j_{\alpha\beta}(z_\beta) = \partial_{z_\beta} f_{\alpha\beta}(z_\beta)$, we have in $U_{\alpha\beta}$

$$dz_\alpha = j_{\alpha\beta}(z_\beta)\left\{1 + t\partial_{z_\beta} f_{\alpha\beta 1}(z_\beta) j_{\alpha\beta}^{-1}(z_\beta) + \frac{t^2}{2}\partial_{z_\beta} f_{\alpha\beta 2}(z_\beta) j_{\alpha\beta}^{-1}(z_\beta) + \cdots\right\} dz_\beta \mod dt .$$

If we take the difference of the box coordinate liftings of $\partial/\partial z_\alpha$ in U_β and U_α we have

$$e_\beta\left(\partial/\partial z_\alpha\right) - e_\alpha\left(\partial/\partial z_\alpha\right) = -\left(t\partial_{z_\alpha} f_{\alpha\beta 1}(z_\beta) + \frac{t^2}{2}\partial_{z_\alpha} f_{\alpha\beta 2}(z_\beta) + \cdots\right)\partial/\partial z_\alpha$$
$$= \left[\left(t f_{\alpha\beta 1}(z_\beta) + \frac{t^2}{2} f_{\alpha\beta 2}(z_\beta) + \cdots\right)\partial/\partial z_\alpha, \partial/\partial z_\alpha\right].$$

It follows that for v, w sections of Θ_X over $U_{\alpha\beta}$

$$\begin{cases} e_{1\alpha\beta}(v) = \left[\tilde{\theta}_{1\alpha\beta}, v\right] \\ e_{2\alpha\beta}(v+tw) = \left[\tilde{\theta}_{2\alpha\beta}, w\right] + \frac{1}{2}\left[\tilde{\theta}_{1\alpha\beta}, v\right] \end{cases} \tag{1.5}$$

using the notation (3.1.1). The fact that the extension classes are in

$$\mathrm{Der}_{\mathbb{C}}\left(\Theta_{m-1}, \Theta_X\right)$$

is a consequence of the Jacobi identity. □

Comparing (3.1.4) and (3.1.5) we see that

$$\theta_m \;\textit{is constructed from}\; \theta_{m-1} \;\textit{and}\; e_m \,.$$

Thus, for $m = 1$ we have

$$0 \longrightarrow \Theta_X \longrightarrow \Theta_1 \longrightarrow \Theta_X \longrightarrow 0$$

and the extension class

$$e_1 = \theta_1 \,.$$

From

$$e_1(\theta_1) = [\theta_1, \theta_1] = 0$$

we have

$$[\theta_1, \theta_1] = \delta\tilde{\theta}_2$$

and by (3.1.4)

$$\theta_{2\alpha\beta} = \tilde{\theta}_{1\alpha\beta} + t\tilde{\theta}_{2\alpha\beta} \,.$$

In general, we will have

$$e_m = \theta_{m-1} \in H^1\left(\mathrm{Der}_{\mathbb{C}}\left(\Theta_{m-1}, \Theta_X\right)\right)$$

and

$$e(\theta_{m-1}) = [\theta_{m-1}, \theta_{m-1}] = \delta\tilde{\theta}_m$$

leading to

$$\theta_{m\alpha\beta} = \theta_{m-1\alpha\beta} + t^m \tilde{\theta}_{m\alpha\beta} .$$

In addition to the sequence (2.1.1) we now consider the sequences analogous to (2.1.3)

$$\begin{cases} \text{(i)} \quad 0 \longrightarrow \mathcal{O}_X \longrightarrow \mathcal{O}_{\mathcal{X}_m} \longrightarrow \mathcal{O}_{\mathcal{X}_{m-1}} \longrightarrow 0 & m \geqq 1 \\ \text{(ii)}_1 \; 0 \longrightarrow \mathcal{O}_X \oplus \Omega^1_{X/\mathbb{C}} \longrightarrow \Omega^1_{\mathcal{X}_m/\mathbb{C}} \longrightarrow \Omega^1_{\mathcal{X}_{m-1}/\mathbb{C}} \longrightarrow 0 & m \geqq 1 \\ \text{(iii)}_1 \; 0 \longrightarrow \mathcal{O}_X \oplus \Omega^1_{X/\mathbb{Q}} \longrightarrow \Omega^1_{\mathcal{X}_m/\mathbb{Q}} \longrightarrow \Omega^1_{\mathcal{X}_{m-1}/\mathbb{Q}} \longrightarrow 0 & m \geqq 1 . \end{cases} \tag{1.6}$$

Proposition 1.7. *The extension classes of (3.1.6) are all induced from θ_m. Specifically, there are maps*

$$\begin{cases} \Theta_{m-1} \longrightarrow \mathrm{Der}_{\mathbb{C}}\left(\mathcal{O}_{\mathcal{X}_{m-1}}, \mathcal{O}_X\right) \\ \Theta_{m-1} \longrightarrow \mathrm{Der}_{\mathbb{C}}\left(\Omega^1_{\mathcal{X}_{m-1}/k}, \mathcal{O}_X \oplus \Omega^1_{X/k}\right) & k = \mathbb{C}, \mathbb{Q} \end{cases}$$

which induce maps

$$\begin{cases} H^1\left(\Theta_{m-1}\right) \longrightarrow H^1\left(\mathrm{Der}_{\mathbb{C}}\left(\mathcal{O}_{\mathcal{X}_{m-1}}, \mathcal{O}_X\right)\right) \\ H^1(\Theta_{m-1}) \longrightarrow H^1\left(\mathrm{Der}_{\mathbb{C}}\left(\Omega^1_{\mathcal{X}_{m-1}/k}, \mathcal{O}_X \oplus \Omega^1_{X/k}\right)\right) \end{cases}$$

under which θ_m goes to corresponding extension classes.

Proof. The inclusion

$$\Theta_{X/B} \subset \Theta_{\mathcal{X}}$$

and pairing

$$\Theta_{\mathcal{X}} \otimes_{\mathbb{C}} \mathcal{O}_{\mathcal{X}} \longrightarrow \mathcal{O}_{\mathcal{X}}$$

induce

$$\Theta_{m-1} \otimes_{\mathbb{C}} \mathcal{O}_{\mathcal{X}_{m-1}} \longrightarrow \mathcal{O}_{\mathcal{X}_{m-1}} .$$

The case $m = 1$ of the proposition is given in the proof of Proposition 2.1.5. For the case $m = 2$ we consider a function $g \in \mathcal{O}_{\mathcal{X}_{m-1},x}$. If $x \in U_\alpha$ we write

$$g = g_0(z_\alpha) + t g_1(z_\alpha)$$

and extend g to $e_\alpha(g) \in \mathcal{O}_{\mathcal{X}_m,x}$ using the (z_α, t) box coordinates. If $x \in U_\beta$ then we write as in (2.1.6)

$$z_\alpha = f_{\alpha\beta}(z_\beta) + t f_{\alpha\beta 1}(z_\beta) + \frac{t^2}{2} f_{\alpha\beta 2}(z_\beta) + \cdots$$

and look at the t^2 term in $e_\beta(g) - e_\alpha(g)$; i.e.,

$$e_\beta(g) - e_\alpha(g) = \frac{1}{2} f_{\alpha\beta 2} \partial_{z_\alpha} g_0 + f_{\alpha\beta 1} \partial_{z_\alpha} g_1$$

which was to be proved. We will write the RHS of this equation as $\mathcal{L}_{\theta_2} g$ where

$$\left(\mathcal{L}_{\theta_2} g\right)_{\alpha\beta} = \frac{1}{2} \tilde{\theta}_{2\alpha\beta} g_0 + \tilde{\theta}_{1\alpha\beta} g_1 \; .$$

A similar calculation for

$$\omega = g_0(z_\alpha) dz_\alpha + t g_1(z_\alpha) dz_\alpha$$

gives

$$e_\beta(\omega) - e_2(\omega) = \theta_2 \rfloor \omega \oplus \mathcal{L}_{\theta_2} \omega$$

where the cocycle representatives are

$$\begin{cases} (\theta_2 \rfloor \omega)_{\alpha\beta} = \tilde{\theta}_{2\alpha\beta} \rfloor g_0(z_\alpha) dz_\alpha + \tilde{\theta}_{1\alpha\beta} \rfloor g_1(z_\alpha) dz_\alpha \\ \left(\mathcal{L}_{\theta_2} \omega\right)_{\alpha\beta} = \mathcal{L}_{\tilde{\theta}_{2\alpha\beta}} (g_0(z_\alpha) dz_\alpha) + \mathcal{L}_{\tilde{\theta}_{1\alpha\beta}} (g_1(z_\alpha) dz_\alpha) \; . \end{cases}$$ □

Definition. *We define*

$$\mathcal{L}_{\theta_m} : H^q\left(\Omega^p_{\mathfrak{X}_{m-1}/k}\right) \longrightarrow H^{q+1}\left(\Omega^p_{X/k}\right) \qquad k = \mathbb{Q}, \mathbb{C} \tag{1.8}$$

to be the $\Omega^p_{X/k}$-component of the coboundary map in the exact cohomology sequence of

$$0 \longrightarrow \Omega^{p-1}_{X/k} \oplus \Omega^p_{X/k} \longrightarrow \Omega^p_{\mathfrak{X}_m/k} \longrightarrow \Omega^p_{\mathfrak{X}_{m-1}/k} \longrightarrow 0 \, , \qquad k \geqq 1 \, . \tag{1.9}$$

As in the case $m = 0$ (cf. Proposition 2.1.8) we have

Proposition 1.10. *The maps (3.1.8) are all zero.*

Proof. As in the proof of (2.1.8) and (2.1.11) it will suffice to do the case $k = \mathbb{C}$. The case $m = 2$ will illustrate how the general pattern goes. The coboundary maps in the cohomology sequences of

$$\begin{array}{ccccccccc} 0 & \longrightarrow & \Omega^{p-1}_{X/\mathbb{C}} \oplus \Omega^p_{X/\mathbb{C}} & \longrightarrow & \Omega^p_{\mathfrak{X}_2/\mathbb{C}} & \longrightarrow & \Omega^p_{\mathfrak{X}_1/\mathbb{C}} & \longrightarrow & 0 \\ & & \downarrow & & \downarrow & & \downarrow & & \\ 0 & \longrightarrow & \Omega^p_{X/\mathbb{C}} & \longrightarrow & \Omega^p_{\mathfrak{X}_2/B} & \longrightarrow & \Omega^p_{\mathfrak{X}_1/B} & \longrightarrow & 0 \end{array}$$

give a commutative diagram

$$\begin{array}{ccccc} H^q\left(\Omega^p_{\mathcal{X}_2/\mathbb{C}}\right) & \longrightarrow & H^q\left(\Omega^p_{\mathcal{X}_1/\mathbb{C}}\right) & \xrightarrow{\mathcal{L}_{\theta_2}} & H^{q+1}\left(\Omega^p_{X/\mathbb{C}}\right) \\ \downarrow & & \downarrow & & \| \\ H^q\left(\Omega^p_{\mathcal{X}_2/B}\right) & \longrightarrow & H^q\left(\Omega^p_{\mathcal{X}_1/B}\right) & \xrightarrow{\delta''} & H^{q+1}\left(\Omega^p_{X/\mathbb{C}}\right) . \end{array}$$

Now $H^q(\Omega^p_{\mathcal{X}_m/B})$ represents m-jets of sections of the vector bundle $R^q\Omega^p_{\mathcal{X}/B} \to B$ (the fact that this is a locally free sheaf being a consequence of dim $H^q(\Omega^p_{X_t/\mathbb{C}}) =$ constant). It follows that $\delta'' = 0$, which implies the result. □

The remaining coboundary map in the exact cohomology sequence of (3.1.9) will be denoted by

$$H^q\left(\Omega^p_{\mathcal{X}_{m-1}/k}\right) \xrightarrow{\theta_m} H^{q+1}\left(\Omega^{p-1}_{X/k}\right) . \tag{1.11}$$

It has the following geometric interpretation: An element in $H^q(\Omega^p_{\mathcal{X}_{m-1}/\mathbb{C}})$ represents an $(m-1)$-jet of section of $R^q\Omega^p_{\mathcal{X}/\mathbb{C}} \to B$, and the coboundary

$$\delta(\varphi) =: \delta'(\varphi) \oplus \delta''(\varphi) \in H^{q+1}\left(\Omega^{p-1}_{X/\mathbb{C}}\right) \oplus H^{q+1}\left(\Omega^p_{X/\mathbb{C}}\right)$$

represents the obstruction to lifting φ to an m-jet. By (3.1.8) and (3.1.10)

$$\delta''(\varphi) = \mathcal{L}_{\theta_m}\varphi = 0 .$$

We now denote by $\Phi \in (R^{p+q}_\pi\mathbb{C})_{t_0}$ the unique local (in the analytic topology) extension of $\varphi_0 \in H^q(\Omega^p_{X/\mathbb{C}}) \subset H^{p+1}(X, \mathbb{C})$ to a section of $R^{p+q}_\pi\mathbb{C}$. By our assumption, to order $m-1$ the section corresponding Φ of $R^{p+q}_\pi\mathbb{C} \otimes \mathcal{O}_B$ lies in the sub-bundle

$$F^p\left(R^{p+q}_\pi \otimes \mathcal{O}_B\right) \subset R^{p+q}_\pi\mathbb{C} \otimes \mathcal{O}_B . \tag{1.12}$$

Then

$$\theta_m \rfloor \varphi$$

represents the deviation of Φ remaining in the sub-bundle (3.1.12) to order m. If we write everything out in terms of Čech cocycles as above we see that $\theta_m \rfloor \varphi$ is expressed by an algebraic contraction operator of φ with the mth order Kodaira–Spencer class θ_m.

We conclude this section with a higher order analogue of Proposition 2.1.13. The proof will be given elsewhere.

Proposition 1.13. *Let* $\eta \in H^q\left(\Omega^p_{\mathcal{X}_{m-1}/\mathbb{Q}}\right)$ *have image denoted by* $\eta_{\mathbb{C}} \in H^q\left(\Omega^p_{\mathcal{X}_{m-1}/\mathbb{C}}\right)$. *Suppose that*

$$\theta_m \rfloor \eta_{\mathbb{C}} = 0$$

in $H^{q+1}(\Omega^{p-1}_{X/\mathbb{C}})$. *Then there exists* $\eta' \in F^q H^q(\Omega^p_{\mathcal{X}_{m-1}/\mathbb{C}})$ *such that*

$$\theta \rfloor (\eta + \eta') = 0$$

in $H^{q+1}(\Omega^{p-1}_{X/\mathbb{Q}})$.

Corollary 1.14. *Under the assumptions of the proposition, we may modify* η *by* $\eta' \in F^1 H^q(\Omega^p_{\mathcal{X}_{m-1}/\mathbb{Q}})$ *such that* $\eta + \eta'$ *lifts to* $H^q(\Omega^p_{\mathcal{X}_m/\mathbb{Q}})$.

3.2 Higher order obstructions to lifting Chow groups; applications. This section is essentially an amalgam of Sects. 2.2 and 3.1. Referring to (2.1.1) and (2.1.2), the basic sequence for the higher order obstruction theory is

$$0 \longrightarrow \Omega^{p-1}_{X/\mathbb{Q}} \longrightarrow \mathcal{K}^M_p(\mathcal{O}_{\mathcal{X}_m}) \longrightarrow \mathcal{K}^M_p\left(\mathcal{O}_{\mathcal{X}_{m-1}}\right) \longrightarrow 0\,. \qquad (2.1)$$

Referring to (2.2.3), the first map is induced by

$$f\frac{dg_1}{g_1} \wedge \cdots \wedge \frac{dg_{p-1}}{g_{p-1}} \longrightarrow \left\{1 + t^m \tilde{f}, \tilde{g}_1, \ldots, \tilde{g}_{p-1}\right\}\,, \qquad t^{m+1} = 0\,. \qquad (2.2)$$

The exactness of (3.2.1) follows from the obvious general case of the following result, for simplicity of notation stated here in the case $p = 2$.

In the following proposition and its proof, it is convenient to replace t by ϵ, so that $\epsilon^{m+1} = 0$.

Proposition 2.3. *Every element of* $\mathcal{K}_2(\mathcal{O}_{\mathcal{X}_{m,x}})$ *can be written as a product of elements of the form*

$$\left\{f, g_0 + g_1\epsilon + \cdots + g_m\epsilon^m\right\} \qquad f, g_i \in \mathcal{O}_{X,x}\,.$$

Proof. We will say that a ring R is *spanned by powers* if for every $d > 0$ every $r \in R$ is a sum of dth powers

$$r = p_1^d + \cdots p_l^d\,, \qquad p_i \in R\,.$$

Claim 1. $\mathcal{O}_{X,x}$ is spanned by powers.

Proof. We first note that a homogeneous polynomial $r \in \mathbb{C}[x_0, \ldots, x_k]$ of degree $k \cdot d$ is a sum of dth powers; this is just the fact that for V a vector space the image of the Veronese map

$$\begin{array}{ccc} \mathbb{P}V & \longrightarrow & \mathbb{P}\,\mathrm{Sym}^d\, V \\ \Psi & & \Psi \\ v & \longrightarrow & v^d \end{array}$$

spans $\mathbb{P}\,\mathrm{Sym}^d\, V$.

Next we see that $\mathcal{O}_{\mathbb{P}^l,x}$ is spanned by powers: Given $f \in \mathcal{O}_{\mathbb{P}^l,x}$ we write

$$f = P/Q^d \qquad \text{(restricted to } X\text{)}$$

where P, Q are homogeneous polynomials with $Q(x) \neq 0$ and apply the first observation.

Claim 2. Every element of $\mathcal{K}_x(\mathcal{O}_{\mathcal{X}_m,x})$ is a product of elements of the form

$$\left\{f, g_0 + g_1\epsilon + \cdots + g_m\epsilon^m\right\}, \qquad \left\{1 + u\epsilon^i, 1 + v\epsilon^j\right\}$$

where $f, g_i, u, v \in \mathcal{O}_{X,x}$.

Proof. We will show that for any k with $0 < k \leqq m$

$$1 + f_k\epsilon^k + f_{k+1}\epsilon^{k+1} + \cdots + f_m\epsilon^m = \left(1 + f_k\epsilon^k\right) \times \left(1 + f_{k+1}\epsilon^{k+1} + \tilde{f}_{k+2}\epsilon^{k+2} + \cdots + \tilde{f}_m\epsilon^m\right) . \tag{2.4}$$

Assuming this we have

$$f_0 + f_1\epsilon + \cdots + f_m\epsilon^m = f_0(1 + u_1\epsilon)\left(1 + u_2\epsilon^2\right)\cdots(1 + u_m\epsilon^m)$$

for some $u_1, \ldots, u_m$, from which it follows that

$$\begin{aligned}
&\left\{f_0 + f_1\epsilon + \cdots + f_m\epsilon^m, g_0 + g_1\epsilon + \cdots + g_m\epsilon^m\right\} \\
&= \left\{f_0(1 + u_1\epsilon)\left(1 + u_2\epsilon^2\right)\cdots\left(1 + u_m\epsilon^m\right), g_0(1 + v_1\epsilon)\left(1 + v_2\epsilon^2\right)\cdots(1 + v_m\epsilon^m)\right\} \\
&= \{f_0, g_0\}\prod_i\left\{f_0, 1 + v_i\epsilon^i\right\}\prod_j\left\{1 + u_j\epsilon^j, g_0\right\}\prod_{k,l}\left\{1 + u_k\epsilon^k, 1 + v_l\epsilon^l\right\}
\end{aligned}$$

which proves the claim.

As for (3.2.4) we have

$$\left(1 + f_k\epsilon^k\right)^{-1} = 1 - f_k\epsilon^k + f_k^2\epsilon^{2k} - \cdots$$

so that

$$\left(1 + f_k\epsilon^k\right)^{-1}\left(1 + f_k\epsilon^k + f_{k+1}\epsilon^{k+1} + \cdots\right) = 1 + f_{k+1}\epsilon^{k+2} \mod \epsilon^{k+2}$$

which gives (3.2.4).

Claim 3. This is the same as Claim 2, but where now we only use elements of the form

$$\left\{1 + u^i\epsilon^i, 1 + v^j\epsilon^j\right\} \qquad u, v \in \mathcal{O}_{X,x}$$

in the second factors.

Proof. Using Claim 1 to write

$$\begin{cases} u = p_1^i + \cdots + p_k^i \\ v = q_1^j + \cdots + q_l^j \end{cases}$$

we have

$$\prod_{\alpha,\beta} \left\{1 + p_\alpha^i \epsilon^i, 1 + q_\beta^j \epsilon^j\right\} = \left\{1 + u^i \epsilon^i, 1 + v^j \epsilon^j\right\} \cdot T$$

where T is a product of terms of the form

$$\left\{1 + u' \epsilon^{i'}, 1 + v' \epsilon^{j'}\right\}$$

where

$$i' + j' > i + j .$$

Now we use descending induction on $i + j$, the case $i + j \geqq 2m + 1$ being automatic.

Claim 4. If $\omega_1, \ldots, \omega_i$ are the ith roots of -1 and $\eta_1, \ldots, \eta_j$ the jth roots of -1, then

$$\prod_{\alpha,\beta} \left\{1 + \omega_\alpha u \epsilon, 1 + \eta_\beta v \epsilon\right\} = \left\{1 + u^i \epsilon^i, 1 + v^j \epsilon^j\right\} .$$

Proof. This follows from

$$\prod_{\alpha=2}^{i} (1 + \omega_\alpha u \epsilon) = 1 + u^i \epsilon^i$$

$$\prod_{\beta=1}^{j} (1 + \eta_\beta v \epsilon) = 1 + v^j \epsilon^j .$$

Claim 5. If $u(x) \neq v(x)$, then

$$\{1 + u\epsilon, 1 + v\epsilon\} \left\{\frac{v}{v - u}, 1 + v\epsilon\right\} \left\{1 + u\epsilon, \frac{u}{u - v}\right\} = 1 .$$

Proof. If A and $A - 1 \in \mathcal{O}_{X,x}^*$, then $A + B\epsilon \in \mathcal{O}_{\mathcal{X}_1,x}^*$ and by the Steinberg relation

$$\{A + B\epsilon, 1 - A - B\epsilon\} = 1 .$$

If follows that

$$1 = \{A, 1 - A\} \left\{A, 1 - \frac{B\epsilon}{1 - A}\right\} \left\{1 + \frac{B\epsilon}{A}, 1 - A\right\} \left\{1 + \frac{B\epsilon}{A}, 1 - \frac{B\epsilon}{1 - A}\right\} . \quad \square$$

Setting

$$u = B/A\ , \qquad v = B/A - 1\ ,$$

i.e.

$$A = v/(v-u)\ , \qquad 1 - A = u/(u-v)$$

then

$$1 = \left\{\frac{v}{v-u}, 1 + v\epsilon\right\}\left\{1 + u\epsilon, \frac{u}{u-v}\right\}\{1 + u\epsilon, 1 + v\epsilon\}\ . \qquad \square$$

In our arguments we will always be able to arrange that $u(x) \neq v(x)$.

Proof (of Proposition 3.2.3). We first write

$$u = p_1^i + \cdots + p_k^i\ , \qquad v = q_1^j + \cdots q_l^j$$

where, using the notations from Claim 3,

$$\omega_\alpha p_\lambda(x) \neq \eta_\beta q_\mu(x) \qquad \text{for all } \alpha, \beta, \lambda, \mu\ .$$

This is easy to arrange. Then

$$\left\{1 + u\epsilon^i, 1 + v\epsilon^j\right\} \equiv \prod_{\alpha,\beta,\lambda,\mu} \left\{1 + \omega_\alpha p_\lambda \epsilon, 1 + \eta_\beta q_\mu \epsilon\right\}$$

modulo terms of the form $\{1 + u'\epsilon^{i'}, 1 + v'\epsilon^{j'}\}$ with $i' + j' > i + j$, and

$$\begin{aligned}&\left\{1 + \omega_\alpha p_\lambda \epsilon, 1 + \eta_\beta q_\mu \epsilon\right\}\\ &\quad = \left\{1 + \eta_\beta q_\mu \epsilon, \frac{\eta_\beta q_\mu}{\eta_\beta q_\mu - \omega_\alpha p_\lambda}\right\}\left\{\frac{\omega_\alpha p_\lambda}{\omega_\alpha p_\lambda - \eta_\beta q_\mu}, 1 + \omega_\alpha p_\lambda \epsilon\right\}\ .\end{aligned}$$

This completes the proof: Claim 2 gets us down to elements in the statement of the proposition and terms of the forms $\{1 + u\epsilon^i, 1 + \epsilon v^j\}$. Claim 3 gets us down to elements we want plus terms $\{1 + u^i\epsilon^i, 1 + v^j\epsilon^j\}$. Claim 4 reduces these to products of $\{1 + A\epsilon, 1 + B\epsilon\}$ terms. Finally, the $\{1 + A\epsilon, 1 + B\epsilon\}$ terms are written as products of the form

$$\{f, 1 + A\epsilon\} \ \text{ and } \ \{1 + A\epsilon, f\}$$

which have the desired form. $\square$

From (3.1.6) (iii)$_p$ and (3.2.1) we infer the commutative diagram

$$\begin{array}{ccccccccc} 0 & \longrightarrow & \Omega^{p-1}_{X/\mathbb{Q}} & \longrightarrow & \mathcal{K}^M_p(\mathcal{O}_{\mathcal{X}_m}) & \longrightarrow & \mathcal{K}^M_p(\mathcal{O}_{\mathcal{X}_{m-1}}) & \longrightarrow & 0 \\ & & \downarrow & & \downarrow{\scriptstyle \eta} & & \downarrow{\scriptstyle \eta} & & \\ 0 & \longrightarrow & \Omega^{p-1}_{X/\mathbb{Q}} \oplus \Omega^p_{X/\mathbb{Q}} & \longrightarrow & \Omega^p_{\mathcal{X}_m/\mathbb{Q}} & \longrightarrow & \Omega^p_{\mathcal{X}_{m-1}/\mathbb{Q}} & \longrightarrow & 0\ . \end{array} \qquad (2.5)$$

Here the left hand vertical arrow is

$$\varphi \longrightarrow \varphi \oplus d\varphi \,,$$

and the other two vertical arrows are the $\wedge^p d\log$ maps. We recall that the bottom left horizontal arrow is

$$\gamma \oplus \omega \to m t^{m-1} dt \wedge \gamma + t^m \omega \,,$$

and the commutativity of (3.2.5) follows from (3.2.2). From Proposition 3.1.10 we deduce the

Proposition 2.6. *The obstruction to lifting* $\xi \in H^p(\mathcal{K}_p^M(\mathcal{O}_{\mathcal{X}_{m-1}}))$ *to* $H^p(\mathcal{K}_p^M(\mathcal{O}_{\mathcal{X}_m}))$ *is given by*

$$\delta(\xi) = \theta_m \rfloor \eta(\xi)$$

where θ_m *is the* m^{th} *order Kodaira–Spencer class and* $\eta(\xi)$ *is the arithmetic cycle class of* ξ.

Corollary 2.7. *Let* $\xi \in CH^p(X)$ *and suppose that the fundamental class* $\psi_0(\xi)$ *remains of Hodge type* (p, p) *in an analytic neighborhood of* $t_0 \in B$. *Assume also that*

$$H^{p+1}\left(\Omega^{p-2}_{X/\mathbb{C}}\right) = \cdots = H^{p+1}(\mathcal{O}_X) = 0 \,. \tag{2.8}$$

Then ξ *lifts formally to* $H^p(\mathcal{K}_p^M(\mathcal{O}_{\hat{\mathcal{X}}}))$.

Remark. If (3.2.8) is not satisfied, then the Hodge Conjecture implies that we may find $\xi' \in F^1CH^p(X)$ such that $\xi + \xi'$ lifts to $H^p(\mathcal{K}_p^M(\mathcal{O}_{\mathcal{X}}))$ (and hence to $H^p(\mathcal{K}_p^M(\mathcal{O}_{\hat{\mathcal{X}}}))$). From (3.1.14) we know the corresponding statement for the arithmetic cycle class $\eta(\xi)$; *we do not know that we can choose*

$$\eta' = \eta(\xi')$$

for some $\xi' \in F^1CH^p(X)$; in fact, this is a geometric existence theorem that would follow from the Hodge Conjecture applied to the spread of X and ξ (cf. (2.2.23) – there are evident higher order analogues of this statement).

Corollary 2.9. *Any* $\xi \in F^pCH^p(X)$ *formally lifts to* $H^p(\mathcal{K}_p^M(\mathcal{O}_{\hat{\mathcal{X}}}))$.

Proof. This is just the higher order extension of (2.2.24). □

We now turn to the more arithmetic aspects and shall prove the following

Proposition 2.10. *Suppose that* X *is defined over* $\mathbb{Q}$ *and that* $\xi \in CH^p(X(\mathbb{Q}))$. *Suppose further that the fundamental class* $\psi_0(\xi)$ *remains of Hodge type* (p, p) *in all* $H^{2p}(X_t, \mathbb{C})$ *for* t *in an analytic neighborhood of* t. *Then* ξ *lifts formally to* $H^p(\mathcal{K}_p^M(\mathcal{O}_{\hat{\mathcal{X}}}))$.

Proof. It will simplify the notation to take the case $p = 2$; the general argument is the same. Since X is defined over $\mathbb{Q}$, the sequence

$$0 \longrightarrow \Omega^1_{\mathbb{C}/\mathbb{Q}} \otimes \mathcal{O}_X \longrightarrow \Omega^1_{X/\mathbb{Q}} \longrightarrow \Omega^1_{X/\mathbb{C}} \longrightarrow 0 \tag{2.11}$$

splits as a sequence of $\mathcal{O}_X$-modules (see below). Making as usual the identification $CH^2(X) \cong H^2(\mathcal{K}_2(\mathcal{O}_X))$, referring to (2.2.11) the obstruction $\delta(\xi) \in H^3(\Omega^1_{X/\mathbb{Q}})$ to lifting ξ to $H^2(\mathcal{K}_2(\mathcal{O}_{\mathcal{X}_1}))$ may, using the splitting of (3.2.11), be written as

$$\delta(\xi) = \mathcal{O}_1(\xi) \oplus \mathcal{O}_2(\xi) \in H^3\left(\Omega^1_{X/\mathbb{C}}\right) \oplus \left(\Omega^1_{\mathbb{C}/\mathbb{Q}} \otimes H^3(\mathcal{O}_X)\right) .$$

The primary 1st order obstruction $\mathcal{O}_1(\xi)$ vanishes by our assumption on $\psi_0(\xi)$, and therefore we have to show that the secondary 1st order obstruction

$$\mathcal{O}_2(\xi) = 0 . \tag{2.12}$$

Geometrically, this is clear from the interpretation of $\mathcal{O}_2(\xi)$, the reason being that the spread of the pair (X, ξ) has no continuous parameters. However, we want to give a computational proof that will extend to higher order.

Proof. We begin with some background remarks. Let $x_1, \ldots x_n \in \mathbb{Q}(X)$ give local uniformizing parameters in a Zariski neighborhood of x in X. Denote by $df \in \Omega^1_{X/\mathbb{Q},x}$ the absolute Kähler differential of $f \in \mathcal{O}_{X,x}$, and by $d_{\mathbb{C}} f$ the image of df in $\Omega^1_{X/\mathbb{C},x}$. The map

$$d_{\mathbb{C}} x_i \longrightarrow dx_i \tag{2.13}$$

extended by $\mathcal{O}_{X,x}$-linearity splits the sequence (3.2.11).

Locally in the analytic topology, any $f \in \mathcal{O}_{X,x}$ may be expanded as a power series in the x_i

$$f = \sum_I a_I x^I \qquad \left(x^I = (x_1)^{i_1} \cdots (x_n)^{i_n}\right) . \tag{2.14}$$

If $f \in \mathcal{O}_{X(k),x}$ for some field k with $\mathbb{Q} \subset k \subset \mathbb{C}$, then the $a_I \in k$. As was proved in [11], we may compute the absolute differential df by differentiating the series (3.2.14) term by term, where $da_I \in \Omega^1_{k/\mathbb{Q}}$ is the absolute Kähler differential of $a_I \in k$. The resultant series converges and represents the power series expansion of df. If $k = \bar{\mathbb{Q}}$, then the $da_I = 0$ and df is given by termwise differentiation of (3.2.14) in the usual sense. As a consequence, if $f \in \mathcal{O}_{X,x} \cap \mathbb{Q}(X)$ and

$$d_{\mathbb{C}} f = \sum_i f_i d_{\mathbb{C}} x_i$$

where $f_i \in \mathcal{O}_{X,x} \cap \mathbb{Q}(X)$, then it follows that

$$df = \sum_i f_i dx_i .$$

This implies that the splitting of (3.2.11) given by (3.2.13) is independent of the choice of local uniformizing parameters.

Now let $\{U_\alpha\}$ be a Zariski covering of X and let

$$\begin{cases} \xi \in H^2\left(\mathcal{K}_2\left(\mathcal{O}_{X(k)}\right)\right) \text{ be represented by } \{g_{\alpha\beta\gamma}, h_{\alpha\beta\gamma}\} \\ \theta \in H^1\left(\Theta_{X/\mathbb{C}}\right) \text{ be represented by } \theta_{\alpha\beta}\partial/\partial x_\alpha \ . \end{cases}$$

Then

$$\delta(\xi)_{\alpha\beta\gamma\delta} = \theta_{\alpha\beta}\rfloor\left(\frac{dg_{\beta\gamma\lambda}}{g_{\beta\gamma\lambda}} \wedge \frac{dh_{\beta\gamma\lambda}}{h_{\beta\gamma\lambda}}\right) . \tag{2.15}$$

If $g_{\beta\gamma\lambda}$ and $h_{\beta\gamma\lambda}$ are represented by series as in (3.2.14) above, then the $\mathcal{O}_2(\xi)_{\alpha\beta\gamma\lambda}$ component of $\delta(\xi)_{\alpha\beta\gamma\lambda}$ is represented by the series

$$\theta_{\alpha\beta}\rfloor\left\{\frac{\bar{d}g_{\beta\gamma\lambda}}{g_{\beta\gamma\lambda}} \wedge \frac{d_{\mathbb{C}}h_{\beta\gamma\lambda}}{h_{\beta\gamma\lambda}} + \frac{d_{\mathbb{C}}g_{\beta\gamma\lambda}}{g_{\beta\gamma\lambda}} \wedge \frac{\bar{d}h_{\beta\gamma\lambda}}{h_{\beta\gamma\lambda}}\right\}$$

where in (3.2.14) we have set

$$\bar{d}f = \sum_I da_I x^I \in \Omega^1_{k/\mathbb{Q}} \otimes \mathcal{O}_{X,x} \ .$$

In particular, if $k = \mathbb{Q}$ then $\bar{d}g_{\beta\gamma\lambda} = \bar{d}h_{\beta\gamma\lambda} = 0$ and so $\mathcal{O}_2(\xi) = 0$, thereby establishing (3.2.12).

If $\xi \in H^2(\mathcal{K}_2(\mathcal{O}_{X(\bar{\mathbb{Q}})})$ and $\mathcal{O}_1(\xi) = 0$, then $\delta(\xi) = 0$ and we may write the obstruction cocycle (3.2.15) as a coboundary of a cochain *defined over* $\mathbb{Q}$. Thus we may lift ξ to

$$\left\{g_{\alpha\beta\gamma} + tg'_{\alpha\beta\gamma}, h_{\beta\gamma\lambda} + th'_{\beta\gamma\lambda}\right\}, \qquad t^2 = 0 , \tag{2.16}$$

where $g'_{\alpha\beta\gamma}\mathcal{O}_{X(\mathbb{Q})}(U_{\alpha\beta\gamma})$ and $h'_{\beta\gamma\lambda} \in \mathcal{O}_{X(\mathbb{Q})}(U_{\beta\gamma\lambda})$. The fact that we may keep things defined over $\mathbb{Q}$ through 1st order implies that when we calculate the 2nd order obstruction to lifting ξ only the deviation of $\psi_0(\xi)$ remaining of Hodge type (2, 2) to 2nd order will appear. This process, when iterated, will lead to a proof of (3.2.10).

We will discuss the next case of extending $\xi_1 \in H^2(\mathcal{K}_2(\mathcal{O}_{\mathcal{X}_1}))$ given by (3.2.16) to $\xi_2 \in H^2(\mathcal{K}_2(\mathcal{O}_{\mathcal{X}_2}))$. By Proposition 3.2.6 the obstruction is given by

$$\delta(\xi_1) = \theta_2\rfloor\eta(\xi_1) \in H^3\left(\Omega^1_{X/\mathbb{Q}}\right) . \tag{2.17}$$

Using the splitting of (3.2.11) we may write

$$\delta(\xi_1) = \mathcal{O}_1(\xi_1) \oplus \mathcal{O}_2(\xi_1)$$

where

$$\begin{cases} \mathcal{O}_1(\xi_1) = \theta_2\rfloor\eta_{\mathbb{C}}(\xi_1) \in H^3\left(\Omega^1_{X/\mathbb{C}}\right) \\ \mathcal{O}_2(\xi_1) \in \Omega^1_{\mathbb{C}/\mathbb{Q}} \otimes H^3(\mathcal{O}_X) \ . \end{cases}$$

Now $\mathcal{O}_1(\xi_1)$ represents the deviation of $\psi_0(\xi)$ remaining of Hodge type (2, 2) to 2nd order, and by assumption it vanishes. As for $\mathcal{O}_2(\xi_1)$, it is computed from terms of the type

$$\bar{d}\log\left(g_{\alpha\beta\gamma}+tf'_{\alpha\beta\gamma}\right)\wedge d\log\left(h_{\beta\gamma\lambda}+th'_{\beta\gamma\lambda}\right)\ . \tag{2.18}$$

More precisely, by (3.2.17) a Čech representative of $\mathcal{O}_2(\xi_1)$ will be given by the contraction of (3.2.18) with the vector field

$$\left(\frac{1}{t}\right)\tilde{\theta}_{1\alpha\beta}\partial/\partial x_\alpha+\tilde{\theta}_{2\alpha\beta}\partial/\partial x_\alpha \tag{2.19}$$

where the first term contracts the coefficient of t in (3.2.8) and the second contracts the other term. Since $g_{\alpha\beta\gamma}$ and $g'_{\alpha\beta\gamma}$ are defined over $\mathbb{Q}$ this contraction vanishes.

Since the cocycles representing the cohomology classes that arise as in (3.2.17) are only defined up to coboundaries, the conclusion that we may draw is that the Čech cocycle representing $\delta(\xi_1)$ is a coboundary, and as before since the vector field (3.2.19) is defined over $\mathbb{Q}$ this coboundary may be taken to also be defined over $\mathbb{Q}$.

The argument now proceeds inductively over the infinitesimal neighborhoods $\mathcal{X}_m$, taking care to ensure at each state that the coefficients in the formal power series remain in $\mathbb{Q}$. □

4 Failures of the Formal Theory

In classical algebraic geometry – where "classical" in the study of cycles refers to codimension one – it is a general principle that *a construction that can be made formally can be made geometrically*; we abbreviate this by saying that

$$\textit{formal} \Rightarrow \textit{actual}\ . \tag{4.1}$$

For example, a result from the earliest days of deformation theory – valid in either the algebraic or analytic setting – states that if we are given a family

$$\{X_t\}_{t\in B}$$

where $h^2(\mathcal{O}_X)=0$, and given $[Z]\in CH^1(X)$ then

(4.2) *there is a family* $[Z_t]\in CH^1(X_t)$ *which specializes to* $[Z]$ *at* $t=t_0$.

Here there are two caveats: First, we are not saying that the cycle Z deforms; only that its class in $CH^1(X)$ does. The second is that in the algebraic case we may have to pass an étalé covering of B. Finally, the assumption $h^2(\mathcal{O}_X)=0$ may be replaced by the assumption that the fundamental class $\psi_0(Z)$ remains of Hodge type (1, 1) in $H^2(X_t,\mathbb{C})$ for t in an analytic neighborhood of t_0.

In the analytic setting, the proof of this result may be done by showing inductively that the successive terms in the power series construction of $\xi_t\in H^1(\mathcal{K}_1(\mathcal{O}_{X_t}))$ may be chosen in a "bounded" fashion so as to insure convergence. Such arguments are common in the works of Kodaira–Spencer, Grauert and others during the 1950's and 1960's.

However, in order to use the Bloch–Quillen result, we need our formal deformation to lift to a rational/algebraic function on B. Here, there is no GAGA principle to rely on. There has been considerable progress made in recent years in approaching issues of rationality and algebraicity; see the Bourbaki talk of [6].

An even more basic illustration of (4.1) is the statement

(4.3) *every tangent vector $\tau \in TCH^1(X)$ is tangent to an arc in $CH^1(X)$.*

Here, by $TCH^1(X)$ we mean the $p = 1$ case of the formal tangent space

$$TCH^p(X) = H^p\left(\Omega^{p-1}_{X/\mathbb{Q}}\right) \tag{4.4}$$

to the Chow groups defined by Bloch [4]. The result (4.3) is more commonly expressed by saying that *$CH^1(X)$ is representable and reduced.* For reasons that will appear shortly, we have expressed it in the form (4.3).

The central point of this section is to explain the following:

(4.5) *For $p \geqq 2$ the principle (4.1) is false. In fact, already for $p = 2$ the analogues of (4.2) and (4.3) are false.*

This will be done by a series of examples, based mostly on our paper [11] in which we defined for $p = 1, n$ the tangent space $TZ^p(X)$ to the space $Z^p(X)$ of codimension-p cycles on X, the subspace $TZ^p_{\text{rat}}(X)$ to the subgroup of cycles rationally equivalent to zero, and then we showed in some cases that

$$TCH^p(X) \cong TZ^p(X)/TZ^p_{\text{rat}}(X)\ . \tag{4.6}$$

As discussed in loc. cit we expect that these definitions can be extended for all p and that (4.6) will hold in general. Also, as discussed in §10 we expect that – essentially for the geometric reasons stated there – it may be the case that

$$\textit{Both } TZ^p(X) \textit{ and } TZ^p_{\text{rat}}(X) \textit{ are formally reduced}\ . \tag{4.7}$$

This means that every tangent vector is tangent to a formal arc in $Z^p(X)$ and $Z^p_{\text{rat}}(X)$.

We shall show that

(4.8) *The principle (4.1) is valid for $TZ^1(X)$ and $TZ^1_{\text{rat}}(X)$, but is in general false for $p \geqq 2$.*

In the following examples we will use the notations and terminology of [11].

Example (i). In $CH^2(\mathbb{P}^2, T)$ we consider the arc

$$z(t) = (b, t) - (b, 1) - (1, t) + (1, 1)$$

where $b \in \bar{Q}^*$, b not a root of unity. This arc is non-constant in $CH^2(\mathbb{P}^2, T)$, but its velocity

$$z'(t) = \frac{db}{b} \in \Omega^1_{\mathbb{C}/\mathbb{Q}} \cong TGr^2CH^2(\mathbb{P}^2, T)$$

is identically zero. In this, building a formal power series $F(t)$ in

$$(Y, Y \cap T) \bigoplus_{Y \subset (\mathbb{P}^2, T)} \mathbb{C}(Y, Y \cap T)^* \tag{4.9}$$

which maps to $z(t)$ under the map

$$(Y, f) \mapsto \operatorname{div} f$$

involves inductively solving in $\mathbb{C}^* \otimes_{\mathbb{Z}} \mathbb{C}^*$ equations of the type

$$\prod_{i=1}^{m} a_i \otimes b_i = \prod_{\nu=1}^{n} (1 - c_\nu) \otimes c_\nu, \qquad c_\nu \neq 1 .$$

Here, the LHS is given and the RHS is to be solved for.

The assumption that b is not a root of unity implies that $z(t)$ is non-constant. We are grateful to A. Merkurjev for explaining this. To solve for the terms in the series, at the first step the assumption that $b \in \bar{Q}^*$ gives

$$db = 0 \qquad (d = d_{\mathbb{C}/\mathbb{Q}}) .$$

We can then solve the equation

$$(1+t) \otimes b = \prod_{\nu=1}^{n} (a_\nu + tb_\nu) \otimes (1 - a_\nu - tb_\nu), \quad t^2 = 0 \text{ and } a_\nu \in \mathbb{Q}\backslash\{0, 1\}, b_\nu \in \mathbb{Q},$$

where the length n of the Steinberg relation on the right depends on the arithmetic complexity of b. This gives the t-coefficient in the formal series for $F(t)$. For the t^2-coefficient we again have to solve an equation of this type where the length of the Steinberg relation depends on the arithmetic complexity of the a_ν and the b_ν. There is *no* inductive choice of the solution to these equations that will yield a convergent answer. Although we don't know how to make this precise, we feel that the underlying reason is that there is no way to bound the arithmetic complexity of the successive choices. In any case, the above is an example of an arc in $Z^2(\mathbb{P}^2, T)$ where $z'(t) \in Z^2_{\text{rat}}(\mathbb{P}^2, T)$ but where there can only be formal arcs in (4.9) that map to $z(t)$.

Example (ii). Let X be a regular surface with $H^2(\mathcal{O}_X) \neq 0$, $Y \subset X$ a smooth curve and suppose that X and Y are both defined over $\mathbb{Q}$ (or over a number field). Then the map

$$CH^1(Y) \longrightarrow CH^2(X)$$

is in general non-constant, while its differential

$$\begin{array}{ccc} TCH^1(Y) & \longrightarrow & TCH^2(X) \\ \wr\Vert & & \wr\Vert \\ H^1(\mathcal{O}_Y) & \longrightarrow & \Omega^1_{\mathbb{C}/\mathbb{Q}} \otimes H^2(\mathcal{O}_X) \end{array}$$

is identically zero. Although not as explicit, the reason seems to be analogous to that in example (i).

Example (iii). Let X be a threefold defined over $\mathbb{Q}$ having $H^2(\mathcal{O}_X) = 0$ and for which the abelian part $J^2_{\text{ab}}(X)$ of the intermediate Jacobian $J^2(X)$ is zero – i.e., there are no sub-Hodge structures in $H^{2,1}(X) \oplus H^{1,2}(X)$ (for the existence of such see example (iv) below). Then (cf. [11])

$$TGr^2CH^2(X) \cong H^2\left(\Omega^1_{X/\mathbb{C}}\right)$$

may be non-zero but, since $J^2_{\text{ab}}(X) = 0$ there are no actual families. In this case, for $\varphi \in H^2(\Omega^1_{X/\mathbb{C}})$ there is a tangent vector

$$\tau \in TZ^2(X)$$

which maps to φ under the isomorphism (4.6). Any arc in $Z^2(X)$ with tangent τ can only be formal; if the heurestic (4.7) is true, $Z^2(X)$ can only be formally reduced.

In concluding we would like to speculate on one possible way of understanding the convergence issues that arise above and in [11].

In the classical case of deformation theory – e.g., the deformation of classes in $CH^1(X)$ – the objects encountered are finite dimensional schemes (more formally, the functors are representable), and by results such as the Artin approximation theorem any construction that can be made formally can be modified at sufficiently high order to be convergent. Analytically we are inductively solving equations of the type

$$\varphi_i = d\psi_i \tag{4.10}$$

with sup norm estimates of the sort

$$\|\psi_i\| \leqq C\|\varphi_i\| \tag{4.11}$$

that are sufficient to obtain convergence in the analytic category.

In the non-classical case where the objects are not finite dimensional and arithmetic/geometric considerations arise – e.g., the deformation of classes in $CH^p(X)$ for $p \geqq 2$ – the issues of convergence are more subtle. One way of thinking of it is that we have to be able to construct convergent series expansions of algebraic functions defined over, say, $\mathbb{Q}$. To do so we have to be able to solve (4.10) with estimates (4.11) where we think of $\|\ \|$ as $|\ |_\infty$, together with estimates

$$|\psi_i|_p \leqq C|\varphi_i|_p \tag{4.12}$$

for each prime p. We must bound "size" both in the usual analytic sense and in the sense of arithmetic complexity.

References

[1] Y. André, G-functions and geometry, *Aspects of Mathematics*, vol. E13 (1989).

[2] S. Bloch, Semi-regularity and deRham cohomology, *Invent. Math.* **17** (1972), 51–56.

[3] S. Bloch, K_2 and algebraic cycles, *Ann. of Math.* **99** (2) (1974), 349–379.

[4] S. Bloch, Lectures on algebraic cycles, *Duke Univ. Math Ser.* IV (1980), 182 pp.

[5] H. Clemens, Cohomology and obstruction I: On the geometry of formal Kuranishi theory, to appear at math. AG/9901084 (rev 2002).

[6] A. Chambert-Loir, Théorèmes d'algébricité en géométrie diophantienne, *Sem. Bourbaki* n° 886 (2000–2001).

[7] H. Esnault and K. Paranjape, Remarks on absolute deRham and absolute Hodge cycles, *C. R. Acad. Sci. Paris* Sér. I Math. *319* (1) (194), 67–72.

[8] H. Esnault and E. Viehweg, Higher Kodaira–Spencer classes, *Math. Ann.* **299** (1994), 491–527.

[9] M. Green, Griffiths' infinitesimal invariant and the Abel–Jacobi map, *J. Differential Geom.* **29** (1989), 545–555.

[10] M. Green and P. Griffiths, Hodge-theoretic invariants for algebraic cycles, IMRN No. 9 (2003).

[11] M. Green and P. Griffiths, On the tangent space to the space of algebraic cycles on a smooth algebraic variety, to appear, Annals of Mathematics Studies, Princeton University Press.

[12] Z. Ran, Derivatives of moduli, *Internat. Math. Res. Notices* no. 4 (1993), 93–106.

[13] C. Soulé, , Opérations en SKS-théorie algébrique. [Operations in algebraic SKS-theory] *Canad. J. Math.* **37** (1985) no. 3, 488–550.

[14] V. Srinivas, Gysin maps and cycle classes for Hodge cohomology, *Proc. Indian Acad. Sci. Math. Sci.* **103** (1993), no. 3, 209–247.

An Analogue of Abel's Theorem

Herbert Clemens*

Abstract. This work makes a parallel construction for curves on threefolds to a "current-theoretic" proof of Abel's theorem giving the rational equivalence of divisors P and Q on a Riemann surface when $Q - P$ is (equivalent to) zero in the Jacobian variety of the Riemann surface. The parallel construction is made for homologous "sub-canonical" curves P and Q on a general class of threefolds. If P and Q are algebraically equivalent and $Q - P$ is zero in the (intermediate) Jacobian of a threefold, the construction "almost" gives rational equivalence.

1 Introduction

This work establishes a parallel between

1) a proof of the classical theorem of Abel deriving rational equivalence classes of divisors P and Q of degree d on a Riemann surface X from equality of their images in the Jacobian variety $J(X)$,

and

2) a construction for certain cohomologous curves P and Q on a threefold X for which the image of $Q - P$ in the (intermediate) Jacobian $J(X)$ is zero.

This paper was motivated by work of Richard Thomas (see [2]). Indeed the point of view is in large measure due to him. To explain the analogy with Abel's theorem, one must recast the classical proof of that theorem in the language of forms with values in distributions, that is, currents (see [4]). In that language, the classical proof of Abel's theorem goes something like this.

* Partially supported by NSF grant DMS-0200895

Take two effective divisors P and Q of degree d on a Riemann surface X. Consider a one-current Γ such that

$$Q - P = \partial(\Gamma),$$

that is, the operator which assigns to a C^∞-one-form on X its integral over some path Γ from P to Q. This current is the pull-back of a current on the multiplicative group $\mathbb{C}^*$ of complex numbers as follows:

For the unique topological line bundle L_∞ of degree d on X, there are two complex structures, one giving $\mathcal{O}(P)$ and the other giving $\mathcal{O}(Q)$. Thus the holomorphic sections s_P and s_Q with respect to these complex structures give two different C^∞-sections of L_∞ which we will continue to call s_P and s_Q. The quotient

$$g := \frac{s_Q}{s_P}$$

then gives a C^∞-map

$$X - (|P| \cup |Q|) \to \mathbb{C}^*.$$

The current Γ is just the pull-back of the current on $\mathbb{C}^*$ given by the positive real axis, that is

$$\Gamma = g^*((0, +\infty)).$$

Furthermore, on the compactification $\mathbb{P}^1$ of $\mathbb{C}^*$ we have the cohomological relation

$$d \log z \sim 2\pi i \cdot (0, +\infty)$$

of currents. Pulling back via g^*, we have the equality

$$\int_\Gamma = \frac{1}{2\pi i}\int_X \alpha \wedge$$

as functionals on $H^{1,0}(X)$, where

$$\alpha := d \log g.$$

Also we note that the $(0,1)$-component of $d \log z$, that is, the linear operator obtained by integrating $(1,0)$ -forms against $d \log z$, is actually bounded, that is, is given by integrating against a C^∞-form $\alpha^{0,1}$ on X of type $(0,1)$.

Now if P is Jacobian-equivalent to Q, the $(0,1)$-summand of the current $\alpha^{0,1}$ (appropriately normalized by the $(0,1)$-summand of an integral cocycle) is $\overline{\partial} b$ for some C^∞-function b. So in this case,

$$\psi := g^*(d \log z) - db$$

is a form of type $(1, 0)$ which is d-closed on $X - (|P| \cup |Q|)$. So ψ is $\bar{\partial}$-closed and of type (1, 0) and therefore meromorphic, having poles with residues which are integral multiples of $2\pi i$. We complete the proof of Abel's theorem by remarking that

$$f = e^{\int \psi}$$

is the rational function giving the rational equivalence of P and Q.

After first giving in detail the distribution-theoretic proof of the classical Abel's theorem, we will make an analogous construction in the case in which P and Q are certain types of effective algebraic one-cycles on a threefold X. We first produce a rank-2 vector bundle E_∞ whose first Chern class is trivial and whose second Chern class is represented by P or by Q. We note that a choice of a metric μ on E_∞ is equivalent to giving E_∞ the structure of a quaternionic line bundle. Then we mimic, for the quaternionic line bundle E_∞, all but the "trivial" last step of the proof of Abel's theorem for divisors on curves. Namely for each metric μ on E_∞ we use Chern–Simons theory to produce a three-form $\alpha_{\mu, PQ}$ on X giving the normal function $Q - P$ and such that:

1) If P and Q are algebraically equivalent, then $\alpha_{\mu, PQ}^{(1,2)+(0,3)}$ is $\bar{\partial}$ -closed as a form on X.

2) If in addition P and Q are Abel–Jacobi equivalent, then $\alpha_{\mu, PQ}^{(1,2)+(0,3)}$ (normalized by the $(1, 2) + (0, 3)$ summand of an integral cycle) is $\bar{\partial}$-exact as a form on X; so there is an associated form ψ_μ on X of type $(3, 0) + (2, 1)$ such that

$$\psi_\mu - \alpha_{\mu, PQ} \in dA_X^2.$$

Recall that P and Q are *algebraically equivalent* if they are homologous on some (possibly reducible) divisor on X.

Finally, for the quaternionic line bundle E_∞ we will use Chern–Simons theory to produce a 3-current α_{PQ} giving the normal function $Q - P$ such that α_{PQ} is d-closed on $X' := X - (|P| \cup |Q|)$ and:

1) If P and Q are algebraically equivalent, then $\alpha_{PQ}^{(1,2)+(0,3)}$ is $\bar{\partial}$-closed as a current on X.

2) If in addition P and Q are Abel–Jacobi equivalent, then $\alpha_{PQ}^{(1,2)+(0,3)}$ (normalized by the $(1, 2) + (0, 3)$ summand of an integral cycle) is $\bar{\partial}$-exact as a current on X; so there is a canonically associated current ψ on X of type $(3, 0) + (2, 1)$ such that

$$(\psi - \alpha_{PQ}) \in d\,\{2 - \text{currents on } X\}.$$

Furthermore $\psi|_{X'}$ is a d-closed 3-current, that is, its integral against the coboundary of compactly supported 2-forms on X' is zero.

To understand a potential significance of ψ, suppose that $\psi|_{X'}$ turns out to be a 3-form and the algebraic equivalence of P and Q is given by an algebraic family

$$\begin{array}{ccc} S & \xrightarrow{s} & X \\ \downarrow^{r} & & \\ C & & \end{array}$$

such that

1) C is a smooth irreducible curve,
2) S is a smooth surface proper and flat over C,
3) for two points p and q in C and the corresponding fibers S_p and S_q of r give P and Q respectively.

Then $r_*s^*\psi$ is a d-closed form of type $(1,0)$ on $C' := C - \{p,q\}$ and

$$f = e^{\int r_*s^*\psi}$$

would therefore be the natural candidate for a meromorphic function on C giving the rational equivalence of $Q - P$ and a divisor supported at points parametrizing curves incident to $|P| \cup |Q|$

2 Classical Theorem of Abel

Let L_∞ denote the unique C^∞ complex-line bundle on X with

$$c_1(L_\infty) = d\ .$$

If P is an effective divisor of degree d on X, then there is a holomorphic structure

$$\overline{\partial}_P : A_X^0(L_\infty) \to A_X^{0,1}(L_\infty)$$

on L_∞ and a section s_P of L_∞ such that $\overline{\partial}_P(s_P) = 0$ and $\operatorname{div}(s_P) = P$.

Given any metric μ on L_∞, there is a unique metric-$(1,0)$ connection $D_{\mu,P}$ on L_∞ such that

$$D_{\mu,P}^{0,1} = \overline{\partial}_P\ .$$

Alternatively, if we restrict our consideration to $X - |P|$, there is a unique connection D_P on $L_\infty|_{X-|P|}$ such that s_P is flat. Again for this connection

$$D_P^{0,1} = \overline{\partial}_P\ .$$

If Q is another effective divisor of degree d, we consider both s_P and s_Q as C^∞-sections of the C^∞-line bundle L_∞ and define as above the map

$$g = \frac{s_Q}{s_P} : X - (|P| \cup |Q|) \to \mathbb{C}^*,$$

assuming that g is meromorphic in a small analytic neighborhood of $(|P| \cup |Q|)$. We compute

$$\begin{aligned} 0 = D_Q(g \cdot s_P) &= dg \cdot s_P + g \cdot D_Q s_P \\ D_Q s_P &= -g^{-1} \cdot dg \cdot s_P\ . \end{aligned}$$

So, for $\alpha_{PQ} = D_Q - D_P$, we have

$$\alpha \cdot s_P = \left(D_Q - D_P\right) s_P = D_Q s_P = -g^{-1} \cdot dg \cdot s_Q$$

so that

$$\begin{aligned} \alpha_{PQ} &= -g^{-1} \cdot dg \\ \alpha_{PQ}^{(0,1)} &= \overline{\partial}_Q - \overline{\partial}_P = -g^{-1} \cdot \overline{\partial} g \,. \end{aligned}$$

Now via residue and the fact that $\lim_{r \to 0} r \log r = 0$, we have cohomologous currents

$$2\pi i \int_0^\infty \sim \int_X d\theta \wedge \sim \int_X \left(z^{-1} \cdot dz\right) \wedge$$

on $\overline{\mathbb{C}^*} = \mathbb{P}^1$. We therefore obtain the equality

$$\int_P^Q \eta = \frac{1}{2\pi i} \int_X \alpha_{PQ} \wedge \eta \qquad (1)$$

for $\eta \in H^{1,0}(X)$ by pulling back the above cohomology between currents via g. Also, since

$$\begin{aligned} D_P^{0,1} &= D_{\mu,P}^{0,1} \\ D_Q^{0,1} &= D_{\mu,Q}^{0,1} \end{aligned}$$

we have for $\alpha_{\mu,PQ} := D_{\mu,Q} - D_{\mu,P}$ that $\alpha_{\mu,PQ}^{0,1} = \alpha_{PQ}^{0,1}$, so that the (0, 1)-summand of $-g^{-1} \cdot dg$ is bounded.

The equality (1) shows that if, for some

$$\varepsilon \in H_1(X; \mathbb{Z}) \,,$$

one has

$$\int_P^Q = \int_\varepsilon : H^{1,0}(X) \to \mathbb{C} \,,$$

then the de Rham class

$$A = A^{1,0} \,(\text{with poles}) + A^{0,1} \,(\text{bounded})$$

such that

$$\left\{A^{0,1}\right\} \in 2\pi i \cdot H^1(X; \mathbb{Z}) + H^{1,0}(X) \,.$$

So the Poincaré dual of ε is a de Rham class

$$\{\xi\} \in H^1(X; \mathbb{Z})$$

such that

$$A^{0,1} - \xi^{0,1} = \overline{\partial}\gamma$$

for some C^{∞}-function $\gamma \in A_X^0$. Then the form

$$\psi := g^{-1} \cdot dg + \xi + d\gamma$$

on $X - (|P| \cup |Q|)$ is d-closed and of type $(1, 0)$ and therefore meromorphic on X with poles at the divisor $Q - P$.

As mentioned above, the "second part" of of the proof of Abel's theorem consists in defining the meromorphic function

$$f = e^{\int \psi}$$

and noticing that div $(f) = Q - P$. Thus $f : X \to \mathbb{P}^1$ gives the rational equivalence between P and Q.

2.1 Analogue for one-cycles on threefolds. In the remainder of this paper we present the analogue of Abel's theorem in the case in which P and Q are cohomologous (sums of) smooth sub-canonical curves on a threefold X and E_∞ is a C^∞ rank-2 vector bundle on X whose first Chern class is trivial and whose second Chern class is represented by P (and so also by Q).

3 Quaternionic Connections

3.1 Serre's construction. Let X be a smooth projective threefold with

$$H^1(\mathcal{O}_X) = H^2(\mathcal{O}_X) = 0$$

and assume that P is a smooth (but possibly reducible) Riemann surface lying inside X. We wish to consider situations in which

$$\omega_X|_P = \omega_P, \tag{2}$$

that is, the line bundle

$$\omega_X^{-1} \otimes \omega_P = \omega_X^{-1} \otimes Ext_X^2(\mathcal{O}_P, \omega_X) = \omega_X^{-1} \otimes Ext_X^2(\mathcal{O}_P, \mathcal{O}_P)$$

on P is the trivial bundle. Of course this is often not the case. However if we assume, more generally, that there exists an *effective* divisor

$$\sum_j x_j$$

on P representing the line bundle $\omega_X^{-1} \otimes \omega_P$ with x_j distinct, and denote by

$$\tilde{X} \to X \tag{3}$$

the blow-up of X at the points x_j containing the proper transform $\tilde{P}$ of P, then

$$\omega_{\tilde{X}}\big|_{\tilde{P}} = \omega_{\tilde{P}} \, .$$

Assuming (2) from now on, we obtain

$$\mathcal{O}_P = \mathcal{E}xt^2_X\left(\mathcal{O}_P, \mathcal{O}_X\right) = \mathcal{E}xt^1_X\left(\mathcal{I}_P, \mathcal{O}_X\right). \tag{4}$$

Since, for $k = 1, 2$,

$$H^k\left(\mathcal{H}om_X\left(\mathcal{I}_P, \mathcal{O}_X\right)\right) = H^k\left(\mathcal{O}_X\right) = 0\,,$$

we have

$$\operatorname{Ext}^1\left(\mathcal{I}_P, \mathcal{O}_X\right) = H^0\left(\mathcal{E}xt^1_X\left(\mathcal{I}_P, \mathcal{O}_X\right)\right)$$

and a nowhere-vanishing section of $\mathcal{E}xt^1_X\left(\mathcal{I}_P, \mathcal{O}_X\right)$ gives, via the Serre construction, an exact sequence of vector bundles

$$0 \to \mathcal{O}_X \to E \to \mathcal{I}_P \to 0. \tag{5}$$

Also E has a distinguished global section s_P vanishing exactly at P. Also, since $\det(E) = \mathcal{O}_X$, we have

$$E^\vee = E\,. \tag{6}$$

Tensor the surjection

$$E^\vee \to \mathcal{I}_P \tag{7}$$

induced by s with $\mathcal{O}_P$ to obtain a surjection of rank-2 bundles on P which is therefore an isomorphism. Dualize to obtain an isomorphism

$$N_{P/X} \to E|_P\,.$$

Thus we have the isomorphism

$$H^{\cdot}\left(N_{P/X}\right) = H^{\cdot+1}\left(\mathcal{I}_P \otimes E \to E\right). \tag{8}$$

On the other hand, (5) and (6) give the exact sequence

$$0 \to E \to E^\vee \otimes E \to \mathcal{I}_P \otimes E \to 0 \tag{9}$$

from which comes

$$H^{\cdot}\left(E^\vee \otimes E\right) \to H^{\cdot}\left(\mathcal{I}_P \otimes E\right). \tag{10}$$

Together (8) and (10) relate the deformation functor of E to that of P.

3.2 Quaternionic line bundles. Now $\det(E) = \mathcal{O}_X$, allowing us to choose a non-vanishing holomorphic section

$$1 \in H^0(\det(E)).$$

We next chose a hermitian metric μ on E such that 1 has norm 1 with respect to the metric on $\det(E)$ induced by μ. We have an associated structure of a *quaternionic line bundle* on E by defining

$$(\mathbf{j} \cdot s')$$

for any (locally defined) C^∞-section s' of E as the unique element such that we have an equality of linear operators

$$\frac{s \wedge (\mathbf{j} \cdot s')}{1} = \mu(s, s')$$

on sections s of E. Thus

$$\begin{aligned} (\mathbf{j} \cdot s) &\perp s \\ s \wedge (\mathbf{j} \cdot s) &= \|s\|^2 \cdot 1 . \end{aligned} \tag{11}$$

Since

$$s \wedge (\mathbf{j} \cdot i \cdot s') = \mu(s, i \cdot s') = -i \cdot \mu(s, s') = s \wedge (-i \cdot \mathbf{j} \cdot s'),$$

we have a well-defined (left) action on E by the group $\mathbb{H}^*$ of non-zero quaternions and the action of $\mathbf{j}\cdot$ is conjugate linear. Since

$$\begin{aligned} s \wedge (\mathbf{j} \cdot s) &= -(\mathbf{j} \cdot \mathbf{j} \cdot s) \wedge (\mathbf{j} \cdot s) \\ &= (\mathbf{j} \cdot s) \wedge (\mathbf{j} \cdot \mathbf{j} \cdot s) \\ &= \|\mathbf{j} \cdot s\|^2 \cdot 1 \end{aligned}$$

$\mathbf{j}\cdot$ acts as a (conjugate-linear) isometry. Also, for any two sections s and s', we have that

$$\mathbf{j} \cdot s \wedge \mathbf{j} \cdot s' = \mu(\mathbf{j} \cdot s, s') = \overline{\mu(s', \mathbf{j} \cdot s)} = \overline{s' \wedge \mathbf{j} \cdot \mathbf{j} \cdot s} = \overline{s \wedge s'}.$$

For a fixed non-zero local section s_0 of E, we have the framing $(s_0, \mathbf{j} \cdot s_0)$ as a complex vector bundle, and, for any section s, we have the formula

$$\begin{aligned} s &= \frac{\mu(s, s_0)}{\|s_0\|^2} \cdot s_0 + \frac{\mu(s, \mathbf{j} \cdot s_0)}{\|s_0\|^2} \cdot (\mathbf{j} \cdot s_0) && (12) \\ &= \frac{s \wedge (\mathbf{j} \cdot s_0)}{\|s_0\|^2} \cdot s_0 - \frac{s \wedge s_0}{\|s_0\|^2} \cdot (\mathbf{j} \cdot s_0) && (13) \end{aligned}$$

with respect to the unitary basis

$$\left(\frac{s_0}{\|s_0\|}, \mathbf{j} \cdot \frac{s_0}{\|s_0\|} \right).$$

Now suppose that $D_{\mu,P}$ denotes the metric $(1,0)$ -connection on E with respect to the metric μ. Then we have that

$$d\left(\mu\left(s,s'\right)\right)=\mu\left(D_{\mu.P}\left(s\right),s'\right)+\mu\left(s,D_{\mu.P}\left(s'\right)\right)$$

so that

$$d\left(s\wedge\mathbf{j}\cdot s'\right)=D_{\mu,P}\left(s\right)\wedge\mathbf{j}\cdot s'+s\wedge\mathbf{j}\cdot D_{\mu,P}\left(s'\right)$$

from which follows that

$$d\left(s\wedge s'\right)=D_{\mu,P}\left(s\right)\wedge s'-s\wedge\mathbf{j}\cdot D_{\mu,P}\left(\mathbf{j}\cdot s'\right).$$

Since

$$d\left(s\wedge s'\right)=D_{\mu,P}\left(s\right)\wedge s'+s\wedge D_{\mu,P}\left(s'\right)$$

it follows that $\mathbf{j}\cdot$ commutes with $D_{\mu,P}$ so that $D_{\mu,P}$ is a quaternionic connection. Since $-\mathbf{j}\cdot a\cdot\mathbf{j}=\overline{a}$,

$$\begin{aligned}-\mathbf{j}\cdot D^{1,0}_{\mu,P}\cdot\mathbf{j}\cdot &= D^{0,1}_{\mu,P}\\ -\mathbf{j}\cdot D^{0,1}_{\mu,P}\cdot\mathbf{j}\cdot &= D^{1,0}_{\mu,P}\,.\end{aligned}$$

Thus

$$D^{1,0}_{\mu,P}=-\left(\mathbf{j}\cdot\right)\circ\overline{\partial}\circ\left(\mathbf{j}\cdot\right).$$

Now let $s=s_P$, the holomorphic section of E. If we write

$$\begin{aligned}D_{\mu,P}\left(s_P\right)=D^{1,0}_{\mu,P}\left(s_P\right)&=-\mathbf{j}\cdot\overline{\partial}\cdot\mathbf{j}\cdot s_P\\ &=\alpha_P\cdot s_P+\beta_P\cdot\mathbf{j}\cdot s_P\end{aligned}$$

for forms α_P and β_P of type $(1,0)$, then

$$\begin{aligned}D_{\mu,P}\left(\mathbf{j}\cdot s_P\right)=\mathbf{j}\cdot D_{\mu,P}s_P&=\mathbf{j}\cdot\left(\alpha_P\cdot s_P+\beta_P\cdot\mathbf{j}\cdot s_P\right)\\ &=-\overline{\beta_P}\cdot s_P+\overline{\alpha_P}\cdot\mathbf{j}\cdot s_P\,.\end{aligned}$$

And since

$$\begin{aligned}\left(D_{\mu,P}\left(\mathbf{j}\cdot s_P\right)\right)^{0,1}\wedge s_P&=\overline{\partial}\left(\mathbf{j}\cdot s_P\right)\wedge s_P\\ &=-\overline{\partial}\left(\|s_P\|^2\right)\end{aligned}$$

we can write

$$D_{\mu,P}\begin{pmatrix}s_P\\ \mathbf{j}\cdot s_P\end{pmatrix}=\begin{pmatrix}\partial\log\left(\|s_P\|^2\right) & \beta_P\\ -\overline{\beta_P} & \overline{\partial}\log\left(\|s_P\|^2\right)\end{pmatrix}\begin{pmatrix}s_P\\ \mathbf{j}\cdot s_P\end{pmatrix}. \tag{14}$$

With respect to the basis $(s_P,\mathbf{j}\cdot s_P)$, the curvature of the connection $D_{\mu,P}$ then becomes

$$\begin{pmatrix} \overline{\partial}\partial \log\left(\|s_P\|^2\right) & d\beta_P \\ -d\overline{\beta_P} & \partial\overline{\partial}\log\left(\|s_P\|^2\right) \end{pmatrix}$$

$$+\begin{pmatrix} -\beta_P \wedge \overline{\beta_P} & \left(\partial-\overline{\partial}\right)\log\left(\|s_P\|^2\right)\wedge\beta_P \\ \overline{\beta_P}\wedge\left(\overline{\partial}-\partial\right)\log\left(\|s_P\|^2\right) & \beta_P\wedge\overline{\beta_P} \end{pmatrix}.$$

Since this matrix must be of type (1, 1) we conclude

$$\partial\beta_P = \beta_P \wedge \partial\log\left(\|s_P\|^2\right) \tag{15}$$

and the curvature matrix becomes

$$R_{\mu,P} = \begin{pmatrix} \overline{\partial}\partial\log\left(\|s_P\|^2\right) - \beta_P\wedge\overline{\beta_P} & \overline{\partial}\beta_P + \beta_P\wedge\overline{\partial}\log\left(\|s_P\|^2\right) \\ -\partial\overline{\beta_P} - \overline{\beta_P}\wedge\partial\log\left(\|s_P\|^2\right) & \partial\overline{\partial}\log\left(\|s_P\|^2\right) + \beta_P\wedge\overline{\beta_P} \end{pmatrix}. \tag{16}$$

3.3 Chern–Simons functional. Following [3] define the holomorphic Chern–Simons functional as follows. For any two connections D_0, D_1 on a C^∞-vector bundle E, form the connection

$$\tilde{D} := D_0 + t\left(D_1 - D_0\right) + d_t$$

on $X \times [0, 1]$, and let

$$\tilde{R} = \tilde{D}^2$$

denote its curvature form. Let $A_X^{\cdot}$ denote the $\mathbb{C}$-deRham complex in X and let $F^{\cdot}$ denote the Hodge filtration on $A_X^{\cdot}$.

Definition 3.1. *The holomorphic Chern–Simons current* $CS_{D_0}\left(D_1\right)$ *is the current of type* $(1, 2) + (0, 3)$ *given by the functional*

$$\begin{aligned} CS_{D_0}\left(D_1\right) \ &: \ F^2A_X^3 \to \mathbb{C} \\ \tau &\mapsto \int_{X\times[0,1]} \tau\wedge tr(\tilde{R}\wedge\tilde{R})\,. \end{aligned}$$

(This is an extension of the standard definition of holomorphic Chern–Simons functional which refers to the restriction of the above current to $F^3H^3\left(X\right)$*.)*

Since $\mathrm{tr}(\tilde{R}\wedge\tilde{R})$ pulls back to an exact form on the associated principal bundle of frames of E (see §3 of [1]), one has for three connections D_0, D_1, D_2 that

$$CS_{D_0}\left(D_1\right) + CS_{D_1}\left(D_2\right) = CS_{D_0}\left(D_2\right). \tag{17}$$

Let

$$A := D_1 - D_0$$

We compute $\tilde{R}$ as follows

$$(D_0 + tA + d_t) \circ (D_0 + tA + d_t) = R_0 + t \cdot D_0 A + t^2 \cdot A \wedge A + dt \wedge A$$

so that

$$\tau \wedge \tilde{R} \wedge \tilde{R} = \tau \wedge dt \wedge 2A \wedge \left(R_0 + t \cdot D_0 A + t^2 \cdot A \wedge A\right)$$

and finally

$$CS_{D_0}(D_1)(\tau) = \int_X \tau \wedge \operatorname{tr}\left(A \wedge \left(R_0 + D_0 A + \frac{2}{3} A \wedge A\right)\right). \tag{18}$$

Suppose that $D_0^{0,1}$ and $D_1^{0,1}$ both give complex structures on E, that is

$$R_0^{0,2} = \left(D_0^{0,1}\right)^2 = 0 = \left(D_1^{0,1}\right)^2 .$$

Then

$$0 = \left(D_0^{0,1} + A^{0,1}\right)^2 = D_0^{0,1} A^{0,1} + A^{0,1} \wedge A^{0,1} .$$

Thus in this case

$$\operatorname{tr}\left(A \wedge (R_0 + D_0 A + \frac{2}{3} A \wedge A)\right)^{0,3} = -\frac{1}{3} \operatorname{tr}\left(\left(A^{0,1}\right)^{\wedge 3}\right). \tag{19}$$

On the other hand, suppose that D_0 and D_1 are both flat. Then $R_0 = 0$ and

$$0 = (D_0 + A)^2 = D_0 A + A \wedge A .$$

so that (18) becomes

$$CS_{D_0}(D_1)(\tau) = -\frac{1}{3} \int_X \tau \wedge \operatorname{tr}(A \wedge A \wedge A) . \tag{20}$$

3.4 Comparing connections via Chern–Simons theory. Let D_P denote the unique $\mathbb{H}$-connection on the restriction E' of E to

$$X' := X - |P|$$

such that $D_P(s_P) = 0$. (D_P is of course flat.) We first apply the Chern–Simons theory to the two connections

$$D_0 = D_P = D_P^{1,0} + D_P^{0,1} \text{ and } D_1 = D_P^{1,0} + D_{\mu,P}^{0,1} =: D'_P \tag{21}$$

where, as above, $D_{\mu,P}$ is the metric $(1,0)$-connection associated to μ and the complex structure on E. Thus D'_P is a $(1,0)$-connection for the complex structure on E. Let

$$A_P := D'_P - D_P .$$

Then by (14)

$$A_P \begin{pmatrix} s_P \\ \mathbf{j} \cdot s_P \end{pmatrix} = \begin{pmatrix} 0 & 0 \\ -\overline{\beta_P} & \bar{\partial} \log \|s_P\|^2 \end{pmatrix} \begin{pmatrix} s_P \\ \mathbf{j} \cdot s_P \end{pmatrix}.$$

Also, in terms of the basis $(s_P, \mathbf{j} \cdot s_P)$ we can compute the curvature

$$\begin{aligned} R'_P &= \left(D'_P\right)^2 = (D_P + A_P)^2 \\ &= \begin{pmatrix} 0 & 0 \\ -d\overline{\beta_P} - \bar{\partial} \log \|s_P\|^2 \wedge \overline{\beta_P} & \partial\bar{\partial} \log \|s_P\|^2 \end{pmatrix} \\ &= \begin{pmatrix} 0 & 0 \\ -\partial\overline{\beta_P} & \partial\bar{\partial} \log \|s_P\|^2 \end{pmatrix} \end{aligned} \tag{22}$$

using (15).

By (18) and (21), for any (3, 0) + (2, 1)-form τ, the Chern–Simons functional $CS_{D_P}\left(D'_P\right)(\tau)$ is given by the expression

$$\int_{X'} \operatorname{tr}\left(\bar{\partial} \log \|s_P\|^2 \wedge \partial\bar{\partial} \log \|s_P\|^2\right) \wedge \tau \tag{23}$$

By (18) and Stokes theorem, if τ is d-closed, this reduces to

$$\begin{aligned} CS_{D_P}\left(D'_P\right)(\tau) &= \int_{X'} \left(\bar{\partial} \log \|s_P\|^2 \wedge \partial\bar{\partial} \log \|s_P\|^2\right) \wedge \tau \\ &= \int_{X'} \left(d \log \|s_P\|^2 \wedge \partial\bar{\partial} \log \|s_P\|^2\right) \wedge \tau \\ &= \int_{\partial X'} \left(\log \|s_P\|^2 \wedge \partial\bar{\partial} \log \|s_P\|^2\right) \wedge \tau \, , \end{aligned} \tag{24}$$

where $\int_{\partial X'}$ is computed as

$$\lim_{\varepsilon \to 0} \int_{B_\varepsilon} \left(\log \|s_P\|^2 \wedge \partial\bar{\partial} \log \|s_P\|^2\right) \wedge \tau$$

where B_ε is a tubular neighborhood of P in X, of radius ε in some fixed metric. The integrand is bounded by

$$const. \cdot \frac{(\log(\varepsilon))^2}{\varepsilon^2} \, .$$

So the above integral is bounded by a constant multiple of $\varepsilon \cdot (\log(\varepsilon))^2$, which tends to zero as $\varepsilon \to 0$. Thus

$$CS_{D_P}\left(D'_P\right) \sim 0 \, .$$

Finally, since

$$D_{\mu,P} - D'_P = A^{1,0}$$

is of type $(1,0)$ we again use (18) to conclude by type that for any form τ of type $(3,0)+(2,1)$,

$$CS_{D'_P}\left(D_{\mu,P}\right)(\tau) = \int_X \tau \wedge \operatorname{tr}\left(A^{1,0} \wedge \left(R'_P + dA^{1,0} + \frac{2}{3}A^{1,0} \wedge A^{1,0}\right)\right) \quad (25)$$
$$= 0$$

since R'_P is of type $(1,1)$. So by the additivity formula (17) we conclude

$$CS_{D_P}\left(D_{\mu,P}\right) \sim 0\,, \quad (26)$$

that is, the $(1,2)+(0,3)$ current $CS_{D_P}\left(D_{\mu,P}\right)$ is d-exact in the sense of currents.

4 Analogue of the First Part of Abel's Theorem

4.1 Changing the curve P. Let X and P be as in the previous section. Suppose now we have another irreducible curve $Q \subseteq X$ with

$$\omega_X^{-1} \otimes \omega_Q = \mathcal{O}_Q\,.$$

We suppose there is a fixed C^∞-vector bundle, which we call E_∞ such that, as C^∞-vector bundles, we have isomorphisms

$$E_P \leftrightarrow E_\infty \leftrightarrow E_Q\,. \quad (27)$$

We choose a fixed holomorphic structure on $\det(E_\infty)$ and choose the C^∞-isomorphisms (27) so as to induce holomorphic isomorphisms

$$\det(E_P) \leftrightarrow \det(E_\infty) \leftrightarrow \det\left(E_Q\right)\,.$$

We next choose a hermitian structure μ on E_∞ such that a fixed holomorphic section 1 of $\det(E_\infty)$ has norm 1 in the metric induced by μ. We then proceed as above to make E_∞, and therefore also E_P and E_Q, into quaternionic line bundles such that the above correspondences are C^∞-isomorphisms of quaterionic line bundles. Again considering s_P and s_Q as sections of the quaternionic line bundle E_∞ we have

$$s_Q = g \cdot s_P$$

where, for $X' = X - (|P| \cup |Q|)$,

$$g = \begin{pmatrix} a & b \\ -\overline{b} & \overline{a} \end{pmatrix} : X' \to \mathbb{H}^*\,.$$

That is

$$s_Q = a \cdot s_P + b \cdot (\mathbf{j} \cdot s_P)\,.$$

Then

$$\begin{aligned}\mathbf{j}\cdot s_Q &= \mathbf{j}\cdot a\cdot s_P + \mathbf{j}\cdot b\cdot (\mathbf{j}\cdot s_P)\\ &= -\overline{b}\cdot s_P + \overline{a}\cdot (\mathbf{j}\cdot s_P)\end{aligned}$$

so that the matrix g gives the expression for the $\mathbb{C}$-basis $\left(s_Q, \mathbf{j}\cdot s_Q\right)$ in terms of the $\mathbb{C}$-basis $(s_P, \mathbf{j}\cdot s_P)$, that is,

$$\begin{pmatrix} s_Q \\ \mathbf{j}\cdot s_Q \end{pmatrix} = g\cdot \begin{pmatrix} s_P \\ \mathbf{j}\cdot s_P \end{pmatrix}.$$

We are interested in comparing two connections on E', namely the $\mathbb{H}$-connection D_P with flat section s_P and the $\mathbb{H}$ -connection D_Q with flat section s_Q. Again we compute

$$\begin{aligned}0 &= D_Q\,(g\cdot s_P) = dg\cdot s_P + g\cdot D_Q s_P\\ D_Q s_P &= -g^{-1}\cdot dg\cdot s_P\,.\end{aligned}$$

So

$$\left(D_Q - D_P\right)(s_P) = -g^{-1}\cdot dg\cdot s_P\,.$$

Since $\mathbb{H}$-connections are determined by their values on a single section, then in terms of the basis $(s_P, \mathbf{j}\cdot s_P)$, we have

$$A_{PQ} := D_Q - D_P = -g^{-1}\cdot dg\,.$$

Since D_P and D_Q are both flat,

$$0 = \left(D_P + A_{PQ}\right)^2 = dA_{PQ} + A_{PQ}\wedge A_{PQ}\,.$$

Write

$$h = \begin{bmatrix} a & b \\ -\overline{b} & \overline{a} \end{bmatrix} = r\cdot \begin{bmatrix} u & v \\ -\overline{v} & \overline{u} \end{bmatrix}$$

with $u\overline{u} + v\overline{v} = 1$. So for

$$\varkappa = \begin{bmatrix} u & v \\ -\overline{v} & \overline{u} \end{bmatrix} = u + v\cdot \mathbf{j}$$

we have

$$\varkappa^{-1} = \overline{u} - v\cdot \mathbf{j}$$

since

$$\begin{aligned}(u + v\cdot \mathbf{j})\cdot (\overline{u} - v\cdot \mathbf{j}) &= u\overline{u} + v\cdot \mathbf{j}\cdot \overline{u} - uv\cdot \mathbf{j} - v\cdot \mathbf{j}\cdot v\cdot \mathbf{j}\\ &= u\overline{u} + vu\cdot \mathbf{j} - uv\cdot \mathbf{j} + v\overline{v}\\ &= 1\,.\end{aligned}$$

The form

$$\begin{aligned} h^{-1} \cdot dh &= d \log r + \overline{\varkappa}^t d\varkappa \\ &= r^{-1} dr + (\overline{u} - v \cdot \mathbf{j})\,(du + dv \cdot \mathbf{j}) \\ &= r^{-1} dr + (\overline{u} du + v d\overline{v}) + (\overline{u} dv - v d\overline{u}) \cdot \mathbf{j} \end{aligned}$$

is a left-invariant 1-form on $\mathbb{H}^* = (0, +\infty) \times SU(2)$. Since $u\overline{u} + v\overline{v} = 1$, the form

$$\overline{u} du + v d\overline{v} = -\,(u d\overline{u} + v d\overline{v})$$

is purely imaginary.

Using the rule

$$(A + B \cdot \mathbf{j}) \wedge (C + D \cdot \mathbf{j}) = \left(A \wedge C - B \wedge \overline{D}\right) + \left(A \wedge D + B \wedge \overline{C}\right) \cdot \mathbf{j}$$

we have that, if $\overline{A} = -A$,

$$(A + B \cdot \mathbf{j}) \wedge (A + B \cdot \mathbf{j}) = -B \wedge \overline{B} - 2\left(\overline{A} \wedge B\right) \cdot \mathbf{j}$$

and

$$(A + B \cdot \mathbf{j})^{\wedge^3} = 3\overline{A} \wedge B \wedge \overline{B}\,.$$

We compute

$$\begin{aligned} \left(h^{-1} dh\right)^{\wedge^3} &= r^{-1} dr \wedge ((\overline{u} du + v d\overline{v}) + (\overline{u} dv - v d\overline{u}) \cdot \mathbf{j})^{\wedge^2} \\ &\quad + ((\overline{u} du + v d\overline{v}) + (\overline{u} dv - v d\overline{u}) \cdot \mathbf{j})^{\wedge^3} \end{aligned}$$

and

$$\overline{(\overline{u} du + v d\overline{v})} \wedge (\overline{u} dv - v d\overline{u}) = d\overline{u} dv\,.$$

So

$$\begin{aligned} ((\overline{u} du + v d\overline{v}) + (\overline{u} dv - v d\overline{u}) \cdot \mathbf{j})^{\wedge^3} &= 3 d\overline{u} dv \wedge (u d\overline{v} - \overline{v} du) \\ &= 3\,(u d\overline{u} dv d\overline{v} - \overline{v} du d\overline{u} dv)\,. \end{aligned}$$

On the other hand

$$0 = d\left(h \cdot h^{-1}\right) = \left(dh^{-1}\right) \cdot h + h^{-1} \cdot dh$$

so that

$$d\left(h^{-1} \cdot dh\right) = -h^{-1} \cdot dh \wedge h^{-1} \cdot dh \tag{28}$$

and

$$\left(h^{-1}\cdot dh\right)^{\wedge 3}=-d\log r\wedge d\left(\varkappa^{-1}\cdot d\varkappa\right)+\left(\varkappa^{-1}\cdot d\varkappa\right)^{\wedge 3}\ . \tag{29}$$

So

$$\operatorname{tr}\left(\left(\varkappa^{-1}\cdot d\varkappa\right)^{\wedge 3}\right)=3\left(ud\overline{u}dvd\overline{v}-\overline{v}dud\overline{u}dv\right)+3\overline{\left(ud\overline{u}dvd\overline{v}-\overline{v}dud\overline{u}dv\right)} \tag{30}$$

is real and d-closed on X', as is

$$\operatorname{tr}\left(\left(h^{-1}\cdot dh\right)^{\wedge 3}\right)\ .$$

Since the left-invariant form (30) evaluated at $(u,v)=(1,0)$ is

$$12d\left(\operatorname{Im}a\right)\wedge d\left(\operatorname{Re}b\right)\wedge d\left(\operatorname{Im}b\right)$$

we have

$$\int_{S^3}\operatorname{tr}\left(\left(h^{-1}\cdot dh\right)^{\wedge 3}\right)=12\cdot\operatorname{vol}\left(S^3\right)=24\pi^2\ .$$

We compare these connections on $E_\infty|_{X'}$ via the Chern–Simons functional. Now

$$A_{PQ}=-g^*\left(h^{-1}\cdot dh\right)$$

for the invariant one-form $h^{-1}\cdot dh$ on $\mathbb{H}^*$. Then by (28) and (29) we have the formula

$$\begin{aligned}CS_{D_P}\left(D_Q\right)(\tau)&=\int_X g^*\operatorname{tr}\left(h^{-1}\cdot dh\wedge d\left(h^{-1}\cdot dh\right)-\frac{2}{3}\left(h^{-1}\cdot dh\right)^{\wedge 3}\right)\wedge\tau \qquad (31)\\&=\frac{1}{3}\int_X g^*\operatorname{tr}\left(\left(h^{-1}\cdot dh\right)^{\wedge 3}\right)\wedge\tau\ .\end{aligned}$$

Analogously to the classical Abel theorem on curves, via residue and the fact that

$$\lim_{r\to 0} r\log r=0\ ,$$

we have cohomologous currents

$$\operatorname{tr}\left(\left(h^{-1}\cdot dh\right)^{\wedge 3}\right)\sim\operatorname{tr}\left(\varkappa^{-1}\cdot d\varkappa\right)^{\wedge 3}\sim 24\pi^2\int_{(0,+\infty)}$$

on $\overline{\mathbb{H}^*}=\mathbb{HP}^1$. Pulling back via g we therefore have by (31) that, whenever τ is a d-closed form of type $(3,0)+(2,1)$,

$$CS_{D_P}\left(D_Q\right)(\tau)=\frac{24\pi^2}{3}\int_P^Q\tau\ . \tag{32}$$

Here on the right-hand side we integrate over the 3-chain

$$g^{-1}\left((0,\infty)\right)\ ,$$

which indeed bounds $Q - P$. This is of course completely analogous, for rank-2 vector bundles, to the main step in the above version of the proof of classical Abel's theorem for line bundles on curves. Thus

$$CS_{D_P}\left(D_Q\right) = 8\pi^2 \int_P^Q$$

is d-exact in the sense of currents. Since

$$\operatorname{tr}\left(\left(h^{-1} \cdot dh\right)^{\wedge 3}\right)$$

is d-closed on $\mathbb{H}^*$, $CS_{D_P}\left(D_Q\right)$ is the $(1,2)+(0,3)$ summand of a d-closed form on X', not of a d-closed current on X. Indeed, by what we have just shown,

$$\frac{1}{3}\int_X g^* \operatorname{tr}\left(\left(h^{-1} \cdot dh\right)^{\wedge 3}\right) \wedge d\beta = 8\pi^2 \int_P^Q d\beta = 8\pi^2 \left(\int_Q \beta - \int_P \beta\right).$$

Finally by (32), (26) and the additivity property (17), we have that

$$CS_{D_{\mu,P}}\left(D_{\mu,Q}\right) - CS_{D_P}\left(D_Q\right)$$

is a current coboundary and so, for any d-closed form τ of type $(3,0)+(2,1)$ that

$$CS_{D_{\mu,P}}\left(D_{\mu,Q}\right)(\tau) = 8\pi^2 \int_Q^P \tau\ . \tag{33}$$

4.2 P and Q Abel–Jacobi equivalent. Let $A_{\mu,PQ} = D_{\mu,Q} - D_{\mu,P}$, where we recall that

$$A_{\mu,PQ}^{0,1} = \overline{\partial}_Q - \overline{\partial}_P \text{ and } A_{\mu,PQ}^{1,0} = -\mathbf{j} \cdot \left(\overline{\partial}_Q - \overline{\partial}_P\right) \cdot \mathbf{j}\ .$$

Let

$$\alpha_{\mu,PQ} := \operatorname{tr}\left(A_{\mu,PQ} \wedge \left(R_{\mu,P} + dA_{\mu,PQ} + \frac{2}{3} A_{\mu,PQ} \wedge A_{\mu,PQ}\right)\right) \tag{34}$$

be the C^∞-de Rham form giving $CS_{D_{\mu,P}}\left(D_{\mu,Q}\right)$ so that

$$CS_{D_{\mu,P}}\left(D_{\mu,Q}\right)(\tau) = \int_X \tau \wedge \alpha_{\mu,PQ} = \int_X \tau \wedge \alpha_{\mu,PQ}^{(1,2)+(0,3)}\ .$$

If P and Q are algebraically equivalent, we can arrange that Γ with

$$\partial\Gamma = Q - P$$

be chosen to lie inside a (possibly reducible) algebraic surface on X. Suppose now that

$$\tau = \overline{\partial}\beta^{2,0}\ .$$

Then, since any (3, 0)-form restricts to zero on Γ, we conclude by (33) that

$$\int_X \tau \wedge \alpha_{\mu,PQ}^{(1,2)+(0,3)} = \int_X \overline{\partial}\beta^{2,0} \wedge \alpha_{\mu,PQ} = 8\pi^2 \int_\Gamma \overline{\partial}\beta^{2,0}$$
$$= 8\pi^2 \int_\Gamma d\beta^{2,0} = 8\pi^2 \left(\int_Q \beta^{2,0} - \int_P \beta^{2,0} \right) = 0 .$$

Thus:

Lemma 4.1. *If P and Q are algebraically equivalent, $\alpha_{\mu,PQ}^{(1,2)+(0,3)}$ is $\overline{\partial}$-closed.*

Referring to (34) and writing the distribution-valued differential

$$\alpha_{PQ} := \mathrm{tr}\left(A_{PQ} \wedge \left(dA_{PQ} + \frac{2}{3} A_{PQ} \wedge A_{PQ} \right) \right) = -\frac{1}{3}\mathrm{tr}(A_{PQ} \wedge A_{PQ} \wedge A_{PQ}) ,$$

then

$$CS_{D_P}\left(D_Q\right)(\tau) = \int_X \tau \wedge \alpha_{PQ} = \int_X \tau \wedge \alpha_{PQ}^{(1,2)+(0,3)} .$$

Again if P is algebraically equivalent to Q, by (32), Γ can be chosen so that

$$\int_X \tau \wedge \alpha_{PQ}^{(1,2)+(0,3)} = 0$$

whenever $\tau \in F^2 A_X^3$ is $\overline{\partial}$-exact. Thus:

Lemma 4.2. *If P is algebraically equivalent to Q, then $\alpha_{PQ}^{(1,2)+(0,3)}$ is $\overline{\partial}$-closed.*

The equality (33) shows that, if P is algebraically equivalent to Q and

$$\int_Q^P = \int_\varepsilon : F^2 H^3 (X) \to \mathbb{C}$$

for some $\varepsilon \in H_3 (X; \mathbb{Z})$, then we have the containment

$$\left\{ \alpha_{\mu,PQ}^{(1,2)+(0,3)} \right\} \in H^3 (X; \mathbb{Z}) + F^2 H^3 (X)$$

of the Dolbeault class of $\alpha_{\mu,PQ}^{(1,2)+(0,3)}$. So the Poincaré dual of ε is a de Rham class

$$\{\xi\} \in H^3 (X; \mathbb{Z})$$

with the property that

$$\alpha_{\mu,PQ}^{(1,2)+(0,3)} - \xi^{(1,2)+(0,3)} = \overline{\partial}\gamma_\mu$$

for some C^∞-form

$$\gamma_\mu \in A_X^{(1,1)+(0,2)} .$$

Then the form

$$\psi_\mu := \alpha_{\mu,PQ} - \xi - \overline{\partial}\gamma_\mu$$

on X is of type $(3,0)+(2,1)$.

The equality (32) shows that, if P is algebraically equivalent to Q and

$$\int_Q^P = \int_\varepsilon : F^2H^3(X) \to \mathbb{C}$$

for some $\varepsilon \in H_3(X;\mathbb{Z})$, then we have as above, taking the de Rham dual

$$\{\xi\} \in H^3(X;\mathbb{Z})$$

of ε, we have

$$\int_X (\alpha_{PQ} - \xi) \wedge \tau = 0$$

for any d-closed form τ of type $(3,0)+(2,1)$ and any $\overline{\partial}$-exact form of type $(3,0)+(2,1)$, hence for any $\overline{\partial}$-closed form of type $(3,0)+(2,1)$. This means that the current $(\alpha_{PQ} - \xi)^{(1,2)+(0,3)}$ is $\overline{\partial}$-exact, and so, in particular,

$$\alpha_{PQ}^{0,3} - \xi^{0,3} = \overline{\partial}\gamma^{0,2}$$

for some distribution-valued form $\gamma^{0,2}$ of type $(0,2)$. Then the current

$$\alpha_{PQ} - \xi - d\gamma^{0,2}$$

on X is of type $(3,0)+(2,1)+(1,2)$ and is d-closed on X'. Since $\alpha_{PQ} - \xi - d\gamma$ integrates to zero against any $\overline{\partial}$-closed form of type $(3,0)+(2,1)$ on X, the current $(\alpha_{PQ} - \xi - d\gamma)^{1,2}$ is again $\overline{\partial}$-exact, we can find a current $\gamma^{1,1}$ such that the distribution-valued form

$$\psi : \alpha_{PQ} - \xi - d\gamma^{0,2} - d\gamma^{1,1}$$

on X is a current of type $(3,0)+(2,1)$ and is d-closed on X'.

Of course, in the case of classical Abel's theorem, P and Q are always algebraically equivalent and ψ gives directly the rational equivalence of P and Q. ψ_μ does not enter the picture because $\alpha^{0,1}$ ($= \alpha_\mu^{0,1}$ for any metric μ) is already smooth. In the threefold case, (23) and (25) imply that

$$\alpha_{PQ}^{0,3} = \alpha_{\mu,PQ}^{0,3}$$

for any metric μ but in general the currents $\alpha_{PQ}^{1,2}$ and $\alpha_{\mu,PQ}^{1,2}$ are not equal. However it should be true that, as we vary the metric μ nicely so that it becomes flat on X', we achieve that $\alpha_{\mu,PQ}^{1,2}$ converges to $\alpha_{PQ}^{1,2}$.

References

[1] Chern, S. S., Simons, J., *Characteristic forms and geometrical invariants*, Annals of Math. **99** (1974), 48–69.

[2] Clemens, H., *Cohomology and Obstructions II: Curves on K-trivial threefolds*. Preprint, math.AG/0206219 (2002).

[3] Donaldson, S. K., Thomas, R., *Gauge theory in higher dimensions*. The Geometric Universe: Science, Geometry and the work of Roger Penrose, Edited by S. A. Huggett et al., Oxford University Press, 1998.

[4] King, J. R., *The currents defined by analytic varieties*, Acta Math. **127** (1971), 185–220.

Arithmetic Questions Related to Rationally Connected Varieties

Tom Graber, Joe Harris, Barry Mazur, and Jason Starr

Contents

1 The Classical Theory: Homogeneous Forms over Finite Fields, and over Fields of Transcendence Degree 1 over $\mathbb{C}$

Let $X \subset \mathbb{P}^n$ be a nonempty variety defined over a field K. Under what combination of conditions on the geometry of X and the algebra of K can we be sure that X has a K-rational point?

There are two extreme cases: if $X \subset \mathbb{P}^n$ is a linear space, no condition is needed on K; and of course if K is algebraically closed we need no condition on X beyond being nonempty. But in between is a range of fascinating questions about classes of fields over which certain types of equations are "guaranteed to have points" and, reciprocally, classes of varieties that "should" have points over specific types of fields.

Consider to start with the two remarkable companion theorems, due to Chevalley–Warning and to Tsen, which have been pivotal in suggesting possible answers to our general question. (For an excellent general reference to this theory, see [6]; for the source articles, see [19], [21], and [15].)

Chevalley–Warning: Let k be a finite field. A homogeneous polynomial

$$f(x_1, x_2, \dots, x_n)$$

of degree strictly less than n (its number of variables) with coefficients in k has a *nontrivial* zero over k. Moreover, the number of rational points over k of the hypersurface $X \subset \mathbb{P}^{n-1}$ cut out by the homogeneous form f is congruent to 1 modulo the characteristic p of k.

Tsen: Let K be the function field of a curve over $\mathbb{C}$ (equivalently: a field of transcendence degree 1 over $\mathbb{C}$). A hypersurface $X \subset \mathbb{P}^{n-1}$ over K, of degree d strictly less than n, has a rational point over K.

Let $\mathcal{V}$ be a class of varieties, where we are using the word *class* informally. Let $\mathcal{F}(\mathcal{V})$ be the class of fields such that any variety $V \in \mathcal{V}$ defined over a field $F \in \mathcal{F}(\mathcal{V})$ has an F-rational point (we might call $\mathcal{F}(\mathcal{V})$ the class of fields *cut out* by $\mathcal{V}$). Dually, if $\mathcal{F}$ is a class of fields, $\mathcal{V}(\mathcal{F})$ will denote the class of varieties such that any $V \in \mathcal{V}(\mathcal{F})$ defined over a field $F \in \mathcal{F}$ has an F-rational point.

In the traditional literature, the classes of varieties $\mathcal{V}$ that are used to "cut out classes of interesting fields" in this way have almost always been only hypersurfaces. For example, if $\mathcal{V}$ is the class of hypersurfaces $X \subset \mathbb{P}^{n-1}$ whose degree is less than n, the corresponding class of fields $\mathcal{F}(\mathcal{V})$ is usually denoted C_1, these fields being also called *quasi-algebraically closed.*

Chevalley and Tsen tell us that finite fields, and fields of transcendence degree one over $\mathbb{C}$, respectively, are both quasi-algebraically closed. In his thesis, Serge Lang defines a series of conditions C_r for non-negative integers r: a field k is $\mathbf{C_r}$ if all homogeneous forms with coefficients in k of degree d in n variables with $d^r < n$ have nontrivial rational points over k. Practicing our informal terminology, if $\mathcal{V}_r$ is the class of projective hypersurfaces $X \subset \mathbb{P}^{n-1}$ of degree d with $d^r < n$, $C_r = \mathcal{F}(\mathcal{V}_r)$. Lang proves that finite extensions of C_r fields are C_r and if a field k is C_r then the rational function field in one variable $k(t)$ over k is C_{r+1}.

We might try to go in the other direction: that is, to ask for $\mathcal{V}(C_r)$. For example, what geometric condition on a variety in general will suffice to ensure that it has rational points over a C_1 field? Tsen's theorem (and the fact that the inequality $d < n$ in Tsen's theorem is sharp) gives $d < n$ as the necessary and sufficient condition for a smooth hypersurface to be a member of $\mathcal{V}(C_1)$ but that's not much to go on.

Recent results due to Kollár, Kollár–Miyaoka–Mori, ([11], [13]) and three of the four authors of the present article (Graber–Harris–Starr: [4]) have suggested an answer. They provide something of an extension of this classical theory, and suggest arithmetic questions about classes of (smooth projective) varieties significantly more general than merely hypersurfaces.

Kollár and his co-authors make a very good case for the claim that an appropriate generalization of the notion of rational curves to higher dimensions is the class of varieties that they call *rationally chain connected.* See, for example, [11], [12]. For a beautiful introductory account of some of this theory, see [2]. Some basic results on rational connectivity were also independently developed in [1].

A smooth projective variety over a field F is *rationally chain connected* if over an algebraic closure $\bar{F}$ of F any two points can be connected by a "path" which is a chain of images of the projective line. Over $\mathbb{C}$, any (smooth projective) rationally chain connected variety V enjoys the stronger property that given any finite set of points of V and, if you wish, an assignment of a tangent vector to each of those points, you can find a rationally parametrized curve passing through all of those finite points with tangent vector at each of those points equal to the corresponding assigned tangent vector. The class of rationally chain connected varieties is a very flexible class: it is immediate from the definition that the image of a rationally chain connected variety is again rationally chain connected, that the class is closed under products, and it is almost immediate that membership in the class is a birational invariant. Smooth projective varieties, then, that are *unirational*, i.e., whose field of rational functions over $\mathbb{C}$ is contained in a purely transcendental extension of $\mathbb{C}$ are rationally chain connected. In particular, rational varieties are also rationally chain connected, and for dimension 1 and 2 that is all there are: any rationally chain connected variety is rational.

The two notions diverge for smooth projective varieties in dimensions 3 and higher. It is not expected that rational varieties, at least in dimensions ≥ 3, behave well in families, but rationally chain connected varieties do: by a result of Kollár, Miyaoka, and Mori if B is a smooth irreducible base, and $f_B : X_B \to B$ a smooth proper morphism over $\mathbb{C}$ then all fibers are rationally chain connected if one is.

Rationally chain connected varieties belong to the larger class of $\mathcal{O}$-acyclic varieties, to which we now turn.

Consider the class of projective varieties V such that the higher cohomology of the structure sheaf of V vanishes, i.e.,

$$H^i(V, \mathcal{O}_V) = 0$$

for $i > 0$, and $H^0(V, \mathcal{O}_V)$ is of dimension 1 over the base field. If V is smooth and geometrically irreducible this is equivalent, using Serre duality, to saying that every meromorphic differential form of positive degree on V has a pole, i.e., there are no nontrivial holomorphic differential forms on V of positive degree. Although this is a class of varieties that occurs from time to time in the literature, it doesn't seem to have a standard name; let us just call these varieties *$\mathcal{O}$-acyclic projective varieties*.

By results of Kollár, Miyaoka, and Mori we know that smooth projective rationally chain connected varieties have no global holomorphic differentials of positive degree; i.e., they are all $\mathcal{O}$-acyclic. But there are many $\mathcal{O}$-acyclic surfaces that are not rationally chain connected; e.g., Enriques surfaces (these being surfaces with fundamental group cyclic of order 2 and universal cover a $K3$-surface).

Before we take a closer look at these two notions, *rational chain connectivity* and *$\mathcal{O}$-acyclicity*, let us note that both conditions are at the same time "extensions" to smooth projective varieties of any dimension of the condition (for hypersurfaces) that enters into the theorems of Chevalley–Warning, and of Tsen: it is relatively elementary to prove the

Proposition 1. *Let $X \subset \mathbb{P}^n$ be a smooth projective hypersurface of degree d in projective n-space. These conditions are equivalent:*

(i) $d \leq n$.
(ii) V *is rationally chain connected.*
(iii) V *is* $\mathcal{O}$*-acyclic.*

2 Generalizing Chevalley–Warning to $\mathcal{O}$-acyclic Varieties

An $\mathcal{O}$-acyclic variety V is very pleasant to deal with if you have a "Woods Hole"-type Lefschetz fixed point theorem. For example, any morphism $f : V \to V$ is immediately seen to have a fixed point since the alternating sum of the traces of f computed on the cohomology groups $H^i(V, \mathcal{O}_V)$ is 1. Applying this to the Frobenius endomorphism, if V is defined over a finite field k, one would get, then, that such a variety V has a k-rational point. All this is a consequence of a more general result due to N. Katz (See pp. 410,411 in [10]) which indeed implies:

Generalized Chevalley–Warning. Theorem of Katz. *Any projective* $\mathcal{O}$*-acyclic variety over a finite field k has a rational point over k.*

In view of this result, it would be natural to ask, as Serre did in a letter to Grothendieck (page 152 in [9]) whether $\mathcal{O}$-acyclic varieties over fields K (of transcendence degree 1 over $\mathbb{C}$) are also guaranteed to have K-rational points. In other words, does Tsen's theorem generalize to the class of $\mathcal{O}$-acyclic varieties? As we shall discuss is section 6 below, the answer to this is *no!* even for smooth projective $\mathcal{O}$-acyclic varieties. More specifically, there are Enriques surfaces over fields K (of transcendence degree 1 over $\mathbb{C}$) with no K-rational points; this follows from [5]. One can then retreat a bit, and ask whether the analogous theorem holds for the field $\mathbb{C}((t))$ of Laurent series in one variable over $\mathbb{C}$. (We don't know the answer.)

3 Generalizing Tsen to Rationally Chain Connected Varieties, and a "Converse" Theorem

In this section, all varieties are over $\mathbb{C}$.

Generalized Tsen Theorem (Graber–Harris–Starr). *Any smooth projective rationally connected variety over a field K of transcendence degree* 1 *over* $\mathbb{C}$ *has a rational point over* K.

This was initially conjectured by Kollár and has as (essentially immediate) consequence that if $f : X \to Y$ is a dominant morphism with rationally chain connected fibers and rationally chain connected base Y, then the total space X is again rationally chain connected. See Graber–Harris–Starr [4]. See also [3] for a companion result over fields of positive characteristic.

In [5] we prove a result that we think of as a "converse" to the generalized Tsen Theorem of Graber–Harris–Starr.

To explain what we mean by “converse” we shall generalize somewhat the notion of *section of a mapping* and paraphrase the Graber–Harris–Starr Theorem.

Let

$$\pi : X \to B$$

be a proper faithfully flat morphism of algebraic varieties (over $\mathbb{C}$) and for simplicity, suppose that B is smooth and irreducible (but not necessarily proper). Let F_B denote the field of rational functions on B.

Definition. *By a* (rational) pseudosection *of π let us mean a morphism $S \to X \to B$ where S is an irreducible variety, $S \to B$ is proper, dominating, and the fiber of $S \to B$ over F_B is a smooth rationally connected variety.*

The generalized Tsen Theorem of Graber–Harris–Starr. Second formulation. *If a proper faithfully flat morphism morphism $\pi : X \to B$ admits a (rational) pseudosection, then for any sufficiently general smooth irreducible curve $C \subset B$, the pullback $\pi_C : X_C \to C$ admits a section.*

We prove (in [5]) the following:

Weak converse theorem. *If for every smooth irreducible curve $C \subset B$ the pullback $\pi_C : X_C \to C$ admits a rational section, then the morphism $\pi : X \to B$ admits a (rational) pseudosection.*

In the next section we will try to explain why we regard this as “weak” and possible directions in which one might hope to strengthen it.

Notes: 1. We cannot sharpen the conclusion in the above theorem to state that “$\pi : X \to B$ admits a (rational) *section*” rather than *pseudosection* for we may easily find (smooth, irreducible) varieties B over $\mathbb{C}$ over which there exists conic bundles $X \to B$ which admit no section. All conic bundles over smooth curves over $\mathbb{C}$, however, admit sections. (The conic bundle itself, of course, *is* a pseudosection.)

2. If $X \to B$ is finite flat this converse theorem is an easy consequence of one version of the Lefschetz theorem for the fundamental group π_1.

Putting this weak converse theorem together with the Graber–Harris–Starr result above, we get:

Theorem 3.1. *The morphism $\pi : X \to B$ admits a (rational) pseudosection if and only if for all sufficiently general smooth irreducible curves, the pullback $f_C : X_C \to C$ admits a rational pseudosection (or, equivalently, a rational section).*

Here are some applications (see [5]) of the weak converse theorem.

Proposition 3.2. *There exists a flat family $f : X \to B$ where the generic fiber is an Enriques surface and such that the family does not admit a pseudosection.*

It follows from this proposition, and from the weak converse theorem, that *the generalization of Tsen's theorem to the class of $\mathcal{O}$-acyclic varieties is false:*

Corollary 1. *There exists a field K of transcendence degree 1 over $\mathbb{C}$ and a smooth projective $\mathcal{O}$-acyclic variety over K which does not have a K-rational point.*

Note. Our Proof of the existence of such an example (a family of Enriques surfaces not admitting a section) is nonconstructive. Recently, Guillaume Lafon [14] has constructed explicit examples of such families.

Since curves of genus 1, and more generally, torsors over abelian varieties, contain no rational curves, it follows that for $f : X \to B$ a flat family of (smooth projective) curves of genus 1 (or more generally, a torsor for some abelian scheme over the base B) any rational pseudo-section is necessarily a rational section, and any rational section is (or rather, extends to) a regular section over the smooth base B. We therefore get the following consequence of the weak converse theorem:

Corollary 2. *A torsor for an abelian scheme over B is trivial if and only if its pullback to every smooth irreducible curve mapping to B is trivial. That is, if $A \to B$ is an abelian scheme, the homomorphism*

$$H^1(B, A) \longrightarrow \prod_{C \to B} H^1(C, A)$$

is injective, where the product is taken over all morphisms, $C \to B$, of irreducible smooth curves into B.

The above Corollary is reminiscent of the standard application of the Lefschetz theorem to the Picard functor (cf. Cor 3.6 of Exp XII in [8]): a line bundle over a smooth base B is trivial if (and only) if it is trivial when restricted to all surfaces in B; it would be interesting to consider analogues of this, and/or of Corollary 2, for more general group schemes.

We can use the statements of Corollaries 1 and 2 to illustrate why we call the converse theorem formulated above *weak*. Let us start with Corollary 2.

It would be very useful to have a stronger version of Corollary 2 which pinpointed a *single* irreducible, smooth, curve $C \hookrightarrow B$ such that the restriction homomorphism

$$H^1(B, A) \to H^1(C, A)$$

is injective. More broadly, if $f : X \to B$ is a flat family, let us borrow the word *witness* from the theory of primality testing, and say that a curve in B, $C \subset B$, is a *witness for pseudosections of* f if every section of the pullback $f_C : X_C \to C$ extends to a pseudosection of f. We would like to establish results that would guarantee that *sufficiently general* curves in B are witnesses to pseudosections of *large classes of* morphisms

$$f : X \to B,$$

where, given any f, the criterion for "sufficiently general" is reasonably testable, and demonstrably fulfilled by at least some curves. It is too optimistic to hope for criteria as clear-cut and as general as the criteria that occur in the statement of the classical Lefschetz theorem guaranteeing surjectivity of $\pi_1(C) \to \pi_1(B)$ but we should aim for something better than the current "weak converse theorem." Here, more specifically, is what we can prove at present and what we can't.

Proposition 3.3. *Let B be smooth, irreducible, and quasi-projective. For each positive integer d there is a sufficiently positive projective embedding $B \hookrightarrow \mathbb{P}^N$ such that for any morphism of varieties $\pi : X \to B$ of fiber dimension $\leq d$ the morphism π has a pseudosection (over all of B) if and only if the restriction, $\pi_C : X_C \to C$, of π to every linear smooth curve section $C = B \cap L$ of B with respect to this projective embedding has a pseudosection (equivalently: a section).*

Question 1. Can we find a single (quasi-)projective embedding of B that works independent of fiber dimension d?

We suspect that the answer to this is no.

Question 2. Can one guarantee the existence of a sufficiently positive projective embedding $B \hookrightarrow \mathbb{P}^N$ such that that any pseudosection over a sufficiently general linear smooth curve section actually *lies in* a global pseudosection?

We hope, but cannot prove, that the answer to this is yes.

Our "witness curves" in the above theorem are obtained by avoiding certain countable unions of proper subvarieties of the parameter space of linear smooth curve sections. In some applications we have an existence result (of "witnesses") without being able to write down explicitly the equations of any witness curve. To underscore this, consider

Question 3. Let $\pi : X \to B$ be a morphism defined over a countable algebraically closed field F, possessing no pseudosection over F. Is there a witness curve in B defined over F?

Our ignorance of the answer illustrates our lack of control of the example asserted to exist in Corollary 1. Our method "produces," in some sense, an Enriques surface S over a field K of transcendence degree 1 over $\mathbb{C}$ without a K-rational point, but it is not explicit. Happily, the recent construction of Lafon [14] offers explicit examples, e.g., of Enriques surfaces defined over $\mathbb{Q}(t)$, possessing no section over $\mathbb{C}((t))$.

Problem. Find an explicit example for Corollary 1.

Corollary 2 guarantees that nontriviality of any given torsor for an abelian scheme over a general base is detected by restriction to appropriate curves in the base. It is natural to ask the analogous question about the Mordell–Weil group of an abelian scheme over a general base. Explicitly,

Question 4. If A is an abelian scheme over the smooth, irreducible base B (over $\mathbb{C}$) is there a smooth curve $C \subset B$ such the the restriction map of the groups of global sections

$$A(B) \longrightarrow A(C)$$

is an isomorphism?

Finally, to the extent that we believe the class of rationally connected varieties is the natural dual of the class of C_1 fields, we would like to return to our original formulation of this class of problems, and ask:

Question 5. Is there a geometric characterization of the class $\mathcal{V}(C_r)$ of varieties? For example, can we find a geometric condition on a variety X that will ensure that it has a point over any field of transcendence degree r over $\mathbb{C}$?

4 "Analogous" Questions over Number Fields

In this section, let F be a number field, and let our varieties be defined over F. Let $\pi : X \to B$ be a flat morphism of schemes over F, with B smooth and irreducible. Say that π is *arithmetically surjective* if for every finite degree extension field L/F, the induced mapping on L-rational points,

$$\pi : X(L) \to B(L)$$

is surjective.

Question 6. If $\pi : X \to B$ is a flat, arithmetically surjective, morphism (of finite type) of schemes over F, does π admit a rational pseudo-section (over F)?

That you cannot replace "pseudosection" by "section" in Question 1 (and expect an affirmative answer) is seen by various examples in the literature. We are thankful to J.-L. Colliot-Thélène (for commentary on an early draft of this paper, and) for pointing out to us such explicit examples in the article of R. Parimala and R. Sujatha, [17]. Here, briefly, is a description of their construction.

Let K be a number field, Y a smooth projective curve over K, and $X \to Y$ a conic bundle corresponding to $c \in \mathrm{Br}(Y)[2]$, an element of order two in the Brauer group of the scheme Y/K (cf. pp. 31–188 in [7]) with the further property that the image of c in $\mathrm{Br}(Y \otimes \mathbb{Q}_p)$ is zero for all primes p and is zero in $\mathrm{Br}(Y \otimes \mathbb{R})$ as well.

Since c is assumed nontrivial, $X \to Y$ has no section. Now let L be any number field extension of K, $\lambda : \mathrm{Spec}\, L \to Y$ be any L-valued point of Y, and denote by X_λ the pullback via λ. i.e., the fiber of X over the point λ.

By our hypotheses, it follows that the Brauer–Severi scheme X_λ is "trivial" over $L \otimes \mathbb{Q}_p$ for all p and over $L \otimes \mathbb{R}$; consequently the conic X_λ splits over every completion of L, and hence it splits over L. In particular, $X \to Y$ is arithmetically surjective and yet possesses no section.

One way of producing conic bundles of the above sort is to rely on a result of Artin and Tate that connects the Brauer group to the Shafarevich–Tate group for the jacobians of curves over the rings of integers of number fields (especially if these curves have integral points and geometrically connected fibers (cf. Sect. 4, Prop. 4.3 of [7]). The slight awkwardness of this approach is that the relationship between the Brauer group and the Shafarevich–Tate group is given through mappings that are painful to work with to get the precise equations of the conic bundle X.

In the article of Parimala and Sujatha a (nontrivial) bundle $X \to Y$ is constructed where Y is the elliptic curve over $\mathbb{Q}$ given by the equation $y^2 = x^3 - 113x$ and where the fiber X_η over the generic point $\eta = \text{Spec } \mathbb{Q}(Y)$ of the curve Y is the conic associated to the quaternionic algebra $(-(x+8), 2)$. The authors show that this quaternionic algebra is split over $\mathbb{Q}_p(Y)$ for all primes p. This algebra also splits over $\mathbb{R}(Y)$, since 2 is a square in $\mathbb{R}$. Therefore, by the discussion above, $X \to Y$ is an arithmetically surjective conic bundle over an elliptic curve over $\mathbb{Q}$ and $X \to Y$ does not admit a section.

An unpublished construction due to Mak Trifkovic ([20]) based on the same strategy obtains (via a slightly different route) explicit equations for conic bundles over elliptic curves over number fields K with the weaker property that every K-rational fiber is trivial.

We have an affirmative answer to Question 4 when the fiber dimension of π is zero:

Proposition 4.4. *Let B be an integral, integrally closed scheme over a number field F. A finite flat morphism $\pi : X \to B$ defined over F is arithmetically surjective if and only if it admits a section over F.*

Proof. Suppose, then, that π satisfies the hypotheses of the proposition and yet admits no rational section over F. We will obtain a contradiction. Using Noether Normalization we fix a quasi-finite flat morphism of a nonempty Zariski open in B onto a Zariski open U in affine d-space over F. Replacing U, B and X by appropriate (nonempty) Zariski opens (but retaining the letters U, B and X to denote them) we may assume that $X \to B \to U$ are finite étale morphisms over F (and that X is nonempty). For W a reduced irreducible scheme over F let F_W denote the field of rational functions on W. Fix Y an irreducible component of X and form the finite field extensions

$$F_U \subset F_B \subset F_Y \subset \Phi$$

where Φ is a Galois closure of F_Y/F_U. We have $\Phi = F_Z$ where Z is the normalization of Y in Φ. Restricting U to yet a smaller nonempty Zariski open we can assume that $Z \to U$ is finite étale. Let $G_U := \text{Gal}(F_Z/F_U)$, $G_B := \text{Gal}(F_Z/F_B)$ and $G_Y := \text{Gal}(F_Z/F_Y)$. Since π admits no rational section, we have a strict inclusion $G_Y \subset G_B$. Define

$$H(Y) := \left\{\alpha \in F^d \mid Z \times_U \alpha \text{ is irreducible over } F\right\}.$$

Then $H(Y)$ is a *Hilbertian set* relative to the number field F in the sense of Lang (see his treatment of the classical Hilbert Irreducibility Theorem in Chap. 9 of [16]).

For each $\alpha \in H(Y)$ the morphism

$$Z \times_Y \alpha \to Z \times_B \alpha$$

is a nontrivial morphism of the spectra of fields containing F; denote these fields $F \subset F_B(\alpha) \subset F_Y(\alpha)$.

Lemma 1.

$$\bigcap_Y H(Y) = \emptyset \subset F^d,$$

where Y runs through the (finite) set of irreducible components of X.

Proof. For any $\alpha \in \cap_Y H(Y)$, put $L := F_B(\alpha)$. We have that $X(L)$ is empty yet $B(L)$ is not, contradicting arithmetic surjectivity of π.

Our proposition then follows from the fact that any finite intersection of Hilbertian sets relative to a number field is nonempty contradicting the conclusion of the lemma.

Question 6 is interesting, we think, even in the case of pencils of curves of genus 1 (also: ≥ 1):

Question 7. Let C be a smooth curve of genus ≥ 1 defined over $F(t)$, the field of rational functions in one variable over a number field F. Suppose that for every algebraic extension L/F and every element $t_0 \in L$ such that C_t has a smooth specialization at t_0, the curve C_{t_0} has an L-rational point. Is it the case that C has an $F(t)$-rational point?

We don't know the answer to this question, or to similar ones, even in very explicit contexts. For example, are there pencils of plane cubics (over $\mathbb{Q}$) which have the property that every nonsingular member of the family over $\mathbb{Q}$ has a $\mathbb{Q}$-rational point and yet the family itself has no section?

Question 8. Let B be a scheme of finite type over a number field F. Is it the case that a torsor for an abelian scheme over B is trivial if and only if its pullback to every algebraic point of B is trivial?

The analogue of Question 4 is:

Question 9. Let B be a smooth irreducible scheme of finite type over a number field F. Let A be an abelian scheme over B. Is there a number field extension L/F and an L-valued point $b : \mathrm{Spec}(L) \to B$ such that the homomorphism

$$A(B) \longrightarrow A(L)$$

obtained by composition with b is an isomorphism?

Proposition 4.5. *An affirmative answer to Question 8 implies that for every integer $N \geq 1$ and number field F there is a number field extension L/F and an elliptic curve E over L with Mordell–Weil group $E(L)$ cyclic of order N.*

Proof. It suffices, for each $N \geq 1$, to exhibit a family of elliptic curves over a smooth base over a number field, whose Mordell–Weil group over $\mathbb{C}$ is cyclic of order N. Indeed, for each $N \geq 1$ we will point to a family $\mathcal{E}_N$ of elliptic curves over a smooth base curve over $\mathbb{Q}$ with the property that the Mordell–Weil group of the family over $\mathbb{C}$ is cyclic of order N and such that the family is "truly varying," i.e., the

j-invariant of the fibers is a nonconstant function of the base curve. For $N \geq 5$ we take $\mathcal{E}_N$ to be the universal elliptic curve supplied with point of order N over the base curve $Y_1(N)$ over $\mathbb{Q}$ and we use Shioda's theorem ([18]) which guarantees that the family $\mathcal{E}_N \to Y_1(N)$ has the property that its global sections (over $\mathbb{C}$) are generated by the "tautological" point of order N. For $N = 1, 2, 3, 4$ we take $\mathcal{E}_N$ to be some appropriate quotient of a universal elliptic curve of higher level, again using Shioda's result.

Note. It seems difficult to establish the conclusion of Proposition 5 even for specific values of N. For example, for any number field F is there a number field extension L/F and an elliptic curve over L with trivial Mordell–Weil group over L?

References

[1] Campana, F.: Connexité rationnelle des variétés de Fano, Ann. Sci. École Norm. Sup. (4) **25** (1992), 539–545.

[2] Debarre, O.: *Higher-Dimensional Algebraic Geometry,* Springer (2001).

[3] de Jong, J., Starr, J.: Every rationally connected variety over the function field of a curve has a rational point, Amer. Jour. of Math. **125** (2003), 567–580.

[4] Graber, T., Harris, J., Starr, J.: Families of rationally connected varieties, Jour. of the Amer. Math. Soc. **16** (2003), 69–90.

[5] Graber, T., Harris, J., Mazur, B., Starr, J.: Rational connectivity and sections of families over curves, accepted Ann. Sci. E. N. S.

[6] Greenberg, M.: *Lectures on Forms in Many Variables,* Benjamin, 1969.

[7] Grothendieck, A.: Le groupe de Brauer I,II,III, pp. 31–188 in *Dix exposés sur la cohomologie des schémas,* eds: Giraud, Grothendieck, Kleiman, Raynaud, Tate; Masson, Paris; North-Holland, Amsterdam (1968).

[8] Grothendieck, A.: Application au groupe de Picard, Exp XI, XII, pp. 124–171 in *Cohomologie locale des faisceaux cohérents et Théorèmes de Lefschetz locaux et globaux (SGA 2)* eds: Grothendieck, A., and Kuiper, N.H.; Masson, Paris; North-Holland, Amsterdam (1968).

[9] *Correspondance Grothendieck–Serre,* eds.: Pierre Colmez – Jean-Pierre Serre, Société Mathématique de France, 2001).

[10] Katz, N.: Une formule de congruence pour la fonction ζ, exposé XXII, pp. 401–438 in SGA 7 II *Groupes de Monodromie en Géométrie Algébrique*, **340**, Lecture Notes in Mathematics, Springer (1973).

[11] Kollár, J.: Rational curves on algebraic varieties, Ergebnisse der Math. **32** Springer (1996)

[12] Kollár, J.: Which are the simplest algebraic varieties? Bull. A.M.S. **38** (2000), 409–434.

[13] Kollár, J., Miyaoka, Y., Mori, S.: Rationally connected varieties, J. Alg. Geom. **1** (1992), 429–448.

[14] Lafon, G.: Une surface d'Enriques sans point sur $\mathbb{C}((t))$ (to appear in C. R. Acad. Sci. Paris).

[15] Lang, S.: On quasi-algebraic closure, Ann. of Math, **55** (1952), 373–390.

[16] Lang, S.: *Fundamentals of Diophantine Geometry,* Springer (1983).

[17] Parimala, R., Sujatha, R.: Hasse principle for Witt groups of function fields with special reference to elliptic curves, Duke Math J. **85** (1996), 555–582.

[18] Shioda, T.: Elliptic modular surfaces I, II, Proc. Japan Acad. **45** (1969), 786–790; ibid. **45** (1969), 833–837.
[19] Tsen, C.: Divisionsalgebren über Funktionenkörper. Nachr. Ges. Wiss. Göttingen (1933), 335.
[20] Trifkovic, M.: Conics over elliptic curves (unpublished notes).
[21] Warning, E.: Bemerkung zur vorstehenden Arbeit von Herrn Chevalley, Abh. Math. Sem. Univ. Hamburg, vol. II (1936), 76–83.

Hyperbolicity in Complex Geometry

Yum-Tong Siu*

Summary. A complex manifold is said to be hyperbolic if there exists no nonconstant holomorphic map from the affine complex line to it. We discuss the techniques and methods for the hyperbolicity problems for submanifolds and their complements in abelian varieties and the complex projective space. The discussion is focussed on Bloch's techniques for the abelian variety setting, the recent confirmation of the longstanding conjecture of the hyperbolicity of generic hypersurfaces of high degree in the complex projective space, and McQuillan's techniques for compact complex algebraic surfaces of general type.

0 Introduction

A complex manifold X is said to be hyperbolic if there exists no nonconstant holomorphic map $\mathbf{C} \to X$. The hyperbolicity problem in complex geometry studies the conditions for a given complex manifold X to be hyperbolic. Hyperbolicity problems have a long history and trace back to the small Picard theorem and the hyperbolicity of compact Riemann surfaces of genus ≥ 2 from Liouville's theorem and the uniformization theorem. The hyperbolicity of compact Riemann surfaces of genus ≥ 2 led to the study of compact complex manifolds X with positive canonical line bundle K_X. The small Picard theorem led to the study of the complement of a hypersurface Y in a compact algebraic manifold X with conditions related to the positivity of $Y + K_X$. For a long time a great part of the research on the hyperbolicity problems has been focussed on the following two environments.

* Partially supported by a grant from the National Science Foundation.
AMS 2000 Mathematics Subject Classification: 32H30, 14K20.

(1) The setting of abelian varieties. The hyperbolicity problem concerns a subvariety in an abelian variety or concerns the complement of a hypersurface in an abelian variety.

(2) The setting of the complex projective space $\mathbf{P}_n$. The hyperbolicity problem concerns a generic hypersurface of sufficiently high degree or concerns the complement of a generic hypersurface of sufficiently high degree.

In this survey article we present the essential techniques from function theory developed over the years and especially recently for the investigation of hyperbolicity problems for these two settings. There are three main parts in this survey paper. The first part concerns the techniques of Bloch for the setting of abelian varieties. The second part concerns the most recent methods for the setting of the complex projective space which were developed from the techniques of Clemens–Ein–Voisin from algebraic geometry. The third part concerns the techniques of McQuillan for hyperbolicity problems for surfaces of general type.

The main emphasis of this article is to explain the techniques and methods and not to present a comprehensive list of results. The background and the history of the problems of hyperbolicity together with lists of references can be found in other survey articles such as [18][19].

Notations and terminology. The structure sheaf of a complex space Y is denoted by $\mathcal{O}_Y$. The bundle of holomorphic p-forms on a complex manifold Z is denoted by Ω_Z^p.

For a complex manifold X denote by $J_k(X)$ the space of all k-jets of X. At a point P_0 of X an element of $J_k(X)$ is the k-th jet of a holomorphic map $f : U \to X$ at 0 and is determined by the derivatives of f up to order k at 0 with respect to the coordinate of U, where U is an open neighborhood of 0 in $\mathbf{C}$ and $f(0) = P_0$. For nonnegative integers $\ell < k$ there is a natural projection $J_k(X) \to J_\ell(X)$ and $J_0(X) = X$.

A k-jet differential on X with local coordinates $z_1, \cdots, z_n$ is locally a polynomial in $d^\ell z_j$ ($1 \le \ell \le k$, $1 \le j \le n$) of homogeneous weight, when the weight of $d^\ell z_j$ is defined to be ℓ. For a complex hypersurface Y in X locally defined by a holomorphic function g, a k-jet differential on X with logarithmic poles along Y is locally a polynomial in $d^\ell z_j$, $d^\ell \log g$ ($1 \le \ell \le k$, $1 \le j \le n$).

For a holomorphic map $\varphi : \mathbf{C} \to X$ the k-th differential $d^k\varphi$ of φ is a map from $J^k(\mathbf{C})$ to $J_k(X)$. Most of the time, we just regard $d^k\varphi$ as a map from $\mathbf{C}$ to $J_k(X)$, which assigns to every point of $\mathbf{C}$ the k-jet of the curve in X defined by φ at that point. Another way to interpret it is that we map $\mathbf{C}$ to $J_k(\mathbf{C})$ by taking the k-jet of the curve in $\mathbf{C}$ defined by the identity map of $\mathbf{C}$ and take the composite of it with $d^k\varphi : J_k(\mathbf{C}) \to J_k(X)$ to get a map from $\mathbf{C}$ to $J_k(X)$. When a k-jet differential ω of weight m on X, as a function on $J_k(X)$, is pulled back by $d^k\varphi : J_k(\mathbf{C}) \to J_k(X)$, one gets a k-jet $\varphi^*\omega$ of weight m on $\mathbf{C}$ and a polynomial P in $d^j\zeta$ ($1 \le j \le k$) of weight m. When $d^k\varphi$ is regarded as a map from $\mathbf{C}$ to $J_k(X)$, the pullback $\varphi^*\omega$ is a function on $\mathbf{C}$ which is the coefficient of $(d\zeta)^m$ in the polynomial P. Most of the time we will use the latter interpretation. For statements involving the vanishing of

$\varphi^*\omega$, its vanishing in the second interpretation for all choices of the coordinate ζ in $\mathbf{C}$ implies its vanishing in the first interpretation.

For a holomorphic map $\varphi : \mathbf{C} \to X$ and a real closed $(1, 1)$-current ω on X, the Nevanlinna characteristic function is $T(r, \varphi, \omega) = \int_{\rho=0}^{r} \frac{d\rho}{\rho} \int_{|\zeta|<\rho} \varphi^*\omega$, where ζ is the coordinate of $\mathbf{C}$. When ω is positive definite and the context requires only the growth order of $T(r, \varphi, \omega)$ up to a positive constant factor, we drop ω from $T(r, \varphi, \omega)$ and simply use $T(r, \varphi)$. When Y is a complex hypersurface of X, the counting function is $N(r, \varphi, Y) = \int_{\rho=0}^{r} n(\rho, \varphi, Y) \frac{d\rho}{\rho}$, where $n(\rho, \varphi, Y)$ is the number of points ζ in $\{|\zeta| < \rho\} \cap \varphi^{-1}(Y)$ with multiplicities counted. The truncated counting function $N_k(r, \varphi, Y)$ means $\int_{\rho=0}^{r} n_k(\rho, \varphi, Y) \frac{d\rho}{\rho}$, where $n_k(\rho, \varphi, Y)$ is the number of points ζ in $\{|\zeta| < \rho\} \cap \varphi^{-1}(Y)$ counted under the rule that the multiplicity is replaced by k if it is greater than k. Let σ_Y be a function on X which is locally equal to a smooth positive function times the absolute value of a local holomorphic function defining Y. The proximity function is $m(r, \varphi, Y) = \oint_{|\zeta|=r} \log^+ \frac{1}{(\sigma_Y)\circ\varphi}$, where $\oint$ means the average over the circle $|\zeta| = r$. The growth behavior of the proximity function is determined up to a bounded additive term.

A meromorphic function F on $\mathbf{C}$ defines a holomorphic map $\varphi_F : \mathbf{C} \to \mathbf{P}_1$ and $T(r, F)$ means $T\left(r, \varphi_F, \omega_{\mathbf{P}_1}\right)$, where $\omega_{\mathbf{P}_n}$ is the Fubini–Study form on $\mathbf{P}_n$. The proximity term $m(r, \varphi_F, \infty)$ up to a bounded additive term is equal to $\oint_{|\zeta|=r} \log^+ |F|$, which we also denote by $m(r, F, \infty)$ or simply $m(r, F)$. Another description of $T(r, F)$ up to a bounded additive term is $N(r, F, \infty) + m(r, F, \infty)$. When $\varphi : \mathbf{C} \to \mathbf{P}_n$ is defined by entire holomorphic functions $F_0, \cdots, F_n$ on $\mathbf{C}$ without common zeros, $T\left(r, \frac{F_j}{F_0}\right) \leq T\left(r, \varphi, \omega_{\mathbf{P}_n}\right) + O(1) \leq \sum_{j=1}^{n} T\left(r, \frac{F_j}{F_0}\right) + O(1)$.

For a holomorphic map $\varphi : \mathbf{C} \to X$ and an ample divisor Y whose cohomology class is represented by a smooth closed positive $(1, 1)$-form ω, the First Main Theorem of Nevanlinna states that $T(r, \varphi, \omega) = m(r, \varphi, Y) + N(r, \varphi, Y) + O(1)$. The defect $\delta(\varphi, Y)$ is the limit inferior of $\frac{m(r,\varphi,Y)}{T(r,\varphi,\omega)}$ as $r \to \infty$. The defect $\delta(\varphi, Y)$ can only take value in the closed interval $[0, 1]$. When the defect $\delta(\varphi, Y)$ is < 1, the image of the holomorphic map φ must intersect Y. What is usually referred to as a Second Main Theorem is an estimate of the error when the limit is taken in the definition of the defect, such as $m(r, \varphi, Y) \leq \delta(\varphi, Y)\, T(r, \varphi, \omega) + E(r)$ for r outside a set with finite measure with respect to $\frac{dr}{r}$, where $E(r)$ is an error term, for example, of the order $o(T(r, \varphi, \omega))$ in a weaker version and of the order $O(\log T(r, \varphi, \omega) + \log r)$ in a stronger version. An estimate of the defect or a Second Main Theorem is a more quantitative description of the hyperbolicity of $X - Y$.

For a meromorphic function F on $\mathbf{C}$, Nevanlinna's logarithmic derivative lemma states that $m\left(r, (\log F)'\right) \leq O(\log T(r, F) + \log r)$ for r outside a set with finite measure with respect to $\frac{dr}{r}$, where $(\log F)'$ is the derivative of $\log F$ with respect to the coordinate of $\mathbf{C}$. When we invoke the logarithmic lemma, the condition that r is outside a set with finite measure with respect to $\frac{dr}{r}$ is understood and will not be explicitly mentioned.

The symbol A is reserved to denote an abelian variety of complex dimension n. An algebraic subvariety of $J_k(A) = A \times \mathbf{C}^{nk}$ means a subvariety which is algebraic along the factor $\mathbf{C}^{nk}$. In other words, it is defined by local functions on $J_k(A) =$

$A \times \mathbf{C}^{nk}$ which are holomorphic in the local coordinates of A and are polynomials in the nk global coordinates of $\mathbf{C}^{nk}$. Algebraic Zariski closures in $J_k(A)$ are defined using such algebraic subvarieties.

The position-forgetting map $\pi_k : J_k(A) \to \mathbf{C}^{nk}$ denotes the projection $J_k(A) = A \times \mathbf{C}^{nk} \to \mathbf{C}^{nk}$ onto the second factor. For $\ell > k$ denote by $p_{k,\ell} : \mathbf{C}^{n\ell} \to \mathbf{C}^{nk}$ the natural projection so that $\pi_k = p_{k,\ell} \circ \pi_\ell$. For an ample divisor D of A we denote by θ_D the theta function on $\mathbf{C}^n$ whose divisor is D. For a holomorphic map φ from $\mathbf{C}$ to A, let $W_{k,\varphi}$ be the algebraic Zariski closure of $\pi_k\left(\mathrm{Im}\,(d^k\varphi)\right)$ in $\mathbf{C}^{nk}$. Clearly, $p_{k,\ell}(W_{\ell,\varphi}) = W_{k,\varphi}$ for $\ell > k$. When Z_k is a proper subvariety of $W_{k,\varphi}$, the inverse image $p_{k,\ell}^{-1}(Z_k)$ is a proper subvariety of $W_{\ell,\varphi}$ for $\ell > k$.

Denote by $\lceil u \rceil$ the smallest integer no less than u and denote by $\lfloor u \rfloor$ the greatest integer not exceeding u. Denote by $\mathbf{N}$ the set of all nonnegative integers.

1 The Setting of Abelian Varieties

For hyperbolicity problems in the setting of abelian varieties there are three important techniques from function theory, all introduced by Bloch [2].

(1) *The use and construction of holomorphic jet differentials.* Bloch's construction uses the map from the jet-space induced by the position-forgetting map from the jet-space of an abelian variety and by using the field extension of meromorphic functions from a holomorphic map between compact algebraic manifolds with generically finite fibers.

(2) *The Schwarz lemma on the vanishing of the pullback, to* $\mathbf{C}$*, by a holomorphic map, of a holomorphic jet differential on a compact algebraic manifold which vanishes on some ample divisor.* Bloch's proof uses the logarithmic derivative lemma and its idea agrees with that of nowaday's best prevailing proof.

(3) *Differential equations constructed from position-forgetting maps of different order.* These differential equations link hyperbolicity to natural conjectured geometric properties. In the case treated in Bloch's paper the link is the translational invariance of the Zariski closure of an entire holomorphic curve in an abelian variety, which is nowadays referred to as Bloch's theorem.

The language and the style of Bloch's paper are very different from those of papers published in today's mathematical journals. The description of arguments in his papers is extremely concise and is more conversational in style. He considered only the case of surfaces, but his techniques are easily adaptable to general dimensions. Here I will first present Bloch's three techniques formulated in modern mathematical terms. Then I will explain where the three techniques are located in Bloch's paper and how Bloch originally presented them.

1.1 First Technique of Bloch

Lemma 1 (Construction of holomorphic jet differentials by generically finite maps). *Let $\mathcal{X}$ be an algebraic subvariety of the space $J_k(A)$ of k-jets of an*

n-dimensional abelian variety A such that $\sigma_k : \mathcal{X} \to \mathbf{C}^{nk}$ induced by the position-forgetting map $\pi_k : J_k(A) \to \mathbf{C}^{nk}$ is generically finite onto its image. Let $\tau : \mathcal{X} \to A$ be induced by the natural projection $J_k(A) \to A$. Let D be an ample divisor of A. Then there exists a polynomial P in $d^\ell z_j$ ($1 \leq \ell \leq k, 1 \leq j \leq n$) with constant coefficients which vanishes on $\tau^{-1}(D)$ but does not vanish identically on $\mathcal{X}$. In particular, P defines a holomorphic jet differential on A whose pullback to $\mathcal{X}$ vanishes on the divisor $\tau^{-1}(D)$ of $\mathcal{X}$.

Proof. Take a meromorphic function F on A whose pole divisor is a positive multiple of D. By assumption on τ and σ_k, $F \circ \tau$ belongs to a finite extension of the field of all rational functions of $\mathbf{C}^{nk}$. There exist polynomials P_j ($0 \leq j \leq p$) with constant coefficients in the variables $d^\ell z_\nu$ ($1 \leq \ell \leq k, 1 \leq \nu \leq n$) such that $\sum_{j=0}^{p} (\sigma_k^* P_j)(\tau^* F)^j = 0$ on $\mathcal{X}$ and $\sigma_k^* P_p$ is not identically zero on $\mathcal{X}$. Then P_p must vanish on $\tau^{-1}(D)$ and the holomorphic jet differential P_p on $\mathcal{X}$ must vanish on $\tau^{-1}(D)$. We need only set $P = P_p$.

1.2 Second Technique of Bloch

Lemma 2 (Schwarz Lemma on vanishing of pullback of holomorphic jet differentials vanishing on ample divisors to entire holomorphic curves). *Let X be a compact complex algebraic manifold and ω be a holomorphic jet differential on X vanishing on an ample divisor D of X. Let $\varphi : \mathbf{C} \to X$ be a holomorphic map. Then the pullback $\varphi^*\omega$ of ω by φ is identically zero on $\mathbf{C}$.*

Proof. Assume $\varphi^*\omega$ not identically zero on $\mathbf{C}$. We embed X into some $\mathbf{P}_N$ so that the hyperplane section line bundle of $\mathbf{P}_N$ pulls back to a positive multiple ℓ of D. By applying the defect relation of Ahlfors–Cartan [1][5] to $\mathbf{C} \to \mathbf{P}_N$ induced by φ, for any $0 < \varepsilon < 1$ we can find a hyperplane in $\mathbf{P}_N$ whose intersection E with X satisfies $N(r, \varphi, E) \geq (1 - \varepsilon) T(r, \varphi, E)$. Let s_D (respectively s_E) be the holomorphic section associated to D (respectively E) so that the divisor of s_D (respectively s_E) is D (respectively E). By replacing ω by $s_E (s_D)^{-\ell} \omega$, we can assume that $\ell = 1$ and $D = E$.

Let k and m be respectively the order and weight of ω. Using the logarithms of global meromorphic functions as local coordinates on X, we can find global meromorphic functions $F_{\ell,j}$ on X such that

$$|\omega| \leq C \sum_{1 \leq \ell \leq q, 1 \leq j \leq k} \left| d^k \log F_{\ell,j} \right|^{\frac{m}{k}}$$

as functions on $J_k(X)$ for some $C > 0$. By the logarithmic derivative lemma, $T(r, \varphi^*\omega) = o(T(r, \varphi))$ when $\varphi^*\omega$ is considered as a function on $\mathbf{C}$. On the other hand, the vanishing of ω on D implies that $T(r, \varphi^*\omega)$ dominates the counting function $N(r, \varphi, D)$, which yields the contraction that $(1 - \varepsilon) T(r, \varphi^*\omega) = o(T(r, \varphi))$.

This proof by logarithmic derivative lemma is taken from [23] and is already implicit in a preprint of Min Ru and Pit-Mann Wong [17]. The Schwarz lemma and its proof can straightforwardly be adapted to the case of jet differentials with logarithmic poles along divisors.

Remark 1. The intuitive reason for the Schwarz lemma is that **C** does not admit a metric with curvature bounded from above by a negative number, even when certain degeneracies of the metric are allowed. The statement holds also for jet metrics. A jet metric assigns a value to a k-jet instead of to a tangent vector as is the case with a usual metric. One can use a holomorphic jet differential ω which vanishes on an ample divisor D to define a jet metric on **C** with curvature bounded from above by a negative number to give a contradiction if the pullback of the jet differential is not identically zero. The idea is to use the pullback of $\frac{|\omega|^2}{|s_D|^2}$, where $|s_D|$ is the pointwise norm of the canonical section s_D of the line bundle L associated to D computed with respect to a positively curved metric of L, and $|\omega|^2$ is just the usual absolute-value square without the use of any metric. A rigorous proof by curvature arguments can be found in [23].

Remark 2. The proof of Lemma 2 actually gives the following stronger result. Let X be a compact complex algebraic manifold and Y be a complex hypersurface in X (or the empty set). Let $\varphi : \mathbf{C} \to X - Y$ be a holomorphic map and V be the algebraic Zariski closure of the image of $d^k\varphi$ in $J_k(X)$. Let ω be a locally defined holomorphic k-jet differential on X with logarithmic poles along Y which vanishes on an ample divisor D of X and which is globally defined as a function on V. Then the pullback $\varphi^*\omega$ of ω by φ is identically zero on **C**.

1.3 Third Technique of Bloch. *Bloch's system of differential equations constructed from position-forgetting maps of different order.* Let $\varphi : \mathbf{C} \to A$ be a holomorphic map from **C** to an n-dimensional abelian variety A whose universal cover $\mathbf{C}^n$ has coordinates $z_1, \cdots, z_n$. Let $\mathfrak{X}_k$ be the algebraic closure of the image of $d^k\varphi$ in $J_k(A)$. Let $\Phi_k : \mathfrak{X}_k \to \mathbf{C}^{kn}$ be the map induced by the position-forgetting map $\pi_k : J_k(A) = A \times \mathbf{C}^{kn} \to \mathbf{C}^{kn}$. Bloch introduced a system of differential equations from the properties of the kernel and image of $d\Phi_k$ and their relations to Φ_{k+1}. For the result (known now as Bloch's theorem) treated in Bloch's paper the specific property involving the kernel and image of $d\Phi_k$ is the finiteness of the generic fiber of Φ_k (for $k = 2$ in Bloch's case of a surface in a 3-dimensional abelian variety).

Lemma 3 (Translational invariance from differential equations). *Let $X = \mathfrak{X}_0$ which is the algebraic Zariski closure of the image of φ in A. Assume that X is of complex dimension $m < n$. If the fibers of $\Phi_m : \mathfrak{X}_m \to \mathbf{C}^{mn}$ are not generically finite, then there exists a nontrivial constant tangent vector v of A such that the translation operator $T_v : A \to A$ defined by v maps X to X.*

Proof. We assume that $\varphi(0)$ is a regular point of X where $z_1, \cdots, z_m$ are local coordinates of X. Write $dz_\alpha = \sum_{\nu=1}^m \omega_{\alpha\nu} dz_\nu$ on X for some meromorphic functions $\omega_{\alpha\nu}$ on X which are holomorphic near $\varphi(0)$ for $m + 1 \leq \alpha \leq n$.

Assume the fibers of $\Phi_m : \mathfrak{X}_m \to \mathbf{C}^{mn}$ not generically finite. Since $\mathfrak{X}_m \subset J_m(X)$, for $\zeta_0 \in \mathbf{C}$ close to 0 some nonzero tangent vector T to $J_m(X)$ at $Q := \varphi(\zeta_0)$ is mapped to zero by $J_m(A) \to \mathbf{C}^{mn}$. We can represent T by $\left(\left(\partial_\zeta^j \partial_t (z_\alpha \circ \psi)\right)(\zeta, 0)\right)_{0 \leq j \leq m, 1 \leq \alpha \leq n}$ for some holomorphic map $\psi(\zeta, t)$ from a neighborhood of $(\zeta_0, 0)$ in $\mathbf{C}^2$ to X with $\psi(\zeta, 0) = \varphi(\zeta)$. From the vanishing of $\left(\partial_\zeta^j \partial_t (z_\alpha \circ \psi)\right)(\zeta_0, 0)$ $(1 \leq j \leq m,\ 1 \leq \alpha \leq n)$ and $dz_\alpha = \sum_{\nu=1}^m \omega_{\alpha\nu} dz_\nu$ $(m+1 \leq \alpha \leq n)$ it follows from the chain rule that $\sum_{\nu=1}^m (\partial_\zeta^j (\omega_{\alpha\nu} \circ \varphi)) \partial_t (z_\nu \circ \psi) = 0$ at $(\zeta_0, 0)$ for $1 \leq j \leq m$ and $1 \leq \alpha \leq n$. Since the m complex numbers $(\partial_t (z_\nu \circ \psi))(\zeta_0, 0)$ $(1 \leq \nu \leq m)$ are not all zero due to the nontriviality of T, it follows that the rank of the $((n-m)m) \times m$ matrix $M := ((\partial_\zeta^j (\omega_{\alpha\nu} \circ \varphi))(\zeta_0))$ is less than m when $1 \leq \nu \leq m$ is regarded as the row index and the double index (j, α) with $1 \leq j \leq m$ and $m+1 \leq \alpha \leq n$ is regarded as the column index.

We now apply the usual Wronskian type argument simultaneously to the $(n-m)m$ sets of m functions which are the column vectors of M and conclude that there exist constants c_ν $(1 \leq \nu \leq m)$ not all zero, independent of ζ, such that $\sum_{\nu=1}^m c_\nu (\partial_\zeta^j (\omega_{\alpha\nu} \circ \varphi))(\zeta) \equiv 0$ for $1 \leq j \leq m$ and $m+1 \leq \alpha \leq n$ and for all ζ close to 0. Hence the 1-parameter group action on A defined by the vector field $\sum_{\mu=1}^n c_\mu \frac{\partial}{\partial z_\mu}$ maps X to itself.

Let B be the set of all elements a of A such that X is invariant under translation of A by elements of a. By induction on the dimension of A and replacing A by A/B and replacing $\varphi : \mathbf{C} \to A$ by the composite of φ with the quotient map $A \to A/B$, we get the following theorem of Bloch.

Theorem 1.1. [2] *Let A be an abelian variety and $\varphi : \mathbf{C} \to A$ be a holomorphic map. Let X be the Zariski closure of the image of φ. Then X is the translate of an abelian subvariety of A.*

The techniques for Bloch's theorem of translational invariance can easily be adapted to give the following generalization for the space of jets of entire holomorphic curves.

Theorem 1.2. *Let A an abelian variety and $\varphi : \mathbf{C} \to A$ be a holomorphic map whose image is Zariski dense in A. Let k be a positive integer and $\mathrm{Var}_{\varphi,k}$ be the algebraic Zariski closure of the k-th order jet of φ in the space $J_k(A)$ of k-jets of A. Then there is a finite unramified covering $\pi : \tilde{A} \to A$ of A and a decomposition $\tilde{A} = B_1 \times B_2 \times \cdots \times B_s$ of $\tilde{A}$ as a product of positive-dimensional abelian varieties such that the lift $\pi^* \mathrm{Var}_{\varphi,k}$ is invariant under translation by elements in B_1, and, in general, $\rho_i \left(\pi^* \mathrm{Var}_{\varphi,k}\right)$ is invariant under translation by elements in B_{j+1}, where $\rho_j : J_k(\tilde{A}) \to J_k\left(\tilde{A} / \left(B_1 \times \cdots \times B_j\right)\right) = J_k\left(B_{j+1} \times \cdots \times B_s\right)$ is the natural projection map $(1 \leq j < s)$.*

This theorem is in the "Correction" of [22] and is a modification of Theorem 2.2 of [22]. The original Theorem 2.2 of [22] is stronger than what was demonstrated in its proof, as pointed out to us by Paul Vojta in an e-mail message on July 13, 2000. A trivial counter-example to the original Theorem 2.2 of [22] is that A is the product of two copies of any elliptic curve and φ is induced by the map $\zeta \to (z_1, z_2) = \left(\zeta, \zeta^2\right)$ from $\mathbf{C}$ to the universal cover $\mathbf{C}^2$ of A.

Remark 3. Bloch's theorem (Theorem 1.1) gives the hyperbolicity of any complex submanifold Y in an abelian variety which does not contain any translate of an abelian subvariety. The problem concerning the Schottky radius for the hyperbolicity of Y is still open. The Schottky radius is the largest r such that there exists a holomorphic map from the 1-disk of radius r to Y when the differential of the map is normalized at the origin. The problem is to express or estimate the Schottky radius effectively in terms of numbers computable from the geometry and topology of Y.

1.4 Bloch's Original Presentation of the Three Techniques. To see how the three techniques are originally presented in Bloch's paper, first let us look at the setting of Bloch's paper, which is described on line 15 of page 36, "Soit une surface algébrique S d'irrégularité supérieure à 2, que nous supposons placée dans l'espace à trois dimensions, où elle a pour équation $f(x, y, z) = 0$. Les coordonnées d'un point P de la surface sont supposées fonctions méromorphes, dans tout le plan complexe, d'une variable t. Soient U, V, W trois integrales simples de premiere espèce linéairement indépendantes de la surface."

So the abelian variety A in Bloch's setting has complex dimension 3 and U, V, W are the coordinates of the universal cover $\mathbf{C}^3$ of A. His algebraic surface S is inside A and at the same time inside $\mathbf{P}_3$ defined by $f(x, y, z) = 0$ with inhomogeneous coordinates x, y, z of $\mathbf{P}_3$.

Bloch used the notation gm (r, f) to denote Nevanlinna's characteristic function $T(r, f)$ for a meromorphic function f on $\mathbf{C}$. He considered a holomorphic map φ from $\mathbf{C}$ to a complex manifold X of complex dimension n and considered a point P on X where some global meromorphic functions $g_1, \cdots, g_n$ on X are used to define local coordinates of X at P. He used $\widehat{gm}(r, P)$ to denote the maximum of $gm\left(r; g_j \circ \varphi\right)$ for $1 \leq j \leq n$. So far as the growth order up to sandwiching by two constant multiples is concerned, $\widehat{gm}(r, P)$ is independent of the choice of $g_1, \cdots, g_n$. The use of P in the symbol $\widehat{gm}(r, P)$ simply emphasizes the fact that P is a variable point on a manifold X which is a target of a holomorphic map $\varphi : \mathbf{C} \to X$.

Bloch's Construction of Holomorphic Jet Differentials. On page 37, lines 10–11 from the bottom of [2], he introduced holomorphic jet differentials by forming algebraic functions of $dU, dV, dW, d^2U, d^2V, d^2W$. His statement that "on obtiendra pour x et y des fonctions algébriques de $dU, dV, dW, d^2U, d^2V, d^2W$" means that the meromorphic functions x and y on X satisfies some polynomial equation $\sum_{j=0}^{k} P_j \xi^k = 0$, where $P_\ell = P_\ell\left(dU, dV, dW, d^2U, d^2V, d^2W\right)$ is a polynomial of $dU, dV, dW, d^2U, d^2V, d^2W$. The equation actually is meant only to hold on $\mathcal{X}_2$, as one can see from "Si, donc, cette résolution est possible, comme cela aura lieu *au général*, on sera conduit à une impossibilité en supposant x, y, z fonctions méromorphes de t" on line 10 from the bottom of page 37 of [2].

Bloch's Proof of Schwarz Lemma. From the preceding equation $\sum_{j=0}^{k} P_j \xi^k = 0$ Bloch argued to get the contradiction that $\widehat{gm}(r, P)$ is dominated by $\log \widehat{gm}(r, P)$. He did not explicitly mention that the holomorphic jet differential P_0 vanishes on the ample divisor defined by the zero-set of x. Without much details he simply referred to Lemme III of [2], which is an argument corresponding to the logarithmic derivative lemma. Since in the proof of Lemme III on lines 9–13 on page 33 of [2], when he compared the characteristic functions of e^U and Φ, he used the comparison of the counting functions of the zero-sets of $e^U - j$ and Φ for some judiciously chosen point j, the natural way to fill in the details which he left out is that he would use the same method of comparison of the counting functions of the zero-sets. In other words, he would use the fact that the holomorphic jet differential P_0 vanishes on the zero-divisor of x. By this comparison he would get the domination of $T(r, x \circ \varphi)$ by a positive constant times $T(r, \varphi^* P_0)$. Thus by the use of the logarithmic derivative lemma $T(r, \varphi^* P_0)$ is dominated by a positive constant times $\widehat{gm}(r, P)$.

For this step the proof of the Schwarz lemma (Lemma 2) applies the logarithmic derivative lemma to $\frac{dx}{x}, \frac{dy}{y}, \frac{d^2x}{x}, \frac{d^2y}{y}$ and compare $|P_0|$ to the absolute value of a polynomial of $\frac{dx}{x}, \frac{dy}{y}, \frac{d^2x}{x}, \frac{d^2y}{y}$. Bloch did not apply the logarithmic derivative lemma the same way, but used some essentially equivalent argument. He applied the differentiation of Poisson's formula to $dU = \frac{d\left(e^U\right)}{e^U}$ and $\frac{d^2\left(e^U\right)}{e^U}$ and similar expressions for V, W. Since the function e^U is not a meromorphic function on the abelian variety A, he had to use the argument that the maximum of $T\left(r, e^U \circ \varphi\right)$, $T\left(r, e^V \circ \varphi\right)$, $T\left(r, e^W \circ \varphi\right)$ dominates a positive constant times $T(r, g \circ \varphi)$ for any meromorphic function g on the 3-dimensional abelian variety A. For the case of an elliptic curve C whose universal cover has coordinate U, he argued that there is a covering map $\pi : \mathbf{C} - 0 \to C$ with e^U as the coordinate of $\mathbf{C} - 0$. On lines 9 to 13 on page 33, by comparing the counting function for the zero-set of $e^U - j$ for a judiciously chosen point j of $\mathbf{C} - 0$ and the counting function of the zero-set of $\Phi - \pi(j)$ on C, he concluded that $T\left(r, e^U \circ \psi\right)$ dominates a positive constant times $T(r, \Phi \circ \psi)$ for any holomorphic map $\psi : \mathbf{C} \to \mathbf{C}$. He then adapted the argument for the elliptic curve C to the 3-dimensional abelian variety A.

Bloch's Differential Equations. In Bloch's paper the differential equations are given as follows. Consider the position-forgetting map defined by $dU, dV, dW, d^2U, d^2V, d^2W$. Bloch's paper (in equation (6) on page 38 of [2]) states that if the position-forgetting map from the algebraic Zariski closure $\mathcal{X}_2$ in $J^2(A)$ to $\mathbf{C}^6$ is not generically of finite fiber, then for any local coordinates ξ, η of S one has the vanishing on $\mathcal{X}_2$ of the Jacobian determinant

$$\frac{\partial\left(d^2U, d^2V, d^2W, d^3U, d^3V, d^3W\right)}{\partial\left(d\xi, d\eta, d^2\xi, d^2\eta, d^3\xi, d^3\eta\right)} \tag{1}$$

when $d^2U, d^2V, d^2W, d^3U, d^3V, d^3W$ are regarded only as functions of $d\xi, d\eta, d^2\xi, d^2\eta, d^3\xi, d^3\eta$ with ξ, η regarded as constants for the purpose of forming the Jacobian determinant. In [2] $\xi = x$ and $\eta = y$ are used. Since the vanishing of (1) on $\mathcal{X}_2$ is independent of the choice of local coordinates ξ, η, the notations ξ and η are used here just to avoid confusion in notations later. In the notation of Bloch's third technique in (1.3) with $m = 2$ and $n = 3$, this corresponds to the vanishing of the determinant of the 2×2 matrix $\left((\partial_\zeta^j(\omega_{\alpha\nu} \circ f))(\zeta_0)\right)_{j=1,2;\, \nu=1,2}$ with $\alpha = 3$. When we choose $U = z_1$, $V = z_2$, $W = z_3$ and the local coordinates $\xi = z_1$, $\eta = z_2$, the 2×2 matrix becomes

$$\begin{pmatrix} dW_\xi & dW_\eta \\ d^2W_\xi & d^2W_\eta \end{pmatrix} \tag{2}$$

pulled back to $\mathcal{X}_2$. Straightforward computation shows that on $\mathcal{X}_2$ the vanishing of (1) agrees with the vanishing of (2). Bloch then used the usual Wronskian argument to show that there are constants a, b, c such that $adU + bdV + cdW$ vanishes on S, which is the equation on line 16 on page 39 of [2]. The vanishing of (1) on $\mathcal{X}_2$ is the system of differential equations from the link between $d\Phi_k$ and Φ_{k+1}.

Remark 4. In Bloch's third technique the order of the jet space used is the same as the dimension of X. The zero-dimensionality of the generic fiber of $\Phi_k : \mathcal{X}_k \to \mathbf{C}^{kn}$ corresponds to the following statement used in diophantine approximation (Lemma 5.1 of [26]).

Suppose A is an abelian variety and X is a subvariety of A which is not invariant under the translation of any nonzero element of A. Then for any $m >$ dim X the map $X^{\times m} \to A^{\times\left(\frac{m(m-1)}{2}\right)}$ defined by $(x_j)_{1\le j\le m} \mapsto (x_j - x_k)_{1\le j<k\le m}$ is generically finite onto its image.

The jet space of order $m - 1$ in Bloch's function theory case is replaced by a product of m copies of X in Vojta's case of diophantine approximation. The use of differences in diophantine approximation corresponds to the use of differentials in function theory. McQuillan [12] used this correspondence to give a completely new proof of Bloch's theorem. Vojta's article [27] gives a comprehensive survey on the relation between Nevanlinna theory and diophantine approximation.

1.5 Complement Case in the Setting of Abelian Varieties. Though hyperbolicity for the complement of a hypersurface in an abelian variety was not considered in Bloch's paper [2], Bloch's techniques are also applicable to the complement case. We treat here the Second Main Theorem with truncated multiplicity which is the quantitative version and is the strongest known result for this setting. The result is in the Addendum of [23]. The system of differential equations from Bloch's third technique (1.3) is a bit unwieldy. In the paper proper of [23] we attempted to skirt their use by using the semi-continuity of cohomology groups in deformations. Noguchi, Winkelmann, and Yamanoi [16] drew our attention to a difficulty in that part of [23]. The Addendum of [23] restores the use of Bloch's third technique (1.3) of a system of differential equations.

Theorem 1.3. *Let A be an abelian variety of complex dimension n and D be an ample divisor of A. Inductively let $k_0 = 0$ and $k_1 = 1$ and $k_{\ell+1} = k_\ell + 3^{n-\ell-1}(4(k_\ell+1))^\ell D^n$ for $1 \le \ell < n$. If $\varphi : \mathbf{C} \to A$ is a holomorphic map whose image is not contained in any translate of D, then $m(r, \varphi, D) +$*

$\left(N(r,\varphi,D) - N_{k_n}(r,\varphi,D)\right) = O(\log T(r,\varphi,D) + \log r)$ *for* r *outside some set whose measure with respect to* $\frac{dr}{r}$ *is finite. In particular, the defect* $\delta(\varphi, D)$ *is zero and the complex manifold* $A - D$ *is hyperbolic.*

Proof. We observe that it suffices to show that

$$\pi_{k_n}(J_{k_n}(D)) \text{ does not contain } W_{k_n,\varphi}. \tag{3}$$

The reason is as follows. By (3) there exists a polynomial P in the variables $d^\mu z_\nu$ $(1 \le \mu \le k_n, 1 \le \nu \le n)$ with coefficients in $\mathbf{C}$ such that P is not identically zero on $W_{k_n,\varphi}$ and $\pi^*_{k_n} P$ is identically zero on $J_{k_n}(D)$. Let ∇ be the covariant differential operator for D defined from a metric of D whose curvature form has constant coefficients. We can write P as $\sum_{\mu=0}^{k_n} \tilde{\rho}_\mu (\nabla^\mu \theta_D)$ so that $\tilde{\rho}_\mu$ is a polynomial in $d^\mu z_\nu$ $(1 \le \mu \le k_n, 1 \le \nu \le n)$ whose coefficients are bounded smooth functions on $\mathbf{C}^n$. Let $\tilde{\varphi} : \mathbf{C} \to \mathbf{C}^n$ be the lifting of φ. By the logarithmic derivative lemma, both $m(r, \tilde{\varphi}^* P, \infty)$ and $m\left(r, \tilde{\varphi}^*\left(\frac{P}{\theta_D}\right), \infty\right)$ are of order $O(\log T(r,\varphi,D) + \log r)$. From the expression $P = \sum_{\mu=0}^{k_n} \tilde{\rho}_\mu (\nabla^\mu \theta_D)$ it follows that $N\left(r, \tilde{\varphi}^*\left(\frac{P}{\theta_D}\right), \infty\right)$ is dominated by $N_{k_n}(r,\varphi,D)$. Our conclusion then follows from the domination of $m(r,\varphi,D) + N(r,\varphi,D) = T\left(r, \tilde{\varphi}^*\left(\frac{1}{\theta_D}\right)\right) + O(1)$ by $T\left(r, \tilde{\varphi}^*\left(\frac{P}{\theta_D}\right)\right) + T\left(r, \tilde{\varphi}^*\left(\frac{1}{P}\right)\right) + O(1)$.

To prove (3) we define the following meromorphic connection for D. Choose $\tau_j \in \Gamma(A, jD)$ for $j = 3, 4$ with the property that for $1 \le k \le k_n$ there exists a proper subvariety Z_k of $W_{k,\varphi}$ such that for $\gamma_k \in W_{k,\varphi} - Z_k$ the set $\{\tau_3\tau_4 = 0\}$ $\cap J_k(D) \cap \pi_k^{-1}(\gamma_k)$ is a nowhere dense subvariety of $J_k(D) \cap \pi_k^{-1}(\gamma_k)$. Define $\mathcal{D}^k\theta_D = (\tau_4)^{k+1} d^k\left(\frac{\theta_D \tau_3}{\tau_4}\right)$ for $1 \le k \le k_n$.

Let $0 \le \ell_0 \le n$ be the largest integer such that the following $(A)_\ell$ holds with $\ell = \ell_0$.

$(A)_\ell$. There exists a proper subvariety E_ℓ of $W_{k_\ell,\varphi}$ such that the complex dimension of the common zero-set of $\langle \mathcal{D}^j\theta_D, \gamma\rangle$ $(0 \le j \le k_\ell)$ in $A - \{\tau_3\tau_4 = 0\}$ is no more than $n - 1 - \ell$ for $\gamma \in W_{k_\ell,\varphi} - E_\ell$.

The definition of ℓ_0 implies that $(B)_\ell$ stated below holds with $\ell = \ell_0 + 1$ for the following reason. For a local complex subspace Y defined by holomorphic functions g_j $(0 \le j \le q-1)$ and for a holomorphic germ g_q, if both the dimension and the multiplicity of the subspace defined by g_j $(0 \le j \le q)$ are the same of those of Y at a generic point Q of the highest-dimensional branches of Y, then g belongs to the ideal generated by $g_1, \cdots, g_q$ at Q.

We apply this to $g_j = \langle \mathcal{D}^j\theta_D, \gamma\rangle$ $(0 \le j \le q)$ for a generic $\gamma \in W_{q,\varphi}$ and note that $\langle \mathcal{D}^j\theta_D, \gamma\rangle$ depends only on $p_{j,q}\gamma$. We bound the multiplicity of Y by using $n - \ell_0 - 1$ generic global sections of the very ample line bundle $3D$ over A to get a zero-dimensional dimensional intersection with Y. We end up with $3^{n-\ell_0-1}\left(4\left(k_{\ell_0}+1\right)^{\ell_0}\right) D^n$ for the multiplicity. The factor $4\left(k_{\ell_0}+1\right)$ occurs, because we use ℓ_0 elements of $\Gamma\left(A, \left(4\left(k_{\ell_0}+1\right)\right) D\right)$ to locally define Y

which are generic $\mathbf{C}$-linear combinations of $\sigma_j g_j$ $(0 \le j \le k_{\ell_0})$ for generic $\sigma_0 \in \Gamma\left(A, 4\left(k_{\ell_0}+1\right)\right)$ and $\sigma_j \in \Gamma\left(A, 4\left(k_{\ell_0}-j\right)\right)$ $(1 \le j < k_{\ell_0})$ and $\sigma_{k_{\ell_0}} \equiv 1$. The sections σ_j are introduced so that all $\sigma_j g_j$ are sections of the same line bundle to enable us to take their $\mathbf{C}$-linear combinations.

$(B)_\ell$. There exist $k_{\ell-1} < q \le k_\ell$, a subvariety V_ℓ of $J_q(D) \cap \pi_q^{-1}(W_{q,\varphi})$ containing $J_q(D) \cap \{\tau_3\tau_4 = 0\}$, and a proper subvariety F_ℓ of $W_{q,\varphi}$, such that

(i) $\left(J_q(D) - V_\ell\right) \cap \pi_q^{-1}(\gamma_q)$ is a nonempty $(n-\ell)$-dimensional subvariety of $J_q(A) - V_\ell$ for every $\gamma_q \in W_{q,\varphi} - F_\ell$, and

(ii) on $J_q(A) - \pi_q^{-1}(F_\ell) - V_\ell$ the function $\mathcal{D}^q\theta_D$ locally belongs to the ideal sheaf which is locally generated by the functions $\mathcal{D}^j\theta_D$ $(0 \le j \le q-1)$ and by the pullbacks to $J_q(A)$ of the local holomorphic functions on $\mathbf{C}^{nq}$ vanishing on $W_{q,\varphi}$.

There exists some $\zeta_0 \in \mathbf{C}$ such that $\pi_q\left((d^q\varphi)(\zeta_0)\right) \in W_{q,\varphi} - F_{\ell_0+1}$. By $(B)_{\ell_0+1}$(i) there exists some point $Q_0 \in D$ such that when we define $g(\zeta) = (Q_0 - \varphi(\zeta_0)) + \varphi(\zeta) \in A$ for $\zeta \in \mathbf{C}$, we have $(d^q g)(\zeta_0) \in J_q(D) - V_{\ell_0+1} - \pi_q^{-1}\left(F_{\ell_0+1}\right)$. Since $\pi_q\left((d^q g)(\zeta)\right) = \pi_q\left((d^q\varphi)(\zeta)\right) \in W_{q,\varphi}$ for $\zeta \in \mathbf{C}$, from $(B)_{\ell_0+1}$(ii) it follows that $\theta_D \circ g$ satisfies a differential equation $\left(\frac{d^q}{d\zeta^q}\right)(\theta_D \circ g)(\zeta) = \sum_{j=0}^{q-1} h_j(\zeta)\left(\frac{d^j}{d\zeta^j}\right)(\theta_D \circ g)(\zeta)$ on some open neighborhood U of ζ_0 in $\mathbf{C}$ for some holomorphic functions $h_0(\zeta), \cdots, h_{q-1}(\zeta)$ on U. By the uniqueness part of the fundamental theorem of ordinary differential equations, $(\theta_D \circ g)(\zeta)$ is identically zero for $\zeta \in \mathbf{C}$. Hence the image of φ is contained in the translate of D by $Q_0 - \varphi(\zeta_0)$. This contradicts the assumption of the image of φ not contained in any translate of D and finishes the verification of (3).

Remark 5. The same arguments work for semi-abelian varieties with straightforward modifications.

2 Setting of the Complex Projective Space

What has been guiding the investigations for this setting are the following two conjectures.

Conjecture 1. A generic hypersurface of degree at least $2n-1$ in $\mathbf{P}_n$ is hyperbolic for $n \ge 3$.

Conjecture 2. The complement in $\mathbf{P}_n$ of a generic hypersurface of degree at least $2n+1$ is hyperbolic.

Remark 6. The number $2n+1$ is used in Conjectures 1 and 2, because when one replaces the generic hypersurface in Conjecture 2 by a union of hyperplanes in general position, Green [9] proved that the image of a holomorphic map from $\mathbf{C}$ to $\mathbf{P}_n$ minus $n+k$ hyperplanes in general position is contained in a linear subspace of dimension no more than the integral part $\left\lfloor \frac{n}{k} \right\rfloor$ of $\frac{n}{k}$, where the bound $\left\lfloor \frac{n}{k} \right\rfloor$ is sharp.

Some partial results are known. The set of hyperbolic hypersurfaces in $\mathbf{P}_n$ is open in the Hausdorff topology on the moduli space of hypersurfaces [28]. The complement in $\mathbf{P}_2$ of a generic curve of sufficiently high degree is hyperbolic [21]. A generic surface of degree at least 36 in $\mathbf{P}_3$ is hyperbolic [14]. A generic surface of degree at least 21 in $\mathbf{P}_3$ and the complement of a generic curve of degree at least 21 are both hyperbolic [7]. For the surface case the new important techniques come from [13]. We will discuss some of those techniques in Section 3. Improvements in the degree bound in [7] are mainly due to the use of a more restricted class of jet differentials called Semple jet differentials so that one needs fewer independent holomorphic jet differentials vanishing on ample divisors in this class in order to use the Schwarz lemma to force the image of every holomorphic map to lie in a proper subvariety. The Semple jet differentials are those whose pullbacks to any local complex curve depend only on the first differential of the local coordinate of the curve and not on any of its higher order differentials. Examples of Semple jet differentials are the determinants of the form $\left(d^j f_\ell\right)_{1\leq j,\ell\leq k}$, where $f_1, \cdots, f_k$ are functions. Such jet differentials were already used in [21].

The most recent result for Conjecture 1 for general dimension is the following. We will discuss the main ideas in its proof. The method can also be adapted for Conjecture 2, but we will not discuss its adaptation here.

Theorem 2.4. [20] *There exists a positive integer δ_n such that a generic hypersurface in $\mathbf{P}_n$ of degree $\geq \delta_n$ is hyperbolic.*

The method of proof is motivated by the work of Clemens–Ein–Voisin [6][8][25] and uses the following ingredients: (1) the global generation of the bundle of meromorphic vector fields on the vertical jet space of the universal family of hypersurfaces with low pole order in the fiber direction; (2) the method of counting monomials, motivated by the theorem of Riemann–Roch and the lower bound of the negativity of jet-differential bundles of the projective space, to produce holomorphic families of vertical jet differentials on fibers vanishing on ample divisors; (3) the action of low pole-order vector fields on the vertical jet space of the universal family of of hypersurfaces to produce independent vertical jet differentials on generic hypersurfaces.

The method is generalizable to complete intersections, but is unable to handle the conjecture that entire holomorphic curves in a compact algebraic manifold of general type must be contained in a fixed proper subvariety.

By Schwarz lemma (1.2), the hyperbolicity problem is reduced to the following two steps: (1) the construction of holomorphic jet differentials vanishing on some ample divisor; (2) to make sure that there are enough independent such jet differentials.

Besides Bloch's method (1.1) for the abelian setting, some available methods of construction of jet differentials are as follows: (1) the explicit construction similar to $\frac{P(x,y)dx}{R_y(x,y)} = -\frac{P(x,y)dy}{R_x(x,y)}$ for holomorphic 1-forms on a regular plane curve $R(x, y) = 0$ with deg $P(x, y) \leq$ deg $R(x, y) - 3$; (2) the use of the theorem of Riemann–Roch (somctimes coupled with the stability properties of tangent bundles for more information on Chern classes). The last method is applicable only to complex surfaces because of the inability to handle the terms of the higher cohomology groups in the general dimension case.

One new technique introduced in the proof of Theorem 2.4 is to use a method motivated by the lower bound of the negativity of the jet differential bundle on $\mathbf{P}_n$ to take care of the terms of the higher cohomology groups in the application of the theorem of Riemann–Roch to hypersurfaces in $\mathbf{P}_n$.

Since our method is motivated by the result of Clemens–Ein–Vosin [6] [8] [25], we first discuss their method.

2.1 Clemens's Result and Its Techniques (with Generalizations by Ein and Voisin). Let $S = \mathbf{P}_N$ be the moduli space of all hypersurfaces in $\mathbf{P}_n$ of degree δ, where $N = \binom{\delta+n}{n} - 1$. Let $\mathcal{X} \to S$ be the universal family, defined by $f = \sum_{\nu \in \mathbf{N}^{n+1}, |\nu|=\delta} \alpha_\nu z^\nu$, which is nonsingular, of bidegree $(\delta, 1)$ in $\mathbf{P}_n \times \mathbf{P}_N$. Here for a multi-index $\nu = (\nu_0, \cdots, \nu_n)$ we denote $\nu_0 + \cdots + \nu_n$ by $|\nu|$.

Theorem 2.5 (Clemens [6]). *If X is a generic hypersurface of degree δ in $\mathbf{P}_n$, then there does not exist any rational curve (respectively elliptic curve) in X if $\delta \geq 2n-1$ (respectively $\delta \geq 2n$).*

The key step is the following lemma.

Lemma 4. *The $(1,0)$-twisted tangent bundle of $\mathcal{X}$ is globally generated.*

Proof. For $0 \leq p \leq n$ let e_p denote the $(n+1)$-tuple whose only nonzero entry is the integer 1 in the p-th position. Let L be any homogeneous polynomial L of degree 1 in the set of variables $\{\alpha_\nu\}_{|\nu|=\delta}$. If $0 \leq p \neq q \leq n$ and $\nu, \mu \in \mathbf{N}^{n+1}$ with $\nu + e_p = \mu + e_q$, then the $(1,0)$-twisted vector field $L\left(z_q\left(\frac{\partial}{\partial \alpha_\nu}\right) - z_p\left(\frac{\partial}{\partial \alpha_\mu}\right)\right)$ on $\mathbf{P}_n \times \mathbf{P}_N$ is tangential to $\mathcal{X}$. For any given $a_{j,k}$ $(0 \leq j,k \leq n)$, the vector field $\sum_{j,k=0}^n a_{j,k} z_j \frac{\partial}{\partial z_k} + \sum_{|\nu|=\delta} \beta_\nu \frac{\partial}{\partial \alpha_\nu}$ is tangential to $\mathcal{X}$ when $\beta_\nu = -\sum_{0 \leq j,k \leq n, j \neq k} \alpha_{\Phi_{j,k}(\nu)} a_{j,k}(\nu_k + 1) - \sum_{j=0}^n \alpha_{\nu_0,\nu_1,\cdots,\nu_n} a_{j,j} \nu_j$, where $\Phi_{j,k} : \mathbf{N}^{n+1} \to \mathbf{N}^{n+1}$ is defined by $\left(\Phi_{j,k}(\nu)\right)_\ell = \nu_\ell$ for $\ell \neq j,k$ with $\left(\Phi_{j,k}(\nu)\right)_j = \nu_j - 1$ and $\left(\Phi_{j,k}(\nu)\right)_k = \nu_k + 1$. Thus the $(1,0)$-twisted tangent bundle of $\mathcal{X}$ is globally generated.

Proof (of Theorem 2.5). Suppose there exists a curve C of genus 0 or 1 in a generic X. From such curves C we define a subvariety $\mathcal{C}$ of $\mathcal{X}$ so that for a generic X the restriction to X of the normal bundle $N_{\mathcal{X},\mathcal{C}}$ (of $\mathcal{C}$ in $\mathcal{X}$) equals the normal bundle $N_{X,C}$ (of C in X). By Lemma 4, the $(1,0)$-twisted normal bundle $N_{\mathcal{X},\mathcal{C}}$ is globally generated. Thus $N_{C,X} \otimes \mathcal{O}_{\mathbf{P}_n}(1)$ is globally generated. The adjunction formula implies that the genus of C is no less than $\delta - 2n + 1$.

2.2 Spaces of Vertical Jets and Vector Fields on Their Universal Space. The global generation of the $(1,0)$-twisted tangent bundle of $\mathcal{X}$ in Lemma 4 implies the global generation of all q-jets in the $(q+1,q)$-twisted tangent bundle of $\mathcal{X}$. If there exists a holomorphic jet differential ω on the part of $\mathcal{X}$ over an affine open subset of S which vanishes to sufficiently high order on some ample divisors of the fibers, then the derivatives of ω with respect to low pole-order vector fields on $\mathcal{X}$ would give enough independent holomorphic jet differentials on generic fibers vanishing on ample divisor. However, there is one difficulty with this approach to proving the hyperbolicity of generic fibers.

The difficulty is that one can find holomorphic jet differentials on each fiber of $\mathcal{X}$ but not on the total space $\mathcal{X}$ itself, because a holomorphic jet differential on $\mathcal{X}$ is different from a holomorphic family of fiber-wise jet differentials. The latter can be obtained by sufficient twisting in the moduli space, but there is no way to construct the former. To illustrate the difficulty, consider the situation of a family of plane curves $R(x, y, \alpha) = 0$. On each curve with parameter α, one can get the holomorphic 1-form $\frac{dx}{R_x(x,y,\alpha)} = -\frac{dy}{R_x(x,y,\alpha)}$ from $dR = 0$ when α is regarded as a constant. However, when α is a variable, $d\alpha$ gives us trouble in $dR = R_x dx + R_y dy + R_\alpha d\alpha$ and we cannot use this method to get a holomorphic 1-form on the total space of plane curves even when we confine α to the open unit disk in $\mathbf{C}$.

To overcome this difficulty one introduces the space $J_{n-1}^{\text{vert}}(\mathcal{X})$ of vertical $(n-1)$-jets which is defined by $f = df = \cdots = d^{n-1} f = 0$ in $J_{n-1}(\mathcal{X})$ with the coefficients α_ν of f regarded as constants when forming $d^j f$ $(1 \leq j \leq n-1)$.

The proof of the following proposition is along the lines of the proof of Lemma 4, but the bookkeeping of indices and the algorithms used are very involved. The two constants c_n, c_n' in the proposition can be explicitly written.

Proposition 1 (Existence of low pole-order vector fields). *There exist c_n, $c_n' \in \mathbf{N}$ such that the (c_n, c_n')-twisted tangent bundle of the projectivization of $J_{n-1}^{\text{vert}}(\mathcal{X})$ is globally generated.*

To avoid considering singularities of weighted projective spaces, instead of using the projectivization of $J_{n-1}^{\text{vert}}(\mathcal{X})$, for the proof of Proposition 1 and its application we actually use functions whose restrictions to each fiber are polynomials, of homogeneous weight, in the differentials of the homogeneous coordinates of $\mathbf{P}_n$.

2.3 Construction of Jet Differentials on Hypersurfaces. The motivation for the new method of construction of holomorphic jet differentials on hypersurfaces of high degree is from the theorem of Riemann–Roch and the lower bound of the negativity of jet bundle of hypersurface X. Since the twisted cohomology groups of $\mathbf{P}_n$ can be computed from counting the number of certain monomials, in our construction we use the method of directly counting certain monomials instead of going through the theorem of Riemann–Roch. One key step is the following lemma on the nonvanishing of the restrictions, to a hypersurface of high degree, of jet differentials given by polynomials of low weight in the differentials of the inhomogeneous coordinates of $\mathbf{P}_n$. By using the high degree of the hypersurface and the vanishing of certain sheaf cohomology of the complex projective space minus a certain linear subspace, one proves the lemma by showing that the identical vanishing of the restriction of the jet differential would imply that the low-weight polynomial can be expressed in an impossible way in terms of the differentials of the high-degree polynomial f.

Lemma 5 (Nonvanishing of restrictions of low-weight jet differentials). *Let $1 \leq k \leq n-1$ and let f be a polynomial of degree δ in the inhomogeneous* *coordinates* $x_1, \cdots, x_n$ *of* $\mathbf{P}_n$ *so that the zero-set of* f *defines a complex manifold* X *in* $\mathbf{P}_n$. *Let* Q *be a non identically zero polynomial in the variables* $d^j x_\ell$ $(0 \leq j \leq k, 1 \leq \ell \leq n)$. *Assume that* Q *is of degree* m_0 *in* $x_1, \cdots, x_n$ *and is of homogeneous weight* m *in the variables* $d^j x_\ell$ $(1 \leq j \leq k, 1 \leq \ell \leq n)$ *when the weight of* $d^j x_\ell$ *is assigned to be* j. *If* $m_0 + 2m < \delta$, *then* Q *is not identically zero on the space of* k*-jets of* X.

Lemma 5 enables us to use the method of counting certain monomials to construct holomorphic jet differentials on hypersurfaces of sufficiently high degree which vanish on an ample divisor. The following proposition gives the precise statement on the existence of such jet differentials.

Proposition 2 (Existence of holomorphic jet differentials). *Let* X *be a nonsingular hypersurface of degree* δ *in* $\mathbf{P}_n$ *defined by a polynomial* $f(x_1, \cdots, x_n)$ *of degree* δ *in the affine coordinates* $x_1, \cdots, x_n$ *of* $\mathbf{P}_n$. *Suppose* ϵ, ϵ', θ_0, θ, *and* θ' *are numbers in the open interval* $(0, 1)$ *such that* $n\theta_0 + \theta \geq n + \epsilon$ *and* $\theta' < 1 - \epsilon'$. *Then there exists an explicit positive number* $A = A(n, \epsilon, \epsilon')$ *depending only on* n, ϵ, *and* ϵ' *such that for* $\delta \geq A$ *and any nonsingular hypersurface* X *in* $\mathbf{P}_n$ *of degree* δ *there exists a non identically zero* $\mathcal{O}_{\mathbf{P}_n}(-q)$*-valued holomorphic* $(n-1)$*-jet differential* ω *on* X *of total weight* m *with* $q \geq \delta^{\theta'}$ *and* $m \leq \delta^{\theta}$. *Here, with respect to a local holomorphic coordinate system* $w_1, \cdots, w_{n-1}$ *of* X, *the weight of* ω *is in the variables* $d^j w_\ell$ $(1 \leq j \leq n-1, \quad 1 \leq \ell \leq n-1)$ *with the weight* j *assigned to* $d^j w_\ell$. *Moreover, for any affine coordinates* $x_1, \cdots, x_n$ *of* $\mathbf{P}_n$, *when* $f_{x_1} = 1$ *defines a nonsingular hypersurface in* X, *the* $(n-1)$*-jet differential* ω *can be chosen to be of the form* $\frac{Q}{f_{x_1}-1}$, *where* $q = \left\lfloor \delta^{\theta'} \right\rfloor$ *and* Q *is a polynomial in* $d^j x_1, \cdots, d^j x_n$ $(0 \leq j \leq n-1)$ *which is of degree* $m_0 = \left\lceil \delta^{\theta_0} \right\rceil$ *in* $x_1, \cdots, x_n$ *and is of homogeneous weight* $m = \left\lceil \delta^{\theta} \right\rceil$ *in* $d^j x_1, \cdots, d^j x_n$ $(1 \leq j \leq n-1)$ *when the weight of* $d^j x_\ell$ *is assigned to be* j.

The proof of Proposition 2 consists of counting the number of coefficients of Q and counting the number of equations needed for the jet differential on X defined by Q to vanish on an ample divisor in X of high degree which is defined by the vanishing of a polynomial $g =$ in $\mathbf{P}_n$. For the counting of equations, we use $f = df = \cdots = d^{n-1} f = 0$ (with the coefficients α_ν treated as constants) to eliminate one coordinate and its differentials. One has to use the nonvanishing result from Lemma 5.

2.4 Argument of Genericity and Final Step of Proof of Hyperbolicity. Before we do the last step of the proof of hyperbolicity which is done by applying low-pole order vector fields to produce enough independent holomorphic jet differentials, we first explain where the genericity of the hypersurface comes from in our argument. A part of the genericity comes from the following well-known statement.

Proposition 3. *Let* $\tilde{\pi} : \mathcal{Y} \to S$ *be a flat holomorphic family of compact complex spaces and* $\mathcal{L}$ *be a holomorphic vector bundle over* $\mathcal{Y}$. *Then there exists a proper subvariety* Z *of* S *such that for* $s \in S - Z$ *the restriction map* $\Gamma(U_s, \mathcal{L}) \to \Gamma\left(\tilde{\pi}^{-1}(s), \mathcal{L}|_{\tilde{\pi}^{-1}(s)}\right)$ *is surjective for some open neighborhood* U_s *of* s *in* S.

For our application $S = \mathbf{P}_N$ is the parameter space of hypersurfaces of degree δ in $\mathbf{P}_n$ and the fiber $Y_s = \tilde{\pi}^{-1}(s)$ of the flat holomorphic family $\pi : \mathcal{Y} \to S$ of compact complex spaces is the weighted projective space defined from the space $J_{n-1}(X_s)$ of $(n-1)$-jets of the hypersurface X_s with moduli s, where the weight is chosen to correspond to the weight of $d^j x_\ell$ being j. The restriction to Y_s of the holomorphic vector bundle $\mathcal{L}$ corresponds to the bundle of $(n-1)$-jet differentials of weight m twisted by $\mathcal{O}_{\mathbf{P}_n}(-q)$.

When $s_0 \in S - Z$ the hypersurface satisfies the following weaker form of hyperbolicity (also known as the algebraic degeneracy of entire holomorphic curves): there exists a proper subvariety X'_{s_0} in X_{s_0} such that the image of any nonconstant holomorphic map $\varphi : \mathbf{C} \to X_{s_0}$ must be contained in X'_{s_0}. The reason is as follows.

On a generic hypersurface X_{s_0} the jet differential $\frac{Q}{g}$ can be extended holomorphically to ω on $\bigcup_{s \in U_{s_0}} X_s$ with values in $\mathcal{O}_{\mathbf{P}_n}(-q)$. Consider ω as a function on $J_{n-1}(X_s)$. The vanishing order of this function ω as a function of the variables $d^j x_\ell$ $(1 \leq \ell \leq n-1,\ 1 \leq j \leq n-1)$ is no more than the weight m of ω at a generic point of X_{s_0} outside the zero-section of $J_{n-1}\left(X_{s_0}\right)$. We use meromorphic vector fields $v_1, \cdots, v_p$ on $J_{n-1}^{\text{vert}}(\mathcal{X})$ of low pole order in the fiber direction from Proposition 2. We now differentiate this function ω no more than m times by such meromorphic vector fields $v_1, \cdots, v_p$ on $J_{n-1}^{\text{vert}}(\mathcal{X})$ to form $v_1 \cdots v_p \omega$. The restriction of $v_1 \cdots v_p \omega$ to X_{s_0} defines a holomorphic jet differential on X_{s_0} which vanishes on an ample divisor, because the vanishing order of ω in the fiber direction is far greater than the pole order of v_j in the fiber direction. The collection of holomorphic jet differentials on X_{s_0} defined by $v_1 \cdots v_p \omega$ would have no common zeros on $J_{n-1}\left(X_{s_0}\right)$ over a generic point of X_{s_0} outside the zero-section of $J_{n-1}\left(X_{s_0}\right)$.

For the proof of hyperbolicity in its original sense, we need another part of the definition of genericity which concerns the maximum possible touching order of jet differentials of low degree and weight. More precisely, there exists some proper subvariety Z' of S with the following property.

Lemma 6 (Generic low vanishing order of constructed jet differentials). *Suppose $s \in S - Z'$ and Q is a polynomial in $d^j x_\ell$ $(1 \leq \ell \leq n,\ 0 \leq j \leq n-1)$ of degree no more than m_0 in $x_1, \cdots, x_n$ and of homogeneous weight no more than m in $d^j x_\ell$ $(1 \leq \ell \leq n,\ 1 \leq j \leq n-1)$. Then the vanishing order of the pullback of Q to X_s is no more than $a_{\delta,m_0,m}$ when $\delta \geq \delta_{n,m_0,m}$, where the numbers $a_{\delta,m_0,m}$ and $\delta_{n,m_0,m}$ can be explicitly written and the product of $a_{\delta,m_0,m}$ and the maximum fiber pole order of the meromorphic vector fields from Proposition 1 is far less than the minimum fiber vanishing order of the holomorphic vector field ω from Proposition 2 and from extension by Proposition 3.*

Once we have this additional part of genericity, the collection of holomorphic jet differentials on X_{s_0} defined by $v_1 \cdots v_p \omega$ would have no common zeros on $J_{n-1}\left(X_{s_0}\right)$ at any point of X_{s_0} outside the zero-section of $J_{n-1}\left(X_{s_0}\right)$. Let us know look at the proof of Lemma 6. Low vanishing order is a Zariski open condition. It suffices to prove the existence of one point $s_0 \in S - Z'$. The idea is to embed $\mathbf{P}_n$ generically into $\mathbf{P}_{\tilde{n}}$ by a high degree map of degree d and then pull back the holomorphic jet differential on a hypersurface $\tilde{X}$ in $\mathbf{P}_{\tilde{n}}$ to the intersection X of $\tilde{X}$ with $\mathbf{P}_n$. This method enables us to control the vanishing order of the coefficients at the expense of increasing the order of the jet differential. One drawback is that the degree of the hypersurface X must now contain the factor d. To overcome this drawback, we use a generic embedding of $\mathbf{P}_n$ into $\mathbf{P}_{2n+1} \times \mathbf{P}_{2n+1}$ by a map of appropriate bidegree (d_1, d_2) and construct jet differentials on a hypersurface $\tilde{X}$ in $\mathbf{P}_{2n+1} \times \mathbf{P}_{2n+1}$ and pull them back to the intersection X of $\tilde{X}$ with $\mathbf{P}_n$. This concludes the discussion of the ideas of the proof of Theorem 2.4.

3 McQuillan's Techniques for Hyperbolicity of Surfaces

McQuillan introduced a number of completely new techniques to treat the hyperbolicity problems for compact complex algebraic surfaces of general type [13]. His techniques were motivated by concepts from diophantine approximation. Here we highlight some of his techniques, recast them in more analytical formulations, and sketch a purely function-theoretical approach which uses the familiar tools of Nevanlinna theory.

3.1 Three Techniques of McQuillan. We select three techniques from [13]. The last two are relevant to our sketch of a purely function-theoretical approach which uses the familiar tools of Nevanlinna theory. The first one is an interpretation of Nevanlinna theory in terms of cohomology and intersection theory which may point the way to other possibilities of Nevanlinna theory in other types of intersection theory.

Currents Associated to an Entire Holomorphic Curve. Let X be a compact complex manifold of complex dimension n and $\varphi : \mathbf{C} \to X$ be a holomorphic map. Let H be an ample line bundle over X with positive smooth curvature form Θ_H and R be a sequence of positive real numbers with ∞ as limit. One defines the positive $(n-1, n-1)$-current $[\varphi]$ on X as follows. For a smooth $(1,1)$-form ω on X, the value $[\varphi](\omega)$ at ω is defined as the limit of $\frac{1}{T(r,\varphi,\Theta_H)} \int_{\rho=0}^{r} \frac{d\rho}{\rho} \int_{|\zeta|<\rho} \varphi^* \omega$ as $r \in R$ goes to ∞. McQuillan [13] verified that for a suitable choice of the coordinate ζ of $\mathbf{C}$ and of the sequence R the positive $(n-1, n-1)$-current $[\varphi]$ is closed on X. When Y is a complex hypersurface in X, since the closed positive $(1,1)$-current $[Y]$ defined by (integration over the regular part of) Y is not smooth, it is not meaningful to evaluate the closed positive $(n-1, n-1)$-current $[\varphi]$ at $[Y]$. However, one can still define the value of $[\varphi]$ at $[Y]$ as the limit of $\frac{1}{T(r,\varphi,\Theta_H)} \int_{\rho=0}^{r} n(r, \varphi, Y) \frac{d\rho}{\rho}$ as $r \in R$ goes to ∞. In general this value may be different from the cup product of the cohomology class of $[\varphi]$ and the cohomology class of $[Y]$. As a matter of fact, when the cohomology class $[Y]$ agrees with the first Chern class of some ample line bundle E, the two values agree if and only if the defect $\delta(\varphi, Y)$ for φ and Y is 0. In [13] this technique is on pages 127–132.

Remark 7. This technique gives an enlightening interpretation of Nevanlinna theory in terms of the discrepancy between the intersection number of the current of an entire holomorphic curve and a divisor and the cup product of the two cohomology classes defined by both. It leads naturally to possibilities of analogs of Nevanlinna theory for other products of cohomology classes and the corresponding intersection theory, for example, the quantum cohomology theory (see *e.g.*, [24][11]). We take three subvarieties Y_1, Y_2, Y_3 in a compact complex algebraic manifold X. In quantum cohomology one counts the number of holomorphic maps f from $\mathbf{P}_1$ with three marked points P_1, P_2, P_3 to X so that $P_j \in Y_j$ for $1 \leq j \leq 3$ and f satisfies certain topological conditions. We replace one of the subvarieties by the disk of radius r in an entire holomorphic curve and pass to limit as $r \to \infty$ after normalization to get a number. The analog of Nevanlinna theory is to study how this number varies when the other two subvarieties vary in their respective cohomology classes.

From Parabolic Leaf in Foliation to Defect for Reduced Points. We now consider the case of a compact complex algebraic surface X. Suppose $\mathcal{F}$ is an algebraic foliation in X with possible singularities. Assume that the singularity set Z of $\mathcal{F}$ at worst consist of reduced points. It means that $\mathcal{F}$ is locally defined by $f(x, y)dx+g(x, y)dy = 0$ for local coordinates x, y so that the common zero-set of the two holomorphic functions $f(x, y)$ and $g(x, y)$ consist of points with reduced structure (*i.e.*, every stalk of the ideal sheaf generated by $f(x, y)$ and $g(x, y)$ is either the unit ideal or the maximum ideal). Associated to $\mathcal{F}$ we have an sequence $0 \to N \to \Omega^1_X \to L \to 0$, where N and L are holomorphic line bundles over X, so that the sequence is exact outside Z.

We give X some Kähler metric so that it induces a Hermitian metric along the fibers of the line bundle N on $X - Z$. Let Θ_N be the curvature of N. Now assume that the holomorphic map φ is along the leaves of the algebraic foliation $\mathcal{F}$ in the sense that the local holomorphic 1-form $f(x, y)dx + g(x, y)dy$ defining $\mathcal{F}$ is pulled back to 0 by φ. Since the transition functions for N on $X - Z$ can be chosen to be locally independent of the coordinate along the leaves of $\mathcal{F}$, one can choose a closed $(1, 1)$-form $\tilde{\Theta}_N$ on $X - Z$ representing the Chern class of N on $X - Z$ so that its pullback to the leaves of $\mathcal{F}$ vanishes (*i.e.*, its pullback by φ vanishes). In particular, $T\left(r, \varphi, \tilde{\Theta}_N\right) = 0$. If Z is empty, then the difference of the two smooth $(1, 1)$-forms Θ_N and $\tilde{\Theta}_N$ is exact on X and, by a lemma of Kodaira, is equal to the $\partial\bar{\partial}$ of some smooth function on X. So under the assumption of Z being empty, we have $T(r, \varphi, \Theta_N) = O(1)$. When Z is a nonempty set of reduced points, the same argument done carefully with estimates near the points of Z gives us $T(r, \varphi, \Theta_N) = m(r, \varphi, Z) + O(1)$, where $m(r, \varphi, Z)$ is the proximity function to Z defined as $\oint_{|\zeta|=r} \log^+ \frac{1}{(\mathrm{dist}_Z)\circ\varphi}$ with dist_Z denoting the distance function to Z. In [13] this technique is in the section on residue heights on pages 144–146.

Defect Relation for Reduced Points. In [13] McQuillan introduced his "refined tautological inequality" to give the defect relation for reduced points on page 167. We are going to give the roughly equivalent traditional function-theoretical formulation for the defect relation for points with condition on the cotangent bundle. This is in contrast to the defect relation for hypersurfaces with condition on the canonical line bundle. The idca of a condition on the cotangent bundle for defect for points is compatible with the intuitive geometric expectation in Nevanlinna theory.

Theorem 3.6 (Defect relation for points). *Let H be an ample line bundle over a compact complex manifold Y of complex dimension n. Let Z be a finite subset of Y. Let $\varphi : \mathbf{C} \to Y$ be a holomorphic map. Let α be a positive rational number and ℓ be a positive integer such that $\alpha\ell$ is an integer. Let $\sigma \in \Gamma\left(Y, \mathrm{Sym}^{\ell}\left(\Omega_Y^1\right) \otimes (\alpha\ell H)\right)$ such that $\varphi^*\sigma$ is not identically zero on $\mathbf{C}$. Let W be the zero divisor of σ in the projectivization $\mathbf{P}(T_Y)$ of the tangent bundle T_Y of Y when σ is naturally regarded as the section of a line bundle over $\mathbf{P}(T_Y)$. Then $\frac{1}{\ell} N(r, d\varphi, W) + \oint_{|\zeta|=r} \log^+ \left|\frac{1}{\mathrm{dist}_Z(\varphi(\zeta))}\right| \leqq \alpha\, T(r, \varphi, H) + O(\log T(r, \varphi, H) + \log r)$ for r outside a set with finite measure with respect to $\frac{dr}{r}$, where dist_Z is the distance function to Z with respect to any smooth metric of Y. The conclusion still holds when τ is replaced by a global holomorphic section of $\mathrm{Sym}^{\ell}\left(\Omega_Y^1\right) \otimes (\alpha\ell H)$ over only the algebraic Zariski closure of the image of $d\varphi$.*

For the proof of Theorem 3.6 we need the following simple lemma for the pullback of holomorphic forms by the blow-up map for a point.

Lemma 7. *Let G be an open neighborhood of 0 in $\mathbf{C}^n$ and $\pi : \tilde{G} \to G$ be the blow-up map for blowing up the point 0. Let $D = \pi^{-1}(0)$. Then $\pi^*\left(\mathcal{O}_G\left(\Omega_G^1\right)\right) \subset \mathcal{I}_D\left(\Omega_{\tilde{G}}^1(\log D)\right)$, where $\mathcal{I}_D$ is the ideal sheaf for the divisor D and $\Omega_{\tilde{G}}^1(\log D)$ is the bundle of 1-forms on $\tilde{G}$ with logarithmic poles along D.*

Proof. The blowup $\tilde{G}$ is the complex submanifold of $G \times \mathbf{P}_{n-1}$ defined by $w_j z_k = w_k z_j$ for $1 \leq j \neq k \leq n$, where $(z_1, \cdots, z_n)$ is the coordinate of $\mathbf{C}^n$ and $[w_1, \cdots, w_n]$ is the homogeneous coordinate of $\mathbf{P}_{n-1}$. Let U_k be the affine subset of $\mathbf{P}_{n-1}$ defined by $w_k \neq 0\,(1 \leq k \leq n)$. Let $\widetilde{U_k}$ be $\tilde{G} \cap (G \times U_k)$. On $\widetilde{U_k}$ we choose the coordinate $z_k, \frac{w_j}{w_k}$ $(1 \leq j \leq n,\ j \neq k)$ so that we have $dz_j = z_k d\left(\frac{w_j}{w_k}\right) + \left(\frac{w_j}{w_k}\right) dz_k$ on $\widetilde{U_k}$ for $1 \leq j \leq n$ and $j \neq k$. The subvariety $D \cap \widetilde{U_k}$ of $\widetilde{U_k}$ is defined by the single equation $z_k = 0$. Since for $1 \leq j \neq k \leq n$ on $\widetilde{U_k}$ we can write $dz_k = z_k \left(\frac{dz_k}{z_k}\right)$ and $dz_j = z_k\left(d\left(\frac{w_j}{w_k}\right) + \left(\frac{w_j}{w_k}\right)\left(\frac{dz_k}{z_k}\right)\right)$ for $j \neq k$, it follows that the restriction of $\pi^*(dz_\ell)$ to $\widetilde{U_k}$ as a differential with log-pole along D has coefficients vanishing on E for $1 \leq \ell \leq n$.

Proof (of Theorem 3.6). Let $\pi : \tilde{Y} \to Y$ be the blow-up of Z and let $E = \pi^{-1}(Z)$. Let $\tilde{\varphi} : \mathbf{C} \to \tilde{Y}$ be the lifting of φ and let $\tau = \pi^*\sigma$. By Lemma 7, τ is a holomorphic section of $\mathrm{Sym}^{\ell}\left(\Omega_{\tilde{X}}^1(\log E)\right) \otimes \pi^*(\alpha\ell H)$ over $\tilde{Y}$ and vanishes to order at least ℓ on E. Let $e^{-\chi}$ be a smooth metric for H.

We denote also by E the line bundle on $\tilde{Y}$ associated to the divisor E. Let s_E be the canonical section of E and $e^{-\kappa}$ be a smooth Hermitian metric along the fibers of E. Let $\tilde{\tau} = \frac{\tau}{s_E^{\ell}}$. Then $\tilde{\tau}$ is a holomorphic section of $\mathrm{Sym}^{\ell}\left(\Omega_{\tilde{X}}^1(\log E)\right) \otimes \pi^*(\alpha\ell H) \otimes (-\ell E)$ over $\tilde{Y}$.

By applying the logarithmic derivative lemma and using the logarithms of global meromorphic functions as local coordinates, we conclude that both $\oint_{|\zeta|=r} \log^+ \varphi^*$

$\left(|\tau|^2 e^{-\alpha\ell\chi}\right)$ and $\oint_{|\zeta|=r} \log^+ \tilde{\varphi}^* \left(|\tilde{\tau}|^2 e^{\ell\kappa-\alpha\ell\chi}\right)$ are of the order $O\,(\log T\,(r,\varphi) + \log r)$. Using $\log x = \log^+ x - \log^+ \frac{1}{x}$ for any $x > 0$ and the identity $-\log \Phi(0) + \oint_{|\zeta|=r} \log \Phi = \int_{\rho=0}^{r} \frac{d\rho}{\rho} \int_{|\zeta|<\rho} \frac{\sqrt{-1}}{\pi} \partial\bar{\partial} \log \Phi$, we obtain

$$
\begin{aligned}
2\ell\, m\,(r,\tilde{\varphi},E) &= \oint_{|\zeta|=r} \log^+ \frac{1}{\tilde{\varphi}^*\left(\left|s_E^{\ell}\right|^2 e^{-\ell\kappa}\right)} + O(1) \\
&\leq \oint_{|\zeta|=r} \log^+ \frac{1}{\varphi^*\left(|\tau|^2 e^{-\alpha\ell\chi}\right)} + \oint_{|\zeta|=r} \log^+ \tilde{\varphi}^* \left(|\tilde{\tau}|^2 e^{\ell\kappa-\alpha\ell\chi}\right) + O(1) \\
&= - \oint_{|\zeta|=r} \log \varphi^* \left(|\tau|^2 e^{-\alpha\ell\chi}\right) + \oint_{|\zeta|=r} \log^+ \varphi^* \left(|\tau|^2 e^{-\alpha\ell\chi}\right) \\
&\quad + \oint_{|\zeta|=r} \log^+ \tilde{\varphi}^* \left(|\tilde{\tau}|^2 e^{\ell\kappa-\alpha\ell\chi}\right) + O(1) \\
&= -2\,N\left(r, \operatorname{div}\left(\varphi^*\tau\right)\right) + 2\alpha\ell\, T\,(r,\varphi,H) + O\,(\log T\,(r,\varphi) + \log r) \\
&= -2\,N\,(r, d\varphi, W) + 2\alpha\ell\, T\,(r,\varphi,H) + O\,(\log T\,(r,\varphi) + \log r)\,,
\end{aligned}
$$

where $N\,(r, \operatorname{div}(\varphi^*\tau))$ means the counting function $\int_{\rho=0}^{r} n\,(\rho, \operatorname{div}(\varphi^*\tau)) \frac{d\rho}{\rho}$ of the divisor $\operatorname{div}(\varphi^*\tau)$ of $\varphi^*\tau$, with $n\,(\rho, \operatorname{div}(\varphi^*\tau))$ denoting the number of points of the divisor of $\varphi^*\tau$ in $|\zeta| < \rho$ with multiplicities counted.

Remark 8. We now reconcile our Theorem 3.6 with McQuillan's refined tautological inequality in Theorem 3.3.2 *bis* on page 167 of [13] in the following way. On $\mathbf{P}\,(T_Y)$ global sections of $\ell\, H_{\mathbf{P}(T_Y)} \otimes (p\,H)$ over $\mathbf{P}\,(T_Y)$ are elements of $\Gamma\left(Y, \mathrm{Sym}^{\ell}\left(\Omega_Y^1\right) \otimes (p\,H)\right)$. Here $H_{\mathbf{P}(T_Y)}$ is the line bundle over the projectivization $\mathbf{P}\,(T_Y)$ of the tangent bundle T_Y of Y so that the restriction of $H_{\mathbf{P}(T_Y)}$ to each fiber of $\mathbf{P}\,(T_Y)$ over a point of Y is the hyperplane section line bundle of that fiber.

When we choose α sufficiently large, the line bundle $H_{\mathbf{P}(T_Y)} \otimes (\alpha\, H)$ is ample over $\mathbf{P}\,(T_Y)$. When we choose ℓ sufficiently large, the line bundle $\ell\, H_{\mathbf{P}(T_Y)} \otimes (\ell\,\alpha\, H)$ is very ample over $\mathbf{P}\,(T_Y)$. Then there exists $\sigma \in \Gamma\left(Y, \mathrm{Sym}^{\ell}\left(\Omega_Y^1\right) \otimes (\alpha\ell\, H)\right)$ not identically zero so that the defect for the divisor W of σ on $\mathbf{P}\,(T_Y)$ and for the map $d\varphi : \mathbf{C} \to \mathbf{P}\,(T_Y)$ is zero. From Theorem 3.6 and $T\left(r, d\varphi, H_{\mathbf{P}(T_Y)}\right) = \frac{1}{\ell} N\,(r, d\varphi, W) - \alpha\, T(r,\varphi,H)$ it follows that $T\left(r, d\varphi, H_{\mathbf{P}(T_Y)}\right) + m\,(r,\varphi,Z) = O\,(\log T(r,\varphi,H) + \log r)$. This is the tautological inequality in Theorem 3.3.2 *bis* on page 167 of [13].

3.2 Nonexistence of Zariski Dense Parabolic Leaf in Surface of General Type. We consider the hyperbolicity problem for a compact complex algebraic surface X of general type (for example, for a surface X of sufficiently high degree in $\mathbf{P}_3$) for which the technique of Bogomolov ([3], page 523 of [7]) produces an algebraic multi-foliation $\mathcal{F}$ so that every entire holomorphic curve in X must be along the leaves of $\mathcal{F}$. Here an algebraic multi-foliation means that it is locally defined by some local holomorphic 1-jet differential. This reduces the problem of showing the nonexistence of Zariski dense entire holomorphic curves in X to the nonexistence of Zariski dense parabolic leaves in $\mathcal{F}$.

We now use the last two of the three techniques of McQuillan listed above to sketch a purely function-theoretic proof of the crucial step of the nonexistence of

Zariski dense parabolic leaves in an algebraic multi-foliation in a compact complex algebraic surface of general type. To make the sketch easier, we consider the case of a surface X of sufficiently high degree in $\mathbf{P}_3$. Assume that there is such a leaf given by a holomorphic map $\varphi : \mathbf{C} \to X$ and we are going to derive a contradiction.

The technique of Miyaoka's almost ampleness theorem ([15], [10], Th.0.2.2 on page 143 of [13], Th.2.3 on page 524 of [7]) gives the bigness of the line bundle B of holomorphic 1-jets along an algebraic multi-foliation of a surface with $c_1^2 > 2c_2$. If we add some twisting to B and replace B by $B + \gamma K_X$ for some positive rational number γ, then, under the weaker assumption of $(1 + \beta)\, c_1^2 > c_2$ with $0 \leq \beta < 1$, Miyaoka's technique gives the bigness of $B + \gamma K_X$ when $\gamma^2 + 3\gamma - \frac{1}{2} > \beta$, which is satisfied when $\gamma > \frac{-3+\sqrt{11+\beta}}{2}$, and, in particular, when $\gamma \geq \frac{1}{6}$ and β is sufficiently small.

To make the presentation a little simpler, we assume at this point that the algebraic multi-foliation $\mathcal{F}$ is actually an algebraic foliation whose singular set Z consists only of reduced points. The general case can be handled by taking a branched cover of X and using monoidal transformations. By Theorem 3.6, the proximity function $m\,(r, \varphi, Z)$ for a set Z of reduced points in X is dominated by $\gamma T(r, \varphi, K_X) + O\,(\log T(r, \varphi) + \log r)$.

The Kähler metric of X induces also Hermitian metrics on the line bundle L on $X - Z$ and on the canonical line bundle K_X of X. Let Θ_{K_X} and Θ_L denote the curvature of K_X and L respectively. For any tangent vector ξ of type $(1, 0)$ in $X - Z$, we have $\Theta_{K_X}\left(\xi, \bar{\xi}\right) = \Theta_N\left(\xi, \bar{\xi}\right) + \Theta_L\left(\xi, \bar{\xi}\right)$, because the contributions, from the second fundamental form for a subbundle and that for the quotient bundle, to Θ_N and Θ_L are equal and of opposite sign when both are computed from the curvature of the cotangent bundle Ω_X^1 of X. Thus $T\left(r, \varphi, \Theta_{K_X}\right) = T\,(r, \varphi, \Theta_N) + T\,(r, \varphi, \Theta_L)$ for any holomorphic map $\varphi : \mathbf{C} \to X$ when we confine our consideration only to points of $X - Z$. Since the singularity set Z of $\mathcal{F}$ consists only of reduced points, the difference between $T\,(r, \varphi, \Theta_N)$ and $T\,(r, \varphi, N)$ and also the difference between $T\,(r, \varphi, \Theta_L)$ and $T\,(r, \varphi, L)$ are each no more than $m\,(r, \varphi, Z)$. We conclude that $T\,(r, \varphi, \Theta_L)$ dominates $(1 - 3\gamma - \varepsilon)\, T(r, \varphi, K_X)$ for any $\varepsilon > 0$.

To derive a contradiction, for some sufficiently large integer ℓ with $\ell\gamma$ also an integer, we take a non identically zero global section τ of $\mathrm{Sym}^\ell\left(\Omega_X^1\right) \otimes (\gamma\ell K_X)$ over the algebraic Zariski closure of the image of $d\varphi$ (which is the algebraic foliation $\mathcal{F}$). We can assume that ℓ is chosen sufficiently large so that, for any sufficiently large integer q, every holomorphic section of $\mathrm{Sym}^\ell\left(\Omega_X^1\right) \otimes ((\gamma\ell + q)\, K_X)$ over $\mathcal{F}$ can be extended to a holomorphic section of $\mathrm{Sym}^\ell\left(\Omega_X^1\right) \otimes ((\gamma\ell + q)\, K_X)$ over X. Take a general element $\xi \in \Gamma\,(X, qK_X)$ with divisor E so that the proximity function for φ and E is $O\,(1)$. Let $\tilde{\tau} \in \Gamma\left(X, \mathrm{Sym}^\ell\left(\Omega_X^1\right) \otimes ((\gamma\ell + q)\, K_X)\right)$ be the extension of $\tau\xi$. Let σ be the global section of $\ell L + (\gamma + q)\,\ell K_X$ over X which is induced by $\tilde{\tau}$. We assume that ℓ is sufficiently large so that $\ell\gamma K_X$ is very ample. Let θ be a general element of $\Gamma\,(X, \ell\gamma K_X)$ with divisor D so that the proximity function for φ and D is $O\,(1)$. Let $e^{-\chi}$ be a smooth metric for $\ell\gamma K_X$.

We obtain $\oint_{|\zeta|=r} \log^+ \varphi^*\left(|\tau|^2 e^{-\chi}\right) = O\,(\log T(r, \varphi) + \log r)$ by using the logarithmic derivative lemma and using the logarithms of global meromorphic functions as local coordinates. Let D be the zero-divisor of θ. Then $\oint_{|\zeta|=r} \log^{+} \varphi^{*}\left(\left|\frac{\tau}{\theta}\right|^{2}\right)$ does not exceed

$$\oint_{|\zeta|=r} \log^{+} \varphi^{*}\left(|\tau|^{2} e^{-\chi}\right)+\oint_{|\zeta|=r} \log^{+} \varphi^{*}\left(\frac{1}{|\theta|^{2} e^{-\chi}}\right)+O(1)$$
$$=T\left(r, \varphi, \gamma \ell K_{X}\right)-N(r, \varphi, D)+O(\log T(r, \varphi)+\log r).$$

We compute $T\left(r, \varphi, \ell L+(\gamma \ell+q) K_{X}\right)$ by using the counting function and the proximity function of the zero-set of σ. We compute the Nevanlinna characteristic function $T\left(r, \varphi^{*}\left(\frac{\tau}{\theta}\right)\right)$ of the meromorphic function $\varphi^{*}\left(\frac{\tau}{\theta}\right)$ on $\mathbf{C}$ by using the counting function and the proximity function of its zero-set. By using Brody's result [4] to reparametrize the entire holomorphic curve $\varphi: \mathbf{C} \rightarrow X$, we can assume that the pointwise norm of $d\varphi$ is uniformly bounded on $\mathbf{C}$. We conclude from the above computations of $T\left(r, \varphi, \ell L+(\gamma \ell+q) K_{X}\right)$ and $T\left(r, \varphi^{*}\left(\frac{\tau}{\theta}\right)\right)$ that

$$T\left(r, \varphi, \ell L+(\gamma \ell+q) K_{X}\right)-N(r, \varphi, D)-N(r, \varphi, E) \leq T\left(r, \varphi^{*}\left(\frac{\tau}{\theta}\right)\right)+O(1).$$

Hence $T(r, \varphi, \ell L) \leq T\left(r, \varphi^{*}\left(\frac{\tau}{\theta}\right)\right)$ does not exceed

$$N\left(r, \varphi^{*}\left(\frac{\tau}{\theta}\right), \infty\right)+T\left(r, \varphi, \gamma \ell K_{X}\right)-N(r, \varphi, D)+O(\log T(r, \varphi)+\log r)$$
$$\leq T\left(r, \varphi, \gamma \ell K_{X}\right)+O(\log T(r, \varphi)+\log r),$$

contradicting the domination of $(1-3\gamma-\varepsilon) T(r, \varphi, K_{X})$ by $T(r, \varphi, \ell L)$ for any $\varepsilon>0$.

References

[1] Ahlfors, L.: The Theory of meromorphic curves, Acta Soc. Sci. Fenn. Nova Ser. A **3**(4), 171–183 (1941)

[2] Bloch, A.: Sur les systèmes de fonctions uniformes satisfaisant à l'équation d'une variété algébrique dont l'irrégularité dépasse la dimension. J. de Math. **5**, 19–66 (1926)

[3] Bogomolov, F.: Holomorphic tensors and vector bundles on projective varieties, Math. USSR, Izv. **13**, 499–555 (1979)

[4] Brody, R.: Compact manifolds in hyperbolicity. Trans. Amer. Math. Soc. **235**, 213–219 (1978)

[5] Cartan, H.: Sur les zéros des combinaisons linéaires de p fonctions holomorphes données, Mathematica (Cluf) **7**, 5–31 (1933)

[6] Clemens, H.: Curves on generic hypersurfaces, Ann. Ec. Norm. Sup. **19**, 629–636 (1986)

[7] Demailly, J.-P., El Goul, J.: Hyperbolicity of generic surfaces of high degree in projective 3-space. Amer. J. Math. **122**, no. 3, 515–546 (2000)

[8] Ein, L.: Subvarieties of generic complete intersections, Invent. Math. **94**, 163–169 (1988). II. Math. Ann. **289**, 465–471 (1991)

[9] Green, M.: Holomorphic maps into complex projective space omitting hyperplanes. Trans. Amer. Math. Soc. **169**, 89–103 (1972)

[10] Lu, S., Yau, S.-T.: Holomorphic curves in surfaces of general type. Proc. Nat. Acad. Sci. USA **87**, 80–82 (1990)
[11] McDuff, D., Salamon, D.: J-holomorphic curves and quantum cohomology. University Lecture Series, **6**. American Mathematical Society, Providence, RI, 1994
[12] McQuillan, M.: A new proof of the Bloch conjecture. J. Algebraic Geom. **5**, 107–117 (1996)
[13] McQuillan, M.: Diophantine approximations and foliations. Inst. Hautes Études Sci. Publ. Math. **87**, 121–174 (1998)
[14] McQuillan, M.: Holomorphic curves on hyperplane sections of 3-folds. Geom. Funct. Anal. **9**, 370–392 (1999)
[15] Miyaoka, Y.: Algebraic surfaces with positive indices, Progr. Math. **39**, Birkhäuser, Boston, MA, 1983, pp. 281–301
[16] Noguchi, J., Winkelmann, J., Yamanoi, K.: The value distribution of holomorphic curves into semi-abelian varieties. C. R. Acad. Sci. Paris Sér. I Math. **331**, 235–240 (2000)
[17] Ru, M., Wong, P.-M.: Holomorphic curves in abelian and semi-abelian varieties, Preprint July, 1995
[18] Siu, Y.-T.: Hyperbolicity problems in function theory, in *Five Decades as a Mathematician and Educator — on the 80th birthday of Professor Yung-Chow Wong*, ed. Kai-Yuen Chan and Ming-Chit Liu, World Scientific: Singapore, New Jersey, London, Hong Kong, 1995, pp. 409–514
[19] Siu, Y.-T.: Recent techniques in hyperbolicity problems. In: *Several complex variables*, ed. M. Schneider & Y.-T. Siu, Cambridge University Press, Cambridge, 1999, pp. 429–508
[20] Siu, Y.-T.: Hyperbolicity of generic high-degree hypersurfaces in complex projective spaces, Preprint 2002
[21] Siu, Y.-T., Yeung, S.-K.: Hyperbolicity of the complement of a generic smooth curve of high degree in the complex projective plane, Invent. Math. **124**, 573–618 (1996)
[22] Siu, Y.-T., Yeung, S.-K.: A generalized Bloch's theorem and the hyperbolicity of the complement of an ample divisor in an abelian variety, Math. Ann. **306**, 743–758 (1996). Correction, Math. Ann. **326**, 205–207 (2003)
[23] Siu, Y.-T., Yeung, S.-K.: Defects for ample divisors of abelian varieties, Schwarz lemma, and hyperbolic hypersurfaces of low degrees. Amer. J. Math. **119**, 1139–1172 (1997). Addendum. Amer. J. Math. **125**, 441–448 (2003)
[24] Vafa, C.: Topological mirrors and quantum rings. Essays on mirror manifolds, 96–119, Internat. Press, Hong Kong, 1992
[25] Voisin, C.: On a conjecture of Clemens on rational curves on hypersurfaces, J. Diff. Geom. **44**, 200–213 (1996). A correction: "On a conjecture of Clemens on rational curves on hypersurfaces," J. Diff. Geom. **49**, 601–611 (1998)
[26] Vojta, P.: Integral points on sub-varieties of semi-abelian varieties I, Invent. Math. **126**, 133–181 (1996)
[27] Vojta, P.: Nevanlinna theory and Diophantine approximation. In: Siu, Y.-T., Schneider, M. (ed) Several complex variables (Berkeley, CA, 1995–1996), 535–564, Math. Sci. Res. Inst. Publ. **37**, Cambridge Univ. Press, Cambridge, 1999
[28] Zaidenberg, M.: Stability of hyperbolic embeddedness and construction of examples. Math. USSR Sbornik **63**, 351–361 (1989)

Abel–Radon Transform and Applications

G. M. Henkin

This paper is an extended version of the lecture given at the Abel Conference 2002. It contains an exposition of several recent results in complex and real integral geometry inspired by Abel's addition theorem and some applications to differential and functional equations.

I. Complex Abel–Radon Transform

Let $z = (z_0, z_1, \dots, z_n) \in \mathbb{C}P^n$, $\xi = (\xi_0, \xi_1, \dots, \xi_n) \in (\mathbb{C}P^n)^*$,

$$\xi \mapsto \mathbb{C}P^{n-1}_\xi = \{z \in \mathbb{C}P^n \ : \ \xi \cdot z = 0\}.$$

Let $D \subset \mathbb{C}P^n$ be a lineary concave domaine, i.e., such that for all $z \in D$ there exists a $\mathbb{C}P^{n-1}_\xi$ such that $z \in \mathbb{C}P^{n-1}_\xi \subset D$. The set of ξ with the last property forms the dual domain $D^* \subset (\mathbb{C}P^n)^*$. Let V be a closed one-dimensional analytic subset of D. Then there exists an open and dense subset $D_0^* \subset D^*$ such that for all $\xi \in D_0^*$ the intersection $\mathbb{C}P^{n-1}_\xi \cap V$ is transverse and forms the finite set $\{z^{(1)}(\xi), \dots, z^{(N)}(\xi)\} \subset V$.

Let $\mathcal{M}^{1,0}(V)$ denote the set of meromorphic 1-forms on V.

Definition (Abel transform). *For all $\psi \in \mathcal{M}^{1,0}(V)$ we set*

$$f(\xi) = \mathcal{A}\psi(\xi) = d_\xi \sum_{j=1}^{N} \int^{z^{(j)}(\xi)} \psi, \ \ \xi \in D_0^* \,.$$

This definition of the Abel transform, inspired by Abel's addition theorem ([2], 1826), is different from the definition of Abel's operator ([1], 1823):

$$\mathcal{A}\varphi(s) = \int_s^\infty \frac{\varphi(r)\,dr}{\sqrt{r-s}},$$

often also called Abel's transform.

Let $(1, z_1, \dots, z_n)$ be affine coordinates on $\mathbb{C}^n = \{z \in \mathbb{C}P^n : z_0 \neq 0\}$ such that $z_n \neq const$ on V. Then any form $\psi \in \mathcal{M}^{1,0}(V)$ can be represented on $V \cap \mathbb{C}^n$ by the expression

$$\psi = \tilde{\psi}(z_1, \dots, z_n)\,dz_n, \quad \tilde{\psi} \in \mathcal{M}(V \cap \mathbb{C}^n)$$

and the Abel transform of ψ takes the form

$$\mathcal{A}\psi(\xi) = \sum_{\nu=1}^{N} \tilde{\psi}(z^{(\nu)}(\xi))\,d_\xi z_n^{(\nu)}(\xi), \quad \xi \in D_0^*.$$

The transform $f = \mathcal{A}\psi$ defines meromorphic 1-form f in the domain D^*.

For any $\psi \in \mathcal{M}^{1,0}(V)$ there exists a discrete locally finite subset $S \subset V$ containing Sing V such that ψ is holomorphic 1-form on $V \backslash S$ with at most polynomial growth near S. For such ψ one can define the $(n, n-1)$-current $\psi \wedge [V]$ in D by the formula

$$< \psi \wedge [V], \varphi >= \lim_{\delta \to 0} \int_{\{z \in V \,:\, dist\,(z,S) \geq \delta\}} \psi \wedge \varphi,$$

where φ is any smooth (0,1)-form with compact support in D.

The meromorphic 1-form ψ on V is called (weakly) holomorphic (or a Rosenlicht form, or a section of the Grothendieck dualising sheaf) if $\bar{\partial}(\psi \wedge [V]) = 0$. Let $H^{1,0}(V)$ denote the space of holomorphic 1-forms on V and $H^{n,n-1}(D)$ the Dolbeault space of $\bar{\partial}$-closed $(n, n-1)$-currents in D considered up to $\bar{\partial}$-exact currents. The current $\psi \wedge [V] \in H^{n,n-1}(D)$ defined by $\psi \in H^{1,0}(V)$ will be called an analytical current.

The holomorphic forms on V can be characterized as those which occur as Poincaré–Leray residues of holomorphic forms in the ambient space ([26], [32]):

The 1-form ψ on V is holomorphic on V iff for any neighbourhood $\mathcal{U}$ of V contained in D and any complete intersection $\{z \in \mathcal{U} : g_1(z) = \ldots = g_{n-1}(z) = 0\}$, $g_j \in H(\mathcal{U})$, $j = 1, \dots, n-1$, containing $V \cap \mathcal{U}$, whose Jacobian $dg_1 \wedge \ldots \wedge dg_{n-1}$ does not vanish identically on any component of $V \cap \mathcal{U}$, there is a holomorphic n-form f on $\mathcal{U}$ such that

$$\psi = \operatorname{Res} \frac{f}{g_1 \cdots g_{n-1}}\Big|_{V \cap \mathcal{U}}.$$

Example. Let V be a closed reduced curve in $\mathbb{C}P^2$ and $V \cap \mathbb{C}^2 = \{z \in \mathbb{C}^2 : p(z_1, z_2) = 0\}$ where the polynomial p is such that $\frac{\partial p}{\partial z_1} \not\equiv 0$ on $V \cap \mathbb{C}^2$. Then $\psi \in H^{1,0}(V)$ iff

$$\psi = \frac{q(z_1, z_2)\,dz_2}{\frac{\partial p}{\partial z_1}(z_1, z_2)},$$

where deg $q \leq$ deg $p - 3$.

The following classical result describes some remarkable properties of the kernel and image of the Abel transform.

Abel's theorem (first version) ([2], [36], [51]). *If V is a reduced closed algebraic curve in $\mathbb{C}P^n$ and $\psi \in \mathcal{M}^{1,0}(V)$, then the Abel transform $f = \mathcal{A}\psi$ is a rational 1-form on $(\mathbb{C}P^n)^*$. In addition, $\mathcal{A}\psi = 0$ on $(\mathbb{C}P^n)^*$ iff $\psi \in H^{1,0}(V)$.*

For elliptic curves this result was obtained already by Euler [18].

The different version of this classical result is the following.

Abel's theorem (second version) ([2], [12], [51]). *Let V be a reduced algebraic curve in $\mathbb{C}P^n$. Two 0-cycles Γ and Γ' of* Reg V *are the sets of zeros and poles respectively of a meromorphic function on V, holomorphic at the points of* Sing V *iff there exists a 1-chain γ in* Reg V *such that $b\gamma = \Gamma - \Gamma'$ and*

$$\int_\gamma \psi = 0\,, \quad \forall \psi \in H^{1,0}(V)\,.$$

Here Reg V is the subset of regular points in V, and Sing $V = V \backslash \operatorname{Reg} V$.

The Abel transform can be naturally interpretated as a important special case of the following version of Radon transform.

Definition (Radon–Martineau transform). *For all $\bar\partial$-closed and continuous representative F of an element of $H^{n,n-1}(D)$ we set*

$$f(\xi) = \mathcal{R}F(\xi) = \frac{1}{2\pi i}\sum_j \left(\int_{z\in D} z_j F \wedge \bar\partial_z \frac{1}{\xi\cdot z}\right) d\xi_j\,, \quad \xi \in D^*\,.$$

Proposition 1. *If a $\bar\partial$-closed continuous (n,n-1)-form F is $\bar\partial$-cohomological in D to the analytical current $\psi \wedge [V]$, then*

$$\mathcal{R}F(\xi) = \mathcal{A}\psi(\xi) \;\; \forall \xi \in D^*.$$

Theorem 1 ([43], [23], [30]). *Let D be a linearly concave domain in $\mathbb{C}P^n$ such that any hyperplane section of D^* is connected. Then the transform $F \mapsto \mathcal{R}F$ defines an injective mapping $\mathcal{R} : H^{n,n-1}(D) \to H^{1,0}(D^*)$ with image $\mathcal{R}H^{n,n-1}(D) = \{f \in H^{1,0}(D^*) \;:\; df = 0\}$.*

This result has the following application, which develops [44], [5].

Corollary 1 ([23]). *Let the compact $G \subset \mathbb{C}P^n$ have a linearly concave complement $D = \mathbb{C}P^n \backslash G$ as in Theorem 1. Let $H^*(G)$ be the dual space to the space $H(G)$ of holomorphic functions on G. Then the Fantappie mapping*

$$\mu \mapsto \psi_\mu(\xi) = \sum_{j=0}^{n} < \mu, \frac{z_j}{\xi\cdot z} > d\xi_j\,, \;\; \mu \in H^*(G), \;\; \xi \in D^*$$

realize an isomorphism between $H^*(G)$ and the space $\{\psi \in H^{1,0}(D^*) \;:\; d\psi = 0\}$.

The analytic current $\psi \wedge [V] \in H^{n,n-1}(D)$ will be called semi-algebraic if V is a restriction to D of some algebraic curve $\tilde{V}$ in $\mathbb{C}P^n$. The semi-algebraic current $\psi \wedge [V]$ will be called algebraic if the holomorphic 1-form on V is rational.

Theorem 1 combined with the classical Oka–Cartan interpolation theorem (see [29]) gives the following caracterization of those elements in $H^{n,n-1}(D)$ which can be represented by algebraic or semi-algebraic currents.

Theorem 2 ([31]). *Under the onditions of Theorem 1 the element F of $H^{n,n-1}(D)$ has a representation by a semi-algebraic current $\psi \wedge [V]$ iff its Radon–Martineau transform $f = \mathcal{R}F$ satisfies the differential equations*

$$P_j \left(\frac{d}{d\xi} \right) f(\xi) = 0 \,, \quad j = 1, 2, \ldots, r; \quad \xi \in D^*,$$

where $\{P_j\}$ are homogeneous polynomials such that rank $[dP_1, \ldots, dP_r] = n - 1$ on

$$V = \{z \in D \;:\; P_1(z) = \ldots = P_r(z) = 0\} \,. \tag{$*$}$$

The semi-algebraic current $\psi \wedge [V]$ is, in addition, algebraic iff f is a rational form.

Theorem 2 implies the following precision of the Abel theorem.

Corollary 2. *Let D be a linear concave domain in $\mathbb{C}P^n$ satisfying the condition of Theorem 1, let $\tilde{V}$ be a closed algebraic curve in $\mathbb{C}P^n$ such that the curve $V = \tilde{V} \cap D$ has the form (*), $\psi \in \mathcal{M}^{1,0}(V)$, and $\psi \mapsto \mathcal{A}\psi$ the Abel transform of ψ. Then*

(i)$\psi \in \ker \mathcal{A}$ iff $\psi = \tilde{\psi}\big|_V$, where $\tilde{\psi} \in H^{1,0}(\tilde{V})$

(ii)$f \in \operatorname{Im} \mathcal{A}$ iff $f \in \mathcal{M}^{1,0}(D^*)$ and $P_j \left(\frac{d}{d\xi} \right) f = 0$, $j = 1, 2, \ldots, r$ on some open subset of D^*

(iii)$\psi \in H^{1,0}(V)$ iff $\mathcal{A}\psi$ is holomorphic in D^*.

The statement (ii) can be interpretated as a version (in terms of the Abel transform) of the Ehrenpreis and Palamodov "fundamental principle" for the system of differential equations considered (see, for example, [46]).

In [31] the explicit formulas are obtained for representation of the solutions of the system $\{P_j \left(\frac{d}{d\xi} \right) f = 0\}$ in the form $f = \mathcal{A}\psi$, $\psi \in H^{1,0}(V)$.

For further applications it turned out to be important to caracterize those elements from $H^{n,n-1}(D)$ which can be represented by analytic currents.

Theorem 3 ([30]). *Under the conditions of Theorem 1 the element F from $H^{n,n-1}(D)$ has a representation in the form of an analytic current $\psi \wedge [V]$ in D iff its Radon–Martineau transform $\mathcal{R}F = \sum_{k=0}^{n} f_k d\xi_k = f$ can be represented in a neighbourhood of some point $\eta \in D^*$ by the finite sum of the closed holomorphic forms $f = \sum_{j=1}^{N} \sum_{k=0}^{n} f_k^{(j)}(\xi)\, d\xi_k$, where the functions $g_k^{(j)} = f_k^{(j)} / f_0^{(j)}$ satisfy the "shock waves" equations*

$$g_k^{(j)} \frac{\partial g_l^{(j)}}{\partial \xi_0} = \frac{\partial g_l^{(j)}}{\partial \xi_k} \;; \quad k, l = 1, \ldots, n \;; \; j = 1, 2, \ldots, N \,.$$

Sketch of the proof. The proof of the necessity is based on a direct calculation involving the following important assertion going back to [14].

Let V be a one-dimensional analytic submanifold in the concave domain $D \subset \mathbb{C}P^n$ and let the hyperplane $\mathbb{C}P^{n-1}_\xi$, $\xi \in D^*$, intersect V in the points $z^{(j)}(\xi) = \{1, z_1^{(j)}(\xi), \dots, z_n^{(j)}(\xi)\}$, $j = 1, 2, \dots, N$. Then for any j the following "shock waves" equations are valid

$$z_k^{(j)} \frac{\partial z_l^{(j)}}{\partial \xi_0} = \frac{\partial z_l^{(j)}}{\partial \xi_k}, \quad k, l = 1, 2, \dots, n .$$

The proof of the sufficiency can be reduced by projections to the case $n = 2$. After using the "shock waves" equations above and Hartogs–Levi extension theorem, we obtain the following statement. Let f be a closed holomorphic 1-form in the domain $D^* \subset (\mathbb{C}P^2)^*$ satisfying the condition of the theorem in the ε-neighbourhood $\mathcal{U}_\eta \subset D^*$ of the point $\eta = (\eta_0, \eta_1, 1)$. We consider for any $\xi \in \mathcal{U}_\eta$ in the domain

$$\{z = (1, z_1, z_2) \in D \ : \ |z_2 + \xi_1 z_1 + \xi_0| < \varepsilon\}$$

the following meromorphic function

$$F_{\xi_1}(z_1, z_2) = \sum_{\nu=1}^{N} \frac{-f_0^{(\nu)}(-\xi_1 z_1 - z_2, \xi_1)}{z_1 - g_1^{(\nu)}(-\xi_1 z_1 - z_2, \xi_1)} .$$

Then
i) this function extends to the meromorphic function in the domain

$$D_{\xi_1} = \{z = (1, z_1, z_2) \in D \ : \ \xi = (-\xi_1 z_1 - z_2, \xi_1, 1) \in D^*\} ;$$

ii) the poles and the residues of the function F_{ξ_1} do not depend on ξ_1, and, moreover,

$$\frac{dF_{\xi_1}(z_1, z_2)}{d\xi_1} = \frac{\partial f_0}{\partial \xi_0}(-\xi_1 z_1 - z_2, \xi_1), \ \ z \in D_{\xi_1} .$$

From this statement we obtain that for any ξ_1 with $|\xi_1 - \eta_1| \le \varepsilon$ in the domain D_{ξ_1}, the meromorphic form $F_{\xi_1}(z_1, z_2)\, dz_1 \wedge dz_2$ has a polar set $V_{\xi_1} \subset D_{\xi_1}$ and a residue $\psi_{\xi_1}(z)\, dz_2$ on V_{ξ_1} such that

$$V_{\xi_1} \cap D_{\eta_1} = V_{\eta_1} \cap D_{\xi_1} \ \text{ and } \ \psi_{\xi_1}\big|_{V_{\xi_1} \cap D_{\eta_1}} = \psi_{\eta_1}\big|_{V_{\eta_1} \cap D_{\xi_1}} .$$

Thus in the domain

$$G = \cup_{\{\xi_1 \, : \, |\xi_1 - \eta_1| < \varepsilon\}} D_{\xi_1}$$

we have a well defined one-dimensional submanifold $V = \cup V_{\xi_1}$ and a 1-form on V,

$$\psi = \{\psi_{\xi_1}(z)\, dz_2, \ z_2 \in V_{\xi_1}, \ |\xi_1 - \eta_1| < \varepsilon\} .$$

In other words, we have obtained an analytic current $\psi \wedge [V]$ in the neighbourhood of the line $\mathbb{C}P^1_\eta \subset D$. From this construction, from the definition of the Abel transform, and from Darboux's theorem, it follows that

$$\mathcal{R}(\psi \wedge [V]) = \mathcal{R}F = f$$

in a neighbourhood of the point $\eta = (\eta_0, \eta_1, 1)$. The statement above allows us to extend $\psi \wedge [V]$ as an analytic current in D. Theorem 1 implies that the currents F and $\psi \wedge [V]$ are equivalent in $H^{n,n-1}(D)$.

Using the statements above and Hartogs–Levi type extensions theorems (see [27]) we obtain the following improvement of classical results of Abel [2], Lie [41], and Darboux [14].

Theorem 4. *Let V be a closed one-dimensional submanifold in the linearly concave domain $D \subset \mathbb{C}P^n$ with the connected dual set D^*. Let ψ be a holomorphic 1-form on V, $\psi \not\equiv 0$ on each component of V. Then*

i) $V = \tilde{V} \cap D$, where $\tilde{V}$ is an algebraic curve in $\mathbb{C}P^n$ and $\psi = \tilde{\psi}\big|_V$, where $\tilde{\psi} \in H^{1,0}(\tilde{V})$ iff $\mathcal{A}\psi(\xi) = 0$, $\forall \xi \in D^*$ (see [26]).
ii) $V = \tilde{V} \cap D$, where $\tilde{V}$ is a closed algebraic curve in $\mathbb{C}P^n$ and $\psi = \tilde{\psi}\big|_V$, where $\tilde{\psi} \in \mathcal{M}^{1,0}(\tilde{V})$ iff $\mathcal{A}\psi(\xi)$ is a rational 1-form on D^*. (see [30]).

Sketch of the proof for the case $n = 2$.
i) If $\mathcal{A}\psi(\xi) = 0$, $\xi \in D^*$, then for the analytical current $\psi \wedge [V]$ in view of Proposition 1 we have $\mathcal{R}(\psi \wedge [V]) = 0$. From Theorem 1 it follows that $\psi \wedge [V] = \bar{\partial}X$, where X is a $(2,0)$-current in D. Since $\operatorname{supp}\psi \wedge [V] \subset V$, X defines a meromorphic $(2,0)$-form in D. Due to the Hartogs–Levi theorem X extends to a meromorphic form on all $\mathbb{C}P^2$. Hence the form X is rational. Let $X = \frac{P}{Q}$ be an irreducible representation of X. From the equality

$$\bar{\partial}\left[\frac{P}{Q}\right]\Big|_D = \psi \wedge [V]\big|_D$$

it follows that $\tilde{V} = \{z \in \mathbb{C}P^2 \ : \ Q(z) = 0\}$ is a reduced curve in $\mathbb{C}P^2$ and

$$\tilde{\psi} = \frac{1}{2\pi i}\operatorname{Res}\left[\frac{P}{Q}\right]\Big|_{\tilde{V}}$$

is a holomorphic 1-form on $\tilde{V}$ such that $V = \tilde{V} \cap D$ and $\psi = \tilde{\psi}\big|_V$.
ii) Let $\mathcal{A}\psi = f$ be a rational 1-form. Following the proof of the sufficiency in Theorem 3 we can obtain such an analytic curve V' in $\mathbb{C}P^2$ and a meromorphic form ψ' on V' that $\mathcal{R}(\psi' \wedge [V'])(\xi) = f(\xi)$ for $\xi \in D^*$. Due to Theorem 4 i) the current $\psi \wedge [V] - \psi' \wedge [V']$ has the form $\tilde{\psi} \wedge [\tilde{V}]$, $\tilde{V}$ is an algebraic curve in $\mathbb{C}P^2$, and $\tilde{\psi}'$ is a holomorphic form on $\tilde{V}$.

Let $\tilde{\tilde{V}} = \tilde{V} \cup V'$ and $\tilde{\tilde{\psi}}$ such a meromorphic form on $\tilde{\tilde{V}}$ that $\tilde{\tilde{\psi}} = \tilde{\psi}$ on $\tilde{V}$ and $\tilde{\tilde{\psi}} = \psi'$ on V'. Then we have $V = \tilde{\tilde{V}} \cap D$ and $\psi = \tilde{\tilde{\psi}}\big|_V$.

Conjecture. Combining Theorems 1, 3 with the theory of Coleff–Herrera residue currents (see [11], [15]) one can apparently obtain the statements of Theorems 2, 4 for the case of nonreduced curves and for the appropriately defined holomorphic forms on them. Such a generalization of part i) of Theorem 4 for the case $n = 2$ was deduced directly from Theorem 1 by B. Fabre (see [19]).

II. Real Abel–Radon Transform

Let $x = (x_1, \dots, x_n) \in \mathbb{R}^n$; $(\theta, t) \in S^{n-1} \times \mathbb{R}^1 \simeq (\mathbb{R}^n)^*$, $(\theta, t) \mapsto \mathbb{R}^{n-1}_{\theta,t} = \{x \in \mathbb{R}^n : \theta \cdot x = t\}$. Let $D \subset \mathbb{R}^n$ be a concave domain, i.e., for all $x \in D$ there exists $\mathbb{R}^{n-1}_{\theta,t}$ such that $x \in \mathbb{R}^{n-1}_{\theta,t} \subset D$. The set of (θ, t) with the last property forms the dual domain $D^* \subset (\mathbb{R}^n)^*$.

Definition (Radon transform). *For any distribution of measure type $f \in C^*(D)$ we set*

$$\mathcal{R} f(\theta, t) = \frac{\partial}{\partial t} \int_{\{x \in \mathbb{R}^n \,:\, \theta x \leq t\}} f(x) \,.$$

If $D \neq \mathbb{R}^n$ this transform is also called the incomplete Radon transform.

Let V be a closed rectifiable bounded 1-chain in D such that

$$\theta \cdot dx\big|_V \geq 0 \ \forall \ (\theta, t) \in D^* \,.$$

Let D_0^* be a subset of D^* such that $\mathrm{vol}_n(D^* \backslash D_0^*) = 0$ and for all $(\theta, t) \in D_0^*)$, the hyperplane $\mathbb{R}^{n-1}_{\theta,t}$ intersects V transversally along a finite set $\{x^{(1)}(\theta, t), \dots, x^{(N)}(\theta, t)\} \subset V$, $N = N(\theta, t)$. Let ψ be a continuous 1-form on V and $\psi \wedge [V]$ the distribution in D, defined by the formula

$$< \psi \wedge [V], \varphi >= \int_V \psi\varphi \,,$$

for all smooth functions φ with compact support in D.

Proposition 2. *For the distribution $F = \psi \wedge [V]$ in D the following equality is valid*

$$\mathcal{R} F(\theta, t) = \mathcal{A}\psi(\theta, t) = \frac{\partial}{\partial t} \sum_{j=1}^{N} \int^{x^{(j)}(\theta,t)} \psi \,,$$

where $(\theta, t) \in D_0^*$.

For the incomplete Radon transform we have the following description of its kernel, obtained recently together with S. Gindikin.

Theorem 5. *Let the dual set D^* for a concave domain $D \subset \mathbb{R}^n$ be connected. Let*

$$D^c = \cup_{(\theta,t) \in D^*} \{z \in \mathbb{C}^n \,:\, \theta \cdot z = t\} \,.$$

Let f be a distribution of measure type with bounded support in $\bar{D}$. Then $\mathcal{R} f \equiv 0$ in D^* iff there exists a $\bar{\partial}$-closed (0,n-2)-form $F \in C^{(\infty)}_{0,n-2}(D^c \backslash D)$ such that $|F(z)| = O(|z|^{2-2n})$, $z \to \infty$, $z \in D^c$, and for all θ in the domain $D_\theta = (\cup_t \mathbb{R}^{n-1}_{\theta,t}) \cap D$, the following residue-formula holds,

$$f(x) = \lim_{\varepsilon \to 0,\ \varepsilon > 0} \int_{\{\theta^\perp \in \mathbb{R}^n \,:\, |\theta^\perp| = 1,\ \theta\theta^\perp = 0\}} F(x + i\varepsilon\theta^\perp) \stackrel{\text{def}}{=} \operatorname{Res} F(x), \quad x \in D_\theta \,.$$

Moreover, any distribution $f \in C^*(D)$ with compact support in $\bar{D}$ can be represented as a sum $f = \text{Res } Ef + \kappa\mathcal{R}f$, where Ef is a representative of some cohomology class $F \in H^{0,n-2}(D^c \backslash D)$ and κ is an inversion operator for the incomplete Radon transform.

This result can be considered on the one hand as an improvement of the "microlocal regularity theorem" ([28], [9]), and on the other hand as the development of the "cohomological extension of real functions" [22].

It can be deduced from Theorem 1, appropriately applied to the linearly concave compact $\bar{D}^c \subset \mathbb{C}P^n$.

The statement of Theorem 5 is especially useful in the case $n = 2$ when the residue formula above has the form

$$f\big|_{D_\theta} = \lim_{\varepsilon \to 0}(F_+(x + i\varepsilon\theta^\perp) - F_-(x - i\varepsilon\theta^\perp)) ,$$

where $F_\pm$ are holomorphic functions in the domains

$$D_\pm = \cup_{(\theta,t)\in D^*}\{(x + iy) \in \mathbb{C}^2 \ : \ \theta z = t, \ \pm\theta^\perp y > 0\} .$$

Using the statements above and the "edge of the wedge" type extensions theorems one can obtain the following generalization of classical results of [25], [14] and [7] who studied the case of a sufficiently smooth form ψ on a smooth chain V under condition $\mathcal{R}(\psi \wedge [V]) = 0$.

The following statement is the development of [32] where the case of a continuous form on C^1-chains was considered.

Theorem 6. *Let D be a concave domain in $\mathbb{R}^n$ and for all $x \in D$ let the set $\{(\theta, t) \in D^* \ : \ \theta \cdot x = t\}$ be connected. Let V be a closed rectifiable bounded 1-chain in D such that $\theta \cdot dx\big|_V \geq 0$ for all $(\theta, t) \in D^*$, and let ψ be a continuous 1-form on V such that $\psi \neq 0$ almost everywhere on V. Then*

i) the data $V \subset D$ and $\psi\big|_V$ are real analytic (restriction of complex analytic data) iff the Abel–Radon transform $\mathcal{R}(\psi \wedge [V])(\theta, t)$ is real analytic in D^*.
ii) V is an intersection with $D \subset \mathbb{R}^n$ of a projective algebraic curve $\tilde{V}$ of degree N and ψ is a restriction on V of the rational (resp. holomorphic) 1-form on $\tilde{V}$ iff $\mathcal{R}(\psi \wedge [V])$ is a rational function (resp. equal to zero) in D^*.

Sketch of the proof for the case of a C^1-chain. The necessity follows from Abel's theorem. The proof of sufficiency can be reduced by projections to the case $n = 2$. So, let further $n = 2$.

i) Put $\xi = (\xi_0, \xi_1) = \left(-\frac{t}{\theta_2}, \frac{\theta_1}{\theta_2}\right)$, $\psi = \tilde{\psi}\, dx_n$ and $f_j^{(\nu)} = \tilde{\psi}(x^{(\nu)}(\xi))\frac{\partial x_n^{(\nu)}(\xi)}{\partial \xi_j}$, $j = 0, 1$. By the conditions of the theorem, the form

$$\mathcal{R}(\psi \wedge [V])(\xi) = f_0 d\xi_0 + f_1 d\xi_1 = \sum_{\nu=1}^{N}(f_0^{(\nu)} d\xi_0 + f_1^{(\nu)} d\xi_1) ,$$

defined for $\xi \in D_0^*$, has a real analytic extention on D^*, and $N = N(\xi) \equiv const$ in $D_0^* \subset D^*$.

Let us consider the following function of variables

$$(x_1, x_2) \in \cup_{\xi \in D_0^*} \mathbb{C}P_\xi^1 \subset D \,,$$

depending on a parameter ξ_1:

$$F_{\xi_1}(x_1, x_2) = \sum_{\nu=1}^{N} \frac{f_0^{(\nu)}(-\xi_1 x_1 - x_2, \xi_1)}{x_1 - g_1^{(\nu)}(-\xi_1 x_1 - x_2, \xi_1)} \,,$$

where $g_1^{(\nu)} = f_1^{(\nu)} / f_0^{(\nu)}$. As in the proof of Theorem 3 we obtain the equation

$$\frac{\partial}{\partial \xi_1} F_{\xi_1}(x_1, x_2) = \frac{\partial f_0}{\partial \xi_0}(-\xi_1 x_1 - x_2, \xi_1) \,.$$

Fix $\tilde{x} \in D$. Since the set

$$D_{\tilde{x}}^* = \{\xi \in D^* \ : \ \xi_0 + \xi_1 \tilde{x}_1 + \tilde{x}_2 = 0\}$$

is connected, it follows from the equations above, that for different ξ' and ξ'' belonging to $D_{\tilde{x}}^*$, there is an equality

$$F_{\xi_1''}(x_1, x_2) = F_{\xi_1'}(x_1, x_2) + \int_{\xi_1'}^{\xi_1''} \frac{\partial f_0}{\partial \xi_0}(-\xi_1 x_1 - x_2, \xi_1)\, d\xi_1 \stackrel{\text{def}}{=} F(x_1, x_2) \,,$$

where x belongs to some neighbourhood of $\tilde{x}$. From this equality, from the definition of $F_{\xi_1}(x_1, x_2)$, and from the real analyticity of f_0, it follows that the function $F(x_1, x_2)$ has locally a meromorphic extension to the CR-manifold $\{z \ : \ \operatorname{Im}(\xi_1'' z_1 + z_2) = 0\}$ and also to the CR-manifold $\{z \ : \ \operatorname{Im}(\xi_1' z_1 + z_2) = 0\}$.

From extension results for separately meromorphic functions [38], [39] it follows that the function $F(x_1, x_2)$ has a meromorphic extension in some neighbourhood of D in $\mathbb{C}^2$.

Hence, the poles and residues of F give rise to complex analytic germs V^c and ψ^c whose restriction to $\mathbb{R}^2$ coincides with the original data V, ψ;

ii) this part follows from part i) and Theorem 4.

Remark. Part ii) of Theorem 6 has the following tomographic interpretation. Let an attenuating object, supported on the set with finite length in some concave domain $D \subset \mathbb{R}^2$, produce the same absorbtion of X-rays along all lines in D. Then, by Theorem 6, the support of this object belongs in fact to a real algebraic curve V in D, and the density of this object is the Abel form on V.

Theorem 6 admits the interesting refinement for the case when the 1-form ψ on V is equal to $\varepsilon_j dx_k\big|_{V_j}$ on the components V_j of V where $\varepsilon_j = \pm 1$, $k = 1, 2, \ldots, n$.

More precisely, let V be a closed rectifiable bounded 1-chain in the concave domain $D \subset \mathbb{R}^n$ of the form $V = \sum \varepsilon_j V_j$, where $\varepsilon_j = \pm 1$ and $\{V_j\}$ are closed rectifiable curves in D not necessary all distinct. Then $\mathbb{R}^{n-1}_\xi = \{x \in \mathbb{R}^n : \xi_0 + \xi_1 x_1 + \ldots + \xi_n x_n = 0\}$, $\xi \in D_0^*$ intersects V along a 0-chain

$$V \cap \mathbb{R}^{n-1}_\xi = \sum_\alpha \sigma_\alpha x^{(\alpha)}(\xi) \ ,$$

where $\sigma_\alpha = \pm 1$ and $x^{(\alpha)}(\xi) = \{x_1^{(\alpha)}(\xi), \ldots, x_n^{(\alpha)}(\xi)\}$.

Definition. *By the Abel transforms of a 1-chain V in D we shall mean the family of functions on $D_0^* \subset D^*$:*

$$a_V^{(k)}(\xi) = \sum_\alpha \sigma_\alpha x_k^{(\alpha)}(\xi), \quad k = 1, 2, \ldots, n \ \text{ and } \ d_V^{\pm}(\xi) = \sum_{\alpha\,:\,\pm\sigma_\alpha = 1} |\sigma_\alpha| \ .$$

Theorem 7. *Let D be a concave domain in $\mathbb{R}^n$ as in Theorem 6. Let $V = \sum \varepsilon_j V_j$ be a closed, bounded, rectifiable 1-chain in D such that $\xi \cdot dx\big|_{V_j} \geq 0 \ \forall \xi \in D^*$. Then $d_V^+(\xi)$ and $d_V^-(\xi)$ are constants and the following statements are equivalent:*

a) the chain V is intersection with D of some algebraic 1-chain V^c in $\mathbb{C}P^n$ of degree $(d^+, d^-) = (d_V^+, d_V^-)$
b) all transforms $a_V^{(k)}(\xi)$ are linear with respect to ξ_0, $k = 1, 2, \ldots, n$
c) any of transforms $a_V^{(k)}(\xi)$, for which $dx_k\big|_V \neq 0$ almost everywhere on V, is rational in ξ.

Theorem 7 generalizes, in particular, the results of Reiss ([50]), ..., Wood ([54]), who studied the case of a sufficiently smooth chain $V = \sum \varepsilon_j V_j$ with $\varepsilon_j = 1$, and also of [31] where the case of an analytic chain was considered.

Sketch of the proof for the case of a C^1-chain. The statements b) and c) follow from a) due to the following precision of the Abel theorem. If V be an algebraic closed curve in $\mathbb{C}P^n$ and $\psi = d\frac{z_k}{z_0}$, $k = 1, 2, \ldots, n$, where $(z_0, \ldots, z_n)$ are homogeneous coordinates in $\mathbb{C}P^n$, then $\mathcal{A}\psi(\xi) = d\varphi_k(\xi_0, \ldots, \xi_n)$, where φ_k is rational and linear with respect to ξ_0.

Theorem 6 applied to the $\psi = \{\varepsilon_j dx_k\big|_{V_j}\}$, proves that a) follows from c).

Let us prove that a) follows also from b). By projections one can reduce the proof to the case $n = 2$, $k = 1$. Fixing $\xi_2 = 1$, we consider further the following function of variables $(x_1, x_2) \in \cup_{\xi \in D_0^*} \mathbb{R}^1_\xi \subset D$, depending on a parameter ξ_1

$$F_{\xi_1}(x_1, x_2) = \frac{\prod_{\alpha\,:\,\sigma_\alpha = 1} \Big(x_1 - x_1^{(\alpha)}(-\xi_1 x_1 - x_2, \xi_1)\Big)}{\prod_{\beta\,:\,\sigma_\beta = -1} \Big(x_1 - x_1^{(\beta)}(-\xi_1 x_1 - x_2, \xi_1)\Big)} \ .$$

As in the proof of Theorem 6 one can obtain the equality

$$\frac{\partial}{\partial \xi_1} \ln F_{\xi_1}(x_1, x_2) = \frac{\partial}{\partial \xi_0} a_V^{(1)}(\xi) \ .$$

From b) it follows that $a_\nu^{(1)}(\xi)$ depends only on ξ_1 and there exists a function $\psi(\xi_1) \neq 0$ such that $\frac{\partial}{\partial \xi_0} a_\nu^{(1)} = \frac{\partial \psi(\xi_1)}{\partial \xi_1}$. Hence the function

$$\tilde{F}_{\xi_1}(x_1, x_2) = \psi(\xi_1) F_{\xi_1}(x_1, x_2) = \tilde{F}_{\xi_1}(-\xi_1 x_1 - x_2, \xi_1)$$

does not depend on ξ_1.

Kneser's theorem [39] on separately rational functions implies that the function $\tilde{F}(x_1, x_2)$ is rational in (x_1, x_2). Hence, the poles and zeros of $\tilde{F}(z_1, z_2)$, $z = (z_1, z_2) \in \mathbb{C}^2$, give rise to an algebraic chain V^c in $\mathbb{C}P^2$ of degree $(d^+, d^-) = (d_V^+, d_V^-)$ such that $V = D \cap V^c$.

Corollary 3. *The family of functions of bounded variations $\{f_j(x, t), j = 1, 2, \dots, N\}$ represents a solution in a domain $D \subset \mathbb{R}^2$ of the differential system of the "resonant shock waves" (see [37]):*

$$\frac{\partial f_j}{\partial t} = f_j \frac{\partial f_j}{\partial x}, \quad j = 1, 2, \dots, N; \quad \frac{\partial^2}{\partial t^2} \sum_{j=1}^{N} \varepsilon_j f_j = 0, \quad \varepsilon_j = \pm 1\,,$$

iff there exist polynomials $\mathcal{P}_+$ and $\mathcal{P}_-$ of two variables and of degrees

$$d_+ = \sum_{j\,:\,\varepsilon_j = 1} \varepsilon_j \text{ and } d_- = -\sum_{j\,:\,\varepsilon_j = -1} \varepsilon_j$$

respectively, such that

$$\mathcal{P}_+(t + x \cdot f_j, f_j) = 0, \quad \forall\, j \,:\, \varepsilon_j = 1 \quad \textit{and}$$
$$\mathcal{P}_-(t + x \cdot f_j, f_j) = 0, \quad \forall\, j \,:\, \varepsilon_j = -1\,.$$

III. Some Applications

Linear superpositions and Webs. In connection with Hilbert's 13th problem Kolmogorov has obtained the following remarkable result.

Kolmogorov's theorem ([40], [21]). *For any $n \geq 2$ there exist strictly increasing, Lipschitz continuous, real functions of one variable $u_{jk} \in C(I)$, $k = 1, 2, \dots, n$, $j = 1, 2, \dots, 2n + 1$, such that any continuous function of n-variables $f \in C(I^n)$ can be represented by superposition*

$$f(x) = \textstyle\sum_{j=1}^{2n+1} \varphi_j(\mathcal{U}_j(x)), \quad x \in I^n, \quad \text{where}$$
$$\mathcal{U}_j(x) = \textstyle\sum_{k=1}^{n} u_{jk}(x_k), \quad \varphi_j \in C(\mathcal{U}_j(I^n)), \quad j = 1, \dots, 2n + 1\,.$$

The inner functions $\{\mathcal{U}_j\}$ of Kolmogorov's formula although Lipschitz continuous can not be continuously differentiable, because the following result is valid.

Theorem 8 ([34]). *Let the real functions $\mathcal{U}_j \in C^{(1)}(\mathbb{R}^n)$, $j = 1, 2, \dots, N$, be such that everywhere in some domain $\Omega \subset \mathbb{R}^n$*

$$d\mathcal{U}_{j_1} \wedge \dots \wedge d\mathcal{U}_{j_n} \neq 0 \ \forall\, 1 \leq j_1 < \dots < j_n \leq N \ .$$

Then for any "regular" polyhedral domain $G \subset \Omega$ the transform

$$\{\varphi_j\} \mapsto A\{\varphi_j\} = \sum_{j=1}^{N} \varphi_j\left(\mathcal{U}_j(x)\right), \quad \varphi_j \in C\left(\mathcal{U}_j\left(\bar{G}\right)\right)$$

defines a mapping $A \ : \ \oplus C(\mathcal{U}_j(\bar{G})) \to C(\bar{G})$ with closed infinite codimensional image in $C(\bar{G})$ and finite dimensional kernel in $\oplus C(\mathcal{U}_j(\bar{G}))$.

The domain $G \subset \Omega$ is called polyhedral if its boundary consists of a finite number of pieces of level sets of of functions $\mathcal{U}_j$.

If the functions $\mathcal{U}_j$, $j = 1, 2, \dots, N$, are smooth enough, then part of this statement concerning Ker A follows from classical results of Blaschke, Bol, and Chern, obtained in the 1930's in the framework of Web geometry.

Definition ([8], [10]). *A (non-singular) N-Web on a domain $\Omega \subset \mathbb{R}^n$ is a family of N foliations $\mathcal{F}_j$, $j = 1, 2, \dots, N$, on Ω of codimension 1, defined (locally) by a family of level sets of the real functionsl $\mathcal{U}_j$, $j = 1, 2, \dots, N$, as in Theorem 8.*

The leaves of the foliations $\{\mathcal{F}_j\}$ are called the Web hypersurfaces.

Two N-Webs $\{\mathcal{F}_1, \dots, \mathcal{F}_N\}$ and $\{\mathcal{F}'_1, \dots, \mathcal{F}'_N\}$ are called equivalent if there exists a homeomorphism h of Ω such that $h(\mathcal{F}_j) = \mathcal{F}'_j$, $j = 1, 2, \dots, N$.

For an N-Web, defined by level sets of the functions $\{\mathcal{U}_j\}$, an equation

$$\sum_{j=1}^{N} f_j(\mathcal{U}_j(x))\, d\mathcal{U}_j(x) \equiv 0, \quad f_j \in C(\mathcal{U}_j(\Omega)), \quad x \in \Omega \ ,$$

is called a (homogeneous) abelian equation.

The dimension of the space of solutions of this equation is called the rank of the N-Web.

So, for the mapping $A \ : \ \oplus C(\mathcal{U}_j(\bar{G})) \to C(\bar{G})$ and N-Web W, defined by functions $\{\mathcal{U}_j\}$, we have the equality

$$\dim \ Ker\ A = \text{rank}\ W + N - 1 \ .$$

For a (non-singular) N-Web W defined by sufficiently smooth $\{\mathcal{U}_j\}$, Bol and Chern (see [8], [10]) have proved a precise inequality

$$\text{rank}\ W \leq \pi\,(N, n) = m\,\{N - \frac{(m+1)(n-1)}{2} - 1\} \ ,$$

where $m = \left[\frac{N-1}{n-1}\right]$.

Under minimal smoothness assumptions as in Theorem 8 this upper bound has been proved by Fridman [21], but only for $n = 2$.

Homogeneous and non-homogeneous Abel's functional equations. An interesting problem, going back, in particular cases, to Euler, Cauchy, Abel, ..., is to describe the image and the kernel of the mapping A under natural additional conditions on the functions $\{\mathcal{U}_j\}$.

The refinement of classical Abel [2], Lie [41], Darboux [14], Blaschke [8] constructions permits to obtain rather complete results in this direction for the case of linear Webs.

Definition. *A Web is called linear if all leaves $\{\mathcal{U}_j = const\}$ of the foliations belonging to W are pieces of linear hyperplanes in $\mathbb{R}^n$.*

A Web is called linearizable if it is equivalent to a linear Web under an appropriate change of coordinates.

From Theorems 2 and 6 one can deduce the following result.

Theorem 9. *Let the domain $D \subset \mathbb{R}P^n$ and the C^1-functions $\mathcal{U}_j(x)$, $x \in D$, $j = 1, 2, \dots, N$, be such that*

$$d\mathcal{U}_{j_1} \wedge \dots \wedge d\mathcal{U}_{j_n} \neq 0, \quad 1 \leq j_1 < \dots < j_n \leq N\,,$$

in D and all level sets $\{x \in D \ : \ \mathcal{U}_j(x) = const\}$ are pieces of hyperplanes in $\mathbb{R}P^n$, all hyperplane sections of D are contractible, and there exist functions $\varphi_j \in C(\mathcal{U}_j(D))$, $\varphi'_j \neq 0$ almost everywhere on $\mathcal{U}_j(D)$, $j = 1, 2, \dots, N$, such that $d(\sum_{j=1}^N \varphi_j(\mathcal{U}_j(x))) = R(x)$ is a rational function for $x \in D$. Then

i) the level sets of $\{\mathcal{U}_j\}$ belong in the dual space $(\mathbb{R}P^n)^*$ to a non-degenerate (i.e., not belonging to any hyperplane) algebraic curve

$$V = \{\xi \in (\mathbb{R}P^n)^* \ : \ P_k(\xi) = 0, \ \ k = 1, 2, \dots, l\}, \ \ l \geq n-1,$$

of degree N, and the collection of forms $\{\varphi'_j d\mathcal{U}_j\}$ is the inverse image of some meromorphic differential on V;

ii) the space of vector functions $\{\psi_j \in C(\mathcal{U}_j(D)), \ \ j = 1, 2, \dots, N\}$ such that

$$\sum_{j=1}^{N} \psi_j(\mathcal{U}_j(x)) \equiv const$$

is the inverse image of the space of functional relations for the Abel integrals of holomorphic differentials on V;

iii) the space of functions $\Phi \in C(D)$ of the form $\Phi(x) = \sum_{j=1}^N \varphi_j(\mathcal{U}_j(x))$, $x \in D$, is exactly the space of solutions of the system

$$P_k\left(\frac{\partial}{\partial x}\right)\Phi(x) = 0, \ \ x = (x_0, x_1, \dots, x_n) \in D \subset \mathbb{R}P^n, \ \ k = 1, 2, \dots, l\,.$$

For the case when $R(x) \equiv 0$ and the functions $\{\mathcal{U}_j\}$ are smooth enough the parts i) and ii) of this theorem can be deduced from [41], [14], [8].

Theorem 9 i) shows that the existence of at least one nontrivial Abel relation

$$d\left(\sum_{j=1}^{N}\varphi_j(\mathcal{U}_j(x))\right) = R(x)$$

for an N-Web, defined by the functions $\{\mathcal{U}_j\}$, implies that this Web is algebraic, i.e., that its leaves $\{\mathcal{U}_j = const\}$, $j = 1, 2, \ldots, N$, belong to the non-degenerate algebraic curve V of degree N in the dual space $(\mathbb{R}P^n)^*$.

By a result of Nakai [45] the topological structure of the algebraic Web determines the curve V up to projective equivalence except for the case $N = 3$.

Theorem 9 ii) implies the following statement [10]: for an N-Web W determined by a non-degenerate algebraic curve V of degree N in the dual space $(\mathbb{R}P^n)^*$, the equality

$$\text{rank } W = \text{genus } V$$

holds. So, the cited results of Bol, Chern, and Griffiths imply, in particular, the Castelnuovo (1889) inequality

$$\text{genus } V \leq \pi\,(N, n)$$

for an arbitrary non-degenerate algebraic curve of degree N in $\mathbb{C}P^n$.

Theorem 9 iii) generalises for "hyperbolic" systems $\{P_k(\frac{\partial}{\partial x})\Phi = 0\}$ some results of Fantappie [20].

For nonlinear Webs very few general statements about the kernel and the image of the mapping A are known. Let us formulate here one of them, based on the "edge of the wedge" theorem for CR-functions [4].

Theorem 10. *Under the conditions of Theorem 8, let all level sets of the functions* $\{\mathcal{U}_j\}$ *be real analytic in* Ω, *and let for some* $\varphi_j \in C(\mathcal{U}_j(\Omega))$, $j = 1, 2, \ldots, n$, *the function* $A\{\varphi_j\}$ *be real analytic in* Ω. *Then each of the functions* $\varphi_j(\mathcal{U}_j(x))$ *is also real analytic in* Ω.

For the case when not only the level sets, but also the functions $\{\mathcal{U}_j\}$ are real analytic this statement was obtained firstly in [37] as a consequence of the theory of elliptic systems.

Surfaces of double translations. Theorem 9 implies the following classical result about surfaces of double translations under minimal regularity conditions.

Lie–Poincaré–Tchebotarev–Wirtinger's theorem ([41], [42], [49], [53], [55]). *Let* S *be a germ of a* C^2 *hypersurface in* $\mathbb{R}^{n+1}$, *admitting the double parametric representation of the form*

$$y_k = \sum_{j=1}^{n}\varphi_{k,j}(u_j(y)) = \sum_{j=n+1}^{2n}\varphi_{k,j}(u_j(y)), \quad y \in S, \quad k = 0, 1, \ldots, n, \qquad (*)$$

where $\{u_j(y)\}$ *and* $\{\varphi_{k,j}(u_j)\}$ *are* C^2*-functions such that*

$\forall\, y \in S$ and $\forall\, 1 \le j_1 < \ldots < j_n \le 2n$
the vectors $\{\varphi'_{0,j_\alpha}(u_{j_\alpha}(y), \ldots, \varphi'_{n,j_\alpha}(u_{j_\alpha}(y)\}$, $\alpha = 1, 2, \ldots, n$,
are linearly independent and the vectors
$\{\varphi''_{0,j}(u_j(y)), \ldots, \varphi''_{n,j}(u_j(y))\}$, $j = 1, \ldots, 2n$, are not tangent to S. (**)

Then the $2n$*-curves*

$$u_j \mapsto \{\varphi'_{0,j}(u_j) \,:\, \varphi'_{1,j}(u_j) \,:\, \ldots \,:\, \varphi'_{n,j}(u_j)\}\,,$$

$j = 1, 2, \ldots, 2n$, *are branches in* $\mathbb{R}P^n$ *of a non-degenerate algebraic curve* V *of degree* $2n$*. The vector-functions* $\{\varphi_{k,1}(u_1), \ldots, \varphi_{k,2n}(u_{2n})\}$, $k = 0, 1, \ldots, n$, *satisfying (*) correspond to the Abel relations for the Abel integrals of* $(n+1)$ *independent holomorphic differentials on* V.

Sketch of the proof. Let us consider the points $\xi^{(j)}(y) = \{\varphi'_{0,j}(u_j(y)) \,:\, \ldots \,:\, \varphi'_{n,j}(u_j(y))\}$, $j = 1, \ldots, 2n$ in $\mathbb{R}P^n$. From (*), (**) it follows that for all $y \in S$ the subspace in $\mathbb{R}P^n$, spanned by these points, is some hyperplane

$$\mathbb{R}P^{n-1}_x = \{\xi \in \mathbb{R}P^n \,:\, x \cdot \xi = 0\}\,,$$

and the correspondence $y \mapsto x(y)$ is (locally) a diffeomorphism, called the Poincaré map. Using x as coordinates on S we obtain C^1-functions $\mathcal{U}_j(x) = u_j(y(x))$, $j = 1, \ldots, 2n$, which satisfy the conditions of Theorem 9. The theorem now follows from the statements i) and ii) of Theorem 9.

Exceptional N-Webs and polylogarithms. An N-Web is called exceptional if it is not linearizable but has maximal rank.

The first example of an exceptional plane N-Web was discovered by Bol ([8]) on the basis of the Spence–Abel functional equation for the Euler dilogarithm:

$$\varphi\left(\frac{x}{y}\right) + \varphi\left(\frac{x+y-1}{x}\right) + \varphi\left(\frac{y-x}{1-x}\right) + \varphi\left(\frac{1-x}{y}\right) + \varphi\left(\frac{y(1-y)}{x(1-x)}\right) \equiv 0\,,$$

where

$$\varphi(z) = Li_2(1-z) - Li_2(z);\quad Li_2(z) = \sum_{k=1}^{\infty} \frac{z^k}{k^2}\,.$$

The existence of functional equations of such a type shows that for applications it is important to find an appropriate generalization of Theorem 9 to the case when level sets of functions $\{\mathcal{U}_j\}$ in a domain of $\mathbb{R}^n$ are pieces of algebraic hypersurfaces.

Let us formulate here only one recent and important result in this direction which shows that the Webs related with Spence–Kummer functional equations for trilogarithm $Li_3(z) = \sum_{k=1}^{\infty} \frac{z^k}{k^3}$ are also exceptional.

Theorem ([47]). *Let* $\mathcal{U}_1 = x$, $\mathcal{U}_2 = y$, $\mathcal{U}_3 = \frac{x}{y}$, $\mathcal{U}_4 = \frac{1-y}{1-x}$, $\mathcal{U}_5 = \frac{x(1-y)}{y(1-x)}$, $\mathcal{U}_6 = xy$, $\mathcal{U}_7 = -x\frac{1-y}{1-x}$, $\mathcal{U}_8 = -\frac{1-y}{y(1-x)}$, $\mathcal{U}_9 = \frac{x(1-y)^2}{y(1-x)^2}$ *be functions which participate in the Kummer functional equation for trilogarithm* $Li_3(z)$*. Let* G *be any simplement connexe domain* G *in* $\Omega = \{(x, y) \in \mathbb{R}^2 \,:\, d\mathcal{U}_j \wedge d\mathcal{U}_k \ne 0,\ 1 \le j < k \le 9\}$.

Then transform $\{\varphi_j\} \mapsto A\{\varphi_j\} = \sum_{j=1}^{9} \varphi_j(\mathcal{U}_j(x, y))$, $\varphi_j \in C(\mathcal{U}_j(\bar{G}))$ *has kernel of (maximal possible) dimension* $36 = 28 + 8$, *which can be described explicitely in terms of the rational functions* $Li_k x$, $k = 1, 2, 3$, *and the function* $arcth\sqrt{x}$.

A result of this type was announced by G. Robert in 2001 as an answer to a question of A. Henaut [35]. Its detailed proof was first obtained by L. Pirio [47], who deduced from it a solution of the A. Goncharov problem [24]: does the Kummer functional equation caracterize $Li_3 x$?

Concluding remarks. In this paper we have considered several results related with the classical Abel's addition theorem in the context of the Radon transform for distributions with support on curves in real or complex spaces. Part of the presented results (Theorems 1–4, 6, 7, 9) admits an appropriate generalization for the cases of multidimensional surfaces in complex or real spaces. We give here only several key references where such generalizations have been considered:

– the significant extensions of Theorem 1 were started in [6] and extended in [23], [17], [33], where the analog of Martineau-Radon transform is often called the Radon–Penrose transform

– some extensions of Theorem 2 and Theorem 3 to the cases of multidimensional surfaces were obtained in [31] and [16]

– the multidimensional versions of Theorem 4 and Theorem 6 were started in [26] and extended in [32], [19]

– the multidimensional versions of Theorem 7 and Theorem 9 one can deduce from [54], [31] and [32].

References

[1] N. H. Abel, *Solution de quelques problèmes a l'aide d'intégrales définies*, Magazin for Naturvidenskaberne, Aargang I, Bind 2, Christiania, 1823; Oeuvres Complètes de N. H. Abel, v. I, 11–27.

[2] N. H. Abel, *Mémoire sur une propriété générale d'une classe très étendue de fonctions transcendantes*, Présenté à l'Acad. Sc. Paris, 30 Octobre 1826; Oeuvres Complètes de N. H. Abel, v. I, 145–211.

[3] N. H. Abel, *Démonstration d'une propriété générale d'une certaine classe de fonctions transcendantes*, Journ. fur reine und ang. Math. **4**, 1829, Oeuvres de N. H. Abel, v. I, 515–517.

[4] R. Airapetyan, G. Henkin, *Analytic continuation of CR-functions across the "edge of the wedge,"* Soviet. Math. Dokl. **24**(1), 1981, 128–132.

[5] L. Aizenberg, *The general form of a linear continuous functional in spaces of functions holomorphic in convex domain in* $\mathbb{C}^4$, Soviet. Math. Dokl. **7**, 1966, 198–202.

[6] A. Andreotti, F. Norguet, *La convexité holomorphe dans l'espace analytique des cycles d'une variété algebrique*, Ann. Scuola Norm. Super. Pisa **21**, 1967, 31–82.

[7] W. Blaschke, *Results and problems about n-webs of curves in a plane*, Bull. Amer. Math. Soc. **38**, 1932, 828–830.

[8] W. Blaschke, G. Bol, *Geometrie der Gewebe*, Grundlehren Math. Wiss, v. **49**, Springer-Verlag, Berlin, 1938.
[9] J. Boman, *Holmgren's uniqueness theorem and support theorems for real analytic Radon transforms*, Contemp. Mathematics, v. **140**, 1992, 23–30.
[10] S. S. Chern, P. Griffiths, *Abel's theorem and Webs*, Jahresbericht der Deutschen Math.-Verein **80**, 1978, 13–110; **83**, 1981, 78–83.
[11] N. Coleff, M. Herrera, *Les courants résiduels associés à une forme méromorphe*, Springer Lecture Notes in Math. **633**, 1978.
[12] A. Clebsch, *Ueber die Anwendung der Abelschen Functionen in der Geometrie*, Journ. fur reine und ang. Math. **63**, 1864, 189–243.
[13] S. Colion, *Transformation d'Abel et formes différentielles algébriques*, C. R. Acad. Sci. Paris, v. **323**, 1996, 1237–1242.
[14] L. Darboux, *Théories des surfaces*, 2e éd, Gauthiers-Villars, Paris, v. I, 1914, 151–161.
[15] A. Dickenstein, C. Sessa, *Canonical representatives in moderate cohomology*, Invent. Math. **80**, 1985, 417–434.
[16] P. Dolbeault, G. Henkin, *Chaînes holomorphes de bord donné dans* $\mathbb{C}P^n$, Bull. Soc. Math. France **125**, 1997, 383–445.
[17] M. G. Eastwood, R. Penrose, R. O. Wells, Jr., Cohomology and massless fields, Commun. Math. Phys. 78, 1981, 305–351
[18] L. Euler, *Opera Omnia*, 4 Séries, Teubner, Leipzig–Berlin– Zurich, 1911–1976, I_{20}, 153–200, I_{21}, 39–56.
[19] B. Fabre, *Nouvelles variations sur des théorèmes d'Abel et de Lie*, Thèse, Univ. de Paris VI, 2000.
[20] L. Fantappié, *Sur les méthodes nouvelles d'intégration des équations aux dérivées partielles au moyen des fonctionnelles analytiques*, Coll. Int. du CNRS LXXI, "La théorie des l'équations aux dérivées partielles," Nancy, 9–15 Avril, 1956, 47–62.
[21] B. Friedman, *An estimate of the dimension of the null spaces of linear superpositions*, Math. USSR Sb. **11**, 1970, 101–114.
[22] S. Gindikin, *The Radon transform from cohomological point of view, 75 years of Radon transform*, (S.Gindikin, P.Michor, eds), International Press, 1994, 123–128.
[23] S. Gindikin, G. Henkin, *Integral geometry for* $\bar{\partial}$*-cohomology in q-linearly concave domains in* $\mathbb{C}P^n$, Funct. Anal. Appl. **12**, 1979, 247–261.
[24] A. Goncharov, *Geometry of configurations, Polylogarithms and Motivic Cohomology*, Advances in Math. **114**, 1995, 197–318.
[25] H. Graf, R. Sauer, *Uber dreifache Geradensysteme ...* , Sitzgsber. bayer. Akad. Math. - naturwiss. Abt. 1924, 119–156.
[26] P. Griffiths, *Variations on a theorem of Abel*, Invent. Math. **35**, 1976, 321–390.
[27] P. Griffiths, J. Harris, *Principles of algebraic geometry*, John Wiley and Sons, New York, 1978.
[28] V. Guillemin, S. Sternberg, *Integral geometry and microlocal analysis*, Amer. Math. J. **1**, 101, 1979, 915–955.
[29] R. G. Gunning, H. Rossi, *Analytic functions of several complex variables*, Prentice Hall, N.J., 1965.
[30] G. Henkin, *La transformation de Radon pour la cohomologie de Dolbeault et un théorème d'Abel inverse*, C.R. Acad. Sci Paris **315**, 1992, 973–978.
[31] G. Henkin, *The Abel–Radon transform and several complex variables*, Ann. of Math. Stud. **137**, 1995, 223–275.
[32] G. Henkin, M. Passare, *Holomorphic forms on singular varieties and variations on a theorem of Lie–Griffiths*, Invent. Math. **135**, 1999, 297–328.

[33] G. Henkin, P. Polyakov, *Homotopy formulas for the $\bar{\partial}$-operator on $\mathbb{C}P^n$ and the Radon–Penrose transform*, Math. USSR Izvestija **28**, 1987,3, 555–587.
[34] G. Henkin, A. Vitushkin, *Linear superpositions of functions*, Russ. Math. Survey, XXII, **1**, 1967, 77–124.
[35] A. Henaut, *Analytic Web geometry*, in "Web theory and related topics" (J.Grifone, E.Salem, ed.), World Scientific, 2001, 6–47.
[36] K. Jacobi, *Theoremata nova algebraica circa systema duarum aequatuenum inter duas variables prepositorum*, Journ. für reine und ang. Math. **14**, 1835, 281–288.
[37] J.-L. Joly, G. Métivier, J. Rauch, *Resonant one dimensional nonlinear geometric optics*, J. Funct. Analysis **114**, 1993, 106–231.
[38] M. V. Kazaryan, *Meromorphic continuation with respect to groups of variables*, Math. USSR Sb. **53**, 1986, 385–398.
[39] H. Kneser, *Einfacher Beweis eines Satzes über rationale Funktionen zweier Veränderlichen*, Hamburg Univ. Math. Sem. Abhandl. **9**, 1933, 195–196.
[40] A. N. Kolmogorov, *On the representation of continuous functions of many variables by superposition of continuous functions of one variable and addition*, Amer. Math. Soc. Transl. **28**, 1963, 55–59.
[41] S. Lie, *Bestimmung aller Flächen die in mehrfacher Weise durch Translationsbewegung einer Kurve erzeugt werden*, Archiv fur Mathematik **7**, 1882, 155–176, Ges. Abhandlungen, Bd I, 450–467.
[42] S. Lie, *Das Abelsche theorem und die Translationsmanningfaltigkeiten*, Leipziger Berichte, 1897, 181–248, Abhandlungen Bd II, 580–639.
[43] A. Martineau, *Indicatrices des fonctionnelles analytiques et inversion de la transformation de Fourier–Borel par la transformation de Laplace*, C. R. Acad. Sci. Paris **255**, 1962, 1845–1847, 2888–2890.
[44] A. Martineau, *Sur la topologies des espaces de fonctions holomorphes*, Math. Ann. **163**, 1966, 62–88.
[45] I. Nakai, *Topology of complex webs of codimension one and geometry of projective space curves*, Topology **26**, 1987, N4, 475–504.
[46] V. Palamodov, *Linear differential operators with constant coefficients*, Springer, 1970.
[47] L. Pirio, *Study of a functional equation associated to the Kummer's equation of the trilogarithm*, Applications, Preprint math. CV/0206172, 2002.
[48] H. Poincaré, *Remarques diverses sur fonctions abeliennes*, J. Math. Pure Appl., Serie 5, v. **1**, 1895, 219–314.
[49] H. Poincaré, *Sur les surfaces de translation et les fonctions abeliennes*, Bull. Soc. Math. France **29**, 1901, 61–86.
[50] M. Reiss, *Mémoire sur les propriétés générales des courbes algébriques etc.*, Corresp. Math. et Phys. de Quetelet **9**, 1837, 249–308.
[51] M. Rosenlicht, *Generalized Jacobian varieties*, Ann. Math. **59**, 1954, 505–530.
[52] B. Saint-Donat, *Variétés de translation et théorème de Torelli*, C. R. Acad. Sci. Paris **280**, 1975, 1611–1612.
[53] N. Tschebotarov, *Über Flächen welche Imprimitivitätssysteme in Bezug auf eine gegebene Kontinuirliche Transformations gruppe enthulten*, Recueie Math. (Sbornik) **34**, 1927, 149–204.
[54] J. A. Wood, *A simple criterion for local hypersurfaces to be algebraic*, Duke Math. J. **51**, 1984, N1, 235–237.
[55] W. Wirtinger, *Lies Translationsmannigfaltigkeiten und Abelsche Integrale*, Monatsch. Math. Phys. **46**, 1938, 384–431.

Abel Transform and Integral Geometry*

Simon Gindikin

1 The Abel Transform and the Radon Transform

In 1917 Radon published his famous paper [26] where he defined the operator which we call today the Radon transform. I already had a chance to write about this very significant paper [17]. Let us remind that Radon considers the problem of the reconstruction of a function of 2 variables $f(x, y)$ through its integrals $F(l)$ along all lines l. For this reconstruction he remarks that this operator $f \mapsto F$ (the Radon transform) commutes with all Euclidean motions and therefore it is sufficient to find a way to reconstruct in one point (let say on $(0, 0)$) a function with radial symmetry in this point. If we have such a function $\varphi(r)$, $r \geq 0$, then its Radon transform will also be symmetric relative to rotations and depend only on the distance q of a line l from $(0, 0)$ and we have a transform of functions of one variable $\varphi(r) \mapsto \Phi(q)$. It is simple to compute this transform explicitly:

$$\Phi(q) = 2 \int_0^\infty \frac{\varphi(r)dr}{\sqrt{r^2 - q^2}} .$$

At this point Radon remarks: "*... so kann man sie leicht nach Art der bekannten Abelshen lösen ...*". So he remarks that his transform for radial functions up to a simple replacement of variables coincides with the Abel transform:

* Partly supported by NSF-grant DMS-0070816. The author thanks the Max-Planck-Institut für Mathematik at Bonn for their hospitality.

$$\mathcal{A}\psi(s) = \int_s^\infty \frac{\psi(r)dr}{\sqrt{r-s}}$$

Abel considered this operator (often also called the Abel equation) in one of his earlier notes on a mechanic problem [1] and found its inversion:

$$\psi(r) = \frac{i}{\pi} \int_r^\infty \frac{dA\psi(s)}{\sqrt{s-r}} \, .$$

Abel considered also a more general operator depending on a parameter (cf. below the Riemann–Liouville integrals) and he integrated on $[0, s]$ instead of $[s, \infty]$.

This formula gives immediately the reconstruction of φ through Φ and, in particular,

$$f(0) = -\frac{1}{\pi} \int_0^\infty \frac{dF(q)}{q}.$$

Let us remark that the inverse Abel transform gives much more than we need for the inversion of the Radon transform of f: it gives the formula for the complete reconstruction of the Radon transform of radial functions but we need their reconstruction only in 0. This difference will be essential in our later discussions of generalizations of the Radon and Abel transforms.

Radon considers the similar problem in all dimensions. He finds that in the 3-dimensional case the situation is simpler and he does not need the Abel transform: the corresponding transform on radial functions can be inverted by a differentiation of order 2. The similar situation holds for all odd dimensions, but in even dimensions the inversion combines a multiple differentiation and the inverse Abel transform. Let us remind that it is connected with the Huygens principle which holds for odd dimensions but does not hold for even ones.

The key circumstance is that it is possible to interpret the Abel transform as the integration of order 1/2 and its inversion – the differentiation of order 1/2 – can be composed from the differentiation and again the integration of order 1/2. In this way the Abel transform can be included in the general family of operators of Riemann–Liouville $\mathcal{R}^\lambda$ – the integration of order $\lambda \in \mathbb{C}$ – which is defined for $\Re\lambda > 0$ and $f \in C_0^\infty(\mathbb{R})$ as

$$\mathcal{R}^\lambda f(x) = \frac{1}{\Gamma(\lambda)} \int_x^\infty f(y)(y-x)^{\lambda-1} dy$$

and then holomorphically extends as an entire function of the parameter $\lambda \in \mathbb{C}$. Abel in [1] practically considered such operators and their inversions for real $0 < \lambda < 1$. The operators $\mathcal{R}^\lambda$ form an Abelian group on λ, where $\mathcal{R}^0$ is the unit operator and for integer negative λ we have multiple differentiations. This gives a possibility to invert $\mathcal{R}^\lambda$ for all λ and includes as a partial case the inversion of the Abel transform. The Radon transform in $\mathbb{R}^n$ is reduced to the inversion of the Riemann–Liouville operator of an integer or semi-integer order depending on the dimension n being even or odd.

2 Hyperbolic Radon Transforms

One of the surprises in Radon's paper is that he understood clearly that these transforms are not necessarily related to $\mathbb{R}^n$, but can be considered also for non Euclidean geometries. Apparently, this broad view on the relationship between analysis and geometry was quite unusual at that time. There was an essential circumstance which could help Radon to state such a point of view: he already knew that Minkowsky and Funk had considered earlier a similar transform on the sphere: the reconstruction of even functions on the sphere S^2 through their integrals along big circles. Moreover, if Minkowsky used spherical functions for the inversion of his transform, Funk applied for this aim the Abel transform. Radon discussed this analogy in his paper.

His next observation is the possiblility to consider a similar transform also on the hyperbolic plane. Radon does not reproduce complete calculations writing down only a few lines: *"Aber auch in der hyperbolischen Ebene hat die gestellte Aufgabe die zu (III) analoge Lösung"*: $f(p) = -\frac{1}{\pi}\int_0^\infty \frac{dF_P(q)}{\sinh q}$. The mentioned formula (III) is just the Radon inversion formula above. We do not know why Radon omitted these computations which can be easily produced in a similar way to the Euclidean case. Namely, let us use the Poincaré model of the hyperbolic plane $\mathcal{L}^2$ on the upper half-plane $z = x + iy, y > 0$. We take a function $f(z), z \in \mathcal{L}^2$, and integrate it along hyperbolic lines which are represented either by Euclidean semicircles with the centers on the x-axis:

$$(x-a)(x-b) + y^2 = 0, y > 0 \,,$$

or by half-lines $x = a$, $y > 0$. The half-lines correspond to $b = \infty$. We integrate relative to the hyperbolic length. Since this transform $f \mapsto F(a, b)$ will commutate with hyperbolic motions it is enough to reconstruct in one point (let say $z = i$) radial symmetric functions with the center i (constant on orbits of $SO(2)$). It means that we have a function $f(z) = \varphi(\frac{|z-i|^2}{2y})$. Its transform $F(a, b)$ will also be $SO(2)$-invariant and depend only on the hyperbolic distance of the hyperbolic lines from the point i so $F(a, b) = \Phi(\frac{1+ab}{a-b})$. So we have an integral transform of functions of one variable $\varphi \mapsto \Phi$; for its calculation we can use an arbitrary family of parallel hyperbolic lines, in particular, ones presented by vertical half-lines. We can compute this transform of functions of one variable explicitly:

$$\Phi(s) = \int_{\sqrt{(s^2+1)}}^{\infty} \left(r^{-1}\varphi(r-1)\right) \frac{rdr}{\sqrt{r^2-(s^2+1)}} \,.$$

This transform can be directly inverted using the inverse Abel transform. In particular, for $s = 0$ we have

$$\varphi(0) = f(i) = -\frac{2}{\pi}\int \frac{d\Phi(r)}{r} \,.$$

We used here parameters slightly different from Radon's parameters.

In hyperbolic geometry there are 2 different analogs of Euclidean lines: hyperbolic lines – geodesics and horocycles – "cycles of infinite radius". We considered the geodesic version of hyperbolic Radon transform but it is possible to consider also the horocyclic Radon transform. This completely different integral transform can be inverted by the absolutely similar reduction to the inverse Abel transform. Horocycles in the Poincaré model are either circles tangent to real axes ($|\xi_2 z - \xi_1|^2 = y, \xi \neq (0,0)$) or lines $y = c$. Again, the trick with the commutativity with motions reduces the problem to a one-dimensional transform: the reconstruction in one point ($z = i$) of (hyperbolically) radial symmetric functions. It is sufficient to integrate such a function along parallel lines and the result will be a function $\Phi(r)$ of hyperbolic distances r of these lines from i. The direct computation gives

$$\Phi(s) = 2\int_{2|\sinh(s/2)|}^{\infty} \frac{\varphi(r) r dr}{e^{r/2}\sqrt{r^2 - 4\sinh^2(s/2)}},$$

which can again be inverted by the classical inverse Abel transform and, as a result,

$$\varphi(0) = f(i) = -\frac{1}{2\pi}\int_0^{\infty} \frac{(e^{r/2}\Phi(r))' dr}{\sinh(r/2)}.$$

We can see that the inversion of four versions of the Radon transform – the affine one (in $\mathbb{R}^n$), the Minkowsky–Funk transform on the sphere S^2, the geodesic and horocyclic ones on the hyperbolic plane $\mathcal{L}^2$ – is reduced to the inverse Abel transform. It is interesting to understand a nature of such an universality. One of possible explanations is that it turns out that the affine Radon transform, the Minkowsky–Funk transform and the geodesic hyperbolic Radon transform are equivalent realizations of the projective Radon transform (this circumstance was missed by Radon and it does not often appear in modern books on the Radon transform). In the broader frames of a conformal version of the Radon transform, the hyperbolic horocyclic one is equivalent to the other three [18], [19]. I believe that this equivalence does not give a complete explanation of the phenomenon, since this equivalence does not commute with the averagings in specific geometries. Rather the explanation is the exceptional universal role of the Abel transform as the fractional integration in one-dimensional analysis.

3 The Horospherical Transform and Abel Transform on Symmetric Spaces

The most interesting generalization of the Abel transform to the more general class of homogeneous manifolds, is connected with the horocyclic hyperbolic Radon transform. Gelfand and Graev suggested in [12] the program of integral geometry on homogeneous manifolds with the focus on a conception of horospherical transforms. Up until today it is not clear for which class of homogeneous manifolds it makes sense to consider horospherical transforms, but without any doubts it is natural for symmetric spaces. The first crucial position of Gelfand–Graev's project was that in many cases the harmonic analysis on homogeneous spaces (Plancherel formula – spectral decomposition on irreducible representations) can be reduced by the classical Fourier transform to the inversion of horospherical transforms. The second crucial point was that inverse horospherical transforms have relatively simple standard structures in spite of the fact that the harmonic analysis on such manifolds can differ dramatically. The first illustration of such a phenomenon is the complete similarity of the inversion formulas for the Radon transform in $\mathbb{R}^n$ and the horospherical Radon transform in the hyperbolic space $\mathcal{L}^n$. At that time the inversion of the analogue of the Fourier transform in the hyperbolic case is much more complicated than in the classical case (it is a consequence of the much more complicated structure of the group of hyperbolic motions compared with the structure of the group of affine motions). Gelfand and Graev considered several examples illustrating their project: complex semisimple Lie groups and their Riemann symmetric spaces. Let us remark that in these cases the inversion operators are differential operators which, as should be clear from the examples above, we cannot expect to see in the general situation.

In [14] we considered the horospherical transform on Riemann symmetric spaces of noncompact type, following the ideas of [12]. We shall in the following only consider symmetric spaces of noncompact type. Let $X = G/K$ be such a space, where G is a semisimple Lie group, and K is its maximal compact subgroup. Let us keep as a basic example $G = SL(p; \mathbb{R})$. Then X is the set of positive symmetric matrices of order p with determinant 1, where elements of G act as the change of variables in the corresponding quadratic form: $x \mapsto g^{\top} x g$; the dimension of X is equal to $n = p^2 - 1$ and the rank is equal to $l = p - 1$.

Horospheres are orbits of all maximal unipotent subgroups (of a maximal unipotent subgroup N and all its conjugates). In our example we can let N be the subgroup of upper triangular matrices with the unit diagonal. By integrating functions on X along the horospheres (relative to consistent invariant measures) we define the horospherical transform on X. The first observation in [14] is that Radon's trick with the reduction of the inversion of his transform to a one-dimensional problem (the Abel transform) works well in this situation: using the invariance relative to G, it is sufficient for the inversion of the horospherical transform to find a way to reconstruct the value at e of functions invariant relative to the action of the isotropy subgroup $K(e)$ of e. Such invariant functions on X induce functions on the Cartan subgroup H of X (which parameterizes $K(e)$-orbits). Their horospherical transforms will also be $K(e)$-invariant, and hence will also define functions on H (since H parameterizes "parallel" horospheres). In this way the horospherical transform induces an integral transform on the Cartan subgroup H: $\varphi \mapsto \Phi$, which later has been called the Abel transform on X (since it reduces to the Abel transform in the simplest case of the hyperbolic plane) [21, 22]. As a result we have an essential decreasing of the number of variables: from the dimension of X to its rank (the dimension of H). It turns out that in the case of spaces of rank 1 it is always possible to reduce the inversion of the Abel transform to the classical inverse Abel transform (the Riemann–Liouville operators). There were several papers in this direction; in the most systematic form by Koornwinder [23]. The case of rank greater than 1 is much more complicated.

Before the discussion of this problem let us look at the Abel transform in our basic example. If e be the unit matrix then $K(e) = K = SO(p)$ and the spherical symmetry of a function $f(x)$ means that it depends only on the eigenvalues of x: f defines a function φ on the diagonal subgroup H: $f(k^{-1}hk) = \varphi(h), k \in SO(p)$. As the subgroup N we take the subgroup of upper triangular matrices with the unit diagonal. Then we take the average $\varphi(n^\top hn)$ on $n \in N$ and we obtain again a function on H: $\varphi \mapsto \Phi = \mathcal{A}\varphi$. We can see that this transform has a very classical nature: in an analytic setting it connects 2 classical ways to reduce quadratic forms to a sum of squares: by orthogonal transformations (Lagrange) and by unipotent transformations (Jacobi). This transform of functions on H has a very clear definition which includes some extension of functions on H to functions on X, and integrations of these extensions along some submanifolds in X, but it is difficult to compute its kernel as an operator on H.

It is possible to consider the problem of the inversion of the Abel transform in 2 versions. The weak version is the reconstruction of the value at the unit point $\varphi(e)$ from $\mathcal{A}\,\varphi(h)$. This is sufficient for the inversion of the horospherical transform. A more general problem is to reconstruct $\varphi(h)$ for all points. Let us start with the first problem. As we mentioned the problem of the decomposition of the representation in $L^2(X)$ in irreducible ones (the Plancherel formula) is equivalent to the inversion of the horospherical transform: corresponding operators are intertwined by the classical Fourier transform. This relation for radial functions can be expressed the following way. For such functions the projection on irreducible representations is the operator whose kernel is the zonal spherical function: $\varphi(h) \mapsto \mathcal{F}\varphi(\lambda)$, where λ is a parameter on the set of zonal spherical functions (representations). Then the transform $\mathcal{F}\varphi(\lambda)$ (which is sometimes called the spherical Fourier transform) and the Abel transform $\mathcal{A}\varphi(h)$ are related via the Fourier transform on the Abelian group H. It is a quite simple corollary of the standard integral representation of zonal spherical functions [14]. In the spectral language it means that the zonal spherical function as the kernel can be replaced by the eigenfunction with the same eigenvalue which is constant on N-orbits ("zonal horospherical" function). This observation goes back to the first computations of Gelfand–Najmark for zonal spherical functions on complex groups. As a result there is a similar relationship between distributions $\mu(\lambda)$, $\nu(h)$ reconstructing $\varphi(e)$ from $\mathcal{F}\varphi$, $\mathcal{A}\varphi$ respectively. The original strategy of Gelfand–Graev was to develop some general methods of integral geometry (without an appeal to group actions) which allow, in particular, to invert the horospherical transform (compute ν) and then, using the classical Fourier transform, to compute the Plancherel measure μ . However, since such direct methods had not been developed at the time of [12], they decided in the beginning to move in the opposite direction and to compute ν for complex semisimple groups G and their Riemann symmetric spaces using the well known Plancherel measure μ for these cases. In [14] we decided to use the same methods for arbitrary Riemann symmetric spaces X.

Let us remark that direct methods for the inversion of horospherical transforms were developed relatively late and only for the special case of even root multiplicities, which includes the case of complex groups (in this case the inversion operators are differential) [16]. General methods which work for nonlocal inversion formulas are still unknown, and it remains an extremely challenging problem of integral geometry. By direct methods we mean methods of integral geometry which use only geometry of submanifolds (horospheres) but not group symmetry.

Let us return to [14]. Just about that time the Plancherel measure μ for arbitrary Riemann symmetric spaces was computed explicitly. This was a consequence of Harish-Chandra's expression of μ through his c-function and the product-formula for the c-function expressing it through classical B-functions [13]. So we had to compute the Fourier transform of this product. Sometimes this is simple. For example, if all root multiplicities are even (as for complex groups G) then μ is a polynomial and its Fourier transform is a result of applications of a differential operator to the δ-function. In a typical case μ looks as a product of a polynomial and a product of tanh of some roots. In our example with $SL(p;\mathbb{R})$ we have

$$\mu(\lambda_1,\ldots,\lambda_p) = c \prod_{1\le i<j\le p} (\lambda_i-\lambda_j)\tanh\left(\pi\frac{\lambda_i-\lambda_j}{2}\right), \quad \lambda_1+\cdots+\lambda_p=0\,.$$

The polynomial factor contributes a differential operator:

$$\mathcal{A}\varphi(h) \mapsto L\mathcal{A}\,\varphi(h)\,.$$

In the example,

$$L = \prod_{1\le i<j\le p}\left(\frac{\partial}{\partial t_i}-\frac{\partial}{\partial t_j}\right),$$

where t_i are logarithms of the diagonal elements.

Thus the problem reduces to the computation of the Fourier transform of products of tanh of some roots. If the number of these roots does not exceed the rank l it is a simple problem, since we are reduced to a one-dimensional problem (by taking roots as new independent variables). It does not work if we have a bigger number of factors but then a very simple trick helps. Using the formula

$$\tanh x \tanh y \tanh(x+y) = \tanh x + \tanh y - \tanh(x+y)\,,$$

(equivalent to the addition formula for tanh), we can transform the product to a sum of products of smaller numbers of tanh's. Experiments showed that eventually we obtain terms with numbers of factors less then rank and it is possible to make the direct computations. Moreover, different terms are obtained from one another by actions of the Weyl group W and, since is W-invariant, it is sufficient to compute one term. In the example for even $p=2k$ the typical term after the reduction of the number of factors is

$$c\tanh\left(\pi\frac{\lambda_1-\lambda_2}{2}\right)\tanh\left(\pi\frac{\lambda_3-\lambda_4}{2}\right)\cdots\tanh\left(\pi\frac{\lambda_{2k-1}-\lambda_{2k}}{2}\right).$$

As a result we have for this case the inversion formula

$$\varphi(e) = c\int_{-\infty}^{\infty}\cdots\int_{-\infty}^{\infty}\frac{L\mathcal{A}\varphi(s_1,-s_1,s_2,-s_2,\ldots,s_k,-s_k)}{\sinh s_1\cdots\sinh s_k}ds_1\cdots ds_k,\ p=2k\ .$$

Of course, we used here the fact that the Fourier transform of $\tanh\pi\lambda$ is equal to $\frac{1}{2\pi i}\frac{1}{\sinh u}$. As we already said, we take as variables on H logarithms of diagonal elements. Similar formulas were found in [14] for all classical normal real groups G. In the final formulas there appear, as in the example above, some subsets of simple roots, whose nature remained unclear. More than 20 years later Beerends [7] found an elegant way to explain their nature and proved their existence for all symmetric spaces X.

4 The Global Inversion of the Abel Transform

The global inversion of the Abel transform looks more difficult. Of course, since we have the inversion of horospherical transform (as a consequence of the inversion in the center) and since the Abel transform is just a specialization of horospherical transform for spherically symmetric functions, we have a construction of the inverse operator (using the averaging of such functions in other points). This construction is, however, not satisfactory: the operator includes too many integrations and, if we take the average with a center $h \in H$ different from e, we go outside of H. It is not simple to rewrite this procedure as an operator on H. We already mentioned the similar situation in the definition of the Abel transform, and its explicit presentation as an operator on H is known only in a few cases. Thus we seek an explicit presentation of $\mathcal{A}^{-1}$ as an operator on functions on H.

As usual, the case of complex groups G is the simplest one (then the inverse operator is differential); as we mentioned above [23], there is also a clear picture for the case of rank 1. Explicit formulas are known basically only for some spaces of rank 2 (cf. e. g. [8], [24]). An exception is the case of the complex Grassmannians corresponding to the groups $SU(p,q)$, for which (differential) inverse operators were computed in [25] using the explicit formulas for zonal spherical functions of [9]. Let us recall that complex Grassmannians is an unique class of symmetric spaces outside of spaces of rank 1 and spaces with complex groups, for which zonal spherical functions were computed explicitly (cf. also the computations for $SL(n;\mathbb{R})$ in [30]). There is probably a strong similarity between the explicit solutions of these two problems.

There is an essential difference of the Abel transform on symmetric spaces from the classical one: in the classical case it is possible to include both the Abel transform and its inversion in the group of the Riemann–Liouville operators. Probably, there does not exist a direct way to generalize this construction on symmetric spaces, but there is a remarkable surrogate of such a construction. It is a part of the general project of the transfer of the harmonic analysis on symmetric spaces X to their

Cartan subgroups H, using spherically symmetric functions. We can consider the restrictions of zonal spherical functions, radial parts of invariant differential operators etc. as elements of an analysis on H without any appeal to symmetric spaces X and work only with the root systems. Such a theory depends on a system of roots Σ and multiplicities of roots m_α. Let us recall that different symmetric spaces can have the same root system Σ but different multiplicities m_α. Opdam and Heckman developed a theory of spherical functions on H which makes sense for all multiplicities m_α, including such ones which are not realized by symmetric spaces. An important part of this theory is "shift" differential operators which in a natural sense intertwine radial Laplacians for different multiplicities for a fixed Σ. Beerends [8] suggested to define the Abel transform also for arbitrary multiplicities and then the "shift" operators will intertwine such Abel transforms for different multiplicities as well. It gives a possibility to inverse the Abel transform if an appropriate "shift" operator exists. Since the existence of the "shift" operators is proven in general, it gives, in particular, the possibility to invert the Abel transform for even multiplicities m_α by differential operators. However, the explicit form of such operators continues to be unknown. Nevertheless, after this interpretation the reduction to the "shift" operators started to be the standard way to compute inverse Abel transforms.

Let us make a few remarks. The concept of "shift" operators for Abel transforms with multiplicities as parameters is similar to J. Bernstein's differential operators intertwining distributions P^λ, where P is a polynomial, relative to the parameter λ. Koorwinder in his paper on the Abel transform for symmetric space of rank 1 already considered "unrealizable" multiplicities, using the general Jacobi polynomials. Let us also mention that the results for symmetric spaces of rank 1 admit the extension to Damec–Ricci spaces which generates some new multiplicities [6].

It is natural to ask why it is is interesting to inverse the Abel transform in all points if for the inversion of horospherical transform we need to invert it only in the central point. The reason is that there are several other problems where the inverse Abel transform can work. The central observation is that the Abel transform transforms the Laplace–Beltrami operator Δ on a symmetric space in the usual Laplacian plus a constant. It opens a way for estimates of operators derived from Δ under the condition that we know the inverse Abel transform either explicitly or, at least, have substantial information about it. The best known project in this direction is connected with estimates of the heat kernel on X. In the beginning Anker [2] obtained a very strong upper estimate in the case $SU(p; q)$ using the explicit expression for the inverse Abel transform for this group [25]; he also conjectured that similar estimates hold for arbitrary symmetric spaces. This estimate was proved by Sawyer for symmetric spaces with root system A by a serious analysis of the inverse Abel transform for such spaces [28], [29], [30]. Anker proved a weaker version of his conjecture in the general case [5]. There are also exact results about the lower estimates.

Let us mention other spectacular applications of the Abel transform on symmetric spaces: Anker, using the support theorem for the Abel transform proved a general L_p multipliers theorem for symmetric spaces (a generalization of the Hörmander–

Michlin multiplier theorem) [3] and gave an elegant proof of the Paley–Wiener theorem for spherical Fourier transform [4].

5 The Riemann–Liouville Operators on Homogeneous Cones

As we already mentioned, the important property of the classical Abel transform, which is not inherited by the Abel transform on symmetric spaces, is its inclusion in the group of Riemann–Liouville operators. It turns out that there is another version of the Abel transform which can be defined on a class of symmetric spaces – symmetric cones – and which can be included in a group of generalized Riemann–Liouville operators. Such operators are reduced to the classical ones if the cone is the half-line. Such a group was first constructed by M. Riesz [27] as (complex) "degrees" of the wave operator. Let

$$\square(u) = t^2 - x_1^2 - \cdots - x_n^2, \quad , u = (t, x) \in \mathbb{R}^{n+1}, x \in \mathbb{R}^n ,$$

and $V = \{u = (t, x); \square(t, x) > 0, t > 0\}$ be the future light cone. Then we define

$$\mathcal{R}_V^\lambda f(v) = \frac{1}{\Gamma_V(\lambda)} \int_{(u-v)\in V} f(u)\square^{\lambda-k}(u-v)du, \quad du = dtdx_1 \ldots dx_n .$$

Here we have $\lambda \in \mathbb{C}$; $f \in C_0^\infty(\mathbb{R}^{n+1})$, $k = (n+1)/2$; Γ_V is the Siegel Γ-function of the cone V which admits the expression through one-dimensional Γ-functions:

$$\Gamma_V(\lambda) = \pi^{(n-1)/2}\Gamma(\lambda)\Gamma\left(\lambda - \frac{n-1}{2}\right) .$$

This integral defines operators for $\Re\lambda > 0$ and extends as an entire function of the parameter λ in such a way that the group property holds and $\mathcal{R}_V^0$ is the unit operator. Also $\mathcal{R}_V^{-1}$ is the wave operator. Riesz applied his operators to the construction of the fundamental solution of the wave equation. If (similar to the classical definition) one were to define the Abel transform of the cone V as

$$\mathcal{A} f(v) = \mathcal{R}_V^{1/2} f(v)$$

then we have the inversion formula

$$\mathcal{A}^{-1} = \mathcal{A}\left(\square\left(\frac{\partial}{\partial u}\right)\right),$$

which is the exact generalization of the classical Abel inversion formula.

Gårding generalized Riesz' construction of Riemann–Liouville operators on the cones of positive symmetric and Hermitian matrices [10]. I constructed them for arbitrary convex homogeneous cones (not necessarily symmetric) [15]. These operators have essential applications to some problems of integral geometry, but these problems are different from those we discussed above (Radon type problems). They apply to problems of reconstructions of integrals along degenerate orbits through integrals along generic ones. A typical example of such a problem is the problem of the reconstruction of $f(0)$ through integrals along the hyperboloids ($\square(u) = const \neq 0$). This kind of applications were important for the very spectacular Gelfand–Graev's deduction of the Plancherel formula for semisimple Lie groups [11]. There are examples of applications of the Riemann–Liouville operators of cones to Radon type problems of integral geometry – some problems of integral geometry on Grassmanians [20]. In my opinion such examples are interesting, but they are quite special relative to the possibilities offered by modern integral geometry (the method of the operator κ).

In the estimate of the place of Abel type transforms in the integral geometry it is important to understand that they appear in situations when a very strong group symmetry gives a possibility to decrease substantially the dimension. The crucial advantage of the integral geometry is that it operates today with geometrical structures more general than group invariance and produces explicit formulas in such a broad environment. Nevertheless, generalized Abel transforms in more general aspects of explicit constructions in multidimensional analysis are impressive examples which are well illustrated by applications to the integral geometry.

Acknowledgments. I thank Jean-Philippe Anker and Patrice Sawyer for their help in the preparation of this paper.

References

[1] N. H. Abel, *Auflösen einer mechanischen Aufgabe*, J. Reine Angew. Math. **1** (1826), 153–157.

[2] J.-Ph. Anker, *Le noyau de la chaleur sur les espaces symtriques $U(p, q)/U(p) \times U(q)$*, Harmonic analysis (Luxembourg, 1987) (P. Eymard and J.-P. Pier, eds.), Lect. Notes Math., vol. 1359, Springer, (1988), pp. 60–82.

[3] J.-Ph. Anker, *L^p Fourier multipliers on Riemannian symmetric spaces of the noncompact type*, Ann. of Math. **132** (1990), 597–628.

[4] J.-Ph. Anker, *The spherical Fourier transform of rapidly decreasing functions. A simple proof of a characterization due to Harish-Chandra, Helgason, Trombi, and Varadarajan*, J. Funct. Anal. **96** (1991), 331–349.

[5] J.-Ph. Anker, *Sharp estimates for some functions of the Laplacian on noncompact symmetric spaces*, Duke Math. J. **65** (1992), 257–297.

[6] J.-Ph. Anker, E. Damek and C. Yacoub, *Spherical analysis on harmonic AN groups*, Ann. Scuola Norm. Sup. Pisa Cl. Sci. **(4)23** (1996), 643–679.

[7] R. J. Beerends, *The Fourier transform of Harish-Chandra's c-function and inversion of the Abel transform*, Math. Ann. **277** (1987), 1–23.

[8] R. J. Beerends, *The Abel transform and shift operators*, Comp. Math. **66** (1988), 145–197.

[9] F. A. Berezin and F. I. Karpelevich, *Zonal spherical functions and Laplace operators on some symmetric spaces*, Dokl. Akad. Nauk SSSR (N.S.) **118** (1958), 9–12.

[10] L. Gårding, *The solution of Cauchy's problem for two totally hyperbolic differential equations by means of Riesz integrals*, Ann. of Math. **48** (1947), 785–826.

[11] I. Gelfand and M. Graev, *On a general method for decomposition of the regular representation of a Lie group into irreducible representations*, Dokl. Acad. Nauk SSSR **92** (1952), 221–224.

[12] I. Gelfand and M. Graev, *Geometry of homogeneous spaces, representations of groups in homogeneous spaces and related questions of integral geometry*, Transl., II, Amer. Math. Soc. **37** (1964), 351–429.

[13] S. G. Gindikin and F. I. Karpelevich, *Plancherel measure of Riemann symmetric spaces of non-positive curvature*, Dokl. Acad. Nauk. USSR **145** (1962), 252–255.

[14] S. Gindikin and F. Karpelevich, *One problem of integral geometry*, In: Pamyati N.G. Chebotareva, Izdatelstvo Kazanskogo Univ., 1964; English transl. in Select. Math. Sov. **1** (1981), 169–184.

[15] S. G. Gindikin, *Analysis in homogeneous domains*, Uspekhi Mat.Nauk **19** (1964), no. 4, 3–92; English transl. in Russian Math. Surveys **19** (1964), no. 4, 1–88.

[16] S. G. Gindikin, *Integral geometry on symmetric manifolds*, Amer. Math. Soc. Transl. (2) **148** (1991), 29–37.

[17] S. G. Gindikin, *Some notes on the Radon transform and integral geometry*, Monatsh. Math. **113** (1992), 23–32.

[18] S. Gindikin, *Real integral geometry and complex analysis*, Integral geometry, Radon transform and complex analysis (E. Casadio Tarabusi, M. Picardello, G. Zampieri, eds.), Lecture Notes in Math., vol. 1684, Springer, (1998), pp. 70–98.

[19] S. Gindikin, *Integral geometry on hyperbolic spaces*, Harmonic analysis and integral geometry (M. Picardello, ed.), Chapman & Hall, (2001), 41–46.

[20] E. Grinberg and B. Rubin, *Radon inversion on Grassmanians via Gårding–Gindikin fractional integrals*, preprint (2002).

[21] S. Helgason, *Group and Geometric Analysis*, Academic Press, 1984.

[22] S. Helgason, *Geometric analysis on symmetric spaces*, Mathematical Surveys and Monographs, vol. 39, American Mathematical Society, 1994.

[23] T. H. Koornwinder, *Jacobi functions and analysis on noncompact semisimple Lie groups*, Special functions: group theoretical aspects and applications (R. A. Askey and other, eds.), Reidel, (1984), pp. 1–85.

[24] Zhu Fu Liu, *On the inverse Abel transforms for certain Riemannian symmetric spaces of rank 2*, Math. Ann. **305** (1996), 617–637.

[25] Ch. Meaney, *The inverse Abel transform for $SU(p, q)$*, Ark. Math. **24** (1986), 131–140.

[26] J. Radon, *Über die Bestimmung von Functionen durch ihre Integralwerte längs gewisser, Mannigfaltigkeiten*, Ber. Verh. Sächs. Acad. Wiss. Leipzig, Math.-Nat. Kl. **69** (1917), 262–277.

[27] M. Riesz, *L'intégrale de Riemann–Liouville et le problème de Cauchy*, Acta Math. **81** (1949), 1–223.

[28] P. Sawyer, *The heat equation on spaces of positive definite matrices*, Canadian Journal of Mathematics **44** 3 (1992), 624–651.

[29] P. Sawyer, *On an upper bound for the heat kernel on $SU^\star(2n)/Sp(n)$*, Canadian Bulletin of Mathematics, **37** 3 (1994), 408–418.

[30] P. Sawyer, *Estimates for the heat kernel on $SL(n, \mathbb{R})/SO(n)$*, Canad. J. Math. **49** (1997), 359–372.

Abel's Inverse Problem and Inverse Scattering

V. P. Palamodov

1 Introduction

The problem of *determination of the shape of a hill from travel time* was solved by N. H. Abel in his third paper written in 1823. He gave more complete arguments in a paper of 1826. The problem is to find the form of the hill $y = h(x)$, $0 \le x \le x_0$ from knowledge of the function $t = t(x)$, $0 < x \le x_0$, t being the time a material point spend to move from $(x, y(x))$ to $(0, y(0))$ in the gravitation field.

This problem is a generalization of the *tautochrone problem* solved earlier by Ch. Huygens: *to shape the curve such that the travel time $\psi(a)$ does not depend on a*. The curve is found to be an arc of a cycloid. Taking the evolute of the cycloid (again an arc of a cycloid), Christian Huygens constructed a pendulum of constant period (1673). Abel's method is based on the inversion ψ of the integral transform

$$A : f \mapsto \pi^{-1/2} \int^{x} (x - y)^{-1/2} f(y)\, dy \tag{1}$$

(later called Abel's transform). Abel considered this transform as an operator of fractional integration of order 1/2. The arc length $s = s(x)$ is related to ψ by the equation $\psi = As'$, hence ψ is equal to 1/2-derivative of s. N. Abel then found the function s by applying the 1/2-integration operator A to ψ. In the original notations Abel's inversion looks as follows

$$s(x) = \frac{1}{\sqrt{\pi}} \frac{d^{-1/2} \psi x}{dx^{-1/2}} = \frac{1}{\sqrt{\pi}} \int^{1/2} \psi x dx^{1/2}$$

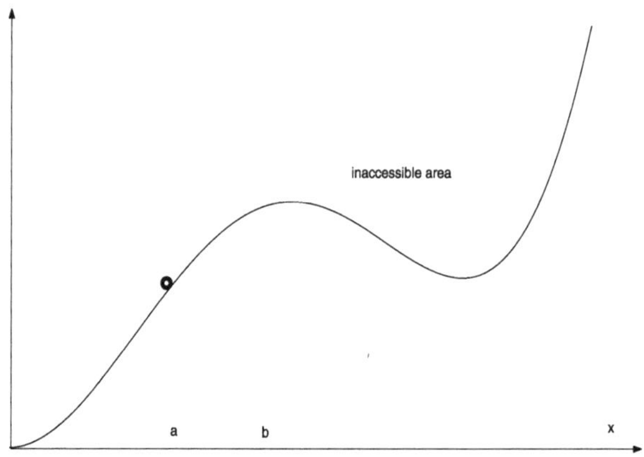

Fig. 1.

where ψx denotes $\psi(x)$, (N. H. Abel, 1823, p. 18). For constant ψ this gives the cycloid and vice versa; this is just Huygens' solution to the tautochrone problem.

Abel's technical idea of *fractional* integration and derivation was applied and generalized in several directions. The modern presentation is done in terms of distribution theory: the operator of fractional integration $I_\lambda,\ \lambda \in \mathbb{C}$ is given by the convolution operator with the kernel $x_\pm^{\lambda-1}/\Gamma(\lambda)$, which has an analytic continuation to the whole complex plane; its values in $\lambda = 0, -1, -2, ...$ are the delta-distribution and its derivatives. The convolution operators $I_\lambda *$ form a commutative group, since $I_\lambda * I_\mu = I_{\lambda+\mu}$.

Abel's method for inversion of the operator A started with a relatively simple problem, appeared very important in more complicated situations:

G. Herglotz (1905) and E. Wiechert (1907) applied the Abel transform for reconstruction of the velocity of elastic waves in the earth from the hodograph.

P. Funk (1916) and J. Radon (1917) studied problems of integral geometry on the sphere and in the Euclidean plane. Inversion of the geodesic transform on the sphere (Minkowski's problem) and on the Euclidean plane is reduced in the final step to inversion of the Abel transform.

J. Hadamard (1932) applied the singular kernel of Abel's type (1) to construction of fundamental solution of a linear second order hyperbolic operator. Hadamard's construction is a key point in the theory of wave propagation.

Huygens's and Abel's results as well as those of Herglotz–Wiechert, Funk, and Radon represent the very beginning of the theory of *Inverse Problems*. The term appears much later at the end of 20th century. There are several features of Abel's solution that are common for another type of inverse problems:

- although the problem was given an explicit solution, there is an inaccessible (or "trapped") area;
- the following similar problem admits no explicit solution: to determine the shape of the segment $y = h(x)$, $0 \leq x \leq b$ of a hill from the passing time, i.e.,

the time $T(E)$ a material point of energy $E \geq h(b)$ spends in passing this segment;

- uniqueness of the solution does not guarantee that it can be found analytically nor numerically.

2 Diffraction Tomography

Let $D \subset X$ be a domain with an unknown refraction coefficient $\mathbf{n}$ =inverse velocity which is supposed to be constant in $X \backslash D$. The harmonic waves with a time frequency $\omega > 0$ satisfy the Helmholtz equation

$$\Delta u + (\omega \mathbf{n})^2 u = 0 \tag{3}$$

The function $u^i = \exp(\iota\omega\mathbf{n}\langle\theta, x\rangle)$ satisfies this equation in $X \backslash D$; it is called the harmonic plane wave with direction θ. Take the solution of (3) that has the form $u = u^i + u^s$, where u^i is the harmonic plane wave and u^s is a function that satisfies the Sommerfeld radiation condition

$$\frac{\partial u^s}{\partial r} - i\omega\mathbf{n}u^s = O\left(r^{-(n+1)/2}\right), \; u^s = O\left(r^{-(n-1)/2}\right), \; r \to \infty$$

It is called the scattered field. Numerical simulation of a plane wave scattered on a velocity perturbation supported by a central disc is shown.

The real (left) and the imaginary (right) part of the scattered field u^s, are plotted in Fig. 2 (courtesy of F. Wuebbeling):

The data for the *inverse scattered problem* is a set of scattered waves from several directions θ measured on the boundary of the domain of interest.

Problem: to evaluate the function $\mathbf{n}$ from boundary measurements of scattered fields for fixed time frequency ω.

Applications: ultrasound imaging technique in medicine, nondestructive evaluation in industry, seismic exploration, etc.

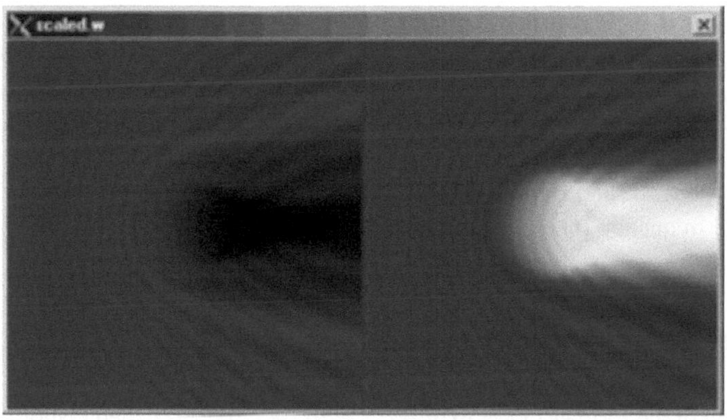

Fig. 2.

3 Resolution Limitation and Born–Wolf Principle

R. Newton (1989): *"the inverse scattering problem is improperly posed, i.e., in any real situation the measured data is not exact and hence a solution does not exist."*

R. Newton's pessimism has strong reasons. Let $\mathbf{n}_*$ be a background refraction coefficient which is an approximation to the unknown one from a priori information. The Helmholtz equation for a background medium is:

$$\Delta u_* + (\mathbf{n}_* \omega)^2 u_* = 0 \tag{3}$$

Suppose for a moment that $\mathbf{n}_*$ is constant and that the perturbation $\boldsymbol{\delta} = \mathbf{n} - \mathbf{n}_*$ is small. In the case $\dim X = 3$ the transmitted field fulfills the Lippman–Schwinger equation

$$u(\theta; x) = \exp(\imath k \theta x) + k^2 \int \frac{\exp(\imath k |x-y|)}{4\pi |x-y|} \boldsymbol{\delta}(y)\, u(\theta; y)\, dy, \ \ k = \mathbf{n}_* \omega$$

In the frame of Born approximation we assume $u(\theta; y) \approx \exp(\imath k \theta y)$ and

$$a(\theta; x) \approx a_B(\theta; x) = k^2 \int \boldsymbol{\delta}(y) \exp(\imath k (\theta - \tilde{x})\, y)\, dy, \ \ \tilde{x} \doteq |x|^{-1} x$$

for the scattering amplitude a. Take a perturbation in the form (Gabor function) $\delta = \varepsilon_0 \exp(-\pi(x-p)^2 + 2\pi\imath\xi x)$. The scattering amplitude is

$$a_B(\theta; x) \doteq \varepsilon_0 k^2 \exp\left(2\pi\imath p\eta - \pi(\eta - \xi)^2\right), \ \ \eta \doteq \mathbf{n}_* \omega (\tilde{x} - \theta)$$

We have $|\eta| \leq 2\mathbf{n}_*\omega$, therefore a_B is exponentially small as $|\xi| \geq (2+\varepsilon)\, \mathbf{n}_*\omega$, $\varepsilon > 0$ and $\omega \to \infty$, since $|\eta - \xi| \geq \varepsilon \mathbf{n}_* \omega$ for arbitrary directions θ and $\tilde{x}$.

Conclusion: *Stable reconstruction of the perturbation $\boldsymbol{\delta}$ is not possible for space frequencies ξ, if $|\xi| \geq (2+\varepsilon)\, \mathbf{n}_*\omega$ and ω is big.*

On the other hand, for small frequency the following positive principle has been claimed: *"If $\mathbf{n}_*$ is a constant, a small perturbation $\boldsymbol{\delta}$ can be stably reconstructed in the frequency ball of radius $2\mathbf{n}_*\omega$."* This principle was called "basic theorem of diffraction tomography" in the book M. Born, E. Wolf, Principles of Optics, 1999. It is proved again in the frame of Born's or Rytov's approximation for the constant background coefficient $\mathbf{n}_*$.

For a general (and rigorous) form of this principle we use localization in both space and frequency spaces, i.e., in the phase space $X \times X^*$.

4 Phase Space Analysis

A function of the form $\mathbf{e}_\lambda(x) = \exp(-\pi(x-p)^2 + 2\pi\imath\xi x)$ is called Gabor function; $\lambda = (p, \theta)$ is a point in the phase space $X \times X^*$. Functions of this form are called

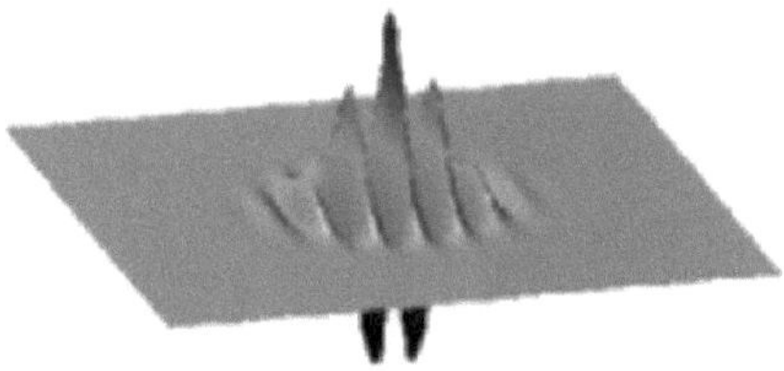

Fig. 3. A Gabor function

the coherent states in quantum field theory; a coherent state is a function that turns the uncertainty inequality to the equation. See Fig. 3 for a graph of Re $\mathbf{e}_\lambda$.

According to D. Gabor (1946): *"the elementary signal* (i.e., $\mathbf{e}_\lambda$) *occupies the smallest possible area in the information diagram,"* i.e., in the phase space. The function $\mathbf{e}_\lambda$ sharply concentrates near p, whereas the Fourier transform $\hat{\mathbf{e}}_\lambda$ focuses to the point $\xi \in X^*$, since

$$\hat{\mathbf{e}}_\lambda(\eta) \doteq \int \exp(-2\pi\imath\eta x)\mathbf{e}_\lambda(x)dx = \exp(2\pi\imath\xi p)\,\mathbf{e}_{\widehat{\lambda}}(\eta)\,, \;\widehat{\lambda} = (\xi, -p)$$

Gabor system versus wavelet system. A wavelet system in a vector space X is obtained from a mother halfdensity by the action of the affine group, i.e., the group that contains any translations and linear transformations of X.

The Gabor system is generated from the gaussian "mother" halfdensity: $\mathbf{e}_0(x) = \exp(-\pi x^2)$ by the action of the Weyl–Heisenberg group that contains translations in X and in X^*. The Gabor system is preserved by the Fourier transform and, moreover, by the action of the orthogonal metaplectic group.

Localization in the phase space. For a function f in an Euclidean space X the Gabor transform is defined as follows $\tilde{f}(\lambda) \doteq \langle f, \mathbf{e}_\lambda\rangle\,, \lambda \in X \times X^*$. The Gabor transform $\tilde{f}(\lambda) = \tilde{f}(p, \theta)$ is a localization of the function f in space and frequency simultaneously. We say that f is stably determined in a domain Ω in the phase space, if its Gabor transform is so for $\lambda \in \Omega$.

Example. Go back to the problem of determination of the shape of a hill from passing time (see Introduction). The function $y = h(x)\,, 0 \le x \le b$ is not stably determined from the data $T(E)\,, E \ge h(b)\,,$ but it is stable in a domain that looks like the one in Fig. 4: This picture shows that high frequencies of the unknown function h can be stably reconstructed only near the point b, whereas only low frequencies of h can be stably recovered far from this point. This implies, in particular, that there is no explicit reconstruction formula for this inverse problem, unlike Abel's case.

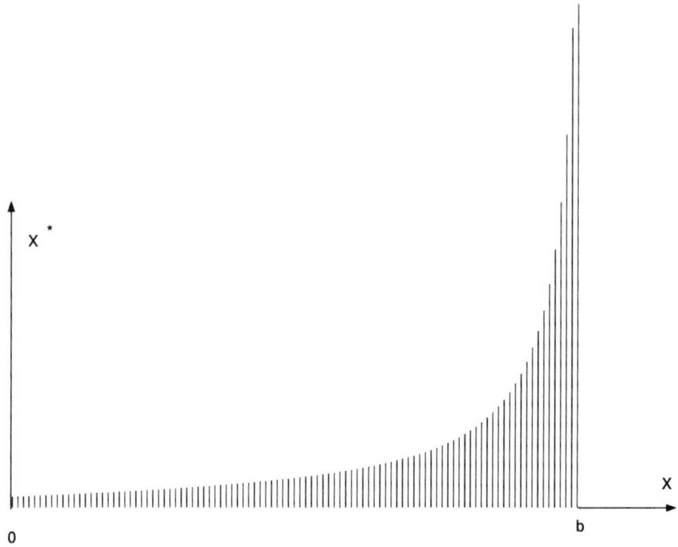

Fig. 4. The stability domain in the phase space

5 "Basic Theorem of Diffraction Tomography"

We give a general form to the Born–Wolf principle. The Riemannian metric $\mathbf{n}ds$ is called non trapping, if no geodesic curve $\boldsymbol{\gamma}$ can stay in a compact set for infinite time. Fix a positive number ε and consider the domain

$$\Omega_\varepsilon(\mathbf{n}, \mathbf{n}_*) \doteq \left\{p \in D,\ (1-\varepsilon)^{-1}\omega\left|\mathbf{n}(p) - \mathbf{n}_*(p)\right| \leq |\xi| \right.$$
$$\left. \leq (1-\varepsilon)\,\omega\,(\mathbf{n}(p) + \mathbf{n}_*(p))\right\}$$

in the phase space $X \times X^*$.

Theorem 1. *Suppose that the metrics $\mathbf{n}ds, \mathbf{n}_*ds \in C^5(X)$ are nontrapping and coincide in $X \backslash D$. Moreover we assume that for an arbitrary point $p \in D$ arbitrary geodesic curves γ, γ_* through p such that $|\sin(\gamma, \gamma_*)| \geq \varepsilon$ have no other common points in D. Then for arbitrary $\omega \geq \omega_0$ the following inequality holds*

$$\max_{\lambda \in \Omega_\varepsilon(\mathbf{n}, \mathbf{n}_*)} |\langle \mathbf{n} - \mathbf{n}_*, \mathbf{e}_\lambda \rangle|^2 \leq C_\varepsilon \omega^{-1} \sup \left| \int_{\partial D} (u_* \partial_\nu u - u \partial_\nu u_*) ds \right|^2 + \omega^{-1/2} R_\varepsilon \quad (4)$$

where the constants C_ε, R_ε do not depend on ω and the supremum is taken over a set of solutions u of (3) and u_* of (3) of unit L_2-norm in D.

We do not assume smallness of the deviation $\mathbf{n} - \mathbf{n}_*$. The supremum in the right side can be taken over only the set of solutions of Fock–Leontovich type: a solution of this type is concentrated to a geodesic curve as $\omega \to \infty$. The set $\Omega_\varepsilon(\mathbf{n}, \mathbf{n}_*)$ looks like that in Fig. 5 It is equal to the union of frequency rings of the exterior radius

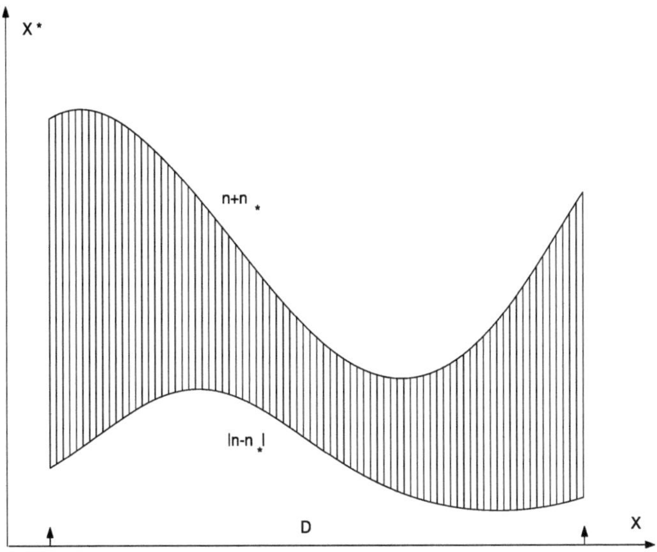

Fig. 5.

$\rho(p) \doteq (\mathbf{n}(p) + \mathbf{n}_*(p))\,\omega$ for $p \in D$. If $\mathbf{n}_*$ is constant and $\mathbf{n}$ is close to $\mathbf{n}_*$, this set is close to the product $D \times \{|\xi| \leq 2\mathbf{n}_*\omega\}$ as in the Born–Wolf principle. Therefore (4) is a rigorous form of the "basic theorem" for an inhomogeneous medium. The condition that the metric $\mathbf{n}_* ds$ is non-trapping in D imposes a limitation on $\mathbf{n}_*$. This situation is similar to that in Abel's result, where the trapped geometry makes a region inaccessible for reconstruction, see Fig. 1.

References

[1] N. H. Abel (1823): *Solution de quelques problèmes à l'aide d'intégrales définies*, Magazin for Naturvidenskaberne, Aargang I, Band 2, Christiania.

[2] N. H. Abel (1826): *Résolution d'un problème de mécanique*, J. Reine Angew. Math. **I**, 13–18, Berlin.

[3] M. Born, E. Wolf (1999): Principles of Optics, Cambridge University Press.

[4] P. Funk (1916): *Über eine geometrische Anwendung der Abelschen Integralgleichung,* Math. Ann. **77**, 129–135.

[5] J. Hadamard (1932): Le problème de Cauchy et les équations aux dérivées partielles linéaires hyperboliques, Hermann, Paris.

[6] V. G. Herglotz (1905): *Über der Elastizität der Erde bei Berücksichtigung ihrer variablen Dichte*, Journal für Math. Phys. **52**, 257–299.

[7] Ch. Huygens (1673): Horologium oscillatorium sive de motu pendulorum, Paris.

[8] J. B. Keller (1976): *Inverse problems,* Amer. Math. Monthly **83**, 107–118.

[9] A. Nachman (1988): *Reconstruction from boundary measurements,* Ann. of Math. **128**, 531–576.

[10] V. P. Palamodov (2002): *Impedance tomography, inverse scattering and phase space analysis,* Partial Differential Equations, Transl. of AMS **206**, Providence RI, 2002 pp.177–192.

[11] J. Radon (1917): *Über Bestimmung von Funktionen durch ihre Integralwerte längs gewisser Mannigfaltigkeiten,* Ber. Verh. Sächs. Akad. Wiss. Leipzig, Math. Nat. Kl. **69**, 262–277.

[12] G. Uhlmann (1992): *Inverse boundary value problems and applications*, Astérisque **207**, 153–221.

[13] E. Wiechert, K. Zöppritz (1907): *Über Erdbebenwellen,* Nachr. Königl. Gesellschaft Wiss. Göttingen **4**, 415–549.

Residues and $\mathcal{D}$-modules

Jan-Erik Björk

Contents

The work *Mémoire sur une propriété génerale d'une classe très étendue de fonctions trancendantes* by Niels Henrik Abel started a new era where geometry, algebra and complex analysis are brought together. It is remarkable that Abel already in 1826 described the process of integrating algebraic functions over cycles – and more generally over chains – which eliminate some variables while the function of the remaining variables has a specific transcendental nature. For example, the conclusive result from 1965 by N. Nilsson in [81] was already suggested in Abel's work. Multi-dimensional residue theory is a subject which also has developed in the spirit of Abel. This article describes how regular holonomic $\mathcal{D}_X$-modules can be realised by distributions satisfying regular holonomic systems on a complex manifold X. Such distributions emerge from Nilsson class functions, i.e., multi-valued analytic functions defined in the complement of a hypersurface T with finite determination and moderate growth along T. Residues and $\mathcal{D}$-modules are treated in Sects. 2–6. Section 1 describes historic examples, foremost related to work by Niels Henrik Abel. Let us finish by citing C. H. Hermite from [9, p. 8]:

> *Abel avait fait la découverte capitale qu'une somme d'un nombre quelconque d'intégrales à limites arbitraires, de la mème fonction algébrique, s'exprime par un nombre fixe d'intégrales semblables auxquelles s'ajoute une quantité algébrique et logarithmique.*

Remark. The work in [50] and [30] provide the source for residue theory in this article. Contributions by the late Miguel Herrera are therefore significant. Prior to this, currents on complex manifolds were put forward in [109], [73] and [33]. See also [34] for a historic account. In the algebraic framework the seminar notes *Residues et Dualité* by Grothendieck from 1963 paved the way to the theory in this article. The GAGA principle in [110] gives a bridge between the algebraic and complex analytic results on quasi-projective manifolds over **C**. The *Desingularisation Theorem* due to Hironaka is essential. The mere existence of principal value currents require Hironaka's result. Chapter 3 and 6 expose work from [48] and [30]. The material in Chaps. 4 and 5 is mainly due to D. Barlet and M. Kashiwara. The text-books [27], [22], [23], [78], [93] and [105] cover results in $\mathcal{D}$-module theory which is used in the chapters.

Acknowledgements. I want to thank M. Andersson, A. Dickenstein, P. Dolbeault, T. Ekedahl, M. Passare, A. Tsikh and A. Yger for discussions which has inspired and improved the presentation.

Related work. The text-books [6], [19] and [113] contain applications of complex residue calculus to division theorems and integral formulas. To avoid confusion we mention that this article does not treat currents which emerge from plurisubharmonic functions in a potential-theoretic framework. Thus, the theory about positive currents in [7], [69], [107] is not discussed here. Among other subjects which are not treated in this article we mention the following. Kernels which yield integral formulas such as the piecewise Cauchy–Fantappie formula in [48] and the Abel–Radon transform in [38], [47] and [49]. Feynman integrals yield regular holonomic systems where residue currents occur. See [66] and [68]. Residues related to Hodge theory is not discussed since this would require extensive discussions. Micro-local analysis does not appear. But let us mention that the residue currents which are constructed in this article yield microfunctions on the cotangent bundle. Inspired by PDE-theory where Fourier integral operators appear, it is tempting to employ micro-localisations in the complex analytic case and study microfunctions satisfying regular holonomic systems with respect to the sheaf $\mathcal{E}_X$ of micro-local differential operators. The major advantage of the micro-local approach is that quantized contact transformations makes it possible to transform systems in a flexible way. See [5] for situations where micro-local methods are used.

1 Historic Examples

1.1 Introduction. Abel's work [4] which proves that the general algebraic equation of degree ≥ 5 cannot be solved by radicals is discussed in 1.3. In [1] integrals of algebraic functions which depend on parameters occur. Even with all sophisticated tools in contemporary mathematics, it appears that Abel's methods still serve as an inspiration for future work.

A theorem of Abel. In the first section of [1] Abel announces the following result:

> *Une fonction dont la derivé est rationelle a, comme on le sait, la propriété qu'on peut exprimer la somme d'un nombre quelconque de semblables fonctions par une fonction algébrique et logarithmique, quelles que soient d'ailleurs les variables de ces fonctions.*

In 1826 the notion of distributions did not exist to express this result. In the context of contemporary mathematics we first consider the following situation: Let $\mu_1, \dots, \mu_n$ be an n-tuple of distributions in $\mathbf{C}^n$ satisfying the integrability condition:

$$\partial\mu_j/\partial x_k = \partial\mu_k/\partial x_j \colon \ 1 \leq j, k \leq n \,. \tag{1.1}$$

Then there exists a distribution ν such that $\partial\nu/\partial x_j = \mu_j$ for every j. To prove this one solves a Pfaffian system. Alternatively, ν is obtained as follows: The holomorphic de Rham complex of distributions in $\mathbf{C}^n$ is a resolution of the anti-holomorphic functions. The Cech Cohomology group of order one with values in the sheaf of anti-holomorphic functions is zero on the Stein manifold $\mathbf{C}^n$. Hence ν exists and is unique up to an anti-holomorphic function.

Abel's case. Here $\mu_1, \dots, \mu_n$ are principal value distributions with poles contained in some algebraic hypersurface $P^{-1}(0)$. Let $P_1, \dots, P_r$ be the irreducible factors of P. Abel's theorem asserts that a distribution solution ν exists where

$$\nu = \mathrm{VP}[R] + \sum_{i=1}^{r} c_i \cdot \mathrm{Log}|P_i|^2 \,.$$

Here R is a rational function with poles contained in $P^{-1}(0)$ and VP[R] its principal value distribution. The $\mathbf{C}$-linear combination of the locally integrable Log-functions expresses the obstruction to solve (1.1) by a principal value distribution. The cohomology group of degree one in the logarithmic subcomplex of the holomorphic de Rham complex of rational functions with poles contained in $P^{-1}(0)$ is therefore described by Abel's discovery!

Divisors and multiplicative forms. We continue to expose Sect. 1 in [1] which illustrates the fertility of Abel's work. Below we follow notations from [1]. Let $(x, y, a_1, \dots, a_m)$ be coordinates in $\mathbf{C}^{m+2}$ and consider an irreducible polynomial

$$P(x, y, a) = y^n + p_{n-1}(x, a)y^{n-1} + \ldots + p_0(x, a) \,.$$

Outside a hypersurface in the (x, a)-space there exists the n-tuple of distinct roots $\{y^{(j)}(x, a)\}$. Let $\Theta = y^{n-1} + q_{n-2}(x, a)y^{n-2} + \ldots + q_0(x, a)$ be another polynomial. The symmetric product below is introduced in [1, p. 147]

$$R(x, a) = \prod_{j=1}^{n} \Theta\left(y^{(j)}(x, a), x, a\right) \,. \tag{1.2}$$

This is a polynomial, i.e., in Abel's words: ... *comme on sait par la théorie des équations algébriques, une fonction rationelle et entière de x et des quantités* $a_1, a_2, \ldots$. Take out the common x-factors, i.e., put $R = F_0(x) \cdot F(x, a)$ where $F_0(x)$ depends on x only. Outside an algebraic hypersurface in the a-space there exists a μ-tuple of distinct root functions $x_1(a), \ldots, x_\mu(a)$ to the equation $F(x, a) = 0$. Then there exists an algebraic hypersurface S in the a-space where the $n \cdot \mu$-tuple $\{y^{(j)}(x_k(a))\}$ yield local branches of multi-valued analytic functions in $\mathbf{C}^m \setminus S$. In a simply connected set $U \subset \mathbf{C}^m \setminus S$ we find single-valued branches of these functions which will be permuted with each other under analytic continuation. Next, consider to each $1 \leq k \leq \mu$ the equation

$$\Theta(y, x_k(a), a) = 0\,. \tag{1.3}$$

By the previous constructions there exists at least one index $1 \leq j \leq n$ such that (1.3) has a root $y^{(j)}(x_k(a), a)$. Following Sect. 2 in [1] we choose indices $j_1, \ldots, j_\mu$ so that $\{y^{(j_k)}(x_k(a))\}$ solve (1.3). Put $\nu_k(a) = y^{(j_k)}(x_k(a))$. These ν-functions belong to $\mathcal{O}(U)$ when $U \subset \mathbf{C}^m \setminus S$ is simply connected. In Sect. 2.8 Abel introduces the differential 1-form:

$$dV = \sum_{k=1}^{\mu} \nu_k(a) \cdot dx_k(a)\,.$$

Notice that single-valued branches exist as above, but multi-valuedness remains in general even though a symmetrization occurs in the family $\{dx_k(a)\}$. Various constructions are suggested in 2.12 in [1]. For example, dV can be integrated over chains or cycles which may be parametrized by rational functions of certain a-variables. The principal aim in [1] is to express functions which arise in this way. Consider as an example the case where $\mathbf{c} = \{0 \leq a_1 \leq 1 \text{ and } a_k = \alpha_k :\ 2 \leq k \leq m\}$ is a 1-chain in $\mathbf{C}^m \setminus S$. Then

$$\phi(a_2, \ldots, a_m) = \int_{\mathbf{c}} dV \tag{1.4}$$

is a germ of an analytic function at $(\alpha_2 \ldots, \alpha_m) \in \mathbf{C}^{m-1}$. This ϕ-function extends to a multi-valued function in $\mathbf{C}^{m-1} \setminus W$ for some algebraic hypersurface W. More precisely, Abel realised that the ϕ-function is of the form

$$\phi = \tau + \gamma \cdot \mathrm{Log}(\rho)\,, \tag{1.5}$$

where τ, γ, ρ are algebraic functions of the variables $a_2, \ldots, a_m$, i.e., they are rational functions of multi-valued root functions to algebraic equations in $a_2, \ldots, a_m$ and W is the union of the discriminant locus of each of these equations. The trancendental logarithm $\mathrm{Log}(\rho)$ appears since $\mathbf{c}$ is a chain. If $\mathbf{c}$ is a cycle in (1.4) then no logarithm would occur in (1.5). In (1.5) we integrated over a 1-chain. More generally, several a-variables can be eliminated by integration over chains of dimension ≥ 2. This process is described in Chap. 5.

Algebraic root functions. Section 6 in [1] gives a construction to eliminate roots in algebraic equations with parameters. On page 170 Abel considers a polynomial

$$F(x, a_1, \dots, a_\alpha) = x^\mu + p_{\mu-1}(a) \cdot x^{\mu-1} + \dots + p_0(a) = \prod_{i=1}^{\mu} (x - x_\nu(a)),$$

where $\{p_\nu(a)\}$ are polynomials with $a \in \mathbf{C}^\alpha$ and $\alpha < \mu$, i.e., the dimension of parameters is strictly smaller than the degree of the x-polynomial. In a suitable open and simply connected set Ω outside the discrimnant locus there exists a k-tuple of roots – say $x_1, \dots, x_\alpha$ – where the Jacobian of the $\alpha \times \alpha$-matrix $(\partial x_\nu / \partial a_j)$ is $\neq 0$. So in Ω every a-coordinate is a holomorphic function of $x_1, \dots, x_\alpha$. To every $\alpha + 1 \leq k \leq \mu$ we then obtain an algebraic function $\{\phi_k(x_1, \dots, x_\alpha)\}$ such that

$$x_k(a) = \phi_k(x_1(a), \dots, x_\alpha(a)).$$

In the final part of [1] Abel investigates properties of differentials which emerge from the construction above. Even with tools from contemporary mathematics it appears to be a challenge to pursue the course of Abel's original constructions.

Logarithmic derivatives. In [2] Abel considers a pair of polynomials ρ, P and studies the primitive function

$$\int \frac{\rho \cdot dx}{\sqrt{P}}. \tag{1.6}$$

The aim is to find conditions in order that this meromorphic differential on the Riemann surface $\{y^2 = P(x)\}$ is a logarithmic differential, i.e., equal to $\frac{\partial f}{f}$ for some meromorphic function f. Actually Abel posed the following problem: Assume that $P(x)$ has some even degree $2+2p$ with simple zeros. When does there exist a polynomial ρ such that (1.6) yields a logarithmic differential. Abel's question is discussed in [9, p. 354–358] since it is related to the problem of finding the minimal number of logarithmic functions expressing an abelian integral, provided such a representation exists. Abel solved the problem about integrals in (1.6) completely. His result is expressed as follows in [9, p. 358]:

Pour qu'il existe un polynome ρ correspondant à la question, il faut et il suffit que le dévelopment de $\sqrt{P}$ en fraction continue algébrique soit périodique

No essential simplification of Abel's original proof exists. See also T. Ekedahl's article in this volume for a study of moduli related to (1.6).

Motion of rigid bodies. Given two polynomials P, Q we consider the integral

$$J(x) = \int Q(x) \cdot \frac{dx}{P^\alpha(x)}, \tag{1.7}$$

where α is a real irrational number or complex with $\mathfrak{Im}(\alpha) \neq 0$. In [53, pp. 82–85] specific expressions of the J-function are found under the hypothesis that $J(x)$ belongs to the algebraic function field generated by x and P^α. The study of (1.7) is the essential ingredient in the proof of the famous theorem – originally due to Liapunov and Poincaré – which asserts that when a rigid body K rotates around a fixed point p different from its center of mass $\mathbf{o}$ and gravity is the sole external force, then the equations of motion cannot be solved by quadrature when the ellipsoid of inertia $\mathcal{E}$ with respect to p has no symmetry, i.e., the quadratic momentum of the principal axes are all different. When $\mathcal{E}$ has a plane of symmetry, properties of J-integrals where α is rational were used in [53] to prove that the rigid body found by S. Kovalevsky in [71] is the only case where the equations of motion can be solved by quadrature – except for the classical cases by Euler and Lagrange. Kovalevsky's condition is that $B = A/2$ where B is the quadratic momentum of the non-symmetric principal axis and A the momentum of the axes in the plane of symmetry while $\mathbf{o}$ can be placed at any point in the plane of symmetry. See [24] for an account about the Kovalevsky gyroscope.

1.2 Integrals on Chains. Let $P(x, t) = t^e + p_{e-1}(x)t^{e-1} + \ldots + p_0(x)$ be a polynomial without multiple factors. Here $x \in \mathbf{C}^n$ are parameters. Let $P_1, \ldots, P_r$ be the irreducible factors of P and $\Delta(x) = \prod_{i \neq \nu} (\alpha_i(x) - \alpha_\nu(x))$ the discriminant where

$$P(x, t) = \prod (t - \alpha_\nu(x))$$

Consider the hypersurface $S = \{P(x, 0) \cdot P(x, 1) \cdot \Delta(x) = 0\}$ in $\mathbf{C}^n$. Let $x_0 \in \mathbf{C}^n \setminus S$ and suppose that $\mathfrak{Re}(P_\nu(x_0, t)) > 0$ for every ν and every $0 \leq t \leq 1$. Let $q_1, \ldots, q_r$ be positive rational numbers. For every $1 \leq \nu \leq r$ we choose a single-valued branch of $P_\nu(x, t)^{-1/q_\nu}$ when x is close to x_0 and $0 \leq t \leq 1$. Then

$$\Phi(x) = \int_0^1 \prod P_\nu(x, t)^{-1/q_\nu} \cdot dt \tag{1.8}$$

is holomorphic in a polydisc centered at x_0 which extends to a Nilsson class function in $\mathbf{C}^n \setminus S$ whose monodromy is quasi-unipotent with index of nilpotency ≤ 1.

The case $n = 1$. Here S is a finite subset of $\mathbf{C}$ and Φ has analytic continuation to $\mathbf{C} \setminus \Sigma$, where the inclusion $\Sigma \subset S$ may be strict. If $\beta \in \Sigma$ the local branches of Φ in a small punctured disc $D^*_\beta(\delta)$ are finite sums

$$\sum g_\nu(x) \cdot (x - \beta)^{r_\nu} + f_\nu(x) \cdot (x - \beta)^{r_\nu} \cdot \mathrm{Log}(x - \beta) \tag{1.9}$$

where $\{0 \leq r_\nu < 1\}$ are rational numbers and $\{g_\nu, f_\nu\}$ single-valued meromorphic functions. No higher powers of the Log-function appear, i.e., the index of nilpotency of monodromy is ≤ 1. When $q_1 = \ldots = q_r = 1$ this follows from Newton's formula:

$$P^{-1} = \sum_{\nu=1}^{\nu=e} 1/(t - \alpha_\nu(x)) \cdot P'_t(x, \alpha_\nu(x)) \tag{1.10}$$

which implies that

$$\Phi(x) = \sum_{\nu=1}^{\nu=e} 1/P_t'(x, \alpha_\nu(x)) \cdot \mathrm{Log}(1/\alpha_\nu(x) - 1)$$

The root functions are permuted under analytic continuation and the local expression in (1.9) follows. Moreover, (1.9) with $n = 1$ and arbitrary rational numbers $r_1, \dots, r_k$, can be reduced to Newton's case, i.e., pass to the Riemann surface Y of $\prod P^{1/r_\nu}$. Under the ramified map with finite fibers from Y onto the x-plane Log-functions on Y yield Log-functions on X and then (1.9) follows for every r-tuple of rational numbers.

Local residues. A superficial glance at Abel's work may give the impression that his intense search of more or less explicit expressions of functions does not bear much relation to contemporary mathematics. However, the situation is rather the contrary thanks to recent advancement such as the Riemann–Hilbert correspondence. Let us give an example starting from another situation in the spirit of Abel: Let $T = P^{-1}(0)$ be an algebraic hypersurface in $\mathbf{C}^n$ and R a rational function with poles contained in T. We seek conditions in order that there exists an n-tuple of rational functions $\phi_1, \dots, \phi_n$ such that

$$R = \sum_{\nu=1}^{n} \frac{\partial \phi_\nu}{\partial x_\nu} \tag{1.11}$$

In contemporary mathematics one settles this as follows. The obstruction to solve (1.11) is the image of $R \cdot dX$ in the cohomology group of degree n in the holomorphic de Rham complex of $\mathbf{C}[x, P^{-1}]$. By Grothendieck's comparison theorem (1.11) has a solution if and only the rational n-form $R \cdot dX$ has zero image in $H^n_{\mathrm{Cech}}(\mathbf{C}^n \setminus P^{-1}(0), \mathbf{C})$. To analyze when this image is zero we consider the finite set of points x_0 which belong to the the support of the local cohomology sheaf $\mathcal{H}^{n+1}_T(\mathbf{C})$. Let $\Sigma \subset T$ be this support. Let $x_0 \in \Sigma$ and $B_\epsilon(x_0)$ be the open ball of radius ϵ centered at x_0. If $\epsilon > 0$ is small, one has:

$$H^n_{\mathrm{Cech}}(B_\epsilon(x_0) \setminus T, \mathbf{C}) \simeq \mathcal{H}^{n+1}_T(\mathbf{C})$$

By Poincaré duality, (1.11) has a solution with ϕ-functions given as germs of meromorphic functions at x_0 if and only if

$$\int_\sigma R = 0 \tag{1.12}$$

for every n-cycle σ in $B_\epsilon(x_0) \setminus T$. Moreover, (1.12) holds at every $x_0 \in \Sigma$ if and only if (1.11) has a solution with rational ϕ-functions. Although this answers the Question, there remains to construct "minimal" complex vector spaces $\mathbf{V}$ in $\mathbf{C}[x, P^{-1}]$ whose images generate $H^n_{\mathrm{Cech}}(\mathbf{C}^n \setminus P^{-1}(0), \mathbf{C})$. No canonical procedure is known, i.e., computations must be made for each given polynomial P. So for problems of this kind one does not exaggerate in saying that Abel's mastery in performing computations would be of help in contemporary mathematics.

1.3 Algebraic Equations. The problem to decide whether an algebraic equation of degree five can be solved by radicals was studied before Abel's work [4]. In 1767 Johann Lambert expressed the solution to the equation

$$y^5 + x \cdot y - 1 = 0 \tag{1.13}$$

where x is a parameter by a series $y(x)$ which later was found to represent a generalised hypergeometric function. The Swedish mathematician Bring proved in 1786 that solutions to the general 5th degree equation can be reduced to (1.13) by radicals and roots. See [29] and F. Klein's account about Bring-transformations of algebraic equations in [70]. The formula below comes from [113]:

Theorem 1.1. *Let $\Gamma(z)$ be the usual Gamma-function. For every real number $0 < x < 1$ the unique real root $0 < y(x) < 1$ of (1.13) is given by*

$$y(x) = \frac{1}{10\pi \cdot i} \cdot \int_{Re(\zeta)=1/2} \frac{\Gamma(\zeta) \cdot \Gamma\left(\frac{1-\zeta}{5}\right)}{\Gamma\left(\frac{1+\zeta}{5} + 1\right)} \cdot x^{-\zeta} \cdot d\zeta$$

The series of Lambert is found when $y(x)$ is expanded around $x = 0$ using the branch with $y(0) = 1$. Here $y(x) = 1 - \frac{x}{5} - \frac{x^2}{25} + \frac{x^3}{125 \cdot 5} + \ldots$ where the coefficients of the series are rational numbers found by an obvious recursion. To obtain the analytic extension of $y(x)$ we notice that the discriminant locus of the equation (1.13) consists of five points; $\{-5 \cdot 4^{\frac{1}{5}} \cdot e^{\frac{2\pi i \nu}{5}} : 0 \leq \nu \leq 4\}$. The integral in Theorem 1.1 represents the root of (1.13) when $y(0) = 1$. In [113] it is shown that the y-function in Theorem 1.1 extends to a single-valued analytic function of $z = u + iv$ in the domain and $-\pi + \frac{\pi}{5} < \arg(u) < \pi - \frac{\pi}{5}$. Of course, $y(z)$ actually extends to a single valued function a bit further, for example around $z = 0$. But eventually it is multi-valued and new local branches occur when we start from the four other branches of root functions $\{y_\nu(x)\}$ with $y_\nu(0) = e^{2\pi i \cdot \nu/5}$ for $1 \leq \nu \leq 4$.

Abel's proof. Abel visited Copenhagen at an early period in his career. It is not unlikely that he knew about the work [29] which was published at the university of Lund. Whether he was aware of a formula similar to Theorem 1.1 is not known. In any case, the proof in [4] uses – at least indirectly – the transformation of Bring since Abel at the end of his article reduces the study to a single p-function. Let us now turn to his work entitled:

> *Beweis der Unmöglichkeit algebraische Gleichungen von höheren Graden als dem vierten allgemein aufzulösen.*

This work contains four paragraphs, where 1–3 contain general results about algebraic equations while the unsolvability of the 5th degree equation is established in the Sect. 4. In Sect. 1 – *Ueber die allgemein Form algebraischen Funktionen* – Abel introduces the basic operations to study algebraic functions, i.e., he defines the order and the degree for algebraic functions which depend on parameters. The conclusive result in [4, p. 70] is:

Wenn ν eine algbraische Function von der Ordnung μ und Grade m ist, so kann man immer setzen:

$$\nu = q_0 + q_1 \cdot p^{\frac{1}{n}} + \ldots + q_{n-1} \cdot p^{\frac{n-1}{n}} ;$$

n ist ein Primzahl, $q_0, \ldots, q_{n-1}$ sind algebraischen Functionen von der Ordnung μ und höchstens grade $m - 1$, und p ist eine algebraische Function von der Ordnung $\mu - 1$ und von der Art, dass sich $p^{\frac{1}{n}}$ nich durch eine rationale Function von $q_0, \ldots, q_{n-1}$ ausdrücken lässt.

In Sect. 2 – *Eigenschaften der algebraischen Functionen welcher einer gegebenen Gleichung genugen* – Abel studies an equation

$$y^r + c_{r-1} y^{r-1} + \ldots + c_0 = 0 \tag{1.14}$$

where $\{c_\nu\}$ are parameters and the roots in (1.14) are supposed to satisfy another equation:

$$y_\nu(x) = q_0(x) + q_1(x) \cdot p(x)^{\frac{1}{n}} + \ldots + q_{n-1}(x) \cdot p^{\frac{n-1}{n}}(x)$$

This *Ansatz* comes from the hypothesis that (1.14) can be solved by radicals. Recall that Gauss used a similar device with $n = 2$ when he proved that the regular polygon with seven sides cannot be constructed by compass and ruler. The material in 2 leads to the conclusive result at the end of this section:

Wenn eine Gleichung algebraisch auflösbar ist, so kann man der Wurzel allezeit eine solche Form geben, dass sich alle algebraische Functionen; aus welchen sie zusammengesetzt ist, durch rationale Functionen der Wurzeln der Gegebenen Gleichung ausdrücken lassen.

In Sect. 3 Abel studies a rational function $R(x_1, \ldots, x_n)$ which generically takes μ different values when the x-variables are permuted. Here $R(x)$ satisfies an algebraic equation whose coefficients are functions of the symmetric polynomials in the x-variables. Various special cases are treated at the end of Sect. 3. One result is that if $n = \mu = 5$, then $R(x)$ is of the form

$$\rho + \rho_1 \xi + \ldots + \rho_4 \cdot \xi^4 \tag{1.15}$$

where the ρ-functions are symmetric functions of the x-variables and ξ a linear form of the x-variables. Of course, by modern algebra this is an exercise in the theory of field extensions but it is of interest to pursue Abel's constructions. Using (1.15) Abel constructs a certain p-function of a single variable x which is multi-valued and defined outside the five points described after Theorem 1.1, where p takes 120 values determined by ordered tuples of the roots of the 5th degree equation. At the same time, using previous results it is shown that if the 5th order equation always can be solved by roots and radicals, then p satisfies an algebraic equation of degree 10. After the formula $(p^5 - u)^2 = (u')^2 \cdot s^2$ in [4, p. 84], Abel concludes:

Diese Gleichung gibt, wie man sieht, p durch eine Gleichung von 10ten *Grade deren Coefficienten sämmtlich symmetrische Funktionen sind, welcher aber zufolge des letzten Lehrsatzes im voriger Paragraph, nicht möglich ist; denn da*

$$p = x_1 + \alpha^4 \cdot x_2 + \alpha^3 \cdot x_3 + \alpha^2 \cdot x_4 + \alpha \cdot x_5$$

ist, so würde p, 120 verschiedene Werthe haben; welches ein Widerspruch ist.

Remark. The proof by E. Galois a few years later was another great discovery. In the case where x is the sole parameter in the equation $y^5 + xy - 1 = 0$, the Galois resolvent admits an explicit representation which is used to determine when this equation can be solved by radicals. Whether this precise result can be recovered from Abel's proof is unclear. But in any case, Abel's general constructions and results are of lasting value. For example, in residue theory trace maps which yield direct images of currents emerge from the constructions in Sect. 1 and Sect. 3 of [1].

Algebraic equations with parameters. We describe two examples and refer to [112] for further material about algebraic equations in parametric form.

The differential system by Mayr. Let $n \geq 2$ and consider an equation

$$p(x) = a_n \cdot x^n + a_{n-1} \cdot x^{n-1} + \ldots + a_0 = 0 \tag{1.16}$$

where $a = (a_0, \ldots, a_n) \in \mathbf{C}^{n+1}$. In the Weyl algebra $A_{n+1}(\mathbf{C})$ of differential operators with polynomial coefficients in the a-space we consider the left ideal $\mathcal{L}$ generated by:

(i) $\partial/\partial a_j \cdot \partial/\partial a_k - \partial/\partial a_s \cdot \partial/\partial a_\ell : \; j + k = s + \ell$
(ii) The Euler field $\sum_{i=0}^n a_i \cdot \partial/\partial a_i$
(iii) The vector field $\sum_{i=0}^n i \cdot \partial/\partial a_i + 1$

Every local solution to (1.16) outside the discriminant locus S is a solution to the homogeneous system defined by $\mathcal{L}$. In [40] it is proved that the left $A_{n+1}(\mathbf{C})$-module $A_{n+1}/\mathcal{L}$ is regular holonomic. This was used in [40] to study hypergeometric functions. Mayr's module $A_{n+1}/\mathcal{L}$ enjoys nice properties. For example, its Fourier transform is again regular holonomic. See [76] for the construction of Fourier transforms of modules over Weyl algebras.

The Horn–Mellin system. Another differential system occurs in [80]. Here one considers the algebraic equation

$$y^{n+1} + x_n \cdot y^n + \ldots + x_1 \cdot y_1 - 1 = 0 \tag{1.17}$$

with $x = (x_1, \ldots, x_n) \in \mathbf{C}^n$. In the Weyl algebra $A_n(\mathbf{C})$ there exists the Fuchsian vector fields $\theta_i = x_i \cdot \partial/\partial x_i$. The Horn–Mellin system is A_n/L where L is the left ideal generated by the following n-tuple:

$$P_i = (-1)^{n-i} \cdot (n+1)^{n+1} \cdot \partial^{n+1}/\partial x_i^{n+1} -$$

$$\prod_{j=0}^{n+1-i} [n\theta_1 + (n-1)\theta_2 + \ldots + \theta_n + 1 + (n+1)j] \prod_{j=0}^{i} [\theta_1 + 2\theta_2 + \ldots + n\theta_n - 1 + (n+1)j]$$

These $(n+1)$th order differential operators commute and generate a left ideal L where A_n/L is regular holonomic. In [80] it was remarked that every solution to (1.16) satisfies the homogeneous system defined by L. The characteristic variety $SS(A_n/L)$ is the union of co-normals to $n+1$ hypersurfaces, one of which is the discriminant locus of the equation (1.16). The remaining hypersurfaces are described in [102].

Epilogue. In his mathematical autobiography [72], S. Lefschetz writes on p. 14: *In the early phase (Abel, Riemann, Weierstrass) algebraic geometry was just a chapter in analytic function-theory*. Abel's early decease means that we can only speculate over the aiming of a work such as [1]. Most likely it was an important issue for him to establish explicit function-theoretic formulas. For example, during the years of Abel's career considerable attention was given to problems in mechanics. A typical problem was to describe force-fields which yield tautochronous motions. J. Lagrange found solutions to this problem around 1780 which rely on periodic behaviour of certain integrals. Inversion formulas to determine the time interval of isochrounous motions was an issue which attracted Abel. His work [3] may be regarded as a first step in this direction. For an account about Lagrange's work we refer to Chap. 10 in [98].

Abel's original constructions remain as a veritable challenge in contemporary mathematics. Volume two of *Théorie des Fonctions Algébriques* by P. Appel and E. Goursat contains material related to Abel's work. Now we describe some results in the spirit of Abel. Most likely he would have appreciated Theorem 1.2 below.

1.4 Jacobi's Integral and the Multiplicity Formula. In [55] Jacobi considered two polynomials P, Q in $\mathbf{C}^2$ in complete intersection and studied the integral

$$\int_\epsilon A(x, y) \cdot dxdy$$

when $A(x, y)$ is a polynomial and the integration is on $\{|P| = |Q| = \epsilon\}$. This is a residue in two variables. More generally, let $F = (f_1, \ldots, f_n)$ be an n-tuple in the local ring $\mathcal{O}_n$ of germs of holomorphic functions at the origin in $\mathbf{C}^n$ in a complete intersection, i.e., the origin is the common zero. Jacobi's residue current of this n-tuple is defined below. But first we study local multiplicities. Then $\mathcal{O}_n/(f_1, \ldots, f_n)$ is a local artinian ring of some length $\ell(F)$. The geometric multiplicity is the number of points in the generic fiber of the map $x \to w$, where $w_i = f_i(x)$ for every i. This positive integer is denoted by $\mu(F)$.

Theorem 1.2. *For every complete intersection one has* $\ell(F) = \mu(F)$.

This result is then used to prove that if $e_1(x), \dots, e_{\mu(F)}(x)$ are germs in $\mathcal{O}_n$ whose images yield a basis in the complex vector space $\mathcal{O}_n/(f_1, \dots, f_n)$, then

$$\mathcal{O}_n = \oplus \mathcal{A} \cdot e_\nu \tag{1.18}$$

where $\mathcal{A}$ is the subalgebra of germs which are expanded in the f-functions, i.e., germs of the form $\sum c_\alpha \cdot f(x)^\alpha$.

Remark. (1.18) without uniqueness with $\oplus$ replaced by Σ was proved by C. Houzel in a Bourbaki seminar in 1960. The equality in Theorem 1.2 is due to Palamodov in [85], who also established (1.18) using division theorems of the Weierstrass type in [84]. A more recent proof of Theorem 1.2 occurs in [113] where the essential ingredient is E. Bishop's integral formula in [21].

Jacobi's residue current. Let $F = (f_1, \dots, f_n)$ be a complete intersection. Consider the linear form defined on holomorphic n-forms $\phi \cdot dX$ by

$$\mathcal{J}_F(\phi \cdot dX) = (2\pi i)^{-n} \cdot \int_{\Gamma(\epsilon_\bullet)} \frac{\phi \cdot dX}{f_1 \cdots f_n} \tag{1.19}$$

where the integral is taken over the set $\Gamma(\epsilon_\bullet) = |f_1| = \epsilon_1 \cap \dots \cap |f_n| = \epsilon_n$. Let us clarify why (1.19) exists and is independent of the n-tuple $\epsilon_\bullet$. The functions $\{f_\nu\}$ are holomorphic in some polydisc D^n. Consider the map $\rho\colon D^n \to \mathbf{R}_+^n$ where $\rho(x) = (|f_1(x)|, \dots, |f_n(x)|)$. The critical values of ρ is some real-analytic set $\Sigma \subset R_+^n$ of positive codimension. If the n-tuple $\epsilon_\bullet$ is outside Σ, then $\Gamma(\epsilon_\bullet)$ is an n-dimensional real manifold whose positive orientation is determined by the n-form $d|f_1|^2 \wedge \dots, \wedge d|f_n|^2$. When $\epsilon_\bullet \in \Sigma$ it is still possible to integrate on $\Gamma(\epsilon)$, i.e., by Lojasiewicz inequality the n-dimensional volume of $\Gamma(\epsilon_\bullet)$ is locally finite and integration is performed on its regular part by the Whitney–Federer theory. Moreover, Stokes theorem is valid by the results in [94]. Since $\phi \cdot dX$ is closed the integral in (1.19) is independent of $\epsilon_\bullet$.

Jacobi's interchange formula. Put $\partial F = \partial f_1 \wedge \dots \wedge \partial f_n$. Residue calculus yields

$$\mathcal{J}_F(\partial F) = \mu_F \tag{1.20}$$

i.e., the geometric multiplicity is recovered. Next, let $H = (h_1, \dots, h_n)$ be another complete intersection whose ideal is contained in F. Then we get a holomorphic matrix $A(x)$ so that $H = A \cdot F$, i.e., $h_\nu = \sum a_{\nu k} \cdot f_k$.

Theorem 1.3. *For every holomorphic n-form $\phi \cdot dX$ one has:*

$$\mathcal{J}_F(\phi \cdot dX) = \mathcal{J}_H(det(A) \cdot \phi \cdot dX)\,.$$

This transformation law is proved in a purely algebraic way in [46]. See also [113, p. 41]. It is then used to compute $\mathcal{J}_F$. Namely, there is a positive integer w_0 such that for every $w \geq w_0$ the ideal $\mathbf{x}^{w+1} = (x_1^{w+1}, \dots, x_n^{w+1})$ is contained in the ideal F.

For every such w we get a holomorphic matrix $A(x)$ such that $\mathbf{x}^{w+1} = A(x) \cdot F$. Cauchy's residue formula gives:

$$\mathcal{G}_F(\phi \cdot dX) = (w!)^{-n} \cdot \partial_1^w \cdots \partial_n^w((\det(A) \cdot \phi))(0) \tag{1.21}$$

Leibniz's rule gives a differential operator with constant coefficients $Q = \sum c_\alpha \cdot \partial^\alpha$ such that

$$\mathcal{G}_F(\phi \cdot dX) = Q(\partial)(\phi)(0)$$

One refers to Q as a noetherian operator of the ideal $(f_1, \ldots, f_n)$. Using Gröbner bases one can obtain Q. See [83]. Integral formulas of the Weil-type can be established using the constructions above. See [113]. The residue $\mathcal{G}_F$ was constructed in a purely algebraic way by Grothendieck. Coleff–Herrera residue currents arise for complete intersections with fewer functions than n which is the main topic in Chap. 4.

1.5 Period Integrals. To clarify the link between the residue theory exposed in this article with differential forms and integrals on algebraic varieties would require an extensive discussion. The branch of algebraic geometry which deals with intersection theory using differential forms and integrals, starting from work by Brill and M. Noether around 1875, followed by the Italian school (Castelnuovo, Enriques around 1890) and by E. Picard (1880–1900), is a subject whose refined results cannot be recovered by contemporary residue theory. However, integrals – say of the first, second or the third kind – which emerge from a meromorphic 1-form on an algebraic surface can be analyzed by residue theory. In fact, many properties of such integrals are related to residue formulas when $\bar{\partial}$-operators are applied to the differential forms. A systematic use of currents and $\mathcal{D}$-module theory may lead to an improved understanding of results presented in a more classical context. We refer to S. Lefschetz in his mathematical autobiography [72] for historic comments and his description of results prior to 1970 which have inspired the residue theory to be exposed in the subsequent chapters.

Example. Recall a construction due to Picard. Let X be a closed Riemann surface, i.e., a compact 1-dimensional complex manifold. Let Σ be a finite subset and consider the integral homology group $H_1(X \setminus \Sigma, \mathbf{Z})$. Let α be a meromorphic 1-form on X with poles $\subset \Sigma$. Let $\mathbf{c}$ be a 1-cycle in $X \setminus \Sigma$ representing an element $\{\mathbf{c}\}$ in this homology group. The integral

$$\int_{\mathbf{c}} \alpha \tag{1.22}$$

depends on $\{\mathbf{c}\}$ only. When $\mathbf{c} = \partial\gamma$ for some 2-chain in X, (1.22) is computed by a residue formula determined by the current $\bar{\partial}\alpha$ which is obtained via the principal value extension of α. See Sect. 3. This construction was also made on surfaces by Picard and extended to higher dimensions by Lefschetz. Here X is a projective manifold of some dimension n and $S \subset X$ a hypersurface. Let α be a meromorphic k-form with poles contained in S. If $\mathbf{c}$ is some k-cycle in $X \setminus S$ representing an element in $H_k(X \setminus S, \mathbf{Z})$ one refers to $\int_{\mathbf{c}} \alpha$ as a period integral.

Prior to Hironka's Desingularisation Theorem, algebraic geometers implicitly supposed that varieties emerged from smooth hyperplane sections. Hironaka's work – inspired by O. Zariski's lectures at Kyoto university 1955 – created a new era where constructions no longer had to be restricted to varieties with modest singularities. Period integrals for general hypersurfaces were studied in [42], [43].

The Gauss–Manin connection. Consider the local analytic case. A holomorphic function $f(x)$ is defined in some neighbourhood of the origin in $\mathbf{C}^n$ with $f(0) = 0$. For pairs $0 < \delta << \epsilon << 1$ the local Milnor fibration exists, i.e., if $B(\epsilon, \delta) = \{|x|^2 < \epsilon \cap 0 < |f| < \delta\}$ then $f\colon B(\epsilon, \delta) \to D^*(\delta)$ is a locally trivial map. To every $1 \leq p \leq n-1$ we consider the homology group $H_p(X(s), \mathbf{C})$ of the Milnor fiber $X(s) = B(\epsilon) \cap f^{-1}(s)$. Let $\mathbf{c}_1(s), \ldots, \mathbf{c}_m(s)$ be a basis. If α is a meromorphic p-form with poles $\subset f^{-1}(0)$ we get the functions

$$\phi_i(s) = \int_{\mathbf{c}_i(s)} \alpha \tag{1.23}$$

One starts to produce this m-tuple at a point $s_0 \in D^*(\delta)$. The ϕ-functions extend to multi-valued analytic functions of finite determination in the punctured disc $D^*(\delta)$ with moderate growth at the origin, i.e., they belong to the Nilsson class. The period integrals determine the monodromy of $s \to H_p(X(s))$. The *Positivity Theorem* asserts that (1.23) tends to zero when α is a holomorphic p-form and $1 \leq p \leq n-1$. Another result relates the roots of the Bernstein–Sato polynomial b_f to the eigenvalues of the local monodromy of the Milnor fibration. More precisely, let $q_1, \ldots, q_s$ be the roots of the b-function. Then the family of roots of unity $\{e^{2\pi i \cdot q_\nu}\}$ is equal to the union of eigenvalues of the mondromy operators in every degree $0 \leq p \leq n-1$. This result is due to B. Malgrange in [74]. See also [23, Theorem 6.3.5] for a proof. Another important result is the invariant cycle theorem established for projective maps in [111]. The proof requires Hodge theory which is essential since the invariant cycle theorem is false in the general case when one allows non-Kählerian complex manifolds. See [42].

Integrals with several functions. The work [45] studies the Gauss–Manin connection for several functions. See also work by Sabbah in [99], [100], [101]. But the case of several variables is more involved. The counter-example by Passare–Tsikh in Chap. 4 shows that asymptotic development of current-valued functions with several functions in general does not exist. However, it is possible to define integrals with several parameters. Consider for example two polynomials P, Q in $\mathbf{C}^n$ in a complete intersection, i.e., $V = P^{-1}(0) \cap Q^{-1}(0)$ has codimension two. Let ϕ be a meromorphic 1-form with polar hypersurface T such that $V \cap T$ has codimension 3. Consider the map $\rho\colon x \mapsto s$ with $s_1 = P(x)$ and $s_2 = Q(x)$. Outside the ρ-image of T and the hypersurface $\partial P \wedge \partial Q = 0$ we construct integrals over 1-cycles, i.e., starting from a 1-cycle γ_0 on a non-critical fiber we get a function

$$g(s_1, s_2) = \int_{\gamma(s_1, s_2)} \phi$$

with moving 1-cycles on fibers $\{P = s_1\} \cap \{Q = s_2\}$. This yields a multi-valued function outside a hypersurface in $\mathbf{C}^2$ which belongs to the Nilsson class. But not much is known about more specific properties of such functions.

Global residues. The subsequent chapters expose results on complex manifolds. By [110] they have an algebraic counterpart on quasi-projective manifolds over $\mathbf{C}$. Let us give an example in the algebraic case. Let $V \subset \mathbf{C}^{n+p}$ be an algebraic set of pure dimension n where the projection $\mathrm{pr} : \mathbf{C}^{n+p} \to \mathbf{C}^n$, mapping $(x, t) \mapsto x$, is proper with finite fibers on V, i.e., $\mathrm{pr}^{-1}(x) \cap V = \{(x, t_1), \ldots, (x, t_p)\}$. Let $\Delta(x)$ be the discriminant and put $S = \Delta^{-1}(0)$. To every rational n-form $\phi(x)$ with poles contained in S and every multi-index α there exists a current on $\mathbf{C}^{n+p}$ of bi-degre $(n + p, p)$ defined by

$$[\phi \cdot \partial_t^\alpha]_V(\Psi^{(0,n)}) = \mathrm{Lim}_{\epsilon\to 0} \int_{V\cap|\Delta(x)|>\epsilon} \phi \wedge \partial_t^\alpha(\Psi)\,. \tag{1.24}$$

The existence of this current follows by desingularisation. We can take finite sums over multi-indices $\alpha = (\alpha_1, \ldots, \alpha_p)$. Let $\mathbf{CH}_V[*S]$ be this space of currents. Let $A_{n+p}(\mathbf{C})$ be the Weyl algebra of differential operators with polynomial coefficients.

Definition 1.4. *The set of $\bar\partial$-closed currents in $\mathbf{CH}_V[*S]$ is denoted by $\mathbf{CH}_V$ and is called the space of Coleff-Herrera currents on V.*

Theorem 1.5. $\mathbf{CH}_V$ *is a right module over A_{n+p} and isomorphic to the regular holonomic module $H^p_V(\Omega)$, where Ω is the free $\mathbf{C}[x, t]$-module of $n + p$-forms.*

Theorem 1.5 is a special case of the result in 6.2. and illuminates the material later on. In the projective closure of V in $\mathbf{P}_{n+p}(\mathbf{C})$ a similar sheaf of Coleff-Herrera currents exists. Above we constructed a right module over the Weyl algebra. By left conversion we get Coleff-Herrera currents of bi-degree $(0, p)$. Then there exists a unique extension of this regular holonomic left A_{n+p}-module to $\mathbf{P}_{n+p}(\mathbf{C})$ which is generated as a $\mathcal{D}$-module by global sections. See [27].

2 Abelian Differentials

2.1 Introduction. Singularities on a variety cause the phenomenon that there exist meromorphic forms whose principal value extensions yield $\bar\partial$-closed currents. The case when $T = P^{-1}(0)$ is a hypersurface in $\mathbf{C}^n$ illustrates this. Namely, construct Leray's $n - 1$-from ϕ such that $\phi \wedge \partial P = dX$. Taking a principal extension there exists a $\bar\partial$-closed current of bi-degree $(n, 1)$ defined by

$$[\phi]_T(\Psi^{0,n-1}) = \mathrm{Lim}_{\delta\to 0} \int_{T_\delta(g)} \phi \wedge \Psi^{0,n-1} \tag{2.1}$$

where g is a polynomial whose zero set contains T_{sing} and $T_\delta(g) = T \cap |g| > \delta$. The current $[\phi]_T$ generates the free $\mathcal{O}_T$-module $\mathcal{A}_T$ whose sections are called abelian differentials on T. More generally, let $V \subset \mathbf{C}^{n+p}$ be as in 1.24. Every rational n-form ϕ on V gives a current

$$[\phi]_V(\Psi) = \operatorname{Lim}_{\delta \to 0} \int_{V_\delta(g)} \phi \wedge \Psi \tag{2.2}$$

If $\bar{\partial}[\phi]_V = 0$ we say that ϕ is an abelian differential. Let $\mathcal{A}_V$ be the space of abelian differentials – also called the dualizing sheaf since one has the isomorphism $\mathcal{A}_V \simeq Ext^p_R(R/I_V, \Omega)$ where $R = \mathbf{C}[x,t]$ and $\Omega = \mathbf{C}[x,t] \cdot dX \wedge dt$. Suppose that $|t| \leq C(1+|x|)$ holds for some constant C when $(x,t) \in V$. Let $\mathbf{L}$ be a linear map from the t-space into the x-space. The projection

$$\rho_{\mathbf{L}} \colon \ (x,t) \mapsto x + \mathbf{L}(t) \tag{2.3}$$

is generically an e-fold covering map, where e is the multiplicity of V. If ϕ is a rational n-form on V the direct image current $(\rho_L)_*([\phi]_V)$ is the trace of ϕ under ρ_L. We can impose the condition

$$\operatorname{Trace}_{\rho_{\mathbf{L}}}(\phi) \in \mathbf{C}[x] \cdot dX \tag{2.4}$$

Direct images preserve $\bar{\partial}$-closedness. Hence (2.4) holds if $\phi \in \mathcal{A}_V$. But (2.4) may also hold for non-abelian differentials. For example, the trace may be zero when $e \geq 2$.

Several projections. Let $\mathbf{L}_1, \dots, \mathbf{L}_e$ be an e-tuple such that $\oplus \rho_{\mathbf{L}_\nu}$ is generically unramified. If (2.4) holds for every $\mathbf{L}_\nu$ one may ask if $\phi \in \mathcal{A}_V$. However, the example at the end of this section shows that this implication does not hold in general. Let us say that the e-tuple $\mathbf{L}_\bullet$ is *non-discrepant* if $\phi \in \mathcal{A}_V$ when (2.4) holds for every $\mathbf{L}_\nu$.

Theorem 2.1. *In the Grassmannian variety the non-discrepant e-tuples is a dense subset whose complement is at most a denumerable union of algebraic hypersurfaces.*

This result appears in the study the Abel–Radon transform. The generic existence of non-discrepant projections was used in [49]. See also [38] for more detailed and intrinsic results. We give a proof of Theorem 2.1 which describes how to avoid discrepancy. Let $\Delta(x)$ define the discriminant locus under the projection $(x,t) \to x$. If $N \geq 1$ then $\mathcal{S}_N$ denotes the rational n-forms ϕ such that $\Delta^N \cdot \phi$ regular. An e-tuple $\mathbf{L}_\bullet$ is non-discrepant of order N if (2.4) above for every $\mathbf{L}_\nu$ gives $\phi \in \mathcal{A}_V$ when $\phi \in \mathcal{S}_N$.

Polarization. By linear change of coordinates in x we may assume that $\Delta(x)$ has x_n as a leading term, i.e., with $x' = (x, \dots, x_{n-1})$,

$$\Delta(x) = \prod_{\nu=1}^{K} (x_n - \beta_\nu(x'))$$

where K is some integer. Fix a point x'_* in $\mathbf{C}^{n-1}$ where the β-roots are all distinct. To every $1 \leq \nu \leq K$ the fiber in V above $(x'_*, \beta_\nu(x'_*))$ consists of distinct points

$\alpha_1(\nu), \ldots, \alpha_{k_\nu}(\nu)$ in the t-space. Notice that this number depends on ν. To every pair ν, j we put $\xi_{\nu,j} = (x'_*, \beta_\nu(x'_*), \alpha_j(\nu))$. Set $S = V \cap \{x' = x'_*\}$ which is a curve in the (x_n, t)-space. Next, consider an e-tuple of projections

$$\rho_i(x', x_n, t) = (x', x_n + \mathbf{L}_i(t)) .$$

So here $\mathbf{L}_i$ map $\mathbf{C}^k$ into the x_n-space. Hartog's theorem applies to $\text{Trace}_{\rho_i}(\phi)$. Together with boundary integrals in (6.19) which define $\bar{\partial}[\phi]_V$ one has:

Proposition 2.2. *The e-tuple $\mathbf{L}_\bullet$ is non-discrepant of degree N if and only if the following holds for every pair (ν, j):*

$$\oplus_{i=1}^{e} \mathbf{C}[x_n - \beta_\nu(x'_*) + \mathbf{L}_i(t - \alpha_j(\nu))] + ((x_n - \beta_\nu(x'_*))^N, I_S(\xi_{(\nu,j)})) = \mathcal{O}_{p+1}(\xi_{(\nu,j)}) \tag{2.5}$$

where $((x_n - \beta_\nu(x'_*))^N, I_S(\xi_{(\nu,j)}))$ is the ideal in the local ring $\mathcal{O}_{p+1}(\xi_{(\nu,j)})$ of germs of holomorphic functions. We say that $\mathbf{L}_\bullet$ yields a polarisation when (2.5) holds.

A study of polarization. There remains to investigate (2.5) at every $\xi_{(\nu,j)}$. Put $x - \beta_\nu(x'_*) = \zeta$ and $\tau = t - \alpha_j$. In the τ-space we can make a linear change of coordinates and assume that τ_1 separates the generic fiber on S under the projection $(\zeta, \tau) \mapsto \zeta$. Then there is an integer m such that $\zeta^m \cdot \tau_i = h_i[\zeta, \tau_1]$ for every $i \geq 2$ where $h_2, \ldots, h_p$ are polynomials. Increasing the integer N in Proposition 2.2 if necessary, the search for the validity of the polarisation condition is reduced to the local study below which explains why Theorem 2.1 holds.

2.2 A Local Study. Let $f(x, t) = t^e + a_1(x) \cdot t^{e-1} + \ldots + a_e(x)$ define a local curve $S = f^{-1}(0)$ in $\mathbf{C}^2$. Here $\{a_\nu(x)\}$ are holomorphic in some disc centered at the origin and each $a_\nu \in \mathfrak{m}^\nu$, where $\mathfrak{m}$ is the maximal ideal in $\mathbf{C}\{x\}$. The discriminant $\delta(x) = \prod_{\nu \neq i} (\alpha_\nu(x) - \alpha_i(x))$ vanishes to some order m at $x = 0$. If $\epsilon > 0$ we set $\sigma_\epsilon = S \cap \{|x| = \epsilon\}$ which yields a 1-cycle on S. Let $\phi = \rho(x, t) \cdot dx$ be a meromorphic 1-form. To every pair of integers $a, b \geq 0$ we compute the residue

$$\int_{\sigma_\epsilon} x^a \cdot t^b \cdot \rho(x, t) \cdot dx . \tag{2.6}$$

Here (2.6) is zero for all pairs a, b if and only if ϕ is an abelian differential. These meromorphic forms yield a sheaf $\mathcal{A}_S$. A classical result using residue calculus shows that $\mathcal{A}_S$ is the free $\mathcal{O}_S$-module generated by $f_t^{-1} \cdot dx$ where $f_t = \partial f / \partial t$.

Trace maps. If $\lambda \in \mathbf{C}$ we get the projection $\pi^\lambda \colon (x, t) \mapsto x + \lambda \cdot t$. The direct image current $\pi_*^\lambda(\rho \cdot dx)$ is $\bar{\partial}$-closed, i.e., a holomorphic form if and only if

$$\int_{\sigma_\epsilon} (x + \lambda \cdot t)^m \cdot \rho(x, t) \cdot dx = 0 . \tag{2.7}$$

for every $m \geq 0$. By Proposition 2.2 an e-tuple $\lambda_\bullet = (\lambda_1, \ldots, \lambda_e)$ is non-discrepant of order N if the following hold: To every meromorphic form such that $x^{N+1} \cdot \rho$ is holomorphic, the vanishing in (2.7) for every λ_ν yields $\rho \cdot dx \in \mathcal{A}_S$. The set of non-discrepant e-tuples is denoted by $D_f^*(N)$.

Polarization. (2.6) and (2.7) give $\lambda_\bullet \in D_f^*(N)$ if and only if

$$\oplus_{\nu=1}^{e} \mathbf{C}[x+\lambda_\nu \cdot t] + \left(x^{N+1}, f\right) = \mathcal{O}_2 \tag{2.8}$$

where (x^{N+1}, f) is the ideal generated by x^{N+1} and f. Consider the homogeneous polynomial

$$f_h(x,t) = t^e + c_1 x \cdot t^{e-1} + \ldots + c_e \cdot x^e \tag{2.9}$$

where $c_\nu = (\nu!)^{-1} \cdot \partial^\nu a_\nu / \partial x^\nu(0)$. It is easily seen that (2.9) holds for f if and only if it holds for f_h, i.e., $D_f^*(N) = D_{f_h}^*(N)$.

The generic hypersurface. Put $H(x,t) = \prod (t - \xi_\nu \cdot x)$ where $\xi_1, \ldots, \xi_e$ are parameters. Let $N \geq e$. If $e \leq m \leq N$ and $\lambda \in \mathbf{C}$ the image of $(x+\lambda \cdot t)^m$ in $\mathcal{O}_2/(x^{N+1}, H)$ is of the form:

$$\sum_{j=0}^{e-1} c_j^m(\lambda) \cdot x^{m-j} \cdot t^j .$$

Here $\{c_j^m(\lambda)\}$ are polynomials in λ whose coefficients are polynomials in the ξ-parameters. The e-tuple $\lambda_\bullet \in D_H^*(N)$ if and only if the determinants of the $e \times e$-matrices $(c_j^m(\lambda_\nu))$ are $\neq 0$ for $e \leq m \leq N$ and to obtain (2.4) when $1 \leq m \leq e-1$ the Vandermonde formula shows that it is necessary and sufficient that the fibers are separated, i.e., $\{\lambda_\nu\}$ are distinct.

Conclusion. To every $N \geq e$ the set $\mathbf{C}^e \setminus D_f^*(N)$ is a hypersurface defined by the determinants above where the specific roots of f_h appear in the generic ξ-variables.

Example. Let $f(t,x) = t^2 - x^2$. A computation shows that a pair $(0, \alpha)$ in $\mathbf{C}^2$ belongs to $D_f^*(N)$ if and only if

$$(\alpha+1)^N \neq (\alpha-1)^N$$

For example, if $N = 3$ discrepancy occurs when $\alpha = i \cdot \sqrt{3}$, i.e., here the meromorphic differential $\frac{dX}{x^3 \cdot t}$ has holomorphic trace under the two projections $(x,t) \mapsto x$ and $(x,t) \mapsto x + i \cdot \sqrt{3} \cdot t$.

3 Principal Value Currents

3.1 Introduction. The existence of the principal value distribution $\mathrm{VP}[\frac{1}{f}]$ and the residue current $\mathcal{R}_f = \bar{\partial}\mathrm{VP}[\frac{1}{f}]$ was established in [50]. More recent work employ meromorphic extensions and asymptotic expansions which give a far more extensive class of currents. Two results from [58] are used to obtain this: The roots of the Bernstein–Sato polynomial of a holomorphic function are strictly negative rational numbers and the functional equation in (4.12). Principal value currents are constructed by mean-values. In this sense Coleff–Herrera residues is a restricted class, but sufficiently ample to express local cohomology and abelian differentials. To get a perspective on the subsequent material we describe how Coleff–Herrera currents are obtained from asymptotic expansions.

Integrals along fibers. Let $f \in \mathcal{O}(X)$ where zero is the sole critical value. Let $n = d_X$ and Ψ is a test-form of bi-degree $(n-1, n-1)$. It is integrated on the oriented complex manifold $f = s$ when $s \neq 0$. Put

$$J_\Psi(s) = \int_{f=s} \Psi \tag{3.1}$$

Desingularisation shows that the current-valued function J_Ψ has an asymptotic expansion

$$\sum \rho_{\nu,k,\alpha,m} \cdot s^\nu \cdot \bar{s}^k \cdot |s|^\alpha \cdot \left[\operatorname{Log}\left(\frac{1}{|s|}\right)\right]^m . \tag{3.2}$$

Here ν, k are non-negative integers, $0 < \alpha < 1$ rational numbers and $0 \leq m \leq M$ for a fixed integer M, i.e., if necessary we restrict to a relatively compact set in X. The indexed ρ-terms are currents supported by $T = f^{-1}(0)$. The constant term ρ_0 is Lelong's integration current on T. Coefficients with positive powers of $\operatorname{Log}(\frac{1}{|s|})$ appear when local monodromy along T has index of nilpotency ≥ 1. The expansion (3.2) has been studied quite extensively since it is related to the Gauss–Manin complex. See [13]. $\mathcal{D}$-module theory is used to find relations between the ρ-currents where the Malgrange–Kashiwara filtration on the regular holonomic $\mathcal{D}_X$-module $\mathcal{O}_X[f^{-1}]$ leads to refined results such as the construction of vanishing cycles. See [75], [41] for original work and also [23, pp. 262–280] for details of proof.

The residue current $\mathcal{R}_f$. Let $X = \mathbf{C}^n$ where dX is the holomorphic n-form. The current $\mathcal{R}_f$ is defined by

$$\mathcal{R}_f\left(dX \wedge \Psi^{0,n-1}\right) = \operatorname{Lim}_{\epsilon \to 0} \int_{|f|=\epsilon} \frac{dX \wedge \Psi^{0,n-1}}{f} \tag{3.3}$$

Let us give another construction using (3.2). Namely, we show that $\mathcal{R}_f$ is obtained from the coefficient ρ_n of the monomial s^n in (3.2). By [28] there is an equation

$$f^n = a_1 \cdot \partial_1(f) + \ldots + a_n \cdot \partial_n(f) ,$$

where $\partial_i(f) = \partial f / \partial x_i$ and $\{a_i \in \mathcal{O}(X)\}$. Residue calculus gives

$$2\pi i \cdot (-1)^{n-1} \cdot \rho_n \left(\sum a_i(x) \cdot dX^{(i)} \wedge \Psi\right) = \mathcal{R}_f(dX \wedge \Psi) \tag{3.4}$$

for every test-form Ψ of bi-degree $(0, n-1)$.

The subsequent material contains results obtained after the pioneering work in [48] and [30]. One may mention the discovery by D. Barlet in [11] concerning the locally square integrable case in the remark after Theorem 3.1. Another new result is that principal value limits exist in Theorem 3.1 when positive C^∞-functions occur in the cut-off functions.

The distribution VP$[\frac{1}{f}]$. Let $f \in \mathcal{O}(X)$. There exists the distribution

$$\mathrm{VP}\left[\frac{1}{f}\right](\Psi) = \mathrm{Lim}_{\epsilon \to 0} \int_{|f|>\epsilon} \frac{\Psi}{f} \tag{3.5}$$

where Ψ are test-forms of bi-degree (n, n) and $n = d_X$. By desingularisation the existence of (3.5) is reduced the proof to the case when X is a polydisc and $f(x) = x^\alpha$ a monomial. In (3.5) the limit is taken over the sets $|f| > \epsilon$. However, one can use a more extensive class of limits. Below $\Phi^{n-1,n}$ is a test-form of bi-degree $(n-1, n)$.

Theorem 3.1. *Let $f \in \mathcal{O}(X)$. For every pair g, ρ where $g \in \mathcal{O}(X)$ is such that $f^{-1}(0) \subset g^{-1}(0)$ and ρ is a positive C^∞-function one has*

$$VP\left[\frac{1}{f}\right](\Psi) = Lim_{\epsilon \to 0} \int_{|\rho \cdot g|>\epsilon} \frac{\Psi}{f} \tag{3.6}$$

$$Lim_{\epsilon \to 0} \int_{|\rho \cdot g|=\epsilon} \frac{\Phi^{n-1,n}}{f} = 0 \tag{3.7}$$

Proof. By desingularisation we may assume that $f(x)$ is a monomial in $x_1, \ldots, x_k$, say $x_1^{\alpha_1} \cdots x_k^{\alpha_k}$ and $g = u \cdot x^\beta$ where u is an invertible holomorphic function and β a multi-index with $\beta_\nu \geq 1$ for every $1 \leq \nu \leq k$. Consider the L^1_{loc}-function $(x_1 \cdot x_k)^{-1}$. Repeated use Stokes formula yield (3.6) and (3.7) if we prove that

$$\mathrm{Lim}_{\epsilon \to 0} \int_{|u \cdot x^\beta| \cdot \rho = \epsilon} \frac{\Phi^{n-1,n}}{x^\gamma} = 0 \tag{3.8}$$

where $1 \leq \gamma_i \leq \alpha_i$ for each i. Decomposing the test-form Φ it suffices to treat the case when $\Phi = \phi \cdot dX^{(1)} \wedge d\bar{X}$ where ϕ is a test-function and $dX^{(1)}$ means that dx_1 is deleted. Put $w = \rho \cdot |u|$ and change coordinates to $\xi_1 = w^{1/\gamma_1} \cdot x_1$ and $\xi_\nu = x_\nu$ for $\nu \geq 2$. Then $d\bar{x}_1 = w^{-1/\gamma_1} \cdot d\bar{\xi}_1 + \bar{\xi}_1 \cdot dw^{-1/\gamma_1}$. Now (3.8) follows if

$$\mathrm{Lim}_{\epsilon \to 0} \int_{|\xi^\beta|=\epsilon} \frac{\Psi}{\xi^\gamma} \tag{3.9}$$

where $\Psi = h \cdot d\xi^{(1)} \wedge d\bar{\xi}^{(1)} \wedge d\bar{\xi}_1 + \bar{\xi}_1 \cdot \chi$ with $h \in C^\infty$ and χ is a test-form of total degree $2n - 1$. At this stage the proof in [50] applies, i.e., use Fourier series expansions in the angular variables of the real hypersurface $|\xi^\beta| = \epsilon$. The point is that it does not matter whether ξ_1 is holomorphic or not, i.e., the limit in (3.9) is zero because Ψ is a sum of differential forms where $d\bar{\xi}_1$ or $\bar{\xi}_1$ appear. ■

Remark. When f is given and m a positive integer such that $m \geq \{\max\{|\alpha_\nu|\}$ in every chart of the desingularisation where f is a monomial x^α, the proof shows that Theorem 3.1 holds when ρ is m-times differentiable. Concerning (3.7) in Theorem 3.1 a precise result occurs in Theorem 3.3.

The L^2-case. The proof in [11] shows that (3.8) from the proof of Theorem 3.1 remains true when $\Phi^{n-1,n}$ is multiplied by a function $|x_1|^{2\lambda_1} \cdots |x_k|^{2\lambda_k}$ where $\mathfrak{Re}(\lambda_\nu) > -1/2$ for every ν. This vanishing plays an essential role in [15] and will also be used in Sect. 5.

The residue current $\mathcal{R}_f$. This is the (0, 1)-current $\bar{\partial}\mathrm{VP}[\frac{1}{f}]$. Let g, ρ be as in Theorem 1.2. Stokes Theorem gives

$$\mathcal{R}_f\left(\Phi^{n,n-1}\right) = \mathrm{Lim}_{\epsilon\to 0} \int_{|\rho\cdot g|=\epsilon} \frac{\Phi^{n,n-1}}{f}.$$

Theorem 3.2. *Let $T = f^{-1}(0)$. Then $\bar{I}_T \cdot \mathcal{R}_f = 0$, i.e., the residue currentis annihilated by anti-holomorphic functions which are zero on T.*

Proof. This is proved in [50] using desingularisation. ∎

Wave front sets. Stokes Theorem and (3.7) in Theorem 3.1 show that if Q is a holomorphic differential operator then

$$Q\left(\mathrm{VP}\left[\frac{1}{f}\right]\right) = \mathrm{VP}\left[Q\left(\frac{1}{f}\right)\right]$$

Hence the cyclic $\mathcal{D}_X$-module generated by $\mathrm{VP}[\frac{1}{f}]$ in the left $\mathcal{D}_X$-module of distributions on the real manifold $X_{\mathbf{R}}$, is isomorphic to the cyclic $\mathcal{D}_X$-module generated by f^{-1} in the sheaf $\mathcal{O}_X[*S]$ of meromorphic functions with poles $\subset f^{-1}(0)$. Put $S = f^{-1}(0)$. A wellknown result in $\mathcal{D}$-module theory shows that the characteristic variety of the regular holonomic module $\mathcal{O}_X[*S]$ is the union of the zero-section $T^*_X(X)$ and the following subset of $T^*(X)$:

$$\{(x, \xi) : x \in S;\ \exists(x_\nu, \lambda_\nu) : x_\nu \to x \,\&\, \lambda_\nu \cdot \partial f(x_\nu) \to \xi\} \tag{3.10}$$

See [23, p. 262]. This complex Lagrangian set is denoted by Λ_f. By Theorem 4.3 the analytic wave-front set of $\mathcal{R}_f$ is contained in Λ_f. Equality holds if $\mathcal{D}_X(f^{-1}) = \mathcal{O}_X[*S]$.

3.2 Meromorphic Extensions. Let λ be a complex parameter. If $m \geq 0$ is an integer and $\mathfrak{Re}(\lambda) > m/2 + 1/2$ then $f^{-1} \cdot |f|^{2\lambda}$ is m-times differentiable. This gives a distribution-valued function of λ holomorphic in $\mathfrak{Re}(\lambda) > m/2 + 1/2$. It extends to a meromorphic function with poles contained in a finite union of arithmetic progressions $\cup\{-q_\nu - \mathbf{N}\}$. Here $q_1, \ldots, q_m$ are positive rational numbers and $\mathbf{N}$ the set of natural numbers. This was proved in [10]. Another way to obtain the meromorphic extension uses the Bernstein–Sato polynomial $b(\lambda)$ which appears in a functional equation

$$b(\lambda) \cdot f^{-1}|f|^{2\lambda} = \sum_{\nu=0}^{\nu=k} \lambda^\nu \cdot \bar{R}_\nu\left(\frac{\bar{f}}{f} \cdot |f|^{2\lambda}\right), \tag{3.11}$$

where $\bar{R}_0 \dots, \bar{R}_k$ are anti-holomorphic differential operators. To be precise, for every relatively compact Stein set in Ω in X we find (1.11) with $\bar{R}_\nu \in \bar{\mathcal{D}}(\Omega)$ and $b(\lambda)$ is the monic polynomial of minimal degree. In [58] it was proved that the roots of b are strictly negative rational numbers. If $-q^*$ is the largest root then $f^{-1} \cdot |f|^{2\lambda}$ is holomorphic in $\mathfrak{Re}(\lambda) > -q^*$. Let μ_0 be the constant term at $\lambda = 0$. One has the equality:

$$\mu_0 = \text{VP}\left[\frac{1}{f}\right] \tag{3.12}$$

The proof uses the functional equation

$$b(\lambda - 1) \cdot f^{-1} \cdot |f|^{2\lambda} = \sum_{\nu=0}^{\nu=\ell} \lambda^\nu \cdot Q_\nu\left(|f|^{2\lambda}\right), \tag{3.13}$$

where $Q_0, \dots, Q_\ell$ are holomorphic differential operators. Stokes Theorem and (3.7) in Theorem 3.1 give $\mu_0 = \text{VP}[\frac{1}{f}]$. If k is the multiplicity of the zero of $b(\lambda)$ at $\lambda = -1$, a local series expansion gives

$$\text{VP}\left[\frac{1}{f}\right] = \sum_{\nu=1}^{\nu=k} P_\nu\left(\left[\text{Log}|f|^2\right]^\nu\right) + g(x),$$

where $g \in \mathcal{O}_X$ and $P_1, \dots, P_k$ are holomorphic differential operators. Since integer powers of $\text{Log}|f|^2$ are locally integrable the right hand side represents a distribution.

Effective contribution of roots. In [14] it is proved that if $-q$ is a root of the b-function of some multiplicity m then there exists some integer $n \geq 0$ such that ν_λ has a pole of order m at $-q - n$. See also [23, pp. 276–279] for the proof.

Vanishing with bounds. Let $-q_*$ be the largest root of the b-function of f and m its multiplicity. Then (3.7) in Theorem 3.1. is sharpened to

Theorem 3.3. *Let g, ρ be as in Theorem 1.1. For every $N \geq 1$ there is a constant C_N and some integer $M \geq 0$ such that*

$$\left|\int_{|\rho \cdot g| = \epsilon} \frac{\Phi^{n-1,n}}{f^N}\right| \leq C_N \cdot \epsilon^{q_*} \cdot [Log(1/\epsilon)]^m \cdot ||\Phi||_M,$$

where $||\Phi||_M$ is the norm of M times differentiable test-forms.

Two limit formulas. Let g, f be a pair of holomorphic functions. No special assumptions is made about the relation between $g^{-1}(0)$ and $f^{-1}(0)$. By the result in (3.10) $\text{WF}(\mathcal{R}_f) \subset \Lambda_f$. Let ρ be a positive C^∞-function. The Morse-Sard Lemma shows that the set of Λ_f-critical values of $\rho \cdot g$ is discrete. Let $\chi_\delta(\rho \cdot g)$ be the characteristic function of $|\rho \cdot g| > \delta$. Then there exists $\delta_0 > 0$ such that the clean Schwarz product $\chi_\delta(\rho \cdot g) \cdot \text{VP}[\frac{1}{f}]$ is defined when $0 < \delta < \delta_0$. See the section about wave front sets in Sect. 4 for details. Using desingularisation one easily proves:

Theorem 3.4. *$Lim_{\delta\to 0}\,\chi_\delta(\rho\cdot g)\cdot VP[\frac{1}{f}] = VP[\frac{1}{f}]$ holds and $|g|^{2\lambda}\cdot VP[\frac{1}{f}]$ is holomorphic in a half-space $\mathfrak{Re}(\lambda) > -q$ where the constant term at $\lambda = 0$ is $VP[\frac{1}{f}]$.*

The next result was proved in [30]. Since it is essential for the theory we supply details of the proof.

Theorem 3.5. *Assume that $T\setminus g^{-1}(0)$ is dense in T. Then*

$$\mathcal{R}_f = Lim_{\delta\to 0}\,\chi_\delta(g)\cdot\mathcal{R}_f$$

Proof. Let $\rho\colon Z\to X$ be a desingularisation. Thus, if $G = g\circ\rho$ and $F = f\circ\rho$ then $G^{-1}(0)$ and $F^{-1}(0)$ are hypersurfaces with normal crossings. Let $S_1,\ldots,S_m$ be the irreducible components of $F^{-1}(0)$. If $\rho(S_\nu)$ is an analytic set of codimension ≥ 2 one says that S_ν is ramified. When S_ν is ramified the pull-back of a test-form in X of bi-degree $(n, n-1)$ is the product of an anti-holomorphic function vanishing on S_ν times a test-form on Z of bi-degree $(n, n-1)$. By Theorem 3.2 a principal residue current along S_ν is annihilated by by anti-holomorphic functions which are zero on S_ν. This means that when the direct image of $\mathcal{R}_F$ is constructed, one can ignore residues along the ramified S-components. At the same time the hypothesis means that common irreducible components of $G^{-1}(0)$ and $F^{-1}(0)$ are ramified. Then Theorem 3.5 follows from (3.7) in Theorem 3.1. ■

Remark. Using the Dolbeault–Lelong representation of $\mathcal{R}_f$ in Theorem 3.6 below it follows that Theorem 3.5 also holds using a cut-off function $\rho\cdot g$ when ρ is a positive C^∞-function.

3.3 The Dolbeault–Lelong Representation. Suppose that f has no multiple factors and put $T = f^{-1}(0)$. Let Φ be a non-vanishing holomorphic n-form, i.e., $\mathcal{O}_X\cdot\Phi = \omega_X$. We get a meromorphic $n-1$-form ϕ where $\phi\wedge\partial f = \Phi$, i.e., this is Leray's construction. Let g be a holomorphic function where $T_{\text{sing}}\subset g^{-1}(0)$ and $T\setminus g^{-1}(0)$ is dense in T. Set $T_\delta(g) = T\cap\{|g| > \delta|\}$. There exists a current of bi-degree $(n, 1)$ defined by:

$$\mathcal{R}_T\left(\Phi\wedge\Psi^{0,n-1}\right) = \text{Lim}_{\delta\to 0}\int_{T_\delta(g)}\phi\wedge\Psi^{0,n-1}. \tag{3.14}$$

The existence of a current in (3.14) follows after a desingularisation $\rho\colon Z\to T$ which reduces the proof to the construction of a principal value distribution on the manifold Z. The right hand side in (3.14) becomes

$$\text{VP}\left[\frac{1}{g\circ\rho}\right]\left(\rho^*\left(\phi\wedge\Psi^{0,n-1}\right)\right) \tag{3.15}$$

Example. Let $f(t, x)$ be a Weierstrass polynomial in t of some degree e without multiple factors. Here $X = D^{n+1}$ is a polydisc in the (x, t)-space and let $\Phi = dX\wedge dt$. Then we can take $\phi = f_t^{-1}\cdot dX$ where $f_t = \partial f/\partial t$. Hence

$$\mathcal{R}_T\left(\Psi^{0,n-1}\right) = \lim_{\delta\to 0}\int_{T_\delta(g)}\frac{dX\wedge\Psi^{0,n-1}}{f_t} \tag{3.16}$$

Theorem 3.6. *One has the equality* $\mathcal{R}_T(\Psi^{0,n-1}) = \mathcal{R}_f(\Phi \wedge \Psi^{0,n-1})$.

Proof. By partitions of the unity it suffices to work locally where f, Φ are as in (3.16). If $\mathrm{Supp}(\Psi) \subset X \setminus g^{-1}(0)$ Cauchy's residue formula and the expression of $\mathcal{R}_T$ in the example gives the equality. The general case follows by Theorem 3.5. ∎

Multiple factors. Let $N \geq 2$ and consider f^N with f as in 2.1. Residue calculus gives

$$\mathcal{R}_{f^N}(dX \wedge \rho \cdot d\bar{X}) = \frac{1}{2\pi i \cdot (N-1)!} \cdot \int_{T_\delta(g)} \partial_t^{N-1} \left(\frac{\rho}{f_t^N} \right) \cdot dX \wedge d\bar{X} . \quad (3.17)$$

This representation indicates that $\mathcal{R}_f$ can be expressed by a meromorphic form on T when $\mathcal{D}_X$-valued sections operate on the text-forms. Theorem 3.7 below gives the conclusive result.

$\mathcal{D}$-modules. Let $\mathcal{O}_X[*T]$ be the sheaf of meromorphic functions with poles in T. This is a regular holonomic $\mathcal{D}_X$-module. Working on a relatively compact set if necessary, we may assume that f^{-k} is a cyclic generator for some $k \geq 1$, i.e., $\mathcal{D}_X(f^{-k}) = \mathcal{O}_X[*T]$. Stokes Theorem and (3.7) in Theorem 3.1 give

$$Q\left(\mathrm{VP}\left[f^{-k}\right]\right) = \mathrm{VP}\left[Q\left(f^{-k}\right)\right] \quad (3.18)$$

when $Q \in \mathcal{D}_X$. Thus $\mathcal{O}_X[*T]$ is isomorphic to the submodule $\mathcal{D}_X(\mathrm{VP}[f^{-k}])$ of $\mathfrak{Db}_X$. Next, consider the local cohomology sheaf $\mathcal{H}^1_{[T]}(\mathcal{O}_X)$. We get isomorphic $\mathcal{D}_X$-modules

$$\mathcal{H}^1_{[T]}(\mathcal{O}_X) \simeq \mathcal{D}_X \cdot \mathcal{R}_{f^k} \quad (3.19)$$

The right $\mathcal{D}_X$-module $[g^{-M}]_T \cdot \mathcal{D}_X$. Let $g \in \mathcal{O}(X)$ where $T_{\mathrm{sing}} \subset g^{-1}(0)$ and $T \setminus g^{-1}(0)$ is dense in T. Put $S = g^{-1}(0)$ and consider the right $\mathcal{D}_X$-module $\mathcal{H}^1_{[T]}(\omega_X)[*S]$. With $\omega_X = \mathcal{O}_X \cdot \Phi$ there exists an integer N such that $g^{-N} \cdot \Phi$ is a cyclic generator. Define the $(n, 1)$-current by

$$[g^{-N} \cdot \Phi]_T(\Psi) = \mathrm{Lim}_{\delta \to 0} \int_{T_\delta(g)} g^{-N} \cdot \Phi \wedge \Psi . \quad (3.20)$$

This current generates a right $\mathcal{D}_X$-module, i.e., to every $Q \in \mathcal{D}_X$ the current $[g^{-N} \cdot \Phi]_V \cdot Q$ is defined by

$$[g^{-N} \cdot \Phi]_T \cdot Q(\Psi) = \mathrm{Lim}_{\delta \to 0} \int_{T_\delta(g)} g^{-N} \cdot \Phi \wedge Q(\Psi) \quad (3.21)$$

The Dolbeault–Malgrange isomorphism gives equal right $\mathcal{D}_X$-modules:

$$[g^{-N} \cdot \Phi]_T \cdot \mathcal{D}_X = \mathcal{H}^1_{[T]}(\omega_X)[*S] \quad (3.22)$$

Using the integer k from (3.19) and right conversion of $\mathcal{D}_X$-modules we obtain:

Theorem 3.7. *There exists* $Q \in \mathcal{D}_X$ *such that* $\mathcal{R}_{f^k}(\Phi \wedge \Psi) = [g^{-N}]_T(Q(\Psi))$

3.4 Integration on Fibers. Let f be a holomorphic function without multiple factors. Choose $\delta_0 > 0$ such that the hypersurface $T(s) = \{f = s\}$ is non-singular when $0 < |s| < \delta_0$. If Φ is a holomorphic n-form we find the $(n-1)$-form ϕ such that $\phi \wedge \partial f = \Phi$. The pull-back of ϕ to $T(s)$ is a holomorphic $n-1$-form for every s in the punctured disc $D^*\delta_0)$. Let Ψ is a test-form of bi-degree $n-1$ and Q a holomorphic differential operator where Ψ has support in $f^{-1}(D(\delta_0))$. Set

$$J_s(Q(\Psi) = \int_{T(s)} \phi \wedge Q(\Psi)\,. \tag{3.23}$$

Theorem 3.8. *One has the equality* $Lim_{s\to 0}\, J_s(Q(\Psi)) = 2\pi i \cdot \mathcal{R}_f(\Phi \wedge Q(\Psi))$.

Proof. Consider the complex Mellin transform

$$\mathcal{M}(\lambda) = \lambda \cdot \int |s|^{2\lambda-2} \cdot J_s(Q(\Psi)) \cdot ds \wedge d\bar{s} \tag{3.24}$$

Theorem 3.8 follows if $\mathcal{M}(\lambda)$ is holomorphic in $Re(\lambda) > -q$ for some $q > 0$ and $\mathcal{M}(0) = \mathcal{R}_f(\Phi \wedge Q(\Psi))$. To prove this one uses Stokes theorem which gives:

$$\mathcal{M}(\lambda) = \int |f|^{2\lambda-2} \cdot \phi \wedge Q(\Psi) \wedge \partial f \wedge \bar{\partial}\bar{f} = \int f^{-1}|f|^{2\lambda} \cdot \Phi \wedge Q(\bar{\partial}\Psi)\,, \tag{3.25}$$

Then (3.7) in Theorem 3.1 gives Theorem 3.8. ∎

$\mathcal{D}_X$-modules. Let $\mu = \mathrm{VP}[\frac{1}{f}]$. In local coordinates the distribution coefficients of $\mathcal{R}_f$ are $\bar{\partial}_j(\mu)$, where $\bar{\partial}_1, \ldots, \bar{\partial}_n$ are sections in the ring $\bar{\mathcal{D}}_X$ of anti-holomorphic differential operators. We can take higher order derivatives, i.e., to every $Q \in \bar{\mathcal{D}}_X$ we get the distribution $Q(\mu)$. It is supported by $T = f^{-1}(0)$ when Q belongs to the left ideal in $\bar{\mathcal{D}}_X$ generated by $\bar{\partial}_1, \ldots, \bar{\partial}_n$. In Theorem 3.5 we proved that $\mathcal{R}_f$ has the standard extension property. The question arises if the standard extension property holds for $Q(\mu)$ when the anti-holomorphic differential operator Q has degree ≥ 2. This is in general not true. Let us explain why this. Recall that the regular holonomic $\mathcal{D}_X$-module $\mathcal{H}^1_{|T|}(\mathcal{O}_X)$ contains the semi-simple module $\mathcal{L}(T)$. When the single degree complex $\mathbf{C}_T[1]$ is not self-dual in $D^b_c(\mathbf{C}_X)$, the inclusion $\mathcal{L}(T) \subset \mathcal{H}^1_{|T|}(\mathcal{O}_X)$ is strict. Assume that f from the start has enough poles along T in order that $\mathcal{D}_X\mu \simeq \mathcal{H}^1_{|T|}(\mathcal{O}_X)$. Theorem 5.1 implies that the cyclic $\bar{\mathcal{D}}_X$-module generated by μ contains a submodule with the same support as the quotient module $\mathcal{H}^1_{[T]}(\mathcal{O}_X)/\mathcal{L}(T)$. This is an analytic set of codimension ≥ 2. Hence there is some $Q \in \bar{\mathcal{D}}_X$ such that $\mathrm{Supp}(Q(\mu))$ has codimension ≥ 2. In particular the standard extension property fails. On the other hand, Theorem 3.1 shows also that if g is a meromorphic function with poles in T and the principal residue current $\mathcal{R}_g$ belongs to $\mathbf{L}(T)$, then $Q(\mathcal{R}_g)$ has the standard extension property for every $Q \in \bar{\mathcal{D}}_X$. This fact extends to higher codimension when residue currents belong to simple submodules of local cohomology. Various constructions in intersection cohomology can therefore be expressed by residue currents on which both $\mathcal{D}_X$ and $\bar{\mathcal{D}}_X$ operate.

4 Regular Holonomic Currents

4.1 Introduction. Let X be a complex manifold and $\mathfrak{Db}_X$ the sheaf of distributions on the underlying real manifold $X_{\mathbf{R}}$. Since $\mathcal{D}_X$-valued sections are differential operators on $X_{\mathbf{R}}$, it follows that $\mathfrak{Db}_X$ is a left $\mathcal{D}_X$-module. A distribution μ is regular holonomic if the cyclic module $\mathcal{D}_X\mu$ is so. Let $\mathrm{RH}(\mathfrak{Db}_X)$ denote the sheaf of regular holonomic distributions. In 4.2. we find a generating class of regular holonomic distributions. Operations on regular holonomic currents occur in 4.3 where meromorphic continuations and limits of non-characteristic restrictions to fibers are studied. Theorem 4.5 is a major result which proves existence of limits along fibers. Let us begin with some basic constructions.

Lelong's integration current. Let $V \subset X$ be a pure-dimensional complex subset of some codimension p. In [73] P. Lelong proved that V_{reg} has locally finite $2k$-dimensional area with $k = \dim(V)$. The complex structure yields the positive orientation on V_{reg} and we get a current of bi-degree (p, p) defined by

$$\mathcal{L}_V(\Psi) = \int_{V_{\mathrm{reg}}} \Psi \tag{4.1}$$

where Ψ are test-forms in $X_{\mathbf{R}}$ of bi-degree (k, k). Since V_{sing} is a complex set of codimension $\geq p + 1$, the real boundary of V_{reg} is negligable and Stokes Theorem shows that $\mathcal{L}_V$ is d-closed, i.e., $\partial\mathcal{L}_V = \bar{\partial}\mathcal{L}_V = 0$.

Regular holonomic currents. If $0 \leq p \leq n = d_X$ the sheaf $\mathfrak{C}_X^{0,p}$ of currents of bi-degree $(0, p)$ is a left $\mathcal{D}_X$-module. We get the sheaf $\mathrm{RH}(\mathfrak{C}_X^{0,p})$ whose sections are regular holonomic currents of bi-degree $(0, p)$. This sheaf is built up from $\mathrm{RH}(\mathfrak{Db}_X)$ since a $(0, p)$-current is regular holonomic if and only if its distribution coefficients are so in every expansion in local coordinates.

Right $\mathcal{D}_X$-modules. A left $\mathcal{D}$-module can be converted to a right $\mathcal{D}_X$-module and vice versa. Namely, let ω_X be the sheaf of holomorphic n-forms and $\mathcal{M}$ a left $\mathcal{D}_X$-module. Then $\mathcal{M}_r = \omega_X \otimes_{\mathcal{O}_X} \mathcal{M}$ is the converted right $\mathcal{D}$-module, and if $\mathcal{N}$ is a right module, then $\mathcal{N}_\ell = \mathcal{H}om_{\mathcal{O}_X}(\omega_X, \mathcal{N})$ is the converted left module. See [23, p. 21]. In particular currents of bi-degree (n, p) are right $\mathcal{D}_X$-modules for every $0 \leq p \leq n$. Let $\mathrm{RH}(\mathfrak{C}_X^{n,p})$ be the sheaf of currents whose right $\mathcal{D}_X$-modules are regular holonomic.

4.2 A Class of Regular Holonomic Currents. To every $0 \leq p \leq n = d_X$ we shall construct a class of regular holonomic currents of bi-degree (n, p) supported by a pure-dimensional analytic sets V in X of codimension p using the following:

1. $V \subset X$ is an analytic set of pure dimension k and $p = d_X - k$.
2. S is a hypersurface in X where $V \setminus S$ is dense on V, $V_{\mathrm{sing}} \subset S$ and $S = g^{-1}(0)$ for some $g \in \mathcal{O}(X)$.
3. α is a holomorphic k-form on V which is $\neq 0$ on $V \setminus S$.

4. $\mathbf{L}$ is a local system in $V \setminus S$
5. A function ϕ on $V \setminus S$, given as a section of the sheaf $\mathcal{O}_{V\setminus S} \otimes_{\mathbf{C}} \bar{\mathcal{O}}_{V\setminus S}$, whose holomorphic part $\mathcal{H}_\phi$ has moderate growth and yields the local system $\mathbf{L}$ and the anti-holomorphic part $\bar{\mathcal{H}}_\phi$ is locally square integrable.

See [15] for condition 5. From 1.–5. we obtain (n, p)-currents as follows:

Theorem 4.1. *To every section Q in $\mathcal{D}_X$ and every non-negative integer N, there exists a current γ defined by*

$$\gamma(\Psi) = Lim_{\delta\to 0} \int_{V_\delta(g)} g^{-N} \cdot \phi \cdot \alpha \wedge Q(\Psi) \tag{4.2}$$

where Ψ are test-forms of bi-degree $(0, k)$ and $V_\delta(g) = \{V \cap |g| > \delta\}$.

Proof. Choose a desingularisation $\rho\colon Z \to V$ where $\rho\colon Z \setminus \rho^{-1}(S) \to V \setminus S$ biholomorphic. Set $G = g \circ \rho$ and $\Phi = \phi \circ \rho$. There remains to prove the existence of a limit:

$$\gamma(\Psi) = \mathrm{Lim}_{\delta\to 0} \int_{Z_\delta(G)} G^{-N} \cdot \Phi \cdot \rho^*(\alpha) \wedge \rho^*(Q(\Psi)) \tag{4.3}$$

Even though inverse images of differential operators do not exist, it is irrelevant above since we use the pull-back of the test-form $Q(\Psi)$. In (4.3) we may assume that $G(x) = x^\beta$ and $\Phi = h \cdot |x|^2 \cdot \lambda$ where $h \in \mathcal{O}_Z[G^{-1}]$ and $\mathfrak{Re}(\lambda - \nu) > -1/2$ for every ν by the L^2-condition on $\bar{\mathcal{H}}_\phi$. Then Barlet's result after Theorem 3.1 gives Theorem 4.1 ∎

Remark. The L^2-condition on $\bar{\mathcal{H}}_\phi$ also gives the vanishing below for every $N \geq 1$ when test-forms W of bi-degree $(k-1, k)$ appear:

$$\mathrm{Lim}_{\delta\to 0} \int_{V\cap\{|g|=\delta\}} g^{-N} \cdot \phi \cdot W^{k-1,k} = 0\,. \tag{4.4}$$

Regular holonomicity. Keeping N fixed while Q varies in $\mathcal{D}_X$ the $(0, k)$-currents in Theorem 4.1. yield a cyclic right $\mathcal{D}_X$-module generated by the current when Q is the identity in $\mathcal{D}_X$. Let us denote it by $\mathcal{M}$. Preservation of regular holonomicity under proper maps shows that $\mathcal{M}$ is regular holonomic. Conversion gives the left $\mathcal{D}_X$-module $\mathcal{M}_\ell$. If N is sufficiently large the solution complex of $\mathcal{M}_\ell$ becomes

$$\mathcal{R}Hom_{\mathcal{D}_X}(\mathcal{M}_\ell, \mathcal{O}_X) = j_!(\mathbf{L}^*)[p]\,. \tag{4.5}$$

Here $\mathbf{L}^*$ is the dual local system on $V \setminus S$ and $j_!(\mathbf{L}^*)$ is the zero extension. In the derived category $D^b_c(\mathbf{C}_X)$ this single degree complex is put in degree p since:

$$\mathcal{E}xt^p_{\mathcal{D}_X}(\mathcal{M}_\ell, \mathcal{O}_X) = j_!(\mathbf{L}^*)\,, \tag{4.6}$$

while $Ext^\nu_{\mathcal{D}_X}(\mathcal{M}_\ell, \mathcal{O}_X) = 0$ when $\nu \neq p$.

Conclusion. $\mathrm{Perv}(\mathbf{C}_X)$ is generated by objects of the form $j_!(\mathbf{L}^*)[p]$ when the triple $V, S, \mathbf{L}$ varies. Then the Riemann-Hilbert correspondence and (4.5) show that $\mathrm{RH}(\mathfrak{Db})$ is sufficiently ample to generate the abelian category of of regular holonomic $\mathcal{D}_X$-modules. To be precise one also needs to relate $\mathcal{D}$-module theoretic images from (4.5) to direct images of currents. This "yoga" is explained in [23, Chap. 7].

Wave front sets. They are defined on distributions in [54, Chap. VIII] and can be used to clarify when products of distributions exist. Let μ and ν be two distributions on a C^∞-manifold whose wave front sets are disjoint, more precisely at least one of the wave front sets have empty intersection with every given line in a cotangent fiber. Then the product $\mu \cdot \nu$ exists in a robust way, i.e., as a limit of C^∞-functions $f_k \cdot g_k$ where $f_k \to \mu$ resp. $g_k \to \nu$ regularize the given distributions. See [54, Theorem 8.2.4]. In this situation we refer to $\mu \cdot \nu$ as a clean Schwarz-product. Leibniz's rule holds for such products, i.e., if δ is a C^∞-vector field then $\delta(\mu \cdot \nu) = \delta(\mu) \cdot \nu + \mu \cdot \delta(\nu)$.

Non-characteristic restrictions. Let $S \subset M$ be a C^∞-submanifold. Let $\mu \in \mathfrak{Db}(M)$ be such that $\mathrm{WF}_\infty(\mu)$ has empty intersection with the conormal bundle $T_S^*(M)$. Then the restriction μ_S to $\mathfrak{Db}(S)$ is obtained as follows: Choose any sequence $\{g_\nu\}$ in $C^\infty(M)$ where $g_\nu \to \mu$. The sequence of restricted $C^\infty(S)$-functions $g_\nu|S$ has a limit in $\mathfrak{Db}(S)$. This limit yields the non-characteristic restriction of μ and is denoted by μ_S.

The analytic case. If M is a real-analytic manifold one defines the analytic wave front set $\mathrm{WF}_A(\mu)$. See [54, Chap. VIII]. Since $\mathrm{WF}_\infty(\mu) \subset \mathrm{WF}_A(\mu)$ the results above apply when we consider analytic wave front sets. In this connection we mention a vanishing theorem in [26] which is used to settle unicity. Let $S \subset M$ be a real-analytic submanifold of some codimension p. Let μ be a distribution defined in some neighbourhood of a point $q \in S$. Assume that $\mathrm{WF}_A(\mu)$ does not intersect the conormal fiber of S above x_0. Let $\delta_1, \dots, \delta_p$ be real-analytic vector fields such that the tangent space $T_q(M)$ is generated by $T_q(S)$ and the tangent vectors $\{\delta_\nu(q)\}$.

Theorem 4.2. *Assume that the non-characteristic restriction of $\delta^\alpha(\mu)$ is zero in a neigbourhood of q for every multi-index $\alpha = (\alpha_1, \dots, \alpha_p)$. Then the distribution μ is identically zero in a neighbourhood of q.*

The complex case. Let X be a complex manifold. The holomorphic cotangent bundle $T^*(X)$ has an underlying real manifold which is naturally identified with the real cotangent bundle $T^*(X_\mathbf{R})$. Hence wave front sets of distributions on X are $\mathbf{R}$-conic subsets of $T^*(X)$. If $V \subset X$ is a complex submanifold then $T_V^*(X)$ is a complex Lagrangian submanfold of $T^*(X)$ and conditions expressed by wave-front sets yield non-characteristic restrictions exactly as in the real case. Thus, if μ is a distribution on $X_\mathbf{R}$ and $\mathrm{WF}_A(\mu) \cap T_V^*(X) = \emptyset$, then we get the distribution μ_V on the real manifold $V_\mathbf{R}$.

The equality $\mathbf{WF}_A(\mu) = \mathbf{SS}(\mathcal{D}_X\mu)$. Theorem 4.3 below is due to the late E. Andronikof in [8]. The proof uses the micro-local extension of Kashiwara's temperate Hom-functor and delicate results about regular holonomic $\mathcal{E}_X$-modules, where $\mathcal{E}_X$ is the sheaf of micro-differential operators on $T^*(X)$. See also [23, Chap 8] for a proof of Theorem 4.3.

Theorem 4.3. *Let μ be a regular holonomic distribution. Then*

$$WF_A(\mu) = WF_\infty(\mu) = SS(\mathcal{D}_X\mu) \tag{4.7}$$

Above $\mathrm{SS}(\mathcal{D}_X\mu)$ is the characteristic variety which by holonomicity is a complex Lagrangian subset of $T^*(X)$. This can be used to obtain non-characteristic restrictions and clean Schwarz products. In general, let μ be a distribution on X whose wave front set is contained in a complex Lagrangian set Λ of $T^*(X)$. The *Morse-Sard Lemma* asserts that the set of Λ-critical values of a real-valued and real-analytic function g on X is a discrete set, i.e.,

$$\{\alpha\colon \exists x\colon (x, \partial g(x)) \in \Lambda\, \alpha = g(x)\} \tag{4.8}$$

is discrete. This means that μ has non-characterstic restrictions to the real hypersurfaces $\{g = s\}$ outside a discrete set on the real s-line.

4.3 Meromorphic Extensions. Let μ be a RH-distribution and $g \in \mathcal{O}(X)$. No special assumptions are made, except that g is not identically zero on $\mathrm{Supp}(\mu)$ in which case the whole construction below disappears, i.e., the meromorphic function is identically zero. Let λ be a complex parameter and N a positive integer. If the real part of λ is sufficiently large we get the distribution

$$\nu_\lambda = g^{-N} \cdot |g|^{2\lambda} \cdot \mu \tag{4.9}$$

In fact, one has only have to ensure that $g^{-N} \cdot |g|^{2\lambda}$ has at least as many derivatives as the order of the distribution μ.

Theorem 4.4. *The distribution-valued function ν_λ extends to a meromorphic function in the complex λ-plane. The poles are contained in a finite union of artithmetic progressions $\{-q_j - \mathbf{N}\}$, where $q_1, \ldots q_k$ are complex numbers and $\mathbf{N}$ the set of non-negative integers.*

Functional equations. To prove Theorem 4.4 and study properties of the meromorphic extension we will use the three basic results below.

Case 1. There exists a polynomial $b(\lambda)$ and $Q_0 \ldots, Q_N$ in $\mathcal{D}_X$ such that:

$$b(\lambda) \cdot \nu_\lambda = \sum_{k=0}^{N} \lambda^k \cdot Q_k(g \cdot \nu_\lambda) \tag{4.10}$$

Case 2. There is a polynomial $\beta(\lambda)$ and anti-holomorphic differential operators $\bar{R}_0, \ldots, \bar{R}_N$ such that:

$$\beta(\lambda) \cdot \nu_\lambda = \sum \lambda^k \cdot \bar{R}_k(\bar{g} \cdot \nu_\lambda) \tag{4.11}$$

Case 3. There exists a positive integer N and $Q_0, \ldots, Q_{N-1}$ in $\mathcal{D}_X$ such that:

$$\lambda^N \cdot \nu_\lambda = \sum_{k=0}^{N-1} \lambda^\nu \cdot Q_k(\nu_\lambda) \tag{4.12}$$

These equations (4.10–12) were established in [66]. The proofs use desingularisation and a careful study in the normal crossing case. See also [23, Chap. 6] for details. We remark that the polynomials $b(\lambda)$ and $\beta(\lambda)$ are not equal in general.

Laurent series. At every point λ_0 one has a Laurent series expansion

$$\nu_{\lambda_0+\zeta} = \sum \rho_k \cdot \zeta^k . \tag{4.13}$$

Every coefficient ρ_k is a regular holonomic distribution. The polar distributions ρ_k with $k < 0$ are supported by the hypersurface $g^{-1}(0)$. See [23, Ch. VII] for details.

Rapid decay. By (4.12) above ν_λ has rapid decay when $\mathfrak{Im}(\lambda) \to \infty$, i.e., for every $A > 0$, every $N \geq 1$ and every test-form ψ one has:

$$\mathrm{Lim}_{|s|\to\infty} |s|^N \cdot \nu_{t+is}(\psi) = 0 \quad -A \leq t \leq A . \tag{4.14}$$

Asymptotic expansions. Consider the real-analytic function $|g|$. If δ is a non-critical value of $|g|$ the oriented real analytic manifold $\{|g|^2 = \delta\}$ is denoted by $S_\delta(g)$ where test-forms of degree $2n-1$ can integrated. Decomposing such test-forms we get the (1,0)-current – resp. the (0,1)-current – defined by

$$\psi^{(n-1,n)} \to \int_{S_\delta(g)} \psi \; ; d\,\phi^{(n,n-1)} \to \int_{S_\delta(g)} \phi . \tag{4.15}$$

The wave front set of these currents are both equal to the conormal bundle of $\{|g|^2 = \delta\}$. So if $\mu \in \mathrm{RH}(\mathfrak{Db}_X)$ there exists $\delta^* > 0$ such that μ has a non-characteristic restriction to $S_\delta(g)$ for every $0 < \delta < \delta^*$. Let μ_δ be the non-characteristic restriction to $S_\delta(g)$. A test-form ϕ of bi-degree $(n, n-1)$ yields the complex-valued function $J_\phi(\delta) = \mu_\delta(\phi)$. Similarly, a test-form ψ of bi-degree $(n-1, n)$ gives the function $J_\psi(\delta) = \mu_\delta(\psi)$. These J-functions have asymptotic expansions as $\delta \to 0$. Namely, Stokes Theorem gives

$$\mathcal{M}_\phi(\lambda) = \lambda \cdot \int |\delta|^{2\lambda-2} \cdot J_\phi(\delta) d\delta = \lambda \cdot \mu \left(|g|^{2\lambda-2} \cdot \phi \wedge g \cdot \bar{\partial}\bar{g} \right) \tag{4.16}$$

where $\phi \wedge d|g|^2 = \phi \wedge g \cdot \bar{\partial}\bar{g}$ is used. If ψ is a test-form of bi-degree $(n-1, n)$ we get

$$\mathcal{M}_\psi(\lambda) = (-1)^n \cdot \lambda \cdot \mu \left(|g|^{2\lambda-2} \cdot \phi \wedge \bar{g} \cdot \partial g \right) \tag{4.17}$$

The meromorphic extensions of the $\mathcal{M}$-functions and their rapid decay means that Mellin's inversion formula can be applied. Hence $J_\phi(\delta)$ and $J_\psi(\delta)$ have asymptotic expansions as $\delta \to 0$. The complex coefficients in the asymptotic expansion depend linearly upon the test-forms and yield currents supported by $g = 0$. Moreover, every such coefficient current is a finite sum of ∂- resp. $\bar{\partial}$-images of distribution coefficients which appear in series expansions of the meromorphic extensions of the $\mathcal{M}$-functions. So the result in 4.13 implies that every current in the asymptotic development is regular holonomic.

Integration along fibers. Let $\mu \in \mathrm{RH}(\mathfrak{Db}_X)$ and $g \in \mathcal{O}(X)$ be given. The Morse-Sard Lemma gives $\delta > 0$ such that the conormal of $\{g = s\}$ does not intersect $\mathrm{WF}(\mu)$ when $0 < |s| < \delta$. Hence non-characteristic restrictions yield (1,1)-currents $\{\mu_s\}$ for every s in the punctured disc $D^*(\delta)$. We also have the meromorphic function defined on test-forms Ψ of bi-degree $(n-1, n-1)$:

$$\mathcal{M}_\Psi(\lambda) = \frac{1}{2\pi i} \cdot \lambda \cdot \mu\left(|g|^{2\lambda-2} \cdot \Psi \wedge \partial g \wedge \bar{\partial}\bar{g}\right) \tag{4.18}$$

Theorem 4.5. *$\mathcal{M}_\Psi(\lambda)$ extends to a meromorphic function in the λ-plane which is holomorphic in $\mathfrak{Re}(\lambda) > -q$ for some $q > 0$. Moreover, for every text-form Ψ one has:*

$$Lim_{s\to 0}\, \mu_s(\Psi) = \mathcal{M}_\Psi(0) \tag{4.19}$$

Barlet distributions. Before Theorem 4.5 is proved we need the following. Let Z be a complex manifold and S a hypersurface. Let ϕ be a real-analytic function in $Z \setminus S$ given by a section in $\mathcal{O}_{Z\setminus S} \otimes \bar{\mathcal{O}}_{Z\setminus S}$ whose anti-holomorphic part is of class L^2. Then, for every holomorphic function f with $S = f^{-1}(0)$ there exists the distribution

$$\mu(\Psi) = \mathrm{Lim} \int_{|f|>\epsilon} \frac{\phi \cdot \Psi}{f} \tag{4.20}$$

called the principal Barlet distribution of the pair (f, ϕ). They provide a generating family of regular holonomic distributions. Namely, by (4.5) every $\mu \in \mathrm{RH}(\mathfrak{Db}_X)$ is locally a finite sum of distributions $\nu_1, \ldots, \nu_k$ where every ν_i is a distribution coefficient of a current given as the direct image of a Barlet distribution (f, ϕ) on a manifold Z in a desingularisation $\rho\colon Z \to V$ where V is the support of ν.

of Theorem 4.5. By the previous discussion we may assume that μ is a Barlet distribution (f, ϕ), and in the desingularisation $f = x^\alpha$ and $g = x^\beta$ are monomials while $\phi = |x|^{2\gamma}$ with $0 \le \gamma_i < 1/2$ for every i. Now Theorem 4.5 follows by straighforward calculations. ∎

5 Kashiwara's Conjugation Functor

5.1 Introduction. In Theorem 5.1 we announce a result from [57]. The fact that a distribution μ on a complex manifold satisfying a regular holonomic system of holomorphic equations, also satisfies a regular holonomic system defined by anti-holomorphic differential operators is remarkable. It is used to study distributions on real manifolds and yields comparison with respect to temperate localisation by **R**-constructible sheaves. See [68]. Since regular holonomicity plays a major role in this article we give comments about major contributions to $\mathcal{D}$-module theory in 5.3 and 5.4 treats Nilsson class functions.

5.2 The Conjugation Functor. Let X be a complex manifold and $\bar{X}$ its conjugate manifold where $\mathcal{O}_{\bar{X}} = \bar{\mathcal{O}}_X$. The sheaf of holomorphic differential operators on $\bar{X}$ is denoted by $\bar{\mathcal{D}}_X$. It is generated by $\bar{\mathcal{O}}_X$ and anti-holomorphic vector fields. Notice that $\mathcal{D}_X$ and $\bar{\mathcal{D}}_X$ commute in the ring $\mathcal{D}_{X_{\mathbf{R}}}$ of differential operators on the real manifold $X_{\mathbf{R}}$. If $\mathcal{M}$ is a left $\mathcal{D}_X$-module we get the left $\bar{\mathcal{D}}_X$-module

$$\mathbf{K}(\mathcal{M}) = \mathcal{H}om_{\mathcal{D}_X}(\mathcal{M}, \mathfrak{Db}_X)\,. \tag{5.1}$$

Similarly there is a functor $\bar{\mathbf{K}}$ from left $\bar{\mathcal{D}}_X$-modules to $\mathcal{D}_X$-modules.

Theorem 5.1. *Let $\mathcal{M} \in RH(\mathcal{D}_X)$. Then $\mathbf{K}(\mathcal{M}) \in RH(\bar{\mathcal{D}}_X)$ and for every $\nu \geq 1$ one has $\mathcal{E}xt^{\nu}_{\mathcal{D}_X}(\mathcal{M}, \mathfrak{Db}_X) = 0$. Moreover, the composed functor $\bar{\mathbf{K}} \circ \mathbf{K}$ is the identity on $RH(\mathcal{D}_X)$.*

Proof. It suffices to consider a generator $\mathcal{M}$ of RH($\mathcal{D}_X$) whose solution complex is $j_!(\mathbf{L})$, where $\mathbf{L}$ is a local system on $V \setminus S$ as in 4.2. Choose a desingularisation $\rho\colon Z \to V$ where $\rho^{-1}(S)$ has normal crossings. Then $\mathcal{M}$ is the direct image $\rho_+(\mathcal{N})$, where $\mathcal{N}$ is the Deligne-extension of $\mathcal{O}_{Z-\rho^{-1}(S)} \otimes_{\mathbf{C}} \rho^{-1}(\mathbf{L}^*)$. In the normal crossing case an induction over the length of the Deligne module reduces the proof to show that if $\{P_\nu = x_\nu \cdot \partial_\nu - \alpha_\nu\}$ is an n-tuple of first order differential operators in a polydisc D^n with $\alpha_\nu \in \mathbf{C}$, then the Koszul complex $\mathcal{K}^\bullet(P_\bullet, \mathfrak{Db})$ only has non-zero cohomology in degree zero whose kernel sheaf $\{\mu \in \mathfrak{Db}\colon\ P_1(\mu) = \ldots = P_n(\mu) = 0\}$ is a regular holonomic $\bar{\mathcal{D}}_X$-module. Next, preservation of regular holonomicity applied to $\bar{\rho}\colon \bar{Z} \to \bar{X}$ yields $\bar{\rho}_+\mathbf{K}(\mathcal{N}) \in \mathrm{RH}(\bar{\mathcal{D}}_X)$. To finish the proof one needs certain formulas for the $\bar{\partial}$-complex in X of the sheaf of extendible distributions on $V \setminus S$, in order to recover direct images of currents via the $\mathcal{D}$-module theoretic direct image functor $\bar{\rho}_+$. See [23, pp. 289–293] for details. ∎

Remark. The $\bar{\partial}$-complex of $\mathfrak{Db}_X$ is a resolution of $\mathcal{O}_X$. Interchange of derived functors gives:

$$\mathrm{DR}_{\bar{X}}(\mathbf{R}\mathcal{H}om_{\mathcal{D}_X}(\mathcal{M}, \mathfrak{Db}_X)) = \mathbf{R}\mathcal{H}om_{\mathcal{D}_X}(\mathcal{M}, \mathcal{O}_X) \tag{5.2}$$

The vanishing of higher $\mathcal{E}xt$-sheaves in Theorem 5.1 means that the derived Hom-functor in the first term is replaced by $\mathbf{K}(\mathcal{M})$. Hence $\mathrm{DR}_{\bar{X}}(\mathbf{K}(\mathcal{M}))$ is the solution complex of $\mathcal{M}$. The Riemann-Hilbert correspondence and duality between the de Rham complex and the solution complex of regular holonomic modules show that $\bar{\mathbf{K}} \circ \mathbf{K}$ is the identity.

Example. Let $\mathcal{D}_X/\mathcal{L}$ be a cyclic RH-module. The RH-module $\mathbf{K}(\mathcal{D}_X/\mathcal{L})$ is always locally cyclic. Choose a distribution μ such that $\bar{\mathcal{D}}_X\mu = \mathbf{K}(\mathcal{D}_X/\mathcal{L})$. Theorem 5.1 implies that $\mathcal{L} = \mathcal{L}_\mu$, where $\mathcal{L}$ is the left ideal of $\mathcal{D}_X$ whose sections annihilate μ. Hence one has:

Theorem 5.2. *Let $\mathcal{L}$ be a left ideal in $\mathcal{D}_X$ such that $\mathcal{D}_X/\mathcal{L}$ is regular holonomic. Then there exists – at least locally – a distribution μ such that $\mathcal{L} = \mathcal{L}_\mu$.*

The previous material shows that if $\mu \in \mathrm{RH}(\mathfrak{D}\mathfrak{b}_X)$ then every distribution ν satisfying $\mathcal{L}_\mu \nu = 0$ is regular holonomic. Moreover, the inclusion $\mathcal{L}_\mu \subset \mathcal{L}_\nu$ holds if and only if $\nu \in \bar{\mathcal{D}}_X \mu$.

Example. Let $X = \mathbf{C}$ and consider the principal value distribution $\mu = [1/x]$. Here $\mathcal{L}_\mu$ is generated by $\partial \cdot x = 1 + x\partial$. On the other hand, $\bar{\partial}(\mu) = 2\pi i \cdot \delta_0$ where δ_0 is the Dirac distribution. It follows that the left annihilator of μ in $\bar{\mathcal{D}}_X$ is generated by $\bar{x} \cdot \bar{\partial}$.

5.3 Regular Holonomic $\mathcal{D}$-modules. The theory about over-determined systems of linear differential systems with analytic coefficients has foremost been developed at RIMS in Kyoto. Prior to this one should mention work by Spencer and D. Quillen [95]. [62] laid the foundations of $\mathcal{D}$-module theory. Micro-differential operators were studied in [104] along with $\mathcal{D}$-modules. A major result from this work is that the characteristic variety of every coherent $\mathcal{D}_X$-module is involutive. The inclusion $\mathrm{DR}_X(\mathcal{M}) \in \mathrm{Perv}(\mathbf{C}_X)$ was established in [64] and an index-formula for holonomic systems in [63]. Kashiwara's lecture at Institute Mittag-Leffler in May 1975 was a historic landmark where the definite version of the Riemann–Hilbert correspondence was formulated. The complete proof was presented by Kashiwara in lectures at Université Paris-Nord in 1979. An independent proof was given in [77]. Moreover, in the algebraic case [79] proves the Riemann–Hilbert correspondence using desingularisation of surfaces only. An inverse functor from $\mathrm{RH}(\mathcal{D}_X)$ to $\mathrm{Perv}(\mathbf{C}_X)$ was constructed in [56]. In addition to this one should mention contributions by J.-P. Schneiders in [108] about images of differential systems in the relative case, by the late E. Andronikof in [8] where the micro-local version of the Riemann–Hilbert correspondence is proved, and by P. Schapira whose sheaf-theoretic notion of micro-support has led to an improved geometric interpretation of the subject. See [68]. Algebraic proofs about the Weyl algebra were established by J. Bernstein in [20] and J-E. Roos in [96]. Results about bidualizing complexes in [97] are used to prove that $A_n(\mathbf{C})$ as well as stalks of $\mathcal{D}_X$ and $\mathcal{E}_X$ are Auslander regular rings. See also [25]. Another important contribution is due to O. Gabber who proved that the set of regularity of every over-determined micro-local system is involutive. See [23, A VII.3.26]. The Decomposition Theorem in [17] is an important result with far-reaching applications. The fundamental discovery in [18] gives a link between $\mathcal{D}$-modules and representations of groups, especially in the equivariant case where regular holonomic systems occur. Regular holonomic $\mathcal{D}$-modules equippped with Hodge structure occur in [103] where the decomposition theorem from [17] is extended to Kähler manifolds.

Coherent $\mathcal{D}$-modules. Let X be a complex manifold. The global dimension of the abelian category of left $\mathcal{D}_X$-modules is $2d_X + 1$. See [23, Theorem 2.1.1]. This facilitates the construction of direct and inverse images of differential systems expressed in the derived category of bounded complexes of $\mathcal{D}$-modules yield transparent formulas. Every stalk of $\mathcal{D}_X$ is isomorphic to the ring $\mathcal{O}_n \otimes_{\mathbf{C}[x]} A_n(\mathbf{C})$ where $\mathcal{O}_n$ is the local ring of convergent power series and $A_n(\mathbf{C})$ the Weyl algebra. The ring $\mathcal{D}_n$ is Auslander regular and has homological dimension n. To every coherent module $\mathcal{M}$ there exists the characteristic variety $\mathrm{SS}(\mathcal{M})$ which is a $\mathbf{C}$-conic analytic set in $T^*(X)$. The *Involutivity Theorem* asserts that the ideal sheaf $I_{\mathrm{SS}(\mathcal{M})}$ is closed under the Poisson product on $\mathcal{O}_{T^*(X)}$.

Regular holonomic modules. The category of holonomic modules consists of coherent modules whose characteristic varieties are n-dimensional. By involutivity, $\mathrm{SS}(\mathcal{M})$ is a Lagrangian set when $\mathcal{M}$ is holonomic. A holonomic module $\mathcal{M}$ is regular if the local Fuchsian condition holds at every point x_0 in $\mathrm{Supp}(\mathcal{M})$, i.e.,

$$\mathrm{Ext}^{\nu}_{\mathcal{D}_X(x_0)}(\mathcal{M}, \hat{\mathcal{O}}(x_0)/\mathcal{O}_X(x_0)) = 0\colon\ 0 \leq \nu \leq n \tag{5.3}$$

where $\hat{\mathcal{O}}(x_0)$ is the local ring of formal power series. Several other regularity conditions exist. See [77], [23, Ch. V] and [27] for the algebraic case.

Micro-local regularity. In PDE-theory Fourier integrals and wave-front sets of distributions are powerful tools. Similarly, refined studies of regular holonomic $\mathcal{D}$-modules require micro-local analysis. Every $\mathcal{D}_X$-module $\mathcal{M}$ has the microlocalisation:

$$\mathcal{E}(\mathcal{M}) = \mathcal{E}_X \otimes_{\pi^{-1}\mathcal{D}_X} \pi^{-1}\mathcal{M} \tag{5.4}$$

where $\pi\colon T^*(X) \to X$ is the projection and $\mathcal{E}_X$ the sheaf of micro-differential operators. See [109], [22, Ch. 4],[60] or [23, Ch. VIII] for material about $\mathcal{E}_X$. We recall only that $\mathcal{E}_X$ is a coherent and noetherian sheaf of rings whose stalks are Auslander regular with homological dimension equal to n. A coherent $\mathcal{E}_X$-module $\mathcal{N}$ is defined on a $\mathbf{C}^*$-conic open subset of $T^*(X)$. If $\dim(\mathrm{Supp}(\mathcal{N})) = n$ one says that $\mathcal{N}$ is holonomic. When $\mathcal{M} \in \mathrm{coh}(\mathcal{D}_X)$ one has $\mathrm{SS}(\mathcal{M}) = \mathrm{Supp}(\mathcal{E}(\mathcal{M}))$.

The category RH($\mathcal{E}_X$). The general theory of regular differential systems in the analytic context was established in [67], where micro-local regularity is defined on coherent $\mathcal{E}_X$-modules, i.e., they are not necessarily holonomic. Regularity is imposed separately on every irreducible component of $\mathrm{Supp}(\mathcal{M})$ when $\mathcal{M} \in \mathrm{coh}(\mathcal{E}_X)$. If micro-local regularity holds along an irreducible component, there exist specialisations and a monodromy operator on the module which gives a canonical good filtration. See also [23, pp. 346–360] for a detailed account of [67]. Micro-local regularity is a generic property, i.e., it suffices to check it on dense open sets in the regular part of the support of any coherent $\mathcal{E}_X$-module $\mathcal{M}$. Moreover, a result due to O. Gabber shows that the support of micro-local regularity is an involutive analytic set in $T^*(X)$. In the regular holonomic case there is a simple local structure theorem. Consider the case when $X = \mathbf{C}^{n+1}$ with base coordinates (x, t) and symplectic co-vector variables ξ, τ in $T^*(\mathbf{C}^{n+1})$. Let $p = (0, 0, 0, dt)$ and let Λ be the Lagrangian $\{\xi = t = 0\}$.

Theorem 5.3. *Let $\mathcal{N} \in RH(\mathcal{E}_X)$ with $Supp(\mathcal{N}) = \Lambda$. Then $\mathcal{N}$ is isomorphic to a finite sum of $\mathcal{E}_X$-modules of the form*

$$\mathcal{E}_X / \left[\sum_{i=1}^{n} \mathcal{E}_X \cdot \partial_{x_i} + \mathcal{E}_X (t\partial_t - \alpha)^m \right] \tag{5.5}$$

where $m \geq 1$ and $\alpha \in \mathbf{C}$.

Remark. By quantized contact transformations this covers the general case. The α-numbers above yield eigenvalues $e^{2\pi i \cdot \alpha}$ of monodromy which are intrinsic since the isomorphism class of $\mathcal{N}$ is determined by the α-images in $\mathbf{C}/\mathbf{Z}$. A fundamental fact – originally proved in [65] – is that

$$\mathcal{E}(\mathcal{M}) \in \mathrm{RH}(\mathcal{E}_X) \tag{5.6}$$

for every $\mathcal{M} \in \mathrm{RH}(\mathcal{D}_X)$. This is used to construct the Kashiwara–Malgrange filtration on every $\mathcal{M} \in \mathrm{RH}(\mathcal{D}_X)$ which is functorial and is reduced in the sense that the annihilating ideal of $\mathrm{gr}(\mathcal{M})$ is equal to its radical. See also [23, Theorem 8.7.3] for the proof of (5.6) which is quite demanding and is not easily obtained by desingularisation.

5.4 Nilsson Class Functions. Let $T \subset X$ be a hypersurface. A multi-valued analytic function F in $X \setminus T$ is called a Nilsson class function if F has finite determination – i.e., the local branches at every point in $x_0 \in X \setminus T$ generate a finite-dimensional subspace $H_F(x_0)$ of $\mathcal{O}_X(x_0)$. Moreover, F has moderate growth along T. Here it suffices to check moderate growth locally at generic points of T_{reg}. See [31] or [22, Ch. 4]. The Nilsson class function yields a local system $\mathcal{H}_F$ in $X \setminus T$ of some rank k whose stalks are k-dimensional vector spaces $H_F(x_0) \subset \mathcal{O}_X(x_0)$. If $\Omega \subset X \setminus T$ is simply connected a trivialisation of $\mathcal{H}_F | \Omega$ arises when we choose $\mathbf{C}$-linearly independent single valued sections $f_1, \ldots, f_k$ in $\mathcal{H}_F(\Omega)$. [31] explains how local systems on Stein manifolds are recaptured by Nilsson class functions. In the algebraic case Nilsson class functions appear in $\mathbf{P}_n(\mathbf{C}) \setminus S$ where S is a hypersurface. An extensive class of Nilsson class functions was constructed in [81] from integrals of rational forms on semi-algebraic chains.

Example. Let $P(x, y)$ be a polynomial in $\mathbf{C}^{n+m}$. Let $x_0 \in \mathbf{C}^n$ and σ is a compact semi-algebraic set in the m-dimensional y-space such that $P(x_0, y) \neq 0$ for every $y \in \sigma$. Suppose that σ is the closure of an m-dimensional skeleton – given by an oriented manifold of real dimension m. Now

$$\phi(x) = \int_\sigma \frac{dy}{P(x, y)} \tag{5.7}$$

is holomorphic close to x_0. In [81] it is proved that ϕ is a local branch of a Nilsson class function F defined in $\mathbf{P}_n(\mathbf{C}) \setminus S$, where the hypersurface S is determined by P and the semi-algebraic set σ. In [82] it is proved that the monodromy F is quasi-unipotent with index of nilpotency $\leq m$. The proof uses classical methods in the spirit of Picard and Goursat, i.e., without desingularisation.

The case $m = 2$. Let $P = u^e + p_1(x,t)\cdot u^{e-1} + \ldots + p_e(x,t)$ be a polynomial in $\mathbf{C}^{n+2}$. Integrate $\frac{1}{P}$ on the 2-chain $\sigma_2 = \{0 \leq u, t \leq 1$. Let $\{\beta_\nu(x,t)\}$ be the roots of the u-polynomial P. Just as in 1.2 we obtain

$$\phi(x) = \sum_{\nu=1}^{\nu=e} \int_0^1 \frac{1}{P'_u(x,t,\beta_\nu(x,t))} \cdot \operatorname{Log}(\frac{1}{\beta_\nu(x,t)} - 1)\cdot dt \tag{5.8}$$

In the (x,t)-space we set $W_1 = \{\prod \beta_\nu(x,t)\cdot(1-\beta_\nu(x,t))\} = 0$ and W_2 is the image of $\{P = P'_u = 0\}$ under the projection $(x,t,u) \mapsto (x,t)$. Then the integrand in (5.8) is a Nilsson class function $F(x,t)$ defined outside $S = W_1 \cup W_2$. Close to $\{x_0 \times [0,1]\}$ the integrand has a single-valued branch $f(x,t)$. Let $Q(x,t) = q_m(x)\cdot t^m + \ldots + q_0(x)$ be the polynomial which defines S. In [81] it is proved that $\phi(x)$ extends to a Nilsson class function in $\mathbf{C}^n \setminus S_*$, where S_* is the union of zeros of $q_m(x)$, $q_0(x)$, $Q(1,x)$ and the discriminant locus of Q.

Construction of distributions. Using the complex conjugate structure we can construct regular holonomic distributions which via $\mathcal{D}$-module theory determine a given Nilsson-class function. This is useful if one wants to perform constructions where multi-valuedness causes technical problems. Especially when direct images with fibers of positive dimension occur. In such situations the direct image of a Nilsson class function can be described via the direct image of its associated distribution. The procedure is as follows. Let F be a Nilsson class function in $X \setminus T$ where T is some hypersurface. If $\Omega \subset X \setminus T$ is a simply connected there exist single valued branches $f_1, \ldots, f_k$ in $\mathcal{O}(\Omega)$ such that $\mathcal{H}_F(\Omega) = \oplus \mathbf{C}\cdot f_\nu$. In $X \setminus$T there exists the left ideal $\mathcal{L} \subset \mathcal{D}_X$ where a stalk $\mathcal{L}(x_0)$ is the left annihilator of $\mathcal{H}_F(x_0)$. An elementary result in $\mathcal{D}$-module theory – see [23, Thm. 1.1.29] – gives the duality formula:

$$\operatorname{Hom}_{\mathbf{C}}(\mathcal{D}_X(x_0), \mathcal{L}(x_0)), \mathcal{O}_X(x_0)) = \mathcal{H}_F(x_0) \tag{5.9}$$

Next, there exists a unique left ideal $\mathcal{L}_F$ defined in the whole of X such that $\mathcal{L}_F|X \setminus T = \mathcal{L}$, $\mathcal{H}^0_T(\mathcal{D}_X/\mathcal{L}_F) = 0$ and $\mathcal{D}_X/\mathcal{L}_F$ is regular holonomic. See [23, Chap. 5] for details. Theorem 5.1 gives a distribution μ such that $\mathcal{L}_\mu = \mathcal{L}_F$. To be precise, this requires that $\mathbf{K}_X(\mathcal{D}_X/\mathcal{L}_F)$ is a cyclic $\bar{\mathcal{D}}_X$-module generated by a global section. This is not always the case and we do not know any procedure to decide when $\mathbf{K}_X(\mathcal{D}_X/\mathcal{L}_F)$ has a global cyclic generator. Concerning existence of global sections we recall a that if $X = \mathbf{P}_n(\mathbf{C})$, then every left $\bar{\mathcal{D}}_X$-module is generated by global sections. See [27]. From now on we assume that μ exists. Then (5.9) implies that $\mu|X \setminus T$ is a section of $\mathcal{O}_X \times_{\mathbf{C}} \bar{\mathcal{O}}_X$. If Ω is a simply connected set and $f_1, \ldots, f_k$ a $\mathbf{C}$-basis of $\mathcal{H}_F(\Omega)$, then (5.9) gives

$$\mu|\Omega = f_1\cdot \bar{g}_1 + \ldots + f_k \cdot \bar{g}_k \tag{5.10}$$

for a unique k-tuple $g_\bullet$ in $\mathcal{O}(\Omega)$. The g-functions are local branches of a new Nilsson class function G. If $T = \phi^{-1}(0)$ for some $\phi \in \mathcal{O}(X)$ one shows that the left annhilating ideal of $\bar{\phi}^m \cdot \mu$ is equal to $\mathcal{L}_\mu$ for every positive integer m. We may replace μ by $\bar{\phi}^m \cdot u$ with m chosen so large that the Nilsson class function

$\phi^m \cdot G$ is locally square integrable along T. See [23, 4.1.7] for the L^2-condition of multi-valued functions with finite determination in $X \setminus T$.

The standard extension. Let μ be as above where its anti-holomorphic part is of class L^2 along T. Then the standard extension of μ exists, i.e., the distribution defined by

$$\mu_{\text{stand}}(\psi) = \text{Lim}_\delta \int_{|h|>\delta} \mu \cdot \psi \tag{5.11}$$

Moreover, $\mathcal{L}_\mu$ is the left annihilator of μ_{stand}. This was originally proved in [11]. See also [23, Chap. 7] for details of proof.

If F also satisfies the local L^2-condition along T, the $\bar{\mathcal{D}}_X$-module generated by μ_{stand} has no torsion and the left $\mathcal{D}_X$-module $\mathcal{D}_X\mu$ is isomorphic to the semi-simple $\mathcal{D}_X$-module given by the minimal Deligne extension of $\mathcal{O}_{X-T} \otimes_{\mathbf{C}} \mathbf{L}^*$. This illustrates how Theorem 5.1 can be used to exhibit specific regular holonomic $\mathcal{D}$-modules. Notice that $\bar{\partial}\mu$ is not supported by T when the g-functions above are not constant. So when a non-trivial local system is present we do not get residues in the sense of Coleff–Herrera which are treated in the next section.

6 Coleff–Herrera Residues

6.1 Introduction. Let $f_1, \ldots, f_p$ be holomorphic functions in a complete intersection, i.e., $\cap f_\nu^{-1}(0)$ has codimension p. The main result in [30] is:

Theorem 6.1. *There exists a current of bi-degree* $(0, p)$ *defined by*

$$\mathcal{R}_F(\Phi) = Lim_{\epsilon_\bullet \to 0} \int_{|F|=\epsilon_\bullet} \frac{\Phi}{f_1 \cdots f_p} \tag{6.1}$$

where $\{|F| = \epsilon_\bullet\} = \{|f_1| = \epsilon_1 \cap \ldots \cap f_p| = \epsilon_p\}$ *and the p-tuple* $\epsilon_\bullet$ *tends to zero in such a way that* $\epsilon_\nu/\epsilon_{\nu+1}^k \to 0$ *for every* $k \geq 2$ *and* $1 \leq \nu \leq p-1$.

Remark. The limit exists since Sard's Lemma gives $\delta > 0$ and a positive integer k such that the level sets $|F| = \epsilon_\bullet$ are real-analytic manifolds when $\epsilon_\nu \leq \epsilon_{\nu+1}^k$ for $1 \leq \nu \leq p-1$ and $\epsilon_p \leq \delta$. A $(2n-p)$-form ω yields the positive orientation on $|F| = \epsilon_\bullet$ when $\omega \wedge d|f_1|^2 \wedge \ldots d|f_p|^2$ is a positive $2n$-form in $X_{\mathbf{R}}$. The current $\mathcal{R}_F$ depends therefore on the ordered p-tuple of functions. In [30] it is shown that one can reverse the limit to recapture $\mathcal{R}_F$, i.e., when $\epsilon_{\nu+1}/\epsilon_\nu^k \to 0$ for every $1 \leq \nu \leq p-1$. The example in 6.4 below shows that the unrestricted limit does not exists in general. Theorem 6.1 is proved in 6.3.

The Bochner–Martinelli kernel $\mathcal{B}_F$. In Theorem 6.1 we replace the limit by mean-values when $\epsilon_1^2 + \ldots + \epsilon_p^2 = \delta^2$, i.e., put

$$\mathcal{B}_F(\delta) = \int_{S^{p-1}(\delta)} [\int_{|\epsilon_\bullet|=\delta} \frac{\Phi}{f_1 \cdot f_p}] \cdot d\sigma_\delta(\epsilon_\bullet)\,, \tag{6.2}$$

where $d\sigma_\delta$ is the normalised Lebesgue measure on the sphere $S^{p-1}(\delta)$. Then

$$\mathrm{Lim}_{\delta\to 0}\,\mathcal{B}_F(\delta) = \mathcal{R}_F\,. \tag{6.3}$$

Here $\mathcal{B}_F(\delta)$ is the $\bar\partial$-image of the Bochner–Martinelli current $\mathbf{B}_F$ which is defined outside the complete intersection as follows: Put

$$||F||^2 = |f_1|^2 + \ldots + |f_p|^2 \;\&\; \bar\partial\bar F^{(\nu)} = (-1)^{\nu-1}\cdot \bar f_\nu \cdot \wedge_{i\neq\nu}\bar\partial\bar f_i \tag{6.4}$$

Then

$$\mathbf{B}_F = ||F||^{-2p}\cdot\sum_{i=1}^{p}\bar\partial\bar F^{(\nu)}\,. \tag{6.5}$$

One verifies that $\mathbf{B}_F$ is $\bar\partial$-closed outside the complete intersection. Using toric charts after desingularisation and Theorem 3.1 one shows that the principal value extension of $\mathbf{B}_F$ exists when $||F||\to 0$ and that $\bar\partial\mathbf{B}_F = \mathcal{R}_F$. See [91].

Abelian differentials. Let V, S, g be as in 4.2. A section ϕ in $\Omega_V^k[*S]$ yields the (n, p)-current

$$[\phi]_V(\Psi) = \mathrm{Lim}\int_{V_\delta(g)}\phi\wedge\Psi\,. \tag{6.6}$$

If $\bar\partial[\phi]_V = 0$ we say that ϕ is an abelian differential on V. The resulting. sheaf is denoted by $\mathcal{A}_V$. The Dolbeault–Malgrange isomorphism shows that the $\mathcal{O}_X$-module $\mathcal{A}_V$ is isomorphic to the dualizing sheaf $\mathcal{E}xt_{\mathcal{O}_X}(\mathcal{O}_V, \omega_X)$. Sect. 6.4 contains material about $\mathcal{A}_V$.

6.2 The Sheaf $\mathbf{CH}_V$. Apply the construction from 4.2 with the local system $\mathbf{L} = \mathbf{C}_{V\setminus S}$, i.e., a pair $\phi\in\Omega_V^k[*S]$ and $Q\in\mathcal{D}_X$ gives the current

$$[\phi\cdot Q]_V(\Psi) = \mathrm{Lim}\int_{V_\delta(g)}\phi\wedge Q(\Psi)\,. \tag{6.7}$$

The family of these currents generate a right $\mathcal{D}_X$-module denoted by $\mathbf{CH}_V[*S]$. From (4.5) we get isomorphic right $\mathcal{D}_X$-modules:

$$\mathcal{H}^p_{[V]}(\omega_X)\simeq\mathbf{CH}_V[*S] \tag{6.8}$$

Let ϕ be a k-form on V which is $\neq 0$ on $V\setminus S$. On a relatively compact Stein open set there is an integer N such that the cyclic module $[g^{-N}\cdot\phi]_V\cdot\mathcal{D}_X = \mathbf{CH}_V[*S]$. Let $\rho\colon Z\to V$ be a desingularisation and put $G = g\circ\rho$. If $\gamma\in\mathbf{CH}_V[*S]$ we find some $Q\in\mathcal{D}_X$ such that

$$\gamma(\Psi) = \mathrm{VP}\left[G^{-N}\right](\rho^*(\phi\wedge Q(\Psi))\,. \tag{6.9}$$

Theorem 3.1 applied to $\mathrm{VP}[G^{-N}]$ gives $\mathrm{Lim}_{\delta\to 0}\,\chi_\delta(\rho\cdot|h|)\cdot\gamma = \gamma$ when ρ is a positive C^∞-function and $h\in\mathcal{O}(X)$ is such that $V\setminus h^{-1}(0)$ is dense in V. We express this by saying that γ is of the standard extension type and write $\gamma = \gamma_{\text{stand}}$. Notice also that $\bar I_V\cdot\gamma = 0$.

The canonical isomorphism $\mathbf{CH}_V \simeq \mathcal{H}^p_{[V]}(\omega_X)$. The subsheaf of $\bar\partial$-closed sections in $\mathbf{CH}_V[*S]$ is denoted by $\mathbf{CH}_V$. Let $\mathcal{C}_V^{n,p}$ be the sheaf of currents of bi-degree (n, p) whose sections are supported by V. Recall the Dolbeault–Malgrange isomorphism

$$\mathcal{H}^p_{[V]}(\omega_X) \simeq \frac{\mathrm{Ker}_{\bar\partial}\left(\mathcal{C}_V^{(n,p)}\right)}{\bar\partial\left(\mathcal{C}_V^{(n,p-1)}\right)} \tag{6.10}$$

Then (6.8) and (6.10) give the canonical isomorphism above. Moreover, there is the direct sum

$$\mathcal{C}_V^{(n,p)} = \bar\partial\left(\mathcal{C}_V^{(n,p-1)}\right) \oplus \mathbf{CH}_V\,. \tag{6.11}$$

The original proof of the canonical isomorphism above was given in [32]. See also [90].

Intrinsic character of $\mathbf{CH}_V$. On the submanifold $V \setminus S$ a $\bar\partial$-closed current annihilitated by $\bar I_{V\setminus S}$ cannot be locally $\bar\partial$-exact, i.e., this follows by elementary distribution theory. Together with the standard extension property a current γ of bi-degree (n, p) yields a section of $\mathbf{CH}_V$ if and only if

$$\bar\partial(\gamma) = \bar I_V \cdot \gamma = 0 \,\&\, \gamma = \gamma_{\mathrm{stand}}\,. \tag{6.12}$$

A Mayer–Vietoris sequence. Put $W = V \cap g^{-1}(0)$. If $\gamma \in \mathbf{CH}_V[*S]$ it follows from (6.9):

$$\bar\partial\gamma(\Psi) = \mathcal{R}_{G^N}(\rho^*(\phi \wedge Q(\Psi)) \tag{6.13}$$

Theorem 3.5 applied to $\mathcal{R}_{G^N}$ shows that $\bar\partial\gamma$ is of the standard extension type. Next, let $h \in I_W$. Then $H = h \circ \rho$ is zero on $G^{-1}(0)$ and then $\bar H \cdot \mathcal{R}_{G^N} = 0$ by Theorem 3.2. It follows that $\bar I_W \cdot \bar\partial(\gamma) = 0$ and (6.12) applied with W gives $\bar\partial(\gamma) \in \mathbf{CH}_W$. Hence we get the exact sequence:

$$0 \to \mathbf{CH}_V \to \mathbf{CH}_V[*S] \xrightarrow{\bar\partial} \mathbf{CH}_W \to \mathcal{H}^{p+1}_{[V]}(\omega_X) \to 0 \tag{6.14}$$

The Cohen–Macaulay case. Assume that V is Cohen–Macaulay. This means precisely that the solution complex of $\mathcal{H}^p_{[V]}(\mathcal{O}_X)$ is the single degree complex $\mathbf{C}_V[p]$ in the derived category of sheaves. Let $h \in \mathcal{O}(X)$ and assume that $\partial h \neq 0$, i.e., $h^{-1}(0)$ is a submanifold. Put $W = V \cap h^{-1}(0)$ and $i\colon W \to V$ is the inclusion map. Since $i^{-1}(\mathbf{C}_V) = \mathbf{C}_W$, the Riemann-Hilbert correspondence gives the $\mathcal{D}$-module theoretic inverse image formula $i^+(\mathbf{CH}_V) = \mathbf{CH}_W$. The inverse image functor can be described in the following explicit fashion:

Theorem 6.2. *To every $\gamma \in \mathbf{CH}_V$ we set $\gamma_* = \bar\partial(h^{-1}\cdot\gamma)$. Then $\gamma \to \gamma_*$ is a surjective and right $\mathcal{D}$-linear map from $\mathbf{CH}_V$ to $\mathbf{CH}_W$ whose kernel is $h \cdot \mathbf{CH}_V$.*

$\mathcal{E}xt$-sheaves. Now V is just pure-dimensional. If $h \in \mathcal{O}(X)$ is such that $V \setminus h^{-1}(0)$ is dense in V, then every $\gamma \in \mathbf{CH}_V$ yields a unique γ_* in $\mathbf{CH}_V$ such that $h \cdot \gamma_* = \gamma$. Put $\gamma_* = h^{-1} \bullet \gamma$ and $W = V \cap h^{-1}(0)$. By (6.13) one has

$$\bar{\partial}(h^{-1} \bullet \gamma) \in \mathbf{CH}_W \tag{6.15}$$

Next, let J be an ideal in $\mathcal{O}_X$ whose radical is I_V. Put

$$\mathcal{M} = \mathcal{E}xt^p_{\mathcal{O}_X}(\mathcal{O}_X/J, \omega_X) \tag{6.16}$$

which by 6.10 can be identified with the subsheaf $\mathbf{CH}_V(J)$ whose sections are annihilated by J. Let $\gamma_1, \ldots, \gamma_k$ generate the coherent $\mathcal{O}_X$-module $\mathcal{M}$. Let h be as above and (J, h) the ideal in $\mathcal{O}_X$ generated by J and h and put $W = V \cap h^{-1}(0)$. As above we get

$$\mathcal{E}xt^{p+1}_{\mathcal{O}_X}(\mathcal{O}_X/J, \omega_X) = \mathbf{CH}_W(J, h) \tag{6.17}$$

With these notations we have the result below whose counterpart in commutative algebra is wellknown.

Theorem 6.3. *Assume that $\mathcal{E}xt^{p+1}_{\mathcal{O}_X}(\mathcal{O}_X/J, \omega_X) = 0$. Then the $\mathcal{O}_X$-module $\mathbf{CH}_W((J, h))$ is generated by the currents $\{\bar{\partial}(h^{-1} \cdot \gamma_\nu)\}$.*

6.3 Complete Intersection. Theorem 6.1 is proved by induction over p. Let $F = (f_1, \ldots, f_p)$ and assume that $\mathcal{R}_F \in \mathbf{CH}_V$. Let g be another function such that (F, g) is a complete intersection. By 6.14 the current $\bar{\partial}(g^{-1} \bullet \mathcal{R}_F)$ belongs to $\mathbf{CH}_W$ where $W = V \cap g^{-1}(0)$. Moreover, one has

$$g^{-1} \bullet \mathcal{R}_F = \mathrm{Lim}_{\delta \to 0}\, g^{-1} \cdot \chi_\delta(g) \cdot \mathcal{R}_F \tag{6.18}$$

It follows that the Coleff–Herrera current $\mathcal{R}_{g,F} = \bar{\partial}(g^{-1} \bullet \mathcal{R}_F)$ and Theorem 6.3 implies the ideal in $\mathcal{O}_X$ which annihilates $\mathcal{R}_F$ is equal to the ideal $(f_1, \ldots, f_p)$. This result was originally proved in [87].

6.4 Abelian Differentials. Let V be of pure dimension p. Put $\mathcal{A}_V = \mathcal{H}om_{\mathcal{O}_X}(\mathcal{O}_V, \mathbf{CH}_V)$ which by 6.10 is the sheaf of abelian differentials on V. Every section in $\mathcal{A}_V$ is given by $[\phi]_V$ for some meromorphic k-form on V. Stokes Theorem shows that $\bar{\partial}[\phi]_V = 0$ if and only if

$$\mathrm{Lim}_{\delta \to 0} \int_{V \cap |\rho \cdot g| = \delta} \phi \wedge \Psi^{0,k-1} = 0 \tag{6.19}$$

when $V_{\mathrm{sing}} \subset g^{-1}(0)$, ρ is a positive C^∞-function and Ψ a test-form of bi-degree $(0, k-1)$.

The sheaf $\mathbf{L}^2(\mathcal{A}_V)$. Let $\rho\colon Z \to V$ be a desingularisation. Every globally defined holomorphic k-form γ on Z yields the $\bar{\partial}$-closed current $\rho_*(\gamma)$ in $\mathcal{A}_V$. These sections are intrinsic. Namely, the ρ_*- image consists of currents $[\phi]_V$ such that $\phi \in \mathbf{L}^2$, i.e., $\phi \wedge \bar{\phi}$ is locally integrable on V. See [44].

The sheaf $\mathbf{L}(\mathcal{A}_V)$. Let $\mathbf{L}(V)$ be the semi-simple $\mathcal{D}_X$-submodule of $\mathbf{CH}_V$. Put $\mathbf{L}(\mathcal{A}_V) = \mathbf{L}(V) \cap \mathcal{A}_V$. The inclusion $\mathbf{L}^2(\mathcal{A}_V) \subset \mathbf{L}(\mathcal{A}_V)$ is proved in [15]. Consider the special case when V is an algebraic set in $\mathbf{C}^n$ given as the image of an injective polynomial map $\mathcal{P}\colon\ \mathbf{C}^k \to V$, i.e., V has a smooth normalisation without branch-points. Then $\mathbf{CH}_V$ is the $\mathcal{D}$-module theoretic direct image $\mathcal{P}_+(\omega_Z)$. It follows that $\mathbf{CH}_V$ is a simple and self-dual $\mathcal{D}_X$-module. Hence, if $p = n - k$ the single degree $\mathbf{C}$-constructible complex $\mathbf{C}_V[p)$ is a simple self-dual object in $\mathrm{Perv}(\mathbf{C}_X)$. In particular $\mathcal{A}_V = \mathbf{L}(\mathcal{A}_V)$. But the description of $\mathcal{A}_V$ with the aid of the polynomial map $\mathcal{P}$ is unclear, i.e., no general procedure is known to exhibit $\mathcal{A}_V$. The case of monodial curves below illustrates the difficulty.

The Grothendieck ring $\mathcal{G}_V$. This is the subring of $\mathcal{D}_X$ whose sections preserve the ideal I_V. Notice that $\mathcal{A}_V$ is a right $\mathcal{G}_V$-module. Very little is known about this module structure. For example, suppose that $\mathbf{L}(\mathcal{A}_V) \neq \mathbf{L}^2(\mathcal{A}_V)$. One may ask if $\mathcal{G}_V$ is ample enough in order that $\mathbf{L}(\mathcal{A}_V) = \mathbf{L}^2(\mathcal{A}_V) \cdot \mathcal{G}_V$. For such a question the full ring structure of $\mathcal{G}_V$ is irrelevant, i.e., it is only essential to know generators of this $\mathbf{C}$-algebra. This restricted consideration appears to be the only reasonable way if one wants to use $\mathcal{G}_V$ for a geometric purpose. The reason is that the ring $\mathcal{G}_V$ in most cases is very complicated to describe. The following result illustrates this. Let V be the union of three hyperplanes $\{t = \alpha_\nu \cdot x\}$ in $\mathbf{C}^2$. Then the $\mathbf{C}$-algebra $\mathcal{G}_V$ requires a minimal set of 15 generators where differential operators of order five occur. See [52].

Monomial curves. Let $a_1, \ldots, a_n$ be integers ≥ 2 with no common divisor. The image of the map

$$\mathcal{P}\colon\ t \mapsto (t^{a_1}, \ldots, t^{a_n}) \tag{6.20}$$

from $\mathbf{C}$ into $\mathbf{C}^n$ is a monomial curve S. The Frobenius number of the n-tuple is the smallest integer M such that $\{M, M+1, \ldots\}$ is contained in the semi-group generated by $a_\bullet$. If $n = 2$ one has $M = (a_2 - 1) \cdot (a_1 - 1)$. But if $n \geq 3$ no closed formula is known, i.e., the Frobenius number has to be computed in every specific case. Residue calculus shows that $\mathcal{A}_S$ is the free $\mathcal{O}_S$-module generated by the direct image current $\mathcal{P}_*(t^{-M} \cdot dt)$. Since no algorithm exists to find M this shows that one cannot expect to establish general formulas to discover the sheaf of abelian differentials. The $\mathbf{L}^2$-lattice $\mathbf{L}^2(\mathcal{A}_S)$ is the free $\mathcal{O}_S$-module generated by $\mathcal{P}_*(dt)$. Since $\mathbf{CH}_S$ is a simple $\mathcal{D}_X$-module the right $\mathcal{G}_S$-module generated by $\mathcal{P}_*(dt)$ is equal to $\mathcal{A}_S$. This equality is confirmed by the description of $\mathcal{G}_S$ in [36] and [37].

Differentials in degree $\neq d_V$. To every $1 \leq q \leq k-1$ there is the sheaf $\mathcal{A}_V^q$ of meromorphic q-forms whose currents of bi-degree $(n-k+q, p)$ are $\bar{\partial}$-closed. The subsheaf $\mathbf{L}^2(\mathcal{A}_V^q)$ was defined in [12]. In [38] an inverse problem is studied: Let $\rho\colon\ W \to V$ be a proper and generically unramified map between analytic sets. A meromorphic k-form ϕ on V has a pull-back $\rho^*(\phi)$. If $\phi \in \mathcal{A}_V$ it is not necessarily true that $\rho^*(\phi) \in \mathcal{A}_W$. Situations where $\rho^*(\mathcal{A}_V) = \mathcal{A}_W$ are found in [38].

6.5 The Passare–Tsikh Example. The unrestricted limit in Theorem 6.1 fails in general. An example occurs already with a pair of functions in $\mathbf{C}^2$, i.e, in the zero-dimensional complete intersection. The first example was found in [90]. We describe another where one function even has non-zero differential at the origin. Let $f = w^3 - z - z^2$ and $g = z^m$ where $m \geq 2$. Put $\Gamma(\epsilon) = \{|f| = \epsilon \cap |z| = \epsilon\}$ and set

$$J(\epsilon) = \int_{\Gamma(\epsilon)} \frac{\bar{w} \cdot dz \wedge dw}{z^m} \tag{6.21}$$

If the unrestricted limit exists then $J(\epsilon) \to 0$ since the antiholomorphic function $\bar{w}$ annihilates the Coleff–Herrera current. However, for suitable large integers m the J-function is even unbounded as $\epsilon \to 0$. To see this we put $z = \epsilon \cdot e^{i\theta}$ and $f = \epsilon \cdot e^{i\phi}$. Then

$$J(\epsilon) = i \cdot \epsilon^{2/3-m+1} \cdot \int_0^{2\pi} e^{-i(m-1)\theta} \cdot \mathbf{W}\left(|1 + \epsilon \cdot e^{i\theta}|\right) \cdot d\theta\,, \tag{6.22}$$

where

$$\mathbf{W}(s) = \int_0^{2\pi} \frac{|e^{i\phi} - s|^{2/3}}{e^{i\phi} - s} \cdot e^{i\phi} \cdot d\phi \tag{6.23}$$

for every real s. The $\mathbf{W}$-function is piecewise real-analytic with a singularity at $s = 1$ which for suitable sufficiently large values of m imply that $J(\epsilon) \to \infty$. See [92] for details.

Remark. Above $f = w^3 - z - z^2$ is even a coordinate since the linear term z appears. The only positive result of a general nature is that the unrestricted limit exists when f, g is a pair of homogeneous polynomials in two variables. See [92]. In [86] it is proved that the limit in Theorem 6.1 exists and yields $\mathcal{R}_F$ when $\epsilon_\bullet \to 0$ outside a null-set in R_+^p, i.e., there exists a certain union of open and conic Passare sectors where the limit exists and yields the Coleff–Herrera current. The absolute case, i.e., when $p = n$ is studied in [114] where the existence of the Coleff–Herrera residue current is proved when $p = n$ without desingularisation including a description of Passare sectors. These results are not trivial since we now consider currents where non-holomorphic numerators occur in the Jacobi integral from Sect. 1.4

References

[1] Abel, N. H., *Mémoire sur une propriété génerale d'une classe très étendue de fonctions trancendantes*, Présenté à L'Académie des sciences à Paris le 30 Octobre 1826. Oeuvres complètes de Niels Henrik Abel. Christiania, 1881, vol. I, 145–211.

[2] Abel, N. H., *Sur l'intégration de la formule différentielle* $\frac{\rho}{\sqrt{R}}$, *R et* ρ *étant des fonctions entières*, 104–144.

[3] Abel, N. H., *Résolution d'un problème de mécanique*, 97–101.

[4] Abel, N. H., *Beweis der Unmöglichkeit algebraischen Gleichungen von Höheren graden als den vierten allgemein auflösen*, J. für die reine und angew. Math. Crelle **1** (1826), 65–84.

[5] D'Agnolo, A., *Sheaves and $\mathcal{D}$-modules in integral geometry*, Contemporary Math. **201** (2000) 141–161.

[6] Ajzenberg, L., Yuzhakov, A. P., *Integral representations and Residues in Multidimensional Complex analysis*, Engl. Transl. Amer. Math. Soc. Providence, R.I. **258** (1983).

[7] Alexander, H., *Holomorphic chains and the support hypothesis conjecture*, J. of Amer. Math. Soc. **10** (1997), 123–138.

[8] Andronikof, E., *The Kashiwara conjugation functor and wave-front sets of regular holonomic distributions on a complex manifold*, Invent. Math. **111** (1993), 35–49.

[9] Appel, P., Goursat, E., *Théorie des fonctions algébriques*, Gauthier-Villars, Paris (1930).

[10] Atiyah, M. F., *Resolutions of singularities and division of distributions*, Comm. Pure appl. math. **23** (1970), 145–150.

[11] Barlet, D., *Fonctions de type trace*, Ann. l'Inst. Fourier **33** (1983), 43–76.

[12] Barlet, D., *Le faisceaux $\omega_V^\bullet$ sur un espace analytique*, Séminaire Norguet 1975, Springer Lecture Notes **670** (1978), 187–204.

[13] Barlet, D., *Développements asymptotiques des fonctions obtenues par intégration sur les fibres*, Invent. Math. **68** (1982), 129–174.

[14] Barlet, D., *Contribution effective de la monodromie aux développements asymptotiques*, Ann. Sci. Ec. Norm. Sup. **4** (1978), 293–315.

[15] Barlet, D., Kashiwara, M., *Le réseau L^2 d'un système holonome égulier*, Invent. Math. **86** (1986), 35–62.

[16] Barlet, D., Maire, H., *Asymptotic expansion of complex integrals via Mellin transforms*, J. Funct. Anal. **83** (1989), no. 2, 233–257.

[17] Beilinson, A. A., Bernstein, J., Deligne, P., *Faisceaux pervers*, Asterisque **100** (1982).

[18] Beilinson, A. A., Bernstein, J., *Localisation des $\mathfrak{g}$-modules*, C. R. Acad. Sci. Paris Ser. I **292** (1981), 15–18.

[19] Berenstein, C. A., Gay, R., Yger, A., *Analytic continuations of currents and division problems*, Forum Math **1** (1989), 15–51.

[20] Bernstein, I. N., *Modules over rings of differential operators. An investigation of the fundamental solution of equations with constant coefficients*, Transl. Functional Anal. appl. **5** (1971), 89–101.

[21] Bishop, E., *Mappings of partially analytic spaces*, Amer. J. Math. **83** (1961), 209–242.

[22] Björk, J-E., *Rings of Differential Operators*, North. Holland Math. Lib. Ser. (1979).

[23] Björk, J-E., *Analytic $\mathcal{D}$-modules and Applications*, Math. and its Appl., Kluwer (1993) **247**.

[24] Björk, J-E., *Rigid bodies and the Bordin Prize*, Operator Theory, Adv. in Math. Birkhäuser **132** (2002), 55–80.

[25] Björk, J-E., *Non-commutative noetherian rings and the use of homological algebra*, J. of Pure and Applied Algebra **18** (1985), 111–119.

[26] Boman, J., *A local vanishing theorem for distributions*, C. R. Acad. Sc. Paris **315** (1992), 1231–1234.

[27] Borel, A. et al., *Algebraic $\mathcal{D}$-modules*, Academic Press (1987).

[28] Briançon, J., Skoda, H., *Sur la cloture intégrale d'un ideal de germes de fonctions holomorphes en un point de* $\mathbf{C}^n$, C. R. Acad. Sc. Paris Ser. A **278** (1974), 949–951.

[29] Bring, L., *Meletama quadem mathematica circa transformationem aequationum algebraicum*, University of Lund (1786).

[30] Coleff, N., Herrera, M., *Les courants residuels associés a une forme meromorphe*, Springer Lect. Notes **633** (1978).

[31] Deligne, P., *Equations différentiels à points singuliers réguliers*, Springer Lect. Notes **163** (1970).

[32] Dickenstein, A., Sessa, C., *Canonical representatives in moderate cohomology*, Invent. Math. (1986), 417–434.

[33] Dolbeault, P., *Formes différentielles et cohomologie sur une variété analytique complexe II*, Ann. of Math. **65** (1957), 282–330.

[34] Dolbeault, P., *General theory of multi-dimensional residues*, Encyclopedia Math. Sciences. Springer vol. 7 (1988).

[35] Dolbeault, P., *Valeur principales et operateurs differentiels semi-holomorphes*, Coll. Int. CNRS. Gauthiers-Villars, Paris (1972).

[36] Eriksen, E., *Graded $\mathcal{D}$-modules on monomial curves*, Thesis. University of Oslo (2000).

[37] Eriksson, A., *The ring of differential operators of a Stanley–Reisner ring*, Comm. Algebra **26** (1998), 4007–4013.

[38] Fabre, B., *On a problem of Griffiths: An inversion of Abel's theorem for families of zero-cycles*, Arkiv för Matematik **41** (2003). To appear.

[39] Gabber, O., *The integrability of the characteristic variety*, Amer. J. Math. **103** (1981), 445–468.

[40] Gelfand, I., Kapranov, M., Zelevinsky, A., *Discriminants, Resultants and Multidimensional Determinants*, Birkhäuser (1994).

[41] Ginsburg, V., *Characteristic varieties and vanishing cycles*, Invent. Math. **84** (1986), 327–403.

[42] Griffiths, P., *Periods of integrals on algebraic manifolds*, Bull. Amer. Math. Soc. (1970) **76**, 228–296.

[43] Griffiths, P., *On the periods of certain rational integrals I*, Ann. of Math. (1969) **90**, 460–495.

[44] Griffiths, P., *Variations on a theorem of Abel*, Invent. Math. **35** (1976), 319–390.

[45] Hamm, H., Zur analytischen und Algebraischen Beschriebung der Picard–Lefschetz Monodromie. Habilitationsschrift. Göttingen, 1974.

[46] Hartshorne, *Residues and duality*, Springer Lect. Notes **20** (1966).

[47] Henkin, G., *La transformation de Radon pour la cohomologie de Dolbeault et un théorème d'Abel inverse*, C. R. Acad. Sci. Paris **315** (1992), 973–978.

[48] Henkin, G., Leiterer, J., *Andreotti–Grauert Theory by integral formulas*, Progress in Math, Birkhäuser **74** (1988).

[49] Henkin, G., Passare, M., *Abelian diferentials on singular varieties and variations on a theorem of Lie–Griffiths*, Invent. Math. **135** (1999), 297–328.

[50] Herrera, M., Liebermann, D., *Residues and principal values on a complex space*, Math. Ann. **194**, (1971), 259–294.

[51] Hironaka, H., *Resolution of singularities on an algebraic variety over a field of characteristic zero*, Ann. of Math. (1954), 109–326.

[52] Holm, P., *Differential operators on arrangements of hyperplanes*, Thesis, Stockholm University (2002).

[53] Husson, E., *Sur un théorème de M. Poincaré, relativement au mouvement d'un solide pesant*, Acta Math. **31** (1908), 71–88.

[54] Hörmander, L., *The analysis of linear partial differential operators* **1**, Springer Verlag Grundlehren (1983).

[55] Jacobi, C. G. J., *Theoremata nova algebraica circa systema duarum aequationium inter duas variabiles prositarum*, Ges. Werke **III**, 285–294.

[56] Kashiwara, M., *The Riemann–Hilbert problem for regular holonomic systems*, Publ. RIMS, Kyoto (1984) **20**, 319–365.

[57] Kashiwara, M., *Distributions and regular holonomic $\mathcal{D}$-modules on complex manifolds*, Adv. Study Pure math **8** (1986), 199–206.

[58] Kashiwara, M., *b-functions and holonomic systems I*, Invent. Math. **39** (1976), 33–53.

[59] Kashiwara, M., *b-functions and holonomic systems II*, Invent. Math. **38** (1976), 121–135.

[60] Kashiwara, M., *Systems of micro-differential operators*, Progress in Math., Birkhäuser (1983).

[61] Kashiwara, M., *Hodge structure and holonomic systems*, Proc. Japan. Acad. 1–4 **62** Ser. A (1985).

[62] Kashiwara, M., *Algebraic study of systems of partial differential equations*, Thesis, Tokyo Univ. (1970).

[63] Kashiwara, M., *Index theorem for maximally over-determined systems*, Proc. Japan. Acad. **49** (1973), 803–804.

[64] Kashiwara, M., *On the maximally over-determined systems of linear differential equations*, Publ. RIMS Kyoto Univ. **10** (1975), 563–579.

[65] Kashiwara, M., Kawai, T., *Second microlocalisation and asymptotic expansions*, Lect. Notes in Physics, Springer **128** (1979), 21–77.

[66] Kashiwara, M., Kawai, T., *Holonomic character and local monodromy of Feynman integrals*, Comm. Math. Physics **54** (1977), 121–134.

[67] Kashiwara, M., Oshima, T., *Systems of differential equations with regular singularities and their boundary value problems*, Ann. of Math. **106** (1977), 145–200.

[68] Kashiwara, M., Schapira, P., *Sheaves on manifolds*, Springer Verlag Grundlehren **292** (1990).

[69] King, J. R., *The currents defined by analytic varieties*, Acta Math. **127** (1971), 185–220.

[70] Klein, F., *Vorlesungen über das Ikosaeder*, Teubner Verlag, Leipzig (1888).

[71] Kovalevsky, S., *Sur le problème de la rotation d'un corps solide autour d'un point fixe*, Acta Math. **12** (1889), 172–232.

[72] Lefschetz, S., *Selected papers*, Chelsea publ. company Bronx, New York (1917).

[73] Lelong, P., *Propriétés métriques des variétés définies par une équation*, Ann. Sci. Ec. Norm. Sup. (1950) **67**, 22–40.

[74] Malgrange, B., *Intégrales asymptotique et monodromie*, Ann. Sci. Ec. Norm. Sup. **4** (1974), 405–430.

[75] Malgrange, B., *Polynome de Bernstein–Sato et cohomologie évanescente*, Astérisque **101–102** (1983), 243–267.

[76] Malgrange, B., *Equations Différentielles à Coéfficients Polynomiaux*, Progress in Math., Birkhäuser **96** (1991).

[77] Mebkhout, Z., *Une équivalence de catégories et Une autre équivalence de catégories*, Compositio Math. **51** (1984), 55–62, 63–69.

[78] Mebkhout, Z., *Le formalisme des six operations de Grothendieck pour les $\mathcal{D}$-modules cohérents*, Hermann, Paris, 1989.

[79] Mebkhout, Z., *Le théorème de comparaison entre cohomologies de de Rham d'une variété algébrique complexe et le théorème d'existence de Riemann*, Publ. Math. IHES **69** (1989), 47–89.

[80] Mellin, H., *Résolution de l'équation algébrique générale à l'aide de la fonction Γ*, C. R. Acad. Sci **172** (1921), 658–661.

[81] Nilsson, N., *Some growth and ramification properties of certain integrals on algebraic manifolds*, Arkiv f. matematik **5** (1965), 463–476.

[82] Nilsson, N., *Monodromy and asymptotic integrals*, Arkiv f. matematik **18** (1980), 181–198.

[83] Oberst, U. *The construction of Noetherian operators*, J. of Algebra **222** (1995), 596–620.

[84] Palamodov, V. P., *Linear differential operators with constant coefficients*, Springer Grundlehren **168**, 1968.
[85] Palamodov, V. P., *The multiplicity of a holomorphic transform*, Functional Prilowzen **1** (1967), 54–65.
[86] Passare, M., *A calculus for meromorphic currents*, J. für die reine und angew. Math. **393** (1988), 37–56.
[87] Passare, M., *Residues, currents, and their relations to ideals of holomorphic functions*, Math. Scand. **62** (1988), 75–152.
[88] Passare, M., *Produit des courants résiduels et règle de Leibniz*, C. R. Acad. Sci. Paris **301** (1985), 727–730.
[89] Passare, M., *Courants méromorphes et égalité de la valeur principale et de la partie finie*, Sém. d'Analyse P. Lelong–P. Dolbeault–H. Skoda, Années 1986–86, Springer Lecture Notes **1295** (1987), 157–166.
[90] Passare, M., Tsikh, A., *Defining the residue current of a complete intersection*, Pitman Res. Notes Math. Series **347**, Longman Harlow, 1996.
[91] Passare, M., Tsikh, A., Yger, A., *Residue currents of the Bochner–Martinelli type*, Publ. Mat. **44** (2000), no. 1, 85–117.
[92] Pavlova, A. A., *On some residue currents for complete intersections*, Mathematical Models. Proc. International Conf. Krasnojarsk **2** (2001), 129–136.
[93] Pham, F., *Singularités des systèmes différentiels de Gauss–Manin*, Progress in Math., Birkhäuser (1979).
[94] Poly, J. B., *Sur l'homologie des courants à support dans un ensemble semi-analytique*, Bull. Soc. Math. France Mémoires **38** (1974), 35–43.
[95] Quillen, D., *Formal properties of over-determined systems of linear partial differential equations*, Thesis, Harvard Univ. (1964).
[96] Roos, J-E., *Détermination de la dimension homologique globale des algèbres de Weyl*, C. R. Acad. Sci. Paris **274** (1972).
[97] Roos, J-E., *Bidualité et structures des foncteurs derivées de Lim dans la catégorie des modules sur un anneaux régulier*, C. R. Acad, Sci. Paris **276** (1962).
[98] Routh, E., *Dynamics of rigid bodies*, Cambridge (1882).
[99] Sabbah, C., *Modules d'Alexander et $\mathcal{D}$-modules*, Duke Math. J. **60** (1989), 729–814.
[100] Sabbah, C., *Proximité évanescente. I. La structure polaire d'un $\mathcal{D}$-module*, Compositio Math. **62** (1987), 283–328.
[101] Sabbah, C., *II. Equations fonctionelles pour plusieurs fonctions analytiques*, Compositio Math. **64** (1987), 213–241.
[102] Sadykov, T. M., *On the Horn system of partial differential equations and series of hypergeometric type*, Math. Scand. **91** (2002), 127–149.
[103] Saito, M., *Modules de Hodge polarisables*, Publ. RIMS Kyoto **24** (1988), 849–995.
[104] Sato, M., Kashiwara, M., Kawai, T., *Hyperfunctions and pseudo-differential equations*, Springer Lecture Notes **287** (1973).
[105] Schapira, P., *Microdifferential systems in the complex domain*, Springer Grundlehren **269** (1985).
[106] Schapira, P., *Vanishing in highest degree for solutions of $\mathcal{D}$-modules and perverse sheaves*, Publ. RIMS, Kyoto **26** (1990), 535–538.
[107] Schiffman, B., *Complete characterization of holomorphic chains of codimension one*, Math. Ann **274** (1986), 233–256.
[108] Schneiders, J-P., *Un théorème de dualité pour les modules différentiels*, C. R. Acad. Sci. Paris **303** (1886), 235–238.
[109] Schwartz, L., *Courants associés a une forme différentielle méromorphe sur une variété analytique complexe*, Coll. Int. CNRS, geom. diff., Strasbourg **52** (1952).

[110] Serre, J. P., *Géométrie algébrique et géométrie analytique*, Ann. l'Inst. Fourier, Grenoble **6** (1955–56), 1–42.

[111] Steenbrink, J., *Limits of Hodge structures*, Invent. Math. **31** (1976), 229–257.

[112] Sturmfels, B., *Solving systems of polynomial equations*, CBMS Regional Conf. Series in Math., **97**, Amer. Math. Soc., Providence, RI, 2002.

[113] Tsikh, A. K., *Multidimensional residues and their applications*, Transl. of Math. Monographs. Amer. Math. Soc. **103** (1992).

[114] Tsikh, A., Yger, A., *Residue currents*, Encyclopedia in Mathematics, Nauka. Moscow (2003).

Algebraic Equations and Hypergeometric Series

Mikael Passare and August Tsikh

Summary. We study the solutions of a general n th order algebraic equation represented by multidimensional hypergeometric series. We provide a detailed description of the domains of convergence of these series in terms of the amoeba and the Horn–Kapranov uniformization of the corresponding discriminant. From a geometric viewpoint this amounts to describing the maximal Reinhardt domains in the complement of the discriminant locus.

1 Introduction

The problem of solving algebraic equations is ubiquitous in Mathematics, and it has a long history. A formidable contribution to the development of the theory of algebraic equations was made by Niels Henrik Abel, proving in 1824 his immortal theorem on the impossibility of solving the general quintic equation by means of radicals. (The complete text [1] was published in 1826.)

This meant that Algebra had to relinquish her monopoly in the field of algebraic equations, and mathematicians turned to the idea, already proposed by Viète over two hundred years earlier, of searching for analytic solutions. The first success in this direction was achieved in 1858 by Hermite and Kronecker, see [8], who were able to express the solution to the quintic equation by means of a modular elliptic function. (This function depends on one variable, connected with the parameter a of the equation $1+a\,x+x^5=0$; it was proved by Bring that any fifth degree equation can be brought into this form by means of Tschirnhaus transformations.)

Around the beginning of the last century another class of special functions made its way into the study of algebraic equations: these were the multivariate *hypergeometric* functions, given either as series [3], as integrals [10], or as solutions to differential equations [9]; see [2] for an extensive account.

Hypergeometric series are among the simplest transcendent functions, for it is possible to apply algebraic procedures to deal with them. This is inherent in their very definition as formulated by Horn in [6]: a power series in m variables is called hypergeometric if all its m ratios of subsequent coefficients are *rational* functions in the summation indices of the series. In the 1920's Birkeland published a series of papers, culminating with [3], where he presented a collection of hypergeometric series in the coefficients of the algebraic equation, which constitute solutions to the equation in certain open subsets of the coefficient space. More precisely, if one considers the general n th degree algebraic equation

$$a_0 + a_1\,x + a_2\,x^2 + \ldots + a_{n-1}\,x^{n-1} + a_n\,x^n = 0 \tag{1}$$

with complex coefficients $a = (a_0, \ldots , a_n)$, one quickly finds, see (2) below, that its root $x(a)$, viewed as a multivalued algebraic function, has a double homogeneity property; hence it may be treated as a function of only $n - 1$ variables. It is therefore enough to be able to solve equations of the type

$$a_0 + a_1\,x + \ldots + x^p + \ldots + x^q + \ldots + a_{n-1}\,x^{n-1} + a_n\,x^n = 0\,, \tag{1$'$}$$

where the coefficients of any two fixed monomials have been frozen. The formulas (6) of Birkeland constitute Taylor series solutions to (1′) in the remaining variables $a_0, a_1, .. [p] .. [q] .., a_n$. By means of a simple monomial substitution corresponding to the bi-homogeneity (2), these Taylor series may at will be turned into Laurent–Puiseux series solutions of the original equation (1).

The aim of our present investigation is to describe the convergence regions of Birkeland's Taylor series solutions to (1′), and consequently, of the Laurent–Puiseux series solutions to (1). The most basic result regarding the convergence of power series of one or several variables is Abel's lemma, which states that if a point a belongs to the domain of convergence D of a power series σ, then σ actually converges in the whole polydisk determined by a. That is, the Cartesian product of the disks $|z_\nu| < |a_\nu|$ is contained in D, and hence the domain of convergence is completely determined by the absolute values of each of the coordinates.

The qualitative structure of the domains of convergence of power series with singularities along algebraic hypersurfaces are closely related to the *amoeba* of this surface, that is, its projection on the space of absolute values in a logarithmic scale. For a power series that represents a rational or meromorphic function the domain of convergence D is given by $\mathrm{Log}^{-1}(E)$, where E is one of the connected components of the complement of the amoeba of the polar hypersurface. Here the principle that "power series converge up to the first singularity" applies in its purest form: "and no further". In our situation, where the power series will be representing the multivalued algebraic function $x\,(a)$ with branching singularities along the discriminant locus ∇ of equation (1), the picture is more complicated: here the series represents one or several of the branches of the full algebraic function $x(a)$, and may therefore fail to "recognize" certain pieces of the discriminant set as being part of its singularity. A combinatorial description of the domains of convergence of the Taylor series solutions to (1′) is given in Theorem 3, which says that the convergence domain D_{pq} of the series (6) projects onto a domain $\mathrm{Log}\,(D_{pq})$ containing a certain collection of the connected components of the complement of the discriminant amoeba.

In order to give a more exact description of the domains of convergence, we introduce the concept of the *contour* of the amoeba, as being the set of critical values of the mapping Log defined on the discriminant locus ∇_{pq}. The contour comprises the boundary of the amoeba, but it also contains the natural continuation of the boundary pieces that enter into the amoeba itself. In this direction we are able to exhibit a parametrization of the boundary of the convergence domains D_{pq} for certain of the series (Theorem 4), and for series depending on just two variables we obtain in Sect. 5.2 an ideal answer: a description of D_{pq} by means of explicit inequalities involving "reflections" of the discriminant. We finally remark that in the course of proving our main results, we also gained new insights regarding the discriminant locus ∇_{pq}, such as a description of its singular set (Theorem 2) and a parametrization (Theorem 1), all highly inspired by the two extraordinary papers by Horn [6] and Kapranov [7].

Both authors were supported by the Göran Gustafsson Foundation, and the second author also by the Russian Fund for Basic Research, grant 02-01-00167.

2 The Solutions as Hypergeometric Series

2.1 Dehomogenization. There are two obvious homogeneity relations associated with equation (1). Firstly, the solutions remain the same if the coefficient vector a is multiplied by any non-zero scalar λ. Secondly, division of the unknown x by a non-zero number λ is compensated by multiplying each coefficient a_ν with λ^ν. In other words, any root $x(a)$ of equation (1), considered as a function of the coefficients a, will have the twofold homogeneity property

$$x(\lambda_0 a_0, \lambda_0\lambda_1 a_1, \ldots, \lambda_0\lambda_1^n a_n) = \lambda_1^{-1}\, x(a)\,, \quad \lambda_0, \lambda_1 \in \mathbf{C}\setminus\{0\}\,. \tag{2}$$

One may thus regard the algebraic multivalued function $x(a)$ as a section of a certain holomorphic line bundle.

The presence of two homogeneities means that the roots x really depend only on $n-1$ variables, and one can actually dispose of any pair of the $1+n$ variable coefficients a_ν. Indeed, let us choose any two integers $0 \le p < q \le n$, that is, we fix an integer subsegment $[p,q] \subset [0,n]$. Then we let the two complex parameters λ_0 and λ_1 satisfy

$$\lambda_0\,\lambda_1^p = a_p^{-1}\,, \quad \text{and} \quad \lambda_0\,\lambda_1^q = a_q^{-1}\,. \tag{3}$$

It is important to observe that there are exactly $q-p$ possible choices for this pair of parameters: first we let λ_1 take any of the $q-p$ values of the radical $(a_p/a_q)^{1/(q-p)}$, and then the other parameter λ_0 is uniquely determined by either of the equations (3).

We shall use the notation $[p]$ to indicate that the index p should be omitted, and similarly for q. In view of the identity (2) this means that if $x\,(a_0, .. [p] .. [q] .., a_n)$ is a solution of the specialized equation $(1')$ then $\lambda_1\, x$ will solve the original equation (1) with coefficients $\tilde{a}_\nu = \lambda_0^{-1}\lambda_1^{-\nu}\, a_\nu$. Conversely, if $x\,(a)$ is a root of (1), then $\lambda_1^{-1}x$ will be a solution of $(1')$ with coefficients $\tilde{a}_\nu = \lambda_0\lambda_1^\nu\, a_\nu$.

Following Sturmfels [14], we also form a matrix out of the two homogeneity relations:

$$A = \begin{pmatrix} 1 & 1 & 1 & \cdots & 1 & 1 \\ 0 & 1 & 2 & \cdots & n-1 & n \end{pmatrix} .$$

The kernel of the corresponding linear mapping is of dimension $n-1$, and for any given choice of integers p and q as above, a basis for this kernel can be composed by the integer vectors

$$(q-p)\, e_\nu + (\nu - q)\, e_p + (p-\nu)\, e_q\,, \quad \nu \neq p, q\,, \tag{4}$$

with e_0, ..., e_n being the standard basis vectors in $\mathbf{R}^{1+n}$. We shall denote by B_{pq} the $(n+1)\times(n-1)$-matrix

$$B_{pq} = \left(\beta_\mu^\nu\right)\,, \tag{5}$$

whose column vectors β^ν are the basis vectors (4). One clearly has $A\,B_{pq} = 0$, and the square matrix obtained by deleting the rows β_p and β_q numbered p and q from B_{pq}, is equal to $q-p$ times the identity matrix of size $n-1$.

2.2 Series Solutions. If we put all coefficients a_ν in equation $(1')$ equal to zero, then we are left with the binomial equation $x^p + x^q = 0$. Disregarding the zero solution, we find that x is equal to any one of the radicals $(-1)^{1/(q-p)}$. It follows, from the implicit function theorem for analytic functions, that for each choice ϵ of this radical there exists a power series solution to $(1')$ which converges for small values of the coefficients a, and equals ϵ at the origin.

There is a classical method of Lagrange that in principle allows one to find the power series expansion of an implicitly given function, and by making clever use of this method, Birkeland was able, see [3], to explicitly write down the $q-p$ solutions to equation $(1')$ as the following power series:

$$\sum_{k\in\mathbf{N}^{n-1}} \frac{\epsilon^{-\langle\beta_q,k\rangle+1}}{(q-p)\,k!}\,\frac{\Gamma\big((-\langle\beta_q,k\rangle+1)/(q-p)\big)}{\Gamma\big(1+(\langle\beta_p,k\rangle+1)/(q-p)\big)}\,a_0^{k_0}a_1^{k_1}\cdot\cdot[p]\cdot\cdot[q]\cdot\cdot a_n^{k_n}\,. \tag{6}$$

Here $k! = k_0!\,k_1!\cdot\cdot[p]\cdot\cdot[q]\cdot\cdot k_n!$ and β_p, β_q are row vectors of the matrix B_{pq} from (5). A word of caution is in order here: the arguments of the Γ functions occurring in (6) will sometimes be negative integers, corresponding to poles of Γ.

However, the quotient of the two Γ functions should be thought of as a meromorphic function with *removable* singularities at the integer points. For instance, we mean that $\Gamma(-1)/\Gamma(-2) = -2\,\Gamma(-2)/\Gamma(-2) = -2$.

We shall *denote this series by* σ_{pq} even though it really does depend also on a choice of radical ϵ, which coincides with its value at the origin. The particular choice of ϵ will not influence the convergence properties of σ_{pq}. Whenever $q-p=1$ the series is truly hypergeometric in the sense of Horn, meaning that the quotient of subsequent coefficients are rational functions of k. As was pointed out by Birkeland, one can regroup (6) as a sum of $(q-p)^{n-1}$ hypergeometric series by simply taking the equivalence classes of each index k_ν modulo $q-p$. We shall here for convenience be referring to σ_{pq} itself as a hypergemetric series.

Our object in this note will be to determine, as exactly as possible, the domains of convergence of the series solutions (6). What is clear from the start is that each σ_{pq} will converge in some sufficiently small polydisk around the origin. In [3] Birkeland also gave an explicit, though rather crude, sufficient condition for the convergence, formulated in terms of a bound on the sum $|a_0| + |a_1| + .. [p] .. [q] .. + |a_n|$, but to the best of our knowledge no exact convergence criteria for σ_{pq} have hitherto been presented.

Let us point out that each series (6) will also give rise to a Laurent–Puiseux series solution of the original equation (1). Indeed, we just have to use the recipe given above, and write

$$\lambda_1\, \sigma_{pq}(\lambda_0\, a_0, \lambda_0\lambda_1\, a_1, .. [p] .. [q] .., \lambda_0\lambda_1^n\, a_n)\,,$$

with λ_0 and λ_1 given by (3). The fact that (6) converges for small enough $|a_\nu|$ will then be equivalent to the corresponding Laurent–Puiseux series converging for

$$|a_\nu|^{q-p}/\big(|a_p^{q-\nu}||a_q^{p-\nu}|\big) < M_\nu\,, \quad \nu = 0, 1, .. [p] .. [q] .., n\,,$$

with the positive constants M_ν being small enough.

3 The Discriminant

3.1 Definition and Basic Properties. When $n > 1$ there is intimately connected with equation (1) a remarkable polynomial Δ in the coefficients a. It is called the discriminant of the equation and it describes the occurrence of multiple roots in the sense that $\Delta\,(a) = 0$ if and only if (1) has at least one multiple root. The discriminant Δ is an irreducible polynomial with integer coefficients and may be expressed in terms of the roots $x_1, x_2, \ldots, x_n$ of the equation by the simple symmetric formula

$$\Delta = \pm\, a_n^{2(n-1)} \prod_{j<k} (x_j - x_k)^2\,.$$

We shall follow the convention of [5], Chap. 12, and choose the sign to be $(-1)^{n(n-1)/2}$.

A more explicit formula for Δ, directly in terms of the coefficients of the equation, is provided by the classical Sylvester determinant

$$\Delta = \frac{1}{a_n}\begin{vmatrix} a_0 & a_1 & a_2 & \dots & a_{n-1} & a_n & 0 & \dots \\ 0 & a_0 & a_1 & \dots & a_{n-2} & a_{n-1} & a_n & \dots \\ \vdots & \dots & & \dots & & \dots & & \vdots \\ a_1 & 2a_2 & \dots & (n-1)a_{n-1} & n\,a_n & 0 & 0 & \dots \\ 0 & a_1 & \dots & (n-2)a_{n-2} & (n-1)a_{n-1} & n\,a_n & 0 & \dots \\ \vdots & \dots & & \dots & & \dots & & \vdots \end{vmatrix}. \tag{7}$$

Here the first $n-1$ rows contain translates of the coefficient vector a, and the remaining n rows involve translates of the coefficients of the differentiated left hand side of (1). For $n = 2$ this gives $\Delta = 4\,a_0a_2 - a_1^2$, and for $n = 3$ one obtains

$$\Delta = 27\,a_0^2a_3^2 + 4\,a_1^3a_3 + 4\,a_0a_2^3 - 18\,a_0a_1a_2a_3 - a_1^2a_2^2\,.$$

The number of terms in Δ grows rapidly with increasing n, and so does the size of its coefficients. For $n = 4$ there are sixteen terms, and for $n = 5$ fifty-nine terms. For any n the monomials $n^n a_0^{n-1}a_n^{n-1}$ and $\pm a_1^2a_2^2\cdots a_{n-1}^2$ occur in Δ, and they are among the 2^{n-1} extremal terms of the discriminant for which a simple combinatorial formula will be given below.

We are going to denote by ∇ the zero locus of the discriminant Δ. Geometrically, this hypersurface ∇ is the dual variety of the Veronese curve $[1 : x : x^2 : \ldots : x^n]$ in $\mathbf{CP}^n$, that is, it consists of those vectors $a \in \mathbf{C}^{1+n}$ for which the corresponding hyperplane $a_0z_0 + a_1z_1 + \ldots + a_nz_n = 0$ is not transversal to the parametrized curve $z_\nu = x^\nu$.

The bi-homogeneity (2) of the roots has the following analog for the discriminant:

$$\Delta\,(\lambda_0\,a_0, \lambda_0\lambda_1 a_1, \ldots, \lambda_0\lambda_1^n a_n) = \lambda_0^{2(n-1)}\lambda_1^{n(n-1)}\Delta\,(a)\,. \tag{8}$$

Now let Δ_{pq} denote the polynomial obtained from the homogeneous discriminant Δ by simply setting a_p and a_q equal to 1. Then Δ_{pq} depends only on the remaining $n-1$ variables; in fact, Δ_{pq} is the discriminant of the equation (1′). Using again the parameters λ_0 and λ_1 that satisfy (3), we deduce from the identity (8) that

$$\Delta_{pq}(\tilde{a}_0, \ldots [p] \ldots [q] \ldots, \tilde{a}_n) = a_p^{s_p} a_q^{s_q}\,\Delta(a)\,, \tag{9}$$

where $\tilde{a}_\nu = a_\nu\, a_p^{(\nu-q)/(q-p)}\, a_q^{(p-\nu)/(q-p)}$ and the exponents s_p and s_q are given by $(n-2q)(n-1)/(q-p)$ and $(2p-n)(n-1)/(q-p)$ respectively.

The equation $\Delta_{pq} = 0$ can be interpreted as the homogeneous equation $\Delta = 0$ expressed in suitable local coordinates where $a_pa_q \neq 0$, and in accordance with the theory of toric manifolds the connection between these equations is given by an identity

$$\Delta_{pq}\big(a^{B_{pq}^{\mathrm{tr}}/(q-p)}\big) = a_p^{s_p}\, a_q^{s_q}\,\Delta(a)\,, \tag{10}$$

where $B^{\mathrm{tr}}_{pq}/(q-p)$ denotes the transpose of the matrix B_{pq} divided by the scalar $q-p$. Since $a_\nu a_p^{(\nu-q)/(q-p)} a_q^{(p-\nu)/(q-p)} = a^{\beta^\nu/(q-p)}$ the equality (10) is in fact the same as equation (9) above. We emphasize once more that in all these identities there are really only radicals of the single quantity a_p/a_q that produce any ambiguity.

3.2 Subdivisions, Polytopes, and Amoebas. A *subdivision* of an integer line segment $[0,n]$ is the collection of adjacent subsegments obtained by dividing the original segment at the points of some subset $I \subset \{1,2,\ldots,n-1\}$.

The *Newton polytope* of a polynomial is the convex hull of the set of exponent vectors of its monomials, with these vectors being considered as integer lattice points in the corresponding real vector space.

The *amoeba* of a polynomial is the image of its zero locus under the mapping Log, which sends each variable to the logarithm of its absolute value.

We shall let the Newton polytope and the amoeba of the discriminant Δ of equation (1) be denoted by $\mathcal{N}$ and $\mathcal{A}$ respectively. We remark in passing that the Newton polytope of the polynomial on the left hand side of (1) is equal to the line segment $[0,n]$.

It will be of importance for our investigations that *there exist natural bijections between the following sets*:

$$\{\text{subdivisions of } [0,n]\} \leftrightarrow \{\text{vertices of } \mathcal{N}\} \leftrightarrow \{\text{components of } \mathbf{R}^{1+n} \setminus \mathcal{A}\}$$

Let us explain these bijections in some detail. The first one, between subdivisions and vertices, is a special instance of a remarkable result from [5] relating triangulations of certain polytopes (in our case the segment $[0,n]$) to extremal monomials in general discriminants (in our case the classical discriminant Δ). To any given subdivision $[0,i_1],[i_1,i_2],\ldots,[i_s,n]$, obtained from an integer subset $I=\{i_1<i_2<\cdots<i_s\}$, one assigns the vector

$$(d_0-1)e_0+(d_0+d_1)e_{i_1}+(d_1+d_2)e_{i_2}+\ldots+(d_{s-1}+d_s)e_{i_s}+(d_s-1)e_n\,, \tag{11}$$

where $e_0,e_1,\ldots,e_n$ are the standard basis vectors in $\mathbf{R}^{1+n}$, and d_ν is the length of the subsegment $[i_\nu,i_{\nu+1}]$; we set $i_0=0$ and $i_{s+1}=n$. This vector is one of the 2^{n-1} vertices of the Newton polytope $\mathcal{N}$, and, since each different subset I gives rise to a different vector in this way, the described assignment does indeed establish a bijection between subdivisions and vertices. In fact, the vertex monomial corresponding to I appears in Δ with the coefficient

$$\prod_{\nu=0}^{s}\left[(-1)^{i_\nu}d_\nu\right]^{d_\nu}.$$

For instance, when $n=2$ the two subdivisions $[0,2]$ and $[0,1],[1,2]$ correspond to the vertex vectors $(1,0,1)$ and $(0,2,0)$ respectively, and the two vertex monomials are $4a_0a_2$ and $-a_1^2$. In the case $n=3$ there are four subdivisions $[0,3]$;

[0, 1], [1, 3]; [0, 2], [2, 3]; [0, 1], [1, 2], [2, 3] with associated vertices (2, 0, 0, 2), (0, 3, 0, 1), (1, 0, 3, 0), (0, 2, 2, 0), and corresponding monomials $27\, a_0^2 a_3^2$, $4\, a_1^3 a_3$, $4\, a_0 a_2^3$, $-a_1^2 a_2^2$.

One may also describe the Newton polytope $\mathcal{N}$ by means of explicit equations and inequalities. Indeed, it is equal to the skew cube of dimension $n-1$ determined by the two homogeneity relations $t_0+t_1+\ldots+t_n = 2(n-1)$ and $t_1+2\,t_2+\ldots+n\,t_n = n(n-1)$, together with the $2(n-1)$ linear inequalities

$$t_k \geq 0\,, \quad \sum_{j=1}^{n-1} \min(j,k)\big[n - \max(j,k)\big]\, t_j \leq nk(n-k)\,, \quad k = 1, 2, \ldots, n-1\,.$$

The other bijection, the one between vertices of the Newton polytope and connected components of the amoeba complement, has to do with normal cones and Laurent series expansions of rational functions.

The *normal cone* C_v at the vertex v of a polytope is the closed convex polyhedral cone, having its apex at the origin, spanned by the outer normals of those faces of the polytope that come together at v. The union of the normal cones at all the vertices of the polytope equals the entire ambient vector space, and the intersection $C_v \cap C_w$ of any two normal cones is a polyhedral cone of codimension equal to the dimension of the smallest face containing the two vertices v and w. The *recession cone* of a convex set E is the maximal cone C with apex at the origin such that the vector sum $E + C$ is contained in E. If E is bounded then $C = \{0\}$.

A multidimensional *Laurent series* is a power series of the form

$$\sum_{k \in \mathbf{Z}^{1+n}} c_k\, a_0^{k_0} a_1^{k_1} \cdots a_n^{k_n}\,, \qquad c_k \in \mathbf{C}\,.$$

Its domain of convergence is given by $\operatorname{Log}^{-1}(U)$ for some convex domain $U \subset \mathbf{R}^{1+n}$. Any holomorphic function in a domain of the type $\operatorname{Log}^{-1}(U)$ can be represented by a unique Laurent series that converges in that domain. For every polynomial f there is a one-to-one correspondence between the Laurent expansions of $1/f$ and the connected components of the amoeba complement $\mathbf{R}^{1+n} \setminus \mathcal{A}_f$.

Let $\{E\}$ denote the family of connected components of the complement of $\mathcal{A}_f$. It was shown in [4] that there is an injective mapping $\nu\colon \{E\} \to \mathcal{N}_f \cap \mathbf{Z}^{1+n}$ with the property that the normal cone to $\mathcal{N}_f$ at $\nu(E)$ is equal to the recession cone of E. In particular, if the recession cone of E has a non-empty interior, then $\nu(E)$ is a vertex of the Newton polytope $\mathcal{N}_f$.

Conversely, given a vertex v of the Newton polytope $\mathcal{N}_f$, one can obtain a convergent Laurent series by identifying the vertex monomial $f_v\, a^v$ in f and forming a geometric progression

$$\frac{1}{f(a)} = \sum_{k=0}^{\infty} \frac{(f_v\, a^v - f(a))^k}{(f_v\, a^v)^{1+k}}\,.$$

The domain of convergence for the Laurent series that arises in this way will have the normal cone C_v as its recession cone. In fact, it coincides with the unique component E of $\mathbf{R}^{1+n} \setminus \mathcal{A}_f$ for which $\nu(E) = v$.

For a general polynomial f the mapping ν provides a bijection between the set of vertices of $\mathcal{N}_f$ and a *subset* of the family of complement components $\{E\}$. However, if we now specialize, and take for f a discriminant polynomial Δ, then the number of complement components is minimal and they all correspond to vertices of the Newton polytope. A proof of this fact is given in [12].

Figure 1 illustrates the amoebas of the discriminants Δ_{03} and Δ_{01} for the case $n = 3$. Notice that the boundary curves cut into the amoebas. These entire curves constitute what we call the contours of the amoebas, and will be carefully studied in connection with the Horn–Kapranov uniformization below.

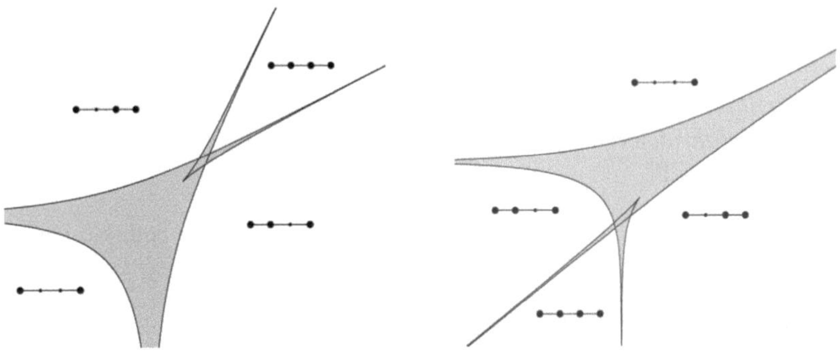

Fig. 1. The cubic [0, 3] and [0, 1] discriminant amoebas (*shaded*), their contours (*solid*), and the subdivisions associated with the complement components

3.3 Uniformization of the Discriminant Locus. By means of the matrix B_{pq} we now define a multivalued algebraic mapping

$$\psi_{pq} \colon \mathbf{CP}^{n-2} \to \mathbf{C}^{n-1}$$

from projective space into the space of coefficients $(a_0, .. [p] .. [q] .., a_n)$ of the equation (1′), by setting

$$\psi_{pq}(s_1 : s_2 : \ \ldots : s_{n-1}) = \left(\prod_{j=0}^{n} \langle \beta_j, s \rangle^{\beta_j^0/(q-p)}, .. [p] .. [q] .., \prod_{j=0}^{n} \langle \beta_j, s \rangle^{\beta_j^n/(q-p)} \right),$$

with the vectors $\beta_0, \ldots, \beta_n$ being the row vectors of the matrix B_{pq}. More precisely, in view of (4) we may write the ν th component of ψ_{pq} as

$$\left[\frac{\langle \beta_p, s \rangle}{\langle \beta_q, s \rangle} \right]^{\frac{\nu-p}{q-p}} \frac{\langle \beta_\nu, s \rangle}{\langle \beta_p, s \rangle},$$

and thus the multi-valuedness arises only from the choice of a radical $\left[\langle\beta_p, s\rangle/\langle\beta_q, s\rangle\right]^{1/(q-p)}$. It follows that ψ_{pq} is a one-to-$(q-p)$ mapping.

The following theorem generalizes (a special instance of) a result of Kapranov, who proved the case $q-p=1$ in [7]. He called ψ_{pq} the Horn uniformization, since it first occurred in the paper [6] in connection with the convergence of hypergeometric series. Recall that the *logarithmic Gauss mapping* of a hypersurface $\nabla=\{f(a)=0\}$ is defined as the mapping γ from ∇ into projective space with components given by $\gamma_\nu(a)=a_\nu\,\partial f(a)/\partial a_\nu$. Geometrically γ is obtained (locally) by taking the image of ∇ under the coordinate-wise mapping $a_\nu\mapsto\log a_\nu$, and then letting $\gamma(a)$ be the normal line to $\log(\nabla)$ at the point $\log a$. We use the same notation γ for the Gauss mapping of any of the discriminant hypersurfaces ∇_{pq}.

Theorem 1. *The mapping ψ_{pq} parametrizes the discriminant hypersurface*

$$\nabla_{pq}=\left\{\Delta_{pq}=0\right\}.$$

Moreover, it is inverse to the logarithmic Gauss mapping $\gamma\colon\nabla_{pq}\to\mathbf{CP}^{n-2}$, which hence realizes the discriminant hypersurface ∇_{pq} as a $(q-p)$-fold covering of the projective space $\mathbf{CP}^{n-2}$.

Proof. Let us first show that $\psi_{pq}(\mathbf{CP}^{n-2})\subset\nabla_{pq}$. To this end, we begin by observing that

$$\Delta(\langle B_{pq}, s\rangle)=0\,,$$

with $\langle B_{pq}, s\rangle$ denoting the vector whose components are

$$\langle\beta_0, s\rangle, \langle\beta_1, s\rangle, \ldots, \langle\beta_n, s\rangle\,.$$

Indeed, in the Sylvester matrix (7) for the polynomial $\Delta(\langle B_{pq}, s\rangle)$ there are $n-1$ rows containing translates of the vector $\langle B_{pq}, s\rangle$ with the remaining entries being equal to zero, and the other n rows similarly all consist of translates of

$$\langle\beta_1, s\rangle, 2\langle\beta_2, s\rangle \ldots, n\langle\beta_n, s\rangle\,.$$

But since $AB_{pq}=0$, this means that the Sylvester matrix annihilates the vector $(1,1,\ldots,1)$, and hence the Sylvester determinant equals zero, which precisely says that $\Delta(\langle B_{pq}, s\rangle)=0$ as claimed. In view of formula (10) we therefore get the desired vanishing

$$\Delta_{pq}\left(\langle B_{pq}, s\rangle^{B^{\mathrm{tr}}/(q-p)}\right)=0\,.$$

Our next step will be to prove that the mapping $\tilde{\psi}_{pq}$, obtained from ψ_{pq} by raising each component to the power $q-p$, is a birational isomorphism from $\mathbf{CP}^{n-2}$ onto some hypersurface $\widetilde{\nabla}_{pq}\subset\mathbf{C}^{n-1}$, and furthermore, that its inverse is equal to the logarithmic Gauss mapping $\tilde{\gamma}$ of $\widetilde{\nabla}_{pq}$. The fact that the image $\widetilde{\nabla}_{pq}$ of the mapping $\tilde{\psi}_{pq}$ is indeed a hypersurface follows by checking that the Jacobian of $\tilde{\psi}_{pq}$ is generically non-zero. This is easily done, for instance by making a computation in the local coordinates $(1, s_1, s_2, ..[p]..[q].., s_n)$ on $\mathbf{CP}^{n-2} \setminus \{s_0 = 0\}$, in terms of which $\tilde{\psi}_{pq}$ has the form

$$\tilde{\psi}_{pq} = \left(\varphi_0, s_1^{q-p}\varphi_1, ..[p]..[q]..s_n^{q-p}\varphi_n\right),$$

with $\varphi_j(0) \neq 0$ because the row vectors β_p and β_q contain no zero coordinates. To see that $\tilde{\psi}_{pq}$ is inverse to the logarithmic Gauss mapping we use the symmetry relations

$$\frac{\partial \log \psi_k}{\partial s_j} = \frac{\partial \log \psi_j}{\partial s_k}, \quad j, k = 0, 1, ..[p]..[q].., n,$$

which hold for both for ψ_{pq} and $\tilde{\psi}_{pq}$. They imply that the Jacobian matrix J of the mapping $\operatorname{Log} \circ \tilde{\psi}_{pq}$ is symmetric, and that it is annihilated by the vector s. Hence the composition $\tilde{\gamma} \circ \tilde{\psi}_{pq}$ is indeed the identity mapping. Similary, we see that also $\gamma \circ \psi_{pq}$ reduces to the identity. This argument is the same as in [7] (the proof of the second half of part *b*) of Theorem 2.1).

All that now remains to be shown is that ψ_{pq} maps $\mathbf{CP}^{n-2}$ onto the discriminant hypersurface ∇_{pq}. Indeed, in analogy with the mapping $\tilde{\psi}_{pq}$ we see that each of the $q-p$ branches of ψ_{pq} is a local immersion. Hence the image of ψ_{pq} is a hypersurface. On the other hand, we know that the discriminant hypersurface $\nabla = \{\Delta = 0\}$ is irreducible, and in view of the identity (10) the hypersurface ∇_{pq} is also irreducible. From this we conclude that the image of ψ_{pq} is equal to ∇_{pq} as required. Theorem 1 is proved. ∎

Let us take a closer look at the relation between the hypersurfaces $\widetilde{\nabla}$ and ∇. To this end we consider the power mapping $F\colon \mathbf{C}^{n-1} \to \mathbf{C}^{n-1}$ given by

$$F_\nu(a) = a_\nu^{q-p}, \quad \nu \neq p, q,$$

and examine the full preimage $F^{-1}(\widetilde{\nabla}_{pq})$. Since $\tilde{\psi}_{pq} = F \circ \psi_{pq}$ we see that $\nabla_{pq} \subset F^{-1}(\widetilde{\nabla}_{pq})$. When $q - p = 1$ there is equality $\nabla_{pq} = F^{-1}(\widetilde{\nabla}_{pq}) = \widetilde{\nabla}_{pq}$, but in general the preimage of $\widetilde{\nabla}_{pq}$ under F will contain a number of "doubles" of the true discriminantal locus ∇_{pq}. This is made precise in the following proposition.

Proposition 1. *The preimage $F^{-1}(\widetilde{\nabla}_{pq})$ consists of $(q-p)^{n-2}$ irreducible components each having the same amoeba $\mathcal{A}_{pq}$ as the true discriminant hypersurface ∇_{pq}.*

Proof. We obtain the full preimage by composing ψ_{pq} with F, and then taking the $(q-p)$ th root of each component separately:

$$\left(\psi_{pq}^{q-p}\right)^{1/(q-p)} = \left\{\left(\omega^{j_0}\alpha\, a_0/a_p, ..[p]..[q].., \omega^{j_n}\alpha\, a_n/a_p\right)\right\}.$$

Here $\omega = \exp 2\pi i/(q-p)$, and α is a fixed value of the radical $(a_p/a_q)^{1/(q-p)}$, while the indices j_k run independently over the set $[0, q-p-1]$. The entire family of such multi-indices $(j_0, .. [p] .. [q] .., j_n)$ splits into $(q-p)^{n-2}$ groups, each consisting of $q-p$ multi-indices, which constitute cycles for the monodromy of the function $\tilde{\psi}^{1/(q-p)}$. Representatives for these cycles can for instance be obtained by fixing some $\nu \neq p, q$, and considering the set J_ν consisting of those multi-indices that satisfy $j_\nu = 0$. This means that we can write

$$F^{-1}(\widetilde{\nabla}_{pq}) = \left\{ \prod_{J_\nu} \Delta_{pq}(\omega^{j_0} a_0, .. [p] .. [q] .., \omega^{j_n} a_n) = 0 \right\} . \tag{12}$$

The factor corresponding to the zero multi-index is the true discriminant Δ_{pq}, and it is clear that all the other factors will also have the same amoeba. ■

We are now going to consider the restriction of the uniformization map ψ_{pq} to the real space $\mathbf{RP}^{n-2}$, and we introduce the following new notion.

Definition. *The set of critical values of the mapping* $\mathrm{Log}\colon \nabla_{pq} \to \mathbf{R}^{n-1}$ *is called the* contour *of the amoeba* $\mathcal{A}_{pq}$, *and it is denoted* $\mathcal{C}_{pq}$.

Observe in particular that the boundary $\partial\mathcal{A}_{pq}$ of the amoeba is contained in the contour $\mathcal{C}_{pq}$.

Proposition 2. *The contour* $\mathcal{C}_{pq}$ *of the amoeba* $\mathcal{A}_{pq}$ *coincides with the image of* $\mathbf{RP}^{n-2}$ *under the composed mapping* $\mathrm{Log} \circ \psi_{pq}$

Proof. We consider first the single-valued mapping $\tilde{\psi}_{pq}$. It is a known fact, see for instance [11] or [15], that the set of critical points for the mapping Log is equal to $\gamma^{-1}(\mathbf{RP}^{n-2})$, and the same relation holds true for corresponding mappings $\widetilde{\mathrm{Log}}$ and $\tilde{\gamma}$ defined on $\widetilde{\nabla}_{pq}$. In fact, since the image of $\mathbf{RP}^{n-2}$ under the mapping $\tilde{\psi}_{pq}$ is contained in the real part of $\widetilde{\nabla}_{pq}$, we actually have that

$$\tilde{\psi}_{pq}\left(\mathbf{RP}^{n-2}\right) = \widetilde{\nabla}_{pq} \cap \mathbf{R}^{n-1} = \mathrm{Crit}\left(\widetilde{\mathrm{Log}}\right) .$$

Since the mappings $\mathrm{Log} \circ F^{-1}$ and $\widetilde{\mathrm{Log}}$ are identical, the proof is complete. ■

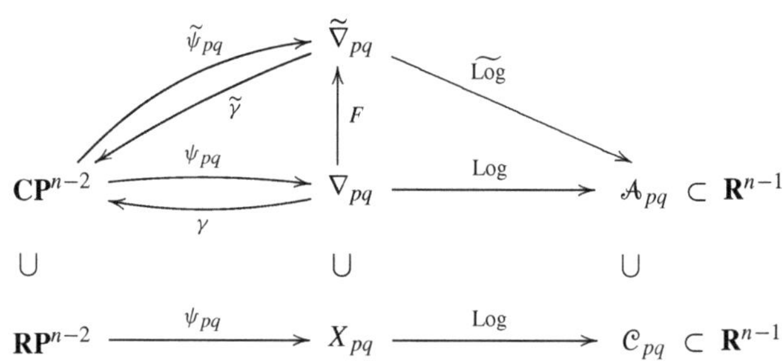

Fig. 2. The Horn–Kapranov parametrization ψ_{pq}, its interaction with the Gauss mapping γ, the Log mapping, and the contour $\mathcal{C}_{pq}$ of the amoeba $\mathcal{A}_{pq}$

3.4 Singularities of the Discriminant Locus. In this section we will show that the singular part of the discriminant hypersurface ∇_{pq} is parametrized by the restriction of the Horn–Kapranov mapping ψ_{pq} to a certain projective hyperplane. Recall that the row vectors β_p and β_q of the matrix B_{pq} have no coordinates equal to zero. They can thus be viewed as elements of the complex torus group $\mathbf{T}^{n-1} = \big(\mathbf{C} \setminus \{0\}\big)^{n-1}$, and we let $\beta_p\beta_q$ denote their product in this group:

$$\beta_p\beta_q = (\beta_{p0}\beta_{q0}, ..[p]..[q].., \beta_{pn}\beta_{qn}) .$$

Theorem 2. *The singular part of the discriminant hypersurface ∇_{pq} consists of the image $\psi_{pq}(L_{pq})$ of the projective hyperplane*

$$L_{pq} = \big\{\langle\beta_p\beta_q, s\rangle = 0\big\} \subset \mathbf{CP}^{n-2} .$$

Consequently, the singular part of the amoeba contour $\mathfrak{C}_{pq}$ consists of the image $\psi_{pq}(L_{pq} \cap \mathbf{RP}^{n-2})$ of the corresponding real projective hyperplane.

Proof. It will be convenient to relabel the components of the mapping ψ_{pq} as $\psi_1, \psi_2, \dots, \psi_{n-1}$, and to consider it as a homogeneous mapping defined on $\mathbf{C}^{n-1} \setminus \{0\}$. We must show that the rank of the Jacobian matrix $\big(\partial\psi_k/\partial s_j\big)$ drops below $n-2$ precisely on the hyperplane L_{pq}, and we begin by observing that it is equivalent to prove that the corresponding rank property holds for the logarithmic Jacobian matrix

$$J = \big(\partial \log \psi_k/\partial s_j\big), \qquad j, k = 1, 2, \dots, n-1 .$$

Note that the homogeneity of $\log \psi_k$ implies that we have an Euler formula $\sum_{j=1}^{n-1} s_j\, \partial\psi_k/\partial s_j = 0$, and that the determinant of the full matrix J therefore vanishes.

What has to be shown is thus that the common zero locus of all the $(n-2)$-minors of J is equal to the hyperplane L_{pq}. This will now be achieved by way of the following proposition, which provides explicit formulas for all these minors. ■

Proposition 3. *The $(n-2)$-minor J_{jk} of the matrix J, obtained by excluding the j th row and the k th column, is equal to*

$$J_{jk} = \frac{(-1)^{j+k}\, s_j s_k \langle\beta_p\beta_q, s\rangle}{s_1 s_2 \cdots s_{n-1}\langle\beta_p, s\rangle\langle\beta_q, s\rangle} . \tag{13}$$

Proof. In order to avoid double indexing we will temporarily rewrite the two rows β_p and β_q of the matrix B_{pq} as $(p_1, \dots, p_{n-1})$ and $(q_1, \dots, q_{n-1})$ respectively. We also write $P(s)$ and $Q(s)$ for the corresponding linear forms $\langle\beta_p, s\rangle$ and $\langle\beta_q, s\rangle$.

The minor $J_{n-1,1}$ equals the coefficient of $ds_1 \wedge \cdots \wedge ds_{n-2}$ in the differential form $d\log\psi_2 \wedge \cdots \wedge d\log\psi_{n-1}$, and keeping in mind the identities $\log\psi_k = \log s_k + p_k \log P(s) + q_k \log Q(s)$ and $dP \wedge dP = dQ \wedge dQ = 0$, it is straightforward to check that this coefficient is equal to

$$(-1)^{n-1}\left[\frac{p_1p_{n-1}}{s_2\cdots s_{n-2}P(s)}+\frac{q_1q_{n-1}}{s_2\cdots s_{n-2}Q(s)}+\sum_{j=2}^{n-2}\frac{\delta_{j1}\delta_{j,n-1}}{s_2\cdots[j]\cdots s_{n-2}P(s)Q(s)}\right],$$

where $\delta_{\mu\nu}$ denotes the determinant $p_\mu q_\nu - p_\nu q_\mu$. We can therefore write $J_{n-1,1}$ as $-N(s)/\big(s_2\cdots s_{n-2}P(s)Q(s)\big)$ with the linear numerator $N(s)$ given by

$$(p_1q_1q_{n-1}+p_1p_{n-1}q_1)\,s_1+\sum_{j=2}^{n-2}(\delta_{j1}\delta_{j,n-1}+p_jq_1q_{n-1}+p_1p_{n-1}q_j)\,s_j \\ +(p_{n-1}q_1q_{n-1}+p_1p_{n-1}q_{n-1})\,s_{n-1}.$$

Using the relations $p_\nu+q_\nu=-1$, which stem from the fact that the elements in each column of B sum to zero, it is now an easy task to rewrite the coefficient of s_ν in $N(s)$ as $-p_\nu q_\nu$, and hence we have verified the identity (13) for the case $(j,k)=(n-1,1)$.

A completely analogous argument will establish (13) for any (j,k) with $j\neq k$, the point being that one still immediately arrives at a linear numerator similar to $N(s)$. In order to settle the remaining diagonal cases J_{kk} we profit from the known vanishing of $\det J$ to get a linear relation

$$\sum_{j=1}^{n-1}(-1)^{j+k}\,\frac{\partial\log\psi_k}{\partial s_j}\,J_{jk}=0\,,$$

which forces (13) to hold true also for J_{kk}. This finishes the proof of Proposition 3. ∎

4 Domains of Convergence

Having gathered the necessary information on the discriminant and its amoeba in the previous section, we are now ready to deal with the problem of finding the domains of convergence of the various hypergeometric series σ_{pq}. We begin with a quite accurate combinatorial description of the convergence domains in terms of the connected components of the amoeba complements $\mathbf{R}^{n-1}\setminus\mathcal{A}_{pq}$.

Theorem 3. *The domain of convergence D_{pq} of the series* (6) *is a complete Reinhardt domain with the property that the corresponding convex domain* Log (D_{pq}) *contains all the connected components of the amoeba complement $\mathbf{R}^{n-1}\setminus\mathcal{A}_{pq}$ that are associated with subdivisions of $[0,n]$ containing the segment $[p,q]$, while it is disjoint from all the other complement components.*

Proof. We know from the implicit function theorem that D_{pq} is not empty, and Abel's lemma tells us that whenever a point a belongs to D_{pq}, then so does the full polydisk determined by a. Therefore D_{pq} is indeed a complete Rienhardt domain, and the corresponding domain $\mathrm{Log}\,(D_{pq})$ will contain the negative orthant $-\mathbf{R}_+^{n-1}$ in its recession cone. In fact, by looking at the restriction of the series (6) to the coordinate axes, and using the fact that any non-confluent hypergeometric series of one variable has a finite radius of convergence, we see that, conversely, the domain $\mathrm{Log}\,(D_{pq})$ must be contained in some translate of the negative orthant. In other words, the recession cone of $\mathrm{Log}\,(D_{pq})$ is exactly equal to $-\mathbf{R}_+^{n-1}$.

Let E be a connected component of $\mathbf{R}^{n-1} \setminus \mathcal{A}_{pq}$ that intersects the domain $\mathrm{Log}\,(D_{pq})$. Then we claim that we actually have an inclusion $E \subset \mathrm{Log}\,(D_{pq})$. Indeed, the holomorphic function $x\,(a)$ is well defined but a priori multi-valued in the domain $\mathrm{Log}^{-1}(E)$. One of its branches is given by the series (6), and hence admits a single-valued continuation to D_{pq}. Now, since the larger domain $D_{pq} \cup \mathrm{Log}^{-1}(E)$ is simply connected, the same branch remains in fact single valued on all of $\mathrm{Log}^{-1}(E)$, and it will thus have a convergent Laurent series expansion in this domain. But since this expansion coincides with the series (6) in the smaller domain $D_{pq} \cap \mathrm{Log}^{-1}(E)$, we conclude that the series (6) is actually convergent in all of $\mathrm{Log}^{-1}(E)$, and hence $E \subset \mathrm{Log}\,(D_{pq})$ as claimed.

It follows, from what we have proved so far, that the domain $\mathrm{Log}\,(D_{pq})$ cannot intersect any component of the amoeba complement $\mathbf{R}^{n-1} \setminus \mathcal{A}_{pq}$ whose recession cone is not contained in the negative orthant. On the other hand, every connected component of $\mathbf{R}^{n-1} \setminus \mathcal{A}_{pq}$ with a recession cone contained in $-\mathbf{R}_+^{n-1}$ will necessarily intersect, and hence be contained in, the domain $\mathrm{Log}\,(D_{pq})$. The following proposition therefore suffices to make the proof of Theorem 3 complete. ∎

Proposition 4. *The normal cone at a vertex of the Newton polytope $\mathcal{N}_{pq}$ is contained in the negative orthant $-\mathbf{R}_+^{n-1}$ if and only if the corresponding subdivision of $[0, n]$ contains the segment $[p, q]$. In fact, the union of such normal cones is equal to $-\mathbf{R}_+^{n-1}$.*

Proof. It will be enough to show that for any vertex v, one can find a good vertex w, that is, a vertex corresponding to a subdivision that includes $[p, q]$, such that the translated cone $w + \mathbf{R}_+^{n-1}$ contains v, and further, that no translate of $\mathbf{R}_+^{n-1}$ contains more than one good vertex.

To see this, we start by letting v be a bad vertex corresponding to a subdivision $[0, i_1], [i_1, i_2], \ldots, [i_s, n]$, which does not include the segment $[p, q]$. Then we can make it into a good subdivision by deleting all i_ν strictly between p and q, and adding p and q, if necessary. Recalling the assignment (11), we see that the new vertex w that we obtain in this way will have all its coordinates less than, or equal to, the ones of the vertex v that we started with. The cone $w + \mathbf{R}_+^{n-1}$ thus contains the vertex v.

On the other hand, we observe that all the good vertices span a face of the polytope $\mathcal{N}_{pq}$. To be precise, this face is given by the intersection of $\mathcal{N}_{pq}$ with the affine space

$$\begin{aligned} p\,t_0 + (p-1)\,t_1 + \ldots + t_{p-1} &= p\,(p-1)\,, \\ t_{p+1} = t_{p+2} = \ldots = t_{q-1} &= 0\,, \\ t_{q+1} + 2\,t_{q+2} + \ldots + (n-q)\,t_n &= (n-q)(n-q-1)\,. \end{aligned} \tag{14}$$

Here the first equation becomes vacuous when $p = 0$, and similarly for the last one when $q = n$. Since all coefficients in the equations (14) are positive, it is clear that no translate of $\mathbf{R}_+^{n-1}$ can contain more than one good vertex. ■

In the cases when $[p, q]$ is equal to $[0, n]$, $[0, 1]$ or $[n-1, n]$ we can determine the domains of convergence completely. This is the content of the following theorem. The result for $[0, n]$ was conjectured in [13], and proved there for $n = 3$.

Theorem 4. *For $[p, q] = [0, n]$, $[0, 1]$ or $[n-1, n]$, the restriction of* $\operatorname{Log}\circ\psi_{pq}$ *to* $\mathbf{RP}_+^{n-2}$ *parametrizes the boundary of the logarithmic convergence domain* $\operatorname{Log}(D_{pq})$ *for the series σ_{pq}.*

Proof. For simplicity we will denote the mapping ψ_{pq} and its components simply by $\psi = (\psi_1, \ldots, \psi_{n-1})$ throughout this proof. We will also consider it as a homogeneous mapping defined on $\mathbf{C}^{n-1} \setminus \{0\}$. In view of Horn's theorem on the convergence of hypergeometric series, it will be enough to prove that the surface $\operatorname{Log}\circ\psi(\mathbf{R}_+^{n-1})$ is convex, that is, that it lies "below" each of its tangent planes. Fix a point $s^0 \in \mathbf{R}_+^{n-1}$. The tangent space at the corresponding point $\operatorname{Log}\circ\psi(s^0)$ is then given by $\langle t, s^0\rangle = 0$. What we shall prove is thus that for any s the scalar λ determined by

$$\langle \lambda\, s + \operatorname{Log}\circ\psi(s) - \operatorname{Log}\circ\psi\left(s^0\right)\,,\, s^0\,\rangle = 0$$

is necessarily positive. In other words, we must show that the mapping $s \mapsto -\langle \operatorname{Log}\circ\psi(s), s^0\rangle$ has a minimum at $s = s^0$. The Hessian matrix of this mapping at the point s^0 is equal to

$$\left[-\sum_i s_i^0\,\frac{\partial^2 \log\psi_i(s^0)}{\partial s_j\,\partial s_k}\right] = \left[-\sum_i s_i^0\,\frac{\partial}{\partial s_i}\left(\frac{\partial \log\psi_k(s^0)}{\partial s_j}\right)\right], \tag{15}$$

and we claim that this is nothing but $J(s^0)$, with J denoting the Jacobian matrix $\left(\partial \log\psi_k/\partial s_j\right)$ that we encountered earlier. Indeed, this follows from (15) and the Euler identity, since the function $\partial \log\psi_k/\partial s_j$ is homogeneous of degree -1.

We know that J has one eigenvalue equal to zero, with the associated eigenvector s, and we want to show that for s in the open positive orthant all the remaining $n-2$ eigenvalues are strictly positive. Since the signature of J will vary continuously with s, it will remain unchanged as long as each $(n-2)$-minor J_{jk} stays non-zero and finite. In view of (13) this is violated only if one of the hyperplanes $\langle\beta_p, s\rangle = 0$, $\langle\beta_q, s\rangle = 0$, or $\langle\beta_p\beta_q, s\rangle = 0$ intersects the open positive orthant. Let us check that this does not happen in our cases. Indeed, for the case $[0, n]$ we get

$$\beta_p = -(n-1, n-2, \ldots, 1)\,, \quad \beta_q = -(1, 2, \ldots, n-1)\,,$$

while in the case [0, 1] we have

$$\beta_p = (1, 2, \dots, n-1), \quad \beta_q = -(2, 3, \dots, n),$$

and similarly for $[n-1, n]$.

All that remains is to verify that for some vector s in the positive orthant, say $s = (1, 1, \dots, 1)$, all the $n-2$ non-zero eigenvalues of $J(s)$ are stricly positive. Letting I denote the diagonal unit matrix, we can compute the characteristic polynomial of $J(s)$ as $\det(J - \lambda I)$. Looking first at the case $[0, n]$, we see that $J - \lambda I$ is equal to

$$(n-\lambda)\, I - \frac{2}{n(n-1)} \beta_p^{\mathrm{tr}} \beta_p - \frac{2}{n(n-1)} \beta_q^{\mathrm{tr}} \beta_q .$$

Since the last two matrices in this expression have rank one, it is a straightforward matter to compute the determinant, and as a result one sees that the characteristic polynomial equals $\pm\lambda(\lambda - n)^{n-3}(\lambda - 2(n+2)/3)$. This shows that all eigenvalues are positive as desired.

Similar computations for the cases [0, 1] and $[n-1, n]$ lead to the characteristic polynomial $\pm\lambda(\lambda-1)^{n-3}(\lambda-4(n+1)/3(n+2))$, and again we see that no eigenvalues are negative. The proof is complete. ■

5 Fewnomial Equations

5.1 Trinomials. In case our equation contains only three non-zero monomials, the procedure of reducing the number of parameters, by choosing a subsegment $[p, q]$ and putting $a_p = a_q = 1$, will leave us with a single remaining variable coefficient. The corresponding series solutions are thus one variable hypergeometric series, and their common domain of convergence is a disk. More precisely, suppose that the equation has the form

$$a_0 + a_m\, x^m + a_n\, x^n = 0,$$

with the two integer exponents $0 < m < n$ being relatively prime. The discriminant of this equation is then equal to the binomial

$$\Delta = n^n a_0^{n-m} a_n^m + (-1)^{m(n-m)} m^m (n-m)^{n-m} a_m^n .$$

It is obtained from the discriminant of the full nth degree equation by making the specialization $a_\nu = 0$ for $\nu \notin \{0, m, n\}$, and dividing out the common factor $a_0^{m-1} a_n^{n-m-1}$. Since m and n are not both even, the sign factor $(-1)^{m(n-m)}$ may be written $(-1)^{1+n}$.

As an illustration of our general approach we now choose the subsegment $[p, q] = [0, m]$. The corresponding inhomogeneous discriminant is given by $\Delta_{0m} = n^n a_n^m - (-1)^n m^m (n-m)^{n-m}$, and its zero locus ∇_{0m} consists of m points which are "parametrized" by the trivial Horn map

$$\psi : 1 \mapsto \left[\frac{m^m (n-m)^{n-m}}{(-n)^n}\right]^{1/m}.$$

The amoeba of Δ_{0m} is equal to the single point $\log|\psi(1)|$, and the connected component $\log|a_n| < \log|\psi(1)|$ of the complement of the amoeba is the one that corresponds to the partition $[0, m, n]$. Setting $a_n = a$, we can write down the m series solutions as

$$x(a) = \sum_{k=0}^{\infty} \epsilon^{1+nk} \frac{\Gamma\big((1+nk)/m)\big)}{\Gamma\big(1+(1+(n-m)k)/m\big)} \frac{a^k}{k!},$$

with ϵ being any m th root of -1. The radius of convergence of these series is $|\psi(1)|$, that is, the exponential of the amoeba point.

In particular, if $m = 1$ and $n = 2$ we recover the solution

$$x(a) = -\sum_{k=0}^{\infty} \frac{(2k)!}{(k+1)!} \frac{a^k}{k!} = \frac{-1+\sqrt{1-4a}}{2a}$$

to the quadratic equation $1 + x + a\,x^2 = 0$, whereas the choice $m = 1$ and $n = 3$ yields the root

$$\begin{aligned} x(a) &= -\sum_{k=0}^{\infty} (-1)^k \frac{(3k)!}{(2k+1)!} \frac{a^k}{k!} \\ &= -3 \Big/ \left\{1 + \left[\sqrt{1+\frac{27a}{4}} + \sqrt{\frac{27a}{4}}\right]^{2/3} + \left[\sqrt{1+\frac{27a}{4}} - \sqrt{\frac{27a}{4}}\right]^{2/3}\right\} \end{aligned}$$

of the cubic equation $1 + x + a\,x^3 = 0$.

5.2 Tetranomials. The next simplest case is the one where the left hand side of the equation consists of four monomials, that is, of a tetranomial. Here the dehomogenization procedure leads to a situation with two varying coefficients a_j and a_k. This means that for each choice of a subsegment $[p, q]$, the corresponding hypergeometric series solutions, as well as the inhomogeneous discriminant and its amoeba will all depend on these two variables a_j and a_k. We are now going to see that in the tetranomial case, which in particular includes the general cubic equation, the domains of convergence of the hypergeometric series solutions admit particularly satisfactory descriptions, not only in terms of the Horn uniformization and the amoeba, but also more directly by means of explicit polynomial inequalities related to the discriminant.

The general tetranomial equation is of the form

$$a_0 + a_\ell\, x^\ell + a_m\, x^m + a_n\, x^n = 0\,, \tag{16}$$

with three relatively prime integer exponents $0 < \ell < m < n$. Its discriminant Δ is homogeneous of degree $n + m - \ell$ and has the further homogeneity property $\Delta(a_0, \lambda^\ell a_\ell, \lambda^m a_m, \lambda^n a_n) = \lambda^{mn}$.

Let us first deal with the cubic case $\ell = 1$, $m = 2$, $n = 3$. If we choose $[p, q] = [0, 1]$ and use the inhomogeneous coordinate $(1, s)$ on $\mathbf{CP}^1$, then we explicitly have

$$B_{01} = \begin{pmatrix} 1 & 2 \\ -2 & -3 \\ 1 & 0 \\ 0 & 1 \end{pmatrix}, \quad \text{and} \quad \psi_{01}(s) = \left[\frac{1+2s}{(2+3s)^2}, -\frac{s\,(1+2s)^2}{(2+3s)^3} \right].$$

We know that $\psi_{01}(\mathbf{RP}^1) = \nabla_{01} \cap \mathbf{R}^2$, and also that $\operatorname{Log} \circ \psi_{01}(\mathbf{RP}^1) = \mathcal{C}_{01}$. By Horn's theorem it follows that the domain of convergence D_{01} for the series σ_{01} is determined by the part of the contour $\mathcal{C}_{01}$ given by $\operatorname{Log} \circ \psi_{01}(\mathbf{RP}^1_+)$. The corresponding part $\psi_{01}(\mathbf{RP}^1_+)$ of the real discriminant curve $\nabla_{01} \cap \mathbf{R}^2$ is located in the fourth quadrant. In fact, each of the four parts of $\nabla_{01} \cap \mathbf{R}^2$ that correspond to the parameter intervals $(-\infty, -2/3)$, $(-2/3, -1/2)$, $(-1/2, 0)$ and $(0, \infty)$ lie in different quadrants. This means that the boundary of D_{01} is given by the equation $\Delta_{01}(|a_2|, -|a_3|) = 0$. Looking more carefully, we can determine on which side of this surface the domain is located, and we find that

$$D_{01} = \left\{ (a_2, a_3) \in \mathbf{C}^2 ;\ 27\,|a_3|^2 - 4\,|a_3| + 4\,|a_2|^3 + 18\,|a_2||a_3| - |a_2|^2 < 0 \right\}.$$

A similar study of the other cases $[p, q]$ reveals the complete list of convergence domains:

$$\begin{aligned}
D_{02} &= \left\{ \Delta_{02}(|a_1|, |a_3|) > 0 \right\} \cap \left\{ \Delta_{02}(|a_1|, -|a_3|) < 0 \right\}, \\
D_{03} &= \left\{ \Delta_{03}(-|a_1|, -|a_2|) > 0 \right\}, \\
D_{12} &= \left\{ \Delta_{12}(|a_0|, -|a_3|) < 0 \right\} \cap \left\{ \Delta_{12}(-|a_0|, |a_3|) < 0 \right\}, \\
D_{13} &= \left\{ \Delta_{13}(|a_0|, |a_2|) > 0 \right\} \cap \left\{ \Delta_{13}(|a_0|, -|a_2|) < 0 \right\}, \\
D_{23} &= \left\{ \Delta_{23}(|a_0|, -|a_1|) < 0 \right\}.
\end{aligned}$$

We can use the same method to treat a general tetranomial equation (16), but in order to obtain precise discriminant inequalities as in the cubic case, we need to assume that ℓ is odd, m is even, and n is odd. Otherwise the four parts of the amoeba contour will not correspond to parts of the real discriminant curve in all the different quadrants. Let us be specific, and consider the quintic equation

$$1 + x + a\,x^2 + b\,x^5 = 0\,. \tag{17}$$

The discriminant of this equation is equal to $3125\,b^2 + 256\,b + 108\,a^5 - 27\,a^4 - 1600\,ab + 2250\,a^2 b$, and the hypergeometric series solution σ_{01} to (17) is given by

$$\sigma_{01} = -\sum_{j,k \geq 0} (-1)^k \frac{(2j+5k)!}{j!k!(j+4k+1)!}\, a^j b^k\,.$$

The domain of convergence D_{01} of this power series is now exactly described, in analogy with the corresponding series in the cubic case, by the single algebraic inequality

$$3125\,|b|^2 - 256\,|b| + 108\,|a|^5 - 27\,|a|^4 + 1600\,|a||b| - 2250\,|a|^2|b| \;<\; 0\,.$$

References

[1] Niels Henrik Abel: Beweis der Unmöglichkeit algebraische Gleichungen von höheren Graden als dem viertem allgemein aufzulösen, *J. Reine Angew. Math. (Crelle)* **1** (1826), 65–84.

[2] Giuseppe Belardinelli: *Fonctions hypergéométriques de plusieurs variables et résolution analytique des équations algébriques générales*, Mémorial des Sciences Mathématiques CXLV, Gauthier–Villars, Paris, 1960.

[3] Richard Birkeland: Über die Auflösung algebraischer Gleichungen durch hypergeometrische Funktionen, *Math. Z.* **26** (1927), 566–578.

[4] Mikael Forsberg, Mikael Passare & August Tsikh: Laurent determinants and arrangements of hyperplane amoebas, *Adv. in Math.* **151** (2000), 45–70.

[5] Israel Gelfand, Mikhail Kapranov & Andrei Zelevinsky: *Discriminants, resultants and multidimensional determinants*, Birkhäuser, Boston, 1994.

[6] Jacob Horn: Über die Convergenz der hypergeometrischen Reihen zweier und dreier Veränderlichen, *Math. Ann.* **34** (1889), 544–600.

[7] Mikhail Kapranov: A characterization of A-discriminantal hypersurfaces in terms of the logarithmic Gauss map, *Math. Ann.* **290** (1991), 277–285.

[8] Felix Klein: *Vorlesungen über das Ikosaeder und die Auflösung der Gleichungen vom fünften Grade*, Teubner, Leipzig, 1884.

[9] Karl Mayr: Über die Lösung algebraischer Gleichungssysteme durch hypergeometrische Funktionen, *Monatsh. Math.* **45** (1937), 280–313.

[10] Hjalmar Mellin: Résolution de l'équation algébrique générale à l'aide de la fonction gamma, *C. R. Acad. Sci. Paris Sér. I Math.* **172** (1921), 658–661.

[11] Grigory Mikhalkin: Real algebraic curves, the moment map and amoebas, *Ann. Math.* **151** (2000), 309–326.

[12] Mikael Passare, Timur Sadykov & August Tsikh: *Singularities of nonconfluent hypergeometric functions in several complex variables*, Preprint 126 (2000), Max-Planck-Institut für Mathematik, Bonn.

[13] Anastasia Semusheva & August Tsikh: *A continuation of Mellin's investigations on solutions to algebraic equations*, pp. 134–146 in "Kompleksnyĭ Analiz i Differentsial'nye Operatory", Publishing Center of the Krasnoyarsk State University, Krasnoyarsk, 2000. (Russian)

[14] Bernd Sturmfels: Solving algebraic equations in terms of A-hypergeometric series, *Discrete Math.* **210** (2000), 171–181.

[15] August Tsikh: Conditions for absolute convergence of the Taylor coefficient series of a meromorphic function of two variables, *Math. USSR-Sb.* **74** (1993), 337–359.

Dirichlet Series and Functional Analysis

Håkan Hedenmalm

1 Introduction

The study of Dirichlet series of the form $\sum_{n=1}^{\infty} a_n n^{-s}$ has a long history beginning in the nineteenth century, and the interest was due mainly to the central role that such series play in analytic number theory. The general theory of Dirichlet series was developed by Hadamard, Landau, Hardy, Riesz, Schnee, and Bohr, to name a few. However, the main results were obtained before the central ideas of Functional Analysis became part of the toolbox of every analyst, and it would seem a good idea to insert this modern way of thinking into the study of Dirichlet series. Some effort has already been spent in this direction; we mention the papers by Helson [20, 21] and Kahane [22, 23]. However, the field did not seem to catch on. It is hoped that this paper can act as a catalyst by pointing at a number of natural open problems, as well as some recent advances. Fairly recently, in [17], Hedenmalm, Lindqvist, and Seip considered a natural Hilbert space $\mathcal{H}^2$ of Dirichlet series and began a systematic study thereof. The elements of $\mathcal{H}^2$ are analytic functions on the half-plane

$$\mathbb{C}_{\frac{1}{2}} = \left\{ s \in \mathbb{C} : \ \mathrm{Re}\, s > \frac{1}{2} \right\}$$

of the form

$$f(s) = \sum_{n=1}^{+\infty} a_n n^{-s} \tag{1.1}$$

where the coefficients $a_1, a_2, a_3, \ldots$ are complex numbers subject to the norm boundedness condition

$$\|f\|_{\mathcal{H}^2} = \left(\sum_{n=1}^{+\infty} |a_n|^2\right)^{\frac{1}{2}} < +\infty .$$

In a natural sense, this is the analogue of the Hardy space H^2 for Dirichlet series. In [17], the pointwise multipliers of $\mathcal{H}^2$ were characterized, and the result was applied to a problem of Beurling concerning 2-periodic dilation bases in $L^2([0, 1])$. The reader is referred to [18] for some historical comments on the topic. We need to introduce the right half plane

$$\mathbb{C}_+ = \left\{s \in \mathbb{C} : \ \mathrm{Re}\, s > 0\right\} ,$$

and the space $\mathcal{H}^\infty$ of bounded analytic functions on $\mathbb{C}_+$ which are given by a convergent Dirichlet series of the form (1.1) in some possibly remote half-plane $\mathrm{Re}\, s > \sigma_0$. By a theorem of Schnee [31], which was later improved by Bohr [7], the Dirichlet series for a function in $\mathcal{H}^\infty$ actually converges on $\mathbb{C}_+$.

This note is an updated version of the survey paper (or problem collection) [16]. Since then, the area has received additional attention, and further progress has been made [2, 3, 4, 12, 24, 26], which is a good reason for making this survey available to a wider audience.

2 Multipliers

We formulate the main result of [17]. We say that an analytic function φ on the half-plane $\mathbb{C}_{\frac{1}{2}}$ is a *multiplier* on $\mathcal{H}^2$ if $\varphi f \in \mathcal{H}^2$ whenever $f \in \mathcal{H}^2$.

Theorem 2.1. *The collection of multipliers on $\mathcal{H}^2$ equals the space $\mathcal{H}^\infty$.*

The above theorem is analogous to the following well-known result for Hardy spaces: the (pointwise) multipliers of H^2 are the functions in H^∞. A noteworthy difference, however, is that the multipliers in the Dirichlet series case are defined as bounded and analytic on a bigger half-plane than the functions in the space. It should be mentioned that the proof of the above theorem in [17] is based on modelling $\mathcal{H}^2$ as the Hardy space on the infinite-dimensional polydisk $\mathbb{D}^\infty$, an idea which goes back to a 1913 paper of Bohr.

3 Convergence Issues

The convergence and analyticity of $f \in \mathcal{H}^2$ given by the series (1.1) in the half-plane $\mathbb{C}_{\frac{1}{2}}$ is a simple consequence of the Cauchy-Schwarz inequality. A deeper fact is that the boundary values of f on the 'critical' line $\partial\mathbb{C}_{\frac{1}{2}} = \{s \in \mathbb{C} : \operatorname{Re} s = \frac{1}{2}\}$ are locally L^2-functions (see [27, formula (29), p. 140] or [17, Theorem 4.11]). It is well-known that functions in $\mathcal{H}^2$ need not have any analytic continuations beyond the half-plane $\mathbb{C}_{\frac{1}{2}}$, and so the Dirichlet series need not converge in any strictly larger open half-plane. The question, then, is what happens precisely on the boundary $\partial\mathbb{C}_{\frac{1}{2}}$. Here, we can compare with Carleson's theorem for Fourier series: given $f \in L^2$ on the unit circle, the corresponding Fourier series converges almost everywhere [10]. Recently, Hedenmalm and Saksman [19] established the validity of the counterpart for Dirichlet series of Carleson's convergence theorem ($\mathbb{R}$ is the set of all real numbers).

Theorem 3.1. *Let* $\sum_{n=1}^{+\infty} |a_n|^2 < +\infty$. *Then the series*

$$\sum_{n=1}^{+\infty} a_n \, n^{-\frac{1}{2}+it}$$

converges for almost every $t \in \mathbb{R}$.

The proof uses an equivalent dual formulation of the strong L^2 maximal function estimate used to prove Carleson's theorem, in the form of a Strong Hilbert inequality. Konyagin and Queffélec later found a really short proof that is based on Carleson's convergence theorem for Fourier integrals [24].

The following problem seems natural.

Problem 1. Suppose the function

$$f(s) = \sum_{n=1}^{+\infty} a_n \, n^{-s}$$

belongs to $\mathcal{H}^\infty$, so that the series converges on $\mathbb{C}_+$. Does the series then also converge almost everywhere on the imaginary axis?

Recently, Bayart [5] obtained an intricate counterexample. His example shows that there exists a function $f \in \mathcal{H}^\infty$ whose Dirichlet series diverges almost everywhere on the imaginary axis.

We mention another type of convergence theorem. Given a function f of the form (1.1), we form the functions

$$f_\chi(s) = \sum_{n=1}^{\infty} a_n \, \chi(n) \, n^{-s}, \tag{3.1}$$

where $\chi(n)$ is a *character*, which means that $\chi(1) = 1$, $\chi(n) \in \mathbb{T}$ for all n, and $\chi(mn) = \chi(m)\,\chi(n)$ for all m and n. The functions f_χ are known as the *vertical limit functions* for f. The terminology is explained by the fact that $f_\chi(s)$ is obtained from f as a limit of a sequence of vertical translates $f(s - it)$, with $t \in \mathbb{R}$. Each character is determined uniquely by its values on the set of primes $\mathcal{P} = \{2, 3, 5, 7, 11, \dots\}$, and the values at different primes may be chosen independently of each other. The set of all characters is denoted by Ξ, and we realize that it can be equated with the infinite-dimensional polycircle $\mathbb{T}^\infty$ by identifying each dimension with a prime number (see [17] for details; see also [30]). The polycircle $\mathbb{T}^\infty$ has a natural product probability measure defined on it, denoted $d\varpi$, the product of the normalized arc length measure $d\sigma$ in each dimension. The set of characters Ξ constitutes the dual group of the multiplicative group of positive rationals $\mathbb{Q}_+$, if the latter is given the discrete topology. The Haar probability measure on the compact group Ξ coincides with $d\varpi$. A natural question arises: given $f \in \mathcal{H}^2$, what is the almost sure convergence behavior of the series (3.1) for $f_\chi(s)$, where s is a point in the complex plane, and χ is a character? It is mentioned in [17] that for almost all χ, $f_\chi(s)$ extends to a holomorphic function on the right half plane $\operatorname{Re} s > 0$, and that this is best possible. The behavior of most of the vertical limit functions is thus in sharp contrast with that of individual functions! As a matter of fact, in [21] (see also [17], Theorem 4.4), Helson shows that for almost all χ, the Dirichlet series (3.1) actually converges in the half-plane $\operatorname{Re} s > 0$. By Theorem 4.1 of [17], the function $f_\chi(it)$ makes sense as a locally L^2 summable function on the real line, for almost all χ. This makes us suspect that we have convergence in (3.1) for almost all s on the imaginary line $\operatorname{Re} s = 0$ and almost all χ. In [19], the following theorem is obtained.

Theorem 3.2. *Let $f \in \mathcal{H}^2$ be of the form (1.1), and let $f_\chi \in \mathcal{H}^2$ be defined by (3.1). Then the series*

$$f_\chi(it) = \sum_{n=1}^{\infty} a_n\,\chi(n)\,n^{-it}$$

converges for almost all characters χ and almost all reals t.

It is possible to use the above theorem to derive estimates of the almost sure growth behavior of partial sums of random characters. More precisely, we have, almost surely,

$$\sum_{n=1}^{N} \chi(n) = O\left(\sqrt{N \log N}\,\left(\log\log N\right)^{1/2+\varepsilon}\right), \qquad \text{as } N \to +\infty\,.$$

Problem 2. Find the best possible growth bound for the almost sure behavior of the above partial sums.

This problem has an unmistakable Erdös-type flavor, in its combination of probability and number theory. And sure enough, in [11, pp. 251–252], Erdös states as a problem to determine the almost sure growth of the analogous sums, where the

$\chi(p)$ for prime indices p are replaced by independent random variables assuming the values ± 1 with equal probabilities $\frac{1}{2}$. Erdös looks to compare the growth of the partial sums with the classical law of the iterated logarithm (see [33]), where all the terms $\chi(n)$ are independent and take values ± 1 with equal probabilities $\frac{1}{2}$. In Erdös' problem, as in ours, the characters have the multiplicative property $\chi(mn) = \chi(m)\chi(n)$, which reduces the randomness and introduces a number-theoretic ingredient. A complete solution should thus shed light on the multiplicative structure of the integers. Some progress on Erdös' problem was obtained by Halász [14].

If we believe that only the random element contributes to the almost sure growth of the partial sums, a natural conjecture would be that

$$\sum_{n=1}^{N} \chi(n) = O\left(\sqrt{\frac{N \log\log N}{\log N}}\right), \qquad \text{as } N \to +\infty ,$$

holds almost surely, and that this is best possible.

4 Composition Operators

Let $f \in \mathcal{H}^2$ be of the form

$$f(s) = \sum_{n=1}^{+\infty} a_n n^{-s}, \qquad s \in \mathbb{C}_{\frac{1}{2}} .$$

Fix a $k = 1, 2, 3, \ldots$. Then

$$f_k(s) = f(ks) = \sum_{n=1}^{+\infty} a_n n^{-ks}, \qquad s \in \mathbb{C}_{\frac{1}{2}} ,$$

is another function in $\mathcal{H}^2$, of the same norm as f. In other words, if $\Phi(s) = ks$, and $\mathcal{C}_\Phi$ is the associated composition operator,

$$\mathcal{C}_\Phi f(s) = f \circ \Phi(s) , \qquad s \in \mathbb{C}_{\frac{1}{2}} ,$$

then $\mathcal{C}_\Phi$ is an isometry on $\mathcal{H}^2$. One would tend to ask what other kinds of composition operators might be around. Recently, Gordon and Hedenmalm found a complete answer to this question. The space $\mathcal{D}$ consists of somewhere convergent Dirichlet series.

Theorem 4.1. *An analytic function $\Phi : \mathbb{C}_{\frac{1}{2}} \to \mathbb{C}_{\frac{1}{2}}$ generates a bounded composition operator $\mathcal{C}_\Phi : \mathcal{H}^2 \to \mathcal{H}^2$ if and only if:*
(a) it is of the form

$$\Phi(s) = ks + \phi(s) ,$$

where $k \in \{0, 1, 2, 3, \ldots\}$ and $\varphi \in \mathcal{D}$; and
(b) Φ has an analytic extension to $\mathbb{C}_+$, also denoted by Φ, such that
(i) $\Phi(\mathbb{C}_+) \subset \mathbb{C}_+$ if $k > 0$, and
(ii) $\Phi(\mathbb{C}_+) \subset \mathbb{C}_{\frac{1}{2}}$ if $k = 0$.

This constitutes a genuine Dirichlet series analogue of Littlewood's subordination principle [25]. Indeed, in case Φ fixes the point $+\infty$, which happens precisely when $k > 0$, the composition operator $\mathcal{C}_\Phi$ is a contraction on $\mathcal{H}^2$.

Note that we again have this dichotomy that sometimes the half-plane $\mathbb{C}_{\frac{1}{2}}$ is relevant, and sometimes we need the whole right half plane $\mathbb{C}_+$ instead.

Problem 3. Suppose $\alpha = \Phi(+\infty) \in \mathbb{C}_{\frac{1}{2}}$, and $\Phi(\mathbb{C}_+) \subset \mathbb{C}_{\frac{1}{2}}$. Try to estimate from above the norm $\|\mathcal{C}_\Phi\|$ in terms of α. Note that it is clear that $\zeta(2\,\mathrm{Re}\,\alpha) \le \|\mathcal{C}_\Phi\|^2$.

Problem 4. Characterize the compact composition operators on $\mathcal{H}^2$.

For the Hardy space H^2 on the unit disk, Shapiro has characterized the compact composition operators [32] in terms of the Nevanlinna counting function. So the question is whether anything similar is possible for the space $\mathcal{H}^2$. Some progress has been made on this problem [3, 4, 12]; however, we are still far from a definitive answer.

5 Integral Means

It is well-known that the norm on $\mathcal{H}^2$ can be expressed in terms of integral means of the function itself, provided the function is "nice". Suppose

$$f(s) = \sum_{n=1}^{+\infty} a_n\, n^{-s}\,,$$

where the sum is finite, that is, all but finitely many of the a_n's are 0. We might call such functions *Dirichlet polynomials*. Then

$$\frac{1}{2T}\int_{-T}^{T} |f(\sigma+it)|^2 dt \to \sum_{n=1}^{+\infty} \frac{|a_n|^2}{n^{2\sigma}} \quad \text{as } T \to +\infty\,, \tag{5.1}$$

for each real σ. We can think of this as a Plancherel formula. However, it is not really useful for calculating the norm of functions in $\mathcal{H}^2$, as such functions need not even be defined along the imaginary axis where the integral mean should then be computed. In fact, functions in $\mathcal{H}^2$ need only be defined in $\mathbb{C}_{\frac{1}{2}}$, which is quite far from the imaginary axis! We shall view (5.1) as a combination of two things:

- a Plancherel formula, and
- an ergodic theorem.

The "genuine" Plancherel formula involves the characters we met earlier:

$$\int_{\Xi} |f_\chi(\sigma)|^2 d\varpi(\chi) = \sum_{n=1}^{+\infty} \frac{|a_n|^2}{n^{2\sigma}}\,,$$

where we recall the notation

$$f_\chi(s) = \sum_{n=1}^{+\infty} a_n\, \chi(n)\, n^{-s}$$

for the vertical limit function associated with the character χ. The characters of the form

$$\chi_t(n) = n^{-it}, \qquad t \in \mathbb{R},$$

constitute a dense "one-dimensional" subset of Ξ; moreover, we can think of them as the result of a motion in Ξ. To make the latter idea precise, just think of the transformation $T_t(\chi) = \chi_t\chi$ which moves the point χ along the time flow parametrized by t. This flow is *ergodic*, because there are not subsets of Ξ of intermediate mass (that is, not equal to 0 or 1) which are preserved by it. The general ergodic theorem then says that the time average along the flow of a continuous function equals the space average, that is, the integral. And the limit

$$\lim_{T\to+\infty} \frac{1}{2T} \int_{-T}^{T} |f(\sigma + it)|^2 dt = \lim_{T\to+\infty} \frac{1}{2T} \int_{-T}^{T} \left|f_{\chi_t}(\sigma)\right|^2 dt$$

is exactly a time average, whereas

$$\int_\Xi |f_\chi(\sigma)|^2 d\varpi(\chi)$$

is the space average. Now, we see that (5.1) holds for more general Dirichlet series f; what is needed is that $f_\chi(\sigma)$ defines a continuous function of $\chi \in \Xi$. For instance, this is true for all σ with $0 < \sigma < +\infty$ if $f \in \mathcal{H}^\infty$.

Problem 5. Suppose $f \in \mathcal{H}^\infty$, so that f has well-defined nontangential boundary values almost everywhere on the imaginary line. Is it true that

$$\frac{1}{2T} \int_{-T}^{T} |f(it)|^2 dt \to \sum_{n=1}^{+\infty} |a_n|^2 \quad \text{as } T \to +\infty\,?$$

6 Hardy Spaces for Dirichlet Series

Suppose f is a Dirichlet polynomial (which means that the Dirichlet series is finite). Fix a p, $1 < p < +\infty$. One can show that the limit

$$\lim_{T\to+\infty} \frac{1}{2T} \int_{-T}^{T} |f(it)|^p dt$$

exists with ergodic methods like in the previous section; it equals the p-th power of the $L^p(\Xi)$ norm of $\chi \mapsto f_\chi(0)$. As a consequence, we can use the above limit to define a norm on the Dirichlet polynomials, and then form the completion of the space with respect to it. The details of this procedure has been worked out by Bayart [3]. The result is the space $\mathcal{H}^p$, the *Hardy space for Dirichlet series*. For each p, the elements of $\mathcal{H}^p$ are Dirichlet series that define analytic functions on $\mathbb{C}_{\frac{1}{2}}$, and generally speaking, not on any other bigger domain.

Problem 6. Find another scale of spaces (perhaps of Orlicz type) which is able to resolve the jump from finite p when the functions are analytic on $\mathbb{C}_{\frac{1}{2}}$, to $p = +\infty$, when the functions are analytic on $\mathbb{C}_+$.

Problem 7. Study the properties of the spaces $\mathcal{H}^p$ in more detail.

Some progress has already been achieved on Problem 7, mainly due to Bayart [2, 3].

7 Weighted Hilbert Spaces of Dirichlet Series

Let ω be a function on the positive integers which takes values in the interval $]0, +\infty[$; we think of it as a weight. We then consider the Hilbert space $\mathcal{H}^2(\omega)$ of Dirichlet series

$$f(s) = \sum_{n=1}^{+\infty} a_n\, n^{-s},$$

subject to the norm boundedness condition

$$\|f\|_{\mathcal{H}^2(\omega)} = \left(\sum_{n=1}^{+\infty} |a_n|^2\, \omega(n)\right)^{\frac{1}{2}} < +\infty\,;$$

here, we only consider weights ω with the property that the norm boundedness implies that the Dirichlet series converges in some half-plane $\operatorname{Re} s > \sigma_0$. These spaces $\mathcal{H}^2(\omega)$ are called *weighted* Hilbert spaces of Dirichlet series. It is then of interest to study the phenomena found for $\mathcal{H}^2$ in this much wider class of spaces $\mathcal{H}^2(\omega)$. A first attempt in this direction has been made by McCarthy in [26], where he studied first Bergman-type spaces of Dirichlet series, and, second, a space with the reproducing kernel

$$K(s, s') = \frac{1}{2 - \zeta(s + \bar{s}')},$$

where ζ refers to Riemann's zeta function. It turns out that this is a complete Nevanlinna-Pick kernel (see [1] for a definition), and that the multiplier space lives on the same half-plane as the space itself, in contrast with the situation for $\mathcal{H}^2$.

The weighted Hilbert spaces of Dirichlet series deserve further attention.

8 Dirichlet Series with General Frequencies

The theorem of Schnee [31] (see also the book of Hardy and Riesz [15]) mentioned earlier says the following: if

$$f(s) = \sum_{n=1}^{+\infty} a_n n^{-s}$$

converges in some (possibly remote) half-plane $\operatorname{Re} s > \sigma_0$, and the function has an analytic continuation to the right half-plane $\mathbb{C}_+$, and satisfies the growth bound for each $\varepsilon > 0$,

$$|f(s)| = O(|s|^{\varepsilon}), \quad \text{as } |s| \to +\infty ,$$

in every half-plane $\operatorname{Re} s > \delta$ with $\delta > 0$, then the Dirichlet series for $f(s)$ converges on $\mathbb{C}_+$. Schnee's theorem also applies to more general Dirichlet series of the form

$$f(s) = \sum_{n=1}^{+\infty} a_n e^{-\lambda_n s} ,$$

where $\lambda_n \in \mathbb{R}$ for all n, and $\lambda_n \to +\infty$ as $n \to +\infty$; the classical case corresponds to having $\lambda_n = \log n$. Schnee's theorem has certain regularity assumptions on the λ_n's. So, for instance, it does not apply when this sequence of frequences "clumps together" too much.

Problem 8. Is it possible to handle the case when we have "clumping together" of the frequencies by enforcing a stronger growth condition on the function?

Problem 9. To what extent are the results mentioned in the previous sections peculiar to $\lambda_n = \log n$?

9 Extremality Methods and Entire Functions

The Hilbert space $\mathcal{H}^2$ has bounded point evaluations in the half-plane $\mathbb{C}_{\frac{1}{2}}$, and therefore a corresponding reproducing kernel function, which is easily found to be

$$K(s,s') = \sum_{n=1}^{+\infty} n^{-s-\bar{s}'} = \zeta(s+\bar{s}') , \qquad s, s' \in \mathbb{C}_{\frac{1}{2}} ,$$

where ζ is Riemann's zeta function. Reproducing kernels have extremal properties; for instance, it follows that if we look for the function $f \in \mathcal{H}^2$ that has unit norm and is the biggest in modulus at a point $s_0 \in \mathbb{C}_{\frac{1}{2}}$, then the solution is the function

$$f(s) = \frac{\zeta(s+\bar{s}_0)}{\sqrt{\zeta(2\operatorname{Re} s_0)}} , \qquad s \in \mathbb{C}_{\frac{1}{2}} .$$

It might be possible to obtain interesting properties of $\zeta(s)$ and especially its zeros by using this fact; however, in practice this approach is problematic, due to the rigidity of $\mathcal{H}^2$ with respect to multiplication, and especially due to the difficulty of finding effective zero divisors for $\mathcal{H}^2$.

We should thus search for other extremal properties of the zeta function. The associated xi function

$$\xi(s) = \frac{1}{2}\, s(s-1)\, \frac{\Gamma\left(\frac{1}{2}s\right)}{\pi^{s/2}}\, \zeta(s)$$

is entire, and it has the symmetry property

$$\xi(1-s) = \xi(s)\,, \qquad s \in \mathbb{C}\,. \tag{9.1}$$

The Riemann hypothesis claims that all the zeros of $\xi(s)$ lie on the critical line $\operatorname{Re} s = \frac{1}{2}$. We proceed as Pólya in his 1926 paper [29], and note that

$$\xi\left(\frac{1}{2} + it\right) = \int_{-\infty}^{+\infty} e^{-itx}\, \Phi(x)\, dx\,, \qquad t \in \mathbb{C}\,,$$

where the function Φ is given by the series expansion

$$\Phi(x) = 2\pi\, e^{5x/2} \sum_{n=1}^{+\infty} \left(2\pi n^2\, e^{2x} - 3\right) \exp\Big(-\pi n^2\, e^{2x}\Big)\,, \qquad x \in \mathbb{R}\,.$$

The symmetry property (9.1) of $\xi(s)$ corresponds to the fact that Φ is even: $\Phi(-x) = \Phi(x)$. This latter property can be derived directly from the Poisson summation formula. We realize that Φ is positive throughout the real line, and that it has the asymptotics

$$\Phi(x) = 4\pi^2\left(e^{9x/2} + e^{-9x/2}\right) \exp\Big(-\pi\, e^{2x} - \pi\, e^{-2x}\Big)\,, \qquad |x| \to +\infty\,.$$

Pólya analyzed the function on the right hand side in place of Φ, and found that its Fourier transform has all its zeros along the real line. We prefer to stick to the original function Φ, and consider

$$\omega(x) = \frac{1}{\Phi(x)}\,, \qquad x \in \mathbb{R}\,,$$

as a weight for the space $L^2(\mathbb{R}, \omega)$ of all locally square summable functions on $\mathbb{R}$ with

$$\|f\|_{L^2(\mathbb{R},\omega)} = \left(\int_{-\infty}^{+\infty} |f(x)|^2\, \omega(x)\, dx\right)^{1/2} < +\infty\,.$$

Let $\mathcal{L}^2(\omega)$ be the image of $L^2(\mathbb{R}, \omega)$ under the Fourier transform:

$$\widehat{f}(t) = \int_{-\infty}^{+\infty} e^{-itx}\, f(x)\, dx\,, \qquad t \in \mathbb{C}\,.$$

This is then a Hilbert space of entire functions, and it has the reproducing kernel

$$K(t,t') = \int_{-\infty}^{+\infty} e^{-it(t+\bar{t}')} \frac{dx}{\omega(x)} = \int_{-\infty}^{+\infty} e^{-it(t+\bar{t}')} \Phi(x)\,dx = \xi\left(\frac{1}{2} + i(t+\bar{t}')\right).$$

The kernel function has, as usual, extremal properties; for instance, the function

$$K(0,0)^{-1/2} K(t,0) = \xi\left(\frac{1}{2}\right)^{-1/2} \xi\left(\frac{1}{2} + it\right)$$

is biggest (in real part, or in modulus) at the origin among all functions in $\mathcal{L}^2(\omega)$ of unit norm.

Problem 10. Let $\lambda \in \mathbb{C}\setminus\mathbb{R}$, and consider the closed subspace $\mathcal{L}^2_e(\omega;\lambda,\bar{\lambda})$ of $\mathcal{L}^2(\omega)$ consisting of all functions that are even and vanish at the four points $\lambda, \bar{\lambda}, -\lambda, -\bar{\lambda}$. Does there exist a meromorphic function $\varphi = \varphi_\lambda$ in the whole plane, with simple poles at these four points only, which multiplies contractively $\mathcal{L}^2_e(\omega;\lambda,\bar{\lambda}) \to \mathcal{L}^2(\omega)$, and has $\varphi(0) = 1$?

If the answer were affirmative, the Riemann hypothesis would follow.

This general idea – to construct appropriate Hilbert spaces of entire functions to treat the Riemann hypothesis – is quite attractive, and has been exploited by de Branges [8, 9]; the idea to use extremality can also be found in the work of Beurling [6, pp. 147–152].

References

[1] J. Agler, J. E. McCarthy, *Pick interpolation and Hilbert function spaces.* (English) Graduate Studies in Mathematics. **44**. Providence, RI: American Mathematical Society (AMS), 2002.

[2] F. Bayart, *De nouveaux espaces de séries de Dirichlet et leurs opérateurs de composition.* (French) [New spaces of Dirichlet series and their composition operators] C. R. Acad. Sci. Paris Sér. I Math. **333** (2001), no. 3, 207–212.

[3] F. Bayart, *Hardy spaces of Dirichlet series and their composition operators.* Monatsh. Math. **136** (2002), no. 3, 203–236.

[4] F. Bayart, *Compact composition operators on a Hilbert space of Dirichlet series,* Illinois J. Math, to appear.

[5] F. Bayart, *Private communication.*

[6] A. Beurling, *The collected works of Arne Beurling. Vol. 2. Harmonic analysis.* Edited by L. Carleson, P. Malliavin, J. Neuberger and J. Wermer. Contemporary Mathematicians. Birkhäuser Boston, Inc., Boston, MA, 1989.

[7] H. Bohr: *Über die Gleichmässige Konvergenz Dirichletscher Reihen,* J. Reine Angew. Math. **143** (1913), 203–211.

[8] L. de Branges, *The convergence of Euler products.* J. Funct. Anal. **107** (1992), 122–210.

[9] L. de Branges, *A conjecture which implies the Riemann hypothesis.* J. Funct. Anal. **121** (1994), 117–184.

[10] L. Carleson, *On convergence and growth of partial sums of Fourier series,* Acta Math. **116** (1966), 135–157.

[11] P. Erdös, *Some unsolved problems,* Magyar Tud. Akad. Mat. Kutató Int. Közl. **6** (1961), 221–254.

[12] C. Finet, H. Queffélec, A. Volberg, *Compactness of composition operators on a Hilbert space of Dirichlet series,* preprint, 2002.

[13] J. Gordon, H. Hedenmalm, *The composition operators on the space of Dirichlet series with square summable coefficients,* Michigan Math. J. **46** (1999), 313–329.

[14] G. Halász, *On random multiplicative functions,* Hubert Delange colloquium (Orsay, 1982), Publ. Math. Orsay, **83-4**, Univ. Paris XI, Orsay, 1983; pp. 74–96.

[15] G. H. Hardy, M. Riesz, *The general theory of Dirichlet's series,* Cambridge Tracts in Mathematics and Mathematical Physics, No. **18** Stechert-Hafner, Inc., New York, 1964 (originally published in 1915).

[16] H. Hedenmalm, *Topics in the theory of Dirichlet series,* Visn. Khark. Univ., Ser. Mat. Prykl. Mat. Mekh. **475** (2000), 195-203.

[17] H. Hedenmalm, P. Lindqvist, K. Seip, *A Hilbert space of Dirichlet series and systems of dilated functions in $L^2(0, 1)$,* Duke Math. J. **86** (1997), 1–37.

[18] H. Hedenmalm, P. Lindqvist, K. Seip, *Addendum to "A Hilbert space of Dirichlet series and systems of dilated functions in $L^2(0, 1)$"*, Duke Math. J. **99** (1999), 175–178.

[19] H. Hedenmalm, E. Saksman, *Carleson's convergence theorem for Dirichlet series,* Pacific J. Math. **208** (2003), 85–109.

[20] H. Helson, *Foundations of the theory of Dirichlet series,* Acta Math. **118** (1967), 61–77.

[21] H. Helson, *Compact groups and Dirichlet series,* Ark. Mat. **8** (1969), 139–143.

[22] J.-P. Kahane, S. Mandelbrojt, *Sur l'équation fonctionnelle de Riemann et la formule sommatoire de Poisson,* Ann. Sci. Ecole Norm. Sup. (3) **75** (1958), 57–80.

[23] J.-P. Kahane, *The last problem of Harald Bohr,* J. Austral. Math. Soc. Ser. A **47** (1989), no. 1, 133–152.

[24] S. V. Konyagin, H. Queffélec, *The translation $\frac{1}{2}$ in the theory of Dirichlet series.* Real Anal. Exchange **27** (2001/02), no. 1, 155–175.

[25] J. E. Littlewood, *On inequalities in the theory of functions,* Proc. London Math. Soc. **23** (1925), 481–519.

[26] J. E. McCarthy, *Hilbert spaces of Dirichlet Series and their multipliers,* Trans. Amer. Math. Soc., to appear.

[27] H. L. Montgomery, *Ten lectures on the interface between analytic number theory and harmonic analysis,* CBMS Regional Conference Series in Mathematics, **84**; published for the Conference Board of the Mathematical Sciences, Washington, DC, by the American Mathematical Society, Providence, RI, 1994.

[28] I. Niven, H. S. Zuckerman, H. L. Montgomery, *An introduction to the theory of numbers,* Fifth edition, John Wiley & Sons, Inc., New York, 1991.

[29] G. Pólya, *Bemerkung über die Integraldarstellung der Riemannschen ξ-Funktion,* Acta Math. **48** (1926), 305–317.

[30] H. Queffélec, *H. Bohr's vision of ordinary Dirichlet series; old and new results.* J. Anal. **3** (1995), 43–60.

[31] W. Schnee, *Zum Konvergenzproblem der Dirichletschen Reihen,* Math Ann. **66** (1909), 337–349.

[32] J. H. Shapiro, *The essential norm of a composition operator,* Ann. of Math. (2) **125** (1987), 375–404.

[33] A. N. Shiryaev, *Probability,* Second edition, Graduate Texts in Mathematics **95**, Springer-Verlag, New York, 1996.

[34] A. Wintner, *Random factorizations and Riemann's hypothesis,* Duke Math. J. **11** (1944), 267–275.

Real Multiplication and Noncommutative Geometry

(ein Alterstraum)

Yu. I. Manin

ἐπεὶ καὶ τὰ γνώριμα ὀλίγοις γνώριμά ἐστιν
... for even subjects that are known are known only to a few.
Aristotle, Poetics IX, 1451b

Preface. Abel's name is associated with a number of key notions of modern mathematics, such as abelian varieties, Abel's integrals, etc. Moreover, it became almost synonymous with the idea of commutativity. Thus, when one speaks about abelian class field theory, one has in mind a description of extensions of a ground (say, number) field with commutative Galois group.

Of course, there is no clear cut boundary between the commutative and the noncommutative worlds. In this paper we sketch a possible approach to the abelian class field theory of real quadratic fields from the vantage point of Connes' noncommutative geometry.

More precisely, classical theory of Complex Multiplication (CM) shows that all abelian extensions of a complex quadratic field K are generated by the values of appropriate modular functions at the points of finite order of elliptic curves whose endomorphism rings are orders in K. For real quadratic fields, a similar description is not known. However, the relevant (still unproved) case of Stark conjectures [69] strongly suggests that such a description must exist. In this paper I propose to use two–dimensional quantum tori corresponding to real quadratic irrationalities as a replacement of elliptic curves with complex multiplication.

Section 1 is dedicated to a general problem of noncommutative geometry Connes style: what are morphisms between noncommutative spaces considered as "spectra" of associative rings (perhaps, with an additional structure)? One natural suggestion is to define morphisms as isomorphism classes of biprojective bimodules as in Morita theory. Slightly extending Rieffel's Morita classification of two–dimensional quantum tori, I present a description of the resulting category (Theorem 1.7.1) in terms of what can be called "period pseudolattices" (Sect. 1.1), by analogy with period lattices of elliptic curves. (In the context of operator algebras, requiring quite sophisticated modification of basic notions, A. Connes calls such morphisms "correspondences", cf. [12], p. 526, and [35]).

Section 2 contains some results on the values and residues of zeta functions of arithmetical progressions in real quadratic fields, in the spirit of earlier work of E. Hecke, continued by G. Herglotz and D. Zagier. Our calculations are strongly motivated by H. M. Stark's conjectures ([69], [70]) proposing very special generators of abelian extensions of such fields.

Section 3 is a contribution to the theory of quantum theta functions (see [45]). It gives a partial answer to the question of A. Schwarz ([66]) about the relationship between quantum thetas and representations of quantum tori. The main Theorem 3.7 of this section generalizes a seminal calculation of F. Boca in [7].

Eventually these results should be considered as pieces of a general picture, which could be called Real Multiplication of two–dimensional quantum tori, by analogy with the classical Complex Multiplication of elliptic curves (Kronecker's *Jugendtraum*).

From this perspective, Sect. 1 outlays basics of the (noncommutative) geometry of Real Multiplication, Sect. 2 presents certain known or conjectural arithmetical facts in the light of this geometric picture, whereas Sect. 3 provides elements of function theory.

Unfortunately, the relations between these parts that I can establish are too sparse yet. An important test for such a theory would be a proof of Stark's conjectures for real quadratic fields. If this plan succeeds, a more ambitious project could address Real Multiplication of multidimensional quantum tori, as an analog and an extension of Shimura–Taniyama multidimensional CM theory.

I tried to facilitate reading this paper for potential readers with varied backgrounds by providing many definitions and introductory explanations, so that large parts of this paper can be read as a review. In particular, the introductory Sect. 0 explains rudiments of the classical Complex Multiplication theory which serves as a guide for the whole enterprise. It discusses as well a possibilty of including this theory in the context of noncommutative geometry. For additional connections, see [46].

Acknowledgements. The crystallization of this project owes much to Matilde Marcolli and our collaboration [47]. Victor Nistor consulted me about the proof of Lemma 1.4.2. Florian Boca's paper [7] and correspondence with him were crucial for recognizing the connection between quantum thetas and modules over quantum tori. Sasha Rosenberg's insights about morphisms between noncommutative spaces developed in [64] helped me to overcome a difficult psychological barrier. Last but not least, I appreciate the proposal of Friedrich Hirzebruch to translate *Alterstraum* in the title as "midlife crisis".

0 Introduction: Lattices, Elliptic Curves, and Complex Multiplication

0.0 An overview. Let K be a field of algebraic numbers of one of the three types: $\mathbf{Q}$, a complex quadratic extension of $\mathbf{Q}$, or a real quadratic extension of $\mathbf{Q}$. Consider the following classical problem: describe the maximal abelian extension K^{ab} of K. Of course, the Galois group of such an extension is known for arbitrary agebraic number fields K: it is the idèle class group of K modulo its connected component. However, explicit generators of K^{ab} and the action of the Galois group on them generally remain a mystery, with exception of two classical cases described below.

According to the Kronecker–Weber theorem (KW), $\mathbf{Q}^{ab}$ is generated by roots of unity, i.e. by the points of finite order of the multiplicative group $\mathbf{G}_m$ considered as an algebraic group over $\mathbf{Q}$. For K imaginary quadratic, the multiplicative group should be replaced by the elliptic curve E_K whose $\mathbf{C}$–points are $\mathbf{C}/O_K$, O_K being the ring of integers in K. To get K^{ab}, one must adjoin to K the values of a power of the Weierstrass function at points of finite order of E_K, and the value of the absolute invariant of E_K. To see that points of finite order generate an abelian extension, one observes that the action of the Galois group on them must commute with the action of algebraic endomorphisms furnished by the power maps $x \mapsto x^m$ in the KW case, resp. the complex multiplication (CM) maps written additively on the universal covering of E_K as $x \mapsto ax$, $a \in O_K$. The commutant of this action suitably completed in profinite topology is abelian, and essentially coincides with the completion of the action itself. The universal idèlic description of the Galois group together with reduction modulo p arguments furnish the rest.

Elliptic curves have a rich analytic theory. Curves admitting a complex multiplication form a subfamily of all elliptic curves. The latter can be parametrized by their period lattices Λ i.e. discrete images of the injective homomorphisms $j : \mathbf{Z}^2 \to \mathbf{C}$ modulo a natural equivalence relation. The moduli space of them is $PGL(2,\mathbf{Z}) \setminus (H^+ \cup H^-)$, $H^{\pm}$ being the upper/lower halfplanes respectively. The curves isogeneous to E_K live over orbits of points $\mathbf{P}^1(K)$. The multiplicative group also appears in this family as the "degenerate elliptic curve" over the cusp, that is the orbit $PGL(2,\mathbf{Z}) \setminus \mathbf{P}^1(\mathbf{Q})$, so that in principle the geometry of the CM and KW cases can be unified.

The cusp corresponds to the very degenerate lattice: j acquires a cyclic kernel. There is an intermediate case of degeneration, invisible in algebraic geometry, where j is still injective, but its image is not discrete. The relevant modular orbit is $PGL(2,\mathbf{Z}) \setminus (\mathbf{P}^1(\mathbf{R}) \setminus \mathbf{P}^1(\mathbf{Q}))$, it contains orbits of $\mathbf{P}^1(K)$ for real quadratic K, but they could not be used in the same way as CM points of the modular curve because of lack of the analog of elliptic curves over this stratum of the moduli space. Hopefully, quantum tori might serve as a substitute.

This introductory section is dedicated to some details of the CM picture and its possible extension to the RM case.

0.1 Category of lattices $\mathcal{L}$. By definition, *a lattice* (of rank 2) is a triple (Λ, V, j), where Λ is a free abelian group of rank two, V is an one–dimensional complex space, and $j : \Lambda \to V$ is an injective homomorphism with discrete image, hence compact quotient.

When no confusion is likely, we will refer to (Λ, V, j) simply as Λ.

A morphism of lattices $(\Lambda', V', j') \to (\Lambda, V, j)$ is a commutative diagram

$$\begin{array}{ccc} \Lambda' & \xrightarrow{j'} & V' \\ \varphi\downarrow & & \downarrow\psi \\ \Lambda & \xrightarrow[j]{} & V \end{array} \tag{0.1}$$

in which φ is a group homomorphism, and ψ is a $\mathbf{C}$–linear map. Clearly, φ is uniquely determined by ψ, and vice versa. Choosing a basis (λ_1, λ_2) in Λ, taking $j(\lambda_2)$ as the base vector of V we see that in any isomorphism class of lattices one can find a representative given by $j : \mathbf{Z}^2 \to \mathbf{C}$ such that $j(0, 1) = 1$, $j(1, 0) := \tau$ is a number in $\mathbf{C} \setminus \mathbf{R}$. Changing the sign of λ_1 if needed we can arrange τ to lie in the upper half–plane H.

Let us denote this lattice Λ_τ. Then any non–zero morphism $\Lambda_{\tau'} \to \Lambda_\tau$ is represented by a non–degenerate matrix

$$g = \begin{pmatrix} a & b \\ c & d \end{pmatrix} \in M\,(2, \mathbf{Z})$$

such that

$$\tau' = \frac{a\tau + b}{c\tau + d}\,. \tag{0.2}$$

This g is obtained by writing φ in (0.1) as the right multiplication of a row by a matrix; the respective ψ is the multiplication by $(c\tau + d)^{-1}$.

Clearly, (0.2) is an isomorphism, iff $g \in GL(2, \mathbf{Z})$. Thus the moduli space of (isomorphism classes of) lattices is

$$PGL(2, \mathbf{Z}) \setminus (\mathbf{P}^1(\mathbf{C}) \setminus \mathbf{P}^1(\mathbf{R})) = PSL(2, \mathbf{Z}) \setminus H\,. \tag{0.3}$$

Endomorphisms of a lattice (Λ, V, j) form a ring, with componentwise addition of (ϕ, ψ) and composition as multiplication. It contains $\mathbf{Z}$ and comes together with its embedding in $\mathbf{C}$:

$$\mathrm{End}\,\Lambda = \{a \in \mathbf{C}\,|\,a j(\Lambda) \subset j(\Lambda)\}\,.$$

Lemma 0.1.1. *(a)* End $\Lambda \neq \mathbf{Z}$ *iff there exists a complex quadratic subfield K of* $\mathbf{C}$ *such that Λ is isomorphic to a lattice contained in K.*

(b) If this is the case, denote by O_K the ring of integers of K. There exists a unique integer $f \geq 1$ (conductor) such that End $\Lambda = \mathbf{Z} + fO_K =: R_f$, *and Λ is a projective module of rank 1 over R_f. Every K, f and a projective module over R_f come from a lattice.*

(c) If lattices Λ and Λ' have the same K and f, they are isomorphic if and only if their classes in the Picard group Pic R_f *coincide.*

Automorphisms of a lattice generally form a group $\mathbf{Z}_2$ (ψ is multiplication by ± 1.) However, integers of two imaginary quadratic fields obtained by adjoining to $\mathbf{Q}$ a primitive root of unity of degree 4 (resp. 6) furnish examples of lattices with automorphism group of order 4 (resp. 6). Only these two fields produce lattices with such extra symmetries.

0.2 Category of elliptic curves $\mathcal{E}$. For any lattice (Λ, V, j) the quotient space $V/j(\Lambda)$ is an one–dimensional complex torus which has a canonical structure of (the set of complex points of) an algebraic curve E_Λ of genus 1 with base point 0. Such curves form a category $\mathcal{E}$ (morphisms should respect base points).

0.3 The functor $P : \mathcal{E} \to \mathcal{L}$. Let E be an elliptic curve. The functor "period lattice" P is defined on objects by the following prescription: $P(E) = (\Lambda_E, V_E, j_E)$ where V_E = the tangent space to E at the base point, considered as its Lie algebra, Λ_E the kernel of the of the exponential map $V_E \to E(\mathbf{C})$, and j_E its canonical embedding. On morphisms, ψ is the induced tangent map and φ its restriction to the period lattices.

Theorem 0.3.1. *P is an equivalence of categories.*

This simple result is crucial for the theory of complex multiplication.

0.4 Abelian extensions of complex quadratic fields. Let now K be a complex quadratic extension of $\mathbf{Q}$. Choose and fix an embedding $K \to \mathbf{C}$. Denote by O_K the ring of integers of K.

There are three related but somewhat different ways to describe the maximal abelian extension K^{ab} of K.

(A) *Approach via elliptic curves.* Here one starts with a single elliptic curve E_K associated to the lattice $O_K \subset \mathbf{C}$. It turns out that its minimal definition field containing K is generated by the value of its absolute invariant $J(E_K)$, and is the maximal unramified extension of K. One can also give a beautiful description of the total set of conjugates of $J(E_K)$ and the action of the Galois group on this set. Namely, any lattice whose endomorphism ring is precisely O_K, is represented by an ideal in O_K, and two lattices are isomorphic iff they lie in the same class. Absolute invariants of the respective elliptic curves are conjugate to each other, and the action of the Galois group is induced by a geometric twisting operation producing from a curve an isogenous curve.

The remaining part of the K^{ab} is generated by the values at points of finite order of E of a special function t. In Weierstrass notation, it is $t = \wp(z, O_K)^u$ where $2u$ is the order of the automorphism group of E, so that our algebraic numbers can be described as the values of a transcendental function

$$\left[\frac{1}{z^2} + \sum_{\lambda \in O_K \setminus \{0\}} \left(\frac{1}{(z+\lambda)^2} - \frac{1}{\lambda^2}\right)\right]^u, \tag{0.4}$$

at $z \in K$. In geometric terms, t is an appropriate coordinate on the projective line

$$\mathbf{P}^1 = E_K / O_K^*, \tag{0.5}$$

to which the points of finite order are mapped.

(B) *Approach via modular curves.* In another approach, one considers an extension of K generated by roots of unity and absolute invariants of all elliptic curves admitting complex multiplication by an order in O_K. Not all of K^{ab} is generated in this way, it remains to produce an additional infinite extension with a Galois group of period two, but in a sense the most essential part of K^{ab} is obtained in this way.

This approach stresses the geometry and arithmetic of *the moduli space (stack)* of elliptic curves rather than that of elliptic curves themselves. This space has special points which can be characterized as fixed points of certain correspondences, and fields of definition of these points are of primary interest.

For a brief introduction to both approaches, see [67] and [71].

(C) *Approach via Stark's numbers.* The general conjectures due to H. M. Stark provide (hypothetical) generators of abelian extensions which are values of zetas (or their derivatives, or the Taylor coefficients next to the residue) similar to (0.4). In the CM case these conjectures are proved in [70], by reducing them to the more classical and geometrical forms of the theory sketched in (A), (B). No independent arguments are known.

To provide the basis for comparison with the RM case, we will briefly describe these numbers.

Let $(\Lambda, \mathbf{C}, j)$ be a lattice in $\mathbf{C}$, $\lambda_0 \in \Lambda \otimes \mathbf{Q}$. Put

$$\zeta(\Lambda, \lambda_0, s) := \sum_{\lambda \in \Lambda} \frac{1}{|j(\lambda_0 + \lambda)|^{2s}} \tag{0.6}$$

where j is extended by $\mathbf{Q}$–linearity. These series admit meromorphic continuation and may have a pole of the first order at $s = 1$ and zero of the first order at $s = 0$. Similar behavior is exhibited in the real case. Zeta-functions for two isomorphic lattices differ by a factor A^s where A is a positive real number. In the CM case we will restrict the choice of Λ in the isomorphism class by considering only lattices with $j(\Lambda) \subset K$. Then A can be modulus squared of any number in K.

We have the following simple lemma.

Lemma 0.4.1. *(a) Assume that $F(s)$ vanishes at $s = 0$. Then for any $A > 0$*

$$\frac{d}{ds}F(s)\bigg|_{s=0} = \frac{d}{ds}(A^s F(s))\bigg|_{s=0}.$$

In particular,

$$S_0(\Lambda, \lambda_0) := e^{\zeta'(\Lambda,\lambda_0,0)} \tag{0.7}$$

is an invariant of the isomorphism class of (Λ, λ_0).

(b) Assume that

$$F(s) = \frac{r}{s-1} + v + O(s-1)\,,\ r \neq 0$$

near $s = 1$. Then the similar formula holds for $A^s F(s)$, with the ratio v/r replaced by $v/r + \log A$. In particular, the following coset

$$S_1(\Lambda, \lambda_0) := e^{v/r} \bmod N_{K/\mathbf{Q}}(K^*) \in \mathbf{C}^*/N_{K/\mathbf{Q}}(K^*) \tag{0.8}$$

is an invariant of the isomorphism class of (Λ, λ_0). Here v, r are calculated via $\zeta(\Lambda, \lambda_0, s)$ for any representative of this class satisfying $j(\Lambda) \subset K$,

The essence of Stark's conjectures consists in the prediction that invariants of the type $S_0(\Lambda, \lambda_0)$ are algebraic units in appropriate abelian extensions of K, and that the action of the Frobenius elements of the Galois group upon them can be explicitly described.

Stark's proof in the CM case is based upon a direct calculation of these invariants, which in turn reduces to the second Kronecker limit formula. A version involving $S_1(\Lambda, \lambda_0)$ might be more feasible from the computational viewpoint, the two versions being essentially equivalent thanks to the classical functional equations. These calculations show that Stark's numbers a priori defined as values of some transcendental functions admit an algebraic geometric interpretation demonstrating their arithmetical nature.

0.5 Real quadratic fields. In this paper we propose some constructions parallel to (A) – (C) above, for the case of real quadratic fields.

(A) *Geometry of real multiplication.* Replacing lattices by pseudolattices and elliptic curves by quantum tori, we develop in Sect. 1 the geometric framework parallel to that of 0.1–0.3 above.

(B) *Geometry of noncommutative modular curves.* The space $PGL(2, \mathbf{Z}) \setminus (\mathbf{P}^1(\mathbf{R}) \setminus \mathbf{P}^1(\mathbf{Q}))$ as an invisible stratum of the classical modular curve was studied in [47]. In particular, it was shown that its K–theory can be written in terms of modular symbols, and that classical modular forms of weight two and their Mellin transforms are represented by interesting densities on this stratum.

For the purposes of real multiplication, however, more relevant might be noncommutative spaces which represent the orbits $PGL(2, \mathbf{Z}) \setminus \mathbf{P}^1(K)$ corresponding to the individual real quadratic K. It seems that the remarkable paper by Bost and Connes [8] and its extensions [1], [29], furnish the right language to describe the arithmetic phenomena that interest us. Provisionally, [8] appears to describe the KW case (cusp) from the noncommutative viewpoint. However, a satisfactory generalization of [8] to more general number fields is not developed as yet (cf. however [29], [1], [10]).

We expect that other noncommutative spaces, besides quantum tori and modular curves, must play an essential role in the future theory. In particular, the projective line (0.5) might be replaced by the crossed product of the algebra of functions on K and its automorphism group of the type $x \mapsto ax + b$ where $a \in O_K^*$, $b \in O_K$. This looks even closer to the spaces studied in [8] and [29].

(C) *Stark's numbers.* For real quadratic fields K, one should consider series of the type (0.6), in which $|a|^{-2s} = (a\overline{a})^{-s}$ is replaced by $N_{K/\mathbf{Q}}(a)^{-2s}$ and furnished with a slight additional twist: the typical term of (0.6) is multiplied by the sign of the conjugate of $j(\lambda_0 + \lambda)$, cf. Sect. 2 below for more details. More important is the following complication: the action of an infinite cyclic group of units makes each term in (0.6) to repeat infinitely often, so to make sense of the whole expression one should only sum over cosets modulo the relevant group.

We show in Sect. 2 that an adaptation of Hecke's calculations leads to formulas for Stark's numbers which are compatible with the general picture of "passing to the quantum limit": cf. 1.8 below.

Our hope, based upon this calculation, is that an appropriate algebraic geometric refinement of (A) and (B) will lead to a proof of Stark's conjectures, in the same way as it worked in the CM case.

Again, a natural question arises: can one see in noncommutative geometry Stark's numbers of the cyclotomic (KW) case? In fact, closely related numbers appear in [8] in their description of the arithmetical symmetry breaking, and in [33], as indices of subfactors of the hyperfinite factor of type II_1. This suggests interesting questions in the framework of our program.

0.6 Elliptic curves as non-commutative spaces. $\mathbf{C}$-points of the elliptic curve E_τ associated with the lattice $\mathbf{Z} + \mathbf{Z}\tau$ can be identified via the exponential map with

$$\mathbf{C}/(\mathbf{Z} + \mathbf{Z}\tau) \cong \mathbf{C}^*/(q^{\mathbf{Z}}),\ q := e^{2\pi i\tau}\,.$$

To treat E_τ as a non-commutative space means to study the appropriate crossed product of an algebra of functions on $\mathbf{C}^*$ with its automorphism group generated by the shift $z \mapsto qz$. The simplest crossed product of this type is $A_q^{alg} := \mathbf{C}[z, z^{-1}][v, v^{-1}]$ where $vz = qzv$, but more sophisticated versions (various completions and their subalgebras) are really interesting.

There is a lot of results in the theory of lattice models and q-deformations that can be interpreted in the light of non-commutative geometry and function theory of elliptic curves. It would be worthwhile to review them systematically for two reasons: first, to get an environment in which elliptic curves and quantum tori could be treated more or less uniformly, and second, because the non-commutative setting considerably enriches even the classical picture.

I will restrict myself to two examples illustrating these points.

0.6.1 Semistable bundles and regular modules. Here we explain the basic reslut of [3], generalized in [4]. The relevant crossed product is $A_q^{\text{form}} := \mathbf{C}((z, z^{-1}))[v, v^{-1}]$ where $\mathbf{C}((z, z^{-1}))$ denotes the field of formal Laurent series finite in negative degrees. Consider the following categories.

Category I. Its objects are left A_q^{form}-modules M, which are finite-dimensional as $\mathbf{C}((z, z^{-1}))$-spaces and which satisfy the following *regularity* condition: there exists a free $\mathbf{C}[[z]]$-submodule $M_0 \subset M$ of maximal rank such that $v^{\pm 1}(M_0) \subset M_0$.

Morphisms are usual module homomorphisms. The tensor product over $\mathbf{C}((z, z^{-1}))$ extends to a structure of rigid tensor category.

Category II. Its objects are semistable degree 0 holomorphic vector bundles over the elliptic curve E_τ. Semistability means that the associated principal bundle admits a global holomorphic connection, which is automatically flat. Morphisms and tensor structure are evident ones.

One of the main results of [3] consists in the construction of an equivalence between the categories I and II. This equivalence is compatible with tensor products.

An interesting comment made in [3] connects this result with a problem in the theory of finite difference equations.

Consider first differential equations of the form

$$z \frac{d\chi}{dz} = m(z)\, \chi(z) \tag{0.9}$$

where $\chi(z)$ is a column of formal series from $\mathbf{C}((z, z^{-1}))$ and $m(z)$ a matrix of such series. We can try to classify such equations by identifying those which can be obtained from each other by a linear transformation of $\chi(z)$. This is equivalent to the gauge transformation of $m(z)$:

$$m(z) \mapsto g(z)\, m(z)\, g(z)^{-1} + z \frac{dg}{dz}\, g(z)^{-1}\,. \tag{0.10}$$

It is known that if (0.9) has a regular (Fuchsian) singularity at $z = 0$, then the formal classification coincides with the analytic one, and the latter is furnished by monodromy around 0.

Now, a finite difference, or q-version of (0.9) is

$$\chi(qz) = m(z)\,\chi(z) \tag{0.11}$$

and the gauge equivalence (0.10) is replaced by

$$m(z) \mapsto g(qz)\,m(z)\,g(z)^{-1}\,. \tag{0.12}$$

We can identify the problem of classification of the equations (0.11) up to gauge equivalence with the problem of classification up to isomorphism of A_q^{form}-modules finite-dimensional over $\mathbf{C}((z, z^{-1}))$. To this end, given $m(z)$, treat it as the matrix of the operator v in a basis. The Baranovski–Ginzburg theorem then implies that for regular modules this classification coincides with the classification of semistable vector bundles over E_τ. But the regularity condition for modules is the standard q-version of Fuchsian regularity. Hence semistable vector bundles over E_τ should be regarded as a q-version of the monodromy data.

It would be important to reconstruct the complete category of coherent sheaves and/or its derived category in terms of an appropriate crossed product. See an interesting discussion in [68], especially 3.3.

0.6.2 Quantum pentagon identity. As above, let $q = e^{2\pi i\tau}$, $\operatorname{Im}\tau > 0$. This time we will consider the elliptic curve $E_{2\tau}$ represented by an appropriate completion of the algebra $\mathbf{C}[u, u^{-1}, v, v^{-1}]$ with $uv = q^2 vu$. Put

$$e_q(t) := \prod_{n\geq 0}\left(1 + q^{2n+1}t\right)\,. \tag{0.13}$$

If t here is understood as a complex number, we get one of the standard classical expressions, for example, occuring in the product formula for the elliptic theta function

$$\theta_q(t) := \sum_n q^{n^2}t^n = e_q(-q)\,e_q(t)\,e_q\left(t^{-1}\right)\,.$$

The following noncommutative identities (with $uv = q^2vu$) are however nonclassical:

$$e_q(u)\,e_q(v) = e_q(u+v)\,, \tag{0.14}$$

$$e_q(v)\,e_q(u) = e_q(u)\,e_q(vu)\,e_q(v)\,. \tag{0.15}$$

In view of (0.14), $e_q(t)$ is sometimes called the q-exponential function.

The second identity (0.15) was proved by Faddeev and Kashaev in [26] and called there a quantum version of the Rogers pentagon identity for the dilogarithm. To explain this, I remind here the classical version of the Rogers identity:

$$L(x) + L(y) - L(xy) = L\left(\frac{x - xy}{1 - xy}\right) + L\left(\frac{y - xy}{1 - xy}\right), \tag{0.16}$$

where

$$L(x) := L_2(x) + \frac{1}{2}\log(1 - x)\log x$$

and

$$L_2(x) := -\int_0^x \log(1 - z)\,\frac{dz}{z} = \sum_{n\geq 1}\frac{x^n}{n^2}.$$

As q tends to the cusp 1, we have a classical asymptotic expansion in τ for the logarithm of $e_q(t)$ for which we write a few first terms in the exponentiated form

$$e_q(t) = \frac{1}{\sqrt{1 + qt}}\exp\left(L_2(-t)/4\pi i\tau\right)(1 + O(\tau)). \tag{0.17}$$

It remains to combine (0.15) and (0.17). This is not quite straightforward. Faddeev and Kashaev argue that an appropriate infinite-dimensional representation of the commutation relations $uv = q^2vu$ and the corresponding notion of the symbol of an operator in this representation produce (0.16).

It is remarkable and promising that (0.15) looks much neater than (0.16) and shows that (0.16) is a boundary reflection of a phenomenon which is both more global and essentially noncommutative.

1 Pseudolattices, Quantum Tori, and Real Multiplication

1.1 Category of pseudolattices $\mathcal{PL}$. By definition, *a pseudolattice* (of rank 2) is a quadruple (L, V, j, s), where L is a free abelian group of rank two, V is an one-dimensional complex space, $j : L \to V$ is an injective homomorphism whose image lies on a real line, and finally s is an orientation of this line. Since this line contains 0, a choice of s defines the notion of positive and negative halves of it. Clearly, this line is the topological closure of $j(L)$.

A strict morphism of pseudolattices $(L', V', j', s') \to (L, V, j, s)$ is a commutative diagram

$$\begin{array}{ccc} L' & \xrightarrow{j'} & V' \\ {\scriptstyle\varphi}\downarrow & & \downarrow{\scriptstyle\psi} \\ L & \xrightarrow[j]{} & V \end{array} \tag{1.1}$$

in which φ is a group homomorphism, and ψ is a $\mathbf{C}$-linear map, which transforms the orientation s' to s. Clearly, ϕ and ψ uniquely determine each other. Moreover, such a strict morphism is a strict isomorphism iff both ϕ and ψ are isomorphisms.

Omitting the condition that ψ respects orientations, we get the notion of *weak morphism.*

As with lattices, several simple observations will help us to clarify the structure of this category.

(i) The orientation s makes L a totally ordered group: by definition, $l > m$ iff $j(l-m)$ lies in the s-positive half-line. Choosing a basis (l_1, l_2) in L and taking $j(l_2)$ as the base vector of V, we see that in any strict isomorphism class of pseudolattices one can find a representative given by $j : \mathbf{Z}^2 \to \mathbf{C}$ such that $j(0,1) = 1$, $j(1,0) := \theta$ is an irrational real number. The remaining piece of data is the sign $\varepsilon = \pm 1$ such that $l > 0$ iff $\varepsilon j(l) > 0$.

Let us denote this pseudolattice (L_θ, ε). Then any non-zero strict morphism $(L_{\theta'}, \varepsilon') \to (L_\theta, \varepsilon)$ is represented by a non-degenerate matrix

$$g = \begin{pmatrix} a & b \\ c & d \end{pmatrix} \in M(2, \mathbf{Z})$$

such that

$$\theta' = \frac{a\theta + b}{c\theta + d}, \quad \operatorname{sgn}(c\theta + d) = \varepsilon\varepsilon' . \tag{1.2}$$

This g is obtained by writing φ in (1.1) as the right multiplication of a row by a matrix.

If $g \in GL(2, \mathbf{Z})$, (1.2) is an isomorphism. Since we can replace g by $-g$ without violating the first condition in (1.2), two lattices are strictly isomorphic iff they are weakly isomorphic.

(ii) We can choose a positive basis in L. This shows that any pseudolattice is isomorphic to some $(L_\theta, \varepsilon = 1)$ with irrational real θ which can be even taken in (0,1). We will denote it simply L_θ. Two such pseudolattices are isomorphic iff their invariants θ lie in the same $PGL(2, \mathbf{Z})$-orbit, that is, their continued fraction expansions coincide starting from some place. Thus set-theoretically, the moduli space of the isomorphism classes of pseudolattices is

$$(PGL(2, \mathbf{Z}) \setminus \mathbf{P}^1(\mathbf{R})) \setminus \{\text{cusp}\} \tag{1.3}$$

where the cusp is the orbit of rational numbers.

(iii) Weak endomorphisms of a pseudolattice L (we omit other structures in notation if there is no danger of confusion) form a ring w-End L (w stands for weak), with componentwise addition of (ϕ, ψ) and composition as multiplication. It contains $\mathbf{Z}$ and comes together with its embedding in $\mathbf{R}$ as $\{a \in \mathbf{R} \mid a\, j(L) \subset j(L)\}$. The non-negative part of this ring is the semiring End L.

Lemma 1.1.1. *(a) w-End $L \neq \mathbf{Z}$ iff there exists a real quadratic subfield K of $\mathbf{R}$ such that L is isomorphic to a pseudolattice contained in K.*

(b) If this is the case, we will say that L is an RM pseudolattice. Denote by O_K the ring of integers of K. There exists a unique integer $f \geq 1$ (conductor) such that $w\text{-End}\, L = \mathbf{Z} + fO_K =: R_f$*, and L is a projective module of rank 1 over R_f. The module L is endowed with a total ordering respected by* End L.

Every K, f and a ordered projective module over R_f come from a lattice.

(c) If pseudolattices L and L_1 have the same K and f, they are isomorphic if and only if their classes in the Picard group Pic R_f *coincide.*

Unlike the case of lattices, the automorphism group of a pseudolattice is always infinite, it is isomorphic to $\mathbf{Z} \times \mathbf{Z}_2$.

For RM pseudolattices embedded in one and the same real quadratic field K, we will say that an isomorphism $L \to L_1 : l \mapsto al_1$, $a \in K$, is an isomorphism *in a narrow sense*, if $N_{k/\mathbf{Q}}(a) > 0$.

1.2 Two-dimensional quantum tori. We now want to define analogs of elliptic curves for pseudolattices, that is, some geometric objects representing quotients $V/j(L)$ where (L, V, j, s) is a pseudolattice.

Choosing L_θ as a representative of the respective isomorphism class, we can naively replace $\mathbf{C}/(\mathbf{Z}+\mathbf{Z}\theta)$ by $\mathbf{C}^*/(e^{2\pi i\theta})$ ("Jacobi uniformization"), and then interpret the last quotient as an "irrational rotation algebra", or two-dimensional quantum torus T_θ. We recall that this torus is (represented by) the universal C^*-algebra A_θ generated by two unitaries U, V with the commutation rule $UV = e^{2\pi i\theta}VU$. A choice of such generating unitaries is called *a frame*; it is not unique.

The next task is to define morphisms between these quantum tori, with properties that would allow us to imitate the framework of 0.3. Already isomorphisms present a problem: we want fractional linear transforms (1.2) to produce isomorphic quantum tori. M. Rieffel's seminal discovery was that to this end we should consider Morita equivalences between appropriate categories of modules as isomorphisms between the tori themselves. Morita equivalences are essentially given by tensor multiplication by a bimodule. Taking this lead, we will formally introduce the general Morita morphisms of associative rings, stressing those traits of the formalism that play a central role in the structure theory of quantum tori (of arbitrary dimension). Our presentation also prepares ground for introducing versions of quantum tori with more algebraic geometric flavor.

1.3 Morita category. Let A, B be two associative rings. A *Morita morphism* $A \to B$ by definition, is the isomorphism class of a bimodule ${}_AM_B$, which is projective and finitely generated separately as module over A and B.

The composition of morphisms is given by the tensor product ${}_AM_B \otimes {}_BM'_C$, or ${}_AM \otimes {}_BM'_C$ for short.

If we associate to ${}_AM_B$ the functor

$$\mathrm{Mod}_A \to \mathrm{Mod}_B : N_A \mapsto N \otimes_A M_B ,$$

the composition of functors will be given by the tensor product, and isomorphisms of functors will correspond to the isomorphisms of bimodules.

We imagine an object A of the (opposite) Morita category as a noncommutative space, right A-modules as sheaves on this space, and the tensor multiplication by ${}_AM_B$ as the pull-back functor, in the spirit of A. Rosenberg's program [64]. We have chosen to work with right modules, but passing to the opposite rings allows one to reverse left and right in all our statements.

Two bimodules ${}_AM_B$ and ${}_BN_A$ supplied with two bimodule isomorphisms ${}_AM\otimes_B N_A \to {}_AA_A$ and ${}_BN\otimes_A M_B \to {}_BB_B$ define mutually inverse Morita isomorphisms (equivalences) between A and B. The basic example of this kind is furnished by $B = \mathrm{Mat}\,(n, A)$, $M = {}_AA^n{}_B$ and $N = {}_BA^n{}_A$.

We will now briefly summarize Morita's theory.

(A) *Characterization of functors* $S : \mathrm{Mod}_A \to \mathrm{Mod}_B$ *of the form* $N_A \mapsto N\otimes_A M_B$. They are precisely functors satisfying any of the two equivalent conditions:

(i) *S is right exact and preserves direct sums.*

(ii) *S admits a right adjoint functor* $T : \mathrm{Mod}_B \to \mathrm{Mod}_A$ *(which is then naturally isomorphic to* $\mathrm{Hom}_B(M_B, *)$*).*

We will call such functors *continuous*.

(B) *Characterization of continuous functors S such that T is also continuous and* $ST \cong 1$. Let S be given by ${}_AM_B$ and T by ${}_BN_A$. Then $M\otimes_B N \cong {}_AA_A$. Moreover, in this case

(iii) *M_B and ${}_BN$ are projective.*

(iv) *${}_AM$ and N_A are generators.*

In particular, equivalences $\mathrm{Mod}_A \to \mathrm{Mod}_B$ are automatically continuous. Hence any pair of mutually quasi-inverse equivalences must be given by a couple of biprojective bigenerators as above.

(C) *Finite generation and balance.* Any right module M_B can be considered as a bimodule ${}_AM_B$ where $A = B' := \mathrm{End}_B(M_B)$. We can then similarly produce the ring $B'' = A' := \mathrm{End}_A({}_AM)$. Module M_B is called *balanced* if $B = B''$. Similarly, one can start with a left module. With this notation, we have:

(v) *M_B is a generator iff ${}_{B'}M$ is balanced and finitely generated projective.*

Properties (i)–(v) can serve as a motivation for our definition of the Morita category above.

1.4 Projective modules, idempotents, traces, and K_0. Projective right A-modules up to isomorphism are exactly ranges of idempotents in various matrix rings $\mathrm{Mat}\,(n, A)$ acting from the left upon (column) vector modules A^n. Morphisms between such modules are also conveniently described in terms of these idempotents. The following (well known) Proposition summarizes the relevant information in the form convenient for us.

We prefer to work with all n simultaneously. So we will denote by $\mathcal{M}A$ the ring of infinite matrices $(a_{ij})\, i, j \geq 1$, $a_{ij} \in A$, $a_{ij} = 0$ for $i + j$ big enough (depending on the matrix in question). Notice that $\mathcal{M}A$ is not unital even if A is. Similarly, denote by A^{∞} the left $\mathcal{M}A$-module of infinite columns (a_i), $i \geq 1$, with coordinates in A and such that $a_i = 0$ for large i.

Denote by Pr_A the category of finitely generated right A-modules. Denote by pr_A the category whose objects are projectors (idempotents) $p \in \mathcal{M}A$, $p^2 = p$, whereas morphisms are defined by

$$\mathrm{Hom}\,(p, q) := q\, \mathcal{M}A\, p$$

with the composition induced by the multiplication in $\mathcal{M}A$.

There is a natural functor $pr_A \to Pr_A$ defined on objects by $p \mapsto pA^{\infty}$. In order to define it on morphisms, we remark that morphisms $pA^{\infty} \to qA^{\infty}$ can be naturally described by matrices in the following way. Clearly, pA^{∞} contains the columns p_k of p which generate pA^{∞} as a right A-module. We can then apply any $\varphi : pA^{\infty} \to qA^{\infty}$ to all p_k and arrange the resulting vectors into a matrix $\Phi \in \mathcal{M}A$ with k-th column $\varphi(p_k)$. One checks that $\Phi p = \Phi$, and since also $q\Phi = \Phi$, we have $\Phi \in q\mathcal{M}Ap$. Conversely, any such matrix determines a unique morphism $pA^{\infty} \to qA^{\infty}$.

Proposition 1.4.1. *(a) The functor $pr_A \to Pr_A$ described above is an equivalence of categories.*

(b) pA^{∞} is isomorphic to qA^{∞} iff there exist $X, Y \in \mathcal{M}A$ such that $p = XY, q = YX$ (von Neumann's equivalence of idempotents). Hence in this case $p - q \in [\mathcal{M}A, \mathcal{M}A]$.

We have already checked most of the statements implicit in (a). As for (b), consider two mutually inverse isomorphisms $pA^{\infty} \to qA^{\infty}$ and $qA^{\infty} \to pA^{\infty}$. Assume that the first one sends columns of p to the columns of qBp whereas the second one sends columns of q to the columns of pCq. Writing that their compositions send p to p and q to q, we see that one can take $X = pCq$, $Y = qBp$. Conversely, if $p = XY, q = YX$, then also $p = (pXq)(qYp)$, $q = (qYp)(pXp)$, so that the matrices pXq and qYp determine mutually inverse isomorphisms of pA^{∞} and qA^{∞}. This completes the proof.

It is also convenient to introduce the parallel formalism for left projective A-modules: here we consider the right $\mathcal{M}A$-module A_{∞} of rows (a_i), $i \geq 1$, $a_i = 0$ for large i, and map a projector p to the left A-module $A_{\infty}p$.

Replacing A by the opposite ring A^{op} switches these two constructions.

A trace of A is any homomorphism of additive groups $t : A \to G$ vanishing on commutators; by definition, it factors through the universal trace $A \to A/[A, A]$. Combining it with the matrix trace, we get its canonical extension to $\mathcal{M}A$. From the Proposition 1.4.1 (b) it follows that $t(p)$ depends only on the isomorphism class of pA^{∞}. The class $p \bmod [A, A]$ is called *the Hattori–Stallings rank* of pA^{∞}.

We define $K_0(A)$ as the Grothendieck group of Pr_A. If $N_A \in Pr_A$, $[N_A]$ denotes its class in $K_0(A)$. If N_A is the range of an idempotent p and t is a trace, $t(p)$ depends only on $[N_A]$ and is additive on exact triples, hence t becomes a homomorphism of $K_0(A)$ (it is called dimension in the theory of von Neumann algebras).

The crucial role of traces in the theory of quantum tori (and in more general functional analytic situations) is explained by the fact that for irrational tori projective modules are exactly classified by the value of the (unique) normalized trace of the respective projector (Rieffel).

The following simple Lemma on traces is a useful technical tool. We assume in it that we work with algebras over a ground field.

Lemma 1.4.2. *Consider two unital algebras A, B and an A–B-bimodule ${}_AM_B$ which is projective as a module over A and over B. Assume that the (dual) space of traces $A/[A,A]$ of A is one-dimensional, whereas that of B is ≥ 1-dimensional, and that $1 \notin [A,A]$. Choose non-zero traces t_A and t_B. Then there exists such a constant c that for any $N_A \in Pr_A$ we have*

$$t_B([N \otimes_A M_B]) = c\, t_A([N_A])) \,. \tag{1.4}$$

The value of this constant is obtained by putting $N_A = A_A$ in (1.4):

$$c = t_B([{}_AM_B])t_A([A_A])^{-1} \,. \tag{1.5}$$

Proof. Let the modules N_A, M_B be given as ranges of idempotents $q \in \mathcal{M}A$, $p \in \mathcal{M}B$ respectively.

Put $A_1 := \mathrm{End}_B(M_B)$ and identify $p\mathcal{M}Bp$ with A_1 as above. The structure of left A-module on M_B is given by a ring homomorphism $\varphi : A \to A_1$. The trace t_B induces a trace on $p\mathcal{M}Bp$ and thus a trace t_1 on A_1. In turn, t_1 induces via φ a trace on A, and since the latter is unique, there exists a constant $c \neq 0$ such that we have identically $t_1(\varphi(a)) = c\, t_A(a)$ for all $a \in A$. By the associativity of tensor multiplication, we have

$$N \otimes_A M_B = (N \otimes_A A_1) \otimes_{A_1} M_B \,.$$

As a right A_1-module, $N \otimes_A A_1$ is isomorphic to $q_1 A_1^\infty$ where $q_1 = \varphi(q)$. We have $q_1^2 = q_1$ and since $q_1 \in p\mathcal{M}Bp$, $q_1 p = pq_1 = q_1$. Hence finally $N \otimes_A M_B$ as a right B-module is isomorphic to $q_1 B^\infty$. Thus

$$t_B([N \otimes_A M_B]) = t_B(q_1) = t_{A_1}(q_1) = c\, t_A(q) = c\, t_A([N_A]) \,.$$

■

1.5 Involutions and scalar products. Assume now that A is endowed with an additive (linear or antilinear in the case of algebras) involution $a \mapsto a^*$, $(ab)^* = b^*a^*$, $a^{**} = a$. It extends to matrix algebras: $(B^*)_{ij} := B^*_{ji}$. Similarly, it extends to $A^\infty \to A_\infty$ and $A_\infty \to A^\infty$, compatibly with the module structures.

In such a context, it makes sense to consider only those projective modules which are ranges of *projections*, that is, $*$-invariant idempotents. In fact, in the case of C^*-algebras the resulting subcategory of projective modules is equivalent to the full category, because *every idempotent is von Neumann equivalent to a projection.* In fact, if p is an idempotent, then

$$P := pp^* \left[1 - (p - p^*)^2\right]^{-1}$$

is an equivalent projection. The C^*-structure is used to ensure the invertibility of $1 - (p - p^*)^2$; the rest is pure algebra: see e. g. [18], IV.1.

Taking into account the involution, we get an additional structure on our modules and bimodules consisting of scalar products and identities relating them. This is a simple but important formalism made explicit by M. Rieffel.

Lemma 1.5.1. *Let M_B be a projective module (over a ring with involution B) isomorphic to $p\mathcal{M}B$ with $p^* = p$. Put $A = \mathrm{End}_B(M_B)$, identify this ring with $p\mathcal{M}Bp$ as above, and consider M as an A–B bimodule. The involution on $p\mathcal{M}Bp$ is induced by that on B.*

Define two scalar products ${}_A\langle *, * \rangle : M \times M \to A$ and $\langle *, * \rangle_B : M \times M \to B$:

$$ {}_A\langle pb, pc\rangle := (pb)(pc)^* = pbc^*p \in p\mathcal{M}Bp = A\,, \tag{1.6}$$

$$\langle pb, pc\rangle_B := (pb)^* pc = b^* pc \in B\,. \tag{1.7}$$

Then the following identities hold, in which $l, m, n \in M$, $a \in A, b \in B$:

$$ {}_A\langle m, n\rangle^* = {}_A\langle n, m\rangle\,, \quad \langle m, n\rangle_B^* = \langle n, m\rangle_B\,, \tag{1.8}$$

$$ {}_A\langle am, n\rangle = a\, {}_A\langle m, n\rangle\,, \quad {}_A\langle m, an\rangle = {}_A\langle m, n\rangle\, a^*\,, \tag{1.9}$$

$$\langle mb, n\rangle_B = b^* \langle m, n\rangle_B\,, \quad \langle m, nb\rangle_B = \langle m, n\rangle_B b\,, \tag{1.10}$$

$$ {}_A\langle l, m\rangle n = l\langle m, n\rangle_B\,. \tag{1.11}$$

We omit the checks which are straightforward.

1.6 Rieffel's projections. As we will see in Sect. 3, over toric C^*-algebras many bimodules ${}_AM_B$ are constructed directly, by inducing them from a Heisenberg representation, and the scalar products with the properties summarized in the Lemma 1.4.2 are introduced by an ad hoc formula.

In this case it is useful to know that, conversely, projections can be produced from such a setup. The following Lemma due to Rieffel [58] furnishes them.

Lemma 1.6.1. *Assume that ${}_AM_B$ is a bimodule over two rings with involution, endowed with two scalar products satisfying the formalism (1.8)–(1.11). Let $m \in {}_AM_B$.*

(a) If $m\langle m, m\rangle_B = m$, then $p := {}_A\langle m, m\rangle$ is a projection in A.

(b) Conversely, assume that from ${}_A\langle n, n\rangle = 0$ it follows that $n = 0$. In this case, if p as above is a projection, then $m\langle m, m\rangle_B = m$.

Proof. (a) Using (1.6) and (1.8), we obtain

$$p^2 = {}_A\langle m, m\rangle {}_A\langle m, m\rangle = {}_A\langle {}_A\langle m, m\rangle m, m\rangle = {}_A\langle m\langle m, m\rangle_B, m\rangle = {}_A\langle m, m\rangle = p.$$

From (1.3) it follows that $p^* = p$.

(b) Conversely, if p is a projection, then we get similarly

$$ {}_A\langle m\langle m, m\rangle_B - m, m\langle m, m\rangle_B - m\rangle = 0.$$

This completes the proof. ■

Rieffel also remarks that if $x \in {}_AM_B$ is such an element that one can construct an invertible $*$-invariant square root $\langle x, x\rangle_B^{1/2}$, then $m := x\,\langle x, x\rangle_B^{-1/2}$ satisfies 1.6.1(a).

F. Boca in [7] takes for x a Gaussian element in the relevant Heisenberg module. Then $\langle x, x\rangle_B$ turns out to be a quantum theta in the sense of [45]. We develop this remark in Sect. 3 for multidimensional case.

The net result is that we have a supply of explicit projections in toric algebras which in the notation of Sect. 3 below are given by the formulas

$$p_T := {}_D\left\langle f_T\, \Theta_{D^!}^{-1/2}, f_T\, \Theta_{D^!}^{-1/2}\right\rangle \in C(D, \alpha).$$

For notation, see (3.20), Theorem 3.7, and Sect. 3.3.

This formula (and its generalizations) relates the representation theory of quantum tori to the theory of quantum thetas which has a distinct flavor of noncommutative *algebraic* geometry. Notice however that the existence of $\Theta_{D^!}^{-1/2}$ is not established in full generality.

The whole formalism sketched in 1.3–1.6.1 is a simple algebraic version of some basic machinery in the theory of von Neumann and C^*-algebras. In particular, see [12] and [34] and original papers by A. Connes and A. Wassermann who overcame some highly nontrivial complications arising in the operator context.

1.6.2 Morita category and two-dimensional quantum tori. By definition, two-dimensional quantum tori are objects of the category $\mathcal{QT}$ whose morphisms are isomorphism classes of projective bimodules ${}_AM_B$ corresponding to projections, so that the formalism of the previous subsections is readily applicable. In particular, A_θ has a unique trace t_A which is normalized by the condition $t_A(1) = 1$ and which vanishes on any frame.

We will see in 1.7 below that there is a functorial correspondence between $\mathcal{QT}$ and pseudolattices which is fairly similar to the correspondence between elliptic curves and lattices. In particular, Real Multiplication of pseudolattices is reflected in $\mathcal{QT}$.

In order to achieve arithmetical applications of Real Multiplication, one has to find still smaller rings and modules, perhaps finitely generated in an algebraic sense and admitting models over rings of algebraic integers. Their definition remains the central unsolved problem in our approach. Since the points of finite order m on an elliptic curve E/K are in fact points of a finite group scheme over K acting upon E, it is conceivable that in the C^*-world the relevant finite objects should be seeked among weak Hopf algebras (or weak quantum groupoids) acting upon C^*-algebras: see recent reports [51], [37], [38], and the references quoted therein.

The famous paper [33] shows how a spectrum of algebraic numbers can be generated from such a setting. Jones's discrete spectrum of indices of subfactors is $\{4\cos^2\frac{\pi}{n} \mid n \geq 3\}$, whereas Stark's numbers in the cyclotomic case are $4\sin^2\frac{\pi m}{n}$. Both generate the maximal real subextension of $\mathbf{Q}^{ab}$.

Is this only a coincidence?

Returning to the C^* (or smooth) context, notice in conclusion that a bimodule ${}_AM_B$ can be treated as an $A \otimes B^{op}$-left module (completed tensor product). If it were projective, we could classify bimodules for toric A, B using the fact that $A \otimes B^{op}$ is again toric: their invariants would come from $K_0(A) \otimes K_0(B^{op})$ (the "trivial part") and from $K_1(A) \otimes K_1(B^{op})$ (the really interesting correspondences). However, intuitively it seems clear that such bimodules are much smaller than projective modules because they are separately A- and B^{op}-projective and hence, like Rieffel's elementary modules, should be realisable in functions of $\dim A = \dim B$ variables, whereas $A \otimes B^{op}$-projective modules are realizable only by functions of the doubled number of variables.

Therefore several questions arise about a possible extension of the classification theory of modules.

Question. Is any Morita morphism bimodule a maximal quotient of a unique projective $A \otimes B^{op}$-module?

More generally, toric projective modules can have nontrivial maximal quotients, like in the situation with highest weight and Verma modules. One should seek for canonical projective resolutions of such modules.

The algebraic machinery might be connected with the fact that $A \otimes B^{op}$ contains large commutative subalgebras, so that a module can be decomposed according to their characters. E. g. if A, B are t-wodimensional quantum tori, $A \otimes B^{op}$ contains two-dimensional classical tori, and prescribing their characters may produce the interesting quotients.

Question. Can one find a description of the derived category of perfect complexes over toric algebras?

1.7 Two functors relating $\mathcal{QT}$ to $\mathcal{PL}$. We start with defining a functor $K: \mathcal{QT} \to \mathcal{PL}$. Let the torus T be represented by an algebra A. On objects, we put:

$$K(T) = (L_A, V_A, j_A, s_A). \tag{1.12}$$

Here $L_A := K_0(A)$, the K_0-group of the category of right projective A-modules (as above, given by projections in finite matrix algebras over A); V_A is the target group of the universal trace on A, that is, the quotient space of A modulo the completed commutator subspace $[A, A]$. Furthermore, $j_A = t_A : K_0(A) \to V_A$ is this universal trace extended to matrix algebras; its value on the class of a module, as we already explained, is its value at the respective projection. Finally s_A is taken in such a way that positive elements in $K_0(A)$ become represented by the classes of actual (not virtual) projective modules.

On morphisms, we define directly the left vertical arrow of the respective diagram (1.1):

$$K({}_AM_B)([N_A]) := [N \otimes_A M_B]. \tag{1.13}$$

The existence of the right vertical arrow follows from the Lemma 1.4.1.

Theorem 1.7.1. *(a) The family of maps (1.12), (1.13) can be uniquely completed to a functor $K: \mathcal{QT} \to \mathcal{PL}$.*

(b) This functor is essentially surjective on objects and (strict) morphisms.

(c) Assume that two bimodules ${}_AM_B$ and ${}_AM'_B$ considered as morphisms in $\mathcal{QT}$ become equal after applying K. Put $A_1 := \mathrm{End}_B(M_B)$ and consider ${}_AM_B$ as an A_1–B bimodule ${}_{A_1}M_B$.

There exist two ring homomorphisms $\varphi, \psi : A \to A_1$ such that if one considers A_1 as an A–A_1 bimodule ${}_\varphi A_{1,A_1}$, (resp. ${}_\psi A_{1,A_1}$) using φ (resp. ψ) to define the left action, and the ring structure of A_1 to define the right action, one obtains

$${}_\psi A_1 \otimes_{A_1} M'_B \cong {}_\varphi A_1 \otimes_{A_1} M_B \tag{1.14}$$

as A–B-bimodules.

In particular, if $\otimes_A M_B$ and $\otimes_A M'_B$ produce Morita equivalences, these functors differ by an automorphism of the category Mod_A *which is induced by an automorphism of the ring A.*

Comments. This result should be compared to the easy Theorem 0.3.1 which provides the geometric basis of the Complex Multiplication. The statement about quantum tori sounds less neat, however in 1.7.2 we will complement it by the construction of a functor in the reverse direction defined only on isomorphisms, which should suffice for the envisioned applications to Real Multiplication.

Proof. (a) Lemma 1.4.2 shows that, after passing to traces, (1.13) becomes the multiplication by a positive number representing a (strict) morphism of pseudolattices $K(m)$. The compatibility with the composition of morphisms is straightforward.

(b) It remains to establish the following three facts.

(i) *Every object of $\mathcal{PL}$ is isomorphic to an object lying in the image of K.*

In fact, the pseudolattice denoted $(L_\theta, 1)$ in 1.1(b) is isomorphic to $K(A_\theta)$ where A_θ is the respective rotation algebra. This is the main result of the theory, due to Connes, Rieffel, Pimsner–Voiculescu, Elliott. It is worth recalling here one of the several known strategies for proving it (cf. [18], Ch. VI).

First, one checks that for any $\alpha \in [0, 1] \cap \mathbf{Z} + \mathbf{Z}\theta$ there exists a projection $p_\alpha \in A_\theta$ with the normalized trace $\tau(p_\alpha) = \alpha$. Using functional calculus, one can directly construct such projections of the form $f(U)V + g(U) + h(U)V^*$ (Rieffel–Powers, see [18], p. 171.) It follows than $\tau(K_0(A_\theta)) \supset \mathbf{Z} + \mathbf{Z}\theta$.

Second, one shows that A_θ can be embedded into an approximately finite algebra $\mathcal{A}_\theta$ which is the completed inductive limit of A_{p_n/q_n}, where p_n/q_n are consecutive convergents to θ. This embedding allows one to calculate $\tau(K_0(A_\theta))$ as the inductive limit of ordered groups $\sigma(K_0(\mathcal{A}_{p_n/q_n}))$, and this inductive limit is explicitly identified with $\mathbf{Z} + \mathbf{Z}\theta$.

This last argument can be read as a weak continuity property of A_θ with respect to θ varying in the set of cusps. In 1.8 below, we will discuss in what sense A_θ can be regarded as a limit of E_τ when τ tends to θ from the upper half-plane.

(ii) *Every morphism $K(T_\theta) \to K(T_{\theta'})$ in $\mathcal{PL}$ is of the form $K(m)$ where m is the tensor multiplication by an appropriate bimodule.*

Clearly, it suffices to choose a generating family of morphisms in $\mathcal{PL}$ (such that any morphism is a composition of members of this family) and to show that each generator can be lifted to $\mathcal{QT}$.

Any morphism of pseudolattices restricted upon L-components is a composition of an injection and an isomorphism (respecting ordering); moreover, this restriction uniquely determines it. Any injection can be decomposed into product of two injections with cyclic quotients.

Isomorphisms between pseudolattices $L_{\theta'} \to L_\theta$ can be decomposed into a sequence of transformations of the form $\theta \mapsto -\theta$, $\theta \mapsto \theta + 1$, $\theta \mapsto \theta^{-1}$. The map $(U, V) \mapsto (V', U')$ produces an isomorphism $T_\theta \to T_{-\theta}$, whereas T_θ and $T_{\theta+n}$ are obviously the same. The only no-ntrivial problem is to find a Morita equivalence $T_\theta \to T_{\theta^{-1}}$. Its solution was given in [13] and generalized to multidimensional tori in a series of works of Rieffel and his collaborators, see [60], [62].

Alternatively, in [17] one can find a direct description of a bimodule furnishing a Morita equivalence between T_θ and $T_{\theta'}$, where θ and θ' are related by a transformation from $PGL(2, \mathbf{Z})$. We will reproduce it in 1.7.2 below.

It remains to treat the case of embedding of pseudolattices. Now choose $n > 0$ and consider the embedding of toric algebras $B := A_{n\theta} \hookrightarrow A := A_\theta$ where in self-evident notation $U_B = U_A^n$, $V_B = V_A$. For ${}_AM_B$ take the bimodule ${}_AA_B$. It is free of rank 1 (resp. n) as A– (resp. B–) module. Denote by t_A, t_B the normalized traces (taking value 1 on 1). Then the constant (1.5) is n, so that the tensor multiplication by ${}_AM_B$ produces the morphism of pseudolattices $L_\theta \mapsto L_{n\theta}$: $\theta \mapsto n\theta$, $1 \mapsto n$. Clearly, any embedding of pseudolattices with cyclic quotient is isomorphic to such one.

(iii) *If* $K(m) = K(m')$, *the respective B-modules* M_B, M'_B *are isomorphic.*

In fact, from (1.4) and (1.5) it follows that the B-traces of them coincide, and for two-dimensional irrational tori this means that they are isomorphic.

The remaining argument is straightforward. Choose and fix an isomorphism of M and M' as B-modules. Actions of A upon M and M' correspond to two different homomorphisms $A \to A_1$. This is the essence of (1.14).

This finishes the proof. ■

1.7.2 The functor $E : \mathcal{PL}_{iso} \to \mathcal{QT}_{iso}$. In this section we rephrase the content of Sect. 2 of the recent preprint [21].

Denote by $\mathcal{PL}_{iso}$ the category whose objects are pseudolattices $L_\theta = \mathbf{Z} + \mathbf{Z}\theta$, $\theta \in \mathbf{R} \setminus \mathbf{Q}$, oriented by their embedding into $\mathbf{R}$, and whose morphisms are strict isomorphisms, that is, multiplications by a positive number identifying two pseudolattices. According to (1.2), such isomorphisms $L_{\theta'} \to L_\theta$ are represented by matrices

$$g = \begin{pmatrix} a & b \\ c & d \end{pmatrix} \in GL\,(2, \mathbf{Z})$$

such that

$$\theta' = g\theta = \frac{a\theta + b}{c\theta + d}, \qquad c\theta + d > 0. \tag{1.15}$$

Denote by $\mathcal{QT}_{iso}$ the category whose objects are irrational toric algebras A_θ and whose morphisms are bimodules inducing Morita equivalences.

Given θ', θ, and g satisfying (1.15), construct an $(A_{\theta'}, A_\theta)$-bimodule ${}_{\theta'}E_\theta(g^{-1})$ (notice the inversion g^{-1}) by the following prescription. The smooth part of ${}_{\theta'}E_\theta(g^{-1})$ consists of functions $f(x, \mu)$ in the Schwartz's space $\mathcal{S}(\mathbf{R} \times \mathbf{Z}_c)$. The generators U, V of A_θ act upon these functions from the right as follows:

$$(fU)(x, \mu) = f\left(x - \frac{c\theta + d}{c}, \mu - 1\right),$$

$$(fV)(x, \mu) = e^{2\pi i (x - \mu d/c)} f(x, \mu).$$

The generators U', V' of $A_{\theta'}$ act from the left:

$$(U'f)(x,\mu) = f\left(x - \frac{1}{c}, \mu - a\right),$$

$$(V'f)(x,\mu) = \exp\left[2\pi i\left(\frac{x}{c\theta + d} - \frac{\mu}{c}\right)\right] f(x,\mu).$$

To become a bimodule over the respective C^* algebras, the Schwartz space must be appropriately completed, cf. Theorem 3.4.1 below.

Theorem 1.7.3. *The map $E : \mathcal{PL}_{iso} \to \mathcal{QT}_{iso}$ defined on objects by $L_\theta \mapsto A_\theta$ and sending the isomorphism (1.15) to the Morita isomorphism $[_{\theta'}E_\theta(g^{-1})]$ is a well defined functor. The composition $K \circ E$ is isomorphic to the identical functor on $\mathcal{PL}_{iso}$.*

This theorem rephrases the main result of [21], Sect. 2, which in our notation establishes an explicit isomorphism of bimodules

$$s_{g,h} : {}_{\theta'}E_\theta\left(g^{-1}\right) \otimes_{A_\theta} {}_\theta E_{\theta''(h^{-1}) \to {}_{\theta'}E_\theta} \left((gh)^{-1}\right)$$

and thus shows that E is multiplicative on isomorphisms of pseudolattices. Here $\theta = g^{-1}(\theta')$ as above, and $\theta“ = h^{-1}(\theta)$ so that $\theta'' = (gh)^{-1}(\theta')$.

This isomorphism is constructed in [21] in the smooth setting. According to [13], extension of rings induces a bijection between the set of isomorphism classes of projective modules of finite type over A_θ and over its smooth subring respectively. Moreover, the trace (dimension) of ${}_{\theta'}E_\theta(g^{-1})$ as a right A_θ-module equals $|c\theta + d|$ ([13], Theorem 7): since ${}_{\theta'}E_\theta(g^{-1})$ is not given as the image of a projection, Connes develops differential geometric methods for calculating this trace. As an exercise, the reader can check that the dimension of the tensor product indeed equals the product of dimensions of factors.

Notice in conclusion that our version of Morita category using isomorphism classes of bimodules as morphisms is a truncation of a finer notion which treats bimodules as functors and leads to the notion of Morita 2-category. A refinement of the Dieng–Schwarz's result in this direction requires an explicitation of the associativity isomorphism connecting $s_{gh,k} \circ (s_{g,h} \otimes \mathrm{id})$ to $s_{g,hk} \circ (\mathrm{id} \otimes s_{h,k})$ which replaces the straightforward associativity of the triple multiplication of morphisms in 1-categories. This looks like a nice exercise.

1.8 Quantum tori as “limits” of elliptic curves. Reading parallelly subsections 0.1 and 1.1, we see that pseudolattices are in a very precise sense limits of lattices, at least, if one forgets orientation; or else one can add orientation to the definition of a lattice, as the choice of a generator of $\wedge^2(\Lambda)$.

Passing to the isomorphism classes of lattices/pseudolattices does not seem to change this impression: compare (0.3) and (1.3).

Comparison of the relevant geometric categories suggests that two-dimensional quantum tori can be thus considered as limits of elliptic curves. More specifically, take a family of Jacobi parametrized curves $E_\tau = \mathbf{C}/(e^{2\pi i\tau})$ with $\operatorname{Im}\tau > 0$ and $\tau \to \theta \in \mathbf{R}$. It is then natural to imagine T_θ as a limit of E_τ.

Fixing a Jacobi uniformization of an elliptic curve (or abelian variety of any dimension) as a part of its structure is necessary, for example, in problems connected with mirror symmetry. In such contexts our intuition seemingly provides a sound picture (cf. a similar discussion in [68], pp. 100, 113–114).

However, limitations of this viewpoint become quite apparent if one has no reason to keep a Jacobi uniformization as a part of the structure, and is interested only in the isomorphim classes of elliptic curves, perhaps somewhat rigidified by a choice of a level structure.

In this case one must contemplate the dynamics of the limiting process not on the closed upper half-plane but on a relevant modular curve X. Letting τ tend to θ along a geodesic, we get a parametrized real curve on X which, when θ is irrational, does not tend to any limiting point. The following lemma shows what can happen.

Lemma 1.8.1. *(a) Let θ be a real quadratic irrationality, θ' its conjugate. Consider the oriented geodesic in H joining θ' to θ. The image of this geodesic on any modular curve X is supported by a closed loop, which we denote $(\theta', \theta)_X$.*

(b) Let θ be as above, and let τ tend to θ along an arbitrary geodesic. Then the image of this geodesic on X has $(\theta', \theta)_X$ as a limit cycle (in positive time).

(c) Each closed geodesic on X is the support of a closed loop $(\theta', \theta)_X$. The union of them is dense in X. It is a strange attractor for the geodesic flow in the following sense. Having chosen a sequence of loops $(\theta_i', \theta_i)_X$, a sequence of integers $n_i \geq 1$, and a sequence of real numbers $\epsilon_i > 0$, $i = 1, 2, \ldots$, one can find an oriented geodesic winding $\geq n_i$ times in the ϵ_i-neighborhood of $(\theta_i', \theta_i)_X$ for each i, before jumping to the next loop.

Proof. We will only sketch a couple of arguments.

For (a), notice that θ' and θ are respectively the attracting and the repelling points of a hyperbolic fractional linear transformation $g \in SL(2, \mathbf{Z})$. This transformation maps into itself the whole geodesic joining θ' to θ and acts upon it as a shift by the distance $\log\varepsilon$ where $\varepsilon > 1$ is a unit in the quadratic field generated by θ (cf. formula (1.16) below). If $X = \Gamma \subset H$, where Γ is a subgroup of finite index of the modular group, then $g^n \in \Gamma$ for an appropriate $n \geq 1$. Therefore the geodesic in question will close to a loop on X.

The distance between two geodesics tending to the same θ in H tends to zero; this shows (b).

Finally, (c) is based upon an elementary argument involving continued fractions and diophantine approximations. The Lemma is proved. ■

Now let us imagine that we have constructed a certain object $R(E_\tau)$ depending on the isomorphism class of E_τ (perhaps, with rigidity). This object can be a number, a function of the lattice, a linear space, a category ... Suppose also that we have constructed a similar object $\mathcal{R}(T_\theta)$ depending on the isomorphism class of T_θ, and that we want to make sense of the intuitive notion that $\mathcal{R}(T_\theta)$ is "a limit of $R(E_\tau)$." Since in the most interesting for us case (a) of the Lemma 1.8.1 E_τ keeps rotating around the same loop, there are two natural possibilities:

(i) *The object $R(E_\tau)$ actually "does not depend on τ", and $\mathcal{R}(T_\theta)$ is its constant value. Here independence generally means a canonical identification of different $R(E_\tau)$, e.g. via a version of flat connection defined along the loop.*

(ii) *The object $R(E_\tau)$ does depend on τ, and $\mathcal{R}(T_\theta)$ is obtained by a kind of integrating or averaging various $R(E_\tau)$ along the loop.*

The second case looks more interesting, however, it is not immediately obvious that such objects occur in nature. Remarkably, they do, and precisely in the context of real multiplication and Stark's conjecture. In fact, this is how we will interpret the beautiful old calculational tricks due to Hecke: see [30], [31], [32], [77]. See also [19] for a similar observation related to what Darmon calls Stark–Heegner points of elliptic curves.

In this section we will only explain the geometric meaning of Hecke's substitution, whereas the (slightly generalized) calculation itself will be treated in the next section.

1.8.2 Hecke's lift of closed geodesics to the space of lattices. Let $K \subset \mathbf{R}$ be a real quadratic subfield of $\mathbf{R}$ and $L \subset K$ an RM pseudolattice. From now on, we denote by $l \mapsto l'$ the nontrivial element of the Galois group of $K/\mathbf{Q}$.

For any real t, consider the following subset of $\mathbf{C}$:

$$\Lambda_t = \Lambda_t(L) := \left\{\lambda_t = \lambda_t(l) := le^{t/2} + il'e^{-t/2} \mid l \in L\right\} \tag{1.16}$$

Lemma 1.8.3. *(a) $\Lambda_t(L)$ is a lattice.*

(b) Any isomorphism $a : L_1 \to L$ in the narrow sense induces isomorphisms $\Lambda_t(L_1) \to \Lambda_{t+c}(L)$ where c is a constant depending only on a and t is arbitrary.

(c) The image of the curve $\{\Lambda_t \mid t \in \mathbf{R}\}$ on the modular curve (0.3) (or any modular curve) is a closed geodesic. The affine coordinate t along this curve is the geodesic length.

Proof. (a) is evident; moreover, if l_1, l_2 form a basis of L, then $\lambda_t(l_1), \lambda_t(l_2)$ form a basis of Λ_t.

For (b), consider an isomorphism $L \mapsto L_1 : l \mapsto al,\ a \in K,\ aa' > 0$. It induces a map $\Lambda_t(L) \to \Lambda_t(L_1)$:

$$\lambda_t(l) \mapsto al\, e^{t/2} + ia'l'e^{-t/2} =$$

$$\sqrt{aa'}\left(\sqrt{\frac{a}{a'}}\, le^{t/2} + \sqrt{\frac{a'}{a}}\, l'e^{-t/2}\right) = \sqrt{aa'}\, \lambda_{t+\log \frac{a}{a'}}(l). \tag{1.17}$$

This produces an isomorphism of $\Lambda_t(L_1)$ with $\Lambda_{t+\log \frac{a}{a'}}(L)$.

For (c), it suffices to consider pseudolattices L generated by 1 and $\theta \in K$ with $\theta' > \theta$. Then $\Lambda_t(L)$ is generated by $e^{t/2} + ie^{-t/2}$ and $\theta\, e^{t/2} + i\theta'\, e^{-t/2}$, and hence isomorphic to the lattice generated by 1 and

$$\tau_t := \frac{\theta\, e^{t/2} + i\theta'\, e^{-t/2}}{e^{t/2} + ie^{-t/2}} = \frac{\theta\, e^t + \theta'\, e^{-t}}{e^t + e^{-t}} + i\,\frac{\theta' - \theta}{e^t + e^{-t}}. \tag{1.18}$$

A straightforward computation shows that

$$\left|\tau_t - \frac{\theta + \theta'}{2}\right|^2 = \left(\frac{\theta' - \theta}{2}\right)^2.$$

Hence τ_t runs over a semicircle in the upper half plane connecting θ' to θ. A further calculation shows that the geodesic length element $\frac{|d\tau|}{\operatorname{Im}\tau}$ restricted to this semicircle coincides with dt. The normalization of t has a simple geometric meaning: $t = 0$ is the upper point of the geodesic semicircle. ■

2 Stark's Numbers and Theta Functions for Real Quadratic Fields

2.1 Stark's numbers at $s = 0$. In this section we fix a real quadratic subfield $K \subset \mathbf{R}$. Denote by $l \mapsto l'$ the action of the nontrivial element of the Galois group of K, and by O_K the ring of integers of K, and put $N(l) = ll'$.

Let L be an arbitrary integral ideal of K which, together with its embedding in $\mathbf{R}$ and the induced ordering, will be considered as a pseudolattice.

Choose also an $l_0 \in O_K$ so that the pair (L, l_0) satisfies the following restrictions:

(i) *The ideals* $\mathfrak{b} := (L, l_0)$ *and* $\mathfrak{a}_0 := (l_0)\mathfrak{b}^{-1}$ *are coprime with* $\mathfrak{f} := L\mathfrak{b}^{-1}$.

(ii) *Let* ε *be a unit of* K *such that* $\varepsilon \equiv 1 \bmod \mathfrak{f}$. *Then* $\varepsilon' > 0$.

Put now

$$\zeta(L, l_0, s) := \operatorname{sgn} l_0'\, N(\mathfrak{b})^s \sum_{l \in L}^{(u)} \frac{\operatorname{sgn}(l_0 + l)'}{|N(l_0 + l)|^s} \tag{2.1}$$

where (u) at the summation sign means that one should take one representative from each coset $(l_0 + l)\varepsilon$ where ε runs over all units $\equiv 1 \bmod \mathfrak{f}$. Notice that $(l_0 + L)\varepsilon = l_0 + L$ precisely for such units.

With these conventions, our $\zeta(L, l_0, s)$ is exactly Stark's function denoted $\zeta(s, \mathfrak{c})$ on the page 65 of [69]: our $\mathfrak{a}_0, \mathfrak{b}, \mathfrak{f}$ have the same meaning in [69], and our l_0 is Stark's γ. The meaning of Stark's $\mathfrak{c}$ is explained below.

The Stark number of (L, l_0) is defined as

$$S_0(L, l_0) := e^{\zeta'(L, l_0, 0)} \tag{2.2}$$

(cf. the general discussion in 0.6).

The simplest examples correspond to the cases when $(L, l_0) = (1)$, $\mathfrak{f} = L$, in particular, $l_0 = 1$.

Notice that pseudolattices which are integral ideals have conductor $f = 1$ in the sense of Lemma 1.1.1.

2.2 Stark's conjecture for real quadratic fields. In [69], Stark conjectures that $S_0(L, l_0)$ are algebraic units generating abelian extensions of K. To be more precise, let us first describe an abelian extension M/K associated with (L, l_0) using the classical language of class field theory. (Our M is Stark's K, whereas our K corresponds to Stark's k.)

In 2.1 above we constructed, starting with (L, l_0), the ideals $\mathfrak{f}$ and $\mathfrak{b}$ in O_K. Let $I(\mathfrak{f})$ be the group of fractional ideals of K generated by the prime ideals of K not dividing $\mathfrak{f}$, and $S(\mathfrak{f})$ be its subgroup called the principal ray class modulo $\mathfrak{f}$. Then Artin's reciprocity map identifies $G(\mathfrak{f}) := I(\mathfrak{f})/S(\mathfrak{f})$ with the Galois group of M/K.

Consider all pairs (L, l_0) as above with fixed $\mathfrak{f}$. It is not difficult to establish that on this set, $S_0(L, l_0)$ in fact depends only on the class $\mathfrak{c}$ of $\mathfrak{a}_0 = (l_0)\mathfrak{b}^{-1}$ in $G(\mathfrak{f})$. Denote the respective number $E(\mathfrak{c})$.

2.2.1 Conjecture. *The numbers $E(\mathfrak{c})$ are units belonging to M and generating M over K. If the Artin isomorphism associates with $\mathfrak{c}$ an automorphism σ, we have $E(1)^\sigma = E(\mathfrak{c})$.*

(We reproduced here the most optimistic form of the Conjecture 1 on page 65 of [69] involving $m = 1$ and Artin's reciprocity map).

2.3 Hecke's formulas. In this subsection we will work out Hecke's approach to the computation of sums of the type (2.1), cf. [31]. It starts with a Mellin transform so that instead of Dirichlet series (2.1) we will be dealing with a version of theta-functions for real quadratic fields. We start with introducing a class of such theta functions more general than strictly needed for dealing with (2.1) (and more general than Hecke's one).

2.4 Theta functions of pseudolattices. Let $K \subset \mathbf{R}$ be as in 2.1. We choose and fix the following data: a pseudolattice $L \subset K$, two numbers $l_0, m_0 \in K$ and a number $\eta = \eta_0 + i\eta_1 \in \mathbf{C}$. A complex variable v will take values in the upper half plane; $\sqrt{-iv}$ is the branch which is positive on the upper part of the imaginary axis.

Finally, choose an infinite cyclic group U of totally positive units in K such that the following conditions hold:

(a) $u(l_0 + L) = l_0 + L$ for all $u \in U$.

(b) $\operatorname{tr} ulm_0 \equiv \operatorname{tr} lm_0 \bmod \mathbf{Z}$, $\operatorname{tr} ul_0m_0 \equiv \operatorname{tr} l_0m_0 \bmod 2\mathbf{Z}$ for all $l \in L$, $u \in U$, where $\operatorname{tr} := \operatorname{tr}_{K/\mathbf{Q}}$.

Put now

$$\Theta^U_{L,\eta}\begin{bmatrix} l_0 \\ m_0 \end{bmatrix}(v) :=$$

$$\sum_{l_0+l \bmod U} (\eta_0 \operatorname{sgn}(l_0' + l') + \eta_1 \operatorname{sgn}(l_0 + l))\, e^{2\pi i\, v|(l_0+l)(l_0'+l')|} e^{-2\pi i \operatorname{tr} lm_0} e^{-\pi i \operatorname{tr} l_0m_0}. \tag{2.3}$$

Notation $l_0 + l \bmod U$ means that we sum over a system of representatives of orbits of U acting upon $l_0 + L$.

Notice that such U always exists, and that if we choose a smaller subgroup $V \subset U$, then

$$\Theta^V_{L,\eta}\begin{bmatrix} l_0 \\ m_0 \end{bmatrix}(v) = [U : V]\, \Theta^U_{L,\eta}\begin{bmatrix} l_0 \\ m_0 \end{bmatrix}(v).$$

In order to relate these thetas to Stark's numbers, consider the function

$$\Theta^U_{L,1}\begin{bmatrix} l_0 \\ 0 \end{bmatrix}(v) = \sum_{l_0+l \bmod U} \operatorname{sgn}(l_0' + l')\, e^{2\pi i\, v|(l_0+l)(l_0'+l')|}. \tag{2.4}$$

Then we have

$$\sum_{l_0+l \bmod U} \frac{\operatorname{sgn}(l_0' + l')}{|N(l_0 + l)|^s} = \frac{(2\pi)^s}{\Gamma(s)} \int_0^{i\infty} (-iv)^s \Theta^U_{L,1}\begin{bmatrix} l_0 \\ 0 \end{bmatrix}(v)\, \frac{dv}{v}. \tag{2.5}$$

We will now show that these RM thetas can be obtained by averaging some theta constants (related to the complex lattices) along the closed geodesics described in 1.8 above.

2.5 Theta constants along geodesics. Starting with the same data as in 2.4, we introduce first of all a family of lattices $\Lambda_t = \Lambda_t(L)$ defined by (1.16). From l_0 which was used to shift L, we will produce a shift of Λ_t:

$$\lambda_{0,t} := l_0\, e^{t/2} + il_0'\, e^{-t/2}.$$

The number m_0 determines a character of L appearing in (2.3): $l \mapsto e^{-2\pi i \operatorname{tr} l m_0}$. Similarly, we will produce a character of Λ_t from

$$\mu_{0,t} := m_0 e^{t/2} + i m_0' e^{-t/2}$$

by using the scalar product on $\mathbf{C}$

$$(x \cdot y) = \operatorname{Im} xy = x_0 y_1 + x_1 y_0 \tag{2.6}$$

where $x = x_0 + ix_1$, $y = y_0 + iy_1$. Since $l_0, m_0 \in L \otimes \mathbf{Q}$, we have similarly $\lambda_{0,t}, \mu_{0,t} \in \Lambda_t \otimes \mathbf{Q}$. Omitting t for brevity, we put:

$$\theta_{\Lambda,\eta} \begin{bmatrix} \lambda_0 \\ \mu_0 \end{bmatrix} (v) := \sum_{\lambda \in \Lambda} ((\lambda_0 + \lambda) \cdot \eta)\, e^{\pi i v |\lambda_0 + \lambda|^2} e^{-2\pi i (\lambda \cdot \mu_0) - \pi i (\lambda_0 \cdot \mu_0)}. \tag{2.7}$$

The two types of thetas are related by Hecke's averaging formula:

Proposition 2.6. *We have*

$$\Theta^U_{L,\eta} \begin{bmatrix} l_0 \\ m_0 \end{bmatrix} (v) = \sqrt{-iv} \int_{-\log \varepsilon}^{\log \varepsilon} \theta_{\Lambda_t,\eta} \begin{bmatrix} \lambda_{0,t} \\ \mu_{0,t} \end{bmatrix} (v)\, dt \tag{2.8}$$

where $\varepsilon > 1$ is a generator of U.

Proof. The following formulas are valid for $\operatorname{Im} v > 0$:

$$e^{2\pi i\, v |mm'|} = \sqrt{-iv}\, |m'| \int_{-\infty}^{\infty} e^{-t/2} e^{\pi i v (m^2 e^t + m'^2 e^{-t})}\, dt =$$

$$\sqrt{-iv}\, |m| \int_{-\infty}^{\infty} e^{t/2} e^{\pi i v (m^2 e^t + m'^2 e^{-t})}\, dt \tag{2.9}$$

(see e. g. [40], pp. 270–271). In the rhs of (2.3), replace the first exponent by its integral versions (2.9), using the first version at η_0 and the second at η_1. We get:

$$\Theta^U_{L,\eta} \begin{bmatrix} l_0 \\ m_0 \end{bmatrix} (v) =$$

$$\sqrt{-iv} \textstyle\int_{-\infty}^{\infty} \sum_{l_0 + l \bmod U} (\eta_0\, (l_0' + l')\, e^{-t/2} + \eta_1\, (l_0 + l)\, e^{t/2}) \times$$

$$e^{\pi i v ((l_0 + l)^2 e^t + (l_0' + l')^2 e^{-t})} e^{-2\pi i \operatorname{tr} l m_0} e^{-\pi i \operatorname{tr} l_0 m_0}\, dt\,. \tag{2.10}$$

In view of (1.16) and (2.6) we have

$$\eta_0\, (l_0' + l')\, e^{-t/2} + \eta_1\, (l_0 + l)\, e^{t/2} = ((\lambda_{0,t} + \lambda_t) \cdot \eta),$$

$$(l_0+l)^2e^t + (l_0'+l')^2e^{-t} = |\lambda_{0,t}+\lambda_t|^2,$$

and similarly

$$\operatorname{tr} lm_0 = (\lambda_t \cdot \mu_{0,t}), \quad \operatorname{tr} l_0m_0 = (\lambda_{0,t} \cdot \mu_{0,t}).$$

Inserting this into (2.10), we obtain

$$\sqrt{-iv}\int_{-\infty}^{\infty} dt \sum_{l_0+l \bmod U} ((\lambda_{0,t}+\lambda_t)\cdot\eta)e^{\pi i\, v|\lambda_{0,t}+\lambda_t|^2}\, e^{-2\pi i\,(\lambda_t\cdot\mu_{0,t})}e^{-\pi i\,(\lambda_{0,t}\cdot\mu_{0,t})}. \tag{2.11}$$

Replacing l_0+l by $\varepsilon(l_0+l)$ is equivalent to replacing t by $t+2\log\varepsilon$. Hence finally the right hand side of (2.11) can be rewritten as

$$\sqrt{-iv}\int_{-\log\varepsilon}^{\log\varepsilon} dt \sum_{\lambda_t\in\Lambda_t} ((\lambda_{0,t}+\lambda_t)\cdot\eta)\, e^{\pi i\, v|\lambda_{0,t}+\lambda_t|^2} e^{-2\pi i\,(\lambda_t\cdot\mu_{0,t})-\pi i\,(\lambda_{0,t}\cdot\mu_{0,t})} \tag{2.12}$$

which is the same as (2.8). ∎

We will now apply Poisson formula in order to derive functional equations for Hecke's thetas.

2.7 Poisson formula. Let V be a real vector space, $\widehat{V}$ its dual. We will denote by $(x\cdot y)\in\mathbf{R}$ the scalar product of $x\in V$ and $y\in\widehat{V}$. Choose a lattice (discrete subgroup of finite covolume) $\Lambda\subset V$ and put

$$\Lambda^! := \left\{\mu\in\widehat{V}\mid\forall\lambda\in\Lambda,\ (\lambda\cdot\mu)\in\mathbf{Z}\right\}. \tag{2.13}$$

Choose also a Haar measure dx on V and define the Fourier transform of a Schwarz function f on V by

$$\widehat{f}(y) := \int_V f(x)\, e^{-2\pi i(x\cdot y)}dx. \tag{2.14}$$

If $f(x)$ in this formula is replaced by $f(x+x_0)\,e^{-2\pi i(x\cdot y_0)-\pi i(x_0\cdot y_0)}$ for some $x_0\in V$, $y_0\in\widehat{V}$, its Fourier transform $\widehat{f}(y)$ gets replaced by $\widehat{f}(y+y_0)\,e^{2\pi i(x_0\cdot y)+\pi i(x_0\cdot y_0)}$.

The Poisson formula reads

$$\sum_{\lambda\in\Lambda} f(\lambda) = \frac{1}{\int_{V/\Lambda}dx}\sum_{\mu\in\Lambda^!}\widehat{f}(\mu), \tag{2.15}$$

and for shifted functions as above

$$\sum_{\lambda\in\Lambda} f(\lambda_0+\lambda)\,e^{-2\pi i(\lambda\cdot\mu_0)-\pi i(\lambda_0\cdot\mu_0)} = \frac{1}{\int_{V/\Lambda}dx}\sum_{\mu\in\Lambda^!}\widehat{f}(\mu_0+\mu)\,e^{2\pi i(\lambda_0\cdot\mu)+\pi i(\lambda_0\cdot\mu_0)}. \tag{2.16}$$

2.8 Functional equations for θ and Θ. In order to transform (2.12) using the Poisson formula, we put

$$V = \mathbf{C} = \{x_0 + ix_1\}, \ \widehat{V} = \mathbf{C} = \{y_0 + iy_1\}, \tag{2.17}$$

and take (2.6) for the scalar product.

Lemma 2.8.1. *Let the lattice $\Lambda_t \subset \mathbf{C}$ be given by (1.15). Then the dual lattice $\Lambda_t^!$ with respect to the pairing (2.6) has the similar structure*

$$\Lambda_t^! = \Lambda_t(M) := \left\{me^{t/2} + im'e^{-t/2} \mid m \in M\right\} \tag{2.18}$$

where we denoted by $M = L^?$ the pseudolattice

$$M := \{m \in K \mid \forall l \in L, \ \mathrm{tr}_{K/\mathbf{Q}}(l'm) \in \mathbf{Z}.\}\,.$$

Proof. Denote by Γ the lattice (2.18). For any $\lambda = le^{t/2} + il'e^{-t/2} \in \Lambda_t$ and $\mu = me^{t/2} + im'e^{-t/2} \in \Gamma$ we have

$$(\lambda \cdot \mu) = \mathrm{Im}\, \lambda\mu = lm' + l'm = \mathrm{tr}_{K/\mathbf{Q}}(lm')\,. \tag{2.19}$$

Therefore this scalar product lies in $\mathbf{Z}$ if $m \in M$ so that $\Gamma \subset \Lambda_t^!$. Clearly, then, Γ must be commensurable with $\Lambda_t^!$, so that the right hand side of (2.19) can be used for computing $(\lambda \cdot \mu)$ on the whole $\Lambda_t^!$. This finishes the proof. ■

For example, $O_K^? = \mathfrak{d}^{-1}$ where $\mathfrak{d}$ is the different. In fact, this is the standard definition of the different.

Now let l_1, l_2 be two generators of the pseudolattice L. Put

$$\Delta(L) := |l_1 l_2' - l_1' l_2|\,. \tag{2.20}$$

Clearly, this number does not depend on the choice of generators.

Lemma 2.8.2. *Let the Haar measure on V be $dx = dx_0\, dx_1$. Choose generators l_1, l_2 of L. Then*

$$\int_{V/\Lambda_t} dx = \Delta(L). \tag{2.21}$$

Proof. If Λ_t is generated by ω_1, ω_2, then the volume (2.21) equals

$$|\mathrm{Re}\, \omega_1\, \mathrm{Im}\, \omega_2 - \mathrm{Re}\, \omega_2\, \mathrm{Im}\, \omega_1|.$$

Taking

$$\omega_1 = l_1 e^{t/2} + il_1' e^{-t/2}\,, \quad \omega_2 = l_2 e^{t/2} + il_2' e^{-t/2}\,,$$

we get (2.20). ■

Lemma 2.8.3. *The Fourier transform of*

$$f_{v,\eta}(x) := (x \cdot \eta)\, e^{\pi i v |x|^2}, \quad \eta = \eta_0 + i\eta_1 \tag{2.22}$$

equals

$$g_{v,\eta}(y) := \frac{i}{v^2}\,(y \cdot i\bar{\eta})\, e^{-\frac{\pi i}{v}|y|^2} \tag{2.23}$$

Proof. Putting $w = -iv$ we have

$$f_{v,\eta}(x) = (x_0\eta_1 + x_1\eta_0)\, e^{-\pi w\left(x_0^2 + x_1^2\right)},$$

so that its Fourier transform by (2.13) and (2.14) is

$$\eta_1 \int_{-\infty}^{\infty} e^{-\pi w x_0^2}\, e^{-2\pi i x_0 y_1}\, x_0\, dx_0 \cdot \int_{-\infty}^{\infty} e^{-\pi w x_1^2}\, e^{-2\pi i x_1 y_0}\, dx_1 +$$

$$\eta_0 \int_{-\infty}^{\infty} e^{-\pi w x_0^2}\, e^{-2\pi i x_0 y_1}\, dx_0 \cdot \int_{-\infty}^{\infty} e^{-\pi w x_1^2}\, e^{-2\pi i x_1 y_0}\, x_1\, dx_1 =$$

$$(\eta_0 y_0 + \eta_1 y_1)\, \frac{1}{i w^2}\, e^{-\pi \frac{y_0^2 + y_1^2}{w}}.$$

This is (2.23). ∎

2.8.4 A functional equation for θ. Let us now write (2.16) for $f = f_{v,\eta}$ and Λ_t:

$$\sum_{\lambda \in \Lambda_t} ((\lambda_{0,t} + \lambda) \cdot \eta)\, e^{\pi i v |\lambda_{0,t} + \lambda|^2}\, e^{-2\pi i (\lambda \cdot \mu_{0,t}) - \pi i (\lambda_{0,t} \cdot \mu_{0,t})} =$$

$$\frac{i}{\Delta(L)\, v^2} \sum_{\mu \in \Lambda_t^!} ((\mu_{0,t} + \mu) \cdot i\bar{\eta})\, e^{-\frac{\pi i}{v}|\mu_0 + \mu|^2}\, e^{2\pi i (\lambda_{0,t} \cdot \mu) + \pi i (\lambda_{0,t} \cdot \mu_{0,t})}.$$

In the notation (2.7) this means:

$$\theta_{\Lambda_t,\eta} \begin{bmatrix} \lambda_{0,t} \\ \mu_{0,t} \end{bmatrix} (v) = \frac{i}{\Delta(L)\, v^2}\, \theta_{\Lambda_t^!, i\bar{\eta}} \begin{bmatrix} \mu_{0,t} \\ -\lambda_{0,t} \end{bmatrix} \left(-\frac{1}{v}\right). \tag{2.24}$$

We now can establish a functional equation for Θ^U as well:

Proposition 2.9. *We have*

$$\Theta^U_{L,\eta} \begin{bmatrix} l_0 \\ m_0 \end{bmatrix} (v) = \frac{1}{\Delta(L)\, v}\, \Theta^U_{L^?, i\bar{\eta}} \begin{bmatrix} m_0 \\ -l_0 \end{bmatrix} \left(-\frac{1}{v}\right). \tag{2.25}$$

Proof. This is a straightforward consequence of (2.8) and (2.24). ∎

3 Heisenberg Groups, Modules over Quantum Tori, and Theta Functions

3.0 Introduction. Most of the constructions of this section are explained for the case of tori of arbitrary dimension. In 3.1–3.5 we remind to the reader the approach to the classical theta functions based upon the theory of Heisenberg groups. We closely follow Mumford's presentation in [50], Sect. 1 and Sect. 2, which ideally suits our goals. The reader can find missing proofs there.

Quantum tori and their representations appear very naturally, when one restricts the basic Heisenberg representation to a lattice. This leads naturally to the emergence of Rieffel's setup as in Lemma 1.5.1, (1.8)–(1.11), although no explicit projections form a part of the picture. A way to remedy this and to construct certain projections starting with theta functions was proposed by F. Boca in [7]. Generalizing his calculation, we prove the Theorem 3.7, which introduces in the context of representation theory of toric algebras *quantum thetas* in the sense of [45]. This is the third type of thetas we meet in this paper (counting Θ^U and θ of Sect. 2 for the first two), and thanks to Boca's theorem, they can be used to construct morphisms of quantum tori.

The initial motivation of [45] was to produce quantized versions of coordinate rings of abelian varieties, generated by the classical theta constants, i. e. the values of theta fuctions at the toric points of finite order. The way they appear here gives a partial answer to the question raised by A. S. Schwarz in [66].

3.1 Heisenberg groups. We start with a locally compact abelian topological group $\mathcal{K}$ and denote its character group $\widehat{\mathcal{K}} = \mathrm{Hom}\,(\mathcal{K}, \mathbf{C}_1^*)$, $\mathbf{C}_1^* = \{z \in \mathbf{C} \mid |z| = 1\}$. We also choose a skew-symmetric pairing $\epsilon : \mathcal{K} \times \mathcal{K} \to \mathbf{C}_1^*$ which is *non-degenerate* in the following sense: it induces an isomorphism $\mathcal{K} \to \widehat{\mathcal{K}}$, and $\epsilon(x,x) \equiv 1$.

Moreover, choose a compatible with ϵ cocycle $\psi : \mathcal{K} \times \mathcal{K} \to \mathbf{C}_1^*$:

$$\psi(x,y)\psi(x+y,z) = \psi(x,y+z)\psi(y,z)\,, \tag{3.1}$$

$$\epsilon(x,y) = \frac{\psi(x,y)}{\psi(y,x)}\,. \tag{3.2}$$

The condition (3.1) holds automatically if ψ is a bicharacter. Hence if one can find a skewsymmetric bicharacter $\epsilon^{1/2}$ which is a square root of ϵ, it can be taken for ψ.

Another useful construction starts with $\mathcal{K}$ which is already represented as $K_0 \times \widehat{K}_0$ for a topological group K_0. Denoting by $\langle *, * \rangle : K_0 \times \widehat{K}_0 \to \mathbf{C}_1^*$ the canonical pairing, we can simultaneously put

$$\psi\left((x,\widehat{x}),(y,\widehat{y})\right) := \langle x, \widehat{y}\rangle\,, \quad \epsilon\left((x,\widehat{x}),(y,\widehat{y})\right) = \frac{\langle x,\widehat{y}\rangle}{\langle y,\widehat{x}\rangle}\,. \tag{3.3}$$

Having chosen $\mathcal{K}$ and ψ, we can construct the following objects:

(i) *The Heisenberg group* $\mathcal{G} = \mathcal{G}(\mathcal{K}, \psi)$. As a set, $\mathcal{G}$ is $\mathbf{C}_1^* \times \mathcal{K}$, and the composition law is given by

$$(\lambda, y)(\mu, z) = (\lambda\mu\psi(y, z), y + z). \tag{3.4}$$

The associativity is assured by (3.1). The group comes as a central extension

$$1 \to \mathbf{C}_1^* \to \mathcal{G} \to K \to 1. \tag{3.5}$$

If $\mathcal{K}$ and ψ split as in (3.3), both subgroups K_0 and $\widehat{K}_0$ of $\mathcal{K}$ come together with their lifts to $\mathcal{G}$: $x \mapsto (1, x)$.

(ii) *Representations of* $\mathcal{G}$ *on functions on* $\mathcal{K}$. Consider a linear space of complex "functions" on $\mathcal{K}$ which is stable with respect to all shifts s_x, $(s_x f)(y) = f(x + y)$, $x, y \in \mathcal{K}$. Here the word "functions" should be understood liberally: completions of spaces of usual functions and distributions will do as long as shifts can be extended in such a way that $s_x s_y = s_{x+y}$, and notation $f(x + y)$ does not imply that we want literally take values at points.

In this case the formula

$$(U_{(\lambda, y)} f)(x) := \lambda\psi(x, y) f(x + y) \tag{3.6}$$

determines a linear representation of $\mathcal{G}$ on this space.

3.2 Basic unitary representation and its various models. In the notation above, a closed subgroup $K_0 \subset \mathcal{K}$ is called *isotropic*, if $\epsilon(x, y) = 1$ for all $x, y \in K_0$, and *maximal isotropic*, if K_0 is maximal with this property. One can then lift K_0 to $\mathcal{G}$, i.e. to find a homomorphism $K_0 \to \mathcal{G} : x \mapsto (\gamma(x), x)$.

Assume that such K_0 and γ are fixed. Consider the subspace $\mathcal{H}(K_0, \gamma) \subset L_2(\mathcal{K})$ consisting of all functions satisfying the condition

$$\forall\, y \in K_0, \quad f(x + y) = \gamma(y)^{-1}\psi(y, x)^{-1} f(x)\,. \tag{3.7}$$

Using (3.6), this can be equivalently written as

$$\forall\, y \in K_0, \quad (U_{(\gamma(y), y)} f)(x) = \epsilon(x, y) f(x)\,. \tag{3.8}$$

A straightforward calculation shows that this space is invariant with respect to the operators (3.6) and therefore determines a unitary representation of $\mathcal{G}$.

In the particular case when $\mathcal{K}$ is $K_0 \times \widehat{K}_0$ and the cocycle is as in (3.3), we can identify $\mathcal{H}(K_0, \gamma)$ with $L_2(\widehat{K}_0)$ because (3.8) allows us to reconstruct any function from its restriction to $\widehat{K}_0$.

This construction plays the central role in the theory of Heisenberg groups because of the following two facts:

Theorem 3.2.1. *(a) $\mathcal{H}(K_0, \gamma)$ is irreducible.*

(b) Any other unitary representation of $\mathcal{G}$ whose restriction on $\mathbf{C}_1^*$ is $U_{(\lambda,0)} = \lambda\,\mathrm{id}$ is isomorphic to the completed tensor product of $\mathcal{H}(K_0, \gamma)$ and a trivial representation. In particular, representations $\mathcal{H}(K_0, \gamma)$ for different choices of (K_0, γ) are all isomorphic.

The non-degeneracy of e is essentially used in the proof of this unicity statement. Everything said in 3.1 holds without any non-degeneracy assumption.

3.3 Heisenberg groups and modules over quantum tori. Since in this section we will be dealing with quantum tori of arbitrary dimension, it is convenient to introduce some invariant notation. Let D be a free abelian group of finite rank and $\alpha : D \times D \to \mathbf{C}_1^*$ a skewsymmetric pairing. The C^* algebra $C(D, \alpha)$ of the quantum torus $T(D, \alpha)$ with the character group D and quantization parameter α is the universal algebra generated by the family of unitaries $e(h) = e_{D,\alpha}(h)$, $h \in D$, satisfying the relations

$$e(g)e(h) = \alpha(g, h)e(g + h)\,. \tag{3.9}$$

(Left) modules over such tori can be obtained by the following construction: choose a Heisenberg group $\mathcal{G}(\mathcal{K}, \psi)$ with a bicharacter cocycle ψ and compatible ϵ. Consider a lattice embedding $l : D \hookrightarrow \mathcal{K}$. Denote by α_D the bicharacter on D induced by ψ. Choose a basic representation U of $\mathcal{G}(\mathcal{K}, \psi)$ in the space $\mathcal{H}$ and define the action of $C(D, \alpha)$ on $\mathcal{H}$ by

$$e_{D,\alpha}(h) f := U_{(1,l(h))} f \tag{3.10}$$

It turns out that an appropriate completion of the subspace of smooth functions is a projective module (see Rieffel's Theorem 3.4.1 below).

3.4 Basic representations as toric bimodules. In the setup of the last paragraph, assume to shorten notation that D is a lattice (discrete subgroup with compact quotient) in $\mathcal{K}$ and denote by $D^!$ the the maximal orthogonal subgroup:

$$D^! := \{x \in \mathcal{K} \mid \forall\, h \in D, \epsilon(h, x) = 1\}\,. \tag{3.11}$$

Let $\alpha^!$ be the pairing induced by ϵ on $D^!$. If $D^!$ is free of finite rank, we get similarly the representation of $C(D^!, \alpha^!)$ on $\mathcal{H}$. Moreover, operators from $C(D, \alpha)$ and $C(D^!, \alpha^!)$ pairwise commute. Identifying $C(D^!, \alpha^!)^{op}$ with $C(D^!, \bar{\alpha}^!)$ in an obvious way, we make of $\mathcal{H}$ an $C(D, \alpha)$–$C(D^!, \bar{\alpha}^!)$ bimodule.

Assuming that we are in the situation of (3.3) and taking the space $L_2(K_0)$ (rather than $L_2(\widehat{K}_0)$) for the basic representation, we will construct the Hermitean scalar products with the properties summarized in the Lemma 1.3.1. For further details, see [58]. We will assume that K_0 is a Lie group of the form $\mathbf{R}^p \times \mathbf{Z}^q \times (\mathbf{R}/\mathbf{Z})^r \times F$ where F is a finite group. Then one can define the Schwartz space $S(K_0)$ consisting of C^∞-functions such that any polynomial times any derivative of the function vanishes at infinity. Rieffel's scalar products are first defined on Schwartz's functions on K_0 and $\mathcal{K}$ and then extended to the appropriate completions. We will write elements of $C(D,\alpha)$ as formal series $F=\sum_{h\in D} a_h e_{D,\alpha}(h)$ where a_h are the (non-commutative) Fourier coefficients defined by $a_h = t(Fe(h)^*)$, t is the normalized trace. If Fourier coefficients form a Schwartz function on D, F will be called smooth.

We start with the standard scalar product on $L_2(K_0)$ (antilinear in the second argument) which will be denoted $\langle *,*\rangle_{L_2}$ and put for $\Phi,\Psi\in S(K_0)$:

$$_D\langle \Phi,\Psi\rangle := \sum_{h\in D}\langle \Phi, e_{D,\alpha}(h)\Psi\rangle_{L_2}\, e_{D,\alpha}(h)\,, \tag{3.12}$$

$$\langle \Phi,\Psi\rangle_{D^!} := \sum_{h\in D^!}\langle e_{D^!,\alpha^!}(h)\Psi, \Phi\rangle_{L_2}\, e_{D^!,\bar\alpha^!}(h)\,. \tag{3.13}$$

(Notice the appearance of both $\alpha^!$ and $\bar\alpha^!$ in the right hand side of (3.13)).

Before summarizing some results due to Rieffel, we have to add a few words about the normalizations of various Haar measures involved. Any Haar measure on K_0 will do; on $\widehat{K_0}$ we take the respective Plancherel measure. For the volumes of the respective fundamental domains we will then have $|\mathcal{K}/D|\cdot|\mathcal{K}/D^!|=1$.

Theorem 3.4.1. *Denote by M the completion of $S(K_0)$ with respect to the operator norm $\|_D\langle\Phi,\Phi\rangle\|^{1/2}$. Put*

$$A=C(D,\alpha),\ \ B=C(D^!,\bar\alpha^!),\ \ _A\langle *,*\rangle = |\mathcal{K}/D|\,_D\langle *,*\rangle,\ \ \langle *,*\rangle_B=\langle *,*\rangle_{D^!}\,.$$

Then we have:

(a) M is a finitely generated projective A–B module isomorphic to the range of a projection (both right and left).

(b) A is the complete endomorphism ring of M_B.

(c) The scalar products defined above satisfy all the identities (1.8)–(1.11).

(d) Let t_B be the normalized trace on B (zeroth Fourier coefficient). Then

$$t_B([M_B]) = |\mathcal{K}/D^!|\,. \tag{3.14}$$

Notice that, contrary to the purely algebraic context of Lemma 1.3.1 where (1.8)–(1.11) followed directly from the definitions (1.6), (1.7), the deduction of (1.11) from (3.12), (3.13) requires application of the Poisson summation formula.

For further details, see [60], Sects. 2 and 3.

3.5 Vector Heisenberg group and classical theta functions. We return now temporarily to the setup of 3.1–3.2, involving no additional lattice D and explain the appearance of the classical theta functions as matrix coefficients of the basic Heisenberg representation. We closely follow [50], Sect. 3.

We choose as $\mathcal{K}$ the real vector space $V=\mathbf{R}^{2N}$. Any element, say, $x\in V$ will be considered as a pair of columns of height N: x_1 consisting of the first N coordinates of x and x_2 consisting of the last N coordinates. Define the standard symplectic form on V and the cocycle ψ (cf. (3.1)) by

$$A(x,y) = x_1^t y_2 - x_2^t y_1, \quad \psi(x,y) = e^{\pi i A(x,y)}, \tag{3.15}$$

so that

$$\epsilon(x,y) = e^{2\pi i A(x,y)}. \tag{3.16}$$

Having chosen a model $\mathcal{H}$ of the basic representation of the resulting Heisenberg group, Mumford defines in $\mathcal{H}$ a finite-dimensional family of vectors $f_T \in \mathcal{H}$ parametrized by the points T in the Siegel upper half space $\mathfrak{H}_N$ consisting of complex symmetric $N \times N$ matrices with positive definite imaginary part.

In abstract terms, this is the space of all flat Kähler structures on $\mathbf{R}^{2N}$ compatible with A. Such a structure can be thought of, for example, as a pair consisting of a complex structure J and a positive definite Hermitean form H with the imaginary part A.

Any given T determines directly the complex structure J_T: it is given by the-complex coordinates $\underline{x}_1, \dots, \underline{x}_N$ on V:

$$\underline{x}_i = \sum_j T_{ij} x_j^{(1)} + x_i^{(2)} \tag{3.17}$$

where now $x_j^{(1)}$ (resp. $x_i^{(2)}$) are the coordinates of x_1 (resp. x_2). The values of the Hermitean form H_T on the basic vectors $e_j^{(2)}$ of the second half of V are

$$H_T\left(e_i^{(2)}, e_j^{(2)}\right) = (\operatorname{Im} T)_{ij}^{-1}. \tag{3.18}$$

We can now define f_T in the Mumford's first realization of the fundamental representation:

$$\mathcal{H} := L_2\left(\mathbf{R}^N\right), \quad (U_{(\lambda,y_1,y_2)} f)(x) = \lambda e^{2\pi i x_1^t y_2 + \pi i y_1^t x_2} f(x_1 + y_1) \tag{3.19}$$

which is a specialization of (3.6) restricted to the subspace (3.8). Namely, we have

$$f_T(x) = e^{\pi i x^t T x}. \tag{3.20}$$

The classical theta function is defined by

$$\theta(\underline{x}, T) := \sum_{n \in \mathbf{Z}^N} e^{\pi i n^t T n + 2\pi i n^t \underline{x}}. \tag{3.21}$$

To express it as a matrix coefficient, Mumford introduces the distribution

$$e_{\mathbf{Z}} := \sum_{n \in \mathbf{Z}^N} \delta_n \tag{3.22}$$

and then checks that

$$\langle U_{(1,x)} f_T, e_{\mathbf{Z}} \rangle = c\, e^{\pi i x_1^t \underline{x}} \theta(\underline{x}, T). \tag{3.23}$$

(See [50], Corollary 2.4).

3.6 Quantum theta functions. In this subsection I give a brief review of the formalism of quantum theta functions introduced in [43] and further studied in [44], [45]. For details and motivation, see [45].

Consider the character group of a quantum torus (D, α). In this subsection we will be interested in the space of formal infinite linear combinations of $e(h) = e_{D,\alpha}(h)$ which we will call formal functions. Theta functions are defined as solutions of functional equations which can be invariantly described in terms of another version of Heisenberg group $\mathcal{G}(D, \alpha)$ acting on this space: it consists of all linear operators on formal functions of the form

$$\Phi \mapsto c\, e(g)\, x^*(\Phi)\, e(h)^{-1}$$

where $c \in \mathbf{C}^*$, $g, h \in D$, $x \in T(D, 1)(\mathbf{C}) := \mathrm{Hom}\,(D, \mathbf{C}^*)$ an arbitrary point of the algebraic torus with the character group D, x^* is the shift automorphism multiplying $e_{D,\alpha}(h)$ by $x(h)$.

Notice that such a shift x^* generally does not respect the unitarity of $e_{D,\alpha}(h)$ and cannot be extended to the automorphisms of $C(D, \alpha)$ unless the values of x belong to $\mathbf{C}_1^*$.

We now define a (formal) theta multiplier for (D, α) as an injective homomorphism $\mathcal{L} : B \to \mathcal{G}(D, \alpha)$ where B is a free abelian group of the same rank as D.

A quantum theta function with multiplier $\mathcal{L}$ is a formal function on $T(D, \alpha)$ invariant with respect to the action of (the image of) B.

$\Gamma(\mathcal{L})$ is the linear space of theta functions with multiplier $\mathcal{L}$.

The theta functions constructed below will have coefficients from the Schwartz space of D and therefore will represent smooth elements of $C(D, \alpha)$. Their multipliers will have the property $\dim \Gamma(\mathcal{L}) = 1$. In other words, as the classical $\theta(\underline{x}, T)$, our thetas will correspond only to the principal polarizations. In order to get more general thetas one should consider more general Heisenberg groups in the sense of 3.1 into which (D, α) could be embedded.

Theorem 3.7. *We have*

$$_D\langle f_T, f_T\rangle = \frac{1}{\sqrt{2^N \det \mathrm{Im}\, T}} \sum_{h \in D} e^{-\frac{\pi}{2}\underline{h}^t (\mathrm{Im}\, T)^{-1} \underline{h}^*} e_{D,\alpha}(h)\,, \tag{3.24}$$

$$\langle f_T, f_T\rangle_{D^!} = \frac{1}{\sqrt{2^N \det \mathrm{Im}\, T}} \sum_{h \in D^!} e^{-\frac{\pi}{2}\underline{h}^t (\mathrm{Im}\, T)^{-1} \underline{h}^*} e_{D^!,\bar{\alpha}^!}(h)\,. \tag{3.25}$$

Here $\underline{h} := Th_1 + h_2$ (cf. (3.17)) and $\underline{h}^* := \overline{T} h_1 + h_2$. These scalar products are quantum theta functions $\Theta_D, \Theta_{D^!}$ satisfying the following functional equations:

$$\forall g \in D, \quad C_g e_{D,\alpha}(g) x_g^*(\Theta_D) = \Theta_D\,, \tag{3.26}$$

$$\forall g \in D^!, \quad C^!_g e_{D^!,\tilde{\alpha}^!}(g) x^{!*}_g(\Theta_{D^!}) = \Theta_{D^!} \tag{3.27}$$

where

$$C_g = e^{-\frac{\pi}{2}\underline{g}^t (\operatorname{Im} T)^{-1} \underline{g}^*}, \quad x^*_g(e_{D,\alpha}(h)) = e^{X_g(h)} e_{D,\alpha}(h),$$

$$X_g(h) = -\pi \operatorname{Re}\left(\underline{g}^t (\operatorname{Im} T)^{-1} \underline{h}^*\right) - \pi i\, A(g,h), \tag{3.28}$$

$$C^!_g = e^{-\frac{\pi}{2}\underline{g}^t (\operatorname{Im} T)^{-1} \underline{g}^*}, \quad x^{!*}_g\left(e_{D^!,\tilde{\alpha}}(h)\right) = e^{X^!_g(h)} e_{D^!,\tilde{\alpha}}(h),$$

$$X^!_g(h) = -\pi \operatorname{Re}(\underline{g}^t (\operatorname{Im} T)^{-1} \underline{h}^*) + \pi i\, A(g,h). \tag{3.29}$$

Proof. We will check (3.24) and (3.26); the other two formulas can be treated similarly.

The general formula (3.12) must be specialized to our case $K_0 = \mathbf{R}^N$, the first half of $\mathcal{K}$. For the L_2-scalar product we take $\int \Phi\overline{\Psi} dx_1$ where dx_1 is the standard Haar measure. From (3.10) and (3.19) it follows that

$$(e_{D,\alpha}(h)\Psi)(x_1) = e^{2\pi i x_1^t h_2 + \pi i h_1^t h_2} \Psi(x_1 + h_1).$$

Hence

$$_D\langle \Phi, \Psi\rangle = \sum_{h\in D} e^{-\pi i h_1^t h_2} \int \Phi(x_1)\overline{\Psi(x_1+h_1)} e^{-2\pi i x_1^t h_2} dx_1 \cdot e_{D,\alpha}(h).$$

Putting here $\Phi = \Psi = f_T$ (see (3.20)), we get:

$$_D\langle f_T, f_T\rangle = \sum_{h\in D} e^{-\pi i h_1^t h_2} \int e^{\pi i [x_1^t T x_1 - (x_1^t + h_1^t)\overline{T}(x_1+h_1) - 2x_1^t h_2]} dx_1 \cdot e_{D,\alpha}(h). \tag{3.30}$$

The exponential expression under the integral sign in (3.30) can be represented as $e^{-(q(x_1)+l_h(x_1)+c_h)}$ where

$$q(x_1) = 2\pi\, x_1^t\, (\operatorname{Im} T)\, x_1,\ l_h(x_1) = 2\pi i\, \left(h_1^t \overline{T} + h_2^t\right) x_1,\ c_h = \pi i h_1^t \overline{T} h_1. \tag{3.31}$$

Denote

$$\lambda_h = \frac{i}{2} (\operatorname{Im}\, T)^{-1} \left[\overline{T} h_1 + h_2\right]. \tag{3.32}$$

Then we have

$$q(x_1 + \lambda_h) - q(\lambda_h) = q(x_1) + l_h(x_1)$$

and therefore

$$\int e^{-(q(x_1)+l_h(x_1)+c_h)}dx_1 = e^{-c_h+q(\lambda_h)}\int e^{-q(x_1+\lambda_h)}dx_1 = e^{-c_h+q(\lambda_h)}\frac{\pi^{N/2}}{\sqrt{\det q}}. \tag{3.33}$$

Putting (3.30)–(3.33) together, we get (3.24).

The equation (3.26) is checked by a straightforward computation: putting $Q(h) = \frac{\pi}{2}\, h_1^t\,(\mathrm{Im}\, T)^{-1}\, h_1$ we have

$$C_g e_{D,\alpha}(g) x_g^*(\Theta_D) = C_g \sum_{h\in D} e^{-Q(h)+X_g(h)+\pi i\, A(g,h)} e_{D,\alpha}(g+h) =$$

$$C_g \sum_{h\in D} e^{-Q(h-g)+X_g(h-g)+\pi i\, A(g,h-g)} e_{D,\alpha}(h) =$$

$$C_g e^{-Q(g)+X_g(-g)} \sum_{h\in D} e^{-Q(h)+Q(g)+X_g(h)+\pi i\, A(g,h)} e_{D,\alpha}(h) = \Theta_D.$$

■

Bibliography

[1] J. Arledge, M. Laca, I. Raeburn. *Crossed products by semigroups of endomorphisms and the Toeplitz algebras of ordered groups.* Documenta Math. **2** (1997), 115–138.

[2] A. Astashkevich, A. S. Schwarz. *Projective modules over non-commutative tori: classification of modules with constant curvature connection.* e-Print QA/9904139.

[3] V. Baranovsky, V. Ginzburg. *Conjugacy classes in loop groups and G-bundles on elliptic curves.* Int. Math. Res. Notices **15** (1996), 733–751.

[4] V. Baranovsky, S. Evens, V. Ginzburg. *Representations of quantum tori and double-affine Hecke algebras.* e-Print math.RT/0005024

[5] B. Blackadar. *K-theory for operator algebras.* Springer Verlag, 1986.

[6] F. Boca. *The structure of higher-dimensional noncommutative tori and metric diophantine approximation.* J. Reine Angew. Math. **492** (1997), 179–219.

[7] F. Boca. *Projections in rotation algebras and theta functions.* Comm. Math. Phys. **202** (1999), 325–357.

[8] J. Bost, A. Connes. *Hecke algebras, type III factors and phase transitions with spontaneous symmetry breaking in number theory.* Selecta Math. **3** (1995), 411–457.

[9] L. Brown, P. Green, M. A. Rieffel. *Stable isomorphism and strong Morita equivalence of C*-algebras.* Pacific J. Math. **71** (1977), 349–363.

[10] P. Cohen. *A C*-dynamical system with Dedekind zeta partition function and spontaneous symmetry breaking.* J. de Th. de Nombres de Bordeaux **11** (1999), 15–30.

[11] P. Cohen. *Quantum statistical mechanics and number theory.* In: Algebraic Geometry: Hirzebruch 70. Contemp. Math. **241**, AMS, Pridence RA, 1999, 121–128.

[12] A. Connes. *Noncommutative geometry.* Academic Press, 1994.

[13] A. Connes. *C*-algebras et géométrie différentielle.* C. R. Ac. Sci. Paris, t. **290** (1980), 599–604.

[14] A. Connes. *Trace formula in noncommutative geometry and the zeros of the Riemann zeta function.* Selecta Math., New Ser., **5** (1999), 29–106.

[15] A. Connes. *Noncommutative geometry and the Riemann zeta function.* In: Mathematics: Frontiers and Perspectives, ed. by V. Arnold et al., AMS, 2000, 35–54.

[16] A. Connes. *Noncommutative Geometry Year 2000.* e-Print math.QA/0011193

[17] A. Connes, M. Douglas, A. Schwarz. *Noncommutative geometry and Matrix theory: compactification on tori.* Journ. of High Energy Physics, **2** (1998).

[18] K. R. Davidson. *C^*-algebras by example.* AMS, 1996.

[19] H. Darmon. *Stark–Heegner points over real quadratic fields.* Contemp. Math. **210** (1998), 41–69.

[20] *The Grothendieck Theory of Dessins d'Enfants (Ed. by L. Schneps).* London MS Lect. Note series **200**, Cambridge University Press, 1994.

[21] M. Dieng, A. Schwarz. *Differential and complex geometry of two-dimensional noncommutative tori.* Preprint, 2002.

[22] V. G. Drinfeld. *On quasi-triangular quasi-Hopf algebras and some groups closely associated with $Gal(\overline{\mathbf{Q}}/\mathbf{Q})$.* Algebra and Analysis **2:4** (1990); Leningrad Math. J. **2:4** (1991), 829–860.

[23] G. A. Elliott. *On the K-theory of the C^*-algebra generated by a projective representation of a torsion-free discrete abelian group.* In: Operator Algebras and Group Representations I, Pitman, London, 1984, 157–184.

[24] G. A. Elliott. *The diffeomorphism groups of the irrational rotation C^*-algebra.* C. R. Math. Rep. Ac. Sci. Canada **8** (1986), 329–334.

[25] G. A. Elliott, D. E. Evans. *The structure of the irrational rotation C^*-algebra.* Ann. Math. **138** (1993), 477–501.

[26] L. D. Faddeev, R. M. Kashaev. *Quantum dilogarithm.* e-Print hep-th/9310070.

[27] C. F. Gauss. *Zur Kreistheilung.* In: Werke, Band 10, p.4. Georg Olms Verlag, Hildsheim–New York, 1981.

[28] F. Goodman, P. de la Harpe, V. Jones. *Coxeter graphs and towers of algebras.* Springer, 1989.

[29] D. Harari, E. Leichtnam. *Extension du phénomène de brisure spontanée de symétrie de Bost–Connes au cas des corps globaux quelconques.* Selecta Math. **3** (1997), 205–243.

[30] E. Hecke. *Über die Kroneckersche Grenzformel für reelle quadratische Körper und die Klassenzahl relativ-abelscher Körper.* Verhandl. d. Naturforschender Gesell. i. Basel **28** (1917), 363–373. (Math. Werke, pp. 198–207, Vandenberg & Ruprecht, Göttingen, 1970).

[31] E. Hecke. *Zur Theorie der elliptischen Modulfunktionen.* Math. Annalen **97** (1926), 210–243. (Math. Werke, pp. 428–460, Vandenberg & Ruprecht, Göttingen, 1970).

[32] G. Herglotz. *Über die Kroneckersche Grenzformel für reelle quadratische Körper I, II.* Berichte über die Verhandl. Sächsischen Akad. der Wiss. zu Leipzig **75** (1923), 3–14, 31–37.

[33] V. Jones. *Index for subfactors.* Invent. Math. **72:1** (1983), 1–25.

[34] V. Jones. *Index for subrings of rings.* Contemp. Math. **43**, AMS(1985), 181–190.

[35] V. Jones. *Fusion en algèbres de von Neumann et groupes de lacets (d'après A. Wassermann).* Séminaire Bourbaki, no. 800 (Juin 1995), 20 pp.

[36] V. Jones, V. Sunder. *Introduction to subfactors.* Lond. Math. Soc. Lecture Note Series **234** (1997).

[37] L. Kadison, D. Nikshych. *Frobenius extensions and weak Hopf algebras.* e-Print math/0102010

[38] L. Kadison, D. Nikshych. *Hopf algebra actions on strongly separable extensions of depth two.* e-Print math/0107064

[39] M. Kontsevich, Y. Soibelman. *Homological mirror symmetry and torus fibrations.* e-Print math.SG/0011041
[40] S. Lang. *Elliptic Functions.* Addison–Wesley, 1973.
[41] Qing Lin. *Cut-down method in the inductive limit decomposition of non-commutative tori, III: A complete answer in 3-dimension.* Comm. Math. Phys. **179** (1996), 555–575.
[42] P. Lochak, L. Schneps. *A cohomological interpretation of the Grothendieck–Teichmüller group.* Invent. Math. **127** (1997), 571–600.
[43] Yu. Manin. *Quantized theta-functions.* In: Common Trends in Mathematics and Quantum Field Theories (Kyoto, 1990), Progress of Theor. Phys. Supplement **102** (1990), 219–228.
[44] Yu. Manin. *Mirror symmetry and quantization of abelian varieties.* In: Moduli of Abelian Varieties, ed. by C. Faber et al., Progress in Math. **195**, Birkhäuser, 2001, 231–254. e-Print math.AG/0005143
[45] Yu. Manin. *Theta functions, quantum tori and Heisenberg groups.* Lett. in Math. Physics **56** (2001), 295–320. e-Print math.AG/001119
[46] Yu. Manin. *Von Zahlen und Figuren.* 27 pp. e-Print math.AG/0201005
[47] Yu. Manin, M. Marcolli. *Continued fractions, modular symbols, and non-commutative geometry.* e-Print math.NT/0102006
[48] D. Mumford. *On the equations defining abelian varieties I.* Invent. Math. **1** (1966), 287–354.
[49] D. Mumford. *An algebro-geometric construction of commuting operators and of solutions to the Toda lattice equation, Korteweg deVries equation and related non-linear equations.* In: Proc. of Int. Symp. on Algebraic Geometry, Kyoto, 1977, 115–153.
[50] D. Mumford (with M. Nori and P. Norman). *Tata Lectures on Theta III.* Progress in Math. **97**, Birkhäuser, 1991.
[51] D. Nikshych, L. Vainerman. *Finite quantum groupoids and their applications.* e-Print math.QA/0006057.
[52] M. Pimsner, D. Voiculescu. *Exact sequences for K-groups and EXT-groups of certain cross-product C^*-algebras.* J. Operator Theory **4** (1980), 93–118.
[53] A. Polishchuk. *Indefinite theta series of signature (1,1) from the point of view of homological mirror symmetry.* e-Print math.AG/0003076.
[54] A. Polishchuk. *A new look at Hecke's indefinite theta series.* e-Print math.AG/0012005.
[55] I. Raeburn, D. Williams. *Morita equivalence and continuous-trace C^*-algebras.* Math. Surveys and Monographs **60**, AMS, 1998.
[56] M. A. Rieffel. *Strong Morita equivalence of certain transformation group C^*-algebras.* Math. Annalen **222** (1976), 7–23.
[57] M. A. Rieffel. *Von Neumann algebras associated with pairs of lattices in Lie groups.* Math. Ann. **257** (1981), 403–418.
[58] M. A. Rieffel. *C^*-algebras associated with irrational rotations.* Pacific J. Math. **93** (1981), 415–429.
[59] M. A. Rieffel. *The cancellation theorem for projective modules over irrational rotation C^*-algebras.* Proc. Lond. Math. Soc. (3) **47** (1983), 285–303.
[60] M. A. Rieffel. *Projective modules over higher-dimensional non-commutative tori.* Can. J. Math., vol. XL, No. 2 (1988), 257–338.
[61] M. A. Rieffel. *Non-commutative tori — a case study of non-commutative differential manifolds.* In: Cont. Math. **105** (1990), 191–211.
[62] M. A. Rieffel, A. Schwarz. *Morita equivalence of multidimensional non-commutative tori.* Int. J. Math. **10** (1999), 289–299. e-Print math.QA/9803057
[63] A. Rosenberg. *Non-commutative algebraic geometry and representations of quantized algebras.* Kluwer Academic Publishers, 1995.

[64] A. Rosenberg. *Non-commutative schemes.* Compositio Math. **112** (1998), 93–125.

[65] A. Schwarz. *Morita equivalence and duality.* Nucl. Phys., B **534** (1998), 720–738.

[66] A. Schwarz. *Theta-functions on non-commutative tori.* e-Print math/0107186

[67] J. P. Serre. *Complex Multiplication.* In: Algebraic Number Fields, ed. by J. Cassels, A. Frölich. Academic Press, NY 1977, 293–296.

[68] Y. Soibelman. *Quantum tori, mirror symmetry and deformation theory.* Lett. in Math. Physics **56** (2001), 99–125. e-Print math.QA/0011162.

[69] H. M. Stark. *L-functions at* $s = 1$*. III. Totally real fields and Hilbert's Twelfth Problem.* Adv. Math. **22** (1976), 64–84.

[70] H. M. Stark. *L-functions at* $s = 1$*. IV. First derivatives at* $s = 0$. Adv. Math. **35** (1980), 197–235.

[71] P. Stevenhagen. *Hilbert's 12th problem, Complex Multiplication and Shimura reciprocity.* In: Class Field Theory — Its Centenary and Prospect. Adv. Studies in Pure Math. **30** (2001), 161–176.

[72] J. Tate. *Les conjectures de Stark sur les fonctions L d'Artin en* $s = 0$. Progress in Math. **47**, Birkhäuser, 1984.

[73] A. Unterberger. *Quantization and non-holomorphic modular forms.* Springer LNM **1742** (2001).

[74] Y. Watatani. *Index for* C^**-subalgebras.* Mem. AMS **83**, Nr. 424, Providence, RA, 1990.

[75] E. Witten. *Overview of K-theory applied to strings.* e-Print hep-th/0007175

[76] D. Zagier. *A Kronecker limit formula for real quadratic field.* Math. Ann. **213** (1975), 153–184.

[77] D. Zagier. *Valeurs des fonctions zeta des corps quadratiques réels aux entiers négatifs.* Astérisque **41–42** (1977), 135–151.

[78] L. Zapponi. *Dessins d'enfants en genre 1.* In: Geometric Galois Actions, ed. L. Schneps, P. Lochak.

On the Quantum Cohomology of Homogeneous Varieties

W. Fulton*

About a decade ago physicists set off something of a "big bang" in the universe of algebraic geometry. A new approach to enumerative geometry, involving mirror symmetry, solved some old questions and raised many new ones. To mention just a few of the areas influenced by these developments, we have new approaches to the study of Calabi-Yau manifolds and orbifolds, the notions of Gromov-Witten and related invariants, and new results about intersection theory on moduli spaces.

From this revolution we now have general and powerful theorems, which produce meaningful Gromov-Witten invariants on every smooth projective complex variety (by means of virtual fundamental cycles), with methods for calculating (localization). On the other hand, a good general understanding of all invariants on a given space has come only recently, in the work first of Kontsevich in the case of a point, and the work of Okounkov and Pandharipande in the case of the projective line.

One of the first situations explained to us by E. Witten [26] was for the Grassmann variety $\mathrm{Gr}(k,n)$ of k-dimensional linear subspaces of an n-dimensional complex vector space, where a deformation of the classical cohomology, called the (small) quantum cohomology, was described. For this variety, and for general homogeneous varieties G/P, the Gromov-Witten invariants are given by naïve counting — no virtual cycles are needed. Some of this story has been extended to other homogeneous varieties. It is somewhat surprising that, after all the progress in this general area, there are still so many open questions about the quantum cohomology of G/P in general, and even for $\mathrm{Gr}(k,n)$. One of the appealing features of this work is the fact that the classical study of these varieties leads to interesting combinatorics. There is already considerable evidence that the quantum versions involve equally interesting combinatorial ideas.

The aim of this lecture is to sketch what is known and what remains open about the quantum cohomology of homogeneous varieties. We will be concerned only with the "small" quantum cohomology, which is formed from the 3-point Gromov-Witten invariants. Unlike much of the algebraic geometry that has come from physics, this

* Partially supported by NSF grant DMS9970435

part of the story is quite accessible to a general audience. Even here, we do not attempt a survey, and apologize to the many whose work is not cited.

We will be concerned with a variety X which is a homogeneous variety: $X = G/P$, where G is a simple complex algebraic Lie group, and P is a parabolic subgroup. These are classified by data from Dynkin diagrams, cf. [15]. When $G = \mathrm{SL}_n(\mathbb{C})$, the varieties G/P are the varieties of partial flags of subspaces of $\mathbb{C}^n$, generalizing the Grassmannian (the one-step flags). There are corresponding manifolds of flags in orthogonal or symplectic vector spaces for the other groups of classical type, together with a few exceptional cases. Those unfamiliar with the general roots and weights story can concentrate on the Grassmannians $\mathrm{Gr}(k, n)$.

The classical cohomology $H^*(X) = H^*(X, \mathbb{Z})$ has a basis (over $\mathbb{Z}$) of classes σ_λ of Schubert varieties, where λ varies over a combinatorial set. In general, this set is the quotient W/W_P of the Weyl group W of G by the subgroup W_P corresponding to P. For $X = \mathrm{Gr}(k, n)$, this combinatorial set can be taken to be the set of partitions whose Young diagram fits in a k by $n - k$ rectangle, i.e., $\lambda = (\lambda_1, \ldots, \lambda_k)$, with $n - k \geq \lambda_1 \geq \lambda_2 \geq \ldots \lambda_k \geq 0$. For example, take $k = 4$, $n = 9$, and $\lambda = (5, 4, 4, 3)$, with Young diagram:

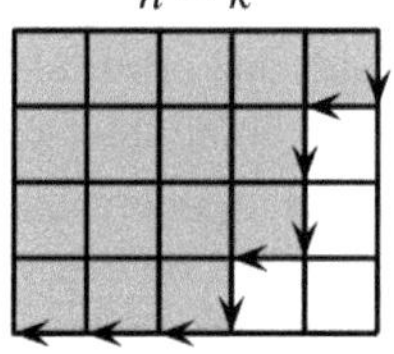

As indicated, we will often regard the partition as a path from the upper right corner of this rectangle to the lower left corner, by a sequence of n steps, each either down or to the left.

If one fixes a complete flag $F_\bullet$ of subspaces of $\mathbb{C}^n$, with $\dim F_i = i$ for $0 \leq i \leq n$, then σ_λ is the cohomology class of the Schubert variety

$$\Omega_\lambda = \Omega_\lambda(F) = \{L \subset \mathbb{C}^n \mid \dim(L \cap F_i) \geq r_i \text{ for } 1 \leq i \leq n\},$$

where r_i is the row one has passed after i steps along the path defining λ. The (complex) codimension of Ω_λ is the number $|\lambda| = \sum \lambda_i$ of boxes in the Young diagram (above the path), so σ_λ is in $H^{2|\lambda|}(X)$.

In the cohomology ring $H^*(X)$, one has

$$\sigma_\lambda \cdot \sigma_\mu = \sum c_{\lambda\mu}^{\nu} \sigma_\nu,$$

the sum over ν with $|\nu| = |\lambda| + |\mu|$. The coefficients are nonnegative integers known as Littlewood-Richardson coefficients, which have several interesting combinatorial descriptions, and play a leading role in algebraic combinatorics. There are dual classes $\sigma_{\lambda^\vee}$, where $\lambda^\vee = (n - k - \lambda_k, \ldots, n - k - \lambda_1)$ is obtained by rotating the diagram of the complement of λ by 180 degrees; the intersection number $\int \sigma_\lambda \cdot \sigma_\mu$ is 1 if $\mu = \lambda^\vee$ and 0 otherwise. With this notation,

$$c_{\lambda\mu}^{\nu} = \int \sigma_\lambda \cdot \sigma_\mu \cdot \sigma_{\nu^\vee}$$

is the number of points in the intersection of three Schubert varieties Ω_λ, Ω'_μ, and $\Omega''_{\nu^\vee}$, using three general flags.

The quantum cohomology $QH^*(X)$ has the same basis σ_λ, but as an algebra over a polynomial ring $\mathbb{Z}[q]$ (in general, the polynomial ring $\mathbb{Z}[q_\beta]$ with a variable for each β in W/W_P with $|\beta| = 1$). Here the (complex) codimension of q is n (or $\int_{\sigma_{\beta^\vee}} c_1(T_X)$ in general). In this ring, with its product denoted by a $*$,

$$\sigma_\lambda * \sigma_\mu = \sum c_{\lambda\mu}^{\nu}(d)\, q^d\, \sigma_\nu,$$

the sum over d and ν with $|\nu| = |\lambda| + |\mu| - dn$. The coefficient $c_{\lambda\mu}^{\nu}(d)$ is the number of maps of degree d from the projective line to X, whose image meets three general Schubert varieties Ω_λ, Ω'_μ, and $\Omega''_{\nu^\vee}$. (In general, the degree is defined by the equation $f_*[\mathbb{P}^1] = \sum d_\beta \sigma_{\beta^\vee}$.) The surprising fact is that this product defines an *associative* and commutative $\mathbb{Z}[q]$-algebra.

These coefficients (the 3-point Gromov-Witten numbers) are known in principle. But many questions about them remain open — very much so for a general G/P, and considerably so for $\mathrm{Gr}(k, n)$. To have a good understanding of these numbers, one wants: (1) a presentation of the quantum cohomology ring:

$$QH^*(X) = \mathbb{Z}[q][x_1, \ldots, x_N]/(\text{relations}).$$

Presentations for quantum cohomology have been given for a general G/B by B. Kim [19]. In type A explicit presentations have been given for the partial flag varieties (see [2] and [12]); the case of Lagrangian and maximal orthogonal Grassmannians can be found in [21]. The presentation for general G/P has been announced in MIT lectures by D. Peterson (unpublished).

One also wants: (2) a "quantum Giambelli" formula, which expresses each σ_λ as a polynomial in the generators x_i and q. Such formulas were worked out first for the Grassmannian by A. Bertram [4] (where the answer is the same as in the classical case — there is no quantum correction), and for complete and partial flag varieties by I. Ciocan-Fontanine [11], [12], and S. Fomin, S. Gelfand, and A. Postnikov [13]; see L. Chen [10] for a concise treatment. A.-L. Mare [22] has recently given an algorithm for arbitrary G/B's.

In addition, one would like: (3) a combinatorial formula for the coefficients $c_{\lambda\mu}^{\nu}(d)$, or at least to know which are nonzero. Although (3) follows in principle from (1) and (2), it is far from obvious how to carry this out explicitly, even for the classical cohomology. In fact, for $H^*(\mathrm{Gr}(k, n))$, there is a recent criterion to tell for which partitions $c_{\lambda\mu}^{\nu}$ is not zero, in terms of a collection of linear inequalities (see [14] for an exposition of this story). In fact, calculations done with A. Buch indicate that there are similar inequalities to describe the positivity of the quantum numbers $c_{\lambda\mu}^{\nu}(d)$. For the classical numbers, the inequalities are determined by the answers to the same questions for smaller Grassmannians $\mathrm{Gr}(r, k)$, $1 \le r \le k$. It is natural to hope that the same is true for the quantum cohomology.[2]

The proof of basic properties of quantum cohomology uses the space of stable maps from rational curves to the given variety [20], cf. [16]. Strangely, proofs of quantum Giambelli and other quantum formulas have used Grothendieck's quot scheme compactifications rather than this space of stable maps. Recently, however, Buch has shown how the basic facts about quantum cohomology of $\mathrm{Gr}(k, n)$ can be proved be entirely elementary methods [6], [7], [8], without any compactifications at all. For a morphism $f : \mathbb{P}^1 \to \mathrm{Gr}(k, n)$ of degree d, consider the intersection K of all the linear spaces $f(t)$ as t varies in $\mathbb{P}^1$, and the span S of all these linear spaces. Buch proves that

$$\dim K \ge k - d \quad \text{and} \quad \dim S \le k + d.$$

With A. Kresch and H. Tamvakis [9], he uses this to show that $c_{\lambda\,\mu}^{\,\nu}(d)$ is equal to a classical intersection number of Schubert classes in the flag variety $\mathrm{Fl}(k-d, k+d; \mathbb{C}^n)$. There is a conjectured combinatorial formula for intersection numbers on two-step flag varieties, from A. Knutson, so one now has at least a conjectured combinatorial formula for the quantum Littlewood-Richardson coefficients.

During the 1996–97 year at the Mittag-Leffler Institute, many tables of products $\sigma_\lambda * \sigma_\mu$ were computed. It was surprising to see that the product was never 0, since, for degree reasons, there are only a finite number of possible nonzero coefficients. This was proved, for the Grassmannian, by S. Agnihotri and C. Woodward [1]. [3]

Note that the classical product $\sigma_\lambda \cdot \sigma_\mu$ is not zero exactly when the diagram of λ is contained in the diagram of $\mu^\vee$. On the basis of some calculations[4], it was natural to conjecture that the smallest power q^d that appears in the quantum product $\sigma_\lambda * \sigma_\mu$ would be the maximum d such that a d by d square fits inside the diagram of λ but outside that of $\mu^\vee$. For example, take $k = 4$, $n = 9$, $\lambda = (5, 4, 4, 3)$ and $\mu = (5, 4, 4, 1)$ (so $\mu^\vee = (4, 1, 1, 0)$:

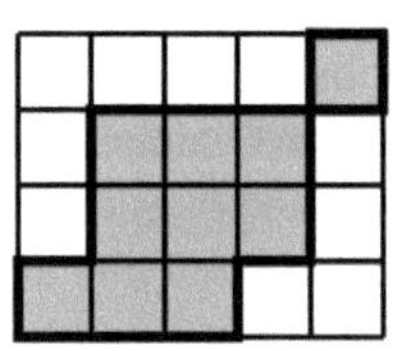

This indicates that $d = 2$, as there are three 2 by 2 squares that can be inserted inside λ and outside $\mu^\vee$. In fact,

[2] (Added later) P. Belkale has done this, in "The quantum Horn conjecture," math.AG/0303013.

[3] Agnihotri gave a very short proof, by observing that, after multiplying by $\sigma_{\lambda^\vee}$ and $\sigma_{\mu^\vee}$ — and noting that by the nonnegativity of all coefficients, there can be no cancelation — it suffices to do this when λ and μ are both the maximal partitions $(n-k, \dots, n-k)$, where it is an easy calculation.

[4] Buch has an efficient computer program for calculating classical and quantum Littlewood-Richardson coefficients, available at http://home.imf.au.dk/abuch/lrcalc/.

$$\sigma_{(5,4,4,3)} * \sigma_{(5,4,4,1)} = q^2(\sigma_{(5,3,2,2)} + \sigma_{(5,3,3,1)} + \sigma_{(5,4,2,1)}) \\ +q^3(\sigma_{(3,0,0,0)} + \sigma_{(2,1,0,0)} + \sigma_{(1,1,1,0)}).$$

Woodward and I [17] have proved this conjecture. Moreover, the product is nonzero for every G/P, and there is a combinatorial formula for the minimum d such that q^d occurs in the product.

For the Grassmannian, other proofs have been given by Buch [6] and Belkale [3], who gives the following refinement. The coefficient of the minimal q^d is equal to a classical product $\sigma_{\lambda'} \cdot \sigma_{\mu'}$, where λ' and μ' are obtained as follows. Pick any maximal square inside λ and outside $\mu^\vee$, and let a be the number of steps one travels down the path of λ to reach the southeast corner of the square, and let b be the number of steps one travels up on the path of $\mu^\vee$ to reach the northwest corner of the square.

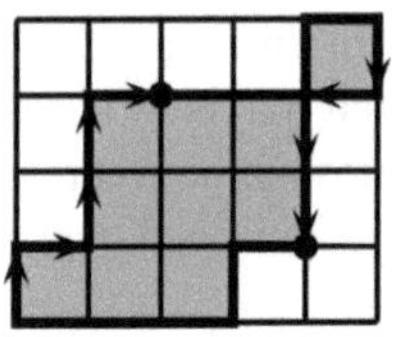

In this example, these corners and paths are marked, so $a = 4$ and $b = 5$. Add a k by a rectangle to the left of λ, and remove n-rims until the result lies inside the original k by $n-k$ rectangle; this result is λ'. Similarly, to find μ', one removes rims from the addition of a k by b rectangle to μ. In the example:

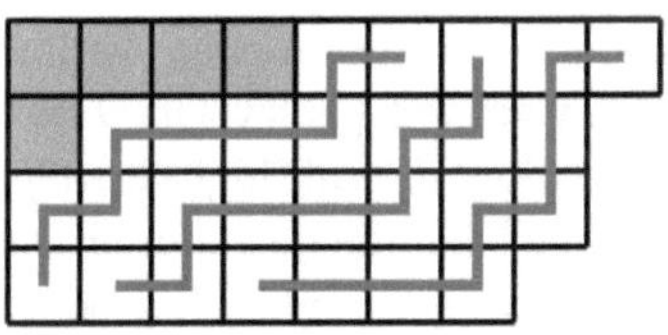

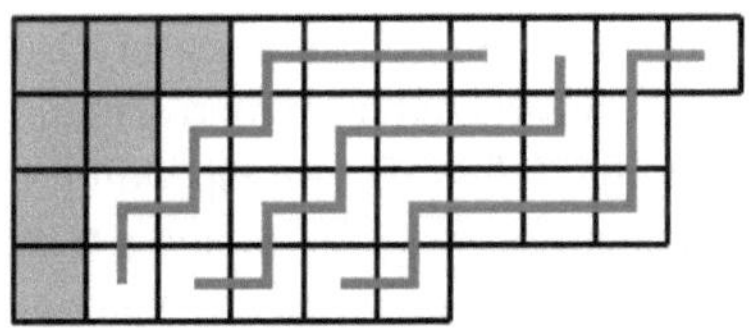

One sees that $\lambda' = (4, 1, 0, 0)$ and $\mu' = (3, 2, 1, 1)$. And, as predicted, the classical product of $\sigma_{(4,1,0,0)}$ and $\sigma_{(3,2,1,1)}$ is $\sigma_{(5,3,2,2)} + \sigma_{(5,3,3,1)} + \sigma_{(5,4,2,1)}$. In general, $\sigma_{(a,\dots,a)} * \sigma_\lambda = q^m \sigma_{\lambda'}$, where m is the number of n-rims removed; this is a special case of the algorithm of [5] for arbitrary multiplication of quantum Schubert classes.

The general formula for the minimal d such that q^d occurs involves the combinatorics of the curves in G/P that are invariant by the action of a maximal torus. It is measured by the length of a chain of such curves needed to join a point of a Schubert variety for λ to an opposite Schubert variety for $\mu^\vee$.

On the Grassmannian, A. Yong [28] found an upper bound for which powers of q can occur, and Postnikov [24] has found exactly which d have q^d occur in a product $\sigma_\lambda * \sigma_\mu$. They are all d between the above $d_{\min}$ and a $d_{\max}$ determined as follows. Slide the path for the diagram of $\mu^\vee$ southeast $d_{\min}$ steps, so there is nothing outside it and inside the path for λ. The ends of this path now extend outside the given k by

$n-k$ rectangle. Think of this rectangle as a torus, with opposite sides identified, and regard the extended path as a loop on the torus.

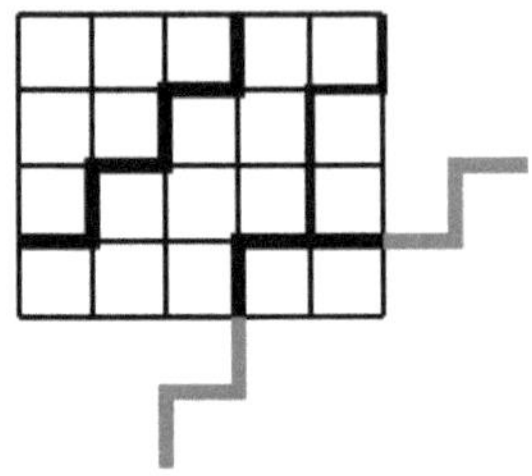

Keep moving this loop to the southeast until it just touches the path for λ again. The total number of steps is $d_{\max}$. In this example, $d_{\max} = 3$. Postnikov has a kind of duality between the minimal and maximal terms, including an explicit formula for the maximal terms, although those in between are not yet well understood. He also shows that, for all G/B, only one minimal degree occurs.

The proof in [17] uses Kleiman's transversality theorem on the Kontsevich-Manin moduli spaces, and the fixed points of the torus action. The work in [5] is combinatorial, as is that in [24].

We conclude with a few remarks and questions.

Postnikov [24] gives a formula for the quantum Littlewood-Richardson numbers in terms of "tableaux on a torus", but not in a way to show their positivity.

The quantum cohomology of $\mathrm{Gr}(k, n)$ is related to the fusion ring (Verlinde algebra) constructed from representations of the unitary group U_k at level $n-k$, with the Schubert class σ_λ corresponding to the representation V_λ with highest weight λ. The fact that $V_\lambda \otimes V_\mu$ is not zero gives another proof that the quantum product $\sigma_\lambda * \sigma_\mu$ is not zero. There are many physics papers studying these tensor products. G. Tudose [25] has a formula for the product when one partition has two parts.

Classically, knowing $H^*(G/B)$ is essentially equivalent to knowing $H^*(G/P)$ for all P, or all maximal parabolic P. In the quantum world, this is far from the case, because of the lack of functoriality of quantum cohomology.

The fact that the coefficients $c_{\lambda\,\mu}^{\,\nu}(d)$ are nonnegative puts very strong restrictions on the quantum cohomology rings. In practise, at least in small examples, one can often use that fact, together with very few simple calculations, to determine the whole quantum cohomology ring. It would be desirable to prove theorems characterizing quantum cohomology along these lines.

Peterson has given a remarkable formula for each $c_{\lambda\,\mu}^{\,\nu}(d)$ on a G/P as a number $c_{\lambda'\,\mu'}^{\,\nu'}(d')$ on the corresponding G/B. This is explained in [27].

Various symmetries have been found in the quantum cohomology of Grassmannians. For these, see [1], [23], [3], [18].

References

[1] S. Agnihotri and C. Woodward, Eigenvalues of products of unitary matrices and quantum Schubert calculus, *Math. Res. Lett.*, 5(6):817–836, 1998, alg-geom/9712013.

[2] A. Astashkevich and V. Sadov, Quantum cohomology of partial flag manifolds $F_{n_1 \cdots n_k}$, *Comm. Math. Phys.*, 170:503–528,1995, hep-th/9401103.

[3] P. Belkale, Transformation formulas in quantum cohomology, math.AG/0210050.

[4] A. Bertram, Quantum schubert calculus, *Adv. Math.*, 128:289–305, 1997, alg-geom/9410024.

[5] A. Bertram, I. Ciocan-Fontanine, and W. Fulton, Quantum multiplication of Schur polynomials, *J. Algebra*, 219(2):728–746, 1999, alg-geom/9705024.

[6] A. Buch, Quantum cohomology of Grassmannians, math.AG/0106268.

[7] A. Buch, Direct proof of the quantum Monk's formula, *Proc. Amer. Math. Soc.*, 131(7):2037–2042, 2003, math.AG/0107202.

[8] A. Buch, Quantum cohomology of partial flag manifolds, math.AG/0303245.

[9] A. Buch, A. Kresch, and H. Tamvakis, Gromov-Witten invariants on Grassmannians, math.AG/0306388.

[10] L. Chen, Quantum cohomology of flag manifolds, *Adv. Math.* 174:1–34, 2003, math.AG/0010080.

[11] I. Ciocan-Fontanine, The quantum cohomology ring of flag varieties, *Trans. Amer. Math. Soc.*, 351(7):2695–2729, 1999, alg-geom/9505002.

[12] I. Ciocan-Fontanine, On quantum cohomology rings of partial flag varieties, *Duke Math. J.*, 98(3):485–524, 1999, math.AG/9710213.

[13] S. Fomin, S. Gelfand, and A. Postnikov, Quantum Schubert polynomials, *J. Amer. Math. Soc.*, 10:565–596, 1997.

[14] W. Fulton, Eigenvalues, invariant factors, highest weights, and Schubert calculus, *Bull. Amer. Math. Soc.*, 37:209–249, 2000, math.AG/0301307.

[15] W. Fulton and J. Harris, *Representation Theory*, Springer-Verlag, 1991, 1999.

[16] W. Fulton and R. Pandharipande, Notes on stable maps and quantum cohomology, In *Algebraic geometry—Santa Cruz 1995*, pages 45–96. Amer. Math. Soc., 1997, alg-geom/9608011.

[17] W. Fulton and C. Woodward, On the quantum product of Schubert classes, *J. Alg.Geom.*, to appear, math.AG/0112183.

[18] H. Hengelbrock, An involution on the quantum cohomology ring of the Grassmannian, math.AG/0205260.

[19] Bumsig Kim, Quantum cohomology of flag manifolds G/B and quantum Toda lattices, *Ann. of Math. (2)*, 149(1):129–148, 1999, alg-geom/9607001.

[20] M. Kontsevich and Yu. Manin, Gromov-Witten classes, quantum cohomology, and enumerative geometry, *Comm. Math. Phys.*, 164(3):525–562, 1994, hep-th/9402147.

[21] A. Kresch and H. Tamvakis, Quantum cohomology of the Lagrangian Grassmannian, math.AG/0306337.

[22] A.-L. Mare, Polynomial representatives of Schubert classes in $QH^*(G/B)$, *Math. Res. Lett.* 9(5–6):757–760, 2002, math.CO/0205309.

[23] A. Postnikov, Symmetries of Gromov-Witten invariants, in *Advances in algebraic geometry motivated by physics (Lowell, MA, 2000)*, pages 251–258, *Contemp. Math.* 276, Amer. Math. Soc., 2001, math.CO/0009174.

[24] Alexander Postnikov, Affine approach to quantum Schubert calculus, math.CO/0205165.

[25] G. Tudose, A special case of sl(n)-fusion coefficients, math.CO/0008034.

[26] E. Witten, The Verlinde algebra and the cohomology of the Grassmannian, in *Geometry, topology, & physics*, Conf. Proc. Lecture Notes Geom. Topology, IV, pages 357–422, Internat. Press, 1995, hep-th/9312104.
[27] C. Woodward, On D. Peterson's comparison formula for Gromov-Witten invariants of G/P, math.AG/0206073.
[28] A. Yong, Degree bounds in quantum Schubert calculus, *Proc. Amer. Math. Soc.* 141(9):2649–2655, 2003, math.CO/0112133.

Quantum Principal Bundles up to Homotopy Equivalence

Christian Kassel

Hopf–Galois extensions are known to be the right generalizations of both Galois field extensions and principal G-bundles in the framework of non-commutative associative algebras. An abundant literature has been devoted to them by Hopf algebra specialists (see [11], [14], [15] and references therein). Recently there has been a surge of interest in Hopf–Galois extensions among mathematicians and theoretical physicists working in non-commutative geometry à la Connes and à la Woronowicz (*cf.* [2], [3], [6], [7], [8], [9]). In their work Hopf–Galois extensions are considered in the setting of "quantum group gauge theory."

In this note we deal with Hopf–Galois extensions in the light of topology. This leads us to simple questions for which we have very few answers, but which ought to be of interest to those working on Hopf algebras and in non-commutative geometry. We derive these questions from certain fundamental properties satisfied by topological principal bundles when we translate them into the setting of Hopf–Galois extensions. The properties we consider are the following.

I. *(Functoriality)* Given a principal G-bundle $X \to Y$ and a map $f : Y' \to Y$, then the pull-back $f^*X \to Y'$ is a principal G-bundle.
II. *(Homotopy)* If $f, g : Y' \to Y$ are homotopy equivalent maps, then f^*X and g^*X are homotopy equivalent bundles.
III. *(Triviality)* Any principal G-bundle over the point is trivial.

In order to translate these properties into algebra, we introduce what we call the homotopy equivalence of Hopf–Galois extensions. This is the main new concept of this note. For our definition of homotopy equivalence we need to restrict to extensions in which the subalgebra of coinvariants is central. In other words, the bases of the quantum principal bundles we consider belong to classical (commutative) geometry. Nevertheless, we impose no restriction on the "structural groups", that is on the Hopf algebras coacting on the quantum principal bundles: they may be non-commutative and non-cocommutative, infinite-dimensional, *etc.*

Despite the self-imposed commutativity of the "base space" the questions we ask on Hopf–Galois extensions do not seem easy to answer. Because they are not the kind of questions usually considered in this subject, it is our hope that they will lead to new developments on Hopf–Galois extensions and encourage research towards their classification over a given base.

We are nevertheless able to answer some of these questions when the Hopf algebra is one of Sweedler or Taft's finite-dimensional Hopf algebras. This is interesting because the latter are neither commutative, nor cocommutative, and are sometimes considered as "toy quantum groups." In particular, we prove that any cleft extension is homotopy equivalent to a trivial one. Cleft extensions form an important class of Hopf–Galois extensions with no non-trivial counterpart in commutative geometry. There are some good reasons to view cleft extensions as the natural non-commutative generalizations of trivial extensions: like the latter, cleft extensions can be constructed from the "structural group" and the "base space"; moreover, in analogy with the local triviality of topological bundles, Rumynin [13] showed that Hopf–Galois extensions over a large class of "base spaces" become cleft after (generic) localization. Our above-mentioned result is of a different, global, nature; we expect it to hold for any Hopf algebra.

The paper is organized as follows. In Sect. 1 we recall the standard terminology related to Hopf–Galois extensions; we define what we call quantum principal bundles as well as their push-forwards along morphisms of commutative algebras (following [13]). The concept of homotopy equivalence is introduced in Sect. 2; we show that up to homotopy equivalence quantum principal bundles over a polynomial algebra $C[t]$ are in bijection with quantum principal bundles over C. In Sect. 3 we consider four notions of triviality for a quantum principal bundle, and ask whether some of them are equivalent. In Sect. 4 we consider the case of Sweedler's and Taft's Hopf algebras for which we prove that any cleft quantum principal bundle is homotopy equivalent to a trivial one. Section 5 is a short conclusion listing three problems dealing with Hopf–Galois extensions over the algebra $k[z, z^{-1}]$ of Laurent polynomials, which we view as the algebraic counterpart of the circle.

We fix a field k. All constructions will be performed in the category of k-vector spaces; in particular, unadorned tensor symbols refer to the tensor product of k-vector spaces. All algebras will be assumed to be associative and unital, and all morphisms of algebras preserve the units.

1 Quantum principal bundles

A principal bundle involves a group G (topological group, group scheme) acting, say on the right, on a space X (topological space, scheme) such that the canonical map

$X \times G \to X \times_Y X$ induced by $(x, g) \mapsto (x, xg)$ is an isomorphism (in the category of spaces under consideration). Here Y represents some version of the quotient space X/G and $X \times_Y X$ the fibre product.

1.1 Hopf–Galois extensions. In a purely algebraic setting the group G is replaced by a Hopf algebra H with coproduct $\Delta : H \to H \otimes H$, coünit $\varepsilon : H \to k$, and antipode $S : H \to H$. We assume that the antipode is bijective. In the sequel we make use of Sweedler's notation: for any $h \in H$ we write

$$\Delta(h) = \sum_{(h)} h_{(1)} \otimes h_{(2)} .$$

The G-space X is replaced by an algebra A carrying the structure of an *H-comodule algebra.* Recall that an algebra A is an H-comodule algebra if it has a right H-comodule structure whose coaction $\Delta_A : A \to A \otimes H$ is a morphism of algebras. We write $\Delta_A(a) = \sum_{(a)} a_{(0)} \otimes a_{(1)}$ for $a \in A$.

The *base* of the H-comodule algebra A is the subspace $C = A^{\mathrm{co}-H}$ of coinvariants defined by

$$A^{\mathrm{co}-H} = \{a \in A \mid \Delta_A(a) = a \otimes 1_H\} ,$$

where 1_H is the unit of H. It is clear that $A^{\mathrm{co}-H}$ is a subalgebra and a subcomodule of A. We say that $C = A^{\mathrm{co}-H} \subset A$ is an *H-extension* or that A is an H-extension of C.

An H-extension $C \subset A$ is said to be *Galois* if the so-called *canonical map* $\beta : A \otimes_C A \to A \otimes H$ defined for $a, b \in A$ by

$$a \otimes b \mapsto (a \otimes 1_H)\, \Delta_A(b) = \sum_{(b)} ab_{(0)} \otimes b_{(1)} \tag{1}$$

is bijective. For more details on Hopf–Galois extensions, see [11], Chap. 8 or [15], Sect. 1.

Following Rumynin [13], we consider *extensions with central invariants,* namely extensions $C \subset A$ such that C sits in the centre of A (implying that C is commutative).

The following are the basic objects investigated in this note.

Definition 1.2. *Given a Hopf algebra H and a commutative algebra C, a quantum principal H-bundle over the base C is a Galois H-extension $C \subset A$ such that C is contained in the centre of A and A is faithfully flat as a C-module.*

The faithful flatness condition imposed in the definition is a crucial standard requirement in the subject (see *e. g.* [15]). Observe that our definition of a quantum principal bundle is different from the one given in [3], [6], [9].

1.3 Functoriality. For any morphism of algebras $f : C \to C'$ and any left C-module A, we define the *push-forward* f_*A as the left C'-module $C' \otimes_C A$. (This is the algebraic counterpart of a topological pull-back.) The most natural way to equip f_*A with a multiplication is to define it by

$$(c'_1 \otimes a_1)(c'_2 \otimes a_2) = c'_1 c'_2 \otimes a_1 a_2$$

for $c'_1, c'_2 \in C'$ and $a_1, a_2 \in A$. Now $c'_1 f(c_1) \otimes a_1 = c'_1 \otimes c_1 a_1$ and $c'_2 f(c_2) \otimes a_2 = c'_2 \otimes c_2 a_2$ in $C' \otimes_C A$ for $c_1, c_2 \in C$. Therefore, in order for the multiplication to be well defined, we must have

$$c'_1 f(c_1) c'_2 f(c_2) \otimes a_1 a_2 = c'_1 c'_2 \otimes c_1 a_1 c_2 a_2 .$$

This forces us to require C to be central in A and the image $f(C)$ to be central in C'. This is why we consider extensions with central invariants.

Theorem 1.4. *Let $C \subset A$ be a quantum principal H-bundle whose base is C. If $f : C \to C'$ is a morphism of commutative algebras, then $A' = f_* A$ is a quantum principal H-bundle whose base is C'.*

This theorem was proved in [13], Theorem 9 under the restriction that H is finite-dimensional over k. We give a proof that dispenses with this restriction.

Proof. As we have seen above, A' is an algebra; it contains C' in its centre. We equip A' with the coaction $\Delta_{A'} = \mathrm{id}_{C'} \otimes_C \Delta_A$, which is well defined because of the condition $C = A^{\mathrm{co}-H}$. In this way A' becomes an H-comodule algebra.

Note that $C' \subset (A')^{\mathrm{co}-H}$. This allows us to define a map $\beta' : A' \otimes_{C'} A' \to A' \otimes H$ by $\beta'(a \otimes b) = (a \otimes 1)\, \Delta_{A'}(b)$ for all $a, b \in A'$. By assumption, the map $\beta : A \otimes_C A \to A \otimes H$ defined by (1) is bijective. Since

$$A' \otimes_{C'} A' = (C' \otimes_C A) \otimes_{C'} (C' \otimes_C A) \cong C' \otimes_C A \otimes_C A$$

and $A' \otimes H = C' \otimes_C A \otimes H$, we have $\beta' = \mathrm{id}_{C'} \otimes_C \beta$, which shows that β' is bijective. By [17], Lemma 4.2, it follows that $C' = (A')^{\mathrm{co}-H}$.

It remains to check that A' is faithfully flat as a C'-module. This is a consequence of [1], Proposition 5. ∎

Observe that, if $C \subset A$ is a quantum principal H-bundle and $f : C \to C'$ and $g : C' \to C''$ are morphisms of commutative algebras, the

$$(g \circ f)_* A = g_*(f_* A) .$$

We also have $\mathrm{id}_* A = A$, where $\mathrm{id} : C \to C$ is the identity map.

2 Homotopy equivalence

The concept of isomorphism of quantum principal H-bundles is clear: two H-extensions $C \subset A$ and $C' \subset A'$ are isomorphic if there exists an isomorphism of H-comodule algebras $A \cong A'$; such an isomorphism will necessarily send C onto C'. We'll use the notation $\cong$ for isomorphic quantum principal bundles.

We now introduce a broader equivalence relation, which is of a global nature. Let $\mathcal{C}_{\text{ét}}$ be the class of *finite étale morphisms* of commutative algebras (for more on étale morphisms, see [12]). We also need the following notation: if K is a commutative algebra and $K[t]$ is the polynomial algebra in one variable t over K, we define $[0], [1] : K[t] \to K$ to be the morphisms of K-algebras sending t to 0 and 1, respectively.

Definition 2.1. *Given two quantum principal H-bundles A and A' over a commutative base C, we write $A \sim A'$ if there exist a morphism $i : C \to \bar{C}$ in $\mathcal{C}_{\text{ét}}$ and a quantum principal H-bundle B with base $\bar{C}[t]$ such that*

$$[0]_* B \cong i_* A \quad \text{and} \quad [1]_* B \cong i_* A' .$$

The homotopy equivalence of quantum principal H-bundles over C is the equivalence relation generated by $\sim$.

Isomorphic quantum principal bundles are clearly homotopy equivalent. In Sect. 4.5 we will show examples of homotopy equivalent quantum principal bundles that are not isomorphic. We also remark that by the very definition of homotopy equivalence two quantum principal bundles A and A' over a base C such that $i_* A \cong i_* A'$ for some morphism $i : C \to \bar{C}$ in $\mathcal{C}_{\text{ét}}$ are homotopy equivalent.

Given a Hopf algebra H and a commutative algebra C, our ultimate goal is to classify quantum principal H-bundles over C up to homotopy equivalence. We denote the set of homotopy equivalence classes of quantum principal H-bundles over C by $\mathcal{P}_H(C)$. The following proposition implies that the push-forward along a morphism $f : C \to C'$ of commutative algebras induces a set-theoretic map $f_* : \mathcal{P}_H(C) \to \mathcal{P}_H(C')$.

Proposition 2.2. *Let $f : C \to C'$ be a morphism of commutative algebras. If A and A' are homotopy equivalent quantum principal H-bundles over C, then so are $f_* A$ and $f_* A'$.*

Proof. It is enough to check that, if $A \sim A'$, then $f_* A \sim f_* A'$. Let B be a quantum principal H-bundle with base $\bar{C}[t]$ ($i : C \to \bar{C}$ is as in Definition 2.1) such that $[0]_* B \cong i_* A$ and $[1]_* B \cong i_* A'$. We extend $f : C \to C'$ to a morphism of algebras $\bar{f} : \bar{C} \to \bar{C'} = C' \otimes_C \bar{C}$. If $i' : C' \to \bar{C'}$ is the natural morphism induced from i, we have $\bar{f} \circ i = i' \circ f$. We next extend $\bar{f} : \bar{C} \to \bar{C'}$ to a morphism $\bar{f}_t : \bar{C}[t] \to \bar{C'}[t]$ of $k[t]$-algebras. By Theorem 1.4 $(\bar{f}_t)_* B$ is a quantum principal H-bundle over $\bar{C'}[t]$. Now, since $\bar{f}(0) = 0$ and $\bar{f}(1) = 1$, we have $[0] \circ \bar{f}_t = \bar{f} \circ [0]$ and $[1] \circ \bar{f}_t = \bar{f} \circ [1]$. Therefore,

$$[0]_* \left(\left(\bar{f}_t \right)_* B \right) = \bar{f}_*([0]_* B) \cong \bar{f}_*(i_* A) = i'_*(f_* A)$$

and

$$[1]_* \left(\left(\bar{f}_t \right)_* B \right) = \bar{f}_*([1]_* B) \cong \bar{f}_*(i_* A') = i'_*(f_* A') .$$

■

We now give a sufficient condition for two morphisms of commutative algebras $f, g : C \to C'$ to induce the same map $\mathcal{P}_H(C) \to \mathcal{P}_H(C')$. We say that f and g are *homotopic* if there exist a morphism $i' : C' \to \bar{C'}$ in $\mathcal{C}_{\text{ét}}$ and a morphism of algebras $\varphi : C \to \bar{C'}[t]$ such that $[0] \circ \varphi = i' \circ f$ and $[1] \circ \varphi = i' \circ g$ (the morphism φ will be called a homotopy between f and g).

Proposition 2.3. *If $f, g : C \to C'$ are homotopic morphisms of commutative algebras and A is a quantum principal H-bundle over C, then the quantum principal H-bundles f_*A and g_*A are homotopy equivalent: $f_*A \sim g_*A$.*

Proof. Let $\varphi : C \to \bar{C}'[t]$ be a homotopy between f and g. By Theorem 1.4, $A' = \varphi_* A$ is a quantum principal H-bundle over $\bar{C}'[t]$. We have

$$[0]_* A' = [0]_*(\varphi_* A) = ([0] \circ \varphi)_* A = i'_*(f_* A)$$

and

$$[1]_* A' = [1]_*(\varphi_* A) = ([1] \circ \varphi)_* A = i'_*(g_* A)\,.$$

■

We give two important applications of Proposition 2.3. For the first one we let $C = \bigoplus_{i \geq 0} C_i$ be a non-negatively graded commutative algebra (then C_0 is a subalgebra of C).

Corollary 2.4. *The inclusion $\iota : C_0 \to C = \bigoplus_{i \geq 0} C_i$ induces a bijection*

$$\mathcal{P}_H(C_0) \cong \mathcal{P}_H(C)\,.$$

Proof. Let $\pi : C \to C_0$ be the canonical projection. We have $\pi \circ \iota = \mathrm{id}_{C_0}$. We claim the identity map id_C of C and $\iota \circ \pi$ are homotopic morphisms. Indeed, let $\varphi : C \to C[t]$ be the morphism of algebras defined by $\varphi(x) = t^i x$ if $x \in C_i$; then $[0] \circ \varphi = i \circ p$ and $[1] \circ \varphi = \mathrm{id}_C$. It follows from Proposition 2.3 that $A \sim i_* p_* A$ for any quantum principal H-bundle A over C. ■

As a consequence of Corollary 2.4, the set of homotopy equivalence classes of quantum principal H-bundles over any polynomial ring $C[x_1, \ldots, x_N]$ with coefficients in a commutative algebra C is in bijection with the set of homotopy equivalence classes of quantum principal H-bundles over C:

$$\mathcal{P}_H(C[x_1, \ldots, x_N]) \cong \mathcal{P}_H(C)\,.$$

Corollary 2.5. *Any morphism $i : C \to \bar{C}$ in $\mathcal{C}_{\text{ét}}$ induces an injection $\mathcal{P}_H(C) \to \mathcal{P}_H(\bar{C})$.*

Proof. It is enough to check that for any two quantum principal H-bundles A and A' over C such that $i_* A \sim i_* A'$, we have $A \sim A'$. Let $\bar{\imath} : \bar{C} \to \bar{C}_1$ be in $\mathcal{C}_{\text{ét}}$ and B be a quantum principal H-bundle with base $\bar{C}_1[t]$ such that

$$[0]_* B \cong \bar{\imath}_*(i_* A) = (\bar{\imath} \circ i)_* A \quad \text{and} \quad [1]_* B \cong \bar{\imath}_*(i_* A') = (\bar{\imath} \circ i)_* A'\,.$$

Since $\bar{\imath} \circ i$ belongs to $\mathcal{C}_{\text{ét}}$ (see [12], Chap. II, Proposition 1), we have $A \sim A'$. ■

3 Triviality

In this section we define four notions of triviality for quantum principal bundles. The first one is the exact algebraic counterpart of the classical concept of a trivial principal bundle. The second one, cleftness, is proper to the world of non-commutative algebras and has no counterpart in the classical theory of principal bundles over a group. The remaining ones, homotopical triviality and homotopical cleftness, are natural byproducts of the concept of homotopy equivalence introduced in the previous section.

3.1. A quantum principal H-bundle over a commutative algebra C is called *trivial* if it is isomorphic to the H-comodule algebra $C \otimes H$, where the coaction is given by $\mathrm{id}_C \otimes \Delta$. (The reader may check that $C \otimes H$ is a quantum principal H-bundle over C.) The push-forward of any trivial quantum principal bundle is trivial.

3.2. An H-extension $C \subset A$ is said to be *cleft* if there exists a morphism $\gamma : H \to A$ of right H-comodules that is invertible under convolution in $\mathrm{Hom}_k(H, A)$, i.e., if there exists another linear map $\gamma' : H \to A$ such that

$$\sum_{(h)} \gamma(h_{(1)})\gamma'(h_{(2)}) = \sum_{(h)} \gamma'(h_{(1)})\gamma(h_{(2)}) = \varepsilon(h)1_A \tag{2}$$

for all $h \in H$. The map γ is called a *cleaving map*. Note that a cleft extension is called a trivial principal bundle in [2], Example 4.2.

By [4] (see also [11], Theorem 7.2.2) an H-extension $C \subset A$ is cleft if and only A is isomorphic to a *crossed product* $C \sharp_\sigma H$, whose underlying vector space is $C \otimes H$ and whose multiplication is given for all $c, d \in C$ and $g, h \in H$ by

$$(c \otimes g)(d \otimes h) = \sum_{(g)(h)} c(g_{(1)} \cdot d)\, \sigma(g_{(2)}, h_{(1)}) \otimes g_{(3)}h_{(2)}\,. \tag{3}$$

Here $\cdot : H \otimes C \to C$ and $\sigma : H \otimes H \to C$ are linear maps defined in terms of the cleaving map γ and its inverse γ' by

$$h \cdot c = \sum_{(h)} \gamma(h_{(1)})\, c\, \gamma'(h_{(2)}) \tag{4}$$

and

$$\sigma(g, h) = \sum_{(g)(h)} \gamma(g_{(1)})\, \gamma(h_{(1)})\, \gamma'(g_{(2)}h_{(2)})$$

for $g, h \in H$ and $c \in C$. The coaction of H on $C \sharp_\sigma H$ is trivial. A cleft H-extension $C \subset A$ is Galois and faithfully flat as a C-module.

For a cleft extension $C \subset C \sharp_\sigma H$ to be a quantum principal bundle in the sense of Definition 1.2, we need C to be central in $C \sharp_\sigma H$. Under this additional condition, (2) and (4) imply $h \cdot c = \varepsilon(h)c$ for all $h \in H$ and $c \in C$. This means that the "quasi-action" $\cdot$ is trivial, which leads to the following simplification of (3):

$$(c \otimes g)(d \otimes h) = \sum_{(g)(h)} cd\, \sigma(g_{(1)}, h_{(1)}) \otimes g_{(2)}h_{(2)} \, .$$

Thus, as already observed in [13], Lemma 7, a cleft quantum principal H-bundle is a *twisted product* in the sense of [11], Example 7.1.5. The push-forward of a cleft quantum principal bundle is cleft by [13], Theorem 9.

3.3. The third and fourth definitions are as follows: a quantum principal H-bundle over a commutative algebra C is called *homotopically trivial* (resp. *homotopically cleft*) if it is homotopy equivalent to a trivial (resp. cleft) quantum principal H-bundle over C. Since push-forwards preserve triviality and cleftness, Proposition 2.2 has the following consequence.

Proposition 3.4. *Let f be a morphism of commutative algebras. If A is a homotopically trivial (resp. homotopically cleft) quantum principal H-bundle, then f_*A is homotopically trivial (resp. homotopically cleft).*

The following provides examples of homotopically trivial bundles.

Proposition 3.5. *If A is a commutative quantum principal H-bundle such that the inclusion $i : C \to A$ belongs to $\mathcal{C}_{\text{ét}}$, then A is homotopically trivial.*

Proof. Since A is commutative, starting from the quantum principal H-bundles A and $C \otimes H$, the latter being trivial, we may apply Theorem 1.4 and define the quantum principal H-bundles $i_*A = A \otimes_C A$ and $i_*(C \otimes H) = A \otimes H$ over A. The bundle $i_*(C \otimes H)$ is trivial. Using the commutativity of A, it is easy to check that the canonical map $\beta : A \otimes_C A \to A \otimes H$ is a morphism of H-comodule algebras. (Observe that H is commutative since β is an isomorphism.) Therefore, $i_*A \cong i_*(C \otimes H)$ for $i \in \mathcal{C}_{\text{ét}}$. This implies $A \sim C \otimes H$. ∎

We end this section with a few questions.

Question 3.6. Is any cleft quantum principal H-bundle over a commutative algebra C homotopically trivial?

An affirmative answer would give a “topological” meaning to the algebraic concept of cleftness and show that homotopical cleftness implies homotopical triviality. We may also wonder whether cleftness is closed under homotopy equivalence.

Question 3.7. Let A and A' be homotopy equivalent quantum principal bundles. Suppose that A' is cleft. Does it imply that A is cleft?

An answer in the affirmative to both Questions 3.6 and 3.7 would imply that the three notions of cleftness, homotopical cleftness, and homotopical triviality coincide. This would give a firm ground to the idea that cleft quantum principal bundles are the right algebraic counterparts of trivial topological principal bundles (compare with the question in [13], Sect. 2.4).

Since a topological principal bundle over a point is trivial, we ask the following question.

Question 3.8. Is any quantum principal H-bundle over the ground field k homotopically trivial?

According to Schneider [16], a Galois H-extension of the ground field k is cleft in the following cases:

(a) H is finite-dimensional (for a proof, see [13], Corollary 17 (3)),
(b) H is pointed (*e.g.*, a group algebra),
(c) H is commutative.

Therefore an affirmative answer to Question 3.6 would imply an affirmative one to Question 3.8 when H satisfies one of the conditions above.

4 The Case of Sweedler's and Taft's Hopf Algebras

The purpose of this section is to answer Questions 3.6 and 3.8 in the affirmative for certain small non-commutative non-cocommutative Hopf algebras.

4.1 Sweedler's four-dimensional Hopf algebra. It is the algebra H_4 generated by X and Y and the relations $X^2 = 1$, $Y^2 = 0$, $XY + YX = 0$. The comultiplication of H_4 is given by

$$\Delta(X) = X \otimes X \quad \text{and} \quad \Delta(Y) = 1 \otimes Y + Y \otimes X\,,$$

the coünit by $\varepsilon(X) = 1$ and $\varepsilon(Y) = 0$, and the antipode by $S(X) = X$ and $S(Y) = XY$.

For any commutative algebra C we construct an H_4-extension $C \subset A$ as follows. Let α, β, and γ be elements of C such that α belongs to the group $C^\times$ of invertible elements of C. We define $A = \left(\frac{\alpha,\beta,\gamma}{C}\right)$ as the C-algebra generated by two variables x and y subject to the relations $x^2 = \alpha$, $y^2 = \beta$, $xy + yx = \gamma$. The algebra A is a H_4-comodule algebra with C-linear coaction $\Delta_A : A \to A \otimes H_4$ defined by

$$\Delta_A(x) = x \otimes X \quad \text{and} \quad \Delta_A(y) = 1 \otimes Y + y \otimes X\,.$$

When $\alpha = 1$, $\beta = \gamma = 0$, then $C \subset \left(\frac{\alpha,\beta,\gamma}{C}\right)$ is trivial in the sense of Sect. 3.1.

Proposition 4.2. *The H_4-comodule algebra $\left(\frac{\alpha,\beta,\gamma}{C}\right)$ is a cleft quantum principal H_4-bundle with base C. Conversely, any cleft quantum principal H_4-bundle with base C is isomorphic to $\left(\frac{\alpha,\beta,\gamma}{C}\right)$ for some $\alpha \in C^\times$ and $\beta, \gamma \in C$.*

Proof. We use Doi and Takeuchi's reformulation of Masuoka's description of cleft Galois H_4-extensions (see [5], [10]). In *loc. cit.* an H_4-cleft datum is defined as a quintuple $(F, D, \alpha, \beta, \gamma)$ where $F, D \in \mathrm{End}_k(C)$, $\alpha \in C^\times$, $\beta, \gamma \in C$ are elements satisfying certain conditions. From any H_4-cleft datum one constructs a cleft H_4-extension $C \subset A$, where A is the free left C-module with basis $\{1, x, y, xy\}$ equipped with an algebra structure such that

$$x^2 = \alpha, \quad y^2 = \beta, \quad xy + yx = \gamma\,,$$

$$xc = F(c)x, \quad yc = cy + D(c)x, \quad xyc = F(c)xy + F(D(c))\alpha$$

for all $c \in C$. Following [5], we denote A by $\left(\frac{F,D,\alpha,\beta,\gamma}{C}\right)$ Conversely, any cleft H_4-extension $C \subset A$ is isomorphic to $\left(\frac{F,D,\alpha,\beta,\gamma}{C}\right)$ for some H_4-cleft datum $(F, D, \alpha, \beta, \gamma)$.

Our definition of a quantum principal bundle requires that in addition C be central in A. This clearly forces F to be the identity map id_C of C and D to be the zero map. The quintuple $(\mathrm{id}_C, 0, \alpha, \beta, \gamma)$ satisfies the requirements of an H_4-cleft datum. It is clear from the definition of $\left(\frac{F,D,\alpha,\beta,\gamma}{C}\right)$ that the algebra associated to the H_4-cleft datum $(\mathrm{id}_C, 0, \alpha, \beta, \gamma)$ is $\left(\frac{\alpha,\beta,\gamma}{C}\right)$. ■

The next result answers Question 3.6 in the affirmative when $H = H_4$.

Theorem 4.3. *Any cleft quantum principal H_4-bundle over a commutative algebra C in which 2 is invertible is homotopically trivial.*

Proof. Let A be a cleft quantum principal H_4-bundle over C. By Proposition 4.2 there exist $\alpha \in C^\times$ and $\beta, \gamma \in C$ such that $A \cong \left(\frac{\alpha,\beta,\gamma}{C}\right)$. Let $i : C \to \bar{C}$ be a finite étale extension containing a square root s of α (by [12], Chap. VI, Proposition 1, we may take $\bar{C} = C[T]/(T^2 - \alpha)$). We have $i_*\left(\frac{\alpha,\beta,\gamma}{C}\right) = \left(\frac{\alpha,\beta,\gamma}{\bar{C}}\right)$. By [5], Theorem 2.7 there is an isomorphism

$$\left(\frac{\alpha, \beta, \gamma}{\bar{C}}\right) \cong \left(\frac{1, \beta, \gamma/s}{\bar{C}}\right).$$

It now suffices to check that any H_4-extension of the form $\left(\frac{1,\beta,\gamma}{\bar{C}}\right)$ is homotopy equivalent to the trivial extension $\left(\frac{1,0,0}{\bar{C}}\right)$. This is proved by considering the cleft quantum principal H_4-bundle $A_t = \left(\frac{1,t\beta,t\gamma}{\bar{C}[t]}\right)$ over the base $\bar{C}[t]$. We have

$$[1]_* A_t = \left(\frac{1, \beta, \gamma}{\bar{C}}\right) \quad \text{and} \quad [0]_* A_t = \left(\frac{1, 0, 0}{\bar{C}}\right).$$

■

This allows us to answer Question 3.8 in the affirmative for H_4.

Corollary 4.4. *If* $\mathrm{char}(k) \neq 2$, *then* $\mathrm{card}\, \mathcal{P}_{H_4}(k) = 1$.

Proof. By [13], Corollary 17 (3), any quantum principal H_4-bundle over the ground field is cleft. We then apply Theorem 4.3. ■

Remark 4.5. By [5], Corollary 2.8, $\left(\frac{\alpha,\beta,\gamma}{k}\right)$ is isomorphic to the trivial quantum principal H_4-bundle $\left(\frac{1,0,0}{k}\right)$ if and only if there exist $s \in k^\times$ and $t \in k$ such that $\alpha = s^2$, $\beta = t^2$, and $\gamma = 2st$. Therefore, $\left(\frac{1,0,\gamma}{k}\right) \not\cong \left(\frac{1,0,0}{k}\right)$ for any $\gamma \neq 0$. In view of Corollary 4.4, this shows the existence of non-isomorphic homotopy equivalent quantum principal bundles.

4.6 Taft's Hopf algebras. Similar results hold for Taft's Hopf algebra H_{N^2}, where N is a fixed integer > 2. Recall that H_{N^2} is the algebra (of dimension N^2) generated by two generators X and Y subject to the relations $X^N = 1$, $Y^N = 0$, and $YX = qXY$, where q is a root of unity of order N. The comultiplication and the coünit are defined by the same formulas as for H_4.

Theorem 4.7. *Any cleft quantum principal H_{N^2}-bundle over a commutative algebra C in which N is invertible is homotopically trivial.*

The proof of Theorem 4.7 relies on [5], Theorems 3.2, 3.5, 3.8 and follows the same lines as the proof of Proposition 4.2 and Theorem 4.3.

Corollary 4.8. *If N is invertible in k, then* card $\mathcal{P}_{H_{N^2}}(k) = 1$.

5 Concluding remarks

So far we have no example of a non-homotopically trivial quantum principal bundle. The search of such an example should be the next step in order to understand better the concept of homotopy equivalence introduced in this note. A good candidate for the non-triviality of $\mathcal{P}_H(C)$ may be the algebra $C = k[z, z^{-1}]$ of Laurent polynomials: it is an algebraic model for the circle, which is the simplest manifold with non-trivial fundamental group.

Problem 5.1. Find examples of quantum principal bundles over $k[z, z^{-1}]$ that are not cleft, not homotopically cleft, or not homotopically trivial.

Problem 5.2. Classify all quantum principal bundles over $k[z, z^{-1}]$ up to isomorphism (for some interesting Hopf algebras). Find the cleft ones.

Problem 5.3. Determine $\mathcal{P}_H(k[z, z^{-1}])$ when $H = H_4$ or $H = H_{N^2}$.

Remark 5.4. The quest for affirmative answers to Questions 3.6 and 3.8 may force us to extend the class $\mathfrak{C}_{\text{ét}}$ to a bigger class $\mathfrak{C}$ of morphisms (closed under composition as is the case with $\mathfrak{C}_{\text{ét}}$). Proposition 3.5 indicates that $\mathfrak{C}$ should not be too big if we want to keep significant non-homotopically trivial quantum principal bundles (see also the example below).

Fix an integer $N \geq 2$ invertible in the ground field k and assume that k contains a root of unity q of order N. Then the extension $k[z, z^{-1}] \subset k[z^{1/N}, z^{-1/N}]$ is a quantum principal H-bundle for the Hopf algebra H dual to the group algebra of the cyclic group C_N acting on $k[z^{1/N}, z^{-1/N}]$ by $\sigma P(z^{1/N}) = P(qz^{1/N})$, where σ is a generator of C_N. Since this extension is finite and étale, it is homotopically trivial by Proposition 3.5.

Acknowledgements. I started working on the problems stated in this note after some conversations with Vladimir Turaev in early 1999.

References

[1] N. Bourbaki, *Algèbre commutative*, Hermann, Paris, 1961.
[2] T. Brzeziński, S. Majid, *Quantum gauge group theory on quantum spaces*, Comm. Math. Phys. **157** (1993), 591–638; Erratum: **167** (1995), 235.
[3] T. Brzeziński, S. Majid, *Line bundles on quantum spheres*, in Particles, fields, and gravitation (Łódź 1998), AIP Conf. Proc., **453**, Amer. Inst. Phys., Woodbury, New York, 1998, 3–8.
[4] Y. Doi, M. Takeuchi, *Cleft comodule algebras for a bialgebra*, Comm. Algebra **14** (1986), 808–818.
[5] Y. Doi, M. Takeuchi, *Quaternion algebras and Hopf crossed products*, Comm. Algebra **23** (1995), 3291–3325.
[6] P. M. Hajac, *Strong connections on quantum principal bundles*, Comm. Math. Phys. **182** (1996), 579–617.
[7] P. M. Hajac, *Bundles over quantum sphere and noncommmutative index theorem*, K-Theory **21** (2000), 141–150.
[8] P. M. Hajac, S. Majid, *Projective module description of the q-monopole*, Comm. Math. Phys. **206** (1999), 247–264.
[9] S. Majid, *Some remarks on quantum and braided group gauge theory*, in Quantum groups and quantum spaces, Banach Center Publications, vol. **40**, Institute of Mathematics, Polish Academy of Sciences, Warszawa, 1997, 335–349.
[10] A. Masuoka, *Cleft extensions for a Hopf algebra generated by a nearly primitive element*, Comm. Algebra **22** (1994), 4537–4559.
[11] S. Montgomery, *Hopf algebras and their actions on rings*, CBMS Conf. Series in Math., vol. **82**, Amer. Math. Soc., Providence, RI, 1993.
[12] M. Raynaud, *Anneaux locaux henséliens*, Lect. Notes in Math. **169**, Springer-Verlag, Berlin, Heidelberg, New York, 1970.
[13] D. Rumynin, *Hopf–Galois extensions with central invariants and their geometric properties*, Algebras and Representation Theory **1** (1998), 353–381.
[14] H.-J. Schneider, *Principal homogeneous spaces for arbitrary Hopf algebras*, Israel J. Math. **72** (1990), 167–195.
[15] H.-J. Schneider, *Hopf Galois extensions, crossed products, and Clifford theory*, in Advances in Hopf algebras (ed. J. Bergen, S. Montgomery), Lecture Notes in Pure and Applied Mathematics, **158**, Marcel Dekker, New York, 1994, 267–297.
[16] H.-J. Schneider, Electronic letter to the author (28 April 1999).
[17] M. Takeuchi, *On extensions of formal groups by μ^A*, Comm. Algebra **5** (1977), 1439–1481.

Non-commutative Crepant Resolutions

Michel van den Bergh

Abstract. We introduce the notion of a "non-commutative crepant" resolution of a singularity and show that it exists in certain cases. We also give some evidence for an extension of a conjecture by Bondal and Orlov, stating that different crepant resolutions of a Gorenstein singularity have the same derived category.

1 Introduction

Let k be an algebraically closed field and let R be an integral Gorenstein k-algebra of dimension n. Put $X = \operatorname{Spec} R$ and let $f : Y \to X$ be a resolution of singularities. f is a crepant resolution of X if $f^*\omega_X = \omega_Y$. Crepant resolutions do not always exist and they are usually not unique. A conjecture of Bondal and Orlov states that if $f_1 : Y_1 \to X$, $f_2 : Y_2 \to X$ are different crepant resolutions of X then $D^b(\operatorname{coh}(Y_1)) \cong D^b(\operatorname{coh}(Y_2))$.

In this paper we study certain non-commutative analogues of crepant resolutions and we call these "non-commutative crepant resolutions". A non-commutative

The author is a director of research at the FWO.

crepant resolution of R is an algebra $A = \mathrm{End}_R(M)$ where M is a reflexive R-module and where A has finite global dimension and is a (maximal) Cohen–Macaulay R-module. For some general discussion of non-commutative resolutions we refer to [9, Sect. 5]

A standard example of a non-commutative crepant resolution is the following:

Example 1.1. Let G be a finite group and let V be a finite dimensional G-representation such that $G \subset \mathrm{Sl}(V)$. Put $S = \mathrm{Sym}(V)$ and $R = S^G$. Then $A = \mathrm{End}_R(S) \cong S * G$ is a non-commutative crepant resolution of R.

In [37] we made the following conjecture.

Conjecture 1.2. If R is three dimensional and has terminal singularities then it has a non-commutative crepant resolution if and only if it has a crepant commutative resolution.

We prove this conjecture below (Theorem 6.6.3). We also show that a version of the Bondal–Orlov conjecture holds in the sense that the two resolutions in Conjecture 1.2 are derived equivalent. One direction in the proof of Conjecture 1.2 is obtained from the beautiful version of the McKay correspondence given in [15] (with virtually the same proof).

In addition to proving Conjecture 1.2 we will give two other instances where a non-commutative resolution exists:

(1) Cones over Del Pezzo surfaces (see Sect. 7).
(2) Invariants of a one dimensional torus acting linearly on a polynomial ring (see Sect. 8).

We would like to thank Paul Smith and Amnon Yekutieli for some useful discussions.

2 Conventions

Throughout k is an algebraically closed field of characteristic zero. All rings are k-algebras. Modules are right modules and Cohen–Macaulay means maximal Cohen–Macaulay. When we say that R is complete local then we mean that R is a (commutative) complete local k-algebra with residue field k.

Below when we say "reduction to the complete case" we usually mean completing the strict Henselization of a commutative noetherian ring at a prime ideal and then lifting the residue field. Replacing k with this lifted field we are in a "complete local" situation, in the sense of the previous paragraph.

We also use it in the following sense: let R be a commutative noetherian ring and let $R \to K$ be a map of R to a field with kernel P. Then a complete local situation may be obtained by replacing R with a suitable flat complete local extension of the completion of R at P which has residue field K and which also contains K.

Below we will sometimes use the theory of dualizing complexes for schemes and for non-commutative coherent sheaves of rings over these. When the underlying scheme is of finite type or when it is the spectrum of a complete local ring then there is a canonical dualizing complex which we will call the "Grothendieck dualizing complex" and which we denote by D. One possible way of characterizing this dualizing complex is through the theory of "rigid dualizing complexes" [39, 41, 42] (and its topological variant). The duality functor associated to D will be denoted by $\mathbb{D}$.

We frequently use the following result by Keller [23]:

Theorem 2.1. *Let $\mathcal{E}$ be a Grothendieck category and assume that $\mathcal{A} = D(\mathcal{E})$ is generated by a compact object E (i.e., $E^{\perp} = 0$ and $\operatorname{Hom}_{\mathcal{A}}(E, -)$ commutes with direct sums). Then $\mathcal{A} = D(\Lambda)$ where Λ is a DG-algebra whose cohomology is given by $\operatorname{Ext}^*(E, E)$.*

In our case we will always have $\operatorname{Ext}^i(E, E) = 0$ for $i \neq 0$ and hence Λ is a true algebra. In that case we refer to E as a *tilting object*.

3 Background on Homologically Homogeneous Rings

It is well-known that for a non-commutative ring the condition of being of finite global dimension is very weak. In order to obtain a good homological theory generalizing the commutative case, a stronger condition is needed. Such a condition was introduced in [17].

Let R be a commutative noetherian k-algebra and let A be a module-finite R-algebra. We will say that A is *homologically homogeneous* if for all $P \in \operatorname{Spec} R$ we have

(H1) $\operatorname{gl\,dim} A_P = \dim R_P$.
(H2) A_P is (maximal) Cohen–Macaulay.

If R is equidimensional of dimension n this is is equivalent to the following condition given in [17].

(H1') All simple A-modules have the same projective dimension n.

Example 3.1. Consider the ring

$$A = \begin{pmatrix} R & m \\ R & R \end{pmatrix}$$

with $R = k[x, y]$, $m = (x, y) \subset R$. One has $\operatorname{gl\,dim} A = 2$ but A is not Cohen–Macaulay. So A is *not* homologically homogeneous.

Zero-dimensional h.h. rings are semi-simple. One-dimensional h.h. rings are classical hereditary orders [28]. Two dimensional h.h. rings over a complete local ring were classified in [2, 29]. The results in [3, 4, 5, 35, 36] may be viewed as a classification of three-dimensional graded local h.h. rings.

Not everything about homologically homogeneous rings is understood. For example it is an interesting question to understand precisely which rings can occur as centers of homologically homogeneous rings. In particular we would like to know the following:

Question 3.2. If A is homologically homogeneous and R is the center of A, is it true that R has rational singularities?

We know one special case where this question has an affirmative answer.

Proposition 3.3. *Assume that A is a finitely generated positively graded homologically homogeneous graded k-algebra with semi-simple degree zero part. Let R be the center of A and assume that R has an isolated singularity. Then R has rational singularities.*

Proof. The part of degree zero of R is a direct sum of fields. Using the resulting central idempotents we may reduce to the case where R is connected (i.e., $R_0 = k$). Put $n = \dim R$. If $n = 0$ then there is nothing to prove so we assume $n > 0$. Put $m = R_{>0}$ and $M = A_{>0}$. Since R has an isolated singularity, it will have a rational singularity if and only if the following condition holds [40]:

$$\begin{aligned} H^i_m(R) &= 0 \qquad \text{for } i < n \\ H^n_m(R)_u &= 0 \qquad \text{for } u \geq 0 \end{aligned} \tag{3.1}$$

Since we are in characteristic zero, R is a direct summand of A and hence it is certainly Cohen–Macaulay. Thus $H^i_m(R) = 0$ for $i < n$.

A regular sequence $x_1, \ldots, x_n$ in R is also a regular sequence in A since A is Cohen–Macaulay. Using this one easily shows that A has the following Gorenstein type property: if S is a simple right A module then $\operatorname{Ext}^i_A(S, A) = n$ for $i < 0$ and $T = \operatorname{Ext}^n_A(S, A)$ is a simple left module. Using a projective resolution of S we find that T lives in purely negative degree.

In [29] it is shown that every two-dimensional homologically homogeneous k-algebra over a complete local ring is obtained as the completion of a graded one. This is unfortunately false in higher dimension (otherwise Proposition 3.3 would yield an affirmative answer to Question 3.2).

Example 3.4. We construct a three-dimensional homologically homogeneous k-algebra which is not the completion of a graded one. Assume that R is a complete local ring with a three-dimensional terminal Gorenstein singularity which is *not* quasi-homogeneous but which is such that $X = \operatorname{Spec} R$ has a crepant resolution. Then it follows from Theorem 5.1 below that there is a reflexive R-module such that $A = \operatorname{End}_R(M)$ is homologically homogeneous. Clearly this A is not the completion of a graded ring since otherwise the same would be true for its center R also.

To make the example explicit consider $R = k[[x, y, z, t]]/(f)$ where

$$f = xy - \left(z - t^2\right)\left(z - t^3\right)\left(z - t^4\right)$$

This is a compound Du Val singularity which has a crepant resolution. Since R has a hypersurface singularity it will be quasi-homogeneous if and only if the Milnor number $\mu(f)$ and the Tyurina number $\tau(f)$ of f are equal [30]. Using the computer algebra program SINGULAR [19] we compute: $\mu(f) = 12$, $\tau(f) = 11$. So R is *not* quasi-homogeneous.

A more sensible approach to Question 3.2 is probably through the following:

Question 3.5. If A is a homologically homogeneous ring over a complete local ring, is there a separated filtration $A = A_0 \supset A_1 \supset \cdots$ on A with $A_1 = \operatorname{rad}(A)$ (maybe the $\operatorname{rad}(A)$-adic filtration(?)) such that $\operatorname{gr} A$ is still homologically homogeneous?

Since a deformation of a rational singularity is rational [18], it not hard to see that an affirmative answer to this question combined with Proposition 3.3 would yield an affirmative answer to Question 3.2.

Remark 3.6. It is tempting to try to develop directly a theory of rational singularities for non-commutative rings. For example (3.1) seems to make perfect sense for non-commutative graded rings. Unfortunately it is easy to see that (3.1) is not invariant under Morita-equivalence, so there are certainly some problems.

A related objection is that there seems to be no obvious non-commutative analogue of the Grauert–Riemenschneider vanishing theorem. It seems unlikely that there is a good theory of rational singularities without the Grauert–Riemenschneider theorem.

4 Non-commutative Crepant Resolutions

Unless otherwise specified, in the rest of this paper R is a normal Gorenstein domain and $X = \operatorname{Spec} R$. Let $f : Y \to X$ be a projective morphism with Y regular. We say that f is *crepant* if $f^*\omega_X = \omega_Y$.

Bondal and Orlov conjecture in [10] that different crepant resolutions of X have equivalent bounded derived categories of coherent sheaves. If X is projective then it is known that they have the same Hodge numbers [7, 25].

We propose a definition of a non-commutative crepant resolution. The motivation is mainly philosophical and the main example is given by the McKay correspondence [15] (see also [21] in dimension two).

Definition 4.1. A non-commutative crepant *resolution of R is an homologically homogeneous R-algebra of the form $A = \operatorname{End}_R(M)$ where M is a reflexive R-module.*

This definition will serve as a vehicle for introducing the various examples we discuss in this paper. Some obvious variations are possible (see Remark 4.3 below).

The following remark will be convenient.

Lemma 4.2. *If $A = \operatorname{End}_R(M)$ for a reflexive R-module M and if* $\operatorname{gl\,dim} A < \infty$ *and A is Cohen–Macaulay then A is homologically homogeneous (and hence A is a non-commutative crepant resolution).*

Proof. We may assume that R is local. Put $n = \dim R$. Using the fact that $\mathrm{Ext}^i_R(A, R) = 0$ for $i > 0$ (since A is Cohen–Macaulay) and $\mathrm{Hom}_A(A, R) \simeq A$ (since $A \simeq \mathrm{End}_R(M)$) we easily deduce for $M \in \mathrm{mod}(A)$: $\mathrm{Ext}^i_A(M, A) = \mathrm{Ext}^i_R(M, R)$. Thus the injective dimension of A_A is $\leq n$. Applying the formula with M a simple A-module obtain that the injective dimension of A_A is precisely n.

Since the global dimension of A equal to the injective dimension of A_A (given that it is finite) we find $\mathrm{gl}\dim A \leq n$. □

The point of Definition 4.1 is that it provides reasonable non-commutative substitutes for "regularity", "birationality" and "crepancy". Obviously regularity correspond to the condition $\mathrm{gl}\dim A < \infty$.

Let K be the function field of R. To have a substitute for birationality we note that in non-commutative geometry it is customary to replace a ring with its module category. So birationality should be expressed by the fact that $A \otimes_R K$ is Morita equivalent to K. I.e., we should have $A \otimes_R K = M_n(K)$.

Assume that R is complete local of dimension n and let ω_R be the dualizing module of R. By the Gorenstein hypotheses ω_R is an invertible R-module. The Grothendieck dualizing complex of R is given by $D_R = \omega_R[n]$. By the adjunction formula A has a dualizing complex given by $D_A = \mathrm{RHom}(A, D_R)$. It follows that A has a dualizing complex concentrated in one degree if an only if $\mathrm{Ext}^i_R(A, \omega_R) = 0$ for $i > 0$, i.e., if and only if A is Cohen–Macaulay. If A is Cohen–Macaulay then in particular it is reflexive.

If A is Cohen–Macaulay and has finite global dimension then it is homologically homogeneous. Hence it follows from [17] that if π is a height one prime in R then A_π is a hereditary order in $M_n(K)$ over the discrete valuation ring by R_π. We denote the ramification index [28] of A_π by $e(\pi)$.

If A is Cohen–Macaulay then the dualizing module of A is given by $\omega_A = \mathrm{Hom}_R(A, \omega_R)$ and it is easy to prove that the latter is equal to the tensor product of ω_R with $(\otimes_\pi \pi^{-1+e(\pi)})^{**}$, where the product runs over all height one primes in R. It follows that that ω_A will be generated by ω_R (a substitute for crepancy) if and only if $e(\pi) = 1$ for all π, i.e., if and only if A is a maximal order. According to [6] all maximal orders in $M_n(K)$ are of the form $\mathrm{End}_R(M)$ for a reflexive R-module M. This finishes our motivating discussion.

Remark 4.3. It is tempting to weaken Definition 4.1 in such a way as to require only that A is an unramified maximal order over R. The proof Theorem 6.3.1 below does not work in this added generality.

Remark 4.4. In all our examples the R-module M may be taken to be itself Cohen–Macaulay but we have no proof that this is always possible. Note that the implication $\mathrm{End}_R(M)$ Cohen–Macaulay $\Rightarrow$ M Cohen–Macaulay is false.

Remark 4.5. If our basescheme X is not affine then we may define a non-commutative crepant resolution of X as a stack of abelian categories which is, locally on an affine open $\mathrm{Spec}\, R$, of the form $\mathrm{Mod}(A)$ where A is a non-commutative crepant resolution of R.

The following conjecture is inspired by the conjecture of Bondal and Orlov.

Conjecture 4.6. All crepant resolutions of X (commutative as well as non-commutative ones) are derived equivalent.

In subsequent sections we will give some examples of non-commutative crepant resolutions and in addition we will give some evidence for Conjecture 4.6. In particular we will prove Conjecture 4.6 for three dimensional terminal Gorenstein singularities.

5 Resolutions with Fibers of Dimension ≤ 1

The following result was proved is [37] (our hypotheses here are slightly more general but this does not affect the proof).

Theorem 5.1. *Assume that there exists a crepant resolution of singularities $f: Y \to X$ such that the dimensions of the fibers of f are ≤ 1 and such that the exceptional locus of f has codimension ≥ 2. Then R has a non-commutative crepant resolution $A = \operatorname{End}_R(M)$ where M is in addition Cohen–Macaulay. Furthermore Y and A are derived equivalent.*

Let us briefly recall how M is constructed. The Grauert–Riemenschneider theorem implies $H^i(Y, \mathcal{O}_Y) = 0$ for $i > 0$ (thus X has rational singularities).

Let $\mathcal{L}$ be an ample line bundle on Y generated by global sections. The hypotheses on the fibers of f imply that $(\mathcal{O}_Y \oplus \mathcal{L})^\perp = 0$ in $D(\operatorname{Qch}(Y))$. Take an extension

$$0 \to \mathcal{O}_Y^r \to \mathcal{M}' \to \mathcal{L} \to 0$$

associated to a set of r generators of $\operatorname{Ext}^1_Y(\mathcal{L}, \mathcal{O}_Y)$ as R-module and put $\mathcal{M} = \mathcal{M}' \oplus \mathcal{O}_Y$. Then $\mathcal{M}^\perp = (\mathcal{O}_Y \oplus \mathcal{L})^\perp = 0$ and furthermore $\operatorname{Ext}^i_A(\mathcal{M}, \mathcal{M}) = 0$ for $i > 0$.

Put $M = \Gamma(Y, \mathcal{M})$. The hypotheses imply that $A = \operatorname{End}_R(M) = \operatorname{End}_Y(\mathcal{M})$. Hence Y is derived equivalent to A. In particular A has finite global dimension. For the fact that A and M are Cohen–Macaulay, we refer to [37].

6 Construction a Crepant Resolution Starting from a Non-commutative One

6.1 Introduction. In this section we assume that $A = \operatorname{End}_R(M)$ is an arbitrary non-commutative crepant resolution of R. We will show that the beautiful approach in [15] to the McKay correspondence generalizes to this situation. In fact an almost literal copy of the proof works, but we will nevertheless give a summary of it, in order to convince the reader (and ourselves!) that nothing specific to the situation of G-equivariant sheaves is used in [15] (see Sects. 6.3–6.6). The only part that is specific to our more general situation is in Sect. 6.3.

An obvious first step in our more general situation is to take for Y some type of A-Hilbert-scheme of M. This would work provided that M is Cohen–Macaulay. Unfortunately we don't know if this is always the case (but as already said, it is true in all examples we know).

Therefore we take a slightly different approach. We construct Y as a moduli-space of certain stable A-representations. It is standard how to do this but since our base ring R is somewhat more general than usual we recall the necessary steps in the next section.

6.2 Moduli spaces of representations. In this section R is a commutative noetherian k-algebra where as usual k is algebraically closed of characteristic zero. Put $X = \operatorname{Spec} R$. Let A be an R-algebra which is finitely generated as R-module. Let $(e_i)_{i=1,\dots,p}$ be pairwise orthogonal idempotents in A such that $1 = \sum_i e_i$ and let $D = \oplus_i Re_i \subset A$ be the corresponding diagonal subalgebra.

For a map $R \to K$ with K a field and V a finite dimensional $A \otimes_R K$ representation we write $\underline{\dim}\, V = (\dim_K Ve_i)_i \in \mathbb{Z}^p$. We put on $\mathbb{Z}^p$ the ordinary Cartesian scalar product.

Pick $\lambda \in \mathbb{Z}^p$ and let $\alpha = \underline{\dim}\, V$. Following [24] We say that V is *(semi-)stable* (with respect to λ) if

$$(\lambda, \alpha) = 0 \tag{6.1}$$

and if for any proper subrepresentation W of V with $\beta = \underline{\dim}\, W$ we have

$$(\lambda, \beta)\ (\geq) > 0 \tag{6.2}$$

Note that if

$$(\lambda, \beta) \neq 0 \qquad \text{for } 0 < \beta < \alpha \tag{6.3}$$

then stability and semi-stability are equivalent. For a fixed α there will exist λ satisfying (6.1) and (6.2) if and only if the greatest common divisor of all α_i is one.

An *affine family* of A representations with dimension vector α is a commutative R-algebra T together with a finitely generated $A \otimes_R T$ module P which is projective as T-module such that Pe_i has constant rank α_i for all i. For such a P we write $\underline{\dim}\, P = \alpha$. This is equivalent to saying that for any map of T to a field K we have $\underline{\dim}(P \otimes_T K) = \alpha$. We say that P is (semi-)stable if for any K we have that $P \otimes_R K$ is (semi-)stable. Non-affine families are defined in the obvious way by gluing affine families. We call families *equivalent* if they are locally isomorphic.

Let $\alpha = (\alpha_i)_{i=1,\dots,p}$ be natural numbers which are relatively prime and pick a λ satisfying (6.1) and (6.3). Consider the functor $\mathbf{R}^s$ which assigns to a commutative R-scheme Z the following set:

$$\{\text{equivalence classes of families of } \lambda\text{-stable } A\text{-representations over } Z \text{ with dimension vector } \alpha\}$$

Proposition 6.2.1. *The functor $\mathbf{R}^s$ is representable by a projective scheme over X.*

We need the following lemma.

Lemma 6.2.2. *Assume that R' is a commutative R-algebra and that T is a commutative R'-algebra. Put $A' = A \otimes_R R'$. Assume that P is a T-family of stable A'-representations. Then P is also stable as a a family of A-representations.*

Proof. This is trivial from the definition.

If we temporarily write $\mathbf{R}^s_R$ for $\mathbf{R}^s$ then this lemma implies that $\mathbf{R}^s_{R'} = \mathbf{R}^s_R \times_{\operatorname{Spec} R} \operatorname{Spec} R'$. Hence if $\mathbf{R}^s_R$ is representable by a projective scheme over R then so is $\mathbf{R}^s_{R'}$.

We use this as follows: we may find a finitely generated subring R_0 of R and a module finite R_0 algebra A_0 such that $A = A_0 \otimes_{R_0} R$. It is then sufficient to prove that $\mathbf{R}^s_{R_0}$ is representable by a projective scheme over R_0. Hence without loss of generality we may (and we will) assume that R is finitely generated over k.

Put $\bar{\alpha} = \sum_i \alpha_i$. Define A' as the centralizer of $M_{\bar{\alpha}}(k)$ in $M_{\bar{\alpha}}(R) *_D A$. So we have

$$M_{\bar{\alpha}}(R) *_D A = M_{\bar{\alpha}}(k) \otimes A' \tag{6.4}$$

where the obvious copy of $M_{\bar{\alpha}}(k)$ on the left and right is the same.

Note that A' is an R-algebra. Since $M_{\bar{\alpha}}(R) *_D A = M_{\bar{\alpha}}(A')$ is finitely generated the same is true for A'.

Put $S' = A'/[A', A']$ and $W' = \operatorname{Spec} S'$. W' is an affine R-scheme of finite type representing the functor [31] which assigns to a commutative R-algebra T:

$$\{A\text{-module structures on } T^{\bar{\alpha}} \text{ which commute with the } T\text{-structure such that } e_i \text{ acts as } \operatorname{diag}(0^{\alpha_1+\cdots+\alpha_{i-1}}, 1^{\alpha_i}, 0^{\alpha_{i+1}+\cdots+\alpha_p})\} \tag{6.5}$$

The corresponding universal bundle is given by $U'_0 = k^{\bar{\alpha}} \otimes_k S$ with the right action of A obtained via the composition $A \to M_{\bar{\alpha}}(R) *_D A = M_{\bar{\alpha}}(A') \to M_{\bar{\alpha}}(S')$.

Let $G = \prod_i \mathrm{GL}_{\alpha_i}(k)$ and let PG be equal to G modulo its center. Conjugation on the first factor induces a G action on $M_{\bar{\alpha}}(R) *_D A$, leaving the elements of A invariant.

The G action on $M_{\bar{\alpha}}(R) *_D A$ leaves $M_{\bar{\alpha}}(k)$ stable and hence it induces rational G-actions on A', S' and W'. The G-action on the righthand side of (6.4) is the diagonal one.

Putting $g(v \otimes \gamma) = vg^{-1} \otimes g\gamma$ defines a rational G-action on U'_0. The center of G acts trivially on S' and via scalar multiplication on U'_0. Let $a \in \mathbb{Z}^p$ be such that $(\alpha, a) = 1$ and define $U' = U'_0 \otimes_S \bigotimes_i \wedge^{\alpha_i}(U'_0 e_i)^{\otimes - a_i}$. U' is still a $A \otimes_R S'$ module which is projective as S'-module and which has dimension vector α. The center of G now acts trivially on U'.

As usual the stability condition (6.2) corresponds to a character $\chi : G \to k^*$ [24]. This character defines a $\mathbb{Z}$-grading on S'.

Let $G_0 = \ker \chi$ and put $S = (S')^{G_0}$, $U = (U')^G$. Then S is still $\mathbb{Z}$-graded and U is a graded S-module. Put $W = \operatorname{Proj} S_{\geq 0}$ and let $f : W \to X$ be the structure map. If W'^s is the open subset of W corresponding to the complement of the closed subscheme defined by $S_{>0}$ then it follows from the Luna slice theorem [26] that $W'^s \to W$ is a principal PG-fiber bundle.

Let $\mathcal{U}$ be the coherent sheaf on W which corresponds to U. $\mathcal{U}$ is a sheaf of right A-modules. It follows from standard descent theory that the pullback of $\mathcal{U}$ to W'^s is the restriction of U' (considered as a coherent sheaf on W'). In particular $\mathcal{U}$ is a vector bundle on W.

It is now standard that W represents the functor $\mathbf{R}^s$ and that $\mathcal{U}$ is the corresponding universal bundle. See [27] for the case of vector bundles.

Lemma 6.2.3. *If $A = M_{\bar{\alpha}}(R)$, then the map $W \to X$ is an isomorphism.*

Proof. It follows from Morita theory that in this case the functor $\mathbf{R}^s$ is represented by X. So this proves what we want. □

6.3 Application to our situation. Now we return to our standard assumption that R is a normal Gorenstein domain and we let $A = \operatorname{End}_R(M)$ be an arbitrary non-commutative crepant resolution of R.

Since A has finite global dimension there is a A-resolution

$$0 \to P \to A^a \to \cdots \to A^b \to M \to 0 \tag{6.6}$$

with P projective.

By the reflexive Morita correspondence we have $P = \operatorname{Hom}_R(M, M_1)$ where

$$M_1 \oplus M_2 \cong M^{\oplus c} \tag{6.7}$$

for some $c \in \mathbb{N}$ and some other reflexive R-module M_2.

From (6.6) we obtain $\operatorname{rk} M = d \operatorname{rk} A \pm \operatorname{rk} P$ for some $d \in \mathbb{Z}$, where "$\operatorname{rk} U$" denotes the rank of U as R-module. Simplifying we find $1 = d \operatorname{rk} M \pm \operatorname{rk} M_1$ and hence the ranks of M and M_1 are coprime. Increasing c if necessary we may also assume that the ranks of M_1 and M_2 are coprime.

We now replace M by $M^{\oplus c}$. This changes A into something Morita equivalent and the decomposition (6.7) becomes $M = M_1 \oplus M_2$.

Let us more generally consider a decomposition $M = \oplus_{i=1}^p M_i$. Denote the rank of M_i by α_i and put $\alpha = (\alpha_i)_i$. Take λ satisfying (6.1) and (6.3) and let $f : W \to X$ and $\mathcal{U}$ be as in the previous section.

If we let $X_1 \subset X = \operatorname{Spec} R$ be the locus where M is locally free then it follows from lemma 6.2.3 that $f^{-1}(X_1) \to X_1$ is an isomorphism. We define Y as the unique irreducible component of W mapping onto X. We will denote the restriction of f to Y also by f. We let $\mathcal{M}$ be the restriction of $\mathcal{U}$ to Y. $\mathcal{M}$ is a sheaf of A-modules on Y.

Following [15] we now define a pair of adjoint functors between $D^b(\operatorname{coh}(Y))$ and $D^b(\operatorname{mod}(A))$.

$$\Phi : D^b(\operatorname{coh}(Y)) \to D^b(\operatorname{mod}(A)) : C \mapsto R\Gamma\left(C \overset{L}{\otimes}_{\mathcal{O}_Y} \mathcal{M}\right)$$

$$\Psi : D^b(\operatorname{mod}(A)) \to D^b(\operatorname{coh}(Y)) : D \mapsto D \overset{L}{\otimes}_A \mathcal{M}^*$$

The following is a straightforward generalization of [15].

Theorem 6.3.1. *Assume that for any point $x \in X$ of codimension n the fiber product $(Y \times_X Y) \times_X \operatorname{Spec} \mathcal{O}_{X,x}$ has dimension $\leq n+1$. Then $f : Y \to X$ is a crepant resolution of X and Φ and Ψ are inverse equivalences.*

To give the proof we need to say something about spanning classes and Serre functors. This is done in the next section.

6.4 Relative Serre duality. Assume that $\mathcal{D} \subset \mathcal{C}$ is a full faithful inclusion of k-linear triangulated categories such that for $C \in \mathcal{C}$, $D \in \mathcal{D}$ we have

$$\sum_i \dim \operatorname{Ext}^i(C, D) < \infty, \quad \sum_i \dim \operatorname{Ext}^i(D, C) < \infty.$$

We say that an auto-equivalence of triangulated categories $S : \mathcal{D} \to \mathcal{D}$ is a relative Serre functor for the pair $(\mathcal{D}, \mathcal{C})$ if the following properties hold:

1. S leaves $\mathcal{D}$ stable.
2. For $C \in \mathcal{C}$, $D \in \mathcal{D}$ there are natural isomorphisms

$$\eta_{D,C} : \operatorname{Hom}(D, C) \to \operatorname{Hom}(C, SD)^* \tag{6.8}$$

Thus in particular S is Serre functor for $\mathcal{D}$.

We will use this in the following situation.

Lemma 6.4.1. *Assume that $X = \operatorname{Spec} R$ where R is a complete local noetherian ring and let $f : Y \to X$ be a projective map. Let $\mathcal{A}$ be a coherent sheaf of $\mathcal{O}_Y$ algebras such that for every $y \in Y$ the stalk of $\mathcal{A}$ at y has finite global dimension. Let $\mathcal{C}$ be the bounded derived category of $\operatorname{coh}(\mathcal{A})$ and let $\mathcal{D}$ be the full subcategory of complexes whose homology has support in $Y_0 = f^{-1}(x)$ where $x \in X$ is the closed point. Then the pair $(\mathcal{D}, \mathcal{C})$ has a relative Serre functor given by tensoring with the Grothendieck dualizing complex $D_{\mathcal{A}} = \mathrm{R}\mathcal{H}om_Y(\mathcal{A}, D_Y)$ of $\mathcal{A}$.*

Proof. We claim first that if $D \in \mathcal{D}$, $C \in \mathcal{C}$ then we have the following identity:

$$\operatorname{Hom}(D, D_Y) \cong \Gamma(Y, D)^* \tag{6.9}$$

Let Y_n be the n'th formal neighborhood of Y_0 and let $j_n : Y_n \to X$ be the inclusion map.

It is easy to see that there is some n and some $D_n \in D^b(\operatorname{coh}(Y_n))$ such that $D = Rj_{n*}D_n$. Then we have $\operatorname{Hom}(D, D_Y) = \operatorname{Hom}(Rj_{n*}D_n, D_Y) = \operatorname{Hom}(D_n, j_n^!(D_Y)) = \operatorname{Hom}(D_n, D_{Y_n})$. Now since Y_n is proper it satisfies classical Serre duality. I.e., there are isomorphisms $\operatorname{Hom}(D_n, D_{Y_n}) \cong \Gamma(Y_n, D_n)^* = \Gamma(Y, D)^*$. It is easy to see that the resulting isomorphism $\operatorname{Hom}(D, D_Y) \cong \Gamma(Y, D)^*$ is independent of n. This finishes the proof of (6.9).

We compute for $D \in \mathcal{D}$, $C \in \mathcal{C}$

$$\begin{aligned} \operatorname{Hom}_{\mathcal{A}}(D, C) &= \Gamma(Y, \mathrm{R}\mathcal{H}om_{\mathcal{A}}(D, C)) \\ &\cong \operatorname{Hom}(\mathrm{R}\mathcal{H}om_{\mathcal{A}}(D, C), D_Y)^* \qquad \text{(by (6.9))} \end{aligned} \tag{6.10}$$

We claim that there are natural isomorphism

$$\mathrm{R}\mathcal{H}om(\mathrm{R}\mathcal{H}om_{\mathcal{A}}(D, C), D_Y)) \cong \mathrm{R}\mathcal{H}om_{\mathcal{A}}\left(C, D \overset{L}{\otimes}_{\mathcal{A}} D_{\mathcal{A}}\right) \tag{6.11}$$

This is easily proved by replacing D_Y with a bounded injective complex and D with a locally projective complex (using the projectivity of f).

So combining (6.10) and (6.11) we obtain an isomorphism

$$\mathrm{Hom}_{\mathcal{A}}(D, C) \cong \mathrm{Hom}_{\mathcal{A}}\left(C, D \overset{L}{\otimes}_{\mathcal{A}} D_{\mathcal{A}}\right)^*$$

which finishes the proof.

6.5 Spanning classes. If $\mathcal{C}$ is a triangulated category then a *spanning class* [13] $\Omega \subset \mathcal{C}$ is a set of objects such that $\Omega^{\perp} = 0$ and ${}^{\perp}\Omega = 0$.

Example 6.5.1. Let $f : Y \to X$ be a *proper* map where X is the spectrum of a noetherian local ring with algebraically closed residue field k. Then $\Omega = \{\mathcal{O}_y \mid y \in Y(k)\}$ is a spanning class for $D^b(\mathrm{coh}(Y))$.

Assume that $F : \mathcal{C} \to \mathcal{E}$ is a functor between pairs of triangulated categories. Assume that F has a left adjoint G and a right adjoint H and that $\Omega \subset \mathcal{C}$ is a spanning class.

Lemma 6.5.2. *[12][13] $F : \mathcal{C} \to \mathcal{E}$ is fully faithful if and only if the map*

$$\mathrm{Ext}^i_{\mathcal{C}}(\omega, \omega') \to \mathrm{Ext}^i_{\mathcal{E}}(F\omega, F\omega') \tag{6.12}$$

is an isomorphism for all $\omega, \omega' \in \Omega$.

Suppose now that F is actually a compatible pair of fully faithful functors $(\mathcal{D}, \mathcal{C}) \to (\mathcal{F}, \mathcal{E})$ between pairs of triangulated categories satisfying the hypotheses given in the beginning of Sect. 6.4 which in addition have relative Serre functors $S_{\mathcal{C}}$ and $S_{\mathcal{E}}$. Assume $\Omega \subset \mathcal{D}$.

Lemma 6.5.3. *Assume that $\mathcal{C}$ is not trivial and that $\mathcal{E}$ is connected. Assume in addition that $S_{\mathcal{E}}F\omega = FS_{\mathcal{C}}\omega$ for $\omega \in \Omega$. Then F is an equivalence of categories.*

Proof. The proof is the same as in [13] with Serre functors being replaced by relative ones. □

6.6 Proof of the main theorem. We now complete the proof of Theorem 6.3.1. We need to show that the canonical natural transformations $\Psi\Phi \to \mathrm{id}$ and $\mathrm{id} \to \Phi\Psi$ are isomorphisms. Since everything is compatible with base change we may reduce to the case where R is a complete local ring containing a copy of its algebraically closed residue field. We will denote this new residue field also by k. We let x be the unique closed point of $X = \operatorname{Spec} R$.

The functors Φ and Ψ have versions for left modules, denoted by Φ° and Ψ° respectively, which are given by the formulas $R\Gamma(Y, \mathcal{M}^* \overset{L}{\otimes}_{\mathcal{O}_Y} -)$ and $\mathcal{M} \overset{L}{\otimes}_A -$. It it easy to see that $\Phi^\circ = \mathbb{D}_A \circ \Phi \circ \mathbb{D}_Y$. Hence Φ also has a right adjoint given by $\mathbb{D}_Y \circ \Psi^\circ \circ \mathbb{D}_A$. Furthermore as in Example 6.5.1 the objects $\mathcal{O}_y$ form a spanning class for $D^b(\mathrm{coh}(Y))$.

For $y \in Y(k)$ denote by $\mathcal{M}_y = \Phi(\mathcal{O}_y)$ the fiber of $\mathcal{M}$ at y. To prove that Φ is fully faithful we need to prove that the canonical maps

$$\mathrm{Ext}^i_Y(\mathcal{O}_y, \mathcal{O}_{y'}) \to \mathrm{Ext}^i_A(\mathcal{M}_y, \mathcal{M}_{y'}) \tag{6.13}$$

are isomorphisms for $y, y' \in Y$.

What do we know already?

1. (6.13) is certainly an isomorphism for $i = 0$.
2. By Serre duality for A and the fact that $D_A = A[n]$ (Lemma (6.4.1)) is an isomorphism if $i = n$ and $y \neq y'$.

There is one more subtle piece of information that may be obtained. Recall that Y is a closed subscheme of the scheme W representing the functor of stable A-representations. This means that there is an injection (the Kodaira–Spencer map):

$$\phi : \mathrm{Ext}^1_Y(\mathcal{O}_y, \mathcal{O}_y) = T_{Y,y} \hookrightarrow T_{W,y} = \mathrm{Ext}^1_W(\mathcal{M}_y, \mathcal{M}_y)$$

where $T_{*,y}$ denotes the tangent space at y. The map ϕ is constructed as follows. An element of $T_{Y,y}$ corresponds to a map $u : \mathrm{Spec}\, k[\epsilon]/(\epsilon^2) \to Y$ and hence to an extension E of $\mathcal{O}_y$ with itself. Then $\phi(u) = u^*(E)$. Thus the Kodaira–Spencer map coincides with (6.13) for $i = 1$ and $y = y'$. Hence we have

3. (6.13) is an injection for $i = 1$ and $y = y'$.

Using an amazing trick based on the intersection theorem in commutative algebra it is shown in [15] that (1),(2) and (3) are sufficient to prove that Φ is fully faithful (under the standing hypothesis $\dim Y \times_X Y \leq n + 1$).

We need a succinct description of $\Psi\Phi$. Let $Y \hat{\times} Y = (Y \times Y)_{X \times X} \mathrm{Spec}\, \hat{\mathcal{O}}_{X \times X,(x,x)}$. This a noetherian scheme proper over $\mathrm{Spec}\, \hat{\mathcal{O}}_{X \times X,(x,x)}$. We denote the projections $Y \hat{\times} Y \to Y$ by $\mathrm{pr}_{1,2}$.

$Y \times_X Y$ may be considered as a closed subscheme of $Y \hat{\times} Y$ (no need to complete). We define $\mathcal{M} \boxtimes_A \mathcal{M}^*$ as the coherent sheaf on $Y \hat{\times} Y$ such that for affine opens $U, V \subset Y$ we have $(\mathcal{M} \boxtimes_A \mathcal{M}^*)(U \hat{\times} V) = \mathcal{M}(U) \otimes_A \mathcal{M}(V)$. $\mathcal{M} \boxtimes_A \mathcal{M}^*$ is clearly supported on $Y \times_X Y$. Using suitable flat resolutions we may define the analogous derived object $\mathcal{Q} = \mathcal{M} \overset{L}{\boxtimes}_A \mathcal{M}^*$ which is also supported on $Y \times_X Y$. Then it is easy to see that

$$\Psi\Phi = R\,\mathrm{pr}_{2*}(L\,\mathrm{pr}_1^*(-) \otimes_{Y \hat{\times} Y} \mathcal{Q})$$

We have $\mathcal{O}_{y,y'} \otimes_{Y \hat{\times} Y} \mathcal{Q} = \mathcal{M}_y \overset{L}{\otimes}_A \mathcal{M}^*_{y'}$. Using Lemma 6.4.1 for A we have $\mathcal{M}^*_{y'} = \mathrm{Hom}_A(A, \mathcal{M}_{y'})^* = \mathrm{Hom}_A(\mathcal{M}_{y'}, A[n])$ and since the righthand side of the last equality has homology only in degree zero, it is equal to $\mathrm{RHom}(\mathcal{M}_{y'}, A[n])$. Thus $\mathcal{M}_y \overset{L}{\otimes}_A \mathcal{M}^*_{y'} = \mathrm{RHom}_A(\mathcal{M}_{y'}, \mathcal{M}_y)[n]$.

Thus if $y \neq y'$ then $H^{-n}(\mathcal{O}_{y,y'} \overset{L}{\otimes}_{Y\hat{\times}Y} \mathcal{Q}) = 0$ and using relative Serre duality for A again we also have $H^0(\mathcal{O}_{y,y'} \overset{L}{\otimes}_{Y\hat{\times}Y} \mathcal{Q}) = \mathrm{Ext}^n_A(\mathcal{M}_y, \mathcal{M}_{y'}) = \mathrm{Hom}_A(\mathcal{M}_{y'}, \mathcal{M}_y) = 0$. So the range of possible non-zero values for $H^i(\mathcal{O}_{y,y'} \overset{L}{\otimes}_{Y\hat{\times}Y} \mathcal{Q}) = 0$ has size $n-1$. But this implies by the intersection theorem (see [15]) that on the complement of the diagonal $\mathcal{Q}$ has support of dimension $\geq \dim(Y\hat{\times}Y) + 1 - (n-1) = n+2$ (if non-empty). But by hypotheses the support of $\mathcal{Q}$ has dimension less than or equal the dimension of $Y \times_X Y$ which is $n+1$. We conclude that $\mathcal{Q}$ is supported on Δ.

Consider $\Psi\Phi\mathcal{O}_y$. By the previous paragraph this is a complex supported on y living in non-positive degree and we have $\mathrm{Ext}^i(\Psi\Phi\mathcal{O}_y, \mathcal{O}_y) = \mathrm{Ext}^i(\mathcal{M}_y, \mathcal{M}_y)$ which is non-zero only in degrees $[0, \ldots, n]$. We claim $H^0(\Psi\Phi\mathcal{O}_y) = \mathcal{O}_y$.

Let $c_y[1]$ be the cone over $\Psi\Phi\mathcal{O}_y \to \mathcal{O}_y$. Thus we have a triangle

$$c_y \to \Psi\Phi\mathcal{O}_y \to \mathcal{O}_y \tag{6.14}$$

By the previous paragraph c_y is supported in y. If is also easy to see that the homology of c_y is concentrated in non-positive degree. There is an exact sequence

$$\begin{aligned}0 \to \mathrm{Hom}(\mathcal{O}_y, \mathcal{O}_y) \to \mathrm{Hom}(\mathcal{M}_y, \mathcal{M}_y) \to \mathrm{Hom}(c_y, \mathcal{O}_y)\\ \to \mathrm{Ext}^1(\mathcal{O}_y, \mathcal{O}_y) \to \mathrm{Ext}^1(\mathcal{M}_y, \mathcal{M}_y)\end{aligned}$$

and using (1)(3) we conclude $\mathrm{Hom}(c_y, \mathcal{O}_y) = 0$ and hence $H^0(c_y) = 0$. The fact that $H^0(\Psi\Phi\mathcal{O}_y) = \mathcal{O}_y$ now follows from (6.14). By the intersection theorem (see [15]) we conclude that Y is regular at y and that $\Psi\Phi\mathcal{O}_y = \mathcal{O}_y$.

Since this is true for all y we now know that Y is regular and that (6.13) holds. Thus Φ is faithful.

To prove that Φ is an equivalence we use lemma 6.5.3 since $D^b(\mathrm{mod}(A))$ is trivially connected. We need that $\Phi S_Y \mathcal{O}_y \cong S_A \mathcal{M}_y$. Now S_A is just shifting n places to the left, and since Y is regular ω_Y is invertible and thus $S_Y\mathcal{O}_y$ is just $\mathcal{O}_y[n]$.

Finally we prove that f is crepant. It is sufficient to prove that $\omega_Y \cong \mathcal{O}_Y$. Indeed if this is the case then $f_*\omega_Y \cong \mathcal{O}_X$ is reflexive and hence it is equal to ω_X. Furthermore it is then also clear that $f^*\omega_X = \omega_Y$.

Let $D^b_x(\mathrm{coh}(Y))$ the full subcategory of $D^b(\mathrm{coh}(Y))$ consisting of complexes supported in $f^{-1}(x)$. Similarly let $D^b_x(\mathrm{mod}(A)) \subset D^b(\mathrm{mod}(A))$ be the complexes supported on x. The functor Φ and Ψ define inverse equivalences between $D^b_x(\mathrm{coh}(Y))$ and $D^b_x(\mathrm{mod}(A))$.

On $D^b_x(\mathrm{coh}(Y))$ we have that $S_Y[-n]$ is isomorphic to the identity functor since the same holds for $D^b_x(\mathrm{mod}(A))$. Thus if $Y_0 = f^{-1}(x)$ then $\omega_Y/\omega_Y(-nY_0) \cong \mathcal{O}_Y/\mathcal{O}_Y(-nY_0)$. Hence if $\hat{Y}$ is the formal scheme associated to Y_0 then $\hat{\omega}_Y \cong \hat{\mathcal{O}}_Y$. But by the Grothendieck existence theorem this implies $\omega_Y \cong \mathcal{O}_Y$.

Remark 6.6.1. As noted in [15] the fact that Y is smooth and the fact that for $y \in Y$ we have $T_{Y,y} = \mathrm{Ext}^1_Y(\mathcal{O}_y, \mathcal{O}_y) \cong \mathrm{Ext}^1_A(\mathcal{M}_y, \mathcal{M}_y) = T_{W,y}$ implies that Y is a connected component of W. If $\dim X = 3$ then it is shown that actually $W = Y$. This result generalizes probably not to our current situation. However if R is complete local and we take a decomposition of M into indecomposables (unique up to non-unique isomorphism by the Krull–Schmidt-theorem) then the proof goes through virtually unmodified.

Remark 6.6.2. Varying λ we get many different crepant resolutions of X. They are all derived equivalent since they are all derived equivalent to A. This gives another instance where the Bondal–Orlov conjecture is true.

Now we may prove.

Theorem 6.6.3. *Assume that R is three-dimensional and has terminal singularities.*

(1) R has a non-commutative crepant resolution if and only if it has a commutative one.
(2) Conjecture 4.6 is true in this case.

Proof. (1) This follows from Theorems 5.1 and 6.3.1.
(2) Let $Y \to X$ be a crepant resolution of singularities and let A be a non-commutative one. By Theorem 6.3.1 there is another crepant resolution $Y' \to X$ of X associated to A. The resulting birational map $Y \dashrightarrow Y'$ is a composition of flops. Hence by [14] Y and Y' are derived equivalent. Since Y' and A are also derived equivalent, we are done. □

7 Cones over Del Pezzo Surfaces

A standard example of a canonical singularity which is not terminal is the cubic cone $w^3 + x^3 + y^3 + z^3 = 0$. This singularity has a crepant resolution obtained by blowing up the origin. Our aim in this section is to show that it also has a non-commutative crepant resolution. The method used applies more generally to cones over del Pezzo surfaces (see Proposition 7.3) below.

Below Z is regular connected projective scheme of dimension $n - 1 > 0$ with an ample line bundle $\mathcal{L}$. We denote the homogeneous coordinate ring $\oplus_i \Gamma(Z, \mathcal{L}^i)$ of Z corresponding to $\mathcal{L}$ by R. R is a finitely generated normal graded ring. Put $X = \operatorname{Spec} R$. We will call R a cone over Z. X has unique singularity at the origin o. This singularity has a standard resolution by $Y = \underline{\operatorname{Spec} S(\mathcal{L})}$. Denote the structure map $Y \to Z$ by π and let f be the canonical map $Y \to X$. Thus we have the following diagram

$$\begin{array}{ccc} Y & \xrightarrow{\pi} & Z \\ {\scriptstyle f}\downarrow & & \\ X & & \end{array}$$

Let $X' \subset X$, $Y' \subset Y$ be the open subsets respectively defined by the ideals $R_{>0}$ and $(S\mathcal{L})_{>0}$. X' is regular and $X - X' = \{o\}$. We also have $Y - Y' = E$ where E is the image of the zero section on π. The map f restricts to an isomorphism $Y' \to X'$ and so the exceptional locus of f is given by E.

We now state a list of properties of X which are well-known and which are easy to prove.

Lemma 7.1. (1) *X is Cohen–Macaulay if and only if*

$$H^i(Z, \mathcal{L}^j) = 0 \qquad \text{for all } 0 < i < n-1,\ j \geq 0$$

(2) *X has rational singularities if and only if*

$$H^i(Z, \mathcal{L}^j) = 0 \qquad \text{for all } i > 0,\ j \geq 0$$

(3) *X has an invertible canonical bundle if and only if $\omega_Z = \mathcal{L}^{-m}$ for $m \in \mathbb{Z}$. In that case $\omega_Y = f^*\omega_X((m-1)E)$.*

(4) *If X has Gorenstein rational singularities then $\omega_Z = \mathcal{L}^{-m}$ for some $m > 0$.*

(5) *If X has Gorenstein rational singularities then f is crepant (or equivalently X is terminal) if and only if $\omega_Z \not\cong \mathcal{L}^{-1}$.*

Now assume that there is a vector bundle $\mathcal{E}_0$ on Z which is a tilting object. Put $\mathcal{E} = \pi^*\mathcal{E}_0$. Then $\mathcal{E}$ is a tilting object on Y if and only if

$$H^i(Z, \mathcal{E}nd(\mathcal{E}_0, \mathcal{E}_0) \otimes \mathcal{L}^j) = 0 \qquad \text{for } i > 0 \text{ and } j \geq 0 \tag{7.1}$$

Thus if (7.1) holds and we put $A = \operatorname{End}(\mathcal{E})$ then A and Y are derived equivalent.

We claim the following result.

Proposition 7.2. *Assume that $\mathcal{E}_0$ is a vector bundle on Z which is a generator of $D(\operatorname{Qch}(Z))$ such that $\mathcal{E} = \pi^*\mathcal{E}_0$ is a tilting object on Y. Then $A = \operatorname{End}(\mathcal{E})$ is a non-commutative crepant resolution of X if and only if the following condition holds:*

$$H^i\left(Z, \mathcal{A}_0 \otimes \mathcal{L}^j\right) = 0 \qquad \text{for } i < n-1 \text{ and } j < 0 \tag{7.2}$$

for $\mathcal{A}_0 = \mathcal{E}nd(\mathcal{E}_0)$.

If $\omega_Z = \mathcal{L}^{-1}$ then (7.2) is always satisfied.

Proof. For $\mathcal{M} \in \operatorname{Qch}(Z)$ put $\underline{\Gamma}(Z, \mathcal{M}) = \oplus_{j\in\mathbb{Z}} \Gamma(Z, \mathcal{M} \otimes \mathcal{L}^j)$ and denote the derived functors of $\underline{\Gamma}$ by $\underline{H}^*$.

If $\mathcal{M}$ is associated to $M \in \operatorname{Gr}(R)$ then it is well-known that we have

$$\underline{H}^i(Z, \mathcal{M}) = H^{i+1}_{R_{>0}}(M) \qquad \text{for } i > 0 \tag{7.3}$$

and there is a long exact sequence

$$0 \to H^0_{R_{>0}}(M) \to M \to \underline{\Gamma}(Z, \mathcal{M}) \to H^1_{R_{>0}}(M) \to 0 \tag{7.4}$$

Now we prove the first part of the proposition. Since A is derived equivalent to Y we already know $\operatorname{gl}\dim A < \infty$, so we only have to be concerned with the Cohen–Macaulayness of A. The latter is equivalent to $H^i_{R>0}(A) = 0$ for $i \leq n-1$ (since the dimension of R is n).

Since we assume (7.1) the condition (7.2) is equivalent to

$$\begin{aligned} \underline{H}^i(Z, \mathcal{A}_0) &= 0 \qquad \text{for } 0 < i < n-1 \\ \underline{\Gamma}(Z, \mathcal{A}_0)_{<0} &= 0 \end{aligned} \tag{7.5}$$

Using (7.3) and (7.4) with $M = A$ we see that A is Cohen–Macaulay if and only if $\underline{H}^i(Z, \mathcal{A}_0) = 0$ for $0 < i < n-1$ and $A = \underline{\Gamma}(Z, \mathcal{A}_0)$. These conditions correspond precisely to the conditions given in (7.5).

Now assume that (7.5) holds and put $E = \underline{\Gamma}(\mathcal{E}_0)$. Then A and $\operatorname{End}_R(E)$ are reflexive R-modules which have the same restriction to X' (when considered as sheaves). Hence they are equal.

If $\omega_Z = \mathcal{L}^{-1}$ then by Serre duality we have $H^i(Z, \mathcal{A}_0 \otimes \mathcal{L}^j) = H^{n-1-i}(Z, \mathcal{A}_0^* \otimes \mathcal{L}^{-j} \otimes \omega_Z)^* = H^{n-1-i}(Z, \mathcal{A}_0 \otimes \mathcal{L}^{-j-1})^*$. If $j < 0$ and $i < n-1$ then $-j-1 \geq 0$ and $i > 0$. I.e., (7.2) follows from (7.1).

Our aim is now to apply this is the case that Z is a surface with ample anti-canonical bundle (i.e., a del Pezzo surface). Recall that by [8] Z is either $\mathbb{P}^1 \times \mathbb{P}^1$ or else is obtained by blowing up $\mathbb{P}^2$ in ≤ 8 points in general position.

Proposition 7.3. *Let Z be a del Pezzo surface and let R be a cone over Z with trivial canonical bundle. Then R has a non-commutative crepant resolution.*

Proof. Let us first discuss the cases where ω_Z is a *proper* multiple of a line bundle. If F is an exceptional curve on a surface Z then $\deg(\omega_Z \mid F) = -1$ and so ω_Z cannot be a proper multiple of a line bundle. The del Pezzo surfaces without exceptional curves are $\mathbb{P}^2$ and $\mathbb{P}^1 \times \mathbb{P}^1$. In the first case $\omega_Z = \mathcal{O}_Z(-3)$ and hence $\mathcal{L} = \mathcal{O}_Z(1)$ and in the second case $\omega_Z = \mathcal{O}_Z(-2,-2)$ and hence $\mathcal{L} = \mathcal{O}_Z(1,1)$.

In the first case the cone over Z is a polynomial ring so this is trivial. In the second case the cone is given by $R = k[u, v, x, y]/(uv - xy)$ which is standard. The non-commutative crepant resolution is given by

$$\begin{pmatrix} R & I \\ I^{-1} & R \end{pmatrix}$$

where $I = (u, x)$.

So from now on we assume $\mathcal{L} = \omega_Z^{-1}$. We will construct a generator for $D(\operatorname{Qch}(Z))$ satisfying condition (7.1). If $Z = \mathbb{P}^1 \times \mathbb{P}^1$ then we take $\mathcal{E}_0 = \bigoplus_{i,j=0,1} \mathcal{O}_{\mathbb{P}^1 \times \mathbb{P}^1}(i, j)$. So assume that Z is obtained by blowing up $\mathbb{P}^2$ in $x_1, \ldots, x_p$ with $p \leq 8$ and denote the corresponding exceptional curves by $F_1, \ldots, F_p$. Let $\alpha : Z \to \mathbb{P}^2$ be the structure map. We have $\omega_Z = \alpha^*(\omega_{\mathbb{P}^2})(F_1 + \ldots + F_p)$.

Put $\mathcal{E}_0 = \alpha^*(\mathcal{O}_{\mathbb{P}^2}) \oplus \alpha^*(\mathcal{O}_{\mathbb{P}^2}(1)) \oplus \alpha^*(\mathcal{O}_{\mathbb{P}^2}(2)) \oplus \mathcal{O}_Z(F_1) \oplus \cdots \oplus \mathcal{O}_Z(F_p)$. It is easy to see that $\mathcal{E}_0$ is a generator for $D(\operatorname{Qch}(Z))$ and that $\pi^*\mathcal{E}_0$ satisfies condition (7.1). □

8 One-dimensional Torus Invariants

Let $T = k^*$ be a one-dimensional torus acting on a finite dimensional vector space V. We may choose a basis $x_1, \ldots, x_n$ for V such that T acts diagonally: $z \cdot x_i = z^{a_i} x_i$ for some $a_i \in \mathbb{Z}$. Put $S = \mathrm{Sym}(V)$ and $R = S^T$. In order to avoid trivialities we assume that there are at least two strictly positive and two strictly negative a_i's and that the greatest common divisor of the a_i's is one. Put $N^+ = \sum_{a_i>0} a_i$, $N^- = -\sum_{a_i<0} a_i$ and $N = \min(N^+, N^-)$.

It will be convenient to use the Artin–Zhang Proj of a graded ring [1]. If T is a noetherian $\mathbb{Z}$-graded ring then $X = \mathrm{Proj}_{\mathrm{AZ}} T$ is the Grothendieck category $\mathrm{Gr}(T)/\mathrm{Tors}(T)$ where $\mathrm{Tors}(T)$ is the localizing subcategory of $\mathrm{Gr}(T)$ given by the graded modules which are limits of right bounded ones. Below we denote the quotient functor $\mathrm{Gr}(T) \to \mathrm{Gr}(T)/\mathrm{Tors}(T)$ by π. We write $\mathcal{O}_X = \pi S$ and $\underline{\Gamma}(X, \mathcal{M}) = \oplus_{n\in\mathbb{Z}} \mathrm{Hom}(\mathcal{O}_X, \mathcal{M}(n))$ where the functor $\mathcal{M} \mapsto \mathcal{M}(n)$ is obtained from the corresponding functor on $\mathrm{Gr}(T)$. The derived functors of $\underline{\Gamma}$ are denoted by $\underline{H}^*$. We write $\mathrm{coh}(X)$ for the category of noetherian objects in X.

It is shown in [1] that $\mathrm{Proj}_{AZ} T = \mathrm{Proj}_{AZ} T_{\geq 0}$ so in principle one may restrict to $\mathbb{N}$-graded rings but below it will be more convenient to work with $\mathbb{Z}$-graded rings.

Putting $\deg x_i = a_i$ defines a $\mathbb{Z}$-grading on S such that $R = S_0$. By [20] R is Cohen–Macaulay. The other homogeneous components S_a of S are finitely generated R-modules. It is known *precisely* when they are Cohen–Macaulay [32, 38]. In particular one has the following result.

Lemma 8.1. *S_a is Cohen–Macaulay for $-N^+ < a < N^-$.*

Let $S^+ = S$ as graded rings and let S^- be the ring S with modified grading given by $S_n^- = S_{-n}$ We define $X^\pm = \mathrm{Proj}_{\mathrm{AZ}} S^\pm$. Thus if I^+, I^- are respectively the ideals generated by $S_{>0}$ and $S_{<0}$ then $X^\pm$ are simply the Grothendieck categories $\mathrm{Gr}(S^\pm)/\mathrm{Tors}(S^\pm)$ where $\mathrm{Tors}(S^\pm)$ consists of the objects in $\mathrm{Gr}(S^\pm)$ which are (elementwise) annihilated by powers of $I^\pm$. Note that I^+, I^- are the graded ideals in S respectively generated by $(x_i)_{a_i>0}$ and $(x_i)_{a_i<0}$. From the description of $\mathrm{Tors}(S^\pm)$ as torsion theories associated to ideals, it follows that $\mathrm{Tors}(S^\pm)$ is stable. I.e., $\mathrm{Tors}(S^\pm)$ is closed under injective hulls [34]. In particular the following lemma follows.

Lemma 8.2. *X^+, X^- have finite global dimension.*

Lemma 8.3. *We have $\underline{H}^i(X^\pm, \mathcal{O}_{X^\pm}) = H^{i+1}_{I^\pm}(S^\pm)$ for $i > 0$ and $S^\pm = \underline{\Gamma}(X^\pm, \mathcal{O}_{X^\pm})$.*

Proof. This is a variant on (7.3)(7.4). Note that $H^{0,1}_{I^\pm}(S^\pm) = 0$ because of the hypotheses on the weights $(a_i)_i$.

Lemma 8.4. *Let $0 \leq i, j < N^\pm$. Then $\mathrm{Ext}^p_{X^\pm}(\mathcal{O}_{X^\pm}(m), \mathcal{O}_{X^\pm}(n)) = 0$ for $p > 0$ and $\mathrm{Hom}_{X^\pm}(\mathcal{O}_{X^\pm}(m), \mathcal{O}_{X^\pm}(n)) = S^\pm_{n-m}$.*

Proof. By the previous lemma we have $\mathrm{Ext}^p_{X^\pm}(\mathcal{O}_{X^\pm}(m), \mathcal{O}_{X^\pm}(n)) = H^{p+1}_{I^\pm}(S^\pm)_{n-m}$ for $p > 0$. A standard computation reveals that $H^u_{I^\pm}(S^\pm)$ is zero in degrees $> -N^\pm$ which is what we want.

Also by the previous lemma we find $\mathrm{Hom}_{X^\pm}(\mathcal{O}_{X^\pm}(m), \mathcal{O}_{X^\pm}(n)) = S^\pm_{n-m}$.

Lemma 8.5. *The objects $\mathcal{O}_{X^\pm}(m)$, $m = 0, \ldots, N^\pm - 1$ form a generating family for $D(X^\pm)$.*

Proof. Let us work with X^+. With a variant of [11, Lemma 4.2.2] we see that $(\mathcal{O}_{X^+}(n))_n$ generates $D(X^+)$. Now we look at the Koszul exact sequence associated to $(x_i)_{a_i>0}$.

$$0 \to S(-N^+) \to \cdots \to \oplus_{a_i>0} S(-a_i) \to S \to S/I^+ \to 0$$

Note that S/I^+ is right bounded. Shifting and applying π we obtain exact sequences

$$0 \to \mathcal{O}_{X^+}(n - N^+) \to \cdots \to \oplus_{a_i>0} \mathcal{O}_{X^+}(n - a_i) \to \mathcal{O}_{X^+}(n) \to 0$$

Hence it follows that all $\mathcal{O}_{X^+}(n)$ may be obtained using triangles from $\mathcal{O}_{X^+}(m)$, $m = 0, \ldots, N^+ - 1$ which finishes the proof. □

To simplify the notations below we will now invert, if necessary, the signs of the a_i's, to insure that $N^+ \le N^-$. Thus $N = N^+$.

Define $\mathcal{E}^\pm = \oplus_{m=0,\ldots,N-1} \mathcal{O}_{X^\pm}(m)$, According to Lemma 8.4 we have $A \stackrel{\mathrm{def}}{=} \mathrm{End}_{X^+}(\mathcal{E}^+) = \mathrm{End}_{X^-}(\mathcal{E}^-)$.

Lemmas 8.4, 8.5 now yield the following

Theorem 8.6. *The functor* $\mathrm{RHom}(\mathcal{E}^+, -)$ *defines an equivalence $D(X^+) \to D(A)$ and the functor $- \otimes_A \mathcal{E}^-$ defines a full faithful embedding $D(A) \to D(X^-)$. In particular there is a full faithful embedding $D(X^+) \to D(X^-)$. All embeddings restrict to embeddings between the corresponding bounded derived categories of coherent objects. If $N^- = N^+$ (or equivalently $\sum_i a_i = 0$) then all embeddings are equivalences.*

The existence of an embedding/equivalence $D(X^+) \to D(X^-)$ was proved by Kawamata in [22] (in a slightly more general situation).

We also have

Proposition 8.7. *A is Cohen–Macaulay.*

Proof. According to Lemma 8.4 we have $A = (S_{n-m})_{0 \le m,n < N}$. Cohen–Macaulauness now follows from Lemma 8.1. □

Now we restrict to the case $N^+ = N^-$, or equivalently $\sum_i a_i = 0$. Under these hypotheses we have

Lemma 8.8. *S_0 is Gorenstein and the S_a are reflexive S_0-modules, satisfying $(S_a S_b)^{**} = S_{a+b}$.*

Proof. S_0 is Gorenstein because of [33, 13.3]. The hypotheses that the greatest common divisor of the $(a_i)_i$ is one implies that that the generic stabilizer of the T-action on $W = \operatorname{Spec}\operatorname{Sym}(V) = V^*$ is trivial.

In general if $(\zeta_i)_i$ is a point in W then the order of its stabilizer is equal to the greatest common divisor of the a_i's such that $\zeta_i \neq 0$. From the relation $\sum_i a_i = 0$ we then deduce that the complement of the locus $W' \subset W$ where T has trivial stabilizer, has codimension at least two. The equality $(S_a S_b)^{**} = S_{a+b}$ is now true on W' and hence it is true on $W' /\!\!/ T$ since $W' \to W' /\!\!/ T$ is a principal T-bundle. Then is is also true on $W / /\!\!/ T$ since $W \to W /\!\!/ T$ contracts no divisor and hence all modules of covariants are reflexive [16, Sect. 1.3]. □

We may now state our final result.

Theorem 8.9. *Let the notations be as above and assume $\sum_i a_i = 0$. Then $R = S^T$ is a Gorenstein ring with a non-commutative crepant resolution given by $A = \operatorname{End}_R(\oplus_{a=0}^{N-1} S_a)$ where $N = \sum_{a_i > 0} a_i = -\sum_{a_i < 0} a_i$.*

Proof. We have

$$A = \oplus_{m,n=0,\dots,N-1}(S_{n-m})$$

which by lemma 8.8 is equal to $\operatorname{End}_R(\oplus_{a=0}^{N-1} S_a)$. □

References

[1] M. Artin and J. J. Zhang, *Noncommutative projective schemes*, Adv. in Math. **109** (1994), no. 2, 228–287.

[2] M. Artin, *Maximal orders of global dimension and Krull dimension two*, Invent. Math. **84** (1986), 195–222.

[3] M. Artin and W. Schelter, *Graded algebras of global dimension 3*, Adv. in Math. **66** (1987), 171–216.

[4] M. Artin, J. Tate, and M. Van den Bergh, *Some algebras associated to automorphisms of elliptic curves*, The Grothendieck Festschrift, vol. 1, Birkhäuser, 1990, pp. 33–85.

[5] ———, *Modules over regular algebras of dimension 3*, Invent. Math. **106** (1991), 335–388.

[6] M. Auslander and O. Goldman, *Maximal orders*, Trans. Amer. Math. Soc. **97** (1960), 1–24.

[7] V. V. Batyrev, *Birational Calabi–Yau n-folds have equal Betti numbers*, New trends in algebraic geometry (Warwick, 1996) (Cambridge), London Math. Soc. Lecture Note Ser., vol. 264, Cambridge Univ. Press, Cambridge, 1999, pp. 1–11.

[8] A. Beauville, *Surface algébriques complexes*, Astérisque, vol. 54, Soc. Math. France, 1978.

[9] A. Bondal and D. Orlov, *Derived categories of coherent sheaves*, available as math.AG/0206295.

[10] ———, *Semi-orthogonal decompositions for algebraic varieties*, available as alg-geom/950601, 1996.

[11] A. Bondal and M. Van den Bergh, *Generators and representability of functors in commutative and non-commutative geometry*, available as math.AG/0204218.

[12] N. Bourbaki, *Algèbre commutative*, Hermann, Paris, 1960–65.

[13] T. Bridgeland, *Equivalences of triangulated categories and Fourier–Mukai transforms*, Bull. London Math. Soc. **31** (1999), no. 1, 25–34.

[14] ______, *Flops and derived categories*, Invent. Math. **147** (2002), 613–632.

[15] T. Bridgeland, A. King, and M. Reid, *The McKay correspondence as an equivalence of derived categories*, J. Amer. Math. Soc. **14** (2001), no. 3, 535–554 (electronic).

[16] M. Brion, *Sur les modules de covariants*, Ann. Sci. École Norm. Sup. (4) **26** (1993), 1–21.

[17] K. A. Brown and C. R. Hajarnavis, *Homologically homogeneous rings*, Trans. Amer. Math. Soc. **281** (1984), 197–208.

[18] R. Elkik, *Singularités rationnelles et déformations*, Invent. Math. **47** (1978), no. 2, 139–147.

[19] G. M. Greuel, G. Pfister, and H. Schönemann, SINGULAR *2.0. A Computer Algebra System for Polynomial Computations*, Centre for Computer Algebra, University of Kaiserslautern, 2001, `http://www.singular.uni-kl.de`.

[20] M. Hochster, *Rings of invariants of tori, Cohen–Macaulay rings generated by monomials, and polytopes*, Ann. of Math. (2) **96** (1972), 318–337.

[21] M. Kapranov and E. Vasserot, *Kleinian singularities, derived categories and Hall algebras*, Math. Ann. **316** (2000), no. 3, 565–576.

[22] Y. Kawamata, *Francia's flip and derived categories*, available as math.AG/0111041.

[23] B. Keller, A_∞ *algebras and triangulated categories*, in preparation.

[24] A. D. King, *Moduli of representations of finite-dimensional algebras*, Quart. J. Math. Oxford Ser. (2) **45** (1994), no. 180, 515–530.

[25] M. Kontsevich, Lecture at Orsay, December 7, 1995.

[26] D. Luna, *Slices étales*, Bull. Soc. Math. France **33** (1973), 81–105.

[27] P. E. Newstead, *Introduction to moduli problems and orbit spaces*, Lectures on Mathematics and Physics, vol. 51, Tata Institute of Fundamental Research, Bombay, 1978.

[28] I. Reiner, *Maximal orders*, Academic Press, New York, 1975.

[29] I. Reiten and M. Van den Bergh, *Tame and maximal orders of finite representation type*, Memoirs of the AMS, vol. 408, Amer. Math. Soc., 1989.

[30] K. Saito, *Quasihomogene isolierte Singularitäten von Hyperflächen*, Invent. Math. **14** (1971), 123–142.

[31] A. Schofield, *Representations of rings over skew fields*, Lecture Note Series, vol. 92, London Mathematical Society, 1985.

[32] R. P. Stanley, *Linear Diophantine equations and local cohomology*, Invent. Math. **68** (1982), 175–193.

[33] ______, *Combinatorics and commutative algebra*, Progress in Mathematics, vol. 41, Birkhäuser Boston Inc., Boston, MA, 1983.

[34] B. Stenström, *Rings of quotients*, Die Grundlehren der mathematischen Wissenschaften in Einzeldarstellungen, vol. 217, Springer Verlag, Berlin, 1975.

[35] D. R. Stephenson, *Artin–Schelter regular algebras of global dimension three*, J. Algebra **183** (1996), 55–73.

[36] ______, *Algebras associated to elliptic curves*, Trans. Amer. Math. Soc. **349** (1997), 2317–2340.

[37] M. Van den Bergh, *Three-dimensional flops and non-commutative rings*, submitted. To appear in Duke Mathematical Journal.

[38] ______, *Cohen–Macaulayness of semi-invariants for tori*, Trans. Amer. Math. Soc. **336** (1993), no. 2, 557–580.

[39] ______, *Existence theorems for dualizing complexes over non-commutative graded and filtered rings*, J. Algebra (1997), 662–679.

[40] K. Watanabe, *Rational singularities with k^*-action*, Commutative algebra (Trento, 1981) (New York), Lecture Notes in Pure and Appl. Math., vol. 84, Dekker, New York, 1983, pp. 339–351.
[41] A. Yekutieli and J. Zhang, *Rings with Auslander dualizing complexes*, J. Algebra **213** (1999), no. 1, 1–51.
[42] ———, *Dualizing complexes and perverse sheaves on noncommutative ringed schemes*, to appear, 2002.

Closed String Operators in Topology Leading to Lie Bialgebras and Higher String Algebra

Moira Chas and Dennis Sullivan

Imagine a collection of closed oriented curves depending on parameters in a smooth d-manifold M. Along a certain locus of configurations strands of the curves may intersect at certain sites in M. At these sites in M the curves may be cut and reconnected in some way. One obtains operators on the set of parametrized collections of closed curves in M. By making the coincidences transversal and compactifying, the operators can be made to act in the algebraic topology of the free loop space of M when M is oriented. The process reveals collapsing sub graph combinatorics like that for removing infinities from Feynman graphs.

Let $\mathbb{H}$ denote the equivariant homology with rational coefficients of the pair (Maps (S^1, M), constant maps (S^1, M)) relative to the S^1 action of rotating the source. $\mathbb{H}$ is called the reduced equivariant homology of the free loop space of M.

Associated to the diagrams I and II of the figure there are operators $c_n : \mathbb{H}^{\otimes n} \to \mathbb{H}$ and $s_n : \mathbb{H} \to \mathbb{H}^{\otimes n}$, $n = 2, 3, 4, ..$ These operators satisfy various relations e.g. c_2

Diagram I (the operator c_n, $n = 5$):

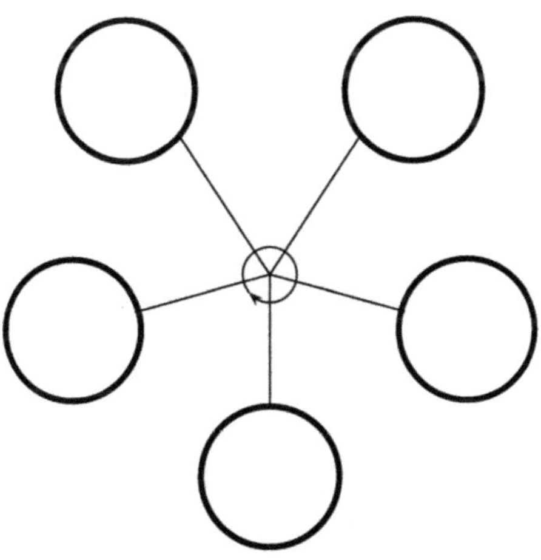

Diagram II (the operator s_n, $n = 6$):

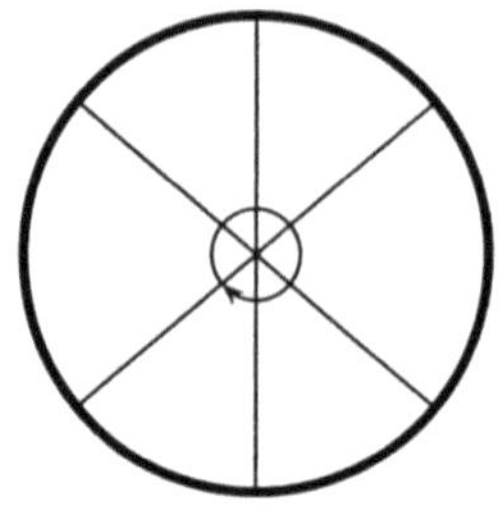

and s_2 satisfy the relations-Jacobi, coJacobi, and Drinfeld compatibility which utilize diagrams III, IV, V, and VI.

Thus we prove the

Theorem: The reduced equivariant homology $\mathbb{H}$ of the free loop space of a smooth oriented d-manifold M has the structure of a Lie bialgebra generated by string operators $c_2 : \mathbb{H} \otimes \mathbb{H} \to \mathbb{H}$ and $s_2 : \mathbb{H} \to \mathbb{H} \otimes \mathbb{H}$ of degree $2 - d$. The string operators $c_n : \mathbb{H}^{\otimes n} \to \mathbb{H}$ and $s_n : \mathbb{H} \to \mathbb{H}^{\otimes n}$ are also defined for $n > 2$ and have degree $n + (1 - n)d$. The relations conjecturally satisfied by c_n and s_n are described below.

Problem and Conjecture: There is evidence (see c_n and s_n Identities paragraph below) that one may prove the c_n and s_n generate an algebraic structure on $\mathbb{H}$ that

Diagram III (to prove the Jacobi identity for c_2):

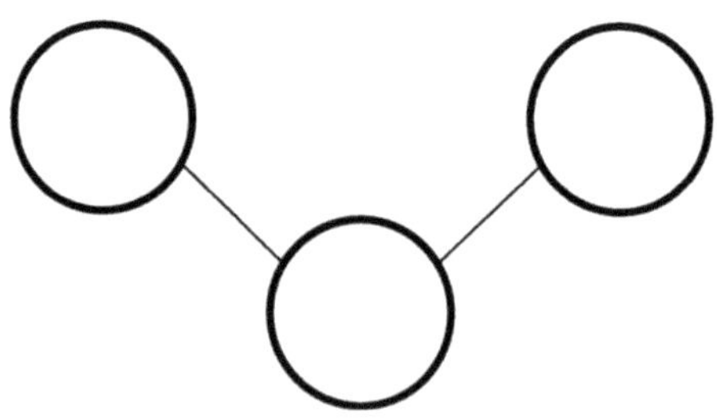

Diagram IV (to prove the coJacobi identity for s_2):

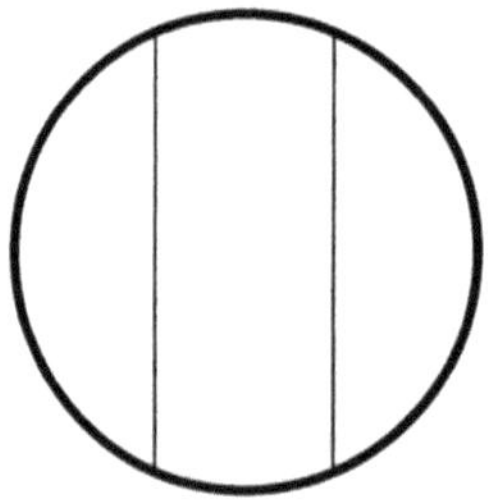

Diagram V (to prove the Drinfeld compatibility for c_2 and s_2):

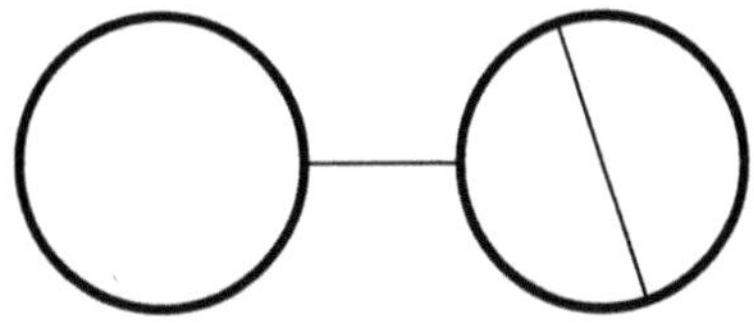

Diagram VI (to prove the Drinfeld compatibility for c_2 and s_2):

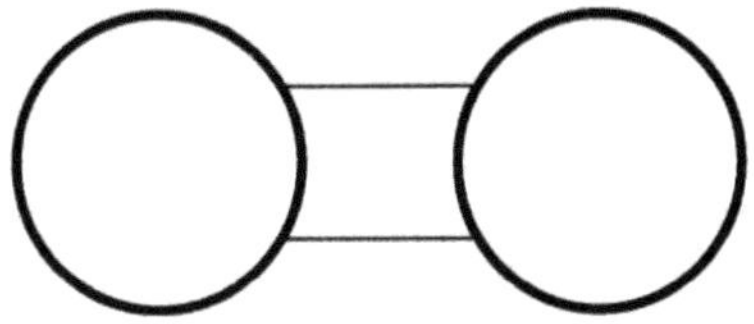

is Koszul dual in the sense of [4] to the positive boundary version (the number of inputs and the number of outputs are both positive) of the algebraic structure (genus zero) of symplectic topology [5].

Remark: This Lie bialgebra is non trivial. For example, when $d = 2$ it is isomorphic to the Lie bialgebra discovered by Goldman and Turaev[10].

General Program: The theorem is part of a more elaborate structure of closed string operators including higher genus acting at the chain level of the spaces of maps of families of closed curves into M. We describe the elements of this general theory and illustrate with the examples indicated by diagrams I, II, III, IV, V, and VI which are required to treat the theorem and diagram VII used to prove the genus one involutive identity.

Diagram VII (Involutive identity $c_2 \cdot s_2 = 0$)

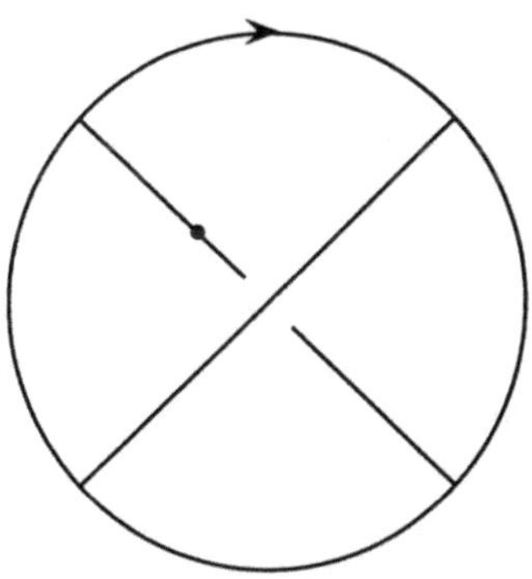

Generalized Chord Diagram: First of all there are operators of order zero, order one, etc. The operators of order zero are associated to diagrams as above. The most general *order zero diagram D* (generalized chord diagram)is specified by the data- a finite subset F of a union C of directed circles, a partition of the subset F into parts of cardinality at least two, and a cyclic order on each part. Higher order operations correspond to general chord diagrams with additional combinatorial data (see paragraphs Diagrams IV, V, VI below). To the order zero diagram one can associate a graph $\Gamma'(D)$ by attaching an n-prong respecting the cyclic order (as in diagram I or II) to each part of the partition. There is also a surface $\Sigma'(D)$ obtained by crossing the collection of circles C with the unit interval [0,1] and attaching along n segments a (thickened n-prong), a disk with n tabs, to Cx$\{1\}$. So $\Sigma'(D)$ equals Cx[0, 1] union (thickened n-prongs) along Cx$\{1\}$.

Ribbon surface and cyclic graph of a diagram: We denote by $\Sigma(D)$ the surface obtained from $\Sigma'(D)$ by collapsing to points the n-prongs slightly extended into the collar to level Cx$\{1/2\}$. The quotient of Cx$\{1/2\}$ by this collapse is called the cyclically ordered graph $\Gamma(D)$ of the diagram D. $\Sigma(D)$ is a *ribbon surface* associated to the *cyclically ordered graph* $\Gamma(D)$, associated to the *generalized chord diagram D*.

Closed String Operator: Let us try to define the chain operator associated to D. The input is a bundle η with fibre C, a labelled union of directed circles and a map of the total space of η into M smooth along the C direction. (These objects can be used to compute tensor products of the equivariant homology reduced or not). We form the associated bundle to η with fibre all isotopic configurations of F embedded in C. We restrict attention to the *D-locus* in the total space of the associated bundle, defined as the locus where the map identifies in M all the points in each part of the partition of F (now considered as a subset of Cx$\{1/2\}$). Along this locus there are induced maps of the quotient graph $\Gamma(D)$ of C into M and thus also of the ribbon surface $\Sigma(D)$ which retracts to $\Gamma(D)$. The restriction of this map on $\Sigma(D)$ to C', the rest of the boundary of $\Sigma(D)$ besides C (actually Cx$\{0\}$) in each fibre is a map of the total space of a C' fibration over the D locus into M. This is the output of the D operation at the level of set theory or topological spaces.

Chain Operator: To have a good object in algebraic topology there are several issues: compactness, transversality, and orientation. We want the output to be a relative chain representing an element in a chain complex computing $\mathbb{H}^{\otimes i}$, $i =$ number of components of C', if the input is. We now discuss these issues in the order mentioned.

Diagram I: Now the compactness property of chains is not a problem for diagram I because its configuration space is already compact being an n-torus. Similarly, compactness is easy to arrange for any cactus diagram (diagram III') generalizing diagram III. (see paragraph Diagram III below). The homological content of this part of theory was discussed in [1] where the non reduced equivariant homology, the Lie bracket c_2 and the higher analogues c_n were considered.

Diagram II: For diagram II the configuration space is non compact and there is an obstacle to overcome. Note however that when two prongs come close enough together (relative to the input C-family) the output C' family has for each parameter at least one component which is small in M. The role of the reduced equivariant homology is to take advantage of this fact. Because families with constant (or small) map components are considered to be null or zero for the reduced discussion we obtain that the configuration space of diagram II is (relatively) compact for the purpose of producing a (relative) chain for the computation of $\mathbb{H}^{\otimes i}$, the reduced equivariant homology.

Relative Compactness: To treat the chain operator for a general diagram we need to complete its configuration space enough so that this relative compactness is achieved. Namely, any strata omitted from a true compactification must correspond to output families which for each parameter have a positive number of tiny components.

Constraint Normal Bundle: Besides this compactness consideration the other "sine qua non" issue is the normal bundle to the constraint locus. Imposing the conditions defining the locus of the diagram in the cases above amounts to taking the preimage of a diagonal with a normal bundle. Then the transversal preimage of a chain will be a chain (the Thom map at the chain level) and we can work in the context of algebraic topology.

We refer to this as the constraint normal bundle issue and we must keep this *constraint normal bundle* as we add pieces to the configuration spaces to obtain (relative) compactness.

Diagram III: Let us consider diagram III. When two different chords on the middle circle coalesce we converge to constraints as in diagram I (n=3). The number of constraints is the same (two) and the normal bundles fit together perfectly. This works for the normal bundle consideration to compactify all the configuration spaces

Diagram III′

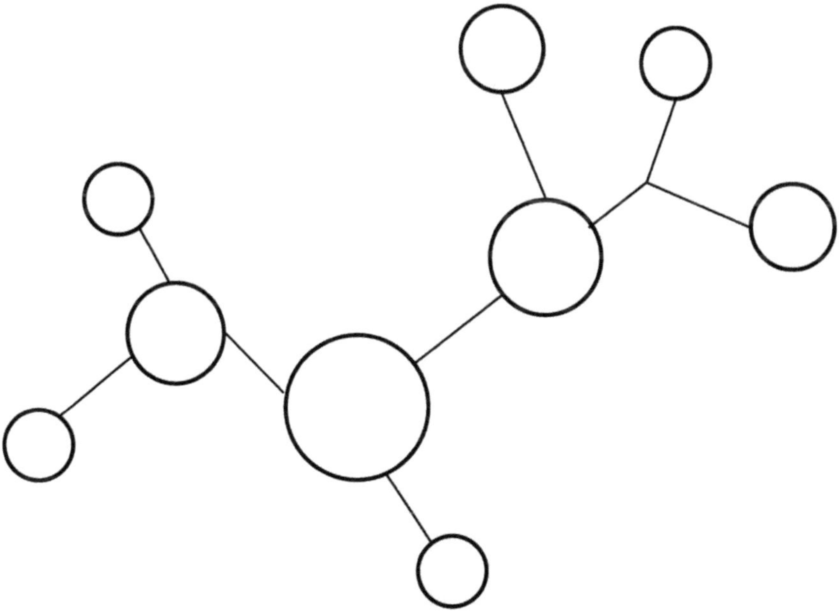

of cactus like diagrams, diagram III' (planar trees with circles inserted at some of the vertices) leading to genus zero n to one operations. The point is that compactness is achieved by allowing *different* parts of the partitions to coalesce where the constraint normal bundle persists. Because of the tree like property it is impossible to let parts self collide. Except for orientations the discussion is conceptually complete in these cases- we have actual compactness and a normal bundle to define a transversal Thom chain map, in the equivariant context. (compare [1]).

Diagram IV: Now consider diagram IV. When different chords come together *at one point* the constraints converge to those of a diagram like II (n=3). The normal bundles fit together perfectly as before. By adding the configuration spaces of diagram II to that of IV we obtain a *good* constraint normal bundle and *relative* compactness for Diagram IV. This also applies to more general internal chord or n-prong diagrams on one circle (disjoint n-prongs in one circle Diagram IV') which are dual to planar cactii with roles of input and output interchanged (see Duality Diagram). We can add strata preserving the constraint normal bundle and achieve relative compactness. Thus dual to all the n to one genus zero operations Diagram III' from the Diagram III paragraph we have one to n genus zero operations Diagram IV' in the reduced theory (however see next paragraph).

Null chains and Degenerate chains: There is one additional caveat about working in the reduced theory even for the diagrams considered up to now. We need to

Diagram IV′

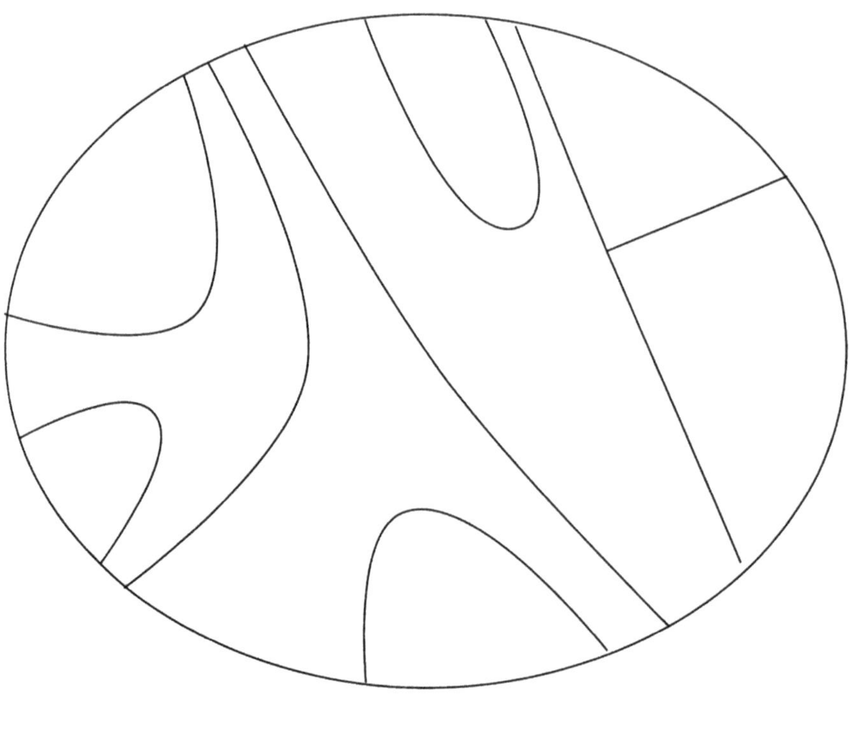

Duality Diagram

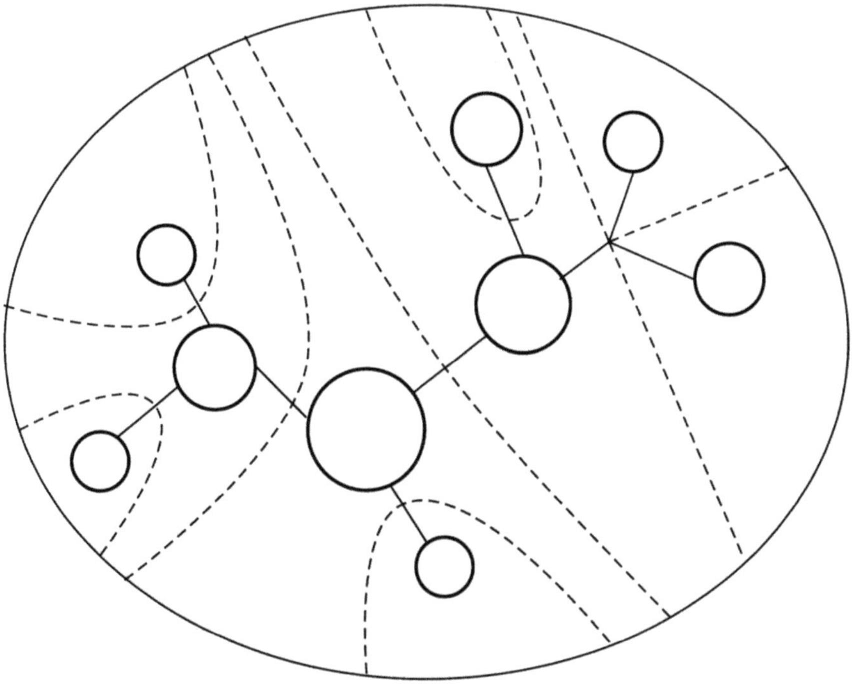

know the null subcomplex consisting of maps where at least one component is constant is invariant by the operations. At first glance this seems problematic but it works out in the end. A constant loop component may be cut up by an operation and mixed into other components. If so we no longer have a constant component. However, the situation is saved because we obtain a degenerate chain-one whose geometric dimension is too low. For as the parameters of the configurations vary in the component which is mapped to a constant loop the image chain is not varying. A transversal pull back will not have the full homological dimension and can be ignored. In fact we can mod out by degenerate chains from the beginning. (This point has to be considered carefully when extraordinary homology theories are studied here). On the other hand if the operation doesn't touch the null component we still have a null component in the output. In summary definable string chain operators act in the relative complex defining the reduced equivariant theory. For example we have all the ingredients now to define the operations c_n and s_n, $n = 2, 3, 4, \ldots$ in the reduced equivariant theory.

One need only add that diagrams I and II correspond to *cycles* since no strata were added to create (relative) compactness. Thus the chain operators corresponding to c_n and s_n commute with the ∂ operators on chains and pass to homology. (Orientations will be discussed in the paragraph below).

To prove the relations of a Lie bialgebra among the compositions of c_2 and s_2 we have to consider diagrams V and VI which bring forth two further considerations.

Diagram V: To achieve relative compactness for the configurations space of diagram V we have to let the two chords touch at one point which is a case already considered above for Diagram III (and IV). We also have to allow the internal chord of Diagram V to collapse to the endpoint of the connecting chord (from opposite sides only- because a one sided approach leads to a tiny output circle and a null chain). This creates in the limit a diagram of type I for $n = 2$ and if we do nothing else the number of constraints goes down and we lose the normal bundle property. However, the collapsing internal chord and the constraint that values at the endpoints of this chord coincide say that in the limit the derivative of the map in the C direction is null at the limit point.

Thus we are led to an order one diagram, a diagram of order zero of type I for $n = 2$ with the additional data that one of the attaching points of the chord is a point of multiplicity two. This means that when the locus of this diagram of order one is defined the condition coincidence of values at the endpoints of the chord is augmented by the condition that the 1st derivative in the C direction is zero at the point of multiplicity two. (In general at a point of multiplicity k the first $(k-1)$ derivatives would be required to be zero). Then the constraint normal bundle extends continuously over the added stratum and we almost have the relative compactness. This treats diagram V.

Diagram VI: One more feature appears in treating diagram VI. To the generic $4D$ configuration space of diagram VI we add three $3D$ strata and two $2D$ strata for relative compactness. One of the $3D$ strata will involve a new kind of consideration similar to that in the Fulton MacPherson compactification of configuration spaces [7]. The other strata will be of the type already considered. Namely, two of the $3D$ strata allow the two chords to touch on one circle or the other. This has already been considered. The new case appears when the two chords approach each other (on opposite sides again) at both endpoints at commensurable distances. We add a $3D$ stratum to our space which records the limiting single chord (two parameters) and a third parameter which can record the signed ratio of the small distances in the approach. We call this an FM-stratum.

We also add two strata for the chords touching first at one endpoint and then at the second endpoint which-like the discussion of diagram V- produces in each case a multiple point. These strata account for the approach of chords at both endpoints with incommensurable distances at the endpoints.

When defining the locus for this completion (to a relatively compact configuration space) we treat the FM stratum in the following way. We have a chord diagram of type I ($n = 2$) with a third ratio parameter λ. We ask first that the map agrees at endpoints of the chord as before and then ask further that derivatives in the C direction at these points be proportional with ratio $1/\lambda$, (the factor $1/\lambda$ because distances appear in the denominator when calculating derivatives).

If the other strata are treated as described above the constraint normal bundle extends continuously over this entire (relative) compactification of the configuration space of diagram VI. This treats diagram VI.

Now we have all the ingredients to define the chain operators of diagrams I through VI up to a question of orientation.

Orientations: It is possible to avoid a nightmare of sign difficulties using a categorial approach to orientations motivated by [6]. First there is the "graded line" functor from finite dimensional real vector spaces to $Z/2$ graded vector spaces. It assigns to V the top exterior power placed in even degree if dimension V is even and in odd degree if dimension V is odd.

An orientation of V is by definition a generator of the graded line of V up to positive multiples. The point of all this is the flip of factors in the direct sum of two different vector spaces is always orientation preserving. Now for any finite family of spaces V_f, $f \epsilon F$ there is a canonical notion of direct sum $\bigoplus_{f \epsilon F} V_f$ and if each V_f is oriented there is a canonical orientation of $\bigoplus_{f \epsilon F} V_f$. Similarly in an exact sequence of spaces $0 \to V \to W \to V' \to 0$ an orientation of any two determines canonically an orientation of the third.

We apply this to the previous constructions as follows. Fix an orientation of M. Then any product $\prod_{f \epsilon F} M$ indexed by a finite set has a canonical orientation by the first fact above. Using the second fact as well, any diagonal corresponding to a part of F and its normal bundle each has a canonical orientation. Further, if the base of a C bundle is oriented the total space η is canonically oriented (since C is oriented) and the total space of the associated bundle of F configurations in C is oriented (if F is ordered up to even permutations). In our examples Diagrams I, II, III, IV, V, VI this ordering on F comes from the ordering of input or output components. In example Diagram VII it comes from the ordering of the chords. Combining all this the transversal pull back is also canonically oriented. The actual definition of the operators c_n and s_n uses these canonical orientations.

Lie bialgebra identities:

Jacobi: The proof of Jacobi for c_2 uses several versions of diagram III, to calculate $[[1,2],3]=c_2\ (c_2(1 \otimes 2) \otimes 3)$ and its cyclic permutations. The arrow on the chords is determined by the order of input arguments and determines, via an ordering of F up to even permutations, the orientation. The cancellation is indicated.

coJacobi: The proof of coJacobi for s_2 uses several versions of diagram IV. Again the direction on the chords is related to the order of arguments and subsequent orientations. Each column represents $(s_2 \otimes 1) \cdot s_2$ or its cyclic permutation. The numbers indicate the output arguments. The particular 4 element subset F pictured with its indicated coincidences could contribute in various ways to the chain operation. A chord could have either direction or be used as first chord or as second chord in the composed operation. Of the 8 possibilities in each case the 4 pictured are the only ones that occur. The cancellation is indicated.

Drinfeld compatibility: The proof of Drinfeld compatibility between c_2 and s_2 uses diagram V and diagram VI. The figure represents $s_2[1, 2]$. The directions on

Jacobi

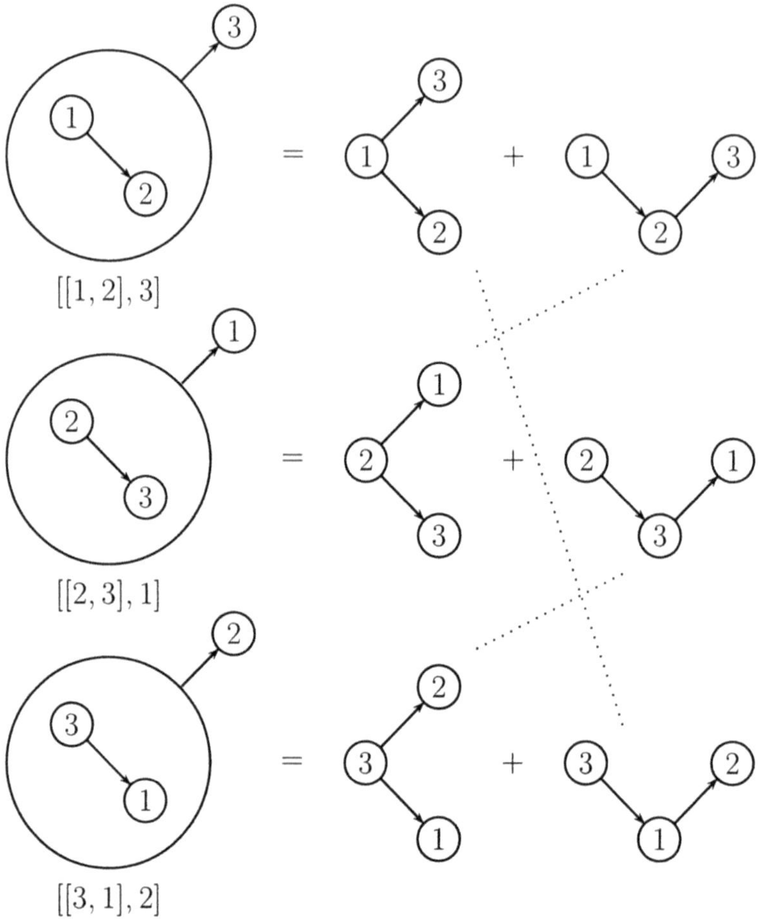

chords indicate ordering of input variables for $s_2[1, 2]$ and subsequent orientations. The dot on the chord indicates it is used first. Cancellation is indicated. The last four terms represent the right hand side of the compatibility equation, $s_2([1, 2]) = [s_2(1), 2] + [1, s_2(2)]$:

Identities for c_n and s_n: Using computations of Getzler [3] one can show the c_n taken together satisfy a generalized Jacobi identity. Some but not all of these were shown at the chain level in [1]. The rest follow as in [3] from the BV homology structure of [1]. The Getzler identities (defining what he calls a "gravity algebra") are Koszul dual [3] to the associative (or commutative) identities in the definition of a "Frobenius manifold" [5] describing the algebraic structure of genus zero Gromov Witten invariants of a closed symplectic manifold.

In the latter case there is a compatible non degenerate inner product- the Poincare duality of a closed symplectic manifold in its Floer homology which is equal to

Cojacobi

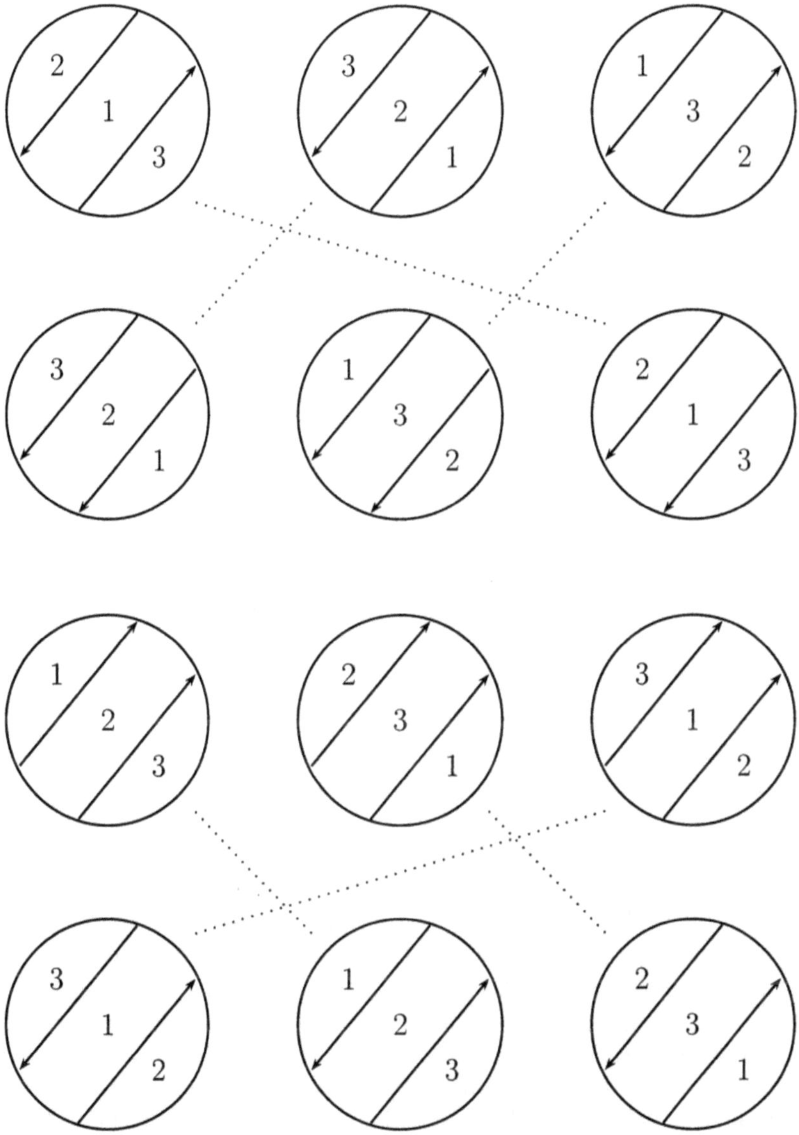

its ordinary homology. One can always form the "positive boundary" version of an algebraic structure by trading in multiplications and inner product for mulitplications and co multiplications satisfying the induced identities. (This idea we learned from David Kahzdan and the terminology "positive boundary" was suggested by Ralph Cohen.)

Drinfeld compatibility

$s_2([1,2]) =$

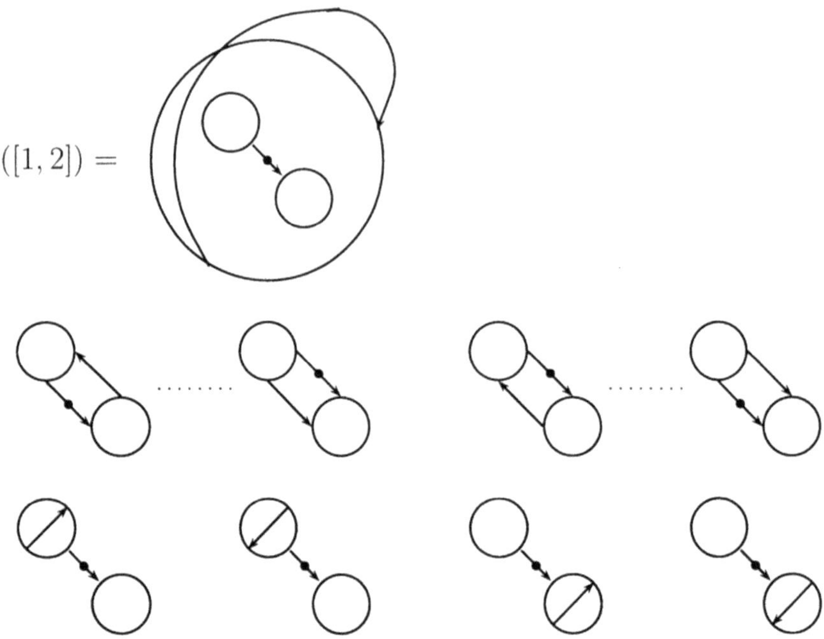

Do this for "Frobenius manifold", form the positive boundary version and then apply Koszul duality [3], [4], [5, p.87]. One obtains an algebraic structure which combines gravity algebra and gravity co algebra with Drinfeld type compatibilities. It contains the notion of Lie bialgebra [4]. At this point one knows the c_n satisfy the gravity algebra identities (generalized Jacobi [3]), the s_n by arrow reversal duality satisfy the gravity coalgebra identities (generalized Cojacobi) and c_2 and s_2 satisfy Drinfeld compatibility (this paper).

This is the evidence for the conjecture and problem mentioned above.

Involutive property of the Lie bialgebra: We may consider the operation $e = c_2 \circ s_2 \colon \mathbb{H} \to \mathbb{H}$. If we think of s_2 and c_2 as associated to pairs of pants pointed in opposite directions then e is associated to a torus with one input circle and one output circle obtained by glueing these two pairs of pants. The diagram for e is one circle with two (ordered) chords whose endpoints are linked (see diagram VII). The (relative) compactification of this Diagram VII uses all of the considerations above-the different parts colliding of Diagrams III and IV, the multiple points of Diagram V, and the FM stratum of Diagram VI. Examining the chain operation shows that there is a complete cancellation because interchanging the ordering of chords is orientation reversing for Diagram VII. Thus the genus one operator e is always zero in the Lie bialgebra.

Now e is the infinitesimal analogue for Lie bialgebras of the square of the canonical antiautomorphism of a Hopf algebra.

When this square is the identity one says the Hopf algebra is involutive so we say in analogy that since e is zero the *Lie bialgebra of the Theorem is involutive.* Note this is a genus one relation. (We are indebted to Pavel Etinghof and Boris Tsygan for this perspective.)

Higher genus operators: Forming the (relative) compactification of a general chord diagram D requires the addition of many strata. When different parts of F collide at one point the normal bundles fit together as indicated above. When the same part self collides at one point we introduce a multiplicity at that point. When several points collide at different rates we can reduce to the previous cases. When several points collide at commensurable rates we may have to introduce FM strata as in paragraph Diagram VI. The constraint bundle issue must be solved by adding relations among derivatives at the coincident points. These relations are associated to cycles in the subgraphs of $\Gamma(D)$ which are collapsing. For example the multiple point of Diagram V (and other examples) corresponds to collapsing a loop of $\Gamma(D)$. The FM stratum of Diagram VI corresponds to collapsing a cycle in $\Gamma(D)$ made out of two edges. More generally when intervals between points of F collapse this determines a collapsing subgraph of $\Gamma(D)$. Cycles on this graph give relations among derivatives. These are required to define the constraint normal bundle.

The theory begins to take on the structure of the collapsing graphs in the renormalization theory of Feynman diagrams appearing in the work of Kreimer et al [8]. This will be discussed elsewhere [9]. See [12] for conjectures related to the fact that the above Lie bialgebra identities hold at the chain level.

References

[1] M. Chas and D. Sullivan "String Topology" GT/ 9911159. Accepted by Annals of Mathematics.

[2] A. Voronov "Notes on Universal Algebra". See author's web page. To appear in proceedings of Stony Brook conference on "Graphs and Patterns in Mathematics and Physics", June 2001.

[3] E. Getzler "Operads and moduli spaces of genus zero Riemann surfaces". In:The Moduli Spaces of Curves ed. by R.Dijkgraaf, C. Faber, G. van der Geer Progress in Math, vol.129 Birkhauser 1995, 199-230.

[4] Wee Liang Gan "Koszul duality for dioperads" preprint University of Chicago 2002. QA/0201074

[5] Yuri I. Manin "Frobenius Manifolds, Quantum cohomology and Moduli spaces". AMS Colloquium Publications Vol.47.

[6] Pierre Deligne "Quantum Fields and Strings: A course for mathematicians" vol. 1 chapter 1 section 1.2 Editors Deligne et al AMS IAS

[7] W. Fulton, R. MacPherson "A compactification of configuration spaces" Annals of Math, 139 (1994) 183-225

[8] D. Kreimer "Structures in Feynman graphs- Hopf algebras and symmetries" See author's web page. To appear in proceedings of Stony Brook conference on "Graphs and Patterns in Mathematics and Physics", June 2001.

[9] D. Sullivan, work in progress "Closed String Operators and Feynman graphs", possible title.

[10] V. Turaev "Skein quantization of Poisson algebra of loops on surfaces" Ann. Sci. Ecole Norm. Sup. (1) 24 (1991), no.6, 635-704.
[11] Moira Chas "Combinatorial Lie Bialgebras of curves on surfaces" GT/0105178, to appear in Topology.
[12] D. Sullivan "Open and closed string field theory interpreted in classical algebraic topology" QA/0302332. To appear in Proceedings of Segalfest.

MIX
Papier aus verantwortungsvollen Quellen
Paper from responsible sources
FSC® C105338

FSC
www.fsc.org

If you have any concerns about our products,
you can contact us on
ProductSafety@springernature.com

In case Publisher is established outside the EU,
the EU authorized representative is:
Springer Nature Customer Service Center GmbH
Europaplatz 3, 69115 Heidelberg, Germany

Printed by Libri Plureos GmbH
in Hamburg, Germany